GW01340171

Flora of Leicestershire

Published 1988.
Leicestershire Museums Publication no.89.

© Leicestershire Museums, Art Galleries and Records Service,
96 New Walk, Leicester LE1 6TD.
All rights reserved. No part of this publication may be reproduced, stored in a retrieval system, or transmitted in any form or by any means, electronic, mechanical, photocopying, recording or otherwise, without the prior permission of Leicestershire Museums, Art Galleries and Records Service.

Designed by Alan Birdsall, Haley Sharpe Associates, Kibworth Beauchamp, Leicestershire.

Typeset and printed by Oldham and Manton Ltd., Leicester.

ISBN 0-85022-230-3

Frontispiece.   *Rosa sherardii*, rough grassland, near Spring Wood Pool, Lount, SK 387190, July 1981. Site destroyed by opencast mining 1987.

Flora of Leicestershire

Edited by
A.L. Primavesi
and
P.A. Evans

Leicestershire Museums, Art Galleries and Records Service
1988

This volume is dedicated to the memory of **Frederick Archibald Sowter (1899-1972)**, who alone contributed both to it and its predecessor. Almost single-handedly at times, he maintained and encouraged the interest of Leicestershire naturalists in the flora of their county over four decades.

CONTENTS

List of figures		5-6
List of plates		7-8
Foreword	C. A. Stace	9
Preface	A. L. Primavesi	11
Acknowledgements		13-15
Topography	I. M. Evans	16-22
Geology and soils	J. G. Martin	23-30
Climate	A. W. Wildig	31-39
Man and the Leicestershire flora	I. M. Evans	41-62
Leicestershire botany and botanists		
1. The period prior to 1900	M. Walpole	63-65
2. The period since 1900	I. M. Evans	65-70
3. Biographical notes	I. M. Evans	71-88
Changes in the flora of the county since 1933	P. A. Evans	89-100
Habitat studies		101-158
Plates		
Recording the flora	A. L. Primavesi	159-165
Plan of the systematic account	A. L. Primavesi	167-171
Systematic account	A. L. Primavesi	173-315
Bibliography	M. Walpole	317-324
Distribution maps M1-M90		
Gazetteer	K. G. Messenger	417-472
Notes on contributors		473-475
Index		477-486

FIGURES

1.	The location of Leicestershire in the British Isles; the relationship of the area surveyed to vice-county 55 and neighbouring vice counties	16
2.	Altitude	18
3.	Drainage	20
4.	Solid geology	24
5.	Schematic geological section through Leicestershire	27
6.	Drift geology	29
7.	Location of weather stations, with their altitudes	32
8.	Mean annual rainfall	33
9.	Mean monthly rainfall	33
10.	Rain days at Loughborough	34
11.	Monthly mean maximum and minimum temperatures	36
12.	Annual mean maximum and minimum temperatures	36
13.	Mean temperatures and ranges	37
14.	Monthly mean temperatures at Nanpantan	37
15.	Frosts and warm days at Nanpantan	38
16.	Daily hours of sunshine at Leicester	38
17.	Centres of population	42
18.	Roads	44
19.	Waterways	46
20.	Railways	50
21.	Major areas of woodland	58
22.	Losses from the flora	98-99
23.	10km. squares which lie wholly or partly in the area surveyed and lettering of the tetrads	160
24.	Areas in sq. km. of the marginal tetrads	161
25.	Number of species recorded from each tetrad (up to the end of 1979)	164
26.	Analysis of the number of species recorded from each tetrad	165

27.	Relationship of the former county to the present districts	465
28.	Parishes and former parishes in the Borough of Charnwood	466
29.	Parishes and former parishes in the Borough of Melton	467
30.	Parishes and former parishes in the eastern part of Harborough District	468
31.	Parishes and former parishes in the western part of Harborough District, the Borough of Oadby and Wigston, Blaby District	469
32.	Parishes and former parishes in the Borough of Hinckley and Bosworth	470
33.	Parishes and former parishes in North-West Leicestershire District	471
34.	Former parishes in the City of Leicester	472

The maps in this book are based upon Ordnance Survey material with the permission of the Controller of Her Majesty's Stationery Office, Crown copyright reserved.

PLATES

Frontispiece. *Rosa sherardii*, rough grassland, near Spring Wood Pool, Lount, SK 387190, July 1981. Site destroyed by opencast mining 1987. — I. M. Evans

1. A. R. Horwood, Sub-Curator, Leicester Museum. Photograph taken on the occasion of the British Association visit to Leicester, July 1907. — Leicestershire Record Office
2. F. A. Sowter, January 1971. — Leicester Mercury
3. Leicestershire Flora Committee, December 1987. Standing (left to right): K. G. Messenger, Dr. F. R. Green, Mrs J. M. Horwood, P. H. Gamble, S. H. Bishop. Seated (left to right): M. Walpole, Rev. A. L. Primavesi, Mrs P. A. Evans, I. M. Evans. — Leicestershire Museums Service
4. Herb-Paris *Paris quadrifolia* in a recently coppiced area of Great Merrible Wood, SP 836962, May 1984. — I. M. Evans
5. Large coppice stool of small-leaved lime *Tilia cordata*, Owston Big Wood, SK 794065, May 1977. — I. M. Evans
6. Bluebell *Hyacinthoides non-scripta* under oak, Spring Wood, Staunton Harold, SK 379227, May 1983. — P. A. Evans
7. Eighteen-year-old planting of larch and sycamore, Out Woods, Loughborough, SK 518158, October 1964. — I. M. Evans
8. Pasture on ridge and furrow with abundant bulbous buttercup *Ranunculus bulbosus*, Birstall, SK 585090, looking north, May 1977. — I. M. Evans
9. Glebe meadow crossed by former brook course, Barkby, SK 654089, July 1974. Ploughed up late 1974. — I. M. Evans
10. Grassland and gorse scrub on marlstone scarp, north-east of Burrough Hill, SK 763122, March 1971. — I. M. Evans
11. Siliceous grassland, habitat for hoary cinquefoil *Potentilla argentea* and subterranean clover *Trifolium subterraneum*, Croft Pasture, SP 509959, June 1975. S. H. Bishop in foreground. — I. M. Evans
12. Roadside verge after summer cut, Lowesby, SK 735082, July 1975. — I. M. Evans
13. Wide roadside verge with invading scrub, A6006 west of Eller's Farm, Wymeswold, SK 636231, looking west, June 1975. — P. A. Candlish
14. Grassland with pignut *Conopodium majus*, churchyard, Packington, SK 358144, June 1980. — I. M. Evans
15. Wet heathland dominated by purple moor-grass *Molinia caerulea*, Charnwood Lodge, SK 466157, looking south, July 1975. — I. M. Evans

16.	Bilberry *Vaccinium myrtillus* on the summit of Timberwood Hill, Charnwood Lodge, SK 470148, looking north-east towards M1, October 1975.	I. M. Evans
17.	Precambrian outcrop with bilberry *Vaccinium myrtillus* and heather *Calluna vulgaris*, High Sharpley, SK 447170, looking south-east, May 1979.	I. M. Evans
18.	Spring-fed marsh, with greater tussock-sedge *Carex paniculata* and common reed *Phragmites australis*, Tilton, SK 765040, April 1968.	I. M. Evans
19.	Marshy grassland with marsh marigold *Caltha palustris* and lady's-smock *Cardamine pratensis*, Tom Long's Meadow, Quorn, SK 557165, May 1975.	P. H. Gamble
20.	Aerial view of Narborough Bog, SP 5497, looking north-west, April 1973.	I. M. Evans
21.	Aerial view of subsidence pools in valley of Saltersford Brook, Oakthorpe and Donisthorpe, SK 3113/3213, looking north, July 1975. Note sinuous course of former Ashby Canal.	I. M. Evans
22.	Greater spearwort *Ranunculus lingua* on margin of Knaptoft Pond, SP 635877, July 1975.	I. M. Evans
23.	Golden dock *Rumex maritimus* on draw-down zone, Charley side of Blackbrook Reservoir, SK 459170, looking north-west, October 1976. Water level rising after drought summer.	P. A. Candlish
24.	Grand Union Canal, Foxton, from bridge at SP 707897, looking west, August 1965. Subsequent increase in use has reduced the aquatic flora.	I. M. Evans
25.	River Eye at Ham Bridge, Stapleford, SK 801186, August 1987.	P. H. Gamble
26.	Deep cutting on dismantled railway line, Thorpe Satchville, SK 728127, looking north, October 1972.	I. M. Evans
27.	Disused railway sidings, Swithland, SK 564131, looking south, July 1975.	P. A. Candlish
28.	Marsh and willow scrub, disused sandpit, Shawell, SP 539795, looking east, July 1970.	I. M. Evans
29.	Long-disused limestone working, habitat for small scabious *Scabiosa columbaria*, Stonesby Quarry, SK 813250, September 1974.	I. M. Evans
30.	Aerial view of arable at Bottesford in the Vale of Belvoir, SK 8037, looking north, September 1975. Grantham Canal in foreground.	I. M. Evans
31.	Common Poppy *Papaver rhoeas* and scentless mayweed *Matricaria perforata* in unsprayed margin of barley field, Ashby Parva, SP 514891, July 1980.	P. A. Evans
32.	Aerial view of clay workings on site of former Moira Reservoir, Ashby Woulds, SK 3016, looking north, April 1973.	I. M. Evans
33.	Waste ground with hoary cress *Cardaria draba*, Humberstone Lane, Leicester, SK 616083, June 1974.	I. M. Evans

FOREWORD

'County Floras' are distributional surveys of the plants occurring in a particular county. Traditionally they do not include descriptions of the plants concerned, but rely for this on various national or international Floras, which in this country are readily available. Nevertheless, the coverage of the counties of the British Isles is very uneven. Many counties have not been sufficiently worked on to merit a single Flora, whereas others are covered by several publications. Leicestershire is particularly well catered for, with three previous Floras nicely spaced at approximately 50 year intervals. The present Flora is in some senses a companion volume to K. G. Messenger's *Flora of Rutland* (1971), for together they cover the whole of the botanical vice-county 55 and there is considerable similarity in the layout of the systematic part of the text.

It is, of course, very important today that authoritative up-to-date county Floras are available for as many regions as possible. The effect of man on the landscape and vegetation has always been a major ecological factor, but is becoming ever-increasingly vital as the rate of change accelerates. There have been many important changes in the plants of such an intensively used county as Leicestershire since the last Flora (1933). Comparison of the two works will amply illustrate this point, and the major changes are summarised in the chapter by P. A. Evans. One devastating change, the virtual annihilation of the elms of Leicestershire by Dutch Elm Disease, has in fact taken place during the compilation of the present Flora. We are very fortunate that the state of play in the period 1967-1987 has been so thoroughly documented by such a highly competent and well organised team of workers.

Whereas county Floras were once rather stereotyped, nowadays they appear in a wide variety of formats dictated partly by the information available and partly by the preferences of the authors. The present *Flora of Leicestershire* seems to follow the best course between the use of recent ideas and the retention of traditional data. Hence we are provided with localities of the less common plants and an indication of the historical trends in abundance and distribution, along with a survey of characteristic habitats and suggested explanations of the distribution patterns; herbarium voucher material is also cited for all species. Moreover, there are 58 pages of detailed habitat studies and 1080 dot-maps based on 2 x 2km. squares. Taxonomically difficult groups have been treated fully, and those such as the dandelions, brambles, elms and roses have received attention from botanists who have made specialist studies of them. In addition to the usual (but essential) introductory chapters on topography, climate, soils, etc., is a most interesting and informative account of the effect of man on the Leicestershire flora by I. M. Evans, and a very helpful and complete gazetteer by K. G. Messenger. The editors, A. L. Primavesi and P. A. Evans, are to be congratulated on having woven together fact and interpretation in such a coherent and readable form.

The very full data, together with the excellent colour photographs, will rank the *Flora of Leicestershire* along with the best available county Floras. The Leicestershire Flora Committee is to be congratulated on its achievement, and the people of Leicestershire and the botanical public of Britain are indebted to their efforts so competently guided by Ian Evans and Tony Primavesi.

C. A. Stace, Ph.D., Sc.D.
Professor of Plant Taxonomy
University of Leicester

PREFACE

Prior to the publication of this present work, there have been three published Floras of Leicestershire. These were *A Flora of Leicestershire* by M. Kirby, published in 1850, *The Flora of Leicestershire* edited by F. T. Mott, published in 1886, and *The Flora of Leicestershire and Rutland* by A. R. Horwood and C. W. F. Noel, 3rd Earl of Gainsborough, published in 1933. The last of these is a monumental work with nearly 300 pages of introductory matter and 680 pages devoted to the species accounts. Though long out of print, it remains a valuable and useful work, and all of us concerned with the production of this present Flora acknowledge our indebtedness to it. However it is now over 50 years since it was published, and many changes have occurred in the county of Leicestershire and in the composition and distribution of its flora.

The Watsonian vice-county 55 includes Leicestershire and Rutland. So now does the administrative county of Leicestershire; Rutland, alas, ceased to be an administrative county in its own right in 1974. However, this Flora of Leicestershire was first mooted in 1967, and at that time work was far advanced on a *Flora of Rutland* by K. G. Messenger, which was published in 1971. This book, therefore, is concerned with the administrative county of Leicestershire as it was before the recent changes. For a complete account of the flora of the present administrative county, or of the Watsonian vice-county 55, it is necessary to consult both this work and Messenger's *Flora of Rutland*.

The original impetus for this Flora arose in 1967 from a field-by-field land-use survey of much of the county, organised by I. M. Evans of what was at that time Leicester Museum. The suggestion for a new Flora came from A. L. Primavesi who was partaking in the scheme, but who at that time had little knowledge of local botanists or means of contact with them. The credit for the initiation of the new Flora project should therefore go to I. M. Evans, whose enthusiasm got the project under way. The first meeting of the Leicestershire Flora Committee was held in September 1967, and field work commenced in 1968. A large number of people have collaborated in the production of the Flora, and detailed acknowledgements will be found elsewhere in this book.

This work is dedicated to F. A. Sowter, who was Secretary and sole surviving member of the previous Flora Committee, and who gave us the benefit of his advice and active encouragement until his death. This also seems a fitting place in which to pay special tribute to E. K. Horwood, son of the author of the previous Flora, who was an active field worker and one of the mainstays of our Committee, but who sadly did not live to see the publication of this work. Biographical notes on both these renowned Leicestershire botanists will be found elsewhere in this book.

It is a well-worn cliché that a local Flora is already out of date at the time of its publication. Fortunately there is no lack of enthusiastic field botanists in Leicestershire; it is hoped that with the help of the Leicestershire Museums Service and various flourishing local societies, records will continue to be kept.

ACKNOWLEDGEMENTS

The members of **Leicestershire Flora Committee** were:

        I. M. Evans (Chairman)
        Rev. A. L. Primavesi (Hon. Secretary)
        M. Walpole (Hon. Treasurer)
        S. H. Bishop
        Mrs P. A. Evans (formerly P. A. Candlish)
        P. H. Gamble
        Dr F. R. Green (left the county 1970)
        E. K. Horwood (deceased 1977)
        Mrs J. M. Horwood
        K. G. Messenger

The Committee was formed in 1967 and met at regular intervals to decide the methods to be employed in the field survey, to discuss progress and any problems which arose, and to decide the content and arrangement of the Flora. Rev. A. L. Primavesi, in his capacity as Hon. Secretary, was responsible for coordinating the efforts of the field workers and transferring the records to a card index as they came in.

The printing of the field and species record cards was made possible by **grants** from the Leicester Museums, the Carnegie United Kingdom Trust and the Leicester Literary and Philosophical Society. **Facilities** for duplication of Instructions to Field Workers, Committee meeting agendas and minutes and the Secretary's circulars were provided free of charge by Ratcliffe College. It is also fitting to acknowledge the very considerable financial burden cheerfully assumed by the field workers and Committee members in petrol, postage and other matters.

The following were responsible for **recording** in large areas of the county assigned to them: S. H. Bishop, O. H. Black, I. M. Evans, Mrs P. A. Evans (formerly P. A. Candlish), P. H. Gamble, H. Handley, Mrs E. Hesselgreaves, Mrs D. M. Hodgkin, E. K. Horwood, Mrs J. M. Horwood, K. G. Messenger, S. D. Musgrove, Rev. A. L. Primavesi.

The following provided records for smaller areas, or assisted the above in recording: N. E. Baker, Miss A. L. Biggs, Miss M. E. Biggs, Dr A. D. Brown, Mrs M. J. Boobyer, Mrs A. Clark, J. E. Dobson, Mrs M. Fishenden, Miss S. M. Fowler, Rev. G. G. Graham, Miss P. A. Henley, Rev. J. M. Hill, Mrs M. D. G. Jones, W. Lemmon, Miss H. Lucking, Mrs N. Maxwell, Dr P. F. Parker, P. C. Powell, M. E. Smith, J. M. Stanley, P. T. Wilkinson.

The following members of the Botanical Society of the British Isles attended one or both of the Society's field meetings held for recording in Leicestershire, together with Leicestershire members acknowledged above: Lady Anne Brewis, Dr and Mrs J. G. Dony, J. M. Fielding, K. Fry, Miss H. Gibbons, Miss R. R. Hadden, H. J. Killick, Miss A. Lewis, Mrs D. K. Lucking, Miss D. Maxey, Miss H. M. Proctor, C. R. Warren, J. R. I. Wood.

Many people provided one or more individual records. The majority of these are acknowledged in the Systematic Account, but unfortunately a few names and records were lost as a result of the fire at Ratcliffe College in 1979.

The field workers could not have achieved the thorough coverage of the areas assigned to them without the generous cooperation of the majority of **land owners**, too numerous to mention individually, who besides allowing access to their property, also frequently gave useful advice and information. A special word of thanks is due to the Divisional Manager of British Rail who kindly provided the field workers with walking permits for recording on the working railways in the county.

The voluntary service provided by the **referees** for the species treated as critical during the eleven years of intensive field work is greatly appreciated. During this time they each determined or confirmed many hundreds of specimens. The names of the referees and the species for which they were responsible are given in the list of critical species on page 162. We are also grateful to A. O. Chater and E. Clements for determination of aliens and casuals.

The following also gave valuable **specialist advice** and encouragement: C. E. A. Andrews, Dr J. E. Dandy, Dr J. G. Dony, E. S. Edees, Rev. G. G. Graham, Dr R. M. Harley, R. D. Meikle, Dr R. Melville, A. Newton, Dr F. H. Perring, Dr A. J. Richards, Dr R. H. Richens, P. D. Sell, F. A. Sowter, Professor C. A. Stace, Professor T. G. Tutin, J. R. I. Wood, Dr P. F. Yeo. Further acknowledgement of specialist services given by many of these will be found in the appropriate places elsewhere in this book.

The **habitat studies** were made in the later stages of the field work programme. The major part of the field work was undertaken by S. H. Bishop, P. H. Gamble, H. Handley and Rev. A. L. Primavesi. These and others who contributed are acknowledged at the head of each of the habitat studies. More studies were made than it was possible to include in the Flora. The ones which were not included have been deposited with Leicestershire Museum Service, together with all other archive material relating to the Flora, for consultation if required. Compilation of the studies selected and the introductory notes on the habitat types were the work of the Editors.

Thanks are due to those field workers who provided **voucher specimens** for the Flora. Mrs D. V. Kolaczec has labelled and accessioned all the material received, latterly on computer. The herbarium material was mounted for the Museums Service by a team including Mrs E. Hesselgreaves, Mrs M. Lewis, Mrs E. Loosmore, Mrs B. M. Nixon, Mrs S. Scott and the Misses K. and E. White.

Authorship of the introductory and other sections is acknowledged on the contents page. Where no such acknowledgement is made, the Editors were responsible for compilation of the text. The distribution maps were compiled progressively during the field work period by Rev. A. L. Primavesi. The art work for the final versions of these maps, which involved the accurate placing of over 150,000 dots, was carried out by Miss J. Leonard. The index was compiled by the Editors.

The Secretary is particularly grateful to Miss S. M. Fowler for **voluntary assistance** in typing all the stencils for the Instructions to Field Workers and the minutes of Committee meetings, and for dealing with his correspondence and other matters at times when he was away.

Production of the Flora has been **sponsored and financed** by the Leicestershire Museums, Art Galleries and Records Service. Its completion has only been possible with the active encouragement and co-operation of the Libraries and Museums Committee of the Leicestershire County Council, the Director of the Museums Service, Dr P. J. Boylan, and many members of his staff.

Publication of the Flora has been the responsibility of I. M. Evans, Assistant Director (Natural Sciences) of the Museums Service. He would like to express his particular thanks to Mrs J. H. Marvin, who has borne the brunt of his clerical work, together with the word-processing of the gazetteer, bibliography and other contributions to the text. To Mrs J. Smith of Oldham and Manton Ltd. must go the credit for skilful typesetting of several million characters of final typescript. Grateful thanks are due to Mr N. A. Arden of Oldham and Manton Ltd. for his long-sustained interest in and supervision of the typesetting, printing and binding and to Mr R. T. Bird of Communitype for specialist assistance. Finally, we should like to thank Mr A. Birdsall of Haley Sharpe Associates for his considerable contribution to the graphics, assembly and overall design of the Flora.

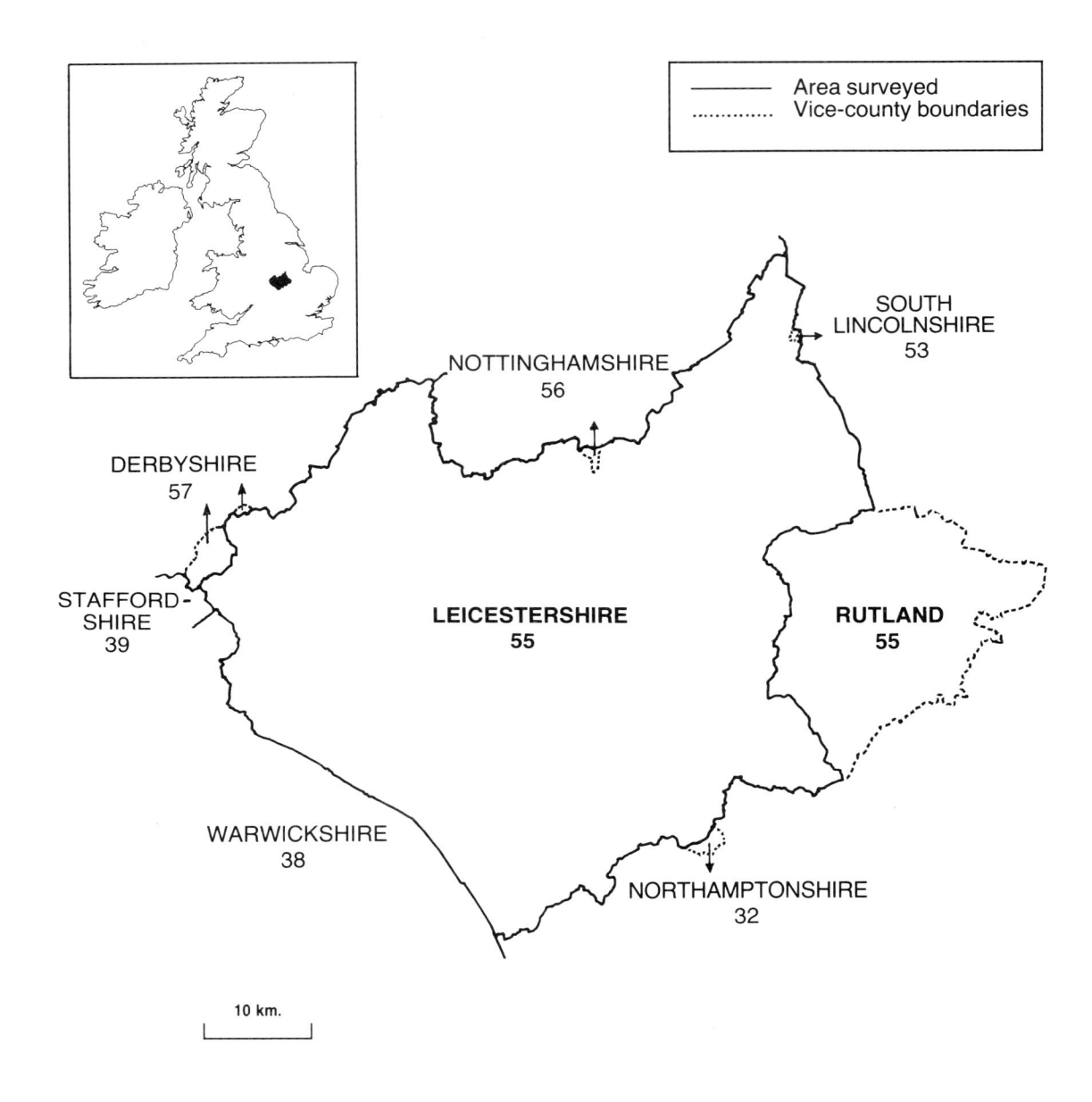

Figure 1. The location of Leicestershire in the British Isles; the relationship of the area surveyed to vice-county 55 and neighbouring vice-counties. Minor boundary changes along the A5 and elsewhere are not shown.

TOPOGRAPHY

Leicestershire is a midland county, situated nearly in the centre of England. It has an area of 2159 square kilometres (834 square miles) or, more precisely, 215928 hectares (533580 acres). Compact in shape, the greatest distance between its boundaries is 72km. (44 miles) north-south and 64km. (39 miles) east-west. Counties or former counties that share a boundary with it are Nottinghamshire, Lincolnshire, Rutland (since 1974 part of the administrative county of Leicestershire), Northamptonshire, Warwickshire, Staffordshire and Derbyshire (figure 1). The county boundary is some 257km. (160 miles) in length, 113km. (70 miles) of which is defined by rivers and 64km. (40 miles) by roads.

Most of Leicestershire lies at altitudes of between 61m. and 183m. (200 and 600ft) above sea level (figure 2). The highest point in the county is the summit of Bardon Hill in Charnwood Forest at 278m. (912ft), and the lowest is at the confluence of the Rivers Soar and Trent at about 27m. (90 ft).

The watersheds of three major drainage systems meet near Knaptoft in south Leicestershire. Most of the county is drained to the north and eventually into the Humber by the River Trent. A small area in the north-east and a larger area in the south-east are drained to the east and into The Wash by the Rivers Witham and Welland respectively and an area in the south-west to the west and into the Bristol Channel by the River Avon (figure 3).

The Trent itself forms the county boundary for a few miles in the vicinity of Castle Donington. Its flood plain broadens eastwards to meet that of the River Soar, its major Leicestershire tributary and the largest river in the county. The Soar originates by the confluence of several small streams near Sharnford on the south-western boundary of the county and flows northwards through Leicester and Loughborough to join the Trent. The valley of the Soar is only really distinguishable from those of its numerous tributaries north of Leicester, where it broadens out. Here there are extensive gravel deposits which have been worked on a large scale, leaving in many places water-filled pits and transforming the landscape. Interspersed with the pits there are still however expanses of flood plain meadows which owe their continued existence to occasional severe floods, when water may extend over an area 2km. (1 mile) wide and 24km. (15 miles) long, from Leicester to the Trent.

The course of the Soar runs parallel to a significant geological boundary just to its east and together the two conveniently divide Leicestershire into western and eastern halves, which differ both in their scenery and botanical interest. Botanically, if not topographically, the western half of the county is dominated by Charnwood Forest, an area of high ground some 117 square km. (45 square miles) in extent, which lies north-west of Leicester and south of Loughborough. The Forest consists of the eroded summits of a Precambrian mountain range. This was originally created by volcanic action and has since been buried and then partially re-exposed, now manifesting itself as hundreds of separate crags. A substantial area lies above the 213m. (700ft) contour and the highest point is at Bardon Hill overlooking Coalville.

Swift-flowing streams drain the Forest north and east into the Soar and south-west into the western River Sence. Notable amongst the tributaries of the Soar are the Grace Dieu and Black Brooks on the northern flank, the Ulverscroft Brook or River Lin rising in the heart of the Forest near Copt Oak and leaving it at Cropston, and the Rothley Brook which drains much of the southern flank of the Forest and the main course of which marks its southern boundary. Groby

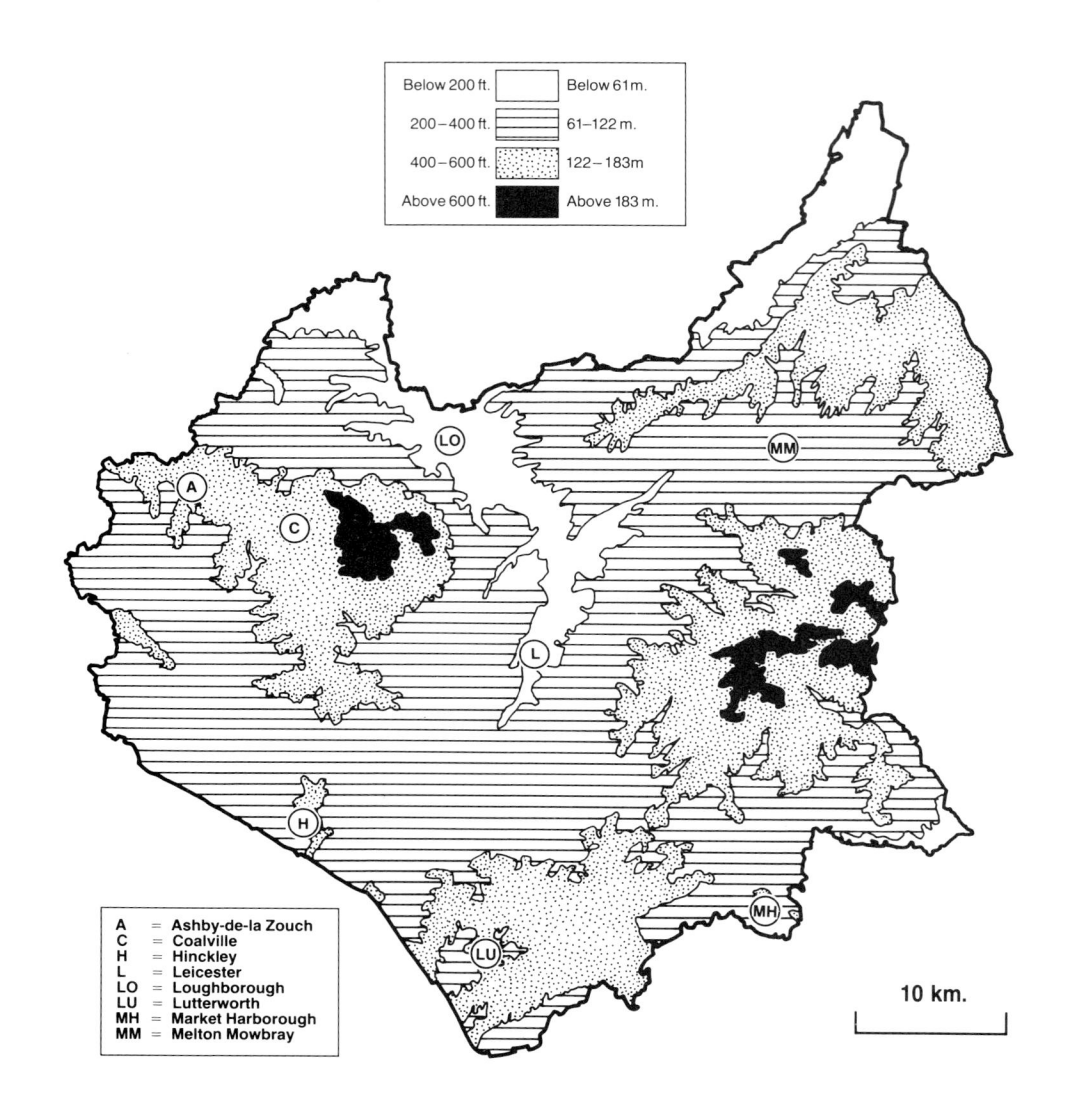

Figure 2. Altitude.

Pool, fed by a small tributary of the Rothley Brook, is probably the only natural lake in the county, though dammed and extended for use as a fishpond, presumably in medieval times. On the north-eastern edge of the Forest, separated from it only by a shallow valley, is an outcrop of hard rock post-dating the Precambrian which gives rise to the twin eminences of Buddon Wood and Broad Hill, Mountsorrel.

Charnwood Forest owes its character, unique at least in a local context, to a combination of geological, topographical and historical factors. For a detailed account of how these have interacted to produce the modern landscape the reader is referred to the Loughborough Naturalists' Club's recent and well illustrated publication (Crocker 1981) but a brief resumé may not be out of place here. Impressions of the area are dominated by craggy outcrops on the skyline, as at Beacon Hill, High Sharpley or Warren Hills, stone walls and alder-fringed watercourses in the valley bottoms, still flanked in places by stretches of marshy meadow. In much of the Forest the shallow slopes of the valley sides are subject to the same agricultural pressures as elsewhere in the county and pasture has been ploughed to give stony arable land. In places however a landscape survives, more or less intact, which predates the early 19th century enclosure of the last 16000 acres of 'unimproved' land on the Forest. The best example is the ancient deerpark at Bradgate but substantial areas of moorland vegetation, dominated by heather, bilberry or bracken, are also to be found for example at Beacon Hill and on the Charnwood Lodge estate, the last now a nature reserve. The woodlands of the Forest have been heavily exploited at various times in the past and many of those which give the area a well-wooded look in places are in fact 19th or 20th century plantings. Others, notably Swithland Wood, are however ancient woodland whose flora testifies to the continuity of this form of land use since prehistoric times (albeit somewhat modified in the case of Swithland Wood by slate quarrying). The scenery of the area is also enhanced, as is its botanical interest, by the presence of four large reservoirs, one, Blackbrook, in the Forest itself and the others, Cropston, Swithland and Thornton, on its periphery.

West of Charnwood Forest, high ground at or around 152m. (500 ft) continues to the county boundary as a broad ridge on which is situated Ashby-de-la-Zouch. This ridge cuts across the Leicestershire coalfield and, although coal only outcrops over a relatively small area to the north-east and south-west of Ashby, the visual and topographical influence of its extraction at depth extends as far away as Desford 16km. (10 miles) to the south-east. The southern edge of the surface coalfield is indicated by the valley of the River Mease, whose headwaters and south-flowing tributaries, the Hooborough, Saltersford and Gilwiskaw Brooks, drain much of the area via the Mease west and north into the Trent. Along the north-eastern edge of the coalfield there are five discrete outcrops of Carboniferous Limestone, one of which, Breedon Hill, crowned by a Saxon church, is a conspicuous landmark. At Dimminsdale, just north of Staunton Harold, the limestone has been largely quarried away but the adjacent Millstone Grit here forms a striking local scarp feature.

The rest of Leicestershire west of the Soar valley is difficult to characterise in topographical terms. Both north of Charnwood Forest and south to Hinckley it is gently rolling country, nearly all lying between the 76m. (250ft) and 122m. (400ft) contours. The Ashby Canal, which traverses the area from Snarestone to Hinckley, follows the 91m. (300ft) contour and is without locks. One unifying feature however is the colour of the soil over much of the area, a particular shade of red derived from iron compounds in the underlying Keuper Marl. This same rock laps the crags of Charnwood Forest, providing most of the younger rock cover which softens the scenery of this area, and it also extends eastwards just to the far side of the Soar valley. Towards the south, around Hinckley, where the Keuper Marl is covered by extensive deposits of glacial sands, gravels and clays, even this colour distinction fails. In many parts of the west of the county the complicated mosaic of rock and soil types provides something of a challenge to the botanist attempting to interpret distribution patterns in terms of the known preferences of species elsewhere. South of Charnwood Forest the area is drained by the many tributaries of the western River Sence, which joins the River Anker on the county boundary, and by the upper reaches of the Soar.

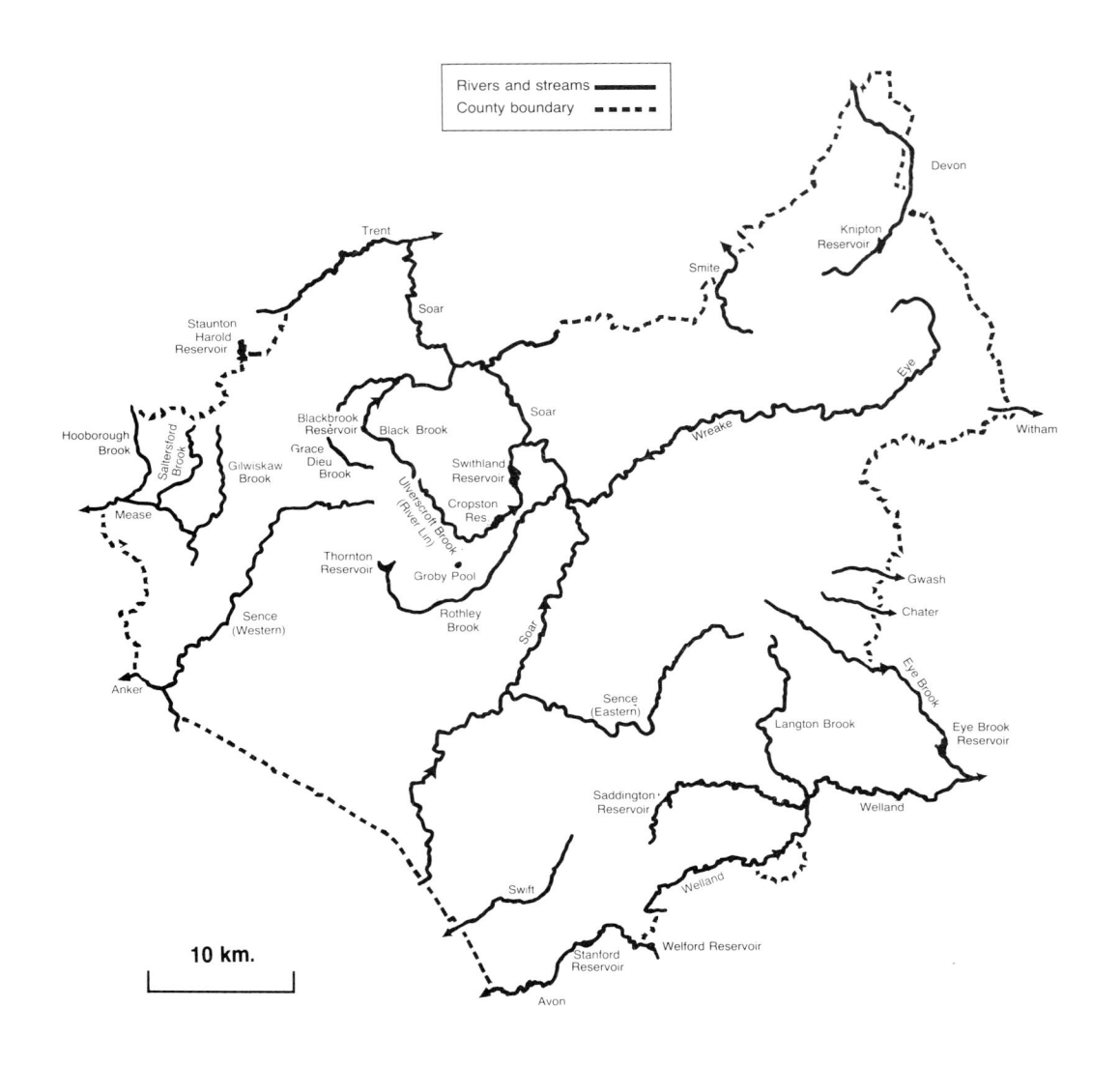

Figure 3. Drainage. Only major watercourses are shown, together with the location of Groby Pool and the reservoirs.

Along the south-eastern edge of the Keuper Marl, in a line running more or less parallel to the upper Soar valley, is an isolated series of outcrops of hard, igneous rocks. They extend in a south-westerly direction for some 11km. (7 miles) from Enderby in the north to Sapcote in the south. One at least, Croft Hill, is a local landmark, rising as it does 60m. (200ft) from the very edge of the Soar. All the outcrops have been quarried in the past and the one remaining quarry, at Croft, is large and active. However, in undisturbed areas the outcrops support a distinctive grassland community which contributes to the locally high diversity of species revealed by the present survey.

East of the Soar valley is a landscape very different from that described hitherto. Its basic form is that of a series of steps shaped by the relative hardnesses of a succession of Jurassic clays, shales, ironstones and limestones. This simple picture is, however, complicated by two factors. The first is that the whole succession dips gently to the east so that the highest ground is not always on the county boundary but some distance to the west of it. The second is that the Jurassic rocks are overlain by more recent superficial deposits of glacial origin, mainly boulder clay, sometimes of considerable thickness. In the event, the land rises, gradually at first and then in places much more suddenly, from the 61m. (200ft) contour on the eastern edge of the Soar valley to the higher ground which levels out at about 152m. (500ft) in the north-east and south of the county, but ascends to over 213m. (700ft) around Tilton, in the east. To the east of the flood plain of the Soar there is a strip some 10-15km. (6-9 miles) wide of whale-backed clay ridges separated by valleys containing westward-flowing tributaries of that river. East of Loughborough these clay ridges fan out in a half circle from a broad central plateau at Six Hills in an area known as the Wolds. The Wolds are drained to the west by tributaries of the Soar, but to the south by those of the River Wreake which joins the Soar at Cossington.

The higher ground east of the Soar valley resolves itself into three separate areas, to the north-east, east and south of Leicester. In the north-east the Wolds are continued through to the county's boundary with Lincolnshire by a ridge at about 152m. (500ft). An abrupt scarp, falling as much as 76m. (250ft) defines its northern boundary. Its steepness has defied cultivation and where it is not wooded it still carries permanent grassland of considerable botanical interest. The scarp runs, though interrupted in places, for some 19km. (12 miles) from just above Old Dalby to Belvoir Castle and it marks the edge of an ironstone, the Marlstone Rockbed, which has been extensively quarried. A spring-line at the base of this ironstone is the source of a number of streams, including the headwaters of the Rivers Smite and Devon, which drain northwards across the flat and rather featureless clay plain of the Vale of Belvoir. The scarp provides some of the most striking views in Leicestershire together with scenery of an unexpectedly intimate nature such as that at Holwell Mouth, where the Smite has cut back into the Marlstone creating a steep-sided valley with wooded sides. The more gentle southern slope of the ironstone ridge is drained by tributaries of the Wreake and of the River Eye, by which name the upper part of the Wreake is known.

The Eye rises at Bescaby, on the western side of a small area of limestone, triangular in overall shape though bisected by a band of boulder clay. This limestone stretches for about 6km. (4 miles) within Leicestershire, from Waltham-on-the-Wolds in the west to The Drift, a green lane that runs along the county's boundary with Lincolnshire, in the east. On its northern edge, as at Lings Hill (173m., 567ft) overlooking the village of Branston, the limestone forms a scarp echoing that of the Marlstone 3 km. (2 miles) to its north. The eastern part of this limestone plateau is the area formerly known as Saltby Heath. Both it and a small area to the south have the distinction of being drained eastwards to The Wash via tributaries of the River Witham, which rises near Wymondham.

The valley of the Eye and Wreake separates the higher ground north-east of Leicester from that to the east of the city. The latter, sometimes known as 'high east Leicestershire', is again delimited, though in an interrupted fashion, by a steep Marlstone scarp along its northern, western and southern edges. The scarp is particularly well developed at Burrough Hill (213m., 690ft) and again at Life Hill, just north of Billesdon (222m., 727ft), but the highest point in the eastern half of the county is in fact at Whatborough Hill (230m., 754ft), to the east of Tilton, which is capped by an ironstone, the Northampton Sands, higher up the Jurassic succession.

A springline at the base of the Marlstone scarp is, again, the source of numerous streams, which drain north, west and south-west into, respectively, the Wreake, Soar and eastern River Sence. To the east and south however 'high east Leicestershire' is drained by tributaries of the River Welland, notably the headwaters of the Rivers Chater and Gwash which flow east across Rutland, and the Eye Brook. These have in places cut steep-sided valleys in the Marlstone, for example the Chater at Launde, on the sides of which are rich spring-fed marshes. In this part of the county there are also preserved the most substantial areas of ancient woodland which it still possesses, in particular Owston Wood, Launde Big Wood, Launde Park Wood and a series down the valley of the Eye Brook stretching from Tilton Wood in the north-west to Great Merrible and Holyoaks Woods in the south-east.

The Welland valley, noted still for its fattening pastures, stretches back along the south-eastern boundary of Leicestershire for some 25km. (15 miles) from Great Easton in the east to a water-divide just to the south of Husbands Bosworth in the west. At this latter point a short distance separates the source of the Welland from the headwaters of the River Avon, which flows westwards along the county boundary and leaves Leicestershire at Catthorpe, the southernmost point of the county.

The Avon and its tributary, the River Swift, drain much of the third area of higher ground east of the Soar Valley, the 'southern uplands', which lie to the south of Leicester and extend westwards from near Market Harborough to Lutterworth. This area is separated from 'high east Leicestershire' by the valleys of the eastern River Sence and the Langton Brook, a tributary of the Welland. In places the Marlstone scarp, though much obscured by overlying glacial deposits, again stamps its mark on the landscape, as at Foxton, Gumley and the Laughton Hills. The last, rising to 171m. (560ft), form quite a striking backdrop to the course of the Welland just 2.5km. (1.5 miles) to the south but 90m. (300ft) below. The rolling hills of this area afford some fine long distance views and its highest point, at Knaptoft (177m., 581ft), is the water divide referred to earlier in this account, from which streams flow their separate ways into the Humber, Wash and Bristol Channel. Those destined for the Humber are tributaries of the eastern Sence and upper reaches of the Soar, which brings us full circle.

Reference

Crocker, J. (ed.), 1981. *Charnwood Forest: A Changing Landscape*. Loughborough: Loughborough Naturalists' Club.

GEOLOGY AND SOILS

To a botanist, geology might seem to be an almost all-embracing science, including as it does the study of rocks, fossils and minerals, their chemistry, biology and structure, landscapes of the past, earthquakes and volcanoes, dating and correlation. However, it is only necessary for a botanist to know the nature of rocks in the area and to understand the relationships and the effect of rocks on the shape of the landscape. Beyond these aspects of geology the reader is referred to the more detailed information available elsewhere.

In the simplest terms, the soil in which plants grow is the product of the decomposition of surface rock *in situ* (no movement is involved) by water, frost and sun. Minerals are selectively dissolved, rain water soaks into cracks to freeze, expand and burst off surface fragments, while the temperature difference between night and day causes contraction and expansion, which breaks the fragments still smaller. Most of the soils of Leicestershire have been developing by these processes of weathering for perhaps 250,000 years, since the area was uncovered by the ice sheets which had overwhelmed it in the last major ice advance to affect the county. However, this apparently simple process of soil development has been modified by three factors.

Firstly, during those 250,000 years it has been both considerably colder and somewhat warmer several times in what has been described as Britain's wildly fluctuating ice age climate; as a result weathering has proceeded at different rates and in different ways. Secondly, as ice crystals grow, snow melts and rivers flow, the products of weathering are moved, either to be deposited somewhere else in the county or removed altogether; so modern soil depths are not representative of 250,000 years of uninterrupted weathering. Lastly, wherever plants grow, organic material is added to the barren debris which would result from the simple weathering of rock. Plants stabilise and protect the soil on which they grow, but, contrarily, they assist in its weathering by root penetration.

Soil chemistry and structure are dependent primarily on the rock from which the soil is formed – the 'parent material'. Even in an area as small as Leicestershire, these parent materials can be as different as granite and clay.

Granite, for example the granodiorite from the Mountsorrel, Buddon Wood and Swithland Reservoir area, is a rock which crystallised from a melt of rock-forming elements such as silicon, aluminium, sodium, potassium and oxygen. The minerals feldspar, quartz and biotite mica of which, among others, the granodiorite consists, crystallised from this molten mixture as a three-dimensional interlocking mosaic with no preferred orientation and no planes of weakness. Physical weathering is therefore very slow. Equally, because of the molecular structure of these minerals, the elements potassium, sodium and calcium, all of which are present in Mountsorrel granodiorite, cannot be easily dissolved by water. The result is that from these rocks a poor soil is produced and very slowly. Only a thin veneer accumulates in rock pockets, and this, because of 'ponding' of peaty organic matter on the impervious rock and the lack of available basic chemicals, is described as 'acid'.

Chemically, soil is acid if water added to it (and thus containing soluble elements of the soil) has more hydrogen ions than hydroxyl ions. This principle is the basis of the pH scale, in which the numbers from 1 to 14 indicate the relative proportions of hydrogen and hydroxyl ions in a solution of soil. A neutral soil has a pH of 7; lower numbers indicate greater concentrations of acid ions while higher numbers represent greater concentrations of basic ions.

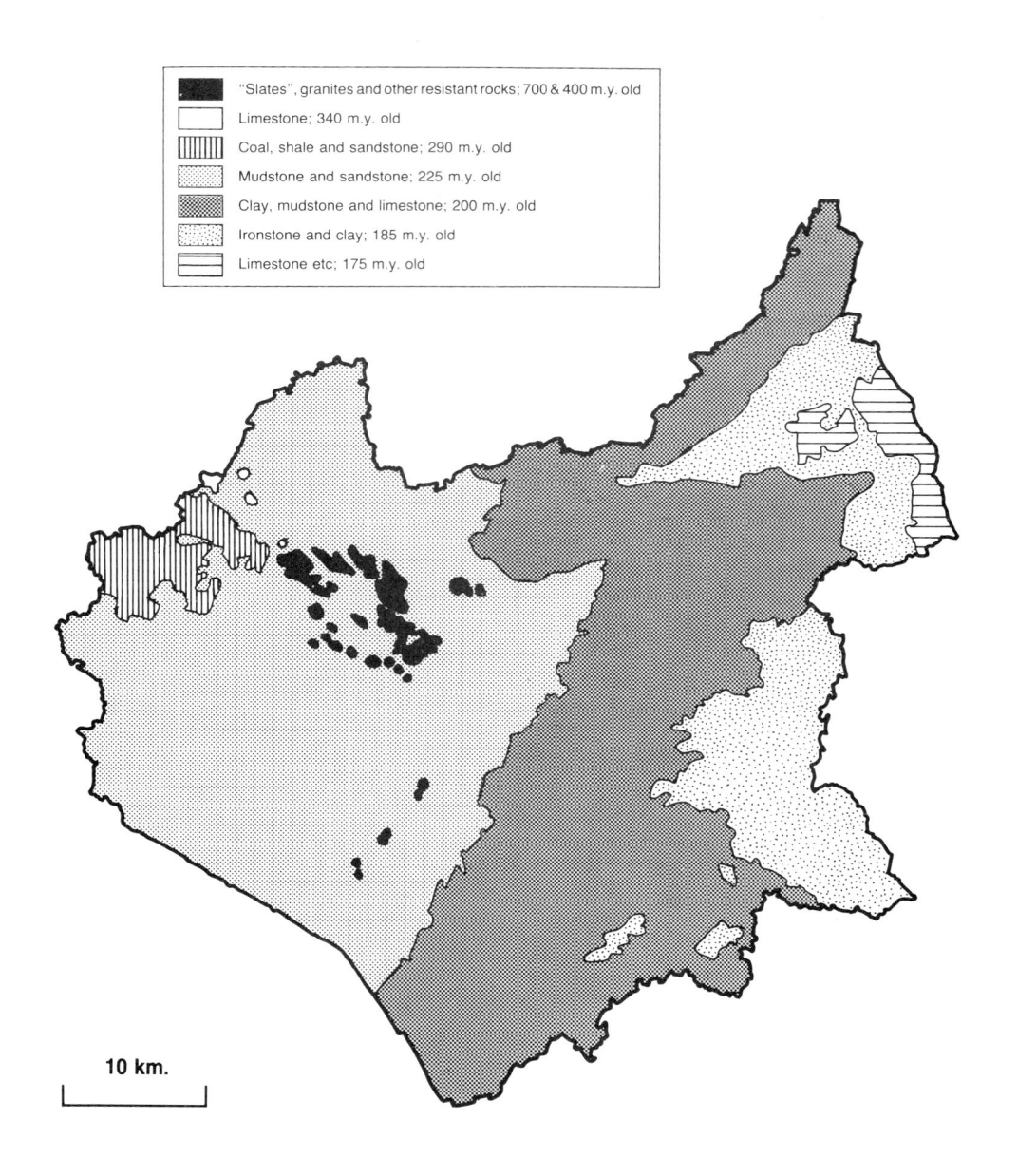

Figure 4. Solid geology. The approximate age of the rock types is given in millions of years.

In contrast to granite, lime-rich clay combines plasticity, which enables physical weathering by sun, frost and water to penetrate deeply, with the base-enriching and texture-improving properties of lime. The soils produced by this parent material are deep, on the basic side of neutral, and potentially very fertile.

As already hinted, parent material is in fact only one of five factors which share in determining the nature of the soil. The others are time, for soils evolve with time, developing new structures as layers, deepening or being eroded; climate, which affects the rates and kinds of weathering, the amount of soil water, evaporation and the deposition of soil minerals; relief, with its effect on drainage, steepness and aspect; and finally plants and animals, of which, in developed countries at least, man has the most profound influence of all on soil chemistry and structure.

It will be apparent that, despite the paramount importance of rock type to soil character and to plant distribution, and although apparently simple soil classifications can be invented, the permutations of time, climate, relief, plants, animals and man mean that soil classification and mapping at the detailed level is actually incredibly complicated. For the Leicestershire botanist a generalised impression gained from this chapter is no substitute for observation, sampling and testing in the field for the prediction or explanation of particular plant occurrences. Reference to some of the Habitat Studies will show how difficult it is to explain plant occurrences solely in terms of an overview of the geology.

In Charnwood Forest are to be found innumerable craggy outcrops of the oldest and most resistant rocks in Leicestershire. These are a complex of metamorphosed rocks, altered by heat and pressure, whose internal structure and mineral constituents show them to have been produced at the bottom of a shallow sea close to one or more volcanoes. There are volcanic ash beds, layers produced by the deposition of clouds of superheated steam and molten rock droplets, and layers shattered by earthquakes, as well as more normal sea floor muds. In addition, sheets of crystalline rock, which may represent the solidified contents of feeders to the volcanic activity, cut through and between the layers. All these rocks are Leicestershire's representatives of Precambrian time, having been laid down almost 700 million years ago. Originally laid in horizontal sheets with the oldest at the bottom, as would be expected, the ancient rocks of Charnwood Forest have since been tilted down towards the south-east, folded and eroded, so that their outcrop is a U shape with the open end pointing north-west. Each rock-type can be traced as a narrow, often interrupted, band around the two sides and the south-east end of the forest. The harder bands stand out as lines of hills topped with crags; the softer layers between, more deeply eroded, are now valleys partially filled with rocks of younger geological ages. Beacon Hill provides a perfect vantage point to see the structure of Charnwood Forest. Here too some of the layered ash beds can be seen, folded, tilted, cleaved and mineralised by intense earth movements.

The rocks of Charnwood Forest are similar in composition, hardness and in the craggy landscape they produce to those of highland Britain. The soils, rather poor, stony and acid, are also similar, and it is this geological peculiarity which gives the area the appearance of 'part of Wales set down in the Midlands'. Floristically the analogy is reinforced, with upland heath communities including heather and bilberry. Only relatively few areas in the Forest remain in this semi-natural state; the actual outcrops of Precambrian rocks are limited in area anyway, while other areas are now farmed, or suffer from the pressures of recreational use.

Scenically continuous with Charnwood Forest is the Mountsorrel–Buddon Wood–Swithland Reservoir area already mentioned as an example of a granite-like rock. The Mountsorrel igneous intrusion, as it is called, cooled deep underground more than 400 million years ago, and like Charnwood Forest has undergone a complicated history of erosion, burial and partial re-excavation. Only the top of what is actually a much more extensive massif is exposed; similar rocks are known to extend under much of east Leicestershire at ever greater depths. Resembling the Mountsorrel intrusion in character and age are a series of isolated outcrops of rocks called tonalites and diorites, some forming low hills, in a narrow band from Narborough to Sapcote.

A further similarity lies in the fact that these south-west Leicestershire intrusions are much more extensive underground; in fact the isolated outcrops seen now are merely the buried and partly re-exposed summits of a range of what were, 200 million years ago, high hills. The best known and most distinctive landscape feature is Croft Hill, with a deep modern roadstone and aggregate quarry alongside.

The summit of Croft Hill has an interesting flora growing precariously in the thin sandy soil or in crevices in the exposed rocks (habitat study 29). At the bottom of the hill is the small glebe meadow through which runs the River Soar and in which outcrops of rock occur. The flora of this meadow illustrates well how dangerous an over-simplification of either geology or botany can be. It contains species characteristic of acid grassland as well as some usually associated with more basic conditions. This unexpected mixture is illustrated in habitat study 28.

How can this small area be both acid and neutral? It is both, and the explanation is that the crystalline rocks, with their acid soil, outcrop through alluvium which has been deposited by the River Soar and which contains pockets of relatively lime-rich sediment. This is a very small-scale mosaic of different soil types and textures; soil chemistry tests within a short distance of each other have yielded pH readings of 3.9 and 7.1, and the plants respond sensitively to these changes of conditions.

Structurally similar in the way they protrude through surrounding younger rocks, but in marked contrast in their rock-type, are a series of isolated outcrops in the north-west of the county. Breedon Hill, Breedon Cloud and Dimminsdale, to name the three main ones, are outcrops of limestone of the same age as that forming the White Peak of Derbyshire. They were laid down in the Carboniferous period, about 380 million years ago. Leicestershire's Carboniferous Limestone differs from that in Derbyshire in its chemistry. Pure limestone consists of calcite (calcium carbonate), but it is possible for this to be replaced, often long after deposition and deep underground, by dolomite (magnesium carbonate). Dolomitised Carboniferous Limestone produces rather unusual well-drained basic soils. A typical area, for example the more open parts as Breedon Hill, is the subject of habitat study 22. Unfortunately much of the hill has been quarried away, as has that part of Cloud Wood which lay on the limestone. The amount of semi-natural vegetation remaining which is influenced by these soils is therefore small.

There is also in north-west Leicestershire an extensive outcrop of more familiar Carboniferous rocks, the Coal Measures. Here, in the exposed part of the Leicestershire Coalfield, the landscape, much of it the result of seven centuries of mining and quarrying, is underlain by a series of sandstones, clays and coals. Although the surface expression of the many layers of these rock-types is complicated in detail by gentle folds, the general effect is to produce either well drained or poorly drained rather acid soils. Documentary evidence and the presence of relict species indicate that the natural flora of the Coalfield was lowland heath. In places, man-made dereliction, the legacy of mining and quarrying, fortuitously creates a complex of rapidly draining hillocks and clay-lined ponds in old spoil heaps. These provide an acceptable alternative for such species, and thus the now fragmented natural habitat is augmented. This is well illustrated by habitat study 41, carried out at Moira.

All the rocks described so far have rather small isolated outcrops. Figure 4 shows them to be surrounded, like islands, by (to continue the analogy) a sea of Triassic rocks. The Precambrian rocks of Charnwood Forest, the igneous intrusions of Mountsorrel and the south-west, and the Carboniferous Limestone and Coal Measures, were buried, folded in earth movements, in some cases refolded, and exposed at the surface as complex upland areas by about 200 million years ago. Then began, first, a short period of erosion while the uplands were dissected and lowered, then a longer period of sedimentation in which they were buried deeper and deeper beneath thousands of metres of younger rocks, deposited as the Earth's crust subsided. Unlike the older rocks, the later ones were not subjected to intensive folding; instead they were gently tilted down towards the south and east. Figure 5, a simplified section across Leicestershire, shows how the dip of these

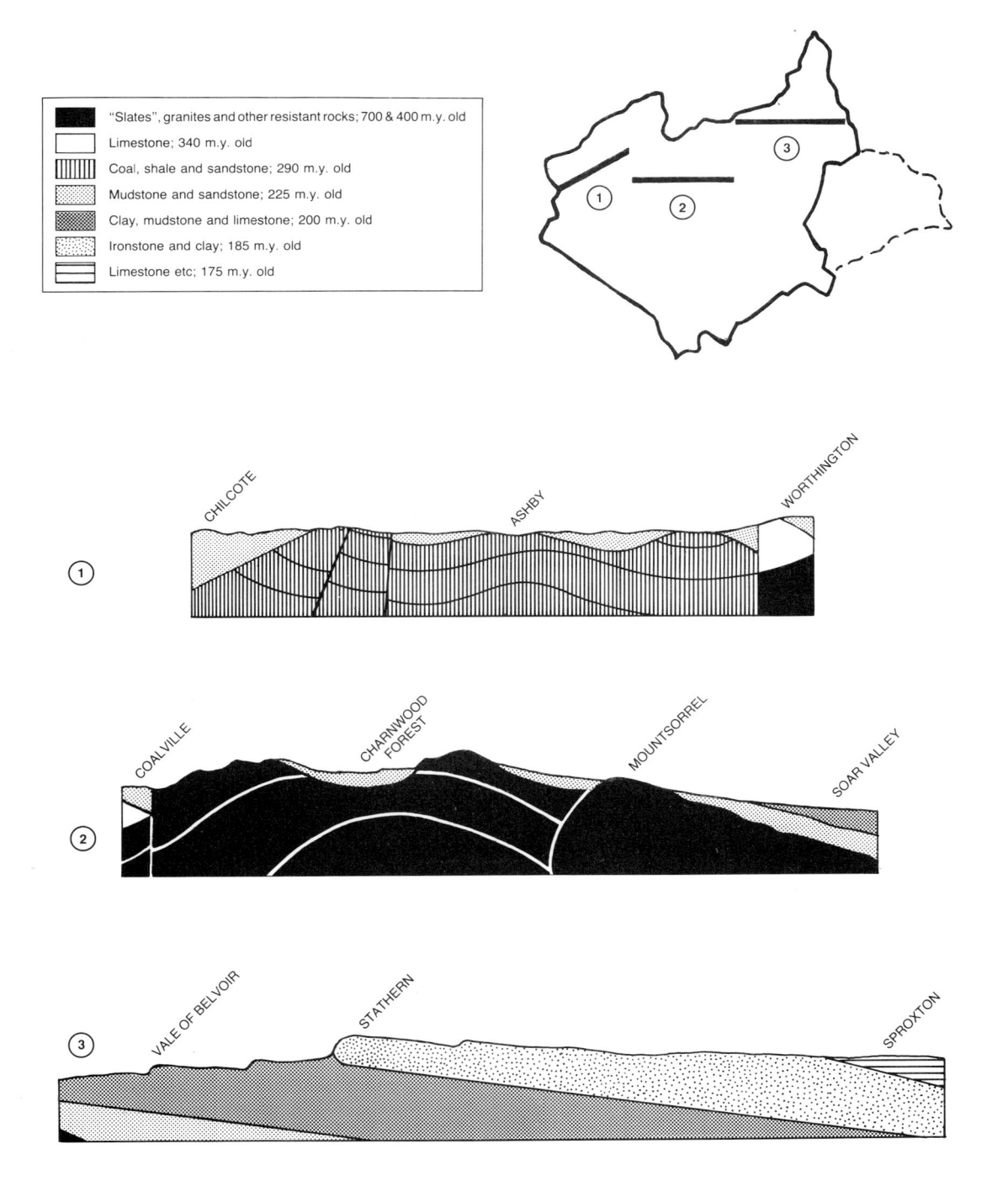

■	"Slates", granites and other resistant rocks; 700 & 400 m.y. old
☐	Limestone; 340 m.y. old
▥	Coal, shale and sandstone; 290 m.y. old
▦	Mudstone and sandstone; 225 m.y. old
▩	Clay, mudstone and limestone; 200 m.y. old
▒	Ironstone and clay; 185 m.y. old
☰	Limestone etc; 175 m.y. old

Figure 5. Schematic geological section through Leicestershire.

27

rocks beneath the modern land surface allows progressively younger layers to be exposed from west to east.

Triassic rocks, the oldest of the relatively undisturbed strata, underlie most of western Leicestershire. They also continue northward into Nottinghamshire and westward across the Midlands to the edge of the Malverns. The rocks are the so-called Bunter sandstones and pebble beds, with only a few outcrops in Leicestershire, and the Keuper 'marls', sandstones and pebble beds, which are widely distributed. The names are German because rocks of the same age and character outcrop on the other side of the North Sea, and were first described and named there. Sandstones and pebble beds within the Triassic strata, being more resistant, form higher ground or steps on hillsides, but their effect on the landscape or flora is small. Much more important are the 'marls' which give the west of the county its gentle rolling brick-red landscape. The rocks are not true marls, which should contain a significant proportion of finely disseminated lime (calcium carbonate), but very fine-grained siltstones and clays. What they do contain in significant proportions, however, are gypsum (calcium sulphate) and iron minerals. In artificial outcrops in quarries the siltstones and clays are seen to be interleaved with bands or lumps of gypsum, while the predominant brick-red colour is mottled or striped with pale green, both colours the result of iron oxide minerals. At the surface, despite these variations, the general result is deep neutral clay soil with no very distinctive flora.

The topmost layers of the Triassic system are thin limestones and clays of the Rhaetian stage (formerly called the Rhaetic). They have little effect on the landscape, forming only slightly elevated land in, for instance, the eastern suburbs of Leicester, but they separate the red Triassic marls of west Leicestershire from the Jurassic rocks which underlie almost the whole of the county east of the River Soar. The oldest major rock units in the Jurassic are the grey-brown clays which form the undulating ground rising towards 'high east Leicestershire'. The clays, the lower and middle Lias, contain thin hard bands of blue-grey limestone which stand up as broken ridges (see figure 5), and, where they outcrop, provide better drained soil than the heavy land which has developed on the Lias clays themselves. Prior to the Second World War much of the Lias clay land was under permanent grass, but with an increasingly intensive agriculture old grassland, even where it survives, is usually 'improved', and fields such as that described in habitat study 37 have become rare.

The highest parts of east Leicestershire, Harby Hills, Belvoir, Burrough Hill, Life Hill and Laughton Hills, are situated on the crest, the westernmost outcrops, of the Marlstone Rock Bed. The Marlstone, to abbreviate its name, is more resistant to erosion than the clays below, and so stands above the lower ground of central Leicestershire as an escarpment. In other places the escarpment is masked by glacial debris or simply has not been formed. The Marlstone is overlain by more clay and then by the Northampton Sand, and the three outcrop in a broad sinuous band from north to south across east Leicestershire. The Marlstone and the Northampton Sand are both ironstones, and the soils they have produced have the characteristic rust-red colour of weathered iron minerals. On level ground, the soils are clay-loams, more open than those of the Lias clays, and contain stone fragments. However, it is in the old quarries, which represent the remnants of more than 700 years of working of the Marlstone for building and ironstone, where the most interesting plant communities are found. Here, on old spoil heaps and disturbed ground, the ironstone produces a thin stony soil with a pH well below 7; however this acidity is offset by the porosity of the broken ground and cracks in the rock beneath, and the plants include species, normally associated with limestone, which are able to grow in these exceptionally well-drained conditions. Grassland on an old quarry site near Terrace Hill Farm, Eaton, the subject of habitat study 36, provides a good example.

Another less spectacular escarpment, seen best between Croxton Kerrial and Waltham on the Wolds, is the western edge of the outcrop of the Lincolnshire Limestone, which occupies the far north-east of the county and crosses the county boundary into Lincolnshire. It is a pure oolitic limestone, consisting of spherical grains (ooliths) of calcium carbonate in a loose cement of

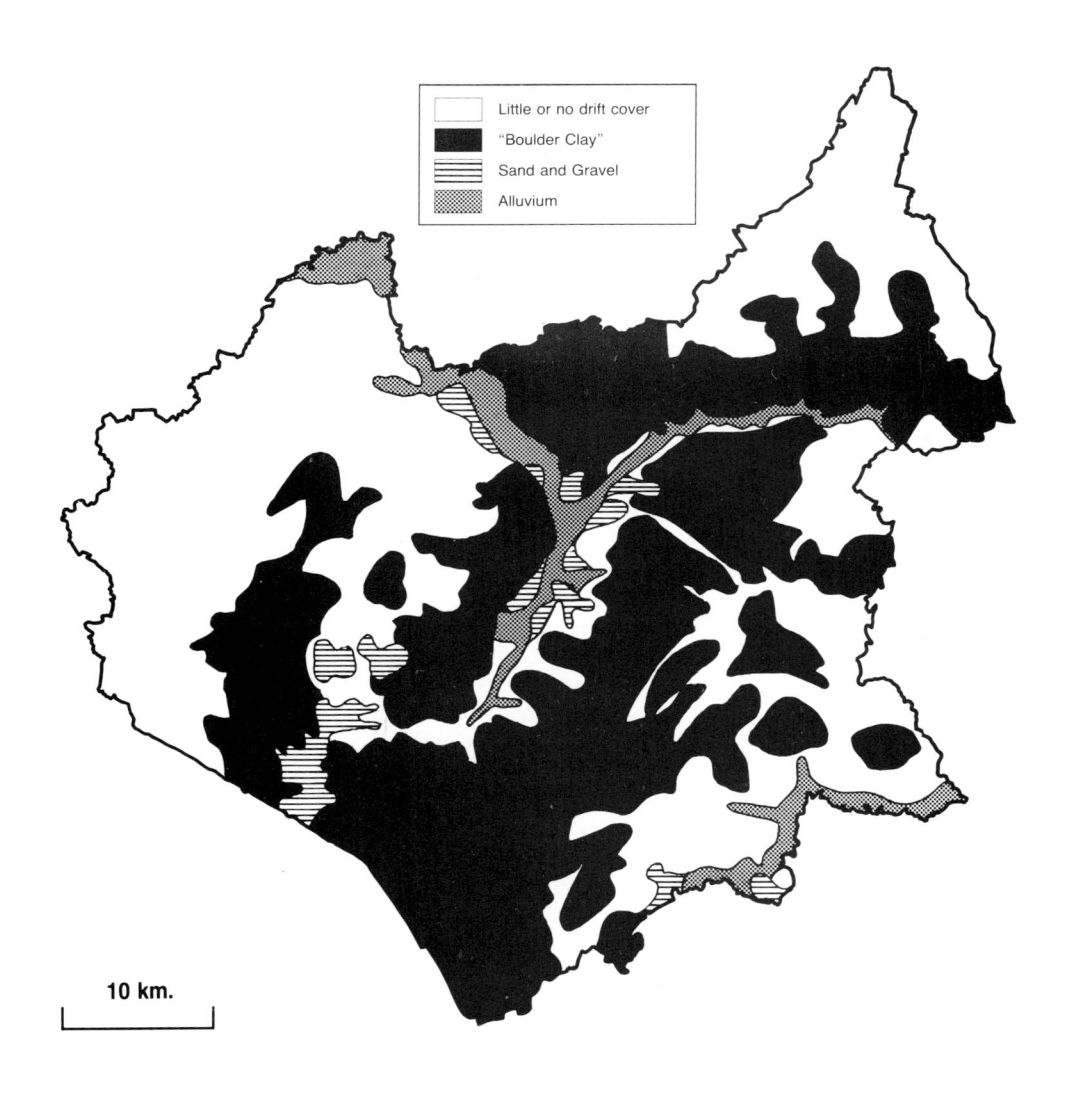

Figure 6. Drift geology. Only the major areas of sand and gravel are shown.

further calcium carbonate. The soil is thin, often less than 30 cm deep. It is calcareous, with a pH above 8, and frequently consists of little more than limestone fragments in the pale brown insoluble residue of the parent material. The Lincolnshire Limestone provides the stations for some county rarities. Examples are illustrated in habitat studies 39, 102 and 103.

Figure 4 shows the general distribution of the rock types described so far. However, the picture is complicated by the occurrence in Leicestershire of thick and extensive spreads of glacial debris, the legacy of the last ice advance 250,000 years ago (figure 6). The debris, boulder clay, is a mixture of all the rock types over which the ice advanced as it spread southward, and so in Leicestershire consists of boulders and pebbles, a few from as far away as Scotland and Scandinavia, more from Derbyshire, Lincolnshire, Yorkshire or Nottinghamshire, and most from further north in Leicestershire, all incorporated in a matrix of clay, mostly of local origin. Two boulder clays have been recognised in Leicestershire on the basis of their colour and origin: a red boulder clay derived from the north-west and consisting predominantly of transported Triassic debris, and chalky boulder clay from the north-east with a Jurassic clay basis and containing chalk as well as other 'boulders'. Because of the fairly local derivation of the greater part of the glacial deposits, the red boulder clay tends to cover the Triassic rocks of the west, while the chalky boulder clay is more or less restricted to the east. Moreover the characteristics of the soils produced are rather similar to those of Triassic marls and Jurassic clays respectively. Chalky boulder clay is, if anything, even heavier, stickier and colder than Jurassic clay, but the chalk and limestone fragments it contains often raise its pH above 7. Grassland on boulder clay is illustrated by habitat study 34. Most of east Leicestershire's oldest woods have survived on outcrops of boulder clay, for example Little Owston Wood, the subject of habitat study 18.

Also of ice age origin, but deposited by water rather than ice, are the beds of gravel which occur in Leicestershire in two situations. One group, those associated intimately with boulder clay, have a ubiquitous but patchy distribution over much of the east and south of the county. These gravels have been important in the past to the people of Leicestershire, providing settlement sites on better drained soil than the surrounding boulder clay, in a prominent situation and with springs for water supply. The thickest deposits of these gravels are found in the south. Their interlayering with boulder clay in places gives rise to some interesting plant community mosaics as in the disused gravel pits at Shawell, the subject of habitat study 100. The second group of gravels, occupying broad swathes and terraces along the Wreake and Soar valleys, have been and are being extensively worked. Where they are not restored to agriculture these valley pits provide valuable wetland sites.

Many details have been omitted from this description of the geology and soils of Leicestershire. It could never account for every subtle nuance of the flora; it should however give clues to the reasons for the main patterns of plant distribution in the county.

CLIMATE

The climate of Leicestershire is to a large extent influenced by its position in the East Midlands plain with the fenlands of Cambridgeshire and Lincolnshire to the east and the higher land of Derbyshire and the West Midlands to the north and west. To the east of Leicester there is a considerable area of land above 120m. which rises to 230m. (754ft) near Tilton, whilst to the north-west lies the high land of Charnwood Forest, somewhat smaller in area but boasting the highest point in the county, Bardon Hill with a height of 278m. (912ft). Thus the east of the county is open to chill winds from the North Sea, whilst the west is more sheltered but somewhat wetter.

Whereas the rolling country of the eastern side of the county is mainly agricultural, industry extends from the central conurbation of Leicester to the smaller towns of Coalville and Loughborough in the north-west and Hinckley and Lutterworth in the south-west. These, and to a lesser extent, other large towns such as Derby and Nottingham, not far across the county boundary, have some effect upon the local climate.

Figure 7 shows the locations of the stations in the county from which records have been taken for this section. They lie roughly on diagonals running north-east to south-west and north-west to south-east across the county. Langham, now in Leicestershire, lies just outside the old county boundary in the district of Rutland. This brief account of the Leicestershire climate can only provide an introduction to those features which may have some bearing on its vegetation. For more detailed information, on the county and the British Isles as a whole, the reader is referred to Pye (1972) and to the Meteorological Office (1952).

Rainfall

Mean annual rainfall increases from a minimum of some 560mm. (22.0 in.) in the south-east to around 700 mm. (27.6 in.) in the north-east and north-west and compares with a general average for England and Wales of 940 mm. (37.0 in.) and for the British Isles as a whole of 1100 mm. (43.3 in.). Thus Leicestershire is one of the drier counties.

Figure 8 compares the mean annual rainfall at five stations in the county, whilst figure 9 shows its distribution throughout the year at three of these. It can be seen that the pattern of rainfall is consistent and is indeed so for the two other stations omitted for reasons of clarity from figure 9. The driest period is from February to June whilst the wettest months are August and November with December and January very close behind.

It does not necessarily follow that the wettest months have the most rainy days as rainfall also depends upon the rate and duration of precipitation in any one day. Figure 10 gives the number of rain days (i.e. with measurable rainfall equal to or greater than 0.2 mm.) recorded at Loughborough Sewage Works for each month over the six-year period from 1974 to 1979 inclusive. Considering average values over the six month period from October to March inclusive, it can be seen that every month has rather more than 15 rain days and that the total for the period is 97.2. The corresponding rainfall is 367.6 mm. so that the mean rainfall per rain day is 3.78 mm. If we now look at the other six month period, April to September, still considering average values, we find that there are 68.8 rain days and that no month has more than 13; with a total mean rainfall over the period of 280.9 mm. this gives a mean rainfall per rain day of 4.08 mm. This is no doubt indicative of the heavier rate of fall of rain, albeit for shorter periods, in the summer months.

Figure 7. Location of weather stations, with their altitudes.

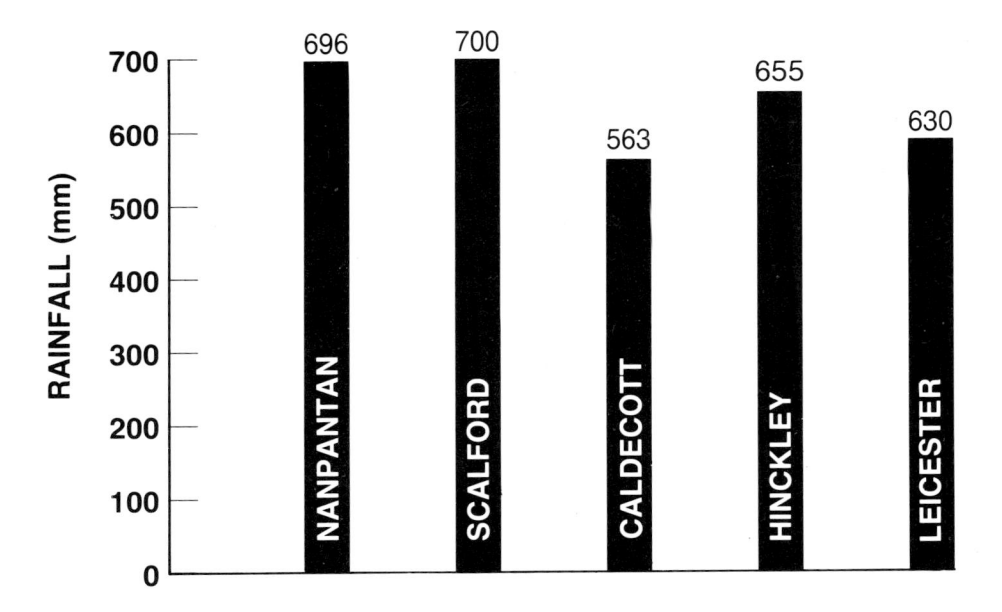

Figure 8. Mean annual rainfall.

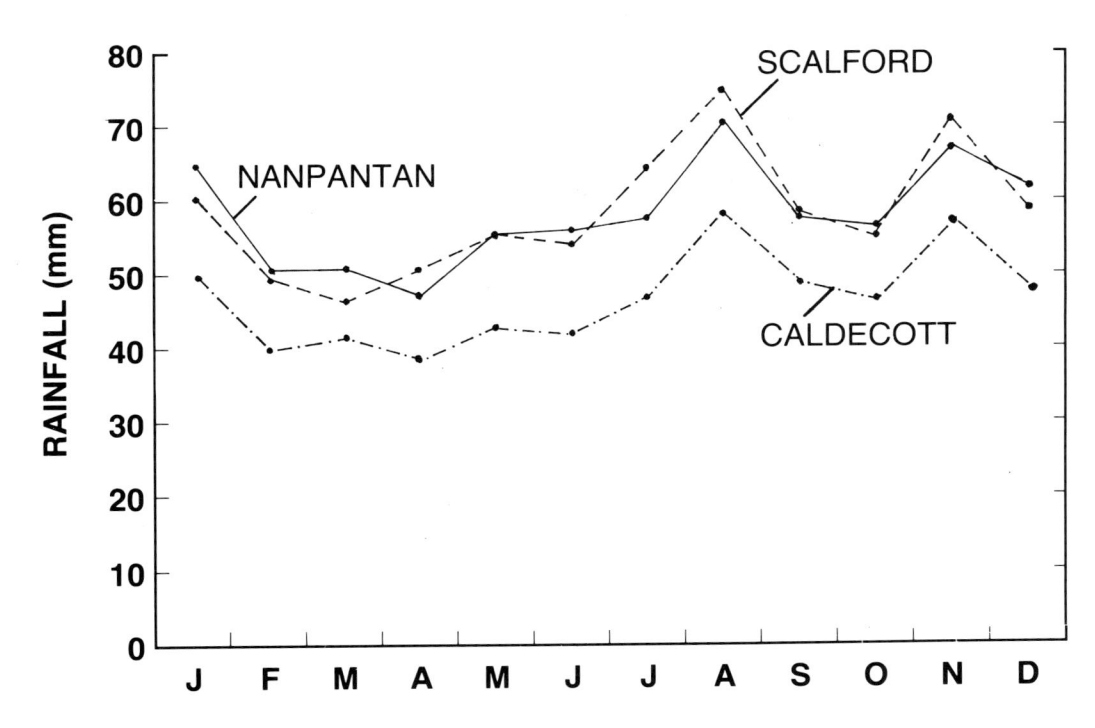

Figure 9. Mean monthly rainfall.

	1974	1975	1976	1977	1978	1979	AVERAGE	MEAN RAINFALL mm	MEAN RAINFALL PER RAIN DAY mm
JANUARY	18	23	13	17	19	20	18.3	64.1	3.50
FEBRUARY	16	7	14	24	22	13	16	58.6	3.66
MARCH	11	20	9	19	16	19	15.7	51.2	3.26
APRIL	4	20	6	15	12	15	12	34.3	2.86
MAY	13	7	17	11	9	19	12.7	49.4	3.89
JUNE	14	6	6	13	10	10	9.8	51.5	5.26
JULY	15	13	6	7	14	4	9.8	37.2	3.80
AUGUST	13	9	4	14	18	14	12	58.0	4.83
SEPTEMBER	23	14	15	8	7	8	12.5	50.5	4.04
OCTOBER	19	9	25	15	6	14	14.7	52.4	3.56
NOVEMBER	19	17	18	18	9	16	16.2	59.0	3.64
DECEMBER	15	8	17	17	24	17	16.3	82.3	5.05
TOTAL	180	153	150	178	166	169	166	648.5	47.35
AVERAGE	15	12.75	12.5	14.8	13.8	14.1	13.8	54.0	3.91

Figure 10. Rain days at Loughborough.

Rainfall is influenced by differences in elevation and this can have a marked effect locally. Readings taken over the six-year period 1974 to 1979 at Nanpantan and at Loughborough Sewage Works, a mere 4.4 km. (2.74 miles) apart but with a difference in elevation of 52.6 m. (172 ft), gave a mean difference in annual rainfall of some 45 mm. (1.78 in.), with Nanpantan (the higher) having the greater rainfall. This difference is substantiated by long term records.

Temperature

Throughout Leicestershire the warmest month is July, with August coming very close to it. The coldest month is January, except in the north-west of the county where February usually has a slightly lower mean temperature. The monthly mean maximum and minimum temperatures are compared in figure 11 for Nanpantan in the north-west and Langham in the east. Apart from the two coldest months, it will be seen that the pattern of the temperature variations is similar (as it is over the rest of the county), but both the maximum temperatures and the range of temperatures are somewhat greater towards the east of the county during the summer months. Figure 12 shows the annual mean maximum and minimum temperatures for four stations around the county; the corresponding average daily mean temperatures taken over the year (i.e. the average of the maximum and minimum temperatures) are 8.8°C, 9.7°C, 9.1°C and 9.5°C respectively, and compare with an average for the British Isles, taken as a whole, of some 9.1°C. Figure 13 shows the daily mean temperature and the mean temperature range at these four stations for the months of January and July. Cloud cover has a blanket effect, tending to reduce the variation in temperature between night and day. This largely explains the reduced range of temperatures obtaining overall in January and to some extent the uniformity of temperature throughout the region. In summer the greater frequency of clear skies and brighter weather results in much greater ranges of temperature between night and day. The smaller range in the north-west reflects the duller conditions that prevail due to the nearby industrial environment, whilst in Leicester there is also the 'heat sink' effect of the urban surroundings.

Figure 14 shows how the monthly mean temperatures at Nanpantan vary throughout the year. On the broad assumption that plant growth occurs at or above minimum mean temperatures of around 6°C, it can be seen that the growing season there is from the beginning of April until the middle of November.

Since the Second World War we have had hard winters in 1947, 1963 and 1979 and hot dry summers in 1959 and 1976. Figure 15 compares the average number of air frosts over the period from 1973 until 1978 inclusive with those in 1979 and also the average number of warm days (maximum temperatures equal to or exceeding 20°C (68°F)) over the period from 1973 to 1979 inclusive with those in the hot summer of 1976. These figures are again based upon readings taken at Nanpantan. For comparison, the average number of air frosts per annum at other stations are 63 at both Leicester and Langham and 89 at Caldecott. Unfortunately these figures are not for identical periods, but they do indicate an increased incidence of frost as one moves south-east across the county. As would be expected, most frosts occur in December, January and February, with no frosts in July and August, nor in June in most places. At Caldecott, however, during the period 1961 to 1970 frosts were recorded at least once in every month of the year.

On average, snow or sleet falls on 17 to 19 days per year but does not always settle and only lies for about 12 to 15 days, the longer period applying to the Charnwood hills. These hills, however, tend to protect the nearby lowlands of the Soar Valley from the worst extremes of the weather.

Sunshine

In spite of being a dry county, Leicestershire is one of the least sunny in England. The number of days with more than 9 hours of sunshine is on average about 40, compared with some 70 on the south coast.

The amount of sunshine does not vary greatly across the county but there is a tendency for it to be sunnier towards the south and east. Typical values for the daily mean (taken over the year) are:

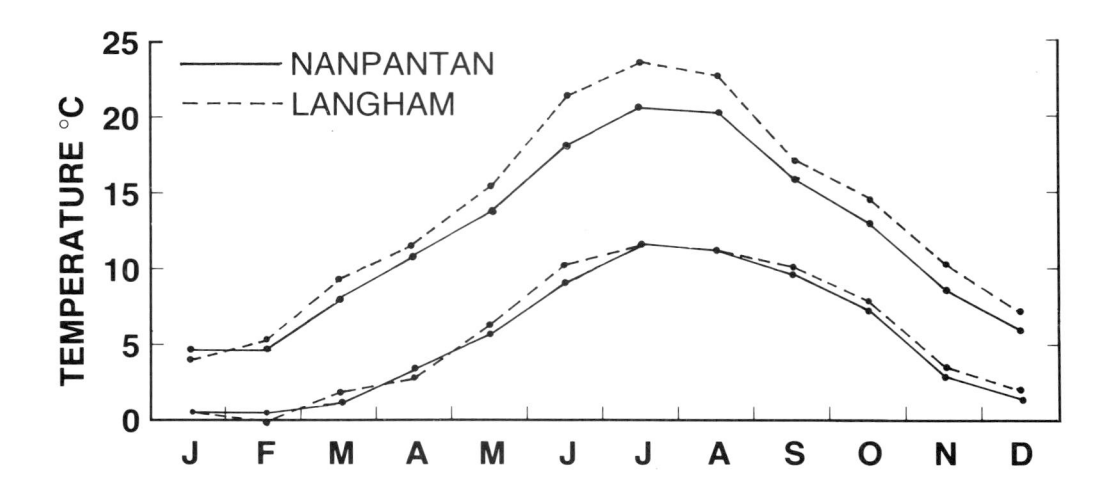

Figure 11. Monthly mean maximum and minimum temperatures.

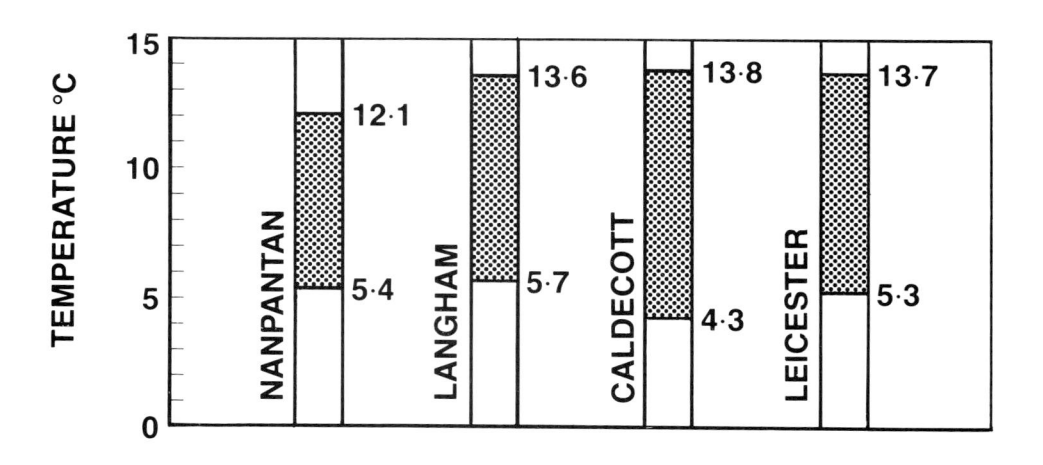

Figure 12. Annual mean maximum and minimum temperatures.

	JANUARY		JULY	
STATION	DAILY MEAN TEMP. °C	MEAN TEMP. RANGE °C	DAILY MEAN TEMP. °C	MEAN TEMP. RANGE °C
NANPANTAN	2.8	4.6	16.6	8.5
LANGHAM	2.6	4.5	17.8	11.1
CALDECOTT	2.7	6.4	15.9	11.5
LEICESTER	2.8	6.0	16.5	9.9

Figure 13.   Mean temperatures and ranges.

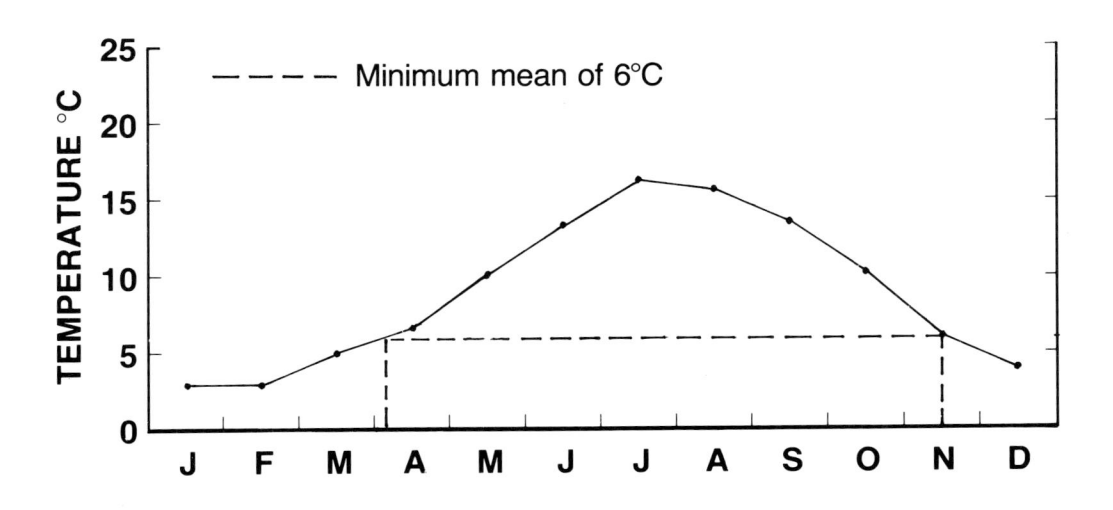

Figure 14.   Monthly mean temperatures at Nanpantan.

		JANUARY	FEBRUARY	MARCH	APRIL	MAY	JUNE	JULY	AUGUST	SEPTEMBER	OCTOBER	NOVEMBER	DECEMBER	YEAR
AIR FROSTS	AVERAGE 1973-78	10.5	12.2	10	6	0.2	—	—	—	0.2	1	7.2	11.7	59
	1979	28	27	16	5	5	—	—	—	—	1	6	11	99
DAYS WITH MAX. TEMP. $\geq 20°C$	AVERAGE 1973-79	—	—	—	0.1	2	8.1	13.9	12.5	2.9	0.3	—	—	39.8
	1976	—	—	—	—	3	17	30	21	2	—	—	—	73

Figure 15.   Frosts and warm days at Nanpantan.

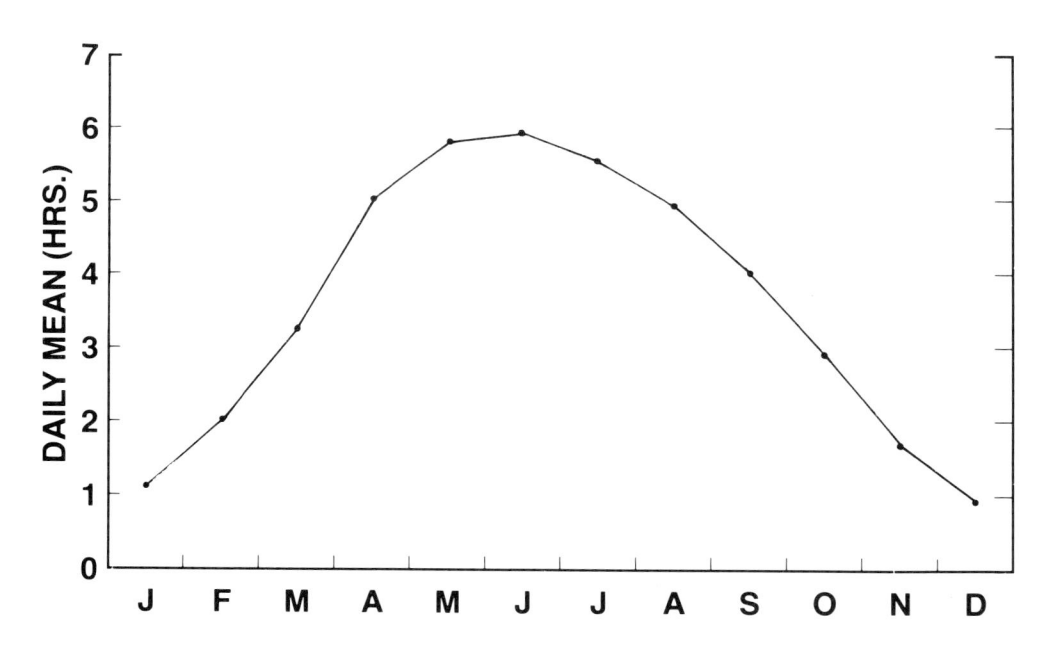

Figure 16.   Daily hours of sunshine at Leicester.

Mount St Bernard Abbey (north-west Charnwood Forest) 3.45 hours, Leicester 3.58 hours, Newtown Linford (south-east Charnwood Forest) and Caldecott 3.71 hours.

The average daily hours of sunshine throughout the year follows, approximately, the hours of daylight. This variation is shown for Leicester in figure 16 and it will be seen that the rate of increase over the spring is greater than the rate of decline during the autumn months.

Fogs and mists

The cleaner air that has resulted from more efficient use of fuels in industry, domestic heating and vehicles, and also the enactment of legislation controlling emissions has meant that fogs and smogs are no longer problems, even in towns. Throughout the county fog can be expected to occur on less than 10 occasions each year.

Mists form on the hills and along the river valleys in both autumn and early spring but usually soon disperse.

Wind

The prevailing wind is from the south-west and winds from between south-west and north-west are predominant. There is an increasing tendency for north-east winds to occur in the summer months often bringing clear skies and sunny weather. Winds from the south-east quarter are not common.

Hail and thunder

These are related to one another and are connected with the formation of cumulo-nimbus clouds. Leicestershire, especially Charnwood Forest, is an area of high incidence of thunderstorms, which can be expected on up to thirteen days per year, whilst hail occurs at something like half the frequency. Both thunder and hail are possible throughout the year but thunder is most likely to occur during the months of May to August, whilst hail can be expected during the first four months of the year.

Acknowledgements

I wish to thank, for assistance and information used in the preparation of this account of the climate of the county: Mr J. J. Eaton of the Rutland Natural History Society; the staff of the Meteorological Office at Bracknell; the staff of the Severn-Trent Water Authority at the Water Reclamation Works, Loughborough, the Leicester Water Centre, Anstey, and the Trent Area Unit, West Bridgford, Nottinghamshire.

References

Meteorological Office, 1952. *Climatological Atlas of the British Isles*. London: H.M.S.O.
Pye, N., 'Weather and climate' *in* Pye, N. (ed.), 1972. *Leicester and its Region*. Leicester: University Press, pp. 84-113.

MAN AND THE LEICESTERSHIRE FLORA

Introduction

When A. R. Horwood started in 1931 to compile the final text of the *Flora* that bears his name, he was drawing on the personal experience and notes of some 20 years fieldwork, undertaken prior to his departure from the county in 1924. The impressions conjured up by his descriptions of localities are those therefore of the Leicestershire landscape as it was at least 60 years ago. The landscape over which the survey team for the present *Flora* ranged is a vastly different one. For thoroughgoing accounts of particular aspects of its human geography the reader is referred to the appropriate chapters in Hoskins and McKinley (1955) and Pye (1972). This section describes in broad outline how the landscape of the county has come to be as it is now and in more detail the changes that have taken place during the last 60 years or so which are likely to have affected the flora. They are, almost without exception, the result of human activity.

Population

At the beginning of the nineteenth century Leicester was a relatively small town. Its population, in 1801, of 17,000 was rather less than that of Melton Mowbray today. At the same time the rest of the county mustered some 113,000 inhabitants. A century later, in 1901, the population of Leicester (albeit with enlarged boundaries) had increased over five-fold, to 95,000, whereas that of the rest of the county stood at only 174,000. Leicester has continued to be the main focus of population growth in the county throughout the twentieth century. As a result the built-up area now stretches virtually unbroken some 13km. (8 miles) from Birstall in the north to Wigston in the south and the same distance from Kirby Fields in the west to Thurnby in the east. This area, sometimes called 'Greater Leicester', contains over 400,000 people, nearly half the population of the county as a whole (figure 17).

The expansion of Leicester has been accomplished in part by the building of houses and factories on what are now known as 'green field' sites. The botanical losses occasioned by these developments are masked by the anonymity of most of the sites. However the city has also engulfed successive whorls of villages and other named areas which do feature in earlier plant lists and Floras. Here we have a better idea of what has been lost. Examples are provided by nineteenth-century records from the former civil parishes annexed in the extension of the borough's boundaries in 1892: Aylestone, Belgrave, Knighton, Leicester Abbey, New Found Pool, North Evington and West Humberstone. Since that time a further ring of once separate settlements have become to all intents and purposes suburbs of Leicester, though they do retain some sense of identity. They are Birstall, Thurmaston, Scraptoft, Thurnby, Evington, Oadby, Wigston, Glen Parva, Braunstone and Glenfield. A little farther out, separated from the built-up area often by only the narrowest strip of agricultural land, is another set of villages which have undergone substantial development in recent years. They include Blaby, Whetstone, Narborough, Enderby, Kirby Muxloe, Ratby, Groby, Anstey and Syston.

The washlands of the River Soar have been a local constraint to development both north and south of the city and the 'green wedges' which have been preserved in this way do retain some areas of botanical interest, as at Aylestone Meadows. However even these are threatened by gravel extraction or tipping and other forms of land reclamation. The only remaining large areas

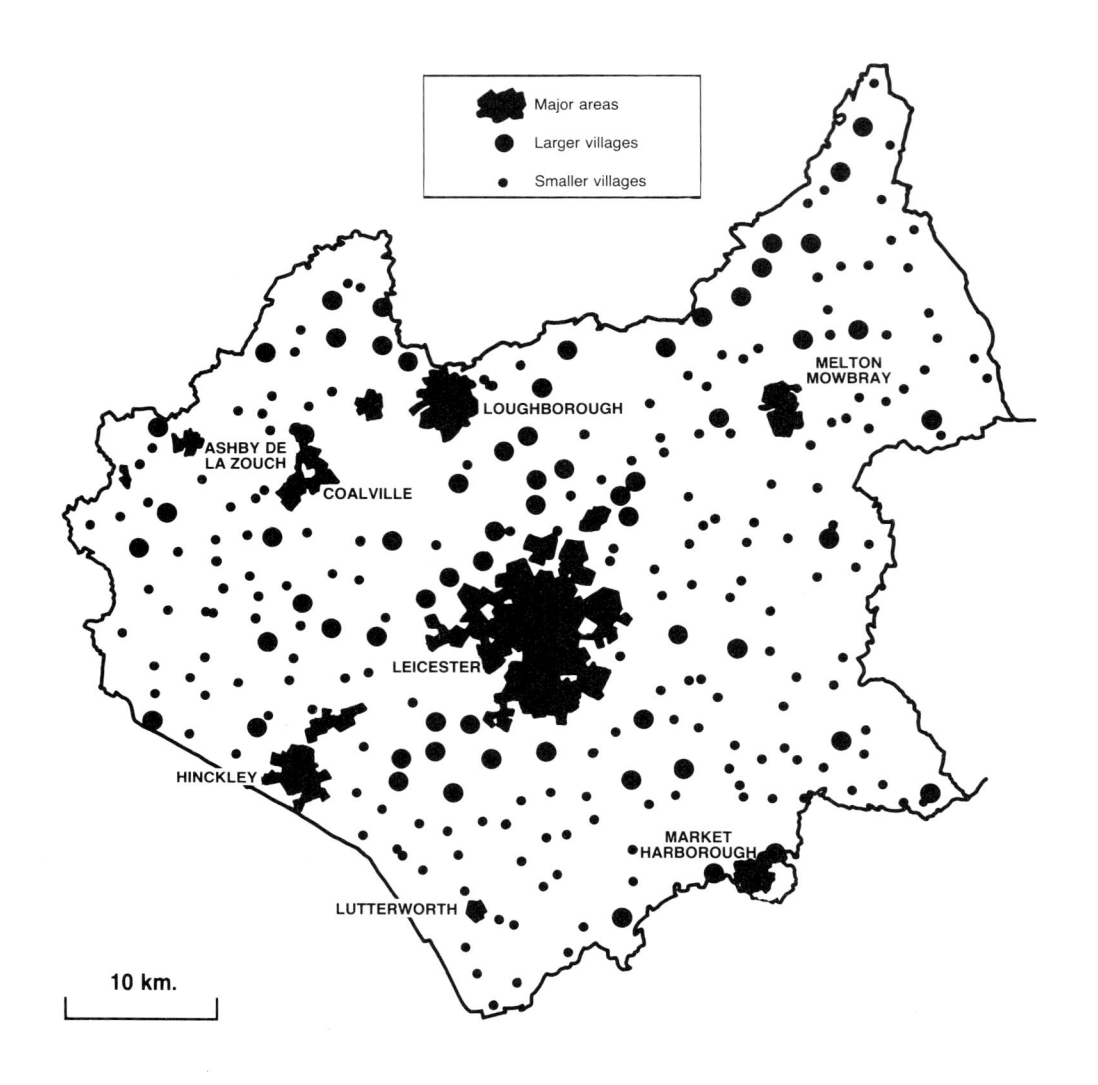

Figure 17. Centres of population.

of agricultural land within the city boundary, at Beaumont Leys and 'Hamilton' north of Humberstone are, respectively, being developed and scheduled for development.

Until well into the nineteenth century it was possible to list the major centres of population in the county, other than Leicester, without any fear of contradiction. They comprised six small market towns spaced around the points of the compass at distances of between 16km. (10 miles) and 32km. (20 miles) from Leicester, namely Loughborough, Melton Mowbray, Market Harborough, Lutterworth, Hinckley and Ashby-de-la-Zouch. Over the last century or so however the picture has become much less clear cut. Loughborough has retained its relative eminence, second only to Leicester, with a population in 1981 of about 48,000. It has absorbed part or whole of the former parishes of Hathern, Dishley, Thorpe Acre, Woodthorpe, Nanpantan and Garendon, and now covers the greater part of some 16 square kilometres (6 square miles) between the flank of Charnwood Forest and the Soar. Melton Mowbray, Market Harborough and Hinckley have all considerably increased in size, particularly the last. However Lutterworth and Ashby have been eclipsed by the expansion of areas such as Coalville, Whitwick, Shepshed, Earl Shilton and Barwell, all, it may be noted, in the western half of the county. Elsewhere there has been a deliberate policy, especially during the last three decades, of concentrating development into some 26 'key settlements', places like Castle Donington, Bottesford, Great Glen, Desford and Rothley and of limiting that of the several hundred other settlements in the county. This policy has accentuated the already high population density along the Soar valley and in parts of west Leicestershire. It has also left a broad swathe, stretching round from Lutterworth through the east of the county and into the north-east, which has largely escaped development, at least in the form of bricks and mortar, and the concomitant increase in population.

New development on the fringes of towns and villages has been accompanied by the demolition and redevelopment of areas in their centres. Such demolition sites often lie vacant for several years and rapidly acquire a diverse and characteristic flora, including many typically ruderal species such as *Sisymbrium altissimum*, *Lepidium ruderale*, *Senecio squalidus* and all four species of *Melilotus*. More difficult to explain, except in terms of dormant seed, are species such as *Atropa belladonna* and *Hyoscyamus niger* which occasionally appear on disturbed ground in built-up areas. The attention which such habitats have received from local botanists in recent years is probably a reflection not only of a generally enhanced interest in aliens, fostered for example by a series of notes in *B.S.B.I. News*, but also of the massive scale of the redevelopment which has taken place, in Leicester and elsewhere, for housing renewal, industry and various road schemes. Horwood mentions urban habitats only in passing (e.g. p.clxxxii) and it is worth noting in this context that he appears to have overlooked Bemrose's useful paper on the adventive flora of Leicester (Bemrose 1927).

Transport and communications: roads

The traditional communication and transport systems in and through Leicestershire, roads, waterways and railways, have each developed over different periods of the county's history and make different contributions to its flora.

There must have been winding tracks between settlements for as long as man has inhabited the Leicestershire landscape, at least 8,000 years. The routes taken by some of these tracks may well have become fossilized in the network of footpaths, bridleways and minor roads which now criss-cross the countryside. The earliest, however, for whose antiquity we have any definite evidence, are prehistoric trackways, which were probably mainly used by the long distance traveller and tended to be more direct and to keep, where they could, to the higher ground. An example is Sewstern Lane or The Drift which may be traced south from Harston, just below the marlstone scarp, for some 32km. (20 miles) to Stamford. The northern half of this ancient trackway is followed by the county boundary with Lincolnshire. For much of its length this stretch is unmetalled and runs across oolitic limestone. The wide green lane is one of the last

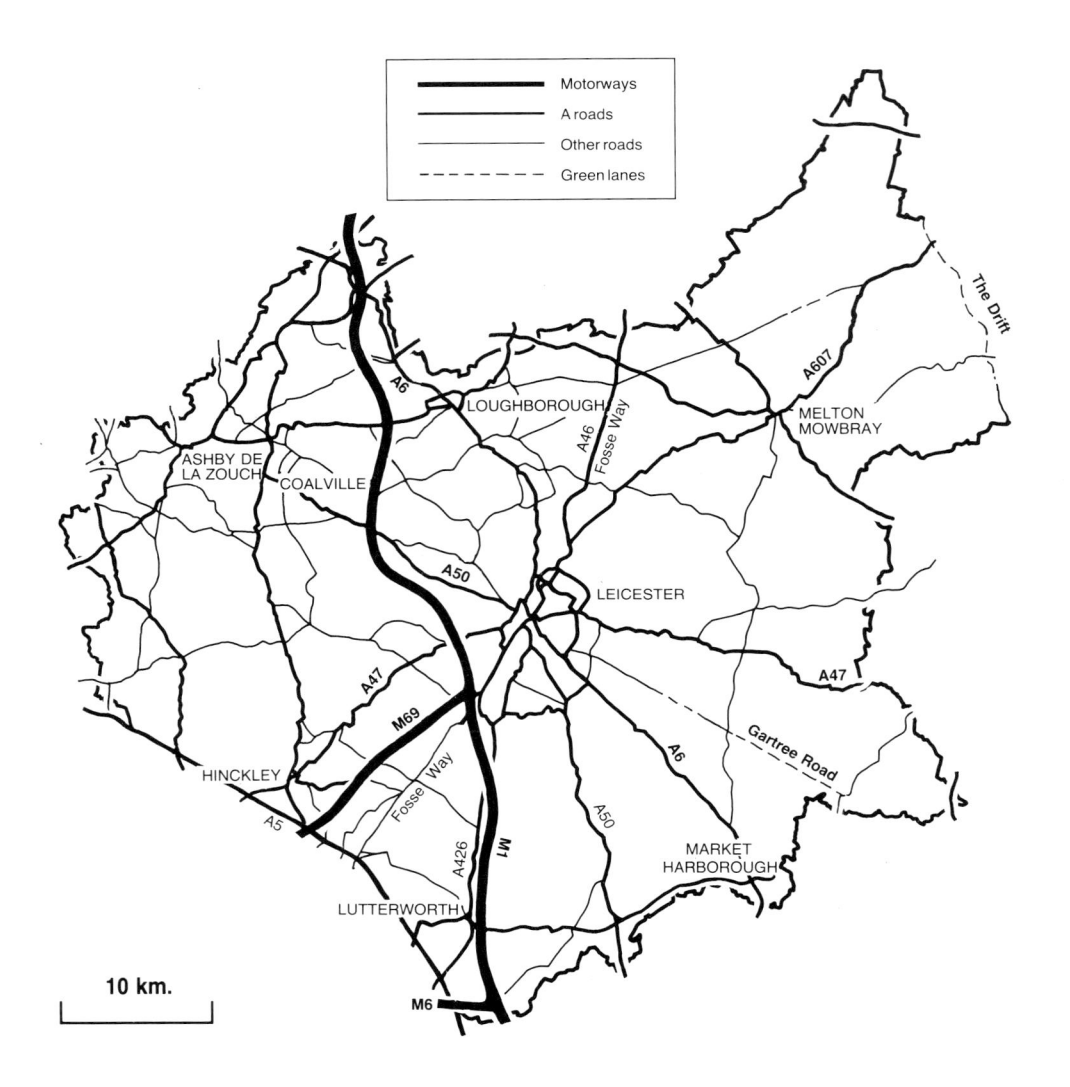

Figure 18. Roads.

refuges, in an intensively cultivated area, of the calcicole flora of the former Saltby Heath.

The Romans superimposed on any pre-existing road pattern a system which largely radiated out from Leicester. Some of their roads have flourished, at least latterly, such as those parts of the Fosse Way which now form the modern A46. Others, like the Gartree Road, which ran south-west from Leicester towards Huntingdon, have largely fallen into disuse and provide opportunities for traffic-free botanising.

During medieval times the development of the county's road network was facilitated by the construction of stone bridges at important river crossings. The shrub diversity of some roadside hedges suggests that they were established during this period. By its close the general pattern of the roads in the county was probably much as it is now, with notable exceptions in the case of large unenclosed areas such as Charnwood Forest. The widespread final enclosure of Leicestershire parishes in the post-medieval period, culminating in the Parliamentary enclosures of the eighteenth and nineteenth centuries, resulted in the realignment of many minor roads and their reconstruction to a standard width. The width seems to us generous until we remember that roads, certainly minor ones, were not in those days metalled. It is to the difficulties of winter travel therefore that we owe the extensive, if often botanically rather dull, verges of many of the straight roads in the east of the county. Contemporaneous with the later enclosures was the turnpiking of most of the county's major roads. The first turnpike trust was set up in 1726 to improve what is now the A6 (Market Harborough – Leicester – Loughborough) and by 1836 there were 288km. (180 miles) of turnpikes including what are now the A426, A46, A50 and A607 (figure 18).

The enormous increase of vehicular traffic on Leicestershire roads during the twentieth century has been accompanied by at least four developments of botanical significance. The first, which has probably been to the benefit of roadside vegetation, is the now almost universal surfacing with tar-macadam. Of the 4990km. (3119 miles) of roads in the county, other than motorways, all but 155km. (97 miles) are metalled (1981 figures). The second, more often destructive than beneficial, has been the continued and extensive widening and straightening of major roads as traffic volume and speed have increased. In places, cut-off bends have become quiet traffic-free refuges, but more often roadworks entail either the actual destruction of verges or the elimination of any long-established communities by the dumping of spoil. Such spoil may have an ephemeral botanical interest because of the occurrence of some of our more striking casuals such as *Cichorium intybus*, but these benefits are fleeting. A third development has been the mechanisation of verge maintenance, a progress from the lengthman with his scythe, through tractor-mounted cutter-bar mowers, to without doubt the most destructive, particularly when it is allowed to scalp the verge, the flail mower. However management agreements do protect the special interest of some 50 roadside verge nature reserves. The fourth development is the new dimension to road building provided since 1960 by the routing through the county of motorways such as the M1 (London – Yorkshire) and M69 (Leicester – Coventry). Their construction has undoubtedly destroyed part or whole of the botanical interest of a number of sites. Examples are the broad swathe carved through Martinshaw Wood by the M1 and the elimination of the last remnants of Potters Marston Bog by the M69. The A42 (Birmingham – Nottingham), which is due to be built through north-west Leicestershire in the second half of the nineteen-eighties, will add to the destruction. On the credit side however, the wide verges created where motorways run in cutting or on embankment offer opportunities for colonisation where seed sources are available and management practices permit, witness *Primula veris,* in quantity where the M1 passes under the A426, and *Genista tinctoria,* which has spread from a small patch between Piper and Oakley Woods to emblazon the sides of a cutting north of Shepshed.

Transport and communications: waterways

The lack of any navigable waterways was a considerably handicap to the economic development of Leicestershire up to the late eighteenth century. Between 1770 and 1820, however, the

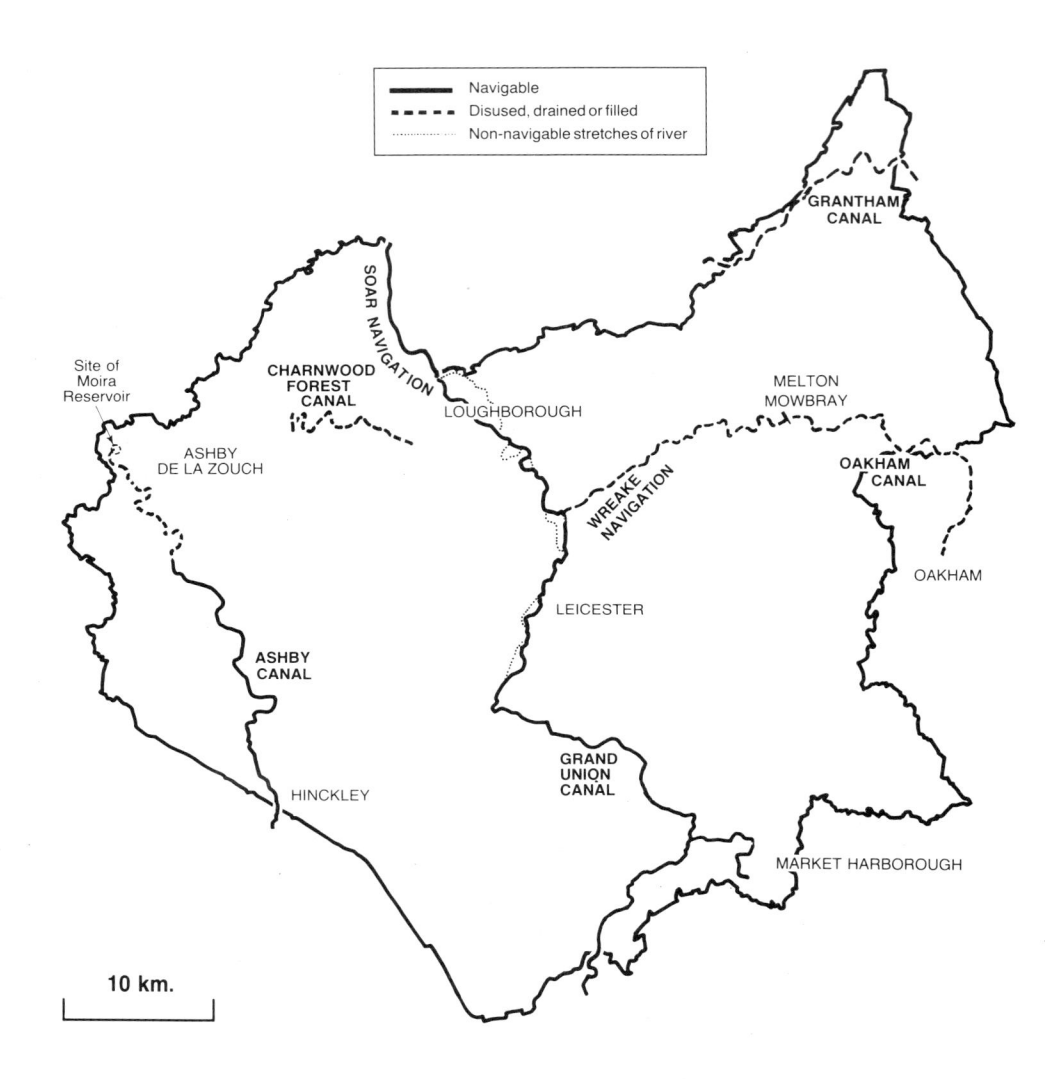

Figure 19. Waterways.

situation dramatically improved. The Soar and Wreake were made navigable up to Leicester and Melton Mowbray respectively and five canals were constructed. The varied history of the navigations and canals since 1820 is reflected in their botanical interest, past and present.

The Soar Navigation was accomplished in two phases, from the Trent to Loughborough (completed in 1778) and from Loughborough to Leicester (completed in 1797). The work entailed by-passing small loops of the former river course at Ratcliffe-on-Soar, Zouch and Barrow, and constructing a substantial length (6km., 4 miles) of new cut from Bishop Meadow, north of Loughborough, to Pillings Lock at Barrow, to serve Loughborough. South of Cossington Mill the navigation followed the first 2km. (1 mile) of the Wreake before diverging south along a new cut which rejoined the Soar at Thurmaston. The original watercourse and new cuts separate and meet again at intervals through Leicester until the final separation of the Soar and what is now the Grand Union Canal, at Aylestone (figure 19).

The Soar Navigation still carries a considerable volume of traffic and it is probably no coincidence that the stretches retaining a well-developed and diverse reedswamp and aquatic flora are mainly those which the navigation bypasses, from Barrow downstream to Stanford-on-Soar for example. However Parliament has recently authorised a massive scheme for the 'improvement' of the Soar downstream of Leicester. This will involve the reshaping of the profile of the river for much of its length. Its purpose is to improve the drainage of the adjacent land and to reduce the frequency of flooding of this land, together with roads and buildings in the flood plain. Although concessions have been made, at the request of the Nature Conservancy Council, which may help to protect areas of flood meadow and associated habitats at Lockington – Hemington, Loughborough and Quorn, the effects, direct or indirect, on the vegetation of most of the rest of the valley cannot help but be drastic.

In this context it is perhaps appropriate to chronicle the 'improvement' of most of the other rivers in the county during the last two or three decades. A recent example is that of the upper reaches of the Soar, from Sharnford downstream to Croft, which appears to have resulted in the extinction of *Oenanthe fluviatile* at its only remaining station in Leicestershire. Other rivers which have received the same destructive treatment, involving straightening, the elimination of pools and riffles, and the grading of one or both banks, are the Welland, eastern Sence, western Sence and upper reaches of the Eye.

Virtually the only stretch of a Leicestershire river of any size that has in fact escaped 'improvement' is the lower part of the Eye, from Ham Bridge downstream to the outskirts of Melton Mowbray. Although the flora includes no rarities, the diversity per unit length is not equalled by any other watercourse within 100km. (62 miles) and the importance of this stretch has been recognised by its designation as a grade 2 N.C.R. site and S.S.S.I. It is nevertheless threatened, not only by the building of a holding reservoir just below Ham Bridge, as part of a proposed improvement scheme for the Wreake which parallels that for the Soar, but also in the long term as the watercourse which would take the drainage from spoil tips at Saltby if mining of the North-east Leicestershire Coalfield goes ahead at that site.

Shortly after the Soar Navigation reached Leicester, the Wreake was opened up to navigation as far as Melton Mowbray (completed in 1797). This scheme involved no substantial lengths of artificial cut, although there were a large number of short cuts which created ox-bows out of former meanders. At the numerous mill sites, the watercourse was doubled or trebled to accommodate the potentially conflicting requirements of the mills and navigation. The Wreake Navigation went into liquidation as long ago as 1877, but a number of the ox-bows and bypass channels have retained their botanical interest over the intervening century, notably two reed-fringed pools which are virtually all that remains of an artificial ox-bow at The Wailes, Frisby-on-the-Wreake.

Following hard on the heels of the navigations, and linking up with them in many cases, was the construction of the five separate canals that Leicestershire briefly boasted. They were the Charnwood Forest Canal (completed in 1794), the Grantham Canal (completed in 1797), the

Oakham Canal (completed 1802-1804), the Ashby Canal (completed in 1804) and the Leicestershire and Northamptonshire Union Canal (later the Grand Union, completed in 1814).

The Charnwood Forest Canal, which ran some 13km. (8 miles) from Osgathorpe to Nanpantan, was the shortest-lived. It was not re-opened after the disastrous bursting of the dam of its feeder reservoir on the Black Brook in 1799 and now survives for most of its length only as an earthwork, although the flora is still of some interest where it crosses the Longcliffe Golf Course. The Oakham Canal was bought out by the Midlands Railway in 1846 and its course crossed in at least seven places by their line from Melton Mowbray to Oakham which was constructed shortly afterwards. It has since deteriorated into a succession of linear pools interspersed with dry stretches, more of the latter than the former. A stretch in Stapleford Park kept its botanical interest, though mainly for the bryologist, until it was drained in the mid nineteen-seventies, but most of the remaining stretches containing water are now solid stands of *Glyceria maxima*. The only substantial length retaining an interesting flora is between Langham and Burley, in Rutland.

The Grantham Canal, which was abandoned for navigation in 1934, is still in water for the whole of its length in Leicestershire. Until the early nineteen-sixties it had an excellent and varied reedswamp which extended right across it in many places. It was at that time dredged; the spoil was dumped on the towpath and it has since been submitted to continued vigorous management in parts, with the result that the botanical interest is now rather patchy, though still considerable in places.

Mining subsidence resulted in the draining, in 1944 and again in 1957 and 1966, of successive lengths of the northernmost 10km. (6 miles) of the Ashby Canal. Much of this stretch was subsequently filled with pulverized fuel ash. The canal now terminates just north of the Snarestone Tunnel, but is navigable from that point south to the county boundary near Hinckley and beyond. The Grand Union Canal south of Leicester is still navigable through its length, including the two branches, to Market Harborough and Welford, although the latter was only reopened to traffic in the nineteen-sixties after a period of disuse.

There has been a marked revival in the use of the navigable waterways of Leicestershire over the last two decades. The lowest point in their decline was probably reached in the early sixties when the Ashby Canal for instance, although still maintained, had lost practically all of its commercial traffic and was a happy hunting ground for the botanist and zoologist alike. Since that time there has been a considerable expansion in the use by pleasure traffic and this has resulted in the active management of the Soar Navigation, Grand Union and Ashby Canals to facilitate the passage of boats and reduce the effects of the resulting erosion of the banks. Long stretches of bank have been protected by sheet-steel piling and on one unhappy occasion in 1963 herbicides were used for 'weed control' over a considerable length of the Grand Union Canal. These and other measures have been responsible for a deterioration in the reedswamp and aquatic floras, although parts of the Ashby Canal still provide a home for species such as *Butomus umbellatus*, *Impatiens capensis* and *Juncus subnodulosus*. Another result of the increase in pleasure boating has been development of areas as 'marinas', notably the lower basin of the Foxton Barge Lift on the Grand Union Canal, where there flourished in the early nineteen-sixties a rich aquatic community dominated in places at the surface by *Hydrocharis morsus-ranae*, now, alas, gone from there.

Associated with the building of the Leicestershire canals was the construction of five feeder reservoirs. The Blackbrook Reservoir, which fed the Charnwood Forest Canal, lasted as we have seen less than a decade. Its site remained of botanical interest however until the construction of a larger drinking-water reservoir, which was completed in 1906. Swainspark Reservoir at Moira, at the head of the Ashby Canal, was famed for its flora, which included *Potentilla palustris* at its only station in the county, until it was destroyed in the course of the opencast mining of coal. The Grantham Canal has two feeder reservoirs still in use, at Knipton and Denton, the latter just over the county boundary in Lincolnshire. The Grand Union Canal also has two, one at Sulby of which only a small part is in Leicestershire and one at Saddington. All have made a local contribution to botanical diversity in a county that almost completely lacks natural lakes and at times of drawdown a specialised although usually inaccessible flora is revealed, including species such as *Apium inundatum* at Saddington and *Limosella aquatica* at Sulby.

Transport and communications: railways

The earliest Leicestershire railways were horse-drawn tramways, the first of which was laid down in 1789 over the 4.5km. (2¾ miles) between Nanpantan and Loughborough. Opened in 1794, it was intended to link the Charnwood Forest Canal with the Soar Navigation. Shortly afterwards a much longer tramway was constructed to link the limestone quarries and limeworks at Ticknall in Derbyshire and Cloud Hill to a wharf on the Ashby Canal. Much of this line was later converted to a railway of the modern type, but the course of the part that remained a tramway may be traced through Old Parks near Ashby. In one place, near where the Cloud Hill branch diverged, some small and inaccessible parcels of rough pasture resulting from its construction harbour the false oxlip *Primula veris* × *vulgaris* and its parents. Further north, in South Wood, the tramway embankment provides a refuge for woodland species that are, curiously, rather sparse in the wood itself.

The first conventional railway employing steam traction constructed in the county was the celebrated line extending 26km. (16 miles) between Leicester and Swannington, which opened in 1832. In the 67 years between the opening of this line and that of the Great Central Railway London to Nottingham route in 1899, over 400km. (250 miles) of railway were built in the county (excluding mineral lines). There were, in 1900, 90 stations and no part of the county was more than 8km. (5 miles) from any one of them. Now only seven remain (figure 20). The construction of this network involved earthmoving on a scale not previously attempted. Over many miles of line, particularly in the more hilly parts of the county, cuttings and embankments follow one another in rapid succession, interspersed with occasional tunnels and viaducts. The resultant earthworks provided new habitats, on a grand scale in places, in a landscape that was still floristically diverse. We know from contemporary photographs that cuttings on the Great Central line south of Leicester were turfed. Whether or not this practice was widespread, it is likely that the sides of cuttings and embankments soon acquired a varied grassland flora, since nearby pastures were a ready seed source. Subsequent maintenance of the lineside vegetation included, as a regular practice, deliberate burning to avoid the risk of accidental fires set off by hot coals. Perhaps because of the efficacy of this technique in controlling the spread of both the coarser grasses and woody perennials such as bramble, the railway flora of the county has both diversified over the decades and in recent years become increasingly important as a refuge for grassland species lost from adjoining pastures as they were ploughed up or 'improved'.

As a means of transport for both people and goods the railways enjoyed ascendancy for some 50 years or so. By the nineteen-twenties, however, the build-up of motor transport had begun to erode their profitability. A few lines were closed, at least to passenger traffic, in the nineteen-thirties, but further cutbacks were postponed by the advent of the Second World War. The post-war rationalisation of the railways culminated in the 'Beeching axe' of 1964, after which the redundant lines were sold off, mainly either to adjacent landowners or local authorities. Leicestershire lines closed at this time included that from Nuneaton to Ashby, the Charnwood Forest line, the Great Central line and most of those in the east of the county. The only ones that remained open for passenger traffic were the Midland line north through Market Harborough, Leicester and Loughborough, its Peterborough branch through Melton Mowbray and the Birmingham route via Hinckley. In addition, lengths of the Ashby-Nuneaton and Great Central lines have been purchased and are worked by railway preservation societies.

The 1933 *Flora* contains only passing references to the railways of the county. A. R. Horwood and his contemporaries may well have found access to working lines difficult. It should also be borne in mind that their botanical interest has, relatively speaking, greatly increased because of

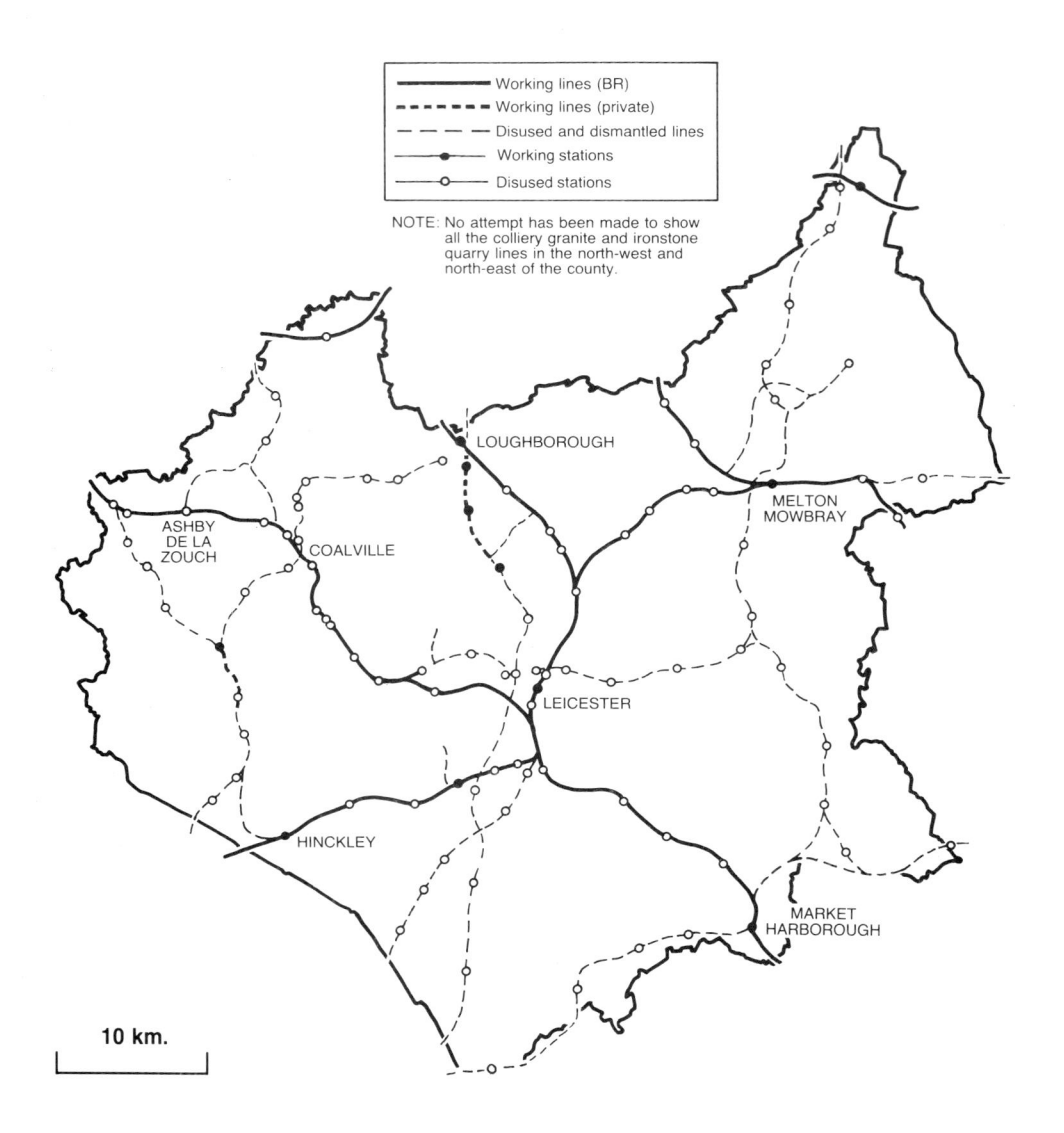

Figure 20. Railways.

the agricultural revolution of the last four decades. Survey workers for the present *Flora* have, in contrast, found both working and dismantled lines a rich source of records. The deep cutting at Red Hill, Thurmaston for example, which is on a working line, retains much of the calcicole flora listed by Horwood for the neighbourhood, including species such as *Blackstonia perfoliata*. The ballast of recently abandoned trackbeds has a characteristic assemblage of species and the distribution maps elsewhere in this volume bear out the importance of the blue brick walls of bridges and abutments for our fern flora, especially those of the Great Central line. Cuttings on dismantled lines in the east of the county have yielded impressive lists of calcicoles, particularly those at Ingarsby, Thorpe Satchville and Scalford, where construction cut deep into Jurassic clays or chalky boulder clays. In contrast, cuttings in the west of the county support a heath grassland flora including *Calluna vulgaris* and *Teucrium scorodonia*. Dismantled stations or goods yards also provide extensive lists, including many casuals, aliens and other species of open habitats.

It was a coincidence that much of the recording for this *Flora* was done in the decade following the 'Beeching axe'. However it is realised that the flora of the dismantled lines may well have reached a peak of diversity and interest in this period. With the sale of the lines and their subsequent use for a variety of purposes many have declined in interest. Cuttings, such as that south of East Norton Station, have been filled with refuse and restored to the level of the adjacent agricultural land. Others have been used for pig farming, effectively eliminating their flora. The sites of stations and goods yards have been developed for industry or housing. At many of the sites which have not otherwise been used, the growth of scrub in the absence of regular burning is gradually shading out grassland species. The end-product of this process, though extending over a much longer period, can be seen at several sites in the east of the county, where the only remaining indicators of a former grassland community are defunct anthills on bare ground under 80 year old hawthorn bushes. In the absence of any form of management, this may well be the fate of most former railway land except in the few areas, as at Shenton, where a dismantled line has been acquired and is managed as nature reserve.

Associated with the main railway lines at various periods in their history have been at least 30, usually short, branch lines serving collieries and mineral workings. The main concentrations were those serving the coalfield in north-west Leicestershire and the iron-ore deposits in the north-east of the county, but the hard rock quarries ringing Charnwood Forest and the diorite quarries between Sapcote and Narborough also had their complement. They were usually single track and in the case of the iron-ore lines the configuration was constantly changing in the immediate vicinity of the quarries so as to serve new faces. Some of the colliery lines are still in use, but most have been dismantled, either because of the phasing-out of ironstone working in the nineteen-sixties, or because of a changeover to road transport. Where the lines were level with adjacent agricultural land, they have often been incorporated into it and little or no trace of their former existence persists. Where they ran on embankment or in cuttings they may still, like the main lines, preserve an element of diversity in an increasingly uniform landscape. The single track mineral line running through Holwell village, which formerly linked most of the ironstone workings in the area with the steelworks at Asfordby, is just such an example. Half a mile was acquired as a county trust nature reserve in 1973 and it contains a rich marsh flora.

Transport and communications: airfields

A new element appeared in the local transportation picture in the second quarter of the twentieth century, the aeroplane. In 1932 (Allen and Potter) there was only one working airfield in Leicestershire, a private one near Ratcliffe-on-the-Wreake, although three more were under consideration, at Braunstone Frith, Desford and Lubenham. At the close of the Second World War in 1945 there were 16, many of them covering 2.5 square kilometres (a square mile) or so, taking into consideration associated features such as camp sites or fuel stores. The effects on the local flora when they were constructed must have been almost wholly destructive, entailing as it did the elimination of all field boundaries, spinneys and other such landscape features within the perimeter of the airfield itself. In places, on level sites, permanent herb-rich grassland survived unscathed between the runways as at the Leicester East Aerodrome at Stoughton, which is one of the three remaining operational airfields. Elsewhere, the difficulty of breaking-up large areas of reinforced concrete prevented until recently the reclamation for agriculture of disused airfields and associated camp sites, and grassland communities have survived in areas where they would otherwise have gone under the plough. An example is on the periphery of Saltby Airfield, in the north-east of the county, where concrete hut bases helped to conserve *Astragalus danicus*, together with other rare calcicoles. However even these marginal areas are now threatened, either by machines which convert the concrete into hardcore or by the hand of Nature in the form of the growth of dense scrub.

Mineral extraction

The use, other than purely casual, by man, of the mineral resources of Leicestershire dates back about 4000 years to the early Bronze Age when a well-attested 'axe factory' was operating in the Charnwood Forest area (Moore and Cummins 1974). Although the type of rock used has been identified, no evidence has yet been found on the ground of the actual site, which may in fact have been destroyed by later working. The first site-specific evidence of mineral extraction in the county is probably that of slag associated with a middle Iron Age site at Sproxton (Liddle 1982). During the Roman occupation use was made of the full range of minerals found in Leicestershire. Since that time the extractive industries have had an increasingly marked effect on the local landscape and hence its botany.

Over the centuries practically every major rock type found in Leicestershire has been quarried or mined for some purpose, be it for building material, ore, flux or fuel. Mineral extraction has, inevitably, an initially destructive effect on local plant communities, whether it is by the removal of the land surface or the location on it of works, machinery, overburden or spoil. In the past, when extraction ceased, quarry and mine sites were often abandoned without any attempt at restoration. If left undisturbed they were rapidly colonised. The constitution of the communities that developed depended on a number of factors, such as the nature of the substrate and the extent of flooding but they were often an intriguing and textbook-defying mix of local species with limited powers of distribution and opportunists like many of the orchids. Today not only are current workings restored to some use, agricultural or otherwise, as quickly as possible, but older sites are in great demand for the disposal of waste. Such has been the unparalleled expansion of the extractive industries over the last 40 years, however, that large numbers of sites remain at least temporarily unrestored, making a considerable contribution to botanical diversity.

The hard rock quarries ringing Charnwood Forest have expanded, eliminating in the process the greater part of a number of famous botanical sites. An example is Spring Hill, Whitwick, in Horwood's time a station for *Empetrum nigrum* and other moorland species. Similarly, the expansion of Bardon Hill Quarry, still continuing, has substantially reduced the area of sessile oakwood and moorland that previously clothed its slopes, though re-afforestation with conifers since the Second World War is also a contributory factor. Bradgate Quarries have eaten deep into Old Wood, Groby, and not far to the east Groby Quarry now stretches right up to the edge of Sheet Hedges Wood, threatening one of the two stations in the county for *Chrysosplenium alternifolium*. Long-disused quarry floors in Cliffe Hill Quarry supported in the nineteen-sixties a flourishing population of *Eriophorum angustifolium*, but this was destroyed by reworking. Further to the east, in the vicinity of Mountsorrel and Quorn, there have been both gains and losses. Abandoned shelves in Main and Hawcliff Quarries support an interesting flora, with *Eriophorum angustifolium* and *Ophrys apifera* close neighbours, but the re-opening and considerable extension of the quarries in Buddon Wood not only destroyed a hybrid *Dactylorhiza* population in Cocklow Wood Quarry but also the greater part of the regenerating sessile oakwood for which this site was famed.

The 20 or so diorite quarries in the south-west of the county present a varied picture both as far as activity and botanical interest are concerned. The southernmost, at Barrow Hill, Earl Shilton, retained much of their interest until recently when the rate of tipping was accelerated, although *Dianthus deltoides* must have become extinct there over 50 years ago. Those in the vicinity of Sapcote and Stoney Stanton, all disused and mostly flooded, preserve elements of a saxicolous flora, including ferns, in an otherwise botanically rather undistinguished area of the county. The quarries at Enderby and Croft have been actively worked in recent decades. Two out of the three at Enderby have been exploited to their economic limits and have since been filled and restored and the third will share this fate in the near future. Croft Quarry is still expanding towards the long-disused Huncote Quarry and the summit of Croft Hill, causing concern for the survival of characteristic open grassland flora of the area which contains an unusually high proportion of annuals such as *Scleranthus annuus*. The rate of change in the vicinity of hard rock quarries today does not leave undisturbed for any appreciable length of time the faces and shelves which provided refuges in the past. However the high totals of species recorded during the present survey from the tetrads including major quarries at Croft, Groby and Mountsorrel shows that their operation even today is not wholly destructive in its effects.

All five of the major outcrops of Carboniferous limestone in the north-west of the county have been quarried in the past. Of the three disused quarries, that at Dimminsdale retains most botanical interest, although this is associated as much with the flooding of the pits and the adjacent Millstone Grit exposure as with a relict limestone flora. The two working quarries, at Breedon and Cloud Hills, have increased substantially in size. At Breedon Hill limestone grassland on the northern and eastern slopes has been engulfed. Cloud Hill Quarry has eaten deep into Cloud Wood and stone-stocking grounds now cover the former site of *Polygonatum multiflorum*.

Coal and fireclay have been mined and quarried in north-west Leicestershire for many centuries. The scale of the operation increased in the early part of the nineteenth century when, for instance, the first pits were sunk in the Ashby Woulds area and again about 40 years ago when the opencast mining of coal commenced. Opencast mining eventually resulted in the destruction of Swainspark Reservoir at Moira and in a marked reduction of habitat diversity, not to say sterilisation, of other areas worked and since restored to agricultural use, as at Ravenstone. The sessile oakwood and acid heath communities of Willesley Wood were totally destroyed in the nineteen-seventies. A compensation however has been the recognition of the botanical interest of adjacent areas of old industrial dereliction, such as that on the site of Newfield Colliery, where dry heathland and bog communities, containing species such as *Calluna vulgaris* and *Carex rostrata* respectively, have developed over the last century or so. Elsewhere on the Leicestershire coalfield deep mining, at places such as Lount, Ellistown and Desford, has resulted in spoil tips and other 'waste ground' areas which do in time acquire a rich adventive flora. These, like the railways, appear not to have been thoroughly investigated by earlier generations of botanists but, again, the causes may be difficulty of access in the past and the increasing degree of dereliction and hence attractiveness at present, rather than any lack of inclination on their part. An interesting feature of some of these areas of mining dereliction is their rapid colonisation by orchids, usually *Dactylorhiza fuchsii,* but in one case, at Coalville, *D. majalis* subsp. *praetermissa*, which appeared in a pit only ten years old. The nearest known seed source for this now uncommon species was several miles away.

Subsidence pools or 'flashes' are a by-product of mining that has locally enhanced the flora of the coalfield. Substantial ones, like those on the Saltersford Brook between Oakthorpe and Donisthorpe, provide a habitat for aquatic and reedswamp species that is otherwise scarce in the vicinity. The species present are usually ones with good powers of colonisation such as *Sparganium erectum* and *Typha latifolia* but if the pools are allowed to remain their flora should diversify.

In 1976 the National Coal Board announced the discovery of substantial reserves of coal underneath north-east Leicestershire, which they proposed to extract through mines at

Asfordby, Hose and Saltby. A lengthy public inquiry, held over the winter of 1979-80, established that the major threats to areas of botanical interest would stem from the tipping of spoil and the effects of subsidence on the Grantham Canal.

The Triassic marls have been widely quarried for brickclay from Measham through Ibstock and Heather to Glen Parva and Gipsy Lane on the outskirts of Leicester and down the Soar valley to Sileby and Loughborough. The older workings are characterised by a sparse grassland containing species such as *Centaurium erythraea*, but the great majority of the worked-out sites have been or are being restored, often by the tipping of domestic refuse. The claypits at Red Hill, Thurmaston, listed by Horwood, have been so restored but fortunately the calcicole flora still survives in the adjacent railway cutting.

East of the Soar valley, there has been exploitation of the Jurassic clays, ironstones and limestones for, as we have seen, nearly two millennia, but again the pace of both extraction and restoration has speeded up during the last 50 years. The relict calcicole floras of the Lower Lias limestone pits at Barrow on Soar, Sileby and Wigston have probably changed little in that time, although one of the sites for *Ophrys apifera* at Kilby Bridge, Wigston, was 'tidied-up' early in 1982 with the almost certain loss of the species.

The quarrying and mining of the Middle Lias Marlstone as iron ore, which continued in the Tilton area until the mid-nineteen-fifties and at Holwell until the sixties, left cuttings, quarry faces and bottoms, and overburden heaps which have been rapidly colonised. At Holwell, for example, overburden only 30-40 years old supports thousands of plants of *Dactylorhiza fuchsii*. A fascinating though regrettably short-lived development in the same area was the formation of some 57 small subsidence ponds by 'crowning-in' at the crossing-points of galleries in the ironstone workings not far underground. Their colonisation was documented by Jones (1971). The huge conical slag-heap of the Holwell Steelworks not far away was, until its lowering in the late sixties, ablaze in the summer with the spikes of *Echium vulgare* and even now the disused sidings and tip base are still rich.

Extensive opencast workings in the Northamptonshire Sand ironstone further to the north and east, around Eastwell and Eaton and near Harston, Sproxton and Buckminster, were, again, abandoned in the early seventies and have largely been restored to agricultural use. However odd corners incapable of restoration do provide a habitat for species such as *Origanum vulgare* and *Carduus nutans* and east of Sproxton insufficiency of fill has left a huge partially water-filled gulley with limestone walls up to 15m. (50ft) high. This has provided a new habitat for the local calcicoles and introduced a freshwater element into an area previously entirely lacking it. The exposure of the Lincolnshire Limestone at Sproxton was in a sense incidental and active quarrying of this rock at Waltham, Stonesby and Nevill Holt appears to have ceased several decades ago. Stonesby Quarry has been affected by tipping but the older parts of the site retain a rich calcicole flora including *Anacamptis pyramidalis* and are now protected by a nature reserve agreement.

To complete this stratigraphical traverse of the extractive industries, mention must be made of the boom in sand and gravel working that has occurred since the Second World War. There were for instance over 20 sites active in 1977. Where plateau gravels are worked, the effect on the local flora may be entirely beneficial, introducing an element of diversity, although the opening up of a gravel pit adjacent to an old Keuper Marl pit at Heather eliminated a colony of *Dactylorhiza majalis* known for a century. At Shawell quarrying of sand stopped in about 1946, when clay and an unworkable layer of conglomerate were reached. The maturing quarry landscape there supported in 1975 *Eriophorum angustifolium* and *Sphagnum* spp. in marshy hollows, aquatics and reedswamp in a large pool and *Ophrys apifera* and *Anacamptis pyramidalis* on clay spoil banks.

The plateau gravel workings are usually relatively small in size but those in the terrace gravels of the Trent, Soar and Wreake valleys are much more extensive, covering hundreds of acres in all. Many of those in the Lockington-Hemington area have been restored with pulverised fuel ash from nearby power stations and some of those in the Wanlip and Thurmaston areas with domestic refuse. However the majority are allowed to flood following cessation of working and these pits and the associated former silt-ponds show textbook hydrosere succession where they remain undisturbed. The oldest have the richest development of aquatic and reedswamp species, whilst stoneworts and *Typha* are features of the most recent. Close behind the herbaceous colonisers are the woody ones, notably *Alnus glutinosa* and *Salix* spp. The variety of the latter may be a reflection of the fact that there were extensive osier beds in parts of the Soar valley until the nineteen-thirties. Areas in and around the pits often preserve elements of the flood meadow flora they have destroyed, such as *Sanguisorba officinalis* and *Saxifraga granulata*, but this was unfortunately not the case with the last remaining locality for *Fritillaria meleagris* at Thurmaston, which was quarried in the late sixties.

Reservoirs

Expanding populations, industrial development and increasing living standards all result in an increasing demand for water. Until the middle of the nineteenth century this demand was satisfied from local sources, whether springs, streams, rivers or wells, without any serious attempt at storage. These sources not only tended to fail in drought years but could also be a risk to health in built-up areas. In the second half of the nineteenth century three water storage reservoirs were constructed, Thornton (completed in 1853), Cropston (completed in 1870) and Swithland (completed in 1896). Early in the twentieth century the Black Brook was once again dammed to provide a more reliable supply for Loughborough (completed in 1906). Work on all four must have eliminated local marsh and streamside communities, but reservoirs soon proved to have botanical interest in their own right as is borne out by T. A. Preston's excellent paper on Cropston Reservoir (Preston 1895), in which he announced the discovery of *Inula britannica* new to Britain and described a 'queer' *Epilobium* which later proved to be the first British record of *E. adenocaulon*.

In the last 60 years three further reservoirs have been constructed, all straddling the boundaries of the former county and two for the benefit of communities outside it. Stanford Reservoir, which lies in the valley of the River Avon, supplies Rugby in Warwickshire and was completed in 1928. Eye Brook Reservoir, whose catchment includes much of 'high east Leicestershire', was built in 1940-41 to supply the expanding steel town of Corby in Northamptonshire. More recently Staunton Harold Reservoir, the greater part of which lies in Derbyshire, was built to augment Leicester's supply by storing water pumped from the River Dove and opened in 1961.

The substantial seasonal fluctuations in the water level of many reservoirs do not favour the development of an extensive reedswamp and aquatic community and in fact this may be actively discouraged by the use of herbicides. In the mid-sixties even more drastic measures had to be taken at Cropston Reservoir, where a luxuriant growth of *Polygonum amphibium* occupied each summer the silted-up shallow southern end, and it was drained and the silt dug out. Episodes like this and severe drought years such as 1976, which emptied the northern half of Swithland Reservoir and drastically lowered the levels of many others, do provide conditions for opportunists like *Rumex* and *Chenopodium* to flourish and spread.

Sewage works and rubbish tips

Complementary, in part at least, to the provision of fresh water is the disposal of sewage. Old-fashioned sewage farms, a 'happy hunting ground' (Bemrose 1927) for the searcher after casuals and aliens, have now virtually disappeared from the Leicestershire landscape. The largest, Beaumont Leys, on the north-western outskirts of Leicester, was closed in 1965 following the construction of the new sewage works at Wanlip, but did until recently afford interesting records. The area is now being developed as a major housing and industrial estate for the city's overspill. A related activity, the disposal of household and other refuse, does provide an alternative if ephemeral habitat for adventives. In recent years the volume of rubbish to be disposed of has steadily increased, despite incineration or composting of a proportion, and suitable sites are now at a premium. There has also been a concentration, by rigorous planning control, on fewer and larger sites, which have a longer working life. Traditional landfill sites, often on washland in river valleys adjacent to built-up areas, as at Aylestone, have now largely been replaced by the use of worked-out pits and quarries, as at Enderby, Syston and Stonesby, or cuttings on dismantled railway lines, as at East Norton. Modern practice is to cover tips with soil as fast as they are filled, but they are nevertheless still briefly of botanical interest as is indicated in the section on changes in the flora. On balance however refuse disposal has destroyed far more interesting sites than can be compensated for by the records of casuals and threatens to do so at an accelerating rate if our consumer society continues to generate more and more waste.

Agriculture

The proportion of the land surface of Leicestershire under the plough has changed continuously and, in the long term, markedly, over the last 1000 years at least. The widespread evidence of ridge and furrow, whether visible now only as crop marks in arable or surviving under permanent grassland, is an indication of how much land was ploughed at some time in the medieval period. The maps, on the other hand, which were produced from the Land Utilisation Survey of Leicestershire in 1933-34 (Auty 1943) show much of the county, particularly the eastern half, as almost uninterrupted permanent grassland, reflecting the position prior to the Second World War, when only about one ninth of the county's farmland was under the plough. In some four years, 1939-43, the proportion of arable rose to almost a half (Turner 1972) and the decline in area occupied by permanent pasture has continued ever since, such that it was only 36% in 1982.

The effects on the flora of the county of this massive change over the last 50 years are described in the section on change in the flora. They are however difficult to quantify with any precision since, unfortunately, the results of very few of the detailed field-by-field and floristic surveys promoted or carried out by A. R Horwood in the first two decades of this century have survived. It is however obvious that many of the species of unimproved permanent grassland have become locally rare or extinct. Many of the remaining small areas of this habitat that do still exist are on small holdings farmed in a traditional manner by an older generation. The future of these areas is not an assured one unless steps are taken in the near future to identify and protect them and the economics of agriculture today are such that this will probably only occur in exceptional circumstances, a bleak picture for the future.

On the credit side however, it should be pointed out that two areas of unimproved permanent grassland in the Wymondham area, previously unknown to botanists and located in the course of field-by-field survey in the late nineteen-sixties, are now county trust nature reserves. A third area near Muston, located as recently as 1979, is now a National Nature Reserve.

Coupled with the trend towards arable farming and away from stock has been an increase in field size in many parts of the county. The average was once about 10 acres, but there are now fields of 300 acres or more without a single hedge or any other internal boundary. Examples of such 'prairie' landscapes may be seen throughout the county, as at Houghton-on-the-Hill, Woodhouse and Upton. In many places the hedges removed were only the comparatively recent product of Parliamentary enclosure in the late eighteenth or early nineteenth century, but elsewhere they were species-rich hedges whose loss is significant in a botanical context.

A further widespread loss of habitat from agricultural land has been that of the field pond. Sowter (1960) records the destruction of just two such ponds at Scraptoft and Thurcaston because of housing development, but much more often changes in farming methods are responsible. Where it is isolated in the middle of an arable field, or even at the edge, a field pond is an inconvenience so far as ploughing and other cultivation techniques are concerned. Where stock is still present on a farm, piped water to a trough is more reliable and less likely to be a source of disease. Some indication of the change in the use of ponds and their condition in the north-east of the county is given by Jones (1971), who reveals that of some 1554 pond sites shown on the prewar 1:2500 Ordnance Survey maps of the 195 square kilometres (75 square miles) of his survey area, over 30% had been filled in and 63% of the remainder were no longer used. The total area of the ponds remaining was about 8ha. (20 acres) dispersed over some 1100 sites. In all 47 species were recorded, giving some idea of the contribution of the field pond to diversity at a local level.

The conversion of pasture to arable is only one facet of the 'agricultural revolution' at present taking place. Another is the removal of scrub from odd corners and steep slopes. This has, for example, rendered *Ulex europaeus*, once widespread if not ubiquitous along the Marlstone scarp, a species to be noted. It may also have favoured grassland species which were being shaded out.

The use of non-selective herbicides along roadside ditches, though happily not yet a widespread practice, is yet another manifestation of modern agricultural technology. Whether or not their application will have a permanent effect on the flora on a wider scale is not yet known, but the local consequences are obvious enough.

Forestry

There are some 1200 woods in Leicestershire exceeding 0.4ha. (one acre) in area. The majority are however so modest in size that the total area amounted to only just over 4850ha. (12000 acres) in 1949 (Chard *et al*. 1952). Of these 1200, it is thought, on the floristic and historical evidence at present available to us, that at least 50 are on sites continuously occupied by woodland since prehistoric times, that is primary woodland sites. These primary woodlands are clustered around the fringes of Charnwood Forest and in a broad band stretching through 'high east Leicestershire' from Tilton and Owston in the north-west to Great Easton in the south east (figure 21).

Other woodlands thought to be of this vintage are scattered very sparsely through the county, for example in the vicinity of Burbage, Shepshed and Breedon-on-the-Hill in the west. Primary woodland is absent from most of the north-east of the county, despite the presence there today, on the Belvoir estates, of nearly 800ha. (2000 acres) of plantations, mostly deciduous. The effect, particularly on the ground flora, of a decline in the management of these woods during the twentieth century, following the removal of their timber in one or both of the World Wars and the felling and replanting of some, often with a high proportion of conifers, has been described in the section on changes in the flora. Fortunately, from a botanical point of view, the post-war 'renewal' of Leicestershire woodlands promoted and carried out by the Forestry Commission came to a halt in the mid-sixties leaving, for instance, about one third of Launde Park Wood, half of Owston Big Wood and the whole of Launde Big Wood unaffected. Even in those woods that were partially or completely replanted, the species used were usually a mixture of conifer and broad-leaf, unlike the case of the rather earlier replantings of woods in eastern Rutland which were largely if not exclusively coniferous.

Woodland is however an economic resource, albeit until recently one neglected for nearly half a century in much of lowland Britain. Apart from the few primary woodlands, which are either in public ownership, that of conservation bodies like the county trust or likely to remain the property of local families who have demonstrated their concern for them, the future of the rest in anything like their present form cannot be guaranteed. Modern management techniques must be to the detriment of a shrub and ground flora which has evolved over half a millenium under a traditional coppicing regime.

Of the 1150 or so woodlands that are probably not primary, the majority are small and a substantial proportion are known to have been planted during the last two or three centuries as cover either for game or foxes or both. The most interesting botanically are the oldest of these secondary woodlands, such as Ambion Wood, which lies on the site of the Battle of Bosworth in 1485. Many of the characteristic herbs of primary woodland appear to have very limited powers of dispersal and do not find their way into secondary woodland unless there is a seed or other source nearby. Other species, such as the orchids *Listera ovata*, *Epipactis helleborine* and *Orchis mascula*, are better colonisers, and are found in woodland lying on ridge and furrow, which provides indubitable evidence of an agricultural interlude in its history.

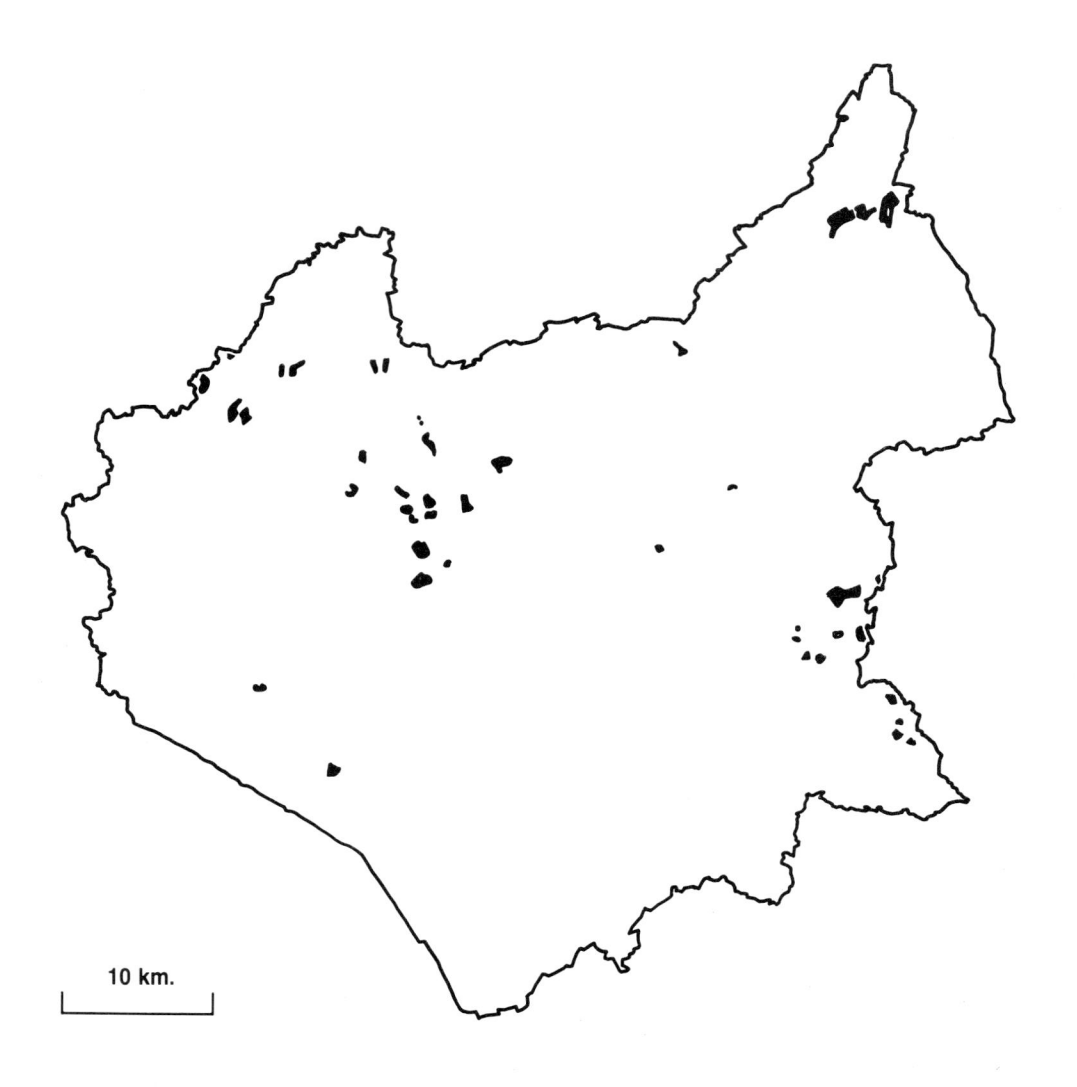

Figure 21. Major areas of woodland.

Recreation

For several millennia the dominant human influences on the flora of Leicestershire were essentially economic, man at work. With increasing affluence, leisure and mobility, phenomena particularly of the last 50 years, the influence of recreational activities, man at play, has become a force to be reckoned with.

This influence extends over large areas of our landscape. Amongst the more obvious effects are those resulting from the sheer numbers of human feet in many so-called 'beauty spots', public open spaces, many famed in the past for their flora. An example is the summit of Beacon Hill, where the vegetation has been virtually eliminated over a substantial area. Even where the effects are not so drastic, there may be marked changes in the composition of the vegetation at viewpoints and along well-trodden paths. The association of *Juncus squarrosus* and *Nardus stricta* which characteristically edges the paths in Bradgate Park must be a very tough one.

Even away from the parts of such public open spaces subject to greatest pressure, management practices, sincerely if short-sightedly carried out for what is considered to be the public benefit, have had a most undesirable effect on their indigenous vegetation. Two such practices are drainage and the planting of trees and shrubs. Parts of both of the Charnwood public open spaces so far mentioned have been drained in the last 30 years, with the result that the bog communities which were one of their special features have been almost eliminated. Thus *Hypericum elodes*, which occurred on Beacon Hill as recently as 1945, is now gone from there and extinct in the county as a result. Even such hardy species as *Erica tetralix* are now virtually confined in parts of Bradgate Park to the edges of the drainage ditches.

The establishment of trees and shrubs, often of exotic species, on public open spaces, an ecologically-unenlightened policy for those in rural areas, has also been responsible for the widespread if localised destruction of their native flora. The long-term effects of this practice are illustrated by the elimination of the ground flora on parts of Beacon Hill by the uncontrolled spread of *Rhododendron*, albeit a planting in this case that long ante-dated the acquisition of the area as a public open space. However both on this area and on the Out Woods nearby, the latter once a fine ancient woodland, there have been post-war plantings of conifers and alien hardwoods which will continue the destructive process.

The twenty public and private golf courses in the county, many of them established during the last 50 years, have on balance made a positive contribution to the conservation of our native flora. The fairways and greens, even if not sown with special grass mixes, are in any case close-mown and often treated with selective herbicides. However they only represent a proportion of the total area of the courses, which amounts to some 800ha. (2000 acres). The rough, often extensive on the older-established courses, is not so regularly or closely mown and may preserve a surprisingly diverse grassland flora, including for instance such species as *Filipendula vulgaris* at Oadby and *Succisa pratensis* on the Melton course. Depending on the topography and original land use of the areas on which courses were established, there may also be substantial areas of heathland, scrub and woodland which receive no regular management. The Longcliffe course has, for instance, large areas of *Calluna*.

The leisure-time exodus to the countryside of a predominantly urban public, whether just to exercise family and pets or for more organised sporting activities, is a comparatively recent phenomenon on the scale on which it now occurs. The rural pursuits of hunting, shooting and fishing have a longer history and continue to have an affect on the vegetation of the county, though changing in its nature. The fox-hunting fraternity must be credited with the establishment over the last three centuries in an essentially agrarian Leicestershire landscape of probably nearly a thousand mainly small areas of scrub or woodland, to provide cover for their quarry. In the establishment of these coverts there is the well-documented planting of species of shrubs such as *Ulex europaeus* which have thereby been given a wider distribution in the county than they might now have had (Ellis 1951). The term 'gorse' which forms part of the name of many such coverts may be a reflection of this practice. Their establishment and maintenance has provided refuges for a characteristic range of hedgerow species such as *Viola odorata* that with the advent of 'prairie' farming have disappeared from considerable areas of the county. In east Leicestershire, coverts often encompassed marshy areas on springlines, thus preserving locally species such as *Caltha palustris*, whose habitat on adjacent farmland has long since been drained.

The development of interest in shooting during the eighteenth and nineteenth centuries resulted in many of these small woods, as well as longer-established larger ones, acquiring a second purpose, that of game preservation, a duality of function not without its problems for gamekeepers. It is to the shooting interests that we owe the widespread occurrence in such woods of alien species of shrubs such as *Mahonia aquifolium* and *Symphoricarpos albus*, both extensively planted as cover and food for game. A less obvious manipulation of the distribution patterns of a native species has resulted in the presence in hedgerows in parts of the county of numerous well-grown trees of *Malus sylvestris*, planted to provide winter food for pheasants. A more ephemeral manifestation of this phenomenon is the planting of strips of species such as *Fagopyrum esculentum* at the edge of arable, obvious enough as game food in the year of planting, but less so if persisting in odd corners in later seasons.

The influence of game preservation on the flora of the county takes another form in the establishment and maintenance by wildfowlers of suitable habitat for the wintering and rearing of waterfowl, a development particularly of the last two decades. Where existing lakes and pools are taken over, there may be little perceptible change in their vegetation. Where, however, a recently created body of water, such as a flooded gravelpit, is developed for this purpose, recommended native aquatic and reedswamp species, such as *Potamogeton natans*, *Polygonum amphibium* and *Sparganium erectum*, are frequently introduced for food and cover (Anon. 1969). That this is not just a recent practice may be reflected in the flora of pools with names such as The Duckery at Market Bosworth (where, incidentally, *Sonchus palustris* was, for whatever purpose, planted in the eighteen-forties and still survives).

Anglers also have a stake in the creation and maintenance of aquatic habitats. Fishing, whether coarse or game, is now the most popular organised sporting pursuit in Britain, and its antecedents, together with those of the related activity of fish farming, stretch back to the monastic period. Stewponds whose creation dates back to that time still exist in some parts of the county and elsewhere as at Launde, new fishing lakes have been created by reflooding ponds long since dried-up. To such fishing lakes, as to other water bodies created in the course of earth moving for some other purpose such as motorway construction, are, again, introduced appropriate species of plants, in this case mainly aquatics. The natural processes of colonisation are thereby greatly speeded-up, perplexing the purist botanist who, since such introductions are rarely documented, does not know what to put down to natural and what to human agency. It should however be said in defence of the anglers and others with recreational interests in aquatic habitats, such as sailing clubs, that without their vociferous protests many such habitats would have been eliminated from our countryside by use as landfill sites.

Gardening is another area of human activity, now predominantly recreational in nature, which has had a substantial effect on the Leicestershire flora. It has a well attested history elsewhere in the British Isles going back to Roman times. In Leicestershire the presence in the neighbourhood of villages of species such as *Armoracia rusticana*, *Ballota nigra*, *Chenopodium bonus-henricus* and *Humulus lupulus* is testimony to their cultivation and use for centuries, whether as a condiment, in herbal medicine, as pot herbs or in brewing. More recently we have the writings of William Hanbury of the Langtons (Hanbury 1770-71), whose landscaping ideas were on the grand scale, or the more modest approach of plantsmen such as the Rev. George Crabbe, who is stated by Horwood (1933) to have been laying out a botanic garden at Muston at about the same time. The last few decades have witnessed an explosive growth of interest in gardens and gardening, if the expansion of the horticultural industry that feeds this interest is anything to go by.

The relationship between gardeners and the Leicestershire flora is such a many-faceted one that a few examples must suffice. The continued existence in accessible habitats of certain attractive native species is still threatened by those who dig them up to remove them to their gardens. Horwood deplored the digging of roots of *Primula vulgaris* and this practice still goes on, though only now in woodland since the species has been virtually extirpated from our hedgerows.

In complete contrast is the practice of dumping on the verges of country lanes surplus material of garden plants of alien origin whose vigour and powers of increase have embarrassed their owners. The more persistent members of this category, characteristically creeping or rhizomatous herbaceous perennials, include species such as *Aster novi-belgii*, *Petasites fragrans*, *Solidago canadensis* and *Tritonia × crocosmiflora*. The reasons for this practice may be those which account for the appearance of bags of hedge clippings and old sinks in these situations. Occasionally however we must allow for worthier, if misguided, motives, as when *Azolla filiculoides* turns up in a regularly visited field pond, or uncommon grassland herbs, whose seeds form part of the 'wildflower mixes' now being marketed by enterprising seedsmen, appear on otherwise botanically undistinguished roadsides.

Nurseries and garden centres now do a flourishing trade in 'containerized' plants and have always distributed much material in media that were not completely sterile. Such places have long been recognised as a major agency in the spread of native weed species such as *Cardamine hirsuta*, but they may have also been the source of some alien species such as *Epilobium nerterioides*, *Montia perfoliata* and, notably, *Galinsoga parviflora* and *G. ciliata*, discovered new to the county in 1945 at a nursery at Anstey. The balance between the native and alien flora of a typical Leicester garden is described in a recent book by Owen (1983), who logged in ten years 148 species in the former and 183 species in the latter category. The introduction into cultivation of showy plants of alien origin has been going on for centuries and is well-documented. What is more likely to provide headaches for future generations of botanists is the less frequently documented introduction of species native to other parts of the British Isles, honourable exceptions being provided by Owen, and the account of the invasive powers of *Hierochloe odorata* in Tutin's fascinating paper on the weeds of his Knighton garden (Tutin 1973).

Finally, a by-product of our current concern for wildlife is the growing list of alien species, often of circum-Mediterranean origin, which have appeared in the vicinity of bird tables, species such as *Bupleurum lancifolium*, being the product of germination of 'wild bird seed'. A further source of these and other aliens are cleanings and spilt seed derived from the keeping of cage birds.

It is perhaps fitting that this section should close with a group of plants that are of patently alien origin. It is a reminder of what a high proportion of the diversity of the present Leicestershire flora is the result of the manipulation, on a increasingly massive and widespread scale, of its landscape by man. Perhaps it is only shortsightedness which leads us to regret the passing of a floristically diverse agricultural landscape, unconsoled by the pleasures of hunting for aliens on waste ground and rubbish tips.

References

Allen and Potter, 1932. *Report of the Leicestershire Regional Town Planning Joint Advisory Committee*. Leicester.

Anon., 1969. *Wildfowl management on inland waters*. Fordingbridge: Eley Game Advisory Station.

Auty, R. M., 1943. *The land of Britain. The report of the Land Utilisation Survey of Britain*. pp.243-327. Part 57. Leicestershire. London: Geographical Publications.

Bemrose, G. J. V., 1927. The adventive flora of Leicester and district. *Trans. Leicester lit. phil. Soc.* **28,** 45-71.

Chard, J. S. R. et al., 1952. *Census of Woodlands 1947-1949*. London: H.M.S.O.

Ellis, C. D. B., 1951. *Leicestershire and the Quorn Hunt*. Leicester: Edgar Backus.

Hanbury, W., 1770-71. *A complete body of planting and gardening . . .* London: Edward and Charles Dilly.

Hoskins, W. G. and McKinley, R. A. (eds), 1955. *A History of the County of Leicester* **3.** London: Oxford University Press.

Jones, R. C., 1971. A survey of the flora, physical characteristics and distribution of field ponds in north east Leicestershire. *Trans. Leicester lit. phil. Soc.* **65,** 12-31.

Liddle, P., 1982. *Leicestershire Archaeology – the present state of knowledge.* **1.** *To the end of the Roman period.* Leicester: Leicestershire Museums Service.

Moore, C. N. and Cummins, W. A., 1974. Petrological identifications of stone implements from Derbyshire and Leicestershire. *Proc. prehist. Soc.* **40,** 59-78.

Owen, J., 1983. *Garden Life.* London: Chatto and Windus/The Hogarth Press.

Preston, T. A., 1895. The flora of Cropstone Reservoir. *Trans. Leicester lit. phil. Soc.* **3,** 430-442.

Pye, N. (ed.), 1972. *Leicester and its Region.* Leicester: University Press.

Sowter, F. A., 1960. Our diminishing flora. *Trans. Leicester lit. phil. Soc.* **54,** 20-27.

Turner, S. E., 'Agriculture' *in* Pye, N. (ed.), 1972. *Leicester and its Region.* Leicester: University Press, pp.325-339.

Tutin, T. G., 1973. Weeds of a Leicester garden. *Watsonia* **9,** 263-267.

LEICESTERSHIRE BOTANY AND BOTANISTS

1. The period prior to 1900

The period covered by this section is already dealt with in considerable detail by Horwood and Gainsborough in the *Flora* of 1933. As this work is not easily obtainable, the more important aspects of early botanical activity relating to the county are outlined below.

Prior to the investigations of Richard Pulteney, Leicestershire would seem to have been a very much neglected county from a botanical point of view. Unlike many other areas of the British Isles, there are no records for the county contained in the early herbals of Gerard, Johnson, Turner and the like, and the only topographical work of any substance relating to the county (Burton 1622) does not contain any reference to the flora. The first published record occurs in James Petiver's work of 1716, *Graminum, muscorum, fungorum, submarinorum &c. Britannicorum Concordia*, which contains a note of the first discovery in the British Isles of *Calamagrostis canescens* by John Scampton, 'a curious botanist who sent it me from Leicestershire'. Although several cryptogams from the county are recorded in the works of John Ray the only vascular plant mentioned (in the third edition of the *Synopsis Methodica* of 1724) is *Ribes rubrum*.

Our early knowledge of the flora of the county is therefore almost exclusively based on the work of Richard Pulteney (1730-1801). Born at Mountsorrel, a small village to the north of Leicester and on the edge of Charnwood Forest, Pulteney was educated at an elementary school in Loughborough and at about 15 years of age apprenticed to an apothecary in the same town, with whom he remained until about 1752. At the end of the apprenticeship he spent a short period working in Nottingham before setting up in practice as an apothecary in Leicester where he stayed until moving to Dorset in 1765.

Pulteney's early botanical interests are evidenced by the many records contained in his manuscript Floras, which were prepared before 1753, and in his correspondence with George Deering of Nottingham and John Blackstone of London, both prominent botanists of the period. The first of these manuscripts is his *Catalogus stirpium circa Lactodorum nascentium*, presented to Leicester Museum in 1861. Although undated, it was probably completed in about 1747 and consists of 186 pages including 51 water colour drawings. Two further manuscript lists (Pulteney 1749 and 1752) contain many additional records. Pulteney actually published very little relating to Leicestershire; a paper on rare plants observed in the county (1756) and, nearly forty years later, his contribution to the first volume of Nichols' History of the county (1795), *A Catalogue of some of the more rare plants found in the neighbourhood of Leicester, Loughborough and in the Charley Forest*. Despite the fact that he had been away from the county for thirty years, this annotated list contains about a hundred records additional to his earlier lists. It includes many species which have since disappeared from the county or are now extremely rare, particularly in the Charnwood Forest area, where extensive drainage has taken place. It is many years indeed since *Pinguicula vulgaris* was 'not uncommon on the bogs' and *Epipactis palustris* 'plentiful in some moist closes at Woodhouse'. In all, Pulteney was responsible for the first published record of over 600 plants within the county. His notes provide a number of interesting clues to early landscape features before the enclosures and subsequent agricultural changes.

The Leicestershire records in the Floras of John Hill (1760), Thomas Martyn (1763) and William Hudson (1798) were communicated to these authors by Pulteney and many of the scattered records in Leicestershire topographical works of the 18th and early 19th century were based on his work, often without acknowledgement. *The Rural Economy of the Midland Counties* by William Marshall (1790) contains a number of first records for Leicestershire, lists of cornfield and hedgerow weeds and those found in various types of meadows, and a good deal of useful observation relating to the state of the landscape. Writing of hedgerow timber in the county he states 'There is not, speaking generally, a young oak in the county. If this error should not be rectified, there may not, in half a century, be a tree left in a township'.

The Rev. George Crabbe, poet and botanist (1754-1832), who spent some twenty years in the north-east of the county between 1782 and 1813, contributed a section on *The natural history of the Vale of Belvoir* to the first volume of Nichols (1795). This includes a list of plants for an area of the county not previously covered in any detail and provides a number of first records. Crabbe also records the introduction of some species, for example *Geranium lucidum* 'brought from Matlock, and . . . scattered about Stathern'. William Pitt's *A general view of the agriculture of the county of Leicester*, published in 1809, contains some useful background information on the natural vegetation. The sections on 'Weeds and Weeding' and 'Pastures' do include a number of first records for the county, although they are mainly unlocalised. Pitt set down a number of proposals to educate the landowners and their tenants in the art of 'improving' their land. These included the recommendation that a library should be established in each parish, in which should be included, in addition to standard works on agriculture, *Curtis's Botanical Magazine* and *Flora Londinensis*, and 'for those who would make any proficiency', Withering's *Botany*.

The arrival in the county in 1826 of the Rev. Andrew Bloxam (1801-78) gave a considerable impetus to the study of the Leicestershire flora. He was for many years Rector of Twycross and is well known for his critical studies of the genus *Rubus*. He contributed a number of papers dealing with the Leicestershire flora to the *Naturalist* and the *Phytologist*. In conjunction with the Rev. Churchill Babington he produced a list of plants found in Charnwood Forest for Potter's *History* of that area (Potter 1842). Further details of his botanical excursions with Prof. Charles Babington are to be found in the *Memorials* of C. C. Babington (AMB 1897), including a record of their visits to the Blackbrook area of Charnwood.

The first complete Flora of the county appeared in 1850, the work of Miss Mary Kirby (1817-1893), encouraged and aided by the Rev. Andrew Bloxam. Preliminary copies, printed on alternate pages, were issued in 1848 and distributed to local botanists in order that they could add their own records and observations for inclusion in the definitive issue. The edition of 1848 is a very rare book and probably only very few copies were printed. The *Flora* lists, with localities, over 900 species, using the nomenclature of the second edition of *The London Catalogue* (1848). The text is interspersed with notes by Mary's sister Elizabeth on the medicinal and other uses of the various families of plants. The section on *Rubus* was prepared by Bloxam and, in the amount of descriptive detail provided for the 35 species dealt with, is a remarkable contribution to a local Flora of this period. Miss Kirby also acknowledges the assistance of the Rev. W. H. Coleman in providing material for the Flora. Joint author with the Rev. R. H. Webb of the *Flora Hertfordiensis* (1849), Coleman (1816-1863) took up residence in Ashby-de-la-Zouch in 1847. By 1852 he had completed a manuscript *Flora Leicestrensis* which was to be the basis of botanical work in the county for the next 30 years. He divided the county into twelve districts based on river basins, a system first used in the Hertfordshire *Flora*, and he visited all the districts to gather records. The whereabouts of the original manuscript is not known, but a typescript copy made by the Rev. T. A. Preston is in the library of Leicestershire Museums (79 L 1937).

Coleman's divisions were used with only slight alteration in the next published Flora of the county. It was based on Coleman's manuscript, which was made available by the then owner, Edwin Brown of Burton-Trent, on condition that Coleman's name appeared on the title page. This work was prepared by a committee of five headed by F. T. Mott and issued in 1886 under the auspices of the Leicester Literary and Philosophical Society. It included the now traditional annotated species lists, in this case following the nomenclature of the 3rd edition of Hooker's *Students' Flora* (1884), with a brief habitat notes and localities for the less common species.

Horwood's subsequent comment that no adequate idea of the flora of the county can be obtained from the 1886 work is perhaps a little harsh but has some justification. One unusual feature was however the inclusion of detailed accounts of most of the cryptogamic groups of plants. The account of the algae was contributed by Frederick Bates who included, for the benefit of the microscopist, the measurements of many species.

During the closing years of the nineteenth century, following the publication of the *Flora*, the *Transactions of the Leicester Literary and Philosophical Society* contained a number of interesting papers, relating to various aspects of Leicestershire botany, by Mott and other society members, notably the Rev. T. A. Preston (1833-1905). The latter's paper on *The Flora of Cropstone Reservoir* (Preston 1895) described the discovery of *Inula britannica*, a plant new to the British Isles. Preston came to live at Thurcaston in 1885 and he and his contemporaries were to form the link between the workers of the nineteenth century and the new impetus engendered by the arrival in the county of A. R. Horwood in 1902.

References

A.M.B.[abington], 1897. *Memorials, Journal and Botanical Correspondence . . . of Charles Cardale Babington.* Cambridge: MacMillan and Bowes.

Burton, W., 1622. *The description of Leicestershire: containing, matters of antiquity, history, armoury, and genealogy.* London: John White.

Hill, J., 1760. *Flora Brittanica, sive Synopsis methodica stirpium Brittanicorum.*

Hudson, W., 1798. *Flora Anglica.* 3rd edition.

Martyn, T., 1763. *Plantae Cantabrigienses or, a catalogue of the plants which grow wild in the County of Cambridge – to which are added Lists of the more rare plants growing in many parts of England and Wales.* London.

Nichols, J., 1795. *The history and antiquities of the county of Leicester.* **1**. London: J. Nichols.

Potter, T. R., 1842. *The history and antiquities of Charnwood Forest.* London: Hamilton, Adams.

Preston, T. A., 1895. The flora of Cropstone Reservoir. *Trans. Leicester lit. phil. Soc.* **3**, 430-442.

Pulteney, R., 1749. *Opusculum Botanicum.* Ms. Linnean Society of London.

Pulteney, R., [1752]. *A catalogue of plants spontaneously growing about Loughborough and adjacent villages.* Ms. Botany Department Library, British Museum (Natural History), ref. no. B.47.5.

Pulteney, R., 1756. On some of the more rare English plants observed in Leicestershire. *Phil. Trans roy. Soc.* **49**, 808-866.

2. The period since 1900

A most valuable feature of Horwood and Gainsborough (1933) is its comprehensive account of the history of botany in Leicestershire, incorporating detailed biographical notes on local botanists. However, A. R. Horwood, conforming to the conventions, did not give as full an account of his contemporaries as he did of his predecessors. To remedy this deficiency, it is necessary for this section to start some decades prior to 1933, a convenient date being 1902, when Horwood took up his appointment at Leicester Museum.

Reconstructing the local botanical scene at such a distance in time, when virtually all the characters have long since died, presents some difficulties. In order that the Flora writers of 2033 shall not find themselves in the same predicament, a liberty has been taken with the conventions and this general account of botany in the county does therefore cover the period up to the present, spanning in all some eight decades.

At the time of Horwood's arrival in Leicester, the study of the county's flora was being actively pursued by members of Section D (Botany) of the Leicester Literary and Philosophical Society. In the report of their meeting on February 19th 1902, the Rev. T. A. Preston records that the Society's herbarium contained 6246 sheets representing 1047 of the 1114 species and varieties of phanerogams then considered to occur in the county. On December 17th of the same year he was able to announce the addition of a further 1000 sheets. Some of the contributors were members of the Section, such as W. Bell, T. Carter, E. F. Cooper, G. B. Dixon, J. E. M. Finch, A. B. Jackson, F. T. Mott and Dr. W. A. Vice, others were correspondents such as T. E. Routh, who sent in specimens from outlying parts of the county. A. B. Jackson was particularly active in the period up to 1907, when he left Leicestershire to take up what proved to be a distinguished career in botany. He published in the *Transactions* of the Society a number of notes on difficult groups such as *Euphrasia* and *Rubus*. In the *Journal of Botany* for 1904 he gave a most useful summary of records which had accumulated since the previous Flora was published in 1886.

Jackson is credited, together with T. A. Preston (in the *Thirteenth Report of the Museum Committee to the Town Council*, published in 1902), with collecting and presenting 'great numbers of specimens' to the Museum herbarium, which comprised at that time 'fifty bound folio volumes of plants of the British Isles'. It would appear that the role of this collection was seen to be complementary to that of the Society's Leicestershire herbarium. This policy was certainly still in operation in 1928, when the Society's *Transactions* record the addition to its herbarium of A. R. Horwood's collection of Leicestershire plants (most or all of his non-Leicestershire material having been purchased by the National Museum of Wales in 1923). Some time later, perhaps during or shortly after the Second World War, the folio volumes containing the Museum herbarium were broken down into their constituent sheets. In 1949 the Society's herbarium was transferred to the Museum on permanent loan. In the same year much of the Museum's non-Leicestershire material was loaned to the Department of Botany at the then University College, being returned in 1975.

At some time in the fifties the Society's herbarium and what remained of the Museum herbarium were amalgamated, losing their separate identities. However the subsequent reorganisation of the Museum's botanical collections, together with the computerised listing which is now taking place, should eventually allow nearly all of the many thousands of herbarium sheets collected in the nineteenth and first half of the twentieth centuries to be confidently assigned to the collections of the individuals or institutions in which they originated.

Horwood was collecting and recording in the county soon after his arrival (witness records for 1902 in Jackson's 1904 paper) but, for reasons that can now only be guessed at, he was not involved in the work of Section D until about 1910. During the intervening eight years he published substantial papers on the cryptogamic flora of the county (Horwood 1909) and the fossil flora of the Leicestershire and South Derbyshire coalfield (Horwood 1907 and 1908). In 1911 he published an update to Jackson's 1904 paper and the following year a Flora Committee was set up under the aegis of Section D, with Horwood as General Editor.

While reviewing botanical activity in the first decade of the century, it is perhaps worth mentioning a piece of research on local grasslands, which, exceptionally, appears to have evaded Horwood's thorough search for publications relating to Leicestershire botany. This is S.F. Armstrong's account, published in 1907, of the botanical and chemical composition of the herbage of named fields in the Market Harborough area. Although this unfortunately groups most of the broad-leaf species as 'various weeds', it does give interesting figures for the contribution by the dominant species of grasses and other plants to the swards of pastures and meadows of various ages and degrees of excellence from an agricultural point of view. It also relates these findings to the management regime.

From 1912, until he left Leicester Museum in 1922, Horwood was the driving force behind work on the county flora. He recruited a number of young enthusiasts known as 'Horwood's boys', many from Wyggeston Boys' School, in whom he instilled a disciplined approach to survey and mapping which he described in one of his more general botanical works (Horwood 1920, p.168). That he used this approach in his own wide-ranging survey work in the county is evident from a number of references in the 1933 *Flora* (see for instance p.xxxvi). Sadly, only three of the detailed ecological surveys with which he was associated were published *in extenso* (Mercer 1914; Measham 1915; Wade 1919). Even more regrettably, virtually none of the maps and manuscript material accumulated by him or his contemporaries in the course of their work on the

Leicestershire flora is still in existence. After Horwood's death in 1937, a few typescripts used by him in the compilation of the *Flora,* together with some of his reference books, reached Leicester Museum. Nothing is known, however, of the fate of the 'important or extensive mss. utilized' by him, including card indices amounting to something like 55000 cards, mentioned on pages cclxxxv-cclxxxvi of the *Flora,* despite lengthy correspondence with institutions where they might have been deposited. It seems unlikely that they still survive. Their loss is all the greater in view of the sweeping changes in the landscape and vegetation of the county that have occurred during the last fifty years.

A happy exception to the general rule is provided by the survival of the notebooks of the Rev. J. A. Cappella of Ratcliffe College, which are in the possession of one of the Editors of this *Flora.* Cappella appears to have been working on the same lines as Horwood, if not with him, and they are known to have botanised together occasionally. The notebooks contain the results of a field by field survey of the land around the College over a period of 45 years (1885-1930), with comments on the changes in flora resulting from changes in management. Numbered sketch maps enable the species recorded to be precisely localised to a field and herbarium material supporting the records has recently been acquired by the Museums Service.

In the decade following Horwood's departure from Leicester Museum in 1922, the close relationship between the Museum and the Botany Section of the Literary and Philosophical Society which had developed was carried on by G. J. V. Bemrose, Assistant Curator at the Museum. His main interest was alien species and his 1927 paper on this subject probably received less than its due recognition in the 1933 *Flora.* In 1930, F. A. Sowter took over from him the posts of Hon. Secretary of the Botany Section and of the Flora Committee. Sowter was one of Horwood's former Wyggeston School protégés and did much of the background research for the final text of the *Flora* in the hectic two years prior to its publication in 1933. To give some idea of his involvement in its preparation, there are amongst his papers thirty-two letters, received from Horwood between January 1932 and August 1933, requesting detailed information on local botany and botanists, copies of herbarium labels, loans of books and 'survey work and special collecting' in areas for which information was not available to Horwood (for further details see Evans 1973).

Other local botanists active during the thirties appear to have been comparatively few. S. A. Taylor, a close friend of Sowter's, was still adding to the personal herbarium which he had begun, as a schoolboy, in 1915, but most of his collecting was from localities beyond the county boundary. He was an enthusiastic member of the Botany and Biology Section of the Literary and Philosophical Society, leading summer field meetings and contributing botanical exhibits to the winter meetings. He was also one of a small committee of its members formed on 15th February 1939 'to consider the matter of getting to work – with a view to publishing a second volume of the Flora'. It is however not clear from the minute whether the purpose was to bring the 1933 *Flora* up to date or to produce the companion volume on cryptogams that A. R. Horwood had intended to write (see his Preface) and the good intentions were in any case overtaken by events.

A. E. Hackett was collecting actively in the south-west of the county in the early thirties, no doubt encouraged by Sowter, with whom he made a number of joint excursions. Following publication of the Flora there appears to have been an understandable if temporary lull in activity, perhaps accentuated by the fact that Sowter's considerable energy and talents had been transferred to the cryptogams, a field in which he was to make a national name for himself. Rev. D. P. Murray returned to England from South Africa in 1937 and spent some years at the Holy Cross Priory in Leicester. However any collections that he may have made in the county at that time have yet to be located.

The early forties were inevitably overshadowed by the Second World War. The Botany and Biology Section of the Literary and Philosophical Society ceased activity for the duration. However by 1943 the thoughts of some people at both national and local level were turning optimistically to post-war reconstruction and in particular the need to conserve our native flora and fauna. The *Report to the Nature Reserves Investigation Committee of the Sub-Committee for Leicestershire and Rutland*, which was circulated in February 1944, provides an assessment of the prime botanical habitats in the two counties. The botanical representatives on the Sub-Committee were Dr. C. T. Ingold, C. J. Lane (Keeper of Botany at Leicester Museum), F. A. Sowter and S. A. Taylor, and the report is obviously based to a large extent on fieldwork carried out during the thirties. Its main recommendations for Leicestershire were the designation of Charnwood Forest and the woodlands in the east of the county as 'National Habitat Reserves', or even in the case of the Forest as a 'National Park'. Meanwhile, S. A. Taylor at least was still botanising in the county, witness his first record for the county in 1940 of *Impatiens capensis*. During the mid-forties he collected a number of *Epilobium* hybrids which are published for the first time in this Flora.

Shortly before the end of the War, in October 1944, a separate Department of Botany was established at the then University College of Leicester. T. G. Tutin, who became in 1947 the first Professor of Botany, started collecting local material in 1945 for what was to become the Department's large and important herbarium. He drew attention to entries in the Flora which needed revision in the light of new information about the taxa concerned and checked some of the doubtful material in the herbarium of the Literary and Philosophical Society. He was ably assisted, both in the revival of interest in Leicestershire botany and the development of the University herbarium, by E. K. Horwood, who was later one of the founder members of the Committee for this Flora. The interests of the Botany Department soon acquired wider horizons, firstly the *Flora of the British Isles* and later *Flora Europaea*, and although the research of both staff and post-graduate students ranged over much wider areas, significant contributions to our knowledge of the Leicestershire flora were made by A. O. Chater, G. Halliday and P. F. Yeo, amongst others.

The activities of the War Agricultural Committees and the continuing changes in farming after the War had an effect on the Leicestershire landscape which must have been depressing for those local botanists who had been active in the thirties. Some measure of the state of the sites known to be of botanical importance is provided in the survey in 1951-52, by E. A. G. Duffey of the Nature Conservancy, of those which were later listed in the first schedule of Sites of Special Scientific Interest notified in 1953. This survey was based on the work of the Nature Reserves Investigation Sub-Committee, but it is sad to note that sites such as Brentingby Field and Saltby Bog had lost in the interim most if not all of their interest, owing to war-time ploughing and drainage.

In 1955 there were two events whose repercussions on Leicestershire botany have continued to the present day. The first was the initiation of the 10km. square mapping of the British flora organised by the Botanical Society of the British Isles, which resulted in the publication of the *Atlas of the British Flora* in 1962. Amongst those involved in mapping the Leicestershire squares were M. K. Hanson, E. K. Horwood, Rev. T. Maloney, Rev. A. L. Primavesi and T. W. Tailby. M. K. Hanson had been actively exploring the county in the company of E. K. Horwood in the early fifties and his photographs of, for instance, *Polygala calcarea* at Saltby in 1951 and *Coeloglossum viride* at Potters Marston in 1953, mark the last occasions on which these rarities were seen in the county. The success of this mapping operation on a national scale led to the idea of mapping the flora of a county at the smaller scale of a tetrad. This was successfully accomplished by Dr. J. G. Dony, whose *Flora of Hertfordshire*, published in 1967, provided a model on which most subsequent local Floras have been based.

The second, less happy, event during 1955 was the destruction by ploughing of two boggy fields to the east of High Sharpley, which were the last station in the county for *Vaccinium vitis-idaea*, despite the scheduling of this site as an S.S.S.I. two years earlier. This was however one of the factors that precipitated the formation in 1956 of the Leicestershire Trust for Nature Conservation (now the Leicestershire and Rutland Trust for Nature Conservation). Since that date as much of the botanical work in the county has been directed to the location and conservation of important habitat as it has to the search for new records, although the two activities often go hand in hand. F. A. Sowter's paper, *Our diminishing flora*, published in 1960, does give a good impression of the concern felt at that time at the rate of loss of habitat, and although with hindsight we may consider that he was a little pessimistic about some sites, there is no doubt that he had correctly interpreted the trend.

The nineteen-sixties witnessed a development of interest in natural history in the county on a scale for which it is difficult to find a previous parallel and botany had a place in this development. The foundation of the Loughborough Naturalists' Club in 1960 provided a focus for the interests of those living in that area, particularly with regard to Charnwood Forest. In Leicester the membership of what is now known as the Natural History Section of the Leicester Literary and Philosophical Society began to increase. A quickening of interest in the flora of the county can be traced through the issues of the newsletters of both societies, notably *Heritage* issued quarterly by the Club. The Club instituted detailed studies of specific parts of the Charnwood Forest of which the first was that on Bradgate Park and Cropston Reservoir Margins published in 1962.

In 1959 the author of this section took up the post of Keeper of Biology at Leicester Museum. With his encouragement the Museum took a renewed interest in the botany of the county during the sixties and was able to acquire the herbaria of a number of local botanists. Notable amongst these was the greater part of the herbarium of William Bell, Hon. Secretary to the Flora Committee in the earliest decades of the century. After Bell's death in 1926 this collection had remained in the family house in Stoneygate. In 1940 the house was bomb-damaged and its contents were transferred to a furniture repository in New Walk. Here the herbarium was found out in the yard by an employee of the City Parks Department, who, fortunately, brought it to the attention of the Parks Superintendent, Mr. J. W. Watson, a local naturalist and bibliophile. He, recognizing its value, rescued as much as could be found and stored it away for safe-keeping in his attic, where he rediscovered it on his retirement in 1964. Much of its interest relates to the fact that it was, apparently, not available to Horwood at the time he was compiling the 1933 *Flora*.

1967 was an important year in the recent annals of Leicestershire botany, since in its twelve months were launched two field-by-field surveys of the county, as well as the flora survey which has resulted in this volume. The field-by-field surveys were that of Charnwood Forest, organised by the Loughborough Naturalists' Club and completed in 1968, and that of the rest of the county, organised by the Natural History Section of the Leicester Literary and Philosophical Society, which extended over three years. In the course of them a number of sites of considerable botanical interest were discovered, which because either of their small size or relative remoteness appear to have been overlooked by previous generations of botanists. Examples are the herb-rich pastures which now constitute the county trust's nature reserves at Cribb's Meadow and Wymondham Rough. At the same time the importance of other botanical sites was brought sharply into focus, for example the marshes in the valley of the western River Sence at Newton Burgoland. A. R. Horwood, who had advocated the use of field-by-field survey 50 or more years earlier, would have been gratified by this proof of the value of his advice.

By a curious coincidence, 1967 was also the year of publication of A. L. Primavesi's paper comparing the flora in the vicinity of Ratcliffe College as it was in the mid-sixties with how it had been forty or more years before, using the lists compiled by J. A. Cappella in the course of his field-by-field survey of the area (Primavesi 1967). It was this publication which brought about the discussions between Primavesi and the author of this section which resulted, as has been mentioned elsewhere, in the setting up of the Flora Committee.

Teamwork was the essence of much of the local botany of the late sixties, but individuals also made important contributions. An example is the survey by R. C. Jones of the floristics of field ponds, which was carried out in 1967-68 (Jones 1971). This paper showed that a high proportion of the field ponds present in the thirties had disappeared by the sixties and also drew some interesting comparisons between the species present in these relatively old-established ponds (where they still existed) and those of a set of recent subsidence ponds in the Holwell area. This work has been further developed in recent years by Jane Beresford at Loughborough University.

The time and efforts of many of the county's botanists have been taken up during the seventies with the flora survey. However, running parallel with this work there has been a separate programme of botanical surveys. These have been undertaken for a variety of purposes, to improve our knowledge of the land-use patterns in poorly known areas of the county, to map the vegetation of sites known to be of interest and to establish the significance on a national level of certain types of site. Most of this work has been carried out by Mrs. P. A. Candlish (later Mrs. P. A. Evans) under the aegis of the Leicestershire Museums Service and financed by that organisation, the county trust and the Nature Conservancy Council. It has included surveys of roadside verges, stand-type analysis of all woodland S.S.S.I., field-by-field surveys of a large number of areas that were the subject of planning applications, including those for the North-East Leicestershire Coalfield and the Improvement Schemes for the Rivers Soar and Wreake. In addition a number of little known parishes have been surveyed field-by-field, yielding in some cases sites of considerable importance that had even escaped the keen eyes of the flora survey workers, such as the *Orchis morio* meadows at Muston. This work is continuing and resulted in 1983 for instance in the discovery of a second station for *Scirpus lacustris* subsp. *tabernaemontani* in the county.

What, then, is the likely pattern for the development of Leicestershire botany in the nineties? There will be a race to locate the remaining isolated unimproved pastures, meadows and marshes before they are improved or ploughed up, so that a few may be conserved. A recent growth of interest in historical ecology may well focus attention on the fifty or so remaining ancient woodlands in the county, which, although neglected, do nevertheless provide fascinating opportunities for a theoretical reconstruction of the floristics of the 'wildwood' from which they were derived. Detailed studies of the flora of single parishes could provide explanations in terms of past human activity for some of the distribution patterns manifested elsewhere in this volume. With disturbance of our landscape showing no sign of diminishing in scale, the hunt for aliens will no doubt continue to provide enjoyment and interest for some. Inevitably however, much of our effort will be directed towards conserving a representative cross-section of the habitats and vegetation types we have inherited, for the enlightenment and enjoyment of future generations.

References

Armstrong, S. F., 1907. The botanical and chemical composition of the herbage of pastures and meadows. *J. agric. Sci.* **2**, 283-304.

Evans, I. M., 1973. F. A. Sowter, F.L.S. – an appreciation. *Trans. Leicester lit. phil. Soc.* **67**, 20-24.

Horwood, A. R., 1907. Palaeontology of the coalfield *in* Fox-Strangways, C. The Geology of the Leicestershire and South Derbyshire Coalfield. *Mem. Geol. Surv.* 114-141.

Horwood, A. R., 1908. The fossil flora of the Leicestershire and South Derbyshire Coalfield, and its bearing on the age of the Coal-Measures. *Trans. Leicester lit. phil. Soc.* **12**, 81-181.

Horwood, A. R., 1909. The cryptogamic flora of Leicestershire. *Trans. Leicester lit. phil. Soc.* **13**, 15-86.

Horwood, A. R., 1920. *The outdoor botanist*. London: T. Fisher Unwin.

Jones, R. C., 1971. A survey of the flora, physical characteristics and distribution of field ponds in North East Leicestershire. *Trans. Leicester lit. phil. Soc.* **65**, 12-31.

Measham, C. E. C., 1915. A botanical survey of some fields near Leicester. *Trans. Leicester lit. phil. Soc.* **19**, 17-25.

Mercer, G. E., 1914. The flora of Belgrave and Birstall. *Trans. Leicester lit. phil. Soc.* **18**, 76-92.

Primavesi, A. L., 1967. Changes in the flora of part of Leicestershire since 1884. *Proc. bot. Soc. Br. Is.* **6**, 343-347.

Wade, A. E., 1919. The flora of Aylestone and Narborough Bogs. *Trans. Leicester lit. phil. Soc.* **20**, 20-46.

3. Biographical notes

The following notes relate to Leicestershire botanists who have died since the publication of the 1933 *Flora*. They therefore include some individuals whose contributions to local botany began more than a century ago, others who have died in the last five years. They are arranged in order of date of birth.

The notes are derived to a large extent from files on Leicestershire naturalists, past and present, held by the Natural Sciences Division of the Leicestershire Museums Service. In some cases they have been written by colleagues or friends still living, in others they draw upon unpublished material gathered from a great variety of sources and people. It is recognized that some Leicestershire botanists deserve a fuller treatment than can be given here, but the research required has so far proved beyond the resources of time that could be allocated to it. Corrections, emendations and additional information, whether factual or anecdotal, would be welcomed.

In general more attention has been given to those essentially of local interest, on whose lives little or nothing has so far been written.

W. A. Vice, M.B, L.D.S., F.E.S. (1852-1937)

William Armston Vice was born on 14th May 1852, son of William Vice, corn miller at Blaby. He graduated in medicine from Aberdeen University in 1875 and after qualifying in dentistry returned to his home county to practice. His practice was in Belvoir Street, Leicester, but he lived for the rest of a long life at Blaby, where he served for many years on the Parish Council. A staunch Free Churchman, he was a member of the Baptist Church at Blaby for over 60 years and for over 40 of those Superintendent of the Sunday School and Treasurer.

His natural history interests were wide, embracing groups such as fungi (Horwood 1909), flowering plants, plant galls, Hymenoptera and Diptera. It is as an entomologist that he is perhaps best known. His copy of Smith (1855) in the Museum Services' library records captures of solitary bees as early as 1873, when he was an undergraduate at Aberdeen, and he made the first substantial contributions to our knowledge of the Leicestershire Hymenoptera and Diptera, his records being summarised in the Victoria County History (Bouskell 1907).

He first joined the Leicester Literary and Philosophical Society in 1880 and was an active member of the various botanical, entomological and zoological sections for about 40 years. He donated specimens to the Society's herbarium during most of this period, but his most noteworthy contribution to Leicestershire botany was a paper describing an assemblage of alien and casual species observed at Blaby Mill in 1903 (Vice 1905). His discovery there of *Myosurus minimus*, in the spring following the closure of the Mill, drew his attention to a flora which was 'certainly not the general herbage of the neighbourhood'. He listed in all 143 species of interest, the origin of many of them being corn from Russian and Turkish ports imported for processing as poultry food. Several of the grasses were sent by A. B. Jackson to E. Hackel in Vienna, where one proved to be identical with *Apera intermedia*, newly described from a volcano in Asia Minor. Its occurrence at Blaby constituted the first British record. His list is a good example of the opportunism required of those interested in alien floras since he records that in the following year, 1904, 'nothing appeared worthy of note except one plant apparently of *Centaurea melitensis* which had no opportunity to bloom, and three plants of chicory'. His records were given a wider audience by his friend A. B. Jackson (Jackson 1904, 1907).

Dr. Vice maintained what might seem to us an unusual breadth of interests until late in life. He served on the Flora Committee until after the First World War and, besides those activities already mentioned, he took a keen interest in archaeology and was an ardent supporter of the temperance movement, being connected with the Band of Hope Union and the Temperance Society. He died on 18th January 1937 in his 85th year leaving a widow, but no children.

References

Anon. (1937). [Obituary] *Leicester Mercury*, 19.1.1937, 6.

Bouskell, F., 1907. *Hymenoptera* and *Diptera* in *The Victoria History of the County of Leicester*. **1**, 64-65, 89-91. London: Archibald Constable.

Horwood, A. R., 1909. The cryptogamic flora of Leicestershire. VIII. Fungi. *Trans. Leicester lit. phil. Soc.* **13**, 47-60.

Jackson, A. B., 1904. Leicestershire plant notes, 1886-1904. *J. Bot.* **42**, 337-349.

Jackson, A. B., 1907. *Apera intermedia* as an alien in Britain. *Ann. Scot. nat. hist.* July 1907, 170-171.

Smith, F., 1855. *Catalogue of the British Hymenoptera in the collection of the British Museum. Apidae.* London: British Museum.

Vice, W. A., 1905. List of casuals gathered at Blaby Mills, 1903. *Trans. Leicester lit. phil. Soc.* **9: 2**, 105-109.

Rev. J. A. Cappella (1858-1943)

The following note has been contributed by Rev. A. L. Primavesi, a confrère of Cappella's.

James Anthony Cappella was born in South Wales in 1858. He was at school at Ratcliffe College from 1868 to 1875, and then joined the Roman Catholic religious order called the Institute of Charity. He went to Italy in 1877 to study for the priesthood, but ill health necessitated his return to England in 1879, and he remained throughout his life a professed brother of the Institute. The rest of his long life was spent at Ratcliffe College, where he taught science for fifty years until his retirement in 1929. He was far too gentle a character to be a good disciplinarian, and the boys ragged him unmercifully, yet surprisingly the examination results of his pupils were always exceedingly good. His scientific knowledge was very wide and thorough, though he was almost entirely self-taught, and he was a very keen observer and collector. When the British Association met at Leicester in 1933 they made a special excursion to Ratcliffe College to examine his geological collection, and he contributed the results of detailed field surveys and numerous records to A. R. Horwood when the 1933 *Flora* was in preparation. His herbarium collection is now housed in the Leicestershire Museums (acc. no. B25.1982), and provides a valuable record of the flora of an area of the county centred on the College from 1884 to 1936. His field notes, written in a neat hand in school exercise books, are still extant at Ratcliffe College, and show that he remained active throughout his long retirement; the last note was added a few months before his death. Of particular interest are comprehensive species lists from fields and copses in the immediate vicinity of the College, coupled with notes on the management of the areas concerned. To these are appended detailed records for the verges of the Fosse Way north from Syston Brook to Six Hills, for other localities such as Twenty Acre in the vicinity of the Fosse Way and for a stretch of the River Wreake from Ratcliffe Mill to Lewin Bridge. To the contemporary botanist familiar with this landscape these make fascinating if depressing reading. Their existence provoked a follow-up study in the nineteen-sixties (Primavesi 1967) which, as related elsewhere, led to the establishment of the Committee responsible for this Flora. Copies of the notebooks have been deposited with the Museums Service. So far as is known, the only publications by Cappella were contributions to the College magazine *The Ratcliffian* in the late eighteen-eighties which included notes on the local flora, the geology of Mountsorrel, 'The history of a boulder' and a short list of the micro-fungi found in the vicinity of the College. Also preserved at the College are some of Cappella's water-colour sketches, which show that he was a competent artist. He had considerable mechanical skill, and made most of his own scientific apparatus. He died at Ratcliffe College in September 1943, aged 85.

Reference

Primavesi, A. L., 1967. Changes in the flora of part of Leicestershire since 1884. *Proc. bot. Soc. Br. Isl.* **6**, 343-347.

G. C. Turner, J.P., F.L.S., F.R.G.S. (1858-1940)

Born in Leicester on 13th August 1858, George Creswell Turner was a pre-eminent example of generations of Victorians who managed to combine a demanding business career with active participation in local affairs, whilst retaining all their lives an enthusiastic interest in natural history. The following notes are based on an obituary by C. T. Ingold (Ingold 1941).

Throughout his life Turner was associated with the city of Leicester. His business was the wholesale boot and shoe trade, where he became in due course Chairman of W. and E. Turner Ltd. He took a keen and practical interest in church affairs, in civic matters and in education. In 1928 he celebrated 50 years as a Sunday School teacher, he was three times President of the Leicester Free Church Council and had also been President of the Free Church Federation. He was elected to the Leicester Town Council in 1892, serving for six years and retaining for some time thereafter a seat on the Education Committee. He was governor of several Leicester schools and Chairman of the College of Art and Technology.

All his life he was a keen observer of nature, being particularly interested in flowering plants and birds. He first joined the Leicester Literary and Philosophical Society in 1878 and was a member of the Committees of the Natural History and Microscopical Sections from 1879 to 1883. His dual interests in botany and microscopy are reflected in the subject of his first paper on 'The reproduction of ferns', read to the Biology Section (as it had become) on 17th October 1883 and 'illustrated by living specimens, microscopic slides, and original drawings'. He gave occasional papers over the next decade or so and became in turn Vice-Chairman and Chairman of the Section, the latter in 1896. He held this office, or latterly that of President, at intervals from 1896 to 1913 and from that year continuously until his death nearly thirty years later. He was elected F.L.S. in 1897, Chairman of the Flora Committee in 1912 and served as President of the Literary and Philosophical Society itself in 1915-1916.

Turner appears to have collected plants for only a relatively brief period in his twenties. His chosen role in later years was one of encouragement and assistance, stimulating work by other local botanists. In 1927 for instance he purchased A. R. Horwood's Leicestershire collections for the Society's herbarium, and paid for their mounting. When the funds required for the publication of the 1933 *Flora* fell short of the total required it was he who 'generously undertook to subscribe the balance'. Horwood acknowledges that without his help the *Flora* would never have appeared, and it contains both a portrait of him and 'several excellent pictures of Charnwood Forest taken by him close to his mountain home'. The last is a reference to his estate at Abbot's Oak on the edge of the Forest near Whitwick where he grew many interesting plants. His love of plants led him to travel extensively in Norway, Switzerland, Palestine and the Mediterranean countries, from which he brought back photographs of the vegetation. His horticultural interests were also reflected in his chairmanship of the Gardens Committee of the University College, Leicester.

He was taken ill at work in his office and died very shortly afterwards on 17th October 1940, at the age of 82. He was married and had at least one daughter who died unmarried, it is believed, in the nineteen-fifties. Prior to moving out to Abbot's Oak he had constructed in 1900-1901 in Elmfield Avenue, Stoneygate, the house known as 'The Beeches' which is a fine example of Art Nouveau style, with natural history motifs in the plaster and metal work, and a conservatory, since demolished, which was large enough to take palms.

Reference

Ingold, C. T., 1941. [Obituary of George Creswell Turner] *Proc. Linn. Soc. (Lond.)* **1940-41**, 301-302.

Miss C. E. C. Measham (1863-1937)

Charlotte Elizabeth Cowper Measham was born in London on 2nd July 1863. She was educated at Devonport High School and entered Newnham College, Cambridge, in 1897, as what we should now call a 'mature student'. She took the Natural Sciences Tripos, getting a First in Part I in 1900 and a Third in Part II (Botany) in 1901. She then took up employment as a teacher at Rochester High School, staying there until 1906, taught at Newcastle-on-Tyne Central High School from 1906 until 1912 and came to Leicester in 1912 where she taught science at the Wyggeston Girls School until her retirement in 1926.

Miss Measham's talents appear to have been recognized soon after she arrived in Leicester, for we find recorded (Turner 1913) that on October 16th 1912 she gave 'a most interesting and instructive lecture on the "Assimilation of Carbon" ' to Section D (Botany) of the Literary and Philosophical Society. Between this date and 1926 when she retired as Vice-Chairman of the Botany Section, she was one of its mainstays, particularly during the war years.

Her other major contributions to local botany were twofold. First was her work on the county herbarium, the state of which she summarised in a paper to Section D on 10th December 1914 (Measham 1914). This opens 'For the past eighteen months I have been able to give a few hours weekly to work in the Herbarium; during that time most of the specimens have passed through my hands. The work has been undertaken in connection with the publication of a new Flora of Leicestershire'. Her continued contribution to the work on the Flora through the war years is acknowledged by A. R. Horwood in the *4th Report of the Flora Committee 1916-1919* (Horwood 1919).

Miss Measham's second major contribution to local botany was her account of a field-by-field survey of an area to the west of Leicester undertaken in the summers of 1913 and 1914. This was read to Section D on 17th February 1915 and published that year with a hand-drawn map (Measham 1915). It gives a detailed and colourful summary of the flora of an area, alas now largely built-over, relating the species present to the management of individual parcels. It describes, for example, damp hayfields 'characterised by an abundance of *Rhinanthus*', some of which contained *Orchis morio* and '*O. maculata*', others *O. mascula*. There are observations on what we now recognise as woodland relict indicator species such as *Anemone*, *Galeobdolon* and *Oxalis* in certain hedges, together with notes on the flora of streams, ponds, spinneys and the weeds of arable fields. Although the species lists for each field, which were forwarded to A. R. Horwood, have not survived, Miss Measham's splendid paper does give us an evocative picture of the flora of an ordinary area of agricultural land in the early part of the century. She ends 'although the vegetation is in no way remarkable, I think that every field I have visited has had some special interest of its own, and I believe work of this kind very amply repays anyone who undertakes it', testimony to the persuasiveness of A. R. Horwood, who induced her to undertake the work which she 'found so enjoyable'.

Whether Miss Measham continued her survey work in further seasons we do not know. Certainly she was kept busy, because besides her normal teaching duties, she took over in November 1917 biology classes at the Museum and in 1921 became the first lecturer in botany at what was to become later the University of Leicester. In this last capacity, she aided A. R. Horwood in the construction and layout of the first university botanic garden (Gornall 1983).

In 1924 she published a novel 'The Sin of Gehazi' and in 1926, several years past normal retirement age, she finally gave up teaching and moved to Teignmouth, where she died on 17th March 1937 in her 74th year. Her papers and the many other references to her in the reports and committee papers of the Literary and Philosophical Society bear eloquent witness to her interest, energy and enthusiasm in the botanical field and a tribute by a former colleague (Mills 1937) indicates the high regard in which she was held as a teacher.

References

Gornall, R. J., 1983. A history of Leicester University Botanic Garden. *Trans. Leicester lit. phil. Soc.* **77**, 42.

Horwood, A. R., 1919. 4th Report of the Flora Committee *in* Report of Section D – Botany. *Trans. Leicester lit. phil. Soc.* **20**, 76.

Measham, C. E. C., 1914. The Society's herbarium *in* Report of Section D – Botany. *Trans. Leicester lit. phil. Soc.* **18**, 68-72.

Measham, C. E. C., 1915. A botanical survey of some fields near Leicester. *Trans. Leicester lit. phil. Soc.* **19**, 17-28.

Mills, N., 1937. [Tribute to Miss C. E. C. Measham]. *The Wyggeston Girls' Gazette.* July 1937, 64.

Turner, G. C., 1913. Report of Section D. Botany. *Trans. Leicester lit. phil. Soc.* **17**, 88.

A. B. Jackson, F.L.S. (1876-1947)

Albert Bruce Jackson was born at Newbury, Berkshire, on 14th February 1876. He grew up at Newbury and after leaving school became a junior reporter on the *Newbury Express*. In 1897 he moved to Leicester to advance his career in journalism, and was soon involved in local botany, being elected a member of the Biology Section of the Literary Philosophical Society on 15th December of that year. From a reference to his 'lamented friend' T. A. Preston (Jackson 1906), we may infer that he was rapidly assimilated into the botanical community of the county.

In the following ten years he made a considerable contribution to knowledge of both the cryptogamic and phanerogamic flora of the county. His 'immense capacity for work and attention to minutiae' (Anon 1947b) is demonstrated both in the meticulous compilations of records that had accumulated since the last Flora, incorporating many of his own (Jackson 1904, 1905), and also by his work on the more taxing groups such as *Rubus* (Jackson 1899, 1906) and *Euphrasia* (Jackson 1904). He delighted in chasing up old records, found amongst herbarium material, as in the case of *Puccinellia distans* (Jackson 1903b) and also had a keen eye for the oddity or hybrid in the field, for example his discovery new to the county of *Alopecurus* × *hybridus* (Jackson 1903a).

In February 1907 he was able to realise a youthful ambition, leaving Leicester to take up a post as a temporary technical assistant in the Herbarium at Kew, the start of a distinguished career as a professional botanist. He left that post soon afterwards to assist Elwes and Henry in the preparation of their great work *The Trees of Great Britain and Ireland* and in 1910 was appointed as technical assistant at the Imperial Institute where he served for 22 years. In the remaining years of his life he was employed as a specialist in the Department of Botany at the British Museum (Natural History).

As a professional botanist he is best remembered for his work on trees, notably his collaboration with W. Dallimore in *A Handbook of Coniferae*, but he retained throughout his life an interest in the British flora, publishing amongst others important papers on *Barbarea* and *Thymus*. His services to botany were recognised by his election as an Associate of the Linnean Society in 1917 and the award of the Veitch Memorial Medal of the Royal Horticultural Society in 1925.

His personal qualities were summarised by J. E. Lousley in an obituary from which much of the above has been taken (Lousley 1947). He wrote 'the outstanding quality of Bruce Jackson was quickness – in speech, in actions and in work. He was overflowing with enthusiasm and his enthusiasm was infectious. His capacity for work was immense and he was always busy. Never idle for a moment and moving about at a speed which would shame many a younger man, he achieved a great deal more in life than might be supposed from a catalogue of his publications. Much of his time was spent in helping others and in tasks which received little or no publicity. Those who had the advantage of his acquaintance have been deprived of a cheery and ever helpful friend.'

He died at his home in Kew on 14th January 1947 in his 71st year, leaving a widow and three sons. We are fortunate that ten years of a productive life were spent in Leicestershire.

References

Anon., 1947a. Obituary. *The Times* 16th January 1947, 7.

Anon., 1947b. Obituary. *The Gardeners' Chronicle* **121**, 46.
Dallimore, W., 1947. Obituary. *Nature (Lond.)* **159**, 156.
Jackson, A. B., 1899. *Rubus kaltenbachii* in Leicestershire. *J. Bot.* **37**, 136.
Jackson, A. B., 1903a. *Alopecurus hybridus* in Leicestershire. *J. Bot.* **41**, 58.
Jackson, A. B., 1903b. *Glyceria distans* var. *obtusa*. *J. Bot.* **41**, 59.
Jackson, A. B., 1904., Leicestershire plant notes, 1886-1904. *J. Bot.* **42**, 337-349.
Jackson, A. B., 1905. Leicestershire mosses. *J. Bot.* **43**, 225-231.
Jackson, A. B., 1906. Charnwood Forest *Rubi*. *J. Bot.* **44**, 261-266.
Lousley, J. E., 1948. Obituary. *Watsonia* **1**, 123-124.
Turrill, W. B., 1947. Obituary. *Proc. Linn. Soc. (Lond.)* **158**, 132-133.

A. R. Horwood, F.L.S. (1879-1937)

Arthur Reginald Horwood was born on 28th May 1879, the son of the Rev. F. E. Horwood, M.A., later Rector of South Croxton, Leicestershire. He was educated at St. John's Foundation School, Leatherhead, and it was intended that he should enter the Indian Civil Service. However he failed the medical examination and after a short period as a private tutor and army coach, his early interest in natural history, especially botany, prevailed and in 1902 he became Sub-Curator at Leicester Museum. He occupied that post for some 20 years, except for a period from August 1916 to March 1919 when he was on home service with the Cheshire Regiment. The annual reports of the Museum chronicle his dedicated development of its natural history collections and displays, in particular those of recent and fossil plants. The museum herbarium was transformed and a reference collection of British species was mounted, bound up and provided in the gallery for consultation by interested members of the public. In May 1908 he initiated labelled displays of living plants from the vicinity of Leicester, the 'wild flower tables', which were still being maintained 50 years later. During this time he wrote a number of popular books on the British flora.

In 1922 he resigned from the Museum. It may be coincidence, but this was also the year of the death of his first wife, Alice Maude, at the comparatively early age of 43. Certainly we know from his correspondence that early in 1923 he was both in poor health and considerable financial difficulties. It was at this time, for instance, that he sold off some of his personal collections. For a short while he took an editorial post with the *Hinckley Journal*, but this did not turn out as expected and in September 1924 he accepted a post as 'temporary botanist' in the Herbarium at Kew, where he remained until his death.

In some respects he was ahead of his time, for in 1907 he read a paper to the Botany Section of the British Association assembled in Leicester, on the disappearance of cryptogams from Charnwood Forest, especially the classic locality of Bardon Hill, due to quarrying and air pollution (Horwood 1907a; see also Hawksworth 1974). He also became Recorder of the Plant Protection Section of the Selborne Society, founded to further efforts to conserve the flora of Britain (Horwood 1913a). The sentiments he expressed and the local bye-laws which followed as the result of the Selborne Society's campaign did not meet with universal approval, as is indicated by the comment in a local newspaper quoted by Sowter (1960), describing him as 'a moss-faced botanist who wished to debar children from picking a few wild violets for their mother's grave'.

He was very interested in the relatively new science of plant ecology and encouraged local botanists to undertake ecological surveys in the county. His book *The Outdoor Botanist* (Horwood 1920) contained extracts from such local surveys, of which those by Mercer (1914), Measham (1915) and Wade (1919) were published in full elsewhere. The cryptogams also claimed his attention and he made considerable collections of fungi, lichens and bryophytes in the county, summarising all the records to date in a substantial paper (Horwood 1909). He was one of those instrumental in the establishment of the Lichen Exchange Club (Horwood 1907b) and Secretary until its closure in 1914. His association with the Leicester Literary and Philosophical Society was mainly on the geological side until 1912, when the Botany Section set up a Sub-Committee to undertake the production of a new Flora. Horwood was appointed General Editor and threw himself energetically into the assembly of information for it (Horwood 1913b, 1914, 1915, 1919). Thereafter he also took a more active role in the general activities of the Section, lecturing and leading field excursions. A member of the Botany Section and noted algologist, Florence Rich, honoured him by naming a new variety of an alga which was first found in Leicestershire *Cosmarium obtusatum* var. *horwoodii*. He was elected a Fellow of the Linnean Society in December 1913.

Always interested in young people, he recruited a group of schoolboys whose botanical studies he encouraged. Mainly from the nearby Wyggeston Boys' School, they became known at the Museum as 'Horwood's Boys'. Amongst those who, under his supervision, assisted in collecting for the herbarium, recording for the Flora and providing material for the flower tables were the following: R. S. Creed, later a chemistry don at Magdalen College, Oxford; T. M. Harris, a noted palaeobotanist and later Professor of Botany at Reading; G. E. Mercer, killed in the First World War; F. A. Sowter; S. A. Taylor; and A. E. Wade, who contributed so much to Welsh botany (and now lives in retirement in New Zealand). All received encouragement and help from him at the onset of their interest in botany.

Had Horwood done no more for Leicestershire botany than the collecting, recording and writing which he carried out during the two decades prior to his leaving the Museum in 1922, then he would be remembered for his energy and industry. It is however his *Flora*, published in 1933, for which local botanists will be eternally grateful. Its preparation occupied him, as he states, characteristically, in the Preface, for 'only eighteen months, and that solely out of my *spare time*.' A massive work, running to 984 pages of small type, it is a superb and fitting monument to his scholarship and wide knowledge of the British as well as the Leicestershire flora.

He died some four years after the publication of the *Flora*, on 21st February 1937, at Brentford in Middlesex. He was buried beside his first wife in the churchyard at Scraptoft, a village he greatly loved and through which he passed often on his botanical excursions. He was survived by his second wife and four sons.

These notes are based on an account prepared by F. A. Sowter, augmented from detailed studies of Horwood's life and works made in 1979-80 by N. J. Moyes and H. E. Stace of the Department of Museum Studies, University of Leicester, the texts of all of which are deposited with the Leicestershire Museums Service.

References

Hawksworth, D. L., 1974. The changing lichen flora of Leicestershire. *Trans. Leicester lit. phil. Soc.* **68**, 32-56.

Horwood, A. R., 1907a. On the disappearance of cryptogamic plants. *J. Bot.* **45**, 334-339.

Horwood, A. R., 1907b. A proposed exchange club for lichens. *J. Bot.* **45**, 412.

Horwood, A. R., 1909. The cryptogamic flora of Leicestershire. *Trans. Leicester lit. phil. Soc.* **13**, 15-86.

Horwood, A. R., 1913a. The State protection of wild plants. *Rept. Trans. Brit. Assoc. Advmt. Sci. (for 1912)*: 764-766.

Horwood, A. R., 1913b. The Leicestershire Flora Committee Report for 1911-1912. *Trans. Leicester lit. phil. Soc.* **17**, 91-92.

Horwood, A. R., 1914. Report of the Flora Committee. *Trans. Leicester lit. phil. Soc.* **18**, 75-76.

Horwood, A. R., 1915. Report of the Flora Committee. *Trans. Leicester lit. phil. Soc.* **19**, 59-61.

Horwood, A. R., 1919. 4th Report of the Flora Committee 1916-19. *Trans. Leicester lit. phil. Soc.* **20**, 76-78.

Horwood, A. R., 1920. *The Outdoor Botanist*. London: T. Fisher Unwin.

Measham, C. E. C., 1915. A botanical survey of some fields near Leicester. *Trans. Leicester lit. phil. Soc.* **19**, 17-25 (with an Appendix by A. R. Horwood, 25-28).

Mercer, G. E., 1914. The flora of Belgrave and Birstall. *Trans. Leicester lit. phil. Soc.* **18**, 76-92.
Sowter, F. A., 1960. Our diminishing flora. *Trans. Leicester lit. phil. Soc.* **54**, 20-27.
Wade, A. E., 1919. The flora of Aylestone and Narborough Bogs. *Trans. Leicester lit. phil. Soc.* **20**, 20-46.

Miss L. E. Cheesman, O.B.E., F.R.E.S. (1881-1969)

Lucy Evelyn Cheesman was born at Westwell, near Ashford, Kent. In her autobiography (Cheesman 1957) she describes her childhood as 'care-free happy days of soaking in wildlife' and it was then that she acquired an abiding interest in natural history. Educated privately, she acquired a good knowledge of French and German which enabled her to take a post as governess with the Murray-Smith family at Gumley Hall in south Leicestershire. She spent about ten years at Gumley, with a break of some eighteen months in 1904-05, when she visited Germany. Despite very restricted time for personal recreation, she recalled these years 'chiefly for the opportunities to continue the study of botany and for the hours snatched at night for watching badgers and foxes'. The latter activity, to which she was introduced by Mr. Adkins the local earth-stopper, takes up much of the section of her autobiography devoted to Gumley, but she nevertheless made some interesting botanical finds during her stay there and on later visits to friends. Notable amongst these was her discovery in 1913 of *Pyrola minor*, new to Leicestershire, beside the pool known as the Big Mott at Gumley, in the company of *Melampyrum pratense*, with *Sphagnum* in the pool itself. Other records credited to her in the 1933 *Flora* include *Botrychium lunaria* in Gumley Wood and *Orchis morio*, found, no doubt in a meadow not far away, on a wartime visit in 1916.

She left Gumley in 1912 when her second charge Elizabeth (later to marry the Highland naturalist Seton Gordon) came of an age to go to school. Unable to obtain entry to the Royal Veterinary College, because of her sex, Miss Cheesman took a post at a dog hospital near Croydon, where she qualified as a canine nurse. She spent the war years as a temporary civil servant at the Admiralty. Shortly afterwards, whilst improving her typing at the Imperial Institute, she obtained an interview with Professor Maxwell Lefroy, who offered her the post of Curator of Insects to the Zoological Society of London. She held this post for six years, but took leave of absence from 1924 to join an expedition to the West Indies, Panama, Galapagos Islands and the South Pacific.

Thus began a unique career as a lone female entomologist and explorer in the islands of the Pacific, which lasted for over thirty years. The experiences of this period are described in her autobiography and a number of books, the results of her work in a long series of scientific papers, mainly on Hymenoptera and Diptera. In recognition of her services to Science she was awarded a Civil List Pension in 1953 and an O.B.E. in 1955.

She died on 15th April 1969 at the ripe age of 88 and a colleague of her later years, K. G. V. Smith of the Department of Entomology at the British Museum (Natural History), wrote of her, 'in common with many others, on first meeting this frail little woman I could hardly believe that she could have survived all the adventures of which I had read . . . Few women can have led a more active, daring and fiercely independent existence and derived such complete satisfaction from it . . . Only time will tell if her like is being bred today; perhaps the social and scientific background will never be quite the same again' (Smith 1969). Her scientific career was not launched until she was in her thirties, but the strength of her interest in wildlife and the indomitability of her character was already manifest in her early years at Gumley, where she is still remembered with affection.

References

Cheesman, L. E., 1957. *Things Worth While*. London: Hutchinson.
Smith, K. G. V., 1969. Obituary. *Ent. mon. Mag.* **105**, 217-219, pl.IV.

The Rev. D. P. Murray, M.A., F.R.E.S., O.P. (1887-1967)

Desmond P. Murray was born in London in 1887. Little is known of his early years, but whilst preparing for the priesthood he studied philosophy and theology at the University of Freiburg, Switzerland. During this period he spent his holidays walking in the mountains, collecting specimens of the Alpine flora and bird-watching.

After completing his studies, he joined the Dominican Order and was sent by them to South Africa, where he served in the missions for a decade or more. His interest in botany was stimulated by studies of the unique flora of that region of the world, but he also developed an interest in the Lepidoptera. He was at one time Honorary Curator in Entomology in the Department of Zoology in Witwatersrand University at Johannesburg, and wrote a number of papers on South African butterflies, notably a monograph on the Lycaenidae (Murray 1935).

In 1937 he was recalled to England and spent several years at the Holy Cross Priory in Leicester. His interest in Lepidoptera continued unabated, witness a note on plume moths in Suffolk (Murray 1942) and further papers on South African butterflies (Murray 1944, 1948). As a 'retiring' post, he was appointed in 1950 Chaplain to the Dominican Sisters of St. Martin's Convent at Stoke Golding near Hinckley. He continued to pursue his natural history interests there, collecting Lepidoptera and, in about 1953, started to put together a herbarium of British plants. This amounts to approximately 1860 specimens and is housed in St. Martin's Catholic High School at Stoke Golding. A microfilm of the herbarium was made in 1977 by the Leicestershire Museums Service and is available for consultation. It inevitably contains a considerable amount of Leicestershire material, but also specimens from as far afield as Kent, Kinross, Kilkenny and Donegal. The Leicestershire specimens provide a number of interesting records for the south-western part of the county, but in view of the fact that most of them date from the nineteen-fifties they do not feature at all conspicuously in the text of the present *Flora*. According to Kent (1958), an earlier herbarium collection is to be found at the University of Cork. A collection of some 2000 specimens of butterflies, mainly South African in origin, with associated manuscripts, is held by the Merseyside Museum Service in Liverpool.

Father Murray's herbarium reveals that he was still actively collecting at Dunvegan in Skye in his 74th year; he died six years later on 2nd March 1967. Sister M. Louis Bertrand, Headmistress of the High School, who knew Father Murray and upon whose notes on his life much of this account is based, wrote in 1978 that his appointment to the Convent 'pleased him well . . . In spite of his many interests, or because of them, he was a deeply religious man carrying out his duties faithfully to the end, which came unexpectedly . . . after having said mass in the Convent Chapel. Returning to his cottage, he collapsed and died at the doorsteps, a botanist, ornithologist, lepidopterist and priest.'

References

Kent, D. H., 1958. *British Herbaria*. London: B.S.B.I.
Murray, D. P., 1935. *South African Butterflies. A monograph of the family Lycaenidae*. London: John Bale, Sons and Danielsson.
Murray, D. P., 1942. On our plume moths. *Trans. Suff. Nat. Soc.* **5:1**, 7-10.
Murray, D. P., 1944. The genus *Cupido* (Lepidoptera: Lycaenidae) in South Africa. *Journ. ent. Soc. S. Africa* **7**, 82-95.
Murray, D. P., 1948. The genitalia of some South African lycaenids (Lepidoptera: Lycaenidae). *Journ. ent. Soc. S. Africa* **10**, 182-?192.

W. P. Powell (1887-1954)

William Percy Powell was born in Hinckley on 22nd September 1887, trained as a teacher and taught for some years in the Boys' School in Hollies Walk, Hinckley. He became in 1931 the first

Principal of the Hinckley Technical College, from which post he retired in 1953. He died the following year on 9th July.

In his youth he made a small collection of plants, mainly from south-west Leicestershire, which was in 1967 donated to the Leicester Museum by his daughter Mrs. D. M. Hodgkin, one of the survey workers for the present Flora. Dates on the specimens and in a notebook accompanying it range from 1905 to 1923, but most are in the seven years 1913-1919. A few of the specimens were evidently collected on his way to or return from the front in France during the First World War. Of particular local interest, however, are detailed records, supported by specimens, of the flora of Burbage Common and Wood, together with that of the diorite outcrops which are a feature of the south-west of the county. He published in *The Hinckley Guardian and South Leicestershire Advertiser* for 2nd and 16th November 1923 a note on 'The Flora of Stoney Stanton, Sapcote and District' which lists 261 species.

Although he was, at one time at least, in correspondence with A. R. Horwood and it is believed that he contributed field notes for the 1933 *Flora*, none of Powell's records appear to be mentioned in that work. This is a pity, since that of *Teesdalia nudicaulis* for example, which he found at Barrow Hill in 1917, predates the record in the Flora by a decade or more. Curiously, amongst 'Recent Correspondents (1902-1933)' to whom Horwood acknowledges his thanks is listed 'J. Powell, Hinckley', presumably John Powell (1865-1954), father of William Percy, who was, according to his grand-daughter, also interested in natural history.

G. J. V. Bemrose, F.M.A. (1896-1972)

Born in Leicester on 12th October 1896, Geoffrey John Bemrose was educated there and on leaving school worked for a short period in the Leicester City Library. He then joined the staff of the Great Eastern Railway Company, but a career with them was interrupted by the First World War. He enlisted in the Leicester Regiment in 1914 and continued to serve, in Aden amongst other places, after the cessation of hostilities in 1918. Following demobilisation, he took up in 1922 the post of Museum Assistant with the Leicester Museum and Art Gallery, where he was employed until his appointment as Curator of the Stoke-on-Trent Museum and Art Gallery in 1930.

Copies of the 1886 *Flora of Leicestershire* annotated by Bemrose have recently been donated by his family to the Leicestershire Museums Service together with other personal papers. Numerous marginal notes in his characteristically neat hand testify to a keen and wide-ranging interest in both phanerogams and fungi. An interest in flowering plants appears to have been kindled in his mid-teens, when he may well have come under the influence of A. R. Horwood. However his main work in the county was done from 1924-1930. In 1924 he became Hon. Secretary of the Botany and Biology Sections of the Literary and Philosophical Society, in the following year Hon. Secretary of the Flora Committee (on the death of W. Bell) and in 1926 Curator of the Society's herbarium. 'Selected Observations', attached to the Botany Section's Reports for 1924-25 and 1925-26, contain evidence of two seasons' active fieldwork, much of it in the company of W. Gardner.

Bemrose's energy and skill in diagnosis bore fruit in a substantial paper in the Society's *Transactions* for 1926-27 on 'The adventive flora of Leicester and district' (Bemrose 1927), containing past observations, but also incorporating many from his own fieldwork. In all over 300 species are listed. A shorter paper, published in the *Transactions* for the following year (Bemrose 1928), summarises the results of four seasons botanising in Rutland, again with W. Gardner. His work towards the new *Flora* is acknowledged by G. C. Turner on the flyleaf of a copy of Horwood and Gainsborough presented by him to Bemrose in 1933. Most of his flowering plant records found their way into either his papers or the *Flora*, although the copies of the 1886 *Flora* annotated by him contain some information that did not and his fungus records have never been published at length.

Following his move to Stoke-on-Trent his interests broadened out into archaeology and ceramics, in the latter of which he became a nationally recognized expert. The *Transactions of the North Staffordshire Field Club* bear witness to his new-found interest in archaeology and a continued interest in botany. He was welcomed to the Staffordshire botanical scene in 1931, attacked that county's flora with the same vigour as he had that of Leicestershire and was still contributing records 20 years later. On his retirement in 1962 he returned to Leicester, where he took up a part-time post as Ceramics Adviser to the Leicester Museums and Art Gallery, which he continued to hold until his death on 2nd August 1972. He does not appear to have amassed a personal herbarium, but much of his Leicestershire material is in the collections of the Museum Service. These notes are based in part on information kindly made available by his son Mr. Paul Bemrose and Mr. G. Halfpenny.

References

Bemrose, G. J. V., 1927. The adventive flora of Leicester and district. *Trans. Leicester lit. phil. Soc.* **28**, 45-71.

Bemrose, G. J. V., 1928. Contribution to the flora of Rutland. *Trans. Leicester lit. phil. Soc.* **29**, 21-25.

F. A. Sowter, F.L.S. (1899-1972)

Frederick Archibald Sowter was born in Leicester on 30th August 1899. He was a pupil of the Wyggeston Boys' School from 1909 to 1914, one of a number of boys from there who came under the influence of A. R. Horwood. He was on active service with the Argyll and Sutherland Highlanders during the latter part of the war years. Returning to Leicester in the early twenties, he set about training himself for a career in textile chemistry and eventually joined Courtaulds Ltd., with which firm he remained until his premature retirement, due to ill-health, in 1958.

His natural history interests were encouraged, not only by Horwood, but also by the Taylor family who were neighbours in Nelson Street, Leicester. It was S. O. Taylor, the father, an accomplished entomologist, who first took Sowter along to meetings of the Leicester Literary and Philosophical Society. Stephen A. Taylor, a fellow pupil at the Wyggeston and a life-long friend, was his boyhood companion in rambles throughout the Leicestershire countryside in search of insects and plants. In 1927, when he began to have some spare time over from establishing himself in his profession, he joined the Botany Section of the Literary and Philosophical Society, of which body and its successors he was a member and served almost continuously as Hon. Secretary, Chairman or President for 45 years.

His interests were not confined to botany. In his youth he played tennis for Leicestershire and he was the President of the Leicester Philatelic Society in 1937. Within botany, he divided his time during the late twenties and early thirties fairly evenly between the cryptogams and phanerogams. His first paper to the Botany Section, in 1929, was one on 'Leicestershire Mosses', but during the next four years he served as Hon. Secretary of the Flora Committee and for eighteen months or more leading up to the publication of the *Flora* he carried out, in his spare time, an enormous amount of work for Horwood, supplying notes and looking up specimens and information. His assistance was acknowledged in the *Flora*, but in correspondence with him Horwood doubted whether this acknowledgement was in any way adequate.

The *Flora* out of the way, he was able to concentrate his efforts on the cryptogams, in which he was encouraged by Horwood. In 1941 he published a revised and extended bryophyte flora of Leicestershire and Rutland (Sowter 1941), which incorporated many of his own records. He then turned his attention to lichens and myxomycetes, which he collected, throughout the two counties and elsewhere, during the nineteen-forties. The second part of 'The Cryptogamic Flora of Leicestershire and Rutland', on lichens, appeared in 1950 (Sowter 1950) and a paper on the myxomycetes in 1958 (Sowter 1958). Although his interest in the cryptogams never flagged, a combination of factors, notably an increasingly severe respiratory condition, restricted his fieldwork during the fifties and sixties. However, after a crisis in his health late in 1966, and with devoted nursing by his second wife Marion, he was able in the last six years of his life to do much more. Four papers, bringing up-to-date the bryophyte and lichen floras of the two counties, were published (Sowter 1969; Hawksworth and Sowter 1969; Sowter and Hawksworth 1970; Sowter 1972) and he had others in preparation at the time of his death.

The above paragraphs give an impression of parochiality in Sowter's interests and endeavours. This would be a false impression and a valuable corrective is provided by the tributes paid on his death by fellow bryologists and lichenologists (Jones 1973; Hawksworth 1973). His home county was always, however, close to his heart and he was one of the founder members of the Leicestershire Trust for Nature Conservation, when it was set up in 1956, and a Council member until his death. His paper 'Our diminishing flora' (Sowter 1960) illustrates his concern over the drastic changes which had taken place in the Leicestershire and Rutland countryside and his dismay at the losses to its flora occasioned by them.

He had been closely associated with the Leicester Museum since his schooldays and was a co-opted member of the Museums and Libraries Committee for 20 years from 1952, becoming in 1971 one of the first two Honorary Associates of the Museum. After his death, which occurred on 16th November 1972, his cryptogam collection, which amounted to over 7000 specimens, together with all his notebooks, manuscripts and part of his extensive library, was bequeathed to the Museum. His collection of flowering plants had been disposed of during the thirties to the Herbarium at Brussels.

When the establishment of a new Flora Committee was canvassed in 1967, Sowter was an enthusiastic supporter of the idea. He made a number of contributions to the Committee's deliberations in its early years, including biographical notes on some of the local botanists active earlier in the century, whom he had known. In recognition of his interest and his very considerable endeavours in keeping Leicestershire botany alive over three decades or more, this *Flora* is dedicated to his memory. The foregoing account is an abbreviated and edited version of an appreciation which appeared in the *Transactions of the Literary and Philosophical Society* (Evans 1973).

References

Evans, I. M., 1973. F. A. Sowter, F.L.S. – an appreciation. *Trans. Leicester lit. phil. Soc.* **67**, 20-24.

Hawksworth, D. L., 1973. Obituary. Frederick Archibald Sowter, 1899-1972. *Lichenologist* **5**, 345-348.

Hawksworth, D. L. and Sowter, F. A., 1969. Leicestershire and Rutland lichens; 1950-1969. *Trans. Leicester lit. phil. Soc.* **63**, 50-61.

Jones, E. W., 1973. Obituary. Frederick Archibald Sowter, 1899-1972. *J. Bryol.* **7**, 465-469.

Sowter, F. A., 1941. *The cryptogamic flora of Leicestershire and Rutland. Bryophytes.* Published for the author by E. Backus, Leicester.

Sowter, F. A., 1950. *The cryptogamic flora of Leicestershire and Rutland. Lichens.* Published for the author by the Leicester Literary and Philosophical Society.

Sowter, F. A., 1958. The mycetozoa of Leicestershire and Rutland. *Trans. Leicester lit. phil. Soc.* **52**, 21-27.

Sowter, F. A., 1960. Our diminishing flora. *Trans. Leicester lit. phil. Soc.* **54**, 20-27.

Sowter, F. A., 1969. Leicestershire and Rutland bryophytes 1945-1969. *Trans. Leicester lit. phil. Soc.* **63**, 40-49.

Sowter, F. A., 1972. Leicestershire and Rutland cryptogamic notes, 2. *Trans. Leicester lit. phil. Soc.* **66**, 21-25.

Sowter, F. A. and Hawksworth, D. L., 1970. Leicestershire and Rutland cryptogamic notes, 1. *Trans. Leicester lit. phil. Soc.* **64**, 89-100.

S. A. Taylor (1900-1954)

Stephen Alfred Taylor was born in Leicester on 13th February 1900, the first of two sons of Stephen Oliver Taylor (1870-1953). The family had an organ building and tuning business at 34, Nelson Street, Leicester, which had been founded in 1866 by Stephen Alfred's grandfather, and both he and his brother Geoffrey followed their father into the business.

His father was a keen field naturalist, primarily interested in beetles, but Stephen Alfred acquired, whilst in his early teens and a pupil at the Wyggeston Boys' School, an interest in flowering plants which lasted all his life. In this he was probably encouraged by friendship with a neighbour at 30, Nelson Street and contemporary at the Wyggeston, F. A. Sowter, and both of them came under the influence of A. R. Horwood at Leicester Museum.

He appears to have collected his first specimens in the summer of 1914, but a small blue notebook now in the possession of the Museums Service records a systematic approach to recording and collecting which commenced early the following year, the first entry reading 'Jan. Euphorbia Helioscopa Thurcaston'. By 6th September 1915, the date of the last entry that year, he had, with youthful energy, visited most of the classic botanical localities within reasonable cycling distance of Leicester and collected specimens of a good proportion of the county's flora. A number of the records from his first year's work found their way into the 1933 *Flora*, for example *Salix pentandra*, which he collected at Knossington on 7th August and *Calamagrostis canescens* at Owston Wood five days later. Others, such as *Centaurea cyanus*, which he recorded on 1st September at Nailstone, but for which there is no specimen in his collection, were not included.

Thereafter his interest seems to have slackened, since there is little material in his herbarium (LSR, acc. no. 286' 1954), which eventually amounted to almost 1800 specimens, collected between 1915 and 1929. His intention seems to have been to collect or acquire by exchange just one specimen of every species found in the British Isles. In the thirties and early forties, with interest rekindled, he concentrated on filling the gaps, where possible with locally collected material, but more often from further afield. Business trips took him to many parts of England, he had friends in Lancashire and elsewhere, but *Carex maritima*, collected at Bettyhill, Sutherland, on 3rd July 1936, must have been acquired whilst he was on holiday. He did however retain an interest in the local flora, making a special study in the late thirties and forties of the species and hybrids of *Epilobium*, and discovering several hybrids new to Leicestershire.

He was an active member of the Botany and Biology Section of the Literary and Philosophical Society from about 1931, when his name appears as leader of an excursion to Moira, cancelled because of bad weather. Thereafter he regularly contributed botanical exhibits to the winter meetings and led summer excursions to many parts of the county. He was a member of the ill-fated committee set up in 1939 to publish a supplement to the 1933 *Flora* and present at the meeting in March 1944 when the Section started up again. He became a committee member the following year, Vice-Chairman in 1947 and Chairman in 1950, the year his father retired after nearly half a century's service to the Section. He was also a member of the Nature Reserves Investigation Sub-Committee for Leicestershire and Rutland, under the chairmanship of Sir Robert Martin, which reported in February 1944 on the nature reserve potential of the two counties.

Stephen Alfred was by all accounts a quiet and unassuming man, who did not feel the need to publish any of his discoveries in the county or elsewhere. The notebook recording his youthful finds, the list of the contents of his herbarium and the many references to him in the minute books of the Botany and Biology Section, do however enable one to conjure up something of his lifetime's interest in botany.

He died, prematurely, on 8th January 1954, following a car accident and left a widow, Margaret. This account of his life is based on notes made in 1979 on the Taylors, father and son, by Susan Ashurst of the Department of Museum Studies, University of Leicester.

J. C. Badcock (1900-1982)

Jack Clement Badcock was born in Kibworth on 9th July 1900, but lived for nearly all his life at Fleckney. He was known to a wide public in Leicestershire as the writer of 'Nature Notes' for the *Leicester Mercury* for twenty years (1955-1975) and further afield as the author of four books based on his childhood and later experiences in the Leicestershire countryside, *Truants* (1953), *Waybent* (1954), *The Four Acre* (1967) and *In the Countryside of South Leicestershire* (1972).

He was an excellent all-round naturalist, with a particular interest in plants and birds. He also made a study of the landscape history of Fleckney and neighbouring parishes, which gave unusual perspective to his natural history observations. His articles and books contain numerous references to the botany of south Leicestershire and some of his records appear in this *Flora*. More useful however, particularly to those with an interest in records at the parish level, is a typescript, amounting to over 200 pages, which resulted from a study made in 1963 by a small group of naturalists based at Fleckney, of which he deposited a copy with the Museum (ref. no. B 1968/394). This group, to whom Jack was guide and mentor, kept a detailed calendar of their observations over twelve months in some six parishes and these are woven into a narrative text, with appendices listing the plants, birds and mammals recorded. He also deposited with the Museum typescripts relating to surveys of individual fields, woods and other sites of natural history interest in Fleckney and elsewhere in the county. The archives of the Natural History Section of the Literary and Philosophical Society, of which he was a member for some years, also contain numerous observations by him.

He died on 13th June 1982, mourned not only by his wife Kit, but also by several generations of younger naturalists whose interest in the subject he had done so much to encourage.

References

Badcock, J. C., 1953. *The Truants*. London: Hutchinson.
Badcock, J. C., 1954. *Waybent*. London: Hutchinson.
Badcock, J. C., 1967. *The Four Acre*. London: Dent.
Badcock, J. C., 1972. *In the Countryside of South Leicestershire*. Leicester: Harvey.

A. E. Hackett (1901-1964)

Arthur Everton Hackett was born in Hinckley on 13th November 1901 and lived there all his life, working in the hosiery industry. The earliest evidence of his interest in plants is that of specimens in his herbarium collected in 1927. Most of his botanising was done in the south-west of the county, where he made a number of interesting discoveries. He was, for instance, probably the original finder in 1929 of *Montia sibirica*, then new to the county, at Cadeby, where it still occurs. Favourite localities, within easy reach of his home were Burbage Common and the adjacent woods, the quarries at Earl Shilton and the bog at Potters Marston nearby. However he made occasional trips to other parts of Leicestershire and to Rutland and also visited adjacent areas in Warwickshire, notably the quarries in the vicinity of Bedworth, Hartshill and Nuneaton. Further afield, holidays took him frequently to Derbyshire, Devon and North Wales and occasionally to other parts of England such as Norfolk and Somerset.

In the late twenties and early thirties he appears to have done some of his botanising in the company of F. A. Sowter, with whom he exchanged specimens. He did not publish any of his records, but a few found their way into the *Transactions of the Leicester Literary and Philosophical Society* (Anon. 1931, 1932) and these and others were communicated by Sowter to The Botanical Society and Exchange Club. As a result Hackett's records tend to be ascribed to Sowter in the 1933 *Flora*. An example is the first record for Leicestershire of *Lactuca serriola*, collected by Hackett, perhaps in the company of Sowter, at 'Earl Shilton Quarry' in August 1930. This record had a chequered history. It was communicated by Sowter to the *B.E.C. Report for 1930* (1931), where it appears as 'East Shelton [*sic*], Leicester, SOWTER'; it was noted in the *Trans. L.L.P.S.*

for 1930-31 as 'Potters Marston. A. E. Hackett and F. A. Sowter.'; it finally appeared in the *Flora* as 'Quarry nr. Earl Shilton, *F. A. Sowter*' and 'First Record (Leics.), *B.E.C. Rep.* (1931)'. Hackett does appear in the list of 'Recent Correspondents (1902-1933)' on p. ccxxxv of the *Flora* but his name is unfortunately mis-spelled, 'A. Hachett'.

After a gap of about a decade in the late thirties and early forties, for the latter part of which he was on war service, Hackett began once again to add to his herbarium. He seems to have confined himself mainly to Leicestershire, exploring the south-west and west of the county in the company of M. K. Hanson and of other local naturalists, but he also contributed records to the computer-mapped flora survey of Warwickshire (Cadbury et al. 1971). During this period he tackled some of the more 'difficult' genera such as *Carex* and *Hieracium*. His last gatherings were in 1962, two years prior to his death on 18th October 1964. His herbarium, which contains some 570 sheets, of which rather less than half are from Leicestershire, passed to his friend M. K. Hanson, by whom it was donated to Leicester Museum in 1966 (acc. nos. 219' 1966 and 223' 1966). A detailed study of the Leicestershire material would provide some useful records from localities in south-west Leicestershire, which either escaped publication in the 1933 *Flora* or, because of the necessary rules for the inclusion of information from the nineteen-forties and fifties, have not been included in this one.

I am grateful to another friend, D. E. Jebbett of Hinckley, and to Hackett's daughter Mrs. J. Mayne, for some personal notes on him. He was a life-long teetotaller, and for many years a member of the Salvation Army. An accomplished musician, he played the tenor horn in the Army band and frequently took solo parts in the male voice choir of which he was a member. He had a keen eye for the unusual and a very good memory for both places and the characteristics of species. His enthusiasm for and thoroughness in the exploration of new localities was such that he sometimes lost all sense of time and missed buses and trains that should have taken him back to Hinckley. His garden, like that of many field botanists, owed little to the seed packet, with *Menyanthes* in a small pond and *Euphorbia amygdaloides* nearby. He was married, but his wife died in 1942 at the early age of 35. He was survived by two married daughters.

References

Anon., 1931. Observations during 1930-31. Plants *in* Report of Section D, Botany, and Section E, Biology. *Trans. Leicester lit. phil. Soc.* **32**, 67-69.

Anon., 1932. Observations during 1931. Phanerogams *in* Report of Section D, Botany, and Section E, Biology. *Trans. Leicester lit. phil. Soc.* **33**, 42-44.

Cadbury, D. A. et al., 1971. *A computer-mapped flora. A study of the county of Warwickshire.* London: Academic Press.

E. K. Horwood (1903-1977)

Edward Kidston Horwood was born on 28th December 1903, the third of four sons of A. R. Horwood. He was educated at the Wyggeston Boys' School where, though a little younger, he was a contemporary of three other boys who later became well-known as botanists, T. M. Harris, F. A. Sowter and A. E. Wade.

An interest in botany was encouraged by his father, whom he accompanied on collecting trips. It was his father's ambition that he might take up a military career, but an early marriage ruled this out. He taught for a while, but in his early thirties found a niche which suited his upbringing and interests and in which he remained for the rest of his life.

He joined the staff of the then University College of Leicester on 5th August 1936, as the only technician in the Botany Department. Under Dr. E. Miles Thomas he developed his skill as a cutter of beautiful free-hand sections, and subsequently, under Dr. (later Professor) C. T. Ingold, became an expert in fungal culture and a good photographer.

In 1945 it was decided to start an Herbarium in the Botany Department and he welcomed the opportunity of displaying yet another skill. He collected a large number of specimens, which can be recognised by the obvious care with which they were pressed and mounted. To begin with the collecting was done mostly within cycling distance of Leicester or on his annual holiday. Later however he and his second wife June, whom he married in 1959, went on collecting holidays to many parts of southern Europe. With a car larger supplies of drying paper could be carried, resulting in many valuable additions to the University Herbarium.

He retired on 31st December 1968 but continued to work part-time for the next five years. After that, until about a week before his death on 16th December 1977, he came into the Herbarium on three mornings a week, giving greatly appreciated voluntary help with mounting and labelling.

The above account of E. K. Horwood's life is based on an appreciation of him written in 1978 by Professor T. G. Tutin. In the *Flora of the British Isles*, first published in 1952, Professor Tutin and his co-authors single out for special mention his 'continuous help' which 'enabled the work to be completed more more rapidly than would otherwise have been possible'. It may also be mentioned that he was a founder member in 1956 of the Leicestershire Trust for Nature Conservation and served on its Council until 1975. He was a Committee member of the Botany and Zoology (later Natural History) Section of the Leicester Literary and Philosophical Society from 1954 until 1966. He and June were founder members in 1967 of the present Flora Committee and together they were responsible for a thoroughly professional survey of the large area of south Leicestershire which they chose to take on. The flora survey herbarium contains a large number of superbly pressed specimens resulting from their joint endeavours. In addition he searched the University Herbarium for pre-1967 Leicestershire material that might provide records useful for the *Flora* and listed it all. Two local herbaria therefore contain abundant evidence of the loving care he devoted to collecting plants and the *Flora* itself is witness to his skills in locating and identifying them. The many botanists who knew him and regarded him with affection will, however, remember him for his unfailing geniality and helpfulness.

R. Wagstaffe (1907-1983)

Reginald Wagstaffe was born on 28th July 1907 and educated at Southport, Lancashire. He received a training in botany and ornithology at the University of Cincinatti and his subsequent post in the University Museum there was the start of a distinguished career in museums, culminating in the post of Keeper of Vertebrate Zoology at Liverpool, which he occupied from 1950 until he resigned because of ill-health in 1970. He died at his home in Cambridge on 11th September 1983 and was survived by his wife and daughter.

He is best known as a taxonomic and field ornithologist with a world-wide reputation and as the co-author of a definitive work on the preservation of natural history specimens, but he was also an extremely competent botanist. He spent nearly three years in Leicester as an Assistant at the Museum, joining the staff in April 1931 and resigning to take up the post of Curator at Stockport Museum in February 1934.

During his stay in Leicester he took an active part in the work of the Botany and Biology Sections of the Literary and Philosophical Society. It is recorded, for instance (Anon. 1932), that on 18th November 1931 he opened the Exhibition Evening 'with a description of the dune flora of the South Lancashire Coast'. He also contributed a number of botanical records to that and the following years' volumes of the *Transactions* of the Society, some of which are to be found in the 1933 *Flora*. Examples are *Stellaria palustris*, which he found near Queniborough in 1931, and what was thought for several decades to have been the only record of the occurrence of *Himantoglossum hircinum* in vice-county 55. This was found in the summer of 1931 by a Mr. Fenton on a roadside verge at Essendine in Rutland and communicated to the Sections by Wagstaffe, thence finding its way into the *Flora*. Unfortunately, careful examination of the evidence by K. G. Messenger has since shown that the spot at which the plant occurred was a few feet on the Lincolnshire side of the county boundary (Messenger 1971).

References

Anon, 1932. Report of Section D, Botany, and Section E, Biology. *Trans. Leicester lit. phil. Soc.* **33**, 41-44.

Messenger, K. G., 1971. *Flora of Rutland*. Leicester: Leicester Museums.

T. W. Tailby, B.Sc., L.R.A.M. (1915-1968)

Trevor William Tailby was born on a farm at Desborough in Northamptonshire on 9th March, 1915. A lonely child (his only sister was eight years younger), he used to wander the countryside of the Desborough area and in his early years acquired a life-long interest in natural history. An equally consuming interest was music. He was taught to play the organ by his father and played in the parish churches at Desborough, Rockingham and Rushden in his late teens and early twenties. He was awarded the L.R.A.M. Diploma in 1952 and was accompanist to the Leicester Male Voice Choir for over twenty years, 1946-1968.

He trained as a teacher at Culham College, Oxford and later completed a B.Sc. On returning from war service with the R.A.S.C., part of which was spent in the West Indies, he took up a teaching post in Coalville. He later transferred to Lancaster Boys' School in Leicester and then, as Deputy Head, to Church Langton and finally to Kibworth High School when Church Langton became a field study centre.

His particular interest in the field of natural history was entomology and he assembled a large and systematically wide-ranging collection of British insects, including moths, beetles, bugs and other groups, which was donated on his death to Leicester Museum. He was also a competent field botanist. His reference collection of British flowering plants (acc. no. 246' 1968) was started in the late thirties, but mainly dates from 1949 and later years. It contains a number of interesting Leicestershire specimens, for example one of *Trifolium micranthum*, from the top of a wall at Groby Pool in 1957, a locality from which it has not otherwise been reported since the early part of this century. Much of his collecting, both of insects and plants, was done in the south and east of the county and the six 10 km. squares he surveyed in 1954-1956 for the *B.S.B.I. Atlas* reflect this, extending as they did from SP 68 (Husbands Bosworth) north-east to SK 70 (Tilton). He was also responsible in 1967 for a field-by-field survey of the Kibworth area, a contribution to a county-wide project organised by the Natural History Section of the Leicester Literary and Philosophical Society.

He was a methodical man and the detailed notes he made for the adult education courses he conducted in natural history show both his enthusiasm for and grasp of his subject and the pains he took in the preparation of talks and field excursions. He had plans for a book on Leicestershire natural history for which he had already drafted a synopsis and some chapters when this and other ambitions were cut short by his premature and unexpected death from a respiratory infection on 9th February 1968. He was survived by his wife Phyllis whom he married in 1942, and daughters Rosemary and Elizabeth. I am grateful to Mrs. Tailby for much of the information on which this note is based.

M. K. Hanson (1918-1981)

Maurice Kimpton Hanson was born at Nuneaton, Warwickshire, on 16th September 1918 and was educated at Nuneaton Grammar School. He moved to London in his late teens, where he shared for a while a flat on the Farringdon Road with Arthur C. Clarke, the science fiction writer. He was amongst those called up early in 1939 and served for some months in France, returning to this country after Dunkirk. The rest of his war service was spent inspecting coastal installations and he used to say that it was during this period in his life that he acquired his interest in botany. In November 1950 he took up a post as Technician in the Physics Department at Leicester University, and shortly afterwards met there E. K. Horwood, who became a frequent companion on botanical excursions and a life-long friend. He was promoted to Senior Technician in 1960 but left in 1966 to move to London. He became, later that year, Librarian Assistant in the Department of Civil Engineering at Imperial College, where he was employed for the rest of his life.

Before leaving Leicester, he deposited with the Biology Department of the Museum a card index of botanical records for 1950-1952 (ref. no. B1966/300) and a copy of his botanical diary for 1957-1964 (ref. no. B1966/302). During the period covered by these he visited, often in the company of A. E. Hackett or E. K. Horwood, many localities of botanical interest in Leicestershire, Rutland and adjacent parts of Northamptonshire. Occasional trips further afield took him to Bedfordshire, Cambridgeshire, Hertfordshire, Huntingdonshire, Lincolnshire, Oxfordshire, Staffordshire, Suffolk, Warwickshire and elsewhere. He was also a contributor to the *Atlas of the British Flora*.

His records are admirably detailed, with a six-figure grid reference or a sketch map to pinpoint the exact location of a find of special interest. They are a fascinating source of information about species and sites in Leicestershire and Rutland for the two decades prior to the establishment of the present Flora Committee in 1967. A number of specimens collected by him are in the Museum collections, among them a specimen of *Aquilegia vulgaris* from Saltby in 1953, the last occasion that the truly wild plant was seen in Leicestershire.

Through the kindness of his friend D. E. Jebbett, the Museums Service has recently acquired copies of a number of photographs of botanical subjects taken by him in the early fifties, many of which are now of considerable historical interest. They include, for example, *Empetrum nigrum* at Spring Hill, Whitwick, in 1952 (site since destroyed by quarrying), *Melampyrum cristatum* at Turnpole Wood in Rutland in 1952 (probably now extinct in vice-county 55) and *Coeloglossum viride* at Potters Marston in 1953 (site since destroyed by construction of the M69).

At the time of his unexpected death in Kettering on 12th May 1981, Maurice Hanson was looking forward to a retirement in which he hoped to develop what had been a hobby for some years, dealing in natural history books, and to pursuing his interests in music and the silent films. He was unmarried, a modest and unassuming man who did not, so far as is known, publish any of his records. In view of the care that he lavished on his botanical diaries and card index, he would perhaps have been glad to know that they are still available for consultation.

G. S. Smith, B.Sc., Dip. Ed. (1946-1969)

Geoffrey Stuart Smith was born in Leicester on 3rd February 1946. He took an interest in natural history from an early age and in their early teens he and his brother Alan were enthusiastic members of a Saturday morning Natural History Club at Leicester Museum.

He took a B.Sc. at the University of East Anglia, graduating in 1967. As part of the course he made in 1965 a collection of plants from waste ground in Leicester; it contains some interesting records and was later presented to Leicester Museum (acc. no. 398' 1966).

Following graduation he went out to Uganda, where he took a Diploma in Education at Makerere University College and was then appointed to a teaching post at Moshi in Tanzania. His life was tragically cut short by a car accident on 26th December 1969 and he is buried at Moshi.

T. G. Tutin, M.A., Sc.D., F.R.S. (1908-1987)

The death occurred on 7th October 1987 of Thomas Gaskell Tutin, Emeritus Professor of Botany at the University of Leicester. As this Flora was then in press, it was impossible to include full biographical notes, for which readers are referred to *Watsonia* 17:2. A reference to Professor Tutin's role in the study of Leicestershire botany can be found on page 68 of this Flora.

CHANGES IN THE FLORA OF THE COUNTY SINCE 1933

Introduction

Leicestershire, although scenically an unspectacular county, was famous fifty years ago for the extensive grasslands which made it some of the best hunting country in Britain. The rather uniform, rolling landscape of fields and hedges was punctuated by a multitude of small spinneys and a few large woods. The outstanding exception to this was Charnwood Forest which provided a few square miles of what was popularly thought of as 'wild' scenery, outcrop rock and bracken clothing relatively steep, though small, hills.

Today large areas of grassland have disappeared beneath the plough. The impression of uniformity has been heightened by two things: the current practice of hedge removal with the consequent increase in field size and the cultivation of many untidy, but ecologically desirable corners. The woodlands have changed little in extent but rather more in character and the area of uncultivated land on Charnwood Forest has shrunk.

Agricultural pressures are acknowledged as posing the greatest threat to wildlife habitats in Britain today and these must be felt particularly in a predominantly farming county such as Leicestershire. It is probably true to say that the past fifty years have seen greater changes in our flora than has any comparable period in the past and there is no indication that the rate of change is diminishing.

There have of course been both increases and decreases in plant populations; some species have become extinct and new county records have been made. There has also been an interesting redistribution of species. Many, whose traditional habitats have been affected by ploughing and grassland improvement, are now found in more recent man-made ones, notably railway verges and sites of mineral extraction.

In considering the changes in the flora since the publication of Horwood and Gainsborough in 1933, it is easy to gain a false impression of the former frequency of any given species by failing to take into account the period of time covered by the records quoted in that work. It should be borne in mind that Horwood's own records were gathered over twenty years and that those of other observers quoted by him go back as far as the mid-eighteenth century.

In the account that follows the main habitats in the county are discussed, together with some of their characteristic species.

Woodland

Leicestershire is one of the most poorly wooded counties in lowland Britain. According to a Forestry Commission Survey in 1947-49 (Chard *et al.* 1952), only 2.3% of the land surface was at that time taken up by woodlands of over five acres; in England and Wales only four counties had less and one had the same. The flora of the county's woodlands, although no doubt slightly poorer than it was fifty years ago, has probably deteriorated less than that of any other semi-natural habitat.

Some of the common and visually attractive species have suffered by the increased mobility of the public, to some of whom the temptation to pick bluebells and dig up primroses is apparently irresistible.

The most destructive influence on the ground flora of our major woodlands has been an increase in shading, brought about in two ways. The felling of broad-leaved trees and their replacement by conifers is a practice which has particularly affected woods on Charnwood Forest. The dense shade of a conifer plantation almost entirely inhibits the shrub and field layer, although bluebells do seem to survive for a time, particularly under larches.

Cessation of the traditional practice of coppicing has also been responsible for increased shading. It is clear from old coppice stools that many woods, particularly those of east Leicestershire, were once managed in this way and this periodic cutting of the shrub layer must greatly have benefited the ground flora. It seems likely that the actual amount of species such as *Paris quadrifolia* and *Platanthera chlorantha* has been reduced somewhat since 1933, but the greater change may lie in the number of plants which flower. It is often possible to find large numbers of 'blind' individuals of both these species with only a few flowering spikes and this may well be due to the now constant shading.

It is encouraging to see, where areas of woodland have been felled and replanted with hardwoods, that much of the ground flora has been able to survive the disturbance. There are few landowners left who do not wish to crop their timber at some time. If they replace with conifers there is little hope for the ground flora, but if hardwoods are used then the future of the field layer depends largely on the extent to which herbicides are used.

An increase has been noted recently in the number of records of *Epipactis purpurata*, a species which is on the edge of its national range in Leicestershire. Only two sites are quoted in Horwood and Gainsborough, both in the east of the county, but it is now known from six, some of them in the north-west. This orchid may well have been under-recorded in the past and indeed may still be so, for it flowers late in the year in a habitat which is most often studied in the spring. Another species for which there are now more records is *Lathraea squamaria*, an easily overlooked plant which again may have been under-recorded.

Calamagrostis canescens was considered locally abundant in 1933, but has only been found once during the present survey. It may have been overlooked in some of its old sites, but this grass typically grows on ride edges and with the general slackening of woodland management could have become overgrown. *Neottia nidus-avis* is represented by one unconfirmed record only, but this species is so difficult to see that it would be unwise to assume that it is therefore on the edge of extinction.

If there have been few losses in the major woodlands, this is far from being the case with scrub. A tendency to clear up rough areas of grass and bushes is a feature of modern farming and the fact that very little such habitat now remains may account for the absence from today's list of *Rosa agrestis, R. micrantha* and *R. rubiginosa*, apart from one record of a garden variety of the last named. All three species were recorded in the previous Flora, although in the light of modern research the list of sites given might be considered a little optimistic for Leicestershire.

Grassland

Although this is the vegetation type for which Leicestershire was once best known, we have little idea what proportion of it was still floristically interesting in 1933. Some indication is given by studies made of Belgrave and Birstall (Mercer 1914), New Parks (Measham 1915) and Aylestone Meadows and Narborough Bog (Wade 1919), but this is hardly sufficient to enable us to draw conclusions about the county as a whole. According to Horwood, many grassland species had already declined and now, over 50 years later, the trend continues. Not only has even more been ploughed, but much of that which remains has been further modified by the use of fertilisers and selective weedkillers, reducing it to a uniform consistency with relatively few broad-leaved herbs. There must now be a number of parishes in the county which can no longer boast a single field of species-rich grassland.

Neutral grassland predominates in Leicestershire and its characteristic species, such as *Ophioglossum vulgatum*, *Rhinanthus minor* and *Silaum silaus*, were once widespread; they are now cause for remark. The population of *Primula veris* in the county has declined very greatly. Fields full of cowslips are now rare, although a number of sites survive in which just a few plants occur.

This suggests either that they do not succumb immediately to cultivation or agricultural chemicals or that they are able to spread in again to such areas. Their ability to colonise is demonstrated by their relatively common occurrence on railway verges, often their remaining stronghold in intensively farmed areas. *Orchis morio* too has dramatically declined in numbers. The impression gained from the 1933 Flora of a widely distributed, locally abundant species, contrasts sharply with the present situation of about a dozen known sites, in only two of which could it be called abundant.

Within the areas of neutral grassland there occur small patches of ground which, judging by their vegetation, are rather more basic in character. Grassland improvement or ploughing on many of these sites has resulted in the relative scarcity of such species as *Cirsium acaule* and *Filipendula vulgaris* compared to 50 years ago.

The picture with regard to limestone grassland is difficult to evaluate. This has never been a common habitat in Leicestershire and is almost entirely confined to small areas on the Oolite in the north-east and on Carboniferous Limestone in the north-west. At the turn of the century much land in the north-east had already been ploughed, but even so the area was considered still to retain a high degree of interest. Since 1933 we have seen the loss of even more of this interesting ground. A now disused aerodrome covers a considerable area and some limestone species such as *Campanula glomerata* and *Cirsium eriophorum* have been able to survive there in small patches of rough grass. Such patches have gradually diminished through ploughing and to a lesser extent through forestry. In 1965 that part of the Leicestershire/Lincolnshire boundary which had previously run down the middle of Mere Road, Saltby (often known as The Drift), was moved for the greater part of its length to the western hedge. Leicestershire lost a small but significant area of limestone grassland in this green lane, which may account for the fact that *Polygala calcarea* and *Hippocrepis comosa* have not been recorded for the county during the present survey. In the north-west, the area of grassland on Breedon Hill has been reduced by the invasion of hawthorn scrub, the extension of the adjoining quarry and ploughing of the lower slopes. The habitat available for calcicoles is largely confined to the spoil heaps of the two quarries in the area, Cloud Hill Quarry and Breedon Hill Quarry. Some of the records given in Horwood and Gainsborough for limestone grassland species are very old; *Asperula cynanchica* had no twentieth century records and *Gentianella campestris* only one, so their absence from today's lists is hardly surprising. A large proportion of our contemporary records for calcicoles comes from railways. *Brachypodium pinnatum*, *Cerastium arvense* and *Origanum vulgare* in particular make use of this habitat.

A distinctive type of open acid grassland occurs occasionally. It has been least affected by grassland improvement and ploughing, perhaps because it occurs on thin soils with rock close to the surface. A particularly species-rich example at Croft supports an interesting community which includes a surprisingly high number of annuals, such as *Aphanes arvensis* s.l., *Trifolium striatum* and *T. subterraneum*, species which are rarely found in closed swards. Also occurring there are *Allium vineale*, *Moenchia erecta* and *Potentilla argentea*, all uncommon in the county. The loss of one example of this type of grassland, at Barrow Hill, Earl Shilton, has further reduced its already limited range.

Heath grassland is found mainly on Charnwood Forest and in the north-west of the county. Fifty years ago there must have been considerably more of this vegetation type, but again, the pressures of modern agriculture have meant that much of it has been improved. *Danthonia decumbens*, *Galium saxatile*, *Nardus stricta* and *Potentilla erecta* have had their once extensive habitats fragmented, whilst the scarcer species such as *Dactylorhiza maculata* ssp. *maculata* are now definitely uncommon. Grassland improvement is nowhere better illustrated than on Charnwood Forest. There it is sometimes possible to see a hillside field, mostly the bright green of improved grassland, but with one or two patches of a quite different colour and texture, comprising small outcrops of rock surrounded by *Agrostis capillaris*, *Festuca ovina*, *Galium saxatile*, *Potentilla erecta* and *Rumex acetosella*.

Heath and moorland

Heath and moorland plants are largely restricted to the acid soils of Charnwood Forest. Some were lost more than a century ago, but others survive, albeit in reduced quantity. *Calluna vulgaris* and *Vaccinium myrtillus*, considered by Horwood to be locally abundant, have certainly decreased in amount and *Erica tetralix* is known only from four sites. Others are even more dangerously low; *Salix repens* has been recorded from only three places and *Genista anglica* from one. Fortunately both are found on a well-protected nature reserve, but the populations are at such an extremely low level that their future is far from assured. The only known Leicestershire station for *Empetrum nigrum* was on this reserve, but the species has not been seen there since 1982. Although *Pinguicula vulgaris* and *Schoenus nigricans* once grew on Charnwood Forest, they were considered extinct there by the time the last Flora was written and were then only known from Saltby Bog in the north-east of the county. With the draining of that bog, *Pinguicula* appears to have been lost, but *Schoenus* was seen as recently as 1970 and its extinction cannot be considered certain.

Marsh

The continuum of habitats from marshy meadows to marshes proper has proved most susceptible to change as modern drainage techniques have brought into cultivation land hitherto regarded as agriculturally valueless. Today, even the once common *Caltha palustris*, *Cardamine pratensis* and *Lychnis flos-cuculi* are quite restricted in their distribution. Some species, which were scarce in 1933, are now verging on extinction; *Epipactis palustris* and *Parnassia palustris* are maintaining a precarious existence at one site only. Actual losses since then include *Potentilla palustris*, whose last remaining station in the county, Moira Reservoir, was lost at about the time of the Second World War.

Horwood states that the flora of Frog Hole, an area of marsh near Groby Pool, was as rich as any in the county and quotes an impressive list of records, including *Epipactis palustris*, *Menyanthes trifoliata* and *Parnassia palustris*. It seems that many of the records were old ones, but no doubt this was still a very rich piece of ground. An estate worker recalls that the area was once managed for snipe, i.e. it was kept wet and the vegetation was cut periodically. This regime presumably suited many small species and on its cessation the latter became crowded out by the alders and reeds which now dominate the site.

Aquatic habitats

There are a number of aquatic species, for example *Baldellia ranunculoides*, *Myriophyllum alternifolium*, *M. verticillatum*, *Pilularia globulifera*, *Utricularia minor* and several species of *Potamogeton*, whose absence from today's flora list should not be taken as proof of extinction in the county, except perhaps in the case of *Pilularia*. Although all were rare in 1933 they are easily overlooked. A full-scale search of aquatic habitats would be needed before any final conclusions were drawn as to their extinction.

Groby Pool is the only stretch of still water in Leicestershire which may be considered natural in origin. It has, for many years, been in the hands of a sympathetic owner and the Pool and its immediate surroundings have retained much of their floral interest.

There are a number of reservoirs, particularly in the west of the county, which enjoy a measure of freedom from disturbance. Most have built up an interesting marginal flora. *Juncus filiformis* was first noted in 1964 at Blackbrook Reservoir. The record was noteworthy, not only because it represented a considerable extension of its range in Britain, but also because of the very large population of the species, which is a national rarity. The number of plants which can be seen at any one time varies from year to year according to the water level, which at this reservoir can fluctuate widely, depending on the current policy of the water authority. However the plant is almost certainly more abundant now than when it was first found. Another new county record was made at this site in 1964 when a small amount of *Juncus tenuis* was discovered. It has since been found at Eyebrook Reservoir and elsewhere in greater quantity. More recently, the margins of Blackbrook Reservoir have been colonised by *Crassula helmsii*, which was first seen there in 1984, since when it has spread around much of the perimeter. *Scirpus fluitans*, which was thought to have been extinct in the county for well over a century, was also discovered on the western bank of the Reservoir in 1984.

Reservoir margins are the habitat in which *Rumex maritimus* has spread so much during the past few decades. Eyebrook Reservoir, which was only completed in 1940, provided the first Leicestershire record for two further docks in 1971. In that year *Rumex palustris* and the hybrid *Rumex crispus* × *palustris* were discovered at the site. Although strictly speaking outside the scope of this chapter, it is interesting to note that reservoir margins provided the habitat for two species new to Britain that were discovered in the county in the last century. *Epilobium adenocaulon* was first recognised in Surrey in 1931 by G. M. Ash who subsequently found it in herbarium material at various museums. The oldest was from Cropston Reservoir and was dated 1891. The now extinct *Inula britannica* was discovered at the same site in 1894.

Moira Reservoir was lost to opencast mining about the time of the Second World War and with it a good aquatic flora including *Pilularia globulifera* and *Myriophyllum alterniflorum*. Today it is thought that the flora of the reservoirs, which are surrounded by agricultural land, is being affected by the run-off of fertilisers, but no specific information is available as yet.

Water-filled gravel pits with their associated marshy areas provide an aquatic habitat which has increased enormously since the nineteen-forties. *Typha latifolia* and *Sparganium erectum* spread quickly into these areas and it seems likely that such efficiently colonising species are now more abundant than they were in 1933. The river valley pits receive seed and plant fragments with every flood, so those which are to remain water-filled could ultimately build up a varied aquatic flora. The development of marginal vegetation is not always possible as the banks of gravel pits tend to be rather sheer.

There is now almost certainly a larger area of freshwater in the county than ever before in historic times. Nevertheless the loss of large numbers of field ponds, consequent upon the increase in arable land and the stringent hygiene requirements for dairy cattle, has considerably restricted the distribution of some plants which seemed particularly suited by this habitat. *Catabrosa aquatica*, *Hippuris vulgaris* and *Ranunculus lingua* have declined to the point where they are now rare in the county and *Hottonia palustris* is known from just two sites, only a matter of yards apart. On the credit side we have a new county record from this habitat. *Mentha* × *gentilis* was recorded in 1979 beside a pond at Whitwick. The pond was subsequently drained, but the mint has survived in the rough marshy ground which remains on the site.

The first decades of the twentieth century saw the beginning of serious pollution of the rivers by sewage and industrial effluent, exacerbated latterly by run-off fertilisers from agricultural land. Although today water quality is beginning to improve, there is a long way to go and it is highly unlikely that the situation will ever be completely reversed.

It is clear after reading Haslam (1978) that we do not yet have sufficiently detailed information about Leicestershire watercourses to make precise statements regarding the effect upon them of pollution. This is a complex subject in which the sensitivity of different species to the type of pollution, flow, water volume and turbidity all have to be taken into account.

It is difficult to quantify the floristic changes in the aquatic and reedswamp flora of the rivers since the publication of Horwood and Gainsborough. The general impression gained from both the text and pictures in that work is of rivers with rich marginal vegetation and a diverse aquatic flora. Today there are few rivers in the county to which this description could be applied. One of them is the River Eye between Ham Bridge and Melton Mowbray. Horwood states that 'the aquatic vegetation is of the usual type . . .'. If we accept that he did know the river well enough to judge, we can only conclude that there were many others of a similar character. Today, far from being the usual type, it is quite exceptional in Leicestershire, as it has recently been classed a Grade 2 N.C.R. Site by the Nature Conservancy Council (Newbold and Palmer 1979). In their opinion it is the relative abundance of a variety of species which is so unusual; there are no uncommon river species apart from *Butomus umbellatus*, which in Leicestershire is almost entirely confined to canals.

It must of course be borne in mind that this information was obtained by an intensive study of the river. The only comparable information available at that time for other watercourses in the county was from the Rivers Wreake (which is the continuation of the River Eye downstream of Melton Mowbray), Devon and Smite and none of these were graded as high as the Eye. However even to a casual observer the Eye appears relatively unspoiled and there are few other stretches of water which give this impression. One which does is the River Mease, from about a mile south of Measham west to the county boundary. This small river can still boast a little *Butomus*. Although no rare species have been seen, there is a good selection of common ones, both aquatic and riparian, including *Cardamine amara*, *Iris pseudacorus*, *Myosotis* spp., *Phalaris arundinacea*, *Ranunculus fluitans*, *Rorippa* spp., *Scirpus lacustris* and *Sparganium emersum*. The water is clear and the impression of richness heightened by an abundance of damselflies and shoals of chub. Perhaps this river and the River Eye may give some idea of what many of the county's small rivers were like fifty years ago.

Some species are pollution tolerant. The River Soar in the heart of industrial Leicester gives every appearance of being extremely dirty but there are good stands of *Nuphar lutea*, *Rorippa amphibia* and *Sparganium erectum*. The presence there of *Sagittaria sagittifolia* is rather more surprising as Haslam does not consider it a species particularly tolerant of sewage and industrial effluent, though more so of eutrophic pollution.

Pollution is not the only factor which has had an adverse effect on the aquatic flora; physical alterations of the river course have also played a part. Many of the smaller water courses have been graded, straightened and had their banks scraped. Dredging alone is said to have only a temporary (2-5 years) effect on the flora but scraping the banks seems to result in the disappearance of marginal plants. Colonisation of the bare banks by invasive species results in the sadly familiar sight of a steep-sided, straight channel, covered with willow-herb, nettle and an occasional teasel. A single strand of barbed wire along the top of the bank often completes the picture. Major flood improvement schemes are proposed for the Rivers Soar, Wreake and Eye, which if carried through will have a most radical effect on the appearance and ecology of our main river valleys.

Canals provide a less polluted aquatic habitat and one which can be particularly suitable for some species, by virtue of a constant water level and slight flow. It is unfortunate that the condition of some of those that run within the county has deteriorated. Several species recorded from the Ashby Canal in Horwood and Gainsborough have not been noted recently. Three factors may account for this. Firstly, some of those records were made in the last century, secondly, a stretch of canal in the north has been drained and filled and, thirdly, that part still in water is used quite intensively by pleasure craft, to the detriment of the flora. The Grantham Canal, no longer navigable, was dredged in the nineteen-sixties, with the predictable result that there was a great increase of invasive species such as *Glyceria maxima* and *Sparganium erectum*. There are however some stretches, particularly in the vicinity of Plungar, which have retained a more varied flora. Unfortunately there are no detailed records in the last Flora with which to compare the state of affairs which exists today. One interesting feature is the occurrence there, sometimes in great quantity, of *Azolla filiculoides*, first recorded in the county from Cossington in 1948. *Nymphoides peltata* was once noted in this canal, but has not been seen during the present survey.

On the credit side, river and canal margins have been colonised by two species of balsam. *Impatiens glandulifera* has spread during the past fifty years and *I. capensis*, first recorded in 1940, now seems firmly established.

Railways

There is very little mention of these in Horwood and Gainsborough, so it is impossible to tell whether our present railway flora represents an actual or only an apparent increase in the populations of some species. *Chaenorhinum minus*, considered rather rare fifty years ago, is now known to be a common plant of railway ballast; *Linaria repens*, with only three records in 1933, is now scattered throughout the county in this habitat, whilst *L. vulgaris*, once sparsely distributed, is now locally abundant.

The outstanding example of a species increasing along railways is *Senecio squalidus*, which in 1933 was regarded as rare, with only two stations in the county. Today it is abundant, not only in this habitat but in waste places generally. *Senecio viscosus*, with thirteen sites quoted in 1933, is now widespread in similar habitats. Consequent no doubt upon these increases was the appearance of a hybrid between the two, *S.* × *londinensis*, which was first recorded on a disused railway line in 1966.

Disused railways, until their verges become overtaken by scrub, are among the richest areas of the county for species which have been squeezed out of their old sites by the demands of agriculture. Large numbers of *Dactylorhiza fuchsii* occur in some cuttings. Once widespread grassland species, such as *Ononis* spp., *Primula veris*, *Sanguisorba minor*, *Stachys officinalis* and *Succisa pratensis*, can thrive there, unaffected by spraying and out of reach of the plough. Abandoned sidings in particular provide good refuge and often present a degree of herb-richness in marked contrast to the surrounding farmland. Those at Rothley have recently been sprayed but until then had *Centaurea scabiosa*, *Knautia arvensis*, *Ornithopus perpusillus* and *Trifolium arvense*, amongst a great number of commoner species. The contrast with the surrounding countryside is often high-lighted by the fact that the railway ballast provides a substrate of a different pH, on which a range of species grows which would not normally be seen in that area.

Railway bridges, particularly those of blue brick, are now the richest habitat in the county for small ferns. The proviso stated at the beginning of the section on railways applies here also; we do not know whether there has been a real or apparent increase in the quantity of these species. *Asplenium adiantum-nigrum*, *Asplenium ruta-muraria* and *Asplenium trichomanes* are today found more often on these bridges than in any other habitat. *Ceterach officinarum* also, although still quite rare, has a third of its sites in such places. The discovery of *Gymnocarpium robertianum* on a railway bridge in 1958 was a new county record and a thriving colony of *Asplenium viride*, considered by Horwood to be extinct on its original single station on Beacon Hill, was found on a bridge north of Loughborough in 1975.

Arable land

The situation with regard to the incidence of arable weeds is complex. As there has been a great increase in the amount of arable land over the past fifty years, there are now almost certainly more arable weeds in the county. However, the increased use of selective weedkillers and higher standards of seed purity mean that, at least in areas of efficient farming, there are fewer of them to the square metre.

Certain species have been effectively controlled in cereal crops; *Papaver dubium*, *P. rhoeas* and *Sinapis arvensis* spring immediately to mind by virtue of their eye-catching flowers, which are no longer a familiar sight in late summer. Other less showy species have also been reduced in numbers; *Mentha arvensis*, for example, is now found almost exclusively on woodland rides and in marshes, whereas at one time it was a common arable weed, hence the name corn mint.

Some of the weeds which were rare in 1933, such as *Anthriscus caucalis*, *Galium tricornutum* and *Torilis arvensis*, have not been recorded at all during the present survey. *Centaurea cyanus* has only been recorded twice from arable land and *Agrostemma githago* not at all. The last two species have however been seen in waste places and on roadsides where they are probably of garden origin.

Surprisingly, some arable weeds have increased in numbers over the past fifty years, relative of course to the area of the county now under plough. The modern practice of growing cereals year after year on the same field, is probably responsible for the increase in certain grass species as weeds. Particularly good examples are *Avena fatua* and *A. ludoviciana* (the latter perhaps under-recorded during the present survey), recorded in 1933 as rare, but now widespread throughout the county. They are difficult and expensive to eliminate by spraying and farmers wishing to sell their crop as good quality seed corn are often reduced to pulling out wild oat by hand. *Bromus sterilis* and *Poa trivialis* are also on the increase and the persistence of many of these grasses is helped by the fact that they drop their seeds before harvest. *Chrysanthemum segetum* and *Matricaria perforata* have proved remarkably spray resistant and there are still fields in which the former seems to be a permanent feature.

Whilst there are very few plants which cannot be controlled by one spray or another, nevertheless the cost of some of the more specialised chemicals puts them out of the reach of many farmers. A further factor which must be considered is that if the numbers of the more easily controlled weeds are reduced, the subsequent lessening of competition may benefit the more spray resistant species, particularly grasses.

Lycopersicon esculentum is now found widely as a transitory arable weed owing to the practice of spreading liquid sewage sludge on the fields.

Taking arable land in its broadest sense, it is perhaps suitable to include here changes in the status of what are usually garden weeds. *Veronica filiformis*, first recorded in 1957, is now widespread in the mown turf of lawns, parks and churchyards. Well established but less persistent members of another community are *Galinsoga ciliata* and *G. parviflora*, species of cultivated ground, allotments and nurseries, which both appear to be on the increase.

Spoil heaps and quarries

The expansion of mineral extraction in the county, with the consequent increase in spoil heaps, has provided a new and suitable habitat which some species, notably orchids, have colonised in large numbers. Thus the populations in Leicestershire of *Dactylorhiza fuchsii* and *Ophrys apifera* are probably larger now than they were fifty years ago, in spite of the fact that many of their original stations have been destroyed. Plants with such diverse requirements as *Anacamptis pyramidalis* and *Solidago virgaurea* are now only to be found in or around disused quarries. Nearly half the current records for *Eriophorum angustifolium* are from this habitat.

Waste ground and rubbish dumps

The spread of *Epilobium angustifolium* since the turn of the century was remarked in 1933. Since that time its population has continued to increase to the point where it is now one of the most successful invasive species of disturbed ground in both rural and urban areas. It is included here as this habitat was the one in which its expansion has been most noticeable, although today there are few niches which it has not penetrated. Another willowherb, *E. adenocaulon*, has also spread phenomenally. In 1933 it was known from only one site in Leicestershire, but today it is locally abundant in waste places throughout the county. Already increasing by 1933 was the introduced *Chamomilla suaveolens*, which is now one of our commonest waste ground species. *Cicerbita macrophylla*, now found in scattered localities, had not been recorded at all fifty years ago. *Reynoutria japonica*, a successful colonist of waste ground roadsides and railways, was then only known from two localities. *Lepidium ruderale* and *Conyza canadensis* are both more frequent today than fifty years ago.

Rubbish dumps and waste places often contain a variety of alien species which appear to be persistent but which are in fact reintroduced annually from various sources. The commonest sources of these aliens are garden rubbish, kitchen waste and seed derived from the sweepings of bird cages. In the first category *Lobularia maritima* and *Calendula officinalis* are apparently more frequent today and there are several records of *Lupinus polyphyllus*, a plant which Horwood does not quote at all. A culinary species which seems to have increased in recent years is *Coriandrum sativum*, especially in those parts of Leicester where there are large Asian communities. Of the bird-seed aliens, *Bupleurum lancifolium*, *Cannabis sativa*, *Panicum mileaceum*, *Phalaris canariensis* and various species of *Setaria* were mentioned in the last Flora. Of these probably only *Cannabis* is rarer today, because hemp seed sold as bird food is now supposed to have been boiled to prevent germination. *Centaurea diluta* and *Guizotia abyssinica* were not mentioned in 1933 but now occur occasionally. There are other casual alien species which appear sporadically and for no apparent reason. One such species is *Datura stramonium*, of which there were few records in the last Flora. This may go unrecorded for a number of years and then suddenly appear in scattered localities all over the county, possibly as a result of a particular set of climatic conditions.

After major road works, verges are sometimes sown with imported grass seed, this being the source of some characteristic alien species. *Hordeum jubatum*, although more frequent than fifty years ago, is almost entirely confined to this habitat. It may persist in one place for a few years before succumbing to competition. *Bromus inermis* is also more widespread today and looks as if it may become permanently established, although it is too early yet to be sure.

Presumably, similar introductions were occurring at the time when the last Flora was written. We have seen some of these, such as *Senecio squalidus*, become permanent additions to our flora. It is therefore important that all such casual aliens which appeared in the county during the survey should be recorded for posterity.

Losses from the flora

The 1933 Flora includes records of plants found in the county as far back as the first half of the 18th century. Of these species, more than 90 have not been seen during the present survey and that figure does not include aliens, obvious introductions, or plants the records for which Horwood regards as of doubtful authenticity. Figure 22 is an attempt to show when those missing species were last recorded.

The first significant list of Leicestershire plant records is that in Richard Pulteney's manuscript of 1747. The time between 1720, the date of his birth, and the commencement of recording for this Flora in 1960, has been divided into periods of 30 years and last known records assigned to one of these periods. In the 1933 Flora, Horwood lists old records giving their author but rarely a date. The notes and cards from which he worked are no longer available and the task of re-tracing the records to source has not been undertaken. It has therefore been necessary to assess the most probable 30-year period to which each record should be assigned.

In the case of some early botanists, the date of their major, relevant publication, if one exists, has been used. When no publication is known, the mid-point of the life-span of the recorder has been used, unless more detailed information is available concerning the dates of his or her work in the county. Many of the last records were made by Horwood himself and these have been assigned to 1900-1929 as most of his work in Leicestershire was carried out prior to 1924, the year he left the county . Records from persons whom he lists as 'recent correspondents, 1902-1933,' have also been assigned to this period. In some recent cases the actual date of an un-published record is known.

Within each habitat, the species have been listed in order of the period of their last record; where several species fall within the same period they are arranged alphabetically.

First records have also been incorporated into the table. The dotted line preceding the first record is an acknowledgement that the species was probably in the county prior to that date.

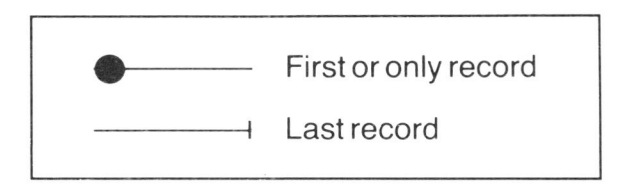

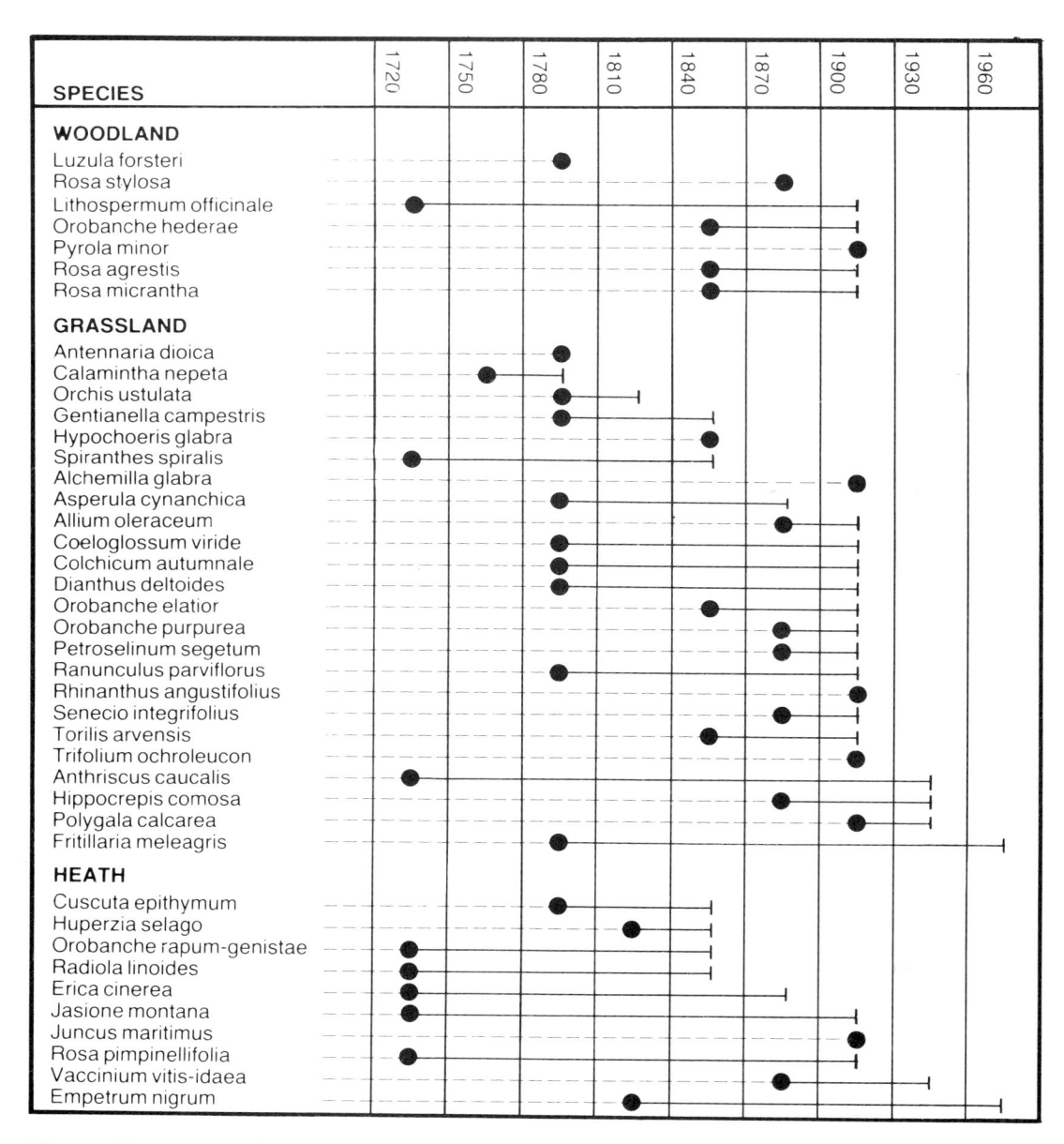

SPECIES	1720	1750	1780	1810	1840	1870	1900	1930	1960
WOODLAND									
Luzula forsteri			●						
Rosa stylosa						●			
Lithospermum officinale	●―――――――――――――――――┤								
Orobanche hederae					●――┤				
Pyrola minor							●		
Rosa agrestis					●―――┤				
Rosa micrantha					●―――┤				
GRASSLAND									
Antennaria dioica			●						
Calamintha nepeta		●――┤							
Orchis ustulata			●――┤						
Gentianella campestris			●――┤						
Hypochoeris glabra					●				
Spiranthes spiralis	●――――――――┤								
Alchemilla glabra							●		
Asperula cynanchica			●―――┤						
Allium oleraceum							●―┤		
Coeloglossum viride			●						
Colchicum autumnale			●						
Dianthus deltoides			●						
Orobanche elatior						●――┤			
Orobanche purpurea							●		
Petroselinum segetum							●		
Ranunculus parviflorus			●						
Rhinanthus angustifolius							●		
Senecio integrifolius						●			
Torilis arvensis						●			
Trifolium ochroleucon							●		
Anthriscus caucalis	●―――――――――――――――――――――――┤								
Hippocrepis comosa							●――┤		
Polygala calcarea							●――┤		
Fritillaria meleagris			●―――――――――――――――――――――┤						
HEATH									
Cuscuta epithymum			●――┤						
Huperzia selago				●――┤					
Orobanche rapum-genistae	●――┤								
Radiola linoides	●――――――――――┤								
Erica cinerea	●――――――――――┤								
Jasione montana	●――――――――――┤								
Juncus maritimus							●		
Rosa pimpinellifolia	●―――――――――――――――┤								
Vaccinium vitis-idaea							●―┤		
Empetrum nigrum				●――――――――――――――――――――┤					

Figure 22. Losses from the flora.

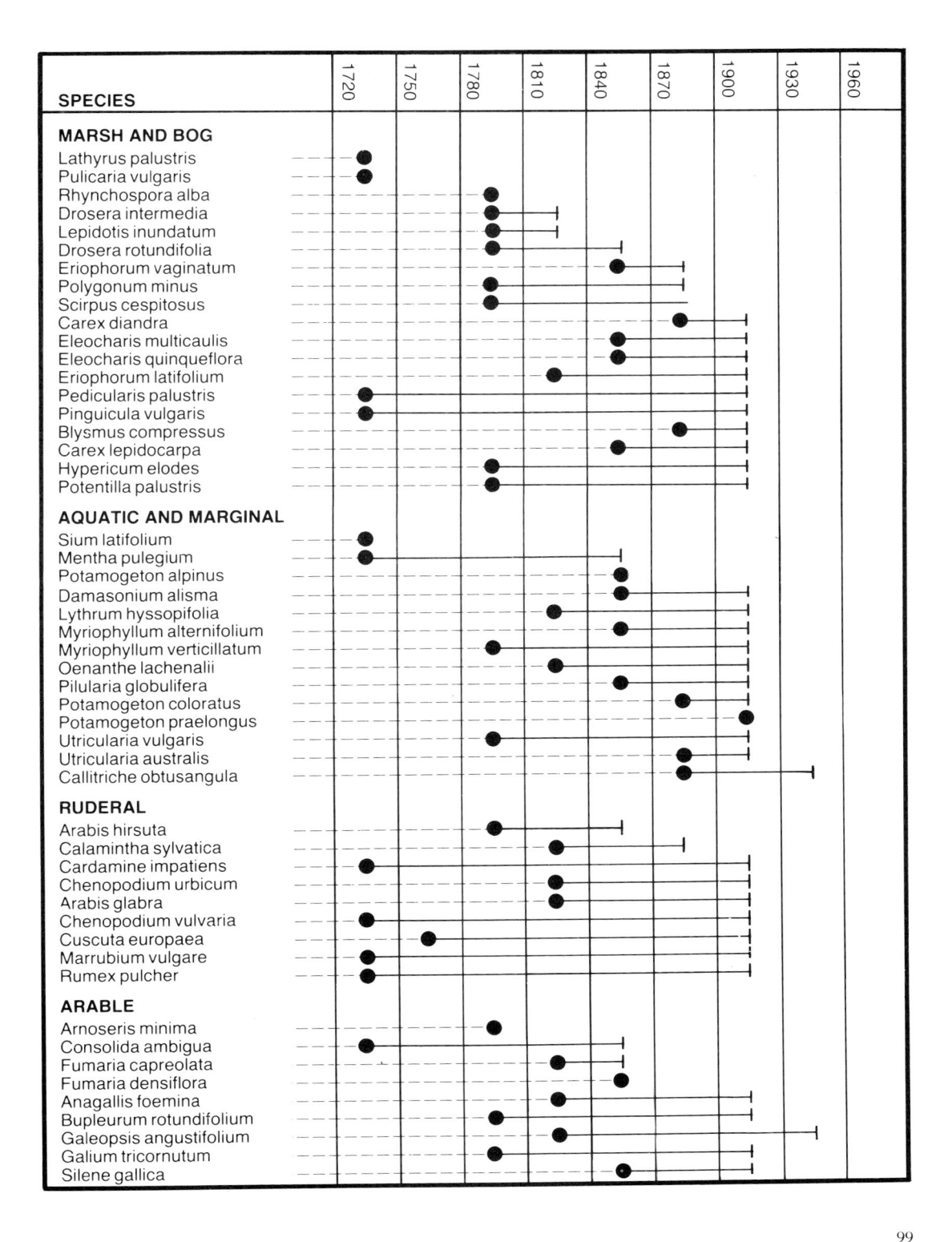

SPECIES	1720	1750	1780	1810	1840	1870	1900	1930	1960
MARSH AND BOG									
Lathyrus palustris	●								
Pulicaria vulgaris	●								
Rhynchospora alba			●						
Drosera intermedia			●——┤						
Lepidotis inundatum			●——┤						
Drosera rotundifolia			●						
Eriophorum vaginatum					●————┤				
Polygonum minus			●						
Scirpus cespitosus			●						
Carex diandra						●——┤			
Eleocharis multicaulis					●————┤				
Eleocharis quinqueflora					●————┤				
Eriophorum latifolium				●————————┤					
Pedicularis palustris	●————————————┤								
Pinguicula vulgaris	●————————————┤								
Blysmus compressus						●——┤			
Carex lepidocarpa					●————┤				
Hypericum elodes			●—┤						
Potentilla palustris			●—┤						
AQUATIC AND MARGINAL									
Sium latifolium	●—┤								
Mentha pulegium	●—┤								
Potamogeton alpinus					●————————————┤				
Damasonium alisma					●————————————┤				
Lythrum hyssopifolia				●					
Myriophyllum alternifolium					●————————————┤				
Myriophyllum verticillatum				●	●				
Oenanthe lachenalii					●————————————┤				
Pilularia globulifera					●————————————┤				
Potamogeton coloratus							●——┤		
Potamogeton praelongus							●		
Utricularia vulgaris				●————————————————————┤					
Utricularia australis							●——————┤		
Callitriche obtusangula							●————————┤		
RUDERAL									
Arabis hirsuta				●————┤					
Calamintha sylvatica				●————┤					
Cardamine impatiens		●							┤
Chenopodium urbicum				●————————————————————————┤					
Arabis glabra				●————————————————————————┤					
Chenopodium vulvaria	●————————————————————————————————————┤								
Cuscuta europaea			●————————————————————————————┤					
Marrubium vulgare	●————————————————————————————————————┤								
Rumex pulcher	●————————————————————————————————————┤								
ARABLE									
Arnoseris minima				●————┤					
Consolida ambigua	●————————————————————————————————————┤								
Fumaria capreolata					●————┤				
Fumaria densiflora					●				
Anagallis foemina					●————————————————————┤				
Bupleurum rotundifolium				●————————————————————————┤					
Galeopsis angustifolium					●————————————————————┤				
Galium tricornutum				●————————————————————————————┤					
Silene gallica						●			

Acknowledgements

I am grateful to Mr. J. C. Dalby, District Quality Control Officer, Severn-Trent Water Authority, for information on water quality in Leicestershire rivers and to Mr. H. A. B. Clements, Agricultural Advisory Officer, Ministry of Agriculture, Fisheries and Food for his help on the subject of agricultural weeds. Mr. D. Wells, Chief Scientist's Team, Nature Conservancy Council, kindly gave his views on the nature of certain grasslands.

References

Chard, J. S. R. et al., 1952. *Census of woodlands, 1947-1949*. London: H.M.S.O.

Haslam, S., 1978. *River plants*. Cambridge: C.U.P.

Measham, C. E. C., 1915. A botanical survey of some fields near Leicester. *Trans. Leicester lit. phil. Soc.* **19**, 17-26.

Mercer, G. E., 1914. The flora of Belgrave and Birstall. *Trans. Leicester lit. phil. Soc.* **18**, 76-92.

Newbold, C. and Palmer, M., 1979. *The River Eye, Leicestershire. A Nature Conservancy Council Report*. Huntingdon: N.C.C.

Wade, A. E., 1919. The flora of Aylestone and Narborough Bogs. *Trans. Leicester lit. phil. Soc.* **20**, 20-46.

HABITAT STUDIES

The introductory chapters of Horwood and Gainsborough (1933) contain detailed species lists from a large number of localities, illustrating the ecological formations represented in the various botanical districts. These species lists are an excellent feature of that Flora, but their usefulness to the present-day botanist is, for several reasons, less than it might be. The precise area listed is not defined, we do not know when the studies were made, and the lists appear to be incomplete. With these limitations in mind, we have tried to see that the habitat studies in this Flora are not lacking essential data.

Selection of sites

When the selection of sites for habitat studies was first discussed, the main object was to provide a good representative cross-section of the habitats in the county. This does however present difficulties. Sites of great botanical interest are obviously not typical of the vast majority of the land in the county as a whole, but these are the sites which most readers of the Flora would wish to know about or visit. An extreme case of this is discussed in the introduction to the studies of arable land; a typical arable field in the county is one where the farmer has done his utmost to eliminate everything of interest to the botanist, and the same is true of most of the grassland in the county.

It must also be admitted that, with hindsight, we are conscious of some unfortunate omissions. There are perhaps two main reasons for this. In the first place, many of the voluntary workers who so generously gave their time to the field work for the Flora, were unable, because of their normal commitments, to do more than the general survey of the part of the county assigned to them. Thus the work for the habitat studies throughout the county devolved upon a few people, and it proved to be a great undertaking. Secondly, the period of the Flora survey coincided with the beginning of the oil crisis, with the consequent great increase in the cost of travel. Here again the generosity of the field workers ensured an adequate coverage of the county in the general survey, but this factor definitely had a limiting effect on the habitat studies.

Selection of study area

There is something to be said for using a small standard area in all cases. However after much thought it was decided that this method has several disadvantages, and can even lead to absurdities such as are found in at least one modern County Flora, where *Fraxinus excelsior* is recorded as frequent in a one metre quadrat. Few habitats, at any rate in Leicestershire, are absolutely uniform throughout, and in fact this diversity adds to their interest. A strictly random selection of a small area for study could give an uninteresting or even atypical picture of the habitat. On the other hand, a subjective selection of area, though perhaps more interesting, could be equally atypical. For example, if one had to select a small area of Botcheston Bog for study, it would not be possible to include *Anagallis tenella, Carex dioica, Epipactis palustris* and *Parnassia palustris* in the list of species, as the selection of a site for one of these would probably exclude one or all of the others which make this site so interesting.

It was therefore decided to record for the whole site, if this was a manageable size which could be searched adequately in one recording session. If the site was too large, a suitable study area was arbitrarily selected.

Subdivision of species lists

The species lists are subdivided in the studies of Woodland, some of the Ponds and Lakes, Canals, Rivers and Hedgerows; in each case the reasons for this subdivision are self-explanatory. Some species may have occurred in both subdivisions, involving a subjective decision as to where to place them. Also, it has sometimes been thought desirable to append a habitat code symbol after a particular species; for example, in the studies of woodland the symbol (Ef) after a species indicates that it is mainly found in the rides or on the edge of the wood.

Designation of frequency

In the absence of a strict definition the use of the frequency designations is inevitably subjective. There are two important factors influencing the frequency of a species on a given site at any one time. Extremes of weather may affect the vegetation both at the time and in subsequent years; the exceptional prolonged drought of 1976 is an example of this. A more gradual change may occur in response to other environmental factors. This is particularly noticeable in unstable habitats such as dismantled railways and ungrazed grassland, where the status and distribution of species may alter markedly during a decade.

Finally, human fallibility must be taken into account. In any survey work there is a degree of under-recording, particularly of small, inconspicuous species and of those not in flower.

Presentation

The studies are grouped by habitat. Within each group the sites selected for study are arranged geographically, working from west to east. A complete list of the studies is appended, together with an index by parish.

Preceding each habitat group is a brief account of the distribution and floristic interest of that habitat in Leicestershire. Full information on the species composition is of course provided by the studies themselves.

Further information concerning recent changes in both habitats and species will be found in the section on 'Changes in the Flora'.

The six figure grid references given are for the approximate centre of the study area, except in the case of some linear features where they indicate the two ends. The areas given for the study areas are approximate, but have all been estimated by the same method. The pH of the surface soil has been given wherever it was possible to determine it.

List of studies

Woodland

1 Spring Wood, Staunton Harold
2 The Smoile, Lount, Worthington
3 Cloud Wood, Breedon on the Hill
4 Cowpastures Spinney, Market Bosworth
5 Asplin Wood, Breedon on the Hill
6 Grace Dieu Wood, Belton
7 Aston Firs, Aston Flamville
8 Holywell Wood, Loughborough
9 Groby Pool, Groby
10 Roecliffe Spinney, Woodhouse
11,12 Swithland Wood, Newtown Linford
13 Narborough Bog, Narborough
14 Cream Gorse, Frisby on the Wreake
15 Tugby Bushes, Tugby and Keythorpe
16 Launde Big Wood, Launde
17 Owston Big Wood, Owston and Newbold
18 Little Owston Wood, Owston and Newbold
19 Allexton Wood, Allexton

Grassland

20 Sheepy Fields, Sheepy
21 Lount Wood, Ashby de la Zouch
22 Breedon Hill, Breedon on the Hill
23 Broad Hill, Whitwick, Coalville
24 Osgathorpe
25 Burbage Common, Hinckley
26 Holly Hayes Wood, Whitwick, Coalville
27 Herbert's Meadow, Ulverscroft
28 Croft
29 Croft Hill, Croft
30 Groby Pool, Groby
31 Swithland Wood, Newtown Linford
32 Loughborough Big Meadow, Loughborough
33 Kinchley Lane, Rothley
34 Berrycott Lane, Seagrave
35 Burrough Hill, Somerby
36 Terrace Hill Farm, Eaton
37 Muston Meadows, Bottesford
38 Wymondham Rough, Wymondham
39 Egypt Plantation, Sproxton
40 Cribb's Meadow, Wymondham

Heathland

41 Moira, Ashby Woulds
42 High Sharpley, Whitwick, Coalville
43,44 Charnwood Lodge, Charley
45,46 Bradgate Park, Newtown Linford
47 Twenty Acre, Burton on the Wolds

Marsh

48 Newton Burgoland, Swepstone
49 Peckleton Fields, Peckleton
50 Botcheston Bog, Desford
51 Lea Wood, Ulverscroft
52 Thurlaston
53 The Wailes, Frisby on the Wreake
54 Saltby Bog, Sproxton

Ponds, lakes and reservoirs

55 Fishpond, Coleorton
56 Pond, Holly Hayes Wood, Whitwick, Coalville
57 Subsidence pool, Bagworth
58 Thornton Reservoir, Bagworth
59 Pond and drainage channel, Beacon Hill, Woodhouse
60 Groby Pool, Groby
61 Swithland Reservoir, Rothley
62 Pond, Sileby
63 Saddington Reservoir, Gumley/Saddington
64 Oxbow, Frisby on the Wreake
65 Pond, Eaton

Canals

66 Ashby Canal, Carlton
67 Grand Union Canal, Loughborough
68 Grand Union Canal, Saddington/Smeeton Westerby
69 Grantham Canal, Stathern

Rivers and streams

70 River Sence, Ratcliffe Culey, Sheepy/Witherley
71 River Trent, Castle Donington
72 Ulverscroft Brook, Ulverscroft
73 River Soar, Croft
74 River Soar, Belgrave, Leicester
75 River Soar, Quorn
76 River Wreake, Hoby
77 River Eye, Freeby

Green lanes, tracks and roadside verges

78 Green lane, Boothorpe, Ashby Woulds
79 Motorway roundabout, Markfield

80	Farm track, Nailstone
81	Roadside verge, Botcheston, Desford
82	Roadside verge, Thurlaston
83	Roadside verge, Park Hill Lane, Seagrave
84	Roadside verge, Withcote

Hedgerows

85	Ratcliffe Culey, Witherley
86	Higham on the Hill
87	Ulverscroft Lane, Newtown Linford
88	Barrow on Soar
89	Little Dalby, Burton and Dalby

Railways

90	Shenton Cutting, Sutton Cheney
91	Swithland Sidings, Swithland
92	Loughborough
93	Scalford Cutting
94	Long Clawson and Hose Station, Clawson and Harby

Arable land

95	Oakthorpe and Donisthorpe
96	Cradock's Ashes, Walton on the Wolds
97	Thrussington Mill
98	Sproxton

Quarries

99	Huncote
100	Cave's Inn Pits, Shawell
101	Swithland Wood, Newtown Linford
102	Waltham
103	Stonesby Quarry, Sproxton

Miscellaneous habitats

104	Refuse tip, Barwell, Hinckley
105	Ruins of Cotes Hall, Cotes
106	Bank by River Soar, Cotes
107	Castle Hill, Mountsorrel

Index of studies by parish

Allexton: 19
Ashby de la Zouch: 21
Ashby Woulds: 41,78
Aston Flamville: 7

Bagworth: 57,58
Barrow on Soar: 88
Belton: 6
Bottesford: 37
Breedon on the Hill: 3,5,22
Burton and Dalby: 89
Burton on the Wolds: 47

Carlton: 66
Castle Donington: 71
Charley: 43,44
Clawson and Harby: 94
Coalville: 23,26,42,56
Coleorton: 55
Cotes: 105,106
Croft: 28,29,73

Desford: 50,81

Eaton: 36,65

Freeby: 77
Frisby on the Wreake: 14,53,64

Groby: 9,30,60
Gumley: 63

Higham on the Hill: 86
Hinckley: 25,104
Hoby with Rotherby: 76
Huncote: 99

Launde: 16
Leicester: 74
Loughborough: 8,32,67,92

Market Bosworth: 4
Markfield: 79
Mountsorrel: 107

Nailstone: 80
Narborough: 13
Newtown Linford: 11,12,31,45,46,87,101

Oakthorpe and Donisthorpe: 95
Osgathorpe: 24
Owston and Newbold: 17,18

Peckleton: 49

Quorn: 75

Rothley: 33,61

Saddington: 63,68
Scalford: 93
Seagrave: 34,83
Shawell: 100
Sheepy: 20,70
Sileby: 62
Smeeton Westerby: 68
Somerby: 35
Sproxton: 39,54,98,103
Stathern: 69
Staunton Harold: 1
Sutton Cheney: 90
Swepstone: 48
Swithland: 91

Thrussington: 97
Thurlaston: 52,82
Tugby and Keythorpe: 15

Ulverscroft: 27,51,72

Waltham: 102
Walton on the Wolds: 96
Withcote: 84
Witherley: 70,85
Woodhouse: 10,59
Worthington: 2
Wymondham: 38,40

Woodland

Apart from those on the more acid soils of parts of Charnwood Forest, and the Coal Measures and Millstone Grit in the north-west, most of the Leicestershire woodlands lie on Boulder Clay, Keuper Marl or Jurassic clays, where in general the soil tends to be neutral in reaction or slightly on the basic side. However, it does not take long for a piece of woodland to produce a deep layer of largely organic soil, the surface layers of which sometimes show a surprisingly low pH value, as can be seen in some of the following studies.

Leicestershire is a sparsely wooded county, in which individual woodlands are on the whole small in area and differing from one another in character. The main factors causing this diversity are as follows: drainage, ranging from extremely wet to comparatively dry; the age of the woodland; the degree of management or other human interference to which the woodland has been subjected, including such practices as coppicing, periodic felling, and the clearance of scrub. A further complication is that many of the larger woods are by no means uniform throughout. A good example of this is Swithland Wood, the subject of two habitat studies (11 and 12). Here there are dry areas, wet areas, and parts affected by the workings and spoil heaps of former slate quarries (study 101).

Woodlands on the more acid soils are characterised by the predominance of *Betula pendula* in the tree layer, and a not very diverse ground flora, the main feature of which is an association of *Pteridium aquilinum*, *Hyacinthoides non-scripta* and *Holcus mollis*. The best example of this was The Smoile (study 2), but this has recently been destroyed in the process of opencast coal mining. Another example is Spring Wood, Staunton Harold, the subject of study 1.

On the more basic soils it is difficult, for reasons stated above, to generalise or classify into clear-cut types, though many of the East Leicestershire woodlands approximate to Peterken's stand type 2A, wet ash-maple (Peterken 1977). The ground flora is much more diverse than in the woodlands on acid soils. Species characteristic of primary woodland are *Galium odoratum*, *Lamiastrum galeobdolon* and *Paris quadrifolia*. Good examples of this type are Tugby Bushes (study 15) and Little Owston Wood (study 18). Secondary woodland, depending on its age and management, can have a diverse ground flora of those woodland species which are capable of colonising such sites, including *Orchis mascula*. An example is the fox covert Cream Gorse (study 14).

Some species of the ground flora show a marked preference for either the east or the west of the county. Bluebells, though present in woodland throughout the county, are at their best on the lighter soils of the western areas; in the east Leicestershire woodlands the impressive sheets of blue to be seen in the spring are in some cases due to *Myosotis sylvatica*, which is rare in the west. *Digitalis purpurea* is a characteristic feature of woodlands on the more acid soils of the west, but here again it is unwise to generalise, because some of the Belvoir woodlands on the Marlstone escarpment in the north-east have been known to produce large stands of this species after clearance. *Campanula latifolia* is frequent in the east Leicestershire woodlands and is much rarer in the west. Other less common species with a geographical preference are *Equisetum sylvaticum* and *Corydalis claviculata* in the west, and *Vicia sylvatica* and *Dipsacus pilosus* in the east.

Wet woodland has its own characteristic species, including alder and willow in the tree canopy, and *Caltha palustris*, *Chrysosplenium oppositifolium*, *Allium ursinum* and *Carex paniculata* in the ground flora. A good example of such wet woodland is the part of Roecliffe Spinney which is the subject of study 10. Other parts of this spinney are dry, showing how conditions in different parts of the same wood may diversify its character.

Reference

Peterken, G. F. (1977). *Woodland survey for nature conservation*. Huntingdon : Nature Conservancy Council.

Habitat Study 1 Spring Wood, Staunton Harold
SK 381228 Area: 3.8 ha (9.4 acres) Alt.: 76 m. (250 ft.)
Recorder: S. H. Bishop. Surveyed: 11th June 1977.
The study is of the northernmost part of this large woodland. The study area is bounded to the north by a green lane which forms the county boundary with Derbyshire,. To the west is rough grassland and to the east and south are overgrown rides. It is dominated by *Betula pendula* and some *Quercus robur* with young standards 30 to 40 years old and much coppice growth especially from the stumps of birch. The shrub layer is very sparse except in some one more open, wetter areas. The wood lies on Millstone Grit.

TREE AND SHRUB LAYERS
Abundant: Betula pendula, Rubus fruticosus agg.
Locally frequent: Lonicera periclymenum, Quercus robur, Salix caprea, Ulex europaeus.
Occasional: Populus tremula, Salix cinerea subsp. oleifolia, Sambucus nigra, Sorbus aucuparia.
Rare: Corylus avellana, Crataegus monogyna, Ilex aquifolium, Rosa canina, Viburnum opulus.

HERB LAYER
Abundant: Pteridium aquilinum.
Locally abundant: Deschampsia flexuosa, Hyacinthoides non-scripta.
Frequent: Holcus mollis, Juncus conglomeratus, Luzula multiflora.
Locally frequent: Agrostis canina, Anthoxanthum odoratum, Cirsium palustre, Dactylis glomerata, Deschampsia cespitosa, Digitalis purpurea, Epilobium angustifolium, Galium saxatile, Juncus acutiflorus, J. effusus, Lysimachia nemorum, Mercurialis perennis, Poa trivialis, Potentilla erecta, Viola palustris.
Occasional: Blechnum spicant, Carex binervis, C. demissa, Cerastium fontanum subsp. triviale, Cynosurus cristatus, Danthonia decumbens, Dryopteris dilatata, Festuca rubra, Hedera helix, Juncus bulbosus, Lotus uliginosus, Ophioglossum vulgatum, Prunella vulgaris, Rumex acetosella, Silene dioica, Stellaria media, Teucrium scorodonia, Trifolium dubium, Urtica dioica.

Rare: Angelica sylvestris, Carex pilulifera, Cirsium vulgare, Dryopteris filix-mas, Epipactis helleborine, Hieracium sp., Hypericum pulchrum, Molinia caerulea, Pedicularis sylvatica, Tamus communis, Taraxacum spp., Trifolium pratense, Veronica officinalis.

Habitat Study 2 The Smoile, Worthington
SK 390195 Area: 12.1 ha (30.0 acres) Alt.: 111 m. (365 ft.)
Recorder: S. H. Bishop. Surveyed: 11th June 1977.
The study covers the northern part of the wood, bounded to the north, west and east by hedges and roads, to the south-west by a hedge and field and to the south by a dismantled mineral railway. The study area is dominated by regenerating *Betula pendula* and some *Quercus robur*, with standards 50 to 100 years old and much coppice growth especially from the stumps of the birch. During the dry summer of 1976 a fire destroyed large parts of the wood, coal seams near the surface continuing to burn for months. In 1979 to 1980 the study area was almost totally felled prior to open-cast coal mining. The wood stands on Coal Measures.

TREE AND SHRUB LAYERS
Abundant: Betula pendula, Rubus fruticosus agg.
Locally abundant: Acer pseudoplatanus, Quercus robur.
Frequent: Crataegus monogyna, Lonicera periclymenum, Sambucus nigra.
Locally frequent: Fraxinus excelsior, Solanum dulcamara.
Occasional: Castanea sativa, Corylus avellana, Ilex aquifolium, Rosa canina, Salix caprea, S. cinerea subsp. oleifolia, Sorbus aucuparia, Ulmus glabra.
Rare: Acer campestre, Alnus glutinosa, Betula pubescens, Cytisus scoparius (Ef), Ligustrum vulgare, Malus sylvestris, Ribes rubrum, R. uva-crispa, Rosa arvensis, Ulex europaeus.

HERB LAYER
Abundant: Dryopteris dilatata, Hyacinthoides non-scripta, Pteridium aquilinum.

Locally abundant: Corydalis claviculata, Deschampsia flexuosa, Dryopteris filix-mas, Holcus mollis, Milium effusum.
Frequent: Circaea lutetiana, Digitalis purpurea, Epilobium angustifolium, Galium saxatile, Geum urbanum, Hedera helix, Poa trivialis, Silene dioica, Teucrium scorodonia, Urtica dioica.
Locally frequent: Arum maculatum, Athyrium filix-femina, Brachypodium sylvaticum, Bromus ramosus, Cardamine flexuosa, Dryopteris carthusiana, Galeopsis tetrahit, Galium aparine, G. palustre (Wp), Glechoma hederacea, Juncus effusus, Mercurialis perennis, Moehringia trinervia, Myosotis arvensis, Polygonum hydropiper, Ranunculus ficaria, Rumex sanguineus, Stachys sylvatica, Stellaria holostea, Tamus communis, Torilis japonica.
Occasional: Alliaria petiolata, Angelica sylvestris, Anthriscus sylvestris, Arctium minus, Callitriche stagnalis, Epilobium montanum, Festuca gigantea, Geranium robertianum, Heracleum sphondylium, Lapsana communis, Polygonum persicaria (Ef), Potentilla erecta, Rumex acetosella (Ef), Sanicula europaea, Scrophularia auriculata, Senecio sylvaticus, Stellaria media, Taraxacum spp., Veronica chamaedrys, Viola riviniana.
Rare: Adoxa moschatellina, Calluna vulgaris, Cerastium fontanum subsp. triviale (Ef), Cirsium arvense, C. vulgare, Dactylis glomerata (Ef), Dactylorhiza fuchsii (Ef), Filaginella uliginosa (Ef), Hypericum hirsutum (Ef), H. pulchrum (Ef), Lamium album, Malva moschata (Ef), Polygonum lapathifolium (Ef), Potentilla anglica, Primula vulgaris, Prunella vulgaris (Ef), Ranunculus auricomus, R. bulbosus, R repens (Ef), Senecio jacobaea (Ef), Veronica montana, V. serpyllifolia (Ef).

Habitat Study 3 Cloud Wood, Breedon on the Hill
SK 415214 Area: 4.7 ha (11.7 acres) Alt.: 91-103 m. (300-340 ft.) S.S.S.I.
Recorders: S. H. Bishop, P. H. Gamble and A. L. Primavesi. Surveyed: 6th June 1977.
The study is of the central part of this ancient woodland. The study area is five-sided, bounded on the west by a large Carboniferous Limestone quarry and on the other four sides by woodland rides. The wood was felled about 40 years ago and has been allowed to regenerate naturally. It has standards 30 to 40 years old of *Quercus robur, Betula pendula, Fraxinus excelsior* and *Ulmus glabra,* with much coppice growth, especially from the stumps of *Ulmus glabra.* The ground slopes gradually down from south-east to north-east. The study area is on Keuper Marl with a small area of Boulder Clay in the eastern part. A sample of surface soil had pH 6.0.

TREE AND SHRUB LAYERS
Abundant: Corylus avellana.
Frequent: Betula pendula, Crataegus laevigata × monogyna, Lonicera periclymenum, Prunus spinosa, Quercus robur, Rosa arvensis, Rubus fruticosus agg., Ulmus glabra.
Locally frequent: Acer campestre, Crataegus monogyna, Fraxinus excelsior, Populus tremula, Salix cinerea.
Occasional: Cornus sanguinea, Rosa canina, Salix caprea, Viburnum opulus.
Rare: Euonymus europaeus, Malus sylvestris, Ribes uva-crispa, Rubus caesius (Ef), R. idaeus, Tilia × vulgaris.

HERB LAYER
Abundant: Hyacinthoides non-scripta, Mercurialis perennis.
Locally abundant: Anemone nemorosa, Campanula latifolia, Dryopteris filix-mas, Hypericum hirsutum.
Frequent: Arum maculatum, Circaea lutetiana, Galium odoratum, Poa trivialis, Potentilla sterilis, Ranunculus ficaria, Sanicula europaea, Veronica chamaedrys.
Locally frequent: Cardamine flexuosa, Deschampsia cespitosa, Filipendula ulmaria (Ef), Galium aparine, Hypericum tetrapterum (Ef), Juncus conglomeratus (Ef), Milium effusum, Myosotis arvensis (Ef), Orchis mascula, Primula vulgaris, Ranunculus repens (Ef), Rumex sanguineus, Silene dioica, Stellaria graminea (Ef), Tamus communis, Urtica dioica, Viola reichenbachiana, V. riviniana.
Occasional: Adoxa moschatellina, Ajuga reptans, Angelica sylvestris, Arctium minus, Brachypodium sylvaticum, Carex pendula, Cirsium arvense (Ef), C. palustre (Ef), Conopodium majus, Fragaria vesca, Geum urbanum, Heracleum sphondylium (Ef), Juncus effusus (Ef), Lamiastrum galeobdolon, Listera ovata, Lysimachia nemorum (Ef), Platanthera chlorantha, Potentilla anserina (Ef), P. reptans (Ef), Prunella vulgaris (Ef), Rumex obtusifolius (Ef), Scrophularia nodosa (Ef), Stachys sylvatica (Ef), Stellaria holostea (Ef), Vicia sepium (Ef).
Rare: Alchemilla filicaulis subsp. vestita, Allium ursinum, Cerastium glomeratum (Ef), C. fontanum subsp. triviale (Ef), Chrysosplenium oppositifolium, Cirsium vulgare (Ef), Dipsacus fullonum (Ef), Dryopteris dilatata, Epilobium hirsutum (Ef), E. montanum (Ef), Epipactis helleborine, Geranium pratense (Ef), Glechoma hederacea, Lapsana communis (Ef), Lotus uliginosus, Lychnis flos-cuculi, Lysimachia nummularia (Ef), Plantago major (Ef), Polystichum aculeatum, Ranunculus auricomus, R. flammula (Ef), Pteridium aquilinum, Senecio jacobaea (Ef), S. vulgaris (Ef), Stachys officinalis, Taraxacum spp. (Ef), Veronica montana, V. serpyllifolia (Ef).

Habitat Study 4 Cowpastures Spinney, Market Bosworth
SK 416037 Area: 2.2 ha (5.6 acres) Alt.: 116 m. (380 ft.)
Recorders: S. H. Bishop, H. I. James and F. T. Smith. Surveyed: 8th June 1977.
The study covers the whole of this small spinney. It is enclosed by hedgerows and on its eastern boundary the hedge is accompanied by a deep ditch and wood bank. *Fraxinus excelsior* and *Quercus robur* are the predominant tree species, with standards between 50 and 100 years old. The spinney stands on Boulder Clay.

TREE AND SHRUB LAYERS
Abundant: Rubus fruticosus agg.
Frequent: Fraxinus excelsior.
Locally Frequent: Crataegus monogyna, Prunus spinosa, Quercus robur.
Occasional: Crataegus laevigata × monogyna, Corylus avellana, Lonicera periclymenum, Rosa canina, Rubus caesius, Sambucus nigra, Solanum dulcamara.
Rare: Acer campestre, Aesculus hippocastanum, Castanea sativa, Cornus sanguinea, Ilex aquifolium, Malus sylvestris, Rosa arvensis, Rubus idaeus, Salix cinerea subsp. oleifolia, S. fragilis, Viburnum opulus.

107

HERB LAYER

Abundant: Hyacinthoides non-scripta, Urtica dioica.
Locally abundant: Anemone nemorosa, Anthriscus sylvestris, Deschampsia cespitosa, Filipendula ulmaria, Mercurialis perennis, Ranunculus ficaria.
Frequent: Angelica sylvestris, Cardamine flexuosa, Circaea lutetiana, Dryopteris dilatata, D. filix-mas, Epilobium hirsutum, Equisetum arvense, Galium aparine, Geum rivale, Glechoma hederacea, Hedera helix, Phalaris arundinacea, Poa nemoralis, Silene dioica, Stachys sylvatica.
Locally frequent: Alopecurus pratensis, Brachypodium sylvaticum, Carex acutiformis, Cirsium arvense, Epilobium angustifolium, Geum rivale × urbanum, Iris pseudacorus, Primula vulgaris, Ranunculus repens, Rumex sanguineus, Viola riviniana.
Occasional: Ajuga reptans, Arum maculatum, Campanula latifolia, Cardamine amara, Conopodium majus, Geranium robertianum, Geum urbanum, Heracleum sphondylium, Lamiastrum galeobdolon, Lapsana communis, Lysimachia nummularia, Stellaria holostea, Vicia cracca, V. sepium.
Rare: Aegopodium podagraria, Dactylis glomerata, Dryopteris borreri, Galeopsis tetrahit, Hypericum hirsutum, Ranunculus auricomus, Rumex obtusifolius, Stellaria media, Tamus communis, Taraxacum spp.

Habitat Study 5   Asplin Wood, Breedon on the Hill SK 431218   Area: 4.3 ha (10.7 acres)   Alt.: 91m. (300 ft.) S.S.S.I.
Recorders: S. H. Bishop, P. H. Gamble and A. L. Primavesi. Surveyed: 6th June 1977.

An area of ash/hazel woodland with a scattering of *Quercus robur*, the standard ash and oak 60 to 80 years old but with some oak more than 100 years old in the extreme north near the road. The part studied was the north-east portion, adjacent to the road and south-east of the more southerly of the two main rides, continuing to a point about half-way into the wood where the ride bends again. Apart from a relatively small area of Keuper Marl in the south-east part, the study area stands on Boulder Clay. A sample of surface soil collected near the middle of the area had pH 4.3.

TREE AND SHRUB LAYERS

Abundant: Fraxinus excelsior, Rubus fruticosus agg.
Locally abundant: Corylus avellana.
Frequent: Crataegus monogyna, Lonicera periclymenum, Rosa arvensis.
Locally frequent: Quercus robur, Tamus communis.
Occasional: Acer campestre, Crataegus laevigata × monogyna, Hedera helix, Ilex aquifolium, Populus tremula, Prunus avium, P. spinosa, Rosa canina.
Rare: Aesculus hippocastanum (Ef), Betula pendula, Cornus sanguinea, Fagus sylvatica (seedling), Malus sylvestris, Ribes rubrum, Sambucus nigra, Sorbus aucuparia, Viburnum opulus.

HERB LAYER

Abundant: Hyacinthoides non-scripta, Milium effusum, Poa trivialis.
Locally abundant: Anemone nemorosa, Deschampsia cespitosa, Dryopteris filix-mas, Holcus mollis, Stellaria holostea.
Frequent: Ajuga reptans, Callitriche stagnalis (Ef), Cardamine flexuosa, Circaea lutetiana, Dryopteris dilatata, Galium odoratum, Geum urbanum, Lonicera periclymenum, Lysimachia nemorum (Ef), Stellaria alsine (Ef), Veronica montana (Ef).
Locally frequent: Arum maculatum, Brachypodium sylvaticum, Carex pendula, Filipendula ulmaria, Galium aparine, Lamiastrum galeobdolon, Melica uniflora, Mercurialis perennis, Moehringia trinervia, Potentilla sterilis, Rumex sanguineus (Ef), Silene dioica, Urtica dioica, Vicia sepium (Ef), Viola riviniana.
Occasional: Angelica sylvestris, Anthriscus sylvestris (Ef), Arctium minus, Carex remota (Ef), C. sylvatica, Festuca gigantea, Galium palustre (Ef), Listera ovata, Mentha arvensis (Ef), Orchis mascula, Oxalis acetosella, Paris quadrifolia, Primula vulgaris, Ranunculus repens (Ef), Stachys sylvatica (Ef), Valeriana officinalis (Ef), Veronica chamaedrys (Ef).
Rare: Alopecurus pratensis (Ef), Arrhenatherum elatius (Ef), Bromus ramosus, Dactylis glomerata, Dactylorhiza fuchsii (Ef), Heracleum sphondylium, Lapsana communis (Ef), Lychnis flos-cuculi, Phalaris arundinacea (Ef), Poa annua (Ef), Ranunculus ficaria, Rumex crispus (Ef), R. obtusifolius (Ef), Scrophularia nodosa (Ef), Senecio vulgaris (Ef), Stellaria media (Ef), Taraxacum spp. (Ef), Veronica beccabunga (Ef).

Habitat Study 6   Grace Dieu Wood, Belton SK 435175   Area: 1.8 ha (4.5 acres)   Alt.: 91-99m. (300-325 ft.) S.S.S.I.
Recorders: S. H. Bishop and P. H. Gamble. Surveyed: 3rd June and 24th August 1976.

An area of wet woodland in the northern part of Grace Dieu Wood, situated at the point of confluence of several spring drains; bounded by the Grace Dieu Brook on its north-west boundary and woodland tracks on its other sides. The Brook here flows in a stony bed more like some northern Pennine streams than most Leicestershire brooks, an impression reinforced by the presence of such characteristic northern plants as *Equisetum hyemale* and *E. sylvaticum*. The area lies on Keuper Sandstone and Marl. The humus-rich soil had pH 3.8 on some of the drier ground and pH 5.2 in a wet location.

TREE AND SHRUB LAYERS

Locally abundant: Acer pseudoplatanus, Rubus fruticosus agg.
Locally frequent: Alnus glutinosa, Corylus avellana, Crataegus laevigata, C. monogyna, Fraxinus excelsior, Hedera helix, Ilex aquifolium, Lonicera periclymenum, Quercus robur, Ulmus glabra.
Occasional: Crataegus laevigata × monogyna, Rhododendron ponticum, Rubus idaeus, Sambucus nigra, Sorbus aucuparia, Viburnum opulus.
Rare: Aesculus hippocastanum, Betula pendula, Cornus sanguinea, Frangula alnus, Malus sylvestris, Prunus avium, P. spinosa, Ribes rubrum, Rosa arvensis, R. canina, Tamus communis, Taxus baccata.

HERB LAYER

Abundant: Anemone nemorosa, Lamiastrum galeobdolon, Ranunculus ficaria.
Locally abundant: Chrysosplenium oppositifolium, Deschampsia cespitosa, Hyacinthoides non-scripta, Petasites hybridus, Poa trivialis.
Frequent: Allium ursinum, Athyrium filix-femina, Cardamine amara, C. flexuosa, Circaea lutetiana,

Dryopteris dilatata, Epilobium adenocaulon, Equisetum sylvaticum, Holcus mollis, Silene dioica, Urtica dioica, Veronica montana.
Locally frequent: Angelica sylvestris, Anthriscus sylvestris, Brachypodium sylvaticum, Carex acutiformis, C. pendula, Conopodium majus, Dactylis glomerata, Dryopteris filix-mas, Elymus caninus, Epilobium montanum, Filipendula ulmaria, Galium aparine, Hedera helix, Heracleum sphondylium, Impatiens glandulifera (Ew), Lonicera periclymenum, Luzula sylvatica, Melica uniflora, Mercurialis perennis, Oxalis acetosella, Phalaris arundinacea, Polygonum hydropiper, Pteridium aquilinum, Ranunculus repens, Rumex sanguineus, Sanicula europaea, Stachys sylvatica, Stellaria holostea.
Occasional: Apium nodiflorum (Ew), Caltha palustris, Equisetum hyemale (Ew), Lapsana communis, Lysimachia nemorum, Milium effusum, Poa annua, P. nemoralis, Polygonum persicaria, Veronica beccabunga.
Rare: Alliaria petiolata, Arctium minus, Arum maculatum, Callitriche stagnalis, Campanula latifolia, Carex remota, C. sylvatica, Cirsium arvense (Rd), Dryopteris carthusiana, Epilobium hirsutum, Equisetum arvense, Galium odoratum, Geranium robertianum, Hypericum tetrapterum (Rd), Juncus bufonius (Rd), Matricaria perforata (Rd), Papaver dubium (Rd), Plantago major, Polygonum aviculare (Rd), P. lapathifolium (Rd), Ranunculus sceleratus (Ew), Rumex obtusifolius, Stellaria media, Taraxacum spp., Teucrium scorodonia, Tussilago farfara, Valeriana officinalis, Veronica chamaedrys, Viola reichenbachiana.

Habitat Study 7 Aston Firs, Aston Flamville
SP 457938 Area: 5.6 ha (13.8 acres) Alt.: 104 m. (340 ft.) S.S.S.I.
Recorder: S. H. Bishop. Surveyed 6th May and 8th September 1979.
The study covers the south-eastern corner of the wood, bounded by Sapcote Road to the south, a trackway and fields to the east and rides to the north and west. *Fraxinus excelsior* and *Quercus robur* are the dominant tree species, with standards up to 100 years old. At the time of the study *Ulmus procera* was locally abundant, but these trees were in an advanced stage of Dutch elm disease with many of them dead or dying. In a large part of the eastern portion of the study area there is much regenerating ash on ridge and furrow. Aston Firs stands on Boulder Clay.

TREE AND SHRUB LAYERS
Abundant: Corylus avellana, Fraxinus excelsior, Rubus fruticosus agg.
Locally abundant: Ulmus procera.
Frequent: Acer campestre, Crataegus monogyna, Lonicera periclymenum, Quercus robur, Rosa arvensis.
Locally frequent: Prunus spinosa, Sambucus nigra.
Occasional: Betula pendula, Ilex aquifolium, Ligustrum vulgare, Salix cinerea subsp. oleifolia, Ribes rubrum, Rosa canina, Rubus idaeus, Solanum dulcamara, Symphoricarpos albus.
Rare: Malus sylvestris, Populus tremula, Prunus avium, Salix caprea, Viburnum opulus.

HERB LAYER
Abundant: Brachypodium sylvaticum, Silene dioica.
Locally abundant: Hyacinthoides non-scripta, Ranunculus ficaria, R. repens (Ef).
Frequent: Ajuga reptans, Arum maculatum, Circaea lutetiana, Deschampsia cespitosa, Dryopteris dilatata, D. filix-mas, Galium aparine, G. odoratum, Geum rivale, G. urbanum, Orchis mascula, Rumex sanguineus, Stellaria holostea, Veronica chamaedrys, Viola riviniana.
Locally frequent: Adoxa moschatellina, Arctium minus, Bromus ramosus, Callitriche stagnalis (Ef), Cardamine flexuosa, Conopodium majus, Dactylis glomerata (Ef), Epilobium angustifolium, E. montanum, Festuca gigantea, Filipendula ulmaria, Glechoma hederacea, Hedera helix, Lapsana communis, Milium effusum, Poa trivialis, Polygonum hydropiper (Ef), Primula veris, Stachys sylvatica, Stellaria media, Urtica dioica, Valeriana officinalis.
Occasional: Angelica sylvestris, Calamagrostis epigejos, Carex pendula, C. remota, C. sylvatica, Cirsium palustre, C. vulgare, Epilobium hirsutum, Geranium robertianum, Heracleum sphondylium, Holcus lanatus (Ef), Juncus effusus, Potentilla anserina (Ef), P. sterilis, Prunella vulgaris (Ef), Tamus communis, Taraxacum spp., Vicia cracca.
Rare: Agrostis capillaris (Ef), Arrhenatherum elatius, Campanula latifolia, Carex hirta (Ef), Cirsium arvense, Epipactis helleborine, Hypericum hirsutum, H. tetrapterum, Lychnis flos-cuculi (Ef), Mentha arvensis (Ef), Plantago major (Ef), Polygonum persicaria (Ef), Rumex obtusifolius, Scrophularia auriculata (Ef), S. nodosa, Sonchus arvensis, Stellaria graminea (Ef), Torilis japonica, Veronica officinalis.

Habitat Study 8 Holywell Wood, Loughborough
SK 507182 Area: 6.5 ha (16.2 acres) Alt.: 55-60 m. (180-220 ft.)
Recorders: S. H. Bishop and P. H. Gamble. Surveyed: 22nd May 1979
This ancient woodland, lying just to the north of Holywell Hall, is three-sided. It is bounded by a stream to the north, a farm track and arable fields to the south-west and pasture to the south-east. It is a composite wood with open to closed canopy and standard trees 40 to 100 years old. It has much *Fraxinus excelsior*, an area of *Betula pendula* in the western part and some *Alnus glutinosa* in the wet portion to the north near the stream. Several colonies of *Ulmus glabra* occur here and there throughout the wood, but at the time of the survey these were mostly in an advanced stage of Dutch elm disease. The wood stands on Keuper Marl except for a small area of Alluvium in the north near the stream. Samples of surface soil varied from pH 4.1 in the dry western area to 5.0 on wet ground near the stream.

TREE AND SHRUB LAYERS
Abundant: Rubus fruticosus agg.
Locally abundant: Betula pendula.
Frequent: Acer campestre, Cornus sanguinea, Corylus avellana, Crataegus monogyna, Fraxinus excelsior, Hedera helix, Malus sylvestris, Prunus spinosa.
Locally frequent: Acer pseudoplatanus, Alnus glutinosa (Fw), Lonicera periclymenum, Populus tremula, Quercus robur, Rubus caesius (Fw), Sambucus nigra,

Solanum dulcamara (Fw), Ulmus glabra.
Occasional: Ilex aquifolium, Ribes rubrum (Fw), Rubus idaeus, Sorbus aucuparia, Rhododendron ponticum, Rosa arvensis, R. canina, Salix caprea, S. cinerea subsp. cinerea (Fw), Viburnum opulus (Fw).
Rare: Larix decidua, Ligustrum vulgare, Picea abies, Prunus avium.

HERB LAYER
Abundant: Hyacinthoides non-scripta, Mercurialis perennis.
Locally abundant: Anemone nemorosa, Deschampsia cespitosa, Holcus mollis, Juncus effusus (Fw), Milium effusum.
Frequent: Cardamine flexuosa (Fw), Dryopteris dilatata, D. filix-mas, Epilobium montanum, Galium aparine, Geum urbanum, Ranunculus ficaria, Scrophularia nodosa, Silene dioica.
Locally frequent: Ajuga reptans, Arum maculatum, Brachypodium sylvaticum, Carex pendula (Fw), Circaea lutetiana, Conopodium majus, Epilobium hirsutum, Filipendula ulmaria (Fw), Glechoma hederacea, Hypericum hirsutum, H. tetrapterum, Lamiastrum galeobdolon, Luzula sylvatica, Moehringia trinervia, Ranunculus repens, Rumex obtusifolius, R. sanguineus, Stachys sylvatica, Urtica dioica, Veronica montana, Vicia sepium (Ef).
Occasional: Angelica sylvestris (Fw), Anthriscus sylvestris, Arctium minus, Athyrium filix-femina, Bromus ramosus, Carex sylvatica, Chrysosplenium oppositifolium (Fw), Cirsium palustre (Fw), C. vulgare (Ef), Dryopteris carthusiana, Elymus caninus (Fw), Epilobium angustifolium, Geranium robertianum, Heracleum sphondylium, Lysimachia nemorum, Oxalis acetosella, Potentilla sterilis, Stellaria holostea, S. media, Taraxacum spp., Valeriana officinalis (Fw), Veronica chamaedrys, Viola riviniana.
Rare: Alliaria petiolata, Cardamine pratensis (Ef), Carex remota, Cerastium fontanum subsp. triviale, Cirsium arvense, Digitalis purpurea, Lamium album (Ef), Listera ovata, Luzula pilosa, Lychnis flos-cuculi, Plantago major (Ef), Polygonum bistorta (Ef), Ranunculus acris, Rumex acetosella (Ef), Sanicula europaea, Scrophularia auriculata (Fw), Senecio jacobaea, Tussilago farfara (Ef), Veronica beccabunga (Ef), V. serpyllifolia, Viola reichenbachiana.

Habitat Study 9 Wet woodland, Groby Pool, Groby SK 520084 Area: 2.1 ha (5.2 acres) Alt.: 99 m. (325 ft.)
S.S.S.I.
Recorders: S. H. Bishop and E. Hesselgreaves. Surveyed: 11th June 1976.
This is an area of wet woodland on the northern shore of Groby Pool. *Alnus glutinosa* is the dominant tree species, with standards aged between 50 and 100 years. *Salix fragilis* is largely confined to the wetter areas, *Fraxinus excelsior* to the drier parts. The area lies in a hollow in the Keuper Marl.

TREE AND SHRUB LAYERS
Abundant: Alnus glutinosa.
Frequent: Rubus fruticosus agg.
Locally frequent: Fraxinus excelsior, Rubus caesius, Salix fragilis.
Occasional: Corylus avellana, Lonicera periclymenum, Quercus robur, Rosa canina, Rubus idaeus, Salix cinerea subsp. oleifolia, Solanum dulcamara.
Rare: Crataegus monogyna, Fagus sylvatica, Ilex aquifolium, Populus × canadensis, Prunus spinosa, Ribes rubrum, Rosa arvensis, Salix purpurea, S. triandra, Sambucus nigra, Sorbus aucuparia, Viburnum opulus.

HERB LAYER
Abundant: Chrysosplenium oppositifolium, Filipendula ulmaria, Glyceria maxima, Phragmites australis.
Locally abundant: Cardamine amara, Carex acutiformis, Mentha aquatica, Poa trivialis, Ranunculus repens, Stellaria alsine.
Frequent: Anemone nemorosa, Athyrium filix-femina, Cardamine flexuosa, Deschampsia cespitosa, Dryopteris dilatata, Galium aparine, Geum urbanum, Iris pseudacorus, Lysimachia nemorum, Lycopus europaeus, Myosotis scorpioides, Poa annua, Rumex conglomeratus, Scutellaria galericulata, Urtica dioica, Valeriana officinalis, Veronica beccabunga.
Locally frequent: Ajuga reptans, Angelica sylvestris, Circaea lutetiana, Cirsium palustre, Epilobium hirsutum, Equisetum fluviatile, Eupatorium cannabinum, Festuca gigantea, Galium palustre, Glechoma hederacea, Glyceria fluitans, Holcus mollis, Hyacinthoides non-scripta, Lamiastrum galeobdolon, Lychnis flos-cuculi, Lysimachia nummularia, Oxalis acetosella, Polygonum aviculare, Prunella vulgaris, Ranunculus flammula, R. sceleratus, Rumex crispus, Scrophularia auriculata, Silene dioica, Stachys sylvatica, Stellaria media, Veronica montana.
Occasional: Apium nodiflorum, Brachypodium sylvaticum, Bromus ramosus, Callitriche stagnalis, Cardamine pratensis, Cerastium fontanum subsp. triviale, Chenopodium rubrum, Dryopteris filix-mas, Epilobium montanum, E. parviflorum, Equisetum arvense, Geranium robertianum, Mercurialis perennis, Moehringia trinervia, Plantago major, Polygonum amphibium, P. persicaria, Ranunculus acris, Rorippa islandica, Sanicula europaea, Senecio aquaticus, S. jacobaea, Tamus communis, Taraxacum spp., Veronica catenata, Viola riviniana.
Rare: Alchemilla filicaulis subsp. vestita, Anthoxanthum odoratum, Arctium minus, Bidens tripartita, Caltha palustris, Carex sylvatica, Cirsium vulgare, Conopodium majus, Digitalis purpurea, Epilobium angustifolium, Filaginella uliginosa, Galeopsis tetrahit, Juncus bufonius, J. conglomeratus, J. inflexus, Primula vulgaris, Scrophularia nodosa, Tussilago farfara, Veronica chamaedrys.

Habitat Study 10 Roecliffe Spinney, Woodhouse SK 531131 Area: 1.3 ha (3.3 acres) Alt.: 107-114 m. (350-375 ft.)
Recorder: P. H. Gamble. Surveyed: 30th May and 1st July 1972.
An area of alder carr at the north end of Roecliffe Spinney, which is being invaded in the south-east by *Acer pseudoplatanus*. The estimated age of the alders is 50 to 100 years. It is intersected by a small trout stream and there are springs on the site. The ground is hummocky and the water table is at or near the surface in the numerous hollows throughout the year. The study area, which slopes gradually from south to north, adjoins the rocky wooded part of Roecliffe Spinney to the southeast. It is

open on its other sides and adjacent to Lingdale Golf Course to the north and west. It lies partly on Alluvium and partly on Keuper Marl. A sample of the humus-rich surface soil had pH 6.5.

TREE AND SHRUB LAYERS
Abundant: Alnus glutinosa.
Frequent: Corylus avellana, Lonicera periclymenum, Rosa arvensis, Rubus fruticosus agg., Solanum dulcamara.
Locally frequent: Acer pseudoplatanus, Salix cinerea subsp. cinerea, Sambucus nigra.
Occasional: Crataegus monogyna, Fraxinus excelsior, Ilex aquifolium, Malus sylvestris, Prunus spinosa, Quercus robur, Ribes rubrum, Salix cinerea subsp. oleifolia.
Rare: Acer platanoides (seedling), Aesculus hippocastanum (seedling), Crataegus laevigata, Frangula alnus, Hedera helix, Ligustrum vulgare, Rosa canina.

HERB LAYER
Abundant: Allium ursinum, Cardamine amara, C. flexuosa, Carex acutiformis, Chrysosplenium oppositifolium, Circaea lutetiana, Deschampsia cespitosa, Geranium robertianum, Lamiastrum galeobdolon, Phalaris arundinacea.
Locally abundant: Anemone nemorosa, Equisetum telmateia, Filipendula ulmaria, Holcus mollis, Ranunculus ficaria, Urtica dioica.
Frequent: Ajuga reptans, Angelica sylvestris, Arum maculatum, Athyrium filix-femina, Caltha palustris, Carex paniculata, Dryopteris dilatata, D. filix-mas, Festuca gigantea, Galium aparine, Hyacinthoides non-scripta, Lysimachia nemorum, Mentha aquatica, Mercurialis perennis, Oxalis acetosella, Poa trivialis, Ranunculus repens, Rumex sanguineus, Stellaria holostea, Valeriana officinalis, Veronica montana.
Locally frequent: Callitriche stagnalis, Nasturtium microphyllum.
Occasional: Anthriscus sylvestris, Apium nodiflorum, Dryopteris carthusiana, Epilobium adenocaulon, Equisetum arvense, Galium palustre, Heracleum sphondylium, Juncus effusus, Moehringia trinervia, Rumex conglomeratus, Silene dioica, Veronica beccabunga.
Rare: Arrhenatherum elatius, Carex sylvatica, Conopodium majus, Elymus caninus, Sanicula europaea.

Habitat Study 11 Swithland Wood, Newtown Linford SK 537129   Area: 2 ha (4.9 acres)   Alt.: 99-114 m. (325-350 ft.) S.S.S.I.
Recorder: P. H. Gamble. Surveyed: 1972.
The study area is the extreme northern part of the Wood, adjacent to The Brand. It is bounded by a dry-stone wall on the north (roadside) and west (trackside), an old slate quarry on the east and an ancient ditch with a raised bank dividing it from the rest of the wood to the south. Standard *Quercus robur* and *Tilia cordata* (150 to 200 years old) are the dominant tree species, producing deep shade with a sparse herb layer under the lime, but a more varied flora under the less heavy canopy of the oak. Some oak regeneration was evident but no young lime was seen. The southern part is being thickly colonized by sycamore seedlings and since 1974 a large part of the study area has been taken for use as a car park with the loss of much ground flora. During the great gale of 1976 numbers of fine mature trees were blown down, thereby opening further areas to invasion by sycamore. Situated on Keuper Marl. Samples of surface soil from different parts of the area had pH 3.7 and 4.2.

TREE AND SHRUB LAYERS
Frequent: Lonicera periclymenum, Quercus robur, Solanum dulcamara, Tilia cordata.
Locally frequent: Acer pseudoplatanus, Hedera helix.
Occasional: Acer campestre, Betula pendula, Corylus avellana, Fraxinus excelsior, Quercus petraea, Rosa arvensis, Sambucus nigra, Sorbus aucuparia.
Rare: Alnus glutinosa, Carpinus betulus, Crataegus monogyna, Fagus sylvatica, Ilex aquifolium, Populus canescens, Ribes uva-crispa, Rubus idaeus, Salix cinerea, Ulmus glabra, Viburnum opulus.

HERB LAYER
Abundant: Circaea lutetiana, Hyacinthoides non-scripta, Oxalis acetosella.
Locally abundant: Anemone nemorosa, Holcus mollis, Mercurialis perennis.
Frequent: Ajuga reptans, Bromus ramosus, Dryopteris dilatata, D. filix-mas, Lonicera periclymenum, Luzula pilosa, Lysimachia nemorum, Pteridium aquilinum, Sanicula europaea.
Locally frequent: Brachypodium sylvaticum, Carex remota, C. sylvatica, Galium aparine, Geranium robertianum, Glechoma hederacea, Hedera helix, Milium effusum, Poa annua, P. trivialis, Urtica dioica, Viola riviniana.
Occasional: Athyrium filix-femina, Arctium minus, Arum maculatum, Cardamine flexuosa, Epilobium adenocaulon, E. montanum, Festuca gigantea, Fragaria vesca, Geum urbanum, Luzula sylvatica, Moehringia trinervia, Poa nemoralis, Potentilla sterilis, Ranunculus ficaria, R. repens, Rumex sanguineus, Scrophularia nodosa, Silene dioica, Stachys sylvatica, Veronica officinalis, Viola reichenbachiana.
Rare: Agrostis capillaris, Dactylis glomerata, Deschampsia flexuosa, Dryopteris carthusiana, Epilobium angustifolium, Epipactis helleborine, Holcus lanatus, Hypericum humifusum, Juncus effusus, Rumex obtusifolius, Stellaria media.

Habitat Study 12 Swithland Wood, Newtown Linford SK 538118   Area: 0.7 ha (1.7 acres)   Alt.: 99-114 m. (325-375 ft.) S.S.S.I.
Recorder: P. H. Gamble. Surveyed: 1976.
This study is of a small part of a large area of mature deciduous woodland. The study area is at the southern end of the wood, about 50 m. in from the main entrance off the B5330 road, and lying between the car park to the south, a wet area with alders and a stream to the north, the wide pipe-line ride to the west and the main track running down the edge of the wood to the east. The dominant tree species is *Quercus robur* with standards 150 to 200 years old, producing a medium to close canopy, whilst *Corylus avellana* is subdominant creating deep shade throughout much of the area. The study area, though largely on Keuper Marl, also has some Alluvium towards the stream to the north. Samples of surface soil from the Keuper Marl varied greatly, from pH 2.6 to 7.6. A sample from the Alluvium had pH 4.2.

TREE AND SHRUB LAYERS
Abundant: Corylus avellana, Quercus robur.
Frequent: Crataegus monogyna, Viburnum opulus.
Locally frequent: Rubus fruticosus agg., Rubus idaeus, Sambucus nigra.
Occasional: Acer pseudoplatanus (seedlings), Alnus glutinosa, Fraxinus excelsior, Hedera helix, Ilex aquifolium, Lonicera periclymenum, Quercus petraea, Ribes rubrum.
Rare: Betula pendula, B. pubescens, Malus sylvestris, Rosa arvensis, R. canina, Salix caprea.

HERB LAYER
Abundant: Hedera helix, Hyacinthoides non-scripta.
Locally abundant: Ajuga reptans, Anemone nemorosa, Circaea lutetiana, Epilobium angustifolium, Galium aparine.
Frequent: Geum urbanum, Lysimachia nemorum, Viola riviniana.
Locally frequent: Arum maculatum, Brachypodium sylvaticum, Bromus ramosus, Carex pendula, Deschampsia cespitosa, Festuca gigantea, Fragaria vesca, Glechoma hederacea, Holcus mollis, Listera ovata, Luzula pilosa, L. sylvatica, Ranunculus repens, Rumex sanguineus, Stellaria holostea, Urtica dioica.
Occasional: Angelica sylvestris (Ef), Arctium minus, Cardamine flexuosa, Carex remota (Ef), C. sylvatica, Conopodium majus, Dryopteris dilatata, Geranium robertianum, Oxalis acetosella, Poa nemoralis, Prunella vulgaris (Ef), Rumex conglomeratus, Scrophularia nodosa, Silene dioica.
Rare: Athyrium filix-femina, Carex pallescens (Ef), Dryopteris carthusiana, D. filix-mas, Epipactis helleborine, Juncus effusus, Ranunculus ficaria (Ef), Sanicula europaea, Scutellaria galericulata, Stachys sylvatica, Taraxacum spp., Torilis japonica, Tussilago farfara.

Habitat Study 13 Wet woodland, Narborough Bog, Narborough
SP 549979   Area: 5.2 ha (13.0 acres)   Alt.: 61 m. (200 ft.) S.S.S.I.; Nature Reserve.
Recorder: S. H. Bishop. Surveyed: 25th May and 18th August 1977.
This study is of an area of wet woodland on the west bank of the River Soar, bounded to the south by the Leicester to Birmingham railway and to the west by a recreation ground and allotments. In the 1960s a flood prevention scheme lowered the level of the River Soar which caused a general drying out of the site. It lies on River Gravels, Peat and Alluvium.

TREE AND SHRUB LAYERS
Abundant: Salix fragilis.
Locally abundant: Crataegus monogyna, Sambucus nigra.
Frequent: Acer pseudoplatanus, Fraxinus excelsior, Ribes rubrum.
Locally frequent: Alnus glutinosa, Prunus spinosa, Quercus robur, Rosa canina, Salix cinerea subsp. oleifolia.
Occasional: Lonicera periclymenum, Rosa arvensis, Rubus fruticosus agg., Solanum dulcamara, Viburnum opulus.
Rare: Acer campestre, Betula pendula, Crataegus laevigata × monogyna, Ilex aquifolium, Ligustrum vulgare, Malus sylvestris, Prunus domestica subsp. domestica, Rhamnus catharticus, Ribes uva-crispa,
Rubus idaeus, Salix alba, S. caprea, S. viminalis, Sorbus aucuparia, Ulmus procera.

HERB LAYER
Abundant: Filipendula ulmaria, Galium aparine, Poa trivialis, Urtica dioica.
Locally abundant: Arum maculatum, Deschampsia cespitosa, Geum urbanum, Phragmites australis, Stachys sylvatica.
Frequent: Circaea lutetiana, Dryopteris filix-mas, Festuca gigantea, Geranium robertianum, Glechoma hederacea, Hedera helix, Lapsana communis, Moehringia trinervia, Ranunculus ficaria, Rumex sanguineus, Silene dioica, Stellaria media, S. neglecta.
Locally frequent: Angelica sylvestris, Arctium minus, Calystegia silvatica, Cardamine flexuosa, Carex acutiformis, C. remota, Cirsium vulgare, Dryopteris dilatata, Epilobium angustifolium, E. hirsutum, Glyceria maxima, Heracleum sphondylium, Juncus effusus, Lychnis flos-cuculi, Phalaris arundinacea, Scrophularia auriculata, Viola riviniana.
Occasional: Alliaria petiolata, Alopecurus pratensis, Arrhenatherum elatius, Athyrium filix-femina, Brachypodium sylvaticum, Bromus ramosus, Carduus acanthoides, Cirsium arvense, C. palustre, Conium maculatum, Elymus repens, Galeopsis tetrahit, Holcus lanatus, Iris pseudacorus, Juncus inflexus, Lamium album, Poa annua, Ranunculus repens, Rumex conglomeratus, Stellaria graminea, Tamus communis, Taraxacum spp., Veronica beccabunga, Vicia cracca, Viola odorata.
Rare: Anthriscus sylvestris, Arctium lappa, Caltha palustris, Calystegia sepium, Capsella bursa-pastoris, Crepis capillaris, Equisetum arvense, Hyacinthoides non-scripta, Impatiens glandulifera, Lolium perenne, Myosotis scorpioides, M. sylvatica, Myosoton aquaticum, Polygonum amphibium, Scutellaria galericulata, Sisymbrium officinale, Sonchus oleraceus, Thalictrum flavum.

Habitat Study 14 Cream Gorse, Frisby on the Wreake
SK 705145   Area: 4.6 ha (11.4 acres)   Alt.: 125 m. (410 ft.)
Recorders: P. H. Gamble and A. L. Primavesi. Surveyed: 9th May 1972, 17th June 1972 and 26th August 1976.
This is a well-managed fox covert with clear rides and a diverse herb layer not obscured by dense scrub. It stands on Boulder Clay.

TREE AND SHRUB LAYERS
Frequent: Acer campestre, Crataegus laevigata × monogyna, C. monogyna, Fraxinus excelsior, Hedera helix, Lonicera periclymenum, Prunus spinosa, Quercus robur, Sambucus nigra, Ulmus glabra.
Locally frequent: Salix caprea.
Occasional: Cornus sanguinea, Corylus avellana, Crataegus laevigata, Ligustrum vulgare, Malus sylvestris, Rhamnus catharticus (Ef), Rosa arvensis, R. canina, Ulex europaeus.
Rare: Viburnum lantana (Ef).

HERB LAYER
Locally abundant: Pteridium aquilinum.
Frequent: Ajuga reptans, Arum maculatum, Brachypodium sylvaticum, Bromus ramosus, Circaea lutetiana, Dryopteris dilatata, D. filix-mas, Epilobium angustifolium, Festuca gigantea, Galium aparine,

Geum urbanum, Glechoma hederacea, Holcus lanatus, Moehringia trinervia, Orchis mascula, Poa trivialis, Potentilla anserina (Ef), Potentilla reptans (Ef), Silene dioica, Stachys sylvatica, Tamus communis, Urtica dioica, Veronica chamaedrys, Viola riviniana.

Occasional: Arctium minus, Centaurea nigra (Ef), Chaerophyllum temulentum (Ef), Cruciata laevipes (Ef), Dactylis glomerata (Ef), Deschampsia cespitosa, Elymus repens (Ef), Filipendula ulmaria (Ef), Geranium robertianum, Heracleum sphondylium, Hyacinthoides non-scripta, Juncus inflexus (Ef), Poa annua (Ef), Potentilla erecta (Ef), Ranunculus acris (Ef), R. auricomus (Ef), R. ficaria, R. repens (Ef), Taraxacum spp. (Ef), Viola reichenbachiana.

Habitat Study 15   Tugby Bushes, Tugby and Keythorpe
SK 770021   Area: 1.5 ha (3.7 acres)   Alt.: 125-146 m. (410-480 ft.) S.S.S.I.
Recorders: S. H. Bishop, P. H. Gamble and A. L. Primavesi. Surveyed: 31st May and 29th August 1976.
The part of the wood studied was the narrow section approximately 350 m. × 50 m. adjacent to the road (on the opposite side to Tugby Wood), running downhill south-west to north-east. A narrow stream in a deep channel with some steep banks cuts through the northern section and joins the Eye Brook. The dominant tree species are *Quercus robur, Fraxinus* and *Ulmus glabra* with some well-grown standards up to 150 years old. The study area is situated on Middle Lias Clays and Sands. Three samples of surface soil collected from different parts of the area had pH of 5.0, 5.4 and 6.7.

TREE AND SHRUB LAYERS
Frequent: Acer campestre, Cornus sanguinea, Corylus avellana, Crataegus laevigata × monogyna, Fraxinus excelsior, Hedera helix, Prunus spinosa, Quercus robur, Rosa canina, Rubus caesius, Rubus fruticosus agg., Ulmus glabra.
Locally frequent: Crataegus laevigata, C. monogyna, Ligustrum vulgare, Lonicera periclymenum, Rosa arvensis.
Occasional: Ulmus minor.
Rare: Acer pseudoplatanus, Alnus glutinosa, Euonymus europaeus, Malus sylvestris, Populus × canadensis, Salix alba, S. caprea, S. fragilis, Sambucus nigra, Solanum dulcamara, Viburnum opulus.

HERB LAYER
Abundant: Anemone nemorosa, Lamiastrum galeobdolon, Mercurialis perennis, Poa trivialis, Urtica dioica.
Locally abundant: Allium ursinum, Circaea lutetiana, Galium aparine, Glechoma hederacea, Hyacinthoides non-scripta, Myosotis sylvatica, Ranunculus ficaria, Silene dioica, Stachys sylvatica, Stellaria holostea.
Frequent: Adoxa moschatellina, Angelica sylvestris, Brachypodium sylvaticum, Dryopteris filix-mas, Festuca gigantea, Galium odoratum, Geranium robertianum, Geum urbanum, Holcus mollis, Hypericum hirsutum, Moehringia trinervia, Rumex sanguineus.

Locally frequent: Ajuga reptans, Bromus ramosus, Campanula latifolia, Cardamine flexuosa, Carex strigosa, C. sylvatica, Chrysosplenium oppositifolium, Elymus caninus (Ef), Epilobium angustifolium, Filipendula ulmaria, Geranium pratense, Heracleum sphondylium, Melica uniflora, Milium effusum, Potentilla sterilis, Pteridium aquilinum, Sanicula europaea, Veronica chamaedrys, V. montana.
Occasional: Arctium minus, Athyrium filix-femina, Campanula trachelium (Ef), Carduus acanthoides, Cirsium palustre, Dryopteris dilatata, Elymus repens (Ef), Epilobium montanum, Oxalis acetosella, Paris quadrifolia, Primula vulgaris, Stellaria alsine, Vicia sylvatica, Viola reichenbachiana.
Rare: Aegopodium podagraria (Ef), Anthriscus sylvestris, Barbarea vulgaris, Cirsium arvense, C. vulgare, Conopodium majus, Cruciata laevipes, Dactylis glomerata, Dipsacus pilosus, Epilobium hirsutum, Humulus lupulus, Lamium album (Ef), Lathraea squamaria, Listera ovata, Luzula pilosa, Mentha aquatica, Myosotis arvensis, Platanthera chlorantha, Poa annua, Ranunculus acris, R. auricomus, R. repens, Rumex obtusifolius, Taraxacum spp., Vicia sepium.

Habitat Study 16   Launde Big Wood, Launde
SK 785035   Area: 3.0 ha (7.4 acres)   Alt.: 167-183 m. (550-600 ft.) S.S.S.I.; Nature Reserve.
Recorders: S. H. Bishop, P. H. Gamble and A. L. Primavesi. Surveyed: 31st May and 29th August 1976.
The study area is situated in the south-west corner of this ancient woodland immediately east of Launde Wood Farm. It is roughly rectangular, 300 m. long by 100 m. wide, with its long axis running north-west to south-east. It is bounded on its south-west and south sides by hedgerows and fields, and on its north-west and north-east sides by woodland rides. Much of the wood was felled during the First World War but has subsequently regenerated naturally. This part of the wood is dominated by coppiced ash, but has a scattering of *Quercus robur* and a localised area of old *Ulmus glabra*. *Acer campestre* is frequent in the understorey and *Corylus avellana* is dominant in the varied shrub layer. The study area lies partly on Upper Lias Clay and partly on Glacial Gravel and Boulder Clay. A sample of surface soil from near its centre had pH 5.5.

TREE AND SHRUB LAYERS
Abundant: Corylus avellana, Fraxinus excelsior.
Frequent: Acer campestre, Cornus sanguinea, Crataegus laevigata, Lonicera periclymenum, Prunus spinosa.
Locally frequent: Rosa arvensis, R. canina, Rubus caesius, Ulmus glabra.
Occasional: Hedera helix, Quercus robur, Salix caprea, Sambucus nigra, Tamus communis, Viburnum opulus.
Rare: Crataegus monogyna, Malus sylvestris.

HERB LAYER
Abundant: Hyacinthoides non-scripta, Mercurialis perennis, Poa trivialis, Ranunculus ficaria, Sanicula europaea.
Locally abundant: Allium ursinum, Urtica dioica, Veronica montana.

Frequent: Circaea lutetiana, Deschampsia cespitosa, Dryopteris filix-mas, Festuca gigantea, Filipendula ulmaria (Ef), Geum urbanum, Glechoma hederacea, Milium effusum, Potentilla sterilis, Ranunculus repens (Ef), Rumex sanguineus, Veronica chamaedrys.

Locally frequent: Ajuga reptans, Anemone nemorosa, Athyrium filix-femina, Bromus ramosus, Campanula latifolia, Dryopteris dilatata, Galium aparine, Lamiastrum galeobdolon, Lotus uliginosus (Ef), Mentha arvensis (Ef), Myosotis sylvatica, Oxalis acetosella, Primula vulgaris, Prunella vulgaris, Silene dioica, Viola reichenbachiana, V. riviniana.

Occasional: Arctium minus, Arum maculatum, Brachypodium sylvaticum, Callitriche stagnalis (Ef), Carex sylvatica, Cirsium palustre (Ef), Conopodium majus, Fragaria vesca, Galium palustre (Ef), Hedera helix, Heracleum sphondylium, Holcus mollis (Ef), Lapsana communis (Ef), Orchis mascula, Polygonum hydropiper, Potentilla anserina (Ef), Scrophularia nodosa (Ef), Stachys sylvatica, Stellaria alsine (Ef), Veronica beccabunga, Vicia sepium.

Rare: Cardamine pratensis (Ef), Carex remota, Cerastium glomeratum (Ef), Cirsium arvense (Ef), Dactylis glomerata (Ef), Dipsacus pilosus, Dryopteris carthusiana, Epilobium adenocaulon (Ef), E. hirsutum (Ef), Galium odoratum (Ef), Geranium robertianum (Ef), Holcus lanatus (Ef), Hypericum hirsutum, Lychnis flos-cuculi (Ef), Lysimachia nemorum (Ef), Platanthera chlorantha, Poa annua (Ef), Potentilla reptans (Ef), Ranunculus acris (Ef), Rumex obtusifolius, Stellaria holostea, Veronica serpyllifolia (Ef).

Habitat Study 17 Owston Big Wood, Owston and Newbold
SK 794065   Area: 6 ha (15 acres)   Alt.: 183 m. (600 ft.) S.S.S.I.
Recorder: P. A. Candlish. Surveyed: May 1976
The study area is in the eastern part of this large ancient woodland and is approximately 250 m. to the west of the Owston Wood Road which separates the Big Wood from Little Owston Wood. The study area is four sided; the north, east and west sides are about 300 m. long and bounded by rides whilst the south side, about 160 m. long, is bounded by fields and overlooks Withcote Hall. The western boundary is a continuation of the line of the avenue leading to the Hall. Owston Big Wood was felled about 50 years ago. The study area has regenerated naturally, although some parts of the wood were planted up by the Forestry Commission in the 1960s. The study area stands on Boulder Clay. A sample of surface soil had pH 5.1.

TREE AND SHRUB LAYERS
Frequent: Betula pendula, Corylus avellana, Crataegus laevigata, Fraxinus excelsior, Lonicera periclymenum, Quercus robur, Rosa arvensis, R. canina, R. fruticosus agg., Salix cinerea, Ulmus glabra.
Locally frequent: Prunus spinosa.
Occasional: Acer campestre, A. pseudoplatanus, Cornus sanguinea, Salix caprea, Tilia cordata.
Rare: Fagus sylvatica, Viburnum opulus.

HERB LAYER
Locally abundant: Filipendula ulmaria (Ef), Holcus lanatus (Ef), Hyacinthoides non-scripta, Juncus bufonius (Ef), Mentha arvensis (Ef).

Frequent: Ajuga reptans, Bromus ramosus,. Carex remota, C. sylvatica, Circaea lutetiana, Dryopteris filix-mas, Galium odoratum, Geum rivale, Glechoma hederacea, Myosotis sylvatica, Ranunculus ficaria, Stachys sylvatica, Urtica dioica, Veronica chamaedrys, Viola reichenbachiana, V. riviniana.

Locally frequent: Agrostis stolonifera (Ef), Anemone nemorosa (Ef), Arrhenatherum elatius (Ef), Callitriche stagnalis (Ef), Carex flacca (Ef), C. hirta (Ef), Chrysosplenium oppositifolium, Cirsium arvense (Ef), Dactylis glomerata (Ef), Deschampsia cespitosa (Ef), Elymus repens (Ef), Equisetum sylvaticum, Glyceria fluitans (Ef), Hypericum hirsutum (Ef), H. tetrapterum (Ef), Juncus acutiflorus (Ef), Lotus uliginosus (Ef), Luzula sylvatica, Mercurialis perennis, Odontites verna (Ef), Oxalis acetosella, Phleum pratense subsp. pratense (Ef), Potentilla anserina (Ef), P. reptans (Ef), Prunella vulgaris (Ef), Rumex sanguineus (Ef), Sanicula europaea, Taraxacum spp. (Ef), Trifolium repens (Ef), Veronica beccabunga (Ef).

Occasional: Alopecurus geniculatus (Ef), Angelica sylvestris (Ef), Athyrium filix-femina, Cardamine flexuosa, Carex otrubae (Ef), Cirsium palustre (Ef), Conopodium majus, Dactylorhiza fuchsii (Ef), Dryopteris dilatata, Epilobium montanum (Ef), Euphrasia nemorosa (Ef), Festuca gigantea, Galium aparine, G. palustre (Ef), Juncus effusus (Ef), J. inflexus (Ef), Lamiastrum galeobdolon, Lathyrus pratensis (Ef), Listera ovata, Luzula pilosa, Lychnis flos-cuculi (Ef), Lysimachia nemorum (Ef), Milium effusum, Moerhingia trinervia, Orchis mascula, Polygonum persicaria (Ef), Potentilla sterilis (Ef), Primula vulgaris, Pteridium aquilinum, Rumex obtusifolius (Ef), Silene dioica, Stellaria alsine (Ef), S. graminea (Ef), S. holostea, Valeriana officinalis (Ef).

Rare: Agrimonia eupatoria (Ef), Arctium minus, Festuca arundinacea (Ef), Platanthera chlorantha, Scrophularia nodosa, Senecio aquaticus (Ef), S. jacobaea (Ef), Tussilago farfara (Ef).

Habitat Study 18 Little Owston Wood, Owston and Newbold
SK 798066   Area: 2.4 ha (6.0 acres)   Alt.: 171 m. (560 ft.) S.S.S.I.
Recorders: S. H. Bishop, P. H. Gamble and A. L. Primavesi. Surveyed: 31st May and 29th August 1976.
Little Owston Wood is on the east side of Owston Wood Road. The study area is near the south-west corner of the wood. It is bounded on the west by a ditch and bank marking the parish boundary and the former county boundary with Rutland. On the north and south sides it is bounded by woodland rides which converge to an apex at the eastern extremity of the study area. The wood was clear-felled a few decades ago, but this part was allowed to undergo natural regeneration, so that it has retained most of its original flora. The study area stands on Boulder Clay. A sample of surface soil had pH 4.9.

TREE AND SHRUB LAYERS
Abundant: Fraxinus excelsior, Rubus fruticosus agg.
Frequent: Acer campestre, Cornus sanguinea, Corylus avellana, Crataegus laevigata × monogyna, Lonicera periclymenum, Prunus spinosa, Quercus robur, Rubus caesius, Viburnum opulus.
Locally frequent: Betula pendula, Crataegus laevigata.

Occasional: Fagus sylvatica, Picea abies, Populus tremula, Rosa arvensis, R. canina, Salix caprea, S. cinerea subsp. oleifolia, Tamus communis, Ulmus glabra.
Rare: Acer pseudoplatanus, Aesculus hippocastanum, Betula pubescens, Malus sylvestris, Populus canescens.

HERB LAYER
Abundant: Anemone nemorosa, Geum rivale, Ranunculus ficaria.
Locally abundant: Chrysosplenium oppositifolium, Dryopteris filix-mas, Galium odoratum, Hyacinthoides non-scripta, Luzula sylvatica, Myosotis sylvatica, Poa trivialis, Sanicula europaea, Urtica dioica.
Frequent: Angelica sylvestris, Anthoxanthum odoratum (Ef), Athyrium filix-femina, Brachypodium sylvaticum, Bromus ramosus, Cardamine flexuosa (Ef), Carex sylvatica, Circaea lutetiana, Deschampsia cespitosa, Festuca gigantea, Filipendula ulmaria, Geum rivale × urbanum, Heracleum sphondylium, Milium effusum, Primula vulgaris, Rumex sanguineus, Stellaria alsine (Ef), S. holostea, Veronica beccabunga (Ef), V. chamaedrys, Vicia sepium.
Locally frequent: Ajuga reptans, Carex flacca (Ef), C. remota, Dryopteris dilatata, Epilobium montanum, Juncus articulatus (Ef), J. effusus (Ef), Lamiastrum galeobdolon, Mentha arvensis (Ef), Phleum pratense subsp. pratense (Ef), Potentilla anserina (Ef), P. sterilis, Ranunculus repens (Ef), Senecio aquaticus (Ef), Valeriana officinalis, Viola reichenbachiana.
Occasional: Agrostis stolonifera (Ef), Alopecurus pratensis (Ef), Arum maculatum, Cerastium glomeratum (Ef), Cirsium palustre, Conopodium majus, Dactylis glomerata (Ef), Epilobium angustifolium, E. hirsutum (Ef), Glechoma hederacea, Glyceria fluitans (Ef), Holcus mollis, Hypericum hirsutum (Ef), Juncus effusus (Ef), Lathyrus pratensis (Ef), Lotus uliginosus (Ef), Omalotheca sylvatica (Ef), Oxalis acetosella, Polygonum aviculare (Ef), P. persicaria (Ef), Stachys officinalis, S. sylvatica, Veronica serpyllifolia.
Rare: Callitriche stagnalis (Ef), Cardamine pratensis (Ef), Carex otrubae (Ef), Epilobium parviflorum (Ef), Epipactis helleborine, Galium palustre (Ef), Geum urbanum, Holcus lanatus (Ef), Hypericum tetrapterum (Ef), Juncus articulatus (Ef), J. bufonius (Ef), J. inflexus (Ef), Lychnis flos-cuculi (Ef), Melica uniflora, Odontites verna (Ef), Plantago major (Ef), Poa annua, Polygonum hydropiper (Ef), Potentilla erecta (Ef), P. reptans (Ef), Prunella vulgaris (Ef), Rumex acetosa (Ef), R. crispus (Ef), Scrophularia nodosa, Sonchus asper (Ef), Stellaria media (Ef).

Habitat Study 19   Allexton Wood, Allexton
SP 820994   Area: 8.8 ha (22 acres)   Alt.: 122-145 m. (400-475 ft.) S.S.S.I.
Recorders: S. H. Bishop and P. H. Gamble. Surveyed: 25th May 1979.
This study covers the northern section of the wood, bounded by an old thorn hedge and fields to the north and west and by a main ride traversing the wood from south-west to north-east. At the time of survey *Fraxinus excelsior, Ulmus glabra* and *U. procera* were the dominant tree species. The ash consisted of standards about 100 years old and numerous older coppiced specimens. The elms were in an advanced stage of disease with many trees dead or dying. This was already showing signs of having a profound effect on the underlying shrubs and herbs because of the increased light. *Quercus robur*, including a scattering of good specimens about 150 years old, occurs throughout the area, and there are also plenty of mature specimens of *Acer campestre*. There is a good shrub layer with *Corylus avellana* and some large colonies of *Cornus sanguinea*. The bryophyte flora is particularly rich, especially on the trunks of the older trees. A stream, in places cutting a deep channel, flows from west to east. A small area of Upper Lias Clay occurs in the north east; the rest of the area stands on Boulder Clay. Samples of surface soil from the Lias Clay had pH 6.3, from the Boulder Clay pH 5.3 and from the streamside on the Boulder Clay pH 4.7.

TREE AND SHRUB LAYERS
Locally abundant: Ulmus procera.
Frequent: Acer campestre, Cornus sanguinea, Corylus avellana, Crataegus laevigata, Fraxinus excelsior, Lonicera periclymenum, Malus sylvestris, Quercus robur, Sambucus nigra, Ulmus glabra.
Locally frequent: Betula pendula, Crataegus monogyna, Prunus spinosa, Rubus caesius (Ef), Rubus fruticosus agg.
Occasional: Ribes rubrum, Rosa arvensis, R. canina, Solanum dulcamara, Viburnum opulus.
Rare: Euonymus europaeus, Salix caprea.

HERB LAYER
Abundant: Anemone nemorosa, Hyacinthoides non-scripta, Mercurialis perennis, Ranunculus ficaria, Urtica dioica.
Locally abundant: Chrysosplenium oppositifolium, Lamiastrum galeobdolon.
Frequent: Arum maculatum, Athyrium filix-femina, Campanula latifolia, Carex sylvatica, Circaea lutetiana, Dryopteris filix-mas, Filipendula ulmaria (Ef), Galium aparine, Poa trivialis, Primula vulgaris, Ranunculus repens (Ef), Rumex sanguineus, Silene dioica.
Locally frequent: Ajuga reptans, Brachypodium sylvaticum, Cardamine flexuosa (Ef), Carex strigosa, Deschampsia cespitosa, Dryopteris dilatata, D. carthusiana, Epilobium montanum (Ef), Glechoma hederacea, Heracleum sphondylium, Hypericum hirsutum (Ef), Juncus effusus (Ef), Lathraea squamaria, Milium effusum, Polygonum hydropiper (Ef), Prunella vulgaris (Ef), Stachys sylvatica, Stellaria holostea, Veronica montana, Vicia sepium (Ef).
Occasional: Anthriscus sylvestris, Arctium minus, Carex remota, Epilobium angustifolium, E. hirsutum (Ef), Geum urbanum, Hypericum tetrapterum (Ef), Poa annua, Potentilla sterilis, Rumex obtusifolius (Ef), Scrophularia nodosa, Taraxacum spp. (Ef), Veronica beccabunga (Ef), V. chamaedrys.
Rare: Cardamine pratensis (Ef), Carex hirta, Cirsium arvense (Ef), C. vulgare, Dactylis glomerata (Ef), Epilobium parviflorum (Ef), Galium odoratum, Lathyrus pratensis (Ef), Lotus uliginosus (Ef), Lychnis flos-cuculi, Melica uniflora, Orchis mascula, Potentilla anserina (Ef), P. reptans (Ef), Quercus cerris, Veronica serpyllifolia.

Grassland

In selecting sites for studies of this particular type of habitat, a difficulty arises, in that there are other habitat categories in which grasses are often the dominant species. For example, on the acid soils in the west of the county one may find a continuum of habitats ranging from heathland, dominated by species such as *Pteridium*, *Calluna* and *Vaccinium myrtillus*, to heath grassland with *Festuca ovina*, *Nardus stricta* and *Danthonia decumbens* as the dominant species. The same applies to wet areas throughout the county, where it is often difficult to decide whether a particular field should be classified as a wet meadow or a marsh. Also there are habitats such as railway verges where the dominant species are grasses, and which form a refuge for many of our rarer grassland species which are declining in numbers in the grasslands properly so-called.

In Leicestershire the majority of grassland lies on Boulder Clay, Keuper Marl and Jurassic clays, all of which tend, under traditional management regimes, to produce a type of neutral grassland. Management such as grazing or cutting for hay is necessary if the grassland is to remain as such and not turn to scrub, but ploughing and the use of artificial fertilizers or selective herbicides destroys the character of a piece of species-rich grassland. Many of our species-rich grasslands are on ridge and furrow and their establishment may date back to the original enclosure of arable land in medieval or later times, but one at least is known to have been under the plough in the mid-19th century. Plainly, then, the species characteristic of these older grasslands are capable of colonisation. How long it would take nowadays for a piece of 'improved' grassland to revert to the species-rich type (if it ever did) we do not know, since the necessary seed sources are now few and far between. Species-rich neutral grassland in Leicestershire continues to decrease in extent, though some of the better sites are now protected as nature reserves. Examples are provided by Muston Meadows (study 37), the grassland adjoining Wymondham Rough (study 38), and Cribb's Meadow (study 40).

Strongly calcareous soils, capable of supporting limestone grassland typical of them, are virtually confined to two areas of Leicestershire. There are a number of small outcrops of Carboniferous Limestone in the north-west of the county. Much of these has been quarried away and the remainder is so subject to human interference that little or no characteristic grassland remains, as is shown by study 22, Breedon Hill. In the north-east of the county there is a somewhat larger exposure of the Lincolnshire Limestone. Here most of the land is under the plough and limestone grassland only survives in places unsuitable for cultivation. There is one such site near Egypt Plantation (study 39), where the concrete bases of huts on the edge of a wartime aerodrome prevent ploughing. Other sites are the grassy floors of former limestone quarries, such as Waltham Quarry (study 102), and the courses of mineral railways. The latter are now all dismantled and have been invaded by scrub or put to agricultural use. In other parts of the county, especially on the Chalky Boulder Clay, there are small areas of somewhat calcareous soils which support a limited number of calcicole species. Examples of this type are the pasture off Berrycott Lane, Seagrave (study 34) and the pasture near Terrace Hill Farm, Eaton (study 36), the site of a former ironstone quarry. Railway verges throughout the county bear a type of grassland, which often contains a number of calcicole species. This will be discussed under the heading of Railways.

Acid grassland or heath grassland is largely confined to the north-west of the county, though there are small areas elsewhere, such as Burbage Common (study 25). Some places on the steep slope of the Marlstone escarpment in the north-east, where the soil is leached, have grassland approximating to this type. A good example of heath grassland on acid soil is the pasture at Broad Hill, Whitwick (study 23). A peculiar type of siliceous grassland is found where the diorite is exposed at Croft (studies 28 and 29), and on the granodiorite at Kinchley Lane (study 33). At these sites there are large numbers of small annual species, possibly favoured by the thinness of the soil layer over the underlying rock, and the consequent open nature of the sward.

The final type of grassland to consider is that occurring on alluvial soils in the flood plains of rivers. Because of periodic flooding these areas are not suitable for arable farming and hence are more likely to be left as permanent grassland. Many have hitherto escaped 'improvement', and the best buttercup meadows in the county can be seen, for instance, in the Soar Valley. Loughborough Big Meadow, part of which is the subject of study 32, is one of the largest areas of flood plain grassland in the Midlands. Another example is the meadow at Sheepy Fields (study 20). The typical species of these fields are *Ranunculus acris, R. bulbosus, Cardamine pratensis, Sanguisorba officinalis*, accompanied sometimes by *Ophioglossum vulgatum* and *Saxifraga granulata*. Loughborough Big Meadow is distinguished by the presence of *Oenanthe silaifolia*. There may also be tussocks of *Deschampsia cespitosa* and *Juncus inflexus*, and parts of these fields may be more properly called marshes.

Habitat Study 20 Meadow, Sheepy Fields, Sheepy SK 332024 Area: 2.3 ha (5.7 acres) Alt.: 76 m. (250 ft.) S.S.S.I.
Recorders: S. H. Bishop, H. I. James and F. T. Smith. Surveyed: 8th June 1977.
This is a long narrow field bounded by high hedges, with the road from Cross Hands to Harris Bridge forming its southern boundary. It is maintained as a hay meadow and is mown in July. It is situated on Alluvium and Keuper Marl.

Abundant: Alopecurus pratensis, Anthoxanthum odoratum, Bromus hordeaceus, Cynosurus cristatus, Holcus lanatus, Lolium perenne, Luzula campestris, Ranunculus acris, R. bulbosus, Rhinanthus minor, Sanguisorba officinalis.
Frequent: Anthriscus sylvestris, Avenula pubescens, Bellis perennis, Centaurea nigra, Cerastium fontanum subsp. triviale, Heracleum sphondylium, Lathyrus pratensis, Plantago lanceolata, Poa pratensis, Rumex acetosa, Trifolium dubium, T. pratense, Trisetum flavescens.
Locally frequent: Conopodium majus, Dactylis glomerata, Leontodon autumnalis, Saxifraga granulata, Taraxacum spp., Vicia cracca.
Occasional: Cardamine pratensis, Filipendula ulmaria, Hypochoeris radicata, Leucanthemum vulgare, Lotus corniculatus, Ophioglossum vulgatum, Primula veris, Silaum silaus.
Rare: Caltha palustris, Cirsium arvense, Equisetum arvense, Lychis flos-cuculi, Orchis morio, Rumex obtusifolius, Veronica chamaedrys.

Locally abundant: Dactylis glomerata, Deschampsia cespitosa, Glyceria fluitans, Juncus acutiflorus, Sanguisorba officinalis.
Frequent: Achillea millefolium, Alchemilla filicaulis subsp. vestita, Briza media, Conopodium majus, Equisetum arvense, Festuca pratensis, Juncus articulatus, Leontodon autumnalis, Lolium perenne, Lotus corniculatus, Potentilla anserina, P. erecta, Prunella vulgaris, Ranunculus acris, Rumex acetosa, Stachys officinalis, Succisa pratensis, Taraxacum spp., Trifolium pratense.
Locally frequent: Alopecurus pratensis, Arrhenatherum elatius, Bellis perennis, Carex flacca, C. pallescens, C. panicea, Cirsium arvense, C. palustre, Holcus mollis, Juncus bufonius, J. conglomeratus, J. effusus, Lotus uliginosus, Phleum pratense subsp. pratense, Ranunculus repens, Scirpus setaceus, Stellaria graminea, Trifolium repens, Vicia cracca.
Occasional: Achillea ptarmica, Angelica sylvestris, Campanula rotundifolia, Carex hirta, C. ovalis, Dactylorhiza fuchsii, Danthonia decumbens, Equisetum palustre, Filipendula ulmaria, Heracleum sphondylium, Juncus inflexus, Lathyrus montanus, L. pratensis, Leontodon hispidus, Potentilla anglica, Rumex conglomeratus, Tussilago farfara, Veronica chamaedrys.
Rare: Agrimonia eupatoria, Cirsium vulgare, Crataegus monogyna (seedling), Epilobium angustifolium, E. hirsutum, Galium palustre, Ophioglossum vulgatum, Plantago major, Primula vulgaris, Quercus robur (seedling), Rosa canina, Rubus fruticosus agg., R. idaeus, Rumex sanguineus, Senecio aquaticus.

Habitat Study 21 Field near Lount Wood, Ashby de la Zouch
SK 377189 Area: 1.1 ha (2.8 acres) Alt.: 122 m. (400 ft.) S.S.S.I.
Recorder: S. H. Bishop. Surveyed: 4th August 1978.
The study is of a long narrow field at the north-west corner of Lount Wood. It is bounded by a dismantled railway to the south and by a shallow marshy ditch to the west and north. Most of it is dry, with a little marshy ground along the northern boundary. It slopes quite steeply, being highest in the south. The drier and less steep parts of the field are cut for hay. The area lies on Coal Measures.

Abundant: Agrostis capillaris, Anthoxanthum odoratum, Centaurea nigra, Cerastium fontanum subsp. triviale, Cynosurus cristatus, Holcus lanatus, Plantago lanceolata.

Habitat Study 22 Breedon Hill, Breedon on the Hill
SK 404233 Area: 3.3 ha (8.1 acres) Alt.: 79-122 m. (260-400 ft.) S.S.S.I.
Recorder: S. H. Bishop. Surveyed: 17th April and 6th August 1977.
The study covers the greater part of the western side of the Hill, bounded by Squirrel Lane to the west, Breedon Church and a large working quarry to the east and north, and Breedon on the Hill village to the south. The study area is limestone grassland with old quarries invaded heavily in parts by scrub. The richest flora is found where there is only a thin layer of soil over the parent rock. The area is common land and has no recent history of grazing. Breedon Hill is Carboniferous Limestone.

Abundant: Achillea millefolium, Anthriscus sylvestris, Crataegus monogyna, Galium aparine, Pimpinella saxifraga, Plantago lanceolata, Poa pratensis,

Ranunculus bulbosus, Stellaria media, Trifolium dubium.

Locally abundant: Aphanes arvensis, Arum maculatum, Avenula pubescens, Campanula rotundifolia, Cardamine hirsuta, Cerastium glomeratum, Cirsium arvense, Koeleria macrantha, Lotus corniculatus, Poa annua, Saxifraga tridactylites, Urtica dioica, Veronica arvensis, Vicia hirsuta, V. sativa.

Frequent: Arenaria serpyllifolia, Bellis perennis, Bromus hordeaceus, Capsella bursa-pastoris, Centaurea nigra, Crepis capillaris, Dactylis glomerata, Epilobium angustifolium, Erophila verna, Festuca ovina, F. rubra, Geranium molle, Hedera helix, Hypericum perforatum, Linaria vulgaris, Lolium perenne, Myosotis ramosissima, Pilosella officinarum, Rumex acetosa, Sisymbrium officinale, Trisetum flavescens, Thymus praecox subsp. arcticus.

Locally frequent: Agrostis stolonifera, Bromus sterilis, Carduus nutans, Cerastium fontanum subsp. triviale, C. semidecandrum, Centaurea debauxii subsp. nemoralis, Clinopodium vulgare, Convolvulus arvensis, Elymus repens, Erodium cicutarium, Galium verum, Glechoma hederacea, Heracleum sphondylium, Lamium album, Lapsana communis, Mercurialis perennis, Moehringia trinervia, Papaver dubium, Phleum pratense subsp. bertolonii, Plantago media, Potentilla reptans, Prunella vulgaris, Prunus spinosa, Ranunculus repens, Reseda luteola, Rosa canina, Rumex crispus, R. obtusifolius, Sedum acre, Senecio jacobaea, S. vulgaris, Sonchus asper, Tamus communis, Taraxacum section Erythrosperma, Taraxacum spp., Torilis japonica, Trifolium repens, Valerianella carinata, Veronica chamaedrys, Viola arvensis.

Occasional: Arabidopsis thaliana, Chaerophyllum temulentum, Desmazeria rigida, Geranium dissectum, G. pratense, G. robertianum, Hordeum murinum, Medicago lupulina, Myosotis arvensis, Plantago coronopus, P. major, Rubus ulmifolius, Sagina apetala, Sambucus nigra, Saxifraga granulata, Senecio squalidus, Sonchus oleraceus, Trifolium pratense, T. striatum, Viola riviniana.

Rare: Acer pseudoplatanus, Agrimonia eupatoria, Alliaria petiolata, Armoracia rusticana, Brachypodium sylvaticum, Bryonia cretica subsp. dioica, Cerastium tomentosum, Chamomilla suaveolens, Conium maculatum,, Cornus sanguinea, Erigeron acer, Euphorbia helioscopia, Fumaria officinalis, Geranium lucidum, Holcus mollis, Lamium hybridum, Ligustrum vulgare, Malus sylvestris, Malva moschata, M. sylvestris, Potentilla sterilis, Primula veris, Prunus avium, Pyrus communis, Rosa obtusifolia, Sanguisorba minor subsp. minor, Scabiosa columbaria, Silene alba, Smyrnium olusatrum, Symphytum officinale, Ulmus glabra, Veronica agrestis, Viola odorata.

Habitat Study 23 Pasture, Broad Hill, Whitwick, Coalville
SK 435171 Area: 5.6 ha (14.0 acres) Alt.: 137-183 m. (450-600 ft.) S.S.S.I.
Recorder: S. H. Bishop. Surveyed: 10th June 1979.
The study is of a large field, bounded by the woodland of Calvary Rock, Broad Hill and Temple Hill on the west and south, by Cademan Wood and a field on the east and by Warren Lane on the north. The field is surrounded by dry stone walls and comprises dry heath grassland with rock outcrops and pockets of wetter ground dominated by rushes. There is local development of scrub. It is grazed by cattle from early summer to autumn. It lies on Beacon Beds and Grimley Porphyroids.

Abundant: Agrostis capillaris, Deschampsia flexuosa,, Festuca ovina, Galium saxatile, Luzula campestris, Potentilla erecta.

Locally abundant: Cardamine pratensis, Conopodium majus, Holcus mollis, Juncus effusus, Plantago lanceolata, Ranunculus repens, Rumex acetosa, Ulex gallii.

Frequent: Anthoxanthum odoratum, Nardus stricta.

Locally frequent: Achillea millefolium, Campanula rotundifolia, Carex ovalis, Cirsium palustre, Cynosurus cristatus, Danthonia decumbens, Hypochoeris radicata, Juncus conglomeratus, J. squarrosus, Lotus corniculatus, L. uliginosus, Poa pratensis, Polygala vulgaris, Prunella vulgaris, Ranunculus acris, R. flammula, R. omiophyllus, Rubus fruticosus agg., Stellaria alsine, Vaccinium myrtillus.

Occasional: Alopecurus pratensis, Betula pendula, Calluna vulgaris, Carex binervis, Centaurea nigra, Cirsium arvense, Crataegus monogyna, Deschampsia cespitosa, Epilobium angustifolium, Glyceria plicata, Juncus bulbosus, Leontodon autumnalis, Ophioglossum vulgatum, Quercus robur, Rumex acetosella, Senecio jacobaea, Sorbus aucuparia, Stellaria graminea, Taraxacum spp., Trifolium pratense, Ulex europaeus, Veronica chamaedrys, V. officinalis.

Rare: Alopecurus geniculatus, Cerastium fontanum subsp. triviale, Dactylis glomerata, Dryopteris dilatata, Heracleum sphondylium, Hyacinthoides non-scripta, Ilex aquifolium, Pilosella officinarum, Plantago major, Prunus spinosa, Pteridium aquilinum, Rumex obtusifolius, Salix cinerea subsp. oleifolia, Stellaria media, Teucrium scorodonia, Vicia sepium.

Habitat Study 24 Pasture, Osgathorpe
SK 427188 Area: 2.6 ha (6.3 acres) Alt.: 84 m. (275 ft.)
Recorder: S. H. Bishop. Surveyed: 29th May 1979.
The study is of a ridge and furrow field north of Junction House, Osgathorpe. It is enclosed by high hedges with a brook following the western boundary. The former Charnwood Forest Canal is adjacent to the eastern and southern boundaries. When surveyed the field was grazed by a small dairy herd. It lies on Keuper Marl and Alluvium.

Abundant: Agrostis capillaris, Anthoxanthum odoratum, Conopodium majus, Plantago lanceolata, Ranunculus bulbosus, Rumex acetosa.

Locally abundant: Ranunculus acris, Sanguisorba officinalis.

Frequent: Achillea millefolium, Alchemilla filicaulis subsp. vestita, Alopecurus pratensis, Centaurea nigra, Cerastium fontanum subsp. triviale, Cynosurus cristatus, Dactylis glomerata, Heracleum sphondylium, Hypochoeris radicata, Leontodon hispidus, Leucanthemum vulgare, Lolium perenne, Lotus corniculatus, Luzula campestris, Primula veris,

Prunella vulgaris, Ranunculus ficaria, Taraxacum spp., Trifolium pratense, Veronica chamaedrys.
Locally frequent: Ajuga reptans, Bellis perennis, Cirsium arvense, Deschampsia cespitosa, Succisa pratensis.
Occasional: Anthriscus sylvestris, Cardamine pratensis, Equisetum palustre, Filipendula ulmaria, Lathyrus pratensis, Saxifraga granulata, Stachys officinalis, Stellaria graminea.
Rare: Anemone nemorosa, Caltha palustris, Carex hirta, Cirsium vulgare, Crataegus monogyna, Equisetum arvense, Hyacinthoides non-scripta, Hypericum tetrapterum, Juncus effusus, Lychnis flos-cuculi, Orchis morio, Plantago major, Poa annua, Quercus robur (seedling), Rumex obtusifolius, Senecio aquaticus, Urtica dioica, Viola riviniana.

Habitat Study 25   Burbage Common, Hinckley
SP 446950   Area: 2.3 ha (5.8 acres)   Alt.: 99 m. (325 ft.)
Recorder: S. H. Bishop. Surveyed: 1st June, 28th June and 10th September 1976.
The study is of the part of the Common immediately to the north of Sheepy Wood. Part of the Common is maintained as a golf course. There is much dry heath grassland with some large tracts of scrub and a little marshy ground. It stands on Glacial Sand and Gravel.

Abundant: Agrostis capillaris, Festuca rubra, Galium saxatile, Holcus lanatus, Luzula campestris, Potentilla erecta.
Locally abundant: Anthoxanthum odoratum, Cynosurus cristatus, Lolium perenne, Poa pratensis, Ulex europaeus.
Frequent: Achillea millefolium, Danthonia decumbens, Deschampsia cespitosa, Festuca pratensis, Leontodon autumnalis, Nardus stricta, Phleum pratense subsp. bertolonii, Ranunculus bulbosus, Rumex acetosella.
Locally frequent: Alopecurus pratensis, Arrhenatherum elatius, Bellis perennis, Carex hirta, Centaurea nigra, Epilobium angustifolium, Festuca ovina, Holcus mollis, Lotus corniculatus, Poa annua, Potentilla reptans, Rubus fruticosus agg., Rumex acetosa, Stachys officinalis, Stellaria graminea, Succisa pratensis, Taraxacum spp., Trifolium repens, Trisetum flavescens.
Occasional: Achillea ptarmica, Crataegus monogyna, Dactylis glomerata, Deschampsia flexuosa, Galium verum, Juncus conglomeratus, Lotus uliginosus, Plantago major, Quercus robur, Ranunculus repens, Rumex obtusifolius, Trifolium pratense.
Rare: Acer campestre, Betula pendula, Calluna vulgaris, Carex ovalis, Cerastium fontanum subsp. triviale, Cirsium arvense, Conopodium majus, Corylus avellana, Dryopteris dilatata, Epilobium hirsutum, Fraxinus excelsior, Hypericum pulchrum, Juncus effusus, J. inflexus, Leontodon hispidus, Pilosella officinarum, Ranunculus acris, R. flammula, Rubus idaeus, Sanguisorba officinalis, Silaum silaus, Sorbus aucuparia, Stellaria media, Trifolium medium.

Habitat Study 26   Meadow near Holly Hayes Wood, Whitwick, Coalville
SK 447152   Area: 1.1 ha (2.7 acres)   Alt.: 160-165 m. (525-540 ft.) S.S.S.I.
Recorders: P. A. Candlish and P. H. Gamble. Surveyed: 13th June 1977.

The study area lies some 300 metres to the south-east of Holly Hayes Wood. It is bounded by a mineral line to its south-east, a bridle track and quarry yard embankment to the north-east, a length of old hedgerow to the north and a ditch with some trees and bushes (formerly a hedgerow) separating it from the remainder of the field to the south. It is highest in the east and slopes gradually to the west. The northern part of the field, now subject to run off from the quarry embankment, is becoming increasingly wet, and the rest of the field has a number of wet hollows (ridge and furrow) supporting marsh type vegetation. This field is probably the best part of an interesting area of heath grassland, agriculturally neglected of late years. It is on Keuper Marl. Samples of surface soil had pH 5.2 in the wet northern part and pH 6.7 in a drier area to the north-west.

Abundant: Centaurea nigra, Conopodium majus, Deschampsia cespitosa, Sanguisorba officinalis, Stachys officinalis, Succisa pratensis.
Locally abundant: Achillea millefolium, Festuca ovina, Holcus mollis.
Frequent: Agrostis capillaris, Carex hirta, Cerastium fontanum subsp. triviale, Cirsium palustre, Dactylis glomerata, Luzula campestris, Plantago lanceolata, Poa pratensis, Potentilla erecta, Rumex acetosa, Serratula tinctoria, Trisetum flavescens.
Locally frequent: Agrostis canina, Ajuga reptans, Alopecurus pratensis, Cardamine pratensis, Cirsium dissectum, Equisetum arvense, Festuca arundinacea, F. pratensis, Galium saxatile, Juncus conglomeratus, J. effusus, Lotus corniculatus, Ranunculus ficaria, R. repens, Stellaria graminea, Trifolium medium.
Occasional: Danthonia decumbens, Festuca rubra, Galium palustre, Nardus stricta, Ranunculus bulbosus, R. flammula, Stellaria media, Trifolium pratense, Veronica chamaedrys.
Rare: Achillea ptarmica, Carex flacca, C. nigra, C. panicea, Cirsium arvense, Crataegus monogyna (seedlings), Dactylorhiza maculata subsp. maculata, Heracleum sphondylium, Hyacinthoides non-scripta, Hypericum tetrapterum, Lathyrus montanus, L. pratensis, Lotus uliginosus, Urtica dioica, Vicia cracca.

Habitat Study 27   Herbert's Meadow, Ulverscroft
SK 494134   Area: 1.8 ha (4.6 acres)   Alt.: 167-175 m. (550-575 ft.) S.S.S.I.; Nature Reserve.
Recorder: P. A. Candlish. Surveyed: June 1972.
This old meadow is situated on sloping ground dipping towards the stream in the south and is bounded on the north, west and east by old hedgerows, the eastern hedgerow being tall and with frequent gaps. A spring-fed drainage channel runs along the western boundary. The southern part is colonised by sallow scrub; the area surveyed was from 3 m. outside the scrub up to the northern boundary. This is a composite type of field with a peculiar assemblage of species. In parts it is marshy and in parts dry, with species indicating a considerable range of soil types. Over a period of 10 years or so many of the species have varied appreciably in frequency, probably because of the extent to which the management has varied from year to year. In some years it was almost completely neglected, in some it was mown for hay, and in others it was grazed. Immediately prior to the survey it had been lightly grazed and the drier parts had been

mown. A few seedlings of *Rosa* and *Crataegus* about 0.5 m. high had become established here and there. The field lies mainly on Keuper Marl but there is also a small area of Boulder Clay in the north-west. Samples of surface soil varied from pH 7.6 on the wetter ground to pH 5.5 on the drier ground.

Abundant: Anthoxanthum odoratum, Carex hirta, Cynosurus cristatus, Festuca rubra, Lotus uliginosus, Plantago lanceolata, Poa annua, Prunella vulgaris, Ranunculus acris, Stachys officinalis.
Locally abundant: Carex panicea, Dactylorhiza maculata subsp. maculata, Equisetum palustre, Galium uliginosum, Rhinanthus minor.
Frequent: Cardamine pratensis, Cerastium fontanum subsp. triviale, Cirsium palustre, Conopodium majus, Juncus conglomeratus, J. effusus, Lathyrus pratensis, Potentilla erecta, Ranunculus ficaria, R. repens, Stellaria graminea, Taraxacum spp., Trifolium pratense.
Locally frequent: Ajuga reptans, Carex flacca, C. nigra, C. ovalis, C. pallescens, Equisetum telmateia, Festuca ovina, Galium palustre, Glyceria fluitans, Gymnadenia conopsea, Holcus lanatus, Juncus inflexus, Lathyrus montanus, Luzula campestris, Lychnis flos-cuculi, Mentha aquatica, Pilosella officinarum, Ranunculus flammula, Rumex acetosa, Senecio aquaticus, Stellaria alsine, Valeriana dioica, Veronica chamaedrys.
Occasional: Alopecurus geniculatus, A. pratensis, Angelica sylvestris, Bellis perennis, Caltha palustris, Carex echinata, C. remota, Dactylorhiza fuchsii, Epilobium palustre, Filipendula ulmaria, Heracleum sphondylium, Leontodon hispidus, Nardus stricta, Polygala vulgaris, Ranunculus bulbosus, Scirpus setaceus, Stellaria media, Triglochin palustris.
Rare: Anagallis tenella, Arrhenatherum elatius, Carex pulicaris, Cirsium vulgare, Dactylis glomerata, Deschampsia cespitosa, Myosotis discolor, Ophioglossum vulgatum, Orchis morio, Pedicularis sylvatica, Primula veris, Serratula tinctoria, Veronica officinalis, V. scutellata.

Habitat Study 28 Pasture, Croft
SP 509959 Area: 0.5 ha (1.2 acres) Alt.: 69 m. (225 ft.) S.S.S.I.; Nature Reserve.
Recorder: S. H. Bishop. Surveyed: 22nd June 1979.
The study is of a small area of sandy grassland interrupted by rock outcrops, situated between the River Soar to the south and Croft Rectory to the north. It is heavily grazed by sheep and cattle for the greater part of the year, which doubtless helps to maintain a thin open sward essential for the survival of the many annual and biennial species found at this site. The area lies on diorite, and a sample of the surface soil had pH 3.9.

Abundant: Achillea millefolium, Aphanes arvensis, Cynosurus cristatus, Lolium perenne, Trifolium dubium, Veronica arvensis, Vulpia bromoides.
Locally abundant: Agrostis capillaris, Aira praecox, Moenchia erecta, Poa annua, Rumex acetosella, Trifolium striatum, T. subterraneum, Veronica hederacea.
Frequent: Alopecurus pratensis, Bellis perennis, Capsella bursa-pastoris, Cardamine hirsuta, Cerastium glomeratum, Dactylis glomerata, Erodium cicutarium, Geranium molle, Montia fontana, Phleum pratense subsp. bertolonii, Plantago lanceolata, Poa pratensis, Ranunculus bulbosus, Taraxacum section Erythrosperma, Trifolium repens, Trisetum flavescens.
Locally frequent: Allium vineale, Arabidopsis thaliana, Arenaria serpyllifolia, Bromus hordeaceus, Campanula rotundifolia, Cerastium semidecandrum, Cirsium arvense, Elymus repens, Erophila verna, Galium verum, Koeleria macrantha, Lamium purpureum, Pilosella officinarum, Ranunculus ficaria, Salvia verbenaca, Scleranthus annus, Stellaria media, Trifolium micranthum.
Occasional: Carduus nutans, Crepis capillaris, Festuca ovina, F. rubra, Geranium pusillum, Hypochoeris radicata, Leontodon taraxacoides, Lotus corniculatus, Myosotis discolor, Poa compressa, P. trivialis, Potentilla argentea, Rumex acetosa, Saxifraga granulata, Sedum acre, S. reflexum, Sisymbrium officinale, Taraxacum spp., Torilis nodosa, Trifolium arvense, Urtica dioica.
Rare: Aira caryophyllea, Anthriscus sylvestris, Ballota nigra, Bromus sterilis, Cerastium fontanum subsp. triviale, Cirsium vulgare, Crataegus monogyna, Galium aparine, Geranium dissectum, Hedera helix, Holcus lanatus, Hordeum murinum, Lamium album, L. amplexicaule, L. hybridum, Malva sylvestris, Montia perfoliata, Pimpinella saxifraga, Plantago major, Polygonum aviculare, Prunus spinosa, Ranunculus repens, Rosa canina, Rumex obtusifolius, Sagina apetala, S. procumbens, Sedum album, Senecio squalidus, Solanum nigrum, Sonchus asper, S. oleraceus, Spergularia rubra, Ulex europaeus, Valerianella locusta, Verbascum thapsus, Veronica agrestis, V. chamaedrys, Vicia angustifolia, V. sativa.

Habitat Study 29 Croft Hill, Croft
SP 510966 Area: 3.4 ha (8.4 acres) Alt.: 128 m. (420 ft.) S.S.S.I.
Recorder: S. H. Bishop. Surveyed: 24th May and 30th July 1976.
This study is of the area surrounding the summit of Croft Hill. It is bounded on the east and south by Croft Quarry, on the west by the Huncote Road and on the north-west and north by a hedgerow and mixed woodland. It is sandy grassland with rock outcrops, some gorse scrub and ground dominated by bracken. It is heavily grazed by cattle and sheep. Croft Hill is composed of diorite. A sample of surface soil had pH 3.8.

Abundant: Achillea millefolium, Agrostis capillaris, Aira praecox, Aphanes microcarpa, Cynosurus cristatus, Festuca ovina, Holcus lanatus, Lolium perenne, Luzula campestris, Pilosella officinarum, Poa annua, Ranunculus bulbosus, Rumex acetosella, Veronica arvensis.
Locally abundant: Conopodium majus, Galium saxatile, Moenchia erecta, Pteridium aquilinum, Ulex europaeus.
Frequent: Anthoxanthum odoratum, Campanula rotundifolia, Cerastium glomeratum, Cirsium arvense, Dactylis glomerata, Festuca rubra, Galium verum, Koeleria macrantha, Lotus corniculatus,

Phleum pratense subsp. bertolonii, Rumex acetosa, Stellaria media, Taraxacum section Erythrosperma, Trifolium dubium, Trisetum flavescens.
Locally frequent: Alopecurus pratensis, Capsella bursa-pastoris, Cerastium semidecandrum, Holcus mollis, Plantago lanceolata, Stellaria pallida, Thymus praecox subsp. articus.
Occasional: Carduus nutans, Carex caryophyllea, Cerastium fontanum subsp. triviale, Cirsium vulgare, Leontodon autumnalis, Montia fontana, Potentilla erecta.
Rare: Bromus hordeaceus, Carex hirta, Crataegus monogyna, Crepis vesicaria, Danthonia decumbens, Geranium molle, Logfia minima, Ornithopus perpusillus, Sanguisorba minor subsp. minor, Rubus fruticosus agg., Trifolium arvense, Veronica chamaedrys.

Habitat Study 30   Pasture near Groby Pool, Groby
SK 518082   Area: 2.0 ha (4.9 acres)   Alt.: 99 m. (325 ft.)
Recorders: S. H. Bishop and E. Hesselgreaves. Surveyed: 11th June and 5th August 1976.
This study is of a small marshy field between Groby Pool and Lady Hay Wood. A wet ditch forms its eastern boundary, with fences on the other sides. When surveyed it was grazed by dairy cattle. An interesting feature is the number of grasses recorded (24 species). The field stands on Alluvium overlying Keuper Marl.

Abundant: Agrostis capillaris, Alopecurus pratensis, Anthoxanthum odoratum, Cynosurus cristatus, Festuca pratensis, Holcus lanatus, Lolium perenne, Lotus corniculatus, Luzula campestris, Ranunculus acris, Rumex acetosa, Sanguisorba officinalis.
Locally abundant: Dactylis glomerata, Glyceria fluitans, Poa trivialis, Ranunculus repens.
Frequent: Achillea millefolium,, Agrostis stolonifera, Bellis perennis, Centaurea nigra, Cerastium fontanum subsp. triviale, Conopodium majus, Juncus effusus, J. inflexus, Phleum pratense subsp. bertolonii, Stachys officinalis, Taraxacum spp., Trifolium pratense, T. repens, Trisetum flavescens.
Locally frequent: Alopecurus geniculatus, Briza media, Cardamine pratensis, Carex hirta, Cirsium arvense, Deschampsia cespitosa, × Festulolium loliaceum, Filipendula ulmaria, Glyceria maxima, Juncus acutiflorus, J. articulatus, Lathyrus pratensis, Leontodon autumnalis, Lotus uliginosus, Lychnis flos-cuculi, Lysimachia nummularia, Polygonum amphibium, Primula veris, Ranunculus flammula, Rumex conglomeratus, Senecio aquaticus, Succisa pratensis, Stellaria graminea.
Occasional: Achillea ptarmica, Ajuga reptans, Alchemilla filicaulis subsp. vestita, Carex disticha, Cirsium palustre, Epilobium parviflorum, E. tetragonum subsp. tetragonum, Equisetum fluviatile, Galium palustre, Glyceria plicata, Heracleum sphondylium, Hypericum tetrapterum, Juncus conglomeratus, Myosotis laxa subsp. caespitosa, Pulicaria dysenterica, Silaum silaus, Stellaria media, Triglochin palustris, Veronica serpyllifolia.
Rare: Alnus glutinosa (seedling), Caltha palustris, Carex otrubae, Capsella bursa-pastoris, Cirsium vulgare, Festuca gigantea, Lolium multiflorum, Phragmites australis, Plantago major, Potentilla anserina, P. erecta, Quercus robur (seedling), Rubus fruticosus agg., Rumex sanguineus, Tussilago farfara, Urtica dioica.

Habitat Study 31   Pasture, Swithland Wood, Newtown Linford
SK 538128   Area: 1.9 ha (4.78 acres)   Alt.: 99-106 m. (325-350 ft.) S.S.S.I.
Recorder: P. H. Gamble. Surveyed: 1975.
Situated within the northern part of Swithland Wood, this area of old pasture has mature deciduous woodland on its western, northern and southern boundaries. To the east, on what was until recently a part of the same field, lies a young mixed plantation. Its north, west and south boundaries consist of a low dry stone wall with a drainage ditch on the field side. The east boundary consists of a post and rail fence on the far side of which runs a deep drainage channel. The field has a few scattered bushes and both wet and dry areas. It is highest in its south-west corner and dips gradually to the north and east. Recent management, including drainage and regular grazing by cattle, has reduced the quantity of certain species, and the colonies of *Epipactis helleborine, Listera ovata* and *Ophioglossum vulgatum* which formerly occurred in association with *Pteridium aquilinum* in the north-west corner of the field have not been seen during the last few years since this bracken was removed. The field stands on Keuper Marl. The soil varies a good deal from place to place, with pH ranging from 5.6 to 7.0, the latter high for a soil derived from Keuper Marl.

Locally abundant: Deschampsia cespitosa, Holcus lanatus, H. mollis, Juncus effusus, Pteridium aquilinum, Ranunculus acris.
Frequent: Achillea millefolium, Agrostis capillaris, Alopecurus pratensis, Anthoxanthum odoratum, Aphanes arvensis, Bellis perennis, Briza media, Cardamine pratensis, Carex caryophyllea, C. nigra, C. panicea, Centaurea nigra, Cerastium fontanum subsp. triviale, Cirsium palustre, Conopodium majus, Cynosurus cristatus, Hypochoeris radicata, Lathyrus pratensis, Lolium perenne, Lotus corniculatus, Luzula campestris, Myosotis arvensis, Pedicularis sylvatica, Plantago lanceolata, Poa annua, Potentilla erecta, Prunella vulgaris, Rumex acetosa, Sanguisorba officinalis, Stachys officinalis, Succisa pratensis, Taraxacum spp., Trifolium pratense, Veronica chamaedrys, Viola riviniana.
Locally frequent: Achillea ptarmica, Ajuga reptans, Arrhenatherum elatius, Brachypodium sylvaticum, Bromus sterilis, Cardamine flexuosa, Carex hirta, Crataegus monogyna, Dactylis glomerata, Dactylorhiza fuchsii, D. maculata subsp. maculata, Deschampsia flexuosa, Epilobium angustifolium, Equisetum palustre, Festuca rubra, Geranium robertianum, Geum urbanum, Glechoma hederacea, Heracleum sphondylium, Juncus acutiflorus, J. articulatus, J. conglomeratus, Leontodon autumnalis, Leucanthemum vulgare, Listera ovata, Lotus uliginosus, Mentha aquatica, Nardus stricta, Ophioglossum vulgatum, Oxalis acetosella, Pilosella officinarum, Pimpinella major, Poa trivialis,

Polygonum hydropiper, Pulicaria dysenterica, Prunus spinosa, Ranunculus ficaria, R. repens, Rumex sanguineus, Scutellaria galericulata, Senecio jacobaea, Silene dioica, Stachys sylvatica, Stellaria alsine, S. graminea, S. holostea, S. media, Trifolium repens, Trisetum flavescens, Tussilago farfara, Urtica dioica, Valeriana dioica, Veronica beccabunga, Vicia cracca.

Occasional: Agrimonia eupatoria, Agrostis canina, Alchemilla filicaulis subsp. vestita, Alopecurus geniculatus, Anemone nemorosa, Angelica sylvestris, Arctium minus, Avenula pubescens, Betula pendula, Carex otrubae, Cirsium vulgare, Danthonia decumbens, Digitalis purpurea, Epilobium adenocaulon, E. palustre, Epipactis helleborine, Festuca arundinacea, F. pratensis, Galium palustre, G. uliginosum, G. verum, Glyceria fluitans, Hyacinthoides non-scripta, Hypericum perforatum, Lathyrus montanus, Luzula multiflora, Lysimachia nemorum, Myosotis laxa subsp. caespitosa, Potentilla sterilis, Primula veris, Quercus robur (seedlings), Rhinanthus minor, Rosa arvensis, Rubus fruticosus agg., Rumex conglomeratus, Scrophularia nodosa, Serratula tinctoria, Solanum dulcamara, Torilis japonica, Valeriana officinalis, Vicia sepium.

Rare: Acer pseudoplatanus (seedling), Galium mollugo, Genista tinctoria, Teucrium scorodonia, Trifolium medium, Vaccinium myrtillus.

Habitat Study 32   Loughborough Big Meadow, Loughborough

SK 540214   Area: 1.4 ha (3.5 acres)   Alt.: 37 m. (120 ft.)
S.S.S.I.
Recorder: P. H. Gamble. Surveyed: 23rd June 1980.
This study is of part of the largest area of unfenced grassland in the county, situated on the flood plain of the River Soar just north of Loughborough. The study area is triangular with an acute angle to the south. It is bounded by the Loughborough to Stanford on Soar road on its west side and a footpath on the east side. The roadside boundary extends for about 260 m. northward to a point directly opposite a field boundary on the other side of the road, and the northern limit of the study area continues this line to the point where it meets the footpath. The site lies on Alluvium. A sample of surface soil had pH 5.7.

Abundant: Ranunculus acris.
Locally abundant: Agrostis stolonifera, Alopecurus geniculatus, A. pratensis, Centaurea nigra, Cynosurus cristatus, Hordeum secalinum, Ranunculus repens.
Frequent: Anthriscus sylvestris, Arrhenatherum elatius, Cardamine pratensis, Lotus corniculatus, Sanguisorba officinalis, Silaum silaus, Tragopogon pratensis, Trisetum flavescens.
Locally frequent: Alliaria petiolata (Ev), Carex hirta, Cerastium fontanum subsp. triviale, Elymus repens, Glyceria fluitans, Heracleum sphondylium, Hypochoeris radicata, Lathyrus pratensis, Leontodon autumnalis, Lolium perenne, Phalaris arundinacea (Ev), Phleum pratense subsp. pratense, Ranunculus ficaria, Rumex crispus, Stellaria graminea, Trifolium repens.
Occasional: Deschampsia cespitosa, Galium verum, Trifolium dubium, T. pratense, Vicia cracca.
Rare: Barbarea stricta (Ev), Oenanthe silaifolia.

Habitat Study 33   Pasture, off Kinchley Lane, Rothley

SK 569145   Area: 0.56 ha (1.4 acres)   Alt.: 85 m. (280 ft.)
Recorder: P. H. Gamble. Surveyed: 1974 and 1975.
This field, situated south-west of Mountsorrel Common, is a close-cropped horse-grazed pasture with small elevated granodiorite outcrops and wet flooded hollows which dry out, or partially dry out, during prolonged dry spells. The part of the field within 2 m. of its boundaries was excluded from the survey. Some of the species occurring on the drier ground against the rock outcrops vary greatly in frequency from year to year; for example, *Erodium cicutarium* was locally abundant during the drought of 1976. Shallow sandy soil with pH 5.2 occurs around the granodiorite outcrops, but heavier soil associated with Boulder Clay also occurs in the area.

Locally abundant: Aira praecox, Aphanes microcarpa, Cerastium glomeratum, Ornithopus perpusillus, Rumex acetosella, Sedum acre, Trifolium dubium, Ulex europaeus.
Frequent: Achillea millefolium, Hypochoeris radicata, Plantago lanceolata.
Locally frequent: Agrostis capillaris, Bellis perennis, Callitriche platycarpa (Wp), Centaurea nigra, Cerastium semidecandrum, Digitalis purpurea, Erophila verna, Festuca ovina, Filaginella uliginosa, Galium palustre (Wp), G. verum, Lemna minor (Wp), Luzula campestris, Lythrum portula, Montia fontana, Poa annua, Ranunculus bulbosus, R. repens, Rumex sanguineus, Senecio jacobaea, Spergularia rubra, Stellaria media, Teucrium scorodonia, Veronica chamaedrys, Vulpia bromoides.
Occasional: Anthriscus sylvestris, Cardamine pratensis, Crataegus monogyna, Crepis capillaris, Erodium cicutarium, Geranium molle, Lamium purpureum, Plantago major, Ranunculus acris, Rhinanthus minor, Rumex acetosa, Stellaria graminea, Taraxacum section Erythrosperma.
Rare: Geranium lucidum, Moenchia erecta, Rorippa islandica, Sorbus aucuparia, Trifolium micranthum, Typha latifolia (Wp).

Habitat Study 34   Pasture, Berrycott Lane, Seagrave

SK 621178   Area: 1.3 ha (3.4 acres)   Alt.: 82 m. (270 ft.)
Recorders: P. H. Gamble, J. M. Horwood and A. L. Primavesi. Surveyed: 15th June 1971 and 20th September 1971.
This narrow field between Berrycott Lane and the Seagrave Brook slopes steeply and irregularly downwards and has probably never been ploughed. It stands on Chalky Boulder Clay.

Abundant: Achillea millefolium, Agrostis capillaris, Bellis perennis, Cerastium fontanum subsp. triviale, Cynosurus cristatus, Poa trivialis, Ranunculus bulbosus, Trifolium repens.
Frequent: Anthoxanthum odoratum, Arrhenatherum elatius, Avenula pratensis, Briza media, Carex flacca, Centaurea nigra, Cirsium arvense, Crepis capillaris, Dactylis glomerata, Elymus repens, Galium verum, Holcus lanatus, Lathyrus pratensis, Leontodon autumnalis, L. hispidus, Leucanthemum vulgare, Lolium perenne, Lotus corniculatus, Luzula campestris, Phleum pratense subsp. bertolonii, P. pratense subsp. pratense, Plantago lanceolata,

Plantago media, Primula veris, Rumex acetosa, Sanguisorba minor subsp. minor, Taraxacum spp., Trifolium pratense, Trisetum flavescens.
Locally frequent: Deschampsia cespitosa, Festuca pratensis, Ononis spinosa, Pilosella officinarum, Polygala vulgaris.
Occasional: Agrimonia eupatoria, Agrostis stolonifera, Ajuga reptans, Alopecurus pratensis, Arum maculatum, Brachypodium pinnatum, Bromus hordeaceus, Cirsium vulgare, Crataegus monogyna, Crepis vesicaria, Festuca rubra, × Festulolium loliaceum, Galium aparine, Hordeum secalinum, Koeleria macrantha, Odontites verna, Potentilla reptans, Prunella vulgaris, Ranunculus acris, R. ficaria, R. repens, Rosa arvensis, R. canina, Rumex crispus, Sambucus nigra, Silaum silaus, Tamus communis, Urtica dioica, Veronica chamaedrys, Viola odorata.
Rare: Anthriscus sylvestris, Brachypodium sylvaticum.

Habitat Study 35 Burrough Hill Camp, Somerby
SK 761119 Area: 2.4 ha (5.9 acres) Alt.: 213 m. (700 ft.)
Recorder: S. H. Bishop. Surveyed: 6th and 22nd August 1977.
The study covers the flat summit of the hill and the banks of this Iron Age encampment. The richest flora is found on the steep banks. The area lies on Marlstone.

Abundant: Agrostis capillaris, Cirsium arvense, Cynosurus cristatus, Holcus lanatus, Phleum pratense subsp. bertolonii, Plantago lanceolata, Ranunculus bulbosus.
Locally abundant: Aphanes arvensis, Bromus hordeaceus, Conopodium majus, Crepis capillaris, Erophila verna, Ranunculus acris, Veronica arvensis.
Frequent: Achillea millefolium, Alopecurus pratensis, Arrhenatherum elatius, Bellis perennis, Campanula rotundifolia, Carduus nutans, Cerastium fontanum subsp. triviale, Dactylis glomerata, Deschampsia cespitosa, Galium verum, Geranium molle, Leontodon autumnalis, Lolium perenne, Lotus corniculatus, Pilosella officinarum, Poa annua, Polygonum aviculare, Rumex acetosa, Sanguisorba minor subsp. minor, Taraxacum section Erythrosperma, Trifolium repens, Trisetum flavescens.
Locally frequent: Cerastium glomeratum, Cirsium vulgare, Elymus repens, Heracleum sphondylium, Koeleria macrantha, Myosotis ramosissima, Phleum pratense subsp. pratense, Plantago media, Potentilla erecta, P. reptans, Ranunculus repens, Saxifraga granulata, Stellaria graminea, S. media, Thymus praecox subsp. arcticus, Torilis nodosa, Veronica chamaedrys.
Occasional: Capsella bursa-pastoris, Cerastium semidecandrum, Crataegus monogyna, Lathyrus pratensis, Pimpinella saxifraga, Plantago major, Prunella vulgaris, Rubus ulmifolius, Rumex acetosella, R. crispus, Stachys officinalis, Trifolium pratense, Urtica dioica.
Rare: Acer pseudoplatanus, Alopecurus geniculatus, Campanula glomerata, Carduus acanthoides, Carex hirta, Cirsium palustre, Fraxinus excelsior, Glechoma hederacea, Matricaria perforata, Medicago lupulina, Sagina apetala, Sambucus nigra, Sisymbrium officinale, Trifolium striatum, Veronica officinalis.

Habitat Study 36 Pasture near Terrace Hill Farm, Eaton
SK 795309 Area: 3.0 ha (7.5 acres) Alt.: 130 m. (425 ft.) S.S.S.I.
Recorder: S. H. Bishop. Surveyed: 13th June 1976.
The study covers part of a large undulating field, much of which was previously quarried and has been left studded with pits and mounds. At the time of the survey the site was grazed by cattle. It is situated on Marlstone.

Abundant: Achillea millefolium, Anthoxanthum odoratum, Bellis perennis, Cirsium arvense, Cynosurus cristatus, Dactylis glomerata, Festuca rubra, Holcus lanatus, Lolium perenne, Lotus corniculatus, Plantago lanceolata, Poa pratensis, Ranunculus bulbosus, Rumex acetosa.
Locally abundant: Cirsium acaule, Galium verum, Koeleria macrantha, Pilosella officinarum, Plantago media, Sanguisorba minor subsp. minor, Trifolium repens.
Frequent: Arrhenatherum elatius, Avenula pubescens, Carduus nutans, Centaurea nigra, Cerastium fontanum subsp. triviale, Conopodium majus, Luzula campestris, Senecio jacobaea, Trifolium pratense, Veronica chamaedrys.
Locally frequent: Aphanes arvensis, Briza media, Carex caryophyllea, Cerastium arvense, Filipendula vulgaris, Knautia arvensis, Myosotis ramosissima, Polygala vulgaris, Primula veris, Rumex acetosella, Saxifraga granulata, Scabiosa columbaria, Stellaria graminea, Taraxacum section Erythrosperma, Thymus praecox subsp. arcticus, Trifolium dubium, Ulex europaeus, Urtica dioica.
Occasional: Aira caryophyllea, Campanula glomerata, C. rotundifolia, Cirsium vulgare, Crataegus monogyna, Poa annua, Ranunculus acris, R. repens, Rubus fruticosus agg., Rumex obtusifolius, Stellaria media, Trifolium striatum, Veronica serpyllifolia.
Rare: Anthyllis vulneraria, Galeopsis tetrahit, Geranium dissectum, Glechoma hederacea, Myosotis arvensis, Ononis repens, Ranunculus ficaria, Rosa canina, Sambucus nigra, Silene alba, Stachys officinalis.

Habitat Study 37 Muston Meadows, Bottesford
SK 824365 Area: 2.43 ha (6.02 acres) Alt.: 46 m. (141 ft.)
Recorders: P. A. Evans and D. A. Wells. Surveyed: 28th May 1979 and 6th June 1980.
S.S.S.I.; National Nature Reserve (no access).
This ridge and furrow field lies almost a mile south-southwest of Muston village. It is separated on its southern boundary from the Grantham Canal by a strip of woodland, and is surrounded by well-maintained hedges. The field to the north is also ridge and furrow hay meadow. On the east and west sides there are arable fields. The good close turf is mown for hay, and occasionally grazed after mowing. This is the best locality for *Orchis morio* in Leicestershire. The following species list was compiled in the course of survey work, not specifically for a habitat study. Some of the frequency designations were therefore appended at a later date. The field lies on Lower Lias Clay.

Abundant: Anthoxanthum odoratum, Festuca rubra, Holcus lanatus, Ranunculus acris, Trifolium repens.
Locally abundant: Lotus corniculatus, Sanguisorba officinalis.

Frequent: Agrostis stolonifera, Centaurea nigra, Conopodium majus, Cynosurus cristatus, Festuca pratensis, Luzula campestris, Ophioglossum vulgatum, Orchis morio, Plantago lanceolata, Primula veris, Prunella vulgaris, Rhinanthus minor, Rumex acetosa, Trifolium pratense, Trisetum flavescens.
Locally frequent: Deschampsia cespitosa, Poa trivialis, Stellaria graminea.
Occasional: Alopecurus pratensis, Briza media, Cardamine pratensis, Cerastium fontanum subsp. triviale, Dactylis glomerata, Galium verum, Heracleum sphondylium, Lathyrus pratensis, Leontodon hispidus, Leucanthemum vulgare, Phleum pratense subsp. pratense, Potentilla reptans, Senecio erucifolius, S. jacobaea, Silaum silaus, Taraxacum spp.
Rare: Juncus conglomeratus, Vicia cracca.

Habitat Study 38 Grassland adjoining Wymondham Rough, Wymondham
SK 833175   Area: 2.6 ha (6.6 acres)   Alt.: 91-99 m. (300-325 ft.) S.S.S.I.; Nature Reserve.
Recorders: S. H. Bishop, P. H. Gamble and A. L. Primavesi. Surveyed: 30th June 1979.
This field, with three long boundaries and one short one, has the woodland of Wymondham Rough adjacent to the south-west, and of Day's Plantation to the south-east, the other two boundaries being hedgerows. The field slopes down from north-west to south-east. The field stands partly on Lower Lias Clays and partly on Chalky Boulder Clay. A sample of surface soil had pH 6.3.

Abundant: Arrhenatherum elatius, Centaurea nigra, Dactylis glomerata.
Locally abundant: Conopodium majus, Filipendula ulmaria, Heracleum sphondylium, Ranunculus acris, Sanguisorba officinalis.
Frequent: Agrostis stolonifera, Anthoxanthum odoratum, Carex flacca, Deschampsia cespitosa, Festuca pratensis, Filipendula vulgaris, Holcus lanatus, Lathyrus pratensis, Lotus corniculatus, Poa trivialis, Primula veris, Rumex acetosa, Taraxacum spp.
Locally frequent: Ajuga reptans, Carex hirta, Cerastium fontanum subsp. triviale, Cirsium arvense, C. palustre, Geum rivale, Hordeum secalinum, Juncus conglomeratus, Phleum pratense subsp. pratense, Potentilla reptans, Ranunculus ficaria, R. repens, Senecio erucifolius, Silaum silaus, Vicia cracca.
Occasional: Alopecurus pratensis, Crataegus monogyna (seedlings), Equisetum arvense, Festuca rubra, Galium verum, Hypochoeris radicata, Leucanthemum vulgare, Poa pratensis, Potentilla erecta, Prunus spinosa (seedlings), Rhinanthus minor, Stellaria graminea.
Rare: Carex spicata, Cirsium vulgare, Cynosurus cristatus, Lysimachia nummularia, Ophioglossum vulgatum, Plantago lanceolata, Prunella vulgaris, Rumex crispus, Tragopogon pratensis.

Habitat Study 39 Limestone grassland near Egypt Plantation, Sproxton
SK 868279   Area: 1.6 ha (4.1 acres)   Alt.: 137 m. (450 ft.) S.S.S.I.
Recorders: S. H. Bishop, P. H. Gamble and A. L. Primavesi. Surveyed: 29th May and 29th October 1978.
This is an oblong area of limestone grassland immediately to the south of Egypt Plantation. It is separated by a post and wire fence from the adjacent arable fields. It is the site of a wartime camp, and scattered throughout the area are the foundations of huts. It is being invaded by scrub but at the time of the survey there were large areas of open grassland. The site stands on Inferior Oolite. A sample of the shallow soil had pH 7.7.

Abundant: Carex flacca.
Locally abundant: Brachypodium pinnatum, Centaurea nigra, Gentianella amarella, Lotus corniculatus, Poa annua, P. pratensis.
Frequent: Achillea millefolium, Acinos arvensis, Arenaria serpyllifolia, Bellis perennis, Centaurea scabiosa, Cerastium fontanum subsp. triviale, C. glomeratum, Cirsium acaule, Crataegus monogyna, Dactylis glomerata, Fraxinus excelsior, Glechoma hederacea, Heracleum sphondylium, Hypericum perforatum, Linum catharticum, Myosotis arvensis, Plantago lanceolata, Rubus fruticosus agg., Torilis japonica, Viola hirta.
Locally frequent: Bromus erectus, Calamagrostis epigejos, Cerastium arvense, Cirsium eriophorum, Clinopodium vulgare, Epilobium angustifolium, Erodium cicutarium, Erophila verna, Galium aparine, Geranium pusillum, Leucanthemum vulgare, Medicago lupulina, Moehringia trinervia, Pilosella officinarum, Primula veris, Sanguisorba minor subsp. minor, Sedum acre.
Occasional: Acer pseudoplatanus, Agrimonia eupatoria, Anthriscus sylvestris, Aphanes arvensis, Arctium minus, Briza media, Campanula rotundifolia, Cirsium arvense, Erigeron acer, Euphrasia nemorosa, Galium verum, Geranium molle, Geum urbanum, Hedera helix, Knautia arvensis, Lamium album, Linaria vulgaris, Luzula campestris, Pimpinella saxifraga, Potentilla reptans, Ranunculus repens, Reseda lutea, Rhamnus catharticus, Rhinanthus minor, Rosa canina, Sagina apetala, Silene alba, S. vulgaris, Stellaria media, Taraxacum section Erythrosperma, Taraxacum spp., Urtica dioica, Veronica persica, Viola arvensis.
Rare: Anagallis arvensis, Betula pendula, Bryonia cretica subsp. dioica, Capsella bursa-pastoris, Chamomilla suaveolens, Cirsium vulgare, Desmazeria rigida, Galeopsis tetrahit, Geranium dissectum, G. robertianum, Lamium purpureum, Ligustrum vulgare, Malus sylvestris, Ononis repens, Polygala vulgaris, Prunella vulgaris, Ranunculus bulbosus, Reseda luteola, Salix caprea, Salix cinerea subsp. cinerea, S. cinerea subsp. oleifolia, Sambucus nigra, Solanum dulcamara, Succisa pratensis, Tamus communis, Tragopogon pratensis, Trifolium pratense, Tussilago farfara.

Habitat Study 40 Cribb's Meadow, Wymondham
SK 899189   Area: 2.4 ha (6 acres)   Alt.: 117 m. (385 ft.) S.S.S.I.; Nature Reserve.
Recorders: S. H. Bishop, P. H. Gamble and A. L. Primavesi. Surveyed: 30th June 1979.
This, the larger of two botanically similar fields owned by the county trust for nature conservation, formerly one but now separated from one another by a dismantled

railway, lies immediately to the north of the railway and to the west of Fosse Lane. It is bounded on the south side by a fence and the scrub-covered railway embankment, and on the other three sides by hedgerows. At the time of the survey, because of an absence of grazing, much of the field was being colonised by small hawthorn bushes. A pond lies near the northern boundary. The survey did not include the plants growing in or on the margins of the pond. The field lies on Boulder Clay. A sample of surface soil had pH 6.5.

Abundant: Anthoxanthum odoratum, Carex flacca, Centaurea nigra, Ranunculus acris, Vicia cracca.
Locally abundant: Arrhenatherum elatius, Carex panicea, Galium verum, Heracleum sphondylium.
Frequent: Agrimonia eupatoria, Briza media, Cerastium fontanum subsp. triviale, Conopodium majus, Dactylis glomerata, Festuca pratensis, Holcus lanatus, Lathyrus pratensis, Lotus corniculatus, Ophioglossum vulgatum, Plantago lanceolata, Poa trivialis, Primula veris, Rumex acetosa, Senecio erucifolius, Taraxacum spp., Trifolium pratense, Trisetum flavescens.
Locally frequent: Agrostis stolonifera, Alopecurus pratensis, Avenula pubescens, Bromus erectus, Carex otrubae, Cirsium arvense, Dactylorhiza fuchsii, Deschampsia cespitosa, Filipendula ulmaria, Potentilla anserina, P. reptans, Ranunculus repens, Rhinanthus minor, Sanguisorba officinalis.
Occasional: Bellis perennis, Bromus hordeaceus, Carex hirta, Crataegus monogyna (seedlings), Cynosurus cristatus, Daucus carota, Elymus repens, Festuca rubra, Hordeum secalinum, Juncus inflexus, Leucanthemum vulgare, Luzula campestris, Orchis morio, Phleum pratense subsp. bertolonii, P. pratense subsp. pratense, Pimpinella major, Poa pratensis, Prunella vulgaris, Rosa canina, Silaum silaus, Succisa pratensis, Trifolium repens, Veronica chamaedrys.
Rare: Cardamine pratensis, Galium album subsp. album, G. palustre, Juncus articulatus, Listera ovata, Lolium perenne, Prunus spinosa (seedlings), Quercus robur (seedlings), Rumex crispus, R. sanguineus, Tragopogon pratensis.

Heathland

The heathlands of Charnwood Forest formerly had a number of species characteristic of upland heath, but some of these, such as *Scirpus cespitosus*, are now almost certainly extinct and others, such as *Empetrum nigrum*, are surviving precariously, if at all. Much of the former Charnwood Forest heathland has been modified or destroyed by various forms of human activity. Those areas which have survived have either been for a long time in enlightened private or public ownership, such as Charnwood Lodge (studies 43 and 44), Bradgate Park (studies 45 and 46), or are so rugged that they cannot be put to any economically useful purpose, such as High Sharpley (study 42). An attempt has been made to provide studies of dry and wet heath, but in most sites on Charnwood Forest there are in close proximity both wet hollows and rocky outcrops where the soil is thin and dry. Nowhere today are there extensive areas of bog, though there are some small wet patches where species of *Sphagnum* are found.

On the Coal Measures in the north-west there were, prior to enclosure, extensive areas of lowland heath, but this region has been so subjected to mining and related human activities that only fragments remain. There are, however, areas of industrial dereliction where the land has tended to revert to lowland heath. One such area, near Moira, is the subject of study 41.

A peculiar piece of ground, which can only be described as a type of heathland, is found at Six Hills. This is the common grazing land known as Twenty Acre, the subject of study 47. This site is on Chalky Boulder Clay, on which one would not expect to find heathland. However freedom from ploughing and impeded drainage has here produced a deep layer of peaty, acid soil supporting *Molinia-Nardus* type heathland. Unfortunately now that the site is now longer grazed it is rapidly becoming invaded by scrub, and the future of this interesting site is problematic.

Habitat Study 41 Heathland near Moira, Ashby Woulds
SK 319154   Area: 1.9 ha (4.7 acres)   Alt.: 107 m. (350 ft.)
Recorder: S. H. Bishop. Surveyed: 5th September 1976.
This site is a relict of the once extensive heathlands of the Ashby Woulds, now one of the major areas of industrial dereliction in Leicestershire. The study embraces an area of heathland which has developed on the site of the former Newfield Colliery. Most of the area is dry, but there are pockets of wet heathland with bog pools. It lies on Coal Measures.

Abundant: Agrostis capillaris, Deschampsia flexuosa.
Locally abundant: Arrhenatherum elatius, Betula pendula, Calluna vulgaris, Centaurea nigra, Crataegus monogyna, Rubus fruticosus agg., Typha angustifolia.
Frequent: Achillea millefolium, Agrostis stolonifera, Cerastium fontanum subsp. triviale, Cirsium arvense, Cynosurus cristatus, Dactylis glomerata, Deschampsia cespitosa, Epilobium angustifolium, Holcus lanatus, Juncus effusus, J. conglomeratus, Lolium perenne, Quercus robur, Rumex acetosella.
Locally frequent: Betula pubescens, Carex rostrata, Dryopteris dilatata, D. filix-mas, Hedera helix, Heracleum sphondylium, Hieracium sp., Holcus mollis, Juncus articulatus, J. bulbosus, Leontodon autumnalis, Lotus corniculatus, Nardus stricta, Plantago lanceolata, Ranunculus repens, Rosa canina, Rumex acetosa, Salix cinerea subsp. oleifolia, Stachys sylvatica, Trifolium pratense, Trisetum flavescens, Urtica dioica.
Occasional: Anthoxanthum odoratum, Avenula pubescens, Crepis capillaris, Danthonia decumbens, Daucus carota, Eloecharis palustris, Lathyrus pratensis, Leontodon hispidus, Linum catharticum, Lonicera periclymenum, Plantago major, Potamogeton natans, Potentilla erecta, Prunella vulgaris, Ranunculus flammula, Rubus idaeus, Rumex crispus, Sambucus nigra, Silene dioica, Solanum dulcamara, Tussilago farfara, Vicia cracca, Viola riviniana.
Rare: Acer pseudoplatanus, Achillea ptarmica, Alisma plantago-aquatica, Artemisia vulgaris, Briza media, Carex demissa, C. flacca, Centaurium erythraea, Circaea lutetiana, Cirsium palustre, C. vulgare, Convolvulus arvensis, Dryopteris carthusiana, Epilobium palustre, Eriophorum angustifolium, Euphrasia nemorosa, Galeopsis tetrahit, Luzula multiflora, Malus sylvestris, Ophioglossum vulgatum, Polygala serpyllifolia, Prunus avium, Pyrus communis, Salix caprea, S. fragilis, Senecio jacobaea, Silene vulgaris, Sorbus aucuparia, Typha latifolia, Veronica chamaedrys.

Habitat Study 42 High Sharpley, Whitwick, Coalville
SK 448171   Area: 12.7 ha (31.4 acres)   Alt.: 160-198 m. (525-650 ft.) S.S.S.I.
Recorder: S. H. Bishop. Surveyed: 28th August 1979.
The study covers the area of heathland which surrounds the rocky outcrop of High Sharpley. It has much rocky, boulder-strewn dry heathland and a little waterlogged ground or wet heath. It is situated on Beacon Beds and Sharpley Porphyroid.

Abundant: Pteridium aquilinum.

Locally abundant: Agrostis capillaris, Calluna vulgaris, Deschampsia flexuosa, Molinia caerulea, Nardus stricta, Vaccinium myrtillus.
Frequent: Betula pendula, Quercus robur.
Locally frequent: Anthoxanthum odoratum, Epilobium angustifolium, Festuca ovina, Galium saxatile, Holcus mollis, Juncus effusus, J. squarrosus, Potentilla erecta, Ulex gallii.
Occasional: Deschampsia cespitosa (Hw), Dryopteris dilatata, Juncus bufonius (Hw), J. conglomeratus, Ranunculus repens, Rubus fruticosus agg., Rumex acetosella, Sorbus aucuparia, Trifolium repens.
Rare: Carex binervis (Hw), C. nigra (Hw), C. ovalis (Hw), Crataegus monogyna, Dactylis glomerata, Erica tetralix, Hydrocotyle vulgaris (Hw), Ilex aquifolium, Leontodon autumnalis, Lotus corniculatus, Luzula multiflora, Malus sylvestris, Plantago lanceolata, Pinus sylvestris, Polygonum persicaria, Ranunculus omiophyllus (Hw), Rumex acetosa, R. obtusifolius, Salix caprea, S. cinerea subsp. oleifolia, Senecio sylvaticus, Urtica dioica.

Habitat Study 43 Wet heathland, Charnwood Lodge, Charley
SK 466154   Area: 0.7 ha (2.65 acres)   Alt.: 192-198 m. (630-650 ft.) S.S.S.I.; Nature Reserve.
Recorder: P. H. Gamble. Surveyed: 20th June 1976.
The study area consists of marshy heathland lying to the north-east of Colony Reservoir and east of Gisborne's Gorse, where it forms part of an extensive area of heath largely dominated by *Pteridium* and *Molinia* on the Charnwood Lodge Nature Reserve. Much of the study area is very wet, with *Sphagnum* spp. and a water table at or near the surface throughout the year. Drainage runnels from a nearby pond in the field to the south help to keep the ground wet. Because of the hummocky nature of the ground and a scattering of large boulders breaking the surface, patches of drier ground occur here and there within the study area. Grazing and trampling by cattle take place intermittently. The site is located on Keuper Marl overlying Hornstone and Grit of the Blackbrook Series, giving rise to a poorly-drained stony loam with pockets of humus-rich soil. Two samples of surface soil had pH 4.80 and 4.85.

Locally abundant: Agrostis capillaris, Galium saxatile, Holcus mollis, Hydrocotyle vulgaris, Juncus bulbosus, J. conglomeratus, Molinia caerulea, Pteridium aquilinum.
Frequent: Achillea ptarmica, Calluna vulgaris, Carex demissa, C. echinata, C. nigra, C. panicea, Cirsium palustre, Epilobium palustre, Erica tetralix, Galium palustre, Luzula multiflora, Mentha aquatica, Montia fontana, Myosotis secunda, Polygonum hydropiper, Potentilla erecta, Ranunculus flammula, R. repens, Stellaria alsine, Viola palustris.
Locally frequent: Anthoxanthum odoratum, Callitriche stagnalis, Cardamine flexuosa, Cynosurus cristatus, Festuca ovina, F. rubra, Glyceria fluitans, Juncus acutiflorus, J. effusus, Lotus uliginosus, Lythrum portula, Poa trivialis, Prunella vulgaris, Ranunculus omiophyllus, Scutellaria minor.
Occasional: Athyrium filix-femina, Betula pendula, Cardamine pratensis, Carex ovalis, Epilobium angustifolium, E. montanum, E. tetragonum subsp. lamyi, Holcus lanatus, Nardus stricta, Plantago major,

Polygala serpyllifolia, Rumex acetosa, Sagina procumbens, Senecio aquaticus, Stellaria graminea, Viola riviniana.
Rare: Anagallis tenella, Briza media, Crataegus monogyna, Dryopteris dilatata, Filaginella uliginosa, Galium uliginosum, Myosotis laxa subsp. caespitosa, Ranunculus ficaria, Rumex obtusifolius, Salix caprea, Veronica scutellata.

Habitat Study 44 Wet heath and bog, Charnwood Lodge, Charley
SK 473149 Area 2.7 ha (6.8 acres) Alt.: 206-230 m. (675-700 ft.) S.S.S.I.; Nature Reserve.
Recorder: P. H. Gamble. Surveyed: 18th July 1976.
This area of wet heath with bog pools lies on the eastern side of Timberwood Hill about midway between the summit and Cat Hill Wood. The study area includes a strip of the open heath, measuring about 350 m. from north to south and between 50 and 75 m. from west to east, bounded on its eastern side by a dry-stone wall. Also included, on the other side of the wall, is a strip 180 m. from north to south and 30 m. from west to east, with the wall forming its western boundary. It is a rocky boulder-strewn area dipping to the north-east. There are numerous wet hollows but it also includes some drier ground. *Sphagnum* is abundant in the wetter areas and *Eriophorum angustifolium,* though rare at the time of the survey, was locally abundant a few years prior to this. The rocks are Beacon Hill Hornstones and Grits and the soil is a poorly drained stony loam with humus-rich pockets. Samples of surface soil had pH 3.7 at the *Eriophorum* site and pH 3.5 at the *Empetrum* site.

Locally abundant: Agrostis canina, Calluna vulgaris, Erica tetralix, Holcus mollis, Hydrocotyle vulgaris, Juncus conglomeratus, Molinia caerulea, Pteridium aquilinum.
Frequent: Anthoxanthum odoratum, Betula pendula, Deschampsia flexuosa, Festuca ovina, Galium saxatile, Juncus squarrosus, Luzula multiflora, Potentilla erecta, Vaccinium myrtillus.
Locally frequent: Carex echinata, Juncus acutiflorus, Rubus fruticosus agg.
Occasional: Betula pubescens, Carex binervis, C. demissa, C. ovalis, C. panicea, Dryopteris dilatata, Juncus effusus, Nardus stricta, Sorbus aucuparia, Ulex europaeus, U. gallii.
Rare: Blechnum spicant, Crataegus monogyna, Empetrum nigrum, Eriophorum angustifolium, Quercus robur, Salix repens.

Habitat Study 45 Pond and wet heath, Bradgate Park, Newtown Linford
SK 529115 Area.: 0.4 ha (1 acre) Alt.: 145 m. (475 ft.) S.S.S.I.
Recorder: P. H. Gamble. Surveyed: 14th July 1976.
Situated near to the northern boundary wall of the Park and about 500 m. north-east of Old John Tower, the study area includes a pond (about 20 m. × 16 m.) with bog mosses, together with some of the neighbouring wet heath and the lower part of ditches draining into the pool. The site supports a good cross-section of the wet heath and bog species of Charnwood Forest. The area stands on Keuper Marl with heavy stony soil and some localised formation of peat. Samples of surface soil from the pool margin had pH 4.3 and from the ditch bank pH 3.5.

Abundant: Molinia caerulea.
Locally abundant: Juncus acutiflorus, J. bulbosus, J. conglomeratus, Pteridium aquilinum.
Frequent: Agrostis canina, A. capillaris, Deschampsia flexuosa, Luzula multiflora.
Locally frequent: Anthoxanthum odoratum, Danthonia decumbens, Eleocharis palustris, Epilobium palustre, Galium palustre, G. saxatile, Holcus mollis, Hydrocotyle vulgaris, Juncus effusus, J. squarrosus, Lotus uliginosus, Montia fontana, Oxalis acetosella, Potamogeton natans, Potentilla erecta, Ranunculus flammula, R. omiophyllus, Scutellaria galericulata, S. minor, Stellaria alsine, Thelypteris limbosperma.
Occasional: Callitriche stagnalis, Campanula rotundifolia, Carex nigra, Cirsium palustre, Deschampsia cespitosa, Digitalis purpurea, Holcus lanatus, Nardus stricta, Teucrium scorodonia.
Rare: Carex demissa, C. ovalis, C. panicea, Dryopteris dilatata, Hypericum humifusum, Scirpus setaceus, Ulex europaeus.

Habitat Study 46 Dry heath, Bradgate Park, Newtown Linford
SK 536110 Area: 3.8 ha (9.4 acres) Alt.: 130-145 m. (425-475 ft.) S.S.S.I.
Recorder: P. H. Gamble. Surveyed: 1976.
The study area lies adjacent to the northern boundary of Dale Spinney and extends some 450 m. along the drystone wall of the wood boundary and some 100 m. out into the open heath. The area is mainly dry heathland without rock outcrops, but there are one or two small wet flushes with *Sphagnum*. A well trodden footpath crosses the area from east to west and is included in the survey. Bradgate and Woodhouse Grits and Hornstones lie just below the surface in the part of the study nearest to the spinney, producing a shallow sandy soil with pH 3.5. The lowest part stands on Keuper Marl, with a heavier and less well-drained soil with pH 3.3 in one of the wetter parts.

Locally abundant: Agrostis capillaris, Galium saxatile, Holcus mollis, Molinia caerulea (Hw), Pteridium aquilinum.
Locally frequent: Agrostis canina (Hw), Anthoxanthum odoratum, Deschampsia flexuosa, Festuca ovina, Juncus acutiflorus, J. conglomeratus (Hw), J. squarrosus, Luzula campestris, Nardus stricta, Potentilla erecta, Rumex acetosella, Spergularia rubra (Rt).
Occasional: Campanula rotundifolia, Carex ovalis, C. pilulifera, Danthonia decumbens (Rt), Erica tetralix (Hw), Hyacinthoides non-scripta, Hypochoeris radicata, Juncus effusus (Hw), Luzula multiflora (Hw), Poa compressa, Polygala serpyllifolia, P. vulgaris.
Rare: Calluna vulgaris, Carex binervis, Epilobium angustifolium, Lotus uliginosus, Salix cinerea, S. repens, Sorbus aucuparia (seedling).

Habitat Study 47 Twenty Acre, Six Hills, Burton on the Wolds
SK 641211 Area: 8.5 ha (21.0 acres) Alt.: 134 m. (440 ft.) S.S.S.I.
Recorders: S. H. Bishop, P. H. Gamble, J. M. Horwood and A. L. Primavesi. Surveyed: 15th June 1971 and 6th June 1977.

This neglected common land is poorly drained and has developed a peaty soil which supports a mixture of wet and dry heath. In recent years it has become heavily invaded by scrub. Fire swept over part of the area in 1967, which probably accounts for the disappearance of *Genista anglica,* last seen in 1960. The common stands on Boulder Clay. Two samples of surface soil had pH 4.2 and 4.4.

Abundant: Centaurea nigra, Potentilla anserina, P. erecta, Prunus spinosa, Salix cinerea, Sanguisorba officinalis, Stellaria graminea.

Locally abundant: Deschampsia cespitosa, Epilobium angustifolium, Festuca ovina, Filipendula ulmaria, Galium aparine, Juncus conglomeratus, Lonicera periclymenum, Molinia caerulea, Nardus stricta, Rubus fruticosus agg., Salix caprea, Stachys sylvatica, Tussilago farfara, Viola riviniana.

Frequent: Achillea ptarmica, Carex panicea, Crataegus monogyna, Dryopteris dilatata, Glechoma hederacea, Holcus lanatus, Luzula campestris, L. multiflora, Quercus robur, Stachys officinalis, Succisa pratensis, Ulex europaeus.

Locally frequent: Ajuga reptans, Anthoxanthum odoratum, Arctium minus, Calamagrostis epigejos, Carex flacca, C. nigra, C. otrubae, Cirsium arvense, C. dissectum, C. palustre, C. vulgare, Dactylorhiza maculata subsp. maculata, Dryopteris filix-mas, Epilobium hirsutum, Equisetum arvense, Galium palustre, G. saxatile, G. uliginosum, Geum urbanum, Heracleum sphondylium, Juncus acutiflorus, J. articulatus, J. bufonius, J. inflexus, Lathyrus pratensis, Lotus uliginosus, Moehringia trinervia, Myosotis arvensis, Poa pratensis, P. trivialis, Populus tremula, Potentilla reptans, Ranunculus ficaria, R. repens, Rumex acetosa, Serratula tinctoria, Silene dioica, Urtica dioica, Viburnum opulus, Vicia cracca, V. sativa.

Occasional: Acer campestre, Alopecurus pratensis, Anthriscus sylvestris, Barbarea vulgaris, Betula pendula, Brachypodium sylvaticum, Cardamine pratensis, Carex hirta, C. pilulifera, Cerastium fontanum subsp. triviale, Crataegus laevigata × monogyna, Dactylorhiza fuchsii, Dryopteris carthusiana, Fraxinus excelsior, Galium verum, Geranium robertianum, Hypericum tetrapterum, Juncus effusus, Lathyrus montanus, Malus sylvestris, Pedicularis sylvatica, Primula veris, Ranunculus bulbosus, Rosa arvensis, R. canina, Rubus idaeus, Rumex acetosella, R. crispus, R. obtusifolius, Sambucus nigra, Scrophularia auriculata, Senecio erucifolius, S. jacobaea, Solanum dulcamara, Sonchus oleraceus, Tamus communis, Taraxacum spp., Trifolium repens, Veronica serpyllifolia.

Rare: Athyrium filix-femina, Bellis perennis, Carex demissa, C. hostiana, Crataegus laevigata, Dactylis glomerata, Festuca gigantea, Geranium dissectum, Lamium album, Lapsana communis, Ligustrum vulgare, Plantago major, Poa annua, Primula veris × vulgaris, Ribes rubrum, Rumex sanguineus, Stellaria alsine, Taraxacum section Celtica, Trifolium dubium, Ulmus glabra, Veronica arvensis.

Marsh

Strictly speaking, a marsh is an area with mineral soil where the water table is at or near the surface throughout the year. A thin layer of peat may accumulate at the surface of such a marsh, but if the soil is mainly of acid peat the area should more correctly be called a bog, if of neutral or basic peat a fen. The plants which normally form acid peat, such as *Sphagnum* and *Eriophorum,* are rare in Leicestershire, and such boggy places as we have in the county have been discussed in the introduction to the studies of heathland. The three sites in Leicestershire known popularly as bogs, namely Botcheston Bog, Narborough Bog and Saltby Bog, though they do have a surface accumulation of peaty soil, do not support a particularly calcifuge flora, and should more correctly be classified as marshes.

The majority of marshes occur in valleys at sites where a high water table results either from impeded drainage or from the proximity of a river or stream. Characteristic species of such valley marshes are *Caltha palustris, Juncus inflexus* and *J. effusus, Carex riparia* and *C. acutiformis* and *Deschampsia cespitosa.* Marshes may also form at a spring line on a hill side. There are several examples of such marshes on the Marlstone escarpment in the east and north-east, in which characteristic species are *Equisetum telmateia, Juncus acutiflorus* and *Carex paniculata.* The marsh at The Wailes, Frisby on the Wreake (study 53) has formed at a spring line, but unfortunately an attempt has been made to drain the upper part where the spring emerges, and the lower portion which remains has more of the character of a valley marsh.

Habitat Study 48 Marshy meadow, Newton Burgoland, Swepstone
SK 381090   Area: 3.2 ha (8 acres)   Alt.: 104 m. (340 ft.)
S.S.S.I.
Recorders: J. E. Dawson and P. H. Gamble. Surveyed: July 1979.

This irregularly shaped horse-grazed field is bounded by a roadside hedge on its long western side, the River Sence on its south-east side, and field hedges to the north and south. The southern hedge is tall with numerous gaps allowing access to a similar but smaller meadow, and there is a wide ditch running alongside it which falls within the study area. At the time of the survey an appreciable area near the roadside gate in the north-west of the field had had hardcore tipped on it, and this area has been further extended since then. The survey excludes those species growing on the tipped area, and also those growing in the hedgerows and along the riverside. The study area stands on Alluvium with a little Boulder Clay in the north-west corner. Two samples of the humus-rich soil had pH 5.5.

Locally abundant: Carex riparia, Cirsium arvense, Glyceria maxima.
Frequent: Alopecurus pratensis, Caltha palustris, Cardamine pratensis, Carex disticha, C. hirta, Centaurea nigra, Cerastium fontanum subsp. triviale, Cirsium palustre, Dactylorhiza majalis subsp. praetermissa, Equisetum fluviatile, Filipendula ulmaria, Galium palustre, Holcus lanatus, Juncus effusus, J. inflexus, Lathyrus pratensis, Plantago lanceolata, Polygonum amphibium, Ranunculus acris, R. ficaria, Rumex acetosa, Sanguisorba officinalis, Trifolium pratense.
Locally frequent: Agrostis stolonifera, Alopecurus geniculatus, Callitriche stagnalis, Carex acuta, C. acutiformis, C. nigra, C. panicea, Cirsium dissectum, Dactylorhiza fuchsii, D. fuchsii × majalis, Eleocharis palustris, Epilobium hirsutum, E. parviflorum, Equisetum arvense, E. palustre, Festuca pratensis, Glyceria fluitans, Juncus articulatus, J. bufonius, Leontodon autumnalis, Lotus uliginosus, Mentha aquatica, Molinia caerulea, Oenanthe fistulosa, Phalaris arundinacea, Poa pratensis, P. trivialis, Polygonum persicaria, Prunella vulgaris, Ranunculus repens, Rumex conglomeratus, R. crispus, R. sanguineus, Scutellaria galericulata, Senecio aquaticus, Stachys sylvatica, Stellaria graminea, Trifolium repens, Valeriana dioica, Veronica beccabunga, V. catenata, Vicia cracca.
Occasional: Achillea ptarmica, Angelica sylvestris, Anthoxanthum odoratum, Briza media, Cirsium vulgare, Conopodium majus, Cynosurus cristatus, Deschampsia cespitosa, Epilobium adenocaulon, E. palustre, Festuca rubra, Hypericum tetrapterum, Juncus acutiflorus, J. conglomeratus, Lotus corniculatus, Myosotis laxa subsp. caespitosa, M. scorpioides, Poa annua, Potentilla erecta, P. reptans, Ranunculus sceleratus, Rumex obtusifolius, Serratula tinctoria, Silaum silaus, Solanum dulcamara, Stachys officinalis, Succisa pratensis, Taraxacum spp., Thalictrum flavum, Valeriana officinalis.
Rare: Ajuga reptans, Barbarea vulgaris, Scirpus setaceus, Scrophularia auriculata.

Habitat Study 49   Marsh, Peckleton Fields, Peckleton
SK 460020   Area: 0.9 ha (2.2 acres)   Alt.: 120 m. (390 ft.)
Recorder: H. Handley. Surveyed: 11th April, 20th May and 4th July 1976.

This marsh is rectangular in shape, approximately 125 m. in length from north-east to south-west and 75 m. wide. The site is divided down its length by a stream and hedgerow. The part on the south-east side is ungrazed and tussocky in contrast to the north-west part which has been grazed. A public footpath from Peckleton to Newbold Verdon crosses the study area from south-east to north-west. The site stands partly on Glacial Sand and Gravel and partly on Boulder Clay. Two samples of soil had pH 5.0 and 5.6.

Abundant: Angelica sylvestris, Apium nodiflorum (Wd), Dactylis glomerata, Dactylorhiza fuchsii, Epilobium hirsutum, Filipendula ulmaria, Juncus inflexus, Poa trivialis, Ranunculus repens, Stellaria alsine, Urtica dioica, Valeriana officinalis.
Frequent: Agrostis stolonifera, Alopecurus pratensis, Anthoxanthum odoratum, Anthriscus sylvestris, Arrhenatherum elatius, Caltha palustris, Cardamine flexuosa, Carduus acanthoides, Carex disticha, C. hirta, C. nigra, Cerastium fontanum subsp. triviale, Cirsium arvense, C. vulgare, Convolvulus arvensis, Deschampsia cespitosa, Elymus repens, Epilobium parviflorum, Equisetum arvense, E. fluviatile (Wd), E. palustre, Festuca rubra, Galium aparine, G. palustre, G. uliginosum, Hedera helix, Holcus lanatus, Hydrocotyle vulgaris, Juncus articulatus, J. bufonius, Lamium album, Lathyrus pratensis, Lemna minor, Ligustrum vulgare (Eh), Lolium multiflorum, Lotus uliginosus, Lychnis flos-cuculi, Mentha aquatica, Phleum pratense subsp. bertolonii, P. pratense subsp. pratense, Plantago lanceolata, Poa annua, Potentilla reptans, Ranunculus ficaria, Rosa arvensis (Eh), Rumex acetosa, R. crispus, Silene dioica, Stachys sylvatica, Stellaria graminea, S. holostea, S. media, Tamus communis (Eh), Veronica beccabunga, Viburnum opulus (Eh), Vicia cracca.
Occasional: Acer campestre (Eh), Arctium minus, Cardamine pratensis, Carex ovalis, C. paniculata, Centaurea debauxii subsp. nemoralis, Corylus avellana (Eh), Crataegus laevigata (Eh), Epilobium obscurum, Filaginella uliginosa, Galeopsis tetrahit, Geranium dissectum, Heracleum sphondylium, Hypericum humifusum, H. tetrapterum, Ilex aquifolium (Eh), Juncus conglomeratus, Mentha arvensis, Polygonum persicaria, Potentilla anserina, Ranunculus acris, Rubus fruticosus agg., Rumex conglomeratus, R. obtusifolius, Salix cinerea subsp. oleifolia (Eh), Sambucus nigra (Eh), Scrophularia nodosa, Sinapis arvensis, Solanum dulcamara, Sonchus asper, Taraxacum section Hamata, Ulmus glabra (Eh), Vicia sepium.
Rare: Barbarea vulgaris, Dryopteris dilatata (Eh), Fraxinus excelsior (Eh), Leucanthemum vulgare, Salix caprea.

Habitat Study 50   Botcheston Bog, Desford
SK 485047   Area: 0.8 ha (2.0 acres)   Alt.: 91 m. (300 ft.)
S.S.S.I.
Recorders: S. H. Bishop, O. H. Black, P. H. Gamble, H. Handley and S. D. Musgrove. Surveyed: 1975, 1976 and 1977.

The area surveyed consists of marshland with a small alder grove and is roughly in the shape of a triangle with its base at the south end of the site. It is bounded on the south by about 80 metres of a stream which formerly served Desford Mill; a hedge forms the western boundary and a drain separates the study area from the larger part of the meadow to the east. Though most of the rare Leicestershire species recorded from this site in the past still survive they are in the main less frequent. *Gymnadenia conopsea* was last seen here in 1965, but heavy grazing and trampling during the period of the survey may have been responsible for this and a few other species being overlooked. This is now the only known site in the county for *Epipactis palustris* and *Parnassia palustris*, though they do not flower every year. The mineral soil is Alluvium overlying Keuper Marl, but because of the high water table and impeded drainage, a thin peat deposit has formed with pH 6.95.

Abundant: Anthoxanthum odoratum, Luzula campestris, Ranunculus bulbosus.

Locally abundant: Agrostis stolonifera, Juncus inflexus, Lysimachia nummularia.

Frequent: Agrostis capillaris, Alopecurus pratensis, Angelica sylvestris, Caltha palustris, Cardamine flexuosa, C. pratensis, Carex acutiformis, C. disticha, Centaurea nigra, Cerastium fontanum subsp. triviale, Cirsium palustre, Cynosurus cristatus, Dactylorhiza majalis subsp. praetermissa, Deschampsia cespitosa, Epilobium parviflorum, Filipendula ulmaria, Glyceria plicata, Holcus lanatus, Hypericum tetrapterum, Juncus articulatus, J. effusus, Lathyrus pratensis, Leontodon autumnalis, Lotus uliginosus, Lychnis flos-cuculi, Mentha aquatica, Myosotis laxa subsp. caespitosa, Myosoton aquaticum, Primula veris, Prunella vulgaris, Ranunculus ficaria, R. flammula, Rumex acetosa, Sanguisorba officinalis, Senecio aquaticus, Stellaria media, Succisa pratensis, Valeriana dioica, Vicia cracca.

Locally frequent: Alnus glutinosa, Apium nodiflorum, Bellis perennis, Berula erecta, Carex panicea, Galium palustre, G. uliginosum, Hordeum secalinum, Hydrocotyle vulgaris, Juncus bufonius, Lotus corniculatus, Myosotis scorpioides, Parnassia palustris, Poa annua, P. trivialis, Pulicaria dysenterica, Ranunculus acris, Rumex conglomeratus, Scirpus setaceus, Trifolium pratense, Trisetum flavescens, Veronica beccabunga.

Occasional: Achillea millefolium, A. ptarmica, Ajuga reptans, Alchemilla filicaulis subsp. vestita, Alopecurus geniculatus, Anagallis tenella, Arrhenatherum elatius, Briza media, Carduus acanthoides, Carex dioica, C. hirta, C. nigra, C. ovalis, Cirsium arvense, Crataegus monogyna, Dactylis glomerata, Eleocharis palustris, Equisetum fluviatile, E. palustre, Festuca pratensis, Glyceria declinata, Hypochoeris radicata, Juncus acutiflorus, J. bulbosus, J. conglomeratus, Lapsana communis, Leontodon hispidus, Leucanthemum vulgare, Molinea caerulea, Oenanthe fistulosa, Phleum pratense subsp. pratense, Plantago lanceolata, P. major, Poa pratensis, Polygonum hydropiper, Potentilla anserina, P. reptans, Ranunculus repens, Rumex crispus, R. sanguineus, Scrophularia auriculata, Silaum silaus, Stachys officinalis, Stellaria alsine, Taraxacum section Hamata, T. nordstedtii, T. spectabile, Trifolium repens, Triglochin palustris,

Valeriana officinalis, Veronica serpyllifolia.

Rare: Carex flacca, Cirsium vulgare, Dactylorhiza fuchsii, Epipactis palustris, Equisetum arvense, Epilobium hirsutum, Eriophorum angustifolium, Festuca arundinacea, F. rubra, Galium aparine, Geranium dissectum, Glyceria fluitans, Hedera helix, Heracleum sphondylium, Holcus mollis, Lolium perenne, Menyanthes trifoliata, Myosotis arvensis, Polygonum persicaria, Prunus spinosa, Quercus robur (seedling), Rosa canina, Rumex obtusifolius, Rubus ulmifolius, Solanum dulcamara, Sonchus asper, Stellaria graminea, Taraxacum haematicum, T. subundulatum, Tragopogon pratensis, Urtica dioica, Viburnum opulus.

Habitat Study 51 Marsh, near Lea Wood, Ulverscroft SK 505114 Area: 2.2 ha (5.4 acres) Alt.: 137-145 m. (450-475 ft.) S.S.S.I.
Recorder: P. H. Gamble. Surveyed: 12th July 1972.

This study is of a mainly marshy corner of a field of heath grassland lying to the east of Lea Wood and adjacent to the Ulverscroft Brook. There is a little dry ground and much ground in which the water table is at or above the surface throughout the summer months. It is grazed and trampled by cattle. It is situated on Keuper Marl and Boulder Clay. Samples of surface soil from four different parts of the study area varied from pH 4.1 on the drier ground to pH 6.2 on the wet ground.

Abundant: Agrostis canina, Anthoxanthum odoratum, Cardamine pratensis, Carex disticha, C. panicea, Cynosurus cristatus, Deschampsia cespitosa, Holcus lanatus, Juncus effusus, J. conglomeratus, Lychnis flos-cuculi, Lysimachia nummularia, Mentha aquatica, Potentilla erecta, Ranunculus flammula, Stachys officinalis, Stellaria graminea, Succisa pratensis.

Locally abundant: Carex hirta, Filipendula ulmaria, Galium saxatile, Glyceria fluitans, Veronica beccabunga.

Frequent: Caltha palustris, Carex nigra, Cerastium fontanum subsp. triviale, Cirsium arvense, C. palustre, Epilobium parviflorum, Equisetum arvense, Juncus acutiflorus, J. articulatus, Myosotis laxa subsp. caespitosa, Poa trivialis, Rumex acetosa, Sanguisorba officinalis, Senecio aquaticus.

Locally frequent: Alopecurus geniculatus, Angelica sylvestris, Apium nodiflorum, Callitriche stagnalis, Carex demissa, C. echinata, C. hostiana, C. ovalis, Dactylorhiza fuchsii, D. maculata subsp. maculata, Eleocharis palustris, Epilobium hirsutum, E. palustre, Festuca pratensis, F. rubra, Galium uliginosum, Juncus inflexus, Lathyrus montanus, Lemna minor, Lotus uliginosus, Luzula multiflora, Molinia caerulea, Nardus stricta, Stellaria alsine, Taraxacum faeroense, Trifolium pratense, T. repens, Triglochin palustris, Valeriana dioica, V. officinalis.

Occasional: Alnus glutinosa (seedlings), Anagallis tenella, Betula sp. (seedlings), Briza media, Carex pallescens, C. pulicaris, C. remota, Danthonia decumbens, Hypericum tetrapterum, Lathyrus pratensis, Pedicularis sylvatica, Salix caprea (seedlings), Scirpus setaceus, Veronica chamaedrys, V. scutellata.

Rare: Alchemilla filicaulis subsp. vestita, Scirpus sylvaticus.

Habitat Study 52   Marsh, Thurlaston
SP 509975   Area: 0.8 ha (2.2 acres)   Alt.: 69 m. (225 ft.)
Recorder: S. H. Bishop. Surveyed: 24th May and 16th June 1976.
This study is of a marshy corner of a neutral grassland field situated in the angle made by the junction of the Feeding Brook with the Thurlaston Brook. There is a little dry grassland and much waterlogged ground. It is bounded by high hedges to the north and east and by the Thurlaston Brook to the south; it is open to the west. A large surface pipeline crosses the area. The field stands on Alluvium.

Abundant: Alopecurus pratensis, Anthoxanthum odoratum, Cynosurus cristatus, Festuca pratensis, Holcus lanatus, Lolium perenne, Poa pratensis, Ranunculus acris.

Locally abundant: Agrostis capillaris, Carex nigra, Glyceria fluitans, Juncus acutiflorus, J. articulatus, Luzula campestris.

Frequent: Bellis perennis, Cerastium fontanum subsp. triviale, Dactylis glomerata, Deschampsia cespitosa, Juncus inflexus, Plantago lanceolata, Ranunculus repens, Rumex acetosa, Sanguisorba officinalis, Taraxacum spp., Trifolium repens.

Locally frequent: Alopecurus geniculatus, Caltha palustris, Cardamine pratensis, Carex hirta, Centaurea nigra, Cirsium palustre, Conopodium majus, Filipendula ulmaria, Glyceria plicata, Hypochoeris radicata, Lychnis flos-cuculi, Poa annua, Scirpus setaceus, Senecio aquaticus, Stachys officinalis, Succisa pratensis, Trifolium pratense, Veronica chamaedrys.

Occasional: Carex disticha, C. ovalis, Cirsium arvense, Juncus conglomeratus, Lathyrus pratensis, Phleum pratense subsp. pratense, Potentilla erecta, Rhinanthus minor, Stellaria graminea, Triglochin palustris, Veronica serpyllifolia.

Rare: Achillea ptarmica, Alchemilla filicaulis subsp. vestita, Carex acuta, Dactylorhiza majalis subsp. praetermissa, Juncus effusus, Phragmites australis, Plantago major, Potentilla anserina, Primula veris, Ranunculus flammula, Saxifraga granulata, Serratula tinctoria, Silaum silaus, Urtica dioica, Valeriana dioica, Veronica beccabunga.

Habitat Study 53   Marsh, The Wailes, Frisby on the Wreake
SK 687173   Area: 0.8 ha (2.1 acres)   Alt.: 61 m. (200 ft.)
S.S.S.I.; Nature Reserve.
Recorders: S. H. Bishop, P. H. Gamble and A. L. Primavesi. Surveyed: 1979.
The study area is the marsh at the bottom of the steeply sloping field north of the Rotherby Lane. The marsh is on Boulder Clay, and is situated just below a spring line where Glacial Sand lies on top of the clay.

Abundant: Juncus inflexus, Lotus uliginosus.

Locally abundant: Carex paniculata, Epilobium hirsutum, Juncus subnodulosus, Petasites hybridus, Ranunculus acris.

Frequent: Holcus lanatus, Juncus effusus, Myosotis laxa subsp. caespitosa, Rumex conglomeratus.

Locally frequent: Agrostis capillaris, Berula erecta, Cardamine pratensis, Carex acutiformis, C. riparia, Cerastium fontanum subsp. triviale, Cirsium arvense, C. palustre, Cynosurus cristatus, Epilobium parviflorum, Equisetum fluviatile, Glyceria fluitans, G. maxima, Juncus articulatus, Lemna minor, Lychnis flos-cuculi, Mentha aquatica, Phalaris arundinacea, Polygonum amphibium, Prunella vulgaris, Ranunculus ficaria, R. repens, Scrophularia auriculata, Stellaria alsine, Taraxacum spp., Trifolium repens, Triglochin palustris.

Occasional: Ajuga reptans, Anthoxanthum odoratum, Apium nodiflorum, Caltha palustris, Cardamine amara, C. flexuosa, Carex hirta, Equisetum palustre, Galium palustre, G. uliginosum, Heracleum sphondylium, Hypericum tetrapterum, Lathyrus pratensis, Leontodon autumnalis, Ranunculus flammula, R. sceleratus, Rumex acetosa, Trifolium pratense, Valeriana dioica, Veronica beccabunga.

Rare: Angelica sylvestris, Carex nigra, Crataegus monogyna, Epilobium angustifolium, Equisetum arvense, Festuca arundinacea, Filipendula ulmaria, Juncus acutiflorus, J. bufonius, J. conglomeratus, Lycopus europaeus, Lysimachia nummularia, Myosotis scorpioides, Phleum pratense subsp. bertolonii, Polygonum aviculare, Rosa canina, Rumex crispus, R. obtusifolius, Sagina procumbens, Salix fragilis, Scutellaria galericulata, Solanum dulcamara, Stachys sylvatica, Urtica dioica.

Habitat Study 54   Saltby Bog, Sproxton
SK 841253   Area: 3.1 ha (7.7 Acres)   Alt.: 128 m. (420 ft.)
Recorders: S. H. Bishop, P. H. Gamble and A. L. Primavesi. Surveyed: 29th May and 19th June 1978.
Lying in a hollow about 1 km to the south-west of Saltby village, at the point of intersection of two deep dykes, the study area consists of marshland partly colonised by scrub. At present it is bounded to the north and south by arable fields and grass ley and to the east and west by more marshland and a good deal of scrub. Because of the lack of grazing during recent years and the deepening of the dykes, much of this area, formerly one of the most important botanical sites in the county, has become dominated by scrub and coarse vegetation, resulting in the loss of many of the rare species recorded here in the 1933 Flora. *Pinguicula vulgaris, Menyanthes trifoliata, Pedicularis palustris, Parnassia palustris, Genista tinctoria* and *Schoenus nigricans* formerly occurred here but were not recorded during the present survey, though the last two species have been seen in recent years. This area of peaty soil supports species typical of fen and bog, suggesting the presence of both alkaline and acidic peat. The site is situated at the junction of Northampton Sands and Upper Lias.

Locally abundant: Epilobium hirsutum, Equisetum arvense, Filipendula ulmaria, Galium uliginosum, Juncus effusus, Valeriana dioica.

Frequent: Angelica sylvestris, Arrhenatherum elatius, Centaurea nigra, Dactylis glomerata, Deschampsia cespitosa, Epilobium parviflorum, Festuca pratensis, Heracleum sphondylium, Lathyrus pratensis, Mentha aquatica, Urtica dioica, Vicia cracca.

Locally frequent: Achillea millefolium, Brachypodium pinnatum, Cardamine pratensis, Cirsium palustre, Crataegus monogyna, Epilobium angustifolium, Galium mollugo, Holcus lanatus, Juncus inflexus,

Moehringia trinervia, Molinea caerulea, Potentilla anserina, Primula veris, Ranunculus ficaria, R. repens, Rubus caesius, Rumex acetosa, Sanguisorba officinalis, Sonchus arvensis, Stachys sylvatica, Trifolium medium, Veronica beccabunga, V. chamaedrys.

Occasional: Ajuga reptans, Alopecurus pratensis, Anthoxanthum odoratum, Anthriscus sylvestris, Apium nodiflorum, Bromus sterilis, Carex flacca, C. lepidocarpa, Cirsium arvense, Conopodium majus, Dryopteris dilatata, Fraxinus excelsior, Galium verum, Hydrocotyle vulgaris, Hypericum tetrapterum, Ligustrum vulgare, Linaria vulgaris, Luzula campestris, Lychnis flos-cuculi, Malus sylvestris, Nasturtium officinale, Ophioglossum vulgatum, Poa pratensis, P. trivialis, Potentilla erecta, P. reptans, Prunus spinosa, Rhamnus catharticus, Rosa canina, Rumex crispus, Salix aurita × cinerea, S. cinerea, Solanum dulcamara, Taraxacum spp., Torilis japonica.

Rare: Achillea ptarmica, Alliaria petiolata, Caltha palustris, Carex hirta, C. hostiana, C. nigra, C. panicea, Cerastium fontanum subsp. triviale, Cirsium vulgare, Crataegus laevigata, Dactylorhiza fuchsii, Galium aparine, Geranium robertianum, Hedera helix, Lamium album, Listera ovata, Lonicera periclymenum, Myosotis laxa subsp. caespitosa, Primula vulgaris, Quercus robur, Ranunculus acris, Salix triandra, Sambucus nigra, Senecio erucifolius, Serratula tinctoria, Silaum silaus, Silene dioica, Sparganium erectum, Stachys officinalis, Tamus communis, Tragopogon pratensis, Trifolium pratense, Tussilago farfara, Typha latifolia, Viola hirta, V. riviniana.

Ponds, Lakes and Reservoirs

The habitats grouped under this heading are virtually all man-made. The only supposedly natural lake in Leicestershire is Groby Pool (study 60), which was managed as a fishpond in medieval times and has a substantial dam along its south-eastern side. The most numerous still water habitats in the county, in excess of 10,000 at one time, are the smallest, field ponds. Many were created for the watering of stock at the time of enclosure and they are exemplified by those at Sileby (study 62) and near Combs Plantation (study 65).

There are about a hundred larger lakes or pools, some created as monastic stewponds, others to beautify the landscape, of which that at Coleorton (study 55) serves as an example. The construction of these involved the damming of a watercourse and even where the dam has since disappeared they may retain their botanical interest, as in the case of that at Grace Dieu (study 56).

In the valleys of the Wreake and Soar are a number of artificial ox-bows cut off by the 18th century navigators, of which that at Frisby on the Wreake (study 64) is a good example. Products of more recent technology are the subsidence ponds resulting from mining, as at Bagworth (study 57).

The largest water bodies in the county are the reservoirs. Oldest are the three created during the relatively brief period of 'canal mania' about 1800, of which Saddington Reservoir (study 63) is a good example. The most notable from a botanical point of view, that at Swains Park, Moira, which fed the Ashby Canal and was the last remaining locality in Leicestershire for *Potentilla palustris*, has unfortunately been destroyed by opencast mining. In the last 140 years seven more large reservoirs have been constructed in or on the periphery of the county to provide water for domestic and industrial purposes. The earliest, as for example that at Thornton (study 58), relied on local catchments for their water, but increasingly they store water pumped from elsewhere. Swithland Reservoir (study 61) now functions as a pumped storage reservoir as does the recently constructed one at Staunton Harold, which takes water from the River Dove. The water supply reservoirs have a more marked fluctuation in level than do the canal feeders and their drawdown zones can have an interesting ephemeral flora, which in the drought summer of 1976 extended right across the bottom of several.

Habitat Study 55 Fishpond, Coleorton
SK 399171 Area: 1.2 ha (3.1 acres) Alt.: 116 m. (380 ft.)
Recorder: S. H. Bishop. Surveyed: 17th August 1979.
This medium-sized fishpond lies within an area of wet woodland on the north side of the Ashby Road between Church Town and Coleorton. A little ornamental planting has been done beside the southern and western edges of the pond next to the gardens of Rose Cottage. The site stands on Boulder Clay.

AQUATIC VEGETATION
Abundant: Nuphar lutea.
Locally abundant: Lemna minor.
Locally frequent: Polygonum amphibium.

RIPARIAN VEGETATION
Abundant: Mentha aquatica, Solanum dulcamara.
Locally abundant: Carex riparia, Phragmites australis, Salix cinerea subsp. oleifolia.
Frequent: Cardamine flexuosa, Deschampsia cespitosa, Juncus effusus, Lycopus europaeus, Myosotis scorpioides, Rumex conglomeratus, Urtica dioica, Veronica beccabunga.
Locally frequent: Alnus glutinosa, Angelica sylvestris, Brachypodium sylvaticum, Carex pseudocyperus, C. rostrata, Epilobium hirsutum, Filipendula ulmaria, Galium palustre, Iris pseudacorus, Ranunculus sceleratus.
Occasional: Alisma plantago-aquatica, Carex remota, Equisetum telmateia, Festuca gigantea, Plantago major, Poa annua, Polygonum hydropiper, Populus canescens, Typha latifolia, Ulmus glabra.
Rare: Acer pseudoplatanus, Bromus ramosus, Cirsium palustre, Dryopteris dilatata, Fraxinus excelsior, Salix fragilis.

Habitat Study 56 Pond, near Holly Hayes Wood, Whitwick, Coalville
SK 444152 Area: 0.4 ha (1 acre) Alt.: 152 m. (500 ft.)
Recorder: S. H. Bishop. Surveyed: 22nd September 1979.
This site is immediately to the south of Holly Hayes Wood. At the time of the survey, it was a fish pond formed by a dam across the Grace Dieu Brook, surrounded on all but its northern side by rough grassland. Since then the dam has been removed and the site of the pond is now rough marshy ground. *Mentha × gentilis* was still present here in 1981. The site is on Keuper Marl.

AQUATIC VEGETATION
Abundant: Apium nodiflorum, Lemna gibba, Nasturtium sp., Potamogeton natans.
Locally abundant: Glyceria fluitans.
Occasional: Elodea canadensis.

RIPARIAN VEGETATION
Locally abundant: Eleocharis palustris.
Frequent: Mentha × gentilis.
Locally frequent: Epilobium hirsutum, Holcus lanatus, Juncus effusus, J. inflexus, Phalaris arundinacea, Polygonum hydropiper, P. persicaria, Veronica beccabunga.
Occasional: Alisma plantago-aquatica, Alopecurus geniculatus, Angelica sylvestris, Cirsium arvense, Galium palustre, Lolium perenne, Myosotis laxa subsp. caespitosa, Plantago lanceolata, Rumex conglomeratus, Stellaria alsine.
Rare: Athyrium filix-femina, Betula pendula, Fraxinus excelsior, Hypericum tetrapterum, Iris pseudacorus, Leontodon autumnalis, Ranunculus acris, Rumex obtusifolius, Salix fragilis, Senecio aquaticus, Solanum dulcamara, Stellaria graminea, Typha latifolia, Tussilago farfara, Urtica dioica.

Habitat Study 57 Mining subsidence pool, Bagworth
SK 450087 Area: 0.9 ha (2.3 acres) Alt.: 137 m. (450 ft.)
Recorder: H. Handley. Surveyed: 31st May and 17th September 1977.
The pool is situated just west of Bagworth Park Farm and is bounded on the south-west side by a railway embankment. It is thought to have been formed during the 1960s on pasture land and is fed from the western side by a small stream. Roughly oval in shape, it is about 250 m. in length and 120 m. at its greatest width. It is partly bisected by a green track which was originally the course of the former Leicester - Swannington Railway. Up to the end of 1976 the two halves of the pool were connected by culvert only but further subsidence early in 1977 flooded about 30 m. of the track and linked the two sections. The site is on Keuper Marl. Two samples of surface soil had pH 5.1 and 6.7.

Abundant: Apium nodiflorum, Ceratophyllum submersum, Nasturtium microphyllum (Ew).
Locally abundant: Myosotis scorpioides (Ew).
Frequent: Eleocharis palustris (Ew), Epilobium hirsutum (Ew), E. adenocaulon (Ew), Glyceria fluitans, Lemna minor, Potamogeton natans, P. pusillus, Ranunculus sceleratus (Ew), Solanum dulcamara (Ew), Typha latifolia.
Occasional: Alisma plantago-aquatica, Juncus articulatus (Ew), J. inflexus (Ew), Polygonum amphibium, Ranunculus aquatilis, Rorippa islandica (Ew), Salix caprea (Ew), Salix cinerea subsp. oleifolia (Ew), S. viminalis (Ew).

Habitat Study 58 Thornton Reservoir, Bagworth
SK 477077 Area: 0.09 ha (0.22 acres) Alt.: 135 m. (440 ft.)
Recorder: O. H. Black. Surveyed: 30th March, 10th May and 1st July 1975.
Situated on the eastern margin of the reservoir 150 m. south-west of Retreat Farm and bounded on the eastern side by a hedgerow and track, the study area is 50 m. in length. The north-east and south-west limits are marked by two parallel field hedgerows which join the track at right angles from the east. The width was 18 m. from 1m. above high water mark down to the then rather low water level. The site lies on Keuper Marl. A sample of surface soil at the high water mark had pH 7.2.

Abundant: Glyceria maxima, Poa trivialis, Ranunculus aquatilis, Rorippa islandica.
Frequent: Agrostis capillaris, Alopecurus aequalis, Arrhenatherum elatius, Atriplex patula, Butomus umbellatus, Capsella bursa-pastoris, Carex acuta, C. hirta, Chamomilla suaveolens, Chenopodium rubrum, Cirsium arvense, Dactylis glomerata, Elymus repens, Epilobium adenocaulon, E. hirsutum, Equisetum arvense, E. fluviatile, E. palustre,

Filaginella uliginosa, Galium aparine, Lapsana communis, Mentha aquatica, Myosoton aquaticum, Juncus compressus, Phalaris arundinacea, Poa annua, Polygonum amphibium, P. aviculare, Ranunculus repens, R. sardous, Rumex crispus, R. obtusifolius, Senecio vulgaris, Stellaria media.
Locally frequent: Eleocharis palustris.
Occasional: Anthriscus sylvestris, Atriplex hastata, Barbarea vulgaris, Bidens tripartita, Calystegia silvatica, Chamomilla recutita, Chenopodium polyspermum, Cirsium vulgare, Convolvulus arvensis, Epilobium obscurum, Galium palustre, Heracleum sphondylium, Holcus lanatus, Juncus articulatus, J. bufonius, J. effusus, Leontodon autumnalis, Matricaria perforata, Myosotis laxa subsp. caespitosa, Plantago major, Polygonum arenastrum, P. lapathifolium, P. persicaria, Potentilla reptans, Ranunculus bulbosus, R. ficaria, R. sceleratus, Rorippa amphibia, Rumex maritimus, Senecio aquaticus, Solanum dulcamara, Sonchus asper, Stachys sylvatica, Stellaria alsine.
Rare: Senecio squalidus.

Habitat Study 59 Pond and drainage channel, Beacon Hill, Woodhouse
SK 519144 Area: not calculated. Alt.: 152 m. (500 ft.)
Recorder: P. H. Gamble. Surveyed: 12th July 1976.
Located about 1000 m. to the east of the summit of Beacon Hill, the pond is in a mature birch wood with open canopy, and has bog mosses including *Sphagnum*. The pond is 18 m. × 14 m. and the drainage channel extends for some 350 m. The channel leaves the pond and flows eastwards for about 90 m., then northwards along the eastern edge of the birch wood until it joins the drainage ditch which runs along the northern edge of the heath. The planting of alien species such as *Rhododendron* and sycamore, and recent management, including deepening of the channel have had a deleterious effect on the site. Keuper Marl overlying Beacon Hill Hornstones and Grits has contributed to a heavy poorly-drained soil with a localised development of peat in parts of the study area. Samples of surface soil from the pond margin had pH 4.5 and from within the drainage channel pH 3.8.

Abundant: Pteridium aquilinum.
Locally abundant: Hydrocotyle vulgaris, Juncus bulbosus (Wp), Potamogeton polygonifolius (Wp).
Frequent: Anthoxanthum odoratum, Betula pendula, Cirsium palustre, Deschampsia flexuosa, Holcus mollis, Juncus effusus, Rubus fruticosus agg.
Locally frequent: Agrostis canina, A. capillaris, Glyceria fluitans (Wp), Juncus acutiflorus, J. conglomeratus, Lotus uliginosus, Potentilla erecta, Ranunculus flammula, Scutellaria minor, Viola palustris.
Occasional: Acer pseudoplatanus, Epilobium angustifolium, Galium saxatile, Holcus lanatus, Molinia caerulea, Nardus stricta, Plantago major, Quercus robur, Teucrium scorodonia, Ulex europaeus.
Rare: Blechnum spicant, Calluna vulgaris, Digitalis purpurea, Erica tetralix, Salix cinerea subsp. cinerea.

Habitat Study 60 Groby Pool, Groby
SK 521082 Area: 4.2 ha (10.5 acres) Alt.: 96 m. (315 ft.) S.S.S.I.
Recorders: S. H. Bishop and E. Hesselgreaves. Surveyed: 17th June and 5th August 1976.
The study covers the eastern part of the pool and includes the whole of the eastern shoreline and a length of the southern shoreline about 150 m. long. The majority of the shoreline studied is open to the road and is bounded by sandy grassland with rock outcrops. It is heavily trampled throughout the year by sightseers. Groby Pool lies in a hollow in the Keuper Marl and diorite.

AQUATIC VEGETATION
Abundant: Potamogeton pectinatus.
Locally abundant: Nuphar lutea, Nymphoides peltata.
Frequent: Polygonum amphibium.
Locally frequent: Lemna minor.

RIPARIAN VEGETATION
Locally abundant: Filipendula ulmaria, Lycopus europaeus, Mentha aquatica, Typha angustifolia.
Frequent: Arrhenatherum elatius, Cirsium arvense, Epilobium hirsutum, Iris pseudacorus, Juncus effusus, J. inflexus, Myosotis scorpioides, Phalaris arundinacea, Plantago lanceolata, Poa annua, P. trivialis, Rorippa amphibia, Rumex conglomeratus, Trifolium repens, Urtica dioica.
Locally frequent: Agrostis capillaris, Carex disticha, C. hirta, Cynosurus cristatus,. Eleocharis palustris, Festuca rubra, Glyceria maxima, Holcus lanatus, Juncus articulatus, J. bufonius, J. compressus, Lolium perenne, Nasturtium sp., Plantago major, Poa pratensis, Polygonum hydropiper, Potentilla anserina, Ranunculus acris, R. repens, Rorippa islandica, Sagina procumbens, Scutellaria galericulata.
Occasional: Alopecurus pratensis, Bellis perennis, Bidens tripartita, Carex acuta, C. ovalis, Crataegus monogyna, Dactylis glomerata, Epilobium angustifolium, E. parviflorum, Galium palustre, Glyceria plicata, Heracleum sphondylium, Luzula campestris, Mentha × verticillata, Oenanthe fistulosa, Potentilla reptans, Ranunculus sceleratus, Rumex acetosa, R. obtusifolius, Senecio jacobaea, Solanum dulcamara, Stachys sylvatica, Stellaria graminea, Tussilago farfara.
Rare: Acer pseudoplatanus, Alnus glutinosa, Alopecurus geniculatus, Angelica sylvestris, Anthoxanthum odoratum, Cardamine pratensis, Carex remota, C. riparia, Centaurea nigra, Cerastium fontanum subsp. triviale, Chamomilla suaveolens, Cirsium vulgare, Deschampsia cespitosa, Fraxinus excelsior, Galium aparine, Hypericum tetrapterum, Lathyrus pratensis, Leontodon autumnalis, Polygonum persicaria, Potentilla erecta, Prunus spinosa, Ranunculus flammula, Rosa canina, Rubus caesius, R. fruticosus agg., R. idaeus, Salix alba, S. cinerea subsp. oleifolia, S. fragilis, Sambucus nigra, Senecio squalidus, S. vulgaris, Sonchus asper, Sparganium erectum.

Habitat Study 61 Swithland Reservoir, Rothley
SK 562142 Area: 1.6 ha (4 acres) Alt.: 58 m. (190 ft.)
S.S.S.I.
Recorder: P. H. Gamble. Surveyed: 1976.
The study area includes approximately 250 m. length of margin and shore to a maximum of 100 m. out from the edge (during a period of exceptional drought). It is situated in a north-east bay of the reservoir between Buddon Wood and Kinchley Hill, each side of a point where a small stream enters the reservoir. The area

includes swamp, sandy shore and part of the exposed muddy bed. Swithland Reservoir was completed in 1896. It lies on Keuper Marl, and a sample of surface soil from the sandy shore had pH 6.9.

Locally abundant: Atriplex hastata, Carex acuta, C. acutiformis, Chenopodium rubrum, Eleocharis acicularis, Epilobium adenocaulon, E. montanum, Filaginella uliginosa, Juncus bufonius, Litorella uniflora, Mentha aquatica, Phalaris arundinacea, Polygonum lapathifolium, Potentilla anserina, P. reptans, Ranunculus sceleratus, Rorippa islandica, Rumex maritimus, Salix cinerea, Scirpus lacustris subsp. lacustris, Typha angustifolia, T. latifolia, Veronica catenata.
Frequent: Juncus acutiflorus, Mentha × verticillata, Myosotis laxa subsp. caespitosa, Myosoton aquaticum, Plantago major, Poa annua, P. trivialis, Polygonum aviculare, Sagina procumbens, Salix viminalis.
Locally frequent: Carex riparia, Eleocharis palustris, Epilobium angustifolium, E. hirsutum, E. parviflorum, Equisetum arvense, Filipendula ulmaria, Iris pseudacorus, Juncus compressus, Montia fontana, Nasturtium microphyllum, Polygonum amphibium, Ranunculus repens, Rorippa amphibia, Urtica dioica, Veronica beccabunga.
Occasional: Carex hirta, Cerastium fontanum subsp. triviale, Chamomilla recutita, C. suaveolens, Cirsium arvense, C. vulgare, Epilobium roseum, Holcus lanatus, Lythrum salicaria, Sagina apetala, Silene dioica, Solanum dulcamara, Sonchus asper, Stellaria alsine, S. media, Trifolium dubium.
Rare: Galeopsis tetrahit, Populus alba, Ranunculus lingua, Samolus valerandi, Senecio jacobaea, S. squalidus.

Habitat Study 62 Field pond, Sileby
SK 592158 Area: 0.2 ha (0.5 acres) Alt.: 43 m. (140 ft.)
Recorder: P. H. Gamble. Surveyed: 3rd July 1976.
Located in a hay meadow on the flood plain of the River Soar, this large pond (approximately 90 m. × 30 m.) was until recent years two separate ponds lying close together. It is in the north-east corner of the field. A 1 m. high bank, part broken to give a muddy or in places gravelly shore, forms the margin of the pond. It is on Alluvium. A sample of soil from the bank had pH 7.1.

AQUATIC VEGETATION
Locally abundant: Butomus umbellatus, Callitriche stagnalis, Elodea canadensis, Equisetum fluviatile, Glyceria fluitans, G. maxima, Potamogeton natans.
Locally frequent: Alisma plantago-aquatica, Apium nodiflorum, Lemna minor, Nasturtium microphyllum, Polygonum amphibium, Potamogeton crispus, Ranunculus circinatus, R. peltatus, Rorippa amphibia, Spirodela polyrhiza.

RIPARIAN VEGETATION
Locally abundant: Eleocharis palustris, Juncus bufonius, Phalaris arundinacea.
Locally frequent: Alopecurus geniculatus, Epilobium adenocaulon, E. hirsutum, Myosotis laxa subsp. caespitosa, Poa trivialis, Ranunculus acris, R. repens, Rorippa islandica, Rumex sanguineus, Veronica catenata.

Occasional: Cardamine pratensis, Galium palustre, Lysimachia nummularia, Plantago major, Polygonum lapathifolium, Ranunculus sceleratus, Rumex crispus.
Rare: Myosotis scorpioides, Saxifraga granulata.

Habitat Study 63 Saddington Reservoir, Saddington and Gumley
SP 664910 Area: not calculated. Alt.: 107 m. (350 ft.)
Recorders: S. H. Bishop and P. H. Gamble. Surveyed: 11th September 1977.
This reservoir was opened in the 1790s as a feeder to what is now the Grand Union Canal. The area surveyed includes 1000 m. of the eastern and western margins which converge to a blunt point at the north, and adjacent areas of shallow water. At the time of the survey the low water level had exposed a considerable area of muddy shore. The reservoir is located on Lower Lias Clays.

Abundant: Glyceria maxima, Polygonum amphibium, Veronica catenata.
Locally abundant: Carex riparia, Mentha aquatica, Oenanthe aquatica, Phalaris arundinacea, Rorippa islandica.
Frequent: Equisetum fluviatile, Galium palustre, Myosotis scorpioides, Ranunculus aquatilis, Salix fragilis, Solanum dulcamara, Stellaria media.
Locally frequent: Atriplex hastata, Bidens tripartita, Carex acuta, C. acutiformis, Filipendula ulmaria, Myosotis laxa subsp. caespitosa, Myosoton aquaticum, Plantago major, Potamogeton crispus, P. pectinatus, Potentilla anserina, Salix purpurea, S. viminalis, Scirpus lacustris subsp. lacustris, Typha latifolia, Urtica dioica.
Occasional: Alisma lanceolatum, Apium inundatum, Atriplex patula, Callitriche intermedia, C. platycarpa, Calystegia sepium, Cardamine amara, Chenopodium rubrum, Cirsium arvense, C. vulgare, Dactylis glomerata, Eleocharis palustris, Epilobium adenocaulon, E. montanum, Festuca gigantea, Galeopsis tetrahit, Galium aparine, Iris pseudacorus, Nasturtium officinale, Polygonum persicaria, Rumex sanguineus, Sonchus oleraceus.
Rare: Bromus ramosus, Crataegus monogyna, Omalotheca sylvatica, Populus tremula, Potentilla reptans, Ranunculus circinatus, Rubus fruticosus agg., Salix alba, S. triandra.

Habitat Study 64 Ox-bow, near The Wailes, Frisby on the Wreake
SK 686174 Area: not calculated. Alt.: 61 m. (200 ft.)
S.S.S.I.; Nature Reserve.
Recorders: S. H. Bishop and P. H. Gamble. Surveyed: 27th August 1976.
Located south of the River Wreake in cattle grazed grassland, the study area is the remains of a loop of the river cut off by the Wreake Navigation in the 1790s. It comprises two oval, reed-fringed pools linked by a strip of swamp vegetation similar in character to that surrounding the pools, with overall dimensions of about 150 m. × 30 m. It is situated on Alluvium. A sample of the surface soil had pH 6.4.

AQUATIC VEGETATION
Abundant: Carex riparia, Phragmites australis.
Locally abundant: Glyceria maxima, Lemna minor, Nuphar lutea.

Locally frequent: Glyceria fluitans, Sparganium erectum, Veronica beccabunga.
Occasional: Alisma plantago-aquatica, Callitriche stagnalis, Carex acutiformis, Equisetum fluviatile.
Rare: Rorippa amphibia.

RIPARIAN VEGETATION
Locally frequent: Agrostis stolonifera, Cirsium arvense, Galium palustre, Juncus articulatus, J. effusus, Mentha aquatica, Myosotis scorpioides, Plantago major, Polygonum aviculare, Ranunculus repens, Rumex conglomeratus.
Occasional: Bellis perennis, Cardamine pratensis, Epilobium parviflorum, Glechoma hederacea, Juncus acutiflorus, J. inflexus, Lathyrus pratensis, Plantago lanceolata, Ranunculus sceleratus, Trifolium pratense, T. repens, Urtica dioica.
Rare: Arctium lappa, Atriplex hastata, Capsella bursa-pastoris, Carex hirta, C. paniculata, Chamomilla suaveolens, Crataegus monogyna, Dactylis glomerata, Deschampsia cespitosa, Epilobium palustre, Equisetum arvense, Filipendula ulmaria, Hypericum tetrapterum, Leontodon autumnalis, Lycopus europaeus, Lysimachia nummularia, Phleum pratense subsp. bertolonii, P. pratense subsp. pratense, Polygonum hydropiper, P. persicaria, Prunella vulgaris, Rorippa islandica, Rosa canina, Rumex obtusifolius, Salix alba, S. cinerea subsp. cinerea, S. fragilis, Scutellaria galericulata, Solanum dulcamara, Taraxacum spp., Veronica chamaedrys.

Habitat Study 65 Pond, near Combs Plantation, Eaton SK 782301 Area: not calculated. Alt.: 152 m. (500 ft.)
Recorder: S. H. Bishop. Surveyed: 13th June 1976.
This small pond, approx. 20 m. × 10 m. in dimensions, lies in a field between Combs Plantation and the Eastwell to Belvoir Road. It is very shallow with little open water. The pond lies on Boulder Clay.

Abundant: Alopecurus geniculatus, Glyceria fluitans, Mentha aquatica, Sparganium erectum.
Locally abundant: Eleocharis palustris, Festuca pratensis (Ew), Juncus inflexus, Lolium perenne (Ew), Poa pratensis (Ew), Veronica scutellata.
Frequent: Bellis perennis (Ew), Cynosurus cristatus (Ew), Dactylis glomerata (Ew), Equisetum fluviatile, Holcus lanatus (Ew), Myosotis scorpioides, Plantago major (Ew), Poa trivialis (Ew), Ranunculus aquatilis, R. repens, Trifolium repens (Ew), Veronica beccabunga.
Locally frequent: Carex otrubae, Centaurea nigra, Cerastium fontanum subsp. triviale, Cirsium arvense (Ew), Epilobium hirsutum (Ew), Equisetum arvense (Ew), Juncus effusus, Lotus corniculatus (Ew), Ranunculus acris (Ew), R. sceleratus, Rumex acetosa (Ew).
Occasional: Apium nodiflorum, Cardamine pratensis, Carex hirta, Epilobium parviflorum, Potentilla reptans (Ew), Prunella vulgaris (Ew), Ranunculus flammula.
Rare: Ajuga reptans, Alisma plantago-aquatica, Alopecurus pratensis (Ew), Crataegus monogyna (Ew), Dactylorhiza fuchsii (Ew), Hypericum tetrapterum (Ew), Lathyrus pratensis (Ew), Rosa canina (Ew), Silaum silaus (Ew), Solanum dulcamara (Ew), Taraxacum spp. (Ew), Urtica dioica (Ew), Vicia cracca (Ew).

Canals

The main features which distinguish canals as a type of habitat from streams and rivers, even when the latter are canalised, are the lack of any appreciable current and a constant water level. The lack of current accounts for the vast quantities of free-floating aquatics such as *Lemna trisulca* in the Ashby Canal, which follows a contour line and is lockless, and of all three species of *Lemna* and at times of *Azolla* in the Grantham Canal. The constant water level seems to favour such marginal species as *Berula erecta*, *Butomus umbellatus* and *Carex pseudocyperus*, which in Leicestershire are typical canal species.

Three Leicestershire canals are still in water for all or the greater part of their lengths. The Grantham Canal, now disused, is shallow over most of its length, and some parts are overgrown with species such as *Sparganium erectum* and *Glyceria maxima*. It has suffered from dredging in recent years, but parts are still of considerable botanical interest. One such part is the subject of study 69. The Ashby Canal (study 66) is now curtailed just north of Snarestone Tunnel, the remainder having been drained and in places filled during the last forty years as a result of mining subsidence. Characteristic of this canal is the occurrence in quantity on its margins of *Cardamine amara* and *Juncus subnodulosus* and, in the water, of a variety of *Potamogeton* species. The Grand Union Canal is navigable throughout a length of some 20 miles in the south-east of the county, before it joins the Soar Navigation at Aylestone. A part of this length of the canal forms the subject of study 68. *Impatiens capensis* has become well-established on its banks since arrival at some time in the late thirties and also occurs in quantity along the Ashby Canal. There are other short lengths of canal further north of Leicester, where artificial cuts have been made to avoid parts of the River Soar which would be awkward for navigation. One such length at Loughborough is the subject of study 67.

Habitat Study 66 Ashby Canal, Carlton Bridge, Carlton
SK 386044   Area: not calculated.   Alt.: 91 m. (300 ft.)
Recorder: H. Handley. Surveyed: 16th April, 22nd May and 6th July 1976.
The study area consists of 100 metres of canal and includes the riparian areas on both sides. It is situated on the west side of Carlton Bridge. The north bank is mainly silt from recent dredging but the south bank is undisturbed. Some of the species recorded, such as *Eleocharis acicularis,* are amphibious; these are listed under the habitat where they are most abundant. The site is on Keuper Marl. Samples of surface soil from the north bank had pH 7.2 and from the south bank pH 7.3.

AQUATIC VEGETATION
Abundant: Lemna trisulca.
Frequent: Alisma plantago-aquatica, Apium nodiflorum, Butomus umbellatus, Carex riparia, Elodea canadensis, Equisetum fluviatile, Glyceria maxima, Nuphar lutea, Potamogeton berchtoldii, P. friesii, P. natans, P. pectinatus, P. perfoliatus, Rumex hydrolapathum, Sagittaria sagittifolia, Sparganium emersum, Veronica beccabunga.
Occasional: Iris pseudacorus, Oenanthe fistulosa, Potamogeton compressus, Scirpus lacustris subsp. lacustris, Typha latifolia.

RIPARIAN VEGETATION
Abundant: Berula erecta, Cardamine amara, Eleocharis acicularis, Mentha aquatica, Ranunculus sceleratus.
Frequent: Bidens tripartita, Carex otrubae, Eleocharis palustris, Epilobium montanum, E. obscurum, E. parviflorum, Equisetum palustre, Filipendula ulmaria, Galium palustre, Impatiens capensis, Juncus inflexus, J. subnodulosus, Lycopus europaeus, Myosotis scorpioides, Phalaris arundinacea, Pulicaria dysenterica, Ranunculus ficaria, Scrophularia auriculata, Stellaria alsine.
Occasional: Alliaria petiolata, Barbarea vulgaris, Caltha palustris, Cardamine pratensis, Carex flacca, C. hirta, Epilobium hirsutum, Equisetum arvense, Juncus articulatus, J. bufonius, J. effusus, Lysimachia nummularia, Myosotis laxa subsp. caespitosa, Myosoton aquaticum, Plantago lanceolata, Polygonum lapathifolium, P. persicaria, Ranunculus repens, Rumex acetosa, Scutellaria galericulata, Senecio aquaticus, Sparganium erectum.
Rare: Carex pseudocyperus, Lythrum salicaria, Sagina nodosa, Solanum dulcamara, Taraxacum section Hamata, T. pannucium.

Habitat Study 67 Grand Union Canal, Loughborough
SK 529213   Area: not calculated.   Alt.: 36 m. (120 ft.)
Recorder: P. H. Gamble. Surveyed: 28th June 1976.
This study is of 570 m. of the canal from the Swingbridge Road bridge to Bishop Meadow Lock. It includes the aquatic flora and the species growing on the east bank up to the tow path. It lies on Alluvium.

AQUATIC VEGETATION
Abundant: Nuphar lutea.
Locally abundant: Potamogeton pectinatus.
Locally frequent: Acorus calamus, Polygonum amphibium, Rorippa amphibia, Sparganium erectum.
Occasional: Alisma plantago-aquatica, Sagittaria sagittifolia.
Rare: Alisma lanceolatum.

RIPARIAN VEGETATION
Abundant: Arrhenatherum elatius.
Locally abundant: Epilobium hirsutum.
Frequent: Alliaria petiolata, Carex riparia, Dactylis glomerata, Lathyrus pratensis, Leontodon autumnalis, Lythrum salicaria, Plantago lanceolata, Rumex hydrolapathum, Scrophularia auriculata.
Locally frequent: Agrostis stolonifera, Anthriscus sylvestris, Elymus repens, Filipendula ulmaria, Galium aparine, Glyceria maxima, Myosotis scorpioides, Phalaris arundinacea, Potentilla reptans, Silene dioica, Tussilago farfara, Urtica dioica, Valeriana officinalis, Vicia cracca.
Occasional: Achillea millefolium, Agrostis capillaris, Armoracia rusticana, Berula erecta, Bromus hordeaceus, Carex otrubae, Cirsium arvense, C. vulgare, Convolvulus arvensis, Crataegus monogyna, Epilobium adenocaulon, E. parviflorum, Festuca arundinacea, Fraxinus excelsior (seedlings), Galium palustre, Hypochoeris radicata, Juncus inflexus, Lamium album, Lycopus europaeus, Phragmites australis, Rosa sp., Rubus fruticosus agg., Rumex conglomeratus, R. obtusifolius, Sambucus nigra, Scutellaria galericulata, Senecio jacobaea, S. squalidus, Sisymbrium officinale, Solanum dulcamara, Stachys palustris, Taraxacum spp., Tragopogon pratensis.
Rare: Angelica sylvestris, Carex acutiformis, Conium maculatum, Humulus lupulus, Impatiens capensis, Lolium multiflorum, Pulicaria dysenterica, Sonchus asper, S. oleraceus.

Habitat Study 68 Grand Union Canal, Saddington/Smeeton Westerby
SP 672921 to 666925   Area: not calculated.   Alt.:107m. (350 ft.) S.S.S.I.
Recorders: S. H. Bishop and P. H. Gamble. Surveyed: 11th September 1977.
An 850 m. length of canal, lying midway between the villages of Saddington and Smeeton Westerby, from the Saddington - Smeeton Westerby road bridge to the field bridge just to the east of the aqueduct sluice. The survey includes the aquatic flora and the riparian species of the north bank up to the tow path. The study area is situated on Lower Lias Clays.

AQUATIC VEGETATION
Locally abundant: Carex riparia, Sparganium erectum.
Frequent: Carex acutiformis, Nuphar lutea, Potamogeton natans, P. perfoliatus.
Locally frequent: Butomus umbellatus, Eleocharis acicularis, Elodea canadensis, Glyceria maxima, Lemna minor, Potamogeton pectinatus, Sagittaria sagittifolia, Sparganium emersum, Typha latifolia.
Occasional: Alisma plantago-aquatica, Apium nodiflorum, Berula erecta, Callitriche platycarpa, Mentha aquatica, Ranunculus circinatus, Typha angustifolia, Veronica beccabunga.
Rare: Alisma lanceolatum, Oenanthe aquatica.

RIPARIAN VEGETATION
Abundant: Arrhenatherum elatius.
Locally abundant: Filipendula ulmaria.
Frequent: Achillea millefolium, Agrostis stolonifera, Anthriscus sylvestris, Carduus acanthoides, Centaurea nigra, Cirsium arvense, C. vulgare, Dactylis glomerata, Elymus repens, Epilobium parviflorum, Holcus lanatus, Impatiens capensis, Lathyrus pratensis, Plantago lanceolata, P. major, Rubus fruticosus agg., Scrophularia auriculata.
Locally frequent: Barbarea vulgaris, Convolvulus arvensis, Crataegus monogyna, Epilobium hirsutum, Equisetum arvense, Galium palustre, Hypericum tetrapterum, Juncus effusus, Lapsana communis, Lolium perenne, Lycopus europaeus, Medicago lupulina, Myosotis arvensis, M. scorpioides, Potentilla reptans, Prunus spinosa, Rumex hydrolapathum, Scutellaria galericulata, Solanum dulcamara, Sonchus asper, Stachys sylvatica, Taraxacum spp., Torilis japonica, Trifolium repens, Tussilago farfara, Urtica dioica, Veronica chamaedrys.
Occasional: Crepis capillaris, Fraxinus excelsior, Galium aparine, G. verum, Geranium dissectum, G. pratense, Juncus inflexus, Pimpinella saxifraga, Potentilla anserina, Ranunculus repens, R. sceleratus, Rosa canina, Rumex conglomeratus, Senecio erucifolius, S. jacobaea, Trifolium pratense, Vicia sepium.
Rare: Bromus ramosus, Carex paniculata, Cerastium fontanum subsp. triviale, Geum urbanum, Lamium album, Lotus corniculatus, Myosotis laxa subsp. caespitosa, Pimpinella major, Plantago media, Rumex sanguineus, Salix fragilis, Sisymbrium officinale, Vicia cracca.

Habitat Study 69 Grantham Canal, Stathern
SK 756328 Area: not calculated. Alt.: 44 m. (145 ft.)
S.S.S.I.
Recorders: S. H. Bishop and P. H. Gamble. Surveyed: 27th August 1979.
A 900 m. length of canal, lying between Stathern Bridge to the south and the dismantled railway track north-west of Stathern Junction to the north. The survey includes the aquatic species and the riparian species of both banks. The towpath runs along the west bank. Most of the study area stands on Upper Lias Clays and Rhaetic Shales, but a small area of Alluvium occurs in the south-east.

AQUATIC VEGETATION
Abundant: Lemna minor, Sparganium erectum.
Locally abundant: Elodea canadensis, Glyceria maxima, Lemna trisulca, Phragmites australis, Typha latifolia.
Locally frequent: Alisma plantago-aquatica, Berula erecta, Eleocharis palustris, Equisetum fluviatile, Polygonum amphibium, Veronica beccabunga.
Occasional: Butomus umbellatus, Nasturtium officinale.
Rare: Carex pseudocyperus, Glyceria fluitans.

RIPARIAN VEGETATION
Frequent: Alopecurus pratensis, Cirsium arvense, Epilobium hirsutum.
Locally frequent: Equisetum arvense, Juncus articulatus, J. inflexus, Lathyrus pratensis, Myosotis scorpioides, Phalaris arundinacea, Ranunculus repens, R. sceleratus, Scrophularia auriculata, Scutellaria galericulata, Tussilago farfara.
Occasional: Calystegia sepium, Carex otrubae, Dactylis glomerata, Deschampsia cespitosa, Epilobium parviflorum, Filipendula ulmaria, Galium palustre, Juncus effusus, Lysimachia nummularia, Mentha aquatica, Potentilla reptans, Pulicaria dysenterica, Rumex conglomeratus, Solanum dulcamara,. Trifolium pratense, T. repens, Urtica dioica, Vicia cracca.
Rare: Conium maculatum, Epilobium angustifolium, Hordeum secalinum, Lycopus europaeus, Mentha × piperita, M. × verticillata, Oenanthe fistulosa, Polygonum aviculare, P. persicaria, Senecio erucifolius, Sonchus arvensis, Stachys sylvatica.

Rivers and Streams

Enough has been said elsewhere in this book concerning the rivers and streams of Leicestershire to obviate the necessity of repeating it here. Suffice it to say that in the following studies examples have been given of the main types of river in the county, and one example of a Charnwood Forest stream.

The western River Sence (study 70), provides a good example of the clean and fast-moving rivers in the west of the county, a characteristic feature of which is the frequency of *Ranunculus fluitans*. Three examples are given of the River Soar. At Croft (study 73) the Soar is still a comparatively small and clean river. At the time when the study was made, a characteristic feature of this stretch of the river was *Oenanthe fluviatilis* in its only Leicestershire station. Unfortunately the upper reaches of the Soar have recently been 'improved', with considerable detriment to the flora. Study 74, of the Soar at Belgrave, shows the river as a navigable waterway in an urban area, where however the quality and variety of the aquatic and riparian vegetation is pleasantly surprising. The third study (75) is of a part of the lower reaches of the river bypassed for navigation by an artificial cut.

The structural, floristic and faunistic diversity of an 8km stretch of the River Eye above Melton Mowbray is such that it has recently been assessed as being of national importance from a conservation point of view. Part of the stretch concerned, near Ham Bridge, forms the subject of study 77. A part of the River Wreake, the lower reaches of the same river under a different name, is the subject of study 76. Finally, the Ulverscroft Brook (study 72) provides an example of a Charnwood Forest stream, and illustrates how poor these streams are in aquatic vegetation, most of the species in the list being recorded from the banks.

In these studies, as also in those of canals, an attempt has been made to distinguish the aquatic and riparian vegetation. However it should be pointed out that some species occur in both habitats, and the decision as to which category they should be placed in is frequently subjective.

Habitat Study 70 River Sence, Ratcliffe Culey, Sheepy/Witherley
SP 325998 to 321996   Area: not calculated.   Alt.: 69 m. (225 ft.)
Recorder: S. H. Bishop. Surveyed: 13th August 1976.
The part of the river surveyed extends to about 800 m. from a point east of Ratcliffe Bridge upstream to the confluence with the Sibson Brook. The river meanders extensively, with small cliffs and islands. The flow is medium fast with a gravel and silt bottom and an extensive riffle and pool structure. The water had pH 7.75.

AQUATIC VEGETATION
Abundant: Lemna gibba.
Locally abundant: Potamogeton pectinatus, Ranunculus fluitans.
Frequent: Elodea canadensis, Nuphar lutea, Scirpus lacustris subsp. lacustris.
Locally frequent: Apium nodiflorum, Sagittaria sagittifolia, Sparganium emersum.
Rare: Butomus umbellatus.

RIPARIAN VEGETATION
Abundant: Epilobium hirsutum, Sparganium erectum.
Locally abundant: Myosotis scorpioides.
Frequent: Barbarea vulgaris, Cardamine flexuosa, Nasturtium sp., Phalaris arundinacea, Plantago major, Polygonum aviculare, P. hydropiper, P. persicaria, Ranunculus repens, Rorippa amphibia, Urtica dioica, Veronica beccabunga.
Locally frequent: Agrostis stolonifera, Alliaria petiolata, Angelica sylvestris, Carex riparia, Cirsium arvense, Dipsacus fullonum, Juncus inflexus, Lycopus europaeus, Myosoton aquaticum, Ranunculus sceleratus, Rumex obtusifolius, Scrophularia auriculata, Solanum dulcamara.
Occasional: Alopecurus geniculatus, Atriplex hastata, A. patula, Cirsium vulgare, Crataegus monogyna, Epilobium montanum, E. parviflorum, Geum urbanum, Polygonum amphibium, P. lapathifolium, Rubus fruticosus agg., Rumex crispus, Salix fragilis, Sonchus asper, Veronica catenata, V. filiformis.
Rare: Acer campestre, Alisma plantago-aquatica, Arctium minus, Bidens tripartita, Carex acuta, Fraxinus excelsior, Hypericum tetrapterum, Juncus effusus, Iris pseudacorus, Ligustrum vulgare, Lythrum salicaria, Populus × canadensis, Prunus spinosa, Quercus robur, Rosa canina, Sambucus nigra, Scirpus sylvaticus, Scutellaria galericulata, Sonchus arvensis.

Habitat Study 71 River Trent, King's Mills, Castle Donington
SK 418278   Area: not calculated.   Alt.: 33 m. (110 ft.)
Recorder: P. H. Gamble. Surveyed: 17th July 1976.
This study is of a 550 m. length of the River Trent north of King's Mills and Quarry Hill Plantation. It includes the aquatic vegetation to a line 5 m. from the bank, and the species growing on the southern (Leicestershire) bank of the river. Much of the bank consists of a cliff up to 1 m. high. Because of the instability of this bank the study area supports a varied weed flora. The deep loamy soil is derived from Alluvium.

AQUATIC VEGETATION
Abundant: Lemna gibba, L. minor.
Locally abundant: Rorippa amphibia.
Frequent: Potamogeton pectinatus, Sparganium emersum.
Locally frequent: Acorus calamus, Apium nodiflorum, Scirpus lacustris subsp. lacustris, Sparganium erectum.
Occasional: Butomus umbellatus.
Rare: Spirodela polyrrhiza.

RIPARIAN VEGETATION
Abundant: Matricaria perforata.

Locally abundant: Brassica napus, Urtica dioica.
Frequent: Artemisia absinthium, Atriplex hastata, A. patula, Barbarea vulgaris, Capsella bursa-pastoris, Carduus acanthoides, Cirsium arvense, Conium maculatum, Epilobium adenocaulon, Leontodon autumnalis, Lolium perenne, Lythrum salicaria, Ranunculus repens, Rorippa islandica, Rumex obtusifolius, Scrophularia auriculata, Scutellaria galericulata, Sonchus asper.
Locally frequent: Agrostis stolonifera, Alliaria petiolata, Anthriscus sylvestris, Arrhenatherum elatius, Epilobium hirsutum, Equisetum arvense, Galium aparine, Glechoma hederacea, Lapsana communis, Myosoton aquaticum, Phalaris arundinacea, Polygonum hydropiper, P. persicaria, Stellaria media.
Occasional: Achillea millefolium, Aethusa cynapium, Angelica sylvestris, Artemisia vulgaris, Barbarea stricta, Chamomilla suaveolens, Cirsium vulgare, Crepis capillaris, Epilobium angustifolium, E. montanum, Lamium album, Leucanthemum vulgare, Linaria vulgaris, Malva sylvestris, Poa trivialis, Polygonum aviculare, Rumex acetosa, Sambucus nigra, Senecio jacobaea, Sisymbrium officinale, Solanum dulcamara.
Rare: Brassica nigra, Symphytum × uplandicum.

Habitat Study 72 Ulverscroft Brook, Ulverscroft
SK 501124   Area: not calculated.   Alt.: 146 m. (480 ft.) S.S.S.I.
Recorder: P. H. Gamble. Surveyed: 14th July 1976.
This study is of a 400 m. length of the brook where it flows south-east across the large field which lies between Ulverscroft Priory ruins and Abbey Lane. The southern part of the study area, running beside some very wet ground, is shady, being overhung by alders. The study includes those species growing in the stream channel and on both banks but does not include species growing in the field away from the bank edge. The site lies on Alluvium and Keuper Marl. A sample of the loamy soil had pH 5.2.

AQUATIC VEGETATION
Abundant: Lemna minor, Veronica beccabunga.
Frequent: Callitriche stagnalis, Nasturtium microphyllum.

RIPARIAN VEGETATION
Abundant: Juncus effusus, Leontodon autumnalis, Ranunculus repens.
Locally abundant: Juncus bufonius.
Frequent: Agrostis capillaris, A. stolonifera, Anthoxanthum odoratum, Cardamine pratensis, Cirsium palustre, Glyceria fluitans, G. declinata, Hypochoeris radicata, Juncus articulatus, Lotus uliginosus, Luzula campestris, Plantago lanceolata, Polygonum hydropiper, Ranunculus acris, R. ficaria, R. flammula, Rumex acetosa, R. sanguineus, Senecio aquaticus, Silene dioica, Stellaria alsine, Trifolium pratense, T. repens, Veronica chamaedrys.
Locally frequent: Ajuga reptans, Alnus glutinosa, Cardamine amara, C. flexuosa, Chrysosplenium oppositifolium, Cynosurus cristatus, Equisetum arvense, Filipendula ulmaria, Galium saxatile, Holcus mollis, Lysimachia nemorum, Moehringia trinervia, Pilosella officinarum, Prunella vulgaris, Tussilago farfara.

Occasional: Achillea millefolium, Angelica sylvestris, Cerastium fontanum subsp. triviale, Cirsium arvense, C. vulgare, Deschampsia cespitosa, Digitalis purpurea, Dryopteris dilatata, Epilobium adenocaulon, E. parviflorum, Galium palustre, Geranium robertianum, Heracleum sphondylium, Holcus lanatus, Hyacinthoides non-scripta, Hypericum tetrapterum, Juncus inflexus, Lathyrus pratensis, Lotus corniculatus, Luzula multiflora, Lychnis flos-cuculi, Myosotis laxa subsp. caespitosa, Plantago major, Polygonum amphibium, P. arenastrum, P. persicaria, Potentilla erecta, P. sterilis, Rubus fruticosus agg., Rumex conglomeratus, R. crispus, R. obtusifolius, Stellaria graminea, S. media, Succisa pratensis, Taraxacum nordstedtii, T. raunkiaerii, Urtica dioica, Valeriana officinalis.
Rare: Arrhenatherum elatius, Betula pendula, B. pubescens, Crataegus monogyna, Dryopteris filix-mas, Epilobium obscurum, Fraxinus excelsior, Geum urbanum, Glechoma hederacea, Hedera helix, Quercus robur, Rosa arvensis, R. canina, Salix cinerea subsp. cinerea, Sambucus nigra, Scrophularia auriculata, Solanum dulcamara, Torilis japonica, Ulex europaeus.

Habitat Study 73 River Soar, Croft
SP 504947 to 511940   Area: not calculated.   Alt.: 72 m. (235 ft.)
Recorder: S. H. Bishop. Surveyed: 30th July 1976.
The stretch of river studied lies between Sopers Bridge and Sutton Hill Bridge and is approximately 1300 m. in length. At the time of the study the river here had a considerable diversity of structure, with low cliffs, spits, shallows, riffles, pools and meanders. There were overhanging trees in places. It was of medium flow, running over a gravel and silt bottom. In some places the river was only a metre wide and as deep, and in others it was 5 metres wide and very shallow. One pool was over 4 metres in diameter and more than 2 metres deep. A Water Authority river improvement scheme carried out during 1979 and 1980 totally changed the nature of the upper River Soar, including this stretch, destroying most of the structural diversity and much of botanical interest.

AQUATIC VEGETATION
Abundant: Nasturtium sp., Sparganium emersum, S. erectum.
Locally abundant: Myosotis scorpioides, Oenanthe fluviatilis, Potamogeton crispus.
Frequent: Lemna minor, Polygonum amphibium, Veronica beccabunga.
Locally frequent: Elodea canadensis, Potamogeton perfoliatus, Scirpus lacustris subsp. lacustris, Zannichellia palustris.
Occasional: Nuphar lutea, Sagittaria sagittifolia.

RIPARIAN VEGETATION
Abundant: Arrhenatherum elatius, Epilobium hirsutum, Urtica dioica.
Locally abundant: Agrostis stolonifera, Holcus lanatus, Phalaris arundinacea.
Frequent: Carduus acanthoides, Cirsium arvense, Dactylis glomerata, Filipendula ulmaria, Galium aparine, Lolium perenne, Myosotis laxa subsp. caespitosa, Phleum pratense subsp. bertolonii, Potentilla anserina, Ranunculus repens, Rumex sanguineus, Sonchus asper.

Locally frequent: Alliaria petiolata, Alopecurus geniculatus, Apium nodiflorum, Cardamine flexuosa, Carex riparia, Cerastium fontanum subsp. triviale, Cirsium vulgare, Crataegus monogyna, Deschampsia cespitosa, Elymus repens, Glyceria plicata, Juncus inflexus, Myosoton aquaticum, Phleum pratense subsp. pratense, Plantago major, Polygonum aviculare, Polygonum hydropiper, P. persicaria, Potentilla reptans, Prunella vulgaris, Rumex conglomeratus, Scrophularia auriculata, Solanum dulcamara, Trifolium repens.
Occasional: Atriplex hastata, Barbarea vulgaris, Capsella bursa-pastoris, Epilobium obscurum, E. parviflorum, Equisetum arvense, Geranium dissectum, Lathyrus pratensis, Rubus fruticosus agg., Torilis japonica.
Rare: Alisma plantago-aquatica, Alnus glutinosa, Arctium minus, Bromus hordeaceus, Chamomilla suaveolens, Conium maculatum, Hypericum tetrapterum, Juncus effusus, Lapsana communis, Ranunculus sceleratus, Rhamnus catharticus, Rorippa islandica, Rosa arvensis, R. canina, Rubus ulmifolius, Salix alba, S. fragilis, Sanguisorba officinalis, Scutellaria galericulata, Veronica catenata.

Habitat Study 74   River Soar, Belgrave
SK 591075   Area: not calculated.   Alt.: 52 m. (170 ft.)
Recorder: S. H. Bishop. Surveyed: 22nd June and 19th August 1976.
The study area is a 250 m. stretch of the River Soar lying between Old Bridge and New Bridge, Belgrave. At this point it is canalised. It is about 12 metres wide and slow flowing. The bottom is silt and the water depth is between 1 and 2 metres.

AQUATIC VEGETATION
Abundant: Lemna gibba, Nuphar lutea.
Locally abundant: Ceratophyllum demersum, Sparganium emersum.
Frequent: Callitriche platycarpa, Potamogeton crispus, Sagittaria sagittifolia, Zannichellia palustris.
Locally frequent: Polygonum amphibium, Potamogeton perfoliatus.
Rare: Alisma plantago-aquatica, Callitriche intermedia, Potamogeton compressus.

RIPARIAN VEGETATION
Abundant: Arrhenatherum elatius, Poa trivialis, Ranunculus repens.
Locally abundant: Cirsium arvense, Epilobium hirsutum, Urtica dioica.
Frequent: Acorus calamus, Bellis perennis, Glyceria maxima, Lolium perenne, Rorippa amphibia, Rumex obtusifolius, Sparganium erectum, Trifolium repens.
Locally frequent: Achillea millefolium, Alliaria petiolata, Barbarea vulgaris, Bromus sterilis, Capsella bursa-pastoris, Dactylis glomerata, Elymus repens, Epilobium montanum, Lycopus europaeus, Nasturtium sp., Phalaris arundinacea, Plantago lanceolata, Ranunculus sceleratus, Scrophularia auriculata, Scutellaria galericulata, Sisymbrium officinale.
Occasional: Artemisia vulgaris, Atriplex hastata, Bidens tripartita, Cardaria draba, Carduus acanthoides, Carex riparia, Cirsium vulgare, Equisetum arvense, Erysimum cheiranthoides, Galium aparine, Myosotis scorpioides, Polygonum aviculare, P. hydropiper, Potentilla anserina, Salix fragilis, Sonchus arvensis, Stellaria media.
Rare: Angelica archangelica, Armoracia rusticana, Ballota nigra, Bryonia cretica subsp. dioica, Calystegia sepium, Conium maculatum, Crataegus monogyna, Fraxinus excelsior, Lamium album, Lapsana communis, Malva sylvestris, Matricaria perforata, Mentha aquatica, Plantago major, Polygonum lapathifolium, P. persicaria, Populus × canadensis, Potentilla reptans, Rorippa sylvestris, Rumex crispus, R. hydrolapathum, Salix cinerea subsp. oleifolia, Sambucus nigra, Senecio jacobaea, S. squalidus, Solanum dulcamara, Sonchus asper, S. oleraceus.

Habitat Study 75   River Soar, north of Pilling's Lock, Quorn
SK 564185   Area: not calculated.   Alt.: 43 m. (140 ft.)
Recorder: P. H. Gamble. Surveyed: 18th July 1976.
This study extends from the point where the railway crosses the river downstream to a point just north of the railway tunnel entrance, a stretch 330 m. in length. The river here flanks a narrow strip of ground next to the railway, with willows and osiers. The survey covers the aquatic vegetation and the vegetation of the west bank, but apart from a narrow strip at the water's edge, does not include those species growing on the top of the bank, much of which is dominated by nettles, brambles and other coarse growth. The study area lies on Alluvium.

AQUATIC VEGETATION
Abundant: Elodea canadensis, Lemna gibba, L. minor, Nuphar lutea.
Locally abundant: Acorus calamus, Rorippa amphibia, Sagittaria sagittifolia.
Locally frequent: Apium nodiflorum, Iris pseudacorus, Polygonum amphibium, Sparganium erectum, Zannichellia palustris.
Occasional: Alisma plantago-aquatica, Callitriche platycarpa.

RIPARIAN VEGETATION
Abundant: Carduus acanthoides, Epilobium hirsutum, Rubus fruticosus agg., Urtica dioica.
Locally abundant: Myosotis scorpioides, Myosoton aquaticum, Phalaris arundinacea.
Frequent: Anthriscus sylvestris, Elymus caninus, Epilobium adenocaulon, Festuca gigantea, Galium aparine, Scrophularia auriculata, Scutellaria galericulata.
Locally frequent: Alliaria petiolata, Bromus sterilis, Calystegia sepium, C. silvatica, Carex riparia, Cirsium arvense, Conium maculatum, Elymus repens, Epilobium parviflorum, Glyceria maxima, Humulus lupulus, Polygonum persicaria, Rumex conglomeratus, Stellaria media.
Occasional: Alnus glutinosa, Angelica sylvestris, Arctium minus, Atriplex hastata, Barbarea stricta, B. vulgaris, Bidens tripartita, Bryonia cretica subsp. dioica, Cardamine flexuosa, Carex otrubae, Cirsium vulgare, Erysimum cheiranthoides, Filipendula ulmaria, Geranium pratense, Geum urbanum, Holcus lanatus, Juncus effusus, Lapsana communis, Lycopus europaeus, Lythrum salicaria, Rumex sanguineus, Salix fragilis, S. viminalis, Sambucus nigra, Solanum dulcamara, Sonchus asper, Stachys palustris.
Rare: Myosotis laxa subsp. caespitosa, Senecio aquaticus.

Habitat Study 76 River Wreake, Hoby
SK 671172   Area: not calculated.   Alt.: 55 m. (180 ft.)
Recorders: S. H. Bishop and P. H. Gamble. Surveyed: 28th June 1980.

A 700 m. stretch of the River Wreake meandering across meadowland south and east of Hoby village. The survey includes the aquatic vegetation of the moderately deep channel and the flora of both banks, between the field bridge south of the village and the site of Hoby Mill. The study area lies on Alluvium.

AQUATIC VEGETATION
Frequent: Nuphar lutea, Rorippa amphibia, Scirpus lacustris subsp. lacustris, Sparganium erectum.
Locally frequent: Carex riparia, Glyceria fluitans, Veronica beccabunga.
Occasional: Carex acuta, Potamogeton pectinatus, Sagittaria sagittifolia.
Rare: Alisma plantago-aquatica, Glyceria maxima, Nasturtium officinale, Polygonum amphibium, Potamogeton perfoliatus, Zannichellia palustris.

RIPARIAN VEGETATION
Abundant: Epilobium hirsutum.
Locally abundant: Conium maculatum.
Frequent: Arrhenatherum elatius, Barbarea vulgaris, Cirsium arvense, Dactylis glomerata, Lolium perenne, Myosotis scorpioides, Poa pratensis, Urtica dioica.
Locally frequent: Alliaria petiolata, Carduus acanthoides, Elymus repens, Myosoton aquaticum, Petasites hybridus, Phalaris arundinacea, Poa annua, P. trivialis, Polygonum persicaria, Stellaria media, Trifolium repens.
Occasional: Alopecurus geniculatus, A. pratensis, Capsella bursa-pastoris, Dipsacus fullonum, Epilobium parviflorum, Equisetum arvense, Festuca pratensis, Galium aparine, Glechoma hederacea, Phleum pratense subsp. pratense, Ranunculus sceleratus, Rumex conglomeratus, R. obtusifolius, R. sanguineus, Salix alba, Scrophularia auriculata, Sisymbrium officinale, Solanum dulcamara, Stachys palustris.
Rare: Atriplex hastata, Arctium minus, Brassica nigra, Cirsium vulgare, Crataegus monogyna, Fraxinus excelsior, Geranium dissectum, Juncus effusus, J. inflexus, Lamium album, Myosotis arvensis, Rorippa islandica, Rosa canina, Salix cinerea subsp. cinerea, S. fragilis, Sambucus nigra, Sinapis arvensis, Sonchus asper, S. oleraceus, Ulmus procera, Veronica catenata.

Habitat Study 77 River Eye, Freeby
SK 800183   Area: not calculated.   Alt.: 81 m. (265 ft.)
S.S.S.I.
Recorders: S. H. Bishop, P. H. Gamble and A. L. Primavesi. Surveyed: 29th August 1976.

The study is of a 1 km. length of the river meandering through fields and meadows downstream from Ham Bridge. It includes the species growing in the water and along both banks. Since the study was carried out the river here suffered short lived but massive pollution by farm effluent but it is not known to what extent this has harmed the aquatic flora. The study area lies on Alluvium.

AQUATIC VEGETATION
Abundant: Glyceria maxima, Scirpus lacustris subsp. lacustris, Sparganium erectum.
Locally abundant: Lemna minor, Myriophyllum spicatum, Rorippa amphibia.
Frequent: Apium nodiflorum, Nuphar lutea, Potamogeton natans, Nasturtium microphyllum × officinale, Sagittaria sagittifolia, Veronica anagallis-aquatica, V. beccabunga, V. catenata.
Locally frequent: Elodea canadensis, Iris pseudacorus, Polygonum amphibium, Potamogeton perfoliatus.
Occasional: Alisma plantago-aquatica, Butomus umbellatus, Callitriche platycarpa, Equisetum fluviatile, Potamogeton crispus, P. pectinatus, Ranunculus circinatus, Zannichellia palustris.

RIPARIAN VEGETATION
Abundant: Cirsium arvense, Urtica dioica.
Locally abundant: Epilobium hirsutum, Myosotis scorpioides, Phalaris arundinacea.
Frequent: Alliaria petiolata, Anthriscus sylvestris, Elymus repens, Ranunculus ficaria, R. repens, Rumex obtusifolius, Senecio aquaticus, Sonchus asper, Stachys palustris.
Locally frequent: Agrostis stolonifera, Arrhenatherum elatius, Barbarea vulgaris, Cirsium vulgare, Conium maculatum, Dactylis glomerata, Equisetum arvense, Lamium album, Matricaria perforata, Myosoton aquaticum, Phleum pratense subsp. pratense, Poa trivialis, Polygonum aviculare, P. hydropiper, P. persicaria, Rumex crispus, Solanum dulcamara.
Occasional: Achillea millefolium, Atriplex hastata, A. patula, Capsella bursa-pastoris, Carduus acanthoides, Epilobium parviflorum, Filipendula ulmaria, Heracleum sphondylium, Lolium perenne, Mentha aquatica, Plantago major, Poa annua, Potentilla anserina, P. reptans, Ranunculus sceleratus, Rumex conglomeratus, R. sanguineus, Scrophularia auriculata, Sisymbrium officinale, Trifolium repens.
Rare: Arctium minus, Calystegia sepium, Carex acuta, Chenopodium album, Cornus sanguinea, Crataegus monogyna, Dipsacus fullonum, Erysimum cheiranthoides, Festuca arundinacea, Galium aparine, Lycopus europaeus, Polygonum lapathifolium, Prunus spinosa, Rorippa islandica, Salix cinerea subsp. oleifolia, S. caprea, S. fragilis, Scutellaria galericulata, Sonchus arvensis.

Green lanes, Tracks and Roadside verges

These studies are grouped together for convenience, but for various reasons which will be discussed, they do not represent a uniform type of habitat, even in any one of the three categories.

Green lanes, where there is no metalled road surface, are subject to very little interference apart from the traffic on them. If the surface is firm and dry throughout the year, the disturbance is usually confined to a worn track up the centre of the lane. At either side there is an undisturbed strip which, if the lane is wide, may provide a refuge for a number of herbaceous species, provided that there is no great invasion of scrub. If the ground tends to be wet, tractors and other vehicles will keep using different parts of the lane as deep ruts are formed and the flora is often disappointing. A feature of such lanes is *Juncus bufonius*, the mucilaginous seeds of which adhere to wheels, boots and feet and are thus able to colonise the wet ruts. The species does not happen to occur in the lane which is the subject of study 78.

Tracks, such as the farm track which is the subject of study 80, where there is some sort of prepared surface on which vehicles can run, differ very little in the character of their verges from small public roads. They may be subjected to a variety of treatments by the owner, even to the extent of treating the verges with a total herbicide. The presence of *Lepidium latifolium* at this particular site is not such an oddity as it may seem, the species being not uncommon in similar sites in this region.

The verges of public roads can be considered under three main headings. There are firstly the majority which receive little or no interference or management, apart from the periodic cutting either of the whole verge or just of a strip next to the road. Such verges retain or take on the character of the vegetation of their surroundings. For instance it is obvious from a perusal of the species list in study 84 that there is woodland adjacent. Verges may in fact provide refuges for species which no doubt once occurred in nearby grassland but are now uncommon in the neighbourhood. An example of this is Park Hill Lane, Seagrave, the subject of study 83.

The mowing of roadside verges, especially with the destructive flail mower, presents many problems from the point of view of conservation. In the past there was a practice of cutting verges three or more times a year, which many species will not tolerate. An agreement was reached whereby certain floristically interesting verges were given a one-swathe wide cut in the spring, and then left until late summer when a complete cut was carried out. However there are wide verges on which any management has now ceased and this can be a disadvantage, since scrub develops and may be cleared later by machine, with very destructive effects.

The second category of verges comprises those which have suffered considerable disturbance, for a variety of reasons. One common example of this is the use of road verges for the dumping of surplus soil from roadworks, which often smothers local grassland species. Another source of disturbance is the laying of pipelines, drains, gas or electricity mains or telephone lines. In these cases, provided that the disturbance has not been too great, the verge will probably soon revert to its former condition. The mechanical cleaning out of ditches causes considerable churning up of the soil by the heavy machinery, and the spoil is usually dumped on the verge. In this case there may be a temporary flourishing of species such as *Barbarea vulgaris* and *Epilobium hirsutum*. In all these cases of drastic disturbance, there is an immediate influx of ruderal species and arable weeds, typical examples being thistles and docks, *Coronopus squamatus*, *Fumaria officinalis*, and various species of *Veronica*. This is well illustrated in study 82.

Finally there are roadside verges which are created in the course of major roadworks. These start as bare soil, but are usually now sown, often with imported grass seed. Their development after work is completed often forms an interesting study. The wide verges of motorway cuttings or embankments are particularly interesting, but unfortunately access to them is restricted. However an example is afforded by the motorway roundabout which is the subject of study 79. On such newly constituted verges alien species such as *Hordeum jubatum* may appear, flourishing for a year or two before they succumb to competition. It is the subsequent colonisation by local species which affords the main interest of these verges, but unfortunately species-rich grassland in the vicinity which could provide seed sources is often scarce and many are probably fated to remain floristically dull.

Habitat Study 78 Green lane, Boothorpe, Ashby Woulds
SK 311166 to 317171 Area: not calculated. Alt.: 122 m. (400 ft.)
Recorder: S. H. Bishop. Surveyed: 26th September 1976.
The study covers a 750 m. length of the lane from its junction with the Albert Village to Moira road eastward towards the hamlet of Boothorpe. It includes the verge and track of the green lane as well as the northern boundary hedgerow. The lane is open to fields on its southern side. The hedgerow has remained unmanaged for many years and the shrubs have attained large proportions. The site stands on Coal Measures.

TREES AND SHRUBS
Locally abundant: Crataegus monogyna.
Frequent: Prunus spinosa, Rubus fruticosus agg.
Locally frequent: Fraxinus excelsior, Quercus robur.
Occasional: Betula pendula, Ilex aquifolium, Rosa arvensis, R. canina, Rubus idaeus, Sambucus nigra.
Rare: Corylus avellana, Cytisus scoparius, Malus sylvestris, Prunus avium, Quercus petraea, Ulex europaeus, U. gallii.

HERBS
Abundant: Agrostis capillaris, Arrhenatherum elatius, Dactylis glomerata, Pteridium aquilinum.
Locally abundant: Achillea millefolium, Epilobium angustifolium, Lolium perenne, Poa annua (Rt), P. pratensis, Urtica dioica.
Frequent: Bromus sterilis, Capsella bursa-pastoris (Rt), Centaurea nigra, Cirsium arvense, Deschampsia cespitosa, Elymus repens, Heracleum sphondylium, Holcus mollis, Hyacinthoides non-scripta, Plantago lanceolata (Rt), P. major (Rt), Rumex acetosella, Teucrium scorodonia.
Locally frequent: Campanula rotundifolia, Centaurea debeauxii subsp. nemoralis, Cerastium fontanum subsp. triviale, Chamomilla suaveolens (Rt), Galium verum, Hedera helix, Lamium album, Leontodon autumnalis, Ranunculus repens, Rumex acetosa, R. obtusifolius, Taraxacum spp., Trifolium repens.
Occasional: Deschampsia flexuosa, Digitalis purpurea, Galium saxatile, Hieracium sp., Matricaria perforata (Rt), Pimpinella saxifraga, Polygonum aviculare (Rt), Rumex crispus, Stellaria graminea, S. media, Vicia cracca.
Rare: Artemisia vulgaris, Lathyrus montanus, Senecio vulgaris (Rt), Serratula tinctoria, Silene vulgaris, Stachys officinalis, Stellaria holostea, Tamus communis, Trifolium medium, Vicia sepium.

Habitat Study 79 Motorway roundabout, Markfield
SK 479107 Area: not calculated. Alt.: 183 m. (600 ft.)
Recorder: S. H. Bishop. Surveyed: 1st July 1979.
The study covers the whole of the roundabout at the junction between the M1 and the A50. The study area is in the main rough grassland with a little invading scrub, none of which was planted at the time of construction of the motorway. There are also some extensive wet flushes. The site is situated on Keuper Marl.

Abundant: Arrhenatherum elatius, Cirsium arvense, C. vulgare, Festuca rubra.
Locally abundant: Aegopodium podagraria, Elymus repens, Equisetum arvense, Juncus conglomeratus, J. effusus, Lotus uliginosus, Lychnis flos-cuculi, Rumex acetosa.
Frequent: Alopecurus pratensis, Dactylis glomerata, Deschampsia cespitosa, Holcus lanatus, Plantago lanceolata.
Locally frequent: Carex hirta, C. ovalis, Centaurea nigra, Cerastium fontanum subsp. triviale, Digitalis purpurea, Epilobium hirsutum, Heracleum sphondylium, Juncus inflexus, Lathyrus pratensis, Leucanthemum vulgare, Plantago major, Rumex obtusifolius, Silene dioica, Sonchus arvensis, Taraxacum spp., Tussilago farfara.
Occasional: Achillea millefolium, Anthriscus sylvestris, Atriplex patula, Bellis perennis, Cardamine pratensis, Conopodium majus, Crataegus monogyna, Cynosurus cristatus, Epilobium angustifolium, E. parviflorum, Holcus mollis, Hypochoeris radicata, Lolium perenne, Lotus corniculatus, Luzula campestris, Poa pratensis, Ranunculus acris, R. repens, Rubus fruticosus agg., Rumex crispus, Sanguisorba officinalis, Solanum dulcamara, Stachys sylvatica, Stellaria graminea, Trifolium repens, Ulex europaeus, Urtica dioica, Vicia cracca.
Rare: Angelica sylvestris, Artemisia vulgaris, Carex flacca, Cirsium palustre, Dryopteris dilatata, Epilobium montanum, Galium saxatile, Hordeum jubatum, Hypericum maculatum, Juncus articulatus, Lamium album, Malus sylvestris, Potentilla erecta, Prunella vulgaris, Quercus robur, Rosa canina, Rubus idaeus, Rumex acetosella, Salix caprea, S. cinerea subsp. oleifolia, Scrophularia auriculata, Senecio erucifolius, S. squalidus, S. vulgaris, Succisa pratensis, Trifolium dubium, Trisetum flavescens, Viburnum opulus.

Habitat Study 80 Farm track, Nailstone Wiggs, Nailstone
SK 423084 Area: not calculated. Alt.: 145 m. (475 ft.)
Recorder: H. Handley. Surveyed: 11th April, 24th July and 19th September 1976.
This study is of a 100 m. length of track from Grange Lane to Nailstone Wiggs Farm, westward from the point where the track turns sharply northwards to skirt Nailstone Wiggs Wood. The western end of the study area is marked by a stile. The verge and the sparse hedgerow on the north side are included in the study. The site is on Glacial Sand and Gravel overlying Boulder Clay. The surface soil is much affected by coal dust from the adjacent colliery. Two samples of it had pH 6.8 and 7.0.

Abundant: Arrhenatherum elatius, Dactylis glomerata, Lepidium latifolium.

Frequent: Agrostis capillaris, A. stolonifera, Anthriscus sylvestris, Bellis perennis, Chamomilla suaveolens, Cirsium arvense, Holcus lanatus, Plantago lanceolata, P. major, Poa annua, Ranunculus repens, Stellaria media, Tragopogon pratensis, Trifolium repens, Urtica dioica.
Occasional: Centaurea nigra, Chaerophyllum temulentum, Cirsium vulgare, Crataegus monogyna, Epilobium angustifolium, Hedera helix, Heracleum sphondylium, Hypochoeris radicata, Juncus bufonius, Leontodon autumnalis, Lolium perenne, Matricaria perforata, Prunella vulgaris, Prunus spinosa, Quercus robur (seedlings), Ranunculus acris, Rosa canina, Rubus fruticosus agg., Rumex obtusifolius, Sambucus nigra, Sonchus asper, Stellaria holostea, Taraxacum cordatum, T. section Hamata, Trifolium pratense, Vicia cracca.
Rare: Rosa arvensis, Sagina apetala.

Habitat Study 81   Roadside verge, Botcheston, Desford
SK 478051 Area: not calculated. Alt.: 105 m. (345 ft.)
Recorder: S. H. Bishop. Surveyed: 29th July 1979.
This study is of the northern verge of the road immediately west of the village of Botcheston. This 250 m. stretch of roadside is cut once annually in mid-August. It lies on Keuper Marl.

Abundant: Anthoxanthum odoratum, Centaurea nigra, Holcus lanatus, Trifolium pratense.
Locally abundant: Leontodon hispidus, Ulex europaeus.
Frequent: Arrhenatherum elatius, Cynosurus cristatus, Dactylis glomerata, Festuca rubra, Lolium perenne, Lotus corniculatus, Medicago lupulina, Pimpinella saxifraga, Plantago lanceolata, Poa pratensis, Potentilla reptans, Ranunculus bulbosus, Trifolium repens, Trisetum flavescens.
Locally frequent: Agrostis capillaris, Carex flacca, Centaurium erythraea, Elymus repens, Heracleum sphondylium, Hieracium sp., Lathyrus pratensis, Ranunculus repens, Rubus fruticosus agg., Rubus ulmifolius, Vicia hirsuta.
Occasional: Achillea millefolium, Anthriscus sylvestris, Agrostis stolonifera, Calystegia sepium, Crataegus monogyna, Galium verum, Hypericum perforatum, Leontodon autumnalis, Leucanthemum vulgare, Phleum pratense subsp. pratense, Pimpinella major, Polygonum aviculare, Prunella vulgaris, Stellaria graminea, Taraxacum spp., Urtica dioica.
Rare: Agrimonia eupatoria, Atriplex hastata, Chamomilla suaveolens, Cirsium arvense, C. vulgare, Crepis capillaris, C. vesicaria, Epilobium angustifolium, Geranium dissectum, Lapsana communis, Matricaria perforata, Pilosella officinarum, Plantago major, Prunus spinosa, Rosa arvensis, Rosa canina, Sambucus nigra, Sonchus asper, Tragopogon pratensis.

Habitat Study 82   Roadside verge, Thurlaston
SP 501977 Area: not calculated Alt.: 76 m. (250 ft.)
Recorder: S. H. Bishop. Surveyed: 16th June 1976.
This study is of a 200 m. stretch of recently disturbed roadside verge on the west side of Croft Road between Yennards Farm and Sandpit Cottages, south of the village of Thurlaston. It is situated on Glacial Sand and Gravel.

Abundant: Chamomilla recutita, Holcus lanatus, Poa annua.
Locally abundant: Arabidopsis thaliana, Poa pratensis, Trifolium dubium.
Frequent: Anchusa arvensis, Capsella bursa-pastoris, Cynosurus cristatus, Lolium perenne, Rumex crispus, Veronica persica, Viola arvensis.
Locally frequent: Aphanes arvensis, Cerastium fontanum subsp. triviale, C. glomeratum, Chenopodium album, Lamium amplexicaule, Papaver dubium, Rumex obtusifolius, Senecio vulgaris, Sonchus asper, Spergula arvensis, Stellaria media, Trifolium repens, Vicia hirsuta.
Occasional: Arrhenatherum elatius, Bilderdykia convolvulus, Bromus sterilis, Chamomilla suaveolens, Cirsium arvense, C. vulgare, Dactylis glomerata, Geranium dissectum, Lolium multiflorum, Papaver argemone, Phleum pratense subsp. pratense, Ranunculus repens, Thlaspi arvense, Trifolium hybridum, T. pratense, Veronica hederifolia.
Rare: Aethusa cynapium, Alopecurus geniculatus, Artemisia vulgaris, Bromus hordeaceus, Crepis vesicaria, Coronopus squamatus, Fumaria officinalis, Geranium molle, Hypochoeris radicata, Lamium album, L. purpureum, Matricaria perforata, Plantago major, Polygonum amphibium, Potentilla reptans, Ranunculus arvensis, Raphanus raphanistrum, Rumex acetosella, Senecio squalidus, Taraxacum spp., Veronica arvensis, Vicia sativa.

Habitat Study 83   Roadside verge, Park Hill Lane, Seagrave
SK 623172 Area: not calculated. Alt.: 70-102 m. (231-334 ft.)
Recorders: P. H. Gamble, J. M. Horwood and A. L. Primavesi. Surveyed: 15th June 1971 and 20th September 1971.
This study is of 500 m. of the southwest verge of the road, from the field entrance just above the brook to the summit of the hill. Most of this verge is a steep bank heavily invaded by scrub, but a strip about 2 m. wide next to the road is cut annually. The site is on Boulder Clay.

Abundant: Achillea millefolium, Arrhenatherum elatius, Crataegus monogyna, Galium aparine, Poa trivialis.
Locally abundant: Alopecurus pratensis.
Frequent: Acer campestre, Agrimonia eupatoria, Anthriscus sylvestris, Arum maculatum, Centaurea nigra, Cerastium fontanum subsp. triviale, Cirsium arvense, Dactylis glomerata, Deschampsia cespitosa, Elymus repens, Equisetum arvense, Galium verum, Glechoma hederacea, Heracleum sphondylium, Lathyrus pratensis, Lotus corniculatus, Phleum pratense subsp. pratense, Plantago lanceolata, Potentilla reptans, Primula veris, Prunella vulgaris, Ranunculus ficaria, R. repens, Rumex acetosa, Sanguisorba minor subsp. minor, Stachys sylvatica, Taraxacum spp., Torilis japonica, Trifolium pratense, T. repens, Urtica dioica.
Locally frequent: Avenula pubescens, Carex flacca, Epilobium hirsutum, Viola hirta.
Occasional: Agrostis capillaris, Anthoxanthum odoratum, Brachypodium pinnatum, B. sylvaticum,

Carduus acanthoides, Cirsium vulgare, Convolvulus arvensis, Dactylorhiza fuchsii, Epilobium angustifolium, Festuca arundinacea, F. pratensis, Fraxinus excelsior, Holcus lanatus, Juncus inflexus, Lamium album, Leontodon hispidus, Listera ovata, Lolium perenne, Malus sylvestris, Ononis spinosa, Plantago media, Polygonum persicaria, Prunus spinosa, Ranunculus acris, Rosa canina, Rumex obtusifolius, R. sanguineus, Senecio erucifolius, Silaum silaus, Solanum dulcamara, Sonchus arvensis, Stellaria media, Succisa pratensis, Tamus communis, Tragopogon pratensis, Trisetum flavescens, Veronica chamaedrys, Vicia sativa.

Rare: Briza media, Festuca rubra, Geranium robertianum, Pimpinella major, Rosa arvensis, Salix caprea, Sonchus asper.

Habitat Study 84 Roadside verge, Withcote SK 801048 Area: not calculated. Alt.: 122-159 m. (400-520 ft.)

Recorders: S. H. Bishop and P. H. Gamble. Surveyed: 28th June 1980.

This length of roadside verge extends from the point where the River Chater passes under the road to a point 400 m. up the hill. Only the western verge of the road was included in the survey, and the study excludes the hedgerow. Part of the verge is a bank up to 2 metres high. A few early-flowering species such as *Adoxa moschatellina* are known to occur here but were not recorded in the survey. The road lies on Middle Lias Clay and Sands near the river and Upper Lias Clay and Boulder Clay on the higher ground. A sample of surface soil from about midway along the length had pH 7.8.

Abundant: Anthriscus sylvestris, Arrhenatherum elatius, Dactylis glomerata.
Locally abundant: Bromus sterilis, Poa trivialis.
Frequent: Alopecurus pratensis, Festuca pratensis, Galium aparine, Heracleum sphondylium, Holcus lanatus, Myosotis sylvatica, Poa pratensis, Potentilla reptans, Rumex sanguineus, Taraxacum spp., Urtica dioica, Veronica chamaedrys, Vicia sepium.
Locally frequent: Brachypodium sylvaticum, Carex hirta, Elymus repens, Equisetum arvense, Geranium dissectum, G. pratense, Lolium perenne, Mercurialis perennis, Plantago major, Potentilla anserina, Ranunculus repens, Silene dioica, Stachys sylvatica.
Occasional: Achillea millefolium, Alliaria petiolata, Bromus hordeaceus, B. ramosus, Cirsium arvense, Cruciata laevipes, Geum urbanum, Glechoma hederacea, Lapsana communis, Lathyrus pratensis, Phleum pratense subsp. pratense, Poa annua, Potentilla sterilis, Tamus communis, Trifolium repens.
Rare: Agrimonia eupatoria, Capsella bursa-pastoris, Carduus acanthoides, Chamomilla suaveolens, Cirsium vulgare, Cornus sanguinea, Deschampsia cespitosa, Equisetum telmateia, Festuca rubra, Fragaria vesca, Hedera helix, Hypericum hirsutum, Lamium album, Polygonum aviculare, Rosa canina, Rubus fruticosus agg., Rumex acetosa, R. obtusifolius, Sonchus asper, Trifolium pratense, Ulmus minor.

Hedgerows

The hedgerows selected for the following studies are all comparatively interesting ones which have not recently been subjected to a great deal of interference, especially by cutting with machinery. The diversity of woody species in a hedgerow can be quantified by making counts of the number present in successive 30m lengths. In some areas of the country it has been shown that this gives an indication of age, since hedges appear to accrete additional species at a rate of one every hundred years or so. This assumes of course that a single species hedge was planted in the first place. However the use of diversity as a reliable means of hedge dating demands considerable knowledge of the previous history of both the hedges and of adjoining land. For example, a hedgerow which once formed the boundary of a now vanished woodland may be expected to have a wider variety of woody species than its neighbours. Such 'woodland ghost' hedges do however often give themselves away, both by a sinuous outline and the presence of woodland herbs such as *Anemone nemorosa*. No comprehensive study has been made of Leicestershire hedgerows, but it can be said that parish boundary or other hedges with an average count of more than five species are noteworthy and those with more than nine exceptional.

Some species have well-marked distribution patterns in the hedges of the county. *Acer campestre* and *Cornus sanguinea* are less common on Charnwood Forest than elsewhere. On the other hand, holly hedges are a common feature of Charnwood Forest and the Coal Measures.

Corylus avellana, a regular constituent of hedgerows in the west of the county, is more or less confined to woodland in the east, though it does occur there in some parish boundary hedges. *Rhamnus catharticus* is characteristic of the valleys of the River Wreake and the lower reaches of the River Soar, whereas *Tamus communis* is unaccountably absent from these valleys.

The now common practice of annual winter trimming of hedgerows means that some shrub species hardly ever produce flowers or fruit. This added considerably to the difficulty of the survey of the county for species of *Rosa*.

It should be understood that in making a study of a hedgerow, the decision as to how far from the hedge to record the herb layer influenced by its presence is arbitrary and subjective. In the lists, climbers and scramblers have been included in the category of trees and shrubs although they are not normally counted as 'woody' species.

Habitat Study 85   Hedgerow, near Ratcliffe Culey, Witherley
SP 329989   Area: not calculated.   Alt.: 76 m. (250 ft.)
Recorder: S. H. Bishop. Surveyed: 13th August 1976.
This hedgerow lies a little to the east of the public footpath between Ratcliffe Culey and Witherley, north-east of Witherley Fields Farm. The hedge extends for some 200 m. and is bounded on its northern side by a shallow ditch and improved pasture and on the southern side by permanent pasture. The site stands on Keuper Marl.

TREES AND SHRUBS
Locally abundant: Crataegus monogyna, Rubus fruticosus agg.
Frequent: Daphne laureola, Ligustrum vulgare, Prunus spinosa.
Locally frequent: Lonicera periclymenum, Rosa arvensis.
Occasional: Acer campestre, Fraxinus excelsior, Rubus ulmifolius, Salix cinerea subsp. oleifolia, Solanum dulcamara.
Rare: Ilex aquifolium, Mahonia aquifolium, Quercus robur, Rhamnus catharticus, Rosa canina, Sambucus nigra.

HERBS
Abundant: Arrhenatherum elatius.
Locally abundant: Agrostis capillaris, Epilobium hirsutum (Wd), Hedera helix, Urtica dioica.
Frequent: Alopecurus pratensis, Cirsium arvense, Dactylis glomerata, Deschampsia cespitosa, Filipendula ulmaria (Wd), Galium verum, Juncus effusus (Wd), Poa trivialis, Ranunculus repens, Rumex sanguineus (Wd).
Locally frequent: Achillea millefolium, Apium nodiflorum (Wd), Centaurea nigra, Elymus repens, Epilobium montanum (Wd), Equisetum arvense, Geranium robertianum, Plantago lanceolata, Scrophularia auriculata (Wd), Veronica beccabunga (Wd).
Occasional: Heracleum sphondylium, Lathyrus pratensis, Potentilla reptans, Sanguisorba officinalis, Serratula tinctoria, Tragopogon pratensis.
Rare: Agrimonia eupatoria, Carex hirta, Cirsium vulgare, Hypericum perforatum, Juncus inflexus (Wd), Tamus communis.

Habitat Study 86   Hedgerow, near Basin Bridge, Higham on the Hill
SP 394958   Area: not calculated.   Alt.: 91 m. (300 ft.)
Recorder: S. H. Bishop. Surveyed: 28th July 1979.

The hedgerow in this study runs southwards for 500 m. from Basin Bridge. It is bounded by a single track road bordering the Ashby Canal on the west and by fields on the east. For much of its length it has remained unmanaged for many years, with the shrubs attaining giant proportions. The area stands on Boulder Clay.

TREES AND SHRUBS
Abundant: Fraxinus excelsior.
Frequent: Crataegus monogyna, Ligustrum vulgare, Prunus spinosa, Quercus robur, Rhamnus catharticus, Rosa arvensis, Rubus fruticosus agg.
Locally frequent: Acer campestre, Lonicera periclymenum, Rosa canina, Solanûm dulcamara.
Occasional: Salix cinerea subsp. oleifolia, Ulex europaeus.
Rare: Ilex aquifolium, Rubus ulmifolius, Salix viminalis, Ulmus glabra.

HERBS
Abundant: Arrhenatherum elatius, Heracleum sphondylium.
Locally abundant: Filipendula ulmaria.
Frequent: Anthriscus sylvestris, Centaurea nigra, Cirsium arvense, Dactylis glomerata, Elymus repens, Lathyrus pratensis, Poa trivialis, Tamus communis, Vicia cracca.
Locally frequent: Agrostis capillaris, Epilobium hirsutum, Galium aparine, Holcus lanatus, Lolium perenne, Pimpinella major, Torilis japonica, Trisetum flavescens, Urtica dioica.
Occasional: Achillea millefolium, Agrimonia eupatoria, Alopecurus pratensis, Cirsium vulgare, Deschampsia cespitosa, Equisetum arvense, Festuca rubra, Holcus mollis, Leontodon autumnalis, Lotus corniculatus, Phleum pratense subsp. pratense, Plantago lanceolata, Potentilla reptans, Ranunculus repens, Rumex acetosa, R. crispus, R. sanguineus, Stachys officinalis, Taraxacum spp., Trifolium medium, T. repens, Veronica chamaedrys.
Rare: Cerastium fontanum subsp. triviale, Juncus inflexus, Lotus uliginosus, Plantago major, Potentilla anserina, Ranunculus acris, Rumex obtusifolius, Senecio erucifolius, Serratula tinctoria, Silaum silaus, Stellaria graminea, Trifolium pratense, Tussilago farfara, Valeriana officinalis.

Habitat Study 87   Hedgerow, Ulverscroft Lane, Newtown Linford
SK 506120   Area: not calculated.   Alt.: 145 m. (475 ft.)
S.S.S.I.

Recorder: P. H. Gamble. Surveyed: 10th May and 12th July 1980.
This hedgerow is on the south-west side of Ulverscroft Lane and extends for 450 m. from the junction with Polly Bott's Lane to a point where the brook veers away from the road and a dry stone wall commences. In its northern part the hedgerow lies on each side of the Ulverscroft Brook and overhangs it. There are frequent standard oak trees about 150 years old and occasional ash trees. All the plants listed were growing under the overhanging foliage of shrubs or trees. The study area lies partly on Alluvium and partly on Keuper Marl. A sample of surface soil from the Alluvium had pH 5.1.

TREES AND SHRUBS
Abundant: Corylus avellana, Crataegus monogyna.
Locally abundant: Hedera helix, Prunus spinosa.
Frequent: Alnus glutinosa, Ilex aquifolium, Lonicera periclymenum, Quercus robur, Rosa arvensis, Rubus fruticosus agg., Sorbus aucuparia.
Occasional: Betula pubescens, Fraxinus excelsior, Sambucus nigra, Ulex europaeus, Viburnum opulus.
Rare: Acer platanoides (seedling), A. pseudoplatanus (seedling), Fagus sylvatica (seedling), Prunus avium (seedling), Salix aurita.

HERBS
Locally abundant: Adoxa moschatellina, Anemone nemorosa, Anthriscus sylvestris, Brachypodium sylvaticum, Geranium robertianum, Holcus mollis, Mercurialis perennis, Pteridium aquilinum, Ranunculus ficaria, R. repens, Stachys sylvatica, Stellaria holostea, Urtica dioica, Veronica montana, Viola riviniana.
Frequent: Arrhenatherum elatius, Dryopteris filix-mas, Poa trivialis, Silene dioica.
Locally frequent: Arum maculatum, Conopodium majus, Dryopteris dilatata, Elymus caninus, Epilobium angustifolium, E. montanum, Galium aparine, Geum urbanum, Hyacinthoides non-scripta, Hypochoeris radicata, Luzula campestris, Oxalis acetosella, Potentilla sterilis, Stellaria neglecta, Taraxacum spp., Trifolium repens, Tussilago farfara, Veronica hederacea, Vicia sepium.
Occasional: Alopecurus pratensis, Angelica sylvestris, Athyrium filix-femina, Bellis perennis, Bromus ramosus, Cardamine flexuosa, Chrysosplenium oppositifolium, Cirsium arvense, C. palustre, Digitalis purpurea, Galeopsis tetrahit, Heracleum sphondylium, Lamium album, L. purpureum, Moehringia trinervia, Plantago lanceolata, P. major, Rumex acetosa, R. obtusifolius, Stachys officinalis, Tamus communis, Teucrium scorodonia, Valeriana officinalis.
Rare: Cardamine pratensis, Chaerophyllum temulentum, Dryopteris carthusiana, Primula vulgaris, Stellaria alsine, Torilis japonica.

Habitat Study 88 Hedgerow, Barrow on Soar
SK 583157 Area: not calculated. Alt.: 43 m. (140 ft.)
Recorder: P. H. Gamble. Surveyed: 12th May and 13th July 1980.
This tall old hedgerow is on the north-west side of a bridle track just to the north of Mountsorrel. Together with a shallow ditch, it forms the south-east boundary of a meadow lying between the bridle track and a mineral railway line. The part of the hedgerow studied extends for 310 m. from the point where the bridle track changes direction from a north-north-west to a north-east course. All the species listed were growing in close association with the hedgerow. It is sited on Alluvium. A sample of the rich loamy soil from the ditch had pH 6.3.

TREES AND SHRUBS
Abundant: Crataegus monogyna, Humulus lupulus.
Frequent: Rhamnus catharticus, Rosa canina, Rubus fruticosus agg.
Occasional: Fraxinus excelsior, Salix fragilis.
Rare: Crataegus laevigata, Sambucus nigra.

HERBS
Locally abundant: Filipendula ulmaria, Phalaris arundinacea, Stachys sylvatica, Urtica dioica.
Frequent: Anthriscus sylvestris, Galium aparine, Glechoma hederacea, Heracleum sphondylium, Myosoton aquaticum, Ranunculus acris, R. repens, Rumex conglomeratus, R. crispus, R. obtusifolius, Stellaria media.
Locally frequent: Alliaria petiolata, Bromus sterilis, Elymus caninus, E. repens, Epilobium hirsutum, E. montanum, E. parviflorum, Poa trivialis, Ranunculus ficaria, Stellaria neglecta, Thalictrum flavum.
Occasional: Angelica sylvestris, Barbarea vulgaris, Bellis perennis, Cardamine pratensis, Cirsium arvense, Geum urbanum, Lamium album, Plantago major, Rorippa amphibia, Rumex acetosa, Stachys palustris, Taraxacum spp.
Rare: Cirsium vulgare, Ranunculus auricomus, Rorippa islandica.

Habitat Study 89 Hedgerow, Little Dalby, Burton and Dalby
SK 763126 Area: not calculated. Alt.: 134 m. (440 ft.)
Recorder: S. H. Bishop. Surveyed: 27th August 1977.
The study is of a 400 m. length of hedgerow running westward from Home Farm on the northern side of the road between Burrough on the Hill and Little Dalby. The presence in the hedgerow of *Sorbus torminalis* is most unusual. It may be significant that a number of herbs typical of ancient woodland occur in the same hedgerow. The area is on Lower Lias Clay.

TREES AND SHRUBS
Abundant: Crataegus monogyna, Prunus spinosa, Cornus sanguinea.
Locally abundant: Rubus fruticosus agg.
Frequent: Acer campestre, Corylus avellana, Fraxinus excelsior.
Locally frequent: Rosa arvensis.
Occasional: Ligustrum vulgare, Sambucus nigra, Solanum dulcamara.
Rare: Lonicera periclymenum, Quercus robur, Rosa canina, Sorbus torminalis.

HERBS
Abundant: Arrhenatherum elatius, Galium aparine, Urtica dioica.
Locally abundant: Stachys sylvatica.
Frequent: Bromus ramosus, Cirsium arvense, Epilobium hirsutum, Filipendula ulmaria, Glechoma hederacea, Rumex sanguineus.
Locally frequent: Anthriscus sylvestris, Arum maculatum, Elymus repens, Galium mollugo, Geranium robertianum, Heracleum sphondylium, Lamiastrum galeobdolon, Silene dioica, Tamus communis.

Occasional: Geranium pratense, Lapsana communis, Mercurialis perennis, Pimpinella major, Potentilla reptans, Rumex obtusifolius, Stellaria holostea, Vicia cracca, V. sepium.

Rare: Anemone nemorosa, Chaerophyllum temulentum, Equisetum arvense, E. telmateia, Hyacinthoides non-scripta, Hypericum hirsutum, Primula veris, Ranunculus auricomus.

Railways

For a variety of reasons railways provide a unique and important type of habitat. In the making of a railway, cuttings and embankments were formed, and layers of ballast were put down on which to lay the track. The cuttings may expose rocks which do not outcrop naturally in the vicinity. This often makes such cuttings botanically interesting, especially in those parts of the county which are otherwise uniformly covered with glacial drift. For example, apart from the small area of Carboniferous Limestone in the north-west, and the places where the Lincolnshire Limestone is exposed in the north-east, the most likely place to find calcicole species in Leicestershire is on railway verges. For this reason species such as *Brachypodium pinnatum*, *Knautia arvensis*, *Linaria vulgaris* and *Linum catharticum* are widespread throughout the county, though not necessarily common. The ballast, and often the part of the verge immediately adjacent to it, also has its characteristic species; the distribution map of *Chaenorhinum minus* is almost a map of the courses of railways, in use and dismantled, in Leicestershire. Other species characteristic of railway verges adjacent to the track, but not confined to this habitat, are *Aira caryophyllea*, *Arabidopsis thaliana*, *Cardamine hirsuta* and *Trifolium campestre*. For a while after a railway has been dismantled, the ballast has a wealth of small and interesting species, such as *Desmazeria rigida*, *Sedum acre*, *Senecio viscosus*, *Vulpia bromoides* and *V. myuros*. The sites at Swithland (study 91) and Long Clawson (study 94), which have large areas of ballast, are particularly interesting in this respect.

Two other factors contribute to the interest and importance of railway verges as a habitat for plants. The first is that when a railway is in working order, access to the track and verges is prohibited, thus providing a refuge for species which otherwise might succumb to human interference. They are also protected from agricultural activities including those of grazing farm animals. The second factor is that the verges of a working railway are usually kept more or less free from scrub. In the days of steam locomotives this was effected by controlled burning, which prevented the occasional accidental fires caused by sparks from the engines from getting out of hand. This kept the scrub down but did not apparently adversely affect the herbaceous plants.

The beginning of the survey for this present Flora followed shortly after the 'Beeching axe' and the resultant closure of numerous unprofitable lines. Working railway lines are now few, and the once extensive railway network in Leicestershire is mostly dismantled. Certainly in the early stages of the survey, and to some extent for the whole of it, these dismantled railways provided a happy hunting ground for the recording botanists. The conditions which have been outlined above persisted for some time, with the added advantage that there was no interference at all, and the whole of the ballast was available for the invasion of such species as were favoured by this habitat. However it is now obvious that this was only a temporary phase. If left entirely to themselves, such lengths of obsolete track soon become taken over by scrub, which eventually colonises the ballast as well, but even in the short term natural succession of species and other factors such as impeded drainage soon alter its character. The stretches of dismantled railway which have been acquired as nature reserves are very difficult to manage so that their original character is retained. Much of the obsolete railway network has now been sold piecemeal and put to other uses, and in many places the ballast has been removed.

Habitat Study 90 Shenton Cutting, Sutton Cheney
SP 397999   Area: not calculated.   Alt.: 94 m. (310 ft.)
Nature Reserve.
Recorder: S. H. Bishop. Surveyed: 24th May and 27th July 1976.
This study covers the 800 m. length of dismantled railway cutting from the site of Shenton Station southwards to the bridge over the Ashby Canal. The banks of the cutting are heavily invaded by scrub. The study area lies on stoneless grey clays known as Brick-earth, derived from Boulder Clay. These clays contain a few sandy and loamy bands and there is a profusion of small calcareous nodules known as 'Race'.

Abundant: Arenaria serpyllifolia, Brachypodium sylvaticum, Crataegus monogyna, Epilobium adenocaulon, Leucanthemum vulgare, Medicago lupulina, Poa annua.

Locally abundant: Arrhenatherum elatius, Brachypodium pinnatum, Cerastium glomeratum, Dactylis glomerata, Equisetum arvense, Fraxinus excelsior, Holcus lanatus, Rubus fruticosus agg., Sagina apetala, Trifolium dubium.

Frequent: Bromus hordeaceus, Cardamine hirsuta, Centaurea debauxii subsp. nemoralis, C. nigra, Daucus carota, Fragaria vesca, Galium aparine, Hedera helix, Heracleum sphondylium, Lathyrus pratensis, Moehringia trinervia, Pilosella officinarum, Plantago lanceolata, Poa trivialis, Potentilla reptans, Primula veris, Rosa canina, Sagina procumbens, Tussilago farfara, Urtica dioica, Veronica chamaedrys, Vicia sativa, Viola hirta, V. riviniana.

Locally frequent: Achillea millefolium, Agrostis capillaris, Arabidopsis thaliana, Cerastium fontanum subsp. triviale, Chaenorhinum minus, Cirsium arvense, Cynosurus cristatus, Deschampsia cespitosa, Erophila verna, Festuca rubra, Filipendula ulmaria, Hypericum perforatum, Juncus inflexus, Lapsana communis, Lotus corniculatus, Luzula campestris, Myosotis arvensis, Pimpinella major, P. saxifraga, Plantago major, Rumex crispus, R. obtusifolius, Solanum dulcamara, Taraxacum spp., Torilis japonica, Trifolium medium, T. repens, Trisetum flavescens, Veronica arvensis, V. serpyllifolia, Vicia cracca, V. hirsuta.

Occasional: Agrimonia eupatoria, Angelica sylvestris, Anthoxanthum odoratum, Anthriscus sylvestris, Bellis perennis, Briza media, Capsella bursa-pastoris, Cardamine flexuosa, C. pratensis, Carex flacca, C. hirta, Cirsium vulgare, Conopodium majus, Convolvulus arvensis, Crepis capillaris, Dactylorhiza fuchsii, Desmazeria rigida, Geranium robertianum, Lamium album, Linum catharticum, Lonicera periclymenum, Lotus uliginosus, Papaver dubium, Potentilla anserina, Prunus spinosa, Ranunculus acris, R. ficaria, R. repens, Rhinanthus minor, Rubus ulmifolius, Rumex acetosa, R. sanguineus, Salix cinerea subsp. oleifolia, Senecio viscosus, Sonchus asper, Stellaria graminea, S. media, Succisa pratensis, Trifolium campestre, Ulex europaeus.

Rare: Acer campestre, A. pseudoplatanus, Anagallis arvensis, Arctium minus, Artemisia vulgaris, Atriplex patula, Barbarea vulgaris, Carex otrubae, Centaurium erythraea, Cirsium palustre, Dryopteris filix-mas, Epilobium hirsutum, E. parviflorum, Filaginella uliginosa, Galium verum, Geranium dissectum, Hyacinthoides non-scripta, Hypericum maculatum, Juncus conglomeratus, J. effusus, Leontodon hispidus, Linaria vulgaris, Ligustrum vulgare, Listera ovata, Malus sylvestris, Matricaria perforata, Odontites verna, Origanum vulgare, Papaver rhoeas, Prunus avium, Pulicaria dysenterica, Ranunculus bulbosus, Rhamnus catharticus, Rubus idaeus, Rumex acetosella, Salix caprea, S. fragilis, Sambucus nigra, Senecio vulgaris, Silaum silaus, Silene dioica, Sonchus arvensis, Tamus communis, Tanacetum vulgare, Taraxacum section Erythrosperma, Teucrium scorodonia, Tragopogon pratensis, Verbascum thapsus, Vicia angustifolia, Viola arvensis.

Habitat Study 91   Swithland Railway Sidings, Swithland
SK 564132   Area: 2.7 ha (6.8 acres)   Alt.: 61 m. (200 ft.)
Recorder: P. H. Gamble. Surveyed: 1976.
The study area comprises what was formerly sidings and a goods yard for the Great Central Railway, and a 400 m. length of the track currently used by the Main Line Steam Trust for a passenger service operating between Loughborough and Rothley. The embankments bordering the road at the north end of the site and along the western boundary are also included, but not the thicket of scrub-covered ground in the north-east. The north-western part of the site lies on Keuper Marl whilst the south-eastern portion is on Glacial Sands and Gravel, but the native soils have been buried beneath ballast throughout much of the study area. Three samples of surface soil had pH 4.6, 4.8 and 5.2.

Abundant: Arabidopsis thaliana, Arenaria serpyllifolia, Hypericum perforatum, Hypochoeris radicata, Lotus corniculatus, Pilosella officinarum.

Locally abundant: Aira caryophyllea, Arrhenatherum elatius, Bromus sterilis, Cerastium glomeratum, Convolvulus arvensis, Dactylis glomerata, Elymus repens, Epilobium angustifolium, Equisetum arvense, Linaria vulgaris, Vulpia bromoides, V. myuros.

Frequent: Achillea millefolium, Agrostis capillaris, Anthriscus sylvestris, Aphanes arvensis, Centaurea nigra, Cerastium fontanum subsp. triviale, Chaenorhinum minus, Crepis capillaris, C. vesicaria, Epilobium montanum, Festuca rubra, Fragaria vesca, Hieracium diaphanum, H. salticola, Plantago lanceolata, Sagina apetala, Taraxacum spp., Taraxacum section Erythrosperma, Trifolium arvense, T. dubium, Vicia hirsuta.

Locally frequent: Aira praecox, Atriplex patula, Betula pendula, Brachypodium sylvaticum, Capsella bursa-pastoris, Cardamine hirsuta, Cardaria draba, Cerastium semidecandrum, Cirsium arvense, Clinopodium vulgare, Crataegus monogyna, Cytisus scoparius, Epilobium hirsutum, Erodium cicutarium, Erophila verna, Fragaria × ananassa, Geranium molle, G. robertianum, Heracleum sphondylium, Holcus lanatus, Knautia arvensis, Leontodon autumnalis, L. hispidus, Leucanthemum vulgare, Logfia minima, Lolium perenne, Luzula campestris, Mercurialis perennis, Myosotis arvensis, Ornithopus perpusillus, Papaver rhoeas, Pimpinella major, Poa annua, P. pratensis, Potentilla reptans, Reseda lutea, Rubus caesius, Rumex acetosa, R. acetosella, Salix cinerea subsp. cinerea, Saxifraga granulata, Senecio viscosus, Silene dioica, S. vulgaris, Stellaria holostea, S. media, Tanacetum vulgare, Trifolium repens, Trisetum

flavescens, Urtica dioica, Valerianella locusta, Vicia angustifolia, V. sativa, Viola hirta.
Occasional: Anthoxanthum odoratum, Arenaria leptoclados, Bromus hordeaceus, Campanula rotundifolia, Centaurium erythraea, Cerastium diffusum subsp. diffusum, Chamomilla sauveolens, Cirsium vulgare, Galium aparine, Geranium pyrenaicum, Lamium album, L. amplexicaule, Lathyrus pratensis, Medicago lupulina, Papaver dubium, Rosa afzeliana, R. canina, R. dumetorum, Rumex obtusifolius, Salix caprea, Senecio jacobaea, S. squalidus, S. vulgaris, Solanum dulcamara, Sonchus asper, Tragopogon pratensis, Trifolium pratense, Tussilago farfara, Ulex europaeus, Veronica arvensis, Vicia sepium, Viola arvensis, V. riviniana.
Rare: Alnus glutinosa, Artemisia vulgaris, Desmazeria rigida, Epilobium lanceolatum, Euphrasia sp., Fraxinus excelsior, Fumaria officinalis, Hypericum hirsutum, Lamium hybridum, Quercus robur, Raphanus raphanistrum, Reseda luteola, Senecio erucifolius, Silene alba, S. alba × dioica, Sambucus nigra.

Habitat Study 92 Railway embankment, Loughborough
SK 540220 Area: 0.74 ha (1.8 acres) Alt.: 43 m. (140 ft.)
Recorder: P. H. Gamble. Surveyed: 1979.
This length of the former Great Central Railway is contained in a loop of the county boundary which follows the River Soar at the north-west end of the study area and crosses the railway 230 m. south-east of the river. The study includes the 9 m. wide track and ballast and the western embankment, but does not include the level area close to the river or the eastern embankment which is largely dominated by scrub. The top part of the embankment is steep and cliff-like in places, with unstable open areas ideal for colonization by ephemeral species. The bank has frequent ant-hills and occasional rabbit holes. The ground here is made up and stands high above the river and nearby meadowland. Samples of surface soil from the ballast area had pH 7.7 and from the bank pH 7.0.

Abundant: Achillea millefolium, Arabidopsis thaliana, Arrhenatherum elatius, Barbarea vulgaris, Galium aparine, Plantago lanceolata.
Locally abundant: Arenaria leptoclados, A. serpyllifolia, Brachypodium sylvaticum, Cardamine hirsuta, Cerastium arvense, Epilobium hirsutum, Leucanthemum vulgare, Myosotis arvensis, Pilosella officinarum, Potentilla reptans, Rumex acetosella, Senecio viscosus, Stellaria media, Urtica dioica, Vicia hirsuta, V. sativa.
Frequent: Cerastium fontanum subsp. triviale, Conium maculatum, Crataegus monogyna (seedlings), Crepis capillaris, Dactylis glomerata, Dipsacus fullonum, Hieracium vagum, Medicago lupulina, Senecio jacobaea, Silene alba, Veronica arvensis.
Locally frequent: Angelica sylvestris, Carduus acanthoides, Cynoglossum officinale, Epilobium montanum, Equisetum arvense, Festuca ovina, F. rubra, Filipendula ulmaria, Glechoma hederacea, Holcus lanatus, Hordeum murinum, Hypochoeris radicata, Inula conyza, Lamium album, L. purpureum, Leontodon autumnalis, Senecio vulgaris, Silene dioica, Stachys sylvatica, Trifolium dubium.

Occasional: Acer pseudoplatanus, Anthriscus sylvestris, Arctium minus, Bellis perennis, Capsella bursa-pastoris, Cirsium arvense, C. vulgare, Fraxinus excelsior (seedlings), Fumaria officinalis, Geranium molle, Heracleum sphondylium, Hieracium spp., Lamium amplexicaule, Lathyrus pratensis, Papaver argemone, P. dubium, Ranunculus acris, Rosa canina, Sambucus nigra (seedlings), Senecio erucifolius, Sisymbrium officinale, Solanum dulcamara, Sonchus asper, Taraxacum spp., Tussilago farfara.
Rare: Acinos arvensis, Geranium pusillum, Ranunculus ficaria, Rumex crispus, Sanguisorba officinalis, Silene alba × dioica, Veronica persica, Viola odorata.

Habitat Study 93 Railway cutting, Scalford
SK 746255 Area: 2.8 ha (7 acres) Alt.: 128 m. (420 ft.)
Recorders: S. H. Bishop and P. H. Gamble. Surveyed: 27th August 1979.
The study area consists of a 370 m. length of dismantled railway track together with the extensive sides of the cutting, between Brock Hill Tunnel to the north and the Landyke Lane road bridge to the south. The area stands on Chalky Boulder Clay, but the ballast of the track and the rubble and stone drainage of the steep sides of the cutting also have a considerable influence on the vegetation; in addition wet flushes account for a variety of marsh species. Samples of surface soil from both sides of the cutting had pH 7.9.

Abundant: Daucus carota, Festuca ovina, Leontodon hispidus, Leucanthemum vulgare, Medicago lupulina.
Locally abundant: Agrostis stolonifera, Brachypodium sylvaticum, Crataegus monogyna, Juncus articulatus, J. bufonius, J. inflexus, Nasturtium officinale, Pteridium aquilinum.
Frequent: Achillea millefolium, Anthoxanthum odoratum, Arenaria serpyllifolia, Arrhenatherum elatius, Cirsium acaule, Equisetum arvense, Fraxinus excelsior, Lathyrus pratensis, Linum catharticum, Lotus corniculatus, Pilosella officinarum, Plantago lanceolata, Poa annua, Rosa canina.
Locally frequent: Agrostis capillaris, Arenaria leptoclados, Bellis perennis, Briza media, Carex flacca, Centaurea nigra, Cerastium fontanum subsp. triviale, Chaenorhinum minus, Chamomilla suaveolens, Cirsium arvense, Cornus sanguinea, Dactylis glomerata, Dactylorhiza fuchsii, Epilobium adenocaulon, E. parviflorum, Festuca rubra, Hieracium sp., Holcus lanatus, Lapsana communis, Myosotis laxa subsp. caespitosa, Plantago major, Poa pratensis, Polygonum arenastrum, Potentilla reptans, Primula veris, Prunella vulgaris, Ranunculus repens, Reseda luteola, Rubus fruticosus agg., Rumex acetosa, Scrophularia auriculata, Senecio erucifolius, S. viscosus, Trifolium pratense, T. repens, Trisetum flavescens, Tussilago farfara, Ulex europaeus, Verbascum thapsus, Veronica beccabunga, Viola hirta.
Occasional: Agrimonia eupatoria, Avenula pratensis, Cardamine hirsuta, Cirsium vulgare, Deschampsia cespitosa, Epilobium angustifolium, E. montanum, Galium aparine, Geranium robertianum, Iberis amara, Leontodon autumnalis, Lolium perenne, Melilotus officinalis, Myosotis arvensis, Polygala vulgaris, Prunus spinosa, Sedum acre, Senecio jacobaea,

Sonchus asper, Sorbus aria, Stellaria media, Taraxacum spp., Torilis japonica, Trifolium campestre, Urtica dioica, Veronica arvensis, V. chamaedrys, Vicia hirsuta, V. sativa, V. tetrasperma.

Rare: Acer campestre, Brachypodium pinnatum, Bromus sterilis, Capsella bursa-pastoris, Carduus acanthoides, Dryopteris filix-mas, Epilobium hirsutum, Festuca arundinacea, Filaginella uliginosa, Galeopsis tetrahit, Heracleum sphondylium, Hypericum hirsutum, Lonicera periclymenum, Malus sylvestris, Matricaria perforata, Moehringia trinervia, Odontites verna, Ononis spinosa, Papaver dubium, Polygonum aviculare, Quercus robur, Reseda lutea, Rosa arvensis, Salix alba, S. caprea, S. caprea × cinerea, S. cinerea, Solanum dulcamara, Sonchus arvensis, Stachys sylvatica, Tamus communis, Tragopogon pratensis, Trifolium dubium, Vicia cracca, Viola riviniana.

Habitat Study 94 Site of Long Clawson and Hose Station, Clawson and Harby
SK 746266 Area: 2.8 ha (7 acres) Alt.: 91-122 m. (300-400 ft.)
Recorders: S. H. Bishop, P. H. Gamble and A. L. Primavesi. Surveyed: 30th June 1979.
The site consists of about 300 metres of dismantled railway track, together with the station yard and embankments, on the former Melton Mowbray to Bottesford railway line. The area surveyed lies immediately to the north of the Brock Hill Tunnel. Much of the ground around the station yard is sparsely vegetated, with ash and limestone clinker. The site is situated on Lower Lias Clays with some sandy Middle Lias Clays in the south-east part near the tunnel entrance.

Abundant: Lotus corniculatus, Medicago lupulina.
Locally abundant: Cerastium fontanum subsp. triviale, Epilobium angustifolium, Equisetum arvense, Hypericum perforatum, Linum catharticum, Poa annua, Pteridium aquilinum, Vulpia bromoides.
Frequent: Achillea millefolium, Anthoxanthum odoratum, Arrhenatherum elatius, Centaurea nigra, Cerastium glomeratum, Crataegus monogyna, Dactylis glomerata, Heracleum sphondylium, Holcus lanatus, Leontodon hispidus, Plantago lanceolata, Poa pratensis, Trifolium campestre, T. pratense, Trisetum flavescens, Veronica arvensis.
Locally frequent: Agrostis capillaris, Alopecurus pratensis, Arenaria serpyllifolia, Bellis perennis, Cardamine hirsuta, Crepis capillaris, Desmazeria rigida, Epilobium adenocaulon, Equisetum telmateia, Festuca ovina, Genista tinctoria, Lathyrus montanus, Pilosella officinarum, Plantago major, Polygala vulgaris, Potentilla reptans, Rubus fruticosus agg., Rumex acetosa, R. acetosella, Salix caprea, Sanguisorba minor subsp. minor, Sedum acre, Trifolium dubium, T. repens, Tussilago farfara, Urtica dioica, Vicia hirsuta.
Occasional: Acer pseudoplatanus, Arabidopsis thaliana, Arenaria leptoclados, Briza media, Cirsium arvense, C. vulgare, Cynosurus cristatus, Deschampsia cespitosa, Dryopteris dilatata, D. filix-mas, Epilobium montanum, Erophila verna, Festuca rubra, Filago vulgaris, Fraxinus excelsior, Galium verum, Hieracium spp., Hypochoeris radicata, Lathyrus pratensis, Leucanthemum vulgare, Linaria vulgaris, Myosotis arvensis, Phleum pratense subsp. bertolonii, Rosa canina, Sagina procumbens, Senecio jacobaea, Sorbus aria, Stellaria holostea, Taraxacum spp., Ulex europaeus, Vicia cracca.
Rare: Acer campestre, Artemisia vulgaris, Athyrium filix-femina, Bromus hordeaceus, B. sterilis, Chaenorhinum minus, Crepis vesicaria, Epilobium hirsutum, E. parviflorum, Galium aparine, Hypericum maculatum, Luzula campestris, Malus sylvestris, Myosotis ramosissima, Moehringia trinervia, Papaver dubium, Pimpinella saxifraga, Polygonum persicaria, Potentilla erecta, Primula veris, Prunella vulgaris, Quercus robur (seedlings), Ranunculus acris, Rumex crispus, Sagina apetala, Salix alba, S. cinerea, Sambucus nigra, Sanguisorba officinalis, Scrophularia auriculata, Senecio viscosus, Silaum silaus, Sonchus asper, Stachys sylvatica, Stellaria media, Tamus communis, Torilis japonica, Trifolium medium, Vicia sativa, Viola riviniana.

Arable land

An arable field is intended to be a monoculture and the aim of the farmer is, as far as this is possible, to eliminate all weeds. On the other hand, the interest of such fields for the botanist lies in the variety of weeds which can be found in particular crops and on particular soils. Farmers spend a large amount of time and money on weed control and they must achieve a measure of success, otherwise the exercise would not be worth their while. Inevitably then the selection of arable fields of botanical interest for habitat studies must be of those which are atypical of arable land in the county as a whole. This is especially true of cereal crops, on which selective herbicides can be employed. Non-cereal crops, which would themselves be affected if such a herbicide were used, usually contain a greater variety of weed species.

Some weed species are considerably more susceptible to selective weed-killers than others. It has been mentioned elsewhere that *Mentha arvensis*, once a common cornfield weed, is now rare in this habitat, and its presence in quantity in the field which is the subject of study 95 is of interest. *Sinapis arvensis* usually germinates in quantity in early spring among cereal crops, but is then virtually eliminated from these crops by spraying. However this spraying must be done early in the year, otherwise the passage of the machinery over the field would damage the crop. There are a number of small annual weed species which are largely unaffected by the spraying because they come up from seed later on. Examples are *Stellaria media*, *Veronica persica*, and to a certain extent the two species of *Kickxia*. Some species, such as *Chrysanthemum segetum* with its thick glaucous leaves, seem to be comparatively resistant to selective herbicides. Grass species are not affected by the herbicides used against broad-leaved weeds, and although there are special sprays which can be used to control them, many farmers do not consider it worth the extra expense.

Many of the arable weeds are widespread throughout the county, but there are some which show a distinct geographical preference. This is probably because soils in parts of the west of the county are in general more acid than those in the east. Thus *Spergula arvensis* and *Raphanus raphanistrum* have a western tendency and the two species of *Kickxia* are almost confined to the east.

Habitat Study 95   Arable field, Oakthorpe and Donisthorpe
SK 334138   Area: 5.0 ha (12.2 acres)   Alt.: 98 m. (320 ft.)
Recorder: S. H. Bishop. Surveyed: 17th August 1979.
The study is of a cereal field opposite Pasture Farm on the north side of Pastures Lane. It is bounded on all four sides by high hedgrows. The site stands on Coal Measures and the soil is sandy.

Abundant: Polygonum aviculare, P. persicaria, Stachys arvensis.
Locally abundant: Agrostis gigantea, Chamomilla recutita, Elymus repens, Mentha arvensis, Stachys palustris.
Frequent: Aethusa cynapium, Anagallis arvensis, Atriplex patula, Bilderdykia convolvulus, Chenopodium album, Matricaria perforata, Sonchus arvensis, Viola arvensis.
Locally frequent: Chrysanthemum segetum, Plantago major, Poa annua, Spergula arvensis.
Occasional: Chamomilla suaveolens, Cirsium arvense, Filaginella uliginosa, Holcus mollis, Juncus bufonius, Polygonum hydropiper, Ranunculus repens, Sinapis arvensis, Sonchus asper, S. oleraceus, Stellaria media, Veronica persica.
Rare: Atriplex hastata, Cirsium vulgare, Euphorbia helioscopia, Galeopsis tetrahit, Galium aparine, Lamium hybridum, Polygonum lapathifolium, Raphanus raphanistrum, Rumex obtusifolius, Senecio vulgaris.

Habitat Study 96   Field next to Cradock's Ashes, Walton on the Wolds
SK 634203   Area: 1.8 ha (4.6 acres)   Alt.: 122 m. (435 ft.)
Recorders: P. H. Gamble and A. L. Primavesi. Surveyed: 3rd October 1971.
This field, lying immediately to the west of Cradock's Ashes, was at the time of survey bearing a crop of sugar beet. The study is of a rectangular area bounded by the road on its south side, the hedgerow continuing the line of the spinney on the north side, the track adjacent to the spinney on the east and an arbitrary line 100 m. from this track on the west. The field lies on Chalky Boulder Clay. A sample of the surface soil had pH 6.8.

Abundant: Euphorbia exigua, Kickxia elatine.
Frequent: Aethusa cynapium, Anagallis arvensis, Aphanes arvensis, Atriplex patula, Lapsana communis, Matricaria perforata, Poa annua, Polygonum aviculare, Ranunculus repens, Senecio vulgaris, Sinapis arvensis, Sonchus asper, Stellaria media, Trifolium repens, Veronica persica, Viola arvensis.
Locally frequent: Bilderdykia convolvulus, Stachys arvensis, Valerianella dentata.
Occasional: Cirsium arvense, C. vulgare, Equisetum arvense, Galium aparine, Geranium dissectum, Myosotis arvensis, Plantago major, Poa trivialis, Rumex crispus.
Rare: Odontites verna, Phleum pratense subsp. bertolonii, Polygonum persicaria, Potentilla reptans, Ranunculus arvensis, Sonchus oleraceus, Trifolium dubium, T. hybridum.

Habitat Study 97   Field near Thrussington Mill, Thrussington
SK 656162   Area: 1.6 ha (4.1 acres)   Alt.: 61 m. (200 ft.)
Recorders: P. H. Gamble and A. L. Primavesi. Surveyed: 3rd October 1971.
This field, immediately south of the Thrussington to Hoby road and east of the lane leading to Thrussington Mill, was at the time of survey bearing a crop of cabbages. The area surveyed is bounded by hedgerows to the east and west, the roadside hedgerow to the north and an arbitrary line 100 m. from the road to the south. A wide headland adjacent to the road was included in the survey. The field lies on River Gravel and Alluvium.

Abundant: Bilderdykia convolvulus, Chamomilla recutita, Chenopodium album, Poa annua, Polygonum aviculare, Stellaria media, Veronica persica.
Locally abundant: Chamomilla suaveolens, Chrysanthemum segetum.

Frequent: Aethusa cynapium, Agrostis gigantea, Cirsium arvense, Elymus repens, Galium aparine, Matricaria perforata, Senecio vulgaris, Sinapis arvensis, Sonchus oleraceus, Viola arvensis.
Locally frequent: Filaginella uliginosa, Fumaria officinalis, Plantago major, Rumex crispus, R. obtusifolius, Urtica urens.
Occasional: Papaver rhoeas, Poa trivialis, Polygonum lapathifolium, Raphanus raphanistrum, Sonchus arvensis, S. asper, Spergula arvensis, Taraxacum spp., Tussilago farfara, Veronica agrestis.
Rare: Anchusa arvensis, Barbarea vulgaris, Lamium amplexicaule, L. purpureum, Papaver dubium, Solanum tuberosum, Thlaspi arvense, Urtica dioica.

Habitat Study 98   Arable field near King Lud's Entrenchments, Sproxton
SK 863278   Area: 5.0 ha (12.3 acres)   Alt.: 122 m. (400 ft.)
Recorder: S. H. Bishop. Surveyed: 30th October 1978.

This study is of a cereal field immediately south of Cooper's Plantation and King Lud's Entrenchments. It is bounded on the east and west by scrub woodland and is open to a road on the south. The field lies on Inferior Oolite. A sample of the stony surface soil had pH 7.4.

Abundant: Anagallis arvensis, Poa annua, Stellaria media, Veronica persica.
Frequent: Kickxia elatine, Silene noctiflora.
Locally frequent: Aphanes arvensis, Arenaria serpyllifolia, Kickxia spuria, Polygonum aviculare, Veronica arvensis, Viola arvensis.
Occasional: Capsella bursa-pastoris, Cerastium arvense, Lamium amplexicaule, Matricaria perforata, Myosotis arvensis.
Rare: Bilderdykia convolvulus, Carduus nutans, Chaenorhinum minus, Chamomilla suaveolens, Cirsium arvense, Euphorbia exigua, Galium mollugo, Lapsana communis, Plantago major, Senecio vulgaris, Sherardia arvensis, Silene vulgaris, Valerianella dentata.

Quarries and Sites of mineral extraction

Almost every kind of rock which occurs at the surface in Leicestershire, and some which occur at considerable depth, have been quarried or mined at some time. Ironstone quarrying has recently ceased after nearly a century, but the mining of brickclay, coal, roadstone, sand and gravel are still major industries in the county. The results of all these activities are the formation of a cavity from which the material was extracted, and the accumulation nearby of heaps of spoil or overburden. These man-made habitats can be of considerable botanical interest. For instance, they often have a large population of orchids. Almost all the Leicestershire records of *Ophrys apifera* are from such sites. An active site of mineral extraction is usually, apart from gravel pits, of little botanical interest. The mineral face is of necessity bare, and the spoil heaps are continually being added to. All the following studies are therefore of disused workings, some of them of great age.

It must be admitted that the range of sites for habitat studies in this section is inadequate. Though the ones selected are of great interest, there are many omissions. One major deficiency is the absence of a study of a typical gravel pit, because these pits are now an important source of wetland habitats in the county. The one selected, at Caves Inn (study 100) is somewhat atypical. It is also a pity that the ironstone quarry at Brown's Hill, Holwell was not made the subject of a study, and that there are no studies of coal mine spoil heaps.

Habitat Study 99   Huncote Quarry, Huncote
SP 512969   Area: 3.4 ha (8.3 acres)   Alt.: 91 m. (300 ft.)
Recorder: S. H. Bishop. Surveyed: 31st May 1980.
The study covers the whole of the disused quarry, which has three distinct layers or floors. The middle floor is particularly interesting, being boulder strewn, dry and stony. Part of the quarry is used as a rifle range. The quarry is in diorite. A sample of the thin surface soil from the middle floor had pH 4.8.

Abundant: Achillea millefolium, Hieracium sp., Hypericum perforatum, Lotus corniculatus, Pilosella officinarum, Plantago lanceolata.
Locally abundant: Aira praecox, Centranthus ruber, Logfia minima, Vulpia bromoides.
Frequent: Aira caryophyllea, Aphanes arvensis, Arabidopsis thaliana, Bromus hordeaceus, Centaurea nigra, Crataegus monogyna, Dactylis glomerata, Galium verum, Geranium molle, Luzula campestris,

Rosa canina, Rubus fruticosus agg., Taraxacum section Erythrosperma, Ulex europaeus, Vicia sativa.

Locally frequent: Betula pendula, Cerastium glomeratum, Festuca ovina, Galium aparine, Hedera helix, Moehringia trinervia, Myosotis discolor, Poa annua, Rumex acetosella, Senecio squalidus, S. sylvaticus, Trifolium arvense, Vicia angustifolia, V. hirsuta.

Occasional: Anthriscus sylvestris, Campanula rotundifolia, Crepis vesicaria, Dryopteris filix-mas, Epilobium angustifolium, Erigeron acer, Erophila verna, Fraxinus excelsior, Lamium purpureum, Salix cinerea subsp. oleifolia, Sambucus nigra, Silene alba, Stachys sylvatica, Stellaria media, Tanacetum parthenium, Taraxacum spp., Urtica dioica, Veronica arvensis.

Rare: Cirsium vulgare, Cytisus scoparius, Dryopteris dilatata, Inula conyza, Malus sylvestris, Pinus sylvestris (seedling), Prunella vulgaris, Salix caprea, Sedum acre, Senecio vulgaris, Silene dioica, Solanum dulcamara, Sonchus asper, S. oleraceus.

Habitat Study 100 Cave's Inn Pits, Shawell
SP 539795 Area: 2.8 ha (7 acres) Alt.: 107 m. (350 ft.) S.S.S.I.
Recorders: E. K. and J. M. Horwood. Surveyed: 6th July 1969.

These old gravel workings are on the county boundary south-east of Shawell Church. They are adjacent to a dismantled railway on the north-west side and Watling Street on the south-west. The study area is a roughly circular section of the northern part of the workings and includes soil and gravel banks, low lying marshy areas and pools with shallow margins. The vegetation includes *Sphagnum* moss as well as calcicole species. The sand and gravel is of glacial origin.

Abundant: Anthoxanthum odoratum, Arrhenatherum elatius, Equisetum arvense, Festuca rubra, Fragaria vesca, Leontodon hispidus, Trifolium dubium, Trisetum flavescens, Tussilago farfara.

Locally abundant: Dactylorhiza fuchsii, Potamogeton obtusifolius.

Frequent: Alopecurus pratensis, Bellis perennis, Bromus sterilis, Cynosurus cristatus, Juncus inflexus, Lathyrus pratensis, Lolium perenne, Phalaris arundinacea, Pilosella officinarum, Poa trivialis, Prunella vulgaris, Ranunculus acris, Rosa canina, Trifolium pratense, T. repens.

Locally frequent: Eleocharis palustris, Juncus acutiflorus, Ophrys apifera.

Occasional: Achillea millefolium, Alisma plantago-aquatica, Alopecurus geniculatus, Bromus × pseudothominii, Cardamine pratensis, Carex flacca, C. pseudocyperus, Chaerophyllum temulentum, Cirsium palustre, Deschampsia cespitosa, Desmazeria rigida, Epilobium angustifolium, Erigeron acer, Festuca ovina, Geranium pratense, Holcus lanatus, Hypochoeris radicata, Juncus effusus, Lemna minor, L. trisulca, Leontodon taraxacoides, Lotus corniculatus, Myosotis arvensis, Ranunculus repens, Rumex conglomeratus, R. obtusifolius, Sagina procumbens, Salix caprea, Senecio jacobaea, Torilis japonica, Veronica arvensis, Vicia sativa, Vulpia bromoides.

Rare: Agrimonia eupatoria, Anacamptis pyramidalis, Artemisia vulgaris, Carlina vulgaris, Crataegus monogyna, Dryopteris filix-mas, Epilobium montanum, E. parviflorum, Hypericum tetrapterum, Rhamnus catharticus, Taxus baccata.

Habitat Study 101 Disused slate quarry, Swithland Wood, Newtown Linford
SK 539122 Area: 1.2 ha (3 acres) Alt.: 107-114 m. (350-375 ft.) S.S.S.I.
Recorder: P. H. Gamble. Surveyed: 1976.

This is a study of a roughly circular area about 130 m. in diameter. It includes a deep flooded slate quarry, last worked over a century ago, with high vertical walls, and surrounding areas of partially vegetated slate spoil. It is bounded by a timber fence. The rock is Swithland Slate of Precambrian age. A sample of surface soil from the cliff area had pH 3.4 and from a spoil heap pH 3.6.

Abundant: Aira praecox, Betula pendula, Deschampsia flexuosa, Rumex acetosella, Teucrium scorodonia.

Locally abundant: Cerastium glomeratum, Holcus mollis, Poa annua, Polygonum amphibium, Pteridium aquilinum.

Frequent: Digitalis purpurea, Dryopteris filix-mas, Hypochoeris radicata, Taraxacum section Erythrosperma.

Locally frequent: Anthoxanthum odoratum, Anthriscus sylvestris, Aphanes microcarpa, Cardamine hirsuta, Cerastium semidecandrum, Cytisus scoparius, Dryopteris borreri, D. dilatata, Epilobium angustifolium, Festuca ovina, Fragaria vesca, Hedera helix, Hieracium diaphanum, H. salticola, H. vulgatum, Juncus effusus, Lonicera periclymenum, Luzula campestris, L. sylvatica, Pilosella officinarum, Poa nemoralis, P. trivialis, Polypodium interjectum, Ranunculus bulbosus, Senecio sylvaticus, Teesdalia nudicaulis, Umbilicus rupestris, Veronica arvensis.

Occasional: Acer pseudoplatanus, Betula pubescens, Corylus avellana, Epilobium montanum, Erophila verna, Fraxinus excelsior, Hypericum pulchrum, Quercus petraea, Q. robur, Ranunculus repens, Sagina apetala, Salix caprea, Scrophularia auriculata, S. nodosa, Sorbus aucuparia, Ulex europaeus.

Rare: Calluna vulgaris, Danthonia decumbens, Eleocharis palustris, Epipactis helleborine, Ilex aquifolium (seedlings), Senecio vulgaris, Taxus baccata (seedlings), Tussilago farfara.

Habitat Study 102 Waltham Quarry, Waltham
SK 800264 Area: 2.5 ha (6.2 acres) Alt.: 153 m. (500 ft.)
Recorders: S. H. Bishop, P. H. Gamble and A. L. Primavesi. Surveyed: 29th May 1978 and 27th August 1979.

The area studied is a disused limestone quarry, mostly open but becoming well colonised and dominated by scrub on its western side. The grassy floor of the quarry is sheltered by the 5 to 7 m. high eroded cliff, and much of the area is covered by low banks of spoil separated by numerous despressions. *Ophrys apifera* was observed here in quantity in 1973, but was not recorded during the survey for this study. This is a Lincolnshire Limestone quarry. Samples of surface soil from the quarry floor had pH 7.6 and from the spoil heaps pH 8.1.

Abundant: Bellis perennis, Lotus corniculatus, Medicago lupulina.
Locally abundant: Arrhenatherum elatius, Crataegus monogyna, Dactylis glomerata, Linum catharticum, Pilosella officinarum, Poa pratensis, Tussilago farfara.
Frequent: Achillea millefolium, Anthriscus sylvestris, Aphanes arvensis, Arenaria serpyllifolia, Bromus hordeaceus, Carex flacca, Centaurea nigra, Cerastium fontanum subsp. triviale, C. glomeratum, Cirsium arvense, C. eriophorum, Convolvulus arvensis, Equisetum arvense, Heracleum sphondylium, Knautia arvensis, Lathyrus pratensis, Leontodon autumnalis, L. hispidus, Plantago lanceolata, Poa annua, Prunella vulgaris, Ranunculus bulbosus, R. repens, Rumex acetosa, Senecio jacobaea, Trifolium dubium, Veronica chamaedrys, Vicia sativa.
Locally frequent: Anthoxanthum odoratum, Cirsium vulgare, Desmazeria rigida, Epilobium angustifolium, Festuca rubra, Luzula campestris, Myosotis arvensis, Potentilla reptans, Rosa canina, Rubus fruticosus agg., Senecio erucifolius, Taraxacum section Erythrosperma, Torilis japonica, Trifolium repens, Trisetum flavescens, Urtica dioica, Veronica arvensis.
Occasional: Agrostis stolonifera, Alliaria petiolata, Astragalus glycyphyllos, Carduus acanthoides, Centaurea scabiosa, Crepis capillaris, C. vesicaria, Cynosurus cristatus, Galium verum, Geranium dissectum, G. molle, G. pusillum, Hypochoeris radicata, Hieracium spp., Lamium album, Phleum pratense subsp. bertolonii, Primula veris, Rumex obtusifolius, Sagina procumbens, Taraxacum spp., Veronica serpyllifolia.
Rare: Agrimonia eupatoria, Alopecurus pratensis, Anagallis arvensis, Artemisia vulgaris, Capsella bursa-pastoris, Cardamine hirsuta, Epilobium montanum, Festuca ovina, Fraxinus excelsior, Glechoma hederacea, Odontites verna, Plantago major, P. media, Quercus robur, Sambucus nigra, Senecio squalidus, Sonchus asper, Stellaria media, Tamus communis, Tanacetum parthenium, Tragopogon pratensis, Trifolium pratense, Vicia sepium.

Habitat Study 103   Stonesby Quarry, Stonesby, Sproxton
SK 814252   Area: 2.5 ha (6.22 acres)   Alt.: 171 m. (560 ft.) S.S.S.I.; Nature Reserve.
Recorders: S. H. Bishop, P. H. Gamble and A. L. Primavesi. Surveyed: 30th June 1979.
This old limestone quarry has become rather overgrown in recent years, much of it now being covered with scrub. A rich assemblage of calcicoles still survives on the small open exposures, on stony ground along the approach road and on the wide grassy margins. The site lies on Lincolnshire Limestone. A sample of the shallow stony soil had pH 7.8.

Locally abundant: Arrhenatherum elatius, Bromus erectus, Dactylis glomerata, Primula veris.
Frequent: Achillea millefolium, Briza media, Carex flacca, Centaurea nigra, Galium verum, Holcus lanatus, Lathyrus pratensis, Lotus corniculatus, Plantago lanceolata, P. media, Trisetum flavescens.
Locally frequent: Avenula pubescens, Bellis perennis, Centaurea scabiosa, Cerastium fontanum subsp. triviale, Cirsium acaule, Clinopodium vulgare, Crataegus monogyna, Euphrasia nemoralis, Gentianella amarella, Linum catharticum, Poa pratensis, P. trivialis, Polygala vulgaris, Rosa canina, Rubus fruticosus agg., Sanguisorba minor subsp. minor, Trifolium campestre, T. dubium, T. repens, Viola odorata.
Occasional: Anacamptis pyramidalis, Anthriscus sylvestris, Anthyllis vulneraria, Cirsium arvense, C. eriophorum, Convolvulus arvensis, Cynosurus cristatus, Deschampsia cespitosa, Erophila verna, Galium aparine, Knautia arvensis, Leucanthemum vulgare, Medicago lupulina, Odontites verna, Ononis repens, Phleum pratense subsp. bertolonii, Potentilla reptans, Prunella vulgaris, Scabiosa columbaria, Senecio jacobaea, Thymus praecox subsp. arcticus, Torilis japonica, Trifolium pratense, Veronica chamaedrys, Vicia cracca.
Rare: Agrimonia eupatoria, Campanula glomerata, Cerastium glomeratum, Chaenorhinum minus, Cornus sanguinea, Dactylorhiza fuchsii, Daucus carota, Dryopteris filix-mas, Festuca ovina, F. pratensis, Fraxinus excelsior, Fumaria officinalis, Geranium pratense, G. robertianum, Glechoma hederacea, Lamium album, Lolium perenne, Myosotis arvensis, Pimpinella major, Plantago major, Prunus spinosa, Ranunculus acris, R. repens, Rhamnus catharticus, Rumex acetosa, R. crispus, R. obtusifolius, Salix caprea, Sambucus nigra, Sherardia arvensis, Silaum silaus, Silene dioica, S. vulgaris, Solanum dulcamara, Sonchus asper, Stachys sylvatica, Stellaria graminea, Tamus communis, Tragopogon pratensis, Tussilago farfara, Urtica dioica, Vicia sativa, V. sepium, Viola hirta.

Miscellaneous habitats

The four studies in this section are of habitats which will not fit into any of the preceding categories. One of them, the refuse tip at Barwell (study 104) is typical of such sites throughout the county, with its assemblage of ruderal species, arable weeds, aliens and adventives. The other three studies in this section all have one thing in common; they are sites which because of the nature of the terrain or the previous history of the site cannot be exploited for any economically useful purpose, and which have acquired over many years an interesting flora.

Habitat Study 104 Refuse tip, Barwell, Hinckley
SP 438973 Area: 10.7 ha (26.6 acres) Alt.: 107 m. (350 ft.)
Recorders: S. H. Bishop and J. E. Dawson. Surveyed: 19th September 1979.

This extensive refuse tip lies to the north-west of Barwell and to the west of Stapleton Lane. It is one of the main domestic refuse tips for the Hinckley and Bosworth Borough Council. The study area is contained by high hedgerows.

Abundant: Atriplex hastata, Chenopodium album, Matricaria perforata, Polygonum aviculare, Sonchus oleraceus.

Locally abundant: Agrostis capillaris, Artemisia vulgaris, Elymus repens, Epilobium angustifolium, Dactylis glomerata, Medicago lupulina, Urtica dioica.

Frequent: Aethusa cynapium, Atriplex patula, Capsella bursa-pastoris, Chamomilla suaveolens, Cirsium arvense, Epilobium montanum, Euphorbia helioscopia, E. peplus, Lapsana communis, Melilotus altissima, Plantago lanceolata, P. major, Poa annua, Polygonum persicaria, Ranunculus repens, Rumex obtusifolius, Senecio squalidus, S. vulgaris, Sisymbrium officinale, Sonchus asper, Stellaria media, Trifolium repens, Tussilago farfara.

Locally frequent: Achillea millefolium, Agrostis stolonifera, Arrhenatherum elatius, Bromus sterilis, Calystegia silvatica, Chamomilla recutita, Crepis capillaris, Deschampsia cespitosa, Dipsacus fullonum, Epilobium adenocaulon, Equisetum arvense, Geranium dissectum, Helianthus annuus, Holcus lanatus, Hordeum murinum, Leontodon autumnalis, Linaria purpurea, Lolium perenne, Lycopersicon esculentum, Phalaris arundinacea, P. canariensis, Phleum pratense subsp. pratense, Senecio viscosus, Solanum tuberosum, Taraxacum spp., Trifolium pratense, Veronica persica.

Occasional: Alliaria petiolata, Anthriscus sylvestris, Antirrhinum majus, Armoracia rusticana, Bilderdykia convolvulus, Cerastium fontanum subsp. triviale, Chenopodium polyspermum, C. rubrum, Cirsium vulgare, Epilobium hirsutum, Fumaria officinalis, Galium aparine, Heracleum sphondylium, Juncus inflexus, Lamium album, Leucanthemum vulgare, Lolium multiflorum, Lotus corniculatus, Malva sylvestris, Papaver somniferum, Phleum pratense subsp. bertolonii, Potentilla reptans, Reseda luteola, Rumex acetosella, R. crispus, Sedum acre, Setaria viridis, Sinapis arvensis, Solanum dulcamara, Solidago altissima, Sonchus arvensis, Tanacetum parthenium, Veronica agrestis.

Rare: Acer pseudoplatanus (seedling), Aegopodium podagraria, Ammi visnaga, Anagallis arvensis, Arctium minus, Artemisia absinthium, Asparagus officinalis, Aster novi-belgii, Atriplex hortensis, Avena fatua, Ballota nigra, Barbarea vulgaris, Bellis perennis, Beta vulgaris, Brassica oleracea, Carduus acanthoides, Centaurea diluta, C. nigra, Chelidonium majus, Coronopus squamatus, Crataegus monogyna, Cytisus scoparius, Daucus carota, Digitalis purpurea, Epilobium parviflorum, Erodium cicutarium, Euphorbia exigua, Fuchsia magellanica, Geranium pusillum, Filaginella uliginosa, Hedera helix, Iberis umbellata, Juncus bufonius, J. effusus, Lamium purpureum, Lepidium ruderale, Lobularia maritima, Lotus tenuis, Lysimachia nummularia, Lythrum junceum, Malus sylvestris, Melilotus alba, M. indica, Mentha × piperita, M. spicata, M. × villosa, Myosotis arvensis, Odontites verna, Papaver dubium, P. rhoeas, Pastinaca sativa, Panicum mileaceum, Picris echioides, Polygonum amphibium, Prunella vulgaris, Quercus robur (seedling), Ranunculus sceleratus, Reynoutria japonica, Ribes nigrum, R. sanguineum, Rosa canina, Rubus fruticosus agg., Sagina apetala, S. procumbens, Salix alba, S. caprea, S. cinerea subsp. oleifolia, S. fragilis, Sambucus nigra, Sanguisorba officinalis, Senecio squalidus × vulgaris, Sherardia arvensis, Silene dioica, Solanum nigrum, Stachys sylvatica, Stellaria graminea, Symphytum officinale, Tanacetum vulgare, Tragopogon pratensis, Trifolium hybridum, Tritonia × crocosmiflora, Typha latifolia, Urtica urens, Verbascum thapsus, Vicia sativa, Viola arvensis.

Habitat Study 105 Ruins of Cotes Hall, Cotes
SK 553209 Area: part of site of area 1.2 ha (3 acres) Alt.: 38 m. (125 ft.) S.S.S.I.
Recorder: P. H. Gamble. Surveyed: 1975.

The study area is on the north-east bank of the River Soar, between the river and Stanford Lane. The study includes those species growing on and against the remains of the old brick and stone walls and on grassy mounds over foundations.

Abundant: Arabidopsis thaliana, Arenaria serpyllifolia, Capsella bursa-pastoris, Senecio vulgaris, Veronica arvensis.

Locally abundant: Bromus sterilis, Medicago arabica, Parietaria diffusa, Poa annua, Reseda luteola, Saxifraga tridactylites, Stellaria media.

Frequent: Carduus nutans, Cerastium glomeratum, Galium aparine, Hedera helix, Taraxacum section Erythrosperma.

Locally frequent: Achillea millefolium, Arenaria leptoclados, Bellis perennis, Cerastium semidecandrum, Crepis capillaris, Cymbalaria muralis, Glechoma hederacea, Lamium purpureum, Myosotis ramosissima, Pilosella officinarum, Poa compressa, Senecio squalidus, Sisymbrium officinale, Stachys sylvatica, Urtica dioica, Veronica chamaedrys, V. hederacea, Vulpia bromoides.

Occasional: Cerastium fontanum subsp. triviale, Chamomilla suaveolens, Crataegus monogyna, Crepis vesicaria, Cynosurus cristatus, Geranium dissectum, G. molle, G. robertianum, Silene alba, Solanum dulcamara, Sonchus asper, Taraxacum spp., Trifolium dubium, Valerianella locusta.

Rare: Arctium minus, Ballota nigra, Bryonia cretica subsp. dioica, Carduus acanthoides, Lamium hybridum, Prunus cerasifera, Rhamnus catharticus, Rosa canina, Salvia verbenaca, Sambucus nigra, Trifolium striatum, Ulmus procera.

Habitat Study 106 Bank by River Soar, Cotes
SK 554208 Area: not calculated. Alt.: 36 m. (120 ft.)
S.S.S.I.
Recorder: P. H. Gamble. Surveyed: 1975.

The study site is an area approximately 200 m. × 18 m. on a raised terrace bank adjacent to and partly flanking the River Soar, located between the former site of Cotes Church and the ruins of Cotes Hall. The bank occupies part of a field frequently grazed and at times heavily trampled by cattle, a factor which helps to keep some of the steeper less stable parts open. It is situated on River Gravels. Two samples of surface soil had pH 5.5 and 6.4.

Abundant: Bellis perennis, Medicago arabica, Plantago lanceolata, Ranunculus bulbosus, Taraxacum lacistophyllum.

Locally abundant: Alopecurus pratensis, Aphanes microcarpa, Cerastium glomeratum, Geranium molle, Salvia verbenaca, Urtica dioica (Ew).

Frequent: Achillea millefolium, Arenaria serpyllifolia, Carduus nutans, Cirsium arvense, Cynosurus cristatus, Dactylis glomerata, Leontodon autumnalis, Lotus corniculatus, Pilosella officinarum, Plantago media, Poa annua, Potentilla reptans, Sisymbrium officinale, Taraxacum pallescens, Trifolium dubium, Veronica arvensis.

Locally frequent: Bromus hordeaceus, Capsella bursa-pastoris, Cerastium semidecandrum, Cirsium vulgare, Conium maculatum (Ew), Erophila verna, Galium verum, Koeleria macrantha, Montia fontana, Poa pratensis, Ranunculus ficaria, Senecio squalidus (Ew), Stellaria media, S. pallida, Trifolium striatum, T. subterraneum, Veronica chamaedrys, Vulpia bromoides.

Occasional: Anthriscus sylvestris, Arrhenatherum elatius, Ballota nigra, Barbarea vulgaris (Ew), Cardamine flexuosa (Ew) Carduus acanthoides (Ew), Cerastium fontanum subsp. triviale, Convolvulus arvensis, Crataegus monogyna, Crepis capillaris, Dipsacus fullonum, Epilobium hirsutum (Ew) E. montanum, Festuca arundinacea, F. ovina, F. rubra, Galium aparine, Geranium dissectum, G. pusillum, Glechoma hederacea, Hordeum murinum, Lamium album, L. purpureum, Lolium perenne, Lycopus europaeus (Ew), Malva sylvestris, Phleum pratense subsp. bertolonii, Poa compressa, P. trivialis (Ew), Ranunculus sceleratus (Ew), Rumex conglomeratus, R. crispus, R. obtusifolius, Salix fragilis, Sambucus nigra, Scrophularia auriculata (Ew), Senecio jacobaea, S. vulgaris, Sonchus arvensis (Ew), S. asper, Torilis nodosa, Trifolium pratense, T. repens, Tussilago farfara, Trisetum flavescens, Veronica hederifolia, V. persica.

Rare: Barbarea stricta (Ew), Coronopus squamatus, Crepis vesicaria, Medicago lupulina, Pimpinella saxifraga, Reseda luteola, Rosa canina, Saxifraga granulata, Ulmus procera, Vicia sepium.

Habitat Study 107 Castle Hill, Mountsorrel
SK 582149 Area: 1.3 ha (3.4 acres) Alt.: 53-69 m. (175-225 ft.)
Recorder: P. H. Gamble. Surveyed: 11th July 1980.

This prominence, standing above the village of Mountsorrel, is the site of a long-vanished castle, and since the First World War the location of the village war memorial. The hill, with rocky outcrops and steep slopes, is largely grassy with a fair amount of scrub. Some of the lower and less steep areas are intermittently grazed by horses. Both granodiorite and Boulder Clay are present, with sandy soil associated with the outcrops of the former and heavy clay soil with the Boulder Clay. Samples of surface soil had pH 4.6 and 6.7.

Abundant: Bellis perennis, Lolium perenne.

Locally abundant: Bromus sterilis, Epilobium angustifolium, Festuca ovina, Geranium pusillum, Phleum pratense subsp. bertolonii, Pilosella officinarum, Plantago lanceolata, Ranunculus bulbosus, Torilis japonica, Trifolium arvense, T. dubium, T. repens, Urtica dioica.

Frequent: Achillea millefolium, Crataegus monogyna, Crepis capillaris, Festuca rubra, Fraxinus excelsior, Galium verum, Hypochoeris radicata, Leontodon autumnalis, Rubus fruticosus agg., Sambucus nigra, Taraxacum spp., Trisetum flavescens.

Locally frequent: Aegopodium podagraria, Agrostis capillaris, Aira praecox (Sr), Anthriscus sylvestris, Aphanes arvensis, Arenaria serpyllifolia, Arrhenatherum elatius, Bromus hordeaceus, Calystegia silvatica, Capsella bursa-pastoris, Carduus nutans, Cerastium tomentosum (Sr), Cirsium arvense, Convolvulus arvensis, Cynosurus cristatus, Dactylis glomerata, Erodium cicutarium, Erophila verna, Galium aparine, Heracleum sphondylium, Hieracium spp. (Sr), Holcus lanatus, Hordeum murinum, Lamium album, Lapsana communis, Lotus corniculatus, Poa pratensis, P. trivialis, Potentilla reptans, Ranunculus repens, Rumex crispus, Sedum reflexum (Sr), Senecio jacobaea, Silene dioica, Tragopogon pratensis, Trifolium pratense, T. striatum,. Veronica chamaedrys, Vicia angustifolia.

Occasional: Ballota nigra, Campanula rotundifolia, Cerastium fontanum subsp. triviale, Cirsium vulgare, Dryopteris dilatata (Sr), Galeopsis tetrahit, Geranium molle, Hieracium perpropinquum, Lonicera periclymenum, Medicago lupulina, Ornithopus perpusillus, Poa annua, Rosa canina, Rubus ulmifolius, Sedum acre (Sr), Senecio squalidus (Sr), Sisymbrium officinale, Sonchus oleraceus, Spergularia rubra, Stellaria graminea, S. media, Tanacetum parthenium, Ulmus procera.

Rare: Acer pseudoplatanus, Carex spicata, Dryopteris filix-mas, Geranium dissectum, Trifolium subterraneum.

Plate 1. A. R. Horwood, Sub-Curator, Leicester Museum. Photograph taken on the occasion of the British Association visit to Leicester, July 1907.

Plate 2. F. A. Sowter, January 1971.

Plate 3. Leicestershire Flora Committee, December 1987. Standing (left to right): K. G. Messenger, Dr. F. R. Green, Mrs J. M. Horwood, P. H. Gamble, S. H. Bishop. Seated (left to right): M. Walpole, Rev. A. L. Primavesi, Mrs P. A. Evans, I. M. Evans.

Plate 4. Herb-Paris *Paris quadrifolia* in a recently coppiced area of Great Merrible Wood, SP 836962, May 1984.

Plate 5. Large coppice stool of small-leaved lime *Tilia cordata*, Owston Big Wood, SK 794065, May 1977.

Plate 6. Bluebell *Hyacinthoides non-scripta* under oak, Spring Wood, Staunton Harold, SK 379227, May 1983.

Plate 7. Eighteen-year-old planting of larch and sycamore, Out Woods, Loughborough, SK 518158, October 1964.

Plate 8. Pasture on ridge and furrow with abundant bulbous buttercup *Ranunculus bulbosus*, Birstall, SK 585090, looking north, May 1977.

Plate 9. Glebe meadow crossed by former brook course, Barkby, SK 654089, July 1974. Ploughed up late 1974.

Plate 10. Grassland and gorse scrub on marlstone scarp, north-east of Burrough Hill, SK 763122, March 1971.

Plate 11. Siliceous grassland, habitat for hoary cinquefoil *Potentilla argentea* and subterranean clover *Trifolium subterraneum*, Croft Pasture, SP 509959, June 1975. S. H. Bishop in foreground.

Plate 12. Roadside verge after summer cut, Lowesby, SK 735082, July 1975.

Plate 13. Wide roadside verge with invading scrub, A6006 west of Eller's Farm, Wymeswold, SK 636231, looking west, June 1975.

Plate 14. Grassland with pignut *Conopodium majus*, churchyard, Packington, SK 358144, June 1980.

Plate 15. Wet heathland dominated by purple moor-grass *Molinia caerulea*, Charnwood Lodge, SK 466157, looking south, July 1975.

Plate 16. Bilberry *Vaccinium myrtillus* on the summit of Timberwood Hill, Charnwood Lodge, SK 470148, looking north-east towards M1, October 1975.

Plate 17. Precambrian outcrop with bilberry *Vaccinium myrtillus* and heather *Calluna vulgaris*, High Sharpley, SK 447170, looking south-east, May 1979.

Plate 18. Spring-fed marsh, with greater tussock-sedge *Carex paniculata* and common reed *Phragmites australis*, Tilton, SK 765040, April 1968.

Plate 19. Marshy grassland with marsh marigold *Caltha palustris* and lady's-smock *Cardamine pratensis*, Tom Long's Meadow, Quorn, SK 557165, May 1975.

Plate 20. Aerial view of Narborough Bog, SP 5497, looking north-west, April 1973.

Plate 21. Aerial view of subsidence pools in valley of Saltersford Brook, Oakthorpe and Donisthorpe, SK 3113/3213, looking north, July 1975. Note sinuous course of former Ashby Canal.

Plate 22. Greater spearwort *Ranunculus lingua* on margin of Knaptoft Pond, SP 635877, July 1975.

Plate 23. Golden dock *Rumex maritimus* on draw-down zone, Charley side of Blackbrook Reservoir, SK 459170, looking north-west, October 1976. Water level rising after drought summer.

Plate 24. Grand Union Canal, Foxton, from bridge at SP 707897, looking west, August 1965. Subsequent increase in use has reduced the aquatic flora.

Plate 25. River Eye at Ham Bridge, Stapleford, SK 801186, August 1987.

Plate 26. Deep cutting on dismantled railway line, Thorpe Satchville, SK 728127, looking north, October 1972.

Plate 27. Disused railway sidings, Swithland, SK 564131, looking south, July 1975.

Plate 28. Marsh and willow scrub, disused sandpit, Shawell, SP 539795, looking east, July 1970.

Plate 29. Long-disused limestone working, habitat for small scabious *Scabiosa columbaria*, Stonesby Quarry, SK 813250, September 1974.

Plate 30. Aerial view of arable at Bottesford in the Vale of Belvoir, SK 8037, looking north, September 1975. Grantham Canal in foreground.

Plate 31. Common Poppy *Papaver rhoeas* and scentless mayweed *Matricaria perforata* in unsprayed margin of barley field, Ashby Parva, SP 514891, July 1980.

Plate 32. Aerial view of clay workings on site of former Moira Reservoir, Ashby Woulds, SK 3016, looking north, April 1973.

Plate 33. Waste ground with hoary cress *Cardaria draba*, Humberstone Lane, Leicester, SK 616083, June 1974.

RECORDING THE FLORA

The numerous local Floras published since the Second World War have provided several excellent examples of what a modern County Flora should contain. All these Floras have their particular good qualities, and comparison would be invidious. However, one of the peculiar attractions of local Floras in general is their individuality, and it was agreed from the outset that for this work it would be a mistake to follow slavishly anyone else's model.

Comparison of modern local Floras with their older counterparts shows the value of a system of recording based upon the National Grid of the Ordnance Survey. In the days when there was no satisfactory system of map references, writers of local Floras were obliged to base the location of their records on parishes or arbitrarily defined botanical districts. It is thus difficult to obtain from these older Floras an adequate idea of the distribution of species. Even when localities are given for individual records, these are often impossible to locate accurately. With the National Grid it is possible to provide a system of references available to anyone who possesses an Ordnance Survey map, and to give an account of distribution which is both detailed and accurate.

Also, as Messenger has pointed out in the *Flora of Rutland* (1971), the National Grid imposes a discipline on the field workers which ensures that, although it is obviously impossible to examine every square metre of the county, at least the whole area has been covered. The older Flora writers were inclined to concentrate upon the botanically interesting regions, and to neglect those which were dull or difficult of access.

The area surveyed

The limit of the area surveyed is the boundary of the administrative county of Leicestershire immediately prior to the incorporation of Rutland in 1974. This corresponds nearly, but not exactly, to the Watsonian vice-county 55 with the exclusion of Rutland (figure 1).

Throughout this work the area defined above is referred to as 'the county' of Leicestershire. This expedient is adopted to avoid much circumlocution and repetition, but it must be clearly understood that the area so described does not correspond to the present administrative county, because Rutland was excluded from the survey.

Division of the area surveyed

The area surveyed is included in the 100 km. squares SP and SK of the National Grid (figure 23). The area selected as a basis for recording was the 4 square kilometre 'tetrad'. There are 25 of these tetrads in each 10 km. square of the National Grid, and a total of 617 of them in the area surveyed, including marginal tetrads where only part of the area lies within the county boundary (figure 24).

The record cards

For recording species in the field, two types of record cards were used. The COMMON SPECIES card was printed with a list of 248 species in alphabetical order, with columns for insertion of code symbols for habitat and frequency. Any species found which were not on this card were written in the blank columns of the OTHER SPECIES card. This second type of card, in addition to the columns for habitat and frequency, had further columns for a six-figure grid reference, and for indicating that a voucher specimen of a critical species had been submitted to

						SK74	SK84	
		SK43				SK73	SK83	
	SK32	SK42	SK52	SK62	SK72	SK82		
SK21	SK31	SK41	SK51	SK61	SK71	SK81	SK91	
SK20	SK30	SK40	SK50	SK60	SK70	SK80		
SP29	SP39	SP49	SP59	SP69	SP79	SP89		
		SP48	SP58	SP68	SP78			
E J P U Z / D I N T Y / C H M S X / B G L R W / A F K Q V			SP57					

Figure 23. 10 km. squares which lie wholly or partly in the area surveyed and lettering of the tetrads.

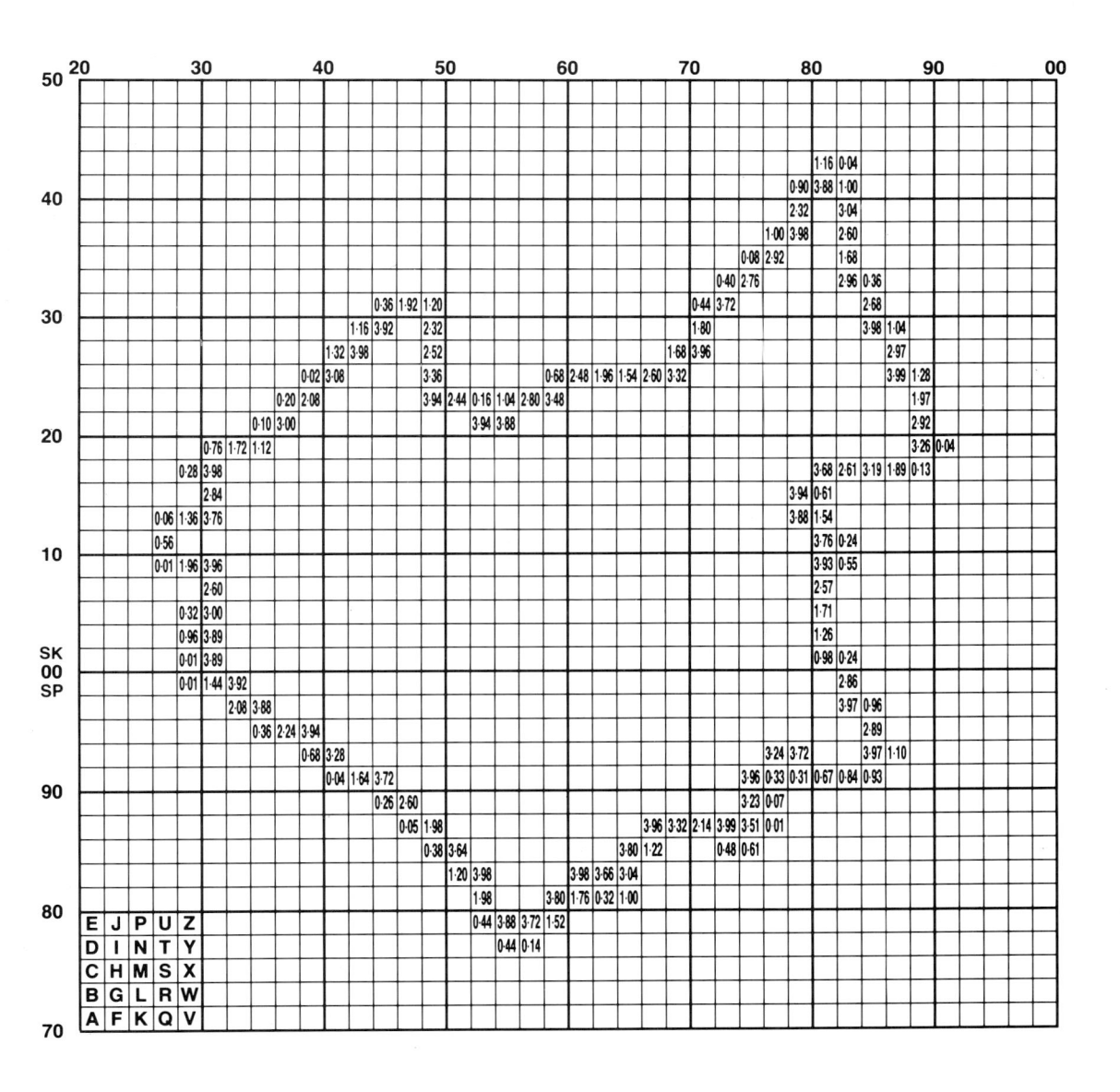

Figure 24. Areas in sq. km. of the marginal tetrads.

the appropriate referee. A pair of these two types of cards was used for each tetrad, and visits were made for recording in spring, summer and autumn.

The species card index

Field workers made an annual return of records to the Flora Committee Secretary, who transferred each year's records to a species card index. At the same time he prepared a dot map for each species. These maps were the basis for the Distribution Maps in this Flora. They were also useful to show the progress of the survey, and they revealed any apparently anomalous gaps in the records which could be dealt with as the survey proceeded. At the end of the survey this species card index provided the information for the account of each species in the Flora. It will be deposited with the Leicestershire Museums Service, together with the annual returns of fair copies of the field record cards, as a permanent record for detailed study.

Critical species

In all local Flora projects where the field work is done by a team of volunteers with varying degrees of skill and experience the problem of the so-called 'critical' species arises. In the survey for this present Flora, three categories of plants were included under this heading. First of all there are the true critical species, which can only be determined with certainty by someone who has specialised in their study. Secondly there are difficult species which anyone with time, patience and sufficient general botanical experience can identify, but for which there is sufficient risk of misidentification to justify their treatment as critical. Finally there are certain groups of species the taxonomy of which is still a matter of debate, and for which taxonomists are not agreed concerning the limits of species and their nomenclature. The list which follows of the species treated as critical for the survey includes species of all three of these categories.

List of species treated as critical

Dryopteris: *D. filix-mas* and *D. borreri*. *D. filix-mas* was also recorded as an aggregate, the referee's confirmation being required to separate the two species. Referee: P. H. Gamble.
Salix: all species except *S. fragilis*, *S. alba*, *S. purpurea*, *S. viminalis*, *S. caprea* and *S. repens*. All hybrids were treated as critical. *S. cinerea* was recorded as an aggregate, the referee's confirmation being required for the segregates. Referee: R. C. L. Howitt.
Betula: *B. pubescens*. Referee: P. H. Gamble.
Quercus: *Q. petraea*. Referee: P. A. Evans.
Ulmus: all taxa except *U. glabra* and *U. procera*. Referee: K. G. Messenger. See also introduction to the genus in the systematic account.
Chenopodium: all species except *C. bonus-henricus*, *C. album* and *C. polyspermum*. Referees: T. G. Tutin and F. R. Green.
Cerastium: all species except *C. fontanum*, *C. glomeratum*, *C. arvense* and *C. tomentosum*. Referee: K. G. Messenger.
Ranunculus subgenus Batrachium: all species. Referee: J. M. Horwood.
Rubus subgenus Rubus: all species except *R. ulmifolius*. Referees: E. S. Edees and A. Newton. See introduction to the subgenus.
Rosa: all taxa except *R. arvensis*. '*R. canina*' was recorded as an aggregate. Referees: G. G. Graham and R. Melville. See introduction to the genus.
Aphanes: both species. Referee: E. K. Horwood.
Viola: all species and hybrids except *V. odorata*, *V. hirta*, *V. riviniana* and *V. arvensis*. Referee: P. H. Gamble.
Epilobium: all species except *E. hirsutum*, *E. parviflorum*, *E. montanum* and *E. angustifolium*. Referee: E. K. Horwood.
Symphytum: all taxa. Referee: I. M. Evans

Callitriche: all species. Referee: P. H. Gamble.
Mentha: all species and hybrids except *M. arvensis* and *M. aquatica.* Referee: P. A. Evans.
Euphrasia: all species. Referee: K. G. Messenger.
Valerianella: all species. Referee: K. G. Messenger.
Taraxacum: all species. The sections were recorded as aggregates. Referees: A. J. Richards and
   C. C. Haworth. See introduction to the genus.
Hieracium: all species. Referees: C. E. A. Andrews and P. D. Sell. See introduction to the genus.
Potamogeton: all species except *P. natans, P. perfoliatus* and *P. crispus.* Referee: J. M. Horwood.
Juncus: J. articulatus, J. acutiflorus and *J. subnodulosus.* Referee: P. A. Evans.
Bromus: all species except *B. sterilis, B. ramosus* and *B. erectus.* Referee: T. G. Tutin.
Agrostis: all species. Referee: T. G. Tutin.
Carex: C. spicata, C. muricata and *C. divulsa; C. lepidocarpa* and *C. demissa.* Referee: P. A. Evans.

For most of these species field workers were required to send voucher specimens to a referee appointed by the Flora Committee. These voucher specimens were deposited, where appropriate, in herb. LSR. The referees were empowered to give certificates of exemption when they were satisfied that an individual could determine species reliably. The genera *Ulmus, Rubus, Rosa, Hieracium, Pilosella* and *Taraxacum* received special treatment during the survey; this is described in an introduction of each of these genera in the systematic account.

Analysis of the records

The species card index contains approximately 160,000 records, including 4984 records for aggregates of species. During the course of the survey, the following numbers of taxa were found.

   Species: 1198
   Additional subspecies: 10
   Hybrids: 71

Also recorded were a number of varieties, forms and other infra-specific taxa of *Ulmus, Rosa* and *Cirsium.* A further 26 species and 10 hybrids are mentioned in the systematic account as having been recorded between 1933 and 1960.

The average number of species per tetrad (excluding marginal tetrads) is 233. It seems likely that this figure and those given on the map (figure 25) showing numbers of records per tetrad represent about 80% coverage by the field workers. This is borne out by the following test case. A field worker's home tetrad is likely to be very thoroughly covered, partly because of its ease of access and partly because at all times the vegetation is immediately under his notice. The writer's home tetrad 61H scored 332 species. Three adjacent tetrads, 61B, 61C and 61G, with similar soil and habitat range, scored 276, 296 and 272. The average of these gives a comparison of 84%.

The average number of species per tetrad for the west of the county is 263, for the east 203, the dividing line between east and west being taken as the 10km. eastings line 6. The county is divided into approximately the same halves by the line of the lower valley of the River Soar, if this line is continued southwards. The difference between west and east is partly due to the nature of the soils, much of the east being covered with glacial drift (boulder clay). However, the main factor would seem to be that most of the eastern part of the county is given over to intensive agriculture, whereas most of the quarries and mines, with their associated man-made habitats, and nearly all the reservoirs, are in the west. There is no significant difference in average numbers of species per tetrad between north and south of the county.

The most important factor in determining the number of species in a tetrad is uniformity or diversity of habitat. Next in importance is the nature and diversity of the underlying geological formations and soils, some of which are botanically richer than others. The habitats most likely to give a large number of species are waterways, lakes and marshes, woodland, villages, railways, quarries, gravel pits and spoil heaps. Thus for example tetrad 71N, on the site of the wartime

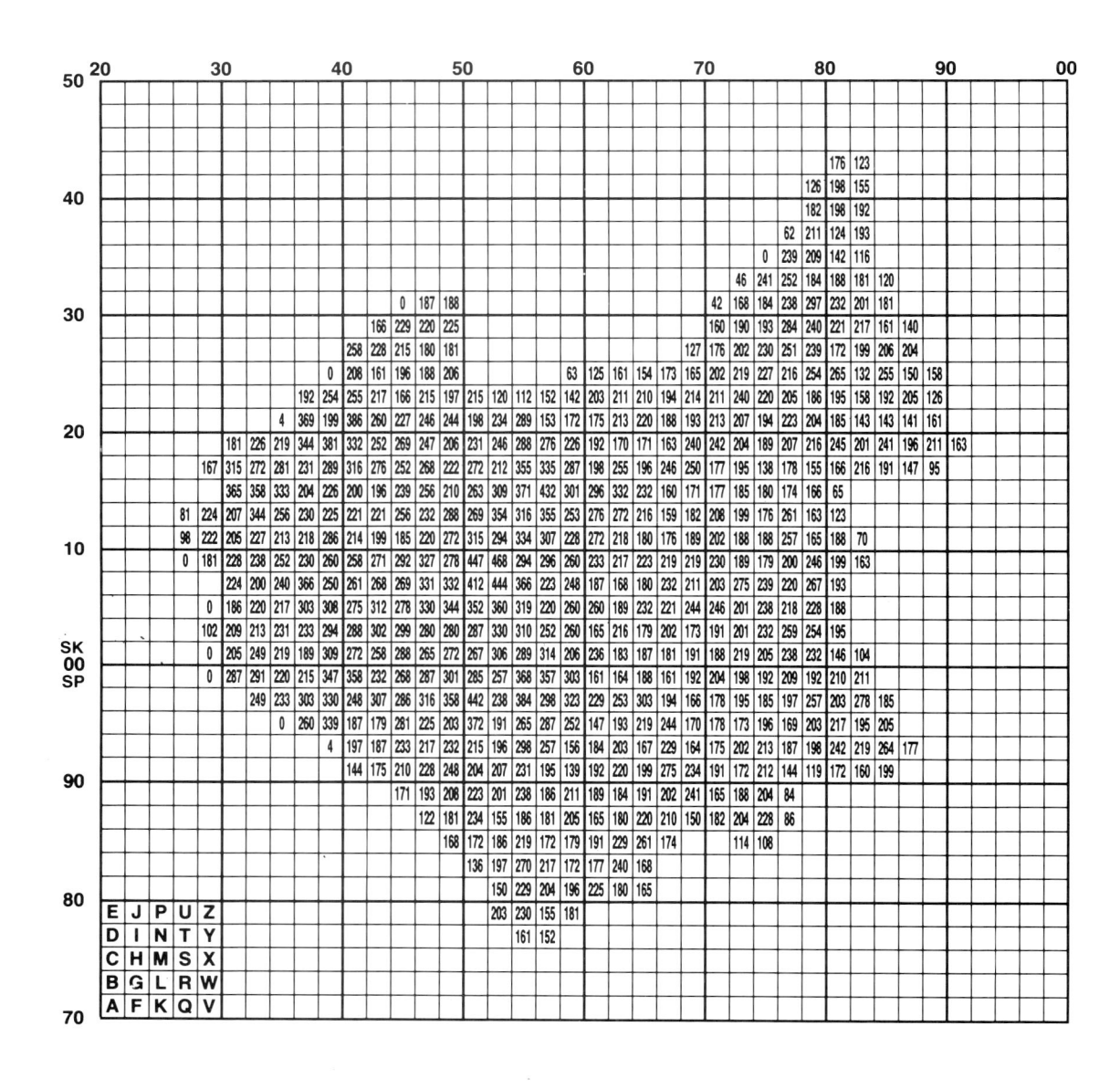

Figure 25. Number of species recorded from each tetrad (up to the end of 1979).

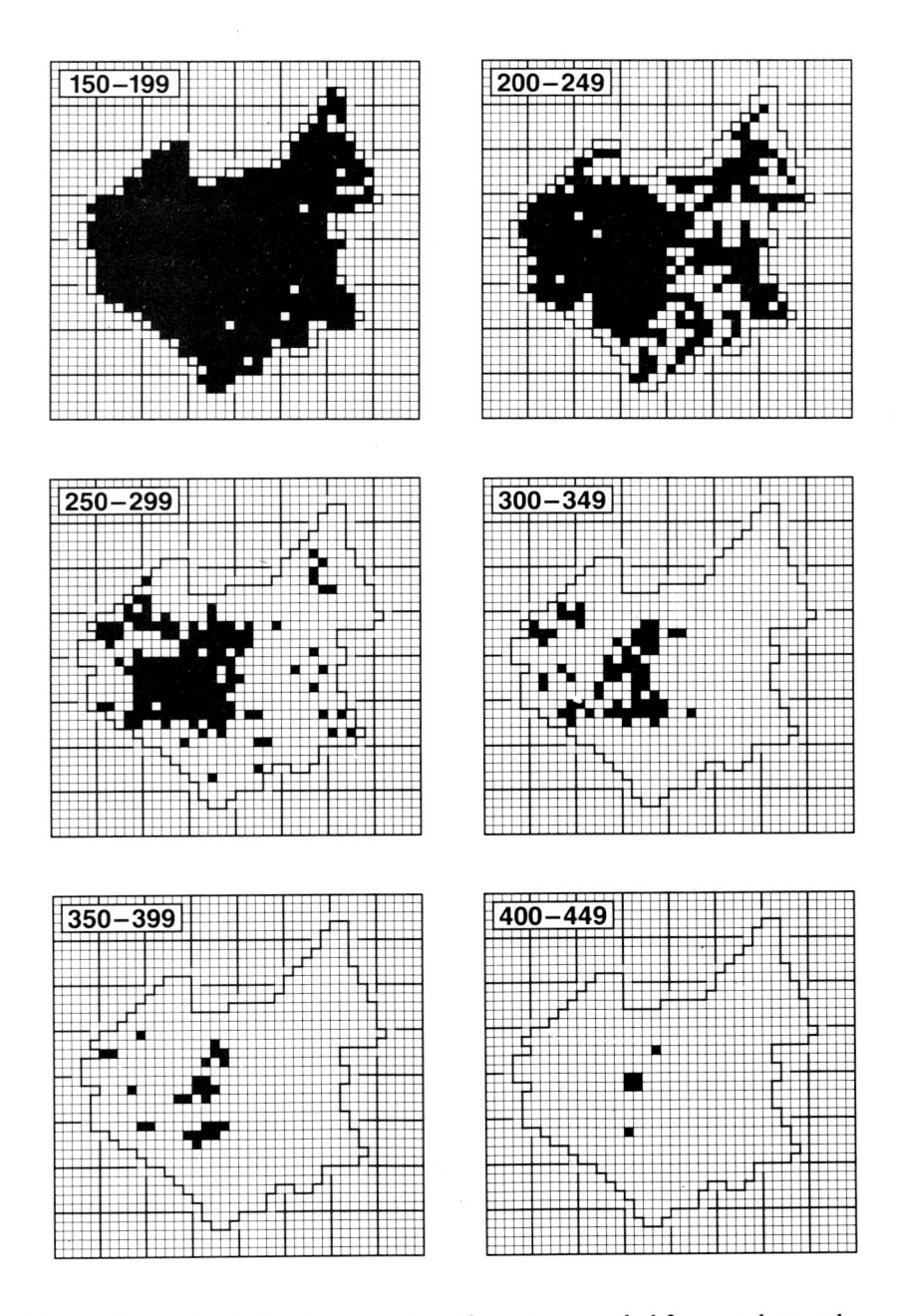

Figure 26. Analysis of the number of species recorded from each tetrad.

Melton aerodrome, has a uniform soil based on glacial drift, all the woodland and nearly all the hedgerows have been removed, and most of the land is now devoted to intensive agriculture. This tetrad is well below the average with 138 species recorded. The three regions with the greatest number of records per tetrad, Croft, Groby and Mountsorrel, are fortunate in combining botanically rich soils with variety of habitats, both natural and man-made.

165

PLAN OF THE SYSTEMATIC ACCOUNT

Authorship

The systematic accounts appearing in this Flora are the work of A. L. Primavesi, except for the accounts of *Ulmus, Rubus, Taraxacum, Hieracium* and *Pilosella* which were written by K. G. Messenger.

Sequence

The sequence of families, genera and species follows the arrangement in *Flora Europaea* (1964-1980).

Nomenclature

The nomenclature used is that of *Flora Europaea*, with the exceptions of the genera *Salix, Ulmus, Rubus, Rosa, Callitriche, Taraxacum* and *Hieracium* where, for reasons stated in the introductory remarks to these genera, it has been thought necessary to depart from *Flora Europaea* nomenclature. The names adopted for the species are given in **bold** type.

The decision to follow the sequence and nomenclature of *Flora Europaea* was made with a certain amount of reluctance. Certainly anyone who is used to the arrangement in Clapham, Tutin and Warburg (1962) will find it confusing at first. Many also will regret the disappearance of some familiar and perhaps cherished names, such as *Gnaphalium, Endymion* and *Agropyron*. British botanists have been accused of being provincial and conservative in their attitude, and the Committee responsible for the production of this Flora therefore thought it best to follow as far as possible the latest authoritative taxonomic opinions of an international body of specialists.

English names

The English names are based on Dony, Perring and Rob (1974). In some cases a second name is given, where there is a local name more commonly used in Leicestershire, or where the name used in Dony, Perring and Rob seems to depart from the commonly accepted usage.

Synonyms

Synonyms, in *italic* type, are given only when the scientific name adopted differs from that used in Clapham, Tutin and Warburg (1962). If the name used in Horwood and Gainsborough (1933) differs from both the above, then it is given in *italic* type immediately after the page reference to that work.

Distribution maps

At the head of the species account, a reference is given to the distribution map, followed by the number of tetrads in which the species was found during the field survey 1968-1980.

Distribution and frequency

In the case of the commoner species, a general assessment is given of their present distribution, relative abundance and habitats in the county. For species with 15 records or less, details are given of habitat, frequency, recorder and date of record at each locality. These details are also included for a small number of species with more than 15 records, e.g. *Paris quadrifolia* and *Orchis morio*, which are considered to be of special interest. The accounts of a few species, mainly casuals and garden escapes, are incomplete because of the loss of a number of record cards in the fire at Ratcliffe College in May 1979. A note to this effect is appended to the accounts of the species concerned.

The following initials are used for the principal contributors of records. The names of others who have contributed records are given in full.

ALP	A. L. Primavesi	KGM	K. G. Messenger
AN	A. Newton	OHB	O. H. Black
EH	E. Hesselgreaves	PAC	P. A. Candlish
EKH	E. K. Horwood	PAE	P. A. Evans
ESE	E. S. Edees		(formerly P. A. Candlish)
HB	H. Bradshaw	PCP	P. C. Powell
HH	H. Handley	PHG	P. H. Gamble
HL	H. Lucking	SHB	S. H. Bishop
IME	I. M. Evans	SM	S. D. Musgrove
JMH	J. M. Horwood	SMF	S. M. Fowler
JRIW	J. R. Ironside Wood		

Status

Following the descriptive account is an assessment of the status of the species in Leicestershire only, based on the recording carried out for this Flora. The categories used for the assessment of status are those devised by K. G. Messenger and used in his *Flora of Rutland* (1971). The following explanation of this system is taken with permission from the *Flora of Rutland*, with slight modifications where the original was applicable to Rutland only.

Sir Edward Salisbury, in a letter to *The Times* of 20 June 1930, dealing with 'Native and Introduced Species of English Wild Flowers' made a statement worth recalling: 'The British Flora as we know it today consists in part of hardy species which survived the glacial epoch, but mainly of species which have been successively introduced by various agents, including man, from that time to the present day. It is therefore as idle as it is unscientific to attempt to draw a hard and fast line between the so-called "native" and the "introduction".'

A county Flora is an account of the contemporary pattern of vegetation in a small area, and of what is known of its recent past history. In the East Midlands there is hardly an acre of land whose vegetation has been allowed to develop without human interference for as much as 100 years. We are concerned with three main types of community, therefore:
  (1) Communities which can be considered to have reached an artificial stability, in spite of, or more commonly because of, human control or human interference.
  (2) Communities which are currently developing towards this type of stability, but are of too recent origin to have reached it.
  (3) Communities which are repeatedly initiated and as repeatedly destroyed before they can acquire any sort of pattern.

In the first of these it is usually possible to distinguish two categories of plants, first, those which by comparison with other similar communities can be called the normal 'Constituent' species of the community, and secondly, those which may invade the community from time to time and perhaps survive for a few generations as 'Intrusive' species before being eliminated by competition of the constituents.

In the second kind of community, where the pattern of constituents is not fully established, a third category can be distinguished, of species surviving from the vegetation of the area before it was last disturbed. These 'Residual' species may in due course become constituents of the new community or they may be eliminated as the community matures. In such a community, though, it will usually be possible to distinguish plants destined for certain to become constituents, and others which equally certainly are no more than intrusive.

The third kind of community is to be found typically on arable land after harvest, on roadsides after widening or the dumping of spoil, and on railway ballast soon after the lifting of track. In the first of these, it will normally be destroyed by ploughing, and when it begins again, the new community will be a mixture of descendants of those plants destroyed on the spot, and of new intrusives from outside. The roadside and railway ballast communities may develop into the more permanent kinds but they, like the ones on arable land, are very often disturbed before they have much chance to do so.

In any of these communities, plants may be found which are certainly not going to survive more than one or two generations. These 'Casuals' are not always easily distinguished from intrusives, but many of them are either of garden origin, or of recent introduction from abroad.

Many of the species in the local flora may therefore have to be considered 'Constituents' of one type of community and 'Residual', 'Intrusive' or even casual elsewhere. In this work, we have used these terms throughout to designate the status of plants in the context of local conditions, feeling that the terms 'Native' and 'Introduction' should be kept for use in designating status in the country as a whole. There are many native British species which in Leicestershire at the present time are certainly introductions.

In addition to the four categories of status mentioned above, we have used the terms 'planted' and 'garden escape' where appropriate. The presence of planted species has a profound effect on the ecology of a community, and we have recorded many localities in which they occur. We have not however recorded garden plants in gardens, or plants grown commercially as crops.

From this it should be clear that the Local Flora writer is only interested in the distinction between native and introduced species when the introduction from abroad has taken place within the period covered by his survey. In Leicestershire this means in effect a period of about 230 years, since very few records were made prior to Pulteney's catalogue of 1747.

Habitat code

Following the assessment of status is a list of symbols for the habitats in which the species has been found in Leicestershire, arranged roughly in order of importance. An explanation of these symbols is given below.

A   Arable : cultivated ground
- Aa   Allotment gardens
- Ac   Cereal crops
- Ag   Gardens
- Ao   Non-cereal crops
- Ap   Public and private parks, churchyards and cemeteries

G   Grassland
- Gl   Leys
- Gp   Permanent grassland
- Gr   Rough grassland

H   Heathland
   Hb   Bog
   Hd   Dry heath
   Hg   Heath grassland
   Hw   Wet heath

F   Forest: woodland
   Fc   Coniferous woodland
   Fd   Deciduous woodland
   Fm   Mixed deciduous and coniferous woodland
   Fs   Scrub woodland
   Fw   Wet or marshy woodland with willows or alders

W   Water: aquatic habitats
   Wc   Canals
   Wd   Ditches
   Wl   Lakes and reservoirs
   Wp   Ponds
   Wr   Rivers and streams

M   Marsh or marshy meadow

E   Edge: marginal habitats
   Ea   Headlands of arable fields
   Ef   Woodland rides and margins
   Eh   Hedgerows
   Er   Railway verges and ballast
   Ev   Roadside verges
   Ew   Banks of waterways, ponds and lakes

R   Ruderal
   Rd   Recently disturbed ground
   Rf   Farmyards
   Rg   Gravel, sand and brick pits
   Rq   Quarries
   Rr   Rubbish dumps
   Rs   Spoil heaps
   Rt   Tracks, pathways and trampled ground
   Rw   Waste ground

S   Saxicole
   Sb   Buildings
   Sr   Rocks
   Sw   Walls

First record

Author and date are given for the first published record unless otherwise stated. Full citation of the references is given in the bibliography following the systematic account.

Herbaria

Herbaria which are known to contain Leicestershire specimens are noted. The following list shows the herbaria searched by K.G.Messenger and E.K.Horwood, and it also serves as a key to the symbols used for these herbaria in the text. It should be understood that it was impossible in the time available to search the major herbaria such as Kew and the British Museum for specimens of all the common species; the search in these herbaria was confined to the rarer or more interesting species.

BM	British Museum (Natural History)
CGE	University of Cambridge
DPM	Herbarium of the Rev. D. P. Murray, Stoke Golding (see below)
K	Kew: The Royal Botanic Gardens
LANC	University of Lancaster
LCR	Ratcliffe College (see below)
LIVU	University of Liverpool
LSR	Leicestershire Museums Service
LTR	University of Leicester
MANCH	University of Manchester
NMW	National Museum of Wales
OXF	University of Oxford
UPP	Uppingham School (now housed at LSR)
WAR	Warwickshire Museums Service

D. P. Murray's herbarium has recently been allocated the symbol STG, but the systematic account had already been typeset when this happened and it has not therefore been revised in this respect. LCR refers, in the context of this Flora, to Leicestershire specimens (mainly *Taraxacum* spp.) at Ratcliffe College which have been collected by A. L. Primavesi during the last decade or so. An earlier Ratcliffe College herbarium, consisting largely of material collected by J. A. Cappella and A. L. Primavesi, was transferred to LSR in 1982.

During the field survey, the field workers were asked to collect herbarium specimens, from each of the 10 km. squares for which they were responsible, of all taxa except those, on a list provided, which because of their rarity it was considered best not to collect. Thus a voucher collection for this Flora was accumulated, referred to in the text by the symbol LSR★.

For further information on the Leicestershire material contained in the herbaria listed the reader is referred to Kent and Allen (1984).

Distribution prior to 1933

Finally for each species, after the symbol HG and in smaller type, a page reference and a brief summary are given of the account of the species in Horwood and Gainsborough (1933). This consists of a quotation of Horwood's summary of frequency and distribution, followed by the number of localities cited in that work for the less common species. Occasionally further comments are added by the author of the species accounts in this present Flora.

References

Clapham, A. R., Tutin, T. G. and Warburg, E. F., 1962. *Flora of the British Isles*. Second edition. Cambridge: University Press.

Dony, J. G., Perring, F. and Rob, C. M., 1974. *English names of wild flowers*. London: Butterworths.

Horwood, A. R. and Noel, C. W. F. (3rd Earl of Gainsborough), 1933. *The Flora of Leicestershire and Rutland*. Oxford: University Press.

Kent, D. H. and Allen, D. E., 1984. *British and Irish Herbaria*. London: B.S.B.I.

Messenger, K. G., 1971. *Flora of Rutland*. Leicester: Leicester Museums.

Tutin, T. G. et al., 1964-1980. *Flora Europaea*. Cambridge: University Press.

SYSTEMATIC ACCOUNT

PTERIDOPHYTA
LYCOPSIDA

LYCOPODIACEAE
Huperzia Bernh.
H. selago (L.) Bernh. ex Schrank & Mart. Fir Clubmoss
Lycopodium selago L.
No recent record.
First record: Bloxam (1837).
HG 671: very rare, apparently extinct; 3 localities cited.

Lepidotis Beauv.
L. inundata (L.) C. Börner   Marsh Clubmoss
Lycopodium inundatum L.
No recent record.
First record: Pulteney *in* Nichols (1795).
HG 671: rare, confined to Charnwood Forest, probably extinct; 3 localities cited.

Lycopodium L.
L. clavatum L.   Stag's-horn Clubmoss
Map 1a; 1 tetrad.
Bardon Hill, dry heath, occasional, PAC 1973.
    Constituent: Hd.
First record: Hands et al. *in* Curtis (1831).
Herb.: LSR★; CGE, LSR.
HG 671: very rare, restricted to Charnwood Forest; 7 localities cited, not including the above.

SPHENOPSIDA

EQUISETACEAE
Equisetum L.
E. hyemale L.   Rough Horsetail, Dutch Rush
Map 1b; 1 tetrad.
Grace Dieu Wood, by brook, locally frequent, SHB 1977.
    Constituent: Ew.
First record: Bloxam (1837).
Herb.: LSR★; CGE, LSR, WAR.
HG 671: rare, restricted in range: 4 localities cited, including the above.

E. fluviatile L.   Water Horsetail
Map 1c; 107 tetrads.
Ponds, lakes, waterways and the wetter marshes, sometimes grows partly submerged in water. Frequent and locally abundant throughout the county wherever there are suitable habitats.
    Constituent: Wp, Wl, Wr, Wc, Wd, M, Fw.
First record: Pulteney (1747).
Herb.: LSR★; DPM, LANC, LSR, LTR.
HG 670 as *E. limosum*: locally abundant generally distributed.

E. palustre L.   Marsh Horsetail
Map 1d; 167 tetrads.
Marshes, ditches, gravel pits and the margins of ponds, lakes and waterways. Frequent throughout the county. Occasionally found in comparatively dry situations, especially on railway ballast.
    Constituent or intrusive: M, Wp, Wl, Wd, Ew, Er, Ea, Rg, Rq, Rw, Hb, Gp.
First record: Pulteney (1752).
Herb.: LSR★; LANC, LSR, LTR.
HG 670: frequent, generally distributed.

E. sylvaticum L.   Wood Horsetail
Map 1e; 6 tetrads.
Grace Dieu Wood, locally frequent, PAC 1969; Ulverscroft, hedgerow, rare, PAC 1969; Staunton Harold, edge of South Wood, locally frequent, KGM 1970; Newtown Linford, hedgerow, rare, PHG 1975; Owston Big Wood, rare, IME 1975; Ashby Old Parks, marsh, rare, SHB 1977; Newtown Linford, hedgerow north of Cover Cloud, PAE 1987.
    Constituent or residual: Fd, Fw, Ef, Eh, M.
First record: Pulteney *in* Nichols (1795).
Herb.: LSR★; BM, LSR, LTR.
HG 669: rare, restricted mainly to Charnwood Forest and NW region; 4 localities cited.

E. arvense L.   Field Horsetail
Map 1f; 555 tetrads.
Throughout the county; locally very abundant on railway verges, frequent in hedgerows and marshes, and sometimes a troublesome weed of arable land and gardens.
    Constituent or intrusive: Er, Ev, Eh, Ew, Ef, Ea, Ac, Ag, Ao, Aa, M, Rg, Rq, Rs, Rw, Rt, Gp, Gr, Wd.
First record: Pulteney (1747).
Herb.: LSR★; BM, LSR, LTR.
HG 669: locally abundant, generally distributed.

E. telmateia Ehrh.   Great Horsetail
Map 1g; 75 tetrads.
Marshes, wet open places in woodland, railway verges, and especially characteristic of springlines below the Marlstone in the east. Occasional throughout the county, sometimes in considerable abundance where it does occur, except on the Boulder Clay, where it is rare or absent. This seems to contradict the statement in Horwood and Gainsborough that it prefers the heavy clay soils, but the localities cited there correspond to the present pattern of distribution.
    Constituent or residual: M, Fw, Fd, Ef, Eh, Er, Ev, Ea, Ew, Wp, Wl, Wd, Gp.
First record: Pulteney (1747).
Herb.: LSR★; LSR, LTR.
Hg 668 as *E. maximum*: local, very sparsely distributed; 42 localities cited.

FILICOPSIDA

OPHIOGLOSSACEAE
Ophioglossum L.

O. vulgatum L.   Adder's-tongue
Map 1h; 70 tetrads.
Occasional throughout the county in old grassland. Inconspicuous and probably often overlooked, but decreasing as suitable habitats are destroyed.
    Constituent, residual or intrusive: Gp, Hg, M, Fw, Fs, Rq, Rg, Ap.
First record: Pulteney (1747).
Herb.: LSR*; LANC, LSR, LTR.
HG 666: Frequent and generally distributed; 62 localities cited.

Botrychium Swartz
B. lunaria Swartz   Moonwort
Map 1i; 3 tetrads.
Charnwood Lodge Nature Reserve, heath grassland, rare, PAC 1971 (has decreased at this site since 1962); Launde, old grassland, rare, IME 1977; Bradgate Park, grassland, rare, PHG 1978.
    Constituent or residual: Gp, Hg.
First record: Pulteney (1749).
Herb.: BM, LSR, LTR.
HG 667: rare, very restricted in range; 29 localities cited.

OSMUNDACEAE
Osmunda L.

O. regalis L.   Royal Fern
No map; 2 tetrads.
Staunton Harold Hall, T. W. Tailby 1960; Woodhouse Eaves, The Brand, naturalised by stream, occasional, PHG 1968.
Planted: Ew.
First record: Pulteney (1756).
Herb.: LSR.
HG 666: very rare, very restricted in range; 4 localities cited, including The Brand.

HYPOLEPIDACEAE
Pteridium Scop.

P. aquilinum (L.) Kuhn   Bracken
Map 1j; 258 tetrads.
Heathland, where it may be dominant over large areas, almost to the exclusion of other vegetation; woodland, hedgerows, roadside and railway verges. Prefers dry acid soils.
Frequent and locally abundant in Charnwood Forest and the west, and has tended to increase with improved drainage of the land, and where there is grazing pressure in grassland from animals to which bracken is unpalatable. Locally frequent elsewhere in the county where there are suitable soils and habitats.
    Constituent: Hd, Hg, Fd, Fc, Fm, Fs, Eh.
    Residual or intrusive: Ev, Er, Rs, Rw, Rd, Sw, Ag, Aa.
First record: Pulteney (1747).
Herb.: CGE, DPM, LSR, LTR.
HG 653 as *Eupteris aquilina*: locally abundant, widely distributed, a very characteristic feature of Charnwood Forest area, less frequent in east Leicestershire; 120 localities cited.

THELIPTERIDACEAE
Thelypteris Schmidel
T. limbosperma (All.) H. P. Fuchs
                                       Lemon-scented Fern
Map 1k; 5 tetrads.
Charnwood Lodge Nature Reserve, drainage channels in moorland and woodland, occasional, PAC 1969; Dimminsdale, edge of water-filled limestone quarry, locally abundant, OHB and HH 1971; Bradgate Park, drainage channels in wet heathland, locally frequent, PHG 1973; Ulverscroft, Poultney Wood, rare, PHG 1987; Beacon Hill, drainage channel, rare, PHG 1987.
    Constituent or residual: Ew, Wd.
First record: Bloxam (1846a).
Herb.: LSR*; LSR, LTR, WAR.
HG 662 as *Dryopteris oreopteris*: rare, restricted in range; 18 localities cited.

ASPLENIACEAE
Asplenium L.
A. trichomanes L.   Maidenhair Spleenwort
Map 1l; 51 tetrads.
Occasional throughout the county on walls, railway bridges, canal locks and similar brickwork or stonework.
    Intrusive: Sw, Sb, Er.
First record: Pulteney (1756).
Herb.: LSR*; LSR.
HG 656: rare, restricted in range; 16 localities cited.

A. viride Hudson   Green Spleenwort
Map 2a; 1 tetrad.
Loughborough, blue brick railway bridge over River Soar, locally frequent, PHG 1975.
    Intrusive: Sw.
First record: Pulteney (1756).
HG 656: very rare, confined to Charnwood Forest; 1 locality cited, not the same as above.

A. adiantum-nigrum L.   Black Spleenwort
Map 2b; 61 tetrads.
Occasional, and almost entirely confined to walls and bridges. Especially characteristic of railway bridges built of blue brick, which accounts for the curious distribution pattern shown on the map.
    Intrusive: Sw, Sb.
First record: Pulteney (1747).
Herb.: LSR*; LSR, LTR.
HG 655: rare, restricted in range; 16 localities cited.

A. ruta-muraria L.   Wall-rue
Map 2c; 99 tetrads.
Locally frequent throughout the county on walls and bridges; rarely on rocks.
    Intrusive or constituent: Sw, Sb, Sr.
First record: Pulteney (1747).
Herb.: LSR*; BM, LSR, LTR.
HG 657: local, sparsely distributed; 32 localities cited.

Ceterach DC.
C. officinarum DC.   Rustyback
Map 2d; 13 tetrads.
Gilmorton, railway bridge, rare, EKH and JMH 1968; Quorndon, railway bridge, locally frequent PHG 1968; Swithland, railway bridge, locally abundant, PHG 1968; East Norton, blue brick retaining wall, dismantled railway, IME 1968 (site since destroyed); Wilson, stone

wall, rare, PAC 1971; Cropston Reservoir, wall of dam, rare, PHG 1973; Croft, wall, rare, SHB 1973; Kibworth Harcourt, canal lock, abundant, EKH and JMH 1973; Mountsorrel, granite wall, rare, PHG 1974; Woodhouse Eaves, wall, rare, PHG 1974; Cotes, wall, rare, PHG 1974.
    Intrusive: Sw.
First record: Brown *in* Mosley (1863).
Herb.: LSR★; LSR.
HG 658: rare, restricted in range; 5 localities cited, including Quorndon.

Phyllitis Hill
P. scolopendrium (L.) Newman   Hart's-tongue
Map 2e; 76 tetrads.
Locally frequent throughout the county on shady old walls and railway bridges; occasional in woodland and on shady stream banks. Characteristic of the brick-lined drainage channels of railway verges.
    Constituent or intrusive: Fd, Fm, Ew, Er, Sr, Sw, Sb, Rq, Rw.
First record: Pulteney *in* Nichols (1795).
Herb.: LSR★; LSR, LTR.
HG 659: rare, and on the decrease; 40 localities cited.

ATHYRIACEAE
Athyrium Roth
Athyrium filix-femina (L.) Roth   Lady-fern
Map 2f; 129 tetrads.
Damp woodland and the banks of rivers and streams. Frequent in the well-wooded parts of the county; occasional elsewhere.
    Constituent: Fw, Fd, Fm, Fs, Ew.
    Residual or intrusive: Eh, Er, Rg, Hw, M, Sw.
First record: Pulteney (1747).
Herb.: LSR★; CGE, DPM, LANC, LSR, LTR.
HG 657: frequent, sparsely distributed; 75 localities cited.

Cystopteris Bernh.
C. fragilis (L.) Bernh.   Brittle Bladder-fern
Map 2g; 1 tetrad.
Quorndon, blue brick railway bridge, rare, PHG 1973.
    Intrusive: Sw.
First record: Pulteney *in* Nichols (1795).
Herb.: LSR★; LSR.
HG 660: very rare, very restricted in range; 6 localities cited, not including the above.

ASPIDIACEAE
Polystichum Roth
P. aculeatum (L.) Roth   Hard Shield-fern
Map 2h; 20 tetrads.
Woodland, especially along the margins of streams and drainage channels. Occasional in the older woodlands throughout the county. Rarely in any quantity where it does occur.
    Constituent, residual or intrusive: Fd, Fm, Eh, Wd, Rq, Sr, Sw.
First record: Pulteney (1752).
Herb.: LSR★; CGE, LANC, LSR, LTR.
HG 660: local, sparsely distributed; 26 localities cited under **P. aculeatum** and 7 localities under *P. lobatum*.

P. setiferum (Forskal) Woynar   Soft Shield-fern
Map 2i; 12 tetrads.
Occasional in the older woodlands and by water. Mainly in the west of the county and the east Leicestershire woodlands.
    Constituent, residual or intrusive: Fd, Fm, Ew, Eh, Sr, Sw, Rq.
First record: Bloxam (1837).
Herb.: LSR★; CGE, LSR.
HG 661: rare, restricted in range; 9 localities cited.

Dryopteris Adanson
D. filix-mas (L.) Schott   Male-fern
Map 2j; 343 tetrads.
Woodland, hedgerows, banks of rivers and streams, and walls. Throughout the county; frequent in the well-wooded parts, occasional elsewhere.
    Constituent: Fd, Fm, Fs, Fw, Eh, Ew.
    Residual or intrusive: Sw, Sb, Sr, Er, Ev, Rq, Rg, Ap, Wd.
First record: Pulteney (1747).
Herb.: LSR★; BM, LANC, LSR, LTR.
HG 662: frequent, but diminishing, generally distributed.

D. borreri Newman   Scaly Male-fern
Map 2k; 24 tetrads.
Woodland. Occasional in Charnwood Forest and the west of the county; rare or absent elsewhere (or possibly overlooked).
    Constituent or residual: Fd, Fc, Fm, Fw, Ew, Wd, Rq.
First record: 1857, R. B. T. Bunch in herb. LTR.
Herb.: LSR★; LANC, LSR, LTR.
HG: not recorded.

D. carthusiana (Vill.) H. P. Fuchs
                       Narrow Buckler-fern
Map 2l; 30 tetrads.
Occasional in woodland in Charnwood Forest and the west of the county; rare or absent elsewhere.
    Constituent or residual: Fd, Fm, Fs, Fc, Fw, Hg, M, Er.
First record: Kirby (1850).
Herb.: LSR★; CGE, LSR, LTR.
HG 663 as *D. spinulosa*: frequent, generally distributed; 60 localities cited.

D. dilatata (Hoffm.) A. Gray   Broad Buckler-fern
Map 3a; 280 tetrads.
Occasional or locally frequent in woodland and hedgerows throughout the county.
    Constituent or residual: Fd, Fm, Fs, Fw, Eh, Er, Ew, Wd, Rq, Rs, Sr.
First record: Pulteney *in* Nichols (1795).
Herb.: LSR★; BM, LANC, LSR, LTR, WAR.
HG 664 as *D. aristata*: frequent, generally distributed; 111 localities cited.

Gymnocarpium Newman
G. robertianum (Hoffm.) Newman   Limestone Fern
Thelypteris robertiana (Hoffm.) Slosson
Map 3b; 2 tetrads.
Shawell, blue brick railway bridge, locally frequent, EKH and JMH 1968 (first found here by EKH in 1958); Newton Harcourt, canal lock, single plant, EKH and JMH 1974.
    Intrusive: Sw.
First record: Gumley Wood, 1915, S.A. Taylor, in herb. LSR.
Herb.: LSR★; LSR, WAR.
HG: not recorded.

BLECHNACEAE
Blechnum L.

B. spicant (L.) Roth　　　　　　　　　Hard Fern
Map 3c; 14 tetrads.
Heathland and woodland. Occurs at Moira, Dimminsdale, Mountsorrel and Launde Big Wood, but otherwise confined to Charnwood Forest, in a number of scattered localities but nowhere in quantity.
　　Constituent: Hg, Hd, Hw, Fm, Fd, Fs, Eh, Ew, Wd.
First record: Pulteney (1747).
Herb.: LSR★; LSR.
HG 655: local, restricted in range; 27 localities cited.

POLYPODIACEAE
Polypodium L.

P. vulgare L. agg.　　　　　　　　　　Polypody
Map 3d; 33 tetrads.
Locally frequent on rocks in Charnwood Forest; occasional on walls elsewhere in the county. Most field workers did not distinguish the segregates, but the following records for **P. interjectum** Shivas were received: Medbourne, wall, rare, KGM 1971; Great Merrible Wood, epiphytic on ash, rare, KGM 1973; Croft, three records from walls, locally frequent, SHB 1973.
　　Constituent or intrusive: Fd, Eh, Sr, Sw, Rq, Rg.
First record: Pulteney (1747).
Herb.: LSR★; LSR, LTR.
HG 665: local, generally distributed; 48 localities cited.

MARSILEACEAE
Pilularia L.

P. globulifera L.　　　　　　　　　　Pillwort
No recent record.
First record: Kirby (1848b).
Herb.: LSR.
HG 672: rare, confined to NW region; 2 localities cited.

AZOLLACEAE
Azolla Lam.

A. filiculoides Lam.　　　　　　　　Water Fern
Map 3e; 8 tetrads.
Cossington, Church Pond and Astill's pond, H. P. Moon 1971; Cossington, artificial pond at Goscote Nurseries, a troublesome weed probably derived from the preceding, ALP 1973; Cotes, River Soar, PHG 1973; Burton Overy, small field pond, EKH and JMH 1973; Elmesthorpe, field pond, SHB 1974; Grantham Canal between Hose and Harby, ALP 1975. In all these localities the quantity varies from year to year and from season to season. For example, in the winter of 1974 to 1975 the Grantham Canal from Hose to Harby was so completely covered that it looked like a red asphalt road, but by the summer of 1975 the **Azolla** had been almost completely ousted by **Lemna gibba**.
　　Intrusive: Wp, Wc, Wr.
First record: 1948, M. Borrill in herb. LTR (from Church Pond, Cossington).
Herb.: LSR★; LTR.
HG: not recorded.

SPERMATOPHYTA
GYMNOSPERMAE
CONIFEROPSIDA
CONIFERALES

Various species of conifers are planted as forestry crops. The ones most frequently found in Leicestershire are **Picea abies, Pinus sylvestris, Pinus nigra, Picea sitchensis** and various species of **Larix**. Less frequently plantations can be found of **Thuja plicata, Pinus contorta, Pseudotsuga menziesii** and **Tsuga heterophylla**. All these and many other alien conifer species are also planted as ornamental trees, and mature deciduous woodland often contains a scatter of such conifers.

In the field survey for this Flora, no attempt was made to record all these conifer species. Sufficient data were however acquired for four of them to give an idea of their distribution in the county, and accounts follow for these.

PINACEAE
Picea A. Dietr.

P. abies (L.) Karsten　　　　　　　Norway Spruce
Map 3f; 64 tetrads.
Locally frequent throughout the county in plantations, sometimes in pure dense stands covering a large area. Also planted for ornament. It is possible that some of the isolated trees in woodland and elsewhere may be the result of natural regeneration.
　　Planted or possibly intrusive: Fc, Fm, Fd, Ef, Ew, Eh, Ap, Ag.
First record: Horwood and Gainsborough (1933).
Herb.: LSR★; LTR.
HG 517 as *Picea excelsa*: frequent; 27 localities cited.

Larix Miller

L. decidua Miller　　　　　　　European Larch
Map 3g; 172 tetrads.
This species was formerly much planted as a forestry crop or as ornamental trees, and old mature specimens of larch in the county usually belong to this species. More recently, however, it has become customary to plant either **L. kaempferi** or **L. × eurolepis**. It must be admitted that some field workers were not aware of the presence in the county of these other taxa, and recorded them all as **L. decidua**. Hence this account and the distribution map should be interpreted as a composite picture of the distribution of all three of these larches in the county. Frequent throughout the county in plantations, woodland, and as ornamental trees. Occasional isolated trees may provide evidence for natural regeneration.
　　Planted or possible intrusive: Fc, Fm, Fd, Fs, Fw, Eh, Er, Ew, Ap, Ag, Gr.
First record: Horwood and Gainsborough (1933).
Herb.: LSR, LTR.
HG 517 as *L. europaea*: locally abundant, generally distributed.

Pinus L.

P. nigra Arnold　　　　　　　　　　Black Pine
Map 3h; 59 tetrads.
Locally frequent throughout the county in plantations, and as ornamental trees in parks and gardens. No evidence of natural regeneration.
　　Planted: Fc, Fm, Fd, Fs, Eh, Ev, Ew, Er, Ap.

First record: Horwood and Gainsborough (1933).
HG 516 as *P. laricio*: occasional; 5 localities cited.

P. sylvestris L.   Scots Pine
Map 3i; 215 tetrads.
Frequent in plantations and woodland in those parts of the county where the soil tends to be acid, often planted but sometimes regenerating naturally; occasional and always planted elsewhere in the county.
    Planted or intrusive: Fc, Fm, Fd, Fs, Ef, Eh, Ev, Ew, Er, Ap, Ag, Rq.
First record: Bloxam (1837).
Herb.: DPM, LSR.
HG 515: frequent, generally distributed; 70 localities cited.

TAXOPSIDA
TAXALES

TAXACEAE
Taxus L.

T. baccata L.   Yew
Map 3j; 138 tetrads.
Locally frequent throughout the county in woodland, parks, churchyards and gardens. In places where it is obviously planted, seedlings arise readily, so it can be assumed that many of the specimens in woodland and elsewhere have arisen naturally from bird-sown seed, probably derived from planted specimens originally.
    Planted or intrusive: Fd, Fm, Fs, Ef, Eh, Ev, Ew, Ap, Ag, Rg, Rq, Sw, Hd.
First record: Power (1807).
Herb.: DPM, LANC, LTR, MANCH.
HG 514: frequent, generally distributed; 28 localities cited.

ANGIOSPERMAE
DICOTYLEDONES
SALICALES

SALICACEAE
Salix L.

During the field survey for this Flora, **Salix cinerea** was considered to be represented in the county by two subspecies. In *Flora Europaea* these are treated as distinct species. In the course of the survey, numerous hybrids of **S. cinerea** were collected, and their parentage was not always determined to the subspecific level. For this reason it seems better not to follow *Flora Europaea* in this particular case, as to do so would present insoluble problems in the treatment of the hybrids.

S. pentandra L.   Bay Willow
Map 3k; 1 tetrad.
Ashby de la Zouch, hedgerow by green lane between Ashby and Old Parks, single bush, SHB 1979.
    Probably intrusive: Eh.
First record: Pulteney *in* Nichols (1795).
Herb.: LSR, NMW.
HG 498: rare, very restricted in range; 13 localities cited, including the above.

S. fragilis L.   Crack Willow
Map 3l; 536 tetrads.
Frequent throughout the county on the banks of rivers and streams, often as rows of pollarded trees along the banks of brooks, and as single specimens at the margins of ponds. No attempt was made during the survey to segregate the numerous varieties and cultivars.
    Constituent or intrusive: Ew, Eh, Ev, Er, Wp, Wl, Wd, Wr, Fw, Fd, Fm, Fs, M, Rg, Rq.
First record: Pulteney *in* Nichols (1795).
Herb.: LSR★; BM, CGE, LANC, LSR, LTR, MANCH, NMW.
HG 500: locally abundant, and generally distributed.

S. alba L.   White Willow
Map 4a; 388 tetrads.
Frequent throughout the county on the banks of rivers and streams, and in hedgerows. Apparently more frequent in wet woodland than **S. fragilis.**
    Constituent or intrusive: Ew, Eh, Ev, Er, Fw, Fd, Fm, Fs, Wp, Wl, Wr, Wd, M, Hw, Rq, Ap.
First record: Pulteney (1747).
Herb.: LSR★; BM, CGE, LSR, LTR, MANCH, NMW.
HG 501: frequent and generally distributed.

S. alba × **fragilis** = **S.** × **rubens** Schrank
Map 4b; 2 tetrads.
Whetstone, hedgerow, rare, EH 1977; Groby, disused quarry, single tree, EH 1983.
    Intrusive: Eh, Rq.
First record: Eng. Bot. xxvi (1808) t. 1808 (see HG).
Herb.: BM, LTR.
HG 502: rather rare, very sparsely distributed, 10 localities cited.

S. triandra L.   Almond Willow
Map 4c; 37 tetrads.
Wet woodland and by water. Occasional in scattered localities throughout the county.
    Constituent or intrusive: Fw, Fd, Fs, Wd, Wp, Ew, Eh, Er, Rw.
First record: Pulteney *in* Nichols (1795).
Herb.: LSR★; LSR, LTR, MANCH, NMW.
HG 499: frequent, and generally distributed; 36 localities cited.

S. triandra × **viminalis** = **S.** × **mollissima** Ehrh.
Map 4d; 3 tetrads.
Kirby Muxloe, hedgerow, rare, EH 1973; Glen Parva, canal bank, rare, EH 1974; Narborough, railway verge, rare, EH 1978.
    Intrusive: Eh, Ew, Er.
First record: 1887, F. T. Mott in herb. MANCH.
Herb.: LSR★; MANCH.
HG 500: 2 localities cited, not including the above.

S. cinerea aggregate   Grey Willow
Map 4e; 423 tetrads.
During the field survey **S. atrocinerea** Brot. was treated as a subspecies of **S. cinerea** L., under the name of **S. cinerea** subsp. **oleifolia** Macreight. The two subspecies were treated as critical, and to segregate them required the submission of a voucher specimen to the referee. At the same time, **S. cinerea** was recorded as an aggregate on the Common Species Card, without the necessity for confirmation. This latter policy was undoubtedly a mistake, because many field workers were content to record the aggregate only, so that the distribution maps for the segregates are manifestly patchy and incomplete.

S. cinerea L. subsp. **cinerea**
S. cinerea L.
Map 4f; 24 tetrads.
Hedgerows, the banks of waterways, and wet places in woodland. Nearly always requires comparatively wet soil. Occasional throughout the county.
    Constituent or intrusive: Eh, Ev, Er, Ew, Wd, Wp, Wc, Fw, Fd, M, Rq, Gr.
First record: Bloxam and Babington *in* Potter (1842).
Herb.: LSR★; LTR.
HG 507: frequent, and generally distributed. No distinction is made between this and the following subspecies. There are, however, seven localities cited for var. *aquatica* Sm., which can be related to subspecies **cinerea**.

S. cinerea L. subsp. **oleifolia** Macreight
S. atrocinerea Brot.
Map 4g; 169 tetrads.
Locally frequent throughout the county in similar situations to the preceding, but tolerant of much drier conditions.
    Constituent or intrusive: Eh, Er, Ev, Ew, Wd, Wr, Wp, M, Fd, Fm, Fs, Fw, Rw, Rq, Hw, Hg.
First record: Kirby (1850).
Herb.: LSR★; DPM, LANC, LSR, LTR.
HG 507: frequent, and generally distributed.

S. cinerea × viminalis = S. × smithiana Willd.
Map 4h; 25 tetrads.
Occasional in hedgerows and by water, probably throughout the county; the apparent southerly distribution may be due to its being overlooked by some field workers.
    Constituent or intrusive: Eh, Ew, Ev, Er, Fw, Fm, Rq, Wc.
First record: Coleman (1852).
Herb.: LSR★; LTR, MANCH, NMW.
HG 507: occasional but generally distributed.

S. aurita L.   Eared Willow
Map 4i; 10 tetrads.
Ulverscroft, hedgerow, rare, PAC 1968; Martinshaw Wood, near marl pit, rare, EH 1969; Markfield, ditch, single bush, OHB, HH and SM 1970; Burbage Wood, hedgerow, rare, SHB 1975; Ambion Wood, overgrown ride next to Ashby Canal, rare, SHB 1976; between Woodville and Blackfordby, roadside hedge, single plant, SHB 1976; Staunton Harold, marsh, rare, SHB 1978; Ulverscroft, roadside, rare, PHG 1978; Bardon, hedgerow, locally frequent, SHB 1979; Markfield, roadside ditch, single shrub, HH 1979.
    Constituent or intrusive: Eh, Ev, Ef, Fm, Wd, M.
First record: Bloxam and Babington *in* Potter (1842).
Herb.: LSR★; LSR, LTR, MANCH, NMW.
HG 506: frequent, sparsely distributed; 39 localities cited.

S. aurita × cinerea = S. × multinervis Doell
Map 4j; 34 tetrads.
Hedgerows, wet woodland, and by water. Locally frequent in the west of the county, and probably underrecorded; rare in the east. It is the formation of this hybrid which is probably causing the decrease of **S. aurita** in the county.
    Constituent or intrusive: Eh, Ev, Er, Ew, Wp, Wd, Wr, Fw, Fm, Fs, M, Rq.
First record: Preston (1900).

Herb.: LSR★; LSR.
HG 508: 9 localities cited.

S. aurita × caprea = S. × capreola J. Kerner ex Anderss.
Map 4k; 4 tetrads.
Martinshaw Wood, roadside hedge at southern edge, rare, EH 1970; Leicester, near Freemen's Lock, railway siding, rare, HB 1971; Thurlaston, roadside hedge, rare, EH 1972; Bagworth, roadside ditch, occasional, OHB and HH 1972.
    Intrusive: Eh, Er, Wd.
First record: Horwood and Gainsborough (1933).
Herb.: LSR★.
HG 507: 4 localities cited.

S. caprea L.   Goat Willow
Map 4l; 391 tetrads.
Locally frequent throughout the county in hedgerows and woodland and on the banks of rivers and streams. Often invades the verges and ballast of dismantled railways.
    Constituent or intrusive: Eh, Ew, Er, Ev, Ef, Fw, Fd, Fm, Fs, Wd, Wl, Wr, Wc, M, Rq, Rg, Rs, Rw, Gr, Gp, Ap.
First record: Pulteney (1747).
Herb.: LSR★; CGE, LANC, LSR, LTR, MANCH, NMW.
HG 504: frequent, widely distributed; 81 localities cited.

S. caprea × cinerea = S. × reichardtii A. Kerner
Map 5a; 2 tetrads.
Groby, roadside hedge, single large shrub, EH 1970; Lubbesthorpe, hedgerow, rare, EH 1979.
    Intrusive: Eh.
First record: Horwood and Gainsborough (1933).
Herb.: LSR★.
HG 506: 1 locality cited.

S. caprea × viminalis = S. × sericans Tausch
Map 5b; 7 tetrads.
Claybrooke Magna, hedgerow, occasional, MH 1972; Kirby Muxloe, stream side, occasional, EH 1972; Upton, roadside hedgerow, rare, SHB 1974; Higham on the Hill, roadside hedgerow, rare, SHB 1974; Leicester, Braunstone Park, waterside, rare, EH 1978.
    Intrusive: Eh, Ew, Ev.
First record: Coleman (1852) if HG is correct, otherwise present work.
Herb.: LSR★; LSR.
HG 505: doubtful because it is given under *S. × mollissima*; if correct, 17 localities cited.

S. caprea × cinerea × viminalis = S. × calodendron Wimmer
Map 5c; 4 tetrads.
Groby, old mineral line, rare, EH 1969 (site since destroyed); Blaby, field hedgerow, rare, EH 1974; Glen Parva, canal bank, rare, EH 1976; Lubbesthorpe, woodland, rare, EH 1978.
    Intrusive: Er, Eh, Ew, Fd.
First record: 1969, E. Hesselgreaves in present work.
Herb.: LSR★.
HG: not recorded.

S. repens L.   Creeping Willow
Map 5d; 3 tetrads.
Charnwood Lodge Nature Reserve, two localities on wet heath, rare, PAC 1969; Bradgate Park, four localities on wet heath dominated by **Molinia**, locally frequent, PHG 1973; Coalville, wet heath grassland, rare, J.E. Dawson 1986.
    Constituent or residual: Hw.
First record: Pulteney (1749).
Herb.: LSR, LTR, MANCH, NMW.
HG 508: occasional, confined to area W. of R. Soar; 8 localities cited, including the first two above.

S. viminalis L.   Osier
Map 5e; 193 tetrads.
By water, in hedgerows, and occasionally in wet woodland. Locally frequent throughout the county. Genuine osier beds where this species is cultivated no longer occur in the county.
    Constituent or intrusive: Ew, Eh, Er, Ev, Ef, Wp, Wl, Wd, Wr, M, Fw, Fd, Fm, Fs, Rg, Rs, Rw, Rf, Gp.
First record: Pulteney (1747).
Herb.: LSR★; DPM, LTR, MANCH, NMW, UPP.
HG 504: frequent, generally distributed.

S. purpurea L.   Purple Willow
Map 5f; 67 tetrads.
In wet woodland and marshes, and by water. Locally frequent in the south-west of the county; occasional and possibly under-recorded elsewhere.
    Constituent or intrusive: Fw, Fs, M. Ew, Eh, Er, Wd, Wp, Wr, Wc.
First record: Pulteney *in* Nichols (1795)
Herb.: LSR★; BM, LSR, MANCH, NMW.
HG 503: occasional, restricted in range; 27 localities cited.

S. purpurea × **viminalis** = **S.** × **rubra** Huds.
Map 5g; 3 tetrads.
Leicester, Belgrave Lock, margin of River Soar, rare, HB 1971; Enderby, wet willow woodland by River Soar, rare, EH 1972; South Wigston, canal bank, rare, EH 1977.
    Intrusive: Ew, Fw.
First record: Pulteney *in* Nichols (1795).
Herb.: LSR★; LSR.
HG 504: rare, restricted in range; 3 localities cited, not including the above.

S. daphnoides Vill.
Map 5h; 2 tetrads.
Bradgate Park, spinney near Cropston Reservoir, rare, probably planted, EH 1973; Croft, scrub woodland by River Soar, rare, SHB 1976.
    Planted or intrusive: Fs.
First record: Linton (1913).
Herb.: LSR★.
HG 509: rare; 2 localities cited, not including the above.

Populus L.

P. alba L.   White Poplar, Abele
Map 5i; 29 tetrads.
Sometimes planted in parks and other public places as ornamental trees, and possibly occasionally arising spontaneously elsewhere. It is suspected that some of the records for this species made early in the survey should be referred to **P. canescens**.
    Planted or intrusive: Ap, Ag, Eh, Ev, Ew, Er, Ef, Fd, Fm, Fs, Fw, Wp, Gp, Gl.
First record: Pulteney (1747).
Herb.: DPM, NMW.
HG 509: occasional, restricted in range; 23 localities cited.

P. canescens (Aiton) Sm.   Grey Poplar
Map 5j; 67 tetrads.
Hedgerows and woodland. Suckers freely, and often forms a long stand of sucker shoots in a hedgerow on either side of a parent tree. Occasional throughout the county, but locally frequent where it does occur.
    Constituent or intrusive: Fd, Fm, Fs, Fw, Eh, Ev, Ew, Er, Ap, Ag, Rq, Wl, Gr, M.
First record: Kirby (1850).
Herb.: LSR★; DPM, LSR, LTR, MANCH, NMW.
HG 509: occasional, generally distributed.

P. tremula L.   Aspen
Map 5k; 102 tetrads.
Woodland, hedgerows and the banks of rivers and streams. Locally frequent in the west of the county; occasional in the east.
    Constituent or intrusive: Fd, Fm, Fs, Fw, Eh, Ew, Ev, Er, Hg.
First record: Pulteney (1747).
Herb.: LSR★; DPM, LSR, LTR, MANCH, NMW.
HG 510: frequent, generally distributed.

P. gileadensis Rouleau   Balsam Poplar
Map 5l; 14 tetrads.
Sometimes planted in woodland and by water. In scattered localities throughout the county, sometimes in fairly large plantations which betray their presence by their scent.
    Planted: Fd, Fm, Ew, Eh, Ap, Wp, Gr.
First record: WBEC Rep. 1916.
Herb.: CGE, LSR, LTR, NMW.
HG 512 as *P. tacamahacca*: rare; 8 localities cited.

P. nigra L.   Black Poplar
No map; 1 tetrad.
Because of suspected confusion of this with **P. × canadensis**, field workers were asked to submit specimens for expert determination, but only one was received: Barlestone, parish boundary hedge, single tree, F.T. Smith 1981 (conf. E. Milne-Redhead).
    Constituent, intrusive or planted: Eh.
P. nigra var. **italica**, the Lombardy Poplar, however, is frequently planted as an ornamental tree or as shelter belts and wind-breaks.
First record: Pulteney (1747).
Herb.: LSR, MANCH, NMW.
HG 510: occasional, very restricted in range; 11 localities cited, specimens in herb. LSR from two of the localities, Edmondthorpe and Osgathorpe, have recently been confirmed by E. Milne-Redhead as being of the type.

P. deltoides Marshall × **P. nigra** L. = **P. × canadensis** Moench   Italian Poplar
Map 6a; 154 tetrads.
Frequently planted in woodland, by water and on roadsides. Often forms large stands in plantations. Since only male trees are said to be present in Britain, and it does not sucker, presumably every specimen in the county has been planted.

Planted: Fd, Fm, Fs, Fw, Ew, Eh, Ev, Er, Ap, Ag, M, Gp, Gr, Rr.
First record: WBEC Rep. 1908
Herb.: LSR★; CGE, DPM, LSR, LTR, MANCH, NMW.
HG 511 as *P. serotina*: frequent, and now generally distribted.

JUGLANDALES

JUGLANDACEAE
Juglans L.
J. regia L.   Walnut
No map; 14 tetrads.
Occurs usually as isolated trees in parkland and gardens or on roadsides, in which localities they have almost certainly been planted. Occasionally it may appear spontaneously in other habitats. It certainly occurs in far more than 14 tetrads, but most field workers have not recorded it, following the rule whereby garden plants and isolated planted specimens are not recorded.
    Planted or intrusive: Ap, Ev, Eh, Er, Rw, Gp.
First record: Loudon (1838).
Herb.: LSR★; LSR, MANCH.
HG 497: rare, save as a planted tree near houses.

FAGALES

BETULACEAE
Betula L.
B. pendula Roth   Silver Birch
Map 6b; 312 tetrads.
Frequent and locally abundant in woodland or heathland in the west of the county. Occasional and probably mostly introduced in the east. It only regenerates on acid soils, though it always produces plenty of viable seeds.
    Constituent: Fd, Fm, Fs, Fw, Hd.
    Residual, intrusive or planted: Eh, Ev, Er, Ew, Rq, Rs, Rw, Ap.
First record: Pulteney (1749).
Herb.: LSR★; DPM, LSR, LTR, NMW.
HG 487 as *B. alba*: frequent, widely distributed, but as native largely confined to Charnwood Forest and the NW. area; 107 localities cited.

B. pubescens Ehrh.   Downy Birch
Map 6c; 26 tetrads.
Woodland and heathland, preferring wetter ground than **B. pendula.** Occasional in the west of the county; rare in the east.
    Constituent: Fd, Fm, Fs, Fw, Hw.
    Intrusive: Er.
First record: Coleman (1852).
Herb.: LSR★; DPM, LSR, LTR.
HG 488: local, restricted in range; 5 localities cited.

Alnus Miller
A. glutinosa (L.) Gaertn.   Alder
Map 6d; 287 tetrads.
The banks of rivers and streams, and wet places in woodland. Forms a characteristic association with **Salix** spp. in woodland on swampy ground. Frequent in the west of the county; occasional in the east.
    Constituent: Ew, Wp, Wl, Wd, Fw, Fd, Fm, Fs, M.
    Residual or intrusive: Eh, Er, Ev, Ap, Rs.
First record: Pulteney (1747).
Herb.: LSR★; DPM, LSR, LTR, MANCH, NMW.
HG 489 as *A. rotundifolia*: locally abundant, especially in Charnwood Forest, generally distributed; 90 localities cited.

A. incana (L.) Moench.   Grey Alder
No map.
Sometimes planted, mainly on roadside verges as amenity trees, usually in rural areas. In general during the survey field workers seem to have overlooked this species or ignored it as planted.
First record: present work.
HG: not recorded.

A. cordata (Loisel.) Loisel.
No map.
Sometimes plated as amenity trees, mainly in suburban areas. As a planted species, this was not recorded by field workers.
First record: present work.
HG: not recorded.

CORYLACEAE
Carpinus L.
C. betulus L.   Hornbeam
Map 6e; 70 tetrads.
Occasional throughout the county in woodland and as planted trees in public and private grounds. It is likely that all trees of this species in Leicestershire have originated from planted stock.
    Planted or intrusive: Fd, Fm, Fs, Ap, Eh, Ev, Er.
First record: Pulteney *in* Nichols (1795).
Herb.: LSR★; CGE, LSR, LTR, NMW.
HG 490: local, generally but sparsely distributed; 33 localities cited.

Corylus L.
C. avellana L.   Hazel
Map 6f; 421 tetrads.
Frequent and locally abundant in woodland throughout the county. Frequent in hedgerows in the west of the county; occasional in this habitat in the east.
    Constituent: Fd, Fm, Fs, Fw, Ef, Eh.
    Residual or intrusive: Er, Ew, Ev, Rq, Ag.
First record: Pulteney (1747).
Herb.: LSR★; DPM, LSR, LTR, MANCH, NMW.
HG 491: Locally abundant, generally distributed.

FAGACEAE
Fagus L.
F. sylvatica L.   Beech
Map 6g; 326 tetrads.
Locally frequent in woodland in the west of the county; occasional in the east, and often as isolated trees. Much planted.
    Constituent, intrusive or planted: Fd, Fm, Fs, Eh, Ev, Er, Ew, Ea, Ap, Ag, Rq, Gp.
First record: Pulteney (1749).
Herb.: LSR★; DPM, LSR, LTR, MANCH, NMW.
HG 497: frequent, generally distributed; 94 localities cited.

Castanea Miller
C. sativa Miller   Sweet Chestnut
Map 6h; 73 tetrads.
Occasional throughout the county in plantations and

woodland, almost certainly planted or descended from introduced trees.

Planted or intrusive: Fd, Fm, Fs, Eh, Ew, Er, Ap, Rq, Gp.
First record: Crabbe *in* Nichols (1795).
Herb.: LSR★; DPM, LSR, LTR, NMW.
HG 496 as *C. vulgaris*: occasional, restricted in range; 34 localities cited.

Quercus L.

Q. ilex L.   Evergreen Oak
No map; 4 tetrads.
Often planted as an ornamental tree, and occurs as such throughout the county in many more than 4 tetrads. May occasionally be self-sown.

Planted or intrusive: Ap, Fm.
First record: Horwood and Gainsborough (1933).
Herb.: LTR, NMW.
HG 495: rare; 2 localities cited.

Q. cerris L.   Turkey Oak
No map; 57 tetrads.
Occasional throughout the county in woodland, hedgerows, parks and gardens. Usually planted, but sometimes self-seeded from planted trees. A distribution map based on the records received would be misleading, because some field workers seem to have recorded nearly all specimens, others hardly any.

Planted or intrusive: Fd, Fm, Eh, Er, Ev, Ew, Ap, Gp.
First record: Horwood and Gainsborough (1933).
Herb.: LSR★; LSR, LTR.
HG 495: fairly frequent, sparsely distributed; 10 localities cited.

Q. petraea (Mattuschka) Liebl.   Sessile Oak, Durmast Oak
Map 6i; 17 tetrads.
Almost entirely confined to Charnwood Forest, where it is locally frequent in woodland on acid soil.

Constituent, residual or intrusive: Fd, Fm, Eh, Ap.
First record: Hands et al. *in* Curtis (1831).
Herb.: LSR★; LSR, LTR, MANCH, NMW.
HG 494 as *Q. robur* var. *sessiliflora*: locally abundant, largely confined to Charnwood Forest; 54 localities cited.

Q. robur L.   Pedunculate Oak
Map 6j; 573 tetrads.
Frequent throughout the county in woodland; locally frequent in hedgerows, except in the extreme north-east of the county where it is rare in certain areas. In districts where there are acid soils, there may be hybrids with **Q. petraea**.

Constituent, residual or intrusive: Fd, Fm, Fs, Fw, Eh, Ev, Er, Ew, Gr, Hg, Ap, Rq, Rs.
First record: Pulteney (1747).
Herb.: LSR★; CGE, DPM, LSR, LTR, MANCH, NMW.
HG 491: (treats **Q. robur** and **Q. petraea** as varieties of an aggregate *Q. robur*, not as separate species); 220 localities cited for the aggregate.

URTICALES

ULMACEAE
Ulmus L.

The genus **Ulmus** is both difficult and critical; difficult because the characters used in the precise determination of specimens can only be recognised with certainty in mature trees and then only at certain times of the year; critical because of the complexities of the breeding biology of the genus, which has produced a very great diversity of hybrid clones, the ancestry of which cannot be confirmed experimentally, and can only be deduced from a critical assessment of the characters referred to above. In addition, there is disagreement among specialists about the taxonomic status which should be accorded to the most distinctive and readily recognisable of the forms which occur in Britain and much of Europe.

Many of these forms still occurred in Leicestershire in 1977, and although by 1984 practically all mature trees and most of those more than five years old had succumbed to Dutch Elm disease, numerous Elm sucker hedges all over the county still survived. Whether these will eventually succumb remains to be seen. If they do not, and if in due course the county becomes repopulated with maturing trees, the records of distribution of species and varieties which we have made may prove of lasting value.

After much discussion and weighing of the alternative schemes of nomenclature available it has been decided to take that used in *Flora Europaea* as the basis of this account, making use of varietal names to distinguish some forms regarded by the leading British authority, Dr. R. Melville, as separate species. This is not the place to argue the relative merits of the different schemes which have been postulated, but it must be stressed that in our experience of elms in the field, the fact that several of the commonest and most widespread types have no acceptable names has been frustrating in the extreme, and we have wished that the rules of taxonomy could be modified to provide for a rational treatment of a genus such as **Ulmus**.

Three taxa of British elms are afforded specific rank in *Flora Europaea*, namely **U. glabra** Hudson, **U. procera** Salisb. and **U. minor** Miller. Of these **U. glabra** and **U. procera** are 'true' species in the sense that precise descriptions of holotypic material exist and populations exist in the field whose members conform closely to these published descriptions. No such description exists for **U. minor**. This name has been adopted, on grounds of priority only, for an assemblage of clones which show almost continuous variation between extremes in a number of characters such as habit, leaf shape, fruit shape, indumentum, bark and seasonal growth of twigs. Certain of these clones are widespread in Britain, and some occur in Leicestershire. Dr. Melville regards four of them as biological species, and describes the remainder in terms of interspecific hybridisation between them. As in the cases of **U. glabra** and **U. procera**, descriptions of holotypic material of these four have been published and the names *U. angustifolia* (Weston) Weston, *U. coritana* Melville, *U. carpinifolia* Gleditsch and *U. plotii* Druce are taxonomically valid if Melville's view is accepted. Otherwise these must be regarded as subspecies or varieties of **U. minor**, a taxon which Melville does not accept.

Dr. R. H. Richens has recently published the names listed below for varieties of **U. minor**. They include two of Melville's four, Cornish Elm and Plot Elm, and also English Elm and Wheatley's Elm, not previously considered as within the **U. minor** complex. Richens does not regard Melville's 'coritana' and 'carpinifolia' as sufficiently distinct from the main mass of '**minor**' forms to warrant the recognition of them under separate varietal names. However, there are populations of 'coritana' in Leicestershire which seem to conform in all respects to Melville's description, whereas with the exception of a now exterminated population of 'carpinifolia' in east Rutland, the latter form has not been reported in vice county 55. It seems that 'coritana' is capable at least of self propagation in isolation, whether sexually or by means of suckers. Richens' proposals are as follows:

U. minor Miller   Field Elm
  var. **lockii** (Druce) Richens = *U. plotii* Druce Plot Elm
  var. **vulgaris** (Aiton) Richens = *U. procera* Salisb.
                                                           English Elm
  var. **sarniensis** (Loud.) Richens = *U. × sarniensis* (Loud.) Bancroft   Wheatley's Elm
  var. **cornubiensis** (Weston) Richens = *U. angustifolia* (Weston) Weston   Cornish Elm
  var. **minor**   Narrow leaved Elm

All except the Cornish Elm are found widely in Leicestershire and their distribution has been mapped. In addition the form 'coritana' has also been recorded and mapped.

Hybrids between **U. glabra** and **U. minor** are also of widespread distribution in Leicestershire. Melville recognised the following combinations among Leicestershire populations: *U. coritana × glabra*, *U. glabra × plotii*, *U. coritana × glabra × plotii*, in addition to Dutch Elm and Huntingdon Elm in the strict sense (respectively clones of the putative parentage *U. carpinifolia × glabra × plotii* and *U. carpinifolia × glabra*). Specimens of these two latter are only known planted ornamentally in Leicestershire, whereas **U. glabra × minor** hybrids conforming to Melville's other diagnoses are common and widespread in the countryside. However, we have not been sufficiently confident of our own diagnoses to map them separately, and have only therefore produced a map for the **U. glabra × minor** aggregate.

Hybridisation within the **U. minor** complex is the most controversial aspect of the Elm problems. In Leicestershire, as far as we have been able to judge, Field Elms act as seed parents only very rarely. Professor T. G. Tutin recalls a year during the Second World War when they did so, but there are no records to show whether all forms did so, nor what proportion of the offspring survived. There are in fact no records known which give experimental confirmation to Melville's view that some forms of Field Elm constitute true biological species while others are the result of natural hybridisation between them. Nor is it known whether elms in the English Midland counties produce viable seed less frequently now than they did 500 or 5000 years ago. We are in fact only really certain that hybridisation can still occur here and hereabouts when **U. glabra** or one of its hybrids acts as the seed parent. Furthermore, even if other elms do still produce viable seed in the Midland countryside, the chance of any seedling surviving under the agricultural regime of the last 150 - 200 years must have been minimal, and any successful hybridisation which has led to the natural formation of the clones we now recognise must have occurred a long time ago, and not necessarily here. We have to admit that the present distribution of elms owes a lot to human introductions of favoured varieties for timber, fodder and amenity. Hence in addition to the comparatively widespread forms referred to above, we find in Leicestershire numerous examples of 'local' clones both of **U. minor** and of **U. glabra × minor**, like those described by Richens for other counties in south-east England. The villages where such clones occur have been recorded, but the clones themselves have not been described in detail. Herbarium material exists of some of them.

The names **U. × hollandica** Miller, **U. × vegeta** (Loud.) A. Ley, **U. × elegantissima** Horwood, **U. × diversifolia** Melville, **U. × sarniensis** (Loud.) Bancroft and **U. × viminalis** Lodd. take on a different taxonomic significance according to whether Melville's classification of elms or that of *Flora Europaea* is accepted. In the former case for example, the name **U. × hollandica** is taxonomically correct for any elm of the parentage *U. carpinifolia × glabra × plotii*; in the latter case however it is correct to use it only for the particular clone known as the Dutch Elm. This raises an interesting problem in connection with the late A. R. Horwood's Midland Elm (Horwood and Gainsborough pp. 482-484). It is no longer possible to be absolutely certain whether the specimen which Horwood used for his description of **U. elegantissima** was in fact the one he noted at Launde and elsewhere. If Melville is correct, this tree was a form of *U. glabra × plotii*, and the name is valid under taxonomic rules for any elm of this parentage. If not then the name is meaningless unless the problem of the identity of Horwood's material can be resolved. In 1977 Melville failed to locate any tree in or about Launde which conformed to Horwood's description, though there is still a tree near Exton in Rutland which Melville determined as **U. × elegantissima** in 1959. A recent critical re-examination of this tree shows that while its habit of growth conforms to Horwood's description its leaf characters are quite different, and the supposition must surely be that Melville named it **U. × elegantissima** only in the broad sense as an *U. glabra × plotii* hybrid, and not in the strict sense. Richens has suggested that Horwood's original **U. elegantissima**, as illustrated in Plate xxxiv (p. 482) of *Flora of Leicestershire and Rutland*, was a clone of **U. minor** and that the attachment of the name to any hybrid of **U. glabra** is an error.

There are therefore several unresolved problems in connection with the genus **Ulmus**, and this account is unlikely to satisfy anyone who has studied it in depth. Thanks are nevertheless due to Dr. R. Melville who led a field meeting of the B.S.B.I. to south and east Leicestershire in 1977, and made his views on the genus known; to Dr. C. A. Stace and Dr. J. G. Dony who contributed valuable advice; and to Dr. R. H. Richens and Dr. J. N. R. Jeffers, whose work on multivariate analysis of the morphological characters of elms led us to a clearer understanding of what we were attempting to do. Dr. Richens criticised the first draft of this account and suggested a number of improvements. The distribution maps are almost entirely the result of an intensive survey carried out by the author in 1977.

                                                                           K. G. Messenger

U. glabra Hudson   Wych Elm
Map 6k; 344 tetrads.
Woods, hedges, parks, avenues etc. Less widely planted than some of its hybrids; its distribution in Leicestershire and Rutland suggests that a significant proportion of specimens are descended from post-glacial native populations. That these have been augmented, both in reafforestation and in ornamental planting, by stocks

derived from elsewhere in the British Isles or from continental Europe seems certain. It is commoner in Charnwood Forest than other elms; elsewhere the distribution pattern is blurred by the presence of obviously introduced populations or individual trees, but some woods contain populations of probably native origin, these including woods north-west of Charnwood Forest as well as those of east Leicestershire.

Constituent, residual or introduced: Ap, Eh, Ev, Fd, Fm, Fs.

First record: Pulteney (1747).

Herb.: LSR*; BM, DPM, LSR, LTR.

HG 478 as *U. montana*: locally abundant, and dominant or subdominant, generally distributed; 148 localities cited.

U. glabra × minor sensu lato

Map 61; 273 tetrads.

Attempts to record hybrids between **U. glabra** and individual segregates of **U. minor** were abandoned when it was realised that distinctions between some of the segregates were themselves suspect. In the end the only useful map to emerge was that showing the distribution of the readily recognisable 'large-leaved' forms. These have a somewhat similar pattern of distribution to that of **U. glabra**, but because of their freely suckering habit are much more frequent in hedgerows, where they are significant contributors to elm-sucker hedges. In east Leicestershire they are sometimes allowed to develop into shelter belts, but they are not common in woodland. Various kinds are planted ornamentally in parkland and around villages, but specimens conforming to the type descriptions of Dutch Elm and Huntingdon Elm are not often seen.

It was recognised during the field survey that other **U. glabra × minor** hybrids are by no means infrequent among hedgerow populations, but these require more critical examination than could be given in the time available, and no maps have been made of their distribution.

We have no clear evidence of hybridisation occurring in situ and leading to the local development of a hybrid clone, though there are places where this is a distinct possibility. There is for example a roadside hedge near Bottesford where a clone of hedgerow hybrids is established close to a population of **U. minor** var. **lockii** and which shows leaf characters reminiscent of the latter; there is no **U. glabra** in the immediate vicinity, whereas the var. **lockii** is close to that part of Lincolnshire in which its English populations may very well have originated.

U. × hollandica Miller, Dutch Elm, has been recorded in avenues at one or two farms in south Leicestershire.

U. × vegeta (Loud.) A. Ley, Huntingdon Elm, occurs in the avenue at Saxelbye Park, and individual specimens provisionally recognised as this type have been seen in former parkland elsewhere in north-east Leicestershire. It is also planted in parks in Leicester.

U. × elegantissima has not been recorded during the course of the recent survey.

Constituent (perhaps), intrusive or introduced: Ap, Eh, Ev, Er, Fd, Fm.

First record (Dutch Elm): Bloxam (1837).

Herb.: LSR, LTR, OXF.

HG 480 as *U. hollandica*: rare, restricted in range; only 6 localities are cited, and it may be that the records under *U. montana* include many that should in fact be here.

U. procera Salisb.   English Elm

U. minor var. *vulgaris* (Aiton) Richens

Map 7a; 525 tetrads.

Hedgerows, particularly by streams and ditches, occasional to frequent; woodland, rare; planted ornamentally in parks and avenues, both rural and suburban; in the outer suburbs of Leicester it is often a relic of former hedgerows, and it is still a frequent constituent of sucker hedges everywhere. Richens believes that in Leicestershire it is near the northern limit of its natural distribution, having been introduced into southern Britain in pre-Roman times from the Iberian Peninsula. Others regard it as having originated in Britain by hybridisation between unknown continental parents.

Constituent, residual or introduced: Ap, Eh, Ev, Ew, Fd, Fm.

First record: Pulteney (1747).

Herb.: LSR*; LSR, LTR.

HG 480 as *U. sativa*: Generally distributed.

U. minor Miller   Field Elm

Includes *U. angustifolia*, *U. coritana*, *U. carpinifolia*, *U. plotii*, *U. × sarniensis* and perhaps *U. × elegantissima*; see under varieties, below.

U. minor Miller var. **minor**   Narrow-leaved Elm

Map 7b; 346 tetrads.

Hedgerows, in much of the county much commoner than **U. procera.** The distribution map includes records of all those specimens not definitely recognisable as named varieties or clones. Many of the records are of sucker hedges which have so far survived Dutch Elm disease.

Constituent, residual or introduced: Ap, Eh, Ev, Ew, Fd, Fm.

First record: Watson, (1837).

HG 481: about 30 localities are cited, from all over the county.

U. minor var. **lockii** (Druce) Richens   Plot Elm

U. plotii Druce

Map 7c; 28 tetrads.

Hedgerows, usually grouped round a farm or estate, occasional in scattered localities, from north-east to south of Leicester; although not recorded by A. R. Horwood, a number of the specimens now known are old enough to have been recognisable in his time; the discontinuous pattern however suggests that the present populations have been introduced, rather than that they are relics of a form originally more common.

Residual and perhaps introduced: Ev, Eh.

First record: Horwood and Gainsborough (1933).

Herb.: BM, CGE, LTR, NMW.

HG 484: 'Leicestershire', G. C. Druce in litt.; not seen by Horwood; no specific locality cited.

U. minor var. **sarniensis** (Loud.) Richens

Wheatley's Elm

U. × sarniensis (Loud.) Bancroft

Map 7d; 81 tetrads.

Extensively planted as an ornamental tree; mainly urban and suburban, but it is also to be seen in many villages, and in a few places it has been introduced into field hedges, as between Queniborough and Barkby; in some places there has been some propagation by means of suckers.

Introduced: Ap, Ev, Eh.
First record: Horwood and Gainsborough (1933).
HG 484: Leicester, and elsewhere.

U. minor var. **cornubiensis** (Weston) Richens
Cornish Elm
U. angustifolia (Weston) Weston
Although this form is recorded as an introduction by A. R. Horwood, it has not been recorded during the recent survey.
First record: B.E.C. Rep. 1916.
HG 484: as *U. minor* Miller: 14 localities cited.

U. minor 'coritana'
U. coritana Melville
Map 7e; 100 tetrads.
Hedgerows and near villages, mainly below 350 foot contour, particularly in river valleys; it replaces the commoner types of *U. minor* almost entirely over a broad swathe of country west of Lutterworth, and also occurs in the Soar and Wreake valleys, and in the Vale of Belvoir. It has been seen on high ground close to Belvoir Castle but here is is almost certainly introduced. Melville has records of **U. coritana** × **procera** in four Leicestershire localities.
 Constituent, residual or introduced: Ap, Eh, Ev, Ew.
First record: 1977, R. Melville (in litt.).
Herb.: LTR.
HG: not recorded.

Local clones of **U. minor** and **U. glabra** × **minor** have been noted at Drayton, Bringhurst, Tilton, Lowesby, Launde, Church Langton, Battle Flat near Bardon, Market Bosworth and Osgathorpe; there are doubtless many others.

CANNABACEAE
Humulus L.
H. lupulus L. Hop
Map 7f; 202 tetrads.
Locally frequent in hedgerows throughout the county. Occasional in woodland.
 Constituent or intrusive: Eh, Er, Ev, Ef, Fd, Fm, Gr. R.W. Rg, Rr. Rd, Sw.
First record: Pulteney (1747).
Herb.: LSR★: DPM, LSR, LTR, NMW.
HG 485: frequent, generally ditributed.

Cannabis L.
C. sativa L.
No map; 3 tetrads.
Beaumont Leys, rubbish tip, rare, EH 1970; Syston, roadside verge adjacent to rubbish tip, several plants with other birdseed aliens, ALP 1971; Sileby, rubbish tip, occasional, ALP 1985.
 Casual: Rr.
First record: 1708, *in* Macaulay (1791).
Herb.: LSR★; LSR, LTR, MANCH, NMW.
HG 485: rare, restricted in range; 8 localities cited.

URTICACEAE
Urtica L.
U. dioica L. Common Nettle
Map 7g; 607 tetrads.
Abundant throughout the county in almost any habitat where it is allowed to grow, except very wet ground.

Prefers nitrogen-rich soils, and is often very luxuriant in secondary woodland.
 Intrusive or constituent: Fd, Fm, Fs, Ef, Ev, Eh, Er, Ew, Ea, Rw, Rf, Rq, Rs, Rd, Rt, Gp, Gr, Gl, Ap, Ac.
First record: Pulteney (1747).
Herb.: LSR★; DPM, LANC, LSR, LTR, NMW.
HG 486: locally abundant, sometimes dominant in ground flora of dry woods, generally distributed.

U. urens L. Small Nettle
Map 7h; 267 tetrads.
A weed of gardens, non-cereal crops and recently disturbed ground. Frequent and locally abundant throughout the county.
 Intrusive: Ag, Ao, Ap, Aa, Ac, Rd, Rf, Rw, Rs, Rt, Ea, Ev, Er, Ew.
First record: Pulteney (1747).
Herb.: LSR★; LANC, DPM, LSR, LTR, MANCH, NMW.
HG 486: frequent, generally distributed.

Parietaria L.
P. diffusa Mert. & Koch Pellitory-of-the-wall
Map 7i; 56 tetrads.
Occasional on walls and buildings throughout the county. Also occasionally on pathways at the bases of walls, and in quarries and waste places.
 Intrusive: Sw, Sb, Rt, Rw, Rq, Ev, Er, Eh, Ap.
First record: Pulteney (1747).
Herb.: LSR★; DPM, LSR, LTR, MANCH, NMW.
HG 487 as *P. ramiflora*: frequent, generally distributed.

Soleirolia Guad.-Beaup.
S. soleirolii (Req.) Dandy Mind-your-own-business
Helxine soleirolii Req.
No map; 4 tetrads.
Often planted in gardens and greenhouses and escaping, becoming established on pathways and walls, and in shady waste places. Some records were lost in the fire at Ratcliffe College, and in any case it is probably more frequent as an escape than the records received seem to indicate.
First record: present work.
HG: not recorded.

SANTALALES

LORANTHACEAE
Viscum L.
V. album L. Mistletoe
Map 7j; 23 tetrads.
In scattered localities throughout the county, almost entirely on apple trees in orchards.
 Intrusive or planted: Ag.
First record: Pulteney *in* Nichols (1795).
Herb.: LSR, DPM.
HG 473: local, sparsely distributed; 9 localities cited.

ARISTOLOCHIALES

ARISTOLOCHIACEAE
Asarum L.
A. europaeum L. Asarabacca
No record since HG.

First record: Kirby (1850)
Herb.: LSR.
HG 471: rare, restricted in range; 3 localities cited.

Aristolochia L.

A. clematitis L.   Birthwort
No record since HG.
First record: Watson (1837).
Herb.: LSR.
HG 472: rare, restricted in range; 2 localities cited.

POLYGONALES

POLYGONACEAE
Polygonum L.

P. aviculare agg.   Knotgrass
Map 7k; 598 tetrads.
Field workers were instructed to record the following three species as an aggregate on the Common Species Cards, and at the same time to record the segregates if possible. The result was an almost complete distribution map for the aggregate, which probably represents the true distribution of **P. aviculare** s.s.

P. aviculare L. s.s.
Map 7l; 369 tetrads.
Arable land, waste places and recently disturbed ground. Abundant throughout the county, and almost certainly more widely distributed than the distribution map shows.
    Intrusive: Ac, Ao, Ag, Ap, Rt, Rd, Rw, Rf, Rs, Ea, Ev, Er, Ew, Gr, Gp, Gl.
First record: Pulteney (1747).
Herb.: LSR★; CGE, DPM, LSR, LTR, MANCH, NMW.
HG 459: locally abundant, generally distributed.

P. rurivagum Jord. ex Bor.
No map; 1 tetrad.
Mountsorrel, arable land, EKH 1961.
    Intrusive: A.
First record: Coleman (1852).
Herb.: LSR, LTR, MANCH, NMW.
HG 460: 19 localities cited.

P. arenastrum Boreau
Map 8a; 147 tetrads.
Trackways, arable land, dry waste places, with a preference for lighter soil. Locally frequent throughout the county; probably under-recorded in some areas.
    Intrusive: Rt, Rq, Rg, Rs, Rd, Rw, Rr, Ev, Er, Ew, Ea, Ao, Ac, Gp, Gl.
First record: Coleman (1852).
Herb.: LSR★; CGE, LSR, LTR, MANCH, NMW.
HG 461: 8 localities cited.

P. minus Huds.   Small Water-pepper
No map; 3 tetrads.
Cropston Reservoir, muddy depression on margin, occasional, PHG 1972; Cropston, marshy meadow, occasional, EH 1974; Blackbrook Reservoir, on mud in draw-down zone, occasional, C. Newbold and D. Wheatley 1984.
    Constituent: Ew, M.
First record: Pulteney in Nichols (1795).

Herb.: LSR★; LSR.
HG 462: Very rare, restricted to Charnwood Forest; 3 localities cited.

P. hydropiper L.   Water-pepper
Map 8b; 166 tetrads.
Locally frequent throughout the county on woodland rides and on the banks of rivers and canals. Occasionally a weed of arable land.
    Constituent: Ew, Ef, Fw, Fd, Fs, M, Wd, Wl, Wp.
    Intrusive: Gp, Hg, Rq, Rs, Rt, Rd, Rw, Rr, Ac, Ao, Ea, Ev, Er.
First record: Pulteney (1747).
Herb.: LSR★; CGE, DPM, LSR, LTR, MANCH, NMW.
HG 461: frequent, generally distributed.

P. persicaria L.   Redshank, Persicaria
Map 8c; 569 tetrads.
Arable land, waste places, and the banks of rivers. Frequent and locally abundant throughout the county.
    Intrusive: Ac, Ao, Ag, Aa, Ap, Ea, Ev, Er, Ew, Rw, Rf, Rd, Rq, Rg, Rs, Rt, Wd, Wp.
First record: Pulteney (1747).
Herb.: LSR★; CGE, DPM, LANC, LSR, LTR, MANCH, NMW.
HG 462: locally abundant, generally distributed.

P. lapathifolium L.   Pale Persicaria
Including *P. nodosum* Pers.
Map 8d; 354 tetrads.
Frequent and locally abundant throughout the county in arable land and waste places, and on the banks of streams.
    Intrusive: Ao, Ac, Aa, Ag, Ea, Ew, Ev, Er, Ef, Rw, Rd, Rr, Rq, Rf, Rs, Gl, Gr, Gp.
First record: Hands et al. *in* Curtis (1831).
Herb.: LSR★; CGE, DPM, LANC, LSR, LTR, MANCH, NMW.
HG 463 as *P. scabrum* and 464 as *P. peteticale*: frequent, widely distributed.

P. amphibium L.   Amphibious Bistort
Map 8e; 200 tetrads.
Frequent throughout the county in marshes, wet places, on river and canal banks, and, in its aquatic form, in shallow parts of ponds, lakes, rivers and canals, sometimes totally covering the water surface over large areas.
    Constituent: M, Ew, Wp, Wl, Wr, Wc, Wd.
    Residual or intrusive: Gp, Gr, Gl, Ev, Er, Ea, Rg, Rq, Rs, Rw, Rd, Ac, Aa, Fs.
First record: Pulteney (1747).
Herb.: LSR★; DPM, LANC, LSR, LTR, MANCH, NMW.
HG 464: locally abundant, generally distributed.

P. bistorta L.   Common Bistort
Map 8f; 9 tetrads.
Shepshed, Tyler Bridge, meadow adjoining the Black Brook, locally abundant, PAC 1969; Kirby Muxloe Castle, long grass, occasional, EH 1970 (not found 1975); Newtown Linford, marshy rough pasture, rare, EH 1971; Groby, pasture, rare, EH 1971; Peckleton, wet roadside verge, occasional, HH 1974; Loughborough, Holywell Wood, old grassland, rare, PHG 1974; Willesley, roadside verge, locally frequent, SHB 1977; Broughton Astley, meadow, abundant, PAE 1981; Sib-

son, edge of spinney, locally frequent, H.I. James 1986.
   Constituent or intrusive: Gp, Gr, M, Ev.
First record: Pulteney (1756).
Herb.: LSR*; LSR, LTR, NMW.
HG 465: occasional, very sparsely distributed; 23 localities cited, including Kirby Muxloe Castle and Shepshed.

P. amplexicaule D. Don   Red Bistort
Map 8g; 1 tetrad.
South Kilworth, rough grassland by outflow of Stanford Reservoir, occasional, EKH and JMH 1972.
   Intrusive: Gr.
First record: 1972, E. K. and J. M. Horwood in present work.
Herb.: LSR*.
HG: not recorded.

P. polystachyum Wall. ex Meissner
Himalayan Knotweed
Map 8h; 2 tetrads.
Blaby, roadside, rare, EH 1974; Gumley, margin of fish pond, locally frequent, EKH and JMH 1974.
   Intrusive or casual: Ev, Ew.
First record: 1974, in present work.
Herb.: LSR*; LTR.
HG: not recorded.

Bilderdykia Dumort

B. convolvulus (L.) Dumort.   Black-bindweed
Polygonum convolvulus L.
Map 8i; 491 tetrads.
Arable land and waste ground. Frequent and locally abundant throughout the county.
   Intrusive: Ac, Ao, Aa, Ag, Ap, Ea, Ev, Er, Eh, Ef, Rd, Rw, Rt, Rf, Rr, Rq, Rg, Rs, Gl, Gp, Gr.
First record: Pulteney (1747).
Herb.: LSR*; DPM, LANC, LSR, LTR, MANCH, NMW.
HG 459: locally abundant, generally distributed.

B. aubertii (Louis Henry) Moldenke   Russian-vine
Polygonum baldschuanicum auct. non Regel
No map.
Often planted in gardens and as a cover for walls and unsightly buildings. Very vigorous in growth, and sometimes escapes, so that it is often difficult to tell whether a particular specimen has been planted or has arisen spontaneously.
   Planted or garden escape: Sw, Sb, Eh, Ev, Rw.
First record: present work.
Herb.: LSR*.
HG: not recorded.

Reynoutria Houtt

R. japonica Houtt.   Japanese Knotweed
Polygonum cuspidatum Siebold & Zucc.
Map 8j; 122 tetrads.
Waste places, roadside and railway verges, and gardens, where it can be a troublesome weed. Locally frequent throughout the county, and appears to be increasing.
   Intrusive: Rw, Rr, Rd, Rf, Rq, Rg, Rs, Ev, Er, Ew, Eh, Ea, Ef, Ag, Ap, Fd, Fs.
First record: as garden plant, 'in and near Leicester', 1881, 1892, in herb. MANCH (Conolly 1977); as escape from cultivation, 'near Leicester', 1908, in herb. LSR (Conolly 1977).

Herb.: LSR*; LSR, LTR, MANCH.
HG 465: rare, 2 localities cited.

R. sachalinensis (Friedrich Schmidt Petrop.) Nakai
Giant Knotweed
Polygonum sachalinense Friedrich Schmidt Petrop.
Map 8k; 8 tetrads.
Gumley, marshy ground by fish pond, locally frequent, EKH and JMH 1974; Bottesford, roadside, rare, HL 1974; Blaby, roadside, rare, EH 1975; Willesley Park, ornamental woodland, locally abundant, SHB 1976; Coleorton, woodland, occasional, SHB 1978; Staunton Harold Hall Park, occasional, SHB 1978; Osgathorpe, wild part of garden, locally frequent, ALP 1978.
   Intrusive or planted: Ev, Ew, Fd, Ap, Ag.
First record: 1956, T. D. Maloney and A. L. Primavesi in herb. LSR.
Herb.: LSR, LTR.
HG: not recorded.

Fagopyrum Miller

F. esculentum Moench   Buckwheat
No map; 2 tetrads.
Freemen's Common, Leicester, waste ground, rare, HB 1968; Shepshed, roadside, occasional, PAC 1976.
   Casual: Rw, Ev.
First record: Mott et al. (1886).
Herb.: DPM, LSR, MANCH, NMW.
HG 465 as *F. sagittatum*: occasional, sparsely distributed; 18 localities cited.

Rumex L.

R. acetosella L. sensu lato   Sheep's Sorrel
Map 8l; 240 tetrads.
Heathland and grassland on acid or sandy soil, and a characteristic feature of railway ballast. Frequent and locally abundant in the west of the county; occasional in the east, except where the soil is acid or on railway ballast, in which situations it may be locally abundant.
   Constituent or residual: Hg, Hd, Gp, Gr, Gl, Fd, Fs, Ev, Er, Ef.
   Intrusive: Er, Ea, Rw, Rq, Rg, Rs, Rt, Rd, Rr, Rf, Sw, Ao, Ag, Ac, Ap.
One record for **R. angiocarpus** Murb. was received: Thurmaston, railway cutting, locally abundant, IME 1969. Otherwise, **R. acetosella** was treated as an aggregate of the three species.
First record: Pulteney (1747).
Herb.: LSR*; DPM, LSR, LTR, MANCH, NMW.
HG 471: locally abundant, generally distributed.

R. acetosa L.   Common Sorrel
Map 9a; 592 tetrads.
Grassland. Frequent and locally abundant throughout the county.
   Constituent, residual or intrusive: Gp, Gr, Gl, M, Hg, Ev, Er, Ew, Ea, Ef, Ap, Ag, Aa, Ao, Rw, Rd, Rt, Rf, Rq, Rs, Fs, Sb.
First record: Pulteney (1747).
Herb.: LSR*; LANC, LSR, LTR, MANCH, NMW.
HG 470: locally abundant, generally distributed.

R. hydrolapathum Hudson   Water Dock
Map 9b; 64 tetrads.
Locally frequent on the banks of the Grand Union and Ashby Canals, but not on the Grantham Canal; occasio-

nal on the banks of rivers. Rarely in ponds and marshes.
    Constituent: Wc, Wr, Wd, Wp, Wl, Ew, M.
First record: Pulteney (1749).
Herb.: LSR★; CGE, DPM, LANC, LSR, LTR, MANCH, NMW.
HG 470: rare, sparsely distributed; 34 localities cited.

R. patientia L.   Patience Dock
No record in the recent survey.
Earl Shilton, old tip heap at Barrow Hill Quarry, E. K. Horwood 1958.
First record: 1958, E. K. Horwood in herb LTR.
Herb.: LTR.
HG: not recorded.

R. crispus L.   Curled Dock
Map 9c; 596 tetrads.
Frequent and locally abundant throughout the county in pastures and leys, on waste ground and roadside verges. Occasional as a weed of arable land.
    Intrusive: Gp, Gr, Gl, Ev, Er, Ew, Ea, Ef, Eh, Rw, Rq, Rg, Rs, Rt, Rd, Rr, Rf, Ao, Ac, Ap, Ag, M.
First record: Pulteney (1749).
Herb.: LSR★; LSR, LTR, MANCH.
HG 469: locally abundant, generally distributed.

R. crispus × **palustris** = **R.** × **areschougii** Beck
Map 9d; 2 tetrads.
Eye Brook Reservoir, western shore between extreme high water mark and normal level, locally frequent, KGM 1971. This was first identified in 1961 by J. E. Lousley from material collected by J. R. I. Wood on the eastern (Rutland) shore of the Reservoir, together with both parents; in 1962 it was found to be widely distributed on both Rutland and Leicestershire shores. The Rutland record, according to Lousley, was the first fully authenticated British record of a hybrid previously only known for certain from the Danube valley. It is apparently almost completely sterile and arises afresh each year. The Reservoir was completed in 1940.
First record: 1962, in Messenger (1971).
HG: not recorded.

R. conglomeratus Murray   Clustered Dock
Map 9e; 384 tetrads.
Frequent throughout the county in marshes and other wet places.
    Constituent or intrusive: M, Wd, Wp, Wr, Wc, Ew, Ev, Er, Ef, Eh, Ea, Rg, Rq, Rw, Rd, Gr, Gp, Ap, Ag, Ac.
First record: Kirby (1850).
Herb.: LSR, LTR, MANCH, NMW.
HG 466: locally abundant, generally distributed.

R. conglomeratus × **crispus** = **R.** × **schulzei** Mausskn.
No map; 1 tetrad.
Groby, Community College, rare, EH 1983 (conf. D.H. Kent).
    Intrusive or casual: Rd.
First record: 1983, E. Hesselgreaves in present work.
Herb.: LSR★.
HG: not recorded.

R. conglomeratus × **obtusifolius** = **R.** × **abortivus** Ruhmer

No map; 2 tetrads.
Donington le Heath, edge of car park, rare, EH 1982; Groby, Community College, disturbed ground, rare, EH 1983 (both records conf. D.H. Kent).
    Instrusive or casual: Ev, Rd.
First record: 1982, E. Hesselgreaves in present work.
Herb.: LSR★.
HG: not recorded.

R. sanguineus L.   Wood Dock
Map 9f; 491 tetrads.
Woodland, hedgerows and shady places. Frequent throughout the county.
    Constituent: Fd, Fm, Fs, Fw, Ef, Eh.
    Residual or intrusive: Ev, Er, Ew, Ea, Wd, Gp, Gr, Gl, M, Rw, Rg, Rs, Rd, Rf.
First record: Pulteney (1747).
Herb.: LSR★; LSR, LTR, MANCH, NMW.
HG 466 as *R. sanguineus* and 467 as *R. viridis*: locally abundant, generally distributed.

R. pulcher L.   Fiddle Dock
No record since HG.
First record: Pulteney (1749).
Herb.: LSR, NMW.
HG 468: rare, very restricted in range; 7 localities cited.

R. obtusifolius L.   Broad-leaved Dock
Map 9g; 605 tetrads.
Frequent and locally abundant throughout the county on roadside verges, in waste places, and in pastures and leys. The commonest of the larger docks.
    Intrusive: Ev, Er, Eh, Ew, Ef, Ea, Gl, Gr, Gp, M, Rw, Rq, Rg, Rs, Rr, Rt, Rf, Rd, Ap, Ac, Ag.
First record: Pulteney (1747).
Herb.: LSR★; DPM, LANC, LSR, LTR, MANCH, NMW.
HG 469: locally abundant, generally distributed.

R. palustris Sm.   Marsh Dock
Map 9h; 2 tetrads.
Eye Brook Reservoir, western shore, locally frequent, KGM 1971.
    Constituent: Ew.
First record: Bloxam (1830).
Herb.: LSR★; BM, LSR, MANCH.
HG 468: rare, restricted in range; 2 localities cited for which Horwood does not express doubt.

R. maritimus L.   Golden Dock
Map 9i; 18 tetrads.
Locally frequent on the margins of Swithland, Cropston, Blackbrook, Thornton and Stanford Reservoirs. Also at Groby Pool, and in marshes north of Loughborough.
    Constituent: Ew, Wl, Wd, M.
First record: Pulteney (1749).
Herb.: LSR★; LTR, MANCH, NMW.
HG 467: rare, very restricted in range; 13 localities cited.

CENTROSPERMAE

CHENOPODIACEAE
Chenopodium L.
C. bonus-henricus L.   Good-King-Henry
Map 9j; 31 tetrads.

Sometimes grown in gardens as a vegetable. Roadside verges, waste places, rubbish dumps. Occasional throughout the county, and rarely present in quantity. At Thrussington Mill it has however been abundant over a large area for a number of years.
    Intrusive: Ev, Rw, Rf, Rq, Rr, Gp, Aa, Ap.
First record: Pulteney (1747).
Herb.: LSR★; DPM, LSR, LTR, MANCH, NMW.
HG 455: frequent, generally distributed.

C. rubrum L.   Red Goosefoot
Map 9k; 119 tetrads.
Locally frequent throughout the county in farmyards and on rubbish dumps, preferring soils with a high nitrogen content. Especially characteristic of manure heaps. Also occurs on reservoir margins, and occasionally in arable land.
    Intrusive: Rf, Rr, Rw, Rg, Rs, Rd, Ew, Ev, Er, Ea, Ag, Ao, Aa.
First record: Pulteney (1747).
Herb.: LSR★; LANC, LSR, LTR, MANCH, NMW.
HG 454: frequent, widely distributed; 55 localities cited.

C. hybridum L.   Maple-leaved Goosefoot
No map; 2 tetrads.
Beaumont Lodge Farm, arable field, single plant, EH 1973; Barwell, rubbish tip, two plants, EH 1979.
    Casual: Ao, Rr.
First record: Preston (1900).
Herb.: LSR★; LSR, LTR, NMW.
HG 454: rare, sparsely distributed; 4 localities cited.

C. polyspermum L.   Many-seeded Goosefoot
Map 9l; 83 tetrads.
Locally frequent throughout the county in arable land, especially non-cereal crops; also on reservoir margins, waste places, spoil heaps and other ruderal habitats.
    Intrusive: Ao, Ac, Ag, Aa, Ea, Ew, Ev, Rw, Rg, Rq, Rs, Rd, Rf, Rr.
First record: Bloxam (1837).
Herb.: LSR★; LANC, LSR, LTR, MANCH, NMW.
HG 450: local, restricted in range; 21 localities cited.

C. vulvaria L.   Stinking Goosefoot
No record since HG.
First record: Pulteney (1749).
Herb.: LSR, MANCH, NMW.
HG 451: rare (on the decrease), very restricted in range; 7 localities cited.

C. urbicum L.   Upright Goosefoot
No record since HG.
First record: Hands et al. *in* Curtis (1831).
Herb.: NMW.
HG 454: very rare, restricted in range; 1 confirmed record.

C. murale L.   Nettle-leaved Goosefoot
Map 10a; 4 tetrads.
Littlethorpe, garden, locally frequent, EH 1974; Newtown Linford, arable land, rare, EH 1978; Groby, garden, rare, EH 1978; Kirby Muxloe, allotment, rare, EH 1979.
    Intrusive or casual: Ag, Aa, Ao.
First record: Bloxam (1848d).
Herb.: LSR★; LSR.
HG 454: very rare, restricted in range; 5 localities cited.

C. ficifolium Sm.   Fig-leaved Goosefoot
Map 10b; 13 tetrads.
Arable land, manure heaps and waste places. Occasional in the southern half of the county; not recorded from the north.
    Intrusive: Ao, Ag, Rf, Rw, Rs, Rd.
First record: Bloxam and Babington *in* Potter (1842).
Herb.: LSR★; CGE, LSR, MANCH, NMW, LTR.
HG 453: occasional, restricted in range; 31 localities cited.

C. opulifolium Schrader ex Koch & Ziz
     Grey Goosefoot
No record from the recent survey.
Leicester, Welford Road, rubbish tip, S. O. Taylor 1939.
First record: 1939, S. O. Taylor in herb. LSR.
Herb.: LSR.
HG: not recorded.

C. album L.   Fat-hen
Map 10c; 527 tetrads.
Arable land, waste places and recently disturbed ground. Locally abundant throughout the county.
    Intrusive: Ac, Ao, Ag, Aa, Ap, Ea, Ev, Er, Ew, Eh, Rd, Rf, Rw, Rg, Rq, Rs, Rr, Gl, Gp, Cr.
First record: Pulteney (1747).
Herb.: LSR★; CGE, DPM, LANC, LSR, LTR, MANCH, NMW.
HG 452: locally abundant, generally distributed.

Atriplex L.
A. hortensis L.   Garden Orache
No map.
Barwell, rubbish dump, rare, SHB 1979. Several other records for this species were lost as a result of the fire at Ratcliffe College. All were from rubbish dumps and waste places.
    Casual: Rr, Rw.
First record: Horwood and Gainsborough (1933).
HG 458: rare; 2 localities cited.

A. patula L.   Common Orache
Map 10d; 516 tetrads.
Frequent throughout the county in arable land, especially non-cereal crops, and in farmyards, manure heaps, waste places and recently disturbed ground.
    Intrusive: Ao, Ag, Ac, Ap, Aa, Rf, Rw, Rd, Rq, Rg, Rs, Rr, Ea, Ev, Er, Gl.
First record: Pulteney (1749).
Herb.: LSR★; DPM, LANC, LSR, LTR, MANCH, NMW.
HG 456: locally abundant, generally distributed.

A. hastata L.   Spear-leaved Orache
Map 10e; 238 tetrads.
Frequent throughout the county in waste places and on roadside verges, especially where the road is treated with salt in the winter. Occasional in arable land and gardens.
    Intrusive: Rw, Rg, Rs, Rt, Rf, Rr, Rd, Ev, Er, Ew, Ea, Ag, Aa, Ac, Ao, Ap.
First record: Pulteney (1749).
Herb.: LSR★; CGE, LANC, LSR, LTR, MANCH, NMW.
HG 457: locally abundant, widely distributed; 29 localities cited.

Salsola L.
S. kali L. subsp. **ruthenica** (Iljin) Soo
　　　　　　　　　　　　　　Spineless Saltwort
S. pestifer A. Nelson
No map. 1 tetrad.
Holwell Iron Works, near Asfordby, flattened top of what was once a very large conical slag heap, locally abundant, ALP 1972. In 1976 the site was bulldozed and the plant has not been seen since.
　　Casual: Rs.
First record: 1972, A. L. Primavesi in present work.
Herb.: LSR★; LTR.
HG: not recorded.

AMARANTHACEAE
Amaranthus L.
A. retroflexus L.
No map; 2 tetrads.
Anstey, manured corner of field, occasional, EH 1970; Quorndon, garden, PHG 1977 and 1979.
　　Casual: Rf, Ag.
First record: Vice (1900).
Herb.: LSR★; LSR.
HG 450: rare, restricted in range; 2 localities cited.

A. albus L.
No map; 3 tetrads.
Gaddesby, building site, single plant, EH 1967; Anstey, manured corner of field, occasional, EH 1970; Barwell, rubbish dump, rare, SHB 1979.
　　Casual: Rw, Rr.
First record: 1967, E. Hesselgreaves in present work.
Herb.: LSR★.
HG: not recorded.

PORTULACACEAE
Montia L.
M. fontana L. subsp. **chondrosperma** (Fenzl) Walters
　　　　　　　　　　　　　　Blinks
Map 10f; 14 tetrads.
Heath grassland, especially where water lies in winter; in marshes, and beside water. In scattered localities throughout the county, sometimes locally frequent.
　　Constituent, residual or intrusive: Hw, Hg, M, Gp, Wd, Wp, Ew, Rw, Rq.
First record: Pulteney (1747).
Herb.: LSR★; LSR, MANCH, NMW.
HG 93 as *M. verna*: local, confined to a few districts; 29 localities cited.

M. perfoliata (Willd.) Howell　　　　Springbeauty
Map 10g; 15 tetrads.
Occasional in scattered localities as a garden weed, especially characteristic of nursery gardens.
　　Intrusive: Ag, Ap, Rd, Eh.
First record: 1885, C. Packe in herb. CGE.
Herb.: LSR★; CGE, LANC, LSR, LTR.
HG 92 as *Claytonia perfoliata*: rare; 1 locality cited.

M. sibirica (L.) Howell　　　　Pink Purslane
Map 10h; 6 tetrads.
Ulverscroft, roadside, locally abundant, PHG 1969; Cadeby, roadside ditch, locally abundant, HH 1972; Cadeby, damp scrub woodland, locally frequent, HH 1973; Cadeby, bank of stream, locally frequent, HH 1975; Burton on the Wolds, western edge of Twenty Acre, on pile of loose soil, rare, ALP 1975; Measham, hedgerow, occasional, SHB 1976; Markfield, hedgerow, locally frequent, E.M. Penn-Smith 1986 (white flowers).
　　Intrusive: Ev, Ew, Eh, Wd, Fs, Rd.
First record: 1886, F. T. Mott in herb. MANCH.
Herb.: LSR★; DPM, LSR, LTR, MANCH.
HG 92 as *Claytonia sibirica*: rare; 1 locality cited (Cadeby).

CARYOPHYLLACEAE
Arenaria L.
A. serpyllifolia L.　　　　Thyme-leaved Sandwort
Map 10i; 118 tetrads.
Frequent and locally abundant throughout the county on railway ballast, 69% of the records received being from this habitat. Occasional on walls, pathways and spoil heaps, and in quarries.
　　Intrusive: Er, Ev, Ea, Sw, Sr, Sb, Rt, Rq, Rs, Rw, Rr, Rf, Rg.
First record: Pulteney (1747).
Herb.: LSR★; LANC, LSR, LTR, MANCH, NMW.
HG 88: abundant, generally distributed.

A. leptoclados (Reichb.) Guss.　　　　Slender Sandwort
Map 10j; 88 tetrads.
Similar to the preceding species in habitat and distribution, but less frequent. 62% of the records for this species are from railway ballast.
　　Intrusive: Er, Ev, Sw, Sb, Rw, Rt, Rg, Rq, Rs, Gr, Gp.
First record: Power (1805).
Herb.: LSR★; LSR, LTR, MANCH.
HG 88: local, range restricted; 21 localities cited.

Moehringia L.
M. trinervia (L.) Clairv.　　　　Three-nerved Sandwort
Map 10k; 426 tetrads.
Frequent throughout the county in woodland, hedgerows and other shady places.
　　Constituent: Fd, Fm, Fs, Fw, Ef.
　　Residual or intrusive: Eh, Ev, Er, Ew, Ea, Wd, Rd, Rt, Sw, Gp.
First record: Pulteney in Nichols (1795).
Herb.: LSR★; DPM, LANC, LSR, LTR, NMW.
HG 88: locally abundant, generally distributed.

Minuartia L.
M. hybrida (Vill.) Schischkin　　　　Fine-leaved Sandwort
Map 10l; 8 tetrads.
Plungar, dismantled railway, locally frequent, Lady Anne Brewis and PAC 1969; Wymondham, railway embankment, occasional, KGM 1971; Glen Parva, dismantled railway, locally frequent, EH 1974; Whetstone, dismantled railway, frequent, EH 1975; Leicester, Chatham Street, brick and concrete rubble, locally abundant, IME 1977; Cosby, dismantled railway, occasional, EH 1977.
First record: 1969, Lady Anne Brewis and P. A. Candlish in present work.
Herb.: LSR★.
HG 88 as *Arenaria tenuifolia*: no Leicestershire record.

Stellaria L.
S. media (L.) Vill.　　　　Common Chickweed
Map 11a; 605 tetrads.
Arable land and gardens, woodland, grassland especially leys, and the edges of pasture near the hedgerow. Abun-

dant throughout the county.
  Constituent, residual or intrusive: Fd, Fs, Ef, Eh, Ac, Ao, Ag, Ap, Aa, Gl, Gp, Gr, Hg, Ev, Ea, Er, Ew, Rd, Rw, Rf, Rt, Rq, Rs.
First record: Pulteney (1747).
Herb.: LSR★; DPM, LANC, LSR, LTR, MANCH, NMW.
HG 84: abundant, generally distributed.

S. neglecta Weihe   Greater Chickweed
Map 11b; 13 tetrads.
Hedgerows, woodland and other shady habitats. Occasional, probably throughout the county and overlooked.
  Constituent or intrusive: Fw, Fd, Ef, Eh, Ew, Wd, M, Hb.
First record: Preston (1900).
Herb.: LSR★; CGE, LANC, LSR, LTR, MANCH, NMW.
HG 85: occasional, sparsely distributed; 13 localities cited.

S. pallida (Dumort.) Piré   Lesser Chickweed
Map 11c; 9 tetrads.
Beaumont Leys, field gateway, occasional, EH 1969; Medbourne, ironstone wall, occasional, KGM 1971; Blaston, ironstone wall, occasional, KGM 1971; Drayton, ironstone wall, occasional, KGM 1971; Groby, bare ground, locally frequent, EH 1972; Leicester, Western Park, allotment garden, rare, EH 1972; Bringhurst, churchyard wall, rare, KGM 1973; Croft Hill, heath grassland, locally frequent, SHB 1975; Cotes, grassland, occasional, PHG and E. G. Webster 1977.
  Intrusive: Sw, Rt, Rd, Aa, Gp, Hg.
First record: Coleman (1852).
Herb.: LANC, LSR, NMW.
HG 84: 8 localities cited.

S. holostea L.   Greater Stitchwort
Map 11d; 221 tetrads.
Woodland, hedgerows and shady roadside and railway verges. Frequent in the well-wooded parts of the county, especially in the west. Occasional elsewhere.
  Constituent: Fd, Fm, Fs, Ef.
  Residual or intrusive: Eh, Ev, Er, Ew, Ea, M, Ag.
First record: Pulteney (1747).
Herb.: LSR★; DPM, LANC, LSR, LTR, MANCH, NMW, UPP.
HG 86: locally abundant, generally distributed.

S. alsine Grimm   Bog Stitchwort
Map 11e; 289 tetrads.
Frequent throughout the county in marshes, ditches and on woodland rides, and on the banks of rivers and canals.
  Constituent: M, Ew, Ef, Wd, Wp.
  Residual or intrusive: Gr, Gp, Gl, Ev, Er.
First record: Pulteney (1747).
Herb.: LSR★; DPM, LANC, LSR, LTR, MANCH, NMW.
HG 87: local, widely distributed; 66 localities cited.

S. palustris Retz.   Marsh Stitchwort
Map 11f; 3 tetrads.
Ratcliffe on the Wreake, near Lewin Bridge, marshy meadow, locally frequent, ALP 1967, the field is now heavily grazed and the plant has not been seen recently; Lockington-Hemington, marshy ground in several sites near the junction of the River Soar and River Trent, locally frequent, PAC 1973 and 1974.
  Constituent or residual: M.
First record: Bloxam (1837).
Herb.: LSR★; CGE, LSR, MANCH, NMW.
HG 86 as *S. dilleniana*: local, restricted in range; 13 localities cited.

S. graminea L.   Lesser Stitchwort
Map 11g; 400 tetrads.
Rough grassland, heath, marsh, roadside and railway verges. Frequent and locally abundant throughout the county.
  Constituent: Gr, Gp, Hg, Hd, M, Ev, Er, Ew, Ef, Fd, Fs.
  Residual or intrusive: Gl, Rw, Rd, Rt, Rg, Ea, Ac, Ap.
First record: Pulteney (1747).
Herb.: LSR★; CGE, DPM, LANC, LSR, LTR, MANCH, NMW.
HG 86: frequent, generally distributed.

Cerastium L.
C. tomentosum L.   Snow-in-summer
Map 11h; 38 tetrads.
A garden escape, becoming established on walls, roadsides and railway verges, and in waste places. Occasional throughout the county. Sometimes covers large areas in railway cuttings.
  Intrusive: Sw, Sr, Rw, Rq, Rs, Rr, Rt, Er, Ev, Gr, Ap.
First record: 1960, E. K. Horwood in herb. LANC.
Herb.: LSR★; LANC, LTR.
HG: not recorded.

C. arvense L.   Field Mouse-ear
Map 11i; 8 tetrads.
Occasional in grassland and the headlands of arable fields on calcareous soils in the north-east of the county; otherwise all the records (67% of the total) are from railway verges, where it is sometimes locally abundant.
  Constituent: Gp.
  Residual or intrusive: Ea, Er.
First record: Crabbe *in* Nichols (1795).
Herb.: LSR★; LSR, LTR, MANCH.
HG 83: rare in Leicestershire, more frequent in Rutland; 14 localities cited.

C. fontanum Baumg. subsp. **triviale** (Link) Jalas
                             Common Mouse-ear
C. holosteoides Fries
Map 11j; 598 tetrads.
Grassland. Abundant throughout the county.
  Constituent: Gp, Gr, M, Ev, Er, Ew.
  Intrusive: Rw, Rq, Rg, Rt, Rf, Rd, Ea, Ac, Ao, Ag, Ap, Aa, Sw, Gl.
First record: Pulteney (1747).
Herb.: LSR★; DPM, LANC, LSR, LTR, MANCH, NMW.
HG 82 as *C. vulgatum*: locally abundant, generally distributed.

C. glomeratum Thuill.   Sticky Mouse-ear
Map 11k; 288 tetrads.
Frequent throughout the county on sandy soils, railway ballast, walls, and dry places such as pathways. More frequent in the west than in the east.
  Constituent or intrusive: Gp, Gr, Gl, Hd, Ev, Er, Ef, Ew, Ea, Ao, Ac, Ag, Ap, Sw, Rw, Rt, Rq, Rg, Rs, Rd, Rf, Rr, M.
First record: Pulteney (1747).

Herb.: LSR*; LSR, LTR, MANCH, NMW.
HG 83 as *C. viscosum*: frequent, generally distributed.

C. semidecandrum L.   Little Mouse-ear
Map 11l; 8 tetrads.
Groby, sandy turf, occasional, EH 1971; Bradgate Park, dry heath grassland, rare, EH 1971; Croft, siliceous grassland, abundant, SHB 1975; Rothley, railway verge, locally abundant, PHG 1976; Croft Hill, siliceous grassland, frequent, SHB 1976; Burrough Hill, pasture, frequent, SHB 1977; Somerby, pasture, frequent, SHB 1977; Newtown Linford, pasture, locally frequent, PHG 1978; Cotes, pasture, locally frequent, PHG 1978.
  Constituent: Hg, Gp.
First record: Pulteney (1747).
Herb.: LSR*; DPM, CGE, LSR, LTR, MANCH.
HG 81: local, restricted to certain soils; 24 localities cited.

C. diffusum Pers. subsp. **diffusum**   Sea Mouse-ear
C. atrovirens Bab.
Map 12a; 13 tetrads.
Thorpe Satchville, two localities on ballast of dismantled railway, locally frequent, HB 1968 (ballast since removed); near Swithland Reservoir, railway verge, rare, PHG 1971; Freeby, cinder ballast of dismantled railway at site of Saxby Station, occasional, KGM 1971; Wymondham, cinder ballast of dismantled railway, occasional, KGM 1971; Groby, dismantled mineral railway, rare, EH 1971 (site since destroyed); Kirby Muxloe, two sites on railway ballast, occasional, EH 1972; Braunstone, railway ballast, rare, EH 1972; Whetstone, two sites on ballast of dismantled railway, occasional, EH 1974 and 1976; Rothley, railway verge, occasional, PHG 1976; Loughborough, railway verge, PHG 1977.
  Intrusive: Er.
First record: Britten *in* White (1877).
Herb.: LSR*; LSR.
HG 80 as *C. tetrandrum*: very rare, restricted in range; 2 localities cited.

Moenchia Ehrh.
M. erecta (L.) P. Gaertner, B. Meyer & Scherb.
                                           Upright Chickweed
Map 12b; 5 tetrads.
Groby, sandy turf near the Pool, occasional, EH 1971; Bradgate Park, dry heath grassland, rare, EH 1972; Croft, siliceous grassland, locally abundant, SHB 1973; Croft Hill, siliceous grassland, locally abundant, SHB 1973; Rothley, heath grassland off Kinchley Lane, occasional, PHG 1974; Groby, near Bradgate House, heath grassland, occasional, EH 1977.
  Constituent: Hg.
First record: Pulteney *in* Nichols (1795).
Herb.: LSR*; LSR, LTR, MANCH, NMW.
HG 83 as *Cerastium erectum*: occasional, very restricted in range; 14 localities cited, including Groby and Croft.

Myosoton Moench
M. aquaticum (L.) Moench   Water Chickweed
Map 12c; 200 tetrads.
Locally frequent on the banks of rivers, canals and streams, and in wet ditches. Occasional in marshes and other wet places, and on roadside verges far from water.
  Constituent: Ew, Wp, Wl, Wd, M, Fw.
  Residual or intrusive: Ev, Er, Rg, Rw, Rd.
First record: Pulteney (1747).

Herb.: LSR*; CGE, LANC, LSR, LTR, MANCH, NMW.
HG 84 as *Stellaria aquatica*: local, but generally distributed.

Sagina L.
S. nodosa (L.) Fenzl   Knotted Pearlwort
Map 12d; 3 tetrads.
Market Bosworth, margin of Ashby Canal by Jackson's Bridge, locally frequent, HB 1971; Carlton, margin of Ashby Canal by Carlton Bridge, occasional, OHB and HH 1971; Sutton Cheney, margin of Ashby Canal by Sutton Wharf, rare, SHB 1972; Carlton, margin of Ashby Canal by Iliffe Bridge, occasional, HH 1975; Dadlington, stonework of canal bridge, rare, SHB 1976.
  Constituent or intrusive: Ew, Sw.
First record: Pulteney (1749).
Herb.: LSR*; LSR, MANCH, WAR.
HG 90: rare, confined to region west of R. Soar; 10 localities cited, including 3 from margin of Ashby Canal.

S. subulata (Swartz) C. Presl   Heath Pearlwort
No record in the recent survey.
Oadby, lawn of Beaumont House, T. G. Tutin 1949.
First record: 1949, T. G. Tutin in herb. LTR.
Herb.: LTR.
HG: not recorded.

S. procumbens L.   Procumbent Pearlwort
Map 12e; 340 tetrads.
Frequent throughout the county on pavements and pathways in towns and villages. Occasional in marshes, in waste places, on railway ballast and walls. Sometimes a troublesome weed of shaded lawns.
  Constituent or intrusive: Rt, Rw, Rq, Rg, Rs, Rd, Sw, Sb, Er, Ev, M, Gp, Gr, Ag, Ap.
First record: Pulteney (1747).
Herb.: LSR*; LSR, LTR, MANCH, NMW.
HG 90: frequent, generally distributed.

S. apetala Ard.   Annual Pearlwort
Including *S. ciliata* Fries
Map 12f; 57 tetrads.
Occasional throughout the county on dry sandy soils, walls, and railway ballast.
  Constituent or intrusive: Hd, Gp, Er, Ev, Sw, Sb, Rt, Rw, Rs, Rr.
First record: Hands et al. *in* Curtis (1831).
Herb.: LSR*; LSR, LTR, MANCH, NMW.
HG 89: frequent, generally distributed.

Scleranthus L.
S. annuus L.   Annual Knawel
Map 12g; 10 tetrads.
Thrussington, cabbage field on river gravel soil, locally frequent, ALP 1967; Woodhouse, arable field on Windmill Hill, locally frequent, PHG 1968; Ab Kettleby, arable field west of Holwell Mouth, locally abundant, PHG 1969; Sutton Cheney, cornfield on sandy soil, locally frequent, OHB and HH 1971; Earl Shilton, Barrow Hill Quarry, occasional, SHB 1972; Osbaston, headland of cornfield, rare, OHB and HH 1973; Croft, siliceous grassland, locally frequent, SHB 1973; Groby, old quarry, rare, EH 1974; Whetstone, disturbed ground by roadside, rare, EH 1975; Cosby, pathway, rare, EH 1975; Potters Marston, pathway, rare, SHB 1978.
  Intrusive: Ao, Ac, Ea, Ev, Rt, Rd, Rq, Hg.

First record: Pulteney (1747).
Herb.: LSR★; LSR, LTR, MANCH, NMW.
HG 449: local, confined to a few districts; 24 localities cited.

Herniaria L.
H. hirsuta L.
No map; 1 tetrad.
Sileby, ballast of dismantled railway goods yard, occasional, ALP 1970, det. T. G. Tutin (had completely disappeared the following year).
Casual: Er.
First record: 1970, A. L. Primavesi in present work.
HG: not recorded.

Spergula L.
S. arvensis L.   Corn Spurrey
Map 12h; 164 tetrads.
Arable land, waste places and recently disturbed ground, preferring acid soils. Frequent and locally abundant in the west of the county, occasional in the east.
Intrusive: Ao, Ac, Ag, Aa, Rw, Rd, Rt, Rq, Rs, Ea, Ev, Er, Ef, Gr, Gl.
First record: Pulteney (1747).
Herb.: LSR★; CGE, DPM, LSR, LTR, MANCH, NMW.
HG 91: frequent, generally distributed.

Spergularia (Pers.) J. & C. Presl
S. rubra (L.) J. & C. Presl   Sand Spurrey
Map 12i; 30 tetrads.
Dry sandy places. Locally frequent in Charnwood Forest; occasional elsewhere, and absent from the east of the county.
Constituent or intrusive: Hg, Hd, Gr, Sr, Rt, Rq, Rs, Rw, Rd, Er, Ev, Ea.
First record: Pulteney (1747).
Herb.: LSR★; LSR, LTR, MANCH, NMW.
HG 91: occasional, restricted in range; 57 localities cited.

Lychnis L.
L. coronaria (L.) Desr.
No map; 2 tetrads.
Newtown Linford, rough grass, rare, EH 1978; Sileby, rubbish tip, occasional, ALP 1986.
Casual: Gr, Rr.
First record: 1978, E. Hesselgreaves in present work.
HG: not recorded.

L. flos-cuculi L.   Ragged Robin
Map 12j; 245 tetrads.
Marshes, ditches, meadows and the banks of rivers, streams and canals. Frequent, but decreasing as suitable habitats are destroyed.
Constituent: M, Ew, Ef, Fw, Wd, Gp, Gr.
Residual or intrusive: Rg, Fd, Fm, Fs, Ev, Er.
First record: Pulteney (1747).
Herb.: LSR★; LSR, DPM, LANC, LTR, MANCH, NMW.
HG 79: frequent, generally distributed.

Agrostemma L.
A. githago L.   Corncockle
No map; 2 tetrads.
Quorndon, casual in garden, rare, PHG 1960; Somerby, rubbish tip, rare, IME 1972.
Casual: Ag, Rr.

First record: Pulteney (1747).
Herb.: LSR★; LSR, LTR, MANCH, NMW.
HG 80 as *Lychnis githago*: occasional, has become rarer save as an alien during the last 30 years; 55 localities cited.

Silene L.
S. nutans L.   Nottingham Catchfly
No recent record.
First record: Crabbe *in* Nichols (1795).
HG 77: reported by Rev. G. Crabbe as introduced at Stathern.

S. vulgaris (Moench) Garcke   Bladder Campion
Map 12k; 58 tetrads.
Frequent on the calcareous soils of the north-east. For the rest of the county, most of the records are from railway verges (74% of records received), where it may be locally frequent. It also occurs occasionally on roadsides and in quarries.
Constituent or intrusive: Gp, Gr, Er, Ev, Rq, Rs, Rw, Rd, Rt.
First record: Pulteney (1747).
Herb.: LSR★; LSR, LTR, MANCH, NMW.
HG 75 as *S. cucubalus*: occasional, generally distributed; 66 localities cited.

S. noctiflora L.   Night-flowering Catchfly
Map 12l; 4 tetrads.
Eastwell, two localities in arable land, occasional, JMS 1968; Ab Kettleby, arable field off Landyke Lane, rare, and cornfield east of Holwell Mouth, rare, PHG and J. Gibbons 1969; Groby, cornfield south-east of Sheet Hedges Wood, rare, EH 1969; Harby Hills, cornfield, rare, JMS 1970; Saltby, cornfield south of King Lud's Entrenchments, frequent, SHB and PHG 1978; Sproxton, sugar beet field north-east of Gallops Plantation, occasional, SHB 1978.
Intrusive: Ac, Ao.
First record: Coleman (1852).
Herb.: LANC, LSR, LTR, MANCH, NMW.
HG 77: rare, rather restricted in range; 29 localities cited.

S. alba (Miller) E. H. L. Krause   White Campion
Map 13a; 410 tetrads.
Frequent throughout the county in waste places, quarries and gravel pits, and on roadsides and railway verges. Occasional as an arable weed.
Intrusive: Rw, Rq, Rg, Rs, Rr, Rd, Rf, Rt, Ev, Er, Eh, Ea, Ew, Gr, Gl, Ac, Ao, Ag.
First record: Pulteney (1747).
Herb.: LSR★; DPM, LSR, LTR, MANCH, NMW.
HG 78 as *Lychnis alba*: locally abundant, widely distributed.

S. alba × dioica
Map 13b; 32 tetrads.
Occurs occasionally in similar habitats to those of the two parents. Almost certainly under-recorded.
Intrusive: Ev, Er, Eh, Ef, Ew, Gr, Rw, Rq, Rd, Ag.
First record: Horwood and Gainsborough (1933).
Herb.: LSR★; LSR.
HG 79: 15 localities cited.

S. dioica (L.) Clairv.   Red Campion
Map 13c; 512 tetrads.
Woodland, hedgerows, roadside and railway verges. Frequent and locally abundant over most of the county; occasional in parts of the south and west.

Constituent: Fd, Fm, Fs, Ef.
Residual or intrusive: Eh, Ev, Er, Ew, Wd, Gr, Gl, Hg, Rw, Rq, Ag.
First record: Pulteney (1747).
Herb.: LSR★; DPM, LANC, LSR, LTR, MANCH, NMW.
HG 78 as *Lychnis dioica*: locally abundant or frequent, generally distributed.

S. gallica L.   Small-flowered Catchfly
No record since HG.
First record: Kirby (1850).
Herb.: LSR.
HG 76: very rare, very restricted in range; 5 localities cited under *S. anglica* and 2 localities under *S. quinquevulnera*.

Saponaria L.
S. officinalis L.   Soapwort
Map 13d; 17 tetrads.
Waste places and roadsides, usually near dwellings. Occasional; nearly always a shortly persistent casual or garden escape.
Intrusive or casual: Rw, Rr, Rs, Ev, Er, Eh, Gr, Aa.
First record: Pulteney (1749).
Herb.: LSR★; DPM, LSR.
HG 74: rare, very restricted in range; 10 localities cited.

Vaccaria Medicus
V. pyramidata Medicus   Cowherb
No map; 3 tetrads.
Croxton Kerrial, disturbed ground at roadside, rare, KGM 1972; Newtown Linford, heath grassland, locally frequent, SHB 1978; Rothley, garden weed, single plant, D. S. Fieldhouse 1978.
Casual: Ev, Hg, Ag.
First record: Preston (1900).
Herb.: LSR, NMW.
HG 74 as *Saponaria vaccaria*: rare, restricted in range; 10 localities cited.

Dianthus L.
D. deltoides L.   Maiden Pink
No record since HG.
First record; Power (1805).
Herb.: LSR.
HG 74: rare, very restricted in range; 5 localities cited.

RANALES

NYMPHAEACEAE
Nymphaea L.
N. alba L.   White Water-lily
Map 13e; 15 tetrads.
Occasional in scattered localities throughout the county in ponds, gravel pits and clay pits; there is one record from the Grand Union Canal.
Planted or intrusive: Wp, Rg, Wc.
First record: Pulteney (1752).
Herb.: LSR.
HG 24 as *Castalia alba*: rather rare, restricted in range; 18 localities cited.

Nuphar Sm.
N. lutea (L.) Sibth. & Sm.   Yellow Water-lily
Map 13f; 134 tetrads.
Frequent throughout the county in rivers and canals. Occasional in the larger brooks and in large ponds and lakes.
Constituent: Wr, Wc.
Intrusive or planted: Wp, Wl, Rq.
First record: Pulteney (1747).
Herb.: LSR★; DPM, LANC, LSR, LTR, MANCH, NMW.
HG 23 as *Nymphaea lutea*: frequent and generally distributed.

CERATOPHYLLACEAE
Ceratophyllum L.
C. demersum L.   Rigid Hornwort
Map 13g; 25 tetrads.
Occasional or sometimes locally abundant in scattered localities throughout the county in ponds, lakes, rivers and canals.
Constituent or intrusive: Wp, Wl, Wr, Wc, Rg.
First record: Pulteney *in* Nichols (1795).
Herb.: LSR★; CGE, DPM, LSR, LTR, MANCH, NMW.
HG 513: rare, restricted in range; 13 localities cited.

C. submersum L.   Soft Hornwort
Map 13h; 22 tetrads.
Mainly in ponds, where it sometimes occurs in great abundance. Occasional in the west of the county; rare in the east.
Constituent: Wp, Wl, Wr, Wd.
First record: Pulteney (1749).
Herb.: LSR★, NMW.
HG 514: rare, restricted in range; 5 localities cited.

RANUNCULACEAE
Helleborus L.
H. foetidus L.   Stinking Hellebore
Map 13i; 2 tetrads.
Ratcliffe College, wild shady parts of grounds, naturalised for at least 30 years and formerly locally frequent, last seen by ALP 1967 (two plants) but now apparently extinct; Woodhouse Eaves, The Brand, mixed woodland, occasional, PHG 1975.
Intrusive: Fm, Ap.
First record: Jackson (1904b).
Herb.: LSR.
HG 20: rare, restricted in range; 3 localities cited, not including the above.

H. viridis L.   Green Hellebore
Map 13j; 1 tetrad.
Woodhouse Eaves, The Brand, mixed woodland, rare, PHG 1968.
Intrusive: Fm.
First record: 1890, F. T. Mott in herb. MANCH.
Herb.: LSR, MANCH, NMW.
HG 20: rare, restricted in range; 4 localities cited.

Eranthis Salisb.
E. hyemalis (L.) Salisb.   Winter Aconite
Map 13k; 18 tetrads.
Frequently planted in gardens, shrubberies and spinneys; occasionally becoming established, thoroughly naturalised and locally abundant. Rarely appears spontaneously in places where it has not been introduced.
Planted or intrusive: Fd, Fm, Fs, Ap, Ev.
First record: Power (1807).

Herb.: LSR★; CGE, DPM, LSR, LTR, NMW.
HG 21 as *Cammarum hyemale*: rare, restricted in range; 6 localities cited.

Caltha L.
C. palustris L.   Marsh-marigold, Kingcup
Map 13l; 224 tetrads.
Marshes, and wet places in woodland. Frequent throughout the county, but decreasing as suitable habitats are destroyed by drainage or agricultural 'improvement'.
    Constituent: M, Ew, Wd, Wp, Fw, Fd.
    Residual: Gp, Gr.
First record: Pulteney (1747).
Herb.: LSR★; LANC, LSR, LTR, MANCH, NMW.
HG 19: locally abundant, generally distributed.

Aconitum L.
A. napellus L.   Monk's-hood
A. anglicum Stapf
Map 14a; 2 tetrads.
Gumley, open woodland by fish pond, occasional, EKH and JMH 1974; Withcote, shrubbery beside lake, rare, IME 1974.
    Intrusive: Ew.
First record: Kirby (1850).
Herb.: LSR.
HG 22: rare, restricted in range; 6 localities cited, not including the above.

Consolida (DC.) S. F. Gray
C. ambigua (L.) P. W. Ball & Heywood   Larkspur
Delphinium ambiguum L.
No record since HG.
First record: Pulteney (1747).
Herb.: LSR, MANCH.
HG 22 as *Delphinium ajacis*: rare, restricted in range; 2 localities cited.

C. regalis S. F. Gray subsp. **regalis**
Delphinium consolida L.
No record since HG.
First record: Kirby (1850).
HG 22: rare, restricted in range; 4 localities cited.

Anemone L.
A. nemorosa L.   Wood Anemone
Map 14b; 137 tetrads.
Frequent and locally abundant in old and long-established woodland; occasional in hedgerows and on shady roadside verges in wooded regions. Rare or occasional in the less wooded parts of the county.
    Constituent: Fd, Fm, Fs, Fw.
    Residual or intrusive: Eh, Ev, Er, Ew, Wd, Gp, Gr, Ap, Ag.
First record: Pulteney (1747).
Herb.: LSR★; DPM, LSR, LTR, MANCH, NMW, UPP.
HG 4: locally abundant and generally distributed.

A. ranunculoides L.   Yellow Anemone
No record in the recent survey.
East Langton, T. W. Tailby 1954.
First record: Crabbe *in* Nichols (1795).

Herb.: LSR.
HG 4: rare, restricted in range; 2 localities cited.

A. apennina L.   Blue Anemone
No record in the recent survey.
East Langton, T. W. Tailby 1954; Grace Dieu Wood, rare, ALP 1957.
First record: Preston (1900).
Herb.: LSR.
HG 5: rare, restricted in range; 2 localities cited.

Pulsatilla Miller
P. vulgaris Miller   Pasqueflower
Anemone pulsatilla L.
No recent record.
HG 3: a single doubtful record in Watson (1832) is cited, which probably relates to Lincolnshire.

Clematis L.
C. vitalba L.   Traveller's-joy
Map 14c; 50 tetrads.
Occasional in hedgerows and on railway verges throughout the county. In Leicestershire it has nearly reached the northern limit of its range, and is not obviously restricted to strongly calcareous soils.
    Constituent or intrusive: Eh, Fd, Fm, Fs, Er, Ev, Rq, Rw.
First record: Marshall (1790).
Herb.: LSR★; DPM, LSR, LTR, NMW.
HG 1: occasional, range restricted; 19 localities cited.

Adonis L.
A. annua L.   Pheasant's-eye
No map; 1 tetrad.
Leicester, Western Park, abandoned allotment, rare, IME 1971.
    Casual: Aa.
First record: Pulteney (1750 - 52).
Herb.: LSR★; BM, LSR.
HG 5: rare, restricted in range; 11 localities cited.

Ranunculus L.
R. repens L.   Creeping Buttercup
Map 14d; 606 tetrads.
Grassland, especially in wet places; marshes, hedgerows and ditches, roadside verges, waste places, and as a weed of arable land. Abundant, the commonest buttercup throughout the county.
    Constituent, residual or intrusive: Gp, Gr, Gl, Hg, M, Ew, Ev, Er, Ea, Fd, Fw, Rw, Rg, Rq, Ao, Ac.
First record: Pulteney (1747).
Herb.: LSR★; DPM, LSR, LTR, MANCH, NMW.
HG 17: common, generally distributed.

R. acris L.   Meadow Buttercup
Map 14e; 594 tetrads.
Abundant in old grassland and water meadows throughout the county. Frequent in leys and other grassy habitats such as roadside verges.
    Constituent, residual or intrusive: Gp, Gr, Gl, Hg, M, Ev, Er, Ea, Ag, Aa, Rw.
First record: Marshall (1790).
Herb.: LSR★; DPM, LANC, LSR, LTR, MANCH, NMW.
HG 16: locally abundant, generally distributed.

R. bulbosus L.   Bulbous Buttercup
Map 14f; 482 tetrads.
Old established grassland, roadside and railway verges. The usual buttercup weed of lawns and sports fields, because with its rosette habit it can tolerate close mowing. Occasional in leys; frequent throughout the county in the other habitats.
    Constituent, residual or intrusive: Gp, Gr, Gl, M, Ev, Er, Ew, Ea, Ag, Ap, Rq.
First record: Pulteney (1747).
Herb.: LSR★; DPM, LSR, LTR, MANCH, NMW.
HG 17: locally abundant, generally distributed.

R. sardous Crantz   Hairy Buttercup
Map 14g; 4 tetrads.
Thornton Reservoir, several sites on shore at about high water mark, locally frequent, OHB and HH 1970; Twycross, disturbed roadside verge, rare, probably casual, HB 1974; Cadeby, recently disturbed roadside verge, occasional, probably casual, HH 1975; Snarestone, recently disturbed roadside verge, single plant, HH 1976.
    Constituent or intrusive: Ew, Ev, Rd.
First record: Hands et al. *in* Curtis (1831).
Herb.: LSR★; LSR, LTR, MANCH, NMW.
HG 17: rare, restricted in range; 16 localities cited, including Thornton Reservoir.

R. arvensis L.   Corn Buttercup
Map 14h; 27 tetrads.
Occasional throughout the county in arable land and recently disturbed ground. This is one of the arable weeds which has decreased with the higher standards of purity of crop seed and the advent of selective weed killers, but it does not appear to be decreasing further, and is sometimes locally abundant.
    Intrusive: Ac, Ao, Ea, Ev, Rd, Rf, Rr, Gl.
First record: Pulteney (1747).
Herb.: LSR★; DPM, LSR, LTR, MANCH, NMW.
HG 18: frequent, and generally distributed.

R. parviflorus L.   Small-flowered Buttercup
No record in the recent survey.
Near Kilby Bridge, D. P. Murray 1953.
First record: Pulteney *in* Nichols (1795).
Herb.: BM, DPM, LSR, WAR.
HG 18: rare, very restricted in range; 10 localities cited, not including the above.

R. auricomus L.   Goldilocks Buttercup
Map 14i; 119 tetrads.
Woodland, hedgerows and roadside verges on basic soils. Usually found on the edges of woodland or in rides, avoiding deep shade. Occasional throughout the county; locally frequent in well-wooded districts where the soil is not acid.
    Constituent: Fd, Fm, Fs, Ef.
    Residual: Eh, Ev, Gp, Ap, Ag, Wd, M.
First record: Pulteney (1747).
Herb.: LSR★; CGE, DPM, LANC, LSR, LTR, MANCH, NMW, UPP.
HG 15: locally abundant, and generally distributed; 77 localities cited.

R. sceleratus L.   Celery-leaved Buttercup
Map 14j; 346 tetrads.
Ponds, marshes and the banks of rivers and canals. Frequent throughout the county.
    Constituent, Wl, Wr, Wc, Wd, Ew, M.
    Residual or intrusive: Rg, Rs.
First record: Pulteney (1749).
Herb.: LSR★; DPM, LANC, LSR, LTR, MANCH, NMW.
HG 13: locally abundant, generally distributed.

R. ficaria L.   Lesser Celandine
Map 14k; 586 tetrads.
Woodland, grassland, grassy verges and hedgerows. Locally abundant throughout the county.
    Constituent or residual: Fd, Fm, Fs, Gp, Gr, Gl, M, Ew, Eh, Ev, Er, Wd, Ap, Ag, Rw.
First record: Pulteney (1747).
Herb.: LSR★; DPM, LANC, LSR, LTR, MANCH, NMW, UPP.
HG 18: locally abundant, generally distributed.

R. flammula L.   Lesser Spearwort
Map 14l; 79 tetrads.
Marshes and the edges of ponds, lakes and rivers, preferring the more acid soils. Frequent in Charnwood Forest and on the Marlstone. Occasional elsewhere.
    Constituent or residual: M, Wp, Wl, Ew, Ef, Wd, Fw, Fs, Hw, Hb.
First record: Pulteney (1747).
Herb.: LSR★; CGE, LSR, LTR, MANCH, NMW.
HG 13: frequent and widely distributed; 76 localities cited.

R. lingua L.   Greater Spearwort
Map 15a; 10 tetrads.
Knaptoft, large pond, locally abundant, EKH and JMH 1968; Aston Flamville, pond, rare, MH 1969; Groby, field pond, locally frequent, EH 1969 (pond drained 1977); Swithland Reservoir, reedswamp, rare, PHG 1970; Mowsley, pond, locally frequent, EKH and JMH 1970; Groby Pool outfall, marshy ground, locally abundant, PHG 1971; Leicester, Braunstone Park, stream side, naturalised introduction, occasional, EH 1971; Cropston Reservoir, reedswamp, locally abundant, PHG 1972; Kirkby Mallory, small pond, rare, HH 1975; Quorn, marshy meadow, occasional, PHG 1983.
    Constituent or intrusive: Wp, Wl, Ew, Hw, M.
First record: Pulteney (1747).
Herb.: LSR★; CGE, LANC, LSR, LTR, MANCH, NMW.
HG 14: rare or very local, sparsely distributed; 22 localities cited, including Groby Pool, Cropston Reservoir and Swithland Reservoir.

R. hederaceus L.   Ivy-leaved Crowfoot
Map 15b; 19 tetrads.
Ponds, ditches, marshes and the edge of rivers. In scattered localities throughout the county; often locally frequent where it does occur.
    Constituent: Wp, Wd, Wr, M.
First record: Pulteney (1747).
Herb.: LSR★; LANC, LSR, LTR, MANCH, NMW.
HG 12: local, widely distributed; 27 localities cited.

R. omiophyllus Ten. Round-leaved Crowfoot
Map 15c; 8 tetrads.
Grace Dieu, ponds between Grace Dieu Wood and Cademan Wood, locally frequent, PAC 1968; High Sharpley, Gun Hill, pond in heathland, locally frequent, PAC 1968; Blackbrook Reservoir, stream at infall, rare, PAC 1968; Markfield, near Cliffe Hill Quarry, pond, rare, PAC 1969; Charnwood Lodge Nature Reserve, pools and ponds, locally frequent, PAC 1969; Bradgate Park, pools and drainage channels, locally frequent, PHG 1972; Copt Oak, pond, occasional, PHG 1973; Ulverscroft Nature Reserve, pond, rare, PHG 1974; Newtown Linford, Foxley Hay, boggy ditch by ride, occasional, PHG 1975; Bardon, Old Rise Rocks Farm, wet ditch and marsh, locally frequent, PHG 1982.
    Constituent or residual: Wp, Wr, Wd.
First record: Babington (1847a).
Herb.: CGE, LANC, LSR, LTR, MANCH, NMW.
HG 12 as *R. lenormandi*: local, almost confined to Charnwood Forest; 12 localities cited, including most of the above.

R. peltatus Schrank
Map 15d; 66 tetrads.
Ponds and lakes, sometimes terrestrial on margins liable to flooding. Occasional throughout the county.
    Constituent or intrusive: Wp, Wl, Wd, Wc, Rg.
First record: Pulteney (1747).
Herb.: LSR, LTR.
HG 9: frequent, widely distributed; 50 localities cited.

R. pseudofluitans (Syme) Newbould ex Baker & Foggitt
R. peltatus Schrank subsp. *pseudofluitans* (Syme) C. Cook
Map 15e; 4 tetrads.
Aylestone, River Soar, frequent, HB 1970; Narborough, River Soar, locally frequent, EH 1974; Blaby, River Sence, locally frequent, EH 1974.
    Constituent: Wr.
First record: Mott et al. (1886).
Herb: DPM, LSR, LTR.
HG 11: rather rare, restricted in range; 7 localities cited, including River Soar at Aylestone.

R. aquatilis L. Common Water Crowfoot
Map 15f; 75 tetrads.
Ponds, ditches and streams. Sometimes terrestrial in wet places liable to periodic submersion. Occasional throughout the county.
    Constituent or intrusive: Wp, Wl, Wr, Wd, Ew, Rg.
First record: Mott et al. (1886).
Herb.: LSR★; CGE, DPM, LSR, LTR, UPP.
HG 8 as *R. heterophyllus*: frequent, generally distributed; 56 localities cited.

R. trichophyllus Chaix Thread-leaved Water Crowfoot
Map 15g; 43 tetrads.
Ponds and ditches. Occasional throughout the county.
    Constituent: Wp, Wd, Wr, Wc.
First record: Pulteney (1747).
Herb.: LSR★; DPM, LSR, LTR, MANCH, NMW.
HG 7: locally abundant, sparsely distributed; 21 localities cited.

R. circinatus Sibth. Fan-leaved Water Crowfoot
Map 15h; 16 tetrads.
Rivers and canals, less frequently in ponds and ditches. In scattered localities throughout the county, often locally abundant where it does occur.
    Constituent or intrusive: Wr, Wc, Wl, Wp, Wd, M, Rg.
First record: 1837, *in* Babington (1897).
Herb.: LSR★; CGE, DPM, LSR, LTR, MANCH, NMW.
HG 5: frequent, widely distributed; 33 localities cited.

R. fluitans Lam. River Water Crowfoot
Map 15i; 23 tetrads.
In the more swiftly flowing and cleaner rivers. Frequent and locally abundant in the west of the county. Rare or absent elsewhere.
    Constituent: Wr.
First record: Bloxam and Babington *in* Potter (1842).
Herb.: LSR★; DPM, LSR, LTR, NMW.
HG 6: local, restricted in range mainly to region W. of R. Soar; 19 localities cited.

Myosurus L.
M. minimus L. Mousetail
Map 15j; 1 tetrad.
Mountsorrel, field gateway in periodically flooded meadow adjacent to River Soar, locally frequent, PHG 1969. The ground is often trampled into a morass by cattle, so that sometimes the plants are destroyed before they can flower. This species appears to thrive on such treatment, but the number of plants varies considerably from year to year.
    Constituent: Rt.
First record: Pulteney (1747).
Herb.: LSR★; BM, CGE, LSR, MANCH, WAR.
HG 5: rare, very restricted in range; 10 localities cited, not including the above.

Aquilegia L.
A. vulgaris L. Columbine
No map; 10 tetrads.
In scattered localities throughout the county, mainly on roadside and railway verges. Never in any quantity, and almost certainly in every case a garden escape. Last recorded as a native by M. K. Hanson in 1951 in a small area of scrub north of King Lud's Entrenchments, specimen in herb. LSR (locality since destroyed). Some records were lost as a result of the fire at Ratcliffe College.
    Garden escape: Er, Ev, Eh, Rq, Gr.
First record: Pulteney (1747).
Herb.: DPM, LSR, LTR, MANCH.
HG 21: rare; 14 localities cited.

Thalictrum L.
T. flavum L. Common Meadow-rue
Map 15k; 31 tetrads.
Banks of rivers and streams, ditches and wet roadside verges. Rare or absent in the east of the county; occasional elsewhere.
    Constituent or residual: Ew, Wd, Fw, Ev, Er, Eh.
First record: Pulteney (1747).
Herb.: LSR★; DPM, LSR, LTR, MANCH, NMW.
HG 2: local, sparsely distributed; 32 localities cited.

BERBERIDACEAE
Berberis L.

B. vulgaris L. — Barberry
Map 15l; 20 tetrads.
In scattered localities throughout the county, usually as single isolated bushes in hedgerows, and rarely in any quantity.
    Constituent or intrusive: Eh, Fs, Ev, Ew.
First record: Pulteney (1747).
Herb.: LSR★; BM, DPM, LANC, LSR, LTR, UPP.
HG 23: rare, generally distributed.

Mahonia Nutt.

M. aquifolium (Pursh) Nutt. — Oregon-grape
Map 16a; 82 tetrads.
Occasional in woodland and hedgerows throughout the county. Sometimes planted in gardens and as an undershrub in coverts, but often becoming established spontaneously, probably from bird-sown seed.
    Intrusive or planted: Fd, Fm, Fs, Eh, Ev, Ew, Er, Ap, Rs, Sw, Sr.
First record: Bemrose (1927).
Herb.: LSR★; LSR, LTR.
HG 23 as *Berberis aquifolium*: probably frequent, but not noted; 4 localities cited.

RHOEADALES

PAPAVERACEAE
Papaver L.

P. somniferum L. — Opium Poppy
Map 16b; 83 tetrads.
Frequent on rubbish dumps. Occasional in waste places, on recently disturbed ground, and elsewhere as a garden escape.
    Intrusive or garden escape: Rr, Rw, Rd, Rs, Rg, Ev, Er, Ew, Ea, Ao, Ag.
First record: Pulteney (1747).
Herb.: LSR★; LANC, LSR, LTR, MANCH, NMW.
HG 24: rare, restricted in range; 19 localities cited.

P. rhoeas L. — Common Poppy
Map 16c; 328 tetrads.
Waste places, recently disturbed ground and railway verges. Frequent throughout the county in these habitats. Less frequently a weed of arable land.
    Intrusive: Rw, Rd, Rs, Rq, Rr, Er, Ev, Ew, Ea, Ef, Ac, Ao, Ag, Ap, Gl, Gr.
First record: Pulteney (1747).
Herb.: LSR★; LANC, DPM, LSR, LTR, MANCH, NMW.
HG 25: frequent, generally distributed.

P. dubium L. — Long-headed Poppy
Map 16d; 327 tetrads.
Very similar in habitat, frequency and distribution to the preceding species. **P. rhoeas** is slightly more frequent in the county as a whole, especially in the east; **P. dubium** appears to be more tolerant of the acid soils in the west.
    Intrusive: Rw, Rd, Rs, Rq, Rg, Rf, Ev, Er, Ew, Ea, Ac, Ao, Aa, Ap, Ag, Gl, Gr, Sw.
First record: Pulteney (1747).
Herb.: LSR★; LANC, LSR, LTR, MANCH, NMW.
HG 25: frequent, widely distributed.

P. lecoqii Lamotte — Yellow-juiced Poppy
Map 16e; 23 tetrads.
Occasional throughout the county as a ruderal and in arable land. Possibly overlooked because of its similarity to **P. dubium**, but certainly less frequent than that species.
    Intrusive: Ag, Ao, Ac, Aa, Er, Ev, Rd, Rs, Rw, Sw.
First record: Mott et al. (1886).
Herb.: LSR★; LSR, MANCH, NMW.
HG 26: local, sparsely distributed; 25 localities cited.

P. argemone L. — Prickly Poppy
Map 16f; 27 tetrads.
Occasional throughout the county in waste places, on recently disturbed ground, on railway verges and in arable land. Seldom occurs in any quantity and is often non-persistent.
    Intrusive: Rw, Rd, Rf, Rg, Rs, Er, Ev, Ao, Ac, Ag.
First record: probably Pulteney (1752); if not, then Hands et al. *in* Curtis (1831).
Herb.: LSR★; LSR, MANCH, NMW.
HG 26: occasional, widely distributed; 33 localities cited.

P. hybridum L. — Rough Poppy
No record since HG.
First record: Britten (1877).
HG 27: rare; a single unsubstantiated record.

Meconopsis Vig.

M. cambrica (L.) Vig. — Welsh Poppy
No map; 13 tetrads.
Occurs occasionally as a casual or garden escape on roadsides and in waste places.
    Casual or garden escape: Ev, Ew, Rw, Rs.
First record: present work.
HG: not recorded.

Chelidonium L.

C. majus L. — Greater Celandine
Map 16g; 186 tetrads.
Hedgerows, roadside and railway verges, and waste places, usually not far from human habitations. Occasional throughout the county; locally frequent in and around villages.
    Intrusive: Eh, Ev, Er, Ew, Rw, Rq, Rf, Rr, Rd, Rs, Ag, Ap, Sw, Sb.
First record: Pulteney (1747).
Herb.: LSR★; DPM, LANC, LSR, LTR, MANCH, NMW, UPP, WAR.
HG 28: frequent, generally distributed.

Eschscholzia Cham.

E. californica Cham. — Californian Poppy
No map.
Has been recorded for a few localities as a casual or garden escape, but most of the records were lost as a result of the fire at Ratcliffe College.
First record: c.1915, A. E. Wade in herb. NMW.
Herb.: NMW.
HG 28: rare; 1 locality cited.

Corydalis Vent.

C. claviculata (L.) DC. — Climbing Corydalis, White Climbing Fumitory
Map 16h; 11 tetrads.
Confined to Charnwood Forest and the Coal Measures in

the north-west of the county, where it is occasional or locally frequent in woodland.
   Constituent: Fd, Fm, Fc, Fs, Ef.
   Intrusive: Sw, Sr.
First record: Pulteney (1747).
Herb.: LSR★; LSR, LTR, NMW, WAR.
HG 28 as *Capnoides claviculata*: local, restricted to Charnwood Forest and Ashby coalfield; 21 localities cited.

C. lutea (L.) DC.   Yellow Corydalis
Map 16i; 46 tetrads.
Of garden origin, now thoroughly established and naturalised on walls and roadsides in and around towns and villages. Locally frequent throughout the county.
   Intrusive; Sw, Ev, Rw, Rs, Ap.
First record: Kirby (1850).
Herb.: LSR★; DPM, LSR, LTR, NMW.
HG 28 as *Capnoides lutea*: occasional, sparsely distributed; 15 localities cited.

C. solida (L.) Swartz
No map; 3 tetrads.
Sibson, churchyard approach, EKH 1960; Eastwell, churchyard, EKH 1962; Leicester, Belgrave House, neglected garden, rare, SHB 1977.
   Casual: Ap, Ag.
First record: Vice (1902).
Herb.: LSR, LTR, NMW.
HG 29 as *Capnoides solida*: rare; 5 localities cited.

Fumaria L.
F. capreolata L.   White Ramping-fumitory
No recent record.
First record: Hands et al. in Curtis (1831).
HG 29: very rare, very restricted in range, not observed during the last fifty years, and possibly extinct; 2 localities cited.

F. muralis Sonder ex Koch subsp. **boraei** (Jordan) Pugsley   Common Ramping-fumitory
No map; 1 tetrad.
Groby, old quarry, rare, EH 1972 (site since destroyed).
   Casual: Rq.
First record: Ardington (1911).
Herb.: LSR★; CGE, LSR, MANCH.
HG 29: rare, very restricted in range; 1 locality cited, not the same as above.

F. densiflora DC.   Dense-flowered Fumitory
F. micrantha Lag.
No recent record.
First record: Kirby (1849).
HG 30: rare, very restricted in range; 1 locality cited, not found there in 1885.

F. officinalis L.   Common Fumitory
Map 16j; 205 tetrads.
Frequent throughout the county on recently disturbed ground and among non-cereal crops. Occasional among cereals.
   Intrusive: Rd, Rw, Rq, Rg, Rs, Rf, Ao, Ag, Ac, Aa, Ap, Ev, Er, Ea, Ew.
First record: Pulteney (1747).
Herb.: LSR★; CGE, DPM, LANC, LSR, LTR, MANCH, NMW.
HG 29: common, generally distributed.

CRUCIFERAE
Sisymbrium L.
S. irio L.   London-rocket
No map; 1 tetrad.
Leicester, Central Street roundabout, W. W. Herrington 1964.
   Casual: Ev.
First record: Brown in Mosley (1863).
Herb.: LSR.
HG 43: rare; 1 locality cited, not the same as above.

S. altissimum L.   Tall Rocket
Map 16k; 51 tetrads.
Waste places, gravel pits, railway verges, mainly in or near large towns. Often locally abundant where it does occur.
   Intrusive: Rw, Rg, Rq, Rs, Rr, Rd, Er, Ev, Ag, Gr.
First record: 1896, F. T. Mott in herb. MANCH.
Herb.: LSR★; CGE, LSR, LTR, MANCH, NMW.
HG 42: occasional, sparsely distributed; 15 localities cited.

S. orientale L.   Eastern Rocket
Map 16l; 27 tetrads.
Waste places, gravel pits and railway verges, in scattered localities throughout the county, where it may be locally frequent.
   Intrusive: Rw, Rg, Rs, Rr, Rd, Er, Ev.
First record: Jackson (1904b).
Herb.: LSR★; LSR, NMW.
HG 44: rare; 2 localities cited.

S. officinale (L.) Scop.   Hedge Mustard
Map 17a; 474 tetrads.
Waste places, gravel pits and quarries, recently disturbed ground and roadsides. Frequent and locally abundant throughout the county.
   Intrusive: Rw, Rf, Rg, Rq, Rs, Rr, Rd, Rt, Ev, Er, Ew, Eh, Ea, Ao, Ac, Ag, Aa, Ap, Gr, Gl, Gp, Sw, Sb.
First record: Pulteney (1747).
Herb.: LSR★; DPM, LANC, LSR, LTR, MANCH, NMW.
HG 42: abundant, generally distributed.

Descurainia Webb & Berth.
D. sophia (L.) Webb ex Prantl   Flixweed
No map; 7 tetrads.
Leicester, Thornton Lane, G. Smith 1965; Leicester, St. Nicholas Street, G. Smith 1965; Fleckney, old clay pit, locally abundant, EKH and JMH 1973; Blaby Mill, rough grass, rare, EH 1974; Enderby, rubbish dump in old quarry, rare, EH 1974; Groby, quarry spoil heap, rare, EH 1976; Leicester, Forest Road, waste ground, rare, SHB 1977.
   Casual: Rw, Rr, Rs, Rg, Gr.
First record: Pulteney (1747).
Herb.: LSR★; CGE, LANC, LSR, NMW.
HG 43 as *Sisymbrium sophia*: local; 31 localities cited.

Alliaria Scop.
A. petiolata (Bieb.) Cavara & Grande   Garlic Mustard
Map 17b; 547 tetrads.
Hedgerows, roadside and railway verges, woodland and the banks of rivers. Frequent throughout the county, but with a curiously discontinuous distribution, being abundant in some areas and unaccountably rare or absent in adjacent ones.

Constituent: Fd, Fw, Eh, Ef.
Residual or intrusive: Ev, Er, Ew, Wd, Rw, Rs, Ap, Gr.
First record: Pulteney (1747).
Herb.: LSR★; DPM, LANC, LSR, LTR, NMW, UPP.
HG 43 as *Sisymbrium alliaria*: locally abundant, generally distributed.

Arabidopsis (DC.) Heynh.
A. thaliana (L.) Heynh.   Thale Cress.
Map 17c; 183 tetrads.
Frequent and locally abundant on railway verges throughout the county (59% of the records are from this habitat). Occasional in waste places, on walls, in quarries and gravel pits. Rarely an arable or garden weed. Prefers dry sandy soils, and is especially characteristic of ant-hills.
   Intrusive: Er, Ev, Ew, Eh, Rw, Rq, Rg, Rs, Rd, Rt, Rf, Sw, Sr, Ag, Aa, Ac, Ap.
First record: Pulteney (1747).
Herb.: LSR★; LANC, LSR, LTR, MANCH, NMW.
HG 41 as *Sisymbrium thalianum*: locally abundant, widely distributed; 40 localities cited.

Isatis L.
I. tinctoria L.   Woad
No record in the recent survey.
Knighton, clay pit, E. K. Horwood 1954 (site since built on).
First record: Blith (1653).
Herb.: LSR, LTR.
HG 56: garden escape, rare; 2 localities cited.

Bunias L.
B. orientalis L.   Warty Cabbage
No map; 2 tetrads.
South Wigston, waste ground, occasional, EH 1976; Leicester, Belgrave, roadside, occasional, HB 1977.
   Casual: Rw, Ev.
First record: 1923, H. P. Reader in herb. BM.
Herb.: LSR★; BM.
HG: not recorded.

Erysimum L.
E. cheiranthoides L.   Treacle Mustard
Map 17d; 37 tetrads.
Occasional throughout the county in waste places and on verges. Rarely as a weed of arable land. Usually non-persistent and of casual status only in this county.
   Casual or intrusive: Rw, Rd, Rr, Rg, Rs, Rt, Ev, Er, Ea, Ew, Ao, Ac, Gp, Gl.
First record: Kirby (1850).
Herb.: LSR★; LSR, LTR, MANCH, NMW.
HG 44: occasional, on the increase, sparsely distributed; 22 localities cited.

Hesperis L.
H. matronalis L.   Dame's-violet
Map 17e; 54 tetrads.
A casual or garden escape, found throughout the county in waste places and on roadside or railway verges. Rarely persists and is seldom found in large numbers.
   Intrusive or casual: Ev, Er, Ew, Eh, Rd, Rw, Rs, Rr, Gr.
First record: Kirby (1848b).
Herb.: LSR★; LSR, LTR, NMW.
HG 41: rare, very restricted in range; 13 localities cited.

Malcolmia R. Br.
M. maritima (L.) R. Br.   Virginia Stock
No map; 2 tetrads.
Occurs occasionally as a casual or garden escape.
   Casual or escape: Rw, Er.
First record: Wade (1919).
Herb.: NMW.
HG 41: rare; 3 localities cited.

Cheiranthus L.
C. cheiri L.   Wallflower
Map 17f; 23 tetrads.
Occasional on walls throughout the county, usually as an established garden escape. In a few sites, e.g. the walls of Leicester Abbey and of Grace Dieu Priory ruins, it is known to have been present for a very long time, and the flowers have the yellow colour of the 'wild' type.
   Intrusive: Sw, Sb, Sr, Rs, Rw.
First record: Pulteney (1747).
Herb.: CGE, LSR, LTR, NMW.
HG 30: occasional; 14 localities cited.

Barbarea R.Br.
B. vulgaris R.Br.   Winter-cress
Map 17g; 393 tetrads.
Frequent throughout the county on the banks of rivers, streams and lakes, and in ditches. Occasional on roadsides and as an arable weed.
   Constituent: Ew, Wd, M.
   Residual or intrusive: Ev, Eh, Er, Ea, Rg, Rw, Rs, Rd, Rr, Ao, Ac, Ag, Gp, Gr, Gl, Sw.
First record: Pulteney (1747).
Herb.: LSR★; DPM, LANC, LSR, LTR, MANCH, NMW, UPP.
HG 33: locally abundant, generally distributed.

B. stricta Andrz.   Small-flowered Winter-cress
Map 17h; 11 tetrads.
By water and in gravel pits and waste places; locally frequent in the lower Soar valley, to which it appears to be almost confined, though it may have been overlooked elsewhere.
   Intrusive or constituent: Wr, Wd, Ew, Ev, Eh, Er, Ea, Rg, Rq, Rw.
First record: Horwood and Gainsborough (1933).
Herb.: LSR★.
HG 34: 1 locality cited.

B. verna (Miller) Ascherson   American Winter-cress
Map 17i; 9 tetrads.
Wigston, railway verge and waste ground, occasional, HB 1969; Groby, quarry, railway ballast, occasional, EH 1969; Glenfield, railway ballast, occasional, EH 1970; Groby village, dismantled railway, ballast, rare, EH 1972 (site since destroyed); Kirby Muxloe, railway ballast, occasional, EH 1972; Croft, quarry spoil heap, rare, SHB 1974; Enderby, roadside verge, rare, EH 1974; Narborough, waste ground, rare, EH 1976.
   Casual or intrusive: Er, Ev, Rw, Rs.
First record: Bloxam (1837).
Herb.: LSR★; LSR, LTR.
HG 34: rare, very restricted in range; 5 localities cited.

B. intermedia Boreau
Medium-flowered Winter-cress
Map 17j; 8 tetrads.
Leicester, Freemen's Common, disused allotments, rare, HB 1968; Lowesby, site of John O'Gaunt Station, dismantled railway, rare, HB 1969; Bagworth, headland of cornfield, occasional, HH 1971; Groby village, dismantled railway, ballast, rare, EH 1971 (site since destroyed); Earl Shilton, Barrow Hill, quarry, waste ground, SHB 1972; Enderby, quarry, mineral railway, locally frequent, EH 1974; Ashby Woulds, waste ground, locally frequent, SHB 1976; Ashby Woulds, near Albert Village, dismantled railway, ballast, occasional, SHB 1976.
    Casual or intrusive: Er, Ea, Rw.
First record: 1885, F. T. Mott in herb. MANCH.
Herb.: LSR★; MANCH.
HG 34: rare, very restricted in range; 1 locality cited, not any of the above.

Rorippa Scop.

R. austriaca (Crantz) Besser   Austrian Yellow-cress
Map 17k; 1 tetrad.
Potters Marston, former sand pit used as a rubbish dump, locally frequent, SHB 1973 (first recorded at this site by E. K. Horwood in 1958).
    Intrusive: Rg, Rr.
First record: 1958, E. K. Horwood in herb. LTR.
Herb.: LSR★; LTR.
HG: not recorded.

R. amphibia (L.) Besser   Great Yellow-cress
Map 17l; 106 tetrads.
Frequent throughout the county at the margins of rivers, canals and reservoirs; occasional in ponds and marshes.
    Constituent or residual: Wr, Wc, Wl, Wp, Wd, Ew, M, Rg.
First record: Pulteney *in* Nichols (1795).
Herb.: LSR★; DPM, LSR, LTR.
HG 32: frequent, sparsely distributed; 47 localities cited.

R. amphibia × sylvestris = R. × anceps (Wahlenb.) Reichenb.
No record in the recent survey.
Groby Pool, margin, T. G. Tutin 1945.
First record: 1945, T. G. Tutin in herb. LTR.
Herb.: LTR.
HG: not recorded.

R. sylvestris (L.) Besser   Creeping Yellow-cress
Map 18a; 23 tetrads.
Gardens, especially nursery gardens where it may be locally abundant and a troublesome weed; in waste places and by water. In scattered localities throughout the county, but rare in the east.
    Intrusive: Ag, Ap, Rw, Rr.
    Possibly constituent: Ew, Wd, M.
First record: Pulteney (1747).
Herb.: LSR★; LSR, LTR, MANCH, NMW.
HG 31: rather local, sparsely distributed; 19 localities cited.

R. islandica (Oeder) Borbás   Marsh Yellow-cress
Map 18b; 112 tetrads.
Frequent throughout the county on the banks of rivers, streams and canals, by ponds, and in marshes. Occasional in waste places and where water lies in winter.
    Constituent: Ew, Wp, Wl, Wd, M.
    Residual or intrusive: Rd, Rw, Rr, Rg, Rq, Rs, Rt, Ac, Ao, Ap, Gp, Er.
First record: Crabbe *in* Nichols (1795).
Herb.: LSR★; CGE, DPM, LANC, LSR, LTR, MANCH, NMW.
HG 32 as *R. palustris*: frequent, generally distributed.

Armoracia Gilib.
A. rusticana P. Gaertner, B. Meyer & Scherb.
Horse-radish
Map 18c; 380 tetrads.
Roadsides, railway verges and waste places; frequent throughout the county, usually but not always near to dwellings. Also frequent in allotment gardens as a relic of former cultivation. Since it is said never to set seed in this country, it is interesting to speculate on how it becomes established in places some distance from sites of former cultivation.
    Intrusive: Ev, Er, Ew, Ea, Rw, Rf, Rd, Rs, Rr, Aa, Ag, Ap, Gr, Gp.
First record: Pulteney (1747).
Herb.: LSR★; LANC, LSR, MANCH, NMW.
HG 41 as *Cochlearia armoracia*: fairly frequent, generally distributed.

Nasturtium R.Br.
N. officinale R.Br.   Water-cress
Rorippa nasturtium-aquaticum (L.) Hayek
Map 18d; 91 tetrads.
Ponds, ditches, rivers, canals and marshes. Occasional throughout the county.
    Constituent: Wp, Wd, Wr, Wc, M.
    Residual or intrusive: Rq.
First record; Pulteney (1747).
Herb.: LSR★; DPM, LANC, LSR, LTR, MANCH, NMW.
HG 31: locally abundant, generally distributed.

N. microphyllum (Boenn.) Reichenb.
Rorippa microphylla (Boenn.) Hyland.
Map 18 e; 180 tetrads.
Ponds, ditches, rivers, canals and marshes. Frequent throughout the county.
    Constituent: Wp, Wd, Wr, Wc, M.
    Residual or intrusive: Rg.
First record: 1875, F. T. Mott in herb. MANCH.
Herb.: LSR★; LTR, MANCH, NMW.
HG 31 as *Rorippa nasturtium-aquaticum* var. *microphyllum*: 7 localities cited.

N. microphyllum × officinale
Rorippa × sterilis Airy Shaw
Map 18f; 35 tetrads.
Ponds, ditches, rivers, canals and marshes. Occasional throughout the county. Of vigorous growth and spreads vegetatively. It often fails to flower, and it cannot then be identified for certain, hence it may be under-recorded.
    Constituent: Wp, Wd, Wr, Wc, M.
First record: 1958, E. K. Horwood in herb. LTR.
Herb.: LSR★; LTR.
HG: not recorded.

Cardamine L.

C. bulbifera (L.) Crantz   Coralroot
Dentaria bulbifera L.
Map 18g; 1 tetrad.
Glenfield, shady border in garden of The Gynsills, occasional, EH 1971.
    Intrusive or planted: Ag.
First record: 1971, E. Hesselgreaves in present work.
HG: not recorded.

C. amara L.   Large Bitter-cress
Map 18h; 65 tetrads.
Marshes, canal and river banks, and wet places in woodland. Locally frequent in such habitats, but absent from the extreme eastern part of the county.
    Constituent: Wr, Wc, Wd, Ew, M, Fw, Fd.
    Residual: Gr.
First record: Pulteney *in* Nichols (1795).
Herb.: LSR★; DPM, LANC, LSR, LTR, MANCH, NMW.
HG 35: occasional, restricted in range; 41 localities cited.

C. pratensis L.   Cuckooflower, Lady's-smock
Map 18i; 463 tetrads.
Marshes, meadows, woodland rides, and the banks of rivers, streams and canals. Frequent throughout the county, but decreasing as suitable habitats are destroyed.
    Constituent: M, Gp, Gr, Ew, Ef, Wd.
    Residual or intrusive: Ev, Ea, Gl, Ap, Ag.
First record: Pulteney (1747).
Herb.: LSR★; DPM, LANC, LSR, LTR, MANCH, NMW, UPP.
HG 36: frequent, generally distributed.

C. impatiens L.   Narrow-leaved Bitter-cress
No record since HG.
First record: Pulteney (1749).
Herb.: LSR, MANCH.
HG 37: rare, restricted in range; 6 localities cited.

C. flexuosa With.   Wavy Bitter-cress
Map 18j; 350 tetrads.
Woodland, marshes, river banks and damp shady places. Frequent throughout the county.
    Constituent: M, Ew, Ef, Fw, Fd, Fs, Wd.
    Residual or intrusive: Rg, Rd, Rw, Ev, Er, Ap, Ag.
First record: Kirby (1850).
Herb.: LSR★; CGE, DPM, LSR, LTR, MANCH, NMW, UPP.
HG 37: frequent, widely distributed.

C. hirsuta L.   Hairy Bitter-cress
Map 18k; 269 tetrads.
Frequent on railway verges throughout the county. Frequent also as a garden weed, but it varies considerably in abundance in this habitat from year to year. Occasional on roadsides and walls, and in waste places.
    Intrusive: Er, Ev, Ef, Sw, Ag, Ap, Rw, Rd, Rt, Rq, Rs.
First record: Pulteney (1749).
Herb.: LSR★; CGE, DPM, LANC, LSR, LTR, MANCH, NMW.
HG 36: frequent, generally distributed.

Arabis L.

A. glabra (L.) Bernh.
No recent record.
First record: Bloxam (1838).
Herb.: LSR.
HG 35: very rare, restricted in range, perhaps extinct in Leicestershire; 4 localities cited.

A. hirsuta (L.) Scop.   Hairy Rock-cress
No recent record.
First record: Pulteney *in* Nichols (1795).
HG 34: very rare, possibly extinct in Leicestershire, confined to Charnwood Forest; 2 localities cited.

A. turrita L.   Tower Cress
No record since HG.
First record: Power (1807).
HG 35: rare, restricted in range; 1 locality cited.

A. caucasica Schlecht.   Garden Arabis
No map.
Often grown in gardens, and sometimes becoming established on old walls.
    Intrusive or garden escape: Sw.
First record: 1962, E. K. Horwood in herb. LTR.
Herb.: LTR.
HG: not recorded.

Aubrieta Adanson

A. deltoidea (L.) DC.
No map.
Often grown in gardens, and frequently escaping to become established at least for a while on walls and in similar places.
    Intrusive or garden escape: Sw.
First record: 1976, P. H. Gamble in present work.
HG: not recorded.

Lunaria L.

L. annua L.   Honesty
Map 18l; 57 tetrads.
A garden escape, becoming established in hedgerows and waste places, and on roadside and railway verges. Occasional throughout the county, with seldom more than a few isolated plants in any one locality.
    Intrusive or garden escape: Eh, Ev, Er, Ea, Rw, Rq, Rs, Rd, Rr, Gr, Sw, Ap, Fm.
First record: 1948, E. K. Horwood in herb. LTR.
Herb.: LSR★; DPM, LTR.
HG: not recorded.

Lobularia Desv.

L. maritima (L.) Desv.   Sweet Alison
No map; 18 tetrads.
Frequently planted in gardens and often escaping on to waste ground. Usually only of casual status, or shortly persistent; on rubbish dumps its persistence is probably due to repeated introduction.
    Intrusive or garden escape: Rw, Rr, Rt, Rs, Rd, Rf, Er, Gr, Sw.
First record: Horwood and Gainsborough (1933).
Herb.: LANC, LSR,, NMW.
HG 38 as *Alyssum maritimum*: rare; 3 localities cited.

Draba L.
D. muralis L.   Wall Whitlowgrass
Map 19a; 1 tetrad.
Thorpe Langton, dismantled railway sidings, grassy embankment with cinder ballast, a colony of about 30 plants, KGM 1972.
    Intrusive: Er.
First record: 1972, K. G. Messenger in present work.
Herb.: LSR★.
HG: not recorded.

Erophila DC.
This is a taxonomically difficult genus, and there appears to be some doubt as to whether or not all our British material consists of variants of **E. verna** subsp. **verna**. However, in the following account of the genus, the records are given as named during the recent survey. A map is given for the distribution of subspecies **verna**, and localities are given for the records of subspecies **spathulata** and **praecox**, without maps.

E. verna (L.) Chavall.   Whitlowgrass
Subsp. **verna**
Map 19b; 143 tetrads.
Railway verges, short dry grassland, rocky places and tops of walls. Locally frequent throughout the county.
    Constituent or intrusive: Er, Ev, Ea, Sw, Sr, Sb, Gp, Gr, Hd, Hg, Rt, Rq, Rw, Rd, Rg, Ag, Ap, Ac.
First record: Pulteney (1747).
Herb.: LSR★; CGE, LSR, LTR, MANCH, NMW.
HG 39: Locally abundant, widely distributed.

Subsp. **spathulata** (A. F. Láng) Walters
No map; 4 tetrads.
Quorndon, railway verge, locally frequent, PHG 1971; Stathern, dismantled mineral railway, occasional, SHB 1973; Mountsorrel, rocks, locally frequent, PHG 1975; Bradgate Park, wall, locally frequent, PHG 1975.
    Intrusive or constituent: Er, Sw, Sr.
First record: Preston (1900).
Herb.: CGE.
HG 40 as *E. boerhavii*: 5 localities cited.

Subsp. **praecox** (Steven) Walters
No map; 1 tetrad.
Glen Parva, railway ballast on bridge, locally frequent, EH 1975.
    Intrusive: Er.
First record: A. B. Jackson (1904b).
HG 40: occasional, sparsely distributed; 15 localities cited.

Cochlearia L.
C. danica L.   Danish Scurvygrass
Map 19c; 4 tetrads.
Between Syston and Cossington, railway cutting, occasional, ALP 1969; Sileby, railway cutting, occasional, ALP 1971; North Kilworth, dismantled railway track, rubble, frequent, EKH and JMH 1971; Leicester, soil heap, rare, J. Daws 1984. The plants in these situations are very tiny, and inconspicuous when not in flower, and it is likely that the species has been overlooked in similar situations.
    Intrusive: Er.
First record: 1950, E. K. Horwood in herb. LTR.
Herb.: LSR★; LTR.
HG: not recorded.

C. officinalis L.   Common Scurvygrass
No record since HG.
First record: Kirby (1850).
Herb.: LSR.
HG 40: very rare, restricted to Soar valley; 2 localities cited.

Camelina Crantz
C. sativa (L.) Crantz   Gold-of-pleasure
No map: 1 tetrad.
Sileby, council rubbish dump, occasional, ALP 1970.
    Casual: Rr.
First record: Pulteney (1749).
Herb.: LSR, LTR, MANCH, NMW.
HG 45: rare, area restricted; 15 localities cited.

Capsella Medicus
C. bursa-pastoris (L.) Medicus   Shepherd's-purse
Map 19d; 597 tetrads.
Arable land, gardens, recently disturbed ground, and the trampled ground of field gateways. Abundant throughout the county.
    Intrusive: Ag, Ao, Ac, Aa, Ap, Rt, Rd, Rw, Rf, Rq, Rs, Ev, Er, Ew, Ea, Eh, Gl.
First record: Pulteney (1747).
Herb.: LSR★; CGE, DPM, LANC, LSR, LTR, MANCH, NMW.
HG 50 as *Bursa pastoris*: abundant, generally distributed.

Teesdalia R.Br.
T. nudicaulis (L.) R.Br.   Shepherd's Cress
Map 19e; 1 tetrad.
Swithland Wood, spoil heap of old slate quarry, PHG 1968. This species is frequent in some years and scarce in others, but it has been recorded periodically from this site for over 200 years.
    Constituent or residual: Rs.
First record: Pulteney (1747).
Herb.: CGE, LSR, LTR, MANCH, NMW.
HG 56: frequent, but locally distributed; 15 localities cited.

Thlaspi L.
T. arvense L.   Field Penny-cress
Map 19f: 102 tetrads.
Occasional throughout the county in arable land and waste places. Among non-cereal crops it is sometimes locally abundant.
    Intrusive: Ao, Ac, Ag, Aa, Rw, Rd, Rt, Rs, Rr, Rg, Rq, Rf, Ea, Ev, Er, Ew, Gr, Gp.
First record: Marshall (1790).
Herb.: LSR★; LSR, LTR, MANCH, NMW.
HG 55: frequent, sparsely distributed; 36 localities cited.

Iberis L.
I. amara L.   Wild Candytuft
Map 19g; 1 tetrad.
Scalford, deep cutting on dismantled railway, south end of Brock Hill tunnel, in considerable quantity, IME 1967; the future of this site is uncertain. Other records, occasional throughout the county, are almost certainly garden escapes, or more likely the following species.
    Residual or intrusive: Er.
First record: Horwood and Gainsborough (1933).
Herb.: LSR★; LSR, LTR.
HG 56: very rare; 1 locality cited, not the same as above.

I. umbellata L.
No map; 7 tetrads.
An occasional garden escape, sometimes becoming established for a short while on roadsides and in waste places.
Garden escape: Ev, Rw, Rd, Rs.
First record: present work.
HG: not recorded.

Lepidium L.

L. campestre (L.) R.Br.　　　　Field Pepperwort
Map 19h; 10 tetrads.
Loughborough, Midland Station, bank beside goods yard, rare, PHG 1971; Earl Shilton, Barrow Hill, quarry, waste ground, rare, SHB 1972; Ratby, garden, rare, EH 1972; Leicester, Western Park, allotment, rare, EH 1972; Cosby, dismantled railway, verge, rare, EH 1976; Syston, Pontylue gravel pit, occasional, ALP 1977; Donisthorpe, recently disturbed ground, rare, SHB 1977; Loughborough, trackway, locally frequent, PHG 1977; Enderby, two localities on roadside verges, EH 1979; Barrow on Soar, arable headland, rare, PHG 1979.
Intrusive: Ev, Er, Ea, Rq, Rg, Rd, Rt, Ag, Aa.
First record: Pulteney (1747).
Herb.: LSR★; LANC, LSR, MANCH, NMW.
HG 54: rare, restricted in range; 23 localities cited.

L. heterophyllum Bentham　　　Smith's Pepperwort
Map 19i; 6 tetrads.
Glenfield, dismantled railway track, locally frequent, EH 1970; Beaumont Leys, rubbish tip, rare, EH 1971; Sapcote, roadside, rare, MH 1973; Croft, roadside verge, single plant, SHB 1973; Thurmaston, waste ground, rare, IME 1975.
Intrusive: Er, Ev, Rw, Rr.
First record: Coleman (1852).
Herb.: LSR★; LSR, LTR.
HG 54 as *L. smithii*: very rare, restricted in range; 6 localities cited.

L. sativum L.　　　　　　　　　Garden Cress
No map; 8 tetrads.
Occurs occasionally as a casual on rubbish dumps, sometimes in large quantity, and in other waste places. Probably under-recorded. May appear yearly on rubbish dumps because of repeated reintroduction.
Casual or intrusive: Rr, Rf, Rt, Ao.
First record: Power (1807).
Herb.: LSR★; LSR, LTR, NMW.
HG 53: occasional, range restricted; 10 localities cited.

L. ruderale L.　　　　Narrow-leaved Pepperwort
Map 19j; 39 tetrads.
Waste places, rubbish dumps and gravel pits. Locally frequent in or near large towns; occasional elsewhere.
Intrusive: Rw, Rg, Rq, Rs, Rr, Rt, Rd, Rf, Er, Ev, Ew.
First record: Walpoole (1782).
Herb.: LSR★; LANC, LSR, LTR, MANCH, NMW.
HG 53: occasional, on the increase, restricted in range; 31 localities cited.

L. latifolium L.　　　　　　　　　　Dittander
Map 19k; 18 tetrads.
Roadside and railway verges and waste places; almost entirely confined to the south-west of the county, where it is locally frequent. Horwood, commenting on the single record (from Blaby Mill), says 'A typical salt marsh plant, of very rare occurrence inland, and then only as a casual.' However, the plants in the localities described above are well established and behave as if thoroughly naturalised.
Intrusive: Ev, Er, Eh, Rw, Rs, Rt, Rf.
First record: Crabbe *in* Nichols (1795).
Herb.: LSR★; LSR, LTR.
HG 53: restricted in range to one area; 1 locality cited.

Cardaria Desv.

C. draba (L.) Desv.　　　　　　　Hoary Cress
Map 19l; 60 tetrads.
Occasional throughout the county in waste places, gravel pits and clay pits, and on railway verges. In these habitats it may be locally abundant. Rarely a weed of arable land.
Intrusive: Rw, Rg, Rq, Rs, Rd, Rr, Er, Ev, Ew, Aa, Ag, Ao, Ac, Ap.
First record: Preston (1900).
Herb.: LSR★; LANC, LSR, LTR, NMW.
HG 54 as *Lepidium draba*: frequent, sparsely distributed; 26 localities cited.

Coronopus Haller

C. squamatus (Forskal) Ascherson　　Swine-cress
Map 20a; 272 tetrads.
The principal habitat for this species is the trampled ground of field gateways and cattle tracks in fields, where it may be locally abundant. Also occasionally in arable land, especially the headlands of non-cereal crops. Frequent on the clay soils, occasional or rare elsewhere.
Intrusive: Rt, Rd, Rw, Rf, Rs, Rr, Rg, Ea, Ev, Ew, Ao, Ag, Ac, Gp, Gl, Gr.
First record: Pulteney (1747).
Herb.: LSR★; LANC, LSR, LTR, MANCH, NMW.
HG 52 as *C. ruellii*: frequent, generally distributed.

C. didymus (L.) Sm.　　　　Lesser Swine-cress
Map 20b; 11 tetrads.
Leicester, New Parks, recently disturbed ground, rare, HB 1972; Earl Shilton, ditch and trampled ground, locally frequent, SHB 1973; Bottesford, sugar beet field, rare, HL 1973; Huncote, recently disturbed roadside, rare, SHB 1974; Ratcliffe College, staff house garden, single plant, ALP 1976; Leicester, margins of temporary car park, occasional, HB 1976; Leicester, Belgrave Hall, garden weed, rare, SHB 1977; Narborough, garden, locally frequent, EH 1978; Higham on the Hill, waste ground, locally frequent, SHB 1979; Sileby, Council rubbish tip, occasional, ALP 1983, Evington, St. Paul's School, waste ground, locally abundant, ALP 1985.
Intrusive or casual: Ag, Ao, Rd, Rw, Rr, Wd.
First record: 1885, J. A. Cappella in herb. LSR.
Herb.: LSR★; LSR, NMW.
HG 52: rare, sparsely distributed; 3 localities cited.

Conringia Adanson

C. orientalis (L.) Dumort.　　Hare's-ear Mustard
No record since HG.
First record: 1893, F. T. Mott in herb. MANCH.
Herb.: LSR, MANCH.
HG 45: rare, restricted in range; 7 localities cited.

Diplotaxis DC.

D. muralis (L.) DC.   Annual Wall-rocket
Map 20c; 24 tetrads.
Occasional throughout the county on railway verges, and especially on the ballast of dismantled railways. Also in waste places and on recently disturbed ground.
    Intrusive: Er, Rd, Rw, Rs, Aa, Ag, Sw.
First record: 1883, F. T. Mott in herb. MANCH.
Herb.: LSR*; LANC, LSR, MANCH, NMW.
HG 49: occasional, sparsely distributed; 6 localities cited.

D. tenuifolia (L.) DC.   Perennial Wall-rocket
No map; 1 tetrad.
Wigston Junction North, ballast of railway sidings, rare, EH 1974.
    Casual: Er.
First record: Vice (1905).
Herb.: LSR*; LSR.
HG 49: rare; 1 locality cited.

Brassica L.

B. oleracea L.   Cabbage
No map.
Occurs occasionally as a casual in waste places and on roadside verges. Some records were lost in the fire at Ratcliffe College, and not all the field workers recorded such relics of cultivation.
    Casual: Ev, Er, Rw, Rd.
First record: Horwood and Gainsborough (1933).
Herb.: LSR*; LSR, NMW.
HG 46: rare; 2 localities cited.

B. napus L.   Rape
Map 20d; 67 tetrads.
Roadside verges, waste places and arable land. Occasional throughout the county; usually a non-persistent casual or escape from cultivation. Probably under-recorded in some parts of the county.
    Casual or intrusive: Ev, Er, Ew, Ea, Rw, Rq, Rs, Rd, Aa, Ag, Ao, Ac.
First record: Pulteney (1747).
Herb.: LSR*; LSR, LTR, MANCH, NMW.
HG 46: frequent, generally distributed.

B. rapa L.   Wild Turnip
Map 20e; 59 tetrads.
Roadside verges, waste places and arable land. Occasional throughout the county; usually a non-persistent casual.
    Casual or intrusive: Ev, Ew, Ef, Er, Rw, Rr, Rq, Rs, Rf, Ao, Ac, Gl, Gr, Fm.
First record: Pulteney (1747).
Herb.: LSR*; CGE, LSR, NMW.
HG 47: occasional, but generally distributed.

B. nigra (L.) Koch   Black Mustard
No map; 6 tetrads.
Witherley, weed of cereal crop, rare, SHB 1974; Barwell, recently disturbed ground, occasional, SHB 1974; Huncote, arable land, rare, SHB 1974; Thurlaston, recently disturbed ground on roadside, rare, SHB 1974; Leicester, Saxby Street, garden weed, rare, HB 1974; Leicester, margins of temporary car park, locally frequent, HB 1976; Sileby, ditch, rare, PHG 1978; Leicester, Spinney Hills, garden weed, rare, HB 1978.
    Casual or intrusive: Ag, Ac, Rd, Rw, Wd.

First record: Pulteney (1749).
Herb.: DPM, LANC, LSR, LTR, MANCH, NMW.
HG 48: local, sparsely distributed; 13 localities cited.

Erucastrum C. Presl.

E. gallicum (Willd.) O. E. Schulz
No record in the recent survey.
Measham, canal bank near colliery, E. K. Horwood 1959.
First record: 1949, E. K. Horwood in herb. LTR.
Herb.: LTR.
HG: not recorded.

Sinapis L.

S. arvensis L.   Charlock, Kedlock
Map 20f; 518 tetrads.
Arable land, waste places and recently disturbed ground. Frequent throughout the county. By no means as common as in the days before selective weed killers; although there is often an abundance of seedlings among cereal crops before spraying, most succumb and the bright yellow cornfields of former days are now seldom seen.
    Intrusive: Ac, Ao, Aa, Ag, Ap, Ea, Ev, Ew, Er, Rw, Rf, Rd, Rr, Rg, Rq, Rs, Gl, Gr.
First record: Pulteney (1747).
Herb.: LSR*; CGE, DPM, LSR, LTR, MANCH, NMW.
HG 48 as *Brassica arvensis*: very abundant, generally distributed, often an agricultural pest.

S. alba L.   White Mustard
No map; 6 tetrads.
Groby, cornfield, rare, EH 1968; Groby, hedgerow, rare, EH 1969; Glenfield, dismantled railway, verge, occasional, EH 1970; Ellistown, rough grassland, rare, PAC 1971; Ratcliffe Culey, arable land, occasional, SHB 1975; Little Dalby, roadside, rare, KGM 1975; Groby, Community College, disturbed ground, occasional, EH 1983.
    Casual or intrusive: Ac, Ev, Er, Eh, Gr, Rd.
First record: Pulteney (1752).
Herb.: LANC, LSR, LTR, NMW.
HG 49 as *Brassica alba*: local, sparsely distributed; 14 localities cited.

Eruca Miller

E. vesicaria (L.) Cav. subsp. **sativa** (Miller) Thell.
E. sativa Miller
No record since HG.
First record: B.E.C. Rep. 1917.
Herb.: LSR, NMW.
HG 50: rare; 1 locality cited.

Hirschfeldia Moench

H. incana (L.) Lagrèze-Fossat   Hoary Mustard
No map; 2 tetrads.
Barwell, rubbish tip, single plant, EH 1981; Glen Parva, rubbish tip, single plant, EH 1983.
    Casual: Rr.
First record: Preston (1900).
Herb.: LSR*
HG 48 as *Brassica incana*: rare, restricted in range; 1 locality cited.

Rapistrum Crantz
R. rugosum (L.) All.   Bastard Cabbage
Subsp. **rugosum**
No map; 2 tetrads.
Burton Overy, railway embankment, rare, EKH and JMH 1973; Groby, quarry spoil heap, occasional, EH 1978.
    Casual: Er, Rs.
First record: B.E.C. Rep. 1919.
Herb.: LSR★; LSR, LTR.
HG 57: rare, very restricted in range; 2 localities cited.

Subsp. **orientale** (L.) Arcangeli
R. orientale (L.) Crantz
No map; 1 tetrad.
Asfordby, roadside at entrance to gravel pits, single plant, ALP 1972.
    Casual: Ev.
First record: 1972, A. L. Primavesi in present work.
Herb.: LTR.
HG: not recorded.

Raphanus L.
R. sativus L.   Garden Radish
No map; 1 tetrad.
Great Glen, farmyard, rare, EH 1975. This species occasionally occurs on rubbish dumps and as a garden throwout, but it seems to have been overlooked in the recent survey.
    Casual or escape: Rf.
First record: Pulteney (1747).
Herb.: LSR★; LSR, NMW.
HG 58: occasional or rare; 7 localities cited.

R. raphanistrum L.   Wild Radish
Map 20g; 249 tetrads.
Arable land, waste places and roadside verges. Frequent and locally abundant, especially in the west of the county. It avoids the heavier clay soils.
    Intrusive: Ao, Ac, Aa, Ag, Ea, Ev, Er, Eh, Ew, Rw, Rr, Rs, Rf, Rd, Rq, Gl, Gr.
First record: Watson (1832).
Herb.: LSR★; LANC, LSR, LTR, MANCH, NMW.
HG 57: locally abundant, generally distributed.

RESEDACEAE
Reseda L.
R. luteola L.   Weld
Map 20h; 140 tetrads.
Locally frequent throughout the county in waste places, gravel pits and quarries, also on spoil heaps and roadside and railway verges.
    Intrusive: Rw, Rg, Rq, Rs, Rd, Rr, Rt, Er, Ev, Ew, Eh, Gr, Sw, Ag.
First record: Pulteney (1747).
Herb.: LSR★; LSR, LTR, MANCH.
HG 59: local, rather restricted in range; 32 localities cited.

R. alba L.   White Mignonette
No map; 1 tetrad.
Ratby, allotment garden, single plant, SHB 1979.
    Casual or escape: Aa.
First record: Wade (1919).
Herb.: LSR, NMW.
HG 58: waste places, rare; 2 localities cited.

R. lutea L.   Wild Mignonette
Map 20i; 70 tetrads.
Forms part of the constituent vegetation of the Oolite district in the north-east. Otherwise most of the records are from railway ballast (69% of the records received). Especially characteristic of the ballast of dismantled railways, where it may be locally abundant.
    Constituent: Gr.
Residual or intrusive: Er, Ev, Rq, Rg, Rs, Rw, Rd.
First record: Pulteney (1749).
Herb.: LSR★; LANC, LSR, LTR, MANCH, NMW.
HG 58: rare, confined save as an alien to N.E. 36 localities cited.

SARRACENIALES
DROSERACEAE
Drosera L.
D. rotundifolia L.   Round-leaved Sundew
No recent record:
First record: Pulteney in Nichols (1795).
HG 235: very rare, originally confined to Charnwood Forest, not seen since Bloxam's time, or 1842; 4 localities cited.

D. intermedia Hayne   Oblong-leaved Sundew
No recent record.
First record: Pulteney in Nichols (1795).
HG 236 as *D. longifolia*: very rare - if ever found - now extinct, originally confined to Charnwood Forest; 2 localities cited.

ROSALES
CRASSULACEAE
Crassula L.
C. helmsii (T. Kirk) Cockayne
No map; 5 tetrads.
Harston, garden pond, occasional (accidentally introduced from Denton Manor, Lincs), E.M. Pearce 1981; Blackbrook Reservoir, stony and muddy margins, locally frequent, PAE 1984; Mountsorrel, small field pond, locally frequent, ALP 1985; Cropston, Puddledyke, occasional, M. Statham 1986.
    Intrusive: Wp, Wl, Rg.
First Record: 1981, E.M. Pearce in present work.
Herb: LSR★; LCR.
HG: not recorded.

Umbilicus DC.
U. rupestris (Salisb.) Dandy   Navelwort
Map 20j; 2 tetrads.
Woodhouse Eaves, The Brand, old slate quarries, locally frequent, PHG 1968; Swithland Wood, old slate quarries, locally frequent, PHG 1968; Swithland churchyard, boundary wall, rare, PHG 1970. The slate quarry sites have been known since Pulteney's time.
    Constituent or residual: Rq.
    Intrusive: Sw.
First record: Pulteney (1747).
Herb.: LSR, MANCH, NMW, WAR.
HG 232 as *Cotyledon umbilicus-veneris*: extremely rare, confined to Charnwood Forest, in one native station; 4 localities cited.

Sempervivum L.
S. tectorum L.   House-leek
No map.
Often found on old roofs and walls, but the majority of the field workers in the recent survey seem to have assumed that it was always planted and did not record it.
   Intrusive or planted: Sw, Sb.
First record: Pulteney (1747).
Herb.: BM, LSR.
HG 235: occasional, but generally distributed; 22 localities cited.

Sedum L.
S. telephium L.   Orpine
No record since HG.
First record: Horwood and Gainsborough (1933).
HG 233*as *S. fabaria*: very rare; 1 locality cited.

S. spurium Bieb.
Map 20k; 10 tetrads.
Hallaton, churchyard wall, occasional, KGM 1971; Freeby, railway near site of Saxby Station, retaining wall, occasional, KGM 1971; Wymondham village, old garden walls, occasional, KGM 1971; Wymondham, former railway station garden, occasional, KGM 1971; Drayton, old garden walls, occasional, KGM 1972; Noseley Hall, ha-ha, rare, KGM 1973; Bringhurst, churchyard wall, locally frequent, KGM 1973; Rothley, railway verge, locally frequent, PHG 1974; South Wigston, grassy railway verge, occasional, EH 1975; Goadby Marwood, stone wall, occasional, ALP 1977; Saltby, near Egypt Plantation, concrete base of former airfield buildings, locally frequent, SHB 1978.
   Intrusive: Sw, Er.
First record: 1971, K. G. Messenger in present work.
HG: not recorded.

S. reflexum L.   Reflexed Stonecrop
Map 20l; 33 tetrads.
Sometimes planted in gardens and occasionally becoming established on walls and in waste places.
   Intrusive or planted: Sw, Sr, Sb, Rw, Rs, Ev, Er.
First record: Pulteney (1747).
Herb.: LSR★; LSR, LTR, NMW.
HG 234: occasional, generally distributed; 35 localities cited.

S. acre L.   Biting Stonecrop
Map 21a; 135 tetrads.
Frequently planted on walls and rockeries, and becoming established on walls and in stony places where it has not been planted. Frequent and locally abundant on railway ballast, especially that of dismantled railways.
   Intrusive or planted: Sw, Sb, Sr, Er, Ev, Rw, Rq, Rs, Rt, Rr, Ap, Ag, Gr.
First record: Pulteney (1747).
Herb.: LSR★; LSR, LTR, MANCH, NMW.
HG 234 as *S. drucei*: frequent, generally distributed.

S. album L.   White Stonecrop
Map 21b; 38 tetrads.
A garden escape, often becoming established on walls and in stony waste places.
   Planted or intrusive: Sw, Sb, Rw, Rt, Rq, Rd, Ap, Er, Ev.
First record: Kirby (1848b).
Herb.: LSR★; DPM, LSR, LTR, MANCH, NMW.
HG 233: rare, restricted in range; 19 localities cited.

S. anglicum Hudson   English Stonecrop
Map 21c; 2 tetrads.
Markfield, waste ground, in cracks of old concrete and tarmacadam, occasional, PAC 1969; Newtown Linford, wall, locally abundant, EH 1972.
   Intrusive or planted: Rt, Sw.
First record: 1969, P. A. Candlish in present work.
HG 234: no Leicestershire record.

S. dasyphyllum L.   Thick-leaved Stonecrop
Map 21d; 3 tetrads.
Nevill Holt, wall, rare, KGM 1971; Wycomb, stone wall, locally frequent, ALP 1977; Chadwell, stone wall, locally frequent, ALP 1977.
   Intrusive or planted: Sw.
First record: Horwood (1911).
HG 233: rare; 2 localities cited, including Nevill Holt.

SAXIFRAGACEAE
Saxifraga L.
S. spathularis Brot. × **umbrosa** L. = **S.** × **urbium** D. A. Webb   Londonpride
No map; 3 tetrads.
Hugglescote, wall, rare, PAC 1970; Willesley, waste ground, occasional, KGM and ALP 1973; Harston, verge of dismantled mineral railway, rare, KGM 1974.
   Garden escape: Rw, Er, Sw.
First record: 1970, P. A. Candlish in present work.
HG: not recorded.

S. cymbalaria L.
Map 21e; 1 tetrad.
Woodhouse Eaves, persistent garden weed, occasional, PAC 1968.
   Intrusive: Ag.
First record: 1968, P. A. Candlish in present work.
Herb.: LSR.
HG: not recorded.

S. tridactylites L.   Rue-leaved Saxifrage
Map 21f; 18 tetrads.
Walls and dry sandy grassland. Occasional, in scattered localities throughout the county.
   Intrusive or constituent: Sw, Gr, Hd, Er, Ap, Ag.
First record: Pulteney (1747).
Herb.: LSR★; CGE, LANC, LSR, LTR, MANCH, NMW.
HG 228: frequent, generally distributed.

S. granulata L.   Meadow Saxifrage
Map 21g; 74 tetrads.
Locally frequent in old grassland throughout the county. Occasional on roadside and railway verges. Probably decreasing as suitable habitats are destroyed.
   Constituent, residual or intrusive: Gp, Gr, M, Hg, Er, Ev, Ew, Ea, Ap, Ag.
First record: Pulteney (1747).
Herb.: LSR★; LANC, LSR, LTR, MANCH, NMW.
HG 228: frequent, and locally distributed; 51 localities cited.

Chrysosplenium L.
C. alternifolium L.   Alternate-leaved Golden-saxifrage
Map 21h; 2 tetrads.
Shepshed, wet woodland south-east of Hookhill Wood, occasional, PHG 1969; Groby, banks of outflow stream

from the Pool, rare, EH 1971.
   Constituent or residual: Fw, Ew.
First record: Pulteney (1749).
Herb.: LSR, MANCH, NMW.
HG 230: rare, practically confined to Charnwood Forest; 8 localities cited, two of which approximate to the above.

C. oppositifolium L.   Opposite-leaved Golden-saxifrage
Map 21i; 55 tetrads.
Wet places in woodland, marshes, and the banks of streams. Locally frequent in Charnwood Forest, on the Coal Measures, and in the East Leicestershire woodlands; rare or absent in the rest of the county.
   Constituent: Fw, Fd, Fm, Ef, Ew, M.
First record: Pulteney (1747).
Herb.: LSR★; LANC, LSR, LTR, MANCH, NMW, UPP.
HG 229: occasional, local in distribution; 48 localities cited.

PARNASSIACEAE
Parnassia L.
P. palustris L.   Grass-of-Parnassus
Map 21j; 1 tetrad.
Botcheston Bog, marsh, locally frequent, SHB 1977.
   Constituent: M.
First record: Pulteney (1747).
Herb.: LSR, LTR, NMW.
HG 230: rather rare, distribution restricted; 16 localities cited, including Botcheston Bog.

GROSSULARIACEAE
Ribes L.
R. rubrum L.   Red Currant
R. sylvestre (Lam.) Mert. & Koch
Map 21k; 111 tetrads.
Occasional throughout the county in wet places, particularly in woodland.
   Constituent or intrusive: Fw, Fd, Fm, Fs, Ef, Ew, Eh, Ev, Er, M, Wd, Wp, Gr, Rq.
First record: Ray (1724).
Herb.: LSR★; DPM, LSR, LTR, MANCH, NMW.
HG 232: local, generally distributed.

R. nigrum L.   Black Currant
Map 21l; 45 tetrads.
Occasional throughout the county in wet places, especially in woodland.
   Constituent or intrusive: Fw, Fd, Fc, Fm, Fs, Ef, Ew, Eh, Er, M, Wd, Wp, Rw, Rg, Rs, Gr, Sw.
First record: Power (1807).
Herb.: LSR★; CGE, LSR.
HG 232: rare, restricted in range; 19 localities cited.

R. sanguineum Pursh
No map; 3 tetrads.
Glenfield, dismantled railway track, rare, EH 1970; Worthington, hedgerow, rare, SHB 1978; Whitwick, Spring Hill, scrub woodland, single plant, SHB 1979.
   Garden escape: Er, Eh, Fs.
First record: 1970, E. Hesselgreaves in present work.
HG: not recorded.

R. uva-crispa L.   Gooseberry
Map 22a; 114 tetrads.
Woodland and hedgerows, usually in small numbers or as single specimens, which seems to imply that it is an escape from cultivation, either bird-sown or from fruits or seeds accidentally scattered by man. Occasional throughout the county. One large specimen was found near Diseworth (PAC 1970) growing eight feet up in the crown of a pollarded willow.
   Intrusive, or possibly constituent: Fd, Fm, Fs, Fw, Eh, Ew, Er, Ev, Wd, Rw, Rq, Sw, Sb, Sr, Gr.
First record: Power (1807).
Herb.: LSR★; DPM, LANC, LSR, LTR, MANCH, NMW.
HG 231: frequent, generally distributed.

R. alpinum L.   Mountain Currant
No map; 2 tetrads.
Misterton Hall park, ornamental copse, occasional, EKH and JMH 1969; Glenfield, hedgerow, rare, EH 1972.
   Planted or garden escape: Fd, Eh.
First record: Bloxam (1837).
Herb.: LTR.
HG 231: rare, restricted in range; 3 localities cited.

ROSACEAE
Spiraea L.
S. salicifolia L.   Bridewort
Map 22b; 20 tetrads.
Planted as an ornamental shrub in gardens, and occasionally becoming established spontaneously by water, in hedgerows and in waste places.
   Intrusive or garden escape: Ew, Eh, Er, Ev, Ef, Fd, Fs, Rw, Rq, Sb, Aa.
First record: Horwood and Gainsborough (1933).
Herb.: LSR★; LSR, LTR, NMW.
HG 155: rare, restricted in range; 2 localities cited.

Filipendula Miller
F. vulgaris Moench   Dropwort
Map 22c; 28 tetrads.
Old grassland, including railway and roadside verges, particularly where the soil is calcareous. Occasional in the east and south of the county, but absent from the more acid soils of Charnwood Forest and the north-west.
   Constituent or residual: Gp, Gr, Er, Ev, Ew.
   Intrusive: Rw.
First record: Pulteney (1747).
Herb.: LSR★; LSR, LTR, MANCH.
HG 156 as *Spiraea filipendula*: rather rare in Leicestershire, more frequent in Rutland, sparsely distributed; 51 localities cited.

F. ulmaria (L.) Maxim.   Meadowsweet
Map 22d; 571 tetrads.
Marshes, ditches, woodland rides, wet woodland, and the banks of rivers, streams and canals. Locally abundant throughout the county.
   Constituent: M, Wd, Ew, Ef, Fw, Fd, Fs.
   Residual or intrusive: Ev, Er, Eh, Gp, Gr, Rw, Rd.
First record: Pulteney (1749).
Herb.: LSR★; DPM, LANC, LSR, LTR, MANCH, NMW.
HG 155 as *Spiraea ulmaria*: locally abundant, generally distributed.

Rubus L.
Subgenus Idaeobatus Focke
R. idaeus L.   Raspberry
Map 22e; 235 tetrads.
Locally abundant in woodland and hedgerows in the west of the county. In the east it is mainly confined to woodland, where it may be locally frequent. Occasional in other habitats as an escape from cultivation.
    Constituent: Fd, Fc, Fm, Fs, Fw, Ef, Eh.
    Residual or intrusive: Er, Ev, Ew, Ea, Gr, Ap, Aa, Rw, Rq, Rg, Rr, Rd.
First record: Pulteney (1747).
Herb.: LSR*; CGE, DPM, LSR, LTR, MANCH, NMW.
HG 157: frequent, generally distributed.

Subgenus Glaucobatus Dumort. pro parte
R. caesius L.   Dewberry
Map 22f; 102 tetrads.
Woods, scrub, roadsides, railway banks and ballast, walls, quarries and other ruderal situations; widespread but with a discontinuous distribution particularly outside the Charnwood Forest area, although in some parts of Charnwood Forest it has probably been overlooked.
    Constituent, residual and intrusive: Fd, Fm, Fs, Fw, Ef, Er, Ev, Eh, Ew, Wd, Rq, Rw, Sw, Ap.
First record: Pulteney (1749).
Herb.: CGE, LANC, LSR, LTR, MANCH, NMW.
HG 190: locally abundant, generally distributed.

R. loganobaccus L. H. Bailey   Loganberry
No map.
This introduced species is widely cultivated and has been reported as a garden escape, e.g. at a car park in Coalville, EH 1976.

Subgenus Rubus   Brambles, Blackberries
This subgenus, referred to in many books as **Rubus fruticosus** agg., consists of a number of groups of mainly apomictic species, of which over 2000 have been described and of which nearly 400 occur in Britain. The subgenus has been regarded as critical during this survey, although it is not difficult for field workers to learn the characters of some of the commoner species in a county such as this, where under fifty species are known to occur.

Nevertheless, it was agreed at the outset of the survey that records of only one of these, namely **R. ulmifolius**, should be accepted from field workers without verification by specialists. This has made it impossible to produce maps of the full distribution of most species, and the maps represent the state of our knowledge as a result of some half dozen visits to the county by specialists, and collecting on a limited scale by some field workers. We are grateful to Mr. E. S. Edees and to Mr. A. Newton both for their personal visits and for their verifications of numerous specimens sent to them. Mr. J. R. I. Wood has also visited the county twice and has allowed us to make use of his records.

Horwood and Gainsborough give an account of 106 taxa (species, subspecies and varieties) of **Rubus** in addition to **R. idaeus** and **R. caesius**, the account being based largely on field observations and taxonomic study made by A. Bloxam, J. Moyle Rogers, E. F. Linton and W. H. Coleman. This account has been thoroughly appraised during the course of the present survey by Newton and Edees and in particular the major British herbaria have been searched for supporting material collected in Leicestershire. The records in Horwood and Gainsborough have, as a result, been divided into four groups.
1. Taxa which can be equated to currently accepted **Rubus** species because a. Leicestershire material dating from c.1850-1933 has been found and the original identifications have been confirmed by Newton and Edees and b. the species has been found in the field during the current survey and determined by Newton, Edees, Miles or Wood.
2. Taxa which can be equated to currently accepted **Rubus** species because a. herbarium material, collected in other counties by Bloxam, Rogers, Linton or Coleman has been accepted by Newton and Edees as reasonable proof that they knew the taxa concerned though no Leicestershire material has been located and b. the species has been found in the field during the current flora survey and determined as above.
3. Taxa which can be equated to currently accepted **Rubus** species on either of the grounds given above, but which have not been found in the current flora survey.
4. Taxa which cannot be equated to any currently accepted **Rubus** species, because no herbarium material from Leicestershire has been located and because taxonomic revisions made since 1933 make it impossible to judge what their modern equivalents might be. About 30 of Horwood and Gainsborough's taxa fall into this group.

In the species accounts which follow, the phrase 'no known herbarium material' is used to indicate taxa in groups 2 and 3 above for which no Leicestershire material has been found dated earlier than the start of the current survey.

    All records made during the current survey, with the exception of those of **R. idaeus, R. caesius, R. ulmifolius** and undetermined populations of the Section *TRIVIALES*, are based on determinations by Newton, Edees, Wood, Miles or, in a few cases, Messenger. Full coverage of the county has been impossible in the time available but a general idea of the distribution of many species found in woodland and heathland can be gained from the distribution maps. There is still a good deal of fieldwork to be done on hedgerow species particularly those of the Section *TRIVIALES*. Map 22g shows the distribution of all records of **Rubus fruticosus** aggregate other than Section *TRIVIALES* and **R. ulmifolius**.
                                                    K. G. Messenger

Section SUBERECTI P. J. Mueller
It is convenient to include these species under the general heading of Subgenus **Rubus**, although they are in fact hybrids between various 'brambles' and **R. idaeus**.

R. nessensis W. Hall
Map 22h; 2 tetrads.
Near Blackbird's Nest, AN 1977-8; Out Woods, AN 1977-8.
First record: Bloxam and Babington *in* Potter (1842).
Herb.: CGE, LSR.
HG 158.

R. scissus W. C. R. Watson
No recent record.
First record: 1850, A. Bloxam in herb. LSR.
Herb.: CGE, LSR, MANCH.
HG 157 as *R. fissus*.

R. bertramii G. Braun
No recent record.
First record: 1846, Bloxam in herb. LSR.
Herb.: LIVU, LSR, MANCH.
HG 158 as *R. plicatus* (in part).

R. pergratus Blanchard
No recent record.
A North American species rarely naturalised in Britain.
First record: 1912, in herb. LSR (from Willesley Park).
Herb.: LSR.
HG: not recorded.

R. plicatus Weihe & Nees
No recent record.
First record: 1853, A. M. Bernard in herb. CGE.
Herb.: CGE, LSR.
HG 158.

R. arrheniformis W. C. R. Watson
No recent record.
First record: Bloxam (1838).
HG 158 as *R. plicatus* (in part).

Section *TRIVIALES* P. J. Mueller
These species, like the *SUBERECTI*, are also considered under the subgenus **Rubus**, though they are hybrids between 'bramble' species and **R. caesius**. There are far more of these than there are of the *SUBERECTI*, and only a very small proportion of them are of sufficiently wide distribution to have been afforded specific names. Many of those found in Leicestershire hedges are unnamed because they are too restricted in distribution, and it is for this reason that it has been necessary to plot them on an aggregate map (Map 22i; 408 tetrads). At the same time it must be admitted that those which have got names have been very much under-recorded.

R. conjungens (Bab.) W. C. R. Watson
Map 22j; 1 tetrad.
Shenton, railway cutting, locally frequent, KGM 1978.
First record: Mott et al. (1886).
Herb.: LSR.
HG 189 as *R. corylifolius* var. *conjungens*.

R. bagnallianus E. S. Edees
Map 22k; 12 tetrads.
Countesthorpe, railway bank, EH 1975; Ashby de la Zouch, South Wood, AN 1977; Gopsall Park, south-west corner, AN 1977; Hawcliff Hill, roadside, AN 1978; Piper Wood, roadside, AN 1978; Holwell Mouth, AN 1978; Thornton Reservoir, EH 1978; Ulverscroft, Lea Lane, EH 1979; Outwoods, EH 1981; Groby, Lawn Hill, EH 1981; Groby, old quarry, EH 1982; Donington le Heath, Blackberry Lane, EH 1982.
First record: Babington (1869), if HG is correct; otherwise 1898, E. F. Linton in herb. LSR.
Herb.: LSR.
HG 188 as *R. dumetorum* var. *fasciculatus*.

R. eboracensis W. C. R. Watson
Map 22l; 15 tetrads.
Probably the most widespread of the Rubi *TRIVIALES* in Leicestershire which have specific names; it has not been widely collected as it occurs mainly in hedgerows, not woodland.
Orton Wood, BAM 1965; Potters Marston, roadside, ESE 1969; Shackerstone, ESE 1969; Swithland Wood, ESE 1969; Twycross, roadside, ESE 1969; Ulverscroft Lane, ESE 1969; Saltby Heath, scrub, KGM and JRIW 1972; Harston, roadside, KGM and JRIW 1974; Cosby, railway verge, EH 1975; Countesthorpe, railway verge, EH 1975; Kirby Muxloe, hedgerow, EH 1975; Anstey, hedgerow, EH 1975; Cosby, hedgerow, EH 1976; Piper Wood, roadside, EH and AN 1978; Groby, two localities, roadside and hedgerow, EH 1979.
First record: Hands et al. *in* Curtis (1831), if HG is correct, otherwise 1891, F. T. Mott in herb. LSR.
Herb.: LSR★; CGE, LSR.
HG 188 as *R. corylifolius*.

R. sublustris Lees
Map 23a; 4 tetrads.
Ashby de la Zouch, BAM 1962; Thurlaston, roadside, ESE 1969; Narborough, railway verge, EH 1979; Donington le Heath, Blackberry Farm, EH 1982.
First record: 1846, A. Bloxam in herb. LSR.
Herb.: LSR★; CGE, LSR.
HG 188 as *R. corylifolius* var. *sublustris*.

R. balfourianus Blox. ex Bab.
No recent record.
First record: Babington (1847c).
Herb.: LSR.
HG 189.

R. warrenii Sudre
Map 23b; 2 tetrads.
Twycross, Appleby Road, AN 1977.
First record: 1904, W. Bell in herb. LSR.
Herb.: LSR.
HG: not recorded.

R. tuberculatus Bab.
Map 23c; 4 tetrads.
Quorndon, roadside, ESE 1969, recorded as **R. scabrosus** but redet. AN 1977; Gopsall Park, woodland margin, AN 1977; Ashby Pastures, scrub and woodland margin, KGM and AN 1978; Thorpe Trussels, scrub and woodland margin, KGM and AN 1978.
First record: 1846, A. Bloxam in herb. LSR.
Herb.: LSR.
HG 186-187 as *R. dumetorum* vars. *ferox, diversifolius* and *tuberculatus*. AN also equates HG's var. *scabrosus* to this species and admits no Leicestershire record of **R. scabrosus.**

R. rubriflorus Purchas
No recent record.
First record: 1906, W. Bell in herb. LSR.
Herb.: BM (this specimen, dated 1930, is not one of those recorded in HG), LSR.
HG 187.

Section *SYLVATICI* P. J. Mueller
R. calvatus P. J. Mueller
Map 23d; 6 tetrads.
Twycross, Appleby Road, ditch, BAM 1961; Twycross, Appleby Road, ditch, ESE 1969; Lount, Spring Wood, ESE 1969; Peckleton, roadside, ESE 1969; Charnwood Lodge Nature Reserve, heath grassland, JRIW 1974; Charnwood Lodge Nature Reserve, woodland, JRIW 1974; Gopsall Park, woodland margin, AN 1977; Martinshaw Wood, woodland margin, EH and AN 1978; Newtown Linford, Hallgate Spinney, EH 1981.
First record: Bloxam (1846b).
Herb.: BM, CGE, LSR, MANCH.
HG 163 as *R. villicaulis* var. *calvatus*.

R. platyacanthus Muell. & Lefev.
Map 23e; 3 tetrads.
Peckleton, roadside, ESE 1969; Swithland Wood, roadside, ESE 1969; Swithland, hedgerow, EH 1983.
First record: Bloxam (1846b).
Herb.: CGE, LSR.
HG 159 as *R. carpinifolius*.

R. nemoralis P. J. Mueller
Map 23f; 7 tetrads.
Ulverscroft, ESE 1969; Abbot's Oak, roadside, JRIW 1974; Martinshaw Wood, EH 1975; Ulverscroft, Lea Lane, AN 1977; Blakeshay Wood, EH 1978; Newtown Linford, Barn Hills, EH 1978; Lingdale Golf Course, scrub, EH 1979; Newtown Linford, Hallgate Spinney, EH 1981.
First record: Bloxam (1848a).
Herb.: LSR★; BM, CGE, LSR, MANCH.
HG 164 as *R. selmeri*.

R. laciniatus Willd.
Map 23g; 1 tetrad.
This cultivated species has been seen by EM as a garden escape on a number of occasions; Out Wood, AN 1978.
First record: 1978, A. Newton in present work.
Herb.: LSR★.
HG: not recorded.

R. lindleianus Lees
Map 23h; 38 tetrads.
One of the more widespread woodland brambles in Leicestershire, occurring also in hedgerows, by roadsides, on derelict mining sites, heathland and railway banks.
First record: Kirby (1850) and 1850, A. Bloxam in herb. LSR.
Herb.: LSR★; CGE, LSR, MANCH.
HG 160.

R. macrophyllus Weihe & Nees
Map 23i; 2 tetrads.
In Piper Wood and on the roadside nearby, EH and AN 1978.
First record: 1902, W. M. Rogers in herb. LSR (from Piper Wood).
Herb.: BM, LSR.
HG 166: AN regards most of the early records of this species as unreliable owing to confusion with *R. amplificatus* and other species.

R. subinermoides Druce
No recent record.
First record: Horwood and Gainsborough (1933).
Herb.: NOT.
HG 166 as *R. pubescens* var. *subinermis*.

R. amplificatus Lees
Map 23j; 2 tetrads.
Newtown Linford, roadside, ESE 1969; Tugby Wood, edge of ride, KGM and AN 1978.
First record: Bloxam (1846b).
Herb.: BM, CGE, LSR, MANCH, OXF.
HG 167 as *R. schlectendalii*, and as *R. macrophyllus* var. *amplificatus* (see **R. macrophyllus** above).

R. pyramidalis Kalt.
Map 23k; 7 tetrads.
Buddon Wood, ESE 1969; Charnwood Lodge Nature Reserve, woodland, heath grassland and roadside, JRIW 1974; Ulverscroft, Lea Lane, woodland margin, JRIW 1974; Martinshaw Wood and woodland margin, EH 1977; Swithland Wood, AN 1977; Out Wood, EH and AN 1978; Croft, edge of churchyard, EH 1982. This is evidently a species of the Charnwood Forest area in particular, with four records from the vicinity of Charnwood Lodge Nature Reserve alone.
First record: Kirby (1850).
Herb.: LSR★; BM, CGE, LSR, MANCH, OXF.
HG 169.

R. leptothyrsos G. Braun
R. danicus Focke
Map 23l; 2 tetrads.
Martinshaw Wood, EH 1977; Newtown Linford, Hallgate Spinney, EH 1981.
First record: 1898, E. F. Linton in herb. LSR.
Herb: LSR.
HG 169 as *R. hirtifolius* var. *danicus* (AN regards these records as erroneous) and var. *mollissimus* (of which specimens in herb. LSR collected by E. F. Linton have been confirmed as this species).

R. polyanthemus Lindb.
Map 24a; 11 tetrads.
Ulverscroft, ESE 1969; Belvoir, Knipton Pasture, KGM 1974; Belvoir, Old Park Wood, KGM 1974; Charnwood Lodge Nature Reserve, woodland, woodland margins and roadside, JRIW 1974; Ulverscroft, Lea Lane, woodland margin, JRIW 1974 and 1977; Martinshaw Wood, EH 1975 and 1977; Ulverscroft, Lea Lane, AN 1977; Swithland Wood, AN 1977; Gopsall Wood, AN 1977; Twycross, Appleby Road, AN 1977; Newtown Linford, Barn Hills, EH 1978; Blakeshay Wood, EH 1978; Nailstone Wiggs, KGM 1978; Newtown Linford, Hallgate Spinney, EH 1981; Ulverscroft, footpath off Lea Lane, EH 1986.
A characteristic Charnwood Forest species also found at Belvoir and in the coalfield area.
First record: 1846, A. Bloxam in herb. LSR.
Herb.: LSR★; BIRM, CGE, LSR, MANCH.
HG 162.

R. rhombifolius Weihe ex Boenn.
No recent record.
First record: Bloxam (1846b).
Herb.: CGE, LSR, MANCH.
HG 159 as *R. nitidus*.

R. cardiophyllus Mueller & Lefev.
Map 24b; 5 tetrads.
Swithland Wood, ESE 1969, AN 1977; near Abbot's Oak, roadside, JRIW 1974; Ulverscroft, Lea Lane, AN 1977; Blackbird's Nest, AN 1977; Newtown Linford, Hallgate Spinney, EH 1981.
First record: Bloxam (1838).
Herb.: LSR, MANCH, OXF.
HG 161 as *R. rhamnifolius*

R. lindebergii P. J. Mueller
Map 24c; 3 tetrads.
Swithland Wood, ESE 1969; Charnwood Lodge Nature Reserve, heath grassland and woodland rides, JRIW 1974; Newtown Linford, Hallgate Spinney, EH 1981.
First record: 1898, E. F. Linton in herb. LSR.
Herb.: BM, LSR.
HG 163

Section *DISCOLORES* P. J. Mueller
R. ulmifolius Schott.
Map 24d; 463 tetrads.
Populations of Section *TRIVIALES* apart, this is the commonest and most widespread bramble in Leicestershire hedgerows, where it is readily recognisable by the silvery indumentum on the underside of the leaflets and its pinkish flowers which open some three weeks later than those of most of the *TRIVIALES*. Apart from hedges it is found occasionally in woodland, quarries, scrub and waste land of various kinds.
First record: Pulteney (1747).
Herb.: LSR★; CGE, LSR MANCH, OXF.
HG 166 as *R. rusticanus*.

R. ulmifolius × vestitus
This hybrid was identified among a collection from Ratby Burroughs made by EH in 1978.
First record: 1978, E. Hesselgreaves in present work.
Herb.: LSR★.
HG: not recorded.

R. armipotens Barton ex A. Newton
Map 24e; 1 tetrad.
Theddingworth, roadside, ESE 1969.
First record: 1969, E. S. Edees in present work.
HG: not recorded.

R. procerus P. J. Mueller
Map 24f; 7 tetrads.
A large-fruited cultivated species frequently found in Britain as an escape. Gumley, ESE 1969; Kirby Muxloe, hedgerow, EH 1974; Groby, hedgerow, EH 1974; Narborough, railway verge, EH 1975; Braunstone, woodland, EH 1975; Beaumont Leys, hedgerow, EH 1975; Whitwick, EH 1976. EH regards it as frequent west of Leicester.
First record: 1969, E. S. Edees in present work.
Herb.: LSR★.
HG: not recorded.

R. anglocandicans A. Newton
R. falcatus Kalt.
Map 24g; 1 tetrad.
Harston, JRIW 1974.
First record: 1846, A. Bloxam in herb. LSR.
Herb.: BM, LSR, MANCH.
HG 165 as *R. thyrsoideus*.

Section *SPRENGELIANI* (Focke) W. C. R. Watson
R. sprengelii Weihe
Map 24h; 2 tetrads.
Ashby de la Zouch, South Wood, AN 1977; Nailstone Wiggs, KGM 1978.
First record: Babington (1847b) and 1847, A. Bloxam in herb. LSR.
Herb.: BM, CGE, LSR, MANCH, NOT, OXF.
HG 168

Section *APPENDICULATI* (Genev.) Sudre
R. vestitus Weihe & Nees
Map 24i; 23 tetrads.
A widespread but not particularly frequent species; records are mainly from Belvoir, east Leicestershire woodland, Charnwood Forest and the Burbage area. Horwood records it from numerous other woods but in roughly the same general areas, except that it was then frequent to the north-west of Charnwood Forest, and present in the south near Market Harborough. It is known to occur occasionally in disused quarries and on railway banks, and rarely on roadsides.
First record: Bloxam (1837).
Herb.: LSR★; CGE, LSR, MANCH, OXF.
HG 170

R. criniger (Linton) Rogers
Map 24j; 2 tetrads.
Copt Oak, roadside, AN 1977; Nailstone Wiggs, KGM 1978 (probable, det. AN).
First record: 1850, A. Bloxam in herb. LSR.
Herb.: BM, CGE, LSR, MANCH.
HG 171.

R. mucronulatus Bor.
Map 24k; 3 tetrads.
Twycross, Appleby Road, AN 1977; Martinshaw Wood, EH 1978.
First record: Kirby (1850) and 1850, A. Bloxam in herb. LSR.
Herb.: BIRM, CGE, LSR, MANCH, OXF.
HG 171 as *R. mucronifer*.

R. radula Weihe ex Boenn.
Map 24l; 3 tetrads.
Belvoir, woodland margin, JRIW 1974; Harston, hedgerow, JRIW 1974; Old Dalby Wood, AN 1978.
First record: 1845, A. Bloxam in herb. LSR.
Herb.: CGE, MANCH, LSR, NOT, OXF.
HG 173.

R. echinatus Lindl.
Map 25a; 12 tetrads.
Thurlaston, roadside, ESE 1969; Kirkby Mallory, roadside, ESE 1969; Martinshaw Wood, ESE 1969; Stockerston, roadside, KGM and JRIW 1974; Allexton, Knob Hill, roadside, KGM and JRIW 1974; Belvoir, Old Park Wood, woodland margin, KGM and JRIW 1974; Belvoir, Granby Wood, woodland margin, KGM and JRIW 1974; Charnwood Lodge Nature Reserve, woodland, JRIW 1974; Swithland Wood, JRIW 1974; Old Dalby Wood, KGM and AN 1978; Tugby Wood, KGM and AN 1978; Groby, old quarry, EH 1978; Lingdale Golf Course, scrub, EH 1979; Newtown Linford, roadside, EH 1982; Ratby, hedgerow, EH 1982.
First record: Bloxam (1846b) and 1846, A. Bloxam in

herb. LSR.
Herb.: LSR*; BIRM, CGE, LSR, MANCH, NOT, OXF.
HG 173.

R. echinatoides (Rogers) Wallman
Map 25b; 2 tetrads.
Swithland Wood, ESE 1965; Launde Park Wood, JRIW 1973.
First record: 1902, W. M. Rogers in herb. LSR.
Herb.: BM, CGE, LSR.
HG 173 as *R. radula* var. *echinatoides*.

R. rudis Weihe & Nees.
Map 25c; 19 tetrads.
Usually in woodland where it may be the dominant species; occasionally in hedgerows near woods. It has a markedly eastern distribution in Leicestershire, only two records coming from west of the River Soar.
Burbage Wood or Aston Firs, ESE 1969; Owston Wood, ESE 1969; Stapleford Park, woodland, KGM and JRIW 1971; Noseley, woodland, KGM and JRIW 1971; Great Merrible Wood, KGM and JRIW 1971; Launde, Park Wood, KGM and JRIW 1971; Bolt Wood, woodland margin and nearby hedgerow, KGM and JRIW 1971; Bescaby Oaks, JRIW 1972; Saltby Heath, scrub, JRIW 1972; King Lud's Entrenchments, scrub, JRIW 1972; Stockerston, Park Wood and nearby hedgerow, KGM and JRIW 1974; Belvoir, Old Park Wood, woodland margin, KGM and JRIW 1974; Swithland Wood, AN 1977; Tugby Wood, KGM and AN 1978; Old Dalby Wood, KGM and AN 1978; Burton on the Wolds, Twenty Acre, scrub, KGM and AN 1978; Tilton Wood, KGM and AN 1978; Skeffington Wood, KGM and AN 1978; Cradock's Ashes, KGM and AN 1978.
First record: Mott et al. (1886).
Herb.: CGE, LSR, MANCH, OXF.
HG 174.

R. cantianus (W. C. R. Watson) Edees & Newton
R. prionodontus Mueller & Lefev.
Map 25d; 1 tetrad.
Gopsall Wood, lane side, BAM 1965.
First record: 1965, B. A. Miles in present work.
Herb.: CGE.
HG: not recorded.

R. bloxamianus Coleman ex Purchas
R. granulatus Mueller & Lefev.
Map 25e; 22 tetrads.
A widespread woodland bramble which is probably still present in many woods where it was recorded in Horwood and Gainsborough but which have not recently been visited; it also occurs frequently in nearby hedgerows.
Burbage Wood or Aston Firs, ESE 1969; Swithland Wood, ESE 1969; Owston Wood, ESE 1969; Ulverscroft, Ulverscroft Lane, ESE 1969; Copt Oak, ESE 1969; Kirkby Mallory, roadside, ESE 1969; Launde, Park Wood, KGM and JRIW 1971; Stockerston, Park Wood, KGM and JRIW 1974; Ulverscroft, Lea Lane, AN 1977; Swithland Wood, AN 1977; Old Dalby Wood, KGM and AN 1978; Piper Wood, EH and AN 1978; Ashby Pastures and Thorpe Trussels, woodland margins, KGM and AN 1978; Skeffington Wood, KGM and AN 1978; Tugby Wood, KGM and AN 1978; Tilton Wood, KGM and AN 1978; Ambion Wood, KGM 1978; Nailstone Wiggs, KGM 1978; Lingdale Golf Course, scrub, EH 1979; near Groby Pool, hedgerow, EH 1979; Newtown Linford, Hallgate Spinney, EH 1981; Newtown Linford, Lea Lane, EH 1981.
First record: 1850, A. Bloxam in herb. LSR.
Herb.: LSR*; BM, CGE, LIVU, LSR, MANCH, OXF.
HG 175 as *R. oigoclados* var. *bloxamianus*.

R. flexuosus Mueller & Lefev.
Map 25f; 5 tetrads.
Buddon Wood, ESE 1969; Ulverscroft, Lea Lane, AN 1977; Blackbird's Nest, AN 1977; Ashby de la Zouch, South Wood, AN 1977; Swithland Wood, AN 1977.
First record: Kirby (1850) and 1850, A. Bloxam in herb. LSR (from Buddon Wood).
Herb.: LSR, MANCH.
HG 179.

R. rubristylus W. C. R. Watson
Map 25g; 1 tetrad.
Bittesswell Hall Park, near entrance, ESE 1969.
First record: Coleman (1852).
Herb.: MANCH.
HG 175 as *R. oigoclados*.

R. bloxamii Lees
Map 25h; 2 tetrads.
Burbage Wood or Aston Firs, ESE 1969; Orton Wood, ESE 1969.
First record: Bloxam (1846b).
Herb.: CGE, LIVU, MANCH, OXF.
HG 178.

R. adamsii Sudre
No recent record.
First record: 1850, A. Bloxam in herb. LSR (from Buddon Wood).
Herb.: LSR.
HG 177 as *R. babingtonii* var. *phyllothyrsus* p.p.

R. watsonii W. K. Mills
Map 25i; 2 tetrads.
Lea Wood, ESE 1965 and 1969; Swithland Wood, ESE 1969.
First record: 1965, E. S. Edees in present work.
Herb.: CGE.
HG: not recorded.

R. pallidus Weihe & Nees
Map 25j; 4 tetrads.
Swithland Wood, ESE 1969, AN 1977; Owston Wood, ESE 1969; Tugby Wood KGM and AN 1978; Martinshaw Wood, EH 1983.
First record: 1897, E. F. Linton in herb. LSR (from Swithland Wood).
Herb.: BM, CGE, LIVU, LSR.
HG 178.

R. euryanthemus W. C. R. Watson
Map 25k; 1 tetrad.
Martinshaw Wood, EH 1977, originally recorded as **R. pallidus** but re-examined in 1978 by AN and determined as this species.
First record: 1977, E. Hesselgreaves in present work.
Herb.: LSR*.
HG: not recorded.

R. insectifolius Mueller & Lefev.
Map 25l; 5 tetrads.
Blakeshay Wood, ESE 1969; Lea Wood, ESE 1969 and JRIW 1974; near Lea Wood, hedgerow, JRIW 1974; Martinshaw Wood, EH 1975; near Lady Hay Wood, EH 1975; Ulverscroft, Lea Lane, AN 1977; Piper Wood, woodland margin, EH and AN 1978.
First record: Kirby (1850).
Herb.: LSR★; LSR.
HG 178 as *R. fuscus* and *R. nuticeps* (equated to this species by AN on the strength of herbarium material from other vice counties).

R. scaber Weihe & Nees
No recent record.
First record: Babington (1847c).
Herb.: CGE, LSR, MANCH, OXF.
HG 179.

R. rufescens Mueller & Lefev.
Map 26a; 10 tetrads.
Orton Wood, ESE 1969; Owston Wood, ESE 1969; Lea Wood, ESE 1969; Swithland Wood, woodland and woodland margin, JRIW 1974; Lea Wood, woodland, woodland margins and nearby hedgerow, JRIW 1974; Gopsall Park, south-west corner, woodland, AN 1977; Twycross, Appleby Road, AN 1977; Ashby de la Zouch, South Wood, AN 1977; Piper Wood, woodland and woodland margin, EH and AN 1978; Nailstone Wiggs, KGM 1978; Newtown Linford, Lea Lane EH 1981; Newtown Linford, Blakeshay Farm, EH 1982.
First record: Bloxam (1846) and 1846, A. Bloxam in herb. LSR.
Herb.: BM, CGE, LSR, MANCH, OXF.
HG 180 as *R. rosaceus* var. *infecundus*.

R. raduloides (Rogers) Sudre
Map 26b; 3 tetrads.
Copt Oak, roadside, AN 1977; Piper Wood, roadside, AN 1978; Martinshaw Wood, EH 1983.
First record: 1898, E. F. Linton in herb. LSR.
Herb.: LSR.
HG: not recorded.

R. griffithianus Rogers
Map 26c; 1 tetrad.
Old Dalby Wood, by bridleway, very local, AN 1978.
First record: 1902, W. M. Rogers in herb. LSR.
Herb.: BM, CGE, LSR.
HG 176.

R. leightonii Lees ex Leighton
Map 26d; 11 tetrads.
Twycross, Appleby Road, BAM 1961 and AN 1977; Gumley, ESE 1969; Orton Wood, woodland and roadside, ESE 1969; Barton in the Beans, roadside, ESE 1969; Nailstone, roadside, ESE 1969; Kirby Muxloe, roadside, ESE 1969; Markfield, roadside, ESE 1969; Buddon Wood, ESE 1969; Blackbird's Nest, AN 1977; Gopsall Park, south-west corner, woodland margin, AN 1977; Out Wood, EH and AN 1978; Hawcliff Hill, EH and AN 1978; Ambion Wood, KGM 1978; Nailstone Wiggs, KGM 1978; Beacon Hill, EH 1982.
First record: 1850, A. Bloxam in herb. LSR.
Herb.: CGE, LSR, MANCH, OXF.
HG 177 as *R. ericetorum*.

R. diversus W. C. R. Watson
Map 26e; 2 tetrads.
Buddon Wood, ESE 1965; Buddon Wood, ESE 1969; Swithland Wood, ESE 1969, JRIW 1974, EH 1974, AN 1977.
First record: 1891, F. T. Mott in herb. LSR.
Herb.: LSR★; BM, CGE, LSR, MANCH, OXF.
HG 185 as *R. hirtus* var *kaltenbachii*.

R. rotundifolius (Bab.) Bloxam
No recent record.
First record: Babington (1848).
Herb.: LSR, MANCH, OXF.
HG 185 as *R. hirtus* var. *rotundifolius*.

Section *GLANDULOSI* P. J. Mueller
R. murrayi Sudre.
Map 26f; 1 tetrad.
Twycross, Appleby Road, AN 1977; Gopsall Park, south-west corner, AN 1977.
First record: 1847, A. Bloxam in herb. LSR.
Herb.: CGE, MANCH, OXF.
HG 181 as *R. rosaceus* subsp. *adornatus*.

R. hylocharis W. C. R. Watson
Map 26g; 2 tetrads.
Rough Park, woodland, ESE 1969; Swithland Wood, JRIW 1974.
First record: Salter (1845).
Herb.: LSR, MANCH, OXF.
HG 179-180 as *R. rosaceus* var. *typica* and var. *hystrix*.

R. dasyphyllus (Rogers) E. S. Marshall
Map 26h; 28 tetrads.
Mainly a woodland species but also found in hedges and scrub. Widely distributed; recent records have come mainly from Charnwood Forest and Belvoir but there scattered records from other parts of the county. It probably still occurs at many other sites which are listed in HG but which have not been visited recently.
First record: Bloxam (1837).
Herb.: LSR★; CGE, LSR, MANCH, NOT, OXF.
HG 182-183, including *R. koehleri*

R. anglohirtus E. S. Edees
R. hirtus Waldst & Kit.
Map 26i; 3 tetrads.
Saltby Heath, scrub, KGM and JRIW 1972; Belvoir, Old Park Wood and Briery Wood, KGM and JRIW 1974.
First record: Kirby (1850).
HG 184.

R. bellardii Weihe & Nees
Map 26j; 2 tetrads.
Bolt Wood, JRIW 1974; Tugby Wood, AN 1978.
First record: Babington (1851).
HG 184.

Rosa L.

Roses

The genus **Rosa** presents peculiar difficulties to the taxonomist. Many of the species are unbalanced polyploids, with some paired and some unpaired chromosomes. The unpaired sets are lost in the formation of the pollen grain, but are retained in the formation of the ovule. Thus the seed parent in an unbalanced pentaploid rose provides four fifths of the genetic material, the pollen parent one fifth. To add to the difficulty, nearly every species of **Rosa** is capable of successfully pollinating any other, producing often complex hybrids with varying degrees of fertility, from fully fertile to completely sterile.

With the enormous number of forms which arise from this situation, it is not surprising that rhodologists find it difficult to agree or even to decide on the limits of a species. In Leicestershire, for example, where northern and southern British species tend to meet at the limits of their geographical ranges, apart from the diploid **Rosa arvensis** and alien diploids such as **R. rugosa**, there are very few rose bushes which correspond exactly to the author's description of a named species. It is not very far from the truth to say that in this county, apart from the exceptions mentioned, every rose bush is different from every other.

The theoretically ideal solution to this problem would be to examine every rose bush in the county and determine its hybrid ancestry. This would present impossible practical difficulties, apart from the enormous magnitude of the task. The alternatives are either to leave most of our roses unrecorded, or to take a wide view of the limits of a species. At the time of this present work going to press, the taxonomy and nomenclature of the British roses is being revised, with a view to devising a system more in line with modern taxonomic principles. Hitherto the majority of British rhodologists have followed the system devised by A.H. Wolley-Dod (1930-31), based on the work of Keller and Gams, with a few changes in nomenclature. Though not entirely satisfactory in the light of modern taxonomic principles, Wolley-Dod's system did succeed in bringing some sort of order out of chaos. It divides the more troublesome species into 'groups' which will accommodate most of the roses found in the field. Using these 'groups' as recording units, it is possible to record the roses on a tetrad basis, and produce significant distribution maps. This is the system used in the following account of **Rosa**. It should be regarded as an interim arrangement, which can be readily correlated with the new system when this is published. The author of this Leicestershire account does not see much significance in the varieties and forms described by Wolley-Dod. Indeed, Wolley-Dod himself said that the majority of British roses found in the field do not correspond exactly or even approximately to the descriptions of these varieties and forms. However, all named varieties and forms which have been found in the recent survey and which have been determined by either G. G. Graham or R. Melville are listed in the following accounts.

Named hybrids between species have only been recorded when the specimen has been determined by either Graham or Melville. The only exceptions to this are the hybrids between **R. arvensis** and **R. canina**, which we considered ourselves competent to identify. In naming hybrids of **Rosa** species it has been necessary to depart from the usual convention of putting the trivial names of the hybrid formula in alphabetical order. Hybrids of the unbalanced polyploid roses are matriclinal in their characters, and reciprocal pairs of hybrids can be very different. In the following accounts, the seed parent is named first in the hybrid formula. For example, **R. arvensis** × **canina**, with **R. arvensis** as the seed parent, will be found immediately after the account of **R. arvensis**; **R. canina** × **arvensis**, the reciprocal hybrid, will be found after **R. canina**.

In spite of the often great difference between these reciprocal pairs of hybrids, the rules of taxonomy only allow one hybrid name for both of them. In the following text, hybrid names, where available, are only given for the combination to which they were originally applied.

The specialist field work on which this account of **Rosa** is based was done mainly by H. Handley and A. L. Primavesi. Other field workers supplied some records and large numbers of specimens, both fresh and pressed, for determination. We acknowledge with gratitude the help given by the Rev. G. G. Graham, who spent some time with us in the field in three successive seasons, suggested the method of treatment of **Rosa** used in this Flora, and determined many specimens. We are also grateful to the late Dr. R. Melville, who determined many specimens, gave much useful advice, and criticised this account of Rosa.

A. L. Primavesi

R. arvensis Hudson   Field Rose
Map 26k; 507 tetrads.
Hedgerows and woodland. Frequent throughout the county; locally abundant in the well-wooded regions. The map shows that this species is absent from parts of the lower Wreake valley and the conurbation of Leicester.
 Constituent or residual: Eh, Ev, Er, Ef, Fd, Fm, Fs, Gr, Gp, Ap.
First record: Pulteney (1747).
Herb.: LSR★; LSR, LTR.
HG 218: locally abundant, generally distributed.

R. arvensis × **canina** = **R.** × **kosinsciana** Besser
Map 26l; 90 tetrads.
Occasional throughout the county. A careful search would probably show it to be more frequent than at present appears. The fruits are usually abortive.
 Intrusive: Eh, Ev, Er, Ef, Gr, Gp.
First record: 1976, J. Chandler in present work.
Herb.: LSR★.
HG: not recorded.

R. arvensis × **afzeliana**
No map; 1 tetrad.
Woodhouse Eaves, ditch, single bush, HH 1982.
 Intrusive: Wd.
First record: 1982, H. Handley in present work.
HG: not recorded.

R. arvensis × **sherardii**
No map; 1 tetrad.
Lount, scrub woodland, single bush, PAE 1986. Plant completely sterile. (Site since destroyed).
 Intrusive: Fs.
First Record: 1986, P.A. Evans in present work.
Herb.: LSR★; LCR.
HG: Not recorded.

R. multiflora Thunb.
No map.
Often cultivated but rarely becoming established outside gardens. Its presence in the county does however occasionally result in the formation of hybrids with other species, but in none of these cases has the other parent been identified with certainty.

R. pimpinellifolia L.   Burnet Rose
No record from the recent survey.
First record: Pulteney (1747).
Herb.: LSR.
HG 203 as *R. spinosissima*: rare, very sparsely distributed; 11 localities cited.

R. pimpinellifolia × **mollis** = **R.** × **sabinii** Woods
Map 27a; 1 tetrad.
Ulverscroft, scrub woodland, rare, PHG 1977. Plant since destroyed.
    Constituent or intrusive: Fs.
First record: Bloxam (1846a).
Herb.: LSR★; LSR.
HG 203: occasional, restricted in range; 2 localities cited, not including the above.

R. rugosa Thunb.   Japanese Rose
Map 27b; 7 tetrads.
Often cultivated in gardens, and occasionally escaping and becoming established, probably from bird-sown seed.
    Intrusive or garden escape: Eh, Rw, Fs.
First record: 1973, E. K. and J. M. Horwood in present work.
Herb.: LSR★.
HG: not recorded.

R. stylosa Desv.
No recent record.
First record: Preston (1900).
HG 218: rare, restricted in range; 2 localities cited.

R. canina L.   Dog Rose
Recent records are described under their respective groups.
First record: Pulteney (1747)
Herb.: LSR★; LSR, LTR.
HG 207 ff.: locally abundant, generally distributed.

Group *LUTETIANAE*
Map 27c; 165 tetrads.
Hedgerows, railway verges and roadsides, and open places in woodland. Occasional throughout the county, seldom in any quantity where it does occur, and often as isolated bushes.
    Constituent or intrusive: Eh, Er, Ev, Ef, Fd, Fs, Gr, Rg, Rq, Sw.
The following varieties have been recorded in the recent survey: var. **lutetiana** (Lem.) Baker; var. **sphaerica** (Gren.) Dum.; var. **flexibilis** (Déségl.) Rouy; var. **senticosa** (Ach.) Baker; var. **senticosa** f. **oxyphylla** (Rip.) W.-Dod; var. **senticosa** f. **mucronulata** (Déségl.) W.-Dod.
First record (for var. **lutetiana**): Mott et al. (1886).
Herb.: LSR★; LSR.
HG 208-213 passim.

Group *TRANSITORIAE*
Map 27d; 581 tetrads.
Hedgerows, roadside and railway verges, open places in woodland, and waste ground such as abandoned quarries and spoil heaps which are becoming invaded by scrub. Frequent and locally abundant. By far the commonest group of roses in the county.
    Constituent or intrusive: Eh, Ev, Er, Ew, Ef, Fd, Fs, Gp, Gr, Rs, Rq, Rg, Rw.

The following varieties have been recorded in the recent survey: var. **spuria** (Pug.) W.-Dod; var. **spuria** f. **syntrichostyla** (Rip.) Rouy; var. **rhynchocarpa** (Rip.) Rouy; var. **globularis** (Franch.) Dum.
First record (for var. **spuria**): c.1830, in Horwood and Gainsborough (1933).
Herb.: LSR★; LSR.
HG 211-212.

Group *DUMALES*
Map 27e; 363 tetrads.
Hedgerows, roadside and railway verges, open places in woodland and scrub. Frequent throughout the county.
    Constituent or intrusive: Eh, Ev, Er, Ew, Ef, Fd, Fs, Fw, Gr, Gp, Rg, Rq.
The following varieties have been recorded in the recent survey: var. **dumalis** (Bechst.) Dum.; var. **dumalis** f. **viridicata** (Pug.) Rouy; var. **dumalis** f. **cladoleia** Rip.; var. **medioxima** (Déségl.) Rouy; var. **biserrata** (Mer.) Baker; var. **biserrata** f. **sphaeroidea** (Rip.) W.-Dod; var **biserrata** f. **eriostyla** (Rip.) W.-Dod; var. **fraxinoides** H.Br. f. **recognita** Rouy; var. **sylvularum** (Rip.) Rouy; var. **sylvularum** f. **adscita** (Déségl.) Rouy.
First record (for var. **dumalis**): Kirby (1850).
Herb.: LSR★; LSR.
HG 210-213 passim.

Group *ANDEGAVENSES*
Map 27f; 13 tetrads.
This group, with glandular pedicels, is believed to consist of hybrids of **R. canina** with other species. Hence many of the records, notably of **R. canina** × **arvensis**, which would otherwise have been included here, are included in the accounts of hybrids. In fact the inclusion of a specimen in this group is really an admission of failure to identify it for what it actually is rather than a successful determination.
The following varieties have been recorded in the recent survey: var. **andegavensis** (Bast.) Desp., Chilcote, stream side, rare, HH 1977; var. **verticillacantha** (Mer.) Baker, Kirby Muxloe, railway verge, rare, EH 1975; Peatling Parva, hedgerow, rare, HH 1978; Sproxton, hedgerow, rare, ALP 1978; var. **aspernata** (Déségl.) Briggs f. **globosa** W.-Dod, Rothley, hedgerow, rare, HH 1977. Also of this group, not determined as to variety by a specialist: Higham on the Hill, dismantled railway verge, rare, HH and ALP 1979.
    Constituent or intrusive: Eh, Er, Ew.
First record: Mott et al. (1886).
Herb.: LSR★; LSR.
HG 211-213.

Group *SCABRATAE*
This group, with subfoliar glands and glandular pedicels is also believed to consist of hybrids of **R. canina** with other species. The few recent records which would have gone into this group have been positively identified as hybrids.
First record: B.E.C. Rep. 1907.
Herb.: LSR.
HG 212: var. *blondaeana* (Rip.) Rouy; 4 localities cited.

R. canina × **arvensis**
Map 27g; 30 tetrads.
Hedgerows and scrub. Occasional throughout the county where the parents occur together.

Intrusive: Eh, Ev, Er, Ew, Rq, Rg.
First record: 1975, E. Hesselgreaves in present work (if records for Group *ANDEGAVENSES* are excluded).
Herb.: LSR★.
HG: not recorded.

R. canina × obtusifolia
No map; 3 tetrads.
Congerstone, hedgerow, occasional, HH 1976; Countesthorpe, hedgerow, rare, EH 1976; Long Clawson, Brock Hill, hedgerow, rare, ALP 1977.
Intrusive: Eh.
First record: 1976, H. Handley in present work.
Herb.: LSR★.
HG: not recorded.

R. canina × tomentosa = R. × curvispina W. -Dod
No map; 3 tetrads.
Barton in the Beans, hedgerow, occasional, HH 1976; Nailstone, streamside, rare, HH 1976; Scalford, near Landyke Lane, hedgerow, single bush, HH 1979.
Intrusive: Eh, Ew.
First record: 1976, H. Handley in present work.
Herb.: LSR★.
HG: not recorded.

R. canina × rubiginosa = R. × latens W. -Dod
No map; 1 tetrad.
Ibstock, hedgerow, single bush, HH 1977.
Intrusive: Eh.
First record: 1977, H. Handley in present work.
Herb.: LSR★.
HG: not recorded.

R. dumetorum Thuill.
Recent records are described under their respective groups.
First record: Kirby (1850).
Herb.: LSR★.
HG 215-217: occasional, restricted in range.

Group *PUBESCENTES*
Map 27h; 205 tetrads.
Hedgerows and woodland; seems to be tolerant of shade and is especially characteristic of the hedgerows on the borders of woodland. Occasional throughout the county.
Constituent, residual or intrusive: Eh, Ev, Er, Ef, Fd, Fm, Fs, Gr, Hg, Rw.
The following varieties have been recorded in the recent survey: var. **typica** W.-Dod f. **urbica** (Lem.) W.-Dod; var. **typica** f. **semiglabra** (Rip.) W.-Dod; var. **ramealis** (Pug.) W.-Dod; var. **gabrielis** (F.Gér.) R. Kell.; var. **calophylla** Rouy; var. **platyphylla** (Rau) W.-Dod; var. **sphaerocarpa** (Pug.) W.-Dod; var. **hemitricha** (Rip.) W.-Dod; var. **erecta** W.-Dod.
First record (for var. **typica**): Kirby (1850).
Herb.: LSR★.
HG 215-217 passim.

Group *DESEGLISEI*
Map 27i; 2 tetrads.
This group most probably consists of hybrids with other species. The following varieties have been recorded in the recent survey: var. **deseglisei** (Bor.) Christ, Newbold Verdon, hedgerow, occasional, HH 1976; var. **incerta** (Déségl.) W.-Dod, Breedon on the Hill, scrub woodland, occasional, HH 1976.
Constituent or intrusive: Eh, Fs.
First record: W.B.E.C. Rep. 1907.
Herb.: LSR★.
HG 216-217.

Group *MERCICAE*
Map 27j; 1 tetrad.
This group most probably consists of hybrids with other species. The following variety has been recorded in the recent survey: var. **fanasensis** R. Kell., Ratby, hedgerow, rare, EH 1975.
Constituent or intrusive: Eh.
First record: 1975, E. Hesselgreaves in present work.
Herb.: LSR★; LSR.
HG: not recorded.

R. dumetorum × tomentosa = R. × aberrans W. -Dod
No map; 1 tetrad.
Newton Burgoland, hedgerow, rare, HH 1976 (det. R. Melville).
Intrusive: Eh.
First record: 1976, H. Handley in present work.
Herb.: LSR★.
HG: not recorded.

R. dumetorum × gallica = R. × collina Jacq.
No recent record.
First record: Mott et al. (1886).
HG 217: 3 localities cited.

R. afzeliana Fr.
Recent records are described under their respective groups.
First record: Preston (1900).
Herb.: LSR★; LSR.
HG 217 as *R. glauca*: local, sparsely distributed; 4 localities cited.

Group *REUTERIANAE*
Map 27k; 135 tetrads.
Hedgerows, roadsides and railway verges. Occasional in scattered localities throughout the county; usually few in numbers or as isolated bushes. In Leicestershire this species is near the southern limit of its range.
Constituent or intrusive: Eh, Ev, Er, Ew, Ef, Fs, Gp, Gr, Rq.
The following varieties have been recorded in the recent survey: var. **reuteri** (God.) Cott.; var. **reuteri** f. **transiens** (Kern.) Gren.; var. **glaucophylla** (Winch) W.-Dod; var. **glaucophylla** f. **jurassica** Rouy; var. **glaucophylla** f. **adenophora** Gren.
First record (for var. **glaucophylla**): Preston (1900).
Herb.: LSR★; LSR.
HG 217: 3 localities cited.

Group *SUBCANINAE*
Map 27l; 349 tetrads.
Hedgerows, roadsides and railway verges. Occasional in the west of the county; locally frequent in the south and east.
Constituent or intrusive: Eh, Er, Ev, Ef, Fs, Gr, Hg, Rq, Rg.
The following variety has been recorded in the recent survey: var. **denticulata** R. Kell. f. **subcomplicata**

Hayek.
First record: Horwood and Gainsborough (1933).
Herb.: LSR★; LSR.
HG 217: 1 locality cited.

R. afzeliana × arvensis.
No map; 1 tetrad.
Newtown Linford, hedgerow, single bush, HH 1982.
Intrusive: EH.
First record: 1982, H. Handley in present work.
HG: not recorded.

R. coriifolia Fr.
Recent records are described under their respective groups.
First record: Kirby (1850).
Herb.: LSR★; LSR.
HG 217 as *R. caesia*: rare, restricted in range; 11 localities cited.

Group *TYPICAE*
Map 28a; 80 tetrads.
Hedgerows, railway verges and roadsides. Occasional in scattered localities throughout the county, usually as single isolated bushes and rarely in any quantity. This species also is near the southern limit of its range in this county.
Constituent or intrusive: Eh, Er, Ev, Ef, Gr, Gp, Rs.
The following varieties have been recorded in the recent survey: var. **typica** Christ; var. **subglabra** R. Kell.; var. **subglabra** f. **subovata** Rouy; var. **watsonii** (Baker) W.-Dod; var. **celerata** (Baker) W.-Dod.
First record (for var. **typica**): Kirby (1850).
Herb.: LSR★; LSR.
HG 218: 8 localities cited.

Group *SUBCOLLINAE*
Map 28b; 287 tetrads.
Hedgerows, roadside and railway verges, and open places in woodland. Occasional throughout the county; slightly less frequent than **R. afzeliana** Group *SUBCANINAE*.
Constituent or intrusive: Eh, Er, Ev, Ew, Fs, Fd, Gp, Gr, Hg.
The following varieties have been recorded in the recent survey: var. **subcollina** Christ; var. **subcollina** f. **dimorphocarpa** (Borb. & Br.) R. Kell.; var. **subcoriifolia** (Barclay) W.-Dod; var. **lintonii** Scheutz; var. **pruinosa** (Baker) W.-Dod; var. **caesia** (Sm.) W.-Dod.
First record (for var. **subcollina**): Horwood and Gainsborough (1933).
Herb.: LSR★; LSR.
HG 218: 1 locality cited.

R. obtusifolia Desv.
Map 28c; 177 tetrads.
Mainly in hedgerows; rarely in open habitats. Except for var. **borreri** it is usually low-growing and somewhat inconspicuous, and may be easily overlooked. Locally frequent in the west of the county, occasional in the east.
Constituent or intrusive: Eh, Er, Ev, Ew, Ef, Fs, Gp, Gr.
The following varieties have been recorded in the recent survey: var. **typica** W.-Dod; var. **tomentella** (Lem.) Baker; var. **tomentella** f. **canescens** (Baker) W.-Dod; var. **borreri** (Woods) W.-Dod; var. **sclerophylla** (Scheutz) W.-Dod; var. **decipiens** Dum.

First record: 1882, F. T. Mott in herb. K.
Herb.: LSR★; K, LSR.
HG 213: rather rare, restricted in range; 38 localities cited.

R. obtusifolia × arvensis
No map; 2 tetrads.
South-west of Bagworth, hedgerow, single bush, HH 1979; Swithland, hedgerow, single bush, HH 1981.
Intrusive: Eh.
First record: 1979, H. Handley in present work.
Herb.: K.
HG: not recorded.

R. obtusifolia × canina = R. × concinnoides W.-Dod
No map; 3 tetrads.
Oakthorpe, hedgerow, rare, SHB 1976; Ullesthorpe, hedgerow, rare, HH 1976; Ratby, hedgerow, rare, HH 1978.
Intrusive: Eh.
First record: 1976, S. H. Bishop in present work.
Herb.: LSR★.
HG: not recorded.

R. obtusifolia × tomentosa
No map; 2 tetrads.
Sheepy Parva, hedgerow, rare, HH 1978; Thurcaston, rough grassland, single bush, HH 1981.
Intrusive: Eh, Gr.
First record: 1978, H. Handley in present work.
Herb.: LSR★; K.
HG: not recorded.

R. tomentosa Sm.
Recent records are described under their respective groups.
First record: Watson (1837).
Herb.: LSR★; LSR.
HG 204: rather local, restricted in range; 36 localities cited.

Group *TYPICAE*
Map 28d; 33 tetrads.
Mainly in hedgerows. Occasional in the west of the county; rare in the east.
Constituent or intrusive: Eh, Ef, Er, Ev, Ew, Gr.
The following varieties have been recorded in the recent survey: var. **typica** W.-Dod; var. **pseudocuspidata** (Crép.) Rouy; var. **pseudocuspidata** f. **cuspidatoides** (Crép.) W.-Dod; var. **dumosa** (Pug.) Rouy; var. **dimorpha** (Bess.) Déségl.
First record: Watson (1837).
Herb.: LSR★; LSR.
HG 204–206 passim.

Group *SCABRIUSCULAE*
This group is believed to consist of hybrids of **R. tomentosa** with other species. If there are any recent records which would have gone into this group, they have been positively identified as hybrids.
First record (for var. **scabriuscula**): Kirby (1850).
Herb.: LSR.
HG 205: 3 localities cited for var. **scabriuscula** Sm. and 3 localities for var. **sylvestris** (Lindl.) Woods.

R. tomentosa × canina
No map; 2 tetrads.
Ratby, hedgerow, rare, EH 1975; Ratby, hedgerow,

rare, HH 1976; Appleby Parva, hedgerow, rare, HH 1976.
 Intrusive: Eh.
First record: 1975, E. Hesselgreaves in present work.
Herb.: LSR★.
HG: not recorded.

R. tomentosa × obtusifolia
No map; 1 tetrad.
Thornton, hedgerow, single bush, ALP 1976.
 Intrusive: Eh.
First record: 1976, A. L. Primavesi in present work.
Herb.: LSR★; K.
HG: not recorded.

R. sherardii Davies
Map 28e; 18 tetrads.
Hedgerows, or as isolated bushes in grassland. Rare, in scattered localities in the west and north-east of the county.
 Constituent or intrusive: Eh, Er, Gp, Gr, Fs.
The following varieties have been recorded in the recent survey: var. **typica** W.-Dod; var. **typica** f. **pseudomollis** (Baker) W.-Dod; var. **typica** f. **uncinata** (Lees) W.-Dodd; var **omissa** (Déségl.)W.-Dod f. **resinosoides** (Crép.) W.-Dod; var. **suberecta** (Ley) W.-Dod.
First record (for var. **typica**): Preston (1900).
Herb.: LSR★, LSR.
HG 204: rare restricted in range; 9 localities cited.

R. sherardii × arvensis
No map; 1 tetrad.
Hose Gorse, roadside hedgerow, rare, HH 1978 (observed and det. in the field by G. G. Graham; conf. R. Melville).
 Intrusive: Eh.
First record: 1978, H. Handley in present work.
Herb.: LSR★; K.
HG: not recorded.

R. sherardii × rubiginosa = R. × burdonii W.-Dod
No map; 1 tetrad.
Ratby, hedgerow, occasional, HH 1976.
 Intrusive: Eh.
First record: 1976, H. Handley in present work.
Herb.: LSR★.
HG: not recorded.

R. mollis Sm.
Map 28f; 1 tetrad.
Ulverscroft, scrub woodland, rare, PHG 1977.
 Constituent or intrusive: Fs.
First record: Crabbe in Nichols (1795).
Herb.: LSR★.
HG 203: rare, restricted in range; 14 localities cited.

R. rubiginosa L.
No map; 1 tetrad.
Whetstone, ballast of dismantled railway station sidings, occasional, EH 1980. A garden variety, probably bird-sown.
 Garden escape: Er.
First record: Pulteney (1747).
Herb.: LSR.
HG 206 as *R. eglanteria*: local, locally distributed; 20 localities cited.

R. rubiginosa × sherardii
No map; 1 tetrad.
Cosby, hedgerow, occasional, HH 1978.
 Intrusive: Eh.
First record: 1978, H. Handley in present work.
Herb.: K.

R. micrantha Borrer ex Sm.
No recent record.
First record: Bloxam (1846a).
Herb.: CGE, LSR, MANCH.
HG 206: local sparsely distributed; 19 localities cited.

R. agrestis Savi
No recent record.
First record: Kirby (1850).
HG 207: rare, restricted in range; 30 localities cited.

Agrimonia L.
A. eupatoria L.   Agrimony
Map 28g; 363 tetrads.
Grassland, roadside and railway verges. Frequent throughout the county, but absent from parts of Charnwood Forest.
 Constituent or intrusive: Gp, Gr, Ev, Er, Eh, Ef, Ew, Ea, Rg, Rq.
First record: Pulteney (1747).
Herb.: LSR★; DPM, LSR, LTR, MANCH, NMW.
HG 199: locally abundant, and widely distributed.

A. procera Wallr.   Fragrant Agrimony
A. odorata auct. non Miller
Map 28h; 1 tetrad.
Newtown Linford, grass verge where adjacent to Blakeshay Wood, occasional, PHG 1971.
 Constituent or residual: Ev.
First record: Coleman (1852).
Herb.: LSR, LTR. MANCH.
HG 200: very rare restricted in range; 12 localities cited, including the above.

Sanguisorba L.
S. officinalis L.   Great Burnet
Map 28i; 353 tetrads.
Marshes, meadows and grassland where the ground is always damp; roadside and railway verges, and gravel pits. Frequent and locally abundant throughout the county, though the type of grassland in which it is found is decreasing.
 Constituent, residual or intrusive: M, Gp, Gr, Gl, Hg, Wd, Ew, Ev, Er, Eh, Ea, Ef, Fs, Fw, Rg, Rs, Rw, Rd, Rt, Ap, Ac, As, Ao.
First record: Pulteney (1747).
Herb.: LSR★; CGE, DPM, LANC, LSR, LTR, MANCH, NMW.
HG 201: frequent, generally distributed; 115 localities cited.

S. minor Scop.
Subsp. **minor**   Salad Burnet
Poterium sanguisorba L.
Map 28j; 121 tetrads.
Calcareous grassland, railway and roadside verges, and limestone quarries. Locally frequent in the east of the county, occasional in the west.
 Constituent, residual or intrusive: Gp, Gr, Gl, Hg, Er, Ev, Ea, Ef, Rq, Rg, Rs, Rw, Rt, Sr.

First record: Pulteney (1747).
Herb.: LSR★; LANC, LSR, LTR, MANCH, NMW.
HG 201: frequent, generally distributed; 115 localities cited.

Subsp. **muricata** Briq.   Fodder Burnet
Poterium polygamum Waldst. & Kit.
Map 28k; 6 tetrads.
Mountsorrel, grass verge in granite quarry, locally frequent, PHG 1968; Swithland, railway embankment, locally frequent, PHG 1968; Husbands Bosworth, embankment of dismantled railway, occasional, EKH and JMH 1969; South Wigston, waste ground by railway sidings, rare, EH 1974; Shackerstone, railway cutting, rare, SHB 1976; Loughborough Big Meadow, locally frequent, PHG 1981.
    Intrusive: Er, Gr, Rw.
First record: W.B.E.C. Rep. 1901.
Herb.: LSR★; CGE, DPM, LSR, LTR, MANCH, NMW.
HG 201: rare, restricted in range; 7 localities cited.

Geum L.
G. rivale L.   Water Avens
Map 28l; 15 tetrads.
Occasional in the older woodlands, old grassland and marshes, and by water. In scattered localities throughout the county, but avoiding the more acid soils of Charnwood Forest and the north-west.
    Constituent or residual: Fd, Ef, Ew, Er, M, Gp, Wd.
First record: Power (1805).
Herb.: LSR★; CGE, DPM, LANC, LSR, LTR, MANCH, NMW.
HG 191: rather local, and sparsely distributed; 23 localities cited.

G. rivale × urbanum = G. × intermedium Ehrh.
Map 29a; 4 tetrads.
Owston Wood, EKH 1963; Wymondham Rough Nature Reserve in several places including meadow, Day's Plantation and former canal bank, also on adjacent railway bank, KGM 1976; Aston Firs, woodland ride, rare, SHB 1976; Launde, Big Wood, ride, rare, IME 1982.
    Constituent or intrusive: Fd, Gp, Ew, Ef, Er.
First record: 1880, F. T. Mott in herb. MANCH.
Herb.: LSR★; LANC, LSR, LTR, MANCH, NMW.
HG 191: local, sparsely distributed; 3 localities cited.

G. urbanum L.   Wood Avens, Herb Bennet
Map 29b; 561 tetrads.
Woodland, hedgerows and shady roadside and railway verges. Frequent and locally abundant throughout the county.
    Constituent: Fd, Fc, Fm, Fs, Fw, Ef.
    Residual or intrusive: Eh, Ev, Er, Ew, Wd, Rq, Rw, Ap, Ag.
First record: Pulteney (1747).
Herb.: LSR★; DPM, LANC, LSR, LTR, MANCH, NMW.
HG 190: locally abundant, generally distributed.

Potentilla L.
P. palustris (L.) Scop.   Marsh Cinquefoil
No record since HG.
First record: Pulteney in Nichols (1795).
Herb.: LSR, MANCH, NMW.
HG 197: rare, very restricted in range, Moira seems to be now the only locality for this plant in the county; 3 localities cited.

P. anserina L.   Silverweed
Map 29c; 592 tetrads.
Roadside and railway verges, marshes, waste places, grassland, particularly where the ground is wet. Locally abundant throughout the county.
    Constituent or intrusive: M, Gp, Gr, Gl, Hg, Ew, Ev, Er, Ea, Ef, Fd, Rw, Rt, Rd, Rf, Rq, Rg, Rs, Ap.
First record: Pulteney (1747).
Herb.: LSR★; DPM, LANC, LSR, LTR, MANCH, NMW.
HG 196: locally abundant and generally distributed.

P. argentea L.   Hoary Cinquefoil
Map 29d; 3 tetrads.
Groby, wall top near the Pool, rare, EH 1968; Groby, old quarry, rare, EH 1970; Croft, siliceous grassland near diorite outcrops, locally frequent, SHB 1973.
    Constituent or intrusive: Hg, Rq, Sw.
First record: Pulteney (1756).
Herb.: LSR★; CGE, LSR, LTR, MANCH, NMW.
HG 196: rare, restricted in range to Charnwood Forest and District IX [i.e. Croft region]; 16 localities cited, including Groby and Croft.

P. recta L.   Sulphur Cinquefoil
No map; 3 tetrads.
Leicester, mound on south end of Knighton Tunnel, occasional, HB 1969; Muston, brook side, rare, HL 1974; Acresford, roadside, rare, HH 1977.
    Casual or garden escape: Ev, Ew, Rw.
First record: Mott et al. (1886).
Herb.: LSR★; LSR.
HG 197: rare, restricted to one locality [wall of Leicester Abbey].

P. erecta (L.) Rauschel   Tormentil
Map 29e; 196 tetrads.
Heathland, grassland, woodland rides and marshes on acid soil, of which it is a clear indicator. Locally abundant on the acid soils of the west of the county, and on the marlstone in the east. Occasional elsewhere when there is suitable soil.
    Constituent or intrusive: Hg, Hd, M, Gp, Gr, Ef, Ev, Er, Ew, Ea, Rw, Rg, Rq, Ap, Ag.
First record: Pulteney (1747).
Herb.: LSR★; DPM, LSR, LTR, MANCH, NMW.
HG 193: locally abundant, generally distributed; 145 localities cited.

P. erecta × reptans = P. × italica Lehm.
Map 29f; 1 tetrad.
Launde, Park Wood, grassy banks by rides, KGM 1971.
    Constituent or intrusive: Ef.
First record: 1954, D. P. Murray in herb. DPM.
Herb.: LSR★; DPM, NMW.
HG: not recorded.

P. anglica Laicharding   Trailing Tormentil
Map 29g; 26 tetrads.
Grassland, roadside and railway verges. Occasional in the west of the county; absent from the east. There are 40 records from various localities in the west. It is very likely that most of these are in fact the hybrid between this species and **P. reptans.** Only one specimen has been confirmed: Whetstone, dismantled railway, occasional, EH 1975 (det. S. M. Walters).
    Constituent or intrusive: Gp, Gr, Hg, Hd, Er, Ev, Eh, Ef, Ea, M, Rt, Rw, Rs.

First records: for **P. anglica**, Pulteney (1756); for the hybrid, W.B.E.C. Rep. 1898.
Herb.: LSR★; LSR, LTR.
HG 195 as *P. procumbens*: local, rather restricted in range; 31 localities cited; in addition 2 localities are cited for **P. anglica** × **reptans**.

P. reptans L.   Creeping Cinquefoil
Map 29h; 599 tetrads.
Roadsides and railway verges, waste places, grassland, preferring drier habitats than **P. anserina**. Locally abundant throughout the county.
    Constituent or intrusive: Ev, Er, Ew, Ef, Ea, Gr, Gp, Hg, Rw, Rt, Rd, Rs, Ap, Ag, Ao, Ac.
First record: Pulteney (1747).
Herb.: LSR★; DPM, LANC, LSR, LTR, MANCH, NMW.
HG 195: locally abundant, generally distributed.

P. sterilis (L.) Garcke   Barren Strawberry
Map 29i; 100 tetrads.
Grassland, heath, hedgerows and woodland, especially woodland rides. Occasional in the east of the county, frequent in the west. Rare or absent on the heavy clay soils.
    Constituent, residual or intrusive: Gp, Gr, Hg, Ef, Eh, Er, Fd, Fs, Rq, Rs, Rt, Sw.
First record: Pulteney (1747).
Herb.: LSR★; CGE, LSR, LTR, MANCH, NMW, UPP.
HG 193: locally abundant, and generally distributed.

Fragaria L.

F. vesca L.   Wild Strawberry
Map 29j; 118 tetrads.
Occasional throughout the county in woodland and hedgerows, on railway and roadside verges, and in quarries. Seems to avoid the heavy clay soils.
    Constituent: Fd, Fm, Fs, Ef.
    Residual or intrusive: Eh, Er, Ev, Gr, Rq, Rw, Ap, Ag, Sw.
First record: Pulteney (1747).
Herb.: LSR★; CGE, DPM, LSR, LTR, MANCH, NMW.
HG 192: locally abundant, widely distributed.

F. × ananassa Duchesne   Garden Strawberry
Map 29k; 55 tetrads.
Frequent on railway verges throughout the county, in which habitat it appears to be thoroughly naturalised. Occasional elsewhere as a garden escape. 85% of the records received are from railway verges.
    Intrusive or garden escape: Er, Ev, Rw, Rs, Ap, Fd, Sw.
First record: Horwood and Gainsborough (1933).
Herb.: LSR★; LTR, NMW.
HG 193: 5 localities cited.

Alchemilla L.

A. xanthochlora Rothm.   Lady's-mantle
Map 29l; 12 tetrads.
Swithland, pasture, locally frequent, PHG 1966; pasture north-east of Tugby Wood, occasional, PGH 1967; Osgathorpe, old damp grassland, occasional, PAC 1968; Bardon, damp meadow next to River Sence, frequent, PAC 1969; Bardon, grazed marshy meadow south of the village, frequent, PAC 1969; Ulverscroft, marshy old grassland, frequent, PAC 1969 (site now probably drained); Stanton under Bardon, waste ground, occasional, PAC 1969; Groby, wet old pasture, occasional, EH 1970; Newtown Linford, marshy meadow, rare, EH 1971; Griffydam, old grassland, occasional, PAC 1972; Battleflat, pasture, frequent, PAC 1972; Ashby Old Parks, pasture, frequent, PAC 1976.
    Constituent or residual: GP, M, Rw.
First record: Jackson (1904b).
HG 198 as *A. vulgaris* var. *pratensis*: occasional, sparsely distributed; 2 localities cited, including Groby.

A. filicaulis Buser   Lady's-mantle
subsp. **vestita** (Buser) M. E. Bradshaw
A. vestita (Buser) Raunk.
Map 30a; 101 tetrads.
Grassland, seemingly indifferent to the amount of water in the soil, but avoiding the more basic soils. Frequent in the west of the county; occasional in the east.
    Constituent: Gp, Gr, M, Hg, Ew, Ef, Fd.
    Residual or intrusive: Gl, Ev, Er, Ea, Rt, Ap.
First record: difficult to determine because of taxonomic changes, but Horwood gives Vice (1901) for *A. alpestris* var. *minor*.
Herb.: LSR★; BM, CGE, DPM, LANC, LSR, LTR, NMW.
HG 198 as *A. alpestris* var. *minor*: 61 localities cited.

A. glabra Neygenf.   Lady's-mantle
No recent record.
First record: 1900, A. B. Jackson in herb. BM (det. S. M. Walters 1953).
Herb.: BM.
HG 198 as *A. alpestris*: rare, very restricted in range; gives no locality.

Aphanes L.

A. arvensis L.   Parsley-piert
Map 30b; 94 tetrads.
Arable land, and less frequently grassland. Occasional throughout the county; more frequent in the west than in the east.
    Intrusive: Ac, Ao, Ag, Ea, Er, Ev, Ew, Gp, Gr, Hd, Rg, Rq, Rw, Rt, Rd.
First record: Pulteney (1747).
Herb.: LSR★; BM, LTR, MANCH.
HG 197 as *Alchemilla arvensis*: locally abundant, generally distributed.

A. microcarpa (Boiss. & Reuter) Rothm.
                                  Slender Parsley-piert
Map 30c; 7 tetrads.
Arable land and heath grassland, requiring more acid soils than the preceding species. This species was treated as critical in the survey, and although it was recorded in numerous localities, only those records which were confirmed by the referee have been mapped. It is certainly less frequent than **A. arvensis**, and is confined to the west of the county.
    Constituent or intrusive: Hg, Hd, Ac, Ap, Rq.
First record: 1884, F. T. Mott in herb. MANCH (det. S. M. Walters 1954).
Herb.: LSR★; LTR, MANCH.
HG: not recorded.

Pyrus L.

P. communis L.   Wild Pear
Map 30d; 13 tetrads.
Hedgerows and scrub woodland, usually near the margin of the latter. Rare in scattered localities throughout the county, usually with only a single specimen at each locality. Some of the trees are spiny and some not, and it is of course possible that some of the records represent escapes from cultivation.
Constituent or intrusive: Eh, Fs, Er.
First record: Pulteney (1747).
Herb.: CGE, LTR.
HG 221: rare, restricted in range; 11 localities cited.

Malus Miller

M. sylvestris Miller   Crab Apple
Map 30e; 494 tetrads.
Frequent throughout the county in hedgerows and woodland.
Constituent: Eh, Ef, Fd, Fm, Fs, Fw.
Residual or intrusive: Ev, Er, Ew, Rw, Rq.
First record: Pulteney (1747).
Herb.: LSR★; DPM, LANC, LSR, LTR, MANCH, NMW.
HG 222 as *Pyrus malus*: frequent, generally distributed.

Sorbus L.

S. aucuparia L.   Rowan
Map 30f; 167 tetrads.
Woodland, hedgerows, shrubberies, public and private parks. A characteristic and constituent feature of the vegetation of Charnwood Forest. Elsewhere in the county it is often planted and seeds readily, so that its status in a particular locality may be difficult to determine. Frequent in the west of the county; occasional in the east.
Constituent, residual, intrusive or planted: Fd, Fm, Fs, Fw, Ef, Eh, Ev, Er, Ew, Hd, Hg, Gr, Ap, Rw, Rq, Rs, Rg, Sr, Sw.
First record: Pulteney (1749).
Herb.: LSR★; DPM, LSR, LTR, MANCH, NMW.
HG 220 as *Pyrus aucuparia*: frequent, rather sparsely distributed; 71 localities cited.

S. torminalis (L.) Crantz   Wild Service-tree
Map 30g; 5 tetrads.
Woodhouse, Beaumanor Park, mixed woodland, single tree, PHG 1976; Whetstone, spinney, occasional, EH 1976; Great Dalby, hedgerow, rare, SHB 1977; Burrough on the Hill, hedgerow, rare, A. Johnson 1983; Sutton Cheney, hedgerow, rare, R. Lockwood 1985.
Constituent or intrusive: Fd, Fm, Eh.
First record: Coleman (1852).
Herb.: LSR★; LSR, LTR, MANCH, NMW.
HG 219 as *Pyrus torminalis*: rare, restricted in range; 3 localities cited, not including any of the above.

S. aria (L.) Crantz   Common Whitebeam
Map 30h; 13 tetrads.
Woodhouse Eaves, The Brand, mixed woodland, rare, PHG 1968; Mountsorrel, floor of disused Hawcliff Quarry, single tree, PHG 1968; Buddon Wood, in woodland regenerating after clear felling, single tree, PHG 1969; Wanlip Hall, parkland, occasional planted trees, PHG 1970; Benscliffe Wood, plantation, occasional, PHG 1970; Oadby, hedgerow, rare, IME 1971; Earl Shilton, plantation, rare, SHB 1972; Swannington, roadside verge, planted, occasional, PAC 1972; Croft, scrub woodland, locally frequent, SHB 1973; Eaton, roadside, planted, rare, SHB 1973; Edmondthorpe, roadside, planted, rare, KGM 1973; Stonesby Quarry, plantation, rare, KGM 1973; Hathern, roadside verge, planted, occasional, PHG 1976; King Lud's Entrenchments, scrub woodland, single tree, SHB and PHG 1978.
Constituent, intrusive or planted: Fd, Fm, Fs, Eh, Ev.
First record; Horwood and Gainsborough (1933).
Herb.: LSR★; DPM, LSR, NMW.
HG 220 as *Pyrus aria*: rare, restricted in range; 6 localities cited.

S. intermedia (Ehrh.) Pers.   Swedish Whitebeam
Map 30i; 6 tetrads.
Woodhouse Eaves, The Brand, old slate quarry, single seedling, PHG 1968; Quorndon, roadside verge, planted, occasional, PHG 1968; Buddon Wood, seedlings probably originating from the preceding, occasional, PHG 1968; Saxelby, roadside, single tree, ALP 1970; Congerstone, hedgerow, rare, PHG and MH 1971; Glen Parva, railway verge, single tree, EH 1975; Croxton Kerrial, plantation, planted, occasional, KGM 1976; Glen Parva, spoil heap of brick pit, single sapling, EH 1976.
Intrusive or planted: Ev, Er, Eh, Fd, Fm, Rq, Rs.
First record: Horwood (1911).
Herb.: LSR★; BM, LSR, LTR.
HG 220: rare; 1 locality cited, planted.

S. latifolia (Lam.) Pers.   Broad-leaved Whitebeam
Map 30j; 1 tetrad.
Long Clawson, cutting of dismantled railway at north end of Brock Hill tunnel, occasional, ALP 1970 (det. T. G. Tutin).
Intrusive: Er.
First record: 1970, A. L. Primavesi in present work.
HG: not recorded.

Crataegus L.

C. laevigata (Poiret) DC.   Midland Hawthorn
C. oxyacanthoides Thuill.
Map 30k; 280 tetrads.
Woodland, usually in fairly deep shade; also in hedgerows, especially those on parish boundaries or near to woodland. Locally frequent throughout the county.
Constituent: Fd, Fm, Fs.
Residual or intrusive: Eh, Ew, Er, Ev, Ap.
First record: Pulteney (1749).
Herb.: LSR★; LSR, LTR, MANCH, NMW.
HG 223: locally abundant, generally distributed; 59 localities cited.

C. laevigata × monogyna = C. × media Bechst.
No map.
Hedgerows and woodland, showing all gradations of form between the two species, the ones nearest to woodland often approximating most to **C. laevigata**, those in open habitats to **C. monogyna**. It is probable that hybrid forms are more frequent throughout the county than either of the pure parents, but few field workers recorded them.
First record: W.B.E.C. Rep. 1906.
Herb.: LSR.
HG 227: 4 localities cited.

C. monogyna Jacq.   Hawthorn
Map 30l; 606 tetrads.
Abundant throughout the county in hedgerows and woodland, but preferring more open habitats than **C. laevigata**. This is the first shrub to invade neglected ground. It is also often found as isolated or scattered specimens in old grassland.
    Constituent or intrusive: Eh, Ev, Er, Ea, Fd, Fm, Fs, Fw, Ap, Rq, Rw.
First record: Pulteney (1747).
Herb.: LSR★; DPM, LSR, LTR, MANCH, NMW.
HG 225: locally abundant, generally distributed.

Prunus L.
P. cerasifera Ehrh.   Cherry Plum
Map 31a; 17 tetrads.
Hedgerows and woodland. Occasional, in scattered localities throughout the county.
    Intrusive: Fd, Fm, Ef, Eh.
First record: B.E.C. Rep. 1925.
Herb.: LSR★; LSR, LTR.
HG 155: rare; 1 locality cited.

P. spinosa L.   Blackthorn, Sloe
Map 31b; 595 tetrads.
Hedgerows and woodland. Frequent and locally abundant throughout the county. Sometimes forms impenetrable thickets in neglected fox-coverts.
    Constituent: Fd, Fm, Fs, Fw, Ef, Eh.
    Residual or intrusive: Ev, Er, Rg.
First record: Pulteney (1747).
Herb.: LSR★; DPM, LSR, LTR, MANCH, NMW, UPP.
HG 152: locally abundant, generally distributed.

P. domestica L.   Wild Plum
Subsp. **domestica**
Map 31c; 52 tetrads.
Occasional in hedgerows and woodland throughout the county.
    Constituent or intrusive: Eh, Ev, Er, Ew, Ef, Fd, Fs, Gr, Gp, Rw.
First record: Kirby (1850).
Herb.: LSR★; LSR, LTR, MANCH, NMW.
HG 153: local, sparsely distributed; 30 localities cited.

Subsp. **insititia** (L.) C. K. Schneider
Map 31d; 65 tetrads.
In similar habitats to the preceding, and apparently slightly more frequent, though both subspecies are almost certainly under-recorded in parts of the county.
    Constituent or intrusive: Eh, Fd, Fm, Fs, Fw.
First record: Pulteney (1747).
Herb.: LSR★; LSR, LTR, MANCH, NMW.
HG 152: generally distributed.

P. avium L.   Wild Cherry, Gean
Map 31e; 123 tetrads.
Frequent in woodland throughout the county. Occasional in hedgerows.
    Constituent, residual or intrusive: Fd, Fm, Fs, Fw, Eh, Er, Ev, Ew, Gr, Gp, Rw, Rq, Ap.
First record: Crabbe *in* Nichols (1795).
Herb.: LSR★; DPM, LSR, LTR, MANCH, NMW.
HG 153: occasional, but generally distributed; 53 localities cited.

P. cerasus L.   Dwarf Cherry
Map 31f; 8 tetrads.
Smeeton Westerby, field hedgerow, occasional, EKH and JMH 1968; Botcheston, roadside hedgerow, occasional, OHB, HH and SM 1969; Markfield, roadside hedgerow, occasional, OHB, HH and SM 1969; Kirby Muxloe, hedgerow, rare, EH 1972; Barlestone, streamside hedgerow, locally frequent, HH 1975; Newtown Linford, hedgerow, locally frequent, EH 1975.
    Intrusive: Eh.
First record: Watson (1832).
Herb.: LSR★; LSR, LTR.
HG 154: very rare, restricted in range; 10 localities cited.

P. padus L.   Bird Cherry
Map 31g; 4 tetrads.
Woodhouse, wet woodland between Swithland Reservoir and the railway, rare, PHG 1968; Markfield, roadside hedgerow, locally frequent, HH 1970; Donington Park, woodland, rare, PAC 1971; Loddington Hall Park, rare, possibly planted, IME 1975.
    Probably intrusive: Fd, Fw, Eh, Ap.
First record: Crabbe *in* Nichols (1795).
Herb.: LSR★; LSR, LTR, NMW.
HG 155: rare, range very restricted; 11 localities cited.

P. lusitanica L.   Portugal Laurel
Map 31h; 9 tetrads.
Ulverscroft, remains of former parkland plantation, occasional, PAC 1969; Stapleford Park, hedgerow, presumably planted, occasional, KGM 1971; Gumley, plantation, occasional, EKH and JMH 1972; Husbands Bosworth Hall, ornamental plantation, occasional, EKH and JMH 1973; Shackerstone, plantation, rare, SHB 1976; Staunton Harold, Spring Wood, rare, SHB 1977; Little Dalby Hall, parkland, rare, SHB 1977; Belvoir, Terrace Hills, woodland, locally frequent, PAC 1977; Buddon Wood, rare, PHG 1979.
    Planted or intrusive: Ap, Fd, Fm, Eh.
First record: 1966, E. K. Horwood in herb. LTR.
Herb.: LTR.
HG: not recorded.

P. laurocerasus L.   Cherry Laurel
Map 31i; 38 tetrads.
Woodland and hedgerows, often planted and occasionally escaping or self-sown.
    Planted or intrusive: Fd, Fm, Fs, Ef, Eh, Er, Ap.
First record: 1957, E. K. Horwood in herb. LTR.
Herb.: LSR★; LANC, LTR.
HG: not recorded.

LEGUMINOSAE
Laburnum Fabr.
L. anagyroides Medicus   Laburnum
Map 31j; 19 tetrads.
Frequently planted as an ornamental tree, and occasionally appearing as isolated self-sown trees in hedgerows, woodland and waste places.
    Intrusive or planted: Eh, Ev, Er, Ew, Fd, Fm, Rw, Rq, Ap.
First record: Horwood and Gainsborough (1933).
Herb.: LTR.
HG 121: rare; 1 locality cited.

Cytisus L.

C. scoparius (L.) Link Broom
Sarothamnus scoparius (L.) Wimmer ex Koch
Map 31k; 124 tetrads.
Heathland, woodland, gravel pits and waste places, usually on sandy or acid soils. Locally frequent in the west of the county; occasional in the east.
  Constituent or residual: Hg, Hd, Gr, Sr, Fd, Fc, Fm, Fs, Ef, Eh.
  Residual or intrusive: Rq, Rg, Rs, Rw, Rr, Er, Ev, Ew.
First record: Pulteney (1747).
Herb.: LSR★; LANC, LSR, LTR, MANCH, NMW.
HG 121: local, but generally distributed; 57 localities cited.

Genista L.

G. tinctoria L. Dyer's Greenweed
Map 31l; 14 tetrads.
Blackbrook Reservoir, rough grassland on margin, locally frequent, PAC 1968; Old Dalby Wood, conifer plantation, rare, ALP 1968; Swithland Wood meadow, old grassland, rare, PHG 1968; Eastwell, grassland, occasional, JMS 1968; Long Clawson, dismantled railway, verge, near site of Clawson and Hose Station, occasional, ALP 1970; Groby, Lawn Hill, stony pasture, occasional, EH 1972; Elmesthorpe, hedgerow, rare, SHB 1972; Stathern, dismantled railway verge, rare, D.K. and H. Lucking 1972; Higham on the Hill, dismantled railway, verge, rare, SHB 1974; Stapleton, hedgerow, rare, SHB 1974; Long Whatton, rough grassland, rare, PAC 1974; Ratby, scrub and rough grassland, occasional, HH 1976; Staunton Harold, pasture, rare, SHB 1978.
  Constituent or intrusive: Gp, Gr, Eh, Er, Fm, Fc, Fs.
First record: Pulteney (1747).
Herb.: LSR★; DPM, LSR, LTR, NMW.
HG 117: local or rare, locally distributed; 67 localities cited.

G. anglica L. Petty Whin
Map 32a; 1 tetrad.
Charnwood Lodge Nature Reserve, **Molinia** heath, rare, PAC 1969. Formerly also at Twenty Acre, Burton on the Wolds, but not seen there since fire swept over the site in 1967.
  Constituent: Hw.
First record: Pulteney (1747).
Herb.: LSR, LTR, MANCH, NMW.
HG 117: very rare, confined to a few uncultivated areas on peaty soil; 14 localities cited.

Ulex L.

U. europaeus L. Gorse, Furze
Map 32b; 301 tetrads.
Heathland, woods and hedgerows, roadside and railway verges; calcifuge. Frequent and locally abundant in those parts of the county where there are suitable soils; occasional elsewhere.
  Constituent: Hg, Hd, Hw, Gr, Gp, Fd, Fc, Fm, Fs, Ef, Eh.
  Residual or intrusive: Ev, Er, Rw, Rq, Rg, Rs, Ap.
First record: Pulteney (1747).
Herb.: LSR★; DPM, LSR, LTR, MANCH, NMW.
HG 118: frequent, generally distributed; 140 localities cited.

U. gallii Planchon Western Gorse
Map 32c; 21 tetrads.
Locally frequent in heathland and among rocks in Charnwood Forest. Occasional on the Coal Measures in the north-west of the county, and the diorite near Enderby and Croft.
  Constituent: Hg, Hd, Gp, Sr, Fs.
  Residual or intrusive: Rq, Rs, Rw, Ev, Er.
First record: Watson (1837).
Herb.: LSR★; LANC, LSR, LTR, MANCH, NMW.
HG 120: frequent on Charnwood Forest, restricted in range elsewhere; 44 localities cited.

Lupinus L.

L. polyphyllus Lindley Lupin
Map 32d; 29 tetrads.
An occasional garden escape throughout the county, becoming locally established on railway verges and in quarries, gravel pits, waste places and rubbish dumps. Sometimes very abundant on the ballast of disused or dismantled railways, especially near the sites of stations.
  Intrusive or garden escape: Er, Rq, Rg, Rw, Rr, Rd, Gr.
First record: 1970, E. K. Horwood and J. M. Horwood in present work.
HG: not recorded.

L. arboreus Sims Tree Lupin
No map; 1 tetrad.
Markfield, roadside verge, locally frequent, SHB 1978.
  Garden escape: Ev.
First record: 1978, S. H. Bishop in present work.
HG: not recorded.

Robinia L.

R. pseudacacia L. Acacia
No map; 4 tetrads.
Often planted as an ornamental tree, and occasionally arising spontaneously in hedgerow or woodland. Probably under-recorded as an escape.
  Intrusive or garden escape: Eh, Fd.
First record: 1972, E. Hesselgreaves in present work.
HG: not recorded.

Galega L.

G. officinalis L. Goat's-rue
Map 32e; 23 tetrads.
Waste places, mainly in the vicinity of Leicester. Usually in small quantity and of casual status only, but occasionally (as in the Pontylue gravel pits near Wanlip) well established and locally abundant.
  Intrusive or casual: Rw, Rg, Rs, Rd, Rr, Er, Ev, Gr.
First record: 1891, F. T. Mott in herb. MANCH.
Herb.: LSR★; LSR, LTR, MANCH.
HG 140: rare; 1 locality cited.

Colutea L.

C. arborescens L. Bladder-senna
Map 32f; 4 tetrads.
Newton Harcourt, railway bank, single bush, EKH and JMH 1973; Enderby, rubbish dump in old quarry, rare, EH 1974; Whetstone, disturbed ground, rare, EH 1974; South Wigston, railway verge, occasional, EH 1975.
  Intrusive or garden escape: Er, Rr, Rd.
First record: 1973, E. K. Horwood and J. M. Horwood in present work.
Herb.: LSR★.
HG: not recorded.

Astragalus L.

A. danicus Retz.   Purple Milk-vetch
Map 32g; 1 tetrad.
King Lud's Entrenchments, rough limestone grassland, occasional, KGM 1972.
  Constituent: Gr.
First record: Hands et al. *in* Curtis (1831).
Herb.: LANC, LSR, LTR, MANCH.
HG 140: rare, *not* extinct as stated in *Flora Leics.* (1886); 3 localities cited.

A. glycyphyllos L.   Wild Liquorice
Map 32h; 6 tetrads.
Croft, three localities in and around diorite quarry, spoil heaps and waste ground, occasional, SHB 1973; Waltham on the Wolds, old Oolite quarry, rare, KGM 1973; Breedon Cloud, Carboniferous Limestone quarry, spoil heap, locally frequent, PAC 1974; Acresford, sand pit, rare, SHB 1979; Ashby de la Zouch, waste ground, occasional, SHB 1979.
  Residual or intrusive: Rq, Rg, Rs, Rw.
First record: Crabbe *in* Nichols (1795).
Herb.: LSR★; LSR, LTR, MANCH, NMW.
HG 141: very rare in Leicestershire, more frequent in Rutland, range restricted; 11 localities cited, including Breedon Cloud.

Vicia L.

V. cracca L.   Tufted Vetch
Map 32i; 474 tetrads.
Hedgerows, rough grassland, roadside and railway verges. Frequent throughout the county.
  Constituent or intrusive: Eh, Ev, Er, Ew, Ef, Ea, Gr, Gp, Fd, Fw, Rq, Rs, Rw, Rd, Ap, Ag.
First record: Pulteney (1747).
Herb.: LSR★; DPM, LSR, LTR, MANCH, NMW.
HG 145: frequent, generally distributed.

V. tenuifolia Roth   Fine-leaved Vetch
No map; 1 tetrad.
Bitteswell, grassy bank of road forming boundary to airfield, locally abundant, EKH and JMH 1970.
  Casual or intrusive: Ev.
First record: B.E.C. Rep. 1910.
Herb.: LSR★; LSR, LTR.
HG 147: rare; 1 locality cited.

V. sylvatica L.   Wood Vetch
Map 32j; 2 tetrads.
Loddington Reddish, woodland, rare, IME 1972; Tugby Bushes, woodland, occasional, IME 1976; Tugby Wood, locally frequent, PAE 1979.
  Constituent: Fd.
First record: Pitt (1809).
Herb.: LSR★; CGE, LSR, LTR, MANCH.
HG 145: rare, and rather restricted in range; 9 localities cited, including the above.

V. villosa Roth   Fodder Vetch
No map; 1 tetrad.
Glenfield, waste ground, rare, EH 1971 (site since built over).
  Casual: Rw.
First record: W.B.E.C. Rep. 1909.
Herb.: LSR★; LSR, LTR.
HG 148: rare; 1 locality cited.

V. benghalensis L.
No map; 1 tetrad.
South Wigston, waste ground on site of former railway station, rare, EH 1974.
  Casual: Rw.
First record: B.E.C. Rep. 1919.
HG 148: rare; 1 locality cited.

V. hirsuta (L.) S. F. Gray   Hairy Tare
Map 32k; 342 tetrads.
Locally abundant on railway verges throughout the county. Occasional as an arable weed and in waste places.
  Intrusive: Er, Ev, Ew, Eh, Ea, Ac, Ao, Ag, Rw, Rq, Rg, Rs, Rd, Rt, Gr, Gp, Gl.
First record: Pulteney (1747).
Herb.: LSR★; DPM, LSR, LTR, MANCH, NMW.
HG 143: frequent and generally distributed.

V. tetrasperma (L.) Schreber   Smooth Tare
Map 32l; 44 tetrads.
Occasional throughout the county on roadside and railway verges, in quarries and waste places.
  Intrusive: Ev, Er, Ew, Ea, Rq, Rw, Rd, Gr.
First record: Marshall (1790).
Herb.: LSR★; CGE, DPM, LANC, LSR, LTR, MANCH, NMW.
HG 144: locally abundant, widely distributed; 43 localities cited.

V. sepium L.   Bush Vetch
Map 33a; 374 tetrads.
Roadside and railway verges, hedgerows and rough grassland. Frequent throughout the county, but avoiding the heavy clay soils.
  Constituent or intrusive: Ev, Er, Eh, Ew, Ea, Ef, Gr, Gp, Fd, Fm, Rw, Rq, Ap, Ag, Aa.
First record: Pulteney (1747).
Herb.: LSR★; DPM, LANC, LSR, LTR, MANCH, NMW.
HG 146: frequent, generally distributed.

V. sativa L.
Subsp. **sativa**   Common Vetch
Map 33b; 458 tetrads.
Frequent throughout the county on roadsides and railway verges and in rough grassland.
  Intrusive: Ev, Er, Eh, Ew, Ea, Ef, Gr, Gp, Gl, Rw, Rq, Rg, Rs, Ao, Ac, Ag, Ap, Aa.
First record: Pulteney (1747).
Herb.: LSR★; DPM, LANC, LSR, LTR, MANCH, NMW.
HG 146: frequent, generally distributed.

Subsp. **nigra** (L.) Ehrh.   Narrow-leaved Vetch
V. sativa L. subsp. *angustifolia* (L.) Gaud.
Map 33c; 31 tetrads.
In similar habitats to subsp. **sativa**. Occasional, and possibly overlooked.
  Intrusive: Er, Ev, Gp, Rw, Rq, Rd.
First record: Pulteney (1747).
Herb.: LSR★; CGE, LSR, LTR, MANCH, NMW.
HG 147 as *V. angustifolia*: frequent, generally distributed.

V. lutea L.   Yellow-vetch
No map; 1 tetrad.
Syston By-pass, roadside verge, rare, M. J. Gillham, 1980.

Intrusive: Ev.
First record: 1948, E. K. Horwood in herb. LTR.
Herb.: LSR★; LTR.
HG 146: doubtful record, Rutland only.

V. faba L. Broad Bean
No map.
Occurs occasionally as a casual in waste places, rubbish dumps etc, and as a non-persistent relic of former cultivation as a fodder crop. Almost certainly in such situations throughout the county, but very few field workers recorded it.
First record: Horwood and Gainsborough (1933).
Herb.: LSR, NMW.
HG 148: rare; 1 locality cited.

Lathyrus L.
L. montanus Bernh. Bitter Vetch
Map 33d; 49 tetrads.
Grassland and open places in woodland on acid soil. Frequent in Charnwood Forest and the west of the county. Occasional at Six Hills and on the Marlstone escarpment in the north-east. Absent elsewhere.
   Constituent: Hg, Gp, Gr, M, Fd, Fs, Fm, Fc.
   Residual or intrusive: Eh, Ev, Er, Rq.
First record: Pulteney (1747).
Herb.: LSR★; CGE, DPM, LSR, LTR, MANCH, NMW.
HG 150: frequent, locally distributed; 48 localities cited.

L. pratensis L. Meadow Vetchling
Map 33e; 584 tetrads.
Grassland, marshes, roadside and railway verges. Frequent and locally abundant throughout the county.
   Constituent, residual or intrusive: Gp, Gr, Gl, Hg, M, Ev, Er, Ef, Ew, Eh, Ea, Wd, Fd, Fs, Rq, Rs, Rw, Rd, Ap.
First record: Pulteney (1747).
Herb.: LSR★; DPM, LANC, LSR, LTR, MANCH, NMW.
HG 149: frequent, generally distributed.

L. palustris L. Marsh Pea
No recent record.
First record: Pulteney (1749).
HG 150: very rare, restricted to Charnwood Forest where now probably extinct; 3 localities cited.

L. sylvestris L. Narrow-leaved Everlasting-pea
No map; 1 tetrad.
Cold Newton, railway bank, frequent, HB 1968.
   Intrusive: Er.
First record: Pulteney (1756).
Herb.: LSR★; DPM, LSR, MANCH.
HG 150: rare, very restricted in range; 9 localities cited.

L. latifolius L. Broad-leaved Everlasting-pea
Map 33f; 19 tetrads.
A garden escape becoming established and locally frequent on railway verges, and occasionally on roadsides. It is sometimes a flamboyant feature of dismantled railways.
   Intrusive or established escape: Er, Ev, Eh.
First record: Bemrose (1927).
Herb.: LSR★.
HG 149: rare; 1 locality cited.

L. nissolia L. Grass Vetchling
Map 33g; 2 tetrads.
Shawell, ballast of dismantled railway track near Caves Inn, about 20 plants, EKH and JMH 1969 (not seen in subsequent years); Nevill Holt, disused limestone quarry, locally frequent, H. Godsmark 1984.
   Intrusive: Er, Rq.
First record: Crabbe in Nichols (1795).
Herb.: LSR★; LTR.
HG 149: rare, locally distributed; 3 localities cited.

L. aphaca L. Yellow Vetchling
Fleckney, Victoria Street, garden, bird seed alien (third season) J. C. Badcock 1969.
   Casual: Ag.
First record: Watson (1832).
Herb.: LSR★; LSR.
HG 149: very rare, very restricted in range; 2 localities cited.

Ononis L.
O. spinosa L. Spiny Restharrow
Map 33h; 95 tetrads.
Grassland, roadside and railway verges. Locally frequent in the east of the county; less frequent in the west.
   Constituent or intrusive: Gp, Gr, Hg, M, Ev, Er, Ew, Eh, Ea, Ap, Ag, Rd, Rt.
First record: Pulteney (1747).
Herb.: LSR★; LANC, LSR, LTR, MANCH, NMW.
HG 123: frequent, generally distributed.

O. repens L. Common Restharrow
Map 33i; 26 tetrads.
Locally frequent in grassland on the Oolite in the north-east. Occasional or rare elsewhere in the county, and then mainly on railway verges.
   Constituent or intrusive: Gp, Gr, Er, Ev, Fs, Sr.
First record: Pulteney (1747).
Herb.: LSR★; DPM, LSR, LTR, MANCH, NMW.
HG 122: frequent, generally distributed; 40 localities cited.

O. mitissima L.
No map; 1 tetrad.
Barwell, rubbish dump, single plant, EH 1979.
   Casual: Rr.
First record: 1979, E. Hesselgreaves in present work.
Herb.: LSR★.
HG: not recorded.

Melilotus Miller
M. altissima Thuill. Tall Melilot
Map 33j; 93 tetrads.
Waste places, gravel pits, quarries and spoil heaps, in which places it may be locally abundant. Also on roadsides and railway verges. Occasional throughout the county.
   Intrusive: Rg, Rq, Rs, Rw, Rr, Rd, Rt, Rf, Er, Ev, Ew, Ef, Ao, Gr, Gp, Gl.
First record: Pulteney (1747).
Herb.: LSR★; DPM, LANC, LSR, LTR, NMW.
HG 126: frequent, generally distributed; 92 localities cited.

M. alba Medicus White Melilot
Map 33k; 22 tetrads.
Waste places, gravel pits, spoil heaps and railway verges. Occasional in the neighbourhood of Leicester; rare elsewhere.

Intrusive: Rw, Rg, Rs, Rd, Er, Ev, Ao.
First record: Mott et al. (1886).
Herb.: LSR*; LANC, LSR, LTR, MANCH, NMW.
HG 127: occasional, sparsely distributed; 15 localities cited.

M. officinalis (L.) Pallas   Ribbed Melilot
Map 33l; 107 tetrads.
Waste places, gravel pits, quarries and spoil heaps, often locally abundant in these habitats. Also on roadsides and railway verges. Occasional throughout the county; slightly more frequent than **M. altissima**.
   Intrusive: Rw, Rq, Rg, Rs, Rr, Rd, Er, Ev, Ew, Eh, Ef, Ea, Gr, Wd.
First record: Pitt (1809).
Herb.: LSR*; LANC, LSR, LTR, MANCH, NMW.
HG 127 as *M. arvensis*: rare, restricted in range; 25 localities cited.

M. indica (L.) All.   Small Melilot
No map; 7 tetrads.
Eastwell, dismantled railway, bank, rare JMS 1968; Melton Mowbray, Lag Lane, pile of loose soil by farm track, rare, ALP 1969; Glenfield, waste ground rare, EH 1972; Quorndon, garden, rare, PHG 1975; Barkby, tipped earth and rubble, rare, IME 1975; Leicester, recently disturbed ground, rare, HB 1978; Groby, waste ground, rare, EH 1978.
   Casual: Rd, Rw, Er.
First record: 1886, F. T. Mott in herb MANCH.
Herb.: 128: Rare, Range restricted; 13 localities

Medicago L.
M. lupulina L.   Black Medick
Map 34a; 538 tetrads.
Frequent and locally abundant throughout the county on railway and roadside verges, in quarries and gravel pits. Less frequent in grassland.
   Intrusive or constituent: Ev, Er, Ea, Ew, Ef, Rq, Rg, Rw, Rs, Rd, Rt, Rr, Gp, Gr, Gl, Ap, Ag, Ao, Sw.
First record: Pulteney (1747).
Herb.: LSR*; CGE, DPM, LANC, LSR, LTR, MANCH, NMW.
HG 124: locally abundant, generally distributed.

M. sativa L.
Subsp. **sativa**
                                Lucerne
Map 34b; 55 tetrads.
Occasional throughout the county in grassland, on roadside and railway verges, in arable land and waste places. Often a non-persistent relic of former cultivation.
   Intrusive or casual: Gp, Gr, Gl, Ev, Er, Ea, Ao, Rw, Rq, Rg, Rf.
First record: Pitt (1809).
Herb.: LSR*; LSR, LTR, MANCH, NMW.
HG 123: local; 48 localities cited.

Subsp. **falcata** (L.) Arcangeli   Sickle Medick
M. falcata L.
No map; 2 tetrads.
New Bridge, Belgrave, Leicester, recently disturbed ground, rare, HB 1971; Barkby, tipped earth and rubble, locally frequent, IME 1976.
   Casual: Rd.
First record: Pitt (1809).
Herb.: LSR*; LSR.
HG 124: rare, restricted in range; 7 localities cited.

M. arabica (L.) Hudson   Spotted Medick
Map 34c; 8 tetrads.
Syston Mill, north bank of River Wreake, rare, ALP 1967. The history of the plant at this site is of interest. A specimen from this exact locality, collected by J. A. Cappella and dated June 1885, is in herb. LSR, and it is recorded in Horwood and Gainsborough from this site under Cappella's name. Manuscript notes of Cappella indicate that quantity and site were the same then as when it was recorded in 1967. In 1968, after work on the river bank and nearby in connection with the construction of the new Syston Bypass, the species increased enormously and spread in quantity to the adjacent railway embankment and the embankment made for the new road. Since then it has decreased again, but is still more plentiful than before. Syston, canal bank near Hope and Anchor, single plant, ALP 1968; Donington Park, crevice of old concrete trackway, rare, PAC 1971; Hoby, roadside verge, occasional, ALP 1973; Cotes, bank of River Soar and site of Cotes church, locally abundant, PHG 1973; Cotes Hall ruins, grassy terrace by River Soar, locally abundant, PHG 1974; Barrow on Soar, bank of river, locally frequent, PHG 1975; Dadlington, pasture, locally abundant, SHB 1978; Sileby, waste ground, PAE 1979.
   Constituent or intrusive: Ew, Ev, Gp, Rw.
First record: Pulteney (1747).
Herb.: LSR*; LSR, LTR, MANCH, NMW.
HG 126: local or rare, sparsely distributed; 15 localities cited, including Syston Mill, Cotes and Barrow on Soar.

M. polymorpha L.   Toothed Medick
No record since HG.
First record: 1886, F. T. Mott in herb. MANCH.
Herb.: LSR, MANCH, NMW.
HG 125 as *M. hispida*: occasional, sparsely distributed; 9 localities cited.

Trifolium L.
T. repens L.   White Clover
Map 34d; 601 tetrads.
Meadows, pastures and all grassy places. Abundant throughout the county.
   Constituent or intrusive: Gp, Gr, Gl, Hg, Ev, Er, Ea, Ew, Ef, Rw, Rt, Rq, Rs, Rd, Rf, M, Ap, Ag, Aa.
First record: Pulteney (1747).
Herb.: LSR*; DPM, LSR, LTR, MANCH, NMW.
HG 134: locally abundant, generally distributed.

T. hybridum L.   Alsike Clover
Map 34e; 301 tetrads.
Occasionally grown as a fodder crop, and in such situations locally abundant. Otherwise occasional throughout the county on roadside and railway verges, and in gravel pits and waste places, usually in small isolated colonies.
   Intrusive: Gl, Gp, Gr, Hg, Ev, Er, Ea, Ew, Rw, Rg, Rs, Rd, Rt, Rf, Rr, Ao, Ac, Ap.
First record: Coleman (1852).
Herb.: LSR*; LSR, LTR, MANCH, NMW.
HG 134: frequent, generally distributed.

T. fragiferum L.   Strawberry Clover
Map 34f; 24 tetrads.
Occasional, being mainly confined to the eastern part of the county, especially the north-east, and apparently

preferring clay soils. Grassland, marshes, roadside verges and tracks. A characteristic habitat is the grassy strip in the centre of a farm track. Seldom in great abundance where it does occur.
Constituent or intrusive: Gp, Gr, M, Ev, Ew, Rt, Rd, Rf.
First record: Pulteney (1747).
Herb.: LSR*; DPM, LANC, LSR, LTR, MANCH, NMW.
HG 135: frequent, widely distributed; 64 localities cited.

T. aureum Pollich
No record since HG.
First record: 1891, T. A. Preston in herb. LSR.
Herb.: LSR, MANCH, NMW.
HG 136 as *T. agrarium*: rare, very restricted in range; 2 localities cited.

T. campestre Schreber   Hop Trefoil
Map 34g; 222 tetrads.
Occasionally in grassland throughout the county where the soil tends to be calcareous. Otherwise on roadsides and railway verges, in waste places and quarries. It is particularly characteristic of railway ballast, where it is frequent and locally abundant.
Constituent or intrusive: Er, Ev, Ef, Gp, Gl, Gr, Rq, Rg, Rs, Rw, Rt, Rd, Ao, Aa.
First record: Pulteney (1747).
Herb.: LSR*; LSR, LTR, MANCH, NMW.
HG 136 as *T. procumbens*: frequent, generally distributed.

T. dubium Sibth.   Lesser Trefoil
Map 34h; 484 tetrads.
Frequent and locally abundant on roadside and railway verges, in quarries, gravel pits and grassland throughout the county.
Constituent or intrusive: Gp, Gr, Gl, M, Hg, Ev, Er, Ew, Ea, Rq, Rg, Rs, Rw, Rt, Rd, Ap, Ag, Aa, Sw.
First record: Pulteney (1747).
Herb.: LSR*; LANC, LSR, LTR, MANCH, NMW.
HG 137: locally abundant, generally distributed.

T. micranthum Viv.   Slender Trefoil
Map 34i; 8 tetrads.
Roundhill near Syston, old sand pit, occasional, ALP 1969 (site since filled with rubbish); Knighton, garden lawn, occasional, IME 1971; Kirby Muxloe, old track, rare, EH 1973; Croft, siliceous grassland, locally abundant, SHB 1974; Rothley, siliceous grassland off Kinchley Lane, rare, PHG 1974; Leicester, The Towers Hospital, lawn, rare, IME 1976; Snarestone, overgrown part of metalled road, locally abundant, HH 1976; Bradgate Park, grassland, occasional, PHG 1978.
Constituent, residual or intrusive: Hg, Gp, Ag, Rg, Rt.
First record: Pulteney (1747).
Herb.: LSR*; LSR, LTR, MANCH, NMW.
HG 137 as *T. filiforme*: rare or local, very locally distributed; 20 localities cited.

T. striatum L.   Knotted Clover
Map 34j; 19 tetrads.
Occasional throughout the county on light soils, mainly in quarries and on railway verges, but also in grassland, where it could easily be overlooked.
Constituent, residual or intrusive: Gp, Hg, Rq, Rw, Rs, Sr, Er, Ev.

First record: Pulteney (1747).
Herb.: LSR*; CGE, LSR, LTR, MANCH, NMW.
HG 132: local, locally distributed; 41 localities cited.

T. arvense L.   Hare's-foot Clover
Map 34k; 43 tetrads.
Occasional throughout the county wherever there are light sandy soils, which it prefers. Also occasionally on railway ballast.
Constituent or intrusive: Hg, Gp, Sr, Er, Ev, Rw, Rq, Rg, Rs, Rt, Rd.
First record: Pulteney (1747).
Herb.: LSR*; CGE, LSR, LTR, MANCH, NMW.
HG 132: local, locally distributed; 28 localities cited.

T. scabrum L.   Rough Clover
No record since HG.
First record: Martyn (1763).
Herb.: LSR.
HG 133: very rare, restricted in range; 7 localities cited.

T. incarnatum L.   Crimson Clover
No map.
Quorndon, recently disturbed ground, single plant, PHG 1971. A few other records for this species were lost as a result of the fire at Ratcliffe College.
First record: 1886, F. T. Mott in herb. MANCH.
Herb.: LSR*; LANC, LSR, MANCH, NMW.
HG 131: occasional, sparsely distributed; 20 localities cited.

T. pratense L.   Red Clover
Map 34l; 599 tetrads.
Often grown as a fodder crop and hence locally abundant in leys and recent grassland. Also frequent throughout the county in old grassland, on roadsides and in other grassy places.
Constituent or intrusive; Gp, Gr, Gl, Ev, Er, Ew, Ea, Ef, M, Rw, Rq, Rd, Rt, Ac, Ao, Ag, Ap, Sw.
First record: Pulteney (1747).
Herb.: LSR*; DPM, LSR, LTR, MANCH, NMW.
HG 128: abundant, generally distributed.

T. medium L.   Zigzag Clover
Map 35a; 112 tetrads.
Grassland, roadside and railway verges, and quarries. Occasional throughout the county.
Constituent or intrusive: Gp, Gr, Gl, Hg, Hd, Ev, Er, Ef, Ew, Ea,.Eh, Rq, Rg, Rs, Rt.
First record: Pitt (1809).
Herb.: LSR*; LSR, LTR, MANCH, NMW.
HG 130: local, generally distributed; 73 localities cited.

T. ochroleucon Hudson   Sulphur Clover
No record since HG.
First record: Horwood (1911).
Herb.: LSR, NMW.
HG 131: rare, very restricted in range; 1 locality cited.

T. pannonicum Jacq.
No recent record.
There is an undated specimen in herb. NMW collected by A. R. Horwood at Mountsorrel, presumably in the period 1902-1922.
Herb.: NMW.
HG: not recorded.

T. subterraneum L.   Subterranean Clover
Map 35b; 3 tetrads.
Croft, siliceous grassland, locally abundant, SHB 1974; Cotes, grassy banks adjacent to River Soar, locally abundant, PHG 1974; Mountsorrel, edge of pathway on rocky hillside, occasional, PHG and ALP 1983.
    Constituent: Hg, Gp.
First record: Pulteney (1749).
Herb.: LSR★; LSR.
HG 128: very rare, very restricted in range, though possibly overlooked; 4 localities cited, including Croft.

Lotus L.
L. tenuis Waldst. & Kit. ex Willd.   Narrow-leaved Bird's-foot-trefoil
No map; 1 tetrad.
Barwell, rubbish dump, occasional, SHB 1979.
    Casual or intrusive; Rr.
First record: Watson (1837).
Herb.: LSR, LTR, MANCH, NMW.
HG 139: rather rare, restricted in range; 24 localities cited.

L. corniculatus L.   Common Bird's-foot-trefoil
Map 35c; 553 tetrads.
Grassland and heathland, roadside and railway verges, quarries and grassy waste ground. Frequent throughout the county.
    Constituent or intrusive: Gp, Gr, Gl, Hg, Hd, Ev, Er, Ef, Ew, Ea, Rq, Rg, Rs, Rt, Rd, Rw, Ap, Ag, Sw.
First record: Pulteney (1747).
Herb.: LSR★; DPM, LANC, LSR, LTR, MANCH, NMW.
HG 138: locally abundant, generally distributed.

L. uliginosus Schkuhr   Greater Bird's-foot-trefoil
L. pedunculatus auct. non Cav.
Map 35d; 308 tetrads.
Marshes, woodland rides and wet places. Locally frequent throughout the county in these habitats.
    Constituent or residual: M, Hw, Hg, Wd, Fw, Ef, Ew, Ev, Er, Eh, Ea, Gp, Gr, Rt, Rq, Rd.
First record: Pulteney (1747).
Herb.: LSR★; LANC, LSR, LTR, MANCH, NMW.
HG 139: frequent, generally distributed.

Anthyllis L.
A. vulneraria L.   Kidney Vetch
Map 35e; 18 tetrads.
Calcareous grassland, mainly in the north-east, occasional. Elsewhere most of the records are from railway verges, with a few from sandy ground and quarries. 57% of the records are from railway verges.
    Constituent or intrusive: Gp, Gr, Er, Rq, Rw.
First record: Pulteney (1749).
Herb.: LSR★; LANC, LSR, LTR, MANCH, NMW.
HG 138: rare, restricted in range; 44 localities cited.

Ornithopus L.
O. perpusillus L.   Bird's-foot
Map 35f; 14 tetrads.
Heathland, sandy and rocky places with acid soil. Confined to Charnwood Forest, the Mountsorrel granodiorite and the diorite at Croft. In 1969 it occurred in the sandpits at Roundhill near Syston, but this site is now filled with rubbish.
    Constituent or intrusive: Hg, Gp, Sr, Rq, Rt, Rd, Er.

First record: Pulteney (1749).
Herb.: LSR★; LSR, LTR, MANCH, NMW.
HG 141: local, very restricted in range; 36 localities cited.

Coronilla L.
C. varia L.   Crown Vetch
No record since HG.
First record: 1881, F. T. Mott in herb. MANCH.
Herb.: LSR, MANCH.
HG 141: rare, but of increasing occurrence, sparsely distributed; 3 localities cited.

Hippocrepis L.
H. comosa L.   Horseshoe Vetch
No record in the recent survey.
Croxton Kerrial, The Drift, E. K. Horwood 1952; Croxton Kerrial to Saltby roadside, T. S. Robertson 1953; Saltby, disused airfield, T. S. Robertson 1953; King Lud's Entrenchments, T. S. Robertson 1953 and H. Gawadi 1962.
First record: Tomkins (1887).
Herb.: LSR, LTR.
HG 142: rare, restricted in range; 6 localities cited, all in Saltby area.

Onobrychis Miller
O. viciifolia Scop.   Sainfoin
No map; 2 tetrads.
Saltby, limestone grassland, EKH 1960; Brooksby, railway verge, a single plant, ALP 1968.
    Casual or intrusive: Gp, Er.
First record: Pulteney (1749).
Herb.: LSR★; LANC, LSR, LTR, MANCH.
HG 142: rare, range restricted; 17 localities cited.

GERANIALES

OXALIDACEAE
Oxalis L.
O. corniculata L.
Map 35g; 34 tetrads.
A garden escape, frequently becoming established on pathways; often a troublesome weed in nursery gardens.
    Intrusive: Rt, Rq, Ag, Ap, Ev, Sw.
First record: 1901, F. T. Mott in herb. MANCH.
Herb.: LSR★; MANCH.
HG 110: no Leicestershire record.

O. stricta L.
No record in the recent survey.
Cadeby, A. Hackett 1957.
First record: Horwood and Gainsborough (1933).
Herb.: LSR.
HG 110: rare; 1 locality cited.

O. europaea Jordan
Map 35h; 5 tetrads.
Quorndon, Farnham Street, crevices in pavement and garden paths, locally frequent, PHG 1968 (present for at least 40 years); Croft, roadside, between paving slabs, locally frequent, SHB 1973; Thurlaston, roadside, rare, SHB 1974; Blaby Mill, garden, EH 1974; Peckleton, neglected garden, rare, SHB 1975; Enderby, garden, locally abundant, EH 1975.
    Established escape or casual: Rt, Ev, Ag.

First record: T. G. Tutin *in* Young (1958).
Herb.: LSR★.
HG: not recorded.

O. articulata Savigny
No map; 1 tetrad.
Newbold Verdon, roadside verge, occasional, HH 1972.
   Casual or escape: Ev.
First record: 1972, H. Handley in present work.
Herb.: LSR★.
HG: not recorded.

O. acetosella L.   Wood-sorrel
Map 35i; 92 tetrads.
Frequent and locally abundant in the larger and older woods; occasional in hedgerows and on shady roadsides in wooded districts. Distribution in the county corresponds to that of the major woodlands, though it is more common on the lighter soils of the west than on the heavier clay soils of the east.
   Constituent: Fd, Fm, Fs, Fw, Ef.
   Residual or intrusive: Eh, Ew, Wd, Ap.
First record: Pulteney (1747).
Herb.: LSR★; DPM, LSR, LTR, MANCH, NMW.
HG 109: locally abundant, widely distributed; 90 localities cited.

GERANIACEAE
Geranium L.

G. sanguineum L.   Bloody Cranesbill
No map; 2 tetrads.
Leicester, Blackfriars, dismantled railway sidings, rare, HB 1972 (site since destroyed); Heather, roadside grass verge subject to soil dumping, SM 1973.
   Casual or garden escape: Er, Ev.
First record: 1953, E. K. Horwood in herb. LTR.
Herb.: LSR★; LTR.
HG: not recorded.

G. pratense L.   Meadow Cranesbill
Map 35j; 254 tetrads.
Roadside and railway verges, the banks of streams, and rough grassland. Frequent in the east of the county, occasional in the west.
   Constituent or intrusive: Ev, Er, Ew, Eh, Ea, Ef, Gr, Gp, M, Fd, Ap, Rw, Rq, Rr.
First record: Pulteney (1747).
Herb.: LSR★; DPM, LANC, LSR, LTR, MANCH, NMW.
HG 104: locally abundant, generally distributed.

G. sylvaticum L.   Wood Cranesbill
No map; 1 tetrad.
Gumley Hall, mixed woodland, occasional (together with other established introductions), EKH and JMH 1974.
   Planted or intrusive: Fm.
First record: Kirby (1850).
HG 104: rare, restricted in range; 1 locality cited.

G. endressii Gay   French Cranesbill
No map; 2 tetrads.
Newtown Linford, roadside verge, garden escape, locally frequent, PHG 1969; Noseley Hall, ornamental woodland, established introduction, occasional, KGM 1971.
   Escape or planted: Ev, Fd.

First record: B.E.C. Rep. 1925.
Herb.: LSR★; LSR.
HG 108: rare; 2 localities cited.

G. endressii × versicolor
No map; 1 tetrad.
Cosby, roadside, rare, EH 1977 (det. E. J. Clement).
   Casual: Ev.
First record: 1977, E. Hesselgreaves in present work.
Herb.: LSR★.
HG: not recorded.

G. versicolor L.   Pencilled Cranesbill
No record since HG.
First record: Kirby (1850).
HG 103: rare, restricted in range; 2 localities cited.

G. phaeum L.   Dusky Cranesbill
No map; 3 tetrads.
Newtown Linford, roadside verge, rare, PHG 1969; Osbaston, shady roadside verge, rare, OHB and HH 1973; Gumley Hall, mixed woodland by lake, occasional (together with other established introductions), IME 1981.
   Garden escape: Ev.
First record: Kirby (1850).
Herb.: LSR★; LSR, LTR.
HG 104: rare, restricted in range; 13 localities cited.

G. pyrenaicum Burm. fil.   Hedgerow Cranesbill
Map 35k; 56 tetrads.
Occasional throughout the county on railway verges and in waste places.
   Intrusive: Er, Ev, Ea, Rw, Rq, Rs, Rd, Rt, Gr, Sw, Sr, Ag.
First record: 1875, F. T. Mott in herb. MANCH.
Herb.: LSR★; DPM, LANC, LSR, LTR, MANCH.
HG 104: occasional – on the increase, sparsely distributed; 23 localities cited.

G. molle L.   Dove's-foot Cranesbill
Map 35l; 243 tetrads.
Roadside and railway verges and waste places; occasionally as a weed of arable land. Frequent throughout the county, but seldom in any quantity in any particular locality.
   Intrusive: Ev, Er, Ea, Eh, Rw, Rq, Rg, Rs, Rt, Rd, Rf, Gp, Gl, Gr, Hg, Ao, Ac, Ag, Ap, Sb.
First record: Pulteney (1747).
Herb.: LSR★; DPM, LANC, LSR, LTR, MANCH, NMW.
HG 105: abundant locally, generally distributed.

G. pusillum L.   Small-flowered Cranesbill
Map 36a; 54 tetrads.
Occasional throughout the county on roadside and railway verges, in waste places, arable land and grassland.
   Intrusive: Er, Ev, Ea, Ew, Rw, Rq, Rd, Rf, Rr, Rt, Ao, Ag, Ac, Gr, Gp, Sr, Sw.
First record: Watson (1832).
Herb.: LSR★; DPM, LSR, LTR, MANCH, NMW.
HG 105: frequent, doubtless overlooked, generally distributed.

G. columbinum L.   Long-stalked Cranesbill
Map 36b; 1 tetrad.
Newton Harcourt, railway verge, rare, EKH and JMH 1973.

Constituent or intrusive: Er.
First record: Horwood and Gainsborough (1933).
Herb.: LSR.
HG 106: rare, restricted in range; 3 localities cited.

G. dissectum L.   Cut-leaved Cranesbill
Map 36c; 472 tetrads.
Frequent throughout the county in arable land, on roadside and railway verges, and in waste places. Occasional in grassland, especially leys.
Intrusive: Ac, Ao, Ag, Aa, Ev, Er, Ew, Rw, Rq, Rg, Rs, Rd, Rt, Rr, Gl, Gr, Gp.
First record: Pulteney (1747).
Herb.: LSR*; DPM, LANC, LSR, LTR, MANCH, NMW.
HG 106: abundant locally, generally distributed.

G. lucidum L.   Shining Cranesbill
Map 36d; 21 tetrads.
Occasional on walls in scattered localities throughout the county. Possibly constituent on rocks at Mountsorrel. Rare in other habitats.
Constituent: Sr.
Intrusive: Sw, Er, Ev, Eh, Rw, Rq, Ap, Ag.
First record: Pulteney (1756).
Herb.: LSR*; LSR, LTR, MANCH, NMW.
HG 106: rare, very restricted in range; 21 localities cited.

G. robertianum L.   Herb-Robert
Map 36e; 543 tetrads.
Frequent and locally abundant throughout the county in woodland and hedgerows, and on railway and roadside verges.
Constituent: Fd, Fm, Fs, Fw, Ef, Eh.
Residual or intrusive: Er, Ev, Ew, Wd, Gr, Gp, Sw, Rw, Rs, Rd, Rt, Ap, Ag, Ac.
First record: Pulteney (1747).
Herb.: LSR*; DPM, LANC, LSR, LTR, MANCH, NMW.
HG 107: locally abundant, generally distributed.

Erodium L'Hér.
E. cicutarium (L.) L'Hér.   Common Stork's-bill
Map 36f; 33 tetrads.
Dry sandy places. Occasional throughout most of the county, but apparently absent from the south-east.
Intrusive or constituent: Hg, Gp, Gr, Sr, Sw, Ev, Er, Rd, Rw, Rf, Ao.
First record: Pulteney (1747).
Herb.: LSR*; LSR, LTR, MANCH, NMW.
HG 108: locally abundant, widely distributed; 33 localities cited.

E. moschatum (L.) L'Hér.   Musk Stork's-bill
Map 36g; 1 tetrad.
Rothley, Swithland Lane, roadside verge, rare, D. S. Fieldhouse 1978.
Intrusive: Er.
First record: Crabbe *in* Nichols (1795).
Herb.: LSR*; BM, LSR.
HG 109: very rare, restricted in range; 5 localities cited.

LINACEAE
Linum L.
L. bienne Miller   Pale Flax
Map 36h; 8 tetrads.
Near Ulverscroft Priory, recently disturbed ground, occasional, PHG 1967; Groby, scrub, occasional, EH 1969; Ratby, recently disturbed roadside verge, rare, EH 1970; Birstall, disused allotments, locally abundant, HB 1971; Knighton, garden, occasional, T. G. Tutin 1973; Whetstone, grassy verge of dismantled railway, rare, EH 1974.
Residual or casual: Hg, Gr, Ev, Er, Aa, Rd.
First record: Mott (1892).
Herb.: LSR*.
HG 102: rare, restricted in range; 3 localities cited.

L. usitatissimum L.   Flax
No map; 10 tetrads.
Occurs occasionally as a non-persistent casual in waste places, on roadsides, etc.
Casual: Rw, Rd, Rf, Rs, Ev, Ap, Ao.
First record: 1708, *in* Macaulay (1791).
Herb.: LSR*; CGE, LANC, LSR, MANCH, NMW.
HG 103: rare, locally distributed; 38 localities cited.

L. catharticum L.   Fairy Flax, Purging Flax
Map 36i; 177 tetrads.
Frequent and locally abundant throughout the county on railway verges, 57% of the records being from this habitat. Also occurs in quarries and grassland, preferring calcareous soils but by no means confined to them.
Constituent or intrusive: Er, Ev, Gp, Gr, Hg, Rq, Rs, Rw, Rd.
First record: Pulteney (1747).
Herb.: LSR*; DPM, LSR, LTR, MANCH, NMW.
HG 102: locally abundant, generally distributed.

Radiola Hill
R. linoides Roth   Allseed
No record since HG.
First record: Pulteney (1749).
Herb.: LSR.
HG 101: rare, not observed since Babington's time, confined to Charnwood Forest; 3 localities cited.

EUPHORBIACEAE
Mercurialis L.
M. annua L.   Annual Mercury
Map 36j; 14 tetrads.
Waste places, usually of casual occurrence and non-persistent, or rarely as a garden weed. Occasional, most of the records being from Leicester or its environs.
Casual or intrusive: Rw, Rd, Rq, Rs, Rf, Ag, Aa.
First record: 1890, F. T. Mott in herb. MANCH.
Herb.: LSR*; LSR, MANCH, NMW.
HG 478: rare, restricted in range; 5 localities cited.

M. perennis L.   Dog's Mercury
Map 36k; 259 tetrads.
Frequent and locally abundant in woodland throughout the county. Occasional in hedgerows and shady places, especially in wooded districts, or as a relict of former woodland.
Constituent: Fd, Fm, Fs, Fw, Ef.
Residual or intrusive: Eh, Ev, Ew, Er, Ap, Ag, Wd, Rw.
First record: Pulteney (1747).
Herb.: LSR*; DPM, LSR, LTR, MANCH, NMW, UPP.
HG 476: locally abundant, and becoming dominant in the ground flora, generally distributed; 140 localities cited.

Euphorbia L.

E. serratula Thuill.   Upright Spurge
E. stricta L. nom. illegit.
No record since HG.
First record: Dixon (1904).
Herb.: NMW.
HG 474: rare; 1 locality cited.

E. helioscopia L.   Sun Spurge
Map 36l; 367 tetrads.
Frequent throughout the county as a garden weed and in recently disturbed ground. Occasional in arable land, especially in non-cereal crops.
    Intrusive: Ag, Ao, Aa, Ac, Ap, Rd, Rw, Rr, Rs, Rt, Rf, Ea, Ev, Er.
First record: Pulteney (1747).
Herb.: LSR★; DPM, LSR, LTR, MANCH, NMW.
HG 473: locally abundant, generally distributed.

E. lathyris L.   Caper Spurge
No map; 4 tetrads.
Glooston, waste ground and roadside, rare, KGM 1971; Kirby Muxloe, rubbish heap in churchyard, rare, EH 1971; Narborough, waste ground and rubble, occasional, EH 1976; Ashby Old Parks, recently disturbed ground, rare, SHB 1977.
    Casual or garden escape: Rw, Rd, Rr, Ev.
First record: Kirby (1850).
Herb.: LSR★; LSR.
HG 475: rare, restricted in range; 14 localities cited.

E. exigua L.   Dwarf Spurge
Map 37a; 103 tetrads.
Occasional throughout the county as a weed of arable land, though it may be locally abundant, especially in non-cereal crops. Rarely as a garden weed, and in waste places.
    Intrusive: Ao, Ac, Ag, Ap, Ea, Er, Ev, Rd, Rw, Rs, Rt, Rr.
First record: Pulteney (1747).
Herb.: LSR★; DPM, LSR, LTR, MANCH, NMW.
HG 475: locally abundant, generally distributed.

E. peplus L.   Petty Spurge
Map 37b; 417 tetrads.
Frequent and locally abundant as a garden weed, in arable land and on recently disturbed ground. The commonest spurge throughout the county.
    Intrusive: Ag, Ao, Aa, Ac, Ap, Rd, Rw, Rf, Rs, Rt, Rr, Ea, Ev, Er, Sw.
First record: Pulteney (1747).
Herb.: LSR★; DPM, LSR, LTR.
HG 475: locally abundant, generally distributed.

E. esula L.   Leafy Spurge
Map 37c; 2 tetrads.
Lutterworth, bank of dismantled railway near site of station, locally abundant, EKH and JMH 1970; Market Bosworth, disused railway verge at site of station, occasional, ALP 1973.
    Intrusive: Er.
First record: Bemrose (1927).
Herb.: LSR★; LTR.
HG 474: rare; 1 locality cited.

E. cyparissias L.   Cypress Spurge
Map 37d; 5 tetrads.
Burbage, railway verge, locally frequent, MH 1969; Stoney Stanton, railway verge, locally abundant, MH 1969; Wymondham, railway cutting, locally frequent, KGM 1971; Moira, waste ground, occasional, SHB 1976; Worthington, railway verge, rare, SHB 1978.
    Intrusive: Er, Rw.
First record: Horwood and Gainsborough (1933).
Herb.: LSR★; CGE, LSR, LTR.
HG 474: rare; 4 localities cited.

E. amygdaloides L.   Wood Spurge
Map 37e; 1 tetrad.
Higham on the Hill, Joykin Spinney, deciduous woodland, rare, SHB 1976.
    Constituent: Fd.
First record: Watson (1837).
Herb.: LSR.
HG 474: rare, restricted in range; 6 localities cited.

RUTALES

POLYGALACEAE
Polygala L.

P. vulgaris L.   Common Milkwort
Map 37f; 43 tetrads.
Grassland, heath and railway verges. In scattered localities throughout the county, where it may be locally frequent.
    Constituent or intrusive: Gp, Gr, Hg, Hd, Er, Rq.
First record: Pulteney (1747).
Herb.: LSR★; CGE, LANC, LSR, LTR, MANCH, NMW.
HG 71: locally abundant, widely distributed; 82 localities cited.

P. serpyllifolia J. A. C. Hose   Heath Milkwort
Map 37g; 9 tetrads.
Heath grassland, where it may be locally frequent. Confined to Charnwood Forest, except for a single record from Ashby Woulds.
    Constituent: Hg, Hd, Gp.
    Residual or intrusive: Rq, Rt, Ag.
First record: Pulteney (1749).
Herb.: LSR★; LSR, LTR, MANCH, NMW.
HG 72: frequent, rather restricted in range; 36 localities cited.

P. calcarea F. W. Schultz   Chalk Milkwort
No record in the recent survey.
Sproxton to Skillington roadside, M. K. Hanson 1950; Saltby, calcareous grassland near The Drift, EKH 1956.
First record: B.E.C. Rep. 1915.
Herb.: LSR, LTR.
HG 73: rare, restricted in range; 2 localities cited, more or less corresponding to the above.

SAPINDALES

ACERACEAE
Acer L.

A. platanoides L.   Norway Maple
Map 37h; 30 tetrads.
Occasional throughout the county in woodland and hedgerows. Often planted and sometimes self-seeding.

Intrusive or planted: Fd, Fm, Fs, Eh, Ev, Er, Ew, Ap, Rr.
First record: 1965, E. K. Horwood in herb. LTR.
Herb.: LSR★; LTR.
HG: not recorded.

A. campestre L.   Field Maple
Map 37i; 562 tetrads.
Hedgerows and woodland. Frequent and locally abundant throughout the county, but absent from parts of Charnwood Forest.
    Constituent: Eh, Ef, Fd, Fm, Fs.
    Residual or intrusive: Ev, Er, Ew, Ap, Rq, Gp.
First record: Pulteney (1747).
Herb.: LSR★; CGE, DPM, LSR, LTR, MANCH, NMW.
HG 116: frequent, generally distributed.

A. pseudoplatanus L.   Sycamore
Map 37j; 546 tetrads.
Hedgerows, woodland, public and private parks. Frequent throughout the county. Often planted, and because of its wind-borne fruits, spreading along hedgerows from the source of introduction. In woods it can be a troublesome weed.
    Intrusive or planted: Eh, Ev, Er, Ew, Fd, Fm, Fs, Fw, Ap, Ag, Rw, Rr.
First record: Pulteney (1747).
Herb.: LSR★; DPM, LSR, LTR, MANCH, NMW.
HG 116: frequent, generally distributed.

HIPPOCASTANACEAE
Aesculus L.
A. hippocastanum L.   Horse-chestnut
Map 37k; 417 tetrads.
Hedgerows, woodland, public and private parks and gardens, often planted and readily self-sown. Sometimes found as isolated trees in grassland. Locally frequent throughout the county.
    Planted or intrusive: Ap, Ag, Eh, Ev, Ew, Er, Fd, Fm, Fs, Gp, Rq, Rf.
First record: Horwood and Gainsborough (1933).
Herb.: LSR★; DPM, LTR, NMW.
HG 115: locally abundant, probably generally distributed; 9 localities cited.

BALSAMINACEAE
Impatiens L.
I. capensis Meerb.   Orange Balsam
Map 37l; 42 tetrads.
Frequent and locally abundant on the banks of the Grand Union Canal in the south-east of the county, and on the banks of the Ashby Canal and rivers in the south-west; also in a few isolated localities on the River Soar. A species such as this, only comparatively recently introduced into the country, raises interesting problems concerning the assignment of status. Most of our 'native' species have arrived in the country by one means or another since the last glaciation. A thoroughly naturalised introduction such as this species would therefore seem to have as much claim to be regarded as a constituent of the flora of its habitat as other species which happen to have arrived at a much earlier date.
    Constituent or intrusive: Ew, Wc, Wr, Wd, M, Fw, Er.
First record: 1940, S. A. Taylor in herb. LSR.

Herb.: LSR★; DPM, LANC, LSR, LTR.
HG: not recorded.

L. parviflora DC.   Small Balsam
Map 38a; 7 tetrads.
Barkby, dry ditch at base of wall of Barkby Hall, occasional, ALP 1967; verge of Barkby to Queniborough road, rare, ALP 1971; Kirkby Mallory, shady roadside verge, locally frequent, HH 1971; Loughborough, shady corner of timber yard, locally frequent, PHG 1971; Knipton, roadside verge, occasional, HL 1972; Bottesford, roadside verge, rare, HL 1973; Belvoir, rough ground by River Devon, occasional, KGM 1974; Harston, wooded cutting of dismantled mineral railway, locally frequent, KGM 1974.
    Intrusive: Ev, Er, Rw, Wd.
First record: Horwood and Gainsborough (1933).
Herb.: LSR★; LSR, LTR, NMW.
HG 111: rare, restricted in range; 3 localities cited, including Belvoir and Knipton.

I. glandulifera Royle   Indian Balsam, Policeman's-helmet
Map 38b; 46 tetrads.
Occasional throughout the county on the banks of rivers and streams, in ditches, hedgerows and woodland. Also as a garden escape in villages.
    Intrusive: Ew, Eh, Ef, Ev, Wd, Fw, Fm, Rw, Rd, Ag.
First record: 1886, F. T. Mott in herb. MANCH.
Herb.: LSR★; LTR, MANCH.
HG 111: no Leicestershire record.

AQUIFOLIACEAE
Ilex L.
I. aquifolium L.   Holly
Map 38c; 366 tetrads.
Hedgerows and woodland. Frequent and locally abundant in Charnwood Forest, where it forms a conspicuous and characteristic part of the natural vegetation, and in the west of the county. Occasional and often an escape from cultivation in the east.
    Constituent, residual or intrusive: Fd, Fm, Fs, Fw, Eh, Er, Rq, Ap, Ag.
First record: Pulteney (1747).
Herb.: LSR★; DPM, LSR, LTR, MANCH, NMW.
HG 111: frequent or local, generally distributed; 105 localities cited.

CELASTRALES

CELASTRACEAE
Euonymus L.
E. europaeus L.   Spindle
Map 38d; 17 tetrads.
Woodland and hedgerows in a few scattered localities, mainly in the south of the county. Rare.
    Constituent or residual: Fd, Fw, Eh, Gr.
First record; Pulteney (1749).
Herb.: LSR★; LSR, MANCH, NMW.
HG 113: rare in Leicestershire, frequent in Rutland, restricted in range in Leicestershire; 26 localities cited.

BUXACEAE
Buxus L.
B. sempervirens L. Box
Map 38e; 44 tetrads.
Often planted in public and private parks and gardens, and escaping into woodland or hedgerow.
  Planted or intrusive: Fd, Fm, Fs, Eh, Ap, Rq.
First record: Kirby (1850).
Herb.: DPM, LSR, LTR, NMW.
HG 476: frequent, widely distributed; 16 localities cited.

RHAMNALES

RHAMNACEAE
Rhamnus L.
R. catharticus L. Purging Buckthorn
Map 38f; 233 tetrads.
Mainly in hedgerows, less frequently in woodland. Rare or absent in Charnwood Forest and the north-west of the county; occasional or locally frequent elsewhere. Often found in hedgerows as single isolated shrubs, rarely in any quantity. An exception is the neighbourhood of Eye Kettleby, where many of the hedgerows contain it in abundance.
  Constituent or intrusive: Eh, Ew, Er, Ev, Ea, Ef, Fd, Fs, Fw, Rq, Sw.
First record: Pulteney (1747).
Herb.: LSR★; CGE, DPM, LSR, LTR, MANCH, NMW.
HG 113: frequent, generally distributed; 95 localities cited.

Frangula Miller
F. alnus Miller Alder Buckthorn
Map 38g; 4 tetrads.
Woodhouse, Roecliffe Spinney, two old shrubs, PHG 1968; Woodhouse Eaves, The Brand, stream side, one large old specimen, PHG 1969; Swithland Wood, one large old tree, PHG 1970 (died in drought of 1976); Newtown Linford, landscaped grounds of waterworks, one specimen, probably planted, PHG 1975; Gracedieu Wood, single specimen, SHB 1977.
  Constituent or residual: Fd, Ew, Ap.
First record: Pulteney (1749).
Herb.: LSR★; DPM, LSR, MANCH.
HG 115 as *Rhamnus frangula*: rare, restricted in range; 21 localities cited, including Swithland.

MALVALES

TILIACEAE
Tilia L.
T. platyphyllos Scop. Large-leaved Lime
No map; 13 tetrads.
Occasional throughout the county, mostly as planted trees, which the majority of the field workers did not record. There may be spontaneous occurrences in woodland and hedgerows.
  Planted or intrusive: Ap, Ev, Eh, Fd, Fm, Fs.
First record: Bloxam (1837).
Herb.: CGE, LSR, NMW.
HG 100: rare, restricted in range; 12 localities cited.

T. cordata Miller Small-leaved Lime
Map 38h; 22 tetrads.
Locally frequent in woodland in Charnwood Forest. Rare in other parts of the county. Almost certainly constituent in the Charnwood Forest localities.
  Constituent, intrusive or planted: Fd, Fm, Eh, Ap.
First record: Pulteney (1749).
Herb.: LSR★; CGE, DPM, LSR, LTR, MANCH, NMW.
HG 101: local, restricted in range chiefly to Charnwood Forest and region W. of R. Soar; 17 localities cited.

T. cordata × platyphyllos = T. × vulgaris Hayne Lime
Map 38i; 297 tetrads.
Frequent throughout the county as planted trees in avenues, public and private parks, churchyards etc., and in hedgerows and woodland.
  Planted or intrusive: Ap, Ev, Eh, Ew, Fd, Fm, Fs.
First record: Crabbe *in* Nichols (1795).
Herb.: CGE, LSR, LTR, MANCH, NMW.
HG 100: frequent, generally distributed.

MALVACEAE
Malva L.
M. moschata L. Musk Mallow
Map 38j; 38 tetrads.
Quarries, waste places, railway verges, and occasionally in grassland. Occasional in the eastern part of Charnwood Forest and the south-east of the county. Rare elsewhere.
  Intrusive or constituent: Rq, Rs, Rw, Er, Ev, Ew, Eh, Ef, Gp, Gr, Hg, Ap, Wd.
First record: Pulteney (1747).
Herb.: LSR★; LSR, LTR, MANCH, NMW.
HG 99: local, generally distributed.

M. sylvestris L. Common Mallow
Map 38k; 368 tetrads.
Roadside verges and waste places. Frequent throughout the county, but with a somewhat discontinuous distribution.
  Constituent or intrusive: Ev, Er, Ew, Eh, Ea, Rw, Rq, Rg, Rs, Rt, Rd, Rr, Rf, Gr, Gp, Gl, Ag, Ap, Hd, Sw.
First record: Pulteney (1747).
Herb.: LSR★; DPM, LANC, LSR, LTR, MANCH, NMW.
HG 99: locally abundant, generally distributed.

M. nicaeensis All.
No map; 1 tetrad.
Glenfield, waste ground, rare, EH 1971 (site now built over).
  Casual: Rw.
First record: B.E.C. Rep. 1916.
HG 100: rare; 1 locality cited.

M. parviflora L. Least Mallow
No record since HG.
First record: Horwood and Gainsborough (1933).
Herb.: LSR.
HG 100: rare; 3 localities cited.

M. pusilla Sm. Small Mallow
No record since HG.
First record: Jackson (1904b).
Herb.: LANC, LSR, NMW.
HG 100: rare; 10 localities cited.

M. neglecta Wallr. Dwarf Mallow
Map 38l; 102 tetrads.
Occasional throughout the county on roadside verges, in waste places, and on trackways where it can tolerate some trampling.
 Intrusive: Ev, Er, Eh, Rt, Rw, Rd, Rf, Rr, Ag, Aa, Ap, Gp, Sw.
First record: Pulteney (1747).
Herb.: LSR★; DPM, LSR, LTR, MANCH, NMW.
HG 99 as *M. rotundifolia*: locally abundant, generally distributed.

Lavatera L.

L. trimestris L.
No map; 1 tetrad.
Countesthorpe, disturbed soil on roadside, rare, EH 1975.
 Casual or escape: Ev.
First record: 1975, E. Hesselgreaves in present work.
Herb.: LSR★.
HG: not recorded.

THYMELAEALES

THYMELAEACEAE
Daphne L.

D. laureola L. Spurge-laurel
Map 39a; 17 tetrads.
Hedgerows and woodland in scattered localities throughout the county. Sometimes locally frequent where it does occur.
 Constituent, residual or intrusive: Fd, Fm, Eh, Ap.
First record: Pulteney (1749).
Herb.: LSR★; LSR, LTR, MANCH, NMW.
HG 472: rare, sparsely distributed; 34 localities cited.

GUTTIFERALES

GUTTIFERAE
Hypericum L.

H. calycinum L. Rose-of-Sharon
No map; 1 tetrad.
Leicester, stonework at side of canal, rare, HB 1970. Frequently planted and often escaping, but it seems to have been overlooked in the recent survey.
 Garden escape or intrusive: Ew.
First record: Kirby (1850).
Herb.: DPM, LSR, LTR.
HG 94: rare, restricted in range; 2 localities cited.

H. androsaemum L. Tutsan
No map; 3 tetrads.
Anstey, waste ground, rare, EH 1970; Stapleford Park, rough grass by lake, rare, KGM 1972; Shackerstone, roadside verge, single plant, SHB 1978.
 Garden escape or intrusive: Ev, Ew, Rw.
First record: Kirby (1850).
HG 94: rare, restricted in range; 3 localities cited.

H. hirsutum L. Hairy St. John's-wort
Map 39b; 108 tetrads.
Rough grassland, roadside and railway verges and woodland. Occasional throughout the county, but avoiding the more acid soils.
 Constituent or intrusive: Gr, Gp, Er, Ev, Eh, Ew, Ef, Fd, Fm, Fs, Fw, Wd, Rq, Sw.
First record: Crabbe *in* Nichols (1795).
Herb.: LSR★; DPM, LSR, LTR, MANCH, NMW.
HG 97: locally abundant, widely distributed; 82 localities cited.

H. pulchrum L. Slender St. John's-wort
Map 39c; 16 tetrads.
Locally frequent in woodland and heath at Mountsorrel and on the eastern edge of Charnwood Forest. Occasional in a few scattered localities in the rest of the county.
 Constituent: Fd, Fm, Ef, Hg, Gr.
 Residual or intrusive: Rq.
First record: Pulteney (1747).
Herb.: LSR, LTR, MANCH, NMW.
HG 96: local, widely distributed; 57 localities cited.

H. elodes L. Marsh St. John's-wort
No record in the recent survey.
Beacon Hill, with **Scutellaria minor**, F. A. Sowter 1947.
First record: Pulteney *in* Nichols (1795).
HG 98: rare, confined to Charnwood Forest, probably extinct; 1 locality cited, Beacon Hill [where it was recorded by Pulteney, who said that it 'could not be found a few years afterwards'].

H. humifusum L. Trailing St. John's-wort
Map 39d; 22 tetrads.
Grassland. Occasional in Charnwood Forest, rare and sporadic in its occurrence elsewhere.
 Constituent: Gp, Gr, Hg, Hd, Ew, Ef.
 Residual or intrusive: Rq, Rs, Rw, Rd, Rt, Wd, Ao, Ag.
First record: Pulteney (1747).
Herb.: LSR★; CGE, LANC, LSR, LTR, MANCH, NMW, WAR.
HG 95: occasional, widely distributed; 39 localities cited.

H. tetrapterum Fries Square-stalked St. John's-wort
Map 39e; 310 tetrads.
Frequent throughout the county in marshes and ditches, and on the banks of rivers, canals and lakes.
 Constituent: M, Hw, Wp, Wl, Wd, Ew, Ef.
 Residual or intrusive: Fd, Ev, Eh, Er, Gr, Gp, Rq, Rg.
First record: Pulteney (1747).
Herb.: LSR★; DPM, LANC, LSR, LTR, MANCH, NMW.
HG 95 as *H. acutum*: frequent, generally distributed.

H. maculatum Crantz Imperforate St. John's-wort
Map 39f; 36 tetrads.
Grass verges and rough grassland. Occasional in the south and west of the county. Rare or absent elsewhere.
 Constituent or intrusive: Er, Ev, Ew, Eh, Ef, Fs, Gr, Rw, Rd.
First record: Kirby (1850).
Herb.: LSR★; LSR, LTR, MANCH, NMW.
HG 95 as *H. quadrangulum*: rare, restricted in range; 14 localities cited.

H. perforatum L. Perforate St. John's-wort
Map 39g; 305 tetrads.
Frequent throughout the county on roadsides, in quarries and rough grassland. Locally abundant on railway verges, especially the ballast of dismantled railways.
  Constituent or intrusive: Gr, Gp, Hg, Ef, Er, Ev, Eh, Ew, Ea, Rq, Rw, Rd, Fd, Wd, Sw, Ag.
First record: Pulteney (1747).
Herb.: LSR★; LANC, LSR, LTR, MANCH, NMW.
HG 94: frequent, generally distributed.

VIOLALES

VIOLACEAE
Viola L.

V. odorata L. Sweet Violet
Map 39h; 296 tetrads.
Hedgerows, woodland, and shady roadside verges. Occasional in Charnwood Forest; frequent elsewhere. Characteristic of secondary woodland, rarely found in ancient woods.
  Constituent: Fd, Fm, Fw, Fs.
  Residual or intrusive: Eh, Ev, Er, Ew, Rq, Gr, Gp, Ap, Ag.
First record: Pulteney (1747).
Herb.: LSR★; CGE, LANC, LSR, LTR, MANCH, NMW.
HG 61: frequent, generally distributed.

V. hirta L. Hairy Violet
Map 39i; 31 tetrads.
Occasional in rough grassland, and in scattered localities throughout the county on railway verges, where it may be locally frequent. 56% of the records received are from railway verges.
  Constituent or intrusive: Er, Gr, Eh, Ef, Ev.
First record: Pulteney in Nichols (1795).
Herb.: LSR★; CGE, LSR, MANCH, NMW.
HG 62: rather occasional, sparsely distributed; 83 localities cited.

V. hirta × odorata = V. × permixta Jord.
Map 39j; 1 tetrad.
Braunstone, Kirby Spinney, rare, EH 1976.
First record: Trans. L.L.P.S. (1907).
Herb.: LSR★.
HG 63: 4 localities cited.

V. reichenbachiana Jordan ex Boreau
  Early Dog-violet
Map 39k; 27 tetrads.
Woodland, tolerating fairly deep shade, and occasionally in hedgerows. In scattered localities throughout the county; possibly under-recorded.
  Constituent: Fd, Fs, Fw, Ef.
  Residual or intrusive: Eh, Ew, Er, Gp, Sw.
First record: 1843, F. T. Mott in herb. MANCH.
Herb.: LSR; CGE, DPM, LSR, LTR, MANCH, NMW.
HG 64 as *V. sylvestris*: local, widely distributed; 21 localities cited.

V. reichenbachiana × riviniana
Map 39l; 1 tetrad.
Asfordby Hill, railway cutting at west end of tunnel, locally abundant, ALP 1972. The plants were luxuriant, showing typical hybrid vigour.
  Intrusive: Er.

First record: 1972, A. L. Primavesi in present work.
HG: not recorded.

V. riviniana Reichenb. Common Dog-violet
Map 40a; 299 tetrads.
Woodland, hedgerows, roadsides and railway verges. Frequent throughout the county; more common in the west than in the east.
  Constituent, residual or intrusive: Fd, Fm, Fs, Ef, Eh, Er, Ev, Ew, Wd, Gp, Gr, Ap, A.
First record: Pulteney (1747).
Herb.: LSR★; DPM, LANC, LSR, LTR, MANCH, NMW.
HG 64: locally abundant, generally distributed.

V. canina L. Heath Dog-violet
Map 40b; 4 tetrads.
Elmesthorpe, sandy hedgebank bounding rough pasture, occasional, SHB 1972; Elmesthorpe, grassy top of deep stream ditch, occasional, SHB 1972 (site destroyed 1975 in construction of M69); Swithland Wood meadow, old grassland, rare, PHG 1975; Braunstone Frith, golf course, small deturfed area, rare, EH 1976; Muston, rough grassland, abundant, PAC 1977.
  Constituent or residual: Gp, Gr, Wd, Rd.
First record: 1832, R. H. Cresswell in herb. WAR.
Herb.: LSR★; LSR, WAR.
HG 66: local, rather restricted in range; 12 localities cited.

V. canina × riviniana
Map 40c; 1 tetrad.
Elmesthorpe, grassy top of deep stream ditch, locally frequent SHB 1972 (site destroyed 1975 in construction of M69).
  Intrusive: Ew.
First record: B.E.C. Rep. 1917.
HG 67: 2 localities cited.

V. palustris L. Marsh Violet
Map 40d; 9 tetrads.
Ulverscroft, near Ulverscroft Cottage Farm, marsh, occasional, PAC 1968; Ulverscroft, marshland by brook south of Priory, rare, PHG 1968; Charnwood Lodge Nature Reserve, wet areas of moorland along banks of stream, locally frequent, PAC 1969; Beacon Hill, lower slopes, relict bog, locally frequent, PHG 1969; Ulverscroft, marshy meadow adjoining Nature Reserve, locally frequent, PAC 1974; Staunton Harold, Spring Wood, scrub woodland, locally abundant, SHB 1977; Charley, Gun Hill, ditch, occasional, PHG 1982.
  Constituent: Hw, Hb, M, Fs, Wd.
First record: Power (1805).
Herb.: LANC, LSR, MANCH, NMW.
HG 61: rare, almost entirely confined to region north west of River Soar; 19 localities cited.

V. tricolor L. Wild Pansy
Map 40e; 13 tetrads.
Rearsby, railway cutting, rare, ALP 1969; Anstey, soil tip, occasional, EH 1969; Hoby, corn field, occasional, ALP 1971; Groby, verge of mineral railway in quarry, rare, EH 1971; Quorndon, farmyard, rare, PHG 1971; Groby, soil tip, rare, EH 1972; Sibson, roadside verge A444, rare, HB 1972; Newtown Linford, allotment garden, locally frequent, EH 1973; Groby, soil tip in old sand quarry, rare, EH 1973; Ratcliffe College, newly

made lawn, occasional, ALP 1974 (lawn regularly mown since); Glen Parva, soil heap near brick works, rare, EH 1974; Snarestone, overgrown portion of metalled road, rare, HH 1976; Moira, waste ground, rare, SHB 1977.
    Intrusive: Ac, Aa, Ag, Rd, Rf, Rq, Rt, Er, Ev.
First record: Pulteney (1747).
Herb.: LSR★; LANC, LSR, LTR, MANCH.
HG 67. It is not clear what Horwood includes under this taxon. There seems to be some confusion with **V. arvensis.**

V. arvensis Murray   Field Pansy
Map 40f; 361 tetrads.
Arable land, waste places, roadside and railway verges. Frequent throughout the county; more frequent in the west than in the east.
    Intrusive: Ac, Ao, Ag, Aa, Ap, Rw, Rd, Rs, Rg, Rf, Rt, Ea, Er, Ev, Gl.
First record: Coleman (1852).
Herb.: LSR★; CGE, DPM, LSR, LTR, MANCH, NMW.
HG 68: see note under **V. tricolor.**

CISTACEAE
Helianthemum Miller
H. nummularium (L.) Miller   Common Rock-rose
H. chamaecistus Miller
Map 40g; 3 tetrads.
Croxton Kerrial, King Lud's Entrenchments, limestone grassland, F. A. and M. G. Sowter 1969; Croxton Kerrial, west side of The Drift, limestone grassland, locally frequent, KGM 1973; Sproxton, mineral railway bank, rare, KGM 1973; Saltby Heath, limestone grassland near Egypt Plantation, occasional, KGM 1974.
    Constituent: Gp, Er.
First record: Pulteney *in* Nichols (1795).
Herb.: :LSR, LTR, MANCH, NMW.
HG 60 as *H. vulgare*: rare, very restricted in range; 13 localities, including the above, but not confined to that region.

[ELATINACEAE
Elatine L.
E. hydropiper L.   Eight-stamened Waterwort
There is a specimen in herb. BM labelled as collected by D. P. Murray (though the label is not in Murray's own handwriting) from a pond at Church Langton in August 1923. There is no reference to this specimen in any of Murray's ms. catalogues or in his own collections. There is, however, a specimen in his herbarium at Stoke Golding from Slitting Mill pond, Rugeley, Staffs. dated August 1923. It is most likely that the BM specimen is from the same site, and that the labelling is a mistake.
HG 94: no Leicestershire record.]

CUCURBITALES

CUCURBITACEAE
Bryonia L.
B. cretica L. subsp. **dioica** (Jacq.) Tutin
                                      White Bryony
B. dioica Jacq.
Map 40h; 203 tetrads.
Hedgerows. Locally frequent, especially on the alluvial soils of the Soar and lower Wreake valleys.
    Constituent or intrusive: Eh, Ev, Er, Ew, Ef, Fd, Fs, Wd, Rw, Rf, Ap.
First record: Pulteney (1747).
Herb.: LSR★; DPM, LSR, LTR, NMW.
HG 250: frequent but of local distribution; 66 localities cited.

MYRTALES

LYTHRACEAE
Lythrum L.
L. salicaria L.   Purple-loosestrife
Map 40i; 41 tetrads.
Locally frequent on the banks of the River Soar in its lower reaches; occasional on the banks of the Ashby Canal; rare or absent elsewhere in the county.
    Constituent: Ew, Wl, M.
    Intrusive: Rg, Wd.
First record: Pulteney (1747).
Herb.: LSR★; LANC, LSR, LTR, MANCH, NMW.
HG 241: frequent, but of local distribution; 54 localities cited.

L. junceum Banks & Solander
No map; 1 tetrad.
Barwell, rubbish dump, single plant, J. E. Dawson 1979 (det. E. J. Clement).
    Casual: Rr.
First record: 1979, J. E. Dawson in present work.
Herb.: LSR★.
HG: not recorded.

L. hyssopifolia L.   Grass-poly
No record since HG.
First record: B.E.C. Rep. 1921.
HG 242: very rare, restricted in range; 1 locality cited, Market Bosworth; (the other locality cited, near Calke, is certainly in Derbyshire).

L. portula (L.) D. A. Webb   Water-purslane
Peplis portula L.
Map 40j; 9 tetrads.
Bagworth, marshy area by stream, locally frequent, SM 1969; Cropston Reservoir, muddy margins, locally frequent, PHG 1969; Martinshaw Wood, wet rides, locally frequent, EH 1970; Bradgate Park, wet heath, locally abundant, two localities, PHG 1972 and EH 1972; Swepstone, small pond, locally abundant, HH 1973; Rothley, muddy margins of pond off Kinchley Lane, locally abundant, PHG 1974; Charnwood Lodge Nature Reserve, wet heath, locally frequent, PHG 1979.
    Constituent or residual: Hw, Ew, Ef.
First record: Pulteney (1747).
Herb.: LSR★; LSR, LTR, MANCH, NMW.
HG 241: frequent, but of local distribution; 54 localities cited.

ONAGRACEAE
Circaea L.
C. lutetiana L.   Enchanter's-nightshade
Map 40k; 246 tetrads.
Frequent throughout the county in woodland. Occasional as a garden weed.
    Constituent: Fd, Fm, Fs, Fw, Ef.
    Residual or intrusive: Eh, Ew, Ev, Er, Ag, Ap, Rw.
First record: Pulteney (1747).
Herb.: LSR★; DMP, LANC, LSR, LTR, MANCH, NMW.
HG 249: frequent, generally distributed.

Oenothera L.
O. biennis L.   Common Evening-primrose
No map; 13 tetrads.
Waste ground, roadside and railway verges, usually a non-persistent casual or garden escape. Occasional throughout the county.
    Casual or garden escape: Rw, Rd, Rr, Ev, Er, Ew, Aa.
First record: Power (1807).
Herb.: LTR.
HG 249: rare, restricted in range; 9 localities cited.

O. erythrosepala Borbás
         Large-flowered Evening-primrose
No map; 22 tetrads.
Waste places, rubbish dumps, roadside and railway verges, usually a non-persistent casual or garden escape. In scattered localities throughout the county; slightly more frequent than the preceding species.
    Casual or garden escape: Rw, Rr, Rd, Rq, Rs, Rt, Rg, Ev, Er.
First record: Horwood and Gainsborough (1933).
Herb.: LSR*.
HG 249 erroneously as *O. grandiflora*: rare; 1 locality cited.

Epilobium L.
No hybrids of **Epilobium** species were determined in the recent survey. Records are given for any hybrids which are recorded in Horwood and Gainsborough or for which there are known herbarium specimens. Most of S.A. Taylor's specimens in herb. LSR were determined or confirmed by G. M. Ash.

E. angustifolium L.   Rosebay Willowherb, Fireweed
Chamaenerion angustifolium (L.) Scop.
Map 40l; 594 tetrads.
Frequent and locally abundant in woodland, hedgerows and waste places. Often completely covers an area of recently felled woodland or the site of a fire. Now so universally common that it is hard to believe that it was comparatively rare before the First World War.
    Constituent or intrusive: Fd, Fm, Fs, Ef, Eh, Ev, Er, Ew, Ea, Rw, Rq, Rg, Rs, Rr, Rf, Hg, Gr, Gp, Wd, Ap, Ag, Sw, Sb.
First record: Bloxam (1838).
Herb.: LSR*; DPM, LANC, LSR, LTR, MANCH, NMW.
HG 243: locally abundant, on the increase, generally distributed.

E. hirsutum L.   Great Willowherb
Map 41a; 600 tetrads.
Ditches, marshes, wet woodland rides, the banks of waterways and streams, and damp waste ground. Abundant throughout the county.
    Constituent, residual or intrusive: Wd, Wp, Wl, M, Ew, Ef, Eh, Ev, Er, Rw, Rq, Rg, Rs, Rr, Rd, Gr, Gp, Ap, Ao, Ag.
First record: Pulteney (1747).
Herb.: LSR*; DPM, LANC, LSR, LTR, MANCH.
HG 243: frequent, generally distributed.

E. hirsutum × parviflorum = E. × intermedium Ruhmer
No record in the recent survey.
Debdale Wharf, S. A. Taylor 1940.
First record: Preston (1895).
Herb.: LSR.
HG 244: 2 localities cited.

E. hirsutum × montanum = E. × erroneum Hausskn.
No record in the recent survey.
Groby Quarry, S. A. Taylor 1941.
First record: Preston (1900).
Herb.: LSR.
HG 244: 1 locality cited.

E. hirsutum × tetragonum = E. × brevipilum Hausskn.
No record in the recent survey.
Groby Quarry, S. A. Taylor 1943.
First record: 1943, S. A. Taylor in herb. LSR.
Herb.: LSR.
HG: not recorded.

E. parviflorum Schreber   Hoary Willowherb
Map 41b; 385 tetrads.
Marshes, ditches and the banks of rivers, canals, streams and ponds. Frequent throughout the county.
    Constituent, residual or intrusive: M, Wd, Wl, Wp, Ew, Er, Ev, Ef, Fw, Fd, Fs, Rg, Rq, Gr, Gp, Ag, Ap.
First record: Pulteney (1747).
Herb.: LSR*; DPM, LANC, LSR, LTR, MANCH, NMW.
HG 244: frequent, generally distributed.

E. parviflorum × tetragonum = E. × weissenburgense F. W. Schultz
No record in the recent survey.
Groby Quarry, S. A. Taylor 1941.
First record: 1941, S. A. Taylor in herb. LSR.
Herb.: LSR.
HG: not recorded.

E. parviflorum × roseum = E. × persicinum Reichb.
No record in the recent survey.
Swithland Reservoir, S. A. Taylor 1941.
First record: 1880, T. A. Preston in herb. LSR.
Herb.: LSR.
HG 244: 1 locality cited (Cropston Reservoir).

E. montanum L.   Broad-leaved Willowherb
Map 41c; 431 tetrads.
Frequent throughout the county in woodland and hedgerows, as a garden weed, and in waste places.
    Constituent, residual or intrusive: Fd, Fm, Fs, Fw, Ef, Eh, Ev, Er, Ew, Ag, Ap, Aa, Ao, Rw, Rq, Rg, Rs, Rd, Rf, Rt, Wd, Wp, Wl, Sw.
First record: Pulteney (1747).
Herb.: LSR*; CGE, DPM, LANC, LSR, LTR, MANCH, NMW.
HG 245: locally abundant, generally distributed.

E. montanum × parviflorum = E. × limosum Schur
No record in the recent survey.
Swithland Reservoir, S. A. Taylor 1941.
First record: 1899, T. A. Preston in herb. LSR.
Herb.: LSR.
HG 245: 2 localities cited (Melton Spinney; Cropston Reservoir).

E. montanum × tetragonum = E. × beckhausii Hausskn.
No record in the recent survey.
Groby Quarry, S. A. Taylor 1942.
First record: 1942, S. A. Taylor in herb. LSR.
Herb.: LSR.
HG: not recorded.

E. montanum × obscurum = E. × aggregatum Celak.
No record since HG.
First record: 1897, E. F. Linton in herb. LSR.
Herb.: LSR.
HG 245: 1 locality cited (Cropston).

E. montanum × roseum = E. × mutabile Boiss. & Reut.
No record in the recent survey.
Husbands Bosworth, S. A. Taylor 1947.
First record: 1882, F. T. Mott in herb. LSR.
Herb.: LSR.
HG 245: 2 localities cited (Hinckley; between Swannington and Coalville).

E. montanum × palustre = E. × montaniforme Knaf ex Celak.
No record in the recent survey.
Groby Quarry, S. A. Taylor 1944.
First record: 1944, S. A. Taylor in herb. LSR.
Herb.: LSR.
HG: not recorded.

E. lanceolatum Sebastiani & Mauri   Spear-leaved Willowherb
Map 41d; 7 tetrads.
Quorndon, chapel graveyard, rare, PHG 1974; Swithland, railway verge, occasional, PHG 1976; Quorndon, garden weed, rare, PHG 1979; Frisby on the Wreake, garden weed, occasional, ALP 1979; Billesdon, churchyard, occasional, KGM 1979; Stoughton, churchyard, occasional, KGM 1979; East Norton, garden, occasional, KGM 1979; Tur Langton, churchyard, occasional KGM 1979.
 Intrusive: Ap, Ag, Er.
First record: 1974, P. H. Gamble in present work.
HG: not recorded.

E. tetragonum L.   Square-stalked Willowherb
Subsp. **tetragonum**
Map 41e; 53 tetrads.
Wet places, woodland and waste ground. Occasional throughout the county.
 Constituent or intrusive: M, Ew, Er, Ev, Eh, Ea, Fd, Fm, Fs, Hw, Rw, Rd, Rq.
First record: Pulteney (1747).
Herb.: LSR★; LSR, LTR, MANCH, NMW.
HG 246: frequent, widely distributed; 43 localities cited.

Subsp. **lamyi** (F.W. Schultz) Nyman
Map 41f; 4 tetrads.
Leicester, Freemen's Common, disused allotments, locally frequent, HB 1969 (site since built over); Ratcliffe College, herbaceous border, rare, ALP 1971; Enderby, rough pasture, rare, EH 1974; Markfield, rough marshy grassland, locally frequent, HH 1974.
 Constituent or intrusive: Gr, M, Aa, Ag.

First record: 1969, H. Bradshaw in present work.
Herb.: LSR★.
HG 248 as *E. lamyi*: no Leicestershire record.

E. obscurum Schreber   Short-fruited Willowherb
Map 41g; 76 tetrads.
Wet places, roadside and railway verges, and occasionally as a garden weed. Locally frequent; probably under-recorded in parts of the county.
 Constituent or intrusive: M, Hw, Wd, Wp, Ew, Er, Ev, Ea, Fw, Fd, Ag, Rt, Rd, Rs, Sw.
First record: Coleman (1852).
Herb.: LSR★; CGE, LSR, MANCH, NMW.
HG 247: local, sparsely distributed; 37 localities cited.

E. obscurum × parviflorum = E. × dacicum Borbás
No record in the recent survey.
Holwell Mouth, S. A. Taylor 1942; Thurlaston, sandpit, S. A. Taylor 1946.
First record: 1841, in herb. LSR.
Herb.: LSR.
HG 248: 2 localities cited (Blaby; River Wreake, Syston).

[**E. obscurum × roseum = E. × brachiatum** Celak
All known herbarium material from Leicestershire originally referred to this hybrid has been subsequently redetermined by G. M. Ash as **E. adenocaulon**, q.v. Hence the record in Horwood and Gainsborough p.248, which is based on this material, is erroneous.]

E. obscurum × palustris = E. × schmidtianum Rostk.
No record in the recent survey.
Holwell Mouth, S. A. Taylor 1941.
First record: 1941, S. A. Taylor in herb. LSR.
Herb.: LSR.
HG: not recorded.

E. roseum Schreber   Pale Willowherb
Map 41h; 31 tetrads.
Banks of rivers and streams and other wet places. Occasional throughout the county; probably under-recorded in some areas.
 Constituent or intrusive: Ew, Ev, Er, Ef, Fd, Fm, M, Rq, Rs, Rt, Ag.
First record: Hands et al. *in* Curtis (1831).
Herb.: LSR★; LSR, LTR, MANCH, NMW.
HG 245: rather rare, restricted in range; 26 localities cited.

E. palustre L.   Marsh Willowherb
Map 41i; 30 tetrads.
Marshes and wet places. Occasional in the west of the county; rare in the east.
 Constituent: M, Ew, Fw, Wp.
 Residual or intrusive: Wd, Er, Gp.
First record: Pulteney *in* Nichols (1795).
Herb.: LSR★; DPM, LANC, LSR, LTR, MANCH, NMW.
HG 248: frequent, fairly widely distributed; 40 localities cited.

E. palustre × parviflorum = E. × rivulare Wahlenb.
No record in the recent survey.
Holwell Mouth, S. A. Taylor 1941.
First record: 1941, S. A. Taylor in herb. LSR.
Herb.: LSR.
HG: not recorded.

E. adenocaulon Hausskn.   American Willowherb
Map 41j; 191 tetrads.
Waste places, railway and roadside verges, and as a garden weed. Frequent throughout the county; locally abundant in the neighbourhood of the larger towns. This species has increased phenomenally in the past 30 years.
   Intrusive: Rw, Rq, Rg, Rs, Rd, Rt, Rf, Rr, Er, Ev, Ew, Ef, Eh, Ea, Ag, Aa, Ap, M, Wp, Wd, Fd, Fm, Sw, Sb.
First record: 1891, T. A. Preston in herb. BM and LSR.
Herb.: LSR★; BM, CGE, DPM, LSR, LTR, MANCH, OXF.
HG: not recorded. In fact there are specimens in herb. BM and LSR which were collected by T. A. Preston in July 1891 at Cropston Reservoir and identified by E. S. Marshall, not without protest from Preston, as *E. obscurum* × *roseum*. The species was first recognised in Surrey in 1933 by G. M. Ash, who subsequently found older material in a number of herbaria. Ash visited Cropston in 1938, and deposited a specimen in herb. BM. Preston's specimens from Cropston Reservoir are the earliest known British specimens.

E. adenocaulon × hirsutum
No record in the recent survey.
Leicester, St. Mary's Road, S. A. Taylor 1946.
First record: 1946, S. A. Taylor in herb. LSR.
Herb.: LSR.
HG: not recorded.

E. adenocaulon × parviflorum
No record in the recent survey.
Groby Quarry, S. A. Taylor 1942 and 1946.
First record: 1942, S. A. Taylor in herb. LSR.
Herb.: LSR.
HG: not recorded.

E. adenocaulon × montanum
No record in the recent survey.
Loughborough, Outwoods Drive, S. A. Taylor 1946.
First record: 1946, S. A. Taylor in herb. LSR.
Herb.: LSR.
HG: not recorded.

E. nerterioides A. Cunn.   New Zealand Willowherb
Map 41k; 5 tetrads.
Bardon Hill, bare stony ground on summit, rare, PAC 1970; Mountsorrel, Cocklow Quarry floor, locally frequent, PHG 1970 (site since quarried); Groby, ant hill in pasture, rare, EH 1973; Cossington, nursery garden weed, occasional, ALP 1973; Grace Dieu, Manor Farm garden, occasional, ALP 1974.
   Intrusive: Hg, Sr, Gp, Ag.
First record: 1948, T. G. Tutin in herb. LTR.
Herb.: LSR★; LSR, LTR.
HG: not recorded.

HALORAGACEAE
Myriophyllum L.
M. verticillatum L.   Whorled Water-milfoil
No record since HG.
First record: Pulteney in Nichols (1795).
Herb.: LSR, MANCH, NMW.
HG 257: rare, very restricted in range; 5 localities cited.

M. spicatum L.   Spiked Water-milfoil
Map 41l; 25 tetrads.
Rivers, canals, lakes, ponds and water-filled gravel or clay pits. Locally abundant in scattered localities throughout the county.
   Constituent or intrusive: Wc, Wr, Wl, Wp, Rg, Rq.
First record: Pulteney (1749).
Herb.: LSR★; DPM, LANC, LSR, MANCH, NMW.
HG 237: frequent, but sparsely distributed.

M. alterniflorum DC.   Alternate Water-milfoil
No record since HG.
First record: 1841, Bloxam in herb. MANCH.
Herb.: LSR, MANCH.
HG 238: rare, restricted in range, still grows in Moira Reservoir; 6 localities cited.

HIPPURIDACEAE
Hippuris L.
H. vulgaris L.   Mare's-tail
Map 42a; 10 tetrads.
Oakthorpe, disused canal, EKH 1960 (since filled in); Measham, canal, T. W. Tailby 1960 (since filled in); Eye Brook Reservoir, locally abundant, KGM 1971; Ravenstone, subsidence pool, locally frequent, PAC 1972; Goadby Marwood Hall, middle lake, locally abundant, almost covering large shallow lake, ALP and PCP 1972; Eastwell, marshy verge of large field pond, occasional, ALP and SMF 1974; Eaton, brook, locally abundant, JMS 1975; Staunton Harold, lake, locally frequent, SHB 1978; Garthorpe, ponds, locally frequent, PAE 1979.
   Constituent or intrusive: Wp, Wl, Wc, Wr.
First record: Pulteney (1749).
Herb.: LSR★; CGE, LANC, LSR, MANCH, NMW.
HG 236: occasional, sparsely distributed; 26 localities cited, including Oakthorpe, Goadby Marwood and Staunton Harold.

UMBELLIFLORAE

CORNACEAE
Cornus L.
C. sanguinea L.   Dogwood
Thelycrania sanguinea (L.) Fourr.
Map 42b; 416 tetrads.
Frequent throughout the county in hedgerows and woodland, with a preference for basic soils.
   Constituent: Fd, Fm, Fs, Fw, Ef, Eh.
   Residual or intrusive: Ev, Er, Ew, M, Gr, Rq, Ap.
First record: Pulteney (1747).
Herb.: LSR★; DPM, LTR, MANCH, NMW.
HG 272: frequent, generally distributed.

C. mas L.
No map; 1 tetrad.
Stapleford Park, shrubbery, occasional, KGM 1972.
   Planted: Fs.
First record: Horwood and Gainsborough (1933).
HG 272: rare; 3 localities cited.

ARALIACEAE
Hedera L.
H. helix L.   Ivy
Map 42c; 591 tetrads.
Woodland, hedgerows, walls and buildings, frequent and locally abundant throughout the county. Frequently

forms dense carpets on the floor of woodland, and occasionally also on shady railway verges, sometimes dominant to the exclusion of all other vegetation.
  Constituent or intrusive: Fd, Fm, Fs, Ef, Eh, Er, Ev, Sw, Sb, Wd, Ap, Rq, Rw.
First record: Pulteney (1747).
Herb.: LSR*; CGE, DPM, LANC, LSR, LTR, MANCH, NMW.
HG 271: locally abundant, generally distributed.

UMBELLIFERAE
Hydrocotyle L.

H. vulgaris L.   Marsh Pennywort
Map 42d; 14 tetrads.
Marshes and wet heath. Locally frequent in Charnwood Forest and the west of the county; rare or absent elsewhere.
  Constituent: M, Hw, Hb, Wd, Wl, Ew.
First record: Pulteney (1747).
Herb.: LSR*; LSR, LTR, MANCH, NMW.
HG 351: local, restricted in distribution; 38 localities cited.

Sanicula L.

S. europaea L.   Sanicle
Map 42e; 71 tetrads.
Old established woodlands. Frequent in the well-wooded parts of the county. Occasional or absent elsewhere.
  Constituent: Fd, Fc, Fm, Fs, Fw, Ef.
  Residual: Eh, Ew, Ev, Wd.
First record: Pulteney (1747).
Herb.: LSR*; LSR, LTR, MANCH, NMW.
HG 252: frequent, generally distributed.

Astrantia L.

A. major L.   Astrantia
No record since HG.
First record: Bemrose (1927).
Herb.: LSR.
HG 252: rare; 1 locality cited.

Chaerophyllum L.

C. temulentum L.   Rough Chervil
Map 42f; 366 tetrads.
Frequent throughout the county in hedgerows and woodland, but avoids the more acid soils.
  Constituent, residual or intrusive: Fd, Fm, Fs, Ef, Eh, Ev, Ew, Er, Ea, Gr, Gp, Gl, Rg, Rs, Rd.
First record: Pulteney (1747).
Herb.: LSR*; CGE, DPM, LANC, LSR, LTR, NMW.
HG 260: locally abundant, generally distributed.

Anthriscus Pers.

A. sylvestris (L.) Hoffm.   Cow Parsley, Keck
Map 42g; 605 tetrads.
Woodland, hedgerows, roadside verges, often spreading into grassland. Abundant, the commonest umbellifer throughout the county.
  Constituent or intrusive: Eh, Ev, Er, Ea, Ew, Ef, Fd, Fm, Gp, Gr, Gl, Wd, Ap, Rw, Rq.
First record: Pulteney (1747).
Herb.: LSR*; DPM, LANC, LSR, LTR, MANCH, NMW.
HG 262 as *Chaerefolium sylvestre*: locally abundant, generally distributed.

A. cerefolium (L.) Hoffm.   Garden Chervil
No record since HG.
First record: Mott (1888).
Herb.: LSR, MANCH.
HG 262: rare; 1 locality cited.

A. caucalis Bieb.   Bur Chervil
No record in the recent survey.
Burrough Hill, T. G. Tutin 1945 and M. K. Hanson 1953.
First record: Pulteney (1747).
Herb.: LSR, LTR, MANCH, NMW.
HG 261 as *Chaerefolium anthriscus*: rare, very sparsely distributed; 14 localities cited.

Scandix L.

S. pecten-veneris L.   Shepherd's-needle
Map 42h; 4 tetrads.
Saddington, garden, C. W. Holt 1967; Ratby, cornfield, rare, EH 1970; Stoney Stanton, cornfield, rare, MH 1973; Groby, barley field, rare, EH 1973; Higham on the Hill, cornfield, locally frequent, SHB 1974.
  Intrusive: Ac, Ag.
First record: Pulteney (1747).
Herb.: LSR*; DPM, LSR, LTR, MANCH, NMW.
HG 261: frequent, generally distributed.

Myrrhis Miller

M. odorata (L.) Scop.   Sweet Cicely
Map 42i; 3 tetrads.
Oadby, waste ground off Manor Road, G. Halliday 1961; Ashby Woulds, in 5 sites, hedgerow, roadside and waste ground, rare, SHB 1976; Ashby Woulds, rough grass, locally frequent, SHB 1976; Leicester, dismantled railway track, rare, E.J.W. Venable 1984.
  Intrusive or possibly constituent: Gr, Ev, Eh, Rw.
First record: Crabbe *in* Nichols (1795).
Herb.: LSR*; LTR.
HG 260: rare, very restricted in range; 4 localities cited, not including above.

Coriandrum L.

C. sativum L.   Coriander
No map; 3 tetrads.
Leicester, several localities on waste ground and as a garden weed, HB 1971. Cultivated by the Asian community as a culinary herb and escaping.
  Casual or escape: Rw, Ag.
First record: Horwood (1911).
Herb.: LSR*.
HG 269: rare, 5 localities cited.

Smyrnium L.

S. olusatrum L.   Alexanders
No map; 4 tetrads.
Freeby, ruins of Saxby Station platform, rare, KGM 1971 (rubble since cleared away); Breedon Hill, limestone grassland, single plant, PAC 1971; Bittesby, roadside, rare, MH 1973; Thurlaston, waste ground near church, rare, SHB 1976.
  Casual: Er, Ev, Rw, Gp.
First record: Coleman (1852).
Herb.: LSR*; LSR, LTR, MANCH, NMW.
HG 253: rare, restricted in range; 3 localities cited.

Conopodium Koch

C. majus (Gouan) Loret   Pignut
Map 42j; 362 tetrads.
Old grassland, heath and woodland. Frequent and locally abundant throughout the county, except in those parts where because of intensive agriculture there are no suitable habitats; becoming less frequent as old grassland is destroyed.
    Constituent or residual: Gp, Gr, Gl, Hg, Fd, Fm, Fs, Ef, Ev, Er, Eh, Ew, Ea, Rg, Rs, M, Ap, Ag.
First record: Pulteney (1747).
Herb.: LSR★; DPM, LANC, LSR, LTR, MANCH, NMW.
HG 260: locally abundant, generally distributed.

Pimpinella L.

P. major (L.) Hudson   Greater Burnet-saxifrage
Map 42k; 241 tetrads.
Roadside and railway verges and hedgerows. Locally frequent, but with a discontinuous distribution which is not easily explicable in terms of soil types.
    Constituent or intrusive: Ev, Er, Eh, Ew, Ef, Gp, Gr, Wd, M, Fd, Rw, Rs.
First record: Pulteney (1747).
Herb.: LSR★; CGE, DPM, LANC, LSR, LTR, MANCH, NMW.
HG 259: locally abundant, generally distributed.

P. saxifraga L.   Burnet-saxifrage
Map 42l; 198 tetrads.
Locally frequent throughout the county in grassland and on roadside and railway verges.
    Constituent, residual or intrusive: Gp, Gr, Gl, Hg, Hd, Ev, Er, Ew, Eh, Ea, Rq, Rw, Rd, Rt, M, Fd, Fs, Ap, Sr, Sb.
First record: Pulteney (1747).
Herb.: LSR★; CGE, DPM, LANC, LSR, LTR, MANCH, NMW.
HG 258: frequent, generally distributed.

Aegopodium L.

A. podagraria L.   Ground-elder
Map 43a; 459 tetrads.
Roadside verges and waste places, often a persistent and troublesome garden weed. Frequent throughout the county, especially in or near towns and villages.
    Intrusive: Ev, Er, Ew, Eh, Ef, Ag, Aa, Ap, Rw, Rt, Rq, Rs, Rd, Rr, Fm.
First record: Pulteney (1747).
Herb.: LSR★; DPM, LANC, LSR, LTR, NMW.
HG 258: locally abundant, generally distributed.

Sium L.

S. latifolium L.   Greater Water-parsnip
No recent record.
First record: Pulteney (1749).
HG 257: rare, restricted in range to River Soar; 5 localities cited. All the records in HG are very old ones, and Mott, in the 1886 *Flora*, doubted their accuracy.

Berula Koch

B. erecta (Hudson) Coville   Lesser Water-parsnip
Map 43b; 110 tetrads.
Marshes, and the banks of rivers, streams and canals where there is very shallow water. Especially characteristic of canal margins, probably because the water level is fairly constant. Locally frequent on canal margins; occasional elsewhere throughout the county.
    Constituent: M, Wc, Wr, Wp, Wd, Ew.
First record: Pulteney (1747).
Herb.: LSR★; DPM, LANC, LSR, LTR, MANCH, NMW.
HG 258 as *Sium erectum*: occasional, generally distributed.

Oenanthe L.

O. fistulosa L.   Tubular Water-dropwort
Map 43c; 45 tetrads.
Marshes and the margins of canals and rivers, especially where there is shallow water. Locally frequent by the River Soar and the Ashby and Grantham Canals; occasional elsewhere in the county.
    Constituent: M, Ew, Wc, Wp, Wl, Wd.
    Residual: Gp.
First record: Pulteney (1747).
Herb.: LSR★; DMP, LANC, LSR, LTR, MANCH, NMW.
HG 262: frequent, widely distributed; 54 localities cited.

O. silaifolia Bieb.   Narrow-leaved Water-dropwort
Map 43d; 2 tetrads.
Loughborough Big Meadow, rare, IME 1974 and PHG 1980.
    Constituent: M.
First record: Bloxam (1830).
Herb.: LSR★; BM, LSR.
HG 263: rare, very restricted in range; 5 localities cited, including Loughborough Meadows.

O. lachenalii C. C. Gmelin   Parsley Water-dropwort
No record since HG.
First record: Bloxam (1837).
Herb.: CGE, LSR, MANCH.
HG 264: rare, restricted in range; 11 localities cited.

O. fluviatilis (Bab.) Coleman   River Water-Dropwort
Map 43e; 1 tetrad.
Croft, approx. 2 km. stretch of River Soar between Sutton Hill Bridge and Croft village, locally frequent, SHB 1973 (since much reduced by river improvement).
    Constituent: Wr.
First record: Pulteney *in* Nichols (1795).
Herb.: LSR, MANCH, NMW.
HG 265: occasional, in R. Soar, Eye, Sence, Welland, widely dispersed; 26 localities cited, including the above.

O. aquatica (L.) Poiret   Fine-leaved Water-dropwort
Map 43f; 2 tetrads.
Saddington Reservoir, margins and outflow, locally frequent, EKH and JMH 1968; Smeeton Westerby, canal, rare, SHB and PHG 1977.
    Constituent: Wl, Wc.
First record: Crabbe *in* Nichols (1795).
Herb.: LANC, LSR, LTR, MANCH, NMW.
HG 264: local, and rather restricted in range; 18 localities cited, including Saddington Reservoir.

Aethusa L.

A. cynapium L.   Fool's Parsley
Map 43g; 457 tetrads.
Frequent throughout the county in arable land, gardens

and recently disturbed ground.
   Intrusive: Ac, Ao, Ag, Aa, Ap, Rd, Rw, Rq, Rs, Rr, Rt, Ea, Ev, Er.
First record: Pulteney (1749).
Herb.: LSR★; DPM, LANC, LSR, LTR, MANCH, NMW.
HG 266: locally abundant, generally distributed.

Foeniculum Miller
F. vulgare Miller   Fennel
No map; 17 tetrads.
Waste places, railway and roadside verges, as a casual or an established escape. Occasional throughout the county.
   Casual or intrusive: Rw, Rr, Rd, Er, Ev, Eh, Aa.
First record: Preston (1900).
Herb.: LSR★; DPM, LSR, LTR, NMW.
HG 262: rare, restricted in range; 8 localities cited.

Silaum Miller
S. silaus (L.) Schinz & Thell.   Pepper-saxifrage
Map 43h; 151 tetrads.
Grassland, roadside and railway verges. Occasional throughout the county, and seldom in any quantity in one locality.
   Constituent or intrusive: Gp, Gr, Gl, Hg, M, Ev, Er, Ew, Ea, Rw, Rt, Fs.
First record: Pulteney (1747).
Herb.: LSR★; DPM, LANC, LSR, LTR, NMW.
HG 266: frequent, generally distributed.

Conium L.
C. maculatum L.   Hemlock
Map 43i; 258 tetrads.
Frequent throughout the county on the banks of rivers, streams and canals. Occasionally spreads from these habitats into hedgerows, roadsides and the headlands of arable fields.
   Constituent or intrusive: Ew, Eh, Ev, Er, Ea, Wd, Gr, Gp, M, Rw, Rq, Rs, Rd, Rr, Fd, Fm, Fs.
First record: Pulteney (1747).
Herb.: LSR★; DPM, LSR, LTR, NMW.
HG 252: fairly frequent, and generally distributed.

Bupleurum L.
B. rotundifolium L.   Thorow-wax
Of the few records received for this species in the recent survey, none were confirmed, and it is likely that they should be referred to **B. lancifolium.**
First record: Walpoole (1782).
Herb.: NMW.
HG 253: rare, very restricted in range; 16 localities cited.

B. lancifolium Hornem.   False Thorow-wax
No map.
Groby, edge of cornfield, rare, E. Hesselgreaves 1969; Fleckney, garden, J. C. Badcock 1965; Enderby, garden, rare, EH 1977. Some other records were lost as a result of the fire at Ratcliffe College.
   Casual: Ag, Ac.
First record: Horwood and Gainsborough (1933).
Herb.: LSR★; NMW.
HG 253: rare; 1 locality cited.

Apium L.
A. graveolens L.   Celery
No map; 1 tetrad.
Glen Parva, canal bank, rare, EH 1975.
First record: Pulteney (1747).
   Casual: Ew.
Herb.: LSR★; CGE, LSR, MANCH, NMW.
HG 254: rare, sparsely distributed; 10 localities cited.

A. nodiflorum (L.) Lag.   Fool's Water-cress
Map 43j; 461 tetrads.
Streams, rivers, canals, ditches, ponds, lakes and marshes, always in very wet places and often partly submerged. Frequent and locally abundant throughout the county.
   Constituent: Wr, Wc, Wp, Wl, Wd, Ew, Ef, M.
First record: Bloxam *in* Combe (1829).
Herb.: LSR★; DPM, LANC, LSR, LTR, MANCH, NMW.
HG 254: localy abundant and generally distributed.

A. inundatum (L.) Reichenb. fil.   Lesser Marshwort
Map 43k; 3 tetrads.
Blackbrook Reservoir, locally abundant, PAC 1977; Saddington Reservoir, locally frequent, SHB and PHG 1977; Loughborough Big Meadow, wet hollow, locally abundant, PHG and E.G. Webster 1981.
   Constituent: Wl, M.
First record: Kirby (1850).
Herb.: LSR, MANCH, NMW.
HG 255: rare, restricted in range; 19 localities cited, including Saddington Reservoir.

Petroselinum Hill
P. crispum (Miller) A. W. Hill   Garden Parsley
No map; 5 tetrads.
Groby, garden casual, EH 1972; Shepshed, recently made roadside verge, occasional, PAC 1976; Ashby Woulds, waste ground, rare, SHB 1976; Breedon on the Hill, roadside, rare, SHB 1977; Croft, waste ground, rare, SHB 1973.
   Casual or escape: Ev, Rw, Rd, Ag.
First record: Hands et al. *in* Curtis (1831).
Herb.: LSR★; LSR, LTR, MANCH.
HG 256 as *Carum petroselinum*: rare, restricted in range; 12 localities cited.

P. segetum (L.) Koch   Corn Parsley
No record since HG.
First record: Pulteney (1747).
Herb.: CGE, LSR, MANCH, NMW.
HG 256 as *Carum segetum*: rare, restricted in range; 5 localities cited.

Sison L.
S. amomum L.   Stone Parsley
Map 43l; 25 tetrads.
Hedgerows, roadsides and railway verges. Distribution in the county is discontinuous; occasional in the north-east and south-west; rare or absent elsewhere.
   Constituent or intrusive: Eh, Ev, Er, Ea, Gr, Wd, Ef, Ac, Ao.
First record: Pulteney (1749).
Herb.: LSR★; DPM, LSR, LTR, MANCH, NMW.
HG 257: local, generally distributed.

Ammi L.

A. visnaga (L.) Lam.
No map; 2 tetrads.
Groby, builder's spoil tip, one large plant, EH 1973; Barwell, rubbish dump, rare, EH 1979.
    Casual: Rd, Rr.
First record: B.E.C. Rep. 1920.
Herb.: LSR★.
HG 256: rare; 1 locality cited.

A. majus L.   Bullwort
No record since HG.
First record: Jackson (1904).
Herb.: LSR.
HG 255: rare; 4 localities cited.

Carum L.

C. carvi L.   Caraway
No record since HG.
First record: Watson (1832).
Herb.: LSR.
HG 256: very rare, restricted in range; 4 localities cited.

C. verticillatum (L.) Koch   Whorled Caraway
No recent record.
There is an undated specimen in herb. NMW collected by A. R. Horwood near Thurmaston, presumably in the period 1902-1922.
First record: see above.
Herb.: NMW.
HG: not recorded.

Angelica L.

A. sylvestris L.   Wild Angelica
Map 44a; 328 tetrads.
Woodland, marshes, banks of rivers and streams, roadside and railway verges. Frequent and locally abundant in the west of the county; occasional or locally frequent in the east.
    Constituent or intrusive: Fd, Fm, Fw, Fs, Ef, Ew, Ev, Er, Eh, M, Wd, Wl, Gr, Gp, Rg, Rs, Rw, Rd.
First record: Pulteney (1747).
Herb.: LSR★; DPM, LANC, LSR, LTR, NMW.
HG 267: frequent, generally distributed.

A. archangelica L.   Garden Angelica
Map 44b; 5 tetrads.
Leicester, London Road station, railway embankment, occasional, EKH 1972; Barrow on Soar, roadside of Slash Lane, rare, PHG 1974; Cotes, bank of River Soar, rare, PHG 1975; Thurmaston, river bank, occasional, IME 1976; Leicester, Belgrave, bank of River Soar, occasional, SHB 1977; several other records for this species were lost as a result of the fire at Ratcliffe College.
    Casual or intrusive: Ew, Ev, Er.
First record: Bemrose (1927).
Herb.: DPM, LTR.
HG 267 as *Archangelica officinalis*: doubtless overlooked; 2 localities cited.

Pastinaca L.

P. sativa L.   Wild Parsnip
Map 44c; 61 tetrads.
Occasional in calcareous grassland, probably only truly constituent in this habitat in the north-east of the county. Elsewhere it occurs mainly on railway verges and in waste places.
    Constituent: Gp, Gr.
    Intrusive: Er, Ev, Ea, Ef, Rw, Rq, Rg, Rs, Aa.
First record: Pulteney (1749).
Herb.: LSR★; CGE, DPM, LSR, LTR, MANCH.
HG 267 as *Peucedanum sativum*: locally abundant, generally distributed.

Heracleum L.

H. sphondylium L.   Hogweed
Map 44d; 605 tetrads.
Abundant throughout the county on roadsides and railway verges, in woodland, hedgerows and rough grassland.
    Constituent or intrusive: Ev, Eh, Er, Ef, Ew, Ea, Fd, Fm, Gr, Gp, Gl, M, Rw, Rq, Rf, Ac, Ap.
First record: Pulteney (1747).
Herb.: LSR★; CGE, DPM, LANC, LSR, LTR, MANCH, NMW.
HG 268: frequent, generally distributed.

H. mantegazzianum Sommier & Levier   Giant Hogweed
Map 44e; 16 tetrads.
An escape from cultivation, occasionally becoming established on roadside and railway verges and the banks of rivers. Occasional in scattered localities throughout the county; locally frequent in the neighbourhood of Market Bosworth.
    Intrusive: Ev, Er, Ew, Fd, Fs, Gr.
First record: Wade (1919).
Herb.: DPM, LSR, LTR.
HG 268: rare; 1 locality cited.

Torilis Adanson

T. nodosa (L.) Gaertner   Knotted Hedge-parsley
Map 44f; 5 tetrads.
Husbands Bosworth, brickwork of canal bridge, locally frequent, EKH and JMH 1968; Leicester, Gwendolen Road, edge of pavement, rare, HB 1968; Cotes, old church site, gravel bank above River Soar, locally frequent, PHG 1974; Burrough Hill, pasture, locally frequent, SHB 1977; Croft, siliceous grassland, rare, SHB 1979.
    Intrusive: Gp, Sw, Rt.
First record: Pulteney (1747).
Herb.: LSR★; LANC, LSR, LTR, MANCH, NMW.
HG 270 as *Caucalis nodosa*: frequent, generally distributed; 47 localities cited.

T. arvensis (Hudson) Link   Spreading Hedge-parsley
No record since HG.
First record: Bloxam and Babington *in* Potter (1842).
Herb.: LSR, MANCH.
HG 269 as *Caucalis arvensis*: local and decreasing, restricted in range; 6 localities cited.

T. japonica (Houtt.) DC.   Upright Hedge-parsley
Map 44g; 529 tetrads.
Frequent throughout the county in hedgerows, on roadside and railway verges, and in rough grassland.
    Constituent or intrusive: Eh, Ev, Er, Ew, Ea, Ef, Gr, Gp, Rq, Rg, Rd, Fd, Fc, Ac.
First record: Pulteney (1749).

Herb.: LSR★; DPM, LANC, LSR, LTR, MANCH, NMW.
HG 270 as *Caucalis anthriscus*: abundant, generally distributed.

Daucus L.
D. carota L.   Wild Carrot
Map 44h; 178 tetrads.
Frequent and locally abundant throughout the county on railway verges, which account for 65% of the records received. Occasional on roadsides, in quarries and waste places, and in calcareous grassland.
    Constituent: Gp, Gr.
    Residual or intrusive: Er, Ev, Ew, Ea, Rq, Rs, Rw, Rr, Gl, Ap.
First record: Pulteney (1747).
Herb.: LSR★; CGE, DPM, LANC, LSR, LTR, NMW.
HG 269: locally abundant, frequent, generally distributed.

ERICALES

PYROLACEAE
Pyrola L.
P. minor L.   Common Wintergreen
No record since HG.
First record: Horwood and Gainsborough (1933).
Herb.: LSR, NMW.
HG 360: rare, restricted in range; 1 locality cited (Great Mott, Gumley).

ERICACEAE
Erica L.
E. tetralix L.   Cross-leaved Heath
Map 44i; 5 tetrads.
High Sharpley, heathland, occasional, PAC 1968; Charnwood Lodge Nature Reserve, heath grassland and boggy ground, occasional, PAC 1969; Beacon Hill, heathland, occasional, PHG 1969; Bradgate Park, wet heath, locally frequent, PHG 1977.
    Constituent: Hw, Hb.
First record: Pulteney (1747).
Herb.: LANC, LSR, MANCH, NMW.
HG 359: frequent in Callunetum, almost confined to Charnwood Forest; 19 localities cited, including the above.

E. cinerea L.   Bell Heather
No recent record.
First record: Pulteney (1747).
Herb.: LSR.
HG 360: rare, apparently extinct, original range restricted to Charnwood Forest and Moira Wolds, from both of which it has disappeared owing to deforestation, drainage etc., last seen by Bates at Swannymote Rock in 1884; 7 localities cited.

Calluna Salisb.
C. vulgaris (L.) Hull   Heather, Ling
Map 44j; 37 tetrads.
Locally frequent in Charnwood Forest; occasional elsewhere in the west of the county; absent from the east except for a single record from a railway verge.
    Constituent, residual or intrusive: Hg, Hd, Hw, Fd, Fc, Fs, Ef, Er, Rq, Rg, Rs, Rt, Sr, Gr, Gp, Ap.
First record: Pulteney (1747).
Herb.: LSR★; LANC, LSR, LTR, MANCH, NMW.
HG 358: locally abundant, forming extensive Callunetum, or heather moor, on Charnwood Forest, restricted in range; 59 localities cited.

Rhododendron L.
R. ponticum L.   Rhododendron
Map 44k; 66 tetrads.
Often planted and sometimes escaping. Occasional in woodland throughout the county, especially on the more acid soils of the west, where it may become a serious pest, difficult of eradication and shading out all ground flora.
    Planted or intrusive: Fd, Fm, Fs, Ap.
First record: there is an undated specimen in herb. NMW collected by A. E. Wade, presumably in the period 1900-1920.
Herb.: LSR★; NMW.
HG: not recorded.

Gaultheria L.
G. shallon Pursh   Shallon
No map; 1 tetrad.
Market Bosworth, game covert, occasional, HB 1972.
    Planted or intrusive: Fs.
First record: 1972, H. Bradshaw in present work.
HG: not recorded.

Vaccinium L.
V. vitis-idaea L.   Cowberry
Occurred at High Sharpley until 1956 when its habitat was destroyed by drainage and ploughing.
First record: Tomkins (1887).
Herb.: LSR, NMW.
HG 357: very rare, confined to a single station on Charnwood Forest; 1 locality cited (High Sharpley).

V. myrtillus L.   Bilberry
Map 44l; 20 tetrads.
Heath and woodland. Almost entirely confined to Charnwood Forest, where it is locally frequent.
    Constituent: Hg, Hd, Hw, Fd, Fm, Fs, Gp, Sr.
    Residual or intrusive: Ev, Eh, Sw, Wd, Ag.
First record: Pulteney (1747).
Herb.: LSR★; LSR, LTR, NMW.
HG 357: locally abundant, forming Vaccinium associations, practically confined to Charnwood Forest; 46 localities cited.

EMPETRACEAE
Empetrum L.
E. nigrum L.   Crowberry
Map 45a; 1 tetrad.
Spring Hill, Whitwick, MKH 1952 (this locality now destroyed). Charnwood Lodge Nature Reserve, wet heath, one very poor straggly plant, PHG 1969.
    Constituent: Hw.
First record: Bloxam (1837).
Herb.: LSR, LTR, MANCH, NMW.
HG 513: very rare, restricted to Charnwood Forest; 3 localities cited, not including the above.

PRIMULACEAE
Primula L.
P. vulgaris Hudson   Primrose
Map 45b; 134 tetrads.
Locally frequent in woodland in Charnwood Forest and the west of the county, and in the east Leicestershire woodlands. Occasional elsewhere in the county, often as established introductions or escapes from cultivation.
    Constituent, residual or planted: Fd, Fc, Fm, Fs, Fw, Ef, Eh, Ew, Er, Ev, Ap, Gp, M, Hg, Wd, Rw, Rt.
First record: Pulteney (1747).

Herb.: LSR★; DPM, LSR, LTR, MANCH, NMW.
HG 361: locally abundant, generally distributed.

P. veris L. Cowslip
Map 45c; 288 tetrads.
Old grassland, open parts of woodland, roadside and railway verges. Frequent throughout the county, mainly because of its ability to colonise the grassland of roadside and railway verges, and to survive at the base of hedgerows in land which has been long under the plough. It is, however, decreasing as suitable old grassland is destroyed, and modern machinery allows ploughing right up to the hedge.
Constituent: Gp, Gr, M, Hg, Eh, Ew, Ef, Fd, Fm, Fs, Fw.
Residual or intrusive: Ev, Er, Ea, Gl, Ap, Rq, Rw, Rt.
First record: Pulteney (1747).
Herb.: LSR★; DPM, LANC, LSR, LTR, MANCH, NMW.
HG 362: locally abundant, generally distributed.

P. veris × vulgaris = P. × tommasinii Gren. & Godron
False Oxlip
Map 45d; 8 tetrads.
Breedon on the Hill, grassland in Pasture Wood, on banks of drainage channel, rare, PAC 1970; Cotesbach, roadside verge, single plant, EKH and JMH 1970; Glenfield, grounds of the Gynsills, rare, EH 1972; Swithland Reservoir, under trees on margin, rare, PHG 1973; Ulverscroft, spinney near Poultney Wood, rare, PAC 1972; Shackerstone Mill, rough grassland by River Sence, rare, SHB 1976; Donisthorpe, rough grass in cemetery, rare, SHB 1976; Dimminsdale, woodland, rare, SHB 1978.
Intrusive: Gr, Gp, Fd, Ew, Ev.
First record: Bloxam *in* Combe (1829).
Herb.: LSR, MANCH, NMW.
HG 363: occasional, restricted in range; 13 localities cited.

Hottonia L.
H. palustris L. Water-violet
Map 45e; 1 tetrad.
Castle Donington, large pond and drainage ditch to the east of Back Lane, locally frequent, PAC 1973; Castle Donington, roadside pond and ditch, Back Lane, occasional, D. G. Goddard 1981.
Constituent: Wp, Wd.
First record: Pulteney (1756).
Herb.: LSR★; LSR, MANCH.
HG 361: very rare, restricted in range to R. Soar and Trent, and a few ponds; 5 localities cited, including Castle Donington.

Lysimachia L.
L. nemorum L. Yellow Pimpernel
Map 45f; 37 tetrads.
Frequent and locally abundant in woodland on Charnwood Forest. Locally frequent in some of the East Leicestershire woodlands. Rare or absent in the rest of the county.
Constituent: Fd, Fc, Fm, Fw, Ef.
Residual: Ew, Wd, M, Rq.
First record: Pulteney (1747).
Herb.: LSR★; LSR, LTR, MANCH, NMW.
HG 365: locally abundant, widely distributed; 49 localities cited.

L. vulgaris L. Yellow Loosestrife
Map 45g; 9 tetrads.
Newtown Linford, roadside by wall of Bradgate Park, rare, EH 1969; Groby, rubbish dump in quarry, rare, EH 1970; Loughborough, Out Wood, relict bog, locally abundant, PHG 1970; Groby Pool, marshy margins, occasional, PHG 1971; Anstey, soil and rubbish tip, rare, EH 1971; Beaumont Leys, rubbish tip, rare, EH 1971 (site since built over); Loddington, bank of Eye Brook, rare, IME 1972; Norris Hill, small marsh by stream, locally frequent, SHB 1976; Swannington, ditch, occasional, SHB 1978.
Constituent or intrusive: Ew, Hb, M, Wd, Rr, Ev.
First record: Pulteney (1747).
Herb.: CGE, LSR, LTR, NMW.
HG 363: very rare, restricted in range; 17 localities cited, including Outwoods and Swannington.

L. nummularia L. Creeping-Jenny
Map 45h; 127 tetrads.
Marshes, wet meadows, the banks of rivers and streams, and woodland rides. Locally frequent throughout the county in these habitats. Often grown in gardens, and occasionally escaping on to verges and waste places.
Constituent: M, Ew, Ef, Fd, Fm, Fw, Wd, Wl, Wp.
Residual or intrusive: Gp, Gr, Eh, Er, Rw, Rq, Rd, Ap, Ag, Sw.
First record: Pulteney (1747).
Herb.: LSR★; DPM, LANC, LSR, LTR, MANCH, NMW, UPP.
HG 364: frequent, generally distributed; 76 localities cited.

L. punctata L. Dotted Loosestrife
Map 45i; 16 tetrads.
A garden escape, sometimes becoming established in waste places and on roadside and railway verges. Occasional in scattered localities throughout the county.
Intrusive: Rw, Rr, Rs, Rf, Ev, Er, Gr, Wd.
First record: Horwood and Gainsborough (1933).
Herb.: LSR★.
HG 364: rare; 1 locality cited.

Anagallis L.
A. tenella (L.) L. Bog Pimpernel
Map 45j; 7 tetrads.
Botcheston Bog, rare, SM 1968; Grace Dieu, damp heath grassland, rare, PAC 1968; Woodhouse, marshy ground near stream on Lingdale Golf Course, locally frequent, PAC 1968; Charnwood Lodge Nature Reserve, wet heath, rare, PHG 1969; Ulverscroft, meadow near Poultney Cottage, old pasture, rare, PAC 1969; Ulverscroft, marshy meadow east of Lea Wood, rare, PHG 1969; Groby, pasture north of the Pool, locally frequent, EH 1975.
Constituent or residual: Hw, M, Gp.
First record: Pulteney (1747).
Herb.: LSR, MANCH, NMW.
HG 366: rare, very restricted in range; 14 localities cited, including Botcheston Bog, Woodhouse, Ulverscroft, Gracedieu and Groby Pool.

A. arvensis L. Scarlet Pimpernel
Map 45k; 365 tetrads.
Arable land and recently disturbed ground. Locally frequent throughout the county; more common in the west

than in the east.
    Intrusive: Ac, Ao, Aa, Ag, Rd, Rt, Rw, Rq, Rs, Rr, Ea, Ev, Er, Ew.
First record: Pulteney (1747).
Herb.: LSR★; DPM, LSR, LTR, MANCH, NMW.
HG 365: frequent, generally distributed.

A. foemina Miller   Blue Pimpernel
No record since HG.
First record: Bloxam (1837).
Herb.: LSR, MANCH.
HG 366: rare, very restricted in range; 15 localities cited.

Samolus L.
S. valerandi L.   Brookweed
Map 45l; 3 tetrads.
Redmile, ditch in disused station yard, occasional, IME 1969; Kibworth Harcourt, marsh at canal margin near Pywell's Lock, occasional, EKH and JMH 1972; Swithland Reservoir, sandy shore, rare, PHG 1972.
    Constituent or intrusive: Ew, Wd, M.
First record: Pulteney (1749).
Herb.: LSR★.
HG 367: very rare, very restricted in range; 8 localities cited, not including the above.

PLUMBAGINALES

PLUMBAGINACEAE
Armeria Willd.
A. maritima (Miller) Willd.   Thrift
Subsp. **elongata** (Hoffm.)Bonnier
No recent record.
First record: Pulteney (1749).
HG 361 as *Statice maritima*, in square brackets; of Pulteney's record from a heath near Saltby, Horwood says 'This can only be a mistake'. He was evidently unaware of this subspecies, which grows on sandy heaths inland. It still occurs just over the county boundary in Lincolnshire.

OLEALES

OLEACEAE
Fraxinus L.
F. excelsior L.   Ash
Map 46a; 606 tetrads.
Frequent throughout the county in hedgerow and woodland. The commonest hedgerow tree in Leicestershire.
    Constituent or intrusive: Fd, Fm, Fs, Fw, Eh, Ev, Er, Ew, Ap, Gp, Rw.
First record: Pulteney (1747).
Herb.: LSR★'; DPM, LSR, LTR, MANCH, NMW.
HG 368: locally abundant, generally distributed; 222 localities cited.

Syringa L.
S. vulgaris L.   Lilac
Map 46b; 34 tetrads.
Often planted in gardens and shrubberies, and sometimes escaping and becoming established in hedgerows, and particularly on railway verges.
    Intrusive: Er, Eh, Ev, Ef, Fd, Fs, Rq, Rw.
First record: 1969, H. Bradshaw in present work.
HG: not recorded.

Ligustrum L.
L. vulgare L.   Wild Privet
Map 46c; 394 tetrads.
Hedgerows and woodland. Frequent throughout most of the county, but rare or absent in parts of Charnwood Forest.
    Constituent: Fd, Fm, Fs, Fw, E.h.
    Residual or intrusive: Ev, Er. Rq, Rw, Ap.
First record: Pulteney (1747).
Herb.: LSR★; DPM, LSR, LTR, MANCH, NMW.
HG 370: locally abundant, generally distributed.

L. ovalifolium Hassk.   Garden Privet
No map; 19 tetrads.
Often planted as a hedgerow shrub, and occasionally self-sown. In hedgerows it may be a relic of a former cultivation at the site of a demolished cottage; in woodland it is probably bird-sown.
    Planted or escape: Eh, Fd, Fm, Fs, Rw, Rs.
First record: 1970, E. K. and J. M. Horwood in present work.
Herb.: LSR★.
HG: not recorded.

GENTIANALES

GENTIANACEAE
Blackstonia Hudson
B. perfoliata (L.) Hudson   Yellow-wort
Map 46d; 3 tetrads.
Thurnaston, railway cutting, occasional, IME 1969; King Lud's Entrenchments, limestone grassland, single plant, D. V. Kolaczec 1972; Saltby Heath, former airfield, limestone grassland, occasional, KGM 1973.
    Constituent or intrusive: Gp, Er.
First record: Crabbe *in* Nichols (1795).
Herb.: LSR★; LSR, MANCH.
HG 371: very rare in Leicestershire, range restricted; 3 localities cited, including Saltby and the railway cutting at Thurnaston.

Centaurium Hill
C. erythraea Rafn   Common Centaury
Map 46e; 63 tetrads.
Occasional throughout the county in heath and dry grassland, in dry sandy waste places, and on railway verges. Often locally abundant where it does occur.
    Constituent or intrusive: Hg, Hd, Hw, Gp, Gr, M, Er, Ev, Ew, Ef, Ea, Rq, Rg, Rs, Rt, Rw, Rd.
First record: Pulteney (1747).
Herb.: LSR★; LANC, LSR, LTR, MANCH, NMW.
HG 372 as *C. umbellatum*: local, generally distributed.

Gentianella Moench
G. campestris (L.) Borner   Field Gentian
No record since HG.
First record: Pulteney *in* Nichols (1795).
Herb.: MANCH.
HG 373 as *Gentiana campestris*: very rare in Leicestershire, restricted in range; 7 localities cited.

G. amarella (L.) Borner   Autumn Gentian
Map 46f; 4 tetrads.
Breedon Hill, old pasture, locally frequent, PHG and MW 1966; King Lud's Entrenchments, rough limestone

grassland, occasional, KGM 1972; Stonesby Quarry, oolitic limestone spoil heaps, occasional, ALP 1972; Saltby Heath, former airfield, limestone grassland, occasional, KGM 1973.
    Constituent or residual: Gp, Gr, Rs.
First record: Pulteney (1747).
Herb.: LSR★; LSR, MANCH.
HG 372 as *Gentiana amarella*: rare, very restricted in range; 17 localities cited, including Stonesby and Saltby.

MENYANTHACEAE
Menyanthes L.
M. trifoliata L.   Bogbean
Map 46g; 3 tetrads.
Botcheston Bog, rare, OHB 1968; Loughborough, flooded clay pit called Ingle Pingle, locally abundant, PHG 1969 (disappeared by 1977 because of shading by trees); Shackerstone, marshy edge of Ashby Canal, locally frequent, SHB 1976.
    Constituent: M, Ew.
First record: Pulteney (1747).
Herb.: LSR, MANCH, NMW.
HG 374: rare, restricted in range; 19 localities cited.

Nymphoides Séguier
N. peltata (S. G. Gmelin) O. Kuntze
                                        Fringed Water-lily
Map 46h; 5 tetrads.
Groby Pool, locally frequent, EKH and JMH 1968; Newton Burgoland, flooded clay pit, locally abundant, ALP 1973; Barkby, brook, rare, IME 1976; Sileby, old quarry, locally frequent, PHG 1978.
    Intrusive or introduced: Wl, Wr, Rg, Rq.
First record: 1922, D. P. Murray in herb. DPM (from Groby Pool).
Herb.: LSR★; DPM, LSR, LTR.
HG 374: rare, restricted in range; 2 localities cited, one of which is in Lincolnshire, the other Groby Pool.

APOCYNACEAE
Vinca L.
V. minor L.   Lesser Periwinkle
Map 46i; 26 tetrads.
Woodland, roadsides and railway verges. Probably always an escape from cultivation. Occasional throughout the county.
    Intrusive: Fd, Fm, Fs, Ef, Eh, Ev, Er, Ap, Rw.
First record: Walpoole (1782).
Herb.: LSR, LTR, NMW.
HG 371: rare, and rather restricted in range; 19 localities cited.

V. major L.   Greater Periwinkle
Map 46j; 29 tetrads.
A garden escape, occasionally becoming established in hedgerows and on roadside and railway verges.
    Intrusive or garden escape: Eh, Ev, Er, Ew, Rf, Rt, Rw, Sw, Ap.
First record: Crabbe *in* Nichols (1795).
Herb.: LSR★; CGE, DPM, LSR, LTR, MANCH, NMW, UPP.
HG 370: occasional, and but sparsely distributed; 17 localities cited.

RUBIACEAE
Sherardia L.
S. arvensis L.   Field Madder
Map 46k; 28 tetrads.
Occasional throughout the county on railway verges, in waste places and grassland, or as a garden weed. Rarely in arable land.
    Intrusive: Er, Ev, Ew, Ea, Ag, Gr, Gp, Gl, Rw, Rd, Rq, Rs, Rt, Rr.
First record: Pulteney *in* Nichols (1795).
Herb.: LSR★; LSR, LTR, MANCH, NMW.
HG 286: locally abundant, generally distributed.

Asperula L.
A. cynanchica L.   Squinancywort
No record in the recent survey.
Swithland, T. W. Tailby 1949.
First record: Pulteney *in* Nichols (1795).
Herb.: LSR, MANCH.
HG 285: rare in Leicestershire, confined to the N.E.; 2 localities cited.

Galium L.
G. odoratum (L.) Scop.   Woodruff
Map 46l; 42 tetrads.
Locally frequent in woodland on basic soils. Often planted in gardens, and occasionally escaping into nearby hedgerows.
    Constituent: Fd, Fm, Fs, Fw, Ef.
    Residual or garden escape: Eh, Ev, Ap, Ag.
First record: Pulteney (1747).
Herb.: LSR★; DPM, LSR, MANCH, NMW.
HG 284: locally abundant, widely distributed; 51 localities cited.

G. uliginosum L.   Fen Bedstraw
Map 47a; 38 tetrads.
Occasional throughout the county in the older and wetter marshes.
    Constituent or residual: M, Hg, Ew, Ev, Wd, Wr, Fw, Ag.
First record: Hands et al. *in* Curtis (1831).
Herb.: LSR★; LSR, LTR, MANCH, NMW.
HG 283: local, restricted in range, probably overlooked; 37 localities cited.

G. palustre L.   Common Marsh-bedstraw
Map 47b; 326 tetrads.
Frequent throughout the county in marshes and at the edges of rivers, streams and canals.
    Constituent or residual: M, Wp, Wl, Wd, Wr, Wc, Ew, Ef, Fw, Fd, Fm, Fs, Gr, Gp, Eh, Er, Rs.
First record: Pulteney (1747).
Herb.: LSR★; DPM, LANC, LSR, LTR, MANCH, NMW.
HG 281: locally abundant, generally distributed.

G. verum L.   Lady's Bedstraw
Map 47c; 520 tetrads.
Frequent throughout the county in old grassland and heathland, and on roadside and railway verges.
    Constituent or intrusive: Gp, Gr, Gl, Hg, Hd, Hw, M, Ev, Er, Ew, Eh, Ea, Ef, Ap, Aa, Rq, Rg, Rs, Rt, Sw, Fc.
First record: Pulteney (1747).
Herb.: LSR★; DPM, LSR, LTR, NMW.
HG 278: locally abundant, generally distributed.

G. mollugo L.   Hedge Bedstraw
Map 47d; 86 tetrads.
Hedgerows, roadsides and railway verges on basic soils. Locally frequent in those parts of the county where the soil is suitable.
    Constituent or intrusive: Eh, Ev, Er, Ew, Ea, Ef, Fs, Gr, Gp.
First record: Pulteney (1747).
Herb.: LSR★; DPM, LSR, LTR.
HG 278: frequent, widely distributed; 71 localities cited.

G. album Miller subsp. **album**   Upright Hedge Bedstraw
G. mollugo L. subsp. *erectum* Syme
Map 47e; 12 tetrads.
Occasional in rough grassland, mainly on railway verges, which account for 72% of the records received.
    Intrusive: Er, Ev, Gp, Ap.
First record: Kirby (1849).
Herb.: LSR★; LANC, LSR, MANCH, NMW.
HG 278 as *G. erectum*: rare, restricted in range; 11 localities cited.

G. album × **verum** = **G.** × **pomeranicum** Retz.
No map; 1 tetrad.
Cosby, grass verge of dismantled railway, single plant, EH 1975.
    Intrusive: Er.
First record: 1952, M. K. Hanson in herb. LSR.
Herb.: LSR.
HG 280: 3 localities cited for **G. mollugo** × **verum**.

G. pumilum Murray   Slender Bedstraw
No record since HG.
First record: Coleman (1852).
HG 281: very rare, restricted in range; 4 localities cited.

G. saxatile L.   Heath Bedstraw
Map 47f; 65 tetrads.
Heathland, open woodland and grassland on acid soil. Locally abundant in Charnwood Forest and the north-west of the county. Occasional elsewhere in the county where there are suitable soils.
    Constituent or intrusive: Hg, Hd, Gr, Gp, Fd, Fm, Fs, Ef, Ev, Er, Eh, Sr, Sw, Ap, Rq, Rs.
First record: Pulteney (1747).
Herb.: LSR★; DPM, LSR, LTR, NMW.
HG 280 as *G. hercynicum*: locally abundant or frequent, widely distributed on sandy or peaty soils; 103 localities cited.

G. aparine L.   Goose Grass, Cleavers, Heriff
Map 47g; 604 tetrads.
Abundant throughout the county in hedgerows, woodland and waste places, and as a weed of arable land.
    Constituent or intrusive: Eh, Ev, Er, Ew, Ea, Ef, Fd, Fm, Fs, Ap, Ac, Gr, Rw, Rq, Rs, Rt, Rd, Rf, M.
First record: Pulteney (1747).
Herb.: LSR★; DPM, LSR, LTR, MANCH, NMW.
HG 283: abundant, generally distributed.

G. tricornutum Dandy   Corn Cleavers
No record since HG.
First record: Pitt (1809).
Herb.: LSR, MANCH, NMW.
HG 284: rare, restricted in range; 14 localities cited.

Cruciata Miller
C. laevipes Opiz   Crosswort
Galium cruciata (L.) Scop.
Map 47h; 181 tetrads.
Woodland, hedgerows, roadsides and other verges. Locally frequent on basic soils; occasional elsewhere.
    Constituent, residual or intrusive: Fd, Fm, Fs, Ef, Eh, Ev, Ew, Er, Gr, Gp, Hg, M, Wd, Rt, Rq, Rw.
First record: Pulteney (1747).
Herb.: LSR★; DPM, LANC, LSR, LTR, MANCH, NMW.
HG 277: locally abundant, generally distributed.

TUBIFLORAE

POLEMONIACEAE
Polemonium L.
P. caeruleum L.   Jacob's-ladder
No map; 3 tetrads.
Freeby, Saxby Station, among rubble of demolished platform, rare, KGM 1971; Normanton le Heath, edge of track, single plant, HH 1973; Ashby de la Zouch, waste ground, rare, SHB 1977.
    Casual or escape: Ev, Rw.
First record: Bloxam (1838).
Herb.: LSR★; BM, LSR.
HG 375: rare, restricted in range; 6 localities cited.

CONVOLVULACEAE
Cuscuta L.
C. campestris Yuncker
No map; 1 tetrad.
Sheepy Magna, in garden on carrots, locally abundant, E. Knight 1984.
    Intrusive: Ag.
First record: 1984, E. Knight in present work.
Herb.: LSR★.
HG: not recorded.

C. europaea L.   Greater Dodder
No record since HG.
First record: Pulteney (1756).
Herb.: LSR.
HG 387: rare, very restricted in range; 5 localities cited.

C. epilinum Weihe   Flax Dodder
No record since HG.
First record: Coleman (1852).
HG 388: very rare, restricted in range; 1 locality cited.

C. epithymum (L.) L.   Dodder
No record since HG.
First record: Pulteney *in* Nichols (1795).
Herb.: BM, LSR.
HG 388: rare, apparently confined to Charnwood Forest; 2 localities cited.

Calystegia R. Br.
C. sepium (L.) R.Br.   Hedge Bindweed
Map 47i; 429 tetrads.
Frequent in hedgerows throughout the county.
    Constituent or intrusive: Eh, Ev, Er, Ew, Fd, Fs, Fw, Rw, Rq, Ag, Ap, Aa.
First record: Pulteney (1747).

Herb.: LSR★; DPM, LANC, LSR, LTR, MANCH, NMW.
HG 386 as *Volvulus sepium*: frequent, generally distributed.

C. silvatica (Kit.) Griseb. Large Bindweed
Map 47j; 323 tetrads.
Hedgerows and waste ground. Locally frequent, with a distribution which would seem to indicate that it has spread from the neighbourhood of towns and villages where it had been introduced. It is now thoroughly naturalised.
    Constituent or intrusive: Eh, Ev, Er, Ew, Rw, Rg, Rq, Rs, Rd, Rr, Ag, Ap, Gr.
First record: 1946, T. G. Tutin in herb. LTR.
Herb.: LSR★; LSR, LTR, NMW.
HG: not recorded.

C. pulchra Brummitt & Heywood Hairy Bindweed
Map 47k; 11 tetrads.
Groby, rough grassland bordering old railway track, locally frequent, EH 1969; Martinshaw Wood, hedgerow, rare, EH 1970; Leicester Forest East, hedgerow, rare, EH 1971; Kirby Muxloe, hedgerow, rare, EH 1972; Kirkby Mallory, roadside hedgerow, single plant, OHB 1971; Nailstone, hedgerow, single plant, HH 1971; Thorpe Arnold, hedgerow of cottage garden, occasional, ALP and PCP 1972; Seagrave, hedgerow, locally frequent, ALP 1973; Blaby, roadside verge, rare, EH 1975; Barkby, hedgerow, rare, IME 1976; Blackfordby, site of demolished house, locally frequent, SHB 1976; Seagrave, hedgerow near Quorn Hunt kennels, rare, PHG 1978.
    Intrusive: Eh, Rw, Gr.
First record: 1945, T. G. Tutin in herb. LTR.
Herb.: LSR★; CGE, LTR.
HG: not recorded.

Convolvulus L.
C. arvensis L. Field Bindweed
Map 47l; 543 tetrads.
Arable land, waste places and roadside verges, and a characteristic plant of railway verges. Frequent and locally abundant throughout most of the county, but apparently rare or absent in parts of Charnwood Forest.
    Intrusive: Ac, Ao, Aa, Ag, Ap, Ea, Er, Ev, Eh, Ew, Ef, Gr, Gp, Gl, Rw, Rd, Rf, Rq, Rs, Rt.
First record: Pulteney (1747).
Herb.: LSR★; DPM, LANC, LSR, LTR, MANCH, NMW.
HG 387: locally abundant, generally distributed.

BORAGINACEAE
Lithospermum L.
Lithospermum officinale L. Common Gromwell
No map; 1 tetrad.
Saltby, King Lud's Entrenchments, disturbed ground, single plant, PHG 1983.
    Intrusive: Rd.
First record: Pulteney (1747).
Herb.: LSR, MANCH.
HG 384: rare, restricted in range; 23 localities cited.

Buglossoides Moench
B. purpurocaerulea (L.) I. M. Johnston Purple Gromwell
Lithospermum purpurocaeruleum L.
No recent record.
First record: 1903, F. T. Mott in herb. MANCH (from Rearsby).
Herb.: MANCH.
HG: not recorded.

B. arvensis (L.) I. M. Johnston Field Gromwell
Lithospermum arvense L.
Map 48a; 2 tetrads.
Eastwell, a few plants in farmyard of West End Farm, and a few in arable field, JMS 1968; edge of arable field east of Six Hills, PAC 1977.
    Intrusive: Ac, Ea, Rf.
First record: Pulteney (1747).
Herb.: LSR, LTR, MANCH, NMW.
HG 385: rare, restricted in range; 34 localities cited.

Echium L.
E. vulgare L. Viper's-bugloss
Map 48b; 6 tetrads.
Saltby, disused airfield, limestone grassland, IME 1966; Thorpe Langton, dismantled railway sidings, locally frequent, KGM 1972; Holwell Iron Works, base of spoil heap, single plant, ALP and SMF 1972; Kegworth, rough ground at edge of arable field, rare, PAC 1973; Ashby Woulds, railway sidings near Rawdon Mine, locally frequent, SHB 1976; Ashby Magna, dismantled railway, ballast, single plant, EKH and JMH 1976.
    Constituent: Gp.
    Intrusive: Er, Rw, Rs.
First record: Pulteney (1747).
Herb.: LSR★; LANC, LSR, MANCH, NMW.
HG 385: rare, sparsely distributed; 36 localities cited.

Pulmonaria L.
P. officinalis L. Lungwort
No map; 6 tetrads.
Cranoe, neglected grass in churchyard, occasional, KGM 1971; Wymondham, roadside, occasional, KGM 1971; Cold Overton, churchyard, occasional, KGM 1972; Noseley Hall, rough grass, rare, KGM 1973; Fenny Drayton, hedgerow near church, rare, SHB 1975; Somerby, roadside, occasional, KGM 1975.
    Garden escape: Ev, Eh, Gr, Ap.
First record; Power (1807).
Herb.: LSR★; LSR, LTR, NMW.
HG 380: rare, restricted in range; 7 localities cited.

Symphytum L.
S. officinale L. Common Comfrey
Map 48c; 80 tetrads.
Occasional throughout the county by water and on roadsides.
    Possibly constituent: Wl, Wd, Ew, Fw, M.
    Intrusive: Ev, Er, Eh, Rw, Rd, Gr, Gp.
First record: Pulteney (1749).
Herb.: LSR★; DPM, LANC, LSR, LTR, NMW.
HG 377: occasional, generally distributed.

S. asperum Lepechin × **officinale** L. = **S.** × **uplandicum** Nyman Russian Comfrey
Map 48d; 137 tetrads.
Locally frequent throughout the county on roadside and railway verges, on the banks of rivers and streams, and in waste places.
  Intrusive: Ev, Er, Eh, Ew, Wd, Wp, Rw, Rg, Rf, Rr, Rd, Rq, Fm, Fw, Ap, Aa, Gr.
First record: B.E.C. Rep. 1929.
Herb.: LSR★; LSR, LTR, NMW.
HG 378 as *S. peregrinum*: rare; 1 locality cited (but the taxonomy of this genus was then in some confusion, and no doubt many of the records for **S. officinale** were really for this hybrid).

S. tuberosum L. Tuberous Comfrey
Map 48e; 2 tetrads.
Bescaby Oaks, partly cleared scrub, rare, KGM 1972; Lubbesthorpe, woodland, locally abundant, EH 1975.
  First record: Hands et al. *in* Curtis (1831).
Herb.: LSR★.
HG 377: very rare, very restricted in range; 2 localities cited, not as above.

S. ibiricum Steven Creeping Comfrey
S. grandiflorum auct. non DC.
No record in the recent survey.
Ulverscroft Priory, EKH 1948.
First record: 1948, E. K. Horwood in herb. LTR.
Herb.: LTR.
HG: not recorded.

S. orientale L. White Comfrey
No map; 1 tetrad.
Peatling Parva, churchyard, occasional, EKH and JMH 1973.
  Casual or intrusive: Ap.
First record: 1933, S. A. Taylor in herb. LSR.
Herb.: LSR.
HG 377: no Leicestershire record.

Anchusa L.

A. arvensis (L.) Bieb. Bugloss
Map 48f; 39 tetrads.
Occasional throughout the county in arable land and waste places, usually on sandy soils.
  Intrusive: Ao, Ac, Ea, Er, Ev, Rw, Rd, Rq, Rg, Rf.
First record: Pulteney (1747).
Herb.: LSR★; LSR, LTR, MANCH, NMW.
HG 379: local, sparsely distributed; 29 localities cited.

Pentaglottis Tausch

P. sempervirens (L.) Tausch Green Alkanet
Map 48g; 50 tetrads.
Occasional throughout the county as an established escape from cultivation, on roadsides, in hedgerows and in waste places.
  Intrusive: Ev, Eh, Er, Rd, Rw, Ag, Ap, Gr, Fd, Fm, Sw.
First record: Power (1807).
Herb.: LSR★; LSR, LTR, NMW.
HG 379 as *Anchusa sempervirens*: rare, restricted in range; 12 localities cited.

Borago L.

B. officinalis L. Borage
No map.
Rolleston, roadside, rare, IME 1976. Some records for this species were lost as a result of the fire at Ratcliffe College.
  Casual: Ev.
First record: Power (1807).
Herb.: LSR★; LSR, LTR, MANCH.
HG 378: rare, restricted in range; 16 localities cited.

Amsinckia Lehm.

A. lycopsoides (Lehm.) Lehm.
No record since HG.
First record: Jackson (1904) if correctly identified.
HG 376: rather rare; 6 localities cited, but Horwood suggests that all these records should probably be transferred to the next species.

A. intermedia Fischer & C. A. Meyer
No record since HG.
First record: B.E.C. Rep. 1917 (but see above).
HG 376: rare; 1 locality cited.

Asperugo L.

A. procumbens L. Madwort
No map; 1 tetrad.
Saddington, garden, beneath bird table, C. W. Holt 1967.
  Casual: Ag.
First record: 1967, C. W. Holt in herb. LSR.
Herb.: LSR.
HG: not recorded.

Myosotis L.

M. arvensis (L.) Hill Field Forget-me-not
Map 48h; 559 tetrads.
Frequent and locally abundant throughout the county in open woodland, arable land and waste places.
  Constituent; Fd, Fm, Fs, Fw, Ef.
  Residual or intrusive: Ac, Ao, Ag, Ap, Ea, Ev, Er, Ew, Eh, Gp, Gr, Gl, Rq, Rs, Rw, Rt, Rd, Rf.
First record: Pulteney (1747).
Herb.: LSR★; DPM, LANC, LSR, LTR, MANCH, NMW.
HG 382: locally abundant, generally distributed.

M. ramosissima Rochel Early Forget-me-not
Map 48i; 50 tetrads.
Dry grassland, particularly on railway verges, which account for nearly half the records received. Occasional throughout the county.
  Constituent or intrusive: Hd, Gr, Gp, Er, Eh, Ea, Ac, Rq, Rs, Rw, Rd, Rt, Sw, Sr.
First record: Watson (1832).
Herb.: LSR★; LSR, LTR, MANCH, NMW.
HG 382 as *M. collina*: rather rare, restricted in range; 26 localities cited.

M. discolor Pers. Changing Forget-me-not
Map 48j; 35 tetrads.
Heathland and sandy places. Occasional throughout the county where there are suitable habitats.
  Constituent, residual or intrusive: Hg, Hd, Gp, Gr, M, Er, Ew, Ef, Ea, Rq, Rs, Rw, Rt, Ac.
First record: Pulteney (1747).

Herb.: LSR★; CGE, LANC, LSR, LTR, MANCH, NMW.
HG 383 as *M. versicolor*: frequent, sparsely distributed; 57 localities cited.

M. sylvatica Hoffm.   Wood Forget-me-not
Map 48k; 70 tetrads.
Locally abundant in woodland, but avoids the more acid soils. Especially characteristic of the east Leicestershire woodlands, where it makes a display in late spring which is worth a visit to see.
    Constituent: Fd, Fm, Fs, Fw, Ef.
    Residual or intrusive: Eh, Ev, Ew, Er, Wd, Ap, Gr, Rw, Rd, Rr.
First record: Watson (1832).
Herb.: LSR★; DPM, LSR, LTR, MANCH.
HG 381: locally abundant, restricted in range; 30 localities cited.

M. secunda A. Murray   Creeping Forget-me-not
Map 48l; 3 tetrads.
Charnwood Lodge Nature Reserve, side of small stream, locally frequent, PAC 1971; Staunton Harold, marsh, occasional, SHB 1978; Newtown Linford, small stream north of Bradgate Park boundary, rare, PHG 1981.
    Constituent: M, Ew.
First record: Kirby (1850).
Herb.: LSR, NMW.
HG 381: rare, restricted in range; 8 localities cited, including two near to the above.

M. laxa Lehm. subsp. **caespitosa** (C. F. Schultz) Hyl. ex Nordh.   Tufted Forget-me-not
M. caespitosa C. F. Schultz
Map 49a; 251 tetrads.
Frequent throughout the county in marshes and wet places, and on the margins of ponds, lakes, rivers and canals. This is the species which one finds in marshes more commonly than **M. scorpioides**.
    Constituent or intrusive: M, Ew, Wp, Wl, Wr, Wc, Wd, Fw, Rg.
First record: Watson 1832.
Herb.: LSR★; LSR, LTR, MANCH, NMW.
HG 380: locally abundant, generally distributed.

M. scorpioides L.   Water Forget-me-not
Map 49b; 272 tetrads.
Frequent throughout the county on the margins of rivers, streams, canals and ponds, in wet marshes and in ditches.
    Constituent: Ew, Wr, Wc, Wd, Wp, Wl, M.
First record: Pulteney (1747).
Herb.: LSR★; DPM, LANC, LSR, LTR.
HG 380 as *M. palustris*: frequent, generally distributed.

Lappula Gilib.
L. squarrosa (Retz.) Dumort.
No map; 1 tetrad.
Ratcliffe College, newly sown lawn, single plant, ALP 1970 (det. A. O. Chater).
    Casual: Rd.
First record: 1970, A. L. Primavesi in present work.
Herb.: LSR★.
HG: not recorded.

Cynoglossum L.
C. officinale L.   Hound's-tongue
Map 49c; 3 tetrads.
Bradgate Park, recently disturbed ground, occasional, IME 1970; Thorpe Langton, dismantled railway sidings, rare, J. C. Badcock and E. Harrison 1971; Loughborough, disused railway track and bank of River Soar adjacent, locally frequent, PHG 1976; Loughborough Meadow, roadside verge, single plant, PHG 1982.
    Intrusive: Er, Ew, Rd.
First record: Pulteney (1747).
Herb.: LSR, MANCH, NMW.
HG 375: rare, restricted in range; 19 localities cited.

VERBENACEAE
Verbena L.
V. officinalis L.   Vervain
No map; 1 tetrad.
Rothley, garden, single plant, PHG 1979.
    Casual: Ag.
First record: Pulteney (1747).
Herb.: LSR.
HG 420: rare, restricted in range; 17 localities cited.

CALLITRICHACEAE
Callitriche L.
In the following account of this genus, the nomenclature is that of Clapham, Tutin and Warburg (1962) and not that of *Flora Europaea*. Identification of plants of this genus is difficult, for various reasons. The shape of the leaves varies much with movement and depth of water, and also depends on whether the leaves are floating or submerged. It is necessary to examine small details of the fruit, but plants are frequently sterile. In the recent survey, this genus was treated as critical, and the majority of records are based on determination of fresh specimens by the referee. Very little of this material was preserved as voucher specimens. The survey began in 1968, and Volume 3 of *Flora Europaea* was not published until 1972, so that many of the records could not be redetermined.

C. truncata Guss.   Short-leaved Water-starwort
Map 49d; 2 tetrads.
Barrow on Soar, Proctor's Pleasure Park, old flooded gravel pits, locally abundant, PHG 1972; Groby Pool, locally frequent, EH 1982.
    Intrusive: Rg, Wl.
First record: 1972, P. H. Gamble in present work.
Herb.: LSR★.
HG: not recorded.

C. stagnalis Scop.   Common Water-starwort
Map 49e; 193 tetrads.
Ponds, ditches, reservoirs, canals, rivers and streams. Also terrestrial on wet mud, especially on woodland rides. Frequent and locally abundant throughout the county. Almost certainly under-recorded.
    Constituent or intrusive: Wp, Wl, Wd, Wr, Wc, M, Ef, Ew, Ea, Er, Fm, Rg, Rt.
First record: Bloxam *in* Combe (1829).
Herb.: LSR, MANCH.
HG 238: frequent, widely distributed.

C. obtusangula Le Gall   Blunt-fruited Water-starwort
No record in the recent survey.
Brooksby Hall, pond, S. A. Taylor 1937.
First record: Preston (1900).
Herb.: LSR, MANCH.
HG 240: rare, restricted in range; 5 localities cited.

C. platycarpa Kutz.   Various-leaved Water-starwort
Map 49f; 43 tetrads.
Ponds, ditches, rivers and canals. Occasional throughout the county.
   Constituent or intrusive: Wp, Wr, Wc, Wd, M, Ew, Rg.
First record: Power (1807).
Herb.: LSR, MANCH.
HG 238: as *C. palustris*: rare, restricted in range; 22 localities cited. As *C. stagnalis* var. *platycarpa*: frequent, widely distributed; 22 localities cited.

C. intermedia Hoffm. subsp. **hamulata** Kutz. ex Koch
Intermediate Water-starwort
Map 49g; 13 tetrads.
Nevill Holt, pond, locally frequent, KGM 1971; Bringhurst, drainage ditch near River Welland, locally frequent, KGM 1971; Martinshaw Wood, marl pit, rare, EH 1971; Ashby Magna, field pond, locally frequent, EKH 1971 and JMH 1971; Saddington Reservoir outflow, occasional, EKH and JMH 1971; Wymondham, mill stream, occasional, KGM 1972; Normanton le Heath, small roadside pond, locally abundant, HH 1973; Knossington, pond by roadside, rare, KGM 1973; Nanpantan, old Forest Canal, rare, PHG 1974; Cropston Reservoir, pool on south-east edge (not in reservoir itself), locally frequent, PHG 1975; Leicester, Belgrave, River Soar, rare, SHB 1977; Bradgate Park, pond, locally abundant, PHG 1978; Swithland, marsh, occasional, SHB 1979.
   Constituent or intrusive: Wl, Wp, Wr, Wd, Wc, M, Rg.
First record: Brown *in* Mosley (1863).
Herb.: LSR★; MANCH.
HG 239: rare, restricted in range; 21 localities cited.

LABIATAE
Ajuga L.

A. reptans L.   Bugle
Map 49h; 246 tetrads.
Woodland, especially woodland rides, marshes and grassland. Locally frequent throughout the county, except in built-up areas and regions of intensive agriculture where there are no suitable habitats.
   Constituent or intrusive: Fd, Fm, Fs, Fw, Ef, Eh, Ew, Er, Ev, Ea, M, Gr, Gp, Gl, Hg, Wd, Wp, Rg, Rw, Ap, Sw.
First record: Pulteney (1747).
Herb.: LSR★; DPM, LSR, LTR, MANCH, NMW.
HG 445: locally abundant, generally distributed.

Teucrium L.

T. scorodonia L.   Wood Sage
Map 49i; 64 tetrads.
Heath, open woodland, and verges on acid soil. Almost confined to Charnwood Forest and acid soils in the west of the county. Elsewhere it is probably an introduction; e.g. near Ratcliffe College, where it has persisted for a number of years together with **Veronica officinalis** on Mountsorrel 'granite' waste used to build up a roadside verge.
   Constituent, residual or intrusive: Hg, Hd, Gr, Sr, Fd, Fm, Ef, Eh, Er, Ev, Rq, Rs, Rw.
First record: Pulteney (1747).
Herb.: LSR★; LSR, LTR, NMW.
HG 444: local and restricted in range; 84 localities cited.

Scutellaria L.

S. galericulata L.   Skullcap
Map 49j; 147 tetrads.
Locally frequent throughout the county on the banks of rivers, streams and canals, and in the wetter marshes.
   Constituent: Wr, Wc, Wl, Wp, Wd, Ew, Ef, M, Fw.
First record: Pulteney (1747).
Herb.: LSR★; DPM, LANC, LSR, LTR, MANCH, NMW.
HG 434: frequent, generally distributed.

S. minor Hudson   Lesser Skullcap
Map 49k; 4 tetrads.
Bradgate Park, wet heath, locally abundant, PHG 1971; Charnwood Lodge Nature Reserve, marshy areas and by stream, locally frequent, PAC 1971; Beacon Hill, lower slopes, bog pool and drainage channels, locally frequent, PHG 1969.
   Constituent or residual: Hw, Hb, Wd, Ew.
First record: Bloxam *in* Combe (1829).
Herb.: LSR, NMW.
HG 434: rare, restricted in range to region W. of R. Soar; 13 localities cited, including Bradgate Park and Beacon Hill.

Marrubium L.

M. vulgare L.   White Horehound
No record since HG.
First record: Pulteney (1747).
Herb.: LSR, NMW.
HG 435: local, restricted in range; 20 localities cited.

Sideritis L.

S. montana L.
No record since HG.
First record: Vice (1905).
HG 435: rare; 1 locality cited.

Galeopsis L.

G. angustifolia Ehrh. ex Hoffm.   Red Hemp-nettle
No record in the recent survey.
King Lud's Entrenchments, arable land, EKH 1952.
First record: Bloxam (1835).
Herb.: LTR, MANCH.
HG 438 as *G. ladanum*: rare, restricted in range; 15 localities cited.

G. speciosa Miller   Large-flowered Hemp-nettle
Map 49l; 13 tetrads.
Arable land, waste places and roadside verges. Rare, in scattered localities throughout the county, probably only of casual occurrence in most cases.
   Casual or intrusive: Ao, Ac, Aa, Ag, Ev, Eh, Rw, Rs, Wd, Gr.
First record: Hands et al. *in* Curtis (1831).
Herb.: LSR★; DPM, LSR.
HG 439: rare, restricted in range; 3 localities cited.

G. tetrahit L.   Common Hemp-nettle
Map 50b; 199 tetrads.
Arable land, roadside and railway verges, hedgerows, woodland and waste places. Locally frequent in the west of the county; occasional in the east.
    Intrusive: Ac, Ao, Ag, Ap, Ea, Ev, Er, Ew, Eh, Ef, Fd, Fm, Fs, Fw, Rw, Rf, Rd, Rt, Rq, Rs, Rr, Wd, Gr, Gp, Hg.
This species and the next were recorded both separately and as an aggregate in the recent survey. In the experience of many field workers the plants were often found when the flowers were over, so that it was not possible to distinguish the species. An aggregate distribution map is therefore given (Map 50a; 353 tetrads) as well as the maps for the two species.
First record: Pulteney (1747).
Herb.: LSR★; DPM, LANC, LSR, LTR, NMW.
HG 439: locally abundant, generally distributed.

G. bifida Boenn.
Map 50c; 26 tetrads.
Arable land, waste places and roadside verges. Apparently, from the records, occasional in the southern half of the county, rare in the north, but it may have been overlooked for reasons stated above.
    Intrusive: Ac, Ag, Ap, Ea, Ev, Rw, Rs, Rd, Rt, Gr, Wd, Fc.
First record: 1969, E. Hesselgreaves in present work.
HG: not recorded.

Lamium L.

L. maculatum L.   Spotted Dead-nettle
Map 50d; 21 tetrads.
An occasional garden escape, becoming established on verges and in waste places.
    Intrusive: Ev, Eh, Ew, Gr, Rs, Rd, Rt, Sw.
First record: Preston (1900).
Herb.: LSR★; LSR, NMW.
HG 442: rare, restricted in range; 12 localities cited.

L. album L.   White Dead-nettle
Map 50e; 601 tetrads.
Roadside and railway verges and hedgerows, and sometimes a weed of arable land. Locally abundant throughout the county.
    Constituent or intrusive: Ev, Er, Eh, Ea, Ew, Ef, Gr, Gp, Ag, Ap, Aa, Ao, Rw, Rq, Rr, Rt, Rd, Sw, Sb.
First record: Pulteney (1747).
Herb.: LSR★; DPM, LANC, LSR, LTR, MANCH, NMW, UPP.
HG 442: locally abundant, generally distributed.

L. purpureum L.   Red Dead-nettle
Map 50f; 493 tetrads.
Locally frequent throughout the county in arable land and on waste ground.
    Intrusive: Ag, Ao, Ac, Aa, Ap, Ea, Ev, Er, Ew, Gl, Gr, Gp, Rw, Rd, Rs, Rf, Sw.
First record: Pulteney (1747).
Herb.: LSR★; DPM, LANC, LSR, LTR, MANCH, NMW, UPP.
HG 442: locally abundant, generally distributed.

L. hybridum Vill.   Cut-leaved Dead-nettle
Map 50g; 112 tetrads.
Gardens, arable fields and waste places. Occasional throughout the county. Seldom in any great quantity where it does occur, but more frequent than **L. amplexicaule**.
    Intrusive: Ag, Ac, Ao, Aa, Ap, Ea, Ev, Er, Rw, Rd, Rs, Rt, Sw.
First record: Pulteney (1747).
Herb.: LSR★; BM, LANC, LSR, LTR, MANCH, NMW.
HG 441: rare, restricted in range; 42 localities cited.

L. amplexicaule L.   Henbit Dead-nettle
Map 50h; 46 tetrads.
Arable land and waste places. Occasional throughout the county; seldom in any great quantity where it does occur.
    Intrusive: Ao, Ac, Ag, Aa, Ea, Ev, Er, Rw, Rd, Rt, Rf, Sw.
First record: Pulteney (1747).
Herb.: LSR★; LSR, LTR, MANCH, NMW.
HG 440: local, widely distributed; 53 localities cited.

Lamiastrum Heister ex Fabr.

L. galeobdolon (L.) Ehrend. & Polatschek
                                        Yellow Archangel
Galeobdolon luteum Hudson
Map 50i; 72 tetrads.
Woodland, and hedgerows near woodland. Locally frequent in the well-wooded parts of the county; rare elsewhere.
    Constituent or residual: Fd, Fc, Fm, Fs, Fw, Ef, Eh, Ev, Ew, Wd, Ap.
First record: Pulteney (1747).
Herb.: LSR★; LANC, LSR, LTR, MANCH, NMW, UPP.
HG 443 as *Lamium galeobdolon*: frequent, generally distributed; 85 localities cited.

Leonurus L.

L. cardiaca L.   Motherwort
No record since HG.
First record: Pulteney *in* Nichols (1795).
Herb.: LSR, NMW.
HG 440: very rare, very restricted in range; 9 localities cited.

Ballota L.

B. nigra L.   Black Horehound
Map 50j; 247 tetrads.
Roadsides and hedgerows. Locally frequent throughout the county, especially near villages, but rare in Charnwood Forest.
    Constituent or intrusive: Eh, Ev, Er, Ea, Ap, Ac, Rw, Rq, Rg, Rs, Rd, Rr, Gr, Sw.
First record: Pulteney (1749).
Herb.: LSR★; DPM, LANC, LSR, LTR, NMW.
HG 444: frequent and generally distributed.

Stachys L.

S. officinalis (L.) Trevisan   Betony
Betonica officinalis L.
Map 50k; 172 tetrads.
Heathland and old grassland. Frequent in the west of the

county, occasional in the east. Decreasing as suitable habitats are destroyed.
    Constituent or residual: Hg, Hd, Gp, Gr, M, Ev, Er, Eh, Ew, Ea, Ef, Fd, Fc, Fs, Rt, Wd.
First record: Pulteney (1747).
Herb.: LSR★; DPM, LSR, LTR.
HG 436: frequent, generally distributed.

S. sylvatica L.   Hedge Woundwort
Map 50l; 601 tetrads.
Abundant throughout the county in hedgerows and woodland, and on roadsides and railway verges.
    Constituent, residual or intrusive: Eh, Ev, Er, Ew, Ea, Ef, Fd, Fm, Fs, Wd, Gr, Ap, Rw, Rs.
First record: Pulteney (1747).
Herb.: LSR★; DPM, LSR, LTR, MANCH, NMW.
HG 437: locally abundant, generally distributed.

S. palustris L.   Marsh Woundwort
Map 51a; 63 tetrads.
Locally frequent throughout the county on the banks of rivers and canals. Occasional in marshes.
    Constituent: Ew, Wr, Wc, Wd, Wp, M, Fw.
    Residual or intrusive: Rg, Rw, Er.
First record: Pulteney (1747).
Herb.: LSR★; DPM, LANC, LSR, LTR, MANCH, NMW.
HG 436: frequent, widely distributed; 35 localities cited.

S. palustris × sylvatica = S. × ambigua Sm.
Map 51b; 2 tetrads.
Newton Harcourt, canal bank, locally frequent, EKH and JMH 1969; Oakthorpe, cereal crop, locally abundant, SHB 1976.
    Intrusive: Ew, Ac.
First record: Hands et al. in Curtis (1831).
Herb.: LSR★; LTR, NMW.
HG 437: rare, restricted in range; 12 localities cited.

S. arvensis (L.) L.   Field Woundwort
Map 51c; 16 tetrads.
Occasional in scattered localities throughout the county as a weed of arable land. In most localities it does not occur in any great quantity.
    Intrusive: Ac, Ao, Ea, Rt.
First record: Bloxam (1837).
Herb.: LSR★; LSR, LTR, MANCH, NMW.
HG 437: rare, restricted in range; 16 localities cited.

Nepeta L.
N. cataria L.   Cat-mint
No map; 1 tetrad.
Belvoir, hedgerow, rare, S.S. Jackson and R. Tetley 1986. The garden hybrids also known as Cat-mint occasionally occur as escapes and have been recorded.
    Constituent or intrusive: Eh.
First record: Pulteney (1747).
Herb.: LSR★; MANCH, NMW.
HG 432: rare, restricted in range; 20 localities cited.

Glechoma L.
G. hederacea L.   Ground-ivy
Map 51d; 588 tetrads.
Frequent and locally abundant throughout the county in woodland and hedgerows, and on roadside and railway verges.
    Constituent, residual or intrusive: Fd, Fm, Fs, Fw, Ef, Eh, Ev, Er, Ew, Wd, Gr, Gp, Gl, Ap, Ac, Ag, Rw, Rd, Rr, Rf.
First record: Pulteney (1747).
Herb.: LSR★; DPM, LSR, LTR, MANCH, NMW.
HG 433 as *Nepeta hederacea*: locally abundant, generally distributed.

Prunella L.
P. vulgaris L.   Selfheal
Map 51e; 554 tetrads.
Frequent throughout the county in grassland and grassy places.
    Constituent or intrusive: Gp, Gr, Gl, Hg, Hd, M, Ev, Er, Ew, Ea, Eh, Ef, Fd, Ap, Ag, Rq, Rs, Rw, Rt, Rf, Rd.
First record: Pulteney (1747).
Herb.: LSR★; DPM, LANC, LSR, LTR, MANCH, NMW.
HG 434: locally abundant, generally distributed.

Melissa L.
M. officinalis L.   Balm
No map; 3 tetrads.
Leicester, Freemen's Common, disused allotments, rare, HB 1968 (site subsequently built over); Thurnby, dismantled railway, rare, IME 1969 (site subsequently built over); Birstall, waste ground, rare, HB 1971.
    Casual or escape: Rw, Aa, Er.
First record: Kirby (1850).
Herb.: LSR★; LSR.
HG 431: rare, restricted in range; 3 localities cited.

Acinos Miller
A. arvensis (Lam.) Dandy   Basil Thyme
Map 51f; 2 tetrads.
Saltby, former airfield, rough limestone grassland, occasional, KGM 1972; Ashby Magna, ballast of dismantled railway, locally frequent, ALP 1975.
    Constituent: Gr.
    Intrusive: Er.
First record: Pulteney (1749).
Herb.: LSR★; LSR, LTR, MANCH.
HG 430 as *Satureia acinos*: very rare, restricted in range; 7 localities cited, including Saltby Heath.

Calamintha Miller
C. sylvatica Bromf. subsp. **ascendens** (Jordan) P. W. Ball   Common Calamint
C. ascendens Jordan
No record since HG.
First record: Hands et al. in Curtis (1831).
Herb.: LSR, MANCH.
HG 431 as *Satureia ascendens*, rare, restricted in range; 6 localities cited.

C. nepeta (L.) Savi   Lesser Calamint
No record since HG.
First record: Pulteney (1752).
HG 430: very rare, restricted in range; 4 localities cited.

Clinopodium L.

C. vulgare L.   Wild Basil
Map 51g; 40 tetrads.
Occasional throughout the county on roadside and railway verges, in rough grassland, and in limestone quarries. Not apparently confined to calcareous soil.
  Constituent or intrusive: Ev, Er, Eh, Gr, Gp, Rq, Rg, Rs.
First record: Pulteney (1749).
Herb.: LSR★; LSR, LTR, MANCH, NMW.
HG 429: frequent, generally distributed.

Origanum L.

O. vulgare L.   Marjoram
Map 51h; 24 tetrads.
Locally frequent in grassy habitats on the Carboniferous Limestone near Breedon. Otherwise most of the records (55% of the total) are from railway verges, where it is locally frequent in scattered localities throughout the county. Surprisingly it does not seem to occur on the Oolite in the north-east.
  Constituent or intrusive: Gr, Er, Ev, Eh, Rw, Rq.
First record: Pulteney in Nichols (1795).
Herb.: LSR★; LSR, LTR.
HG 427: rare, restricted in range; 20 localities cited.

Thymus L.

T. praecox Opiz subsp. **arcticus** (E. Durand) Jalas
                                                     Wild Thyme
T. drucei Ronniger
Map 51i; 15 tetrads.
Occasional in scattered localities throughout the county, usually on calcareous or sandy soil. More frequent in the east than in the west.
  Constituent or intrusive: Gp, Gr, Hg, Sr, Rq.
First record: Pulteney (1747).
Herb.: LSR★; LSR, LTR, MANCH, NMW.
HG 428 as *T. serpyllum*: local, sparsely distributed; 43 localities cited.

T. pulegioides L.   Large Thyme
Map 51j; 2 tetrads.
Lowesby, dismantled railway, cutting, occasional, HB 1968; Cold Newton, Springfield Hill, old gravel workings, rare, HB 1968; Hallaton, castle mound, occasional, S. Ford 1986.
  Residual or intrusive: Er, Rg, Gp.
First record: Pulteney (1749).
Herb.: LSR★; LSR, MANCH, NMW.
HG 429 as *T. glaber*: frequent, generally distributed. Horwood's accounts of this and the preceding species appear to be in need of reassessment in the light of modern taxonomic opinion.

Lycopus L.

L. europaeus L.   Gipsywort
Map 51k; 148 tetrads.
Locally frequent throughout the county on the banks of rivers, streams and canals. Occasional in marshes.
  Constituent: Wr, Wc, Wl, Wp, Wd, Ew, M.
First record: Pulteney (1747).
Herb.: LSR★; DPM, LANC, LSR, LTR, MANCH, NMW.
HG 427: locally abundant, generally distributed.

Mentha L.

M. requienii Bentham   Corsican Mint
No recent record.
Belton, C. Babington, 1841.
First record: 1841, C. Babington in herb. CGE.
Herb.: CGE.
HG: not recorded.

M. pulegium L.   Pennyroyal
No recent record.
First record: Pulteney (1747).
Herb.: LSR.
HG 427: very rare, probably extinct, extremely restricted in range, not seen in recent years; 5 localities cited.

M. arvensis L.   Corn Mint
Map 51l; 70 tetrads.
Occasional throughout the county in marshes, on woodland rides, and near water. Only very occasionally found as an arable weed. This species has decreased drastically in the past 30 years. It was formerly frequent in arable land, but is appararently unable to tolerate modern agricultural methods and selective weed killers.
  Constituent or intrusive: M, Wp, Wd, Wl, Ew, Ef, Ea, Ev, Er, Fw, Ac, Ao, Gr, Gp, Rd.
First record: Pulteney (1749).
Herb.: LSR★; CGE, LSR, LTR, MANCH, NMW.
HG 426: locally abundant, generally distributed.

M. arvensis × spicata = M. × gentilis L.
No map; 1 tetrad.
Whitwick, pond near Holly Hayes Wood, locally frequent, SHB 1979. The dam across the Grace Dieu Brook which formed this pond has since been removed, but the mint has persisted there at least until 1981.
  Constituent or intrusive: Wp.
First record: Pulteney (1756).
Herb.: LSR★; BM (Swithland Reservoir, A.B. Jackson 1900; not one of the localities in Horwood and Gainsborough).
HG 423: very rare, very restricted in range; 2 localities cited.

M. aquatica L.   Water Mint
Map 52a; 325 tetrads.
Frequent throughout the county in marshes, at the margins of rivers, streams and canals, and in wet woodland.
  Constituent: Wp, Wl, Wr, Wc, Wd, Ew, Ef, Fw, Fd, M.
  Residual or intrusive: Ev, Rg, Rs, Rw, Rd.
First record: Pulteney (1747).
Herb.: LSR★; DPM, LANC, LSR, LTR, MANCH, NMW.
HG 423: locally abundant, generally distributed.

M. aquatica × arvensis = M. × verticillata L.
Map 52b; 12 tetrads.
Groby Pool, marshy borders, locally frequent, EH 1970; Pegg's Green, reedswamp surrounding Kidger's Pond, locally frequent, PAC 1972; Castle Donington, old river channel north of power station, locally frequent, PAC 1973; Anstey, stream side, rare, EH 1973; Swithland Reservoir, muddy shore, locally frequent, PHG 1973; Cropston Reservoir, muddy margins, locally frequent, PHG 1973; Stathern, canal margin, rare, SHB 1979;

Grace Dieu, marsh, locally frequent, PAE and ALP 1982.
   Constituent or intrusive: Ew, M.
First record: Kirby (1850).
Herb.: LSR★; DPM, LSR, MANCH, NMW.
HG 423: locally abundant, sparsely distributed; 29 localities cited, including Groby Pool, Cropston Reservoir and Castle Donington.

M. aquatica × arvensis × spicata = M. × smithiana
R. A. Graham
Map 52c; 2 tetrads.
Huncote, stream bank, occasional, SHB 1974; Packington, Gilwiskaw Brook, margins and marsh, locally frequent, SHB 1976.
   Intrusive or constituent: Ew, M.
First record: Crabbe in Nichols (1795).
Herb.: LSR.
HG 424 as *M. rubra*: rare, very restricted in range; 7 localities cited, including Packington.

M. aquatica × spicata = M. × piperita L. Peppermint
Map 52d; 11 tetrads.
Hemington, small stream in village street, occasional, PAC 1973; Earl Shilton, Heath Lane, roadside bank, locally frequent (nm. **citrata**), SHB 1973; Potters Marston, waste ground, rare, (nm. **citrata**), SHB 1973; Barwell, rubbish dump, locally frequent (nm. **citrata**), SHB 1974; Appleby Parva, damp grass verge, occasional, HH 1974; Blackfordby, Shell Brook, margins, locally frequent, SHB 1976; Willesley, margin of lake, rare, SHB 1977; Swannington, waste ground, occasional (nm. **citrata**), SHB 1978; Pegg's Green, marsh, locally frequent, SHB 1978; Stathern, canal margin, rare, SHB 1979.
   Constituent or intrusive: Ew, Ev, M, Rw, Rr.
First record: Bloxam (1830).
Herb.: LSR, LTR, NMW.
HG 422: rare, restricted in range; 10 localities cited.

M. longifolia (L.) Hudson   Horse Mint
No map.
The consensus of recent opinion is that this species does not occur in Britain, and all records for it in N.W. Europe are probably hairy variants of **M. spicata**, or cultivars of the hybrid **M. longifolia × spicata** (Harley 1972, 1975 and pers. comm. 1979). However, the following records were received during the recent survey:
Beaumont Leys, rubbish dump, rare, EH 1970 (site since built over); Great Easton, ditch at foot of railway bank by level crossing, rare, KGM 1971; Sutton Cheney, dumped earth by field pond, locally frequent, SHB 1973 (pond subsequently filled and ploughed); Sproxton, damp road verge, rare, KGM 1973 (site subsequently cleared); Barwell, rubbish dump, locally frequent, SHB 1974; Enderby, Froane's Hill Quarry, rubble tip, rare, EH 1974.
First record: Pulteney (1749).
Herb.: LSR★; LANC, LSR.
HG 421: rare, very restricted in range; 9 localities cited.

M. spicata L.   Spear Mint
Map 52e; 28 tetrads.
In waste places as a garden throw-out or escape. Locally frequent throughout the county, and almost certainly under-recorded.
   Intrusive: Rw, Rq, Rs, Rr, Rf, Er, Gr, M, Sw.

First record: Walpoole (1782).
Herb.: LSR★; LSR, NMW.
HG 422: rare, restricted in range; 7 localities cited.

M. spicata × suaveolens = M. × villosa Hudson
M. × niliaca auct.
Map 52f; 15 tetrads.
Peckleton, roadside verge, occasional, HH 1971; Leicester, disused allotments, rare, HB 1971; Aylestone, railway bank, rare, HB 1971; Glenfield, rough grass, rare, EH 1972; Barwell, disused allotment gardens, locally frequent, SHB 1973; Earl Shilton, neglected allotments, rare, SHB 1973; Burton on the Wolds, verge of Six Hills Road, soil heap, occasional, ALP 1974; Witherley, recently disturbed ground, rare, SHB 1975; Ashby Woulds, waste ground, rare, SHB 1976; Moira, waste ground, locally frequent, SHB 1976; Ashby de la Zouch, footpath to Blackfordby, disused allotment gardens, occasional, SHB 1976; Ashby de la Zouch, Cliftonthorpe, dismantled railway, rare, SHB 1976; Leicester Forest West, wet road verge and ditch, locally frequent, HH 1976; Swannington, waste ground, rare, SHB 1978.
   Intrusive: Ev, Er, Aa, Rw, Rd, Wd.
First record: 1898, F. T. Mott in herb. BM.
Herb.: LSR★; BM.
HG 422: 1 locality cited.

Salvia L.
S. verbenaca L.   Wild Clary
S. horminoides Pourret
Map 52g; 3 tetrads.
Croft, rocky outcrops in siliceous grassland and old wall, locally frequent, SHB 1973; Cotes, old church site, gravel bank by River Soar, locally frequent, PHG 1973.
   Constituent and intrusive: Hg, Sw, Ew.
First record: Pulteney (1747).
Herb.: LSR★; LSR, LTR, NMW.
HG 431: rare, very restricted in range; 13 localities cited, including Croft.

S. verticillata L.   Whorled Clary
No record since HG.
First record: 1908, W. Bell in herb. LSR.
Herb.: LSR.
HG 432: rare; 4 localities cited.

SOLANACEAE
Nicandra Adanson
N. physalodes (L.) Gaertner
No map; 8 tetrads.
Ratcliffe College, neglected herbaceous border, occasional, ALP 1967 (set much seed but did not reappear); Queniborough, garden (in place where guinea pig cage cleanings were dumped), B. J. Grodding 1967; Beaumont Leys, rubbish dump, rare, EH 1970; Leicester, De Montfort Square, newly seeded soil, occasional, W. Herrington 1970; Coleorton Hall, flower bed, locally frequent, PAC 1976; East Norton, rubbish dump, rare, IME 1976; Moira, rubbish dump, rare, SHB 1977; Sileby, rubbish tip, locally frequent, ALP 1983.
   Casual: Ag, Rr, Rd.
First record: 1949, R. Rawlinson in herb. LTR.
Herb.: LSR★; LSR, LTR.
HG: not recorded.

Lycium L.

L. barbarum L. Duke of Argyll's Teaplant
L. halimifolium Miller
Map 52h; 53 tetrads.
Often planted in hedgerows and gardens, and escaping or persisting long after cultivation has ceased; its presence in a hedgerow sometimes marks the site of a cottage long since demolished. Occasional throughout the county.
Planted or intrusive: Eh, Ev, Er, Sw, Rw, Rq, Aa.
First record: Power (1807).
Herb.: LSR*; DMP, LSR, LTR, NMW.
HG 390 (as *L. chinense*, but most of the records should almost certainly be referred to **L. barbarum**): frequent, restricted in range; 30 localities cited.

L. chinense Miller China Teaplant
Map 52i; 1 tetrad.
Kibworth Harcourt, hedgerow near Kibworth Bridge, locally frequent, EKH and JMH 1974.
Intrusive: Eh.
First record (i.e. first known authenticated record): 1974, E. K. and J. M. Horwood in present work.
Herb.: LSR*.
HG: see note for preceding species.

Atropa L.

A. bella-donna L. Deadly Nightshade
Map 52j; 10 tetrads.
Stoke Golding, farmyard, D. P. Murray 1966; Newtown Linford, Bradgate Park, car park, small gully, rare, EH 1968 (site since destroyed); Leicester, Abbey Meadow, margins of canals, locally frequent, HB 1970; Eastwell, West End Farm, gravel path, rare, JMS 1970 (plants destroyed as a danger to children); Leicester, New Parks, disturbed ground, rare, HB 1971; Bradgate House ruins, old walls and mortar rubble, locally frequent, PHG 1971; Harby, dismantled railway junction, locally frequent, ALP and PCP 1972 (site since destroyed); Loughborough, neglected garden to rear of shops in High Street, locally frequent, PHG 1975; Leicester, Abbey walls, rare, HB 1976; Harby Hills, woodland at base of marlstone escarpment, occasional, ALP 1976; Keyham, hedgerow, rare, J. Otter 1976; Leicester, garden, rare, J.E. Dawson 1982.
Constituent or instrusive: Fd, Er, Ev, Ew, Ag, Rw, Rt, Sw.
First record: Pulteney (1749).
Herb.: LSR*; DPM, LSR, LTR.
HG 390: very rare, restricted in Leicestershire to six districts; 12 localities cited, including Abbey Park and Abbey Meadow.

Hyoscyamus L.

H. niger L. Henbane
No map; 11 tetrads.
Stoke Golding, allotments, D. P. Murray 1961; Leicester, Vaughan Way, waste ground, T. G. Tutin 1962; Stoke Golding, convent garden, D. P. Murray 1963; Bradgate Park, disturbed ground round Deer Barn and ruins of House, rare, PHG 1968; Leicester, Belgrave House, neglected garden, rare, V. Stevens 1968; Beaumont Leys, rubbish dump, rare, EH 1969 (site since built over); Walcote, old gravel pits, rare, EKH and JMH 1969; Worthington, kitchen garden, rare, PAC 1971; Loughborough, railway verge, rare, PHG 1971; Barkby, tipped earth and rubble, rare, IME 1975; Shackerstone, dismantled railway and waste ground, SHB 1976; Leicester, recently disturbed ground, occasional, HB 1979; Leicester, Waterloo Way, recently disturbed ground, single plant, SHB 1979.
Casual: Ag, Aa, Rd, Rw, Rg, Rr, Er.
First record: Pulteney (1747).
Herb.: LSR*; DPM, LANC, LSR, LTR, MANCH, NMW.
HG 391: very local, widely distributed; 50 localities cited.

Physalis L.

P. alkegengi L. Cape-gooseberry
No map.
Ratcliffe on the Wreake, farmyard, single plant, ALP 1969. Other records for this species were lost in the fire at Ratcliffe College. It is often planted.
Casual: Rf.
First record: 1969, A. L. Primavesi in present work.
Herb.: LSR*.
HG: not recorded.

Solanum L.

S. nigrum L. Black Nightshade
Map 52k; 84 tetrads.
Mainly a plant of waste places and disturbed ground, locally frequent in or near large towns, where it is usually of casual occurrence or shortly persistent. Less frequently a weed of arable land.
Intrusive or casual: Rw, Rd, Rf, Rr, Rs, Rt, Ag, Aa, Ao, Ac, Ea, Ev, Er.
First record: Pulteney (1747).
Herb.: LSR*; LSR, LTR, NMW.
HG 389: rare, restricted in range; 27 localities cited.

S. dulcamara L. Bittersweet, Woody Nightshade
Map 52l; 600 tetrads.
Frequent throughout the county in two quite distinct types of habitat: the margins of ponds, rivers and canals, and as a hedgerow and woodland climbing plant.
Constituent or intrusive: Wp, Wd, Wr, Wc, Ew, Eh, Ev, Er, Ef, Fd, Fm, Fs, Fw, Rw, Rr, Rd, Rq, Ap, Gr, Sb.
First record: Pulteney (1747).
Herb.: LSR*; DPM, LANC, LSR, LTR, MANCH, NMW.
HG 388: locally abundant, generally distributed.

S. tuberosum L. Potato
No map.
Often occurs as an outcast on rubbish dumps, in farmyards and waste places, sometimes shortly persistent, and will appear in arable fields and headlands from tubers left in the ground from a previous year's crop. Though widespread in these habitats, most of the field workers in the recent survey overlooked it and did not record it.
First record: Horwood and Gainsborough (1933).
HG 389: occasional.

S. cornutum Lam.
No map; 1 tetrad.
Sileby, Council rubbish tip, single plant, ALP 1983.
Casual: Rr.
First record: 1983, A.L. Primavesi in present work.
Herb.: LCR, LTR.
HG: not recorded.

Lycopersicon Miller
L. esculentum Miller   Tomato
No map; 31 tetrads.
Waste places and arable land. Locally abundant on the sludge drying beds of sewage farms, and becoming increasingly frequent as a non-persistent weed of arable land now that sewage sludge is being pumped on to fields as a fertiliser.
  Intrusive or casual: Rr, Rw, Rf, Rd, Rq, Rs, Ao, Ea, Ew, Gr.
First record: Preston (1900).
HG 389 as *Solanum lycopersicum*: occasional, sparsely distributed; 5 localities cited.

Datura L.
D. stramonium L.   Thorn-apple
No map; 14 tetrads.
Ratcliffe on the Wreake, garden, single plant, ALP 1960; Braunstone, Amy Street, garden, rare, IME 1960; Dadlington, waste ground, D. P. Murray 1961; Barkby, arable land, two plants, IME 1969; Beaumont Leys, rubbish dump, single plant, EH 1970; Anstey, dung heap, rare, EH 1970; Aylestone, recently disturbed ground, rare, HB 1970; Theddingworth, garden, rare, EKH and JMH 1973; Rothley, garden, locally frequent, PHG 1975; Leicester, Belgrave Hall, garden, occasional, SHB 1977; Leicester, Spinney Hills, recently disturbed ground, occasional, HB 1978; Leicester, New Walk, newly planted shrubbery, rare, SHB 1978; Oadby, on imported topsoil, C. Collinson 1981; Syston, derelict building site, occasional, ALP 1986.
  Casual: Rd, Rf, Rw, Ag, Ac.
First record: Crabbe *in* Nichols (1795).
Herb.: LSR★; DPM, LSR, NMW.
HG 391: rare, sparsely distributed; 15 localities cited.

BUDDLEJACEAE
Buddleja L.
B. davidii Franchet   Butterfly-bush
No map; 17 tetrads.
Often planted in gardens, and occasionally self-seeded, appearing as isolated bushes in waste places.
  Intrusive: Rw, Rd, Er, Fd, Sw, Sb.
First record: 1969, E. Hesselgreaves in present work.
HG: not recorded.

SCROPHULARIACEAE
Limosella L.
L. aquatica L.   Mudwort
Map 53a; 2 tetrads.
Eye Brook Reservoir, ox-bow type ponds at outflow end, rare, KGM 1971; Sulby Reservoir, Leicestershire margin, muddy shore, locally abundant, EKH and JMH 1972.
  Constituent: Ew.
First record: Crabbe *in* Nichols (1795).
Herb.: BM, LSR, LTR, WAR.
HG 401: very rare, very restricted in range; 4 localities cited.

Mimulus L.
M. guttatus DC.   Monkeyflower
Map 53b; 12 tetrads.
Shallow streams and wet marshes. Locally frequent in Charnwood Forest, rare or absent elsewhere in the county.
  Intrusive: Wr, Ew, M, Rw.
First record: Preston (1900).
Herb.: LSR, LTR, NMW.
HG 400: occasional, and sparsely distributed; 14 localities cited.

M. luteus L.   Blood-drop-emlets
No record since HG.
First record: Bemrose (1930).
Herb.: LSR, NMW.
HG 401 as *M. guttatus* var. *concolor*: rare; 1 locality cited.

M. moschatus Douglas ex Lindley   Musk
Map 53c; 2 tetrads.
Quorn House Park, marshy bottom of wooded old sand pit, rare, PHG 1968; Staunton Harold, marsh, locally frequent, SHB 1977.
  Intrusive: M.
First record: Horwood and Gainsborough (1933).
HG 401: rare, 1 locality cited.

Verbascum L.
V. blattaria L.   Moth Mullein
No map; 1 tetrad.
Sutton Cheney, Ambion Hall Farm, by entrance track, rare, SHB 1975.
  Casual: Rt.
First record: Preston (1900).
Herb.: LSR.
HG 394: rare, restricted in range; 4 localities cited.

V. virgatum Stokes   Twiggy Mullein
No map; 3 tetrads.
East Norton, dismantled railway, rare, IME 1975; Glen Parva, dismantled railway, occasional, EH 1978; Sileby, rubbish tip, occasional, ALP 1983.
  Casual: Er, Rr.
First record: 1883, F. T. Mott in herb. MANCH.
Herb.: LSR★; MANCH.
HG 393: very rare, restricted in range; 3 localities cited.

V. pyramidatum Bieb.
No map; 1 tetrad.
Melton Mowbray, railway verge, occasional, R. C. Palmer 1977.
  Garden escape: Er.
First record: 1977, R. C. Palmer in present work.
Herb.: OXF.
HG: not recorded.

V. phlomoides L.   Orange Mullein
No map; 6 tetrads.
Anstey, waste ground, rare, EH 1971; Kibworth Harcourt, railway verge, rare, EKH and JMH 1973; Great Glen, railway yard, occasional, EKH and JMH 1973; Enderby, rubbish dump, occasional, EH 1974; South Wigston, old quarry filled with builder's rubble, rare, EH 1974; Leicester, Abbey wall, rare, HB 1976.
  Casual: Rw, Rr, Er, Sw.
First record: 1971, E. Hesselgreaves in present work.
Herb.: LSR★.
HG: not recorded.

V. thapsus L.   Great Mullein
Map 53d; 132 tetrads.
Occasional throughout the county on roadside and railway verges, in waste places and in gardens. A characteristic feature of the ballast of dismantled railways.
   Intrusive: Er, Ev, Rw, Rq, Rg, Rs, Rd, Rr, Rt, Ag, Ap, Sw, Gr.
First record: Pulteney (1749).
Herb.: LSR★; LSR, MANCH.
HG 392: very local, generally distributed.

V. lychnitis L.   White Mullein
No map; 1 tetrad.
Saltby, King Lud's Entrenchments, single plant, IME and ALP 1983. Apart from a record from Waltham Quarry (E.K. Horwood 1952), this species has not been seen in the county since A.R. Horwood's time.
First record: Horwood and Gainsborough (1933).
Herb.: LSR, LTR.
HG 393: rare, restricted in range; 1 locality cited (King Lud's Entrenchments).

V. nigrum L.   Dark Mullein
No map; 6 tetrads.
Leicester, St. Matthew's Estate, disturbed ground, rare, HB 1969 (site since built over); Loughborough, weedy soil heap, single plant, PHG 1970; Freeby, rubble of demolished Saxby Station, rare, KGM 1971; Glooston, rough ground by pathway, rare, KGM 1971; Bottesford, waste ground opposite church, rare, HL 1974; Barkby, tipped soil and rubble, occasional, IME 1976.
   Casual: Rd, Rw.
First record: Bloxam (1830).
Herb.: LANC, LSR.
HG 393: very rare, restricted to limited area; 4 localities cited.

Scrophularia L.
S. nodosa L.   Common Figwort
Map 53e; 152 tetrads.
Woodland, roadside and railway verges and waste places. Frequent in the west of the county; occasional in the east.
   Constituent, residual or intrusive: Fd, Fm, Fs, Fw, Ef, Eh, Er, Ev, Ew, Ea, Rq, Rs, Rw, Rd, Wd, Gr, Hg, M, Sw, Ap.
First record: Pulteney (1747).
Herb.: LSR★; DPM, LSR, LTR, MANCH, NMW.
HG 399: frequent, generally distributed; 112 localities cited.

S. auriculata L.   Water Figwort
S. aquatica auct. non L.
Map 53f; 476 tetrads.
Frequent throughout the county on the banks of rivers, streams and canals, in ditches and in wet places.
   Constituent or intrusive: Wr, Wc, Wd, Wp, Wl, M, Ew, Ev, Er, Eh, Ef, Fw, Rg, Rs, Rw.
First record: Pulteney (1747).
Herb.: LSR★; DPM, LANC, LSR, LTR, MANCH, NMW.
HG 398: frequent, generally distributed.

S. umbrosa Dumort.   Green Figwort
Map 53g; 4 tetrads.
This species was not recorded during the recent survey by any of the field workers other than Dr. P. F. Parker, who is currently engaged on a study of the genus. As this species is conspicuously different from **S. auriculata** and could hardly have been overlooked by the majority of the field workers if it occurred in any quantity, it would appear to be genuinely rare in Leicestershire, and confined to the eastern part of the county. In all cases it is growing in shallow water. Dr. Parker has supplied the following records of its present occurrence: Halstead, banks of small stream, small cluster of 8-10 plants; Sauvey Castle, wet marshy land in moat, rare; Loddington, banks of stream, one or two plants only; Stockerston, banks of the Eye Brook, 4-5 plants only.
   Constituent: Ew, M.
First record: B.E.C. Rep. 1917.
Herb.: LSR.
HG 399 as *S. neesii*: local, restricted in range; 4 localities cited.

Antirrhinum L.
A. majus L.   Snapdragon
Map 53h; 48 tetrads.
A garden escape, occasional throughout the county as a non-persistent casual in waste places, or well-established on walls and the rock faces of quarries.
   Intrusive or casual: Sw, Rq, Rs, Rw, Rr, Rd, Rf, Er, Ev.
First record: Pulteney (1747).
Herb.: LSR★; CGE, LANC, LSR, LTR, NMW.
HG 398: rare, range limited by suitable habitat; 10 localities cited.

Misopates Rafin
M. orontium (L.) Rafin.   Lesser Snapdragon
No map; 2 tetrads.
Leicester, Belgrave Hall, garden weed, occasional, SHB 1977; Groby, single plant as a weed in a house-plant pot, EH 1979.
   Casual: Ag.
First record; 1977, S. H. Bishop in present work.
HG 398 as *Antirrhinum orontium*: no Leicestershire record.

Chaenorhinum (DC.) Reichenb.
C. minus (L.) Lange   Small Toadflax
Map 53i; 173 tetrads.
A characteristic plant of railway ballast, in which habitat it is frequent and locally abundant throughout the county. 94% of the records received are from railway ballast, and the presence of this species often indicates the course of an old railway which has been dismantled for many years, or ballast transported to another site. Occasional in arable land and waste places.
   Intrusive: Er, Ev, Ea, Ac, Ao, Ag, Sw, Rw, Rd, Rq, Rs, Rt.
First record: Coleman (1852).
Herb.: LSR★; LSR, LTR, MANCH, NMW.
HG 397 as *Linaria minor*: rather rare, sparsely distributed; 22 localities cited.

Linaria Miller
L. purpurea (L.) Miller   Purple Toadflax
Map 53j; 88 tetrads.
A frequent garden escape, sometimes becoming established on walls, but occurring more frequently as a casual. It may be shortly persistent in waste places, or persistent by continuous re-introduction.
   Intrusive, casual or escape: Sw, Rw, Rd, Rs, Rr, Rt, Ev, Er, Eh, Aa, Ag.
First record: Mott et al. (1886).
Herb.: LSR★; LSR, LTR, NMW.
HG 395: rare, restricted in range; 6 localities cited.

Linaria purpurea × repens = L. × dominii Druce
No map; 1 tetrad.
Glenfield, dismantled railway, cutting near tunnel, a hybrid swarm, EH 1973.
First record: E. Hesselgreaves in present work.
Herb.: LSR★.
HG: not recorded.

L. repens (L.) Miller   Pale Toadflax
Map 53k; 15 tetrads.
Railway ballast, especially that of dismantled railways, 69% of the records received being from this habitat. Also on roadsides, on walls and in waste places. Occasional in scattered localities throughout the county.
    Intrusive: Er, Ev, Sw, Rw.
First record: 1910, A. R. Horwood in herb. CGE.
Herb.: LSR★; CGE, LSR, LTR, MANCH.
HG 396: rare; 4 localities cited.

L. vulgaris Miller   Common Toadflax
Map 53l; 200 tetrads.
Locally frequent in calcareous grassland, as at Breedon on Carboniferous Limestone, and in the north-east on the Oolite. Otherwise, 74% of the total records received are from railway verges, where it is frequent and locally abundant throughout the county.
    Constituent or intrusive: Gr, Gp, Er, Ev, Eh, Ea, Rq, Rg, Rs, Rw, Rd, Sw, Ag.
First record: Pulteney (1747).
Herb.: LSR★; LSR, LTR, MANCH, NMW.
HG 396: local, and sparsely distributed; 61 localities cited.

Cymbalaria Hill
C. muralis P. Gaertner, B. Meyer and Scherb.
                              Ivy-leaved Toadflax
Map 54a; 131 tetrads.
Locally frequent on walls throughout the country.
    Intrusive: Sw, Sb, Ag, Ap, Rt, Rr, Er.
First record: Power (1807).
Herb.: LSR★; DPM, LSR, MANCH, NMW.
HG 394 as *Linaria cymbalaria*: frequent, but sparsely distributed; 49 localities cited.

Kickxia Dumort.
K. elatine (L.) Dumort.   Sharp-leaved Fluellen
Map 54b; 12 tetrads.
Cossington, Ratcliffe (formerly Carthagena) Farm, arable land, locally abundant, ALP 1967; Ratcliffe on the Wreake, newly made verge of Fosse Way, occasional, ALP 1968; Cossington, Elms Farm, arable land, occasional, ALP 1968; Plungar, cereal crop, rare, PAC 1969; Walton on the Wolds, sugar beet field west of Cradock's Ashes, locally frequent, ALP 1971; Kirby Muxloe, neglected garden, rare, EH 1971; Wymeswold, Highthorn Farm, arable land, occasional, ALP 1972; Thorpe Arnold, broad bean crop, occasional, ALP 1972; Redmile, cornfield next to Jericho Covert, occasional, ALP 1974; Witherley, cereal crop, locally frequent, SHB 1975; Fleckney, southern end of Coal Pit Lane, cornfield, E. Bishop 1977; south of King Lud's Entrenchments, arable field, frequent, PHG and SHB 1978; Coston, arable field, rare, ALP 1978.
    Intrusive: Ac, Ao, Ag, Rd.
First record: Pulteney *in* Nichols (1795).
Herb.: LSR★; LSR, MANCH, NMW.
HG 394 as *Linaria elatine*: occasional, local, very restricted in range; 17 localities cited.

K. spuria (L.) Dumort.   Round-leaved Fluellen
Map 54c; 15 tetrads.
A weed of arable land, locally frequent in scattered localities in the north-east of the county, rare or absent elsewhere. This and the preceding species vary considerably in their frequency from year to year. In some years they may be in great abundance where they do occur, in others, hard to find.
    Intrusive: Ac, Ao, Ea, Er.
First record: Crabbe *in* Nichols (1795).
Herb.: LSR★; LSR.
HG 395 as *Linaria spuria*: rather rare, restricted in range; 8 localities cited.

Digitalis L.
D. purpurea L.   Foxglove
Map 54d; 178 tetrads.
Locally frequent in the west of the county in hedgerows and woodland, preferring acid soils. Occasional and often an escape from cultivation in the east.
    Constituent or intrusive: Fd, Fm, Fs, Ef, Eh, Ev, Er, Ew, Hg, Gr, Rq, Rg, Rs, Rw, Rd, Rt, Rr, Sw, Ap, Ag.
First record: Pulteney (1747).
Herb.: LSR★; LSR, LTR, MANCH, NMW.
HG 401: locally abundant, restricted in range; 93 localities cited.

Veronica L.
V. serpyllifolia L.   Thyme-leaved Speedwell
Map 54e; 311 tetrads.
Frequent throughout the county in short grazed grassland and in woodland rides. Occasional as a weed of gardens and arable land.
    Constituent or intrusive: Gp, Gr, Gl, Hg, M, Er, Ev, Ew, Ea, Ef, Fd, Fs, Rq, Rs, Rt, Ap, Ag, Ac.
First record: Pulteney (1747).
Herb.: LSR★; LANC, LSR, LTR, MANCH, NMW.
HG 406: locally abundant, generally distributed.

V. officinalis L.   Heath Speedwell
Map 54f; 39 tetrads.
Heathland and dry places where there is acid soil. Locally frequent in parts of Charnwood Forest and the west of the county. Occasional on the Marlstone in the east, and elsewhere where there are suitable soils.
    Constituent or intrusive: Hg, Hd, Gr, Gp, Fd, Fm, Er, Ev, Eh, Ef, Rq, Rr, Rd, Rt.
First record: Pulteney (1747).
Herb.: LSR★; LSR, LTR, MANCH, NMW.
HG 406: frequent, generally distributed.

V. chamaedrys L.   Germander Speedwell
Map 54g; 553 tetrads.
Frequent and locally abundant throughout the county in grassland, on roadside and railway verges, and in woodland rides.
    Constituent or intrusive: Gp, Gr, Gl, Hg, Ev, Er, Ew, Eh, Ea, Ef, Fd, Fm, Fs, Rw, Rq, Rs, Rd, Rt, Rr, Sw, M, Ap, Ao, Ac, Ag.
First record: Pulteney (1747).
Herb.: LSR★; DPM, LANC, LSR, LTR, MANCH, NMW.
HG 406: locally abundant, generally distributed.

V. montana L. Wood Speedwell
Map 54h; 60 tetrads.
Woodland. Frequent and locally abundant in the well-wooded parts of the county; occasional or absent elsewhere.
　　Constituent: Fd, Fc, Fm, Fw, Fs, Ef.
　　Residual: Eh, Ew, Ev, Wd, Ap, Rq.
First record: Bloxam *in* Combe (1829).
Herb.: LSR★; LSR, LTR, MANCH, NMW.
HG 407: occasional, sparsely distributed; 32 localities cited.

V. scutellata L. Marsh Speedwell
Map 54i; 12 tetrads.
Wyfordby, marshy field between River Eye and railway, rare, ALP and PCP 1969; Bagworth, Merrylees, marsh, rare, OHB and HH 1969; Charnwood Lodge Nature Reserve, marsh, occasional, PHG 1969; Cropston Reservoir, muddy margins, locally frequent, PHG 1969; Groby, marshy meadow to west of the Pool, occasional, EH 1970; Newtown Linford, marshy ground south of Cropston Reservoir, rare, EH 1970; Stanton under Bardon, pond, locally frequent, PAC 1970; Ulverscroft, marshy field by Lea Wood, occasional, PHG 1970; Great Easton, pond in cricket ground field, rare, KGM 1971; Kegworth, ditches, occasional, PAC 1973; Stathern, pond south of Combs Plantation, locally abundant, SHB 1973; Stanford Park, shallow ornamental lake west of Hall, locally frequent, EKH and JMH 1974.
　　Constituent or residual: M, Wp, Wd.
First record: Pulteney (1747).
Herb.: LSR★; LSR, LTR, MANCH, NMW, UPP.
HG 407: rare, sparsely distributed; 32 localities cited.

V. beccabunga L. Brooklime
Map 54j; 529 tetrads.
Frequent and locally abundant along the margins of rivers, streams and canals, in ditches and marshes, and in woodland rides.
　　Constituent: Wr, Wc, Wp, Wl, Wd, Ew, Ef, Fw, Fd, M.
　　Residual or intrusive: Gp, Rw.
First record: Pulteney (1747).
Herb.: LSR★; DPM, LSR, LTR, MANCH, NMW.
HG 409: locally abundant, generally distributed.

V. anagallis-aquatica L. Blue Water-speedwell
Map 54k; 8 tetrads.
Nanpantan, old Forest Canal, occasional, PHG 1970; Drayton, north bank of River Welland, rare, KGM 1971 (later destroyed by river 'improvement'); Claybrooke Parva, shallow pond, locally frequent, MH 1971 (did not survive the drought of 1976); Stanford Park, muddy margins of lake, occasional, EKH and JMH 1971; Garthorpe, muddy margin of pond, rare, KGM 1973; Lubbesthorpe, margin of pond, occasional, EH 1973; Cosby, stream in village street, locally frequent, EH 1975; Wyfordby, River Eye near Ham Bridge, SHB and PHG 1977.
　　Constituent: Wr, Wp, Wl, Wc.
First record: 1876, F. T. Mott in herb. MANCH.
Herb.: LSR★; LSR, MANCH.
HG 408: frequent, distribution not fully known; 9 localities cited.

V. catenata Pennell Pink Water-speedwell
Map 54l; 104 tetrads.
Locally frequent in mud and shallow water at the margins of rivers, ponds and lakes.
　　Constituent or residual: Wr, Wp, Wl, Wd, Ew, M, Rg.
First record: 1881, F. T. Mott in herb. MANCH.
Herb.: LSR★; DPM, LSR, LTR, MANCH, UPP.
HG 409 as *V. aquatica*: occasional, distribution not fully known; 3 localities cited.

V. arvensis L. Wall Speedwell
Map 55a; 314 tetrads.
Frequent throughout the county in arable land, on railway ballast, on the tops of walls, and in waste places.
　　Intrusive: Ac, Ao, Ag, Aa, Ap, Ea, Er, Ev, Ef, Sw, Rt, Rq, Rg, Rs, Rw, Rd, Rr, Gp, Gr, Gl, Hg.
First record: Pulteney (1749).
Herb.: LSR★; CGE, LANC, LSR, LTR, MANCH, NMW.
HG 405: locally abundant, generally distributed.

V. agrestis L. Green Field-speedwell
Map 55b; 83 tetrads.
Occasional throughout the county in arable land and waste places.
　　Intrusive: Ac, Ao, Aa, Ag, Ap, Ea, Ev, Er, Rw, Rq, Rs, Rd, Rf, Rt, Rr, Gl, Sw, Sr.
First record: Pulteney (1747).
Herb.: LSR★; CGE, DPM, LSR, LTR, MANCH, NMW.
HG 404: rather local, but generally distributed.

V. polita Fries Grey Field-speedwell
Map 55c; 20 tetrads.
Arable land and gardens. Occasional in scattered localities throughout the county.
　　Intrusive: Ag, Ac, Ao, Ap, Rw, Rd, Rt, Sw.
First record: Bloxam (1837).
Herb.: LSR★; CGE, LSR, LTR, MANCH, NMW.
HG 403 as *V. didyma*: frequent, widely distributed; 34 localities cited.

V. persica Poiret Common Field-speedwell
Map 55d; 521 tetrads.
This species, first recorded in Britain in 1829, is now the commonest speedwell in arable land and waste places throughout the county.
　　Intrusive: Ac, Ao, Aa, Ag, Ap, Ea, Ev, Er, Ew, Rw, Rd, Rs, Rf, Rt, Gl, Gr, Gp.
First record: Bloxam (1846a).
Herb.: LSR★; DPM, LANC, LSR, LTR, MANCH, NMW.
HG 405 as *V. tournefortii*: locally abundant, generally distributed.

V. filiformis Sm. Slender Speedwell
Map 55e; 104 tetrads.
Public and private parks, churchyards, lawns, the banks of streams, and roadside and railway verges. Locally frequent throughout the county, and on the increase.
　　Intrusive: Ap, Ag, Gl, Gp, Gr, Ew, Ev, Er, Rw, Rd.
First record: 1957, E. K. Horwood in herb. LTR.
Herb.: LSR★; BM, LSR, LTR.
HG: not recorded.

V. hederifolia L. Ivy-leaved Speedwell
Map 55f; 331 tetrads.
Frequent and locally abundant as a garden weed and in recently disturbed ground. Less frequent as a weed of

261

arable land, and occasional in open woodland. Throughout the county, possibly under-recorded.
Intrusive: Ag, Ap, Ao, Aa, Ac, Rd, Rw, Rt, Rr, Rf, Sw, Ev, Er, Eh, Ew, Ea, Fd, Fm, Fs.
First record: Pulteney (1747).
Herb.: LSR★; DPM, LANC, LSR, LTR, MANCH, NMW.
HG 402: locally abundant, generally distributed; 33 localities cited.

V. longifolia L.
No map; 1 tetrad.
Ashby Woulds, rough grassland, rare, SHB 1977.
Casual: Gr.
First record: 1977, S. H. Bishop in present work.
HG: not recorded.

Melampyrum L.
M. pratense L. Common Cow-wheat
Map 55g; 7 tetrads.
Ratby Burroughs, deciduous woodland, locally frequent HH 1968; Swithland Wood, locally frequent, PHG 1968; Loughborough, Out Wood, locally frequent, PHG 1969; Ulverscroft, Stoneywell Wood, rare, PHG 1969; Ashby Pastures, woodland ride, rare, HB 1975; Buddon Wood, margin, rare, PHG 1981.
Constituent: Fd, Ef.
First record: Pulteney (1747).
Herb.: LSR★; LSR, LTR, MANCH, NMW.
HG 416: local, very sparsely distributed; 15 localities cited.

Euphrasia L.
E. anglica Pugsley Eyebright
Map 55h; 3 tetrads.
Desford, Lindridge, overgrown ancient mine-shaft spoil heap, locally frequent, OHB, HH and SM 1969 (site levelled and destroyed 1975); Moult Hill near Blackbrook Reservoir, heath grassland, occasional, PAC 1974; Newtown Linford, Roecliffe Manor, grass heath lawn, locally frequent, PHG 1976.
Constituent or residual: Hg, Rs.
First record: 1894, F. T. Mott in herb. NMW.
Herb.: BM, CGE, LTR, NMW.
HG 412: occasional, sparsely distributed; 12 localities cited.

E. arctica Lange ex Rostrup subsp. **borealis** (Townsend) Yeo Eyebright
E. brevipila auct. non Burnat & Gremli
Map 55i; 1 tetrad.
Osgathorpe, old pasture, locally abundant, PAC 1968.
Constituent or residual: Gp.
First record: 1886, F. T. Mott in herb MANCH.
Herb.: LSR★; CGE, LSR, MANCH.
HG 412: rare, very restricted in range; 1 locality cited (Ulverscroft), and 6 localities cited under *E. borealis*.

E. nemorosa (Pers.) Wallr. Eyebright
Map 55j; 20 tetrads.
Heath and limestone grassland, quarries, woodland rides and railway verges. Locally frequent in scattered localities throughout the county.
Constituent or residual: Gr, Gp, Hg, Hd, Rq, Ef, Er, Ev.
First record: Jackson (1904a).
Herb.: LSR★; BM, LSR, LTR, MANCH, NMW.
HG 410: frequent, sparsely distributed; 36 localities cited.

Odontites Ludwig
O. verna (Bellardi) Dumort. Red Bartsia
Map 55k; 208 tetrads.
Roadside and railway verges, green lanes, woodland rides, and sometimes in arable land. Locally frequent throughout the county, but apparently decreasing.
Constituent or intrusive: Ev, Er, Ef, Ew, Ea, Gp, Gr, Gl, Hg, M, Rt, Rq, Rs, Rw, Rr, Rd, Ac, Ao.
First record: Pulteney (1747).
Herb.: LSR★; CGE, DPM, LANC, LSR, LTR, MANCH, NMW, WAR.
HG 412 as *Bartsia odontites*: locally abundant, generally distributed.

Pedicularis L.
P. palustris L. Marsh Lousewort
No record since HG.
First record: Pulteney (1747).
Herb.: LSR, MANCH, NMW.
HG 413: occasional, sparsely distributed; 14 localities cited.

P. sylvatica L. Lousewort
Map 55l; 13 tetrads.
Heathland, marshes and old grassland. Occasional in Charnwood Forest, and at Twenty Acre, Burbage Common and Harby Hills. Absent from the rest of the county.
Constituent: Hg, Hd, M, Gp.
Residual or intrusive: Rq.
First record: Pulteney (1747).
Herb.: LSR★; LSR, LTR, NMW.
HG 414: local, widely distributed; 60 localities cited.

Rhinanthus l.
R. minor L. Yellow Rattle
Map 56a; 183 tetrads.
A characteristic plant of old grassland and marshes, often persisting in leys and arable land. Locally abundant throughout the county in these habitats, but decreasing as old grassland is destroyed.
Constituent, residual or intrusive: Gp, Gr, Gl, Hg, M, Ev, Er, Ef, Ea, Rd, Rg, Rt, Rw, Fs.
First record: Pulteney (1747).
Herb.: LSR★; CGE, DPM, LSR, LTR, MANCH, NMW.
HG 415 as *R. crista-galli*: locally abundant, generally distributed.

R. angustifolius C. C. Gmelin subsp. **angustifolius**
R. serotinus (Schonheit) Oborny
No record since HG.
First record: Coleman (1852).
Herb.: NMW.
HG 415 as *R. major*: very rare, restricted in range; 2 localities cited.

Lathraea L.
L. squamaria L. Toothwort
Map 56b; 10 tetrads.
Loddington Reddish, T. W. Tailby 1965; Sauvey Castle, wet spinney by River Chater, occasional, ALP 1969; Launde, spinney by River Chater, on hazel, rare, KGM 1971; Sheet Hedges Wood, locally frequent, EH 1971; Hallaton Wood, on hazel, occasional, KGM 1973; Owston Wood, occasional, IME 1975; Launde, Big Wood, rare, IME 1975; Tugby Bushes, deciduous woodland, occasional, IME 1976; Skeffington Wood, locally

frequent, IME 1976; Launde, Park Wood, northern part, locally abundant, IME 1978; Allexton Wood, frequent, PAE 1978 and 1979; Tryon Spinney, locally frequent, PAE 1979; Hallaton, hedge, occasional, J.E. Dawson 1984.
  Constituent: Fw, Fd.
First record: Bloxam and Babington in Potter (1842).
Herb.: LSR★; CGE, LSR, LTR, NMW.
HG 418: rare, restricted in range; 3 localities cited, including Owston Wood.

OROBANCHACEAE
Orobanche L.

O. purpurea Jacq. — Yarrow Broomrape
No record in the recent survey.
Stonesby Quarry, EKH 1952.
First record: Mott et al. (1886).
Herb.: LSR, LTR.
HG 416: rare, restricted in range; 2 localities cited, including Stonesby Quarry.

O. minor Sm. — Common Broomrape
Map 56c; 1 tetrad.
Thurcaston, steep railway embankment, about 20 plants, PHG 1974.
  Residual or intrusive: Er.
First record: Britten in White (1877).
Herb.: LSR.
HG 418: rare, sparsely distributed; 3 localities cited.

O. hederae Duby — Ivy Broomrape
No record since HG.
First record: Coleman (1852).
Herb.: LSR.
HG 417: rare, restricted in range; 4 localities cited.

O. elatior Sutton — Knapweed Broomrape
No record since HG.
First record: Kirby (1849).
Herb.: BM, LSR.
HG 417: rare, restricted in range; 2 localities cited.

O. rapum-genistae Thuill. — Greater Broomrape
No recent record.
First record: Pulteney (1747).
HG 416: very rare - possibly extinct - not observed within recent years, almost confined to Charnwood Forest; 9 localities cited.

LENTIBULARIACEAE
Pinguicula L.

P. vulgaris L. — Common Butterwort
No record since HG.
First record: Pulteney (1747).
Herb.: LSR.
HG 419: very rare, very restricted in range, not extinct as suggested (in Fl. Leics. 1886) and rediscovered in E. Leicestershire on leached out detritus and peat on limestone soils; 13 localities cited.

Utricularia L.

U. vulgaris L. — Greater Bladderwort
No record since HG.
First record: Pulteney in Nichols (1795).
Herb.: LSR.
HG 419: rare, very restricted in range; 11 localities cited.

U. australis R.Br.
U. neglecta Lehm.
No record since HG.
First record: 1891, F. T. Mott in herb. MANCH.
Herb.: MANCH.
HG 419 as *U. major*: very rare, only in one district; 1 locality cited.

PLANTAGINALES

PLANTAGINACEAE
Plantago L.

P. major L. — Broad-leaved Plantain, Greater Plantain
Map 56d; 603 tetrads.
Frequent and locally abundant throughout the county in grassland and on verges. Frequent in the trampled ground of field gateways. Occasional in arable land.
  Constituent or intrusive: Gp, Gr, Gl, Hg, Ev, Er, Ew, Ea, Ef, Rt, Rw, Rq, Rs, Rf, Rr, Rd, Ap, Ag, Ac, Ao, Sw.
First record: Pulteney (1747).
Herb.: LSR★; CGE, DMP, LANC, LSR, LTR, MANCH, NMW.
HG 446: locally abundant, generally distributed.

P. coronopus L. — Buck's-horn Plantain
Map 56e; 4 tetrads.
Rearsby, Broome Lane, cinder roadway at gate of Ordnance Depot, locally abundant, ALP 1967 (site later destroyed in development of East Goscote village); Breedon Hill, limestone grassland, rock outcrops, occasional, PAC 1972; Mountsorrel, near Hawcliff Quarry, 'granite' waste, rare, PHG 1975; Leicester, Spinney Hill Park, rare, HB 1978.
  Constituent or intrusive: Gp, Rs, Rt, Ap.
First record: Pulteney (1747).
Herb.: LSR★; LSR, MANCH.
HG 448: rare, restricted in range; 18 localities cited, including Breedon Hill.

P. media L. — Hoary Plantain
Map 56f; 205 tetrads.
Locally frequent throughout the county in grassland and on roadside and railway verges, especially where the soil is basic; more frequent in the east of the county than in the west. Often found in the mown grass of churchyards.
  Constituent or intrusive: Gp, Gr, Gl, Ev, Er, Ew, Ea, Ap, Ag, Ao, Rw, Rq, Rt, Rd, Sw, Sb.
First record: Pulteney (1747).
Herb.: LSR★; DPM, LSR, LTR, MANCH, NMW.
HG 447: frequent, generally distributed.

P. lanceolata L. — Ribwort Plantain
Map 56g; 599 tetrads.
Frequent and locally abundant throughout the county in grassland and on verges. Occasional in arable land.
  Constituent or intrusive: Gp, Gr, Gl, Hg, Ev, Er, Ew, Ef, Ea, Ac, Ao, Ap, Ag, Rw, Rt, Rq, Rs, Rd, Rr.
First record: Pulteney (1747).
Herb.: LSR★; DPM, LSR, LTR, MANCH, NMW.
HG 447: locally abundant, generally distributed.

P. arenaria Waldst. & Kit.
P. indica L. nom. illegit.
No record since HG.
First record: Jackson (1904).
Herb.: LSR.
HG 449: rare; 3 localities cited.

Littorella Bergius
L. uniflora (L.) Ascherson   Shoreweed
Map 56h; 7 tetrads.
Blackbrook Reservoir, sandy, rocky and muddy shores, locally abundant, PHG 1969; Cropston Reservoir, several localities, PHG 1969; Swithland Reservoir, rocky and sandy shores, locally abundant, PHG 1972.
   Constituent: Wl, Ew,.
First record: Bloxam *in* Combe (1829).
Herb.: LSR★; LSR, LTR, MANCH.
HG 449: very rare, restricted in range; 5 localities cited, including Cropston Reservoir.

DIPSACALES

CAPRIFOLIACEAE
Sambucus L.
S. ebulus L.   Danewort, Dwarf Elder
Map 56i; 4 tetrads.
Cropston, sand pit and old hedgerow near site of Rothley Station, locally frequent, PHG 1970 (site destroyed 1971); Stockerston, hedgerow and roadside verge, occasional, KGM 1971; Hose, roadside verge just within the county boundary, locally frequent, ALP 1975; King's Norton, hedgerow on parish boundary, locally abundant, IME 1976.
   Constituent, residual or intrusive: Ev, Eh, Rg.
First record: Pulteney (1747).
Herb.: LSR★; NMW.
HG 274: rare, sparsely distributed; 11 localities cited, including Stockerston, Hose and Rothley Plain.

S. nigra L.   Elder
Map 56j; 602 tetrads.
Frequent and locally abundant throughout the county in hedgerows and woodland. It is tolerant of root disturbance, and hence is characteristic of the disturbed ground near rabbit warrens and badger setts.
   Constituent, residual or intrusive: Fd, Fm, Fs, Fw, Eh, Ev, Er, Ew, Ap, Rw, Rd, Rq.
First record: Pulteney (1747).
Herb.: LSR★; DPM, LANC, LSR, LTR, MANCH, NMW.
HG 273: frequent, generally distributed.

S. racemosa L.   Red-berried Elder
No map; 2 tetrads.
Coalville, Green Hill and Abbot's Oak, roadside strip of ornamental woodland, occasional, probably planted, but there are scattered bushes in secondary woodland nearby which may be self-sown, PAC 1969; Leicester Forest West, Old Brake Spinney, probably planted, rare, HH 1975.
   Planted or intrusive: Fd.
First record: 1951, H. J. Lacey in herb. LTR.
Herb.: LTR.
HG: not recorded.

Viburnum L.
V. opulus L.   Guelder-rose
Map 56k; 308 tetrads.
Locally frequent throughout the county in woodland and hedgerows, particularly in damp situations.
   Constituent or residual: Fd, Fm, Fs, Fw, Ef, Eh, Ew, Ev, Er, Wd, Ap, M, Rw.
First record: Pulteney (1747).
Herb.: LSR★; CGE, DPM, LSR, LTR, MANCH, NMW.
HG 274: frequent, generally distributed.

V. lantana L.   Wayfaring-tree
Map 56l; 10 tetrads.
Woodhouse, small area of woodland between Swithland Reservoir Pumping Station and railway, rare, PHG 1968; Cotesbach, roadside hedge, rare, EKH and JMH 1970; Scraptoft, hedgerow, rare, IME 1971; Newton Harcourt, hedge bordering spinney near river bridge, rare, EKH and JMH 1972; Frisby on the Wreake, Cream Gorse, roadside edge of fox covert, rare, ALP 1972; Buddon Wood, rock face on western edge, single shrub, PHG 1972; Wistow, roadside hedge, rare, EKH and JMH 1973; Melton Mowbray, Great Framlands, field hedgerow, occasional, ALP 1973; King's Norton, roadside hedge, occasional, EKH and JMH 1974; Muston, green lane north of Debdale Farm, single bush, HL 1974; King's Norton, parish boundary hedge, rare, R. Gardner 1978.
   Constituent or intrusive: Eh, Ef, Fd.
First record: Pulteney *in* Nichols (1795).
Herb.: LSR★; LSR, LTR, NMW.
HG 275: rare, sparsely distributed; 22 localities cited, including Cotesbach and Wistow.

Symphoricarpos Duh.
S. albus (L.) S. F. Blake   Snowberry
S. rivularis Suksd.
Map 57a; 264 tetrads.
Often planted in hedges and shrubberies or as game cover, and frequently self-seeded, becoming established in woodland and hedgerows. Locally frequent throughout the county.
   Intrusive or planted: Fd, Fm, Fs, Fw, Ef, Eh, Ev, Er, Ew, Ap, Ag, Aa, Rw, Rr, Sw, M.
First record: 1884, F. T. Mott in herb. MANCH.
Herb.: LSR★; DPM, LSR, LTR, MANCH, NMW.
HG 276 as *S. racemosus*: occasional, sparsely distributed; 31 localities cited.

Lonicera L.
L. caerulea L.
No record since HG.
First record: B.E.C. Rep. 1916.
Herb.: NMW.
HG 277: rare; 3 localities cited.

L. xylosteum L.   Fly Honeysuckle
No recent record.
First record: Kirby (1849).
Herb.: LSR.
HG 277: rare, restricted in range; 4 localities cited.

L. caprifolium L.   Perfoliate Honeysuckle
No record since HG.
First record: Horwood (1911)
HG 276: very rare, restricted in range; 1 locality cited.

L. nitida Wilson　　　　　　Japanese Honeysuckle
No map.
Frequently planted in garden hedges. It persists at sites of former cultivation, and discarded hedge clippings sometimes take root. It has been recorded from waste places and railway verges throughout the county, but very few field workers recorded it, either by oversight or because it was assumed to be planted. The full distribution of these spontaneous occurrences is not known, because many of the records were lost after the fire at Ratcliffe College.
　　Intrusive or garden escape: Rw, Er.
First record: Present work.
HG: not recorded.

L. periclymenum L.　　　　　　Honeysuckle
Map 57b; 359 tetrads.
Locally frequent throughout the county in hedgerows and woodland, but with a tendency to avoid the heavier clay soils. In deep shade in woodland it hardly ever flowers.
　　Constituent, residual or intrusive: Fd, Fm, Fs, Fw, Ef, Eh, Ew, Er, Ev, Sw, Rq, Rw, Gr.
First record: Pulteney (1747).
Herb.: LSR★; DPM, LANC, LSR, LTR, MANCH, NMW, UPP.
HG 277: frequent, generally distributed.

ADOXACEAE
Adoxa L.
A. moschatellina L.　　　　　　Moschatel
Map 57c; 63 tetrads.
Woodland, hedgerows and the banks of streams in shady places. Locally frequent in the west and the south-east of the county; rare or absent elsewhere.
　　Constituent or residual: Fd, Fm, Fs, Fw, Ef, Ew, Eh, Ev, M, Wd.
First record: Pulteney (1747).
Herb.: LSR★; LSR, LTR, MANCH, NMW.
HG 272: occasional, sparsely distributed; 50 localities cited.

VALERIANACEAE
Valerianella Miller
V. locusta (L.) Laterrade　　　Lamb's Lettuce,
　　　　　　　　　　　　　　　Common Cornsalad
Map 57d; 23 tetrads.
Occasional throughout the county on railway verges (76% of the records received). Very rarely a weed of arable land. Almost certainly under-recorded.
　　Intrusive: Er, Ev, Ea, Ag, Rw, Sw.
First record: Pulteney (1747).
Herb.: LSR★; DPM, LANC, LSR.
HG 289 as *V. olitoria*: frequent, widely distributed; 34 localities cited.

V. carinata Loisel.　　　　Keel-fruited Cornsalad
Map 57e; 2 tetrads.
Rearsby, grassy railway cutting by Bleak Moor, locally frequent, ALP 1971; Breedon on the Hill, limestone grassland, locally frequent, SHB 1978.
　　Constituent or intrusive: Gp, Er.
First record: Jackson (1904).
Herb.: BM, LSR.
HG 290: rare, restricted in range; 2 localities cited.

V. dentata (L.) Pollich　　Narrow-fruited Cornsalad
Map 57f; 3 tetrads.
Walton on the Wolds, sugar beet field south-west of Cradock's Ashes, occasional, ALP 1971; Glen Parva, grassy verge of dismantled railway, rare, EH 1974; south of King Lud's Entrenchments, cornfield, rare, SHB 1978.
　　Intrusive: Ac, Ao, Er.
First record: Hands et al. *in* Curtis (1831).
Herb.: LSR★; CGE, LSR, MANCH, NMW.
HG 290: local, restricted in range; 26 localities cited.

Valeriana L.
V. officinalis L.　　　　　　Common Valerian
Map 57g; 94 tetrads.
Wet woodland, marshes, and wet places in general. Locally frequent in the west of the county; occasional in the south-east; absent elsewhere, except for a single record from the north-east.
　　Constituent or residual: M, Fw, Fd, Fm, Ef, Eh, Ev, Er, Ew, Wd, Wr, Wc, Sr, Gr.
No attempt was made in the recent survey to segregate the subspecies **officinalis** and **sambucifolia** (Mikan fil.) Celak
First record: for subsp. **officinalis**, Pulteney (1747); for subsp. **sambucifolia**, Coleman (1852).
Herb.: LSR★; CGE, LSR, LTR, MANCH, NMW.
HG 287 as *V. officinalis*: frequent, widely distributed; 56 localities cited.
HG 288 as *V. sambucifolia*: frequent, sparsely distributed; 46 localities cited.

V. dioica L.　　　　　　　　Marsh Valerian
Map 57h; 23 tetrads.
Occasional throughout the county in marshes and wet woodland.
　　Constituent or residual: M, Fw, Fd, Ew.
First record: Pulteney (1747).
Herb.: LSR★; DPM, LANC, LSR, LTR, MANCH, NMW.
HG 286: frequent, widely distributed; 53 localities.

Centranthus DC.
C. ruber (L.) DC.　　　　　　Red Valerian
Map 57i; 22 tetrads.
An escape from cultivation, sometimes becoming established on walls, rocks, quarry faces and railway verges, and in waste places. Occasional throughout the county.
　　Intrusive: Sw, Sr, Rq, Rg, Rw, Er, Ev, Eh.
First record: Horwood (1911).
Herb.: LSR★; LANC, LTR.
HG 289: rare, restricted in range; 2 localities cited.

DIPSACACEAE
Dipsacus L.
D. sativus (L.) Honckeny　　　Fuller's Teasel
No record since HG.
First record: Power (1807).
HG 291: rare, restricted in range; 6 localities cited.

D. fullonum L.　　　　　　　　Teasel
Map 57j; 260 tetrads.
Locally frequent throughout the county near water, on

roadside and railway verges, and in waste places.
  Constituent, residual or intrusive: Ew, Ev, Er, Eh, Ea, Wd, Ag, Aa, Ap, Rw, Rg, Rq, Rs, Rd, Rr, Rt, M, Gr, Gp.
First record: Pulteney (1747).
Herb.: DPM, LANC, LSR, LTR, NMW.
HG 291 as *D. sylvestris*: frequent, generally distributed.

D. pilosus L. Small Teasel
Map 57k; 8 tetrads.
Cranoe, roadside hedgerow, rare, IME 1963; Hoothill Slang, wet spinney by Eye Brook, occasional, ALP 1968; Lowesby, Carr Bridge Spinney, wet woodland, occasional, HB 1969; East Norton, bank of Eye Brook, rare, IME 1975; Launde, Big Wood, woodland ride, locally frequent, IME 1975; Tilton Wood, woodland edge, locally abundant, IME 1976; Loddington Reddish, woodland, rare, IME 1976; Tugby Bushes, bank of Eye Brook, rare, IME 1976; Tugby Wood, rare, IME 1976; Noseley, Coney Hill Plantation, rare, I. Metcalf 1980.
  Constituent or residual: Fw, Fd, Ef, Ew.
First record: Pulteney (1747).
Herb.: LSR★; CGE, LSR, LTR, MANCH.
HG 292: rare, sparsely distributed; 17 localities cited, including Tugby Wood, Lowesby, East Norton, Tilton Wood and Loddington Reddish.

Succisa Haller
S. pratensis Moench Devil's-bit Scabious
Map 57l; 112 tetrads.
Old grassland, marshes and other grassy places. Locally frequent in the west of the county, and on the Marlstone in the east. Occasional elsewhere.
  Constituent, residual or intrusive: Gp, Gr, Hg, Hd, M, Er, Ev, Eh, Ea, Ew, Ef, Fc, Fm, Wd.
First record: Pulteney (1747).
Herb.: LSR★; DPM, LANC, LSR, LTR, NMW.
HG 293 as *Scabiosa succisa*: frequent, generally distributed.

Knautia L.
K. arvensis (L.) Coulter Field Scabious
Map 58a; 62 tetrads.
Calcareous grassland, railway and roadside verges, and quarries. Frequent on the Oolite in the north-east; occasional throughout the rest of the county.
  Constituent or intrusive: Gp, Gr, Er, Ev, Eh, Ew, Ea, Rq, Rs, Rw, Rd.
First record: Pulteney (1747).
Herb.: LSR★; LSR, LTR, MANCH, NMW.
HG 293: frequent, generally distributed.

Scabiosa L.
S. columbaria L. Small Scabious
Map 58b; 6 tetrads.
Stonesby Quarry, locally frequent, ALP 1972; Saltby Heath, disused airfield, in several localities, occasional, KGM 1972; Sproxton, limestone grassland by mineral railway near county boundary, occasional, KGM 1973; Stathern, calcareous pasture on old ironstone workings near Terrace Hill Farm, locally frequent, SHB 1973; Breedon Hill, limestone grassland, rare, SHB 1978.
  Constituent or residual: Gp, Rq, Rs.
First record: Pulteney (1747).
Herb.: LSR★; LSR, LTR, MANCH, NMW.
HG 293: rare, sparsely distributed; 17 localities cited.

CAMPANULALES

CAMPANULACEAE
Campanula L.

C. patula L. Spreading Bellflower
No map; 1 tetrad.
Buddon Wood, 3 plants on edge of quarry, PHG 1985; 2 plants on spoil heap and 3 in grass, PHG 1986.
  Constituent or residual: Fd, Rs, Gr.
First record: Pulteney (1747).
Herb.: LSR, NMW.
HG 355: very rare, confined to Charnwood Forest; 4 localities cited, including Buddon Wood.

C. rapunculus L. Rampion Bellflower
No record since HG.
First record: Horwood and Gainsborough (1933).
HG 355: rare; 1 locality cited.

C. glomerata L. Clustered Bellflower
Map 58c; 6 tetrads.
Saltby, The Drift, disused airfield, IME 1966; Stathern, calcareous pasture on old ironstone workings, locally frequent, SHB 1973; Sproxton, rough limestone grassland bordering mineral railway near Sproxton Lodge, occasional, KGM 1973; Chadwell, old pasture, occasional, ALP 1975; Burrough Hill Encampment, pasture, rare, SHB 1977; Stonesby Quarry, rock floor, occasional, ALP 1978; Belvoir, grassland, occasional, PAE 1986.
  Constituent or residual: Gp, Gr, Rq.
First record: Pulteney (1749).
Herb.: LSR, LTR, MANCH, NMW.
HG 352: rare, almost confined to N.E. Oolite tract; 26 localities cited.

C. latifolia L. Giant Bellflower
Map 58d; 65 tetrads.
Woodlands and hedgerows. Locally frequent in the well-wooded parts of the county, but absent from most of Charnwood Forest and from places with acid soil. Occasional elsewhere in the county.
  Constituent: Fd, Fm, Fs, Fw, Ef.
  Residual or intrusive: Eh, Ev, Ew, Wd, Gr, Rq.
First record: Bloxam (1830).
Herb.: LSR★; LSR, LTR, MANCH.
HG 354: occasional, sparsely distributed; 55 localities cited.

C. trachelium L. Nettle-leaved Bellflower
Map 58e; 5 tetrads.
Tugby Bushes, margin of woodland by roadside, occasional, IME 1976; Owston Wood, rare, PAC 1976; Launde, Big Wood, rare, IME 1976; Launde, Park Wood, rare, J. D. Buchanan 1978; Brown's Wood, rare, PAE 1979.
  Constituent: Fd, Ef.
First record: Coleman (1852).
Herb.: LSR, LTR, MANCH.
HG 353: local, sparsely distributed; 20 localities cited.

C. rapunculoides L. Creeping Bellflower
Map 58f; 17 tetrads.
An escape from gardens, found in scattered localities throughout the county, nearly half the records being from railway verges.
  Intrusive: Er, Ev, Ef, Ap, Rw, Rq.

First record: Kirby (1850).
Herb.: LSR★; LSR.
HG 355: rare, restricted in range; 6 localities cited.

C. rotundifolia L.   Harebell
Map 58g; 180 tetrads.
Heathland, grassland, roadside and railway verges, and hedgerows. Locally frequent in the west of the county and on the Marlstone in the east. Occasional elsewhere.
    Constituent or intrusive: Hg, Hd, Gp, Gr, Ev, Er, Eh, Ea, Ef, Fd, Rq, Rg, Rt, Rw, Sr, Sw.
First record: Pulteney (1747).
Herb.: LSR★; DPM, LANC, LSR, LTR, MANCH, NMW.
HG 355: frequent, generally distributed.

Legousia Durande
L. hybrida (L.) Delarbre   Venus's-looking-glass
Map 58h; 3 tetrads.
Stonesby, arable field near Stonesby Quarry, rare, ALP and PCP 1972; Sproxton, cottage garden, occasional, PHG 1972; Saltby Heath, arable field south of Egypt Plantation, occasional, KGM 1972.
    Intrusive: Ac, Ag.
First record: Pulteney (1747).
Herb.: LSR★; LSR, LTR, MANCH, NMW.
HG 356: rare, very sparsely distributed; 18 localities cited.

L. speculum-veneris (L.) Chaix
No record since HG.
First record: Jackson (1904).
Herb.: LSR.
HG 357: rare; 2 localities cited.

Jasione L.
J. montana L.   Sheep's-bit
No record since HG.
First record: Pulteney (1747).
Herb.: LSR, MANCH, NMW.
HG 352: local, very restricted in range; 20 localities cited.

COMPOSITAE
Eupatorium L.
E. cannabinum L.   Hemp-agrimony
Map 58i; 24 tetrads.
Wet woodland and the banks of streams. Locally frequent in Charnwood Forest and the west of the county; absent from the east, except for a single record from the bank of the Grantham Canal near Muston.
    Constituent or intrusive: Fw, Fd, Wr, Wd, Wl, Ew, Ev, Eh, Er, M, Rq.
First record: Pulteney (1747).
Herb.: LSR, LTR, MANCH, NMW.
HG 293: local, sparsely distributed; 31 localities cited.

Solidago L.
S. virgaurea L.   Golden-rod
Map 58j; 6 tetrads.
Buddon Wood, old quarry, locally frequent, PHG 1968; Peldar Tor Quarry, spoil heaps, locally frequent, PAC 1969; Ashby Woulds, hedgerow west of Union Lodge, rare, M. E. Smith 1972; Loughborough, rocky outcrop between Buck Hill and Beacon Hill, rare, PHG 1974; Ashby Woulds, heath grassland south-east of Union Lodge, rare, SHB 1976; Swithland Wood, rare, S.F. Woodward 1984.
    Constituent or residual: Rq, Hg, Eh, Sr.
First record: Pulteney (1747).
Herb.: LSR★; LSR, LTR, MANCH.
HG 294: rare, restricted in range; 10 localities cited, including Buddon Wood and Ashby Woulds.

S. canadensis L.   Canadian Golden-rod
Map 58 k; 76 tetrads.
A garden escape, becoming thoroughly established in waste places and on railway verges. Locally frequent in these habitats throughout the county.
    Intrusive: Rw, Rq, Rs, Rr, Rd, Er, Ev, Ew, Ef, Fs, Fm, Gr.
First record: Horwood and Gainsborough (1933).
Herb.: LSR.
HG 295: rare; 1 locality cited.

S. gigantea Aiton
No map; 3 tetrads.
Leicester, Belgrave, Checketts Road, G. Smith 1965; Earl Shilton, roadside, locally frequent, SHB 1972; Thurlaston, arable headland, occasional, SHB 1972.
    Casual or escape: Ev, Ea, Rw.
First record: 1965, G. Smith in herb. LSR.
Herb.: LSR★; LSR.
HG: not recorded.

S. graminifolia (L.) Salisb.
No record in the recent survey.
Aylestone, towpath near rubbish tip, T. S. Robertson 1952.
First record: 1952, T. S. Robertson in herb. LSR.
Herb.: LSR.
HG: not recorded.

Bellis L.
B. perennis L.   Daisy
Map 58l; 596 tetrads.
Frequent and locally abundant in short grazed or mown grassland. Less frequent in leys; in some parts of the county where all the fields are arable or temporary leys, daisies may be difficult to find except on lawns or mown verges.
    Constituent, residual or intrusive: Gp, Gr, Gl, Hg, Ev, Er, Ef, Ew, Ea, Ag, Ap, M, Rt, Rs, Rd, Rw.
First record: Pulteney (1747).
Herb.: LSR★; DPM, LANC, LSR, LTR, MANCH, NMW.
HG 295: locally abundant, generally distributed.

Aster L.
A. novi-belgii L.   Michaelmas-daisy
Map 59a; 34 tetrads.
A garden escape, becoming thoroughly established or shortly persistent on waste ground, spoil heaps, railway verges and similar habitats. Locally frequent near large towns. Occasional elsewhere in the county.
    Intrusive: Rw, Rs, Rq, Rg, Rr, Er, Ev, Ew, Gr, Fs, Wd.
First record: 1969, H. Bradshaw and E. Hesselgreaves in present work. There is a specimen in herb. NMW collected by A. E. Wade which must be earlier than this, but it is undated.
HG: not recorded.

A. lanceolatus Willd.
No map; 1 tetrad.
Owston, Furze Hill, roadside verge, locally abundant, IME 1976 (det. T. G. Tutin).
  Casual or escape: Ev.
First record: 1976, I. M. Evans in present work.
Herb.: LSR★.
HG: not recorded.

A. linosyris (L.) Bernh.   Goldilocks Aster
No record since HG.
First record: Horwood and Gainsborough (1933).
HG 295: rare; 1 locality cited.

Erigeron L.
E. acer L.   Blue Fleabane
Map 59b; 30 tetrads.
Occasional throughout the county on railway verges, and on calcareous or sandy soils in quarries, on spoil heaps and in other man-made habitats.
  Residual or intrusive: Er, Ev, Rq, Rg, Rs.
First record: Crabbe in Nichols (1795).
Herb.: LSR★; CGE, LSR, LTR, MANCH.
HG 296: rare, restricted in range; 16 localities cited.

Conyza Less.
C. canadensis (L.) Cronq.   Canadian Fleabane
Map 59c; 36 tetrads.
Locally frequent in waste places, railway sidings, quarries and gravel pits, and occasionally as a garden weed, mostly in the vicinity of the larger towns. Probably increasing.
  Intrusive: Rw, Rq, Rg, Rs, Rr, Rd, Rt, Er, Ag.
First record: Vice (1901).
Herb.: LSR★; LSR, LTR.
HG 295 as *Erigeron canadensis*: occasional; 3 localities cited.

Filago L.
F. vulgaris Lam.   Common Cudweed
F. germanica L. non Hudson
Map 59d; 10 tetrads.
Plungar, ballast of dismantled railway, rare, PAC 1969; Sharnford, cornfield, occasional, MH 1969; Groby, quarry, locally frequent, EH 1970; Long Clawson, dismantled railway, ballast at site of Clawson and Hose Station, locally frequent, ALP and PCP 1970; Quorndon, Cocklow Wood Quarry, locally frequent, PHG 1970 (site since destroyed); Quorndon, 'granite' spoil adjacent to Hawcliff Quarry, locally abundant, PHG 1971; Bardon Hill, quarry, locally abundant, PAC 1972; Croft, waste ground, locally frequent, SHB 1978; Enderby, trackway, occasional, EH 1978.
  Intrusive: Er, Rq, Rs, Rw, Rt.
First record: Pulteney (1747).
Herb.: LSR★; LSR, LTR, MANCH, NMW.
HG 296: frequent, generally distributed.

Logfia Cass.
L. minima (Sm.) Dumort.   Small Cudweed
Filago minima (Sm.) Pers.
Map 59e; 6 tetrads.
Quorndon, Cocklow Wood Quarry, locally abundant, PHG 1970 (site since destroyed); Croft, waste ground, occasional, SHB 1973; Croft Quarry, locally frequent, SHB 1974; Huncote Quarry, locally frequent, SHB 1974; Potters Marston, trackway, occasional, SHB 1974;

Swithland, old railway sidings, locally frequent, PHG 1975; Houghton on the Hill, miniature railway, ballast, rare, IME 1976.
  Intrusive: Rq, Rw, Rt, Er.
First record: Pulteney (1749).
Herb.: LSR★; LSR, LTR, MANCH, NMW.
HG 296: local, mainly restricted to Charnwood Forest and N.W. region and 'S. Syenitic' area; 13 localities cited.

Omalotheca Cass.
O. sylvatica (L.) Schultz Bip. & F. W. Schultz
                                              Heath Cudweed
Gnaphalium sylvaticum L.
Map 59f; 4 tetrads.
Mountsorrel, Nunckley Hill Quarry, locally frequent, PHG 1968; Groby, southern edge of plantation between the Pool and Sheet Hedges Wood, locally frequent, ALP 1969; Owston Little Wood, ride, occasional, SHB, PHG and ALP 1976; Saddington Reservoir margin, rare, SHB and PHG 1977.
  Constituent or residual: Ef, Ew, Rq.
First record: Pulteney (1747).
Herb.: LSR★; LSR, MANCH, NMW.
HG 298: rare, sparsely distributed; 18 localities cited.

Filaginella Opiz
F. uliginosa (L.) Opiz   Marsh Cudweed
Gnaphalium uliginosum L.
Map 59g; 199 tetrads.
Damp places in grassland and arable fields, and especially characteristic of damp trackways. Locally abundant in the west of the county; locally frequent on the Marlstone in the east; occasional elsewhere in the county.
  Intrusive or constituent: M, Gr, Gp, Gl, Hg, Rt, Rw, Rf, Rg, Rq, Rd, Ao, Ac, Ag, Ap, Ea, Ef, Ev, Er, Ew, Wl, Wp, Wd, Fd.
First record: Pulteney (1747).
Herb.: LSR★; DPM, LANC, LSR, LTR, MANCH, NMW.
HG 297: frequent, generally distributed; 65 localities cited.

Antennaria Gaertner
A. dioica (L.) Gaertner   Mountain Everlasting
No recent record.
First record: Crabbe in Nichols (1795).
HG 297: very rare, restricted to N.E. area, not recently seen anywhere in Leicestershire; Crabbe's record, 'In the road from Croxton to Skillington', is the only record for Leicestershire.

Inula L.
I. helenium L.   Elecampane
No map; 2 tetrads.
Witherley, roadside verge, single plant, SHB 1974; Gumley, open woodland bordering fishpond, with other introductions, occasional, EKH and JMH 1974.
  Planted or escape: Ev, Ew.
First record: Crabbe in Nichols (1795).
HG 299: very rare, restricted in range; 3 localities cited.

I. salicina L.   Irish Fleabane
No record since HG.
First record: 1914, in Horwood and Gainsborough (1933).
Herb.: LSR.
HG 299: rare; 1 locality cited.

I. britannica L.
No record since HG.
First record: 1894, T. A. Preston in herb. BM and B.E.C. Rep.
Herb.: BM, LSR.
HG 299: rare, first found at Cropstone in the British Isles, restricted in range; 1 locality cited, Cropston Reservoir.

I. conyza DC. Ploughman's-spikenard
Map 59h; 16 tetrads.
Occurs in scattered localities throughout the county in various man-made habitats, railway verges, quarries, and a wall. No recent records exist for this species in a natural habitat in the county.
  Presumably intrusive: Er, Rq, Hg, Gr, Sw, Ap.
First record: Bloxam (1838).
Herb.: LSR*; LSR, LTR, NMW.
HG 299 as *I. squarrosa*: rare, restricted in range; 4 localities cited.

Pulicaria Gaertner
P. dysenterica (L.) Bernh. Common Fleabane
Map 59i; 96 tetrads.
Marshes and the banks of rivers, streams and canals, and in small dense colonies marking a wet place in other habitats. Locally frequent throughout the county.
  Constituent: M, Ew, Wr, Wc, Wd, Wp.
  Residual or intrusive: Ev, Er, Ef, Ea, Gp, Gr, Rg, Rt.
First record: Pulteney (1747).
Herb.: LSR*; LSR, LTR, MANCH, NMW.
HG 300: locally abundant, generally distributed.

P. vulgaris Gaertner Small Fleabane
No recent record.
First record: Pulteney (1747).
HG 300: very rare – probably extinct, its special habitats in two areas altered and destroyed; 2 localities cited, Leicester and Loughborough.

Guizotia Cass.
G. abyssinica (L.fil.) Cass.
No map.
An occasional casual, derived from bird seed. It was reported several times as occurring on rubbish dumps and in waste places, but some records were lost during the fire at Ratcliffe College.
  Casual: Rr, Rw.
First record: 1969, E. Hesselgreaves in present work.
Herb.: LSR*.
HG: not recorded.

Bidens L.
B. tripartita L. Trifid Bur-marigold
Map 59j; 50 tetrads.
The margins of ponds, rivers and canals, and in marshes and gravel pits. Locally frequent throughout the county.
  Constituent or intrusive: Ew, Wp, Wl, Wr, Wc, M, Rg.
First record: Pulteney (1747).
Herb.: LSR*; LANC, LSR, LTR, MANCH, NMW.
HG 302: frequent, generally distributed.

B. cernua L. Nodding Bur-marigold
Map 59k; 12 tetrads.
Margins of ponds, reservoirs, rivers and canals. Locally frequent in the south-west of the county, and by the Eye Brook Reservoir in the south-east; rare or absent elsewhere.
  Constituent: Ew, Wp, Wl, Wr, Wc, Wd.
First record: Pulteney (1749).
Herb.: LSR*; DPM, LSR, LTR, MANCH, NMW.
HG 301: occasional, sparsely distributed; 17 localities cited.

Helianthus L.
H. annuus L. Sunflower
No map; 6 tetrads.
Often included in bird seed mixtures, and appearing as a casual on rubbish dumps and in waste places. Certainly more frequent in such habitats than would appear from the records received.
  Casual: Rr, Rw, Rf, Rd.
First record: Vice (1905).
HG 301: rare; 3 localities cited.

H. tuberosus L. Jerusalem Artichoke
No map; 3 tetrads.
Beaumont Leys, rubbish dump, occasional, EH 1969; Newton Harcourt, canal bank, rare, EKH and JMH 1969; Kirby Muxloe, waste ground, rare, EH 1973.
  Casual: Rr, Rw, Ew.
First record: 1969, E. Hesselgreaves, E. K. and J. M. Horwood, in present work.
HG: not recorded.

H. rigidus (Cass.) Desf.
No map; 2 tetrads.
Kirby Muxloe, railway verge, rare, EH 1970; Great Easton, railway verge, occasional, KGM 1971.
  Casual or escape: Er.
First record: 1970, E. Hesselgreaves in present work.
HG: not recorded.

Ambrosia L.
A. artemisiifolia L. Ragweed
No map; 2 tetrads.
Quorndon, garden, single plant, PHG 1976; Loughborough, Knightthorpe Road, garden, single plant, PHG 1976.
  Casual: Ag.
First record: 1976, P. H. Gamble in present work.
HG: not recorded.

A. trifida L.
No record since HG.
First record: Jackson (1904).
Herb.: LSR.
HG 301: rare; 1 locality cited.

Galinsoga Ruiz & Pavón
G. parviflora Cav. Gallant Soldier
Map 59l; 12 tetrads.
Sileby, Ratcliffe Road, garden weed, locally frequent, ALP 1967; Loughborough, High Street, waste ground, locally frequent, PHG 1969; Leicester, demolition sites in two localities, occasional, HB 1969; Glenfield, nursery garden, rare, EH 1970; Rearsby, waste ground by brook, occasional, ALP 1971; Cossington, Goscote Nurseries, locally abundant, ALP 1973; Whetstone, allotment gardens, occasional, EH 1974; Enderby, Palmer's Nursery, occasional, EH 1974; Enderby Hall, gardens, occasional, EH 1975; Ratcliffe College, garden of staff house, rare,

ALP 1975; Mountsorrel, arable land, occasional, PHG 1977.
    Intrusive: Ag, Ao, Aa, Rw.
First record: 1945, in Lacey (1947).
Herb.: LSR★; LSR, LTR, NMW.
HG: not recorded.

G. ciliata (Rafin.) S. F. Blake   Shaggy Soldier
Map 60a; 33 tetrads.
Gardens, waste places, roadsides and railway verges. Locally frequent in or near large towns; occasional elsewhere in the county. This species is much more frequent than **G. parviflora**, but both species appear to be increasing.
    Intrusive: Ag, Ap, Rw, Rd, Rt, Rs, Rr, Ev, Er.
First record: 1946, in Lacey (1947)
Herb.: LSR★; LSR, LTR.
HG: not recorded.

Anthemis L.
A. arvensis L.   Corn Chamomile
No map; 1 tetrad.
Oadby, garden weed, G. Halliday 1967.
    Casual or intrusive: Ag.
First record: Pitt (1809).
Herb.: LANC, LSR, MANCH, NMW.
HG 304: local, more frequent in Rutland, sparsely distributed; 25 localities cited.

A. cotula L.   Stinking Chamomile
Map 60b; 82 tetrads.
Occasional throughout the county in arable land and waste places. It would appear from the account in HG that this species has never been very common in Leicestershire. Some farmers do not distinguish this species from **Chamomilla recutita** and **Matricaria perforata**, and apply the name 'Stinking Nanny' to all of them.
    Intrusive: Ac, Ao, Ea, Ev, Er, Rw, Rd, Rt, Rf, Gl, Gr.
First record: Pulteney (1747).
Herb.: LSR★; LANC, LSR, LTR, MANCH, NMW.
HG 304: fairly frequent, widely distributed; 44 localities cited.

A. tinctoria L.   Yellow Chamomile
No map; 3 tetrads.
Quorndon, margin of A6, rare, PHG 1968 (site destroyed by road works 1969); Aylestone, soil heap in cemetery, 2 plants, EH 1974; South Wigston, railway verge, rare, EH 1977.
    Casual: Ev, Er, Rd.
First record: Mercer (1914).
Herb.: LSR★; LSR.
HG 303: rare; 2 localities cited, one doubtful.

Achillea L.
A. ptarmica L.   Sneezewort
Map 60c; 75 tetrads.
Marshes, ditches and wet places in grassland. Locally frequent in the west of the county; occasional in the east.
    Constituent, residual or intrusive: M, Hg, Hd, Gp, Gr, Wd, Ew, Ev, Er, Ea, Eh, Fw, Rw, Rq, Rg, Rt, Ac.
First record: Pulteney (1747).
Herb.: LSR★; LSR, LTR, NMW.
HG 303: frequent, generally distributed.

A. millefolium L.   Yarrow, Milfoil
Map 60d; 602 tetrads.
Frequent and locally abundant in grassland and on grassy verges.
    Constituent or intrusive: Gp, Gr, Gl, Hg, Hd, Ev, Er, Ew, Ef, Eh, Ea, Ag, Ap, Rw, Rt, Rq, Rs, Rd, Rf, Sw.
First record: Pulteney (1747).
Herb.: LSR★; DPM, LANC, LSR, LTR, MANCH, NMW.
HG 303: locally abundant, generally distributed.

Chamaemelum Miller
C. nobile (L.) All.   Chamomile
No record since HG.
First record: Kirby (1850).
HG 305: very rare, probably extinct, restricted in range; 3 localities cited.

Matricaria L.
M. perforata Mérat   Scentless Mayweed
Tripleurospermum maritimum (L.) Koch subsp. *inodorum* (L.) Hyland. ex Vaarama
Map 60e; 600 tetrads.
Frequent and locally abundant throughout the county as a weed of arable land, in waste places and recently disturbed ground.
    Intrusive: Ac, Ao, Aa, Ag, Ea, Ev, Er, Ew, Rd, Rf, Rw, Rt, Rq, Rg, Rs, Rr, Gl, Gr, Gp.
First record: Hands et al. in Curtis (1831).
Herb.: LSR★; DPM, LANC, LSR, LTR, MANCH, NMW.
HG 307 as *M. inodora*: locally abundant, generally distributed.

Chamomilla S. F. Gray
C. recutita (L.) Rauschert   Scented Mayweed
Matricaria recutita L.
Map 60f; 291 tetrads.
Locally frequent throughout the county as a weed of arable land and in waste places.
    Intrusive: Ac, Ao, Ag, Ea, Ev, Er, Rw, Rd, Rt, Rq, Rg, Rs, Rf, Rr, Gl, Gr, Gp.
First record: Pulteney (1747).
Herb.: LSR★; LANC, LSR, LTR, MANCH, NMW.
HG 307 as *Matricaria chamomilla*: frequent, generally distributed.

C. suaveolens (Pursh) Rydb.   Pineappleweed
Matricaria matricarioides (Less.) Porter p.p.
Map 60g; 597 tetrads.
Abundant throughout the county in arable land, waste places, pathways and the trampled ground of field gateways. This species has increased enormously in recent years. It is not mentioned at all in the *Flora* of 1886. Horwood and Gainsborough, though classing it as frequent and widely distributed, and saying that the plant had spread over the Midlands in the previous 20 years, nevertheless cite only 23 localities.
    Intrusive: Rt, Rw, Rf, Rq, Rg, Rs, Rd, Ac, Ao, Aa, Ag, Ap, Ea, Ev, Er, Ew, Gl, Gr, Gp.
First record: Marshall (1790).
Herb.: LSR★; CGE, DPM, LANC, LSR, LTR, MANCH, NMW.
HG 307 as *Matricaria suaveolens*: frequent, widely distributed; 23 localities cited.

Chrysanthemum L.

C. segetum L.   Corn Marigold
Map 60h; 73 tetrads.
Arable land, mainly on the lighter soils. Locally frequent on the Marlstone in the north-east, and on river gravel or alluvial soils. Occasional elsewhere in the county.
    Intrusive: Ao, Ac, Ea, Ev, Rw, Rf, Rd, Rs, Gr, Gl.
First record: Pulteney (1747).
Herb.: LSR★; DPM, LANC, LSR, LTR, MANCH, NMW.
HG 305: frequent, generally distributed; 63 localities cited.

Tanacetum L.

T. vulgare L.   Tansy
Chrysanthemum vulgare (L.) Bernh. non (Lam.) Gatereau
Map 60i; 86 tetrads.
Railway and roadside verges, river banks and waste places. Occasional throughout the county; more frequent in the west than in the east.
    Intrusive: Er, Ev, Ew, Eh, Rw, Rg, Rq, Rs, Rd, Rr, Gr, Wd, Aa, Ag, Sw.
First record: Pulteney (1747).
Herb.: LSR★; DPM, LANC, LSR, LTR, MANCH, NMW.
HG 308: occasional, widely distributed; 50 localities cited.

T. parthenium (L.) Schultz Bip.   Feverfew
Chrysanthemum parthenium (L.) Bernh.
Map 60j; 196 tetrads.
Locally frequent throughout the county in waste places, on walls, in gardens and on roadsides.
    Intrusive: Rw, Rq, Rg, Rs, Rd, Rt, Rf, Rr, Sw, Sb, Ev, Er, Eh, Ew, Ag, Ap, Gr.
First record: Pulteney (1747).
Herb.: LSR★; DPM, LANC, LSR, LTR, MANCH, NMW.
HG 306: occasional, generally distributed.

Leucanthemum Miller

L. vulgare Lam.   Oxeye Daisy
Chrysanthemum leucanthemum L.
Map 60k; 452 tetrads.
Frequent throughout the county in old grassland, on roadside and railway verges, and in other grassy places.
    Constituent, residual or intrusive: Gp, Gr, Gl, Hg, Hd, Er, Ev, Ew, Ea, Eh, Ef, Rq, Rg, Rs, Rw, Rd, Rt, Ap, Ag.
First record: Pulteney (1747).
Herb.: LSR★; DPM, LANC, LTR, MANCH, NMW.
HG 306: locally abundant, generally distributed.

L. maximum (Ramond) DC.   Shasta Daisy
Chrysanthemum maximum Ramond
Map 60l; 13 tetrads.
Theddingworth, site of station, dismantled railway, locally abundant, EKH and JMH 1968; Glenfield, site of station, dismantled railway, occasional, EH 1970; Market Harborough, roadside verge of A6, rare, KGM 1971; Blackfordby, waste ground, rare, M. E. Smith 1972; Donisthorpe, railway verge, rare, ALP and KGM 1973; Groby, soil tip, rare, EH 1973; Great Glen, railway verge by closed station, locally frequent, EKH and JMH 1973; Belvoir, bank of disused mineral railway, rare, KGM 1974; Enderby, rubbish dump, rare, EH 1974; Cosby, dismantled railway, verge, occasional, ALP 1974; Aylestone, soil heap, rare, EH 1974; Moira, hedgerow, rare, SHB 1977; Blackfordby, waste ground, occasional, SHB 1977; Staunton Harold, Spring Wood, edge of ride lined with brick rubble, occasional, PAE and ALP 1982.
    Persistent escape: Er, Ev, Rw, Rr, Rd, Eh.
First record: 1968, E. K. and J. M. Horwood in present work.
Herb.: LSR★; LTR.
HG: not recorded.

Artemisia L.

A. vulgaris L.   Mugwort
Map 61a; 381 tetrads.
Roadside and railway verges and waste places. Frequent throughout the county, but becoming less frequent towards the east.
    Constituent or intrusive: Ev, Er, Eh, Ea, Ew, Rw, Rq, Rg, Rs, Rr, Rd, Gr, Ac, Ag.
First record: Pulteney (1747).
Herb.: LSR★; DPM, LSR, LTR, NMW.
HG 309: frequent, generally distributed.

A. verlotiorum Lamotte   Chinese Mugwort
No map; 1 tetrad.
Wistow, side of field lane near Newton Harcourt, rare, EKH and JMH 1969.
    Casual: Ev.
First record: 1969, E. K. and J. M. Horwood in present work.
Herb.: LSR★.
HG: not recorded.

A. absinthium L.   Wormwood
Map 61b; 142 tetrads.
Waste places, roadsides and railway verges. Locally frequent in and around Leicester and in the west of the county; occasional in the east. Appears to have increased considerably since 1933.
    Intrusive: Er, Ev, Ew, Ef, Rw, Rr, Rq, Rg, Rs, Rd, Gr, Hg, Ap, Ag.
First record: Pulteney (1747).
Herb.: LSR★; LSR, LTR, MANCH, NMW.
HG 309: rare, sparsely distributed; 18 localities cited.

A. pontica L.
No record since HG.
First record: W.B.E.C. Rep. 1908.
Herb.: NMW.
HG 310: rare; 1 locality cited.

Tussilago L.

T. farfara L.   Colt's-foot
Map 61c; 586 tetrads.
Roadside and railway verges, wet places in grassland, and waste ground, being particularly abundant on the spoil heaps of gravel pits and quarries. Also a weed of arable land and gardens, very difficult to eradicate when once established. Frequent and locally abundant throughout the county.
    Constituent or intrusive: M, Gp, Gr, Gl, Ev, Er, Ew, Eh, Ef, Ea, Ac, Ao, Ag, Ap, Rg, Rq, Rs, Rw, Rr, Rf, Rd, Rt, Wd, Fw, Fd, Fs.
First record: Pulteney (1747).
Herb.: LSR★; LANC, LSR, MANCH, NMW.
HG 310: locally abundant, generally distributed.

Petasites Miller
P. albus (L.) Gaertner　　　　　　White Butterbur
Map 61d; 2 tetrads.
Waltham on the Wolds, bed and banks of partly drained fishponds, locally frequent, KGM 1973; Coleorton Hall, path through ornamental woodlands, occasional, PAC 1976.
　　Intrusive or planted: Ew, Ef.
First record: Power (1807).
Herb.: LSR★; LSR, LTR.
HG 312: rare, restricted in range; 3 localities cited, including Waltham.

P. hybridus (L.) P. Gaertner　　　　　　Butterbur
Map 61e; 72 tetrads.
Banks of rivers and streams, and wet places in woodland. Locally frequent in suitable habitats throughout the county.
　　Constituent: Ew, Wr, Wd, M, Fw, Ef.
　　Residual or intrusive: Ev, Er, Rw, Rf, Gr.
First record: Blackstone (1746).
Herb.: LSR★; CGE, DPM, LANC, LSR, LTR, MANCH, NMW.
HG 310: locally abundant, generally distributed.

P. fragrans (Vill.) C. Presl　　　Winter Heliotrope
Map 61f; 10 tetrads.
Kirby Muxloe, roadside hedgerow, locally frequent, EH 1975; Withcote Hall, margin of fish pond, rare, IME 1975; Owston, Furze Hill, roadside verge, locally abundant, IME 1975; South Kilworth, roadside, locally abundant, EKH and JMH 1975; Skeffington, roadside, locally frequent, IME 1976; Glen Parva, rough grassland, rare, EH 1977; Coleorton, roadside, locally frequent, SHB 1978; Staunton Harold Park, locally frequent, SHB 1978.
　　Intrusive: Ev, Ew, Eh, Gr, Ap.
First record: Horwood and Gainsborough (1933).
Herb.: LSR★; LSR.
HG 310: rare; 1 locality cited.

P. japonicus (Siebold & Zucc.) Maxim.
　　　　　　　　　　　　　　　　Giant Butterbur
No map; 2 tetrads.
Baggrave Hall, marshy field, P. A. H. Bromwich 1969; Normanton Turville, bank of stream, locally abundant, J.E. Dawson 1980.
　　Planted or escape: M, Ew.
First record: 1969, P. A. H. Bromwich in herb. LTR.
Herb.: LTR.
HG: not recorded.

Doronicum L.
D. plantagineum L.　Plantain-leaved Leopard's-bane
No record since HG.
First record: Kirby (1850).
HG 312: rare, restricted in range; 1 locality cited.

D. pardalianches L.　　　　　　Leopard's-bane
Map 61g; 8 tetrads.
Charnwood Lodge Nature Reserve, Gisborne's Gorse, woodland, locally abundant, PHG 1968; Great Easton, edge of wild garden, escaping to stream side, rare, KGM 1971; Ashby Magna, roadside verge, locally frequent, EKH and JMH 1971; Market Bosworth Park, margin of Bow Pool, occasional, HH 1972; Markfield, Cottage Lane, roadside verge, locally abundant, SHB 1972; Croft, roadside, rare, SHB 1973; Saddington, roadside hedgerow, rare, EKH and JMH 1973; Cold Overton Hall, stream side in valley to west, occasional, KGM 1973.
　　Established escape: Ev, Eh, Ew, Fd.
First record: Watson (1837).
Herb.: LSR★; LSR, LTR.
HG 312: very rare, restricted in range; 2 localities cited.

Senecio L.
S. integrifolius (L.) Clairv.　　　　Field Fleawort
No record since HG.
First record: Tomkins (1887).
HG 316: rare, restricted in range; 2 localities cited.

S. jacobaea L.　　　　　　　　Common Ragwort
Map 61h; 381 tetrads.
Roadside and railway verges, grassland and waste places. Frequent over the county as a whole, but the distribution is somewhat discontinuous. Locally frequent in some parts of the county, occasional in others.
　　Constituent or intrusive: Gp, Gr, Gl, Hg, Ev, Er, Ef, Ew, Eh, Ea, Ag, Ap, Ac, Rw, Rg, Rq, Rs, Rd, Rt, Rf, Rr, Sw, Sb.
First record: Pulteney (1747).
Herb.: LSR★; CGE, LANC, LSR, LTR, MANCH, NMW.
HG 315: abundant or frequent, widely distributed.

S. aquaticus Hill　　　　　　　　Marsh Ragwort
Map 61i; 124 tetrads.
Marshes, the banks of rivers and streams, and other wet places. Locally frequent in the west of the county; occasional in the east.
　　Constituent or residual: M, Ew, Ef, Ev, Ea, Wd, Wr, Wl, Wp, Ap, Gp, Gr, Rd, Fw.
First record: Pulteney (1747).
Herb.: LSR★; DPM, LANC, LSR, LTR, MANCH, NMW.
HG 316: frequent, generally distributed.

S. erucifolius L.　　　　　　　　Hoary Ragwort
Map 61j; 191 tetrads.
Roadside and railway verges and rough grassland. Locally frequent throughout the county, but absent from Charnwood Forest.
　　Constituent or intrusive: Ev, Er, Ew, Ef, Eh, Ea, Gr, Gp, Gl, M, Rw, Rq, Rg, Rd, Rr, Rt, Aa, Ap, Sw, Fs.
First record: Pulteney (1747).
Herb.: LSR★; LANC, LSR, LTR, MANCH, NMW.
HG 315: frequent, generally distributed.

S. squalidus L.　　　　　　　　Oxford Ragwort
Map 61k; 315 tetrads.
Waste places, roadside and railway verges, and sometimes a weed of gardens and arable land. Locally abundant in and near large towns, especially on demolition sites. Locally frequent in other parts of the county. Has increased phenomenally since the Second World War.
　　Intrusive: Rw, Rq, Rg, Rs, Rd, Rr, Rt, Ev, Er, Ew, Eh, Ea, Ac, Ao, Ag, Ap, Sw, Gr, Gp, Fs.
First record: Bemrose (1927).
Herb.: LSR★; LANC, LSR, LTR.
HG 315: rare; 2 localities cited.

S. squalidus × viscosus = S. × londinensis Lousley
Map 61l; 4 tetrads.
Leicester, Coleman Road, dismantled railway, IME 1966; Leicester, St. Margaret's Baths, car park, IME 1967; Thurnby, dismantled railway, rare, IME 1969; Thorpe Langton, cinder ballast of dismantled railway, occasional, KGM 1973; Croft, quarry, rare, SHB 1973; South Wigston, waste ground, rare, EH 1979.
    Intrusive: Rw, Er, Rq.
First record: 1966, I. M. Evans in herb. LSR.
Herb.: LSR★; LSR.
HG: not recorded.

S. vernalis Waldst. & Kit.
No map; 2 tetrads.
Lubenham, reseeded roadside verge, occasional, KGM 1968; Market Harborough, reseeded verges of A427, KGM 1969; 56 plants in 1969, 25 in 1970, 12 in 1971, 6 in 1972, none in 1973. Odd plants have been seen at intervals since. The grass seed mixture (according to information supplied by Leicestershire County Council) originated in Holland and Denmark.
    Casual: Ev.
First record: 1968, K. G. Messenger in present work.
Herb.: LSR★; BM, CGE, LIVU, UPP.
HG: not recorded.

S. sylvaticus L.   Heath Groundsel
Map 62a; 46 tetrads.
Heathland and grassland on acid or sandy soils. Locally frequent in the west of the county; rare or absent in the east.
    Constituent or residual: Hg, Hd, Gr, Gp, Sr, Fd, Fm, Fs, Er, Ev, Eh, Ew, Rq, Rs, Rt, Rd, Ap, Ac.
First record: Pulteney (1749).
Herb.: LSR★; LSR, LTR, MANCH, NMW.
HG 313: local, rather restricted in range; 51 localities cited.

S. viscosus L.   Sticky Groundsel
Map 62b; 242 tetrads.
A characteristic plant of railway ballast throughout the county, 62% of the records received being from this habitat. Often very abundant on the ballast of dismantled railways. Occasional but sometimes locally frequent in quarries and other ruderal habitats. Horwood's field workers appear to have neglected railway verges, but nevertheless this species has almost certainly increased greatly since 1933.
    Intrusive: Er, Ev, Ew, Ea, Eh, Rq, Rg, Rs, Rw, Rr, Rd, Rt, Ap, Ao, Gr, Gl, Fd, Sw.
First record: Jackson (1904).
Herb.: LSR★; LSR, LTR, NMW, UPP.
HG 314: rare, restricted in range; 13 localities cited.

S. vulgaris L.   Groundsel
Map 62c; 603 tetrads.
Gardens, arable land and disturbed ground. Locally abundant throughout the county.
    Intrusive: Ag, Ao, Ac, Aa, Ap, Ea, Ev, Er, Ew, Rd, Rf, Rt, Rw, Rq, Rg, Rs, Rr, Gp, Gr, Gl, Sw.
First record: Pulteney (1747).
Herb.: LSR★; CGE, DPM, LSR, LTR, MANCH, NMW.
HG 312: abundant, generally distributed.

Calendula L.
C. officinalis L.   Pot Marigold
Map 62d; 23 tetrads.
Often grown in gardens, and occasionally escaping on to roadside verges, rubbish dumps and waste places. Rarely persists.
    Casual or garden escape: Ev, Er, Rw, Rr, Rd, Rq, Rs, Rf, Wd, Gr.
First record: Jackson (1904).
Herb.: LSR.
HG 317: rare; 3 localities cited.

C. arvensis L.   Field Marigold
No record since HG.
First record: Bemrose (1927).
Herb.: LSR.
HG 317: rare; 1 locality cited.

Carlina L.
C. vulgaris L.   Carline Thistle
Map 62e; 6 tetrads.
Shawell, Cave's Inn, disused gravel pits, rare, EKH and JMH 1969; Woodhouse Eaves, The Brand, dry floor of old slate quarry, rare, PHG 1969; Stonesby Quarry, oolitic limestone, occasional, ALP 1973; Saltby Heath, north-east corner of old airfield, limestone grassland, occasional, KGM 1973; Breedon Cloud, Carboniferous Limestone quarry spoil heaps, locally frequent, PAC 1978; Hallaton, castle mound, locally frequent, J.E. Dawson 1984.
    Constituent, residual or intrusive: Gp, Rq, Rg, Rs.
First record: Pulteney (1747).
Herb.: LSR★; LSR, MANCH.
HG 317: Occasional, restricted in range; 19 localities cited, including Stonesby Quarry, The Brand, Saltby and Breedon.

Arctium L.
A. lappa L.   Greater Burdock
Map 62f; 65 tetrads.
Roadside and railway verges, waste places, and sometimes in woodland. Occasional throughout the county.
    Intrusive or constituent: Ev, Er, Eh, Ew, Ea, Ef, Rw, Rq, Rf, Rd, Rt, Gr, Gp, M, Fd, Fs, Fw.
First record: Pulteney (1747).
Herb.: LSR★; CGE, LSR, NMW.
HG 318: frequent, generally distributed.

A. minus aggregate   Lesser Burdock
Map 62g; 473 tetrads.
At the beginning of the survey for this Flora, an attempt was made to record **A. pubens** Bab., **A. minus** Bernh. and **A. nemorosum** Lej. The records received proved to be very inconsistent and unreliable, and it was not possible to arrange for specimens to be determined by a referee. As stated in *Flora Europaea*, specific limits within this genus cannot be clearly defined, and because of the large numbers of records received, it did not seem practicable to seek the assistance of a specialist. The three species have therefore been recorded as an aggregate. Frequent throughout the county in open woodland, on roadside and railway verges, and in waste places.
    Constituent or intrusive: Fd, Fm, Fs, Fw, Ef, Ev, Er, Eh, Ew, Ea, Rw, Rf, Rq, Rg, Rr, Rd, Gr, Gp, Wd, Ap, Aa.
First record: Bloxam (1837).
Herb.: LSR★; DPM, LSR, MANCH, NMW.
HG 319: locally abundant, generally distributed.

Carduus L.

C. nutans L.   Musk Thistle
Map 62h; 70 tetrads.
Locally frequent in grassland on the Oolite and the Marlstone in the east of the county, and the Carboniferous Limestone at Breedon in the west. Elsewhere it occurs occasionally where the soil is suitable, or as a casual on soil brought in for major road works and similar projects.
   Constituent or intrusive: Gp, Gr, Ev, Er, Eh, Ea, Ew, Rg, Rq, Rs, Rw, Rd, Rf, Rr, Sr.
First record: Pulteney (1747).
Herb.: LSR★; DPM, LANC, LSR, LTR, MANCH, NMW.
HG 319: frequent, generally distributed.

C. acanthoides L.   Welted Thistle
Map 62i; 436 tetrads.
Hedgerows, roadsides and the banks of rivers. Frequent in the east of the county; occasional in the West.
   Constituent, residual or intrusive: Eh, Ev, Ew, Er, Ea, Ef, Fd, Fs, Gr, Gp, Gl, M, Wd, Rw, Rq, Rs, Rd, Rf, Ao.
First record: Marshall (1790).
Herb.: LSR★; DPM, LANC, LSR, LTR, MANCH, NMW.
HG 320 as *C. crispus*: abundant, generally distributed.

Cirsium Miller

C. eriophorum (L.) Scop.   Woolly Thistle
Map 62j; 13 tetrads.
Limestone grassland and quarries. Almost entirely confined to the Oolite in the north-east, and apart from records from a hedgerow near Bruntingthorpe and railway verges near Newton Harcourt and Wymondham, absent from the rest of the county.
   Constituent, residual or intrusive: Gr, Gp, Ev, Er, Eh, Ef, Rq.
First record: Pulteney (1747).
Herb.: LSR★; LSR, MANCH, NMW.
HG 321: occasional, widely distributed; 41 localities cited.

C. vulgare (Savi) Ten.   Spear Thistle
Map 62k; 604 tetrads.
Grassland, grassy verges and waste places, and as a weed of arable land. Locally abundant throughout the county.
   Constituent or intrusive: Gp, Gr, Gl, Hg, M, Ev, Er, Eh, Ew, Ef, Ea, Rw, Rq, Rs, Rr, Rd, Ap, Ac, Ao, Ag.
First record: Pulteney (1747).
Herb.: LSR★; DPM, LANC, LSR, LTR, MANCH.
HG 321: Abundant, generally distributed.

C. dissectum (L.) Hill   Meadow Thistle
Map 62l; 7 tetrads.
Twenty Acre, **Molinia - Nardus** heath grassland, locally frequent, ALP 1968; Ulverscroft, marsh and meadow adjoining Ulverscroft Pond, occasional, PAC 1968; Holly Hayes Farm, old pasture, locally frequent, PAC 1969; Potters Marston, marsh, locally frequent, SHB 1972 (site destroyed in construction of M69, 1975); Newton Burgoland, marsh at Newton Barn, locally frequent, HB 1973; Markfield, Cottage Lane, pasture, occasional, G.N. Thurlow 1987.
   Constituent or residual: Hw, M, Gp.
First record: Pulteney (1749).

Herb.: LSR★; LANC, LSR, LTR, MANCH, NMW.
HG 323: rare, restricted in range; 14 localities cited, including Burton on the Wolds (as 'Six Hills') Ulverscroft and Potters Marston.

C. acaule Scop.   Dwarf Thistle
Map 63a; 62 tetrads.
Grassland and grassy roadside and railway verges. Occasional in the east of the county; rare or absent in the west. In Leicestershire it is approaching the northern limit of its range, and is tending to decrease as suitable habitats are ploughed or destroyed.
   Constituent or residual: Gp, Gr, Er, Ev, Ef, Fs, Rq.
First record: Pulteney *in* Nichols (1795).
Herb.: LSR★; LSR, LTR, MANCH, NMW.
HG 324: fairly frequent, but sparsely distributed; 47 localities cited.

C. palustre (L.) Scop.   Marsh Thistle
Map 63b; 379 tetrads.
Marshes and wet places. Locally frequent throughout the county.
   Constituent or residual: M, Gp, Gr, Hg, Wd, Ew, Ev, Er, Eh, Ea, Ef, Fw, Fd, Fm, Fs, Rg, Rs, Rd.
First record: Pulteney (1747).
Herb.: LSR★; DPM, LSR, LTR, NMW.
HG 322: locally abundant, generally distributed.

C. arvense (L.) Scop.   Creeping Thistle
Map 63c; 606 tetrads.
Grassland, often becoming a serious pest in pastures; grassy verges, waste places, and a persistent and troublesome weed of arable land and gardens. Abundant throughout the county.
   Intrusive or constituent: Gp, Gr, Gl, Ev, Ef, Ew, Eh, Er, Ea, Ac, Ao, Ap, Rw, Rq, Rg, Rs, Rf, Rt, Rr, Rd, M, Fd.
First record: Pulteney (1747).
Herb.: LSR★; CGE, DPM, LANC, LSR, LTR, MANCH, NMW.
HG 325: abundant, generally distributed.

C. arvense var. **incanum** (Fisch.) Ledeb.
No map; 2 tetrads.
Leicester, Waterloo Way, waste ground, rare, M. Hider and IME 1978; Leicester, South Fields, railway bank, locally abundant, K.W. Bindley 1985; Leicester, site of former West Bridge Station, waste ground, occasional, K.W. Bindley 1985. The third locality is that at which this subspecies was last recorded, in about 1926, by G.J.V. Bemrose.
   Casual: Er, Rw.
First record: B.E.C. Rep. 1907.
HG 325: three localities cited, all in the immediate vicinity of West Bridge Station.

Onopordum L.

O. acanthium L.   Cotton Thistle
No map; 5 tetrads.
Beaumont Leys, rubbish dump, rare, EH 1971; Kibworth Beauchamp, rubbish dump, occasional, EKH and JMH 1974; Oadby, rubbish dump, rare, IME 1976;

Groby, waste ground, single plant, EH 1977; Quorndon, recently disturbed ground, PHG 1978.
  Casual or escape: Rr, Rw, Rd.
First record: Pulteney (1747).
Herb.: LANC, LSR, LTR.
HG 326: rare, restricted in range; 9 localities cited.

Silybum Adanson
S. marianum (L.) Gaertner   Milk Thistle
No map; 2 tetrads.
Leicester, Abbey Park, bare ground round Abbey ruins, rare, HB 1969; Holwell Iron Works, base of old slag heap, single plant, ALP 1972 (whole area bulldozed clear 1976).
  Casual: Rs, Rd.
First record: Power (1807).
Herb.: LSR★; LSR, NMW.
HG 326: rare, restricted in range; 8 localities cited.

Serratula L.
S. tinctoria L.   Saw-wort
Map 63d; 38 tetrads.
Heathland and grassland on acid or sandy soils. Locally frequent in the west of the county; occasional in the north-east; absent from the south-east.
  Constituent or residual: Hg, Gp, Gr, M, Eh, Er, Ev, Ea, Fs.
First record: Pulteney (1747).
Herb.: LSR★; CGE, LANC, LSR, LTR, MANCH, NMW.
HG 326: local, sparsely distributed; 71 localities cited.

Centaurea L.
C. scabiosa L.   Greater Knapweed
Map 63e; 26 tetrads.
Locally frequent in grassland on the Oolite in the north-east. Occasional elsewhere in the county, mainly on railway verges or in quarries.
  Constituent: Gp, Gr, Ef.
  Residual or intrusive: Er, Ev, Rq, Rw, Rt.
First record: Pulteney (1747).
Herb.: LSR★; DPM, LSR, LTR, MANCH, NMW.
HG 330: local, generally distributed; 50 localities cited.

C. solstitialis L.   Yellow Star-thistle
No map; 1 tetrad.
Owston, garden, single plant, J. Buchanan 1979.
  Casual: Ag.
First record: Preston (1900).
Herb.: LSR★; LSR.
HG 332: rare, sparsely distributed; 4 localities cited.

C. melitensis L.
No record since HG.
First record: Jackson (1904).
Herb.: LSR.
HG 332: rare; 2 localities cited.

C. diluta Aiton
No map.
This alien species has been reported from scattered localities throughout the county in rubbish dumps and waste places, probably derived from bird seed. Many of the records were lost as a result of the fire at Ratcliffe College.
  Casual: Rr, Rw.
First record: 1970, E. Hesselgreaves in present work.
Herb.: LSR★.
HG: not recorded.

C. nigra aggregate
Map 63f; 561 tetrads.
During the survey for this Flora, **C. debauxii** subsp. **nemoralis** and **C. nigra** were treated as subspecies of **C. nigra**. Field workers were instructed to record the species as an aggregate, and to record also the two subspecies separately. This policy was evidently a mistake, because some of the field workers were content to record the aggregate only, so that the distribution maps for the two segregates are incomplete.

C. debauxii Gren. & Godron subsp. **nemoralis** (Jordan) Dostál   Slender Knapweed
C. nigra L. subsp. *nemoralis* (Jordan) Gugler
Map 63g; 107 tetrads.
Grassland and grassy verges. Frequent throughout the county.
  Constituent or intrusive: Gp, Gr, Ev, Er, Ew, Eh, Ea, Ef, Rq, Rs, Rd, Rt.
First record: 1922, A. E. Ellis in herb. LANC.
Herb.: LSR★; LANC, NMW.
HG 329 as *C. nemoralis*: locally abundant, widely distributed.

C. nigra L.   Common Knapweed
Map 63h; 248 tetrads.
Grassland and grassy verges. Frequent throughout the county.
  Constituent or intrusive: Gp, Gr, Hg, Hd, M, Ev, Er, Ew, Ea, Ef, Rq, Rs, Rt, Rw.
First record: Pulteney (1747).
Herb.: LSR★; DPM, LSR, LTR, MANCH, NMW.
HG 327: abundant, generally distributed.

C. montana L.   Perennial Cornflower
No record in the recent survey.
Saltby, disused airfield, widespread, EKH 1956; Stoughton, roadside, EKH 1956; Groby, hedgerow, EKH 1956.
First record: 1956, E. K. Horwood in herb. LTR.
Herb.: LTR.
HG: not recorded.

C. cyanus L.   Cornflower
Map 63i; 6 tetrads.
Rearsby, East Goscote building site, HB 1965; Quorndon, Bull in the Hollow, disturbed ground on verge of A6, rare, PHG 1970; Potters Marston, recently disturbed roadside, rare, SHB 1974; Potters Marston, old sand pit, rare, SHB 1974; Croft, recently disturbed roadside, rare, SHB 1974; Littlethorpe, hedgerow, rare, EH 1976; Ingarsby, cornfield, undersown with clover seed from Belgium, rare, PAC, 1977; Enderby, recently disturbed ground, single plant, SHB 1977; Walcote, cornfield, rare, PAE 1979.
  Intrusive or casual: Ac, Rd, Rw, Rg, Eh.
First record: Pulteney (1747).
Herb.: LSR★; LSR, MANCH, NMW.
HG 331: rare, rather sparsely distributed; 30 localities cited.

Cichorium L.

C. intybus L.   Chicory
Map 63j; 16 tetrads.
Waste ground, roadside and railway verges, in scattered localities throughout the county, usually only of casual status.
    Casual or intrusive: Rw, Rd, Rf, Ev, Er, Gp.
First record: Pulteney (1747).
Herb.: LSR★; DPM, LTR.
HG 332: rare, restricted in range; 32 localities cited.

Arnoseris Gaertner

A. minima (L.) Schweigger & Koerte   Lamb's Succory
No recent record.
First record: Pulteney *in* Nichols (1795).
HG 333: rare, perhaps extinct, restricted in range; 2 localities cited, both records by Pulteney.

Hypochoeris L.

H. glabra L.   Smooth Cat's-ear
No recent record.
First record: Coleman (1852).
Herb.: LSR.
HG 344 rare, restricted in range; 1 record only, prior to 1852.

H. radicata L.   Cat's-ear
Map 63k; 468 tetrads.
Grassland and grassy places. Frequent throughout the county.
    Constituent or residual: Gp, Gr, Gl, Hg, M, Ev, Er, Ew, Ea, Ap, Ag, Ac, Aa, Rw, Rq, Rs, Fs, Sw.
First record: Pulteney (1749).
Herb.: LSR★; DPM, LSR, LTR, MANCH, NMW.
HG 344: locally abundant, generally distributed.

Leontodon L.

L. autumnalis L.   Autumn Hawkbit
Map 63l; 526 tetrads.
Grassland and grassy places. Frequent throughout the county.
    Constituent or residual: Gp, Gr, Gl, Hg, Ev, Er, Ew, Ef, Ap, Ag, Rw, Rq, Rs, Rf, Rt.
First record: Watson (1832).
Herb.: LSR★; CGE, LANC, LSR, LTR, MANCH, NMW.
HG 345: locally abundant, generally distributed.

L. hispidus L.   Rough Hawkbit
Map 64a; 324 tetrads.
A characteristic plant of railway verges, in which habitat it is locally frequent throughout the county. Occasional in grassland elsewhere, preferring basic soils.
    Constituent or intrusive: Gp, Gr, Gl, Hg, Er, Ev, Ew, Ef, Ea, Ap, Ac, Rw, Rq, Rs, Rd.
First record: Hands et al *in* Curtis (1831).
Herb.: LSR★; CGE, LANC, LSR, LTR, MANCH, NMW.
HG 345: frequent, generally distributed.

L. taraxacoides (Vill.) Mérat   Lesser Hawkbit
Map 64b; 54 tetrads.
Grassland and other grassy places. Occasional, but possibly overlooked and under-recorded.
    Constituent or intrusive: Gp, Gr, Gl, Hg, Ev, Er, Ew, Ap, Rg, Rs.
First record: Hands et al *in* Curtis (1831).
Herb.: LSR★; LANC, LSR, LTR, MANCH, NMW.
HG 344 as *L. nudicaulis*: frequent, generally distributed.

Picris L.

P. echioides L.   Bristly Oxtongue
Map 64c; 18 tetrads.
Locally frequent as an arable weed in the extreme north-east of the county. Occasional and often of only casual occurrence elsewhere.
    Intrusive: Ao, Aa, Ea, Ev, Er, Ew, Gp, Rw, Rd, Rt.
First record: Pulteney (1747).
Herb.: LSR★; LANC, LSR, LTR, MANCH, NMW.
HG 335: local, sparsely distributed; 22 localities cited.

P. hieracioides L.   Hawkweed Oxtongue
Map 64d; 19 tetrads.
Locally frequent on railway verges in scattered localities throughout the county, 68% of the records received being from this habitat. Also occasionally in quarries and waste places.
    Intrusive: Er, Ev, Eh, Rq, Rs, Rw.
First record: Hands et al. *in* Curtis (1831).
Herb.: LSR★; LSR, LTR, MANCH, NMW.
HG 334: local, restricted in range; 43 localities cited.

Tragopogon L.

T. porrifolius L.   Salsify
No record since HG.
First record: Kirby (1850).
Herb.: LSR.
HG 351: rare, restricted in range; 4 localities cited.

T. pratensis L.   Goat's-beard
Map 64e; 450 tetrads.
Locally frequent on roadside and railway verges throughout the county. Occasional in grassland.
    Constituent, residual or intrusive: Ev, Er, Ew, Ea, Ef, Gp, Gr, Gl, Rw, Rg, Rq, Rd, Ap, Ao.
First record: Pulteney (1747).
Herb.: LSR★; DPM, LANC, LSR, LTR, MANCH.
HG 350: frequent, and generally distributed.

Sonchus L.

S. asper (L.) Hill   Prickly Sow-thistle
Map 64f; 583 tetrads.
Arable land, gardens, waste places, roadside and railway verges; frequent and locally abundant throughout the county.
    Intrusive: Ao, Ac, Ag, Ap, Ea, Ev, Er, Eh, Ew, Ef, Rw, Rf, Rq, Rg, Rs, Rt, Rd, Rr, Gr, Gl, Gp, Sw, Wd.
First record: Pulteney (1747).
Herb.: LSR★; DPM, LANC, LSR, LTR, NMW.
HG 349: frequent, generally distributed.

S. oleraceus L.   Smooth Sow-thistle
Map 64g; 530 tetrads.
Frequent throughout the county in gardens, allotments, non-cereal crops and recently disturbed ground. Occasional in waste places, in cereal crops, on roadsides and railway verges.
    Intrusive: Ag, Ao, Aa, Ac, Ap, Ea, Ev, Er, Eh, Ew, Rw, Rd, Rf, Rq, Rg, Rs, Rr, Rt, Wd, Gl, Gp, Gr, Sw.
First record: Pulteney (1747).
Herb.: LSR★; CGE, LANC, LSR, MANCH, NMW.
HG 349: locally abundant, generally distributed.

S. palustris L.   Marsh Sow-thistle
Map 64h; 1 tetrad.
Market Bosworth, wet woodland surrounding lake in former park, locally frequent, HH 1977 (originally planted here by the Rev. N. P. Small c. 1840 - 1850).
   Established introduction: Fw.
First record: Pulteney (1752) if correct.
Herb.: LSR★; LSR, LTR.
HG 350: very rare, restricted in range to the River Soar before alteration, if it ever occurred as a native plant; 2 localities cited, i.e. Pulteney's records for the River Soar, and the introduction at Market Bosworth. Horwood places the whole account for this species in square brackets, but, in spite of his comment quoted above, says that he is willing to accept Pulteney's record. Mott, in the *Flora* of 1886, considers Pulteney's record an error.

S. arvensis L.   Perennial Sow-thistle
Map 64i; 551 tetrads.
Arable land, roadside verges and waste places. Frequent and locally abundant throughout the county.
   Intrusive: Ao, Ac, Aa, Ag, Ap, Ea, Ev, Er, Eh, Ef, Rw, Rf, Rq, Rg, Rs, Rd, Rt, Wd, Gl, Gr, Gp, M, Fd.
First record: Pulteney (1747).
Herb.: LSR★; DPM, LANC, LSR, LTR, NMW.
HG 350: locally abundant, generally distributed.

Lactuca L.
L. serriola L.   Prickly Lettuce
Map 64j; 25 tetrads.
Waste ground, gravel pits and quarries. Occasional in scattered localities throughout the county, especially in the vicinity of Leicester.
   Intrusive: Rw, Rg, Rq, Rs, Rr, Rd, Ev, Er, Eh, Ew, Ag.
First record: B.E.C. Rep. 1931.
Herb.: LSR★; LSR, LTR.
HG 348: rare, restricted in range; 1 locality cited.

L. virosa L.   Great Lettuce
Map 64k; 14 tetrads.
Railway verges (more than half the records received), waste places, quarries and spoil heaps. In scattered localities throughout the county; less frequent than the preceding species.
   Intrusive: Er, Rw, Rq, Rs, Gr, Ev.
First record: Pulteney (1749).
Herb.: LSR★; LSR, LTR.
HG 347: rare, very restricted in range; 17 localities cited.

Cicerbita Wallr.
C. macrophylla (Willd.) Wallr.   Blue Sow-thistle
Map 64l; 11 tetrads.
Leicester, De Montfort Hall, garden weed, rare, HB 1968; Belton, Woods Lane, rough grassland, rare, PAC 1968; Quorndon, bridle track off Chaveney Road, grass verge, locally frequent, PHG 1968; Market Harborough, track side by allotments, rare, KGM 1971; Shangton, roadside verge outside wild garden, occasional, KGM 1971; Blaston, verge of field road and headland of adjacent arable, occasional, KGM 1971; Donington Park, rough ground near the Hall, rare, PAC 1971; Wymondham, roadside verge, locally frequent, KGM 1972; Great Glen, Station Road, roadside ditch, locally abundant, EKH and JMH 1973; Sileby, Hobbs Wick, waste ground by brook, occasional, ALP 1974; Hungarton, roadside, rare, IME 1975.
   Casual or intrusive: Ev, Gr, Wd, Rw, Ag.
First Record: 1964, J. C. Badcock in herb. LSR.
Herb: LSR★, LSR,LTR.
HG: not recorded.

Mycelis Cass.
M. muralis (L.) Cass.   Wall Lettuce
Map 65a; 11 tetrads.
Leicester, Welford Road Cemetery, walls, occasional, HB 1968; Cold Overton, churchyard wall, occasional, KGM 1971; Castle Donington Park, several places on walls, occasional, PAC 1971; Sauvey Castle, rocks, occasional, IME 1972; Orton on the Hill, steep roadside verge, occasional, HB 1973; Goadby Marwood, top of low stone wall, occasional, ALP 1973; Rolleston Hall grounds, occasional, IME 1976; Leicester, Belgrave, churchyard, occasional, SHB 1977; Braunstone, scrub woodland, rare, EH 1977; Grace Dieu, bank of brook near Priory ruins, rare, SHB 1979; Leicester, London Road, garden, occasional, IME 1984.
   Constituent or intrusive: Sw, Sr, Ew, Ap.
First record: Pulteney (1747).
Herb: LSR★; LSR, LTR.
HG 348 as *Lactuca muralis*: rare, sparsely distributed; 23 localities cited, including Orton and Sauvey Castle.

Taraxacum Weber
Dandelions

The present state of our knowledge of this genus in Leicestershire is due largely to the collections made in the field by H. Handley from 1974 onwards. Until 1980 these were submitted to Dr. A. J. Richards for determination, and thereafter to C. C. Haworth. Collections have also been made on a smaller scale in parts of the county not visited by Handley, by A. L. Primavesi, K. G. Messenger and E. Hesselgreaves. Very little herbarium material collected before 1974 has been found; much of what there is has proved indeterminable by modern criteria and there are very few first records of species earlier than 1974. Records from before 1960 which could be verified are represented on the distribution maps by hollow dots. There is, however, no map for section PALUSTRIA, an important specimen of a species belonging to which was discovered in herb. LSR in 1983.

Many of the species recorded are of widespread distribution in Britain and Europe, but some are rare and local, and their discovery in Leicestershire has been of some importance in broadening our knowledge of their range. In general, knowledge both of the taxonomy and of the distribution of **Taraxacum** species has been increasing so rapidly during the period of the survey, that an attempt to include an up-to-date statement of the British and European distribution and frequency of each species has had to be abandoned.

The earlier history of the study of the genus is dealt with in Richards (1972). As in genera such as **Rubus, Hieracium, Alchemilla, Sorbus** and others, the large number of genetic units meriting the rank of species results from their specialised reproductive behaviour. At the present time the great majority of **Taraxacum** species are agamospermous or nearly so (i.e. seeds develop from unfertilised ovules), and although it seems likely that Europe and Asia contained vast hybrid swarms of sexually reproducing species as recently as the end of the last glaciation, these may have been able to hybridise with Arctic agamospermous species as the ice receded, and so conferred agamospermy on the majority of their descendents, with a consequent loss of genetic variability and the 'freezing' of large numbers of true breeding groups of individuals differing from one another constantly but only in comparatively minor respects.

When Handley began to work in 1974 in Leicestershire, Richards' publication *The Taraxacum Flora of the British Isles* was only two years old and 132 species were recognised in the British Isles, divided into five sections, of which three were known to be represented in the county. So much work has been done on the genus since, that now nearly 180 British species have been named and it has been found that a division into nine sections is more rational than into five. The process by which this has been achieved has gone on in several stages involving not only reassessment of the species determinations themselves, but also changes in the order in which they are arranged. The HAMATA, a section well represented in Leicestershire, proved particularly full of taxonomic problems which were not resolved until late in our survey, and we were forced to send some 200 specimens for redetermination.

After much debate, and against the advice of the specialists, we decided to follow *Flora Europaea* in arranging the species accounts and maps in alphabetical order of specific names within their sections to facilitate cross referencing between the Flora and the Herbarium (LSR) and to make it easier for the herbarium keepers to insert new species into the collection.

An unfortunate consequence of these taxonomic changes has been that much of the early recording by field workers of species aggregates has had to be discarded, and the only aggregate distribution map which now has any value is that showing the tetrads from which specialist collections were made. From the records of field workers, we know that **Taraxacum** species are present in virtually every tetrad (over 600), but material firmly determined by the specialists comes from only 197, and collections of material which proved to be indeterminable were made in about 50 more. A brief explanation of this seems necessary. The characters used in determination include not only those of flowers and fruit but also those of foliage. The leaf features are sensitive to seasonal changes (no specialist will undertake to identify material gathered after the end of May), but also to environmental factors, such as the lushness of adjacent vegetation, grazing and treading. Inevitably, successful determinations are concentrated in tetrads to the north, north-west and west of Leicester, where three of the four collectors live, and where some tetrads are known to contain 10 or more species. Equally inevitably hardly any determinations have been made from the area of the city or the larger towns where the factors adversely affecting the normal growth of leaves act most strongly.

In conclusion, just over half the **Taraxacum** species on the British list at the time of publication have been found in Leicestershire, and we believe that a sufficient number of records has been made to give a reasonably accurate idea of their relative frequency, and some idea of their distribution within the county. We also believe that this is the first time an account of the **Taraxacum** flora of a county has been published in such detail.

K. G. Messenger.

Section ERYTHROSPERMA (H. Lindb. f.) Dahlst.
Apart from the addition of species new to the British list, this section has remained unchanged during the period of the survey, and field workers' aggregate records are more meaningful than those of other sections. Specimens belonging to species in this section have been recognised in 62 tetrads, and firm determinations have been made of 8 species during the present survey from 17 tetrads. 3 other species have been found in old herbarium collections.

Dry places, including quarries, dry pastures, heaths, derelict mining sites, ballast of dismantled railways, walls, gravel paths etc. Most frequent in the mining area in the west of the county and in Charnwood Forest; in other parts of the county most records are from railways.
First record: as *T. laevigatum* (Willd.) DC., 1873, F. T. Mott in herb. MANCH.
HG 346 – 347 as *T. lacistophyllum, T. erythrospermum* and *T. brachyglossum*.

T. argutum Dahlst.
No map; 1 tetrad.
Waltham on the Wolds, old Oolite quarry spoil heap, ALP 1986.
   Residual or intrusive: Rq.
First Record: 1986, A. L. Primavesi in present work.
Herb.: LCR.

T. brachyglossum (Dahlst.) Dahlst.
Map 65b; 7 tetrads.
Mountsorrel, roadside near quarries, two localities, HH 1975 and 1976; Groby, roadside near the Pool, HH 1977; Holwell, dismantled mineral railway, HH 1977; Kibworth, canal towpath, ALP 1978; Croft, roadside, HH 1978; Croft, dry pasture, HH 1978; Groby, roadside bank, HH 1980; Ratcliffe on the Wreake, steep dry bank by fish pond, ALP 1985.
   Constituent, residual or intrusive: Ev, Er, Hd, Rt, Gp.
First record: 1907, Druce in herb. OXF.
Herb.: LSR★; LTR, NMW, OXF.

T. canulum Hagl.
Map 65c.
No recent record.
First record: 1916, A. R. Horwood in herb. OXF (Lockington).
Herb.: OXF.

T. fulviforme Dahlst.
Map 65d; 1 tetrad.
Swithland Reservoir, roadside near dam, HH 1975.
   Intrusive: Ev.
First record: 1916, A. R. Horwood in herb. LSR.
Herb.: LSR★; LSR, LTR.

T. fulvum Raunk.
Map 65e; 4 tetrads.
Sutton Cheney, roadside, HH 1976; Ingarsby, dismantled railway track in cutting, HH 1977; Mountsorrel, roadside near quarry, HH 1978; Sileby, garden bed, HH 1981.
   Residual or intrusive: Ev, Er, Ag.
First record: 1905, A. B. Jackson in herb. BM.
Herb.: LSR★; BM, NMW.

T. lacistophyllum (Dahlst.) Raunk.
Map 65f; 8 tetrads.
Sutton Cheney, old brick pit, HH 1976; Cotes, siliceous grassland, HH 1976; Breedon Hill, limestone grassland, HH 1977; Ingarsby, dismantled railway track in cutting, HH 1977; Shenton, dismantled railway track in cutting, HH 1977; Croft meadow, R. J. Pankhurst 1979; Tilton, disused quarry, HH 1980; Waltham on the Wolds, old Oolite quarry spoil heap, ALP 1986.
   Constituent, residual or intrusive: Gp, Er, Rg, Rq.
First record: 1905, A. B. Jackson in herb. BM (Breedon Hill).
Herb.: LSR★; BM.

T. laetum (Dahlst.) Dahlst.
Map 65g.
No record in the recent survey.
Groby, roadside near the Pool, E. K. Horwood 1947.
First record: 1947, E. K. Horwood in herb. LTR.
Herb.: LTR.

T. oxoniense Dahlst.
Map 65h; 8 tetrads.
Sutton Cheney, floor of old brick pit, HH 1976; Staunton Harold, Spring Wood, roadside, HH 1976; Cotes, old brick wall, HH 1976; Holwell, track of dismantled mineral railway, HH 1977; Medbourne, wall top, KGM 1978; Bringhurst, dismantled railway ballast, KGM 1978; Waltham on the Wolds, old Oolite quarry, ALP 1978; King Lud's Entrenchments, calcareous grassland, HH 1980.
   Residual or intrusive: Er, Ev, Sw, Rq, Rg, Gr.
First record: 1875, G. Robson in herb. LSR.
Herb.: LSR★; LSR, NMW.

T. pseudolacistophyllum van Soest
Map 65i; 1 tetrad.
Ingarsby, dismantled railway track in cutting, HH 1977.
   Residual or intrusive: Er.
First record: 1905, A. B. Jackson in herb. BM.
Herb.: LSR★; BM.

T. rubicundum (Dahlst.) Dahlst.
Map 65j; 1 tetrad.
Breedon Hill, limestone grassland, HH 1977.
   Constituent: Gp.
First record: 1945, T. G. Tutin in herb. LTR.
Herb.: LSR★; LTR.

T. silesiacum Dahlst. ex Hagl.
Map 65k; 1 tetrad.
Saltby Heath, limestone grassland, HH 1975.
   Constituent: Gr.
First record: 1945, T. G. Tutin in herb. LTR.
Herb.: LSR★; LTR.

T. simile Raunk.
Map 65l; 1 tetrad.
Freeby, rough grassland, ALP 1987.
   Intrusive: Gr.
First record: 1905, A. B. Jackson in herb. BM (Hemington).
Herb.: BM.

Section PALUSTRIA Dahlst.
The taxonomy of this section was only imperfectly known to A. R. Horwood, and most of the old herbarium material filed under the names *T. palustre* and *T. paludosum* turns out to have been wrongly named, even by the standards of his time. However one specimen in herb. LSR was determined in 1954 by D. E. Allen and confirmed in 1984 by C. C. Haworth as belonging to this section as now understood.

T. palustre (Lyons) Symons
No recent record.
First record: 1896, H. P. Reader in herb. LSR (Burbage Common).
Herb.: LSR.
HG 347 as *T. paludosum.*

Section SPECTABILIA Dahlst.
Shortly before the publication of this work, British and European specialists agreed on the formation of a number of new sections and the removal of most of the species formerly in section SPECTABILIA. Only two species belonging to the section as newly constituted have been found in Leicestershire, in only 12 tetrads mostly in the area of Charnwood Forest. No criteria have yet been published whereby the species in this section can readily be distinguished in the field by non-specialists, and the aggregate records collected by our field workers earlier in the survey have therefore had to be discarded.

T. faeroense (Dahlst.) Dahlst.
Map 66a; 7 tetrads.
Ulverscroft, marshy meadow near Lea Wood; PHG 1972; Thornton, marshy meadow, HH 1976; Ulverscroft Nature Reserve, three localities in marsh, marshy meadow and wet pasture, HH 1976; Frisby on the Wreake, The Wailes, marsh, HH 1976; Groby, marshy grassland, three localities, EH 1981; Gaulby, stream bank, HH 1983.
   Constituent or residual: M, Gp, Ew.
First record: 1972, P. H. Gamble in present work.
Herb.: LSR★

T. spectabile Dahlst.
Map 66b; 8 tetrads.
Narborough Bog Nature Reserve, marsh, EH 1974; Market Bosworth, marsh, two localities, HH 1974; Osbaston, marsh, HH 1975; Ratby, marsh, HH 1975; Ulverscroft, five localities, marsh, marshy meadow and marshy verge of lane, HH 1976; Swithland Reservoir, marsh, HH 1976; Botcheston Bog, marsh, HH 1976.
   Constituent or residual: M, Ev.
First record: 1974, H. Handley and E. Hesselgreaves in present work.
Herb.: LSR★.

Section NAEVOSA Dahlst.
Shortly before this work was published, a number of species having affinities with **T. naevosum** Dahlst. was separated from the rest of those in section SPECTABILIA into a section on their own. No criteria have yet been published by which non-specialists can distinguish all species in the new section from those in SPECTABILIA, and it seems likely that an aggregate map of the section would be of little value. In the recent survey, 4 species of section NAEVOSA have been found in 15 tetrads, widely scattered through the county in a variety of habitats, the commonest species being **T. euryphyllum** (13 records).

T. euryphyllum (Dahlst.) M. P. Chr.
Map 66c; 13 tetrads.
Osbaston, roadside verge, HH 1975; Nailstone, pasture, HH 1975; Nailstone roadside verge, HH 1975; Narborough Bog Nature Reserve, marsh, HH 1975; Cribb's Meadow Nature Reserve, marshy pasture, HH 1975; Ulverscroft, marshy meadow, HH 1976; Blackbirds Nest, roadside verge, HH 1976; Holwell, dismantled railway track, HH 1977; Wymondham Rough Nature Reserve, marshy meadow, HH 1977; south east of Bruntingthorpe, roadside verge, HH 1978; west of Barwell, roadside, HH 1979; Knossington, bank of ditch near Lady Wood, HH 1980; Newtown Linford, Benscliffe Lane, roadside verge, HH 1982; Leire, dismantled railway, HH 1983.
   Constituent or residual: M, Ev, Er, Wd.
First record: 1975, H. Handley in present work.
Herb.: LSR★.

T. fulvicarpum Dahlst.
Map 66d; 1 tetrad.
Medbourne, stone wall, KGM 1978.
   Intrusive: Sw.
First record: 1978, K. G. Messenger in present work.
Herb.: LSR★.

T. maculosum A. J. Richards
Including *T. maculigerum* auct. non Lindb.
Map 66e; 1 tetrad.
Charnwood Lodge Nature Reserve, heath grassland, HH 1977.
   Constituent: Hw.
First record: 1977, H. Handley in present work.
Herb.: LSR★.

T. naevosiforme Dahlst.
Map 66f; 1 tetrad.
Saltby Bog, marsh, ALP 1978.
   Constituent: M.
First record: 1978, A. L. Primavesi in present work.
Herb.: LSR★.

Section CELTICA Sahlin
Shortly before this work was published, European specialists agreed that a number of species formerly included in sections SPECTABILIA and VULGARIA had sufficient affinities to justify their being placed together in a single section. Criteria which could be used by non-specialists to distinguish species belonging to the new section from all others have not yet been published, and an aggregate map of their distribution in Leicestershire would seem to have little value. During the course of the present survey, 10 species have been identified, spread widely throughout the county in a total of 52 tetrads. They occur mainly in wet grassland and marshy ground, and less commonly in ruderal situations, roadside verges and trodden ground; some have been recorded from heathy parts of Charnwood Forest. Specimens previously determined as *T. praestans* H. Lindb. f., a species which is now considered to be very rare in Britain, have been redetermined by C. C. Haworth as either **T. explanatum** H. Øllgaard or **T. bracteatum** Dahlst.

T. explanatum H. Øllgaard.
Map 66h; 1 tetrad.
Saltby Bog, marsh, ALP 1978.
   Constituent: M.
First record: 1978, A. L. Primavesi in present work.
Herb.: LSR★.

T. fulgidum Hagl.
Map 66i; 3 tetrads.
Sutton Cheney, wet floor of brick pit, HH 1976; Holwell, marshy track of dismantled mineral railway, HH 1977; Gaulby, wet pasture, HH 1963.
   Constituent or intrusive: GP, Rg, Er.
First record: 1976, H. Handley in present work.
Herb.: LSR★; LTR.

T. gelertii Raunk.
T. adamii auct.
Map 66g (as *T. adamii*); 10 tetrads.
Market Bosworth, marsh, HH 1974; Heather, roadside verge, HH 1975; Carlton, marshy meadow, HH 1976; Ulverscroft, roadside verge, HH 1976; Woodhouse Eaves, roadside verge, HH 1976; Holwell, Browns Hill Quarry, spoil heap, HH 1976; Appleby Magna, north-west of No Man's Heath, hedgerow, HH 1977; Market Bosworth, roadside verge, HH 1977; Ratcliffe College, rough grass, ALP 1986; Waltham on the Wolds, old Oolite quarry, ALP 1986.
   Constituent, residual or intrusive: M, Ev, Eh, Rs, Rq.
First record: 1885, unsigned, in herb. LSR; 1898, F. T. Mott in herb. LSR may be this species.
Herb.: LSR★; LSR.

T. haematicum Hagl.
Map 66j; 4 tetrads.
Botcheston Bog, marsh, HH 1974; Lindridge, pasture, HH 1976; Thornton, marshy meadow, HH 1976; Frisby on the Wreake, The Wailes, marsh, HH 1976; Frisby on the Wreake, damp meadow, HH 1976.
   Constituent: M, Gp.
First record: 1974, H. Handley in present work.
Herb.: LSR★.

T. laetifrons Dahlst.
Map 66k; 2 tetrads.
Osbaston, marsh, HH 1974; Shackerstone, canal bank, HH 1979.
   Constituent or residual: M, Ew.
First record: 1974, H. Handley in present work.
Herb.: LSR★.

T. landmarkii Dahlst.
Map 66l; 1 tetrad.
Congerstone, canal bank, HH 1977.
   Constituent or residual: Ew.
First record: 1977, H. Handley in present work; material in herb. LTR coll. E. K. Horwood in 1955 has been queried.
Herb.: LSR★.

T. nordstedtii Dahlst.
Map 67a; 19 tetrads.
Market Bosworth, marsh, HH 1974; Botcheston Bog, marsh, HH 1974; Narborough Bog Nature Reserve, marsh, EH 1974 and HH 1975; Ratby, pasture, HH 1975; near Abbots Oak, roadside, HH 1975; Ulverscroft, marsh on Nature Reserve, marsh near Priory, two localities in marsh near Lea Wood, HH 1975, on bank of stream near Lea Wood, HH and PHG 1976; Ratby, marsh, HH 1976; Dimminsdale Nature Reserve, calcareous grassland, HH 1976; Shenton, dismantled railway track in cutting, HH 1977; Ingarsby, dismantled railway track in cutting, HH 1977; Charnwood Lodge Nature Reserve, wet heath, HH 1978; south east of Bruntingthorpe, side of ditch, HH 1978; Loddington, ballast of dismantled railway, HH 1980; Groby, near Lady Hay Wood, marshy meadow, EH 1981; Gracedieu Manor, two localities, marsh and heath grassland, ALP 1985.
   Constituent, residual or intrusive: M, Hg, Hw, Ew, Wd, Gp, Er.
First record: 1948, T. G. Tutin in herb. LTR.
Herb.: LSR★; LTR.

T. raunkiaerii Wiinst.
T. duplidentifrons Dahlst. p.p.
Map 67b; 19 tetrads.
Markfield, roadside verge, HH 1974; Dimminsdale, woodland ride, HH 1976; Staunton Harold, near Spring Wood, roadside verge, HH 1976; Sutton Cheney, roadside verge, HH 1976; Dadlington, roadside, HH 1976; Swithland, damp pasture, HH 1976; Cossington, roadside verge, HH 1976; Ulverscroft, by brook, PHG 1976; East Norton, roadside verge, KGM 1978; Launde, pasture, KGM 1978; Tilton, dismantled railway track, ALP 1978; Walton on the Wolds, roadside verge, ALP 1978; Measham, roadside verge, HH 1979; Ratcliffe Culey, roadside verge, two localities, HH 1979; Fenny Drayton, roadside verge, HH 1979; Kirkby Mallory, roadside verge, HH 1979; Gaulby, wet pasture, HH 1983; Houghton on the Hill, roadside verge, HH 1983; Frisby on the Wreake, The Wailes, marsh, ALP 1985.
   Constituent or intrusive: Ev, Er, Ef, Ew, Gp, M.
First record: 1974, H. Handley in present work.
Herb.: LSR★.

T. subbracteatum A. J. Richards.
Includes records previously determined as *T. crispifolium* H. Lindb. f. Map 67c; 11 tetrads.
Heather, marsh, HH 1975; near Ives Head, roadside verge, HH 1975; Thornton Reservoir, marsh, HH 1975; Ulverscroft, roadside verge, HH 1976; Markfield, roadside verge, HH 1976; Ibstock, roadside verge, HH 1976; Kirby Muxloe, roadside verge, HH 1977; near Beacon Hill crossroads, roadside verge, HH 1977;
Redmile, ditch bank, HH 1980; Gaulby, wet pasture, HH 1983.
   Constituent or intrusive: M, Gp, Ev, Wd.
First record: 1975, H. Handley in present work.
Herb.: LSR★.

T. tamesense A. J. Richards.
Map 67d; 1 tetrad.
Markfield, damp pasture, HH 1978.
   Constituent: Gp.
First record: 1978, H. Handley in present work.
Herb.: LSR★.

T. tenebricans (Dahlst.) Dahlst.
Map 67e; 2 tetrads.
Cossington, roadside verge, HH 1976; Launde, Brook Farm, gateway, KGM 1978.
   Intrusive: Ev, Rt.
First record: 1976, H. Handley in present work.
Herb.: LSR★.

Section HAMATA Øllgaard.
During the early years of this survey the relationships of a group of species with affinities with **T. hamatum** Raunk. were still the subject of intensive research, and the collections made in Leicestershire between 1974 and 1979 added considerably to the British material available for study. Shortly before this work was published, it was agreed by the specialists that the species concerned should be removed from section VULGARIA to form a section on their own, and on the advice of C. C. Haworth we submitted all the available specimens which had been determined originally by A. J. Richards for redetermination. As expected, a large proportion (nearly 30%) of the 170 specimens had to be renamed, and our ideas of the frequency and distribution of some species had to be drastically revised. The section as now understood includes 12 species which occur in Leicestershire, and these include two of the most widespread and frequent in the county. Altogether they have been found in 123 tetrads, three more in fact than the 40 species of section VULGARIA. They occur in a wide variety of habitats, being most frequent in damp disturbed ground and grassland, and less common in damp places.
   Material originally determined under the names of *T. atrovirens* Dahlst., *T. latisectum* H. Lindb. f. and *T. maculatum* Jordan has all been renamed, and these species have not been found in Leicestershire.

T. atactum Sahl.
Map 67f; 3 tetrads.
Woodhouse, pasture, HH 1976; Barrow on Soar, marshy meadow, ALP 1978; Redmile, ditch bank, HH 1980.
   Constituent: Gp, Wd, M.
First record: 1976, H. Handley in present work.
Herb.: LSR★.

T. boekmanii Hagl.
Map 67g; 3 tetrads.
Staunton Harold, Dimminsdale, scrub woodland, HH 1976; Shenton, roadside verge, HH 1979; Ratcliffe on the Wreake, steep dry bank by fish pond, ALP 1985.
   Constituent or intrusive: Ev, Fs, Gp.
First record: 1976, H. Handley in present work.
Herb.: LSR★; LCR.

T. bracteatum Dahlst.
Map 67h; 17 tetrads.
Carlton, damp pasture, HH 1976; Newbold Verdon, grassy track, HH 1976; Desford, stream bank, HH 1976; Walton on the Wolds, roadside verge, HH 1977; Shenton, King Richard's Field, marshy ground, HH 1977; Market Bosworth, roadside verge, HH 1977; Barrow on Soar, marshy meadow, ALP 1978; Launde, roadside verge, KGM 1978; Wymondham Rough Nature Reserve, pasture, KGM 1978; Stockerston, roadside verge, KGM 1978; Bagworth, marshy meadow south west of Thornton, HH 1978; Saltby Bog, marsh, ALP 1978; Barrow upon Soar, pasture, HH 1979; Thorpe Satchville, ditch bank, HH 1982; Twenty Acre, wet heathland, ALP 1978; Grace Dieu Manor, marsh and heath grassland, ALP 1985.
 Constituent or intrusive: Gp, M, Hg, Hw, Ev, Ew, Rt.
First record: 1976, H. Handley in present work.
Herb.: LSR★; LCR.

T. hamatiforme Dahlst.
Map 67i; 2 tetrads.
Shenton Cutting Nature Reserve, track of dismantled railway, HH 1977; Ratcliffe on the Wreake, permanent pasture, ALP 1986. This was the most completely misunderstood species in the section, and some 28 specimens originally given this name have been renamed or are indeterminate.
 Intrusive: Er, Gp.
First record: 1977, H. Handley in present work.
Herb.: LSR★.

T. hamatulum Hagdk.
Map 67j; 2 tetrads.
Asfordby, pasture, HH 1975; Oaks in Charnwood, roadside verge, HH 1976. A formerly misunderstood species; 15 of the specimens originally given this name have been redetermined.
 Constituent or intrusive: Gp, Ev.
First record: 1975, H. Handley in present work.

T. hamatum Raunk.
Map 67k; 42 tetrads.
The most widespread **Taraxacum** species so far recorded in Leicestershire; although 13 of the specimens originally named as this species have now been redetermined, there are still more than 40 certain records. Mainly a species of roadsides and disturbed ground. It is also found in wet and marshy grassland, woodland rides and old pasture.
 Constituent, residual and intrusive: M, Gp, Gr, Ap, Er, Ev, Ew, Ef, Eh, Rs.
First record: 1974, H. Handley in present work.
Herb.: LSR★; LCR.

T. hamiferum Dahlst.
Including records of *T. atrovirens* Dahlst.
Map 67l; 7 tetrads.
Oaks in Charnwood, roadside verge, HH 1976; Swithland Reservoir, roadside verge, HH 1976; Rothley, roadside verge, HH 1976; Carlton Bridge, marshy meadow, HH 1976, Cold Overton, roadside verge, HH 1980; Bottesford, roadside verge, HH 1980; Groby, track of dismantled railway, HH 1980. Material from Farm Town, Coleorton, HH 1979 was not available for redetermination.
 Constituent or intrusive: M, Ev, Er.
First record: 1976, H. Handley in present work.
Herb.: LSR★.

T. lamprophyllum M. P. Chr.
Map 68a; 6 tetrads.
Great Bowden, roadside verge, HH 1975; Thurcaston, marshy meadow, HH 1975; near No Man's Heath, roadside verge, HH 1975; Sutton Cheney, roadside verge, HH 1976; Saxelby, roadside verge, HH 1976; Kimcote, damp roadside verge, HH 1978.
 Constituent or intrusive: M, Ev.
First record: 1975, H. Handley in present work.
Herb.: LSR★.

T. marklundii Palmgr.
No map; 3 tetrads.
Newbold Verdon, roadside verge, HH 1976 (det. H. Øllgaard, 1984); Charnwood Lodge Nature Reserve, heath grassland, HH 1977; Asfordby, old gravel pit, recently disturbed grassland, ALP 1985. Until the determination of the first two of these specimens in 1984, this species was supposedly endemic in Finland.
 Constituent or intrusive: Hg, Ev, Gr.
First record: 1976, H. Handley in present work.
Herb.: LSR★; LCR.

T. polyhamatum Øllgaard
Map 68b; 2 tetrads.
Sproxton, roadside verge, HH 1975 (2nd British record); Ashby de la Zouch, roadside verge, HH 1979 (material not available for redetermination).
 Intrusive: Ev.
First record: 1975, H. Handley in present work.
Herb.: LSR★.

T. pseudohamatum Dahlst.
Map 68c; 39 tetrads.
Roadsides, frequent; pastures and marshy ground, occasional; often in disturbed ground. The second most widespread species in the county; 28 specimens previously determined under other names proved on re-examination to belong here. Material from 30L, 31Z, 41D, 51V, 61A and 61G was not available for redetermination.
 Constituent or intrusive: M, Gp, Ev, Rg, Ag.
First record: 1975, H. Handley in present work.
Herb.: LSR★; LCR.

T. quadrans Øllgaard
Includes records of *T. latisectum* H. Lindb. f.
Map 68d; 12 tetrads.
Seagrave, roadside verge, HH 1975; Peckleton, marsh, HH 1976; Bagworth, trackway, HH 1976; Norton juxta Twycross, hedge bank, HH 1976; Woodhouse, trodden ground, HH 1976; Walton on the Wolds, roadside verges, 3 different sites, HH 1976, 1977, 1978; Ratcliffe on the Wreake, roadside verge, HH 1976; Groby, roadside verge, HH 1976; Barlestone, roadside verge, HH 1976; Stathern, roadside near Easton, HH 1980. Material from Wymeswold, ALP 1979, was not available

for redetermination but was probably this species.
Constituent or intrusive: M, Ev, Eh, Rt.
First record (as *T. latisectum*): 1975, H. Handley in present work.
Herb.: LSR★.

T. subditivum HvSZ.
No map; 1 tetrad.
Cossington, marshy pasture, ALP 1986.
Constituent: M.
First record: 1986, A. L. Primavesi in present work.
Herb.: LCR.

T. subhamatum M. P. Chr.
Map 68e; 10 tetrads.
Material originally determined as *T. marklundii* or *T. subhamatum* by A. J. Richards has been redetermined by him or by C. C. Haworth, and 3 of the 28 specimens have proved to belong to this species as now understood, while 2 others are believed to belong to a species new to Britain and not yet determined. A further 6 specimens have been found among other material collected in 1975 and 1976, and the present position is as follows: Cropston, pasture, HH 1975; Frisby on the Wreake, pasture, HH 1975; Thurcaston, marsh, HH 1975; Staunton Harold, Dimminsdale Nature Reserve, stream bank, HH 1976; Market Bosworth, damp pasture, HH 1976; Peckleton, marshy meadow, HH 1976; Peckleton, roadside verge, HH 1976; Swithland, hedge bank, HH 1976; Normanton le Heath, roadside verge, HH 1979; Asfordby, old gravel pit, recently disturbed grassland, ALP 1985.
Constituent or intrusive: M, Gp, Ev, Eh, Rd.
First record: 1975, H. Handley in present work.
Herb.: LSR★; LCR.

Section VULGARIA Dahlst.
Taraxacum officinale Weber p.p.
Shortly before this work was published, British and European specialists agreed on the separation from this section of two groups of species to form a new section, HAMATA, and part of another, CELTICA. The species remaining in section VULGARIA are by no means the commonest or most widespread in the county, and as a result of the separation, the aggregate map of *T. officinale* prepared from the records of field workers has become meaningless. It has become virtually impossible to give non-specialists a set of criteria by which to distinguish species of this section in the field. 52 species of section VULGARIA have been identified from material collected during the recent survey, from 120 tetrads, but several of the most widespread and frequent species have now been removed to section HAMATA, and we can no longer be confident that species of section VULGARIA would have been found in every tetrad if it had been possible for specialist collectors to visit them all in the time available. The most widespread species is **T. cordatum**, found in 22 tetrads.

T. aequilobum Dahlst.
Map 68f; 1 tetrad.
Stathern, roadside verge, HH 1980.
Intrusive: Ev.
First record: 1980, H. Handley in present work.
Herb.: LSR★.

T. alatum H. Lindb. f.
Map 68g; 7 tetrads.
Botcheston, roadside verge, HH 1975; Mountsorrel, garden weed, HH 1975; near Lubcloud, roadside verge, HH 1976; Burton Bandalls, roadside verge, HH 1976; north of Chilcote, roadside verge, HH 1977; Six Hills, roadside verge, ALP 1978; Ratcliffe College, rough grass, ALP 1984.
Intrusive: Ev, Ag, Gr.
First record: 1975, H. Handley in present work.
Herb.: LSR★; LCR.

T. ancistrolobum Dahlst.
Map 68h; 6 tetrads.
Thornton, rough grassland, HH 1975; Ratby, pasture, HH 1975; Narborough Bog Nature Reserve, marsh, HH 1975; Mountsorrel, marsh, HH 1976; Frisby on the Wreake, The Wailes, marsh, HH 1976; Bagworth, roadside verge, HH 1978; Frisby on the Wreake, kitchen garden, ALP 1985.
Constituent or intrusive: M, Gp, Gr, Ev, Ag.
First record: 1975, H. Handley in present work.
Herb.: LSR★; LCR.

T. angustisquameum H. Lindb. f.
No map; 1 tetrad.
Ratcliffe on the Wreake, permanent pasture, ALP 1986.
Constituent: Gp.
First record: 1986, A. L. Primavesi in present work.
Herb.: LCR.

T. cophocentrum Dahlst.
Map 68i; 8 tetrads.
Loddington, roadside verge, KGM 1978; Church Langton, brick wall, KGM 1978; Glenfield, Leicester Road, roadside verge, HH 1979; Wanlip, roadside verge, ALP 1979; Sileby, roadside verge, HH 1979; Normanton, bank of ditch, HH 1980; Foxton, canal bank, HH 1981; near Stanford Hall, stream bank, HH 1981.
Intrusive: Ev, Ew, Sw, Wd.
First record: 1978, K. G. Messenger in present work.
Herb.: LSR★.

T. cordatum Palmgr.
Map 68j; 22 tetrads.
Roadside verges, wet and marshy grassland, improved pasture, water side, woodland, quarry spoil heaps and dismantled railways. Widespread.
Constituent or intrusive: M, Gp, Gr, Gl, Ev, Ew, Er, Rs, Fd.
First record: 1974; H. Handley in present work.
Herb.: LSR★; LCR.

T. croceiflorum Dahlst.
Map 68k; 11 tetrads.
Carlton, roadside verge, HH 1976; Nailstone, grassy track, HH 1976; Walton on the Wolds, roadside verge, HH 1977; Thorpe Langton, roadside verge, KGM 1978; Hallaton, roadside verge, KGM 1978; Allexton, roadside verge, KGM 1978; Kibworth, canal towpath, ALP 1978; Lount, colliery waste, HH 1979; Beeby, stream bank,

HH 1979; Stathern, canal bank, HH 1980; Easthorpe, canal bank, HH 1980.
  Constituent or intrusive: Ev, Ew, Rt, Rs.
First record: 1976, H. Handley in present work.
Herb.: LSR★.

T. cyanolepis Dahlst.
Map 68l; 3 tetrads.
Cropston, roadside verge, HH 1976; Croft, roadside verge, HH 1978; Swithland, roadside verge, HH 1978.
  Intrusive: Ev.
First record: 1976, H. Handley in present work.
Herb.: LSR★

T. dahlstedtii H. Lindb. f.
Map 69a; 5 tetrads.
Wymondham Rough Nature Reserve, rough grassland, KGM 1978; Bringhurst, dismantled railway track, KGM 1978; Thrussington Wolds, pasture, ALP 1979; Muston, canal towpath, HH 1980; Normanton, roadside verge and grass of disused airfield, HH 1980.
  Constituent or intrusive: Gp, Gr, Er.
First record: 1978, K. G. Messenger in present work.
Herb.: LSR★.

T. ekmanii Dahlst.
Map 69b; 12 tetrads.
Sileby, roadside verge, HH 1975; Hoby, roadside verge, HH 1975; South Croxton, roadside verge, HH 1975; near Ratcliffe College, roadside verge, ALP 1976; Loddington, track of dismantled railway, KGM 1978; Kibworth Bridge, canal towpath, ALP 1978; Six Hills, roadside verge, ALP 1978; Lount Waste, rough grassland, ALP 1979; Wanlip, roadside verge, ALP 1979; Seagrave, roadside verge, ALP 1979; Thorpe Satchville, ditch, HH 1982; Ratcliffe College, rough grass, ALP 1984.
  Constituent or intrusive: M, Gr, Ev, Er, Rt, Wd.
First record: 1915, A. E. Wade in herb. NMW.
Herb.: LSR★; NMW, LCR.

T. exacutum Markl.
No map; 1 tetrad.
Barrow on Soar, marshy pasture, ALP 1985.
  Constituent: M.
First record: 1985, A. L. Primavesi in present work.
Herb.: LSR★; LCR.

T. excellens Dahlst.
Map 69c; 2 tetrads.
Holwell, dismantled mineral railway, HH 1977; Muston, canal bank, HH 1980.
  Constituent or intrusive: Er, Ew.
First record: 1977, H. Handley in present work.
Herb.: LSR★.

T. expallidiforme Dahlst.
Map 69d; 11 tetrads.
Cropston, roadside verge, HH 1976; Cribbs Meadow Nature Reserve, marshy pasture, HH 1978; Kibworth, canal tow path, ALP 1978; Coleorton, Farm Town, hedgerow, HH 1979; Shenton, roadside verge, HH 1979; Claybrooke Magna, roadside verge, HH 1979; Nanpantan, garden weed, HH 1979; Seagrave, pasture,

ALP 1979; Foxton, canal bank, HH 1981; Ratcliffe College, rough grass under trees, ALP 1986.
  Constituent or intrusive: Ev, Ew, Gp, Ag, Rt.
First record: 1976, H. Handley in present work.
Herb.: LSR★; LCR.

T. fasciatum Dahlst.
Map 69e; 3 tetrads.
Desford, headland of arable field, HH 1976; Great Bowden, pasture, KGM 1978; Wigston Magna, canal bank, HH 1983.
  Constituent or intrusive: Gp, Ea, Ew.
First record: 1976, H. Handley in present work.
Herb.: LSR★.

T. grossum van Soest
Records previously determined as *T. sundbergii* Dahlst.
Map 69f; 2 tetrads.
Ratcliffe on the Wreake, Fosse Way verge, ALP 1976; Wymondham, roadside verge, KGM 1978.
  Intrusive: Ev.
First record: 1976, A. L. Primavesi in present work.
Herb.: LSR★.

T. hemicyclum Hagl.
Map 69g; 2 tetrads.
Great Bowden, former railway yard, KGM 1978; Wymondham Rough Nature Reserve, improved pasture, KGM 1978.
  Intrusive: Er, Gl.
First record: 1978, K. G. Messenger in present work.
Herb.: LSR★.

T. hemipolyodon Dahlst.
Map 69h; 1 tetrad.
Near Ratcliffe College, hard shoulder of Fosse Way, ALP 1976.
  Intrusive: Ev.
First record: 1945, E. K. Horwood in herb. LTR.
Herb.: LSR★; LTR.

T. horridifrons Rail.
Map 69i; 1 tetrad.
Woodhouse Eaves, herbaceous border, HH 1983.
  Intrusive: Ag.
First record: 1983, H. Handley in present work.
Herb.: LSR★.

T. huelphersianum Dahlst.
Map 69j; 2 tetrads.
Ratcliffe Culey, roadside verge, HH 1979; Sapcote, roadside verge, HH 1979.
  Intrusive: Ev.
First record: 1979, H. Handley in present work.

T. insigne Ekman ex Raunk.
Map 69k; 6 tetrads.
Roecliffe Hill, roadside verge, HH 1976; near Swithland Reservoir, roadside verge, HH 1976; Bagworth, roadside verge, two localities, HH 1976; north of Chilcote, roadside verge, HH 1977; Willesley, hedgerow, HH 1979; Battle Flat, roadside verge, HH 1979.
  Intrusive: Ev, Eh.
First record: 1976, H. Handley in present work.
Herb.: LSR★.

T. laciniosifrons Wiinst.
Previously determined as *T. sinuatum* Dahlst.
Map 69l; 1 tetrad.
Sheepy Parva, roadside verge, HH 1979.
   Intrusive: Ev.
First record: 1979, H. Handley in present work.

T. lacinulatum Markl.
Map 70a; 1 tetrad.
Anstey, roadside verge, HH 1976.
   Intrusive: Ev.
First record: 1976, H. Handley in present work.
Herb.: LSR*.

T. laticordatum Hagl.
No map; 1 tetrad.
Ratcliffe College, rough grass, ALP 1984.
   Intrusive: Gr.
First record: 1984, A. L. Primavesi in present work.
Herb.: LSR*; LCR.

T. latissimum Palmgr.
Map 70b; 3 tetrads.
Loddington, roadside verge, KGM 1978; Hallaton, churchyard, KGM 1978; Nevill Holt, roadside verge, KGM 1978.
   Intrusive: Ev, Ap.
First record: 1978, K. G. Messenger in present work.
Herb.: LSR*.

T. lingulatum Markl.
Map 70c; 9 tetrads.
Narborough Bog Nature Reserve, marsh, HH 1975; Woodhouse Eaves, roadside verge, HH 1975; Hoby, roadside verge, HH 1975; Owston, roadside verge, KGM 1978; Somerby, churchyard, KGM 1978; south east of Thornton, recently disturbed roadside verge, HH 1978; Stathern, canal bank, HH 1980; Frolesworth, roadside verge, HH 1983; Waltham on the Wolds, old Oolite quarry, ALP 1986.
   Constituent or intrusive: M, Ev, Ew, Ap, Rd, Rq.
First record: 1975, H. Handley in present work.
Herb.: LSR*.

T. longisquameum H. Lindb. f.
Map 70d; 7 tetrads.
Chilcote, roadside verge, HH 1975; Heather, marsh, HH 1975; Heather, wall, HH 1975; Sileby, roadside verge, ALP 1976; Sutton Cheney, roadside verge, HH 1976; near Branston, roadside verge, HH 1976; Saltby, roadside verge, HH 1976.
   Constituent or intrusive: Ev, M, Sw.
First record: 1914, A. E. Wade in herb. NMW.
Herb.: LSR*; NMW.

T. melanthoides Dahlst.
Map 70e; 2 tetrads.
Cribbs Meadow Nature Reserve, marsh, HH 1975; Frisby on the Wreake, damp meadow, HH 1976.
   Constituent: M, Gp.
First record: 1975, H. Handley in present work.
Herb.: LSR*.

T. obliquilobum Dahlst.
Map 70f; 2 tetrads.
Prestwold, roadside verge, ALP 1978; Kirby Bellars, roadside verge, ALP 1978.
   Intrusive: Ev.
First record: 1978, A. L. Primavesi in present work.
Herb.: LSR*.

T. oblongatum Dahlst.
Map 70g; 8 tetrads.
Barrow on Soar, pasture, HH 1976; Barrow on Soar, swamp, HH 1976; Frisby on the Wreake, ditch, HH 1976; Holwell, Browns Hill Quarry, spoil heap, HH 1976; Wymondham Rough Nature Reserve, pasture, HH 1977; Drayton, roadside verge, KGM 1978; Walcote, roadside verge, HH 1978; Lount Waste, rough grass, ALP 1979; Ratcliffe College, pasture, ALP 1984; Frisby on the Wreake, The Wailes, marsh, ALP 1985.
   Constituent or intrusive: Gp, Gr, M, Ev, Wd, Rs.
First record: 1976, H. Handley in present work.
Herb.: LSR*; LCR.

T. obscuratum Dahlst.
Map 70h; 1 tetrad.
Walton on the Wolds, roadside verge, ALP 1978 (det. A. J. Richards; in the opinion of C. C. Haworth the specimen was of an immature plant and indeterminate.)
   Intrusive: Ev.
First record: 1978, A. L. Primavesi in present work.
Herb.: LSR*; LCR.

T. obtusifrons Markl.
No map; 1 tetrad.
Sileby, rough grass on dismantled railway ballast, ALP 1986.
   Intrusive: Er.
First record: 1986, A. L. Primavesi in present work.
Herb.: LCR.

T. pachymerum Hagl.
Map 70i; 3 tetrads.
Wymondham Rough Nature Reserve, pasture, HH 1977; Barrow on Soar, marshy meadow, ALP 1978; south east of Bruntingthorpe, roadside verge, HH 1978.
   Constituent or intrusive: Gp, M, Ev.
First record: 1977, H. Handley in present work.
Herb.: LSR*.

T. pallescens Dahlst.
Map 70j; 3 tetrads.
Cotes, siliceous grassland, HH 1976; Cossington, water meadow, HH 1977; Ratcliffe College, rough grass, ALP 1984.
   Constituent or intrusive: Gp, Gr, M.
First record: 1976, H. Handley in present work.
Herb.: LSR*.

T. pannucium Dahlst.
Map 70k; 4 tetrads.
Carlton, canal bank, HH 1976; Wymondham, roadside verge, KGM 1978; Prestwold, roadside verge, ALP 1978; Thrussington, roadside verge, ALP 1979.
   Residual or intrusive: Ew, Ev.
First record: 1945, E. K. Horwood in herb. LTR, A. J. Wilmott's specimens in herb. BM are undated.
Herb.: LSR*; BM, LTR.

T. piceatum Dahlst.
Map 70l; 2 tetrads.
Sileby, marshy ground near River Soar, HH 1975 (first British record); Syston, Pontylue Gravel Pits, rough grass, ALP 1986.
   Constituent: Gr, M.
First record: 1975, H. Handley in present work.
Herb.: LSR★.

T. polyodon Dahlst.
Map 71a; 21 tetrads.
Recorded mainly from roadside verges and tracks in scattered localities in the west, north and east of the county; there are single records from pasture and marshy ground, and three records from dismantled railways.
   Constituent or intrusive: Ev, Er, Rt, Gp, M.
First record: 1975, H. Handley in present work.
Herb.: LSR★.

T. porrectidens Dahlst.
Map 71b; 3 tetrads.
Ratcliffe College, pasture, ALP 1975; near site of Cossington Gorse, verge of track, ALP 1975; Mountsorrel, quarry spoil heap, HH 1976.
   Constituent or intrusive: Gp, Ev, Rs.
First record: 1975, A. L. Primavesi in present work.
Herb.: LSR★.

T. praeradians Dahlst.
Map 71c; 1 tetrad.
Sapcote, Fosse Way, roadside verge, HH 1983.
   Intrusive: Ev.
First record: 1983, H. Handley in present work.
Herb.: LSR★.

T. privum Dahlst.
Map 71d; 1 tetrad.
Thrussington, roadside verge, ALP 1979.
   Intrusive: Ev.
First record: 1979, A. L. Primavesi in present work.

T. procerisquameum Øllgaard.
Includes records of *T. procerum* Hagl.
Map 71e; 3 tetrads.
Near Groby Pool, roadside verge, HH 1976; Stockerston, roadside verges, 2 localities, KGM 1978; Wanlip, roadside verge, ALP 1979.
   Intrusive: Ev.
First record: 1976, H. Handley in present work.
Herb.: LSR★.

T. reflexilobum H. Lindb. f.
Map 71f; 2 tetrads.
Wanlip, roadside verge, ALP 1979; Broughton Astley, roadside verge, HH 1983.
   Intrusive: Ev.
First record: 1948, T. G. Tutin in herb. LTR.
Herb.: LSR★; LTR.

T. rhamphodes Hagl.
Map 71g; 1 tetrad.
Bilstone, roadside verge, HH 1979.
   Intrusive: Ev.
First record: 1979, H. Handley in present work.

T. sagittipotens Dahlst.
Map 71h; 1 tetrad.
Seagrave, roadside verge, ALP 1979.
   Intrusive: Ev.
First record: 1979, A. L. Primavesi in present work.

T. sellandii Dahlst.
Includes 1 record previously determined as *T. monochroum* Hagl.
Map 71i; 11 tetrads.
Barrow on Soar, pasture, HH 1975; Ulverscroft, marsh, HH 1975; Mountsorrel, roadside verge, HH 1976; Bringhurst, roadside verge, HH 1978; Church Langton, roadside verge, HH 1978; Hallaton, roadside verge, HH 1978; Loddington, roadside verge, HH 1978; Wymondham Rough Nature Reserve, pasture, HH 1978; Markfield, rough grassland, HH 1978; Kirby Bellars, roadside verge, ALP 1978; west of Beeby, roadside verge, HH 1979; Mountsorrel, water meadow, HH 1975; Ratcliffe College, one locality in pasture and two localities in rough grass, ALP 1984.
   Constituent or intrusive: Gp, Gr, Ev.
First record: 1975, H. Handley in present work.
Herb.: LSR★; LCR.

T. stenacrum Dahlst.
Map 71j; 2 tetrads.
Groby, margin of the Pool, HH 1975; Barrow on Soar, marsh, HH 1975.
   Constituent: M, Ew.
First record: 1975, H. Handley in present work.
Herb.: LSR★.

T. subcyanolepis M. P. Chr.
Map 71k; 1 tetrad.
Leire, dismantled railway, HH 1983.
   Intrusive: Er.
First record: 1983, H. Handley in present work.
Herb.: LSR★.

T. sublaeticolor Dahlst.
Map 71l; 4 tetrads.
Narborough Bog Nature Reserve, marsh, HH 1975; Sileby, marsh, HH 1975; Shenton, wet track of dismantled railway in cutting, HH 1977; Barrow on Soar, marshy pasture, ALP 1986.
   Constituent or intrusive: M, Er.
First record: 1975, H. Handley in present work.
Herb.: LSR★.

T. subundulatum Dahlst.
Map 72a; 2 tetrads.
Botcheston Bog, marsh, HH 1974; Bottesford, roadside verge and bank of ditch, HH 1980.
   Constituent or intrusive: M, Ev, Wd.
First record: 1974, H. Handley in present work.
Herb.: LSR★.

T. trigonum M. P. Chr.
Map 72b; 2 tetrads.
Groby, roadside bank, HH 1980; Stathern, roadside verge, HH 1980.
   Intrusive: Ev.
First record: 1980, H. Handley in present work.
Herb.: LSR★.

T. tumentilobum Markl.
Previously determined as *T. sinuatum*.
Map 72c; 1 tetrad.
Mountsorrel, water meadow, HH 1975.
    Constituent or intrusive: Gp.
First record: 1975, H. Handley in present work.
Herb.: LSR★.

T. uncosum Hagl.
Map 72d; 3 tetrads.
Higham on the Hill, roadside verge, HH 1979; South Kilworth, roadside verge, HH 1981; Houghton on the Hill, ditch bank, HH 1983.
    Intrusive: Ev, Wd.
First record: 1979, H. Handley in present work.
Herb.: LSR★.

T. undulatiflorum Chr.
Map 72e; 7 tetrads.
Glooston, roadside verge, KGM 1978; Edmonthorpe, roadside verge, KGM 1978; Glenfield, waste ground, HH 1979; Thrussington, roadside verge, ALP 1979; Thrussington Wolds, roadside verge, ALP 1979; Wigston Magna, canal bank, HH 1983; Ratcliffe College, rough grass, two localities, ALP 1984.
    Constituent or intrusive: Ev, Ew, Rw, Gr.
First record: 1978, K. G. Messenger in present work.
Herb.: LSR★; LCR.

T. valdedentatum Dahlst.
Map 72f; 2 tetrads.
Frisby on the Wreake, river bank, HH 1976; Newbold Verdon, grassy track, HH 1976.
    Constituent or intrusive: Ew, Rt.
First record: 1976, H. Handley in present work.
Herb.: LSR★.

T. vastisectum Markl.
Map 72g; 2 tetrads.
Wigston Magna, canal bank, HH 1983; Sileby, rough grass on dismantled railway ballast, ALP 1986.
    Constituent or intrusive: Ew, Gr.
First record: 1983, H. Handley in present work.
Herb.: LSR★.

T. xanthostigma H. Lindb. f.
Map 72h; 1 tetrad.
Near Sileby Mill, wet ditch, HH 1975.
    Constituent or intrusive: Wd.
First record: 1975, H. Handley in present work.
Herb.: LSR★.

Lapsana L.

L. communis L.   Nipplewort
Subsp. **communis**
Map 72i; 591 tetrads.
Frequent and locally abundant throughout the county in hedgerows and on roadsides, in waste places and as a weed of arable land.
    Constituent or intrusive: Ev, Er, Eh, Ew, Ef, Ea, Ac, Ao, Ag, Aa, Ap, Rw, Rq, Rs, Rd, Rf, Rt, Wd, Gr, Fd, Fm, Fs.
First record: Pulteney (1747).
Herb.: LSR★; DPM, LANC, LCR, LSR, LTR.
HG 333: abundant, generally distributed.

Subsp. **intermedia** (Bieb.) Hayek
L. intermedia Bieb.
Map 72j; 1 tetrad.
Knighton, garden, introduced but established and aggressive, T. G. Tutin 1973.
    Intrusive: Ag.
First record: Tutin (1973).
HG: not recorded.

Crepis L.

Crepis paludosa (L.) Moench   Marsh Hawk's-beard
No record in recent survey.
There is a single leaf, confirmed as being of this species, in the collection of J. A Cappella now at herb. LSR, from a specimen collected by H.R. Hackshaw in Grace Dieu Wood in June 1934. The rest of the specimen, which was seen by A.R. Horwood, is supposed to have been deposited in herb. LSR, but cannot now be found.
First record: Watson (1837).
Herb.: LSR.
HG 337: very rare, perhaps extinct, restricted in range; 2 localities cited, including Grace Dieu. Horwood also states 'Not seen since 1883 when Mr. Thomas Carter refound it at Gracedieu'.

C. biennis L.   Rough Hawk's-beard.
Map 72k; 2 tetrads.
Launde, Park Wood, woodland ride, single plant, PHG 1965; Barrow on Soar, margins of mineral line to concrete works, two plants only, PHG 1969; Barrow on Soar, Paudy Lane, roadside verge, locally frequent, PHG 1972; Barrow on Soar, old disused lime pit, locally frequent, PHG 1972.
    Constituent or intrusive: Ev, Er, Ef, Rq.
First record: Bloxam *in* Combe (1829).
Herb.: LSR★; CGE, DPM, LSR.
HG 337: rare, restricted in range; 15 localities cited, not including the above.

C. tectorum L.
No map; 5 tetrads.
Kirby Muxloe, waste ground, rare, EH 1973; Glen Parva, roadside, EH 1974; Braunstone, Western Park, disturbed soil on building site, single plant, EH 1974; Blaby, bypass, EH 1974; Ratby, roadside, rare, EH 1975; Groby, disturbed soil on roadside near the Pool, single plant, EH 1975.
    Casual: Ev, Rw, Rd.
First record: 1973, E. Hesselgreaves *in* Clement (1976).
Herb.: LSR★; LTR
HG: not recorded.

C. capillaris (L.) Wallr.   Smooth Hawk's-beard.
Map 72l; 461 tetrads.
Grassland, grassy verges and grassy waste places. Frequent throughout the county.
    Constituent or intrusive: Gp, Gr, Gl, Ev, Er, Ew, Ea, Rw, Rg, Rq, Rs, Rd, Rt, Ap, Ag, Ac, Ao, Sw, Sb.
First record: Pulteney (1749).
Herb.: LSR★; DPM, LANC, LSR, LTR, MANCH, NMW.
HG 336: locally abundant, generally distributed.

C. vesicaria L. subsp. **haenseleri** (Boiss. ex DC) P. D. Sell   Beaked Hawk's-beard
C. vesicaria subsp. *taraxacifolia* (Thuill.) Tholl.
Map 73a; 234 tetrads.
Waste places, railway and roadside verges. Locally frequent in these habitats, especially in the neighbourhood of the larger towns. Occasional in grassland and arable land.
    Intrusive: Rw, Rq, Rg, Rs, rd, Rt, Rr, Er, Ev, Ew, Ea, Ag, Ap, Ac, Ao, Aa, Gr, Gp, Gl, Wd.
First record: Preston (1900).
Herb.: LSR*; DPM, LSR, LTR.
HG 335 as *C. taraxacifolia*: frequent, sparsely distributed; 26 localities cited.

C. setosa Haller fil.   Bristly Hawk's-beard.
Map 73b; 11 tetrads.
Bagworth, roadside verge, single plant, HH 1971; Croft, Thurlaston Lane, recently disturbed verge, locally frequent, SHB 1974; Huncote, recently disturbed roadside verge, single plant, SHB 1974; North Kilworth, roadside verge, rare, EKH and JMH 1974; Stapleton, recently disturbed roadside verge, rare, SHB 1975; Market Bosworth Park, grassy footpath south of Bow Pool, rare, HH 1976; Shenton, hedgerow, locally frequent, HH 1977; Ashby Woulds, recently disturbed ground, rare, SHB 1977; Leicester Forest East, rough grassland, 6 plants, EH 1978.
    Intrusive: Ev, Eh, Gr, Rd, Rt.
First record: Kirby (1850).
Herb.: LSR*; LSR.
HG 336: very rare, restricted in range; 7 localities cited.

Hieracium L.
                                                     Hawkweeds
On the advice of C. E. A. Andrews, who has determined the whole of the material of this genus collected during the recent survey, the nomenclature used by P. D. Sell and C. West in the *Critical Supplement to the Atlas of the British Flora* (Perring, F. and Sell, P. D., 1968) has been followed in this account.

Only a very few specimens of **Hieracium** dating from 1933 or earlier have been found in the various British herbaria searched. These have been determined by Sell and West but it has not been possible to confirm many of the records quoted in Horwood and Gainsborough. Changes too have taken place in the taxonomy and nomenclature of the genus and we have been advised to disregard altogether records of several taxa mentioned in that work, either because they cannot be equated exactly with modern taxa, or because as a result of modern knowledge of the distribution of the taxa named, it is regarded as unlikely that they could have occurred in Leicestershire. This does not mean that all Horwood and Gainsborough records are rejected out of hand, and where it is possible to equate new with old, this has been done.

A certain amount of collecting was done in Leicestershire between 1933 and 1967, and those specimens which were determined by Sell and West have been included in the distribution maps. About 250 specimens were collected during the Flora survey, and the determination of these was done by C. E. A. Andrews in 1979. The majority were collected by E. Hesselgreaves and H. Handley, but seven other field workers also contributed to the total. Only seven specimens proved to be unidentifiable because of their condition. A few specimens of **Pilosella**, now separated from **Hieracium**, were also determined.

Four of the 14 species described in the following species accounts show a marked preference for railway banks and ballast, and these are quite the commonest species in the county. They are **H. diaphanum, H. strumosum, H. salticola** and **H. vagum.** Most of the other species recorded are also found on railways, but there are not enough records of these for a habitat preference to be distinguishable. In general though, the Leicestershire species are mainly found in well drained situations subject to some degree of human interference.

Certain areas of the county have been searched far more thoroughly than others for **Hieracium** and it was not until the specimens determined by C. E. A. Andrews came to be sorted that it was realised that a large area of Charnwood Forest had not been included in the search. This is a serious deficiency since it is here if anywhere that traces of indigenous species might be found. Nevertheless, even without such records, the maps show clearly the association of **Hieracium** species with the parts of the county where there are most mines and quarries, and that elsewhere the records show the course of the former railway network, now largely defunct.
                                                          K.G. Messenger

Section *VULGATA*
H. grandidens Dahlst.
Map 73c; 2 tetrads.
Enderby, blue brick railway bridge, rare, EH 1974; Narborough, railway bank, occasional, EH 1974.
    Intrusive: Er, Sw.
First record: 1974, E. Hesselgreaves in present work.
Herb.: LSR*.
HG: not recorded.

[**H. sublepistoides** (Zahn) Druce
A dot appears in 10km square SP 68 in the map for this species in Perring and Sell (1968). The record has not been traced; about 30% of the square is in Northamptonshire.]

H. vulgatum Fr.
Map 73d; 13 tetrads.
Leicester, waste ground, occasional, HB 1979; Newbold Verdon, roadside verge, frequent, HH 1971; west of Merry Lees, roadside verge, occasional, HH 1971; Snarestone, railway verge, rare, EH 1971; Long Clawson station site, dismantled railway, ballast, frequent, ALP 1972; Swithland Reservoir, woodland margin, occasional, PHG 1972; Heather, old quarry, abundant, HH 1973; Coleorton, railway bridge, brickwork, occasional, ALP 1973; Ashby de la Zouch, stone wall top, frequent, ALP 1974; Holwell, Browns Hill Quarry, frequent, ALP 1974; Saxelby, cutting south of railway tunnel, frequent, ALP 1974; Cosby, dismantled railway, cutting, rare, EH 1975; South Wigston Junction, railway sidings, occasional, EH 1976; Countesthorpe, dismantled railway, occasional, EH 1977.
    Intrusive: Er, Ev, Ef, Rq, Rw, Sw.
First record: 1906, W. Bell in herb. CGE.
Herb.: LSR*; CGE, NMW.
HG 339: records under *H. vulgatum* (agg.) and under *H. acroleucum* are probably of this species; 16 localities cited.

H. maculatum Sm.
Map 73e; 24 tetrads.
Locally frequent or abundant on the ballast and verges of little used and dismantled railway lines in and around Leicester; occasionally recorded from quarries, waste ground and rough grassland.
   Constituent or intrusive: Er, Gr, Rq, Rw, Rd.
First record: Pulteney *in* Nichols (1795); 1843, A. Bloxam in herb. CGE.
Herb.: LSR★; CGE, LTR.
HG 340: local, restricted in range; 20 localities cited, including some where it still occurs.

H. submutabile (Zahn) Pugsl.
Map 73f; 5 tetrads.
Scalford, dismantled railway, ballast, frequent, ALP 1971; Snarestone, dismantled railway, frequent, HH 1974; Enderby, railway banks, occasional, Brockington Ecological Society 1974; Blaby, dismantled railway, occasional, EH 1974; Enderby, wall, occasional, EH 1977.
   Intrusive: Er, Sw.
First record: 1906, F. L. Foord-Kelcey in herb. CGE.
Herb.: LSR★; CGE.
HG 340 as *H. acroleucon* var. *mutabile*: 6 localities cited.

H. diaphanum Fr.
Map 73g; 32 tetrads.
More than half the records for this species are from railways, either in use or dismantled, and it varies in different places from rare to abundant. In addition it has been recorded on walls four times, in quarries on rock or on spoil heaps four times, waste ground three times, twice each from rough grassy places and road verges, and once from a wild garden. Most records are concentrated to the west and south-west of Leicester, but there are scattered records from the mining areas further west, and a few in the east and south-east of the county.
   Intrusive: Er, Ev, Gr, Rq, Rs, Rw, Sw, Sb, Sr, Ap.
First record: C. C. Babington in herb. CGE, undated, but prior to 1895.
Herb.: LSR★; CGE, LTR.
HG 339 and 341 as *H. acroleucon* var. *daedalolepium* and as *H. scanicum*. Records of *H. diaphanoides* may also refer to this species.

H. strumosum (W. R. Linton) A. Ley
Map 73h; 36 tetrads.
With 64 records from 36 tetrads, this appears to be the most frequent species in the county. Like several others it is most often recorded from railways and quarries, but there are also records from walls, dry heath, rough grassland, old pasture and roadsides. There is a single record from a rocky slope in Charnwood Forest. Sell and West (in litt.) regard it as possibly native in Wales and the English uplands. The Charnwood Forest population may perhaps be indigenous; but with so many Leicestershire populations being intrusive in man-made habitats it could equally well be a recolonisation from one of these. There is a concentration of records to the west and south-west of Leicester, with scattered outliers further west and to the north-east and south-east. The Charnwood Forest populations deserve further study.
   Perhaps residual: Hd, Gr, Sr.
   Intrusive: Er, Ev, Eh, Fs, Ap, Hg, Rq, Rw, Sw, Fd.

First records: Preston (1900); 1916, A. R. Horwood in herb. K.
Herb.: LSR★; K, LTR.
HG 341 as *H. lachenalii*.

Section *TRIDENTATA* Fr.
H. eboracense Pugsl.
Map 73i; 6 tetrads.
Newtown Linford, roadside verge, rare, EH 1968; Kirby Muxloe, track and verges of little-used railway, locally frequent, EH 1972; Swithland, railway verge, locally frequent, PHG 1972; Groby, Lawn Wood Quarry, stony slope in old quarry, occcasional, EH 1972; Coleorton, railway verge, frequent, ALP 1973; Heather, roadside verge recently disturbed, frequent, HH 1973.
   Intrusive: Er, Ev, Rq, Rd.
First record: 1968, E. Hesselgreaves in present work.
Herb.: LSR★.
HG 342. Some of the records listed under the name *H. tridentatum* may perhaps have been this species.

H. calcaricola (F. J. Hanb.) Roffey
Map 73j; 12 tetrads.
Whitwick, disused railway, abundant, PAC 1969; Holwell Mouth, scrub, occasional, PHG 1969; Holwell, dismantled mineral railway, abundant, PHG 1969; Martinshaw Wood, rides, occasional, EH 1970; Botcheston, disturbed roadside verge, frequent, HH 1970; Groby, dismantled mineral railway, EH 1970; Kirby Muxloe, railway verge, frequent, EH 1971; Groby, Lawn Wood Quarry, Keuper Marl spoil tip, rare, EH 1972; Ratby cemetery, rough grass, occasional, EH 1973; Measham, colliery shale heaps, frequent, HH 1975; Chaplin's Rough, margin, occasional, EH 1975; Markfield, roadside verge, rare, HH 1976.
   Residual or intrusive: Er, Ev, Ef, Fs, Gr, Rq, Rs, Rd.
First record: 1954, E. K. Horwood in herb. LTR.
Herb.: LSR★; LTR.
HG 342. Some of the records listed under *H. tridentatum* may perhaps refer to this species.

Section *UMBELLATA* Fr.
H. umbellatum L.
Map 73k; 5 tetrads.
North of Stapleton, roadside verge of A447, occasional, HH 1971; near Swithland Wood, hedgerow, locally frequent, PHG 1973; Peckleton Common, roadside verge, frequent, HH 1974. This is the only species for which the number of records in HG seems significantly greater than in the recent survey; it has almost certainly suffered from road widening and improvements, but there are woodland sites, particularly in Charnwood Forest, where it still ought to be found. There are four herbarium specimens dated 1957, from Sutton Cheney, Burbage Wood, Market Bosworth and 'Stapleton - Cadeby'.
   Residual or intrusive: Ev, Eh.
First records: Pulteney (1749); 1845, A. Bloxam in herb. BM.
Herb.: LSR★; BM, CGE, LTR, MANCH, NMW.
HG 343: rare; 20 localities cited.

Section *SABAUDA* Fr.
H. perpropinquum (Zahn) Druce
Map 73l; 15 tetrads.
Leicester, Blackbird Avenue, EH 1965; Ulverscroft, near Priory, EH 1967; Swithland Wood, margin near stream,

EH 1967; Martinshaw Wood, rides, frequent, EH 1968, 1970 and 1973; Groby, quarries, frequent, EH 1969; Swithland, Joe Moore's Lane, F. A. Sowter 1970; Groby, Lawn Wood Quarry, Keuper Marl spoil tip, occasional, EH 1970; Bradgate Park, birch scrub, occasional, EH 1971; Desford, railway verge, frequent, HH 1971; Groby, quarry near Groby Pool, rare, EH 1972; near Desford, hedgerow, frequent, HH 1973; Cosby, railway verge, occasional, EH 1975; Twycross, roadside verge of A444 near Zoo, occasional, HH 1975; Chaplin's Rough, edge, rare, EH 1975.

Constituent, residual or intrusive: Ev, Er, Ef, Fs, Rq, Rs.

First records: probably Pulteney (1747); 1909, W. Bell in herb. LTR.

Herb.: LSR*; CGE, LANC, LTR.

HG 342 as *H. sabaudum*: frequent but sparsely distributed. It is not possible to be certain that all the records under this name in HG were in fact of **H. perpropinquum**, in view of the prevalence now of **H. salticola** and **H. vagum**, neither of which is mentioned by Horwood. 53 localities are cited for *H. sabaudum* but only 4 of these are for subsp. *obliquum* which can certainly be referred to this species.

H. rigens Jord.
Map 74a; 2 tetrads.
No record in the recent survey.
Knighton, garden, T. G. Tutin 1952 and 1953; Swithland Wood, T. G. Tutin 1959.
    Casual or intrusive: Ag, Fd.
First record: W.B.E.C. Rep. 1906.
Herb.: CGE, LTR.
HG 343 as *H. sabaudum* var. *rigens*: 1 locality cited.

H. salticola (Sudre) Sell & West
Map 74b; 27 tetrads.
More than half the records for this species are from railways, in use and dismantled, and there are records also from quarries, roadside verges, waste ground and a hedgerow; there is a single record from a ride in secondary woodland. There is nothing in the distribution to suggest that it might be indigenous in the county, and the pattern resembles that of the commoner members of the Section *VULGATA*, though it is less frequent.

Constituent or intrusive: Er, Ev, Eh, Ef, Fd, Gr, Rq, Rw.

First record: 1946, A. B. Hackett in herb. LSR.
Herb.: LSR*; LSR, LTR.

HG: not recorded; however it seems likely that some of the records of *H. sabaudum* were of this species, especially those from mining areas.

H. vagum Jord.
Map 74c; 33 tetrads.
More than half the records of this species are from railways, in use or dismantled, and there are records also from roadside verges, waste ground and woodland, though surprisingly none from quarries. The distribution resembles that of other common species in the county, but there is a wider scatter of records beyond the immediate environs of Leicester. The pattern does not suggest that it is indigenous, although there is a specimen in herb. LTR from The Brand in 1945, where it was found growing on rock.

Constituent or intrusive: Er, Fd, Fm, Ef, Ev, Gr, Rw.

First record: 1945, H. Gilbert Carter, F. A. Sowter and T. G. Tutin in herb. LTR.
Herb.: LSR*; LTR.

HG: not recorded; however some of the records of *H. sabaudum* may have been of this species.

Pilosella Hill
We follow Sell and West (1974) in separating this genus from Hieracium.

P. officinarum C. H. & F. W. Schultz
                                Mouse-ear Hawkweed
Hieracium pilosella L.
Map 74d; 311 tetrads.
Widespread and frequent throughout the county in well drained situations, particularly railway verges, quarries and dry heathland where it may dominate considerable areas of ground. It is also found in other ruderal situations when these are left undisturbed for any length of time.

Constituent or intrusive: Er, Rq, Rg, Rs, Rw, Rt, Rd, Hg, Hd, Gp, Gr, Ev, Eh, Ef, Ap, Ag, Sw.

First record: Pulteney (1747).
Herb.: LSR*; LANC, LSR, LTR, MANCH, NMW.

HG 338: locally abundant, generally distributed. Horwood cites three records for var. *concinnatum*. The earliest were recorded in B.E.C. Rep. 1917 as var. *nigrescens* by A. E. Wade, and subsequently corrected by Horwood.

The following determinations of subspecies have been made by C. E. A. Andrews of specimens collected during the recent survey and of material in herb. LSR.
Subsp. **officinarum**
Croft, A. Hackett 1957; Groby, quarry, EH 1969; Dimminsdale, wall, HH 1970; Carlton Bridge, railway track, HH 1972.
Subsp. **nigrescens** (Fr.) Sell & West
Asfordby Hill, railway verge, ALP 1972; Swepstone, hedgerow, HH 1975; Heather, gravel pit on spoil heaps, HH 1976.
Subsp. **concinnata** (F. J. Hanb.) Sell & West
Groby, rifle range quarry, EH 1969; Bagworth, pasture, HH 1975; Kirkby Mallory, margin of pool, HH 1976; Dimminsdale, spoil heaps near disused limestone quarry, HH 1976; Market Bosworth, dismantled railway, HH 1977; Kirkby Mallory, churchyard, HH 1977.

P. flagellaris (Willd.) Sell & West subsp. **flagellaris**
Hieracium flagellare Willd.
Map 74e; 1 tetrad.
Cosby, dismantled railway verge, occasional, EH 1977.
    Intrusive or residual: Er.
First record: 1956, A. B. Hackett in herb. LTR from quarry at Stoney Stanton. There is also material from this site in herb. LSR coll. 1957 det. C. E. A. Andrews, and in herb. LTR coll. 1956 det. P. D. Sell, coll. 1958 det. A. O. Chater.
Herb.: LSR*; LSR, LTR.
HG: not recorded.

P. aurantiaca (L.) C. H. & F. W. Schultz
                                        Fox and Cubs
Hieracium aurantiacum L.
Map 74f; 19 tetrads.
Gardens, railway verges, churchyards, roadside verges, grassland and waste ground. Usually few specimens are found together, suggesting casual status; occasionally

more frequent and obviously naturalised and spreading, though there is no real evidence that it is indigenous. Of the 19 tetrad records five are supported by herbarium specimens in herb. LSR★ determined by C. E. A. Andrews as subsp. **brunneocrocea** (Pugsl.) Sell & West. There are several older specimens also determined as this subspecies by P. D. Sell (unless otherwise stated), namely: Leicester Abbey, W. Bell, 1909, in herb. CGE; Swithland, F. L. Foord-Kelcey, undated but pre-1914, in herb. CGE; Husbands Bosworth and Market Harborough, A. E. Ellis, c. 1920, in herb. LANC; Ulverscroft, collector unknown, 1934, in herb. LSR; Ulverscroft, D. Brummitt, 1956, in herb. LIVU (det. C. E. A. Andrews). In *Flora Europaea* it is noted that subsp. **aurantiaca** is native in the Alps and Carpathians while subsp. **carpathicola** Naeg. & Peter (=subsp. *brunneocrocea*) is native throughout the range of the species. All British populations are regarded as naturalised outside the range of either subspecies and although it is likely that they all belong to subsp. **brunneocrocea** this cannot be assumed without confirmation by specialists.

First record: of the species, Preston (1900); of subsp. **brunneocrocea**, 1909, W. Bell in herb. CGE.
Herb.: LSR★; CGE, LANC, LIVU, LSR.
HG 338 as *Hieracium aurantiacum*: rare, restricted in range; 16 localities cited.

MONOCOTYLEDONES

HELOBIAE

ALISMATACEAE
Sagittaria L.
S. sagittifolia L. Arrowhead
Map 74g; 92 tetrads.
Locally frequent throughout the county in rivers and canals. Almost entirely confined to these habitats, so that the distribution map shows the courses of the major waterways in the county.
Constituent: Wr, Wc, Ew.
First record: Pulteney (1747).
Herb.: LSR★; DPM, LANC, LSR, LTR, MANCH, NMW.
HG 567: frequent, generally distributed.

Baldellia Parl.
B. ranunculoides (L.) Parl. Lesser Water-plantain
No record since HG.
First record: Pulteney *in* Nichols (1795).
Herb.: LSR.
HG 566: rare, almost confined to Charnwood Forest and two other districts; 6 localities cited.

Luronium Raf.
L. natans (L.) Raf. Floating Water-plantain
No record in the recent survey.
Kilby Bridge, canal, G. Halliday 1962.
First record: B.E.C. Rep. 1916.
Herb.: LTR.
HG 566 as *Elisma natans*: very rare, restricted in range; 3 localities cited, not including the above.

Alisma L.
A. plantago-aquatica L. Water-plantain
Map 74h; 306 tetrads.
Ponds, lakes, ditches, the shallow margins of rivers, streams and canals, and in gravel pits. Frequent throughout the county in these habitats.
Constituent or residual: Wp, Wl, Wr, Wc, Wd, Ew, M, Rg, Rs.
First record: Pulteney (1747).
Herb.: LSR★; DPM, LANC, LSR, LTR, MANCH, NMW.
HG 565: locally abundant, generally distributed.

A. lanceolatum With. Narrow-leaved Water-plantain
Map 74i; 18 tetrads.
Occasional in the eastern part of the county, being almost confined to the Grand Union Canal, the River Soar, the Eye Brook Reservoir, and streams or culverts by railway embankments at Slawston and Whissendine. The only record from the west of the county is from Staunton Harold Reservoir.
Constituent or intrusive: Wc, Wr, Wl, Wp.
First record: 1883, F. T. Mott in herb. MANCH.
Herb.: LSR★; CGE, LANC, LTR, MANCH, NMW.
HG 565: occasional, restricted in range; 15 localities cited.

Damasonium Miller
D. alisma Miller Starfruit
No record since HG.
First record: Jackson (1903b).
HG 567: rare, very restricted in range; 2 localities cited.

BUTOMACEAE
Butomus L.
B. umbellatus L. Flowering-rush
Map 74j; 55 tetrads.
Locally frequent at the margins of the Grand Union, Ashby and Grantham Canals and Thornton Reservoir. Occasional in rivers and other reservoirs; it would seem that it prefers the more constant water level of the canals. Rarely in ponds.
Constituent: Wc, Wr, Wl, Wd, Wp, Ew.
First record: Pulteney (1747).
Herb.: LSR★; DPM, LANC, LSR, LTR, MANCH, NMW, UPP.
HG 567: frequent and generally distributed.

HYDROCHARITACEAE
Hydrocharis L.
H. morsus-ranae L. Frog-bit
Map 74k; 3 tetrads.
Foxton, backwater at bottom of locks, locally abundant, IME and PHG 1964 (site since developed as marina); near Kilby Bridge, canal, rare, S. Walker 1964; Donisthorpe, subsidence pool where Saltersford Brook crosses the Measham road, locally abundant, SHB 1976.
Constituent or intrusive: Wc, Wp.
First record: Pulteney *in* Nichols (1795).
Herb.: DPM, LSR, LTR, MANCH.
HG 520: rare, very restricted in range; 5 localities cited.

Stratiotes L.
S. aloides L.   Water-soldier
Map 74l; 6 tetrads.
Loughborough, University Campus, old clay pit called Ingle Pingle, locally frequent, PHG 1969; Newton Burgoland, flooded brick pit, locally frequent, ALP 1973 (site since destroyed); Redmile, Grantham Canal, locally frequent, G.J. Dalglish 1984; Barkestone, Grantham Canal, locally abundant, H.J. Mousley 1986; Smeeton Westerby, pond, rare, J.E. Dawson 1986; Groby, plants appeared in newly excavated pool on site of medieval fishpond, R.J. Burrows, 1986.
    Intrusive: Rg.
First record: Bemrose (1927).
Herb.: LSR★; LSR.
HG 520: rare, restricted in range; 2 localities cited, not as above.

Egeria Planch.
E. densa Planch.
No map; 1 tetrad.
Enderby, ornamental pond in old quarry, occasional, certainly introduced, EH 1975.
    Introduced: Wp.
First record: 1975, E. Hesselgreaves in present work.
HG: not recorded.

Elodea Michx.
E. canadensis. Michx.   Canadian Waterweed
Map 75a; 122 tetrads.
Locally frequent or abundant in canals, rivers, ponds, lakes and gravel pits throughout the county.
    Constituent or intrusive: Wc, Wr, Wp, Wl, Wd, Rg.
First record: Kirby (1849).
Herb.: LSR★; DPM, LANC, LSR, LTR, MANCH, NMW.
HG 518: locally abundant, but on the decrease, generally distributed; 68 localities cited.

E. nuttallii (Planch.) St. John
Map 75b; 1 tetrad.
Sheepy Magna, River Sence, locally frequent, M. Palmer 1979. This species has increased enormously in East Anglia in recent years, and it may well have been overlooked in other parts of the county in the recent survey.
    Intrusive: Wr.
First record: 1979, M. Palmer in present work.
HG: not recorded.

JUNCAGINACEAE
Triglochin L.
T. palustris L.   Marsh Arrowgrass
Map 75c; 47 tetrads.
Marshes and the margins of canals and streams where there is shallow water. Occasional throughout the county.
    Constituent: M, Ew, Wc.
First record: Pulteney (1749).
Herb.: LSR★; LSR, LTR, MANCH, NMW.
HG 568: occasional, sparsely distributed; 44 localities cited.

POTAMOGETONACEAE
Potamogeton L.
P. natans L.   Broad-leaved Pondweed
Map 75d; 238 tetrads.
Ponds, lakes, canals, rivers and gravel pits. Frequent throughout the county.
    Constituent or intrusive: Wp, Wl, Wc, Wr, Wd, Rg, Rq, Hb.
First record: Pulteney (1749).
Herb.: LSR★; BM, DPM, LSR, LTR, MANCH, NMW.
HG 568: locally abundant, generally distributed.

P. polygonifolius Pourr.   Bog Pondweed
Map 75e; 4 tetrads.
Beacon Hill, two small ponds, locally frequent, PHG 1969; Charnwood Lodge Nature Reserve, pond, locally frequent, PAC 1969; Appleby Parva, small shallow pond, locally frequent, HH 1974; Bradgate Park, wet heathland and pool between Old John and Hallgate Hill Spinney, locally abundant, SHB 1977.
    Constituent or residual: Wp, Hw.
First record: Kirby (1850).
Herb.: LSR, MANCH, NMW.
HG 569 as *P. oblongus*: rare, restricted to Charnwood Forest and Western region; 13 localities cited, including Beacon Hill.

P. coloratus Hornem.   Fen Pondweed
No record since HG.
First record: Britten *in* White (1877).
HG 569: rare, restricted in range; 2 localities cited.

P. lucens L.   Shining Pondweed
Map 75f; 13 tetrads.
Husbands Bosworth, Grand Union Canal, occasional, EKH and JMH 1969; Shawell, old gravel pits at Cave's Inn, rare, EKH and JMH 1969; Thornton Reservoir, locally abundant, HH 1971; Eye Brook Reservoir, occasional, KGM 1971; Melton Mowbray, old canal, locally abundant, ALP 1971; between Enderby and Glen Parva, River Soar, occasional, EH 1974; Glen Parva, Grand Union Canal, rare, EH 1974; South Wigston, canal, rare, EH 1974; Birstall, disused gravel pit, locally abundant, SHB 1977; River Eye between Swan's Nest and Brentingby, occasional, C. Newbold and M. Palmer 1979; Saddington Reservoir, locally frequent, ALP 1983.
    Constituent or intrusive: Wr, Wc, Wl, Rg.
First record: Bloxam (1837).
Herb.: LSR★; BM, LANC, LSR, LTR, MANCH, NMW.
HG 570: frequent, widely distributed; 34 localities cited.

P. lucens × perfoliatus = P. × salicifolius Wolfg.
Map 75g; 4 tetrads.
Wigston, flooded limestone quarry, G. Halliday 1967; Newton Harcourt, Grand Union Canal, locally frequent, EKH and JMH 1969; Foxton, top lock, H. P. Moon 1970; Leicester, Grand Union Canal, King's Lock, occasional, HB 1970 (canal dredged 1976); Aylestone Mill Lock, Grand Union Canal, locally frequent, HB 1971; Lubenham, Grand Union Canal, locally frequent, EKH and JMH 1971.
    Intrusive: Wc.
First record: 1914, G. Chester in herb. BM and herb. CGE.
Herb.: BM, CGE, LSR, LTR, NMW.
HG 571: 1 locality cited.

P. gramineus L.   Various-leaved Pondweed
Map 75h; 1 tetrad.
Cropston Reservoir, locally abundant, PHG 1969.
    Constituent or intrusive: Wl.
First record: Pulteney *in* Nichols (1795).
Herb.: BM, CGE, LSR, LTR.
HG 570: rare, very restricted in range; 10 localities cited, not including the above.

P. gramineus × lucens = P. × zizii Koch ex Roth
Map 75i; 1 tetrad.
Knipton Reservoir, north end, J. R. I. Wood 1969.
    Intrusive: Wl.
First record: 1891, F. T. Mott in herb. MANCH.
Herb.: BM, MANCH, NMW.
HG 572: rare, restricted in range; 3 localities cited, not including the above.

P. alpinus Balb.   Red Pondweed
No record since HG.
First record: 1843, Bloxam in herb. CGE.
Herb.: CGE, LSR.
HG 570: rare, restricted in range; 7 localities cited.

P. praelongus Wulf.   Long-stalked Pondweed
No record since HG.
First record: Fryer (1898-1915).
HG 572: rare; 2 localities cited.

P. perfoliatus L.   Perfoliate Pondweed
Map 75j; 50 tetrads.
Locally frequent in the Grand Union, Ashby and Grantham Canals. Occasional in rivers and reservoirs. Rarely in ponds.
    Constituent: Wc, Wr, Wl, Wp.
First record: Pulteney (1749).
Herb.: LSR★; BM, DPM, LSR, LTR, MANCH, NMW.
HG 572: locally abundant, generally distributed.

P. friesii Rupr.   Flat-stalked Pondweed
Map 75k; 14 tetrads.
Locally frequent in the Ashby Canal, and occasional in the Grand Union Canal. Apart from these, there are records from a gravel pit at Wanlip (since filled in), a stream at Bradgate Park, and an unconfirmed record from a pond at Enderby.
    Constituent or intrusive: Wc, Wr, Wp, Rg.
First record: Mott et al. (1886).
Herb.: LSR★; BM, LSR, LTR, NMW.
HG 576 as *P. mucronatus*: rare, restricted in range; 12 localities cited, including several from the Ashby and Grand Union Canals.

P. pusillus L.   Lesser Pondweed
Map 75l; 8 tetrads.
Foxton Locks, pound pond, occasional, EKH and JMH 1968; Foxton, canal by Bridge 60, H. P. Moon 1970; Barrow on Soar, Proctor's Pleasure Park, old gravel pits, locally frequent, HH 1973; Sutton Cheney, Ashby Canal by Sutton Wharf Bridge, locally abundant, SHB 1975; Shackerstone, Ashby Canal between Turn Bridge and Bates Wharf Bridge, locally frequent, SHB 1976; Shackerstone, Ashby Canal by Hill's Bridge, locally frequent, SHB 1976; Ulverscroft, pool near Nature Reserve, locally abundant, PHG 1976; Ratcliffe on the Wreake, fish pond, occasional, ALP 1983; Acresford, disused sand quarry, locally abundant, IME 1983.
    Constituent or intrusive: Wc, Wp.
First record: 1892, E. F. Cooper in herb. BM.
Herb.: BM, LTR, NMW.
HG 576. Records for this species almost certainly refer to **P. berchtoldii** (see note under that species).

P. obtusifolius Mert. & Koch
           Blunt-leaved Pondweed
Map 76a; 4 tetrads.
Shawell, old gravel pit at Cave's Inn, locally abundant, EKH and JMH 1969; Husbands Bosworth, gravel pit, locally frequent, EKH and JMH 1971; Appleby Magna, pond, locally abundant, HH 1974; Higham on the Hill, pond near Rowdens Gorse, locally abundant, SHB 1974.
    Intrusive: Wp, Rg.
First record: 1846, Bloxam in herb. CGE.
Herb.: LSR★; BM, CGE, DPM, MANCH, NMW.
HG 575: rare, restricted in range; 2 localities cited, not including the above.

P. berchtoldii Fieb   Small Pondweed
Map 76b; 13 tetrads.
Ponds, reservoirs, gravel pits, rivers and ditches. In scattered localities throughout the county; often locally frequent or abundant where it does occur.
    Constituent or intrusive: Wp, Wl, Wc, Wr, Wd, Rg.
First record: Bloxam (1837).
Herb.: LSR★; CGE.
HG 576 as *P. pusillus*: local and sparsely distributed; 21 localities cited. It seems likely that this species is meant because (a) *P. pusillus* auct. mult. non L. is given as a synonym for **P. berchtoldii** in the standard Floras; (b) Horwood states 'This is one of the commoner grass-leaved Pond Weeds'; (c) he cites Bloxam as the author of the first record, and there is a specimen of **P. berchtoldii** collected by Bloxam in herb. CGE.

P. trichoides Cham. & Schlecht.   Hairlike Pondweed
Map 76c; 5 tetrads.
Groby, quarry pool, rare, EH 1972; Leicester, River Soar, occasional, HH 1976; Grand Union Canal between Debdale wharf and Kilby Bridge, frequent, M. Palmer and D. Wheatley 1984.
    Constituent or intrusive: Wr, Rq.
First record: 1972, E. Hesselgreaves in present work.
HG: not recorded.

P. compressus L.   Grass-wrack Pondweed
Map 76d; 19 tetrads.
Locally frequent in the Grand Union Canal. Occasional in the Ashby Canal and the River Soar in Leicester.
    Constituent: Wc, Wr.
First record: 1835, Bloxam in herb. BM and herb. CGE.
Herb.: LSR★; BM, CGE, LANC, LTR, MANCH, NMW.
HG 575: fairly frequent, widely distributed; 29 localities cited, including the Ashby Canal, the Grand Union Canal and the River Soar in Leicester.

P. acutifolius Link   Sharp-leaved Pondweed
Map 76e; 2 tetrads.
Sutton Cheney, Ashby Canal west of Poplar's Farm, occasional, SHB 1975; Ashby Canal near Sutton Wharf Bridge, rare, SHB 1975; Shackerstone, Ashby Canal near Hill's Bridge, rare, SHB 1976.
    Constituent or intrusive: Wc.
First record: 1975, S. H. Bishop in present work.
HG: not recorded.

P. crispus L.   Curled Pondweed
Map 76f; 112 tetrads.
Locally frequent throughout the county in ponds, lakes, rivers and canals.
   Constituent or intrusive: Wp, Wl, Wc, Wr, Wd, Rg, Rq.
First record: Pulteney (1749).
Herb.: LSR★; BM, CGE, DPM, LANC, LSR, LTR, MANCH, NMW.
HG 573: locally abundant, generally distributed.

P. crispus × perfoliatus = P. × cooperi (Fryer) Fryer
No record since HG.
First record: 1880, F. T. Mott in herb. CGE (Canal, Loughborough; first British record).
Herb.: BM, CGE, LSR, MANCH, NMW.
HG 573: rare, restricted in range; 2 localities cited.

P. crispus × friesii = P. × lintonii Fryer
Map 76g; 1 tetrad.
Leicester, junction of canal and River Soar, past Freemen's Lock, rare, HB 1968.
   Intrusive: Wr.
First record: 1906, W. Bell in herb. BM.
Herb.: BM, CGE.
HG 574: 1 locality cited, not the same as above.

P. pectinatus L.   Fennel Pondweed
Map 76h; 67 tetrads.
Locally abundant throughout the county in rivers and canals. Occasional in ponds, lakes, gravel pits and water-filled brick pits. Often very conspicuous as mats of long waving plants in clear rapidly flowing stretches of river.
   Constituent or intrusive: Wr, Wc, Wp, Wl, Wd, Rg.
First record: Pulteney (1749).
Herb.: LSR★; BM, CGE, DPM, LANC, LSR, LTR, MANCH, NMW.
HG 577: locally abundant, widely distributed; 41 localities cited.

Groenlandia Gay
G. densa (L.) Fourr.   Opposite-leaved Pondweed
Map 76i; 8 tetrads.
Cranoe Lodge, concrete tank, H. P. Moon 1965; Ashby Folville, pond, H. P. Moon 1967; Seagrave, field pond, locally frequent, ALP 1967; Croxton Kerrial, field pond, occasional, D. Maxey and A. Lewis 1969; Wymondham, shallow slow-flowing stream, rare, KGM 1971; Wymeswold, 3 field ponds in part of parish formerly in Willoughby on the Wolds, locally frequent in one, occasional in two, P.M. Wade and J.E. Beresford 1979; Thrussington, field pond, abundant, P.M. Wade and J.E. Beresford 1979; Seagrave field pond, occasional, P.M. Wade and J.E. Beresford 1979; Croxton Kerrial, pond, dominant, H.J. Mousley 1987.
   Constituent or intrusive: Wp, Wr.
First record: Bloxam (1837).
Herb.: LSR★; BM, LSR, LTR, MANCH, NMW.
HG 574 as *Potamogeton densus*: local, sparsely distributed; 22 localities cited.

ZANNICHELLIACEAE
Zannichellia L.
Z. palustris L.   Horned Pondweed
Map 76j; 50 tetrads.
Ponds, rivers, canals and gravel pits. Occasional throughout the county.
   Constituent or intrusive: Wp, Wr, Wc, Wl, Wd, Rg.
First record: Pulteney in Nichols (1795).
Herb.: LSR★; LSR, LTR, MANCH, NMW.
HG 578: frequent, generally distributed.

LILIIFLORAE

LILIACEAE
Tofieldia Hudson
T. pusilla (Michx.) Pers.   Scottish Asphodel
No recent record.
First record: Kirby (1850).
Herb.: LSR.
HG 547 as *T. palustris*: very rare, now extinct, restricted to Moira Reservoir and Wolds. 1 locality cited.

Hemerocallis L.
H. fulva (L.) L.   Day Lily
No map; 1 tetrad.
Ratcliffe College, scrub woodland, occasional, ALP 1983 (established at this site for at least 20 years).
   Garden escape: Fs.
First record: 1983, A.L. Primavesi in present work.
HG: not recorded.

Colchicum L.
C. autumnale L.   Meadow Saffron
No record since HG (except as planted in gardens).
First record: Power (1805).
Herb.: LSR.
HG 546: very rare, restricted in range; 9 localities cited. Horwood states that it was then on the verge of disappearance because of eradication by the agriculturist.

Gagea Salisb.
G. lutea (L.) Ker-Gawler   Yellow Star-of-Bethlehem
Map 76k; 1 tetrad.
Withcote, small spinney near Sauvey Castle, occasional, ALP 1969 (first recorded from this site by F. A. Sowter 1950).
   Constituent or residual: Fd.
First record: Power (1805).
Herb.: LSR★.
HG 546: very rare, confined to north-west Leicestershire; 1 locality cited (Breedon Cloud Wood).

Erythronium L.
E. dens-canis L.   Dog's-tooth-violet
No map; 1 tetrad.
Grace Dieu Wood, rare, PAC 1969.
   Planted or intrusive: Fd.
First record: 1969, P. A. Candlish in present work.
HG: not recorded.

Tulipa L.

T. sylvestris L. Wild Tulip
No record since HG.
First record: Power (1805).
Herb.: BM.
HG 545: very rare, very restricted in range; 3 localities cited.

Fritillaria L.

F. meleagris L. Fritillary
No map; 2 tetrads.
Keythorpe, parkland north of Hall, occasional, IME 1975; Ratcliffe College, deciduous woodland, rare, ALP 1978 (introduced 1934, and persisted until 1979 when site was destroyed by building operations). Reported by W. H. Barrow to have survived until about 1965 in a riverside meadow at Thurmaston; site destroyed by gravel extraction 1968-1973. It is occasionally planted, and may persist for many years, but only the above two records were received.
  Planted: Ap, Fd.
First record: Withering (1801).
Herb.: CGE, LSR, NMW (all pre-1933 specimens of indigenous plants).
HG 545: very rare, very restricted in range; 5 localities cited, including Thurmaston.

Lilium L.

L. martagon L. Martagon Lily
No map; 4 tetrads.
Stanford Park, mixed woodland, rare, EKH and JMH 1971; Saddington, small spinney in grounds of Saddington Hall, occasional, EKH and JMH 1973; Gumley, open woodland, locally frequent, EKH and JMH 1974; Sharnford, Bumblebee Spinney, rare, IME 1974.
  Planted or intrusive: Fd, Fm.
First record: Watson (1837).
Herb.: LSR.
HG 544: rare, restricted in range; 5 localities cited.

Ornithogalum L.

O. umbellatum L. Star-of-Bethlehem
Map 76l; 8 tetrads.
Ratcliffe College, wild parts of grounds, occasional, naturalised for at least 30 years, ALP 1967; Thrussington, grassy verge of Hoby Road, rare, ALP 1968; Stanford Park, The Rookery, mixed plantation, occasional, EKH and JMH 1971; Sutton Cheney, roadside verge at Brook Spinney, rare, SHB 1973; Mountsorrel, Slash Lane, roadside verge, locally frequent, PHG 1974; Quorndon, derelict garden site, rare, PHG 1974; Donisthorpe, rough grass in cemetery, locally frequent, SHB 1976; Coleorton, roadside verge, locally frequent, SHB 1977.
  Intrusive or introduction: Fd, Fm, Ev, Gr, Ap.
First record: Kirby (1850).
Herb.: LSR, NMW.
HG 544: rare, restricted in range; 7 localities cited.

O. nutans L. Drooping Star-of-Bethlehem
No record since HG.
First record: Kirby (1850).
Herb.: LSR, NMW.
HG 544: rare, restricted in range; 6 localities cited.

Hyacinthoides Rothm.

H. non-scripta (L.) Rothm. Bluebell
Endymion non-scriptus (L.) Garcke
Map 77a; 283 tetrads.
Locally frequent in the larger woodlands in the east of the county; frequent and locally abundant in woodland, hedgerows and shady roadside verges in the west. The difference in frequency between east and west is probably because it prefers the lighter soils, and hence is not confined in the west to the larger and older woodlands. One of the few species which can compete with **Pteridium aquilinum**, since they occupy different levels in the soil and the bluebells are over before the bracken forms its cover of deep shade.
  Constituent: Fd, Fm, Fs, Fw.
  Residual or intrusive: Eh, Ev, Ew, Er, Wd, Rw, Rq, Gr, Ap, Hd.
First record: Pulteney (1747).
Herb.: LSR*; DPM, LANC, LSR, LTR, MANCH, NMW, UPP.
HG 543 as *Scilla non-scripta*: locally abundant forming an association, complementary with *Pteris* and *Holcus mollis*, widely distributed; 145 localities cited.

H. hispanica (Mill.) Rothm. Spanish Bluebell
Endymion hispanicus (Mill.) Chouard
No map.
Often planted in gardens, and sometimes becoming thoroughly established in woodland and on verges. Probably occasional and locally frequent throughout the county, but many field workers did not record it.
  Intrusive or planted: Fd, Ev, Er, Gr, Rq, Ap.
First record: 1971, K. G. Messenger in present work.
HG: not recorded.

Muscari Miller

M. neglectum Guss. Grape Hyacinth
M. atlanticum Boiss. & Reut.
No map; 1 tetrad.
Markfield, roadside verge, occasional, HH 1976.
  Casual or intrusive: Ev.
First record: Kirby (1850).
HG 542 as *M. racemosum*: rare, restricted in range; 1 locality cited.

Allium L.

A. paradoxum (Bieb.) G. Don Few-flowered Leek
Map 77b; 2 tetrads.
Knipton, roadside verge, E. M. Pearce 1971; Somerby, roadside verge, locally frequent, SHB 1977.
  Intrusive: Ev.
First record: 1971, E. M. Pearce in present work.
Herb.: LSR*.
HG: not recorded.

A. ursinum L. Ramsons, Wild Garlic
Map 77c; 56 tetrads.
Woodland, preferring wet places and the banks of streams. Locally frequent in Charnwood Forest and the north-west of the county, and in the east Leicestershire woodlands. Occasional elsewhere in the county.
  Constituent or residual: Fw, Fd, Fm, Fs, Ef, Ew, Eh, Ev, Wd, M, Ap, Rq.
First record: Pulteney (1747).
Herb.: LSR*; LANC, LSR, LTR, MANCH, NMW.
HG 541: local, sparsely distributed; 34 localities cited.

A. oleraceum L.   Field Garlic
No record since HG.
First record: Preston (1900).
Herb.: LSR.
HG 541: rare, restricted in range; 2 localities cited.

A. vineale L.   Wild Onion, Crow Garlic
Map 77d; 9 tetrads.
Groby, wall top and base of wall by the Pool, locally frequent, EH 1968; Brooksby, grassy railway embankment, occasional, ALP 1968; Potters Marston, Barrow Hill Quarry, locally frequent, SHB 1972; Groby, rough grassland, rare, EH 1972; Glenfield, garden paths at the Gynsills, occasional, EH 1972; Croft, siliceous grassland, locally frequent, SHB 1973; Croft, Soper's Bridge, grass-topped bridge, locally frequent, SHB 1973; Narborough, waste ground, rare, Brockington Ecological Society 1974; Braunstone, top of covered reservoir, locally frequent, EH 1977; Wanlip, roadside verge, rare, PHG 1977.
    Constituent or intrusive: Hg, Gr, Ev, Er, Sw, Rt, Rw.
First record: Pulteney (1747).
Herb.: LSR★; LSR, LTR, MANCH, NMW.
HG 541: rare, confined chiefly to area W. of R. Soar; 19 localities cited, including Groby, Barrow Hill and Soper's Bridge.

Convallaria L.
C. majalis L.   Lily-of-the-valley
Map 77e; 6 tetrads.
Quorndon, Rowhele Wood, locally frequent, PHG 1968; Groby, woodland near quarry, occasional, EH 1968; Martinshaw Wood, rare, EH 1969 (planted approx. 1903); Oakley Wood, ride, rare, PAC 1969; Croft, mixed woodland, occasional, SHB 1973; Moira, waste ground on demolished housing site, locally frequent, SHB 1976.
    Intrusive or planted: Fd, Fm, Ef, Rw.
First record: Pulteney (1747).
Herb.: LSR, MANCH.
HG 540: rare, confined chiefly to area W. of R. Soar; 15 localities cited. The Quorn and Oakley Wood sites are ancient ones, dating back to Pulteney's time.

Polygonatum Miller
P. multiflorum (L.) All.   Solomon's-seal
Map 77f; 11 tetrads.
Occasional in woodland and hedgerows throughout the county; probably almost always introduced or a garden escape.
    Planted, garden escape, or possibly constituent: Fd, Fm, Fs, Ev, Er, Eh, Rq.
First record: Bloxam and Babington *in* Potter (1842).
Herb.: LSR★; LSR, LTR.
HG 540: rare, restricted in range; 8 localities cited.

Paris L.
P. quadrifolia L.   Herb-Paris
Map 77g; 12 tetrads.
Asplin and Pasture Woods, occasional, PAC 1970; Stockerston, Bolt Wood, occasional, KGM 1971; Great Merrible Wood, locally abundant, KGM 1971; Breedon on the Hill, Cloud Wood, locally frequent, PAC 1971; Glooston Wood, rare, KGM 1974; Little Owston Wood, occasional, PAC 1976; Skeffington Wood, occasional, IME 1976; Loddington Reddish, locally frequent, IME 1976; Tugby Bushes, deciduous woodland, occasional, IME 1976; Launde, Big Wood, occasional, PAE 1978; Tugby Wood, occasional, PAE 1979; Skeffington, Crow Wood, rare, PAE 1979; Skeffington, Brown's Wood, locally frequent, PAE 1979; Tilton Wood, locally frequent, PAE 1979; Skeffington, Priest Hill Wood, occasional, PAE 1979.
    Constituent: Fd, Ef.
First record: Pulteney (1747).
Herb.: LSR★; CGE, LSR, LTR, MANCH, NMW.
HG 547: local, sparsely distributed; 34 localities cited.

Asparagus L.
A. officinalis L.   Asparagus
No map; 17 tetrads.
An occasional escape from cultivation, occurring as a casual or shortly persistent on rubbish dumps or in other waste places, never in any quantity.
    Casual or escape: Rr, Rw, Rq, Er, Ev, Sw, Ap.
First record: Horwood and Gainsborough (1933).
HG 540: rare; 1 locality cited.

Ruscus L.
R. aculeatus L.   Butcher's-broom
No map; 2 tetrads.
Cranoe, hedgerow by church, rare, obviously introduced, KGM 1971; Croft, ornamental mixed woodland, occasional, SHB 1974.
    Planted: Eh, Fm.
First record: Pulteney *in* Nichols (1795).
Herb.: LSR.
HG 539: rare, almost confined in Leicestershire to Charnwood Forest; 3 localities cited.

AMARYLLIDACEAE
Galanthus L.
G. nivalis L.   Snowdrop
Map 77h; 41 tetrads.
Often planted in gardens, shrubberies and woodland, and becoming thoroughly naturalised and established.
    Planted or intrusive: Ap, Fd, Fm, Fs, Fw, Ef, Ew, Eh, Ev, Wd, Gp.
First record: Power (1807).
Herb.: LSR★; LSR, LTR, MANCH, NMW.
HG 538: rare, restricted in range; 16 localities cited.

Narcissus L.
N. pseudonarcissus L.   Daffodil
Map 77i; 6 tetrads.
Cotesbach, churchyard, locally abundant, wild type but presumably introduced, EKH and JMH 1970; Harby Hills, mixed woodland, occasional, wild type, known from here for a very long time, EKH and JMH 1973; Broughton Hill, scrub woodland, locally abundant, a garden cultivar in a site where it seems unlikely that it was deliberately introduced, ALP 1973; Sutton Cheney, Brook Spinney, locally abundant, SHB 1973.
    Residual, possibly constituent, or planted: Fd, Fm, Fs, Ap.
First record: Pulteney (1747).
Herb.: CGE, LSR, MANCH.
HG 538: rare, very restricted in range; 14 localities cited, including Harby Hills.

DIOSCOREACEAE
Tamus L.

T. communis L.   Black Bryony
Map 77j; 507 tetrads.
Hedgerows and open places in woodland. Frequent and locally abundant throughout the county, but apparently absent from the river gravel soils of the Wreake valley and part of the valley of the River Soar.
    Constituent or residual: Eh, Ev, Ew, Er, Ef, Fd, Fs, Fw, Wd, Ap, Aa.
First record: Pulteney (1747).
Herb.: LSR★; CGE, DPM, LANC, LSR, LTR, MANCH, NMW.
HG 539: locally abundant, generally distributed.

IRIDACEAE
Iris L.

I. foetidissima L.   Stinking Iris, Gladdon
Map 77k; 4 tetrads.
Carlton Curlieu, side of path from church to Carlton Hall, rare, EKH and JMH 1974; Foston, rough grass in churchyard, rare, EKH and JMH 1974; Kirkby Mallory, shady roadside verge, locally frequent, HH 1976; Willesley, mixed woodland, rare, SHB 1977.
    Intrusive or planted: Ev, Ap, Fm.
First record: Crabbe *in* Nichols (1795).
Herb.: LSR★; LTR.
HG 536: rare, very restricted in range; 3 localities cited, including Willesley Park.

I. pseudacorus L.   Yellow Iris, Yellow Flag
Map 77l; 166 tetrads.
Shallow margins of rivers, canals, lakes and ponds; wet places in woodland, wet ditches, and occasionally marshes. Locally frequent throughout the county.
    Constituent or intrusive: Wr, Wc, Wp, Wl, Wd, Ew, M, Fw, Fm, Fd, Rq, Rw.
First record: Pulteney (1747).
Herb.: LSR★; DPM, LSR, LTR, MANCH, NMW.
HG 537: frequent and locally abundant, generally distributed.

I. germanica L.
No map; 1 tetrad.
South Wigston, site of abandoned railway station, occasional, EH 1975.
    Intrusive or planted: Er.
First record: 1975, E. Hesselgreaves in present work.
HG: not recorded.

Tritonia Ker-Gawler
T. × crocosmiflora (Lemoine) Nicholson
                           Montbretia
Crocosmia × *crocosmiflora* (Lemoine) N.E.Br.
Map 78a; 13 tetrads.
A garden escape or outcast, sometimes becoming established in waste places and on verges. It is certainly more frequent in these situations than the records show, and it appears that some field workers ignored its presence or overlooked it.
    Garden escape: Rw, Rr, Rs, Ev, Er, Wd.
First record: Horwood and Gainsborough (1933).
HG 538: rare; 1 locality cited.

JUNCALES

JUNCACEAE
Juncus L.

J. maritimus Lam.   Sea Rush
No record since HG.
First record: Horwood and Gainsborough (1933).
HG 552: rare, restricted in range; 1 locality cited (bank of Ashby Canal near Shackerstone).

J. filiformis L.   Thread Rush
Map 78b; 2 tetrads.
Blackbrook Reservoir, stony margins between normal summer high water level and maximum level, locally abundant and apparently increasing, PAC 1971.
    Intrusive: Ew.
First record: Blackbrook Reservoir, 1964, P. A. Candlish in herb. LSR.
Herb.: LSR★; LSR.
HG: not recorded.

J. inflexus L.   Hard Rush
Map 78c; 576 tetrads.
Marshes, poorly drained grassland, roadside and railway verges, margins of ponds, lakes and waterways, and wet places in waste ground, quarries and gravel pits. Frequent and locally abundant throughout the county.
    Constituent, residual or intrusive: M, Hw, Gp, Gr, Gl, Wp, Wl, Wd, Wr, Wc, Ew, Ev, Er, Ef, Fd, Fm, Fw, Rq, Rg, Rs, Rw, Rd, Rt.
First record: Pulteney (1747).
Herb.: LSR★; LSR, LTR, MANCH, NMW.
HG 550: locally abundant, generally distributed.

J. effusus L.   Soft Rush
Map 78d; 561 tetrads.
Marshes, ditches, the margins of ponds and waterways, and other wet places. Frequent throughout the county. This is seldom found with **J. inflexus**, though the distribution of both throughout the county is similar. It would seem that this species prefers wet soils with a high organic content, whereas **J. inflexus** is tolerant of drier conditions on mineral soils.
    Constituent, residual or intrusive: M, Wd, Wp, Wr, Wc, Ew, Ev, Er, Ef, Fw, Fd, Fm, Fs, Rg, Rq, Rs, Rw, Rd, Rt, Gp, Gr.
First record: Pulteney (1747).
Herb.: LSR★; DPM, LSR, LTR, MANCH, NMW.
HG 550: locally abundant, generally distributed.

J. effusus × inflexus = J. × diffusus Hoppe
No record in the recent survey.
Owston Wood, F. A. Sowter 1953.
First record: Bloxam (1848c).
Herb.: CGE, LSR, LTR, MANCH, NMW.
HG 551: occasional, restricted in range; 41 localities cited.

J. conglomeratus L.   Compact Rush
Map 78e; 278 tetrads.
Marshes and wet heath, preferring acid soil. Frequent in the west of the county; locally frequent on suitable soils in the east.
    Constituent, residual or intrusive: M, Hw, Hb, Hg, Wd, Wp, Wl, Wr, Ew, Ev, Er, Ef, Fw, Fd, Fm, Fs, Gr, Gp, Rg, Rq, Rs, Rw, Rd.

First record: Hands et al. *in* Curtis (1831).
Herb.: LSR★; DPM, LSR, LTR.
HG 552: locally abundant, generally distributed.

J. squarrosus L.   Heath Rush
Map 78f; 11 tetrads.
Heathland. Almost confined to Charnwood Forest, where it is locally frequent.
    Constituent: Hg, Hd, M, Fw.
First record: Pulteney *in* Nichols (1795).
Herb.: LSR, LTR, MANCH, NMW.
HG 548: locally abundant, restricted in range; 31 localities cited.

J. compressus Jacq.   Round-fruited Rush
Map 78g; 15 tetrads.
Marshes and the margins of reservoirs and streams. Occasional in scattered localities throughout the county.
    Constituent or intrusive: M, Ew, Wl, Gp, Rq, Rt.
First record: Watson (1832).
Herb.: LSR★; CGE, LANC, LSR, LTR, MANCH, NMW, UPP.
HG 549: rare, restricted in range; 25 localities cited.

J. tenuis Willd.   Slender Rush
Map 78h; 6 tetrads.
Blackbrook Reservoir, margins, PAC and PHG 1964; Barrow on Soar, meadow, occasional, PHG 1965; Coalville, disused claypit, locally frequent, PAC 1976 (site since used as a tip); Eye Brook Reservoir, distinct band about 1 metre below extreme high water level, locally frequent, KGM 1976; Leicester, grounds of Technology Museum, rare, SHB 1977; near Cossington Mill, gravel pits, occasional, SHB 1978; Coalville, dismantled railway sidings, rare, IME 1984.
    Intrusive: Ew, Rg, Rw, Gp.
First record: 1964, P. A. Candlish and P. H. Gamble in herb. LSR.
Herb.: LSR★; LSR.
HG: not recorded.

J. bufonius L.   Toad Rush
Map 78i; 319 tetrads.
Marshes, woodland rides and other wet places. Especially characteristic of damp trackways, because its mucilage-coated seeds adhere to footwear, hooves and wheels. Frequent in the west of the county; locally frequent in the east.
    Constituent or intrusive: M, Ew, Ev, Er, Ea, Ef, Fd, Fm, Rt, Rg, Rq, Rs, Rw, Rd, Wd, Wl, Wp, Hw, Ac, Gp, Gr.
First record: Pulteney (1747).
Herb.: LSR★; DPM, LANC, LSR, NMW, UPP.
HG 548: locally abundant, generally distributed.

J. subnodulosus Schrank   Blunt-flowered Rush
Map 78j; 19 tetrads.
Locally abundant on the margins of the Ashby Canal. Occasional in wet marshes and on canal banks elsewhere in the county, preferring basic soils.
    Constituent: M, Ew, Wc, Wp.
    Residual or intrusive: Gr, Rd.
First record: Watson (1837).
Herb.: LSR★; LSR, LTR, NMW.
HG 553: rather rare, restricted in range; 19 localities cited.

J. bulbosus L.   Bulbous Rush
Map 78k; 26 tetrads.
Almost entirely confined to Charnwood Forest and the north-west of the county, where it is locally frequent in marshes, heathland and woodland.
    Constituent: M, Hw, Hb, Hg, Wp, Wl, Ew, Ef, Fw, Fd, Fm, Fs.
    Residual or intrusive: Rg, Rq, Rt.
First record: Crabbe *in* Nichols (1795).
Herb.: LSR★; LSR, MANCH, NMW.
HG 552: local, restricted in range; 25 localities cited.

J. acutiflorus Ehrh. ex Hoffm.   Sharp-flowered Rush
Map 78l; 116 tetrads.
Marshes and wet heathland. Particularly characteristic of the marshy region around a spring. Locally frequent in the west of the county and at the springline beneath the Marlstone in the east. Occasional elsewhere in the county.
    Constituent: M, Hw, Hb, Ew, Wd, Wp, Wl, Wc, Wr.
    Residual or intrusive: Gr, Gp, Rq, Fm, Fs, Er.
First record: Hands et al. *in* Curtis (1831).
Herb.: LSR★; DPM, LSR, LTR, NMW.
HG 554: frequent, generally distributed.

J. articulatus L.   Jointed Rush
Map 79a; 296 tetrads.
Marshes and other wet places. Frequent and locally abundant throughout the county.
    Constituent: M, Hw, Wd, Wp, Wl, Wr, Wc, Ew, Ef, Fw, Fd, Fm.
    Residual or intrusive: Gp, Gr, Ev, Er, Rg, Rq, Rs, Rt, Rw, Rd.
First record: Pulteney (1749).
Herb.: LSR★; DPM, LSR, LTR, MANCH, NMW.
HG 554: locally abundant, generally distributed.

Luzula DC.
L. campestris (L.) DC.   Field Wood-rush
Map 79b; 421 tetrads.
Frequent throughout the county in old grassland and in heathland.
    Constituent or residual: Gp, Gr, Gl, Hg, Hd, M, Er, Ev, Ew, Ef, Ap, Ag, Rq, Rs, Rw, Rd, Rt.
First record: Hands et al. *in* Curtis (1831).
Herb.: LSR★; DPM, LSR, LTR, MANCH, NMW.
HG 556 as *Juncoides campestris*: locally abundant, generally distributed.

L. multiflora (Retz.) Lejeune   Heath Wood-rush
Map 79c; 34 tetrads.
Heathland, grassland and woodland on acid soils. Locally frequent in Charnwood Forest and the west of the county, and at Six Hills. Rare or absent in the rest of the county.
    Constituent or residual: Hg, Hd, Hw, Hb, M, Gp, Gr, Fc, Fd, Fm, Fs, Fw, Ef, Ew, Er, Ev, Rg, Rs.
First record: Kirby (1850).
Herb.: LSR★; DPM, LSR, LTR, MANCH, NMW.
HG 556 as *Juncoides multiflora*: locally abundant, sparsely distributed; 47 localities cited.

L. sylvatica (Hudson) Gaudin   Great Wood-rush
Map 79d; 20 tetrads.
Locally frequent in woodland on Charnwood Forest.

Occasional in the east Leicestershire woodlands. Absent elsewhere in the county.
  Constituent: Fd, Fm, Fs, Ef.
First record: Pulteney (1749).
Herb.: LSR★; CGE, LANC, LSR, LTR, MANCH, NMW.
HG 555 as *Juncoides sylvatica*: occasional, very sparsely distributed; 25 localities cited.

L. pilosa (L.) Willd.   Hairy Wood-rush
Map 79e; 21 tetrads.
Woodland. Locally frequent in Charnwood Forest and in the east Leicestershire woodlands. Rare or absent elsewhere in the county.
  Constituent: Fd, Fm, Fw, Ef.
First record: Hands et al. *in* Curtis (1831).
Herb.: LSR★; CGE, LSR, LTR, MANCH, NMW.
HG 555 as *Juncoides pilosa*: frequent, sparsely distributed; 25 localities cited.

L. forsteri (Sm.) DC.   Southern Wood-rush
No recent record.
First record: 1791 in Jackson (1904).
HG 554 as *Juncoides forsteri*: rare, restricted in range, not seen in Leicestershire since J. Babington's time; 1 locality cited (Belton Wood).

GRAMINALES

GRAMINEAE
Festuca L.

F. gigantea (L.) Vill.   Giant Fescue
Map 79f; 371 tetrads.
Woodland, hedgerows and shady roadside and railway verges. Locally frequent throughout the county.
  Constituent or residual: Fd, Fm, Fs, Fw, Ef, Eh, Ew, Ev, Er, Ea, Gr, Wd, Ap, Rq, Rw, Rf.
First record: Hands et al. *in* Curtis (1831).
Herb.: LSR★; DPM, LANC, LSR, LTR, MANCH, NMW.
HG 641: locally abundant, generally distributed.

F. pratensis Hudson   Meadow Fescue
Map 79g; 448 tetrads.
Grassland and grassy verges. Sometimes included in grass seed mixtures. Frequent throughout the county.
  Constituent or intrusive: Gp, Gl, Gr, Hg, M, Ev, Er, Ew, Ef, Ea, Ap, Ac, Rs, Rw.
First record: Watson (1832).
Herb.: LSR★; CGE, DPM, LSR, LTR, NMW.
HG 640 as *F. elatior* var. *pratensis*: frequent, generally distributed.

F. arundinacea Schreb.   Tall Fescue
Map 79h; 272 tetrads.
Marshes, the banks of rivers and streams, roadside and railway verges, preferring damp situations. Locally frequent throughout the county.
  Constituent or intrusive: M, Ew, Ev, Er, Eh, Ea, Ef, Fw, Wd, Gp, Gr, Gl, Rw, Rq, Rg, Rs, Rd, Rt, Rr.
First record: Marshall (1790).
Herb.: LSR★; LANC, LSR, LTR, MANCH, NMW, UPP.
HG 640 as *F. elatior*: local, rather sparsely distributed.

F. heterophylla Lam.   Various-leaved Fescue
No map; 1 tetrad.
Enderby, old quarry, rare, EH 1977.
  Casual or intrusive: Rq.
First record: 1977, E. Hesselgreaves in present work.
HG: not recorded.

F. rubra L.   Red Fescue
Map 79i; 542 tetrads.
Often included in mixtures of grass seeds (much of the seed coming from abroad), and hence frequent in lawns and on seeded roadside verges. Frequent and locally abundant in these habitats, and in grassland and grassy verges throughout the county. Occasional on the tops of walls.
  Constituent, intrusive or planted: Ev, Er, Eh, Ea, Ef, Gp, Gr, Gl, Hg, Hd, M, Sw, Sb, Ap, Ag, Ac, Rw, Rq, Rt, Rd.
First record: Marshall (1790).
Herb.: LSR★; CGE, DPM, LANC, LSR, LTR, MANCH, NMW.
HG 639 as *F. ovina* var. *duriuscula*: frequent, generally distributed.
As **F. rubra**: local, sparsely distributed; 40 localities cited.

F. tenuifolia Sibth.   Fine-leaved Sheep's-fescue
Map 79j; 3 tetrads.
Ashby Parva, deep railway cutting, locally frequent, EKH and JMH 1968; Groby, rocky heath grassland, occasional, EH 1970.
  Constituent or intrusive: Hg, Er.
First record: Horwood and Gainsborough (1933).
Herb.: LSR★; LSR, LTR.
HG 639 as *F. ovina* var. *capillata*: 2 localities cited, including Groby.

F. ovina L.   Sheep's-fescue
Map 79k; 123 tetrads.
Heath grassland, railway and roadside verges. Locally frequent throughout the county.
  Constituent or intrusive: Hg, Hd, Gp, Gr, Ev, Er, Ea, Ew, Ef, Sw, Ap, Rw, Rq, Rg, Rs, Rt.
First record: Pulteney *in* Nichols (1795).
Herb.: LSR★; LSR, LTR, MANCH, NMW.
HG 638: locally abundant, generally distributed.

                 × **Festulolium** Aschers. & Graebn.
Festuca pratensis × **Lolium perenne**
 = × **Festulolium loliaceum** (Hudson) P. Fourn.
Map 79l; 34 tetrads.
Occasional in grassland and grassy places throughout the county. Possibly overlooked.
  Intrusive: Gp, Gr, Gl, M, Ev, Rw.
First record: Bloxam (1837).
Herb.: LSR★; CGE, LTR, MANCH.
HG 641: frequent and generally distributed; 28 localities cited.

Lolium L.
L. perenne L.   Perennial Rye-grass
Map 80a; 604 tetrads.
Abundant throughout the county in all types of grassland except heath grassland, where it is occasional. Often included in mixtures of grass seeds, and frequently forms the main constituent of leys; most of the rye grass in

apparently natural habitats is probably derived from this source.
    Constituent, intrusive or planted: Gl, Gp, Gr, M, Hg, Ev, Er, Eh, Ew, Ea, Ef, Ap, Aa, Ac, Rw, Rd, Rt, Rq, Rf.
First record: Pulteney (1747).
Herb.: LSR★; DPM, LANC, LSR, LTR, MANCH, NMW.
HG 647: abundant, generally distributed.

L. multiflorum Lam.   Italian Rye-grass
Map 80b; 384 tetrads.
Often included in mixtures of grass seeds for leys, and frequently escaping on to roadsides and other grassy places. Locally abundant in leys; frequent throughout the county as an established escape, and in arable land.
    Planted or intrusive: Gl, Gp, Gr, Ac, Ao, Ap, Ag, Aa, Ea, Ev, Er, Eh, Rf, Rw, Rd, Rs, Rt.
First record: Kirby (1849).
Herb.: LSR★; LANC, LSR, LTR, MANCH, NMW.
HG 648: frequent, widely distributed; 50 localities cited.

L. temulentum L.   Darnel
No map; 2 tetrads.
Leicester, Birstall Street, demolition site, rare, G. S. Smith 1966; Leicester, Spinney Hill Park, edge of brook, rare, HH 1968.
    Casual: Rw, Ew.
First record: Pulteney (1749).
Herb.: LSR★; LSR, LTR, NMW.
HG 649: rare, restricted in range; 6 localities cited.

Vulpia C. C. Gmelin
V. bromoides (L.) S. F. Gray   Squirrel-tail Fescue
Map 80c; 87 tetrads.
Railway ballast, quarries, waste places, preferring light dry soil. Locally frequent in these habitats throughout the county.
    Intrusive: Er, Ev, Ew, Rq, Rs, Rw, Rt, Rd, Sw, Gp, Gr, Hg, Ag.
First record: Watson (1837).
Herb.: LSR★; LSR, LTR, MANCH, NMW.
HG 638 as *Festuca bromoides*: occasional, generally distributed.

V. myuros (L.) C. C. Gmelin   Rat's-tail Fescue
Map 80d; 75 tetrads.
Locally frequent on railway ballast, especially that of dismantled railways where it is sometimes found in great abundance for a time, though it does not usually persist. Also occasionally in quarries, on spoil heaps, in waste places and in other situations with light dry soil. 65% of the records received are from railway ballast.
    Intrusive: Er, Ev, Rw. Rq, Rg, Rt, Rd, Ap, Ag.
First record: Pulteney *in* Nichols (1795).
Herb.: LSR★; CGE, LANC, LSR LTR, MANCH.
HG 637 as *Festuca myuros*: rare, very restricted in range; 16 localities cited.

Desmazeria Dumort.
D. rigida (L.) Tutin   Fern-grass
Catapodium rigidum (L.) C. E. Hubbard
Map 80e; 47 tetrads.
Occasional throughout the county on walls and railway ballast, and in other places with light dry soil.
    Intrusive: Sw, Sb, Sr, Er, Ev, Rw, Rq, Rg, Rs, Rt, Gp.

First record: Pulteney *in* Nichols (1795).
Herb.: LSR★; LANC, LSR, LTR, MANCH.
HG 637 as *Festuca rigida*: frequent, sparsely distributed; 32 localities cited.

Poa L.
P. annua L.   Annual Meadow-grass
Map 80f; 604 tetrads.
Grassland, arable land, gardens, waste places, roadside and railway verges. Abundant throughout the county. Can be found in flower all the year round.
    Constituent or intrusive: Gp, Gr, Gl, Ao, Ac, Ag, Ap, Aa, Ea, Ev, Er, Ew, Ef, Fs, Rw, Rq, Rs, Rt, Rd, Rf, Rr, Sw.
First record: Pulteney (1749).
Herb.: LSR★; DPM, LSR, LTR, MANCH, NMW.
HG 631: locally abundant, generally distributed.

P. trivialis L.   Rough Meadow-grass
Map 80g; 568 tetrads.
Frequent and locally abundant throughout the county in grassland, grassy verges, hedgerows and open places in woodland. Tolerant of shade. Often subdominant in marshy ground.
    Constituent or intrusive: Gp, Gr, Gl, M, Ev, Er, Ew, Eh, Ea, Ef, Fd, Fm, Fs, Ap, Ac, Ao, Ag, Rw, Rq, Rs, Rt, Rd.
First record: Marshall (1790).
Herb.: LSR★; LSR, LTR, NMW.
HG 634: locally abundant, generally distributed.

P. pratensis L.   Smooth Meadow-grass
Map 80h; 527 tetrads.
Frequent throughout the county in grassland and on grassy verges.
    Constituent or intrusive: Gp, Gl, Gr, M, Ev, Er, Ew, Ef, Ea, Ac, Ao, Ap, Ag, Aa, Sw, Sb, Rw, Rq, Rs, Rt, Rf, Rd, Rr, Fs.
First record: Pitt (1809).
Herb.: LSR★; LANC, LSR, LTR, MANCH, NMW.
HG 633: abundant, sometimes dominant, generally distributed.

P. angustifolia L.   Narrow-leaved Meadow-grass
No map; 2 tetrads.
Kirby Muxloe, railway ballast, rare, EH 1973; Belvoir Castle, stone wall, occasional, KGM 1974.
    Casual or intrusive: Er, Sw.
First record: Preston (1900).
Herb.: LSR, LTR, NMW.
HG 633 as *P. pratensis* var. *angustifolia*: 28 localities cited.

P. compressa L.   Flattened Meadow-grass
Map 80i; 56 tetrads.
Walls, quarries, spoil heaps and other stony or dry places. Occasional throughout the county.
    Intrusive: Sw, Rq, Rs, Rt, Rw, Rd, Ev, Er, Eh, Ew, Ap, Gl.
First record: Bloxam and Babington *in* Potter (1842).
Herb.: LSR★; LSR, LTR, MANCH, NMW.
HG 632: occasional, local, sparsely distributed; 34 localities cited.

P. nemoralis L.   Wood Meadow-grass
Map 80j; 106 tetrads.
Locally frequent in woodland throughout the county.
    Constituent: Fd, Fm, Fs, Fw.
    Residual: Eh, Ev, Ew, Er, Gp, Gr, Ap, Ag, Rq.

First record: Pulteney in Nichols (1795).
Herb.: LSR★; CGE, DPM, LSR, LTR, MANCH, NMW.
HG 631: local, restricted in range; 39 localities cited.

Puccinellia Parl.

P. distans (L.) Parl.   Reflexed Saltmarsh-grass
Map 80k; 3 tetrads.
Aylestone Meadows, marsh, locally frequent, HB 1978; Ashby de la Zouch, waste ground, locally abundant, SHB 1979; Staunton Harold, Scotland, recently disturbed roadside verge, locally frequent, SHB 1979.
    Constituent or intrusive: M, Rw, Rd.
First record: Bloxam (1837).
Herb.: LSR★; BM, LSR.
HG 636 as *Glyceria distans*: rare, restricted in range; 9 localities cited.

Dactylis L.

D. glomerata L.   Cock's-foot
Map 80l; 606 tetrads.
Frequent and locally abundant throughout the county in grassland and on grassy verges.
    Constituent or intrusive: Gp, Gr, Gl, M, Hg, Ev, Er, Ew, Eh, Ef, Ea, Ap, Rw, Rq, Rs, Rt, Rf.
First record: Pulteney (1749).
Herb.: LSR★; DPM, LANC, LSR, LTR, MANCH, NMW.
HG 629: locally abundant, generally distributed.

Cynosurus L.

C. cristatus L.   Crested Dog's-tail
Map 81a; 558 tetrads.
Frequent and locally abundant in pastures and leys; sometimes sown with mixtures of grass seeds. Occasional in other grassy places.
    Constituent, intrusive or planted: Gp, Gr, Gl, M, Ev, Er, Ew, Ef, Ea, Ac, Ap, Ag, Rw, Rd, Rt, Rr.
First record: Walpoole (1782).
Herb.: LSR★; DPM, LANC, LSR, LTR, MANCH, NMW.
HG 626: locally abundant, generally distributed.

C. echinatus L.   Rough Dog's-tail
No map; 2 tetrads.
Bagworth, verge of bridle track, rare, HH 1971; Wigston, bank of flood basin, newly sown grass, J. M. Court 1972.
    Casual: Rt, Rd.
First record: Kirby (1850).
Herb.: LSR★; LSR, LTR, MANCH, NMW.
HG 626: rare, restricted in range; 6 localities cited.

Catabrosa Beauv.

C. aquatica (L.) Beauv.   Whorl-grass
Map 81b; 6 tetrads.
Stanford Reservoir, ditch at north-east end, locally frequent, EKH and JMH 1970; Shearsby, marshy field, occasional, EKH and JMH 1973; Kimcote and Walton, small field pond near Bridgemere Farm, muddy margin, locally frequent, EKH and JMH 1974; Rolleston, stream, rare, IME 1976; Stretton Magna, pond, rare, IME 1977; Misterton, stream, locally frequent, PAE 1983.
    Constituent or residual: Wp, Wr, Wd, M, Ew.
First record: Pitt (1809).
Herb.: LSR★; LSR, LTR, MANCH, NMW.
HG 628: local, generally distributed.

Briza L.

B. media L.   Quaking-grass
Map 81c; 213 tetrads.
Grassland, preferring calcareous soils, and also in marshes. Locally frequent throughout the county. Especially characteristic of grassy railway verges, particularly deep cuttings.
    Constituent or intrusive: Gp, Gr, Gl, M, Hg, Hd, Hw, Er, Ev, Ef, Ea, Rq.
First record: Pulteney (1747).
Herb.: LSR★; DPM, LSR, LTR, MANCH, NMW.
HG 630: locally abundant, generally distributed.

B. maxima L.   Great Quaking-grass
No record since HG.
First record: Mercer (1914).
Herb.: LSR, NMW.
HG 630: rare; 1 locality cited.

Melica L.

M. uniflora Retz.   Wood Melick
Map 81d; 29 tetrads.
Woodland, with a preference for light soils; usually at the edges of woodland or in open places where the shade is not too dense. Locally frequent in the west of the county; occasional in the east Leicestershire woodlands.
    Constituent: Fd, Fm, Ef.
    Residual: Eh, Ew, Wd, Rq.
First record: Pulteney (1752).
Herb.: LSR★; LANC, LSR, LTR, MANCH, NMW.
HG 629: local, very restricted in range; 33 localities cited.

Glyceria R.Br.

G. maxima (Hartm.) Holmberg   Reed Sweet-grass
Map 81e; 212 tetrads.
Frequent and locally abundant on the margins of rivers and canals. Occasional in marshes, and forming a swamp terminating the shallow parts of large ponds and reservoirs.
    Constituent or residual: Wr, Wc, Wd, Wp, Wl, Ew, M, Fw, Fm, Gr.
First record: Pulteney (1749).
Herb.: LSR★; DPM, LANC, LSR, LTR, MANCH, NMW.
HG 636 as *G. aquatica*: locally abundant and generally distributed.

G. declinata Bréb.   Small Sweet-grass
Map 81f; 83 tetrads.
Ponds, ditches, marshes and streams. Locally frequent in the west of the county; occasional in the east.
    Constituent, residual or intrusive: M, Wp, Wr, Wd, Wl, Ew, Ef, Gp, Hg, Rq, Rw, Rt.
First record: Horwood and Gainsborough (1933).
Herb.: LSR★; DPM, LSR, LTR.
HG 635 as *G. plicata* var. *declinata*: 1 record only cited.

G. fluitans (L.) R.Br.   Floating Sweet-grass
Map 81g; 335 tetrads.
Ponds, streams and marshes. Frequent and locally abundant throughout the county.
    Constituent or residual: Wp, Wl, Wr, Wd, Wc, Ew, Ev, Ea, Ef, Fw, Fs, M, Gp, Rt.
First record: Pulteney (1749).
Herb.: LSR★; CGE, DPM, LSR, LTR, MANCH, NMW.
HG 635: locally abundant, generally distributed.

G. fluitans × plicata = G. × pedicellata Townsend
Map 81h; 66 tetrads.
Locally frequent throughout the county at the margins of ponds and streams and in marshes, not always in proximity to the parents. Probably overlooked and more frequent and generally distributed than appears from the records received. It is easier to find in the autumn, when, unlike those of its parents, the infertile spikelets have failed to disarticulate.
 Constituent, residual or intrusive: Wp, Wr, Wd, Wc, Ew, Ef, M, Rg.
First record: 1844, F. T. Mott in herb. MANCH.
Herb.: LSR★; CGE, DPM, LSR, LTR, MANCH.
HG 635: 37 localities cited.

G. plicata Fr. Plicate Sweet-grass
Map 81i; 249 tetrads.
Ponds, ditches, streams and marshes. Frequent and locally abundant throughout the county, though slightly less common than **G. fluitans**.
 Constituent, residual or intrusive: M, Wr, Wd, Wp, Wc, Ew, Er, Ev, Ea, Fw, Rw, Rt, Rg, Ac.
First record: Kirby (1849).
Herb.: LSR★; LANC, LSR, LTR, MANCH.
HG 635: frequent, sparsely distributed.

Bromus L.

B. sterilis L. Barren Brome
Anisantha sterilis (L.) Nevski
Map 81j; 530 tetrads.
Frequent throughout the county on roadside verges, in hedgerows and in waste places. Rarely in grassland. Sometimes a persistent garden weed.
 Intrusive: Ev, Eh, Er, Ew, Ea, Ac, Ag, Aa, Ap, Rw, Rq, Rs, Rf, Rt, Rd, Rr, Sw, Sb, Gp, Gr, Gl.
First record: Pulteney (1747).
Herb.: LSR★; DPM, LANC, LSR, LTR, MANCH, NMW.
HG 643: abundant, generally distributed.

B. diandrus Roth Great Brome
Anisantha diandra (Roth) Tutin
No record in the recent survey.
Leicester, University College 'order beds', garden weed, A. J. Wilmott 1948.
First record: 1948, A. J. Wilmott in herb. BM.
Herb.: BM.
HG: not recorded.

B. tectorum L. Drooping Brome
Anisantha tectorum (L.) Nevski
No map; 1 tetrad.
Glenfield, roadside verge of A50, rare, EH 1973.
 Casual: Ev.
First record: Jackson (1904b).
Herb.: LSR★; LSR, LTR.
HG 643: rare; 1 locality cited.

B. inermis Leyss. Hungarian Brome
Zerna inermis (Leyss.) Lindm.
Map 81k; 13 tetrads.
Occasional in rough grass, mainly on roadsides but also on railway verges. In some cases only of casual occurrence but in others becoming firmly established and persisting for a number of years in competition with the indigenous grass species. Its present pattern of distribution implies that it is on the increase, and spreading from two centres of introduction, one in the south of the county and the other in the north-east.
 Intrusive: Ev, Er, Gr.
First record: Preston (1900).
Herb.: LSR★; LSR, LTR.
HG 645: rare, restricted in range; 3 localities cited.

B. ramosus Hudson Hairy-brome
Zerna ramosa (Hudson) Lindm.
Map 81l; 421 tetrads.
Woodland, hedgerows and shady grass verges. Frequent throughout the county.
 Constituent or residual: Fd, Fm, Fs, Fw, Ef, Eh, Ew, Er, Ev, Ea, Ap, Wd, Gr, M.
First record: Hands et al. *in* Curtis (1831).
Herb.: LSR★; LSR, LTR, MANCH, NMW.
HG 642: locally abundant, generally distributed.

B. erectus Hudson Upright Brome
Zerna erecta (Hudson) S. F. Grey
Map 82a; 35 tetrads.
Occasional in calcareous grassland in the north-east of the county. Elsewhere, most of the records (61% of the total received) are from railway verges, where it may be locally abundant.
 Constituent or intrusive: Er, Ew, Ev, Gr, Gp, Rq.
First record: Pulteney *in* Nichols (1795).
Herb.: LSR★; CGE, DPM, LANC, LSR, LTR, MANCH, NMW.
HG 462: rare, restricted in range; 22 localities cited.

B. secalinus L. Rye Brome
No map; 1 tetrad.
Ashby de la Zouch, newly made road verge of A453, locally frequent, ALP 1973 (did not persist the following year).
 Casual: Rd.
First record: Hands et al. *in* Curtis (1831).
Herb.: LSR★; LSR, LTR, MANCH, NMW.
HG 643: rare, restricted in range; 11 localities cited.

B. commutatus Schrad. Meadow Brome
Map 82b; 2 tetrads.
Hinckley, hay meadow near Higham Thorns, locally abundant, SHB 1974; Loughborough Big Meadow, rare, PHG 1983.
 Intrusive: Gp.
First record: Kirby (1850).
Herb.: LSR, LTR, MANCH.
HG 644 as *B. pratensis*: occasional, restricted in range; 19 localities cited, not including the above.

B. racemosus L. Smooth Brome
Map 82c; 11 tetrads.
Locally frequent in meadows in the valley of the River Soar north and south of Loughborough. Occurs also in scattered localities elsewhere in the county.
 Constituent or intrusive: Gp, M, Rw, Rt.
First record: Watson (1837).
Herb.: CGE, LSR, LTR, MANCH, NMW.
HG 644: frequent, generally distributed.

B. hordeaceus L. subsp. **hordeaceus** Soft-brome
B. mollis L.
Map 82e; 323 tetrads.
This and the next two taxa were treated as critical in the recent survey, and specimens were required to be submitted to a referee. At the same time the field workers were asked to record them as an aggregate without confirmation. The aggregate map is given (Map 82d) as well as maps for the three taxa. **B. hordeaceus** s.s. is locally frequent throughout the county in grassland, on grassy verges and in waste places. It is almost certainly under-recorded in parts of the county.
    Constituent or intrusive: Gl, Gp, Gr, Hg, M, Ev, Er, Ew, Ef, Ea, Ac, Aa, Ap, Ag, Rw, Rq, Rg, Rs, Rt, Rd, Rf, Sw.
First record: Marshall (1790).
Herb.: LSR*; CGE, DPM, LSR, LTR, MANCH, NMW.
HG 645: abundant, generally distributed.

B. hordeaceus × **lepidus** = **B.** × **pseudothominii**
P. M. Smith Lesser Soft-brome
B. thominii auct. non Hardouin
Map 82f; 25 tetrads.
Grassland, grassy verges and waste places. Occasional throughout the county.
    Intrusive: Gp, Gl, Ev, Er, Ew, Ea, Ac, Ag, Rg, Rw, Rt.
First record: 1948, T. G. Tutin in herb. LTR.
Herb.: LSR*; LANC, LSR, LTR.
HG: not recorded.

B. lepidus Holmberg Slender Soft-brome
Map 82g; 10 tetrads.
Rare, in scattered localities throughout the county, often as only a few plants; in some localities locally frequent but not always persistent. The records are mainly from grassland and roadside verges.
    Constituent or intrusive: Gp, Gr, Gl, Ev, Ea, Rt, Ag.
First record: 1945, T. G. Tutin in herb. LTR.
Herb.: LSR*; LTR, UPP.
HG: not recorded.

B. willdenowii Kunth Rescue Brome
Ceratochloa unioloides (Willd.) Beauv.
No map; 1 tetrad.
Sileby, council rubbish dump, only 3 or 4 plants, ALP 1973.
    Casual: Rr.
First record: Preston (1890).
Herb.: LSR*; CGE, LSR, LTR.
HG 645 as *B. unioloides*: rare, restricted in range; 2 localities cited.

Brachypodium Beauv.
B. sylvaticum (Hudson) Beauv. False Brome
Map 82h; 405 tetrads.
Woodland, hedgerows and shady grass verges. Frequent and locally abundant throughout the county.
    Constituent or residual: Fd, Fc, Fm, Fs, Fw, Ef, Eh, Ev, Er, Ew, Ea, Ap, Ag, Wd, Gr, Gp, Sw.
First record: Pulteney (1749).
Herb.: LSR*; LANC, LSR, LTR, MANCH, NMW.
HG 646: locally abundant, generally distributed.

B. pinnatum (L.) Beauv. Tor-grass
Map 82i; 71 tetrads.
Locally frequent in calcareous grassland and on grassy verges in the north-east of the county. Occasional elsewhere, and mainly confined to railway verges.
    Constituent or intrusive: Gr, Gp, Er, Ev, Eh, Ew, Ef, Rq, Rd, Rt, Ag, Ac.
First record: Pulteney *in* Nichols (1795).
Herb.: LSR*; LSR, LTR, MANCH, NMW.
HG 646: rare, rather restricted in range; 50 localities cited.

Elymus L.
E. caninus (L.) L. Bearded Couch
Agropyron caninum (L.) Beauv.
Map 82j; 105 tetrads.
Woodland, hedgerows, shady grass verges and stream banks. Locally frequent in the west of the county; occasional in the east.
    Constituent or residual: Fd, Fm, Fs, Fw, Ef, Ew, Eh, Ev, Ea, Wd, Gr, Rw, Rt, Rd.
First record: Watson (1832).
Herb.: LSR*; CGE, LSR, LTR, MANCH, NMW.
HG 650: local, sparsely distributed; 35 localities cited.

The awnless variant of **E. caninus**, formerly known as *Agropyron donianum*, which occurs very locally in Scotland, was introduced into a garden in Knighton, and has become established as an aggressive weed (Tutin 1973).

E. repens (L.) Gould Common Couch
Agropyron repens (L.) Beauv.
Map 82k; 596 tetrads.
Arable land and gardens, where it can be a troublesome and persistent weed; hedgerows, grass verges and waste places. Abundant throughout the county.
    Intrusive or constituent: Eh, Ev, Ew, Er, Ef, Ea, Ac, Ag, Ao, Ap, Gp, Gr, Gl, Rw, Rq, Rs, Rt, Rf, Rd, Sw, Sb.
First record: Marshall (1790).
Herb.: LSR*; LANC, LSR, LTR, MANCH, NMW.
HG 650: abundant, generally distributed.

Hordeum L.
H. murinum L. Wall Barley
Map 82l; 367 tetrads.
Roadside verges and waste places, locally frequent throughout the county in or near villages and towns. Rarely in grassland, except occasionally in association with **Urtica dioica** on alluvial soil in meadows adjacent to a river.
    Intrusive: Ev, Er, Ew, Ef, Ea, Ag, Ap, Ao, Ac, Aa, Gp, Gr, Gl, Rw, Rq, Rg, Rs, Rt, Rd, Rf, Rr, Sw, Sb.
First record: Pulteney (1747).
Herb.: LSR*; DPM, LANC, LSR, LTR, MANCH, NMW.
HG 652: abundant, generally distributed.

H. secalinum Schreb. Meadow Barley
Map 83a; 348 tetrads.
Grassland, preferring heavy clay soils. Frequent and locally abundant in the east of the county; rare or absent on the lighter soils in the north-west.
    Constituent or intrusive: Gp, Gr, Gl, M, Ev, Er, Ew, Ea, Ap, Ag, Ao, Rw, Rd, Rt, Sw.
First record: Pulteney (1747).
Herb.: LSR*; LANC, LSR, LTR, MANCH, NMW.
HG 652 as *H. nodosum*: locally abundant, generally distributed.

H. jubatum L. Foxtail Barley
Map 83b; 36 tetrads.
Roadside verges, especially newly made verges of major roads which have been sown with imported grass seed. Occasional in such habitats throughout the county. In most cases it is of casual status only, and though it may persist for a few years, it usually succumbs to competition with other species of grasses.
 Casual or intrusive: Ev, Rd, Wd, Gp.
First record: Jackson (1904b).
Herb.: LSR★; LTR, NMW.
HG 652: rare; 5 localities cited.

Avena L.
A. fatua L. Wild-oat
Map 83c; 462 tetrads.
Frequent and locally abundant throughout the county in arable land, especially cereal crops; occasional in waste places and on verges. This species, which seems to be favoured by modern agricultural methods, has increased enormously in recent years; it is now a serious pest, and it is a common sight to see its tall culms standing above the corn.
 Intrusive: Ac, Ao, Ag, Ea, Ev, Er, Rw, Rf, Rr, Rd, Rg, Rs, Gr, Gp, Gl.
First record: Pulteney (1747).
Herb.: LSR★; LSR, LTR, MANCH, NMW.
HG 623: rare, restricted in range; 26 localities cited.

A. ludoviciana Durieu Winter Wild-oat
No map.
Branston, roadside verge, rare, J. G. Dony 1969. A map showing this single record would have little significance, because the species has almost certainly been overlooked elsewhere in the county.
 Intrusive: Ev.
First record: 1969, J. G. Dony in present work.
HG: not recorded.

Avenula Dumort.
A. pubescens (Hudson) Dumort. Downy Oat-grass
Helictotrichon pubescens (Hudson) Pilg.
Map 83d; 67 tetrads.
Grassland and grass verges on basic soils, or light sandy soils which are not strongly acid. Locally frequent in those parts of the county where the soil is suitable.
 Constituent or intrusive: Gp, Gr, Hg, Hd, M, Er, Ev, Ew, Ef, Fs, Rq.
First record: Pulteney in Nichols (1795).
Herb.: LSR★; CGE, DPM, LANC, LSR, LTR, MANCH, NMW.
HG 622 as *Avena pubescens*: local, generally distributed.

A. pratensis (L.) Dumort. Meadow Oat-grass
Helictotrichon pratense (L.) Pilg.
Map 83e; 16 tetrads.
Grassland and grassy railway verges, preferring calcareous soils. Occasional in scattered localities in central and eastern parts of the county, but absent from the west.
 Constituent or intrusive: Gp, Er, Ew, Ea.
First record: Walpoole (1782).
Herb.: LSR★; LSR, LTR, MANCH, NMW.
HG 623 as *Avena pratensis*: rather rare, restricted in range; 18 localities cited.

Arrhenatherum Beauv.
A. elatius (L.) Beauv. ex J. & C. Presl False Oat-grass
Map 83f; 604 tetrads.
Grass verges and rough grassland. Abundant in these habitats throughout the county.
 Constituent or intrusive: Gr, Gp, Gl, Hg, Ev, Er, Ew, Ea, Eh, Ef, Fd, Fc, Fm, Fs, Rw, Rq, Rs, Rd, Ap, Ac.
First record: Watson (1832).
Herb.: LSR★; LANC, LSR, LTR, MANCH, NMW.
HG 624: locally abundant, generally distributed.

Koeleria Pers.
K. macrantha (Ledeb.) Spreng. Crested Hair-grass
K. cristata (L.) Pers.
Map 83g; 17 tetrads.
Occasional throughout the county in calcareous grassland or on light sandy soil which is not strongly acid.
 Constituent or intrusive: Gp, Gr, Hg, Rq, Rg, Rs.
First record: Pulteney in Nichols (1795).
Herb.: LSR★; CGE, LANC, LSR, LTR, MANCH, NMW.
HG 627 as *K. gracilis*: local and sparsely distributed; 34 localities cited.

Trisetum Pers.
T. flavescens (L.) Beauv. Yellow Oat-grass
Map 83h; 403 tetrads.
Frequent throughout the county in grassland and on grassy verges.
 Constituent or intrusive: Gp, Gr, Gl, Ev, Er, Ew, Ea, Ef, M, Ap, Rw, Rq, Rd.
First record: Marshall (1790).
Herb.: LSR★; LSR, LTR, MANCH, NMW.
HG 622: locally abundant, generally distributed.

Deschampsia Beauv.
D. cespitosa (L.) Beauv. Tufted Hair-grass
Map 83i; 589 tetrads.
Marshes, poorly drained or neglected grassland, grass verges and woodland. Frequent and locally abundant throughout the county.
 Constituent or intrusive: M, Gr, Gp, Gl, Hg, Ev, Er, Ew, Ea, Ef, Fw, Fd, Rw, Rq, Rg, Rr, Rd, Rt, Ap.
First record: Marshall (1790).
Herb.: LSR★; DPM, LANC, LSR, LTR, MANCH, NMW.
HG 619: locally abundant, generally distributed.

D. flexuosa (L.) Trin. Wavy Hair-grass
Map 83j; 49 tetrads.
Heath grassland and open woodland on acid soil. Frequent and locally abundant in Charnwood Forest; occasional elsewhere in the west of the county; absent from the east except for a cutting on the dismantled railway near Tilton and for two small clumps at the bases of trees on the drive of Ratcliffe College. This latter was first recorded in 1916 and has persisted ever since, marking the spot where the geological map shows a small strip of Glacial Sand in the Boulder Clay.
 Constituent or residual: Hg, Hd, Gr, Fd, Fc, Fs, Er, Ev, Eh, Rq, Rg, Rs, Rw, Rt, Ag.
First record: Pulteney in Nichols (1795).
Herb.: LSR★; LSR, LTR, MANCH, NMW.
HG 620: locally abundant, restricted to high ground, Charnwood Forest and N.W. and W. regions; 73 localities cited.

Aira L.

A. praecox L. — Early Hair-grass
Map 83k; 30 tetrads.
Heath grassland, rock outcrops and grassland on sandy soil. Locally frequent in Charnwood Forest; occasional elsewhere in the west of the county; rare in the east.
    Constituent or residual: Hg, Hd, Gp, Sr, Sw, Rq, Rs, Rd, Rt, Er, Ap.
First record: Pulteney (1747).
Herb.: LSR★; CGE, LANC, LSR, LTR, MANCH, NMW.
HG 618: local, restricted in range; 45 localities cited.

A. caryophyllea L. — Silver Hair-grass
Map 83l; 62 tetrads.
Locally frequent on railway ballast throughout the county, 79% of the records received being from this habitat. Occasional in heath grassland and other dry sandy places.
    Constituent or intrusive: Er, Ev, Ew, Eh, Hg, Hd, Gp, Rq, Rw, Rt, Ag.
First record: Watson (1832).
Herb.: LSR★; CGE, LSR, MANCH, NMW.
HG 618: very local, restricted in range; 26 localities cited.

Hierochloe R.Br.

H. odorata (L.) Beauv. — Holy-grass
No map; 1 tetrad.
Knighton, garden, introduced but established and aggressive, T. G. Tutin 1973.
    Introduced: Ag.
First record: Tutin (1973).
HG: not recorded.

Anthoxanthum L.

A. odoratum L. — Sweet Vernal-grass
Map 84a; 541 tetrads.
Frequent and locally abundant throughout the county in grassland.
    Constituent or intrusive: Gp, Gr, Gl, Hg, M, Ev, Er, Ew, Ef, Ea, Ap, Ac, Rt, Rw, Sw.
First record: Pulteney (1749).
Herb.: LSR★; DPM, LANC, LSR, LTR, MANCH, NMW.
HG 609: abundant, generally distributed.

Holcus L.

H. lanatus L. — Yorkshire-fog
Map 84b; 596 tetrads.
Frequent and locally abundant throughout the county in grassland, on grassy verges and in woodland.
    Constituent or intrusive: Gr, Gp, Gl, Hg, M, Ev, Ew, Er, Eh, Ea, Ef, Fd, Rw, Rq, Rs, Rd, Rt, Rr, Ap, Ac, Ao, Ag.
First record: Pulteney (1749).
Herb.: LSR★; DPM, LANC, LSR, LTR, MANCH, NMW.
HG 622: abundant, generally distributed.

H. mollis L. — Creeping Soft-grass
Map 84c; 181 tetrads.
Heathland, woodland, grassland and grassy verges, preferring acid soils. Frequent and locally abundant in the west of the county; occasional in the east.
    Constituent or intrusive: Hg, Gp, Gr, Gl, M, Fd, Fm, Fs, Ef, Eh, Ev, Ea, Er, Ew, Rq, Rg, Rs, Rw, Rt, Ap, Ac.
First record: Marshall (1790).
Herb.: LSR★; CGE, LANC, LSR, LTR, MANCH, NMW.
HG 620: frequent, generally distributed; 111 localities cited.

Agrostis L.

A. canina L. — Brown Bent
Map 84d; 20 tetrads.
Grassland, preferring damp and acid soil. Occasional, and possibly under-recorded.
    Constituent or residual: Hw, Hb, Hg, M, Gp, Fd, Ef, Ew, Ev, Rq, Rt.
First record: Bloxam (1837).
Herb.: LSR★; LSR, LTR, MANCH, NMW.
HG 613: local, restricted in range; 43 localities cited.

A. capillaris L. — Common Bent
A. tenuis Sibth.
Map 84e; 388 tetrads.
Grassland, especially old permanent pasture and heath grassland. Frequent throughout the county.
    Constituent or intrusive: Gp, Gr, Gl, Hg, M, Ev, Er, Ew, Ea, Ef, Fd, Fm, Fs, Rw, Rq, Rg, Rs, Rt, Rd, Ap, Ac, Ag.
First record: Marshall (1790).
Herb.: LSR★; CGE, LANC, LSR, LTR, MANCH, NMW.
HG 615: locally abundant, generally distributed.

A. capillaris × stolonifera = A. × murbeckii
Fouillade ex Fournier
Map 84f; 13 tetrads.
This hybrid, found in similar habitats to those of the parents, is probably fairly frequent throughout the county, but only those records which have been determined by T. G. Tutin have been accepted, and these were all submitted from the south-west of the county.
    Intrusive: Ev, Er, Ew, Ea, Ac, Gr.
First record: 1945, T. G. Tutin in herb. LTR.
Herb.: LSR★; LTR.
HG: not recorded.

A. gigantea Roth — Black Bent
Map 84g; 210 tetrads.
Arable land, roadside and railway verges, and waste places. Locally frequent throughout the county.
    Intrusive: Ac, Ao, Aa, Ap, Ag, Ea, Ev, Er, Ew, Eh, Gr, Gl, Rw, Rt, Rg, Rs, Rd, Rf.
First record: W.B.E.C. Rep. 1900.
Herb.: LSR★; BM, LSR, LTR.
HG 615 as *A. tenuis* var. *nigra*: 28 localities cited.

A. stolonifera L. — Creeping Bent
Map 84h; 407 tetrads.
Grassland and grassy verges. Frequent throughout the county.
    Constituent or intrusive: Gp, Gr, Gl, Hg, M, Ev, Er, Ew, Ef, Ea, Ac, Ao, Ap, Ag, Rw, Rq, Rg, Rs, Rt, Rd, Rr, Fs.
First record: Marshall (1790).
Herb.: LSR★; LANC, LSR, LTR, MANCH, NMW.
HG 614: locally abundant, generally distributed.

Apera Adans.

A. spica-venti (L.) Beauv.   Loose Silky-bent
No record since HG.
First record: Jackson (1904b).
Herb.: LSR.
HG 617: rare, restricted in range; 1 locality cited.

A. intermedia Hackel
No record since HG.
First record: Jackson (1904b).
HG 617: rare (first British record); 1 locality cited (Blaby Mill).

Calamagrostis Adans.

C. epigejos (L.) Roth   Wood Small-reed
Map 84i; 56 tetrads.
Open places in woodland, and heath grassland. Prefers damp ground but avoids the heavy clay soils. Locally frequent in the west of the county; occasional in the east.
  Constituent or residual: Hw, Hg, M, Gr, Fw, Fd, Fc, Fm, Fs, Ef, Eh, Ew, Er, Ev, Rq, Rs, Rw, Rt, Ag.
First record: Watson (1832).
Herb.: LSR★; LANC, LSR, LTR, MANCH, NMW.
HG 616: very local (more frequent in East Leicestershire), rather restricted in range; 31 localities cited.

C. canescens (Weber) Roth   Purple Small-reed
No map; 2 tetrads.
Buddon Wood, marshy ditch on western boundary, PHG 1962 (now extinct at this site); Ratby Burroughs, deciduous woodland, occasional, OHB 1970. There is also an unconfirmed record from Breedon Cloud Wood. A. R. Horwood's statement that it was locally abundant seems to be dubious. In herb. LSR there are seven good specimens collected on various dates between 1888 and 1915, four from Tugby Wood and one each from Breedon Cloud Wood, Oakley Wood and Loddington Wood; there is a further specimen from Tugby Wood collected in 1949 and one from Ratby Burroughs (1970) confirming the record given above.
  Constituent: Fd.
First record: Petiver (1716).
Herb.: LSR★; DPM, LSR, LTR, MANCH, NMW.
HG 617: locally abundant, sparsely distributed; 16 localities cited.

Phleum L.

P. pratense L.
Subsp. **pratense**   Timothy
Map 84j; 578 tetrads.
Grassland, grass verges, and as a weed of arable land, especially cereal crops. Frequent and locally abundant throughout the county.
  Constituent or intrusive: Gp, Gr, Gl, Ev, Er, Ew, Ef, Ea, Ac, Ao, Ag, Ap, Aa, Rq, Rw, Rd, Rt.
First record: Pulteney (1749).
Herb.: LSR★; CGE, LSR, LTR, MANCH, NMW.
HG 612: locally abundant, generally distributed.

Subsp. **bertolonii** (DC.) Bornm.   Smaller Cat's-tail
P. bertolonii DC.
Map 84k; 518 tetrads.
Frequent throughout the county in grassland and on grass verges.
  Constituent or intrusive: Gp, Gr, Gl, Ev, Ew, Er, Ef, Ea, Ac, Ap, Rq, Rt, Rw.
First record: Marshall (1790).
Herb.: LSR★; CGE, LSR, LTR, MANCH, NMW.
HG 612 as *P. pratense* var. *nodosum*: frequent, sparsely distributed; 17 localities cited.

Alopecurus L.

A. pratensis L.   Meadow Foxtail
Map 84l; 595 tetrads.
Abundant throughout the county in grassland and on grass verges.
  Constituent or intrusive: Gp, Gr, Gl, Hg, M, Ev, Er, Ew, Ef, Eh, Ea, Ac, Aa, Ag, Ap, Rw, Rq, Rt.
First record: Pulteney (1749).
Herb.: LSR★; DPM, LANC, LSR, LTR, MANCH, NMW.
HG 611: locally abundant, sometimes dominant, generally distributed.

A. geniculatus L.   Marsh Foxtail
Map 85a; 407 tetrads.
Marshes, margins of ponds, wet muddy gateways and damp places in arable land and grassland. Frequent throughout the county.
  Constituent or intrusive: M, Wp, Wl, Wd, Ew, Ev, Ef, Er, Ea, Ac, Ao, Gp, Gr, Gl, Rt, Rq, Rw.
First record: Pulteney (1749).
Herb.: LSR★; LANC, LSR, LTR, MANCH, NMW.
HG 610: locally abundant, generally distributed.

A. geniculatus × pratensis = A. × hybridus Wimm.
No record since HG.
First record: Jackson (1903a).
Herb.: NMW.
HG 611: 3 localities cited.

A. aequalis Sobol.   Orange Foxtail
Map 85b; 17 tetrads.
Locally frequent on the margins of reservoirs. Occasional elsewhere in the county on the margins of canals and rivers, and in marshes.
  Constituent: Wl, Wp, Wd, Ew, M.
First record: Coleman (1852).
Herb.: LSR★; DPM, LSR, LTR, MANCH, NMW.
HG 610: rare, restricted in range; 8 localities cited, all but one being reservoir margins.

A. myosuroides Hudson   Black-grass, Black Twitch
Map 85c; 108 tetrads.
Locally frequent throughout the county as a weed of arable land, where it can be a serious pest.
  Intrusive: Ac, Ao, Ag, Ap, Aa, Ea, Ev, Er, Gl, Gp, Rd, Rf, Rw, Rr, Rt.
First record: Pulteney (1749).
Herb.: LSR★; CGE, LSR, LTR, MANCH, NMW.
HG 610: frequent, widely distributed.

Parapholis C. E. Hubbard

P. strigosa (Dumort.) C. E. Hubbard   Hard-grass
No map; 1 tetrad.
Ashby Old Parks, waste ground, rare, SHB 1977.
  Casual: Rw.
First record: 1977, S. H. Bishop in present work.
HG: not recorded.

Phalaris L.

P. arundinacea L.   Reed Canary-grass
Map 85d; 491 tetrads.
Locally frequent throughout the county on the margins of ponds, reservoirs, rivers, canals and streams, and in wet ditches. Occasional in wet woodland.
　　Constituent or intrusive: Wp, Wl, Wr, Wc, Wd, Ew, Ev, Er, Eh, Ef, Fw, Fd, Fs, M, Rq, Rg, Rw, Rd.
First record: Pitt (1809).
Herb.: LSR★; DPM, LANC, LSR, LTR, MANCH, NMW.
HG 609: locally abundant, generally distributed.

P. canariensis L.   Canary-grass
Map 85e; 52 tetrads.
Occasional throughout the county on rubbish dumps, roadside verges and waste places. Usually a non-persistent casual derived from discarded bird seed, but sometimes, especially on rubbish dumps, recurring from year to year because of repeated reintroduction.
　　Casual or intrusive: Rr, Rw, Rs, Rd, Rt, Ev, Ew, Er, Eh, Ao, Ap, Gr, Hg.
First record: Bloxam (1837).
Herb.: LSR★; CGE, DPM, LANC, LSR, LTR, MANCH, NMW.
HG 608: rare, restricted in range; 26 localities cited.

P. paradoxa L.
No record since HG.
First record: Jackson (1904b).
Herb.: LSR, NMW.
HG 609: rare; 1 locality cited.

Milium L.

M. effusum L.   Wood Millet
Map 85f; 90 tetrads.
Woodland. Locally frequent in the west of the county and in the east Leicestershire woodlands. Occasional elsewhere.
　　Constituent: Fd, Fc, Fm, Fs, Fw, Ef.
　　Residual: Wd, Eh, Hg.
First record: Pulteney in Nichols (1795).
Herb.: LSR★; LSR, LTR, MANCH, NMW, UPP.
HG 611: frequent, widely distributed; 47 localities cited.

Phragmites Adans.

P. australis (Cav.) Trin. ex Stend.   Common Reed
P. communis Trin.
Map 85g; 102 tetrads.
Locally frequent throughout the county at the margins of rivers, canals and ponds, and occasionally in marshes. However, in spite of its general distribution throughout the county, and its local abundance in many places where it does occur, the reed cannot be said to be a common Leicestershire species, and its sites usually have to be looked for rather than providing a characteristic feature of the landscape.
　　Constituent, residual or intrusive: Wr, Wc, Wl, Wp, Wd, Ew, Er, Ev, Ea, Ef, Fw, Fs, M, Rw, Rq, Rg, Rd.
First record: Pulteney (1747).
Herb.: LSR★; CGE, DPM, LANC, LSR, LTR, NMW.
HG 626: locally abundant, generally distributed.

Danthonia DC.

D. decumbens (L.) DC.   Heath-grass
Sieglingia decumbens (L.) Bernh.
Map 85h; 36 tetrads.
Heath grassland. Locally frequent in Charnwood Forest. Occasional elsewhere in the west of the county, and at Twenty Acre and Hallaton Castle.
　　Constituent or residual: Hg, Hd, Gp, Gr, Ef, Ap, Rq.
First record: Pulteney in Nichols (1795).
Herb.: LSR★; LANC, LSR, LTR, MANCH, NMW.
HG 625: locally abundant, generally distributed; 72 localities cited.

Molinia Schrank

M. caerulea (L.) Moench   Purple Moor-grass
Map 85i; 21 tetrads.
Grassland on wet peaty soil. Locally frequent in Charnwood Forest; occasional elsewhere in the west. In the eastern part of the county it occurs only at Twenty Acre and Saltby Bog.
　　Constituent: Hw, Hb, Hg, M, Gr, Ef, Fd, Fs.
First record: Pulteney in Nichols (1795).
Herb.: LSR★; DPM, LSR, LTR, MANCH, NMW.
- HG 627: occasional, sparsely distributed; 48 localities cited.

Nardus L.

N. stricta L.   Mat-grass
Map 85j; 28 tetrads.
Heath grassland. Almost confined to Charnwood Forest and the north-west of the county, except for Twenty Acre, Burbage Common, and a single anomalous record from a roadside verge near Hinckley.
　　Constituent or residual: Hg, Hd, M, Gp, Gr, Ap, Ag, Ev.
First record: Pulteney (1756).
Herb.: LSR★; DPM, LSR, LTR, MANCH, NMW.
HG 651: locally abundant, sparsely distributed; 59 localities cited.

Panicum L.

P. mileaceum L.
No map; 5 tetrads.
Sileby, council rubbish dump, locally abundant, ALP 1968; Beaumont Leys, rubbish dump, locally frequent, EH 1969 (site since built over); Barwell, rubbish dump, occasional, SHB 1974; Whetstone, refuse disposal plant, rare, EH 1975; Oadby, rubbish dump, occasional, IME 1976.
　　Casual: Rr.
First record: B.E.C. Rep. 1909.
Herb.: LSR★; LTR.
HG 607: rare; 2 localities cited.

Echinochloa Beauv.

E. crus-galli (L.) Beauv.   Cockspur
No map; 2 tetrads.
Beaumont Leys, rubbish tip, rare, EH 1970; Sileby, rubbish tip, rare, ALP 1983.
　　Casual: Rr.
First record: Horwood and Gainsborough (1933).
Herb.: LSR★; LCR, LTR.
HG 607 (as *Panicum crus-galli*): rare; 1 locality cited.

E. utilis Ohwi & Yabuno
No map; 3 tetrads.
Beaumont Leys, rubbish tip, rare, EH 1970; Whetstone, sewage works, rare, EH 1974; Sileby, rubbish tip, rare, ALP 1984.
　　Casual: Rr.
First record: 1970, E. Hesselgreaves in present work.
Herb.: LSR★; LCR.
HG: not recorded.

Digitaria Haller
D. sanguinalis (L.) Scop.   Crab Grass
No map; 1 tetrad.
Hinckley, Upper Bond Street, occasional, D. E. Jebbett 1982.
    Casual: Ev.
First record: Horwood (1911).
Herb.: LSR★.
HG 607 (as *Panicum sanguinale*): rare; 1 locality cited.

Setaria Beauv.
S. pumila (Poiret) Schultes
S. lutescens (Weigel) Hubbard
No map; 1 tetrad.
Thurnby, garden, A. J. Rose 1971.
    Casual: Ag.
First record: B.E.C. Rep. 1917.
Herb.: LSR★; NMW.
HG 608 as *S. glauca*: rare; 1 locality cited.

S. verticillata (L.) Beauv.   Rough Bristle-grass
No map; 1 tetrad.
Stoughton, garden, rare, M. G. Sowter 1975.
    Casual: Ag.
First record: 1975, M. G. Sowter in present work.
Herb.: LSR★.
HG: not recorded.

S. viridis (L.) Beauv.   Green Bristle-grass
No map.
Mountsorrel, Broad Hill, rare, F. A. Sowter 1970; Barwell, rubbish dump, occasional, J. E. Dawson and EH 1979. Some other records for this species were lost as a result of the fire at Ratcliffe College.
    Casual: Rw, Rr.
First record: Mott et al. (1886).
Herb.: LSR★; LSR, MANCH, NMW.
HG 608: rare, restricted in range; 9 localities cited.

S. italica (L.) Beauv.   Millet
No map; 4 tetrads.
Beaumont Leys, rubbish dump, locally frequent, EH 1969 (site since built over); Potters Marston, rubbish dump, rare, SHB 1973; Oadby, rubbish dump, occasional, IME 1976; East Norton, rubbish dump, rare, IME 1976.
    Casual: Rr.
First record: Jackson (1904b).
Herb.: LSR★; LTR.
HG 608: rare; 1 locality cited.

SPATHIFLORAE

ARACEAE
Acorus L.
A. calamus L.   Sweet-flag
Map 85k; 49 tetrads.
Frequent and locally abundant along the margins of the Grand Union Canal and the River Soar. Elsewhere in the county it occurs in a few scattered localities in ponds and marshes, and at one place on the Ashby Canal.
    Constituent or intrusive: Wc, Wr, Wd, Wp, Ew, M, Fw.
First record: Deering (1738).
Herb.: LSR★; LANC, LSR, LTR, MANCH, NMW.
HG 562: locally abundant; 28 localities cited.

Arum L.
A. maculatum L.   Lords-and-ladies
Map 85l; 391 tetrads.
Frequent throughout the county in woodlands and hedgerows, with a preference for basic soils.
    Constituent: Fd, Fm, Fs, Fw.
    Residual or intrusive: Eh, Ev, Er, Ew, Wd, Ap, Ag, Gp, Gr, Gl.
First record: Pulteney (1747).
Herb.: LSR★; LSR, LTR, MANCH, NMW.
HG 561: locally abundant, generally distributed; 141 localities cited.

LEMNACEAE
Lemna L.
L. trisulca L.   Ivy-leaved Duckweed
Map 86a; 108 tetrads.
Frequent throughout the county in ponds, lakes and canals; probably often overlooked because of its submerged habit of growth. Particularly abundant in the Ashby Canal and parts of the Grantham Canal.
    Constituent or intrusive: Wc, Wp, Wl, Wr, Wd, Rg.
First record: Pulteney (1749).
Herb.: LSR★; DPM, LANC, LSR, LTR, MANCH, NMW.
HG 563: frequent, widely distributed; 51 localities cited.

L. gibba L.   Fat Duckweed
Map 86b; 35 tetrads.
Locally frequent throughout the county in canals and slow-moving rivers. Occasional in ponds and ditches.
    Constituent or intrusive: Wc, Wr, Wp, Wd.
First record: Watson (1837).
Herb.: CGE, LANC, LSR, LTR, MANCH, NMW.
HG 564: fairly frequent, sparsely distributed; 21 localities cited.

L. minor L.   Common Duckweed
Map 86c; 405 tetrads.
Frequent and locally abundant throughout the county on the surface of ponds, lakes, canals and slow-moving rivers; occasional in wet places in marshes.
    Constituent or intrusive: Wp, Wl, Wc, Wr, Wd, M, Rq, Rg.
First record: Pulteney (1747).
Herb.: LSR★; DPM, LANC, LSR, LTR, MANCH.
HG 564: locally abundant, generally distributed.

Spirodela Schleiden
S. polyrhiza (L.) Schleiden   Greater Duckweed
Lemna polyrhiza L.
Map 86d; 7 tetrads.
Cossington, Church Pond, locally frequent, ALP 1967; Rotherby, large ox-bow pond near site of Brooksby station, locally frequent, ALP 1967 (also present were the three species of **Lemna**); Sileby, large field pond, locally abundant, PHG 1968; Barrow on Soar, Proctor's Pleasure Park, old gravel workings, locally frequent, PHG 1972; Hemington, pools between Sawley Cut and River Trent, locally frequent, PAC 1976; Lockington, pool south-west of confluence of Rivers Soar and Trent, rare, PAC 1976; Castle Donington, Kings Mills, River Trent, PHG 1976. Several other records, mostly from the region around Loughborough, were lost as a result of the fire at Ratcliffe College.
    Constituent or intrusive: Wp, Wr, Rg.

First record: Pulteney (1747).
Herb.: LSR*; LSR, LTR, MANCH, NMW.
HG 565: frequent, sparsely distributed; 20 localities cited.

PANDANALES

SPARGANIACEAE
Sparganium L.

S. erectum L. Branched Bur-reed
Map 86e; 409 tetrads.
Frequent throughout the county along the shallow margins of rivers, streams, canals, ponds, reservoirs and wet ditches. Occasional in wet marshes.
Constituent or intrusive: Wr, Wc, Wp, Wl, Wd, Ew, M, Rg.
First record: Pulteney (1747).
Herb.: LSR*; DPM, LANC, LSR, LTR, MANCH, NMW.
HG 559 as *S. ramosum*: frequent, generally distributed.

S. emersum Rehm. Unbranched Bur-reed
Map 86f; 49 tetrads.
Rivers and canals, and occasionally ponds. This species often grows totally submerged. In this condition it does not flower and it is very difficult to identify. Hence it is probably more common than would appear from the records received.
Constituent or intrusive: Wr, Wc, Wp, Wd, Rg.
First record: Pulteney in Nichols (1795).
Herb.: LSR*; LANC, LSR, LTR, MANCH, NMW.
HG 560 as *S. simplex*: frequent, sparsely distributed.

TYPHACEAE
Typha L.

T. angustifolia L. Lesser Bulrush, Lesser Reedmace
Map 86g; 26 tetrads.
Locally frequent on the margins of the Grand Union Canal. Occasional elsewhere in the county in ponds, lakes, reservoirs and gravel pits.
Constituent or intrusive: Wc, Wp, Wl, Wd, Ew, Rg, Rq.
First record: Pulteney in Nichols (1795).
Herb.: CGE, LSR, LTR, MANCH, NMW.
HG 558: occasional, sparsely distributed; 23 localities cited.

T. latifolia L. Bulrush, Reedmace
Map 86h; 238 tetrads.
Frequent throughout the county in ponds, lakes, canals, wet ditches, marshes and flooded parts of quarries and gravel pits. Occasional on the margins of rivers and streams.
Constituent or intrusive: Wp, Wl, Wc, Wd, Wr, Ew, M, Hw, Fw, Rq, Rg, Rw.
First record: Pulteney (1747).
Herb.: LSR*; LANC, LSR, LTR, NMW.
HG 558: frequent, generally distributed.

CYPERALES

CYPERACEAE
Scirpus L.

S. sylvaticus L. Wood Club-rush
Map 86i; 20 tetrads.
Marshes, wet woodland and the banks of streams in scattered localities throughout the county. Often in considerable abundance where it does occur.
Constituent or residual: M, Fw, Ew, Wr, Wl, Wd.
First record: Pulteney (1756).
Herb.: LSR*; LANC, LSR, LTR, MANCH, NMW.
HG 582: occasional, widely distributed; 52 localities cited.

S. lacustris L.
Subsp. **lacustris** Common Club-rush, Bulrush
Schoenoplectus lacustris (L.) Palla
Map 86j; 94 tetrads.
Frequent throughout the county in rivers and the larger brooks; occasional in canals, ponds, reservoirs and marshes.
Constituent or intrusive: Wr, Wl, Wc, Wp, Wd, Ew, M, Rg, Rq.
First record: Pulteney (1747).
Herb.: LSR*; DPM, LSR, LTR, MANCH, NMW.
HG 581: locally abundant, occasionally subdominant – though not as completely so as in the East Anglian Broads, generally distributed through the river systems.

Subsp. **tabernaemontanae** (C. C. Gmelin) Syme Grey Club-rush
Schoenoplectus tabernaemontanae (C. C. Gmelin) Palla
Map 86k; 2 tetrads.
Ashby Woulds, Barratt Pool, locally frequent, SHB 1976; Lockington, pond, occasional, PAE 1983.
Intrusive: Wp.
First record: (if correct) Horwood and Gainsborough (1933).
Herb.: LSR*.
HG 582: rare, restricted in range; 1 locality cited (a somewhat doubtful record from Frisby on the Wreake).

S. setaceus L. Bristle Club-rush
Isolepis setacea (L.) R.Br.
Map 86l; 31 tetrads.
Marshes and the margins of rivers and reservoirs. Locally frequent in Charnwood Forest and the west of the county; occasional on the Marlstone in the east; rare or absent elsewhere.
Constituent or residual: M, Ew, Wr, Wl, Wd, Gp, Ap.
First record: Pulteney in Nichols (1795).
Herb.: LSR*; LSR, LTR, MANCH, NMW.
HG 581: occasional, very restricted in range; 33 localities cited.

S. fluitans L. Floating Club-rush
Eleogiton fluitans (L.) Link
No map; 1 tetrad.
Blackbrook Reservoir, on mud in draw-down zone, occasional, C. Newbold and D. Wheatley 1984.
Constituent: Wl.
First record: Pulteney in Nichols (1795).
HG 580: rare, practically confined to Charnwood Forest, now probably extinct; 4 localities cited.

S. cespitosus L. Deer-grass
Trichophorum cespitosum (L.) Hartm.
No recent record.
First record: Pulteney in Nichols (1795).
Herb.: LSR.
HG 580: rare, almost confined to Charnwood Forest, perhaps extinct; 2 localities cited.

Blysmus Panz.
B. compressus (L.) Panz. ex Link　　　Flat-sedge
No record in the recent survey.
Between Peckleton and Desford, marsh, D. P. Murray 1953; Glooston Lodge, bog, M. K. Hanson 1958.
First record: Horwood (1909).
Herb.: CGE, DPM, LSR, LTR, MANCH.
HG 583: rare, restricted in range; 2 localities cited, not the same as above.

Eriophorum L.
E. angustifolium Honck.　　　Common Cottongrass
Map 87a; 8 tetrads.
Wanlip, gravel pits, EKH 1961; Glooston, bog, EKH 1964; Potters Marston, marsh, rare, IME and PHG 1966 (site since destroyed); Ulverscroft, Herbert's Meadow, locally frequent, PHG 1967; Mountsorrel, quarry, locally abundant, PHG 1968; Botcheston Bog, marsh, rare, HH 1969; Shawell, disused gravel pits, rare, IME 1970; Great Bowden pit, occasional, KGM 1971; Charnwood Lodge Nature Reserve, wet heath, locally frequent, PAC 1971; Moira, bog, locally frequent, SHB 1976.
　Constituent, residual or intrusive: Hb, Hw, M, Rg, Rq.
First record: Pulteney (1747).
Herb.: LSR★; CGE, LSR, LTR, MANCH, NMW.
HG 584: rare, decreasing, restricted in range; 20 localities cited, including Botcheston Bog and Moira.

E. latifolium Hoppe　　　Broad-leaved Cottongrass
No record since HG.
First record: 1833, Bloxam in herb. CGE.
Herb.: CGE, LSR, MANCH.
HG 584: rare, restricted in range; 3 localities cited.

E. vaginatum L.　　　Hare's-tail Cottongrass
No record since HG.
First record: Coleman (1852) if correct, or 1883, F. T. Mott in herb. MANCH.
Herb.: MANCH.
HG 583: very rare, restricted in range; 1 locality cited.

Eleocharis R.Br.
E. quinqueflora (F. X. Hartmann) Schwarz
　　　Few-flowered Spike-rush
No record since HG.
First record: Kirby (1850).
Herb.: LSR.
HG 580 as *Scirpus pauciflorus*: rare, restricted in range; 3 localities cited.

E. acicularis (L.) Roem. & Schult.　Needle Spike-rush
Map 87b; 11 tetrads.
Locally frequent at the margins of Swithland and Cropston Reservoirs, the Ashby Canal and the Grand Union Canal near Saddington; also recorded from the River Soar at Barrow when the river was low, and in disused gravel pits near Frisby on the Wreake when the water level had fallen after a dry summer. Not recorded elsewhere in the county, but as this species often grows submerged and is then difficult to detect or identify, it may well have been overlooked in other localities.
　Constituent: Wl, Wc, Wr, Ew, Rg.

First record: Watson (1835).
Herb.: LSR★; CGE, LSR, LTR, MANCH, NMW.
HG 579: occasional, restricted in range; 13 localities cited.

E. palustris (L.) Roem. & Schult. Common Spike-rush
Map 87c; 179 tetrads.
Marshes, and in the shallow water of ponds and other water bodies. Locally frequent throughout the county.
　Constituent, residual and intrusive: M, Wp, Wl, Wc, Wr, Wd, Ew, Gp, Gr, Rq.
First record: Pulteney (1747).
Herb.: LSR★; CGE, DPM, LANC, LSR, LTR, MANCH, NMW.
HG 579: frequent, generally distributed.

E. multicaulis (Sm.) Sm.　Many-stalked Spike-rush
No record since HG.
First record: Kirby (1850).
HG 580: very rare, restricted in range; 3 localities cited.

Rhynchospora Vahl
R. alba (L.) Vahl　　　White Beak-sedge
No recent record.
First record: Pulteney *in* Nichols (1795).
HG 585: rare, confined to Charnwood Forest; 1 locality cited (Beacon Hill). Presumed extinct by Mott et al. (1886).

Schoenus L.
S. nigricans L.　　　Black Bog-rush
Map 87d; 1 tetrad.
Saltby Bog, J. W. Birch 1970.
　Constituent: M.
First record: Pulteney *in* Nichols (1795).
Herb.: LSR, LTR.
HG 585: rare, though considered extinct (*Fl. Leics.* 1886) rediscovered in E. Leicestershire, restricted in range; 2 localities cited, including Saltby Bog.

Carex L.
C. paniculata L.　　　Greater Tussock-sedge
Map 87e; 38 tetrads.
Locally frequent on the margins of the Grand Union Canal, and in marshes and wet woodland on the springline below the Marlstone in the east of the county. Occasional on the margins of the Ashby Canal, and in marshes and wet woodland elsewhere in the county.
　Constituent or residual: M, Fw, Fd, Fm, Fs, Ew, Wc, Wd.
First record: Pulteney *in* Nichols (1795).
Herb.: LSR★; BM, CGE, DPM, LANC, LSR, LTR, MANCH, NMW.
HG 588: rare, restricted in range; 21 localities cited.

C. diandra Schrank　　　Lesser Tussock-sedge
No record since HG
First record: 1883, F. T. Mott in herb. MANCH.
Herb.: BM, LSR, MANCH, NMW.
HG 587: rare, restricted in range; 1 locality cited.

C. otrubae Podp.　　　False Fox-sedge
Map 87f; 260 tetrads.
Margins of ponds, rivers and canals, marshes and damp woodland rides. Frequent throughout the county.
　Constituent, residual or intrusive: M, Wp, Wl, Wd, Wc, Wr, Ew, Ev, Er, Ef, Fw, Fd, Gp, Gr, Rg, Rw, Rd, Rt.

First record: Pulteney *in* Nichols (1795).
Herb.: LSR★; BM, CGE, DPM, LANC, LSR, LTR, MANCH, NMW.
HG 589 as *C. vulpina*: frequent, generally distributed.

C. otrubae × remota = C. × pseudoaxillaris K. Richt.
No record in the recent survey.
Owston Wood, A. Hackett 1957.
First record: 1957, A. Hackett in herb. LTR.
Herb.: LTR.
HG: not recorded.

C. spicata Hudson   Spiked Sedge
Map 87g; 108 tetrads.
Locally frequent throughout the county on roadside and railway verges and in other rough grassy places.
    Constituent or intrusive: Gp, Gr, M, Ev, Er, Eh, Ea, Ef, Ew, Fd, Wd, Rq, Rw, Rt, Ap.
First record: Pulteney *in* Nichols (1795).
Herb.: LSR★; BM, LANC, LTR, NMW.
HG 590 as *C. muricata* frequent, generally distributed.

C. divulsa Stokes   Grey Sedge
No map.
Castle Donington, single plant in crack in concrete track, PAC 1971 (site since destroyed).
    Casual: Rt.
First record: Watson (1835).
Herb.: LSR★; LSR, LTR.
HG 590: occasional, restricted in range; 7 localities cited.

C. muricata L.   Prickly Sedge
Map 87h; 2 tetrads.
Allexton, bank of Eye Brook, F. H. Perring 1961; Charley, rocky dry pasture north of Charley Hall, locally frequent, PAC 1968.
    Constituent: Ew, Hg.
First record: 1959, A. Copping in herb. LTR (first known material determined by T. G. Tutin in the light of recent taxonomic ideas).
Herb.: CGE, LTR, MANCH, NMW.
HG 590 as *C. pairaei* rare, range restricted. Only gives Pulteney's unlocalised records and expresses doubt as to whether these should be referred to **C. spicata**.

C. disticha Hudson   Brown Sedge
Map 87i; 86 tetrads.
Locally frequent in marshes in the west of the county; occasional in the east.
    Constituent or residual: M, Ew, Wp, Wl, Wd, Gp, Ev, Er.
First record: Pulteney *in* Nichols (1795).
Herb.: LSR★; LSR, LTR, MANCH, NMW.
HG 587: local, restricted in range; 37 localities cited.

C. remota L.   Remote Sedge
Map 87j; 70 tetrads.
Mainly confined to old-established woodland, though it is occasionally found in marshes. Locally frequent in those parts of the county where there is suitable woodland; rare or absent elsewhere.
    Constituent or intrusive: Fd, Fm, Fw, Ew, Er, Ew, Wd, Wp, M, Gp, Rg.

First record: Pulteney *in* Nichols (1795).
Herb.: LSR; BM, CGE, DPM, LANC, LSR, LTR, MANCH, NMW.
HG 591: frequent, widely distributed; 64 localities cited.

C. ovalis Gooden.   Oval Sedge
Map 87k; 63 tetrads.
Marshes and grassland, tolerant of a wide range of soil humidity. Locally frequent in Charnwood Forest; occasional elsewhere throughout the county.
    Constituent or intrusive: M, Hg, Hd, Gp, Gr, Ew, Er, Ev, Ea. Ef, Fd, Wl, Rs.
First record: Pulteney *in* Nichols (1795).
Herb.: LSR★; BM, LSR, LTR, MANCH, NMW.
HG 593: local, generally distributed.

C. echinata Murr.   Star Sedge
Map 87l; 9 tetrads.
Wet heathland and marshes on acid soil. Confined to Charnwood Forest, where it is locally frequent.
    Constituent or residual: M, Hb, Gr, Wd, Ef, Ap.
First record: Watson (1832).
Herb.: BM, LSR, LTR, MANCH, NMW.
HG 591 as *C. stellulata*: local, restricted in range; 32 localities cited.

C. dioica L.   Dioecious Sedge
Map 88a; 1 tetrad.
Botcheston Bog, marsh, occasional, HH 1974.
    Constituent: M.
First record: Pulteney *in* Nichols (1795).
Herb.: LSR★; LSR.
HG 586: rare, very restricted in range; 6 localities cited, including Botcheston Bog.

C. hirta L.   Hairy Sedge
Map 88b; 438 tetrads.
Marshes, grassland, roadside and railway verges. Frequent throughout the county; the commonest sedge in Leicestershire.
    Constituent or intrusive: M, Gp, Gr, Gl, Hg, Ev, Er, Ef, Ea, Ew, Wd, Wp, Wl, Rq, Rg, Rw, Rd, Rt, Ac.
First record: Pulteney *in* Nichols (1795).
Herb.: LSR★; CGE, DPM, LANC, LSR, LTR, MANCH, NMW.
HG 604: locally abundant, generally distributed.

C. acutiformis Ehrh.   Lesser Pond-sedge
Map 88c; 122 tetrads.
Locally frequent throughout the county in marshes and wet places in woodland; occasional on the margins of rivers, canals and ponds. Seems to require soils with higher organic content than **C. riparia**.
    Constituent or intrusive: M, Fw, Fm, Wl, Wd, Wp, Wr, Wc, Ew, Ev, Er, Rg, Rq, Gp.
First record: Pulteney *in* Nichols (1795).
Herb.: LSR★; CGE, DPM, LANC, LSR, LTR, MANCH, NMW.
HG 605: locally abundant, widely distributed; 57 localities cited.

C. riparia Curt.   Greater Pond-sedge
Map 88d; 176 tetrads.
Frequent and locally abundant throughout the county along the margins of rivers, canals, reservoirs and ponds,

in marshes, and in wet places in woodland.
 Constituent or intrusive: Wr, Wc, Wl, Wp, Wd, Ew, M, Fw, Fd, Fm, Rq, Rg, Rw.
First record: Pulteney *in* Nichols (1795).
Herb.: LSR★; LANC, LSR, LTR, MANCH, NMW.
HG 606: locally abundant, generally distributed.

C. pseudocyperus L. Cyperus Sedge
Map 88e; 34 tetrads.
Locally frequent on the margins of the Ashby Canal and the Grand Union Canal; occasional on the Grantham Canal. Elsewhere it occurs occasionally in ponds and marshes, mainly in the west of the county.
 Constituent: Wc, Wp, Wl, Wd, Ew, M, Fw.
First record: Pulteney *in* Nichols (1795).
Herb.: LSR★; BM, CGE, DPM, LANC, LSR, LTR, MANCH, NMW.
HG 604: rare, very restricted in range; 27 localities cited.

C. rostrata Stokes Bottle Sedge
Map 88f; 5 tetrads.
Blackbrook Reservoir, marshy area at inflow, locally frequent, PAC 1968; Ulverscroft, marshland by brook near Priory, rare, PHG 1968; Moira, bog, locally abundant, SHB 1976; Coleorton, fish pond, locally frequent, SHB 1979; Cropston Reservoir, inflow end, locally frequent, PAE 1983.
 Constituent: M, Hb, Ew.
First record: Pulteney *in* Nichols (1795).
Herb.: LSR★; BM, DPM, LANC, LSR, LTR, MANCH, NMW.
HG 606: rare, very restricted in range; 18 localities cited, including Blackbrook Reservoir, Ulverscroft and Moira.

C. vesicaria L. Bladder-sedge
Map 88g; 5 tetrads.
Blackbrook Reservoir, margins, locally frequent, PAC 1968; Ulverscroft, edge of pond, occasional, PAC 1968; Cropston Reservoir, reed swamp at inflow, occasional, PHG 1970; Higham on the Hill, field pond north-west of Heath Wood, locally frequent, SHB 1974; Nailstone Wiggs, waterside, locally abundant, HH 1977.
 Constituent or residual: Wl, Wp, Ew.
First record: Pulteney *in* Nichols (1795).
Herb.: LSR★; BM, DPM, LSR, LTR, MANCH, NMW.
HG 607: rare, largely confined to areas W. of R. Soar; 19 localities cited, including Blackbrook and Cropston Reservoirs.

C. pendula Hudson Pendulous Sedge
Map 88h; 26 tetrads.
Woodland, and occasionally by water. Locally frequent in Charnwood Forest and the west of the county; occasional in the east.
 Constituent or residual: Fd, Fm, Fw, Ef, Ew, Wd.
First record: Pulteney *in* Nichols (1795).
Herb.: LSR★; DPM, LANC, LSR, LTR, MANCH, NMW.
HG 599: local, very restricted in range; 24 localities cited.

C. sylvatica Hudson Wood-sedge
Map 88i; 48 tetrads.
Woodland. Locally frequent in the well-wooded parts of the county, but avoiding the more acid soils.
 Constituent: Fd, Fm, Fs, Fw, Ef.
 Residual or intrusive: Gr, Rw.

First record: Pulteney *in* Nichols (1795).
Herb.: LSR★; BM, DPM, LSR, LTR, MANCH, NMW.
HG 599: frequent, widely distributed; 68 localities cited.

C. strigosa Hudson Thin-spiked Wood-sedge
Map 88j; 9 tetrads.
Stockerston, Park Wood, wet ground by ride, occasional, KGM 1971; Allexton Wood, wet ride, rare, KGM 1971; Launde, Park Wood, gullies by stream, occasional, KGM 1971; Little Owston Wood, wet ride in coniferous woodland, occasional, KGM 1974; Bolt Wood, wet rides, occasional, KGM 1974; Tugby Bushes, deciduous woodland, occasional, IME 1976; Owston Big Wood, rides, occasional, IME 1976; Staunton Harold, Spring Wood, locally frequent, SHB 1977; Launde, Big Wood, rare, IME 1978.
 Constituent: Fd, Fc.
First record: Bloxam (1837).
Herb.: LSR★; BM, DPM, LSR, LTR, MANCH, NMW.
HG 599: rare, very restricted in range; 8 localities cited, including Spring, Tugby, Bolt and Owston Woods.

C. flacca Schreb. Glaucous Sedge
Map 88k; 246 tetrads.
Marshes, grassland and grassy verges, especially of railways. Tolerant of a wide range of soil humidity. Frequent throughout the county.
 Constituent or intrusive: M, Gp, Gr, Hg, Er, Ev, Ew, Ea, Ef, Fd, Wd, Rq, Rg, Rw, Rt, Ac, Ap.
First record: Pulteney *in* Nichols (1795).
Herb.: LSR★; DPM, LSR, LTR, MANCH, NMW.
HG 596 as *C. diversicolor*: locally abundant, generally distributed.

C. panicea L. Carnation Sedge
Map 88l; 60 tetrads.
Marshes and heath grassland. Locally frequent in the west of the county; occasional in the east.
 Constituent or residual: M, Hg, Gp, Gr, Gl, Ew, Fd.
First record: Pulteney *in* Nichols (1795).
Herb.: LSR★; BM, LSR, LTR, MANCH, NMW.
HG 598: frequent, generally distributed.

C. laevigata Sm. Smooth-stalked Sedge
Map 89a; 4 tetrads.
Swithland Wood, wet ride, rare, PHG 1968; Ulverscroft, margin of lake, rare, PAC 1968; Ulverscroft, marshland by stream near Priory, rare, PHG 1968; Ulverscroft, wet flush in field east of Stoneywell Wood, locally frequent, PHG 1973; Grace Dieu Wood, occasional, PHG 1976; Ulverscroft, Poultney Wood, wet ride, occasional, PAE and PHG 1983.
 Constituent or residual: M, Ew, Ef, Fd.
First record: Coleman (1852).
Herb.: LSR★; BM, LSR, NMW.
HG 600 as *C. helodes*: rare, restricted in range; 7 localities cited, including Swithland Wood and Ulverscroft.

C. distans L. Distant Sedge
No record in the recent survey.
Botcheston Bog, E. K. Horwood 1950; Kilby Bridge, canal margin, D. P. Murray 1956; Higham on the Hill, railway verge, D. P. Murray 1958.
First record: 1950, E. K. Horwood in herb. LTR.
Herb.: DPM, LTR.
HG: not recorded.

C. binervis Sm. Green-ribbed Sedge
Map 89b; 18 tetrads.
Heath grassland and woodland on acid soil. Almost confined to Charnwood Forest, where it is locally frequent.
   Constituent or residual: Hg, Hd, Hw, M, Gp, Fd, Fc, Fm, Fw, Ef.
First record: Watson (1837).
Herb.: LSR★; BM, CGE, DPM, LSR, LTR, MANCH, NMW.
HG 601: local, restricted in range; 32 localities cited.

C. hostiana DC. Tawny Sedge
Map 89c; 5 tetrads.
Twenty Acre, **Molinia-Nardus** heathland, rare, ALP 1968; Ulverscroft, marsh between Lea Wood and brook, locally frequent, PHG 1972; Saltby, 'dry bog', rare, PHG 1972; Narborough Bog, rare, EH 1974; Benscliffe Wood, edge, rare, SHB 1977.
   Constituent or residual: Hw, M, Ef.
First record: Kirby (1848b).
Herb.: LSR★; BM, LANC, LSR, LTR, MANCH, NMW.
HG 602 as *C. fulva*: occasional, restricted in range; 9 localities cited, including Twenty Acre and Saltby.

C. lepidocarpa Tausch Long-stalked Yellow-sedge
No record in the recent survey.
Breedon on the Hill, Cloud Wood, F. J. Taylor 1950; Saltby Bog, E. K. Horwood 1953; Burbage Wood, D. P. Murray 1955.
First record: Pulteney (1749) as *C. flava*; Coleman (1852) as **C. lepidocarpa.**
Herb.: LSR★; DPM, LSR, LTR, MANCH, NMW.
HG 602 as *C. flava* and 603 as **C. lepidocarpa**: rare, restricted in range; 22 localities cited under *C. flava* and 8 localities under **C. lepidocarpa** (the authors of the present work accept the authority of T. G. Tutin that **C. flava** does not occur in Leicestershire).

C. demissa Hornem. Common Yellow-sedge
Map 89d; 15 tetrads.
Heathland, marshes and woodland rides, preferring acid soil. Locally frequent in Charnwood Forest, to which it is almost confined, apart from Moira in the north-west of the county and Owston Wood in the east.
   Constituent or intrusive: Hg, Hb, M, Gp, Ef, Wd, Rg.
First record: Bloxam (1837) (fide A. R. Horwood; see note below).
Herb.: LSR★; LANC, LSR, LTR.
HG 603 as '*C. oederi*': rare, restricted in range; 8 localities cited. From the accounts in HG it is obvious that at that time there was confusion concerning the **C. flava** group.

C. pallescens L. Pale Sedge
Map 89e; 15 tetrads.
Marshes, heathland, and occasionally woodland, apparently preferring acid soil. Mostly confined to Charnwood Forest, except for records from Kirby Muxloe, Potters Marston, Six Hills, and Old Dalby. Seldom in great quantity where it does occur.
   Constituent or residual: M, Hg, Hw, Hd, Gr, Gp, Fd, Ef.
First record: Pulteney in Nichols (1795).
Herb.: LSR★; BM, CGE, LANC, LSR, LTR, MANCH, NMW.
HG 598: local, restricted in range; 47 localities cited.

C. caryophyllea Latourr. Spring-sedge
Map 89f; 23 tetrads.
Grassland. Locally frequent in the west of the county; occasional in the east.
   Constituent or residual: Gp, Hg, M, Ev, Ap.
First record: Pulteney (1747).
Herb.: LSR★; DPM, LSR, LTR, MANCH, NMW.
HG 597: frequent, and widely distributed; 61 localities cited.

C. pilulifera L. Pill Sedge
Map 89g; 15 tetrads.
Heath grassland and marshes. Confined to Charnwood Forest, except for records from Twenty Acre and Burbage Common. Seldom in any great quantity where it does occur.
   Constituent or residual: Hg, Hd, M, Ef, Fc, Rq.
First record: Pulteney in Nichols (1795).
Herb.: LSR★; CGE, LANC, LSR, LTR, MANCH, NMW.
HG 596: rather rare, restricted in range; 27 localities cited.

C. elata All. Tufted-sedge
Map 89h; 1 tetrad.
Castle Donington, marsh, occasional, PAC 1973.
   Constituent: M.
First record: Brown in Mosley (1863).
Herb.: LSR, NMW.
HG 593: rare, restricted in range; 5 localities cited, not including the above.

C. nigra (L.) Reichard Common Sedge
Map 89i; 73 tetrads.
Marshes and heathland. Locally frequent in the west of the county; occasional in the east.
   Constituent or residual: M, Hg, Hw, Gp, Gr, Wd, Ew, Ev, Ef.
First record: Pulteney in Nichols (1795).
Herb.: LSR★; BM, LSR, LTR, MANCH, NMW.
HG 595 as *C. goodenowii*: locally abundant, fairly widely distributed; 50 localities cited.

C. acuta L. Slender Tufted-sedge
Map 89j; 43 tetrads.
Margins of rivers, canals, reservoirs and ponds, and in marshes. Locally frequent in the west of the county; occasional in the east.
   Constituent or residual: Ew, Wc, Wl, Wd, Wp, M, Fd, Rq.
First record: Pulteney (1749).
Herb.: LSR★; CGE, LSR, LTR, MANCH, NMW.
HG 594 as *C. gracilis*: fairly frequent, generally distributed.

C. acuta × acutiformis = C. × subgracilis Druce.
No map; 1 tetrad.
Margin of Groby Pool, EH 1982.
   Intrusive: Ew.
First record: 1982, E. Hesselgreaves in present work.
Herb.: LSR★.
HG: not recorded.

C. pulicaris L. Flea Sedge
Map 89k; 2 tetrads.
Ulverscroft, Herbert's Meadow, rare, PAC 1968; Ulverscroft, marshland between Lea Wood and brook, locally frequent, PHG 1970.
   Constituent: M.

First record: Pulteney in Nichols (1795).
Herb.: LSR★; LSR, LTR, MANCH, NMW.
HG 586: rare, range restricted; 22 localities cited, including Ulverscroft.

MICROSPERMAE

ORCHIDACEAE
Epipactis Sw.
E. palustris (L.) Crantz Marsh Helleborine
Map 89l; 1 tetrad.
Botcheston Bog, marshland, rare, 3 plants only seen, ground heavily grazed and trampled, HH 1974 (prior to 1968 upwards of 25 plants were seen).
 Constituent: M.
First record: Pulteney (1747).
Herb.: BM, LSR, LTR, MANCH, NMW.
HG 524 as *Helleborine palustris*: rare, confined to Charnwood Forest; 8 localities cited, including Botcheston Bog (which is not in Charnwood Forest!).

E. helleborine (L.) Crantz Broad-leaved Helleborine
Map 90a; 33 tetrads.
Occasional throughout the county in well-established woodland, and usually only in small numbers where it does occur.
 Constituent: Fd, Fm, Fs, Ef.
 Residual: Eh, Ev, Gr.
First record: Pulteney (1747).
Herb.: LSR★; LANC, LSR, LTR, MANCH, NMW.
HG 523 as *Helleborine latifolia*: frequent, widely distributed; 58 localities cited.

E. purpurata Sm. Violet Helleborine
Map 90b; 4 tetrads.
Stockerston, Park Wood, EKH 1967; Great Merrible Wood, rare, IME 1968; Bolt Wood, rare, KGM 1971; Pasture Wood, rare, S. Jackson 1977; Breedon on the Hill, Cloud Wood, single plant, PAE 1978; Sheet Hedges Wood, rare, PHG 1978.
 Constituent: Fd.
First record: Kirby (1850).
Herb.: LSR★; LSR, LTR.
HG 524 as *Helleborine purpurata*: rare, restricted in range; 2 localities cited, including Park Wood at Stockerston.

Neottia Ludw.
N. nidus-avis (L.) Rich. Bird's-nest Orchid
No record in the recent survey.
Tugby Bushes, under hazel, E. K. Horwood 1949; Elmesthorpe Plantation, A. Hackett 1948, M. K. Hanson 1953, 1962 and 1964; Burbage Wood, unconfirmed report, 1977.
First record: Bloxam (1837).
Herb.: BM, LSR, LTR, MANCH, NMW.
HG 521: rare, restricted in range; 8 localities cited.

Listera R.Br.
L. ovata (L.) R.Br. Common Twayblade
Map 90c; 52 tetrads.
Woodland, railway and roadside verges, quarries and old grassland. Occasional throughout the county; sometimes locally frequent where it does occur.
 Constituent or residual: Fd, Fm, Fs, Fw, Ef, Er, Ev, Ew, Gp, Rg, Rq.

First record: Pulteney (1747).
Herb.: LSR★; BM, DPM, LSR, LTR, MANCH, NMW.
HG 521: frequent, generally distributed.

Spiranthes Rich.
S. spiralis (L.) Chevall. Autumn Lady's-tresses
No record since HG.
First record: Pulteney (1749).
HG 522: rare, confined to Charnwood Forest; 6 localities cited.

Platanthera Rich.
P. chlorantha (Custer) Reichb. Greater Butterfly Orchid
Map 90d; 9 tetrads.
Elmesthorpe Plantation, rare, MH 1968; Launde, Park Wood, rare, PHG 1968; Great Merrible Wood Nature Reserve, occasional, KGM 1971; Breedon on the Hill, Cloud Wood, rare, PAC 1971; Little Owston Wood, occasional, PAC 1976; Owston Big Wood, rare, IME 1976; Tugby Bushes, deciduous woodland, rare, IME 1976; Pasture Wood, occasional, PAE 1978; Launde, Big Wood, occasional, PAE 1979; Tugby Wood, rare, PAE 1979.
 Constituent: Fd.
First record: Pulteney (1747).
Herb.: LSR, LTR, MANCH.
HG 535 as *Habenaria virescens*: local, occasional, sparsely distributed; 38 localities cited, including Cloud Wood, Pasture Wood, Great Merrible Wood, Tugby Wood and Launde Big Wood.

Gymnadenia R.Br.
G. conopsea (L.) R.Br. Fragrant Orchid
Map 90e; 3 tetrads.
Botcheston Bog, occasional, PHG and MW 1965; Ulverscroft, Herbert's meadow, occasional, PAC 1968; Dunton Bassett, disused sand quarry, rare, S. Grover 1986.
 Constituent: M, Rg.
First record: Crabbe in Nichols (1795).
Herb.: LSR★; LSR, LTR, MANCH, NMW.
HG 534: rare, restricted in range; 17 localities cited, not including the above.

Coeloglossum Hartm.
C. viride (L.) Hartm. Frog Orchid
No record in recent survey.
Potters Marston, M.K.Hanson 1953. This site is now destroyed.
First record: Pulteney in Nichols (1795).
Herb.: CGE, LSR, MANCH.
HG 536: rare and rather restricted in range and numbers; 19 localities cited.

Dactylorhiza Soó
D. incarnata (L.) Soó Early Marsh Orchid
Dactylorchis incarnata (L.) Vermeul.
No map.
Potters Marston Bog, EKH 1960; Glooston, bog, EKH 1964. Both these sites are now destroyed.
 Constituent: M, Hb.
First record: 1843, F. T. Mott in herb. MANCH.
Herb.: BM, LANC, LSR, LTR, MANCH, NMW.
HG 527 as *Orchis incarnata*: rare, restricted in range; 9 localities cited, including Potters Marston.

D. majalis (Reichenb.) Soó subsp. **praetermissa** (Druce) D. Moresby Moore & Soó   Southern Marsh Orchid
Dactylorchis praetermissa (Druce) Vermeul.
Map 90f; 18 tetrads.
Marshes, and marshy places in quarries and gravel pits. Occasional in the west of the county; rare or absent in the east.
    Constituent, residual or intrusive: M, Gp, Gr, Rq, Rg, Rs, Ew, Er, Fw.
First record: 1837, Churchill Babington in herb. CGE.
Herb.: LSR★; BM, CGE, DPM, LSR, LTR, NMW.
HG 529 as *Orchis praetermissa*: local, sparsely distributed; 13 localities cited.

D. maculata (L.) Soó subsp. **maculata**   Heath Spotted Orchid
Dactylorchis maculata (L.) Vermeul. subsp. *ericetorum* (E. F. Linton) Vermeul.
Map 90g; 15 tetrads.
Heathland, marsh and grassland, preferring acid peaty soil. Locally frequent where suitable soils occur, in scattered localities throughout the county.
    Constituent or residual: Hg, M, Gp, Gr, Ef, Fd.
First record: B.E.C. Rep. 1916 (it is not possible to tell whether Pulteney's record of 1747 refers to this or the following species).
Herb.: LSR, LTR, NMW.
HG 530 as *Orchis maculata*: locally abundant, generally distributed; 61 localities cited, but it would seem from examination of the present-day distribution of the two species that some of these localities should refer to **D. fuchsii**.

D. fuchsii (Druce) Soó   Common Spotted Orchid
Dactylorchis fuchsii (Druce) Vermeul.
Map 90h; 91 tetrads.
Locally frequent throughout the county in grassland, marshes, quarries, open places in woodland, and on railway and roadside verges. The commonest orchid in Leicestershire.
    Constituent, residual or intrusive: Gp, Gr, Hg, M, Fd, Fs, Fw, Ef, Er, Ev, Ew, Ea, Ap, Rq, Rg, Rs.
First record: B.E.C. Rep. 1916 (it is not possible to tell whether Pulteney's record of 1747 refers to this or the preceding species).
Herb.: LSR★; BM, DPM, LSR, LTR, MANCH, NMW.
HG 531 as *Orchis fuchsii*: frequent, and generally distributed; 39 localities cited (but see note concerning localities cited for the preceding species).

Orchis L.
O. morio L.   Green-winged Orchid
Map 90i; 12 tetrads.
Shepshed, old meadow near Black Brook, PHG 1967 (since ploughed); Shearsby, pasture east of village, rare, J. C. Badcock 1967 (ploughed 1972); Shepshed, small disused quarry, occasional, PAC 1968; Thorpe Satchville, dismantled railway, cutting near Thorpe Trussels, occasional, HB 1968; Ulverscroft, Herbert's Meadow, old grassland, locally frequent, PAC 1968; Arnesby, pasture, rare, H. Godsmark and IME 1969; Wymondham, Cribbs Meadow, old grassland, locally frequent, KGM 1971; Shearsby, old pasture west of village, IME 1972; Sileby, Highgate Lodge Farm, old grassland, occasional, ALP 1974; Sheepy, old pasture east of the Cross Hands, rare, SHB 1977; Osgathorpe, old pasture, rare, SHB 1979; Muston, three separate sites, old pasture, locally abundant in one, PAE 1979.
    Constituent or residual: Gp, Er, Rq.
First record: Pulteney (1749).
Herb.: LSR★; CGE, LSR, LTR, MANCH, NMW.
HG 525: locally abundant, generally distributed; 70 localities cited.

O. ustulata L.   Burnt Orchid
No recent record.
First record: Pulteney *in* Nichols (1795).
HG 525: rare; 2 localities cited. Mott et al. (1886) doubted its occurrence in Leicestershire at all, and stated that there was no recent record for it at that time.

O. mascula (L.) L.   Early-purple Orchid
Map 90j; 30 tetrads.
Occasional in woodland throughout the county. Decreasing as woodland is cleared or fox coverts become overgrown through neglect, but in some of its localities it may be in considerable abundance still.
    Constituent: Fd, Fc, Fm, Fs, Fw, Ef.
    Residual: Gp, Rq, Er, Wd.
First record: Pulteney (1747).
Herb.: LSR★; DPM, LSR, LTR, MANCH, NMW.
HG 526: frequent, generally distributed; 67 localities cited.

Anacamptis Rich
A. pyramidalis (L.) Rich.   Pyramidal Orchid
Map 90k; 4 tetrads.
Shawell, old gravel pits at Cave's Inn, rare, EKH and JMH 1969; Stonesby Quarry, oolitic limestone, quarry floor and spoil heaps, occasional, ALP 1972; Enderby, old quarry, rare, probably planted, EH 1974; Dunton Bassett, disused sand quarry, rare, KGM and ALP 1987.
    Residual or intrusive: Rq, Rg, Rs.
First record: Pulteney *in* Nichols (1795).
Herb.: LSR★; BM, LSR, LTR, MANCH.
HG 524 as *Orchis pyramidalis*: rare, range restricted mainly to NE and Dist. IX; 5 localities cited, including Stonesby.

Ophrys L.
O. apifera Hudson   Bee Orchid
Map 90l; 18 tetrads.
Calcareous grassland, mainly in quarries and on spoil heaps or railway verges. Nearly all the localities for this species in Leicestershire are in man-made habitats as described above, which demonstrates the importance of such habitats in maintaining the variety of the flora. In scattered localities throughout the county, but avoiding the more acid soils of the west. Often locally frequent in the places where it occurs.
    Constituent, residual or intrusive: Gr, Rq, Rg, Rs, Rw, Er.
First record: Pulteney *in* Nichols (1795).
Herb.: LSR★; BM, LSR, LTR, MANCH, NMW.
HG 532: rare, restricted in range; 23 localities cited.

BIBLIOGRAPHY

This bibliography comprises:
1. references for the first records of species contained in the Systematic Account. The authors and dates of these are printed in **bold**.
2. references for taxonomic works cited in the Systematic Account.
3. other references to the flora of Leicestershire.

References to works cited in sections other than the Systematic Account are listed in the appropriate places in those sections.

An attempt has been made to cite by author first records published in the *Transactions of the Leicester Literary and Philosophical Society*, which in Horwood and Gainsborough (1933) are all subsumed under the convention *Trans*. A number of these records are taken from the Reports to Council of the variously-named Sections of the Society devoted to the biological sciences. Some of these Reports identify their author, usually the Chairman or Hon. Secretary at the time. In others, the authorship may be inferred, in which case the author's name has been placed in square brackets. In yet others, authorship is uncertain, in which case the convention 'Anon.' has been used. Those interested in tracking down references in the *Transactions* should be wary of the somewhat irregular systems of pagination used at times in the heyday of that journal in the later 19th and early 20th centuries.

Allen, D. E., 1948. The Rugby flora, 1946-47. *Rep. Rugby Sch. nat. Hist. Soc.* **81**, 18-30.
Allen, D. E., 1949. Rugby botany, 1948. *Rep. Rugby Sch. nat. Hist. Soc.* **82**, 16-21.
Allen, D. E., 1950. Notes on Rugby botany, 1949. *Rep. Rugby Sch. nat. Hist. Soc.* **82**, 13-17.
Allen, D. E., 1957. *The flora of the Rugby district*. Rugby: Rugby School Natural History Society.
Anon., 1886. [Report of field excursion to Burbage Wood and Common] *in* Report of Leicester Literary and Philosophical Society. *Midl. Nat.* **9**, 228.
Anon., 1899-1903. The flora of Burton-on-Trent and neighbourhood. *Trans. Burton-on-Trent nat. Hist. archaeol. Soc.* **3**, 177-190, 269, 282; **4**, 75-88, 117-148; **5**, 37.
Anon., 1931. Observations during 1930-31: Plants *in* Report of Section D, Botany, and Section E, Biology. *Trans. Leicester lit. phil. Soc.* **32**, 67-69.
Anon., 1932. Observations during 1931: Phanerogams *in* Report of Section D, Botany, and Section E, Biology. *Trans. Leicester lit. phil. Soc.* **33**, 43-44.
Anon., 1933. Observations during 1932: Phanerogams *in* Report of Section D, Botany, and Section E, Biology. *Trans. Leicester lit. phil. Soc.* **34**, 69.
Ardington, E. E., 1911. Report of Section D – Botany. *Trans. Leicester lit. phil. Soc.* **15**, 108-112.
Armstrong, S. F., 1907. The botanical and chemical composition of the herbage of pastures and meadows. *J. agric. Sci.* **2**, 283-304.
B[abington], A. M., 1897. *Memorials, journals and botanical correspondence . . . of Charles Cardale Babington*. Cambridge: MacMillan and Bowes.
Babington, C. C., 1846. A synopsis of the British *Rubi*. *Ann. Mag. nat. Hist.* **17**, 165-175, 235-247, 314-322.
Babington, C. C., 1847a *in* Sowerby, J. and Smith, J. E., *Supplement to English Botany* **4**, text to pl. 2930.

Babington, C. C., 1847b. *Manual of British Botany* . . . . Second edition. London: Van Voorst.
Babington, C. C., 1847c. A supplement to 'A synopsis of the British *Rubi*', I and II. *Ann. Mag. nat. Hist.* **19**, 17-19, 83-87.
Babington, C. C., 1848. A supplement to 'A synopsis of the British *Rubi*' II. *Ann. Mag. nat. Hist.* (2) **2**, 32-43.
Babington, C. C., 1851. *Manual of British Botany* . . . . Third edition. London: John van Voorst.
Babington, C. C., 1865 *in* Sowerby, J. and Smith, J. E., *Supplement to English Botany* **5**, text to pl.2993.
Babington, C. C., 1869. *The British Rubi: an attempt to discriminate the species of Rubus known to inhabit the British Isles.* London: Van Voorst.
Bagnall, J. E., 1888. [Plants from Loddington Wood] *in* Reports of societies. *Midl. Nat.* **11**, 237-238.
B. E. C. Rep. Botanical Exchange Club Reports (1879-1948).
Bell, W., 1903. Saltby Heath *in* Vice, W. A., Quarterly Report of Section D – Botany. *Trans. Leicester lit. phil. Soc.* **7:3**, 142-145.
[Bell, W.], 1905. The flora of Leicestershire *in* Nuttall, G. C. (ed.), *Guide to Leicester and neighbourhood.* Leicester: British Medical Association. pp.185-187, 191.
Bell, W., 1907. Botany: Phanerogams *in* Nuttall, G. C. (ed.), *A guide to Leicester and district.* Leicester: British Association. pp.334-345.
Bell, W., 1907. Quarterly Report of Section D – Botany. *Trans. Leicester lit. phil. Soc.* **11**, 58-59.
Bell, W., 1908. The Rev. Thomas Arthur Preston, M.A., F.L.S. (Rector of Thurmaston). A reminiscence and an appreciation. *Trans. Leicester lit. phil. Soc.* **12:2**, 211-220.
Bemrose, G. J. V., 1927. The adventive flora of Leicester and district. *Trans. Leicester lit. phil. Soc.* **28**, 45-71.
[Bemrose, G. J. V.], 1929. Selected observations *in* Report of Section D, Botany, and Section E, Biology. *Trans. Leicester lit. phil. Soc.* **30**, 69.
[Bemrose, G. J. V.], 1930. Selected observations *in* Report of Section D, Botany, and Section E, Biology. *Trans. Leicester lit. phil. Soc.* **31**, 51-54.
Beresford, J. E. and Wade, P. M., 1982. Field ponds in north Leicestershire: their characteristics, aquatic flora and decline. *Trans. Leicester lit. phil. Soc.* **76**, 25-34.
Blackstone, J., 1746. *Specimen botanicum, quo plantarum plurium rariorum Angliae indigenarum loci natales illustrantur.* London: Gulielmi Faden.
Blith, W., 1653. *The English Improver Improved; or, the Survey of Husbandry Surveyed . . . With an Additional Discovery of the Several Tooles, and Instruments &c.* Second edition. London: John Wright.
Bloxam, A., [1829]. A list of wild plants found in Charnwood Forest and its neighbourhood *in* Combe, T. (ed.), *A description of Bradgate Park and the adjacent country* . . . . Leicester: T. Combe. pp.72-82.
Bloxam, A., 1830. Plants in Charnwood Forest and its neighbourhood . . . . *Mag. nat. Hist.* **3**, 167-168.
Bloxam, A., 1831. Plants found in Charnwood Forest. *Mag. nat. Hist.* **4**, 162.
Bloxam, A., 1837. Leicestershire Flora. *Naturalist* **2**, 79-83, 132-136.
Bloxam, A., 1838. Addenda to the 'Leicestershire Flora' . . . . *Naturalist* **4**, 55.
Bloxam, A., 1840. *Myriophyllum alterniflorum* new to the British flora. *Ann. nat. Hist.* **5**, 358.
Bloxam, A., 1846a. List of the rarer plants found in the neighbourhood of Twycross, Leicestershire. *Phytologist* **2**, 640-642.
Bloxam, A., 1846b. List of British *Rubi* found in the neighbourhood of Twycross, Leicestershire. *Phytologist* **2**, 675-676.
Bloxam, A., 1848a. Note on the British *Rubi*. *Phytologist* **3**, 181-183.
Bloxam, A., 1848b. Occurrence of *Botrychium lunaria* near Twycross. *Phytologist* **3**, 183.
Bloxam, A., 1848c. Notice of *Juncus diffusus* in Leicestershire. *Phytologist* **3**, 291.

Bloxam, A., 1848d. British *Rubi* in Yorkshire. *Phytologist* **3**, 325-326.
Bloxam, A., 1851. *Trifolium patens* Schreb. near Ashby-de-la-Zouch. *Phytologist* **4**, 1100.
Bloxam, A. and Babington, C., 1842. Botany of Charnwood Forest *in* Potter, T. R., *The history and antiquities of Charnwood Forest*. London: Hamilton, Adams. Appendix, pp.35-62.
Borrill, M., 1958. A biometrical study of some *Glyceria* species in Britain. *Watsonia* **4:2**, 77-78, 89-100.
Britten, J., 1877. Botany *in* White W., *History, gazetteer, and directory of the counties of Leicester and Rutland* . . . . Third edition. London: William White. pp.54-59. [Revision of text by W. H. Coleman in second edition, 1863.]
Britton, J., 1807. *A topographical and historical description of the county of Leicester*. London.
Brown, E., 1863. The flora of the district surrounding Tutbury and Burton-on-Trent *in* Mosley, O., *The natural history of Tutbury*. London: John van Voorst. pp.231-364.

Carter, T., 1889. Phanerogamous parasites of Leicestershire. *Trans. Leicester lit. phil. Soc.* **1:10**, 16-20.
Clement, E. J., 1976. Recent records of *Crepis tectorum* L. in Adventive News 5. *B.S.B.I. News* **13**, 24.
Clement, E. J., 1981. Aliens and adventives *in* Adventive News 21. *B.S.B.I. News* **29**, 8-13.
Clement, E. J., 1985. Selfheals (*Prunella* spp.) in Britain. *B.S.B.I. News* **41**, 20. [Recognises, as possibly the first British record of *P. laciniata* (L.), the description in Mott (1878) of a supposed white variety of *P. vulgaris* from Birstall.]
Coleman, W. H., 1844. Observations on a new species of *Oenanthe. Ann. Mag. nat. Hist.* **13**, 188.
Coleman, W. H., 1852. Part 1. The flowering plants and ferns *in* Bloxam, A. and Coleman, W. H., *Flora Leicestrensis or a Catalogue of the wild plants of the county of Leicester*. Ms. [Leicestershire Museums Service. 81 L 37.]
Coleman, W. H., 1863. Botany *and* A catalogue of the flowering plants and ferns, known or reported to inhabit the county of Leicester *in* White, W., *History, gazetteer, and dictionary of the counties of Leicester and Rutland* . . . . Second edition. London: Simpkin, Marshall. pp. 62-77.
Conolly, A. P., 1977. The distribution and history in the British Isles of some alien species of *Polygonum* and *Reynoutria. Watsonia* **11**, 291-311.
Cooke, G. A., [1809]. *Topographical and statistical description of the county of Leicester*. London: C. Cooke. pp. 137-138.
Cooper, E. F., 1894. Notes on the Leicestershire Potamogetons. *Trans. Leicester lit. phil. Soc.* **3:9**, 391-398.
Crabbe, G., 1795. The natural history of the Vale of Belvoir *in* Nichols, J., *The history and antiquities of the county of Leicester*. **1:1**, cxci-ccviii. London: J. Nichols.
Creaghe-Howard, L. C., 1893. *Orobanche elatior. Sci. Gossip* **39**, 47.
Crocker, J. (ed.), 1981. *Charnwood Forest: A changing landscape*. [Loughborough]: Loughborough Naturalists' Club.

Dale, A. and Elkington, T. T., 1974. Variation within *Cardamine pratensis* L. in England. *Watsonia* **10**, 1-17.
Deering, C., 1738. *Catalogus stirpium . . . or A catalogue of plants naturally growing and commonly cultivated in divers parts of England, more especially about Nottingham*. Nottingham.
[Dixon, G. B.], 1904. Quarterly Report of Section D (Botany). *Trans. Leicester lit. phil. Soc.* **8**, 74-80.
Dunns, S. T., 1898. *Key in simple language to the families of the wild flowers of Warwickshire, Leicestershire, Northamptonshire, Oxfordshire*. Rugby.
Eller, I., 1841. The flora of the Vale of Belvoir. Appendix to *The History of Belvoir Castle*. London. pp.391-410.
Evans, I. M., 1977. *A contribution to the study of the landscape and natural history of the parish of Launde*. Leicester: Leicestershire Museums Service.

Fisher, H., 1907. Botany *in* Page, W. (ed.), *The Victoria History of the County of Leicester* **1**. London: Archibald Constable. pp.27-59.

Fryer, A., 1891. On a new British *Potamogeton* of the *nitens* group. *J. Bot.* **29**, 289-292.

Fryer, A., Bennett, A. and Evans, A. H., 1898-1915. *The Potamogetons (pondweeds) of the British Isles*. London: L. Reeve.

[Gamble, P. H.], 1976. Flowering plants, ferns and horsetails recorded on the Prestwold Estate, 1973-1975 *in* Middleton, R. (ed.), *Prestwold 2000. A report on the future of a large estate*. Appendix 3. Leicester: Leicestershire Rural Community Council.

Gibbons, E. J. and Lousley, J. E., 1958. An inland *Armeria* overlooked in Britain, 1. *Watsonia* **4**, 125-135.

Gregory, H. H. 1952. Botany and Zoology *in* Hadfield, C. N., *Charnwood Forest. A survey*. Leicester: Leicestershire County Council/Edgar Backus.

Hands, T., Paget, J. and Parkinson, W., 1831. Botany *in* Curtis J., *A topographical history of the county of Leicester . . . .* Ashby-de-la-Zouch: W. Hextall. pp.xxxv-xxxix.

Harley, R. M., 1972. *Mentha* L. *in* Tutin, T. G. et al (eds.), *Flora Europaea* **3**, 183-186.

Harley, R. M., 1975. *Mentha* L. *in* Stace, C. A. (ed.), *Hybridization and the flora of the British Isles*. London: Academic Press.

Harris, G. T., 1880. Unfrequent plants around Atherstone. *Midl. Nat.* **3**, 277.

Hill, J., 1760. *Flora Britanica, sive synopsis methodica stirpium Britanicorum*. London: Jacob Waugh.

Horwood, A. R., 1904. Leicestershire records, 1903. *J. Bot.* **42**, 26.

Horwood, A. R., 1907. Classified bibliography of principal works on the geology, botany . . . of Leicestershire: Botany *in* Nuttall, G.C. (ed.), *A guide to Leicester and district*. Leicester: British Association. pp. 399-403.

Horwood, A. R., 1909. Leicestershire plants. *J. Bot.* **47**, 430-431.

Horwood, A. R., 1911. Leicestershire plants (1905-1910). *J. Bot.* **39**, 31-36, 48-59.

Horwood, A. R., 1911. *Digitalis purpurea* in East Leicestershire. *J. Bot.* **49**, 98.

Horwood, A. R., 1913. The Leicestershire Flora Committee. Report for 1911-12. *Trans. Leicester lit. phil. Soc.* **17**, 91-93.

Horwood, A. R., 1914. Report of the Flora Committee. *Trans. Leicester lit. phil. Soc.* **18**, 75-76.

Horwood, A. R., 1915. Report of the Flora Committee. *Trans. Leicester lit. phil. Soc.* **19**, 59-61.

Horwood, A. R., 1919. 4th Report of the Flora Committee. 1916-19. *Trans. Leicester lit. phil. Soc.* **20**, 76-78.

Horwood, A. R., 1933. The flora of Leicestershire *in* Bryan, P. W. (ed.), *A scientific survey of Leicester and district*. London: British Association. pp.25-33.

Horwood, A. R. and Noel, C. W. F. (3rd Earl of Gainsborough), 1933. *The flora of Leicestershire and Rutland*. London: Oxford University Press. [Usually referred to as 'Horwood and Gainsborough', which usage has been followed in this work.]

Howarth, S. E. and Williams J. T., 1968. Biological Flora of the British Isles. *Chrysanthemum leucanthemum* L. *J. Ecol.* **56**, 585-595.

Jackson, A. B., 1899. *Rubus kaltenbachii* in Leicestershire. *J. Bot.* **37**, 136.

Jackson, A. B., 1900. Notes on the botany of the Beaumont Leys Sewage Farm. *Trans. Leicester lit. phil. Soc.* **5:10**, 495-502.

Jackson, A. B., 1902. Notes on the Kirby herbarium. *Trans. Leicester lit. phil. Soc.* **6:4**, 213-215.

Jackson, A. B., 1903a. *Alopecurus hybridus* in Leicestershire. *J. Bot.* **41**, 58.

Jackson, A. B., 1903. *Glyceria distans* var. *obtusa*. *J. Bot.* **41**, 59.

Jackson, A. B., 1903b. *Damasonium stellatum* in Leicestershire. *J. Bot.* **41**, 249.

Jackson, A. B., 1904a. Notes on Leicestershire eyebrights. *Trans. Leicester lit. phil. Soc.* **8:2**, 116-120.

Jackson, A. B., 1904b. Leicestershire plant notes, 1886-1904. *J. Bot.* **42**, 337-349.

Jackson, A. B., 1906. Charnwood Forest *Rubi*. *J. Bot.* **44**, 261-266.

Jackson, A. B., 1907. *Apera intermedia* as an alien in Britain. *Ann. Scot. nat. Hist.* **July 1907**, 170-171.
Jones, R. C., 1971. A survey of the flora, physical characteristics and distribution of field ponds in north east Leicestershire. *Trans. Leicester lit. phil. Soc.* **65**, 12-31.
Kirby, M., 1848. *The flora of Leicestershire, according to the natural orders; arranged from the London catalogue of British plants.* Leicester: J. S. Crossley. [Leicestershire Museums Service. 83 L 37.]
Kirby, M. 1848a. Revivifying property of the Leicestershire *Udora*. *Phytologist* **3**, 30.
Kirby, M., 1848b. Notes on the flora of Leicestershire, with addenda thereto. *Phytologist* **3**, 179-180.
Kirby, M., 1849. Occurrence of *Udora* in the Lene near Nottingham. *Phytologist* **3**, 387-388.
Kirby, M., 1850. *A flora of Leicestershire* . . . . London: Hamilton, Adams.
Kirk, T., 1849. Notice of a new locality for *Anacharis alsinastrum*. *Phytologist* **3**, 389-390.
Lacey, W. S., 1947. *Galinsoga parviflora* Cav. in Leicestershire. *North west. Nat.* **22**, 114-115.
Lacey, W. S., 1948. The genus *Galinsoga* Ruiz and Pav. in Leicestershire (v.-c. 55). *North west. Nat.* **23**, 162-166.
Lakin, C., 1899. Some medicinal plants of Leicestershire. *Trans. Leicester lit. phil. Soc.* **5:4**, 176-182.
Lees, E., 1845. Investigations of the scientific distinctions of *Oenanthe pimpinelloides*, *Oe. peucedanifolia*, and *Oe. lachenalii*. *Phytologist* **2**, 354-365.
Linton, E. F., 1913. The British willows. *J. Bot.* Supplement to **51**.
Loudon, J. C., 1838. *Arboretum et fruticetum Brittanicum*. London. 8 vols.
Loughborough Naturalists' Club, 1962. *Bradgate Park and Cropston Reservoir margins*. Loughborough.
Lowe, R. T., 1868. *A manual flora of Madeira and the adjacent islands of Porto Santo and the Desertas.* London: John van Voorst. 2 vols. [On p.110 of vol. 2 there is a reference to *Nicandra physalodes* observed at Hathern in 1828.]

Macaulay, A., 1791. *The history and antiquities of Claybrook, in the county of Leicester* . . . . London: J. Nichols.
Marshall, W., 1790. *The rural economy of the midland counties; including the management of livestock in Leicestershire and its environs* . . . . London: G. Nicol. 2 vols.
Martyn, T., 1763. *Plantae Cantabrigienses or, a catalogue of the plants which grow wild in the County of Cambridge – to which are added lists of the more rare plants growing in many parts of England and Wales.* London.
Measham, C. E. C., 1914. The Society's herbarium *in* Turner, G. C., Report of Section D – Botany. *Trans. Leicester lit. phil. Soc.* **18**, 68-72.
Measham, C. E. C., 1915. A botanical survey of some fields near Leicester. *Trans. Leicester lit. phil. Soc.* **19**, 17-25.
Mercer, G. E., 1914. The flora of Belgrave and Birstall. *Trans. Leicester lit. phil. Soc.* **18**, 76-92.
Merrett, C., 1666. *Pinax rerum naturalium Britannicarum* . . . . London: C. Pulleyn.
Messenger, K. G., 1970. *Senecio vernalis* Waldst. and Kit in Leicestershire. *Watsonia* **8**, 90.
Messenger, K. G., 1971. *Flora of Rutland*. Leicester: Leicester Museums.
Monk, J., 1794. *General view of the agriculture of the county of Leicester*. London.
Moscrop, W. J., 1866. A report on the farming of Leicestershire. *J. R. agric. Soc.* (2) **2**, 289-337.
Mott, F. T., 1868. *Notes Botanical in Charnwood Forest*. Third edition. London: W. Kent. pp.25-36.
Mott, F. T., 1878. *Prunella vulgaris*, white variety. *Midl. Nat.* **1**, 136.
Mott, F. T., 1885. The Leicestershire forms of *Capsella bursa-pastoris*. *Midl. Nat.* **8**, 217-220.
Mott, F. T., 1886. The campanulas of Leicestershire. *Trans. Leicester lit. phil. Soc.* **1:1**, 24-26.
Mott, F. T., 1887. The wild geraniums of Leicestershire. *Trans. Leicester lit. phil. Soc.* **1:3**, 14-16.
Mott, F. T., 1888. The native trees of Leicestershire. *Trans. Leicester lit. phil. Soc.* **1:8**, 20-23.

Mott, F. T., 1888. Quarterly Report of Section D, for Biology (Zoology and Botany). *Trans. Leicester lit. phil. Soc.* **1:9**, 40-41.
Mott, F. T., 1889. The ferns of Leicestershire. *Trans. Leicester lit. phil. Soc.* **1:12**, 25-28.
Mott, F. T., 1890. The native bulbs of Leicestershire. *Trans. Leicester lit. phil. Soc.* **2:2**, 68-71.
Mott, F. T., 1892. Quarterly Report of Section D, for Biology (Zoology and Botany). *Trans. Leicester lit. phil. Soc.* **2:10**, 459-461.
Mott, F. T., 1895. The Leicestershire brooks. *Trans. Leicester lit. phil. Soc.* **4:1**, 5-9.
Mott, F. T., 1897. Leicestershire roads and lanes. *Trans. Leicester lit. phil. Soc.* **4:10**, 413-417.
Mott, F. T., 1898. The shrubs of Leicestershire. *Trans. Leicester lit. phil. Soc.* **4:11**, 483-484.
Mott, F. T., 1898. The native undershrubs of Leicestershire. *Trans. Leicester lit. phil. Soc.* **5:1**, 39-40.
Mott, F. T., 1899. The wild fruits of Leicestershire. *Trans. Leicester lit. phil. Soc.* **5:5**, 229-236.
Mott, F. T. et al., 1886. *The flora of Leicestershire, including the cryptogams, with maps of the county.* London: Williams and Norgate.

Nature Conservancy Council, 1975. *Wildlife conservation in Charnwood Forest.* Huntingdon: Nature Conservancy Council.
Nature Conservancy Council, 1981. *A botanical survey of selected grassland sites in Leicestershire. England Field Unit Project no. 12.* 59pp.
Newman, E., 1847. Occurrence of *Udora canadensis*, a plant new to Britain and Europe near Market Harborough in Leicestershire. *Phytologist*, **2**, 1044, 1050.

Peacock, E. A. W., 1896. *Limnathemum peltatum* in S. Lincoln. *J. Bot.* **34**, 229.
Perring, F. and Sell, P. D., 1968. *Critical supplement to the Atlas of British Flora.* London: Thomas Nelson and Sons.
Petiver, J., [1716]. *Brittanicorum Concordia Graminum, Muscorum, Fungorum, Submarinorum &c.* London.
Petiver, J., 1767. *Jacobi Petiveri opera, historiam naturalem spectantia . . . gazophylactium.* London: J. Millan.
Pitt, W., 1809. *A general view of the agriculture of the county of Leicester, with observations on the means of its improvement . . . .* London: Richard Phillips.
Pinnock, W., 1821. *The history and topography of Leicestershire.* London.
Powell, W. P., 1923. The flora of Stoney Stanton, Sapcote and District. *The Hinckley Guardian and South Leicestershire Advertiser*, 2nd and 16th November.
Power, J., 1805. [Ms. notes in a copy of Turner and Dillwyn (1805) in the Library of the National Museum of Wales.]
Power, J., 1807. *Calendar of flora at Market Bosworth.* Hinckley. [Not seen].
Preston, T. A., 1890. Quarterly Report of Section D for Biology (Zoology and Botany). *Trans. Leicester lit. phil. Soc.* **2:5**, 204-206.
Preston, T. A., 1895. The flora of Cropstone Reservoir. *Trans. Leicester lit. phil. Soc.* **3**, 430-442.
Preston, T. A., 1896. Quarterly Report of Section D, for Biology (Zoology and Botany). *Trans. Leicester lit. phil. Soc.* **2:5**, 204-206.
Preston, T. A., 1900. Notes respecting the flora of the county since 1885. *Trans. Leicester lit. phil. Soc.* **5**, 411-419.
Preston, T. A., 1901. [Leicestershire Flora] *in* Vice, W. A., Quarterly Report of Section D (for Botany). *Trans. Leicester lit. phil. Soc.* **5:11**, 609-613.
Preston, T. A., 1902. Report of the Herbarium *in* Vice, W. A., Quarterly Report of Section D (for Botany). *Trans. Leicester lit. phil. Soc.* **6:4**, 205-206.
Preston, T. A., 1903. On the progress of the herbarium during the past year. *Trans. Leicester lit. phil. Soc.* **7**, 191-193.
Primavesi, A. L., 1967. Changes in the flora of part of Leicestershire since 1882. *Proc. bot. Soc. Br. Isl.* **6**, 343-347.

Pulteney, R., 1747. *Catalogus stirpium circa Lactodorum nascentium* or *A Catalogue of plants in the neighbourhood of Loughborough by Richd. Pulteney M.D., the original M.S. in his own handwriting at an early time of his life*. Ms. 186pp. [Leicestershire Museums Service. L.R.O., FY10, box 9.]

Pulteney, R., 1749. *Opusculum botanicum locos plantarum natales circa Loughborough et in agris adjacentibus* . . . . Ms. [Linnean Society of London. Ms. collections. Ms. 31.]

Pulteney, R., [1750-1752]. *Pharmaceutical ms. and hortus siccus*. [British Museum (Natural History). Botany Department Ms. 72a. Ms. not found 1981; specimens thought to have been dispersed through the British herbarium.]

Pulteney, R., [1752]. *A catalogue of plants spontaneously growing about Loughborough and the adjacent villages*. Ms. [British Museum (Natural History), Botany Department Library. Banks Ms.90; B.47.5.]

Pulteney, R., 1756. An account of some of the more rare English plants observed in Leicestershire. *Phil. Trans. R. Soc.* **49**, 803-866.

Pulteney, R., 1789. Rare plants found in Leicestershire *in* R. Gough's translation of W. Camden's *Britannia* . . . . First edition. pp.215-217.

Pulteney, R., 1795. A catalogue of some of the more rare plants found in the neighbourhood of Leicester, Loughborough, and in Charley Forest *in* Nichols, J., *The history and antiquities of the county of Leicester* **1:1**, clxxvii-cxc. London: J. Nichols.

Purchas, W. H., 1860. *Agrimonia odorata* from Staunton Harold *in* Report of December meeting of Thirsk Natural History Society. *Phytologist* (New Series) **4**, 59.

Purton, T., 1817-1850. *A botanical description of British plants in the midland counties*. Stratford-upon-Avon.

Quilter, H. E., 1911. A possible explanation of the presence of the dwarf elder, or danewort, *Sambucus ebulus* L., in Leicestershire *in* Ardington, E. E., Report of Section D – Botany. *Trans. Leicester lit. phil. Soc.* **15**, 108-111.

Ratcliffe. D. and Conolly, A. P., 1972. Flora and vegetation *in* Pye N. (ed.), *Leicester and its region*. Leicester: Leicester University Press. pp.133-159.

Ray, J., 1724. *Synopsis methodica stirpium Brittanicarum*. Third edition. London: William and John Innys.

Reader, H. P., 1883. Leicestershire plants. *J. Bot.* **21**, 374.

Richards, A. J., 1972. The *Taraxacum* Flora of the British Isles. *Watsonia*. Supplement to **9**.

Richens, R., 1983. *Elm*. Cambridge: Cambridge University Press.

Rogers, W. M., 1900. *Handbook of British Rubi*. London: Duckworth.

Rogers, W. M., 1909. Supplementary records of British *Rubi* (April, 1900 – December, 1908). *J. Bot.* **47**, 310-318, 340-346.

Salmon, C. E., 1909. *Gagea lutea* in Leicestershire. *J. Bot.* **47**, 31-32.

Salmon, C. E., 1914. *Alchemilla acutidens* Buser and other forms of *Alchemilla vulgaris* L. *J. Bot.* **52**, 281-289.

Salter, T. B., 1845. Remarks on some forms of *Rubus*. *Ann. Mag. nat. Hist.* **16**, 361-372.

Salter, T. B., 1850a. A descriptive table of British brambles. *Bot. Gaz. Lond.* **2**, 113-131.

Salter, T. B., 1850b. Critical notes on British brambles. *Bot. Gaz. Lond.* **2**, 147-156.

Sell, P. D. and West, C., 1974. Materials for a Flora of Turkey. XXX. Compositae, I. *Hieracium*. *Notes R. bot. Gdn. Edinb.* **33**, 241-248.

Smith, J. E., 1800-1804. *Flora Britannica*. London.

Sowerby, J. and Smith, J. E., 1808. *English Botany* . . . . **26**, text to pl. 1808. [First record of *Salix alba* × *fragilis*.]

Sowter, F. A., 1949. Biological Flora of the British Isles. *Arum maculatum* L. *J. Ecol.* **37**, 207-219.

Sowter, F. A., 1950. Report of B.S.B.I. excursion to Leicester, 18th-21st June, 1948. *B.S.B.I. Yearbook for 1950*. pp.43-44.

Sowter, F. A., 1960. Our diminishing flora. *Trans. Leicester lit. phil. Soc.* **54**, 20-27.

Spanton, J., 1858. *A companion to Charnwood Forest.* Loughborough: J. H. Gray. Appendix. Botany. pp.63-65.

Spencer, J. and T., [no date]. *Spencer's new guide to Charnwood Forest.* Appendix. Botany. pp.92-93.

Stukeley, W., 1724. *Itinerarium curiosum.* London. [Not seen].

Taylor, F. J., 1956. Biological Flora of the British Isles. *Carex flacca* Schreb. *J. Ecol.* **44**, 281-290.

Timson, J., 1963. The taxonomy of *Polygonum lapathifolium* L., *P. nodosum* Pers. and *P. tomentosum* Schrank. *Watsonia* **5**, 386-395.

Tomkins, H., 1887. Quarterly Report of Section D, for Biology (Zoology and Botany). *Trans. Leicester lit. phil. Soc.* **1:5**, 22-26.

Turner, D. and Dillwyn, L. W., 1805. *The botanist's guide through England and Wales* **2**, 374-384. London: Phillips and Farndon.

Tutin, T. G., 1973. Weeds of a Leicester garden. *Watsonia* **9**, 263-267.

Velten, E. C. W., 1934. *A survey of the Market Harborough grasslands.* Oxford: Shakespeare Head Press.

Vice, W. A., 1900. Quarterly Report of Section D (for Botany). *Trans. Leicester lit. phil. Soc.* **5:10**, 524-527.

[Vice, W. A.], 1901. Quarterly Report of Section D (for Botany). *Trans. Leicester lit. phil. Soc.* **6:2**, 88-95.

[Vice, W. A.], 1902. Quarterly Report of Section D (for Botany). *Trans. Leicester lit. phil. Soc.* **7:1**, 53-59.

Vice, W. A., 1905. List of casuals and aliens gathered at Blaby Mills, 1903. *Trans. Leicester lit. phil. Soc.* **9:2**, 105-109.

Wade, A. E., 1919. The flora of Aylestone and Narborough bogs. *Trans. Leicester lit. phil. Soc.* **20**, 20-46.

Wade, P. M. and Beresford, J. E., 1982. The use of the aquatic macroflora to describe changes in the environmental quality of the River Soar. *J. Fac. hum. environ. Sci. Loughborough Univ.* **2**, 47-60.

Walford, T., 1818. *The scientific tourist through England, Wales and Scotland.* London.

Walpole, M., 1969. Orchids in Leicestershire. *Ann. Rep. Leics. Trust Nat. Cons., 1968-69.* pp.27-39.

Walpole, M., 1976. Natural history *in* Evans, I. M. (ed.), *Charnwood's Heritage.* Leicester: Leicestershire Museums Service. pp.93-97.

Walpoole, G. A., [1782]. *The new British traveller. An account of curious plants to be found in different parts of this country. Chap. 7 . . . the county of Leicester.* London. pp.136-144. [Now considered to have been published in 1784.]

Watson, H. C., [1832]. *Outlines of the geographical distributions of British plants . . . .* Edinburgh.

Watson, H. C., 1835-1837. *The new botanist's guide to the localities of the rarer plants of Britain.* London.

Watson, H. C., 1849. Notes on certain plants for distribution by the Botanical Society in 1849. *Phytologist* **3**, 478.

Watson, H. C., 1873-1874. *Topographical Botany.* Thames Ditton. 2 vols.

W.B.E.C. Rep. Watsonian Botanical Exchange Club Reports, 1884-1934.

Withering, W., 1801. *A systematic arrangement of British plants . . . .* Fourth edition. 4 vols. London: T. Cadell et al.

Wolley-Dod, A. H., 1930-1931. A revision of the British Roses. *J. Bot.* Supplement to **68**, pp.1-16; supplement to **69**, pp.17-111.

Young, D. P., 1958. *Oxalis* in the British Isles. *Watsonia* **4:2**, 51-69.

DISTRIBUTION MAPS

In the following pages, distribution maps are given for all taxa described in the text the status of which can be placed unequivocally in one or more of the categories 'constituent', 'residual' or 'intrusive' (see page 168), and which were recorded in the recent survey. Maps are not given for species which are considered to be casuals, non-persistent garden escapes or universally planted. The decision to omit a map is essentially an arbitrary one based on subjective judgement, and some inconsistency in applying the norms stated above must be admitted. It was felt that, allowing for inevitable changes in the course of time, the user of the Flora should reasonably expect the map to show the distribution of the plant concerned at the time it was being compiled. This is certainly not the case with casuals, recorded during a survey extending over thirteen years, which do not persist at one site for more than a year or two. However the distribution of some casuals is of interest in itself, and in other cases it is difficult to decide whether the plant should be classified as casual or intrusive. Moreover there is in some cases evidence that a plant which was formerly of casual status is beginning to establish itself as a permanent member of the flora. Some of these species have been given the benefit of the doubt and have a map. So too have some planted species, such as *Picea abies*, which have a significant effect upon landscape and ecology.

A further source of inconsistency in the presentation of maps is the case of those taxa for which the survey has not provided complete coverage of the county. This applies particularly to some *Rosa* hybrids and species of *Rubus* and *Taraxacum*. As explained in the introduction to the genus, our knowledge of *Rosa* hybrids is at present insufficient to construct meaningful maps, and these have been omitted. On the other hand, it was felt that although the survey for *Rubus* and *Taraxacum* was incomplete, the maps were of sufficient interest to be included. The hollow dots on the *Taraxacum* maps represent pre-1960 records which were verified from herbarium material.

It is important to remember that a dot on the map may represent a single specimen in a four square kilometre tetrad or thousands of specimens. In other words, the maps show occurrence of a species but not its frequency. Thus on the map for *Dactylorhiza fuchsii* the dot in tetrad 61H represents a single plant, whereas the dot in tetrad 72L represents a population of many thousand. Similarly, when *Hippuris vulgaris* was recorded at Goadby Marwood, the entire surface of the shallow middle lake in the Hall grounds was covered with it, and the dot in tetrad 72T probably represents over 90% of the entire population of this species in the county.

It should also be remembered that these maps represent the distribution of the species based essentially on records obtained in the 13 year period 1968-1980, together with any significant additions up to mid-1987. Even during this period habitats altered or were destroyed, either by natural succession or by human interference. Thus, dismantled railway tracks became covered with scrub or were converted to agricultural use, worked-out gravel pits and quarries were filled in and derelict urban sites were built over. It is changes such as these over the last 50 years which have made the production of this present Flora of Leicestershire desirable, and which may inspire others to prepare another in the future.

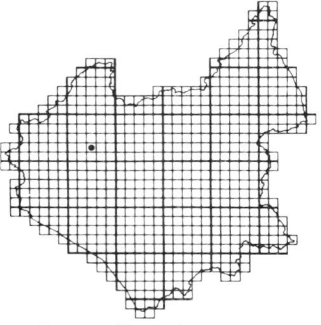

Lycopodiaceae - Aspleniaceae

Maps 1a-l

a. Lycopodium clavatum
b. Equisetum hyemale 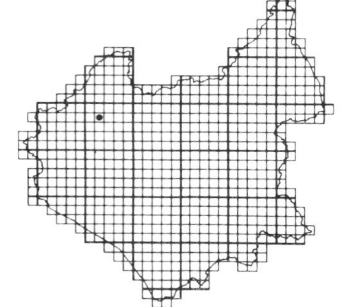
c. Equisetum fluviatile 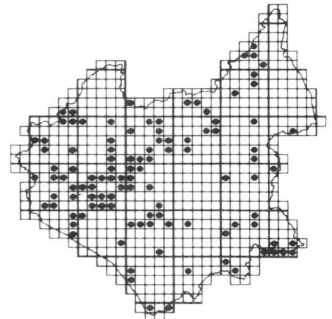
d. Equisetum palustre 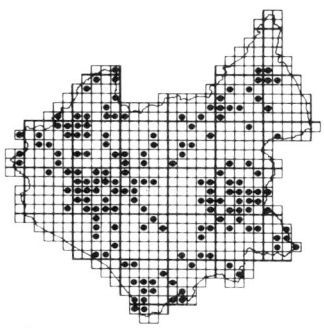
e. Equisetum sylvaticum 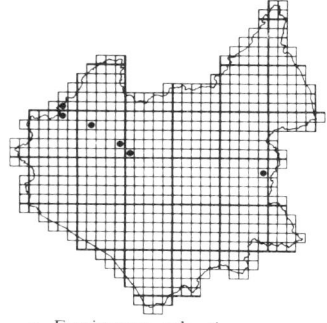
f. Equisetum arvense 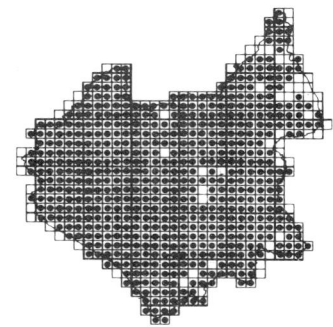
g. Equisetum telmateia 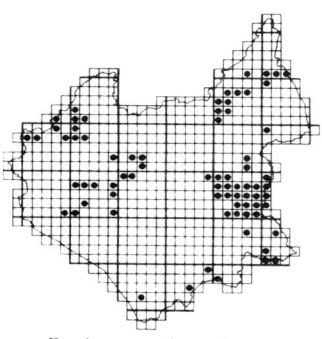
h. Ophioglossum vulgatum 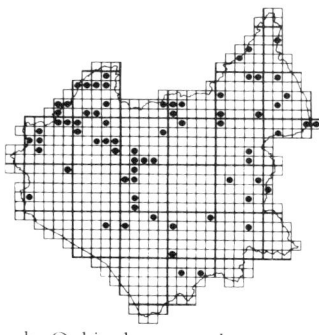
i. Botrychium lunaria 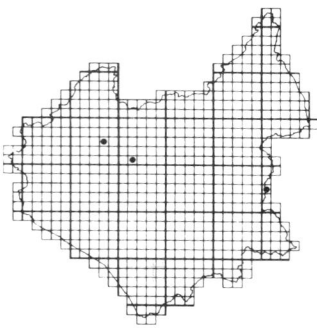
j. Pteridium aquilinum 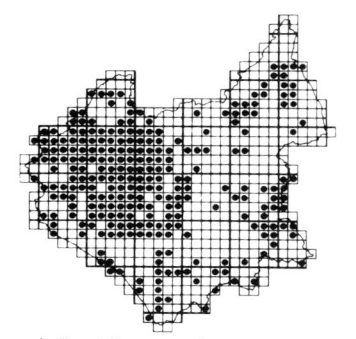
k. Thelypteris limbosperma 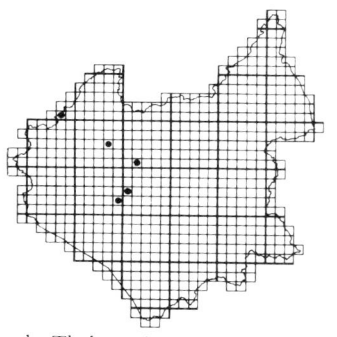
l. Asplenium trichomanes

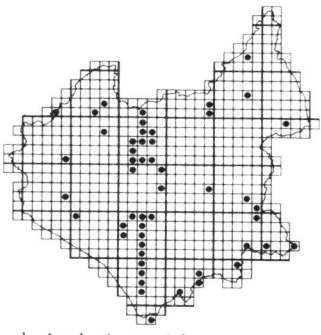

Aspleniaceae - Aspidiaceae

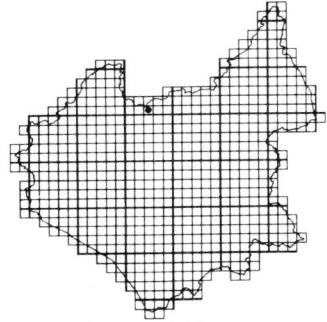

Maps 2a-l

a. Asplenium viride

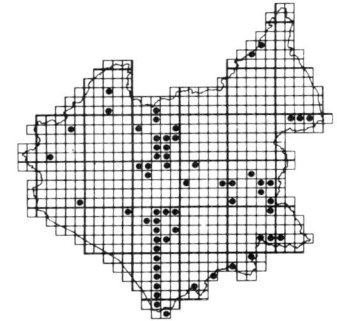

b. Asplenium adiantum-nigrum

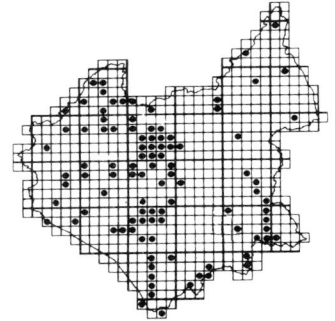

c. Asplenium ruta-muraria

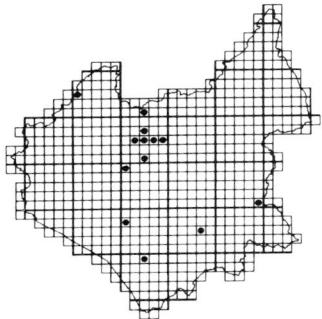

d. Ceterach officinarum

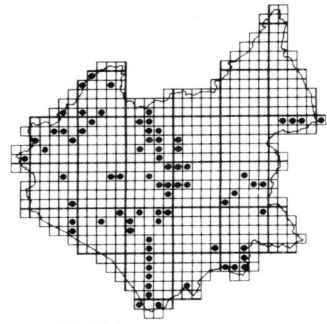

e. Phyllitis scolopendrium

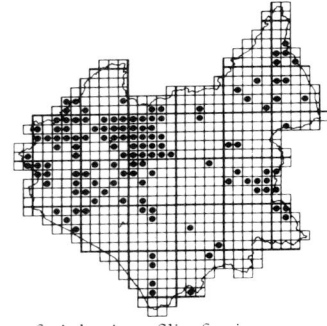

f. Athyrium filix-femina

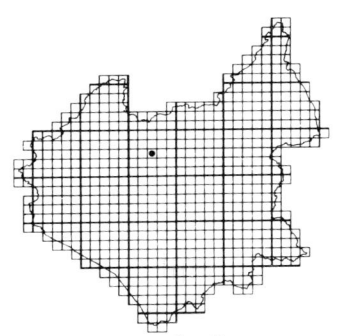

g. Cystopteris fragilis

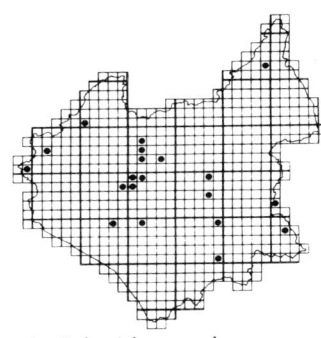

h. Polystichum aculeatum

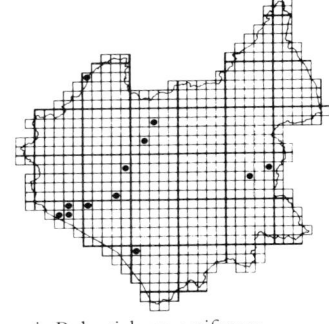

i. Polystichum setiferum

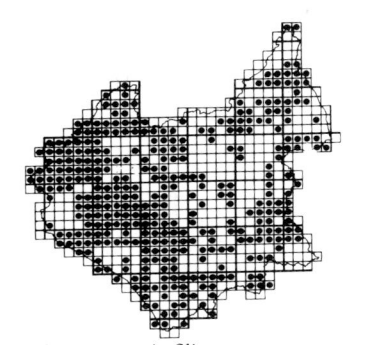

j. Dryopteris filix-mas

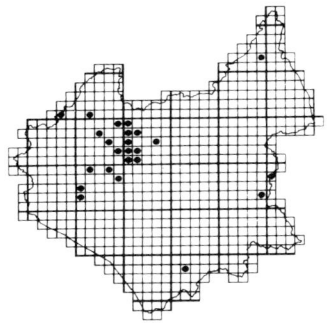

k. Dryopteris borreri

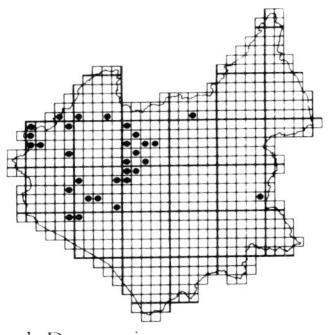

l. Dryopteris carthusiana

Aspidiaceae - Salicaceae

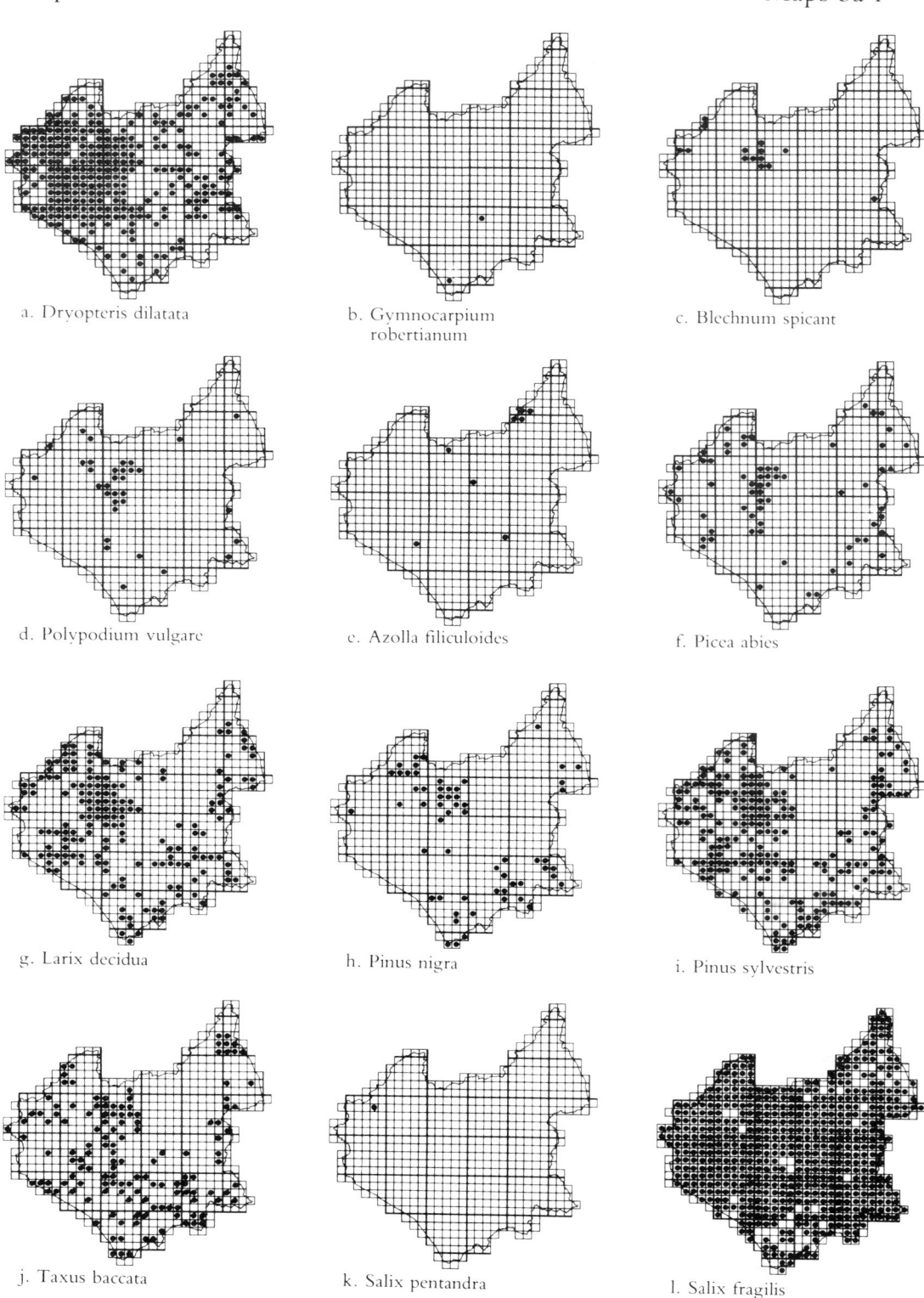

Maps 3a-l

a. Dryopteris dilatata

b. Gymnocarpium robertianum

c. Blechnum spicant

d. Polypodium vulgare

e. Azolla filiculoides

f. Picea abies

g. Larix decidua

h. Pinus nigra

i. Pinus sylvestris

j. Taxus baccata

k. Salix pentandra

l. Salix fragilis

Salicaceae

Maps 4a-l

a. Salix alba

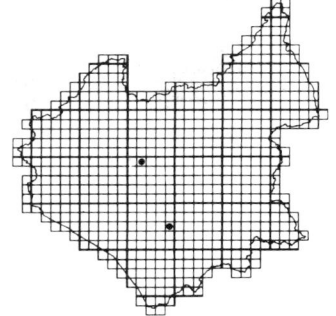

b. Salix alba × fragilis

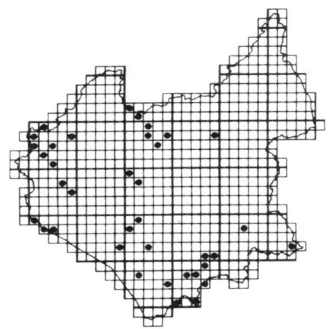

c. Salix triandra

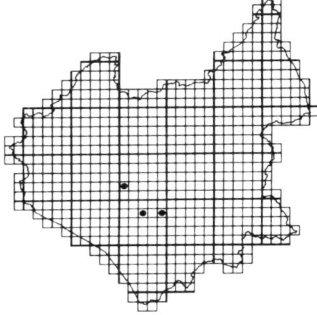

d. Salix triandra × viminalis

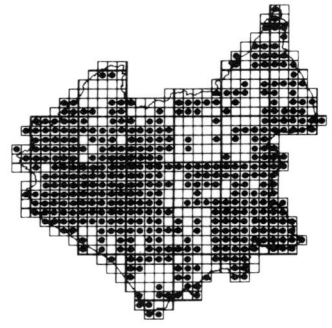

e. Salix cinerea aggregate

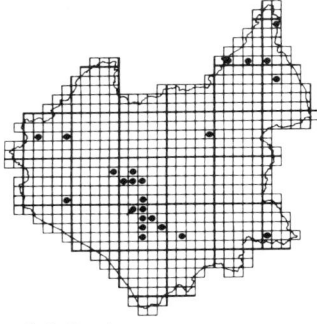

f. Salix cinerea ssp. cinerea

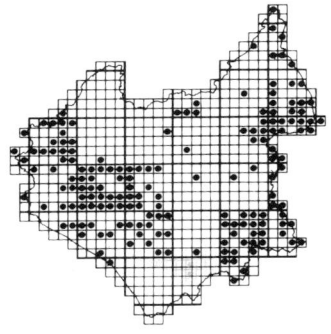

g. Salix cinerea ssp. oleifolia

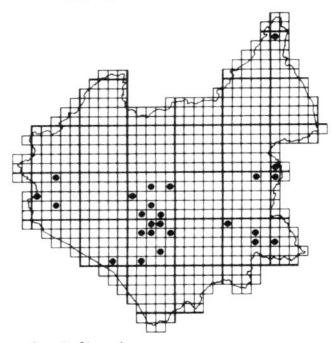

h. Salix cinerea × viminalis

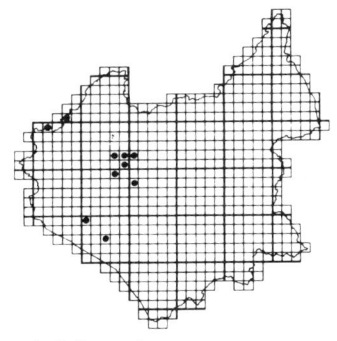

i. Salix aurita

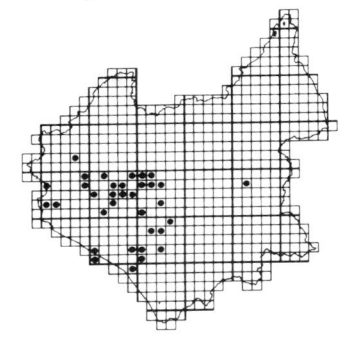

j. Salix aurita × cinerea

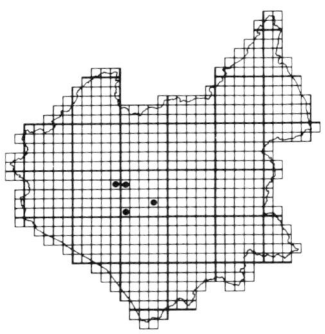

k. Salix aurita × caprea

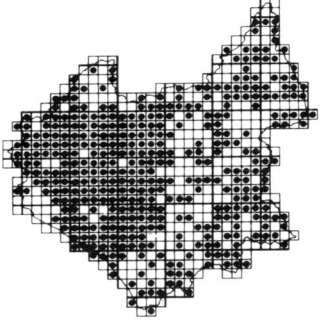

l. Salix caprea

Salicaceae

Maps 5a-l

a. Salix caprea × cinerea

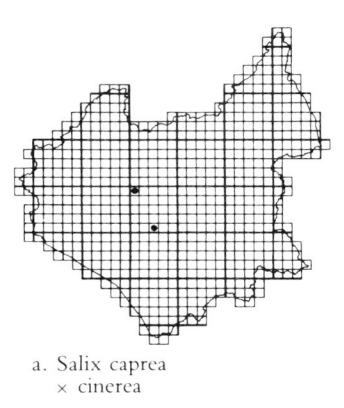

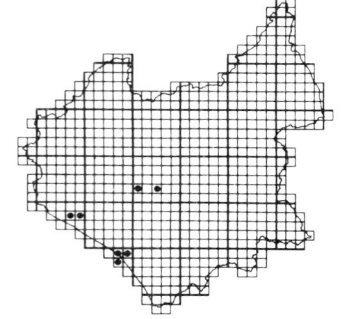

b. Salix caprea × viminalis

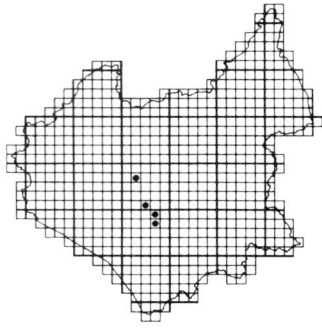

c. Salix caprea × cinerea × viminalis

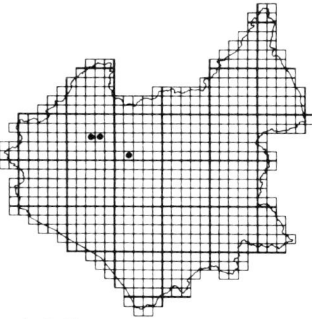

d. Salix repens

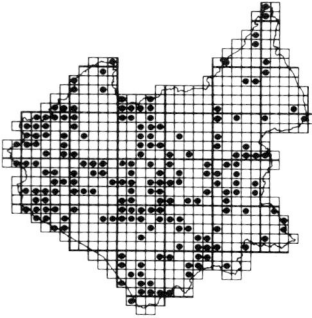

e. Salix viminalis

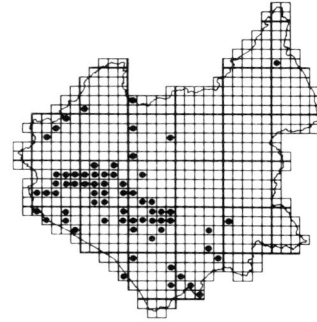

f. Salix purpurea

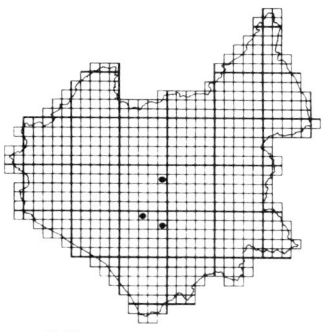

g. Salix purpurea × viminalis

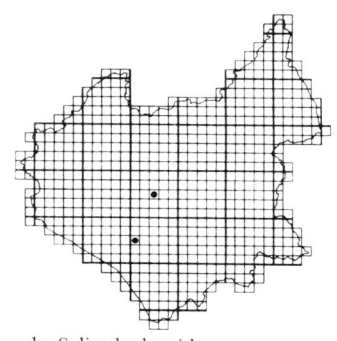

h. Salix daphnoides

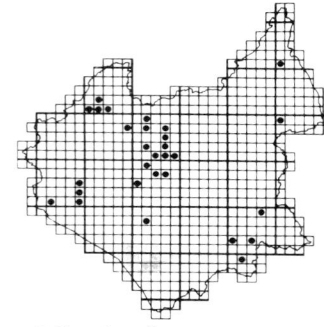

i. Populus alba

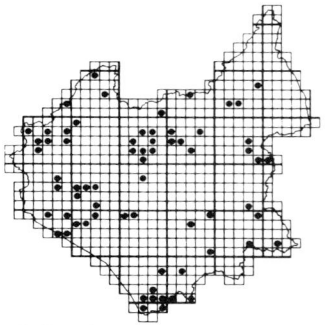

j. Populus canescens

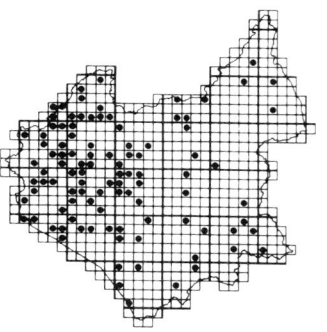

k. Populus tremula

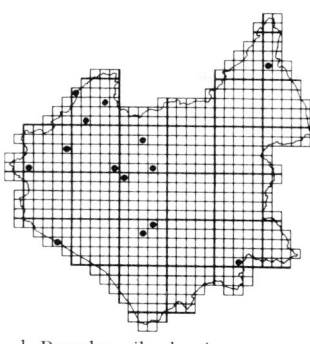

l. Populus gileadensis

Salicaceae – Ulmaceae Maps 6a-l

a. Populus × canadensis
b. Betula pendula
c. Betula pubescens
d. Alnus glutinosa
e. Carpinus betulus
f. Corylus avellana
g. Fagus sylvatica
h. Castanea sativa
i. Quercus petraea
j. Quercus robur
k. Ulmus glabra
l. Ulmus glabra × minor s.l.

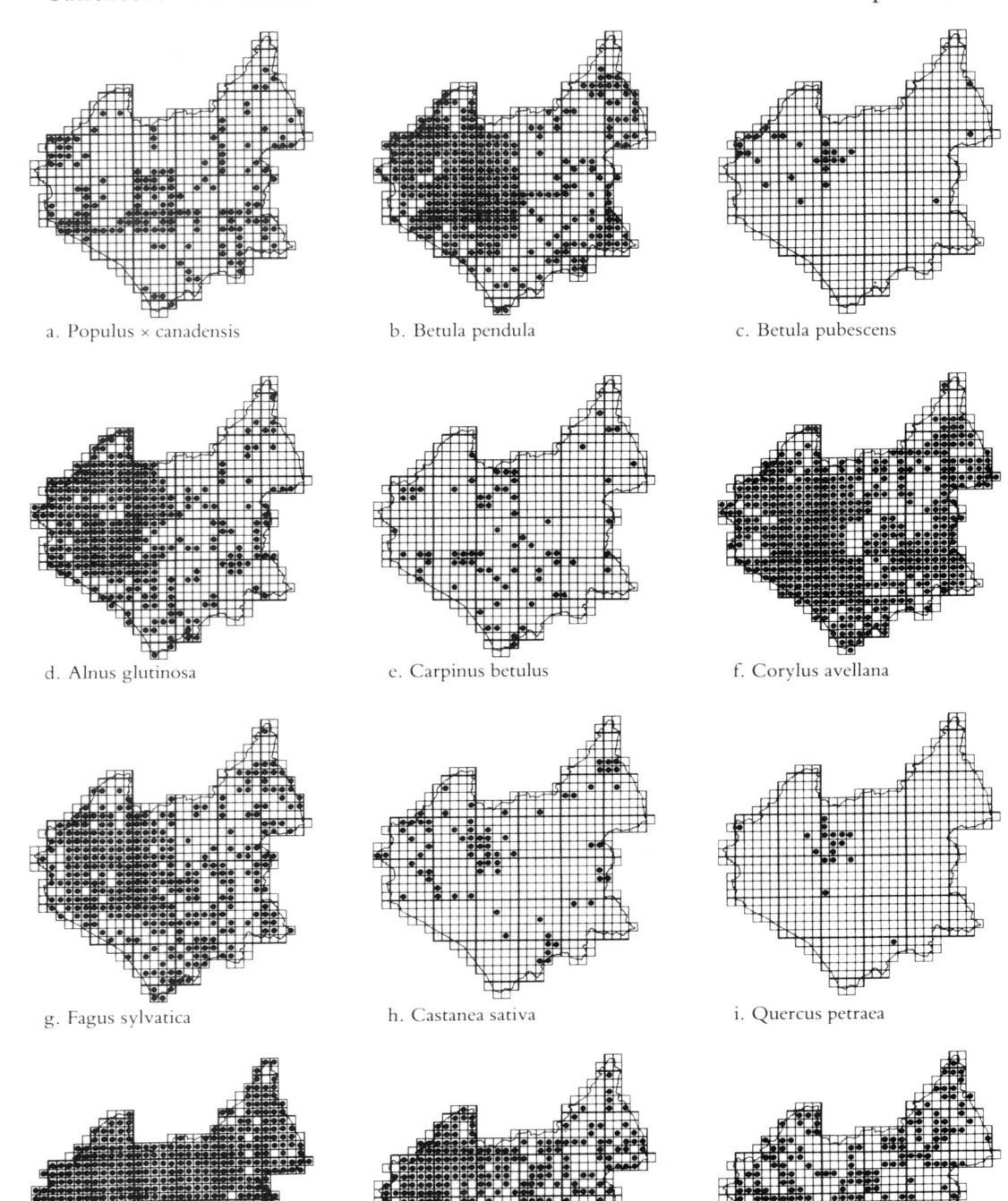

Ulmaceae – Polygonaceae Maps 7a–l

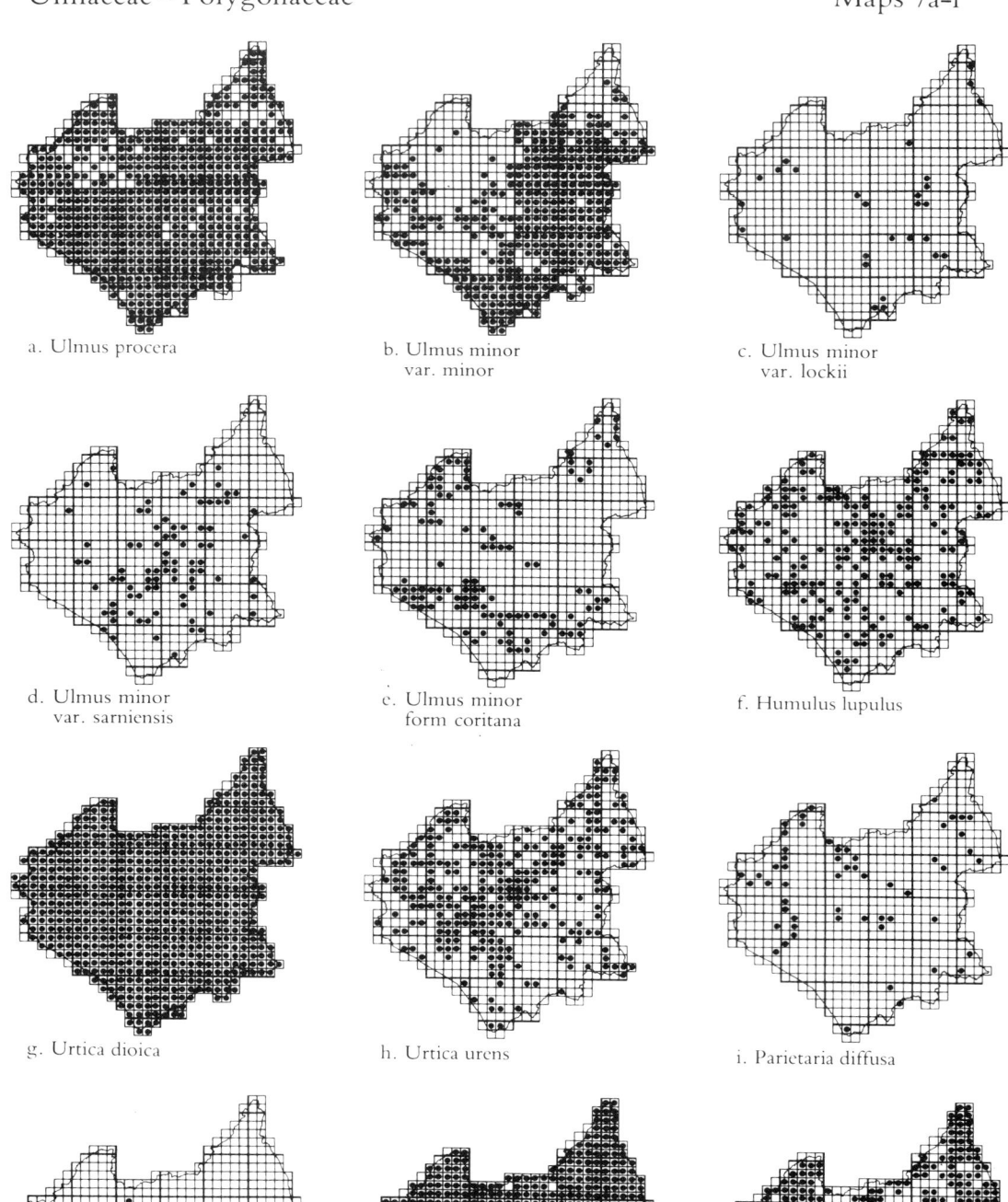

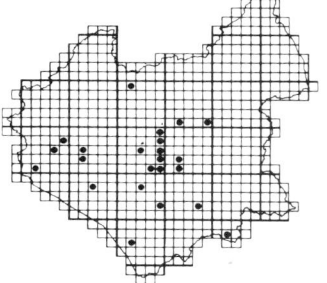

a. Ulmus procera

b. Ulmus minor var. minor

c. Ulmus minor var. lockii

d. Ulmus minor var. sarniensis

e. Ulmus minor form coritana

f. Humulus lupulus

g. Urtica dioica

h. Urtica urens

i. Parietaria diffusa

j. Viscum album

k. Polygonum aviculare aggregate

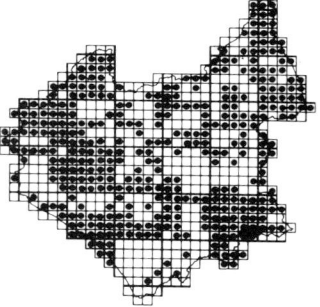

l. Polygonum aviculare s.s.

Polygonaceae

Maps 8a-l

a. Polygonum arenastrum

b. Polygonum hydropiper

c. Polygonum persicaria

d. Polygonum lapathifolium

e. Polygonum amphibium

f. Polygonum bistorta

g. Polygonum amplexicaule

h. Polygonum polystachyum

i. Bilderdykia convolvulus

j. Reynoutria japonica

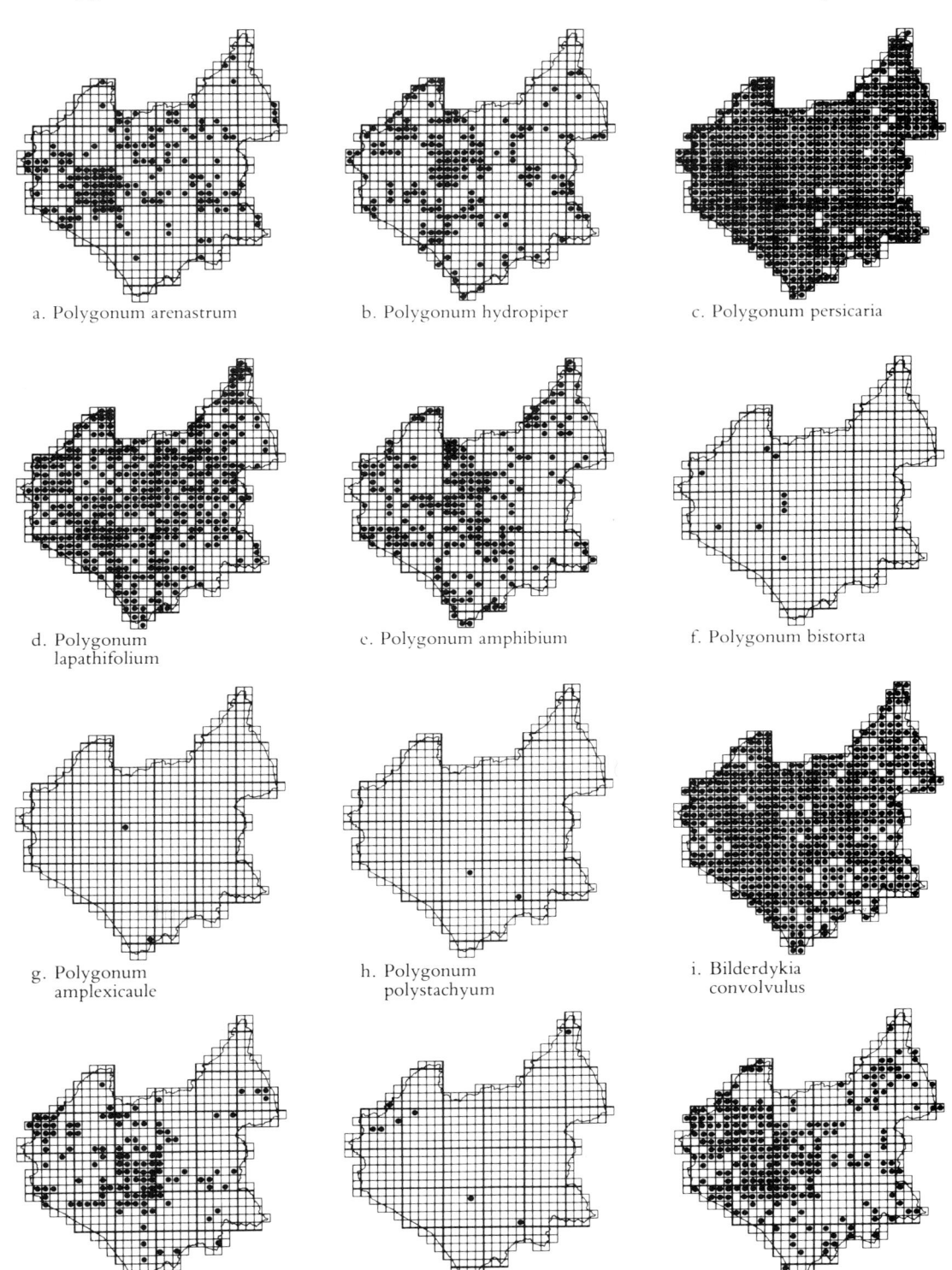

k. Reynoutria sachalinensis

l. Rumex acetosella

Polygonaceae - Chenopodiaceae

Maps 9a-l

a. Rumex acetosa

b. Rumex hydrolapathum

c. Rumex crispus

d. Rumex crispus × palustris

e. Rumex conglomeratus

f. Rumex sanguineus

g. Rumex obtusifolius

h. Rumex palustris

i. Rumex maritimus

j. Chenopodium bonus-henricus

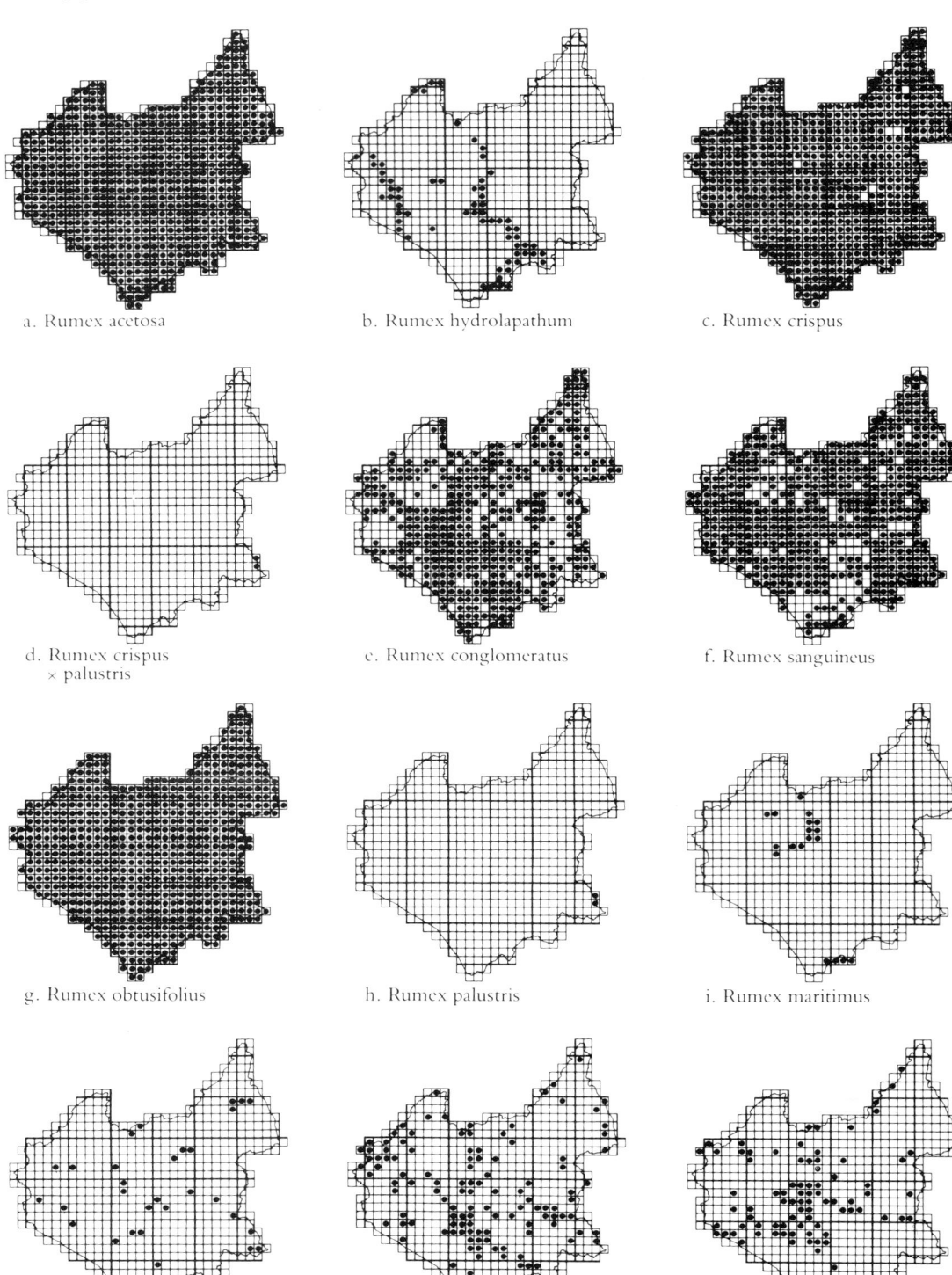

k. Chenopodium rubrum

l. Chenopodium polyspermum

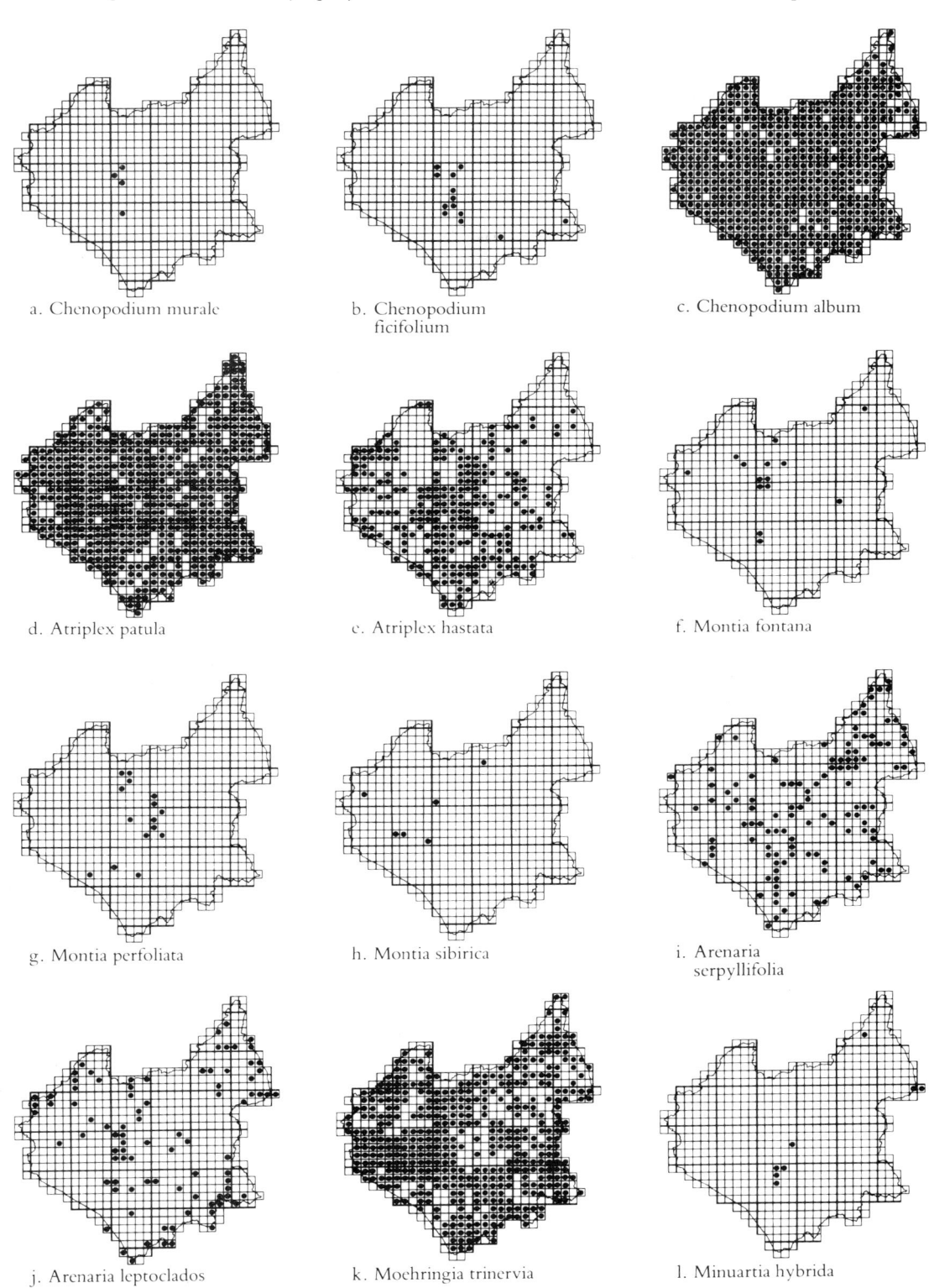

Chenopodiaceae - Caryophyllaceae

Maps 10a-l

a. Chenopodium murale

b. Chenopodium ficifolium

c. Chenopodium album

d. Atriplex patula

e. Atriplex hastata

f. Montia fontana

g. Montia perfoliata

h. Montia sibirica

i. Arenaria serpyllifolia

j. Arenaria leptoclados

k. Moehringia trinervia

l. Minuartia hybrida

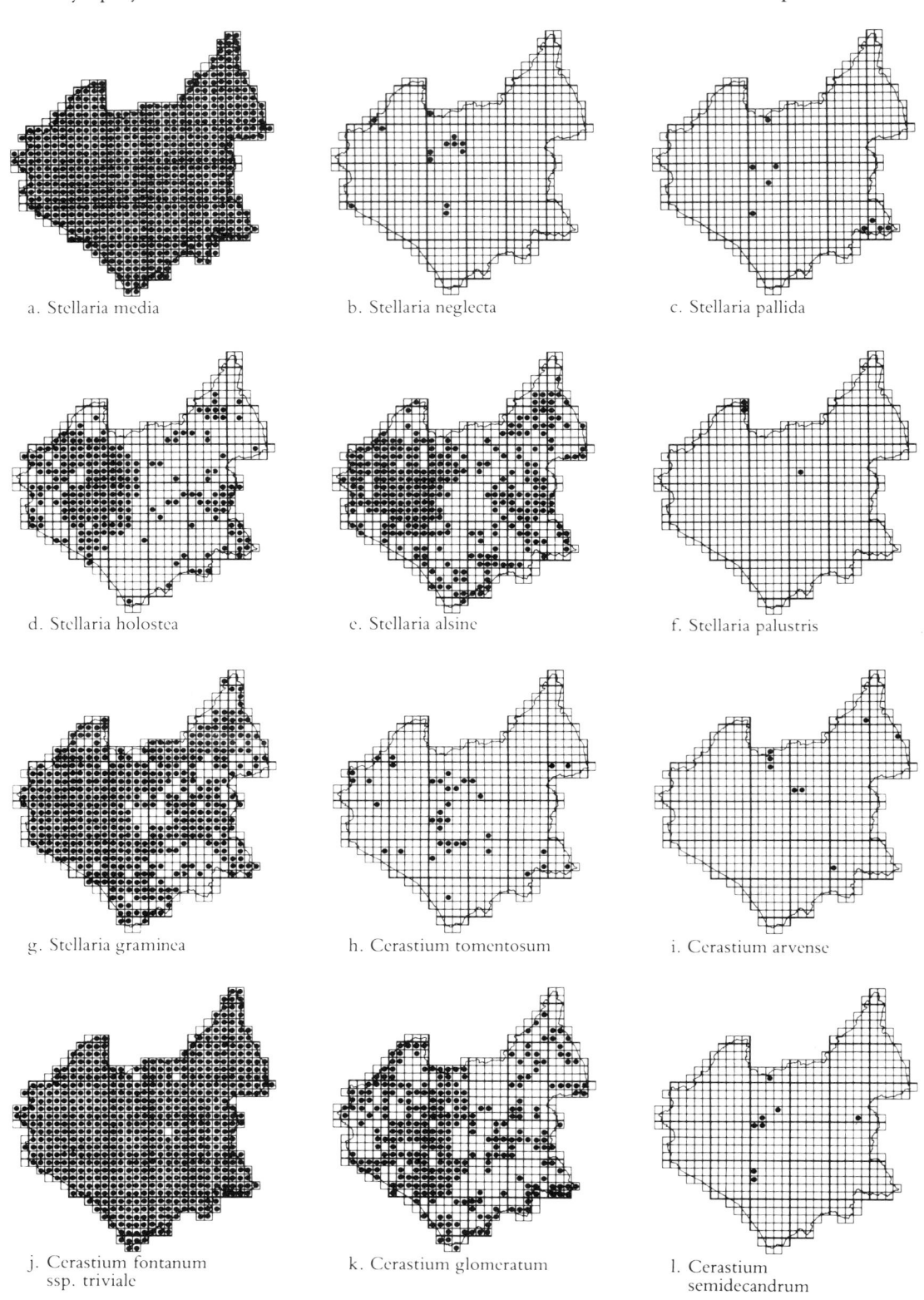

Caryophyllaceae

Maps 11a-l

a. Stellaria media

b. Stellaria neglecta

c. Stellaria pallida

d. Stellaria holostea

e. Stellaria alsine

f. Stellaria palustris

g. Stellaria graminea

h. Cerastium tomentosum

i. Cerastium arvense

j. Cerastium fontanum ssp. triviale

k. Cerastium glomeratum

l. Cerastium semidecandrum

Caryophyllaceae

Maps 12a-l

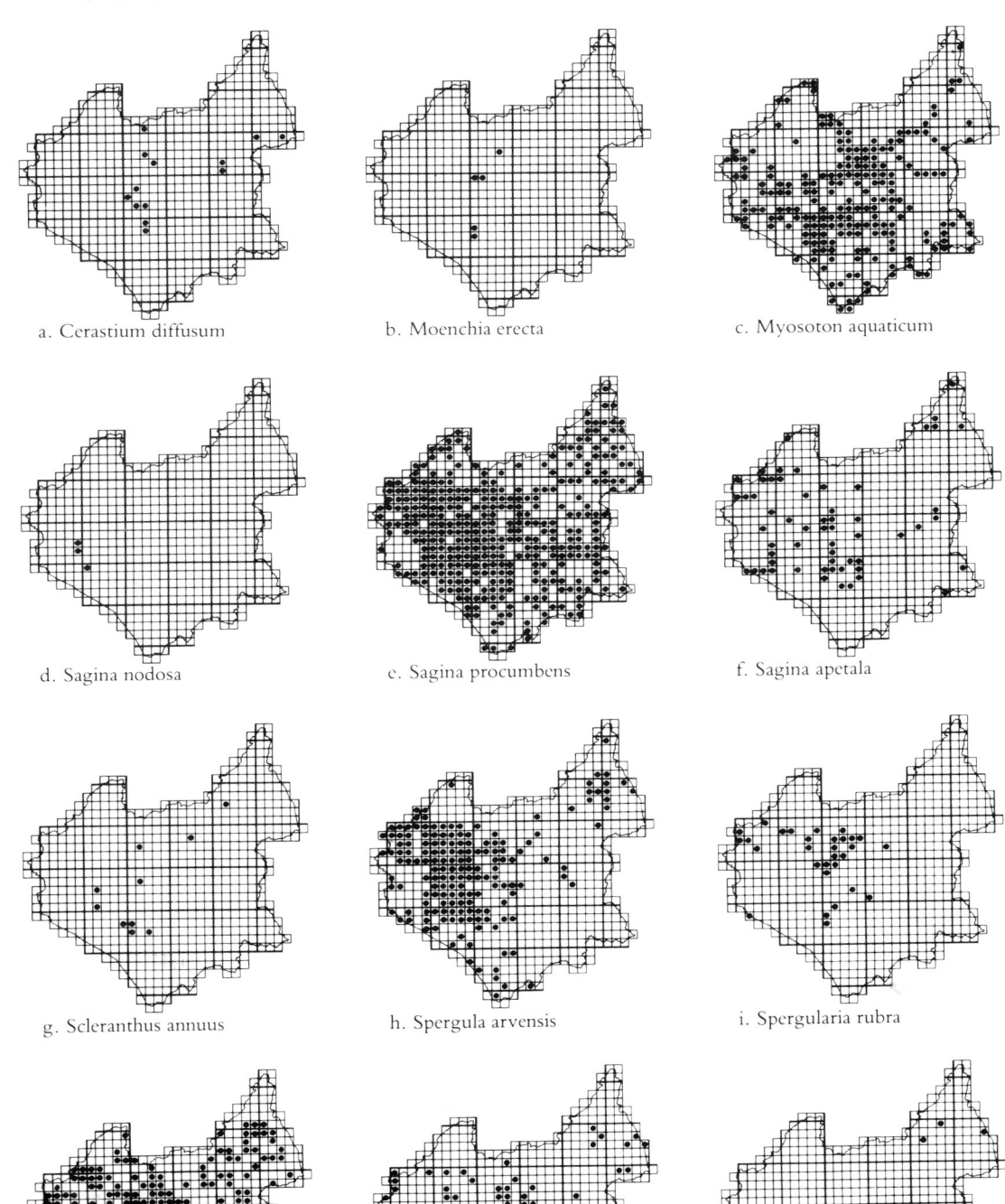

a. Cerastium diffusum

b. Moenchia erecta

c. Myosoton aquaticum

d. Sagina nodosa

e. Sagina procumbens

f. Sagina apetala

g. Scleranthus annuus

h. Spergula arvensis

i. Spergularia rubra

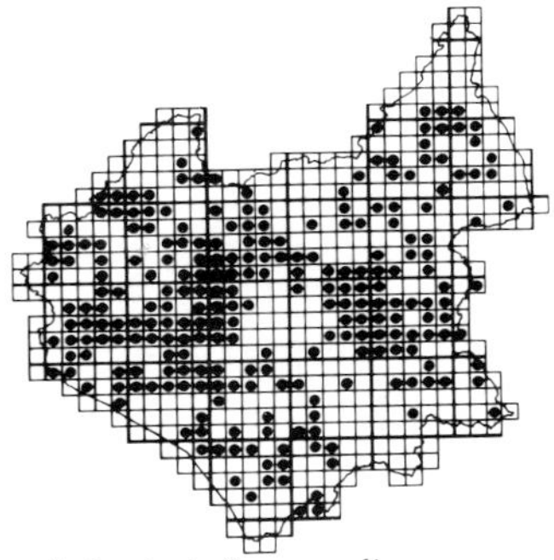

j. Lychnis flos-cuculi

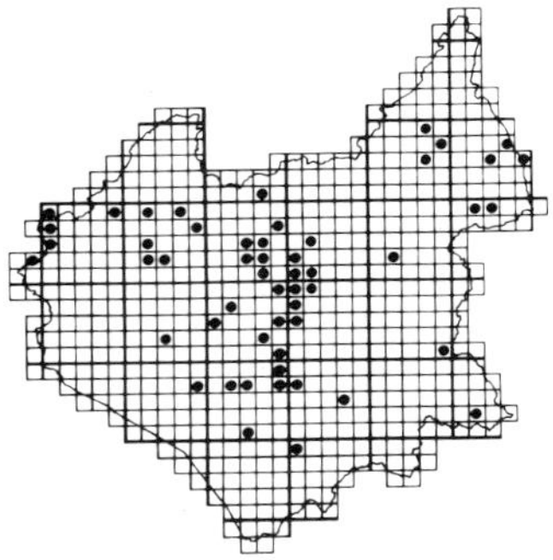

k. Silene vulgaris

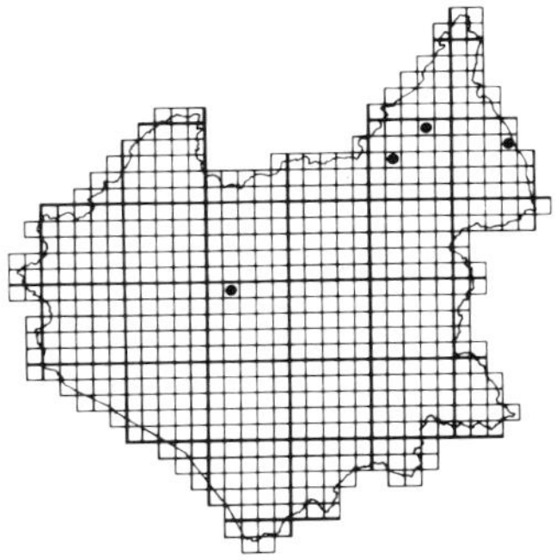

l. Silene noctiflora

Caryophyllaceae - Ranunculaceae

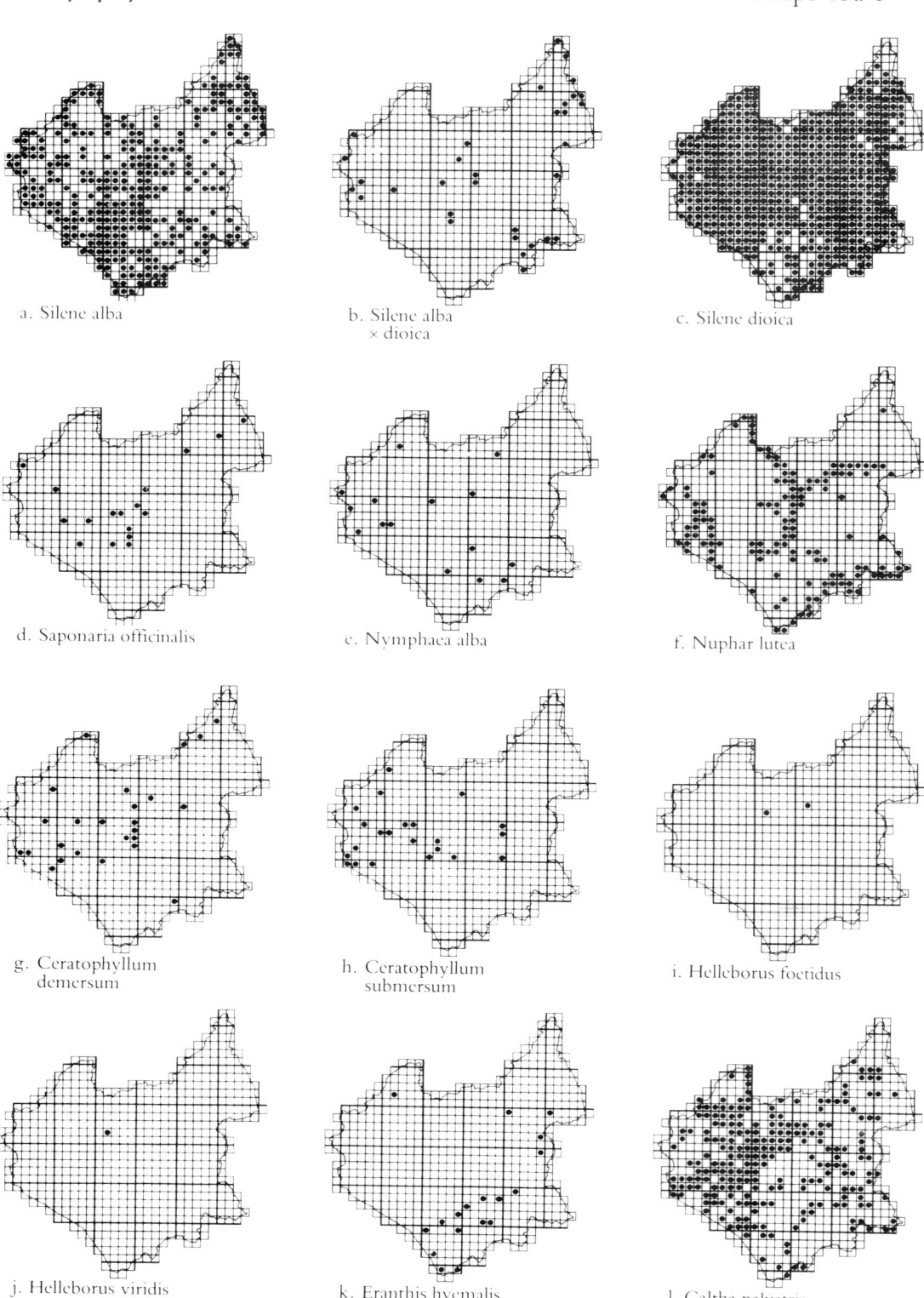

Maps 13a-l

a. Silene alba

b. Silene alba × dioica

c. Silene dioica

d. Saponaria officinalis

e. Nymphaea alba

f. Nuphar lutea

g. Ceratophyllum demersum

h. Ceratophyllum submersum

i. Helleborus foetidus

j. Helleborus viridis

k. Eranthis hyemalis

l. Caltha palustris

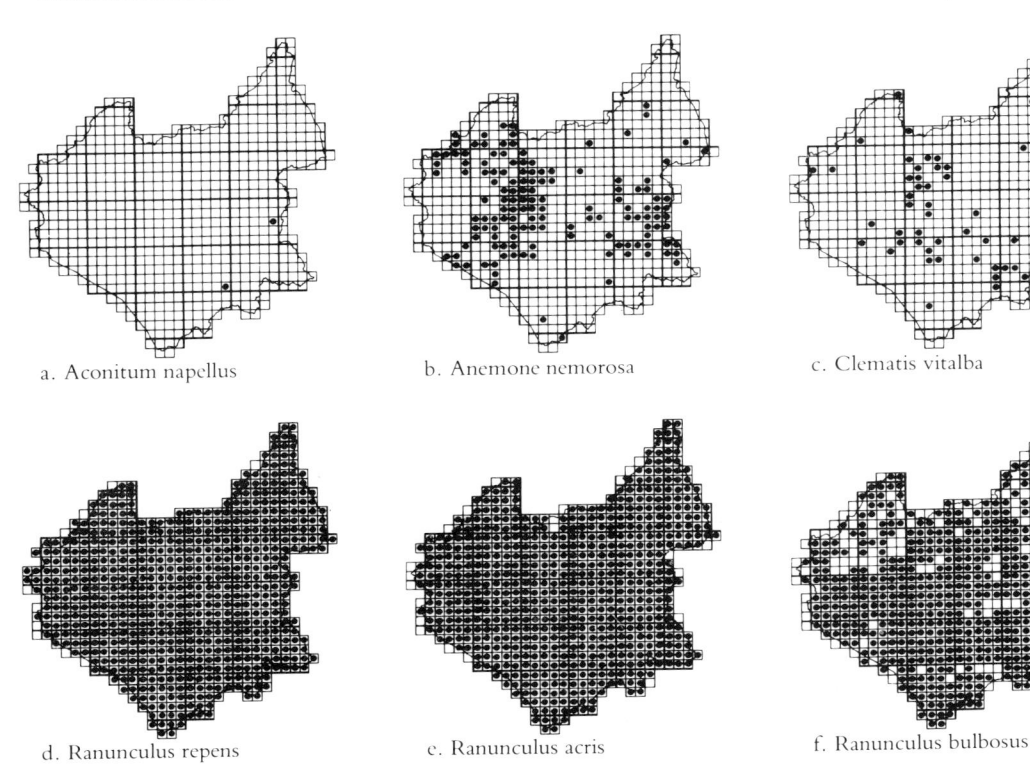

Ranunculaceae

Maps 14a-l

a. Aconitum napellus

b. Anemone nemorosa

c. Clematis vitalba

d. Ranunculus repens

e. Ranunculus acris

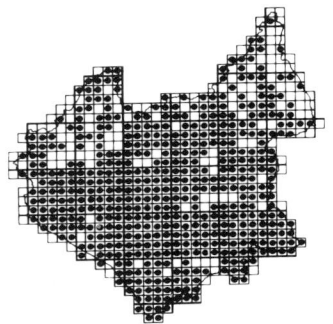

f. Ranunculus bulbosus

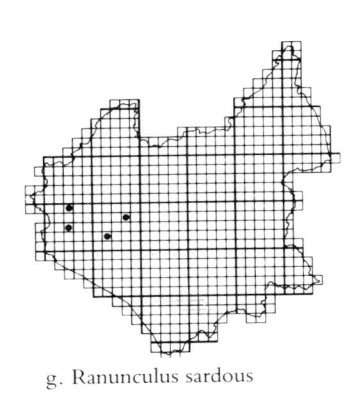

g. Ranunculus sardous

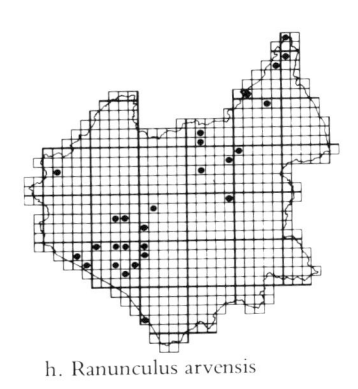

h. Ranunculus arvensis

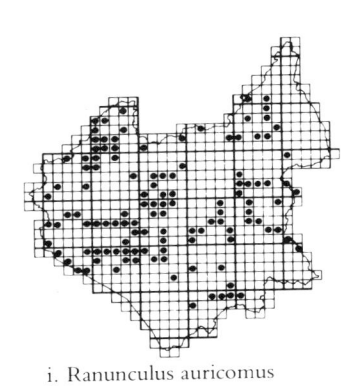

i. Ranunculus auricomus

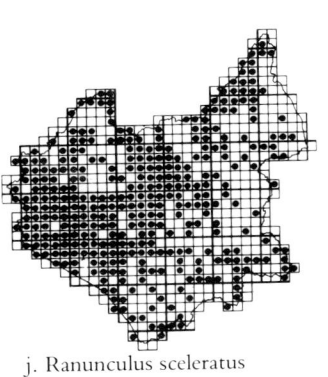

j. Ranunculus sceleratus

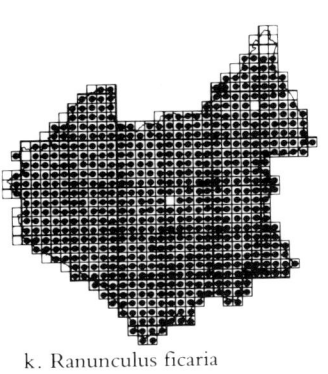

k. Ranunculus ficaria

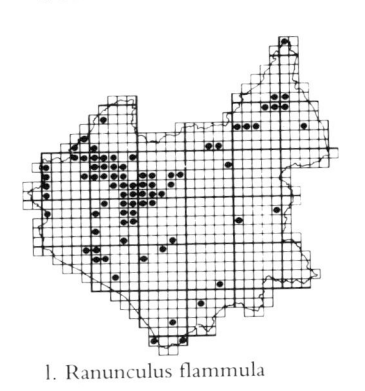

l. Ranunculus flammula

Ranunculaceae - Berberidaceae

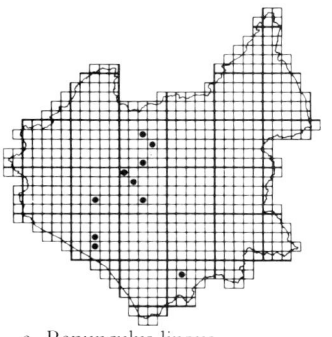

Maps 15a-l

a. Ranunculus lingua

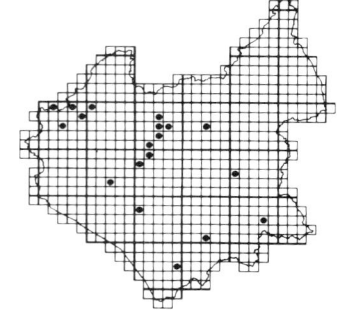

b. Ranunculus hederaceus

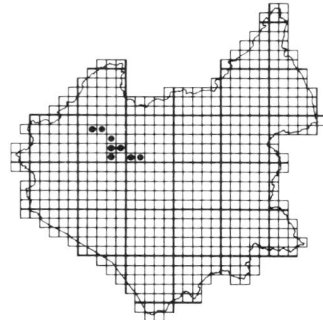

c. Ranunculus omiophyllus

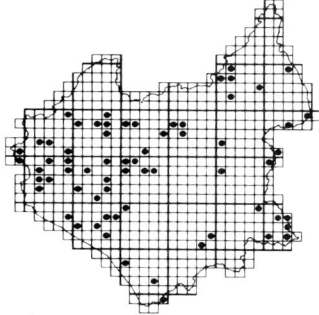

d. Ranunculus peltatus

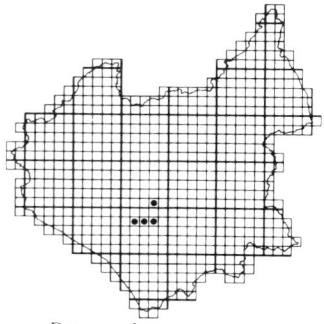

e. Ranunculus pseudofluitans

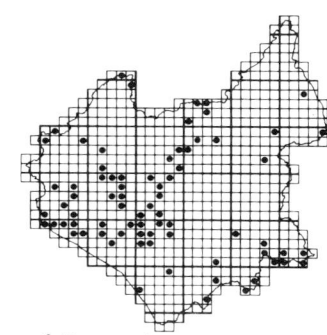

f. Ranunculus aquatilis

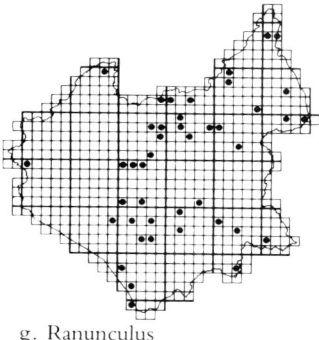

g. Ranunculus trichophyllus

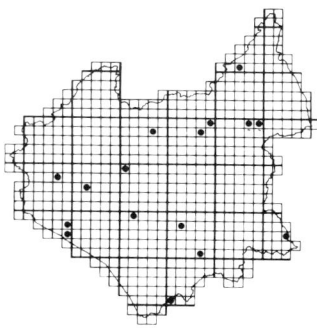

h. Ranunculus circinatus

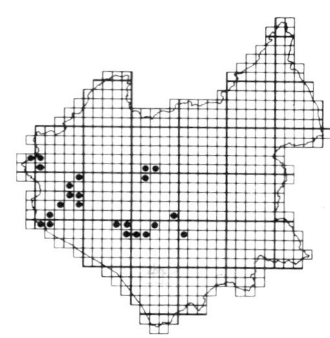

i. Ranunculus fluitans

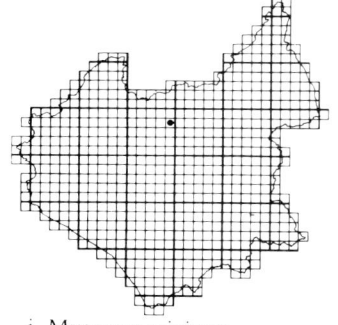

j. Myosurus minimus

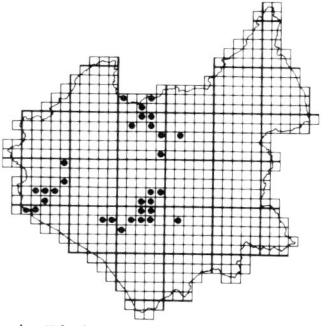

k. Thalictrum flavum

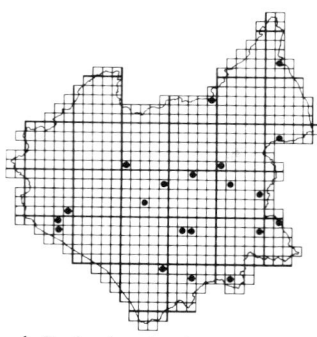

l. Berberis vulgaris

Berberidaceae - Cruciferae

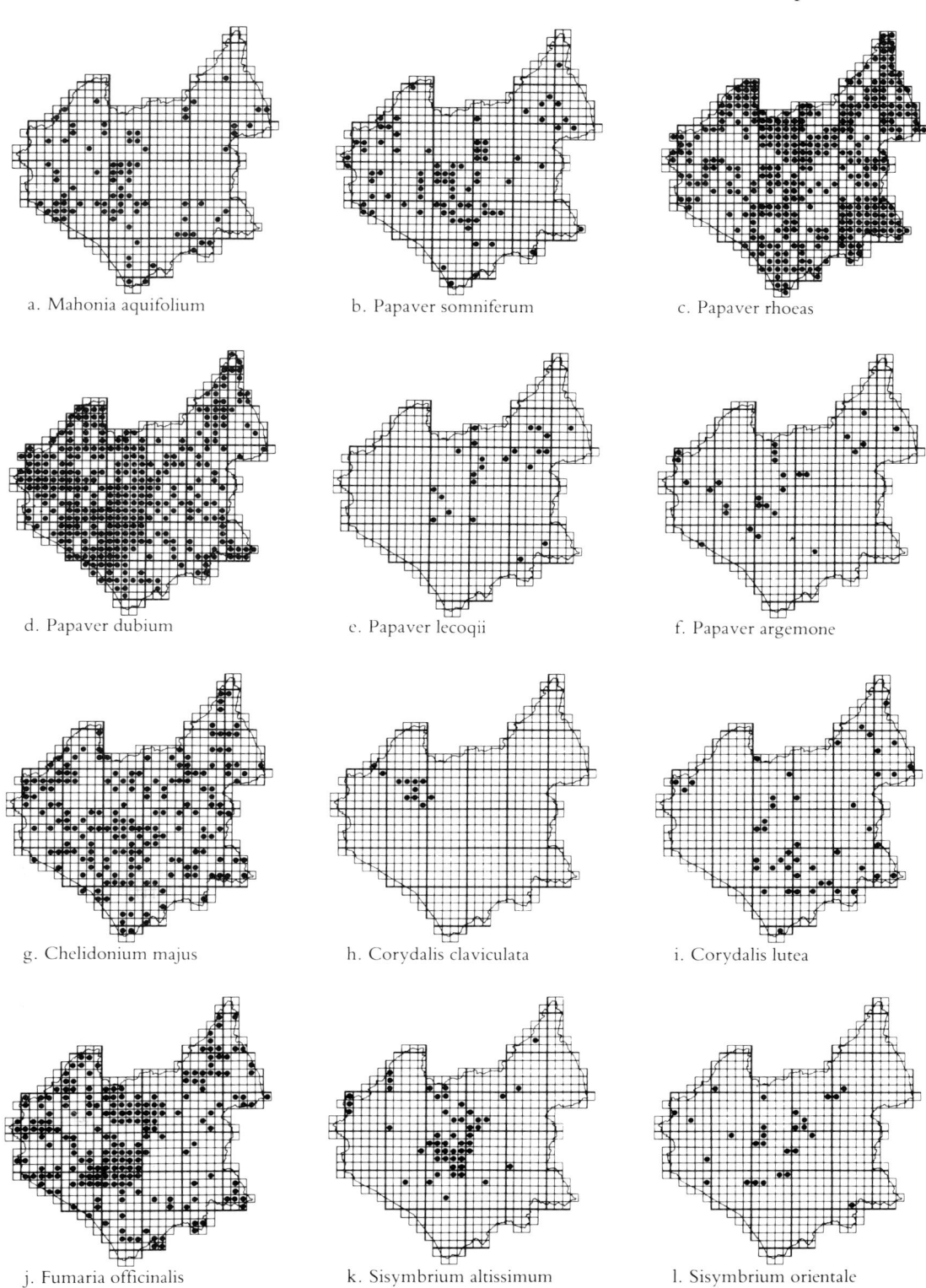

Maps 16a-l

a. Mahonia aquifolium

b. Papaver somniferum

c. Papaver rhoeas

d. Papaver dubium

e. Papaver lecoqii

f. Papaver argemone

g. Chelidonium majus

h. Corydalis claviculata

i. Corydalis lutea

j. Fumaria officinalis

k. Sisymbrium altissimum

l. Sisymbrium orientale

Cruciferae

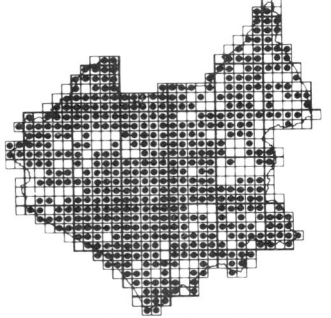

Maps 17a-l

a. Sisymbrium officinale

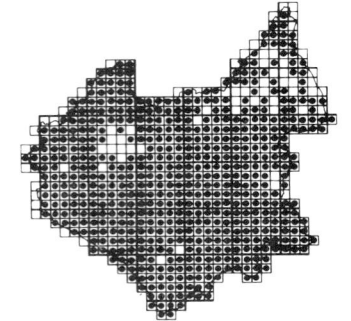

b. Alliaria petiolata

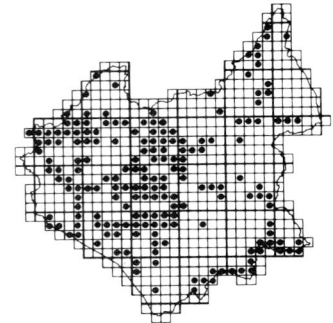

c. Arabidopsis thaliana

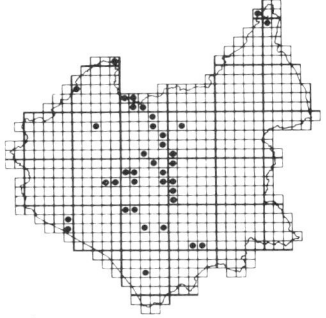

d. Erysimum cheiranthoides

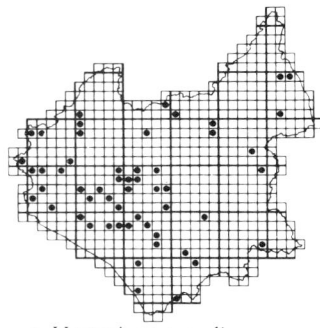

e. Hesperis matronalis

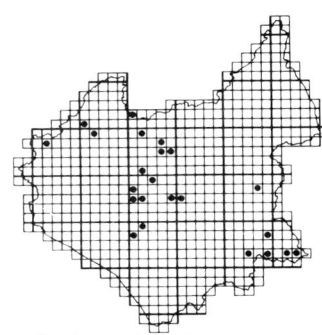

f. Cheiranthus cheiri

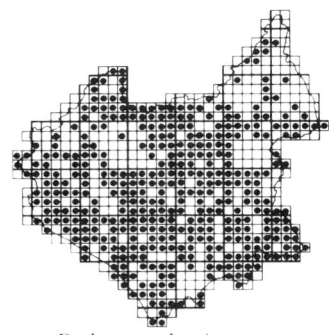

g. Barbarea vulgaris

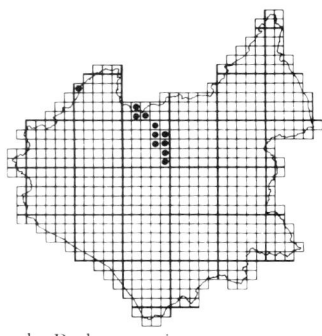

h. Barbarea stricta

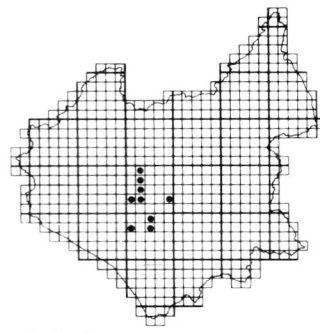

i. Barbarea verna

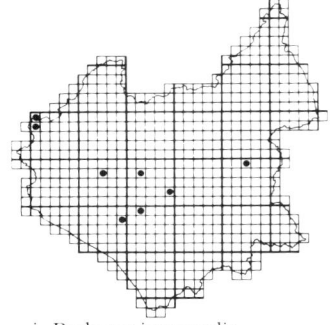

j. Barbarea intermedia

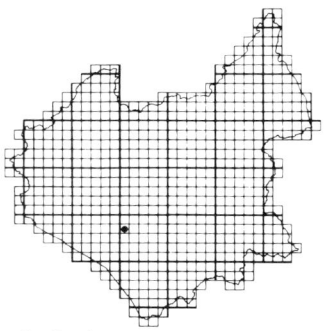

k. Rorippa austriaca

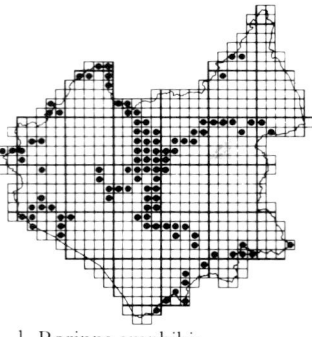

l. Rorippa amphibia

Cruciferae

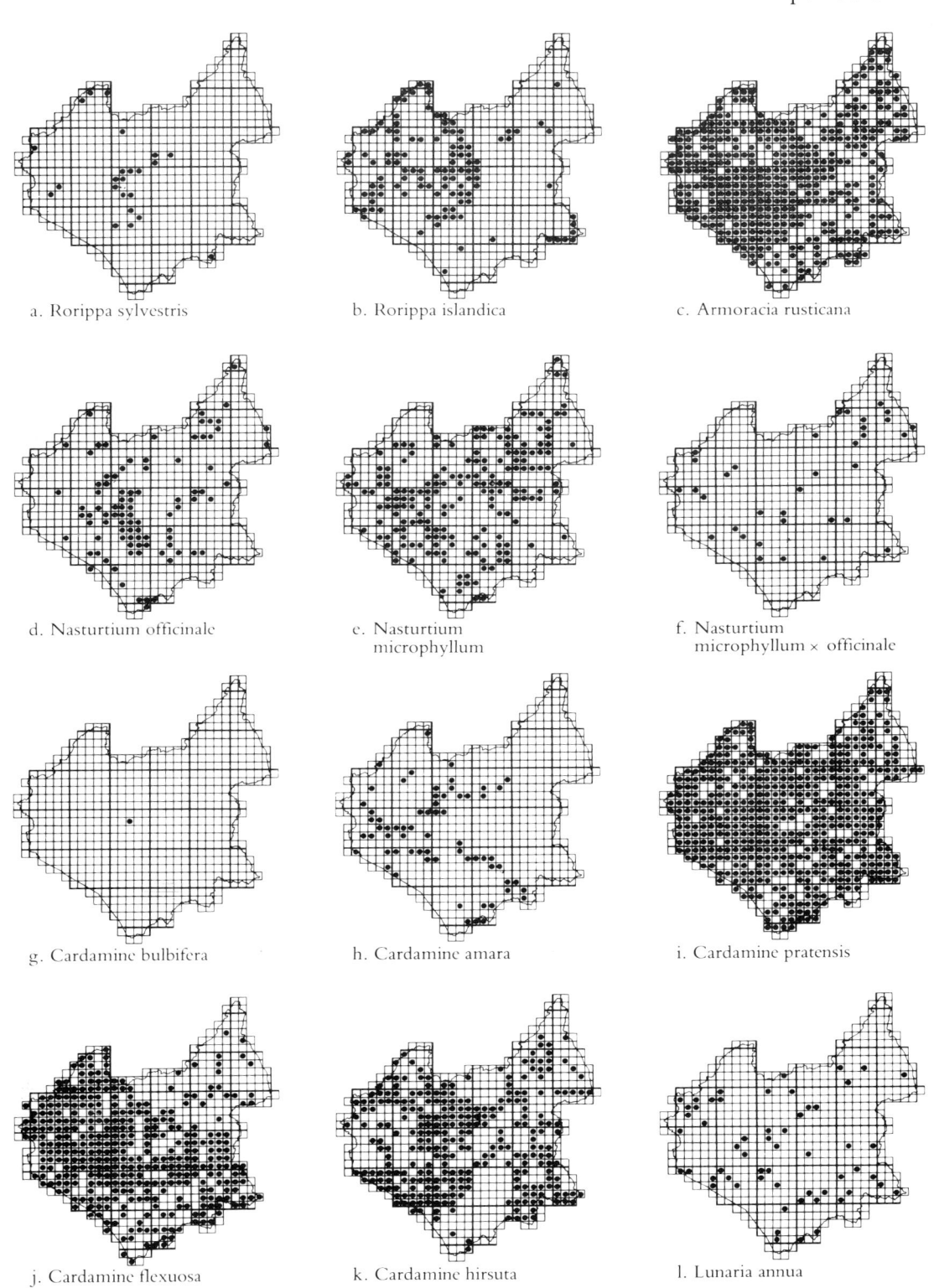

Maps 18a-l

a. Rorippa sylvestris

b. Rorippa islandica

c. Armoracia rusticana

d. Nasturtium officinale

e. Nasturtium microphyllum

f. Nasturtium microphyllum × officinale

g. Cardamine bulbifera

h. Cardamine amara

i. Cardamine pratensis

j. Cardamine flexuosa

k. Cardamine hirsuta

l. Lunaria annua

Cruciferae

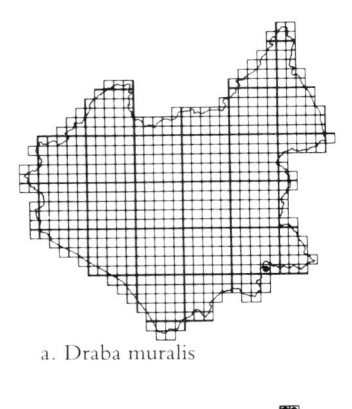

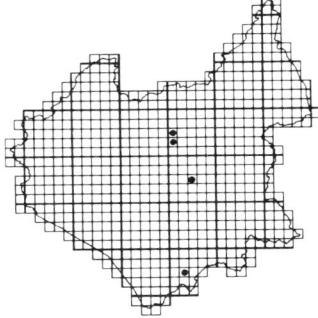

Maps 19a-l

a. Draba muralis

b. Erophila verna ssp. verna

c. Cochlearia danica

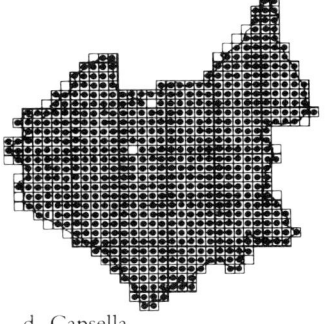

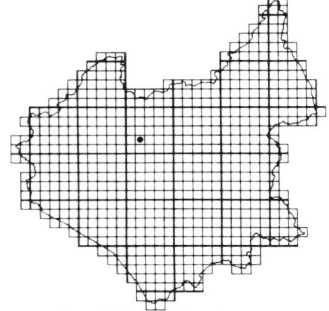

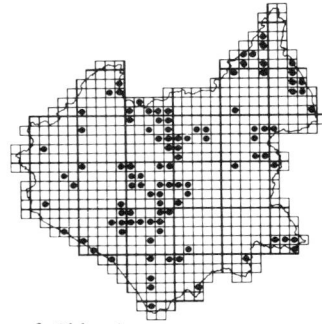

d. Capsella bursa-pastoris

e. Teesdalia nudicaulis

f. Thlaspi arvense

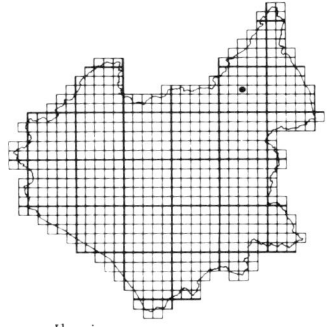

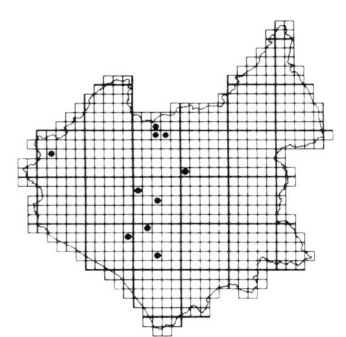

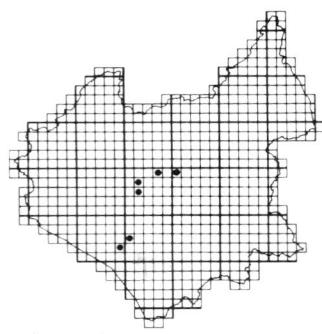

g. Iberis amara

h. Lepidium campestre

i. Lepidium heterophyllum

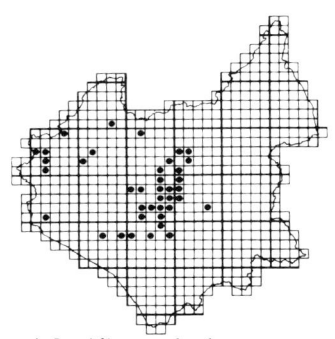

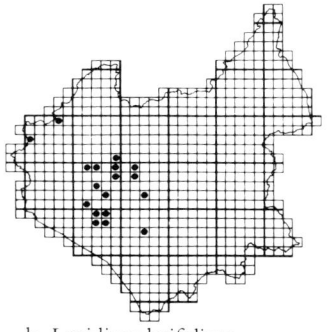

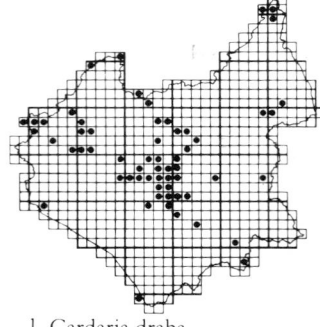

j. Lepidium ruderale

k. Lepidium latifolium

l. Cardaria draba

Cruciferae - Crassulaceae

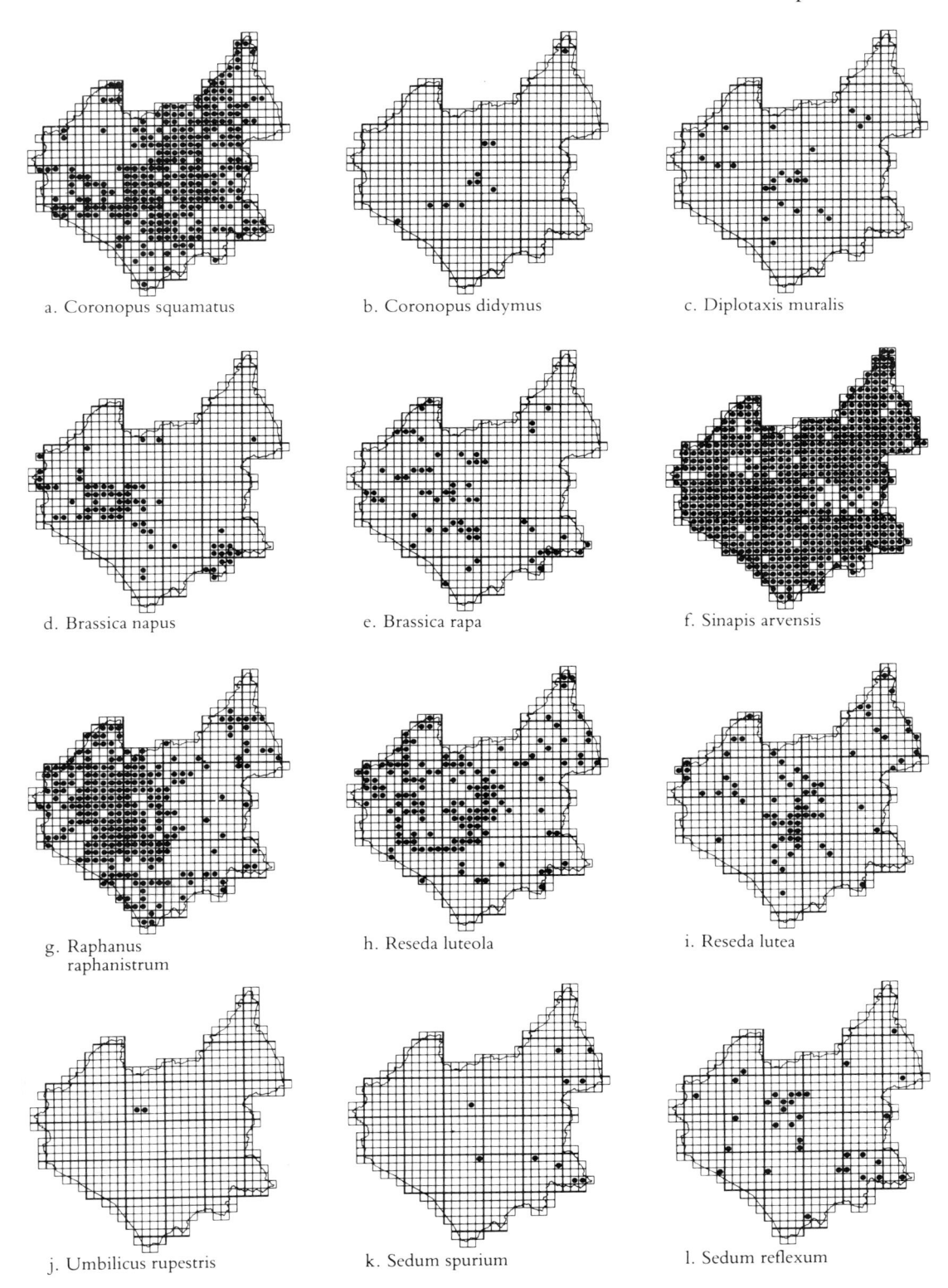

Maps 20a-l

a. Coronopus squamatus

b. Coronopus didymus

c. Diplotaxis muralis

d. Brassica napus

e. Brassica rapa

f. Sinapis arvensis

g. Raphanus raphanistrum

h. Reseda luteola

i. Reseda lutea

j. Umbilicus rupestris

k. Sedum spurium

l. Sedum reflexum

Crassulaceae - Grossulariaceae   Maps 21a-l

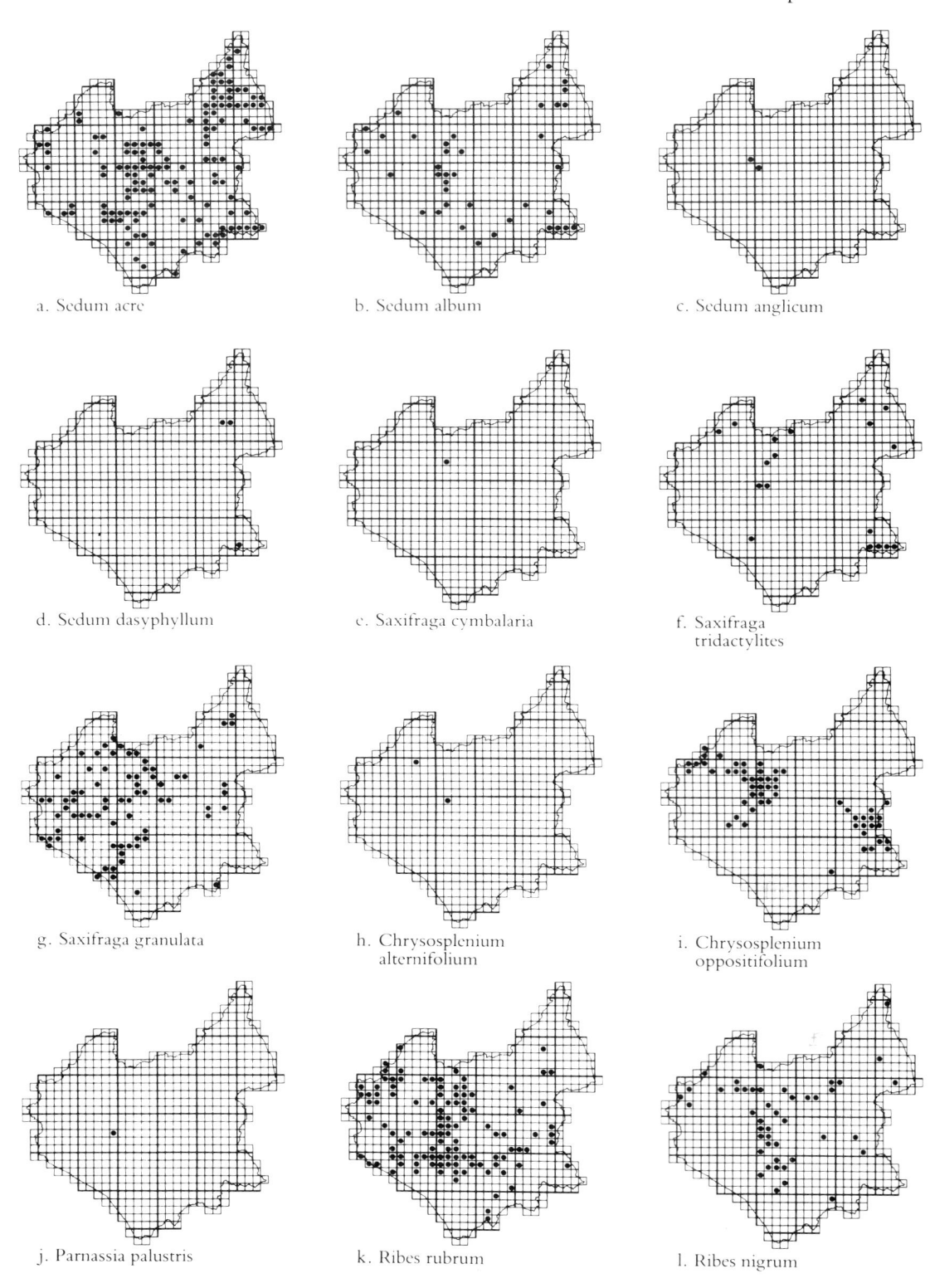

a. Sedum acre

b. Sedum album

c. Sedum anglicum

d. Sedum dasyphyllum

e. Saxifraga cymbalaria

f. Saxifraga tridactylites

g. Saxifraga granulata

h. Chrysosplenium alternifolium

i. Chrysosplenium oppositifolium

j. Parnassia palustris

k. Ribes rubrum

l. Ribes nigrum

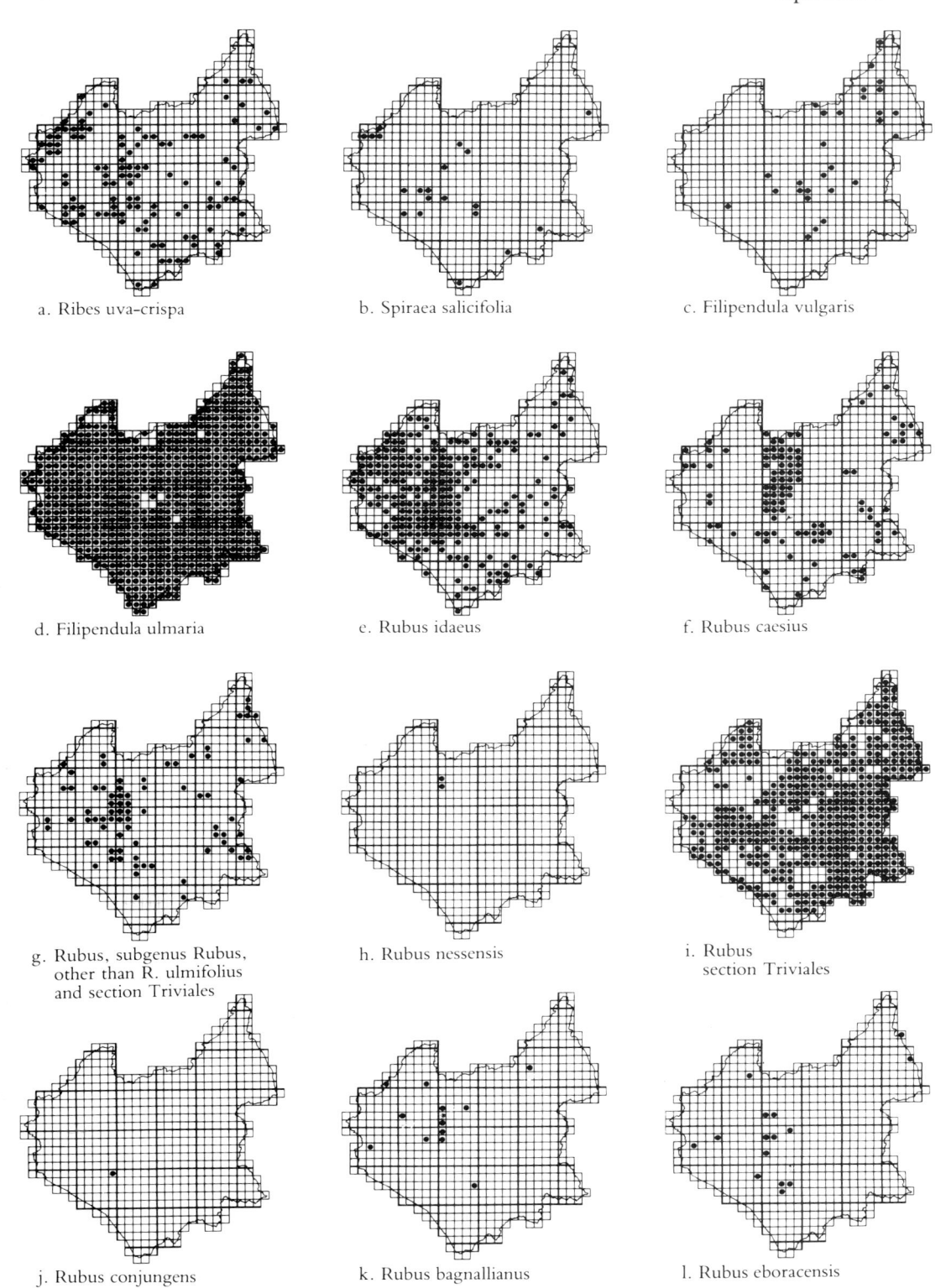

Grossulariaceae - Rosaceae　　　　　　　　　　　　　　　　Maps 22a-l

a. Ribes uva-crispa

b. Spiraea salicifolia

c. Filipendula vulgaris

d. Filipendula ulmaria

e. Rubus idaeus

f. Rubus caesius

g. Rubus, subgenus Rubus, other than R. ulmifolius and section Triviales

h. Rubus nessensis

i. Rubus section Triviales

j. Rubus conjungens

k. Rubus bagnallianus

l. Rubus eboracensis

Rosaceae

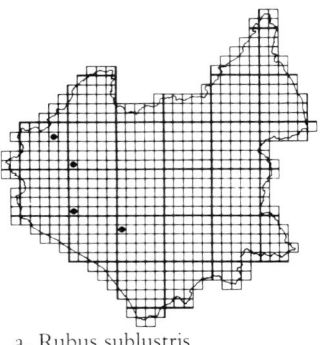

Maps 23a-l

a. Rubus sublustris

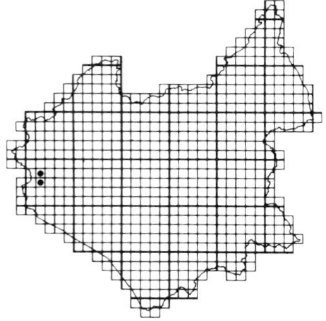

b. Rubus warrenii

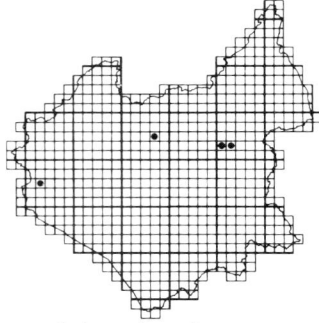

c. Rubus tuberculatus

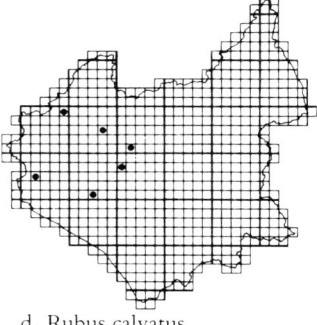

d. Rubus calvatus

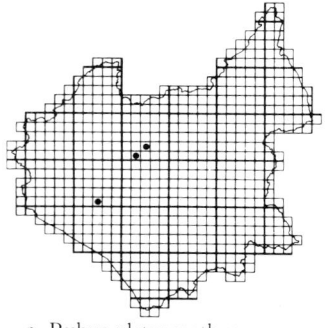

e. Rubus platyacanthus

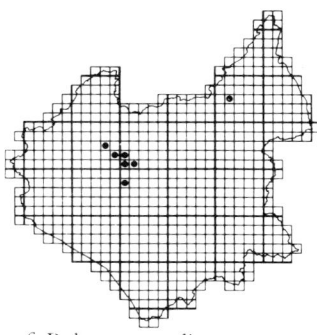

f. Rubus nemoralis

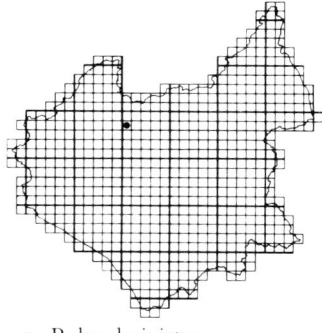

g. Rubus laciniatus

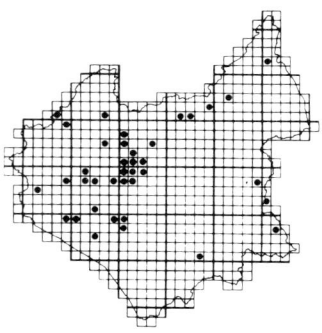

h. Rubus lindleianus

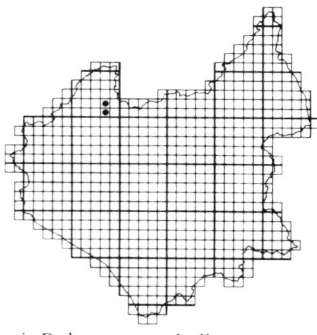

i. Rubus macrophyllus

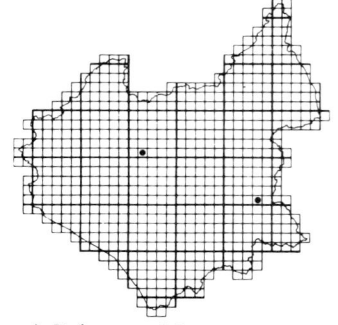

j. Rubus amplificatus

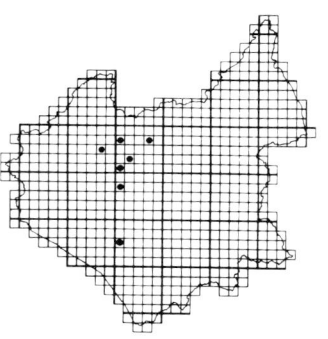

k. Rubus pyramidalis

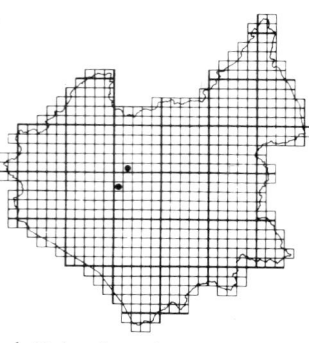

l. Rubus leptothyrsos

Rosaceae

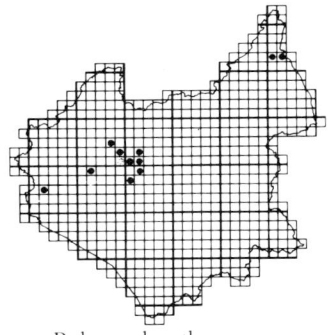

Maps 24a-l

a. Rubus polyanthemus

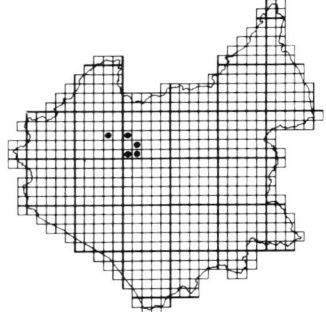

b. Rubus cardiophyllus

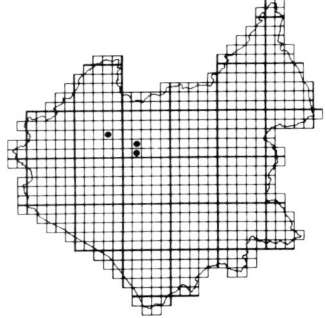

c. Rubus lindebergii

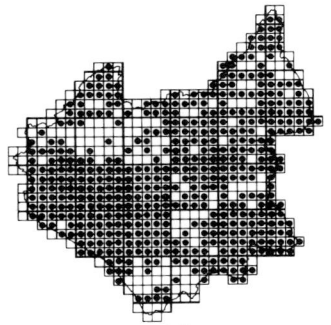

d. Rubus ulmifolius

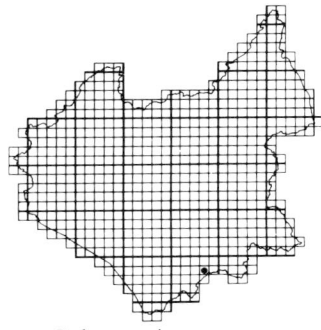

e. Rubus armipotens

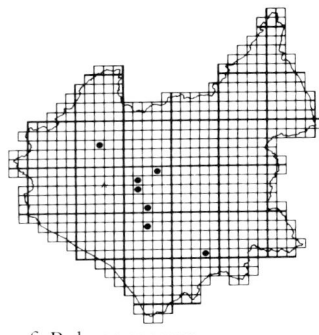

f. Rubus procerus

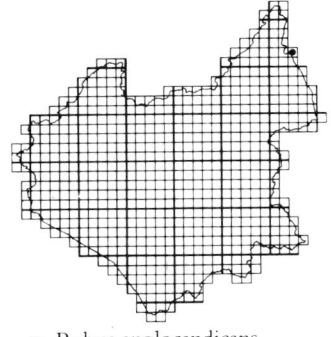

g. Rubus anglocandicans

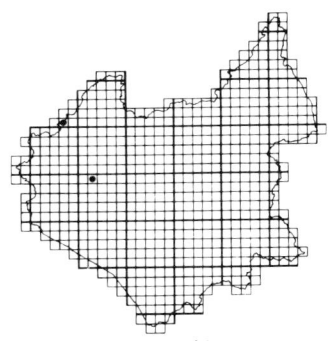

h. Rubus sprengelii

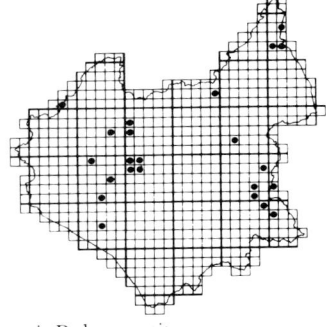

i. Rubus vestitus

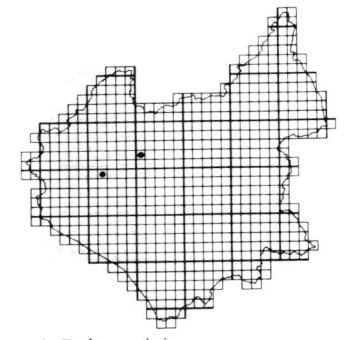

j. Rubus criniger

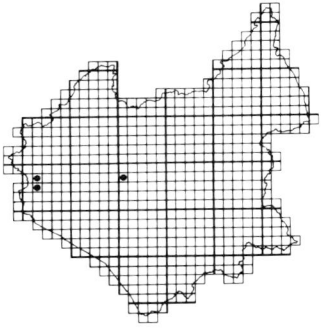

k. Rubus mucronulatus

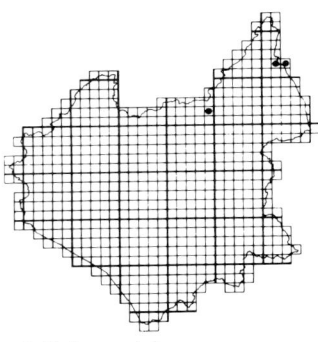

l. Rubus radula

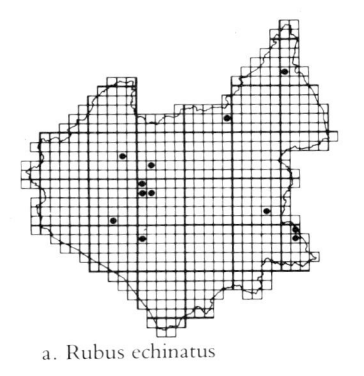

Rosaceae   Maps 25a–l

a. Rubus echinatus
b. Rubus echinatoides 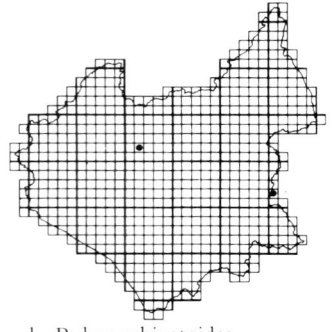
c. Rubus rudis 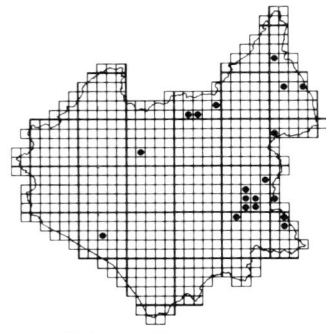
d. Rubus cantianus 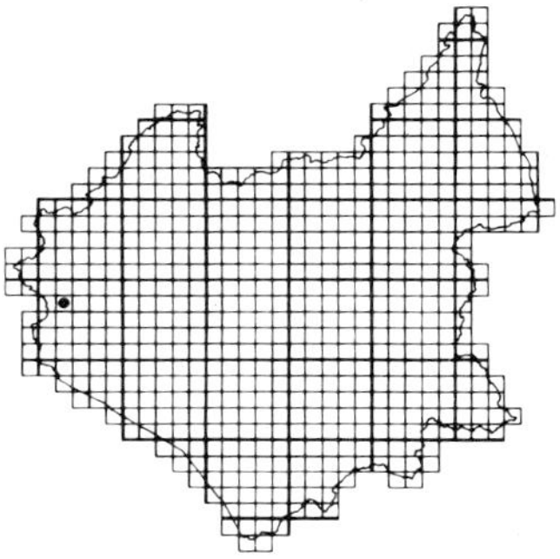
e. Rubus bloxamianus 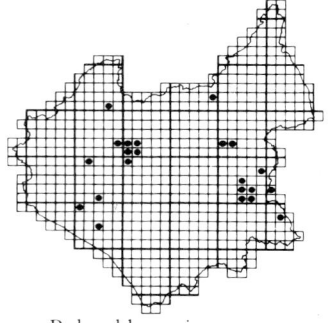
f. Rubus flexuosus 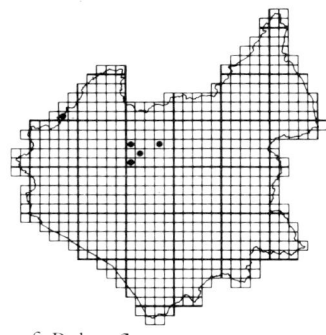
g. Rubus rubristylus 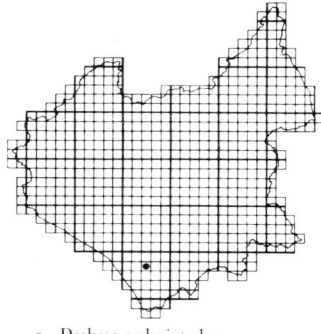
h. Rubus bloxamii 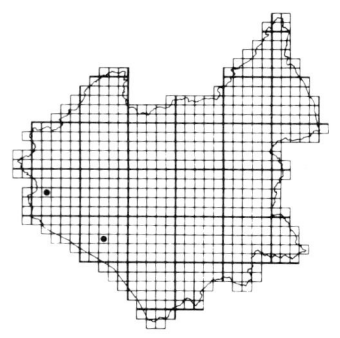
i. Rubus watsonii 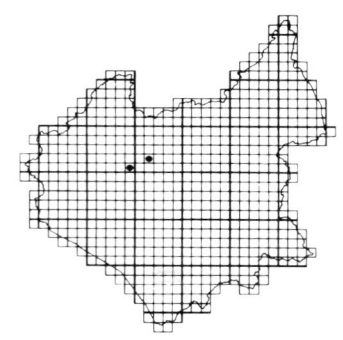
j. Rubus pallidus 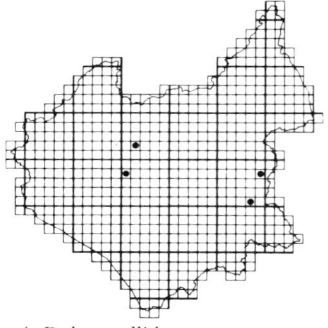
k. Rubus euryanthemus 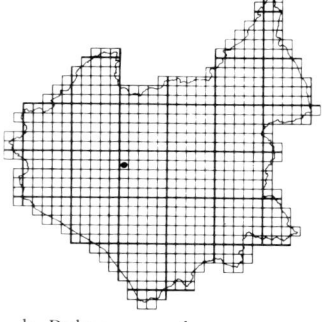
l. Rubus insectifolius

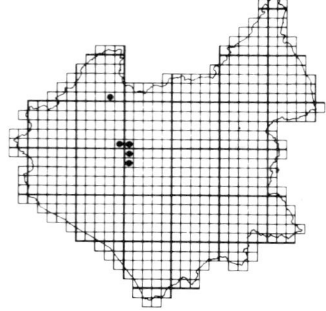

Rosaceae

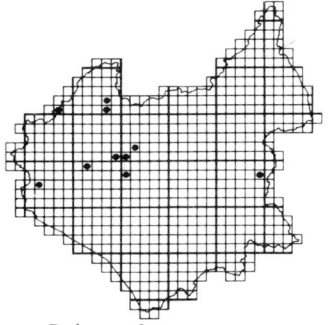

Maps 26a-l

a. Rubus rufescens

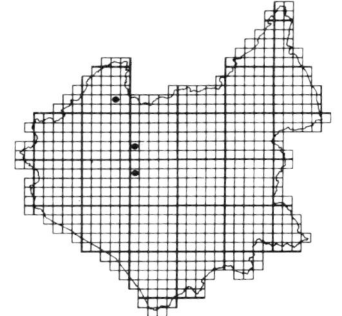

b. Rubus raduloides

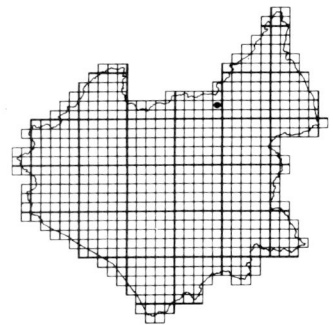

c. Rubus griffithianus

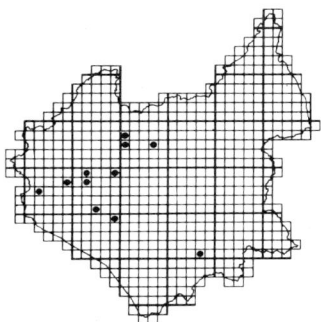

d. Rubus leightonii

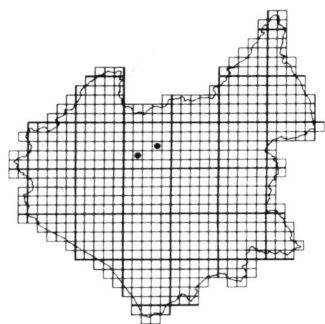

e. Rubus diversus

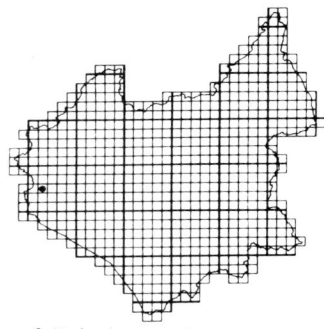

f. Rubus murrayi

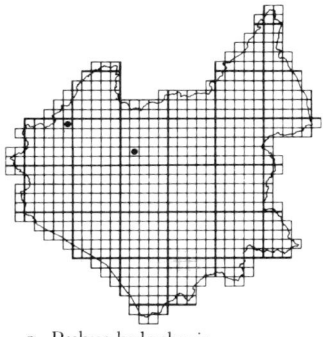

g. Rubus hylocharis

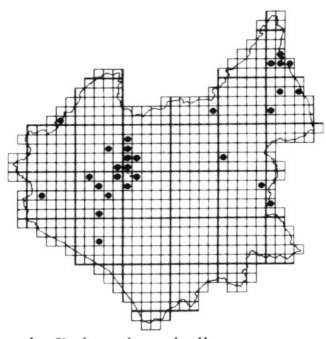

h. Rubus dasyphyllus

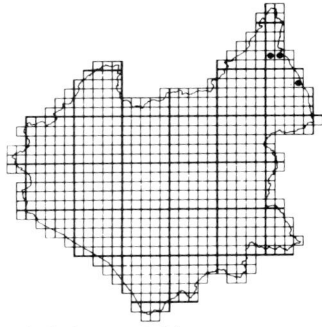

i. Rubus anglohirtus

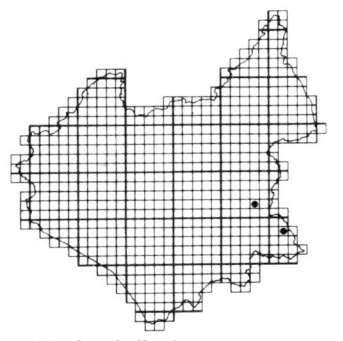

j. Rubus bellardii

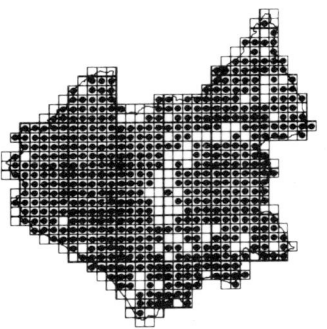

k. Rosa arvensis

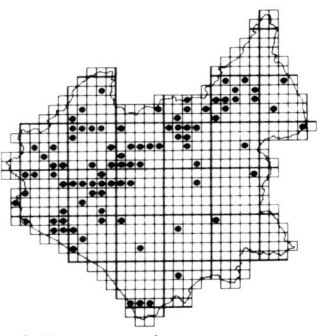

l. Rosa arvensis × canina

Rosaceae   Maps 27a-l

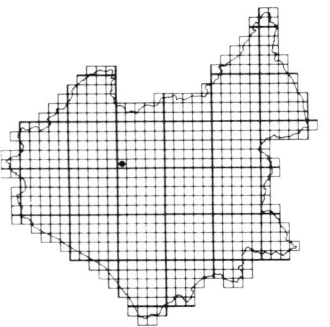

a. Rosa pimpinellifolia × mollis

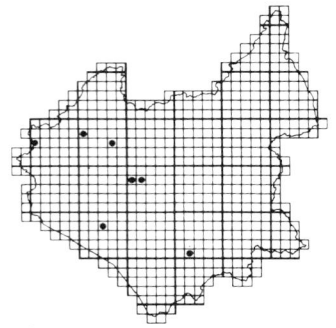

b. Rosa rugosa

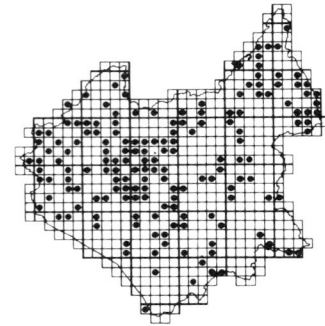

c. Rosa canina group Lutetianae

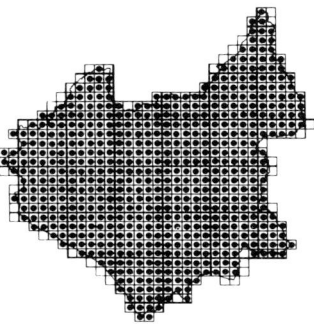

d. Rosa canina group Transitoriae

e. Rosa canina group Dumales

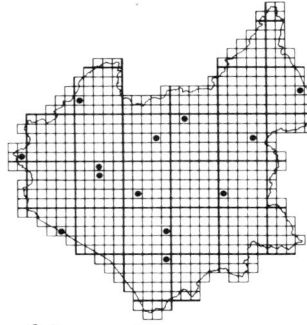

f. Rosa canina group Andegavenses

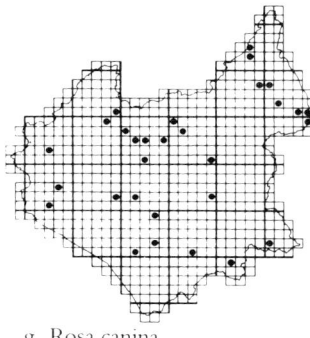

g. Rosa canina × arvensis

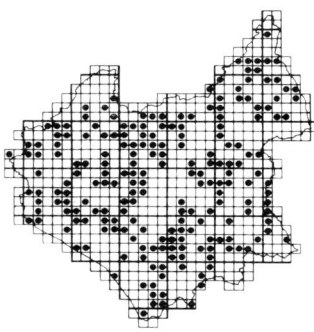

h. Rosa dumetorum group Pubescentes

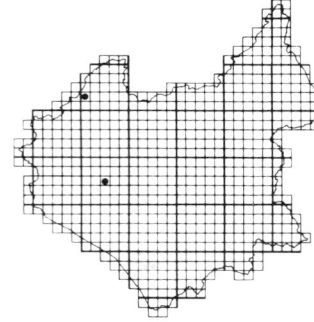

i. Rosa dumetorum group Deseglisei

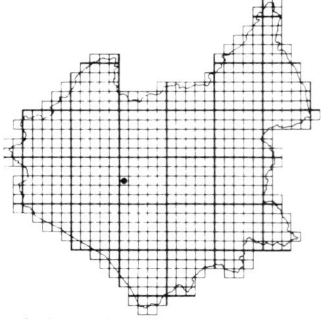

j. Rosa dumetorum group Mercicae

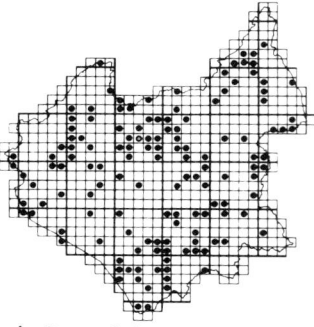

k. Rosa afzeliana group Reuterianae

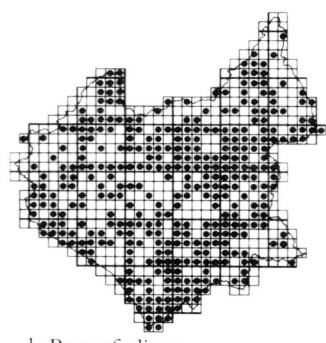

l. Rosa afzeliana group Subcaninae

Rosaceae

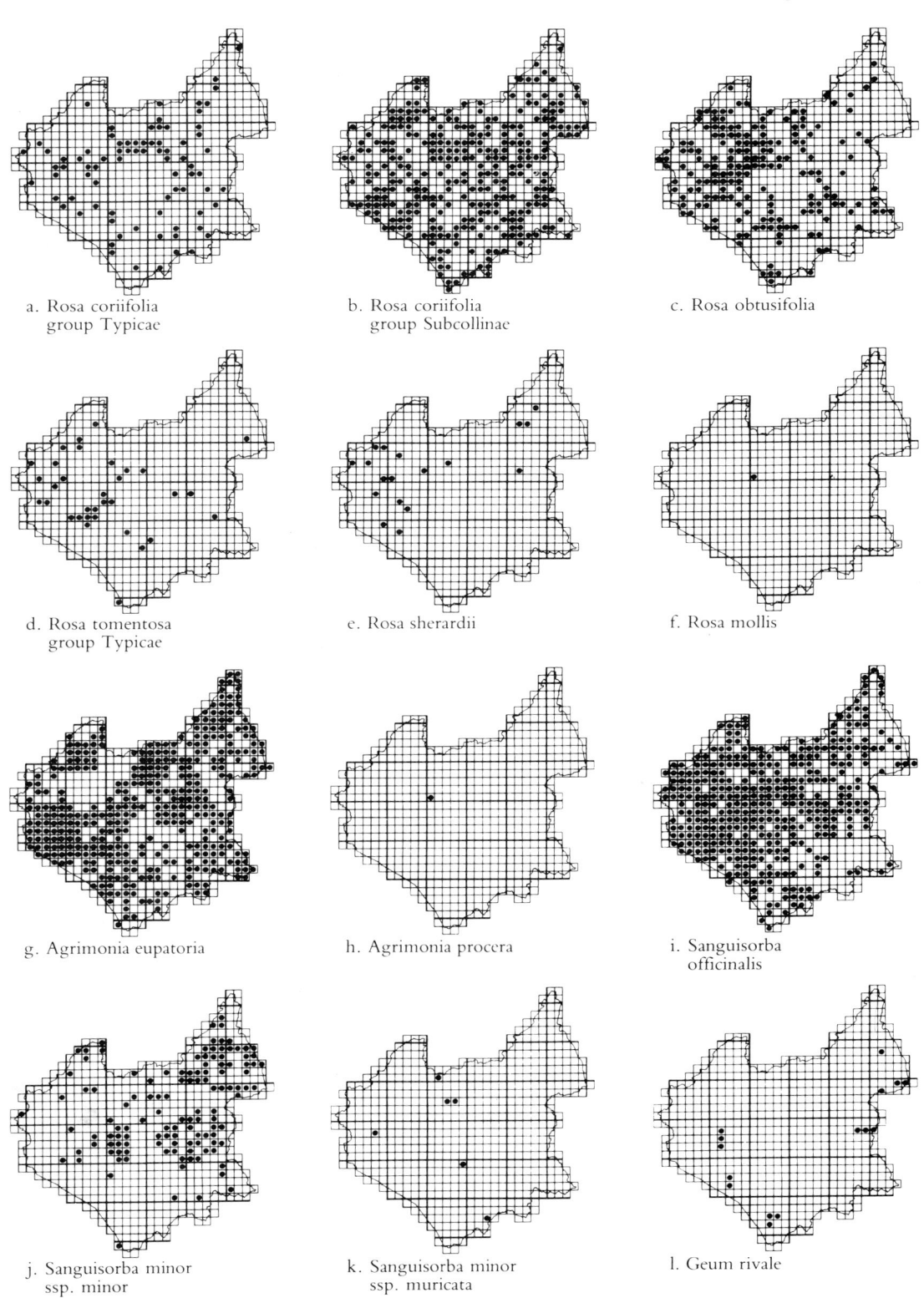

Maps 28a-l

a. Rosa coriifolia group Typicae

b. Rosa coriifolia group Subcollinae

c. Rosa obtusifolia

d. Rosa tomentosa group Typicae

e. Rosa sherardii

f. Rosa mollis

g. Agrimonia eupatoria

h. Agrimonia procera

i. Sanguisorba officinalis

j. Sanguisorba minor ssp. minor

k. Sanguisorba minor ssp. muricata

l. Geum rivale

Rosaceae

Maps 29a-l

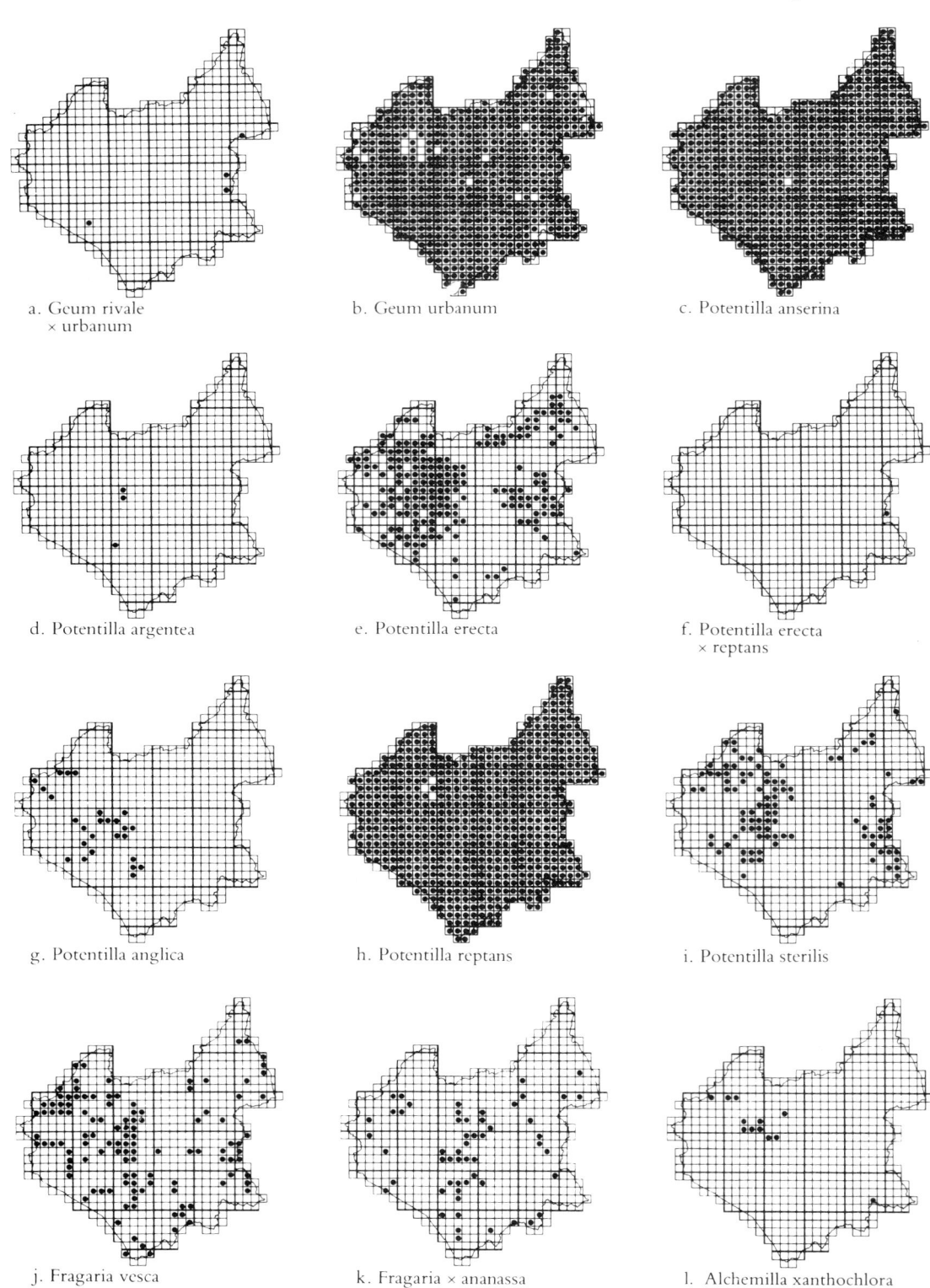

a. Geum rivale × urbanum

b. Geum urbanum

c. Potentilla anserina

d. Potentilla argentea

e. Potentilla erecta

f. Potentilla erecta × reptans

g. Potentilla anglica

h. Potentilla reptans

i. Potentilla sterilis

j. Fragaria vesca

k. Fragaria × ananassa

l. Alchemilla xanthochlora

Rosaceae

Maps 30a-l

a. Alchemilla filicaulis ssp. vestita

b. Aphanes arvensis

c. Aphanes microcarpa

d. Pyrus communis

e. Malus sylvestris

f. Sorbus aucuparia

g. Sorbus torminalis

h. Sorbus aria

i. Sorbus intermedia

j. Sorbus latifolia

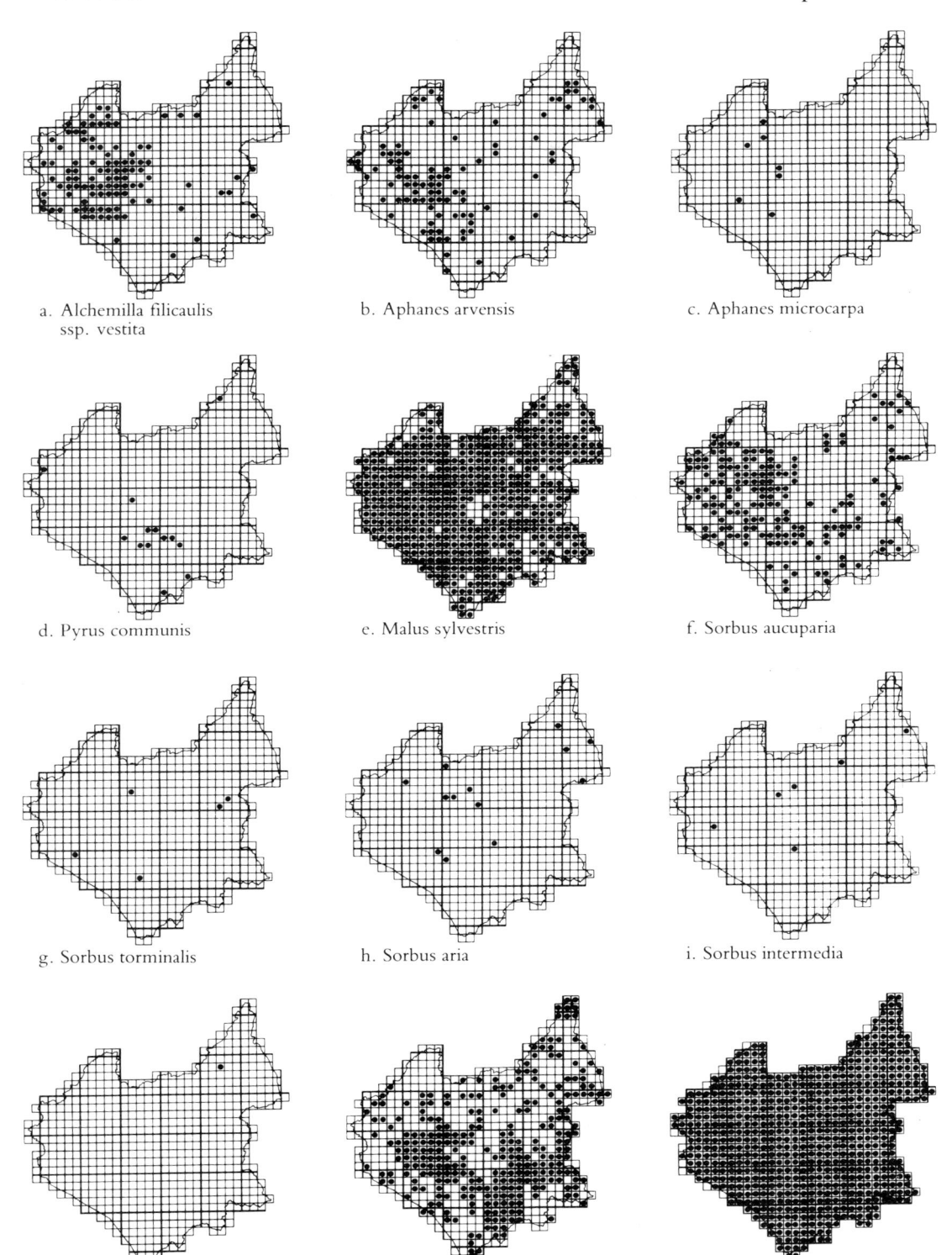

k. Crataegus laevigata

l. Crataegus monogyna

Rosaceae - Leguminosae

Maps 31a-l

a. Prunus cerasifera

b. Prunus spinosa

c. Prunus domestica ssp. domestica

d. Prunus domestica ssp. insititia

e. Prunus avium

f. Prunus cerasus

g. Prunus padus

h. Prunus lusitanica

i. Prunus laurocerasus

j. Laburnum anagyroides

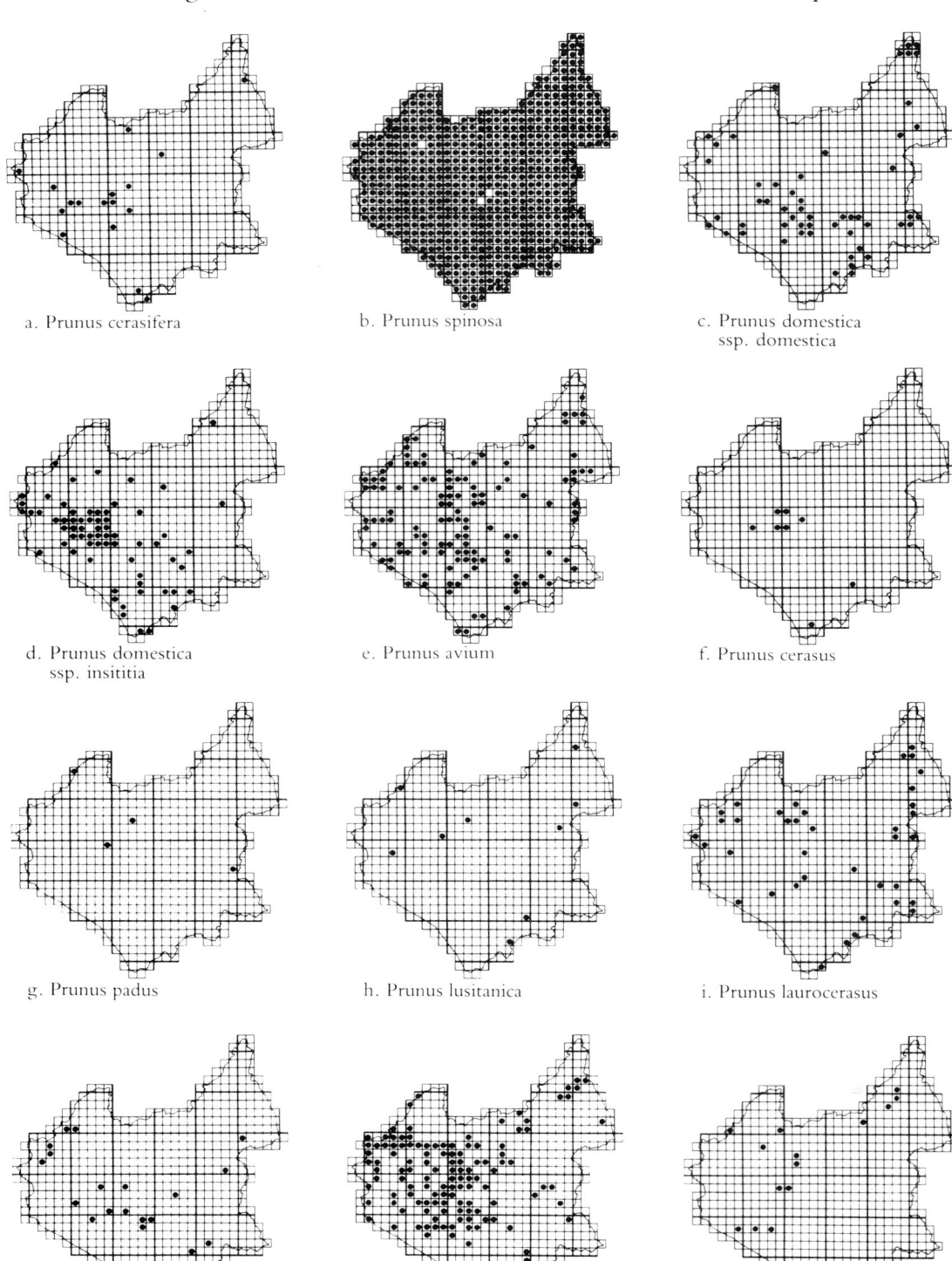

k. Cytisus scoparius

l. Genista tinctoria

Leguminosae

Maps 32a-l

a. Genista anglica

b. Ulex europaeus

c. Ulex gallii

d. Lupinus polyphyllus

e. Galega officinalis

f. Colutea arborescens

g. Astragalus danicus

h. Astragalus glycyphyllos

i. Vicia cracca

j. Vicia sylvatica

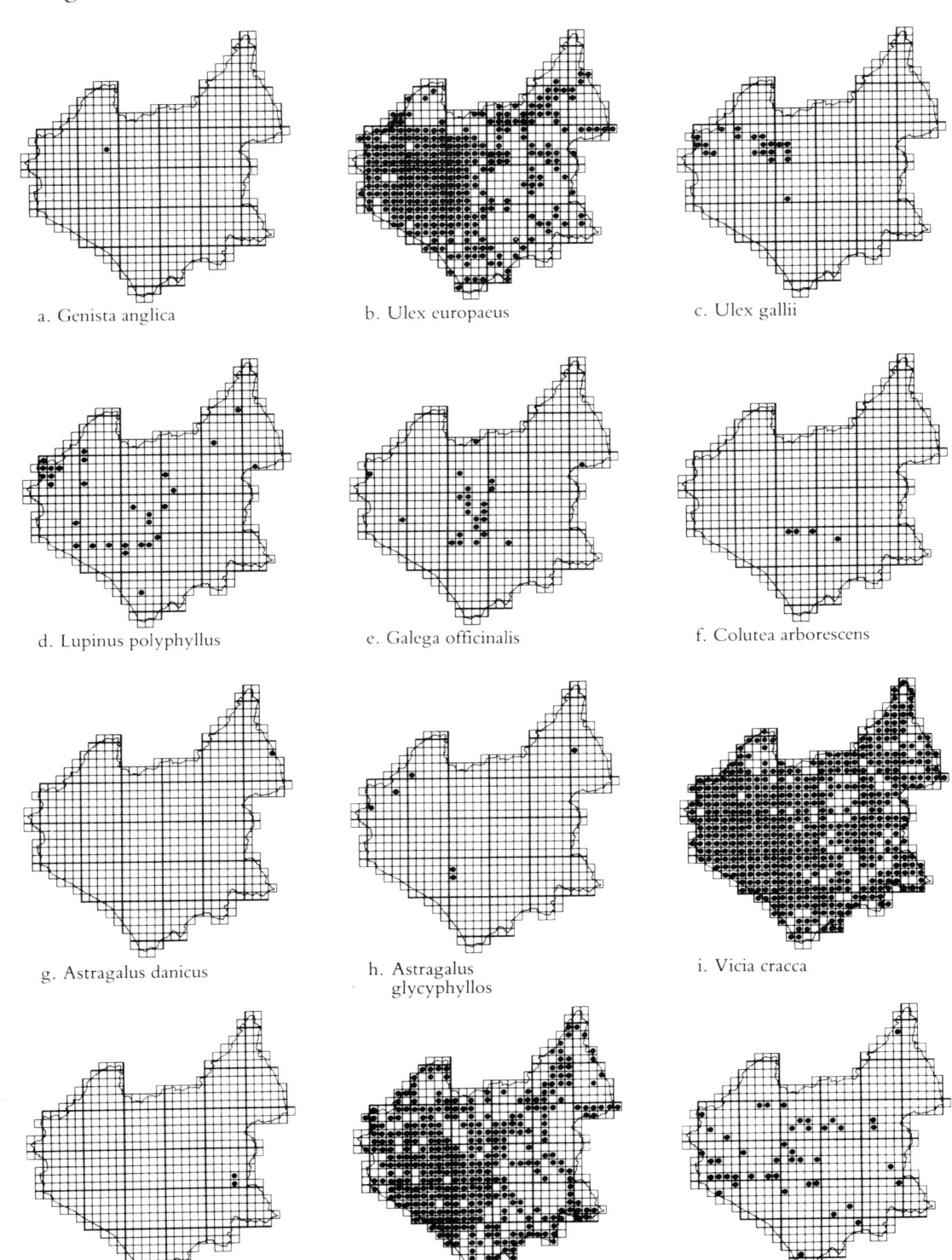

k. Vicia hirsuta

l. Vicia tetrasperma

Leguminosae

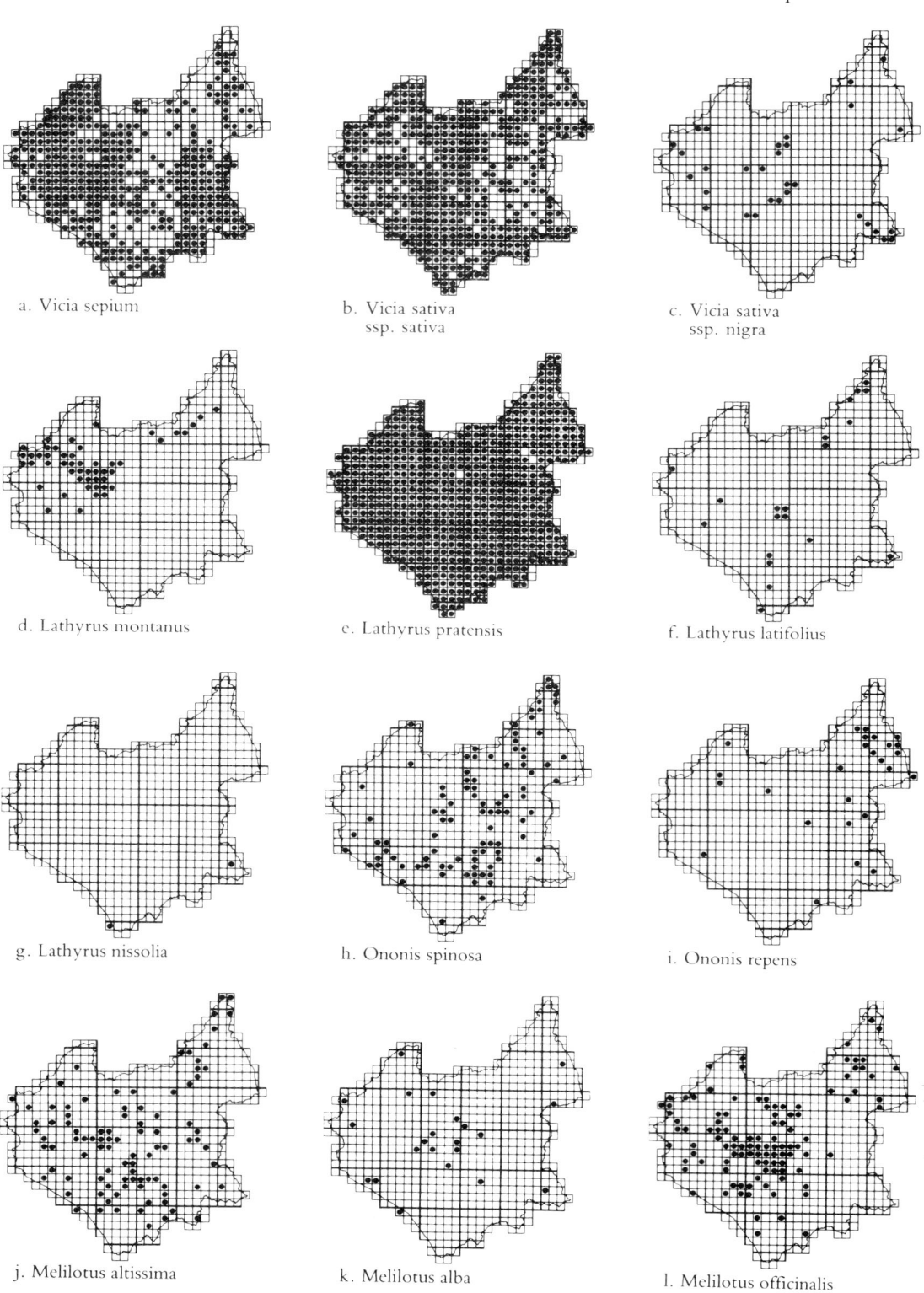

Maps 33a-l

a. Vicia sepium

b. Vicia sativa ssp. sativa

c. Vicia sativa ssp. nigra

d. Lathyrus montanus

e. Lathyrus pratensis

f. Lathyrus latifolius

g. Lathyrus nissolia

h. Ononis spinosa

i. Ononis repens

j. Melilotus altissima

k. Melilotus alba

l. Melilotus officinalis

Leguminosae

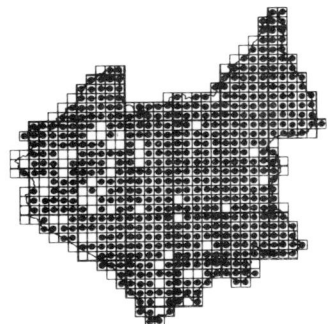

Maps 34a-l

a. Medicago lupulina

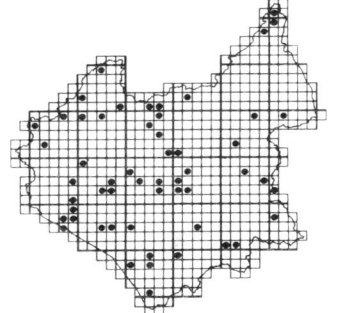

b. Medicago sativa ssp. sativa

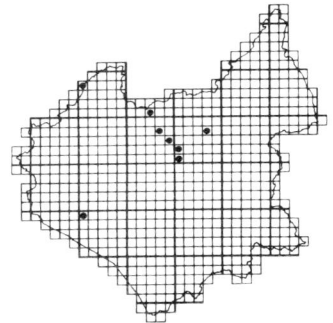

c. Medicago arabica

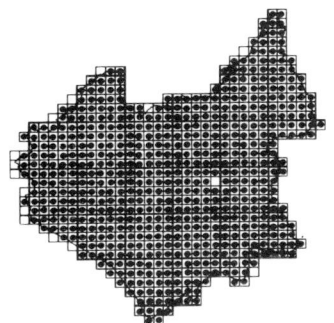

d. Trifolium repens

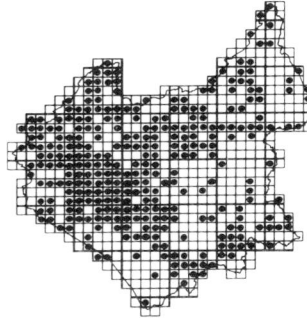

e. Trifolium hybridum

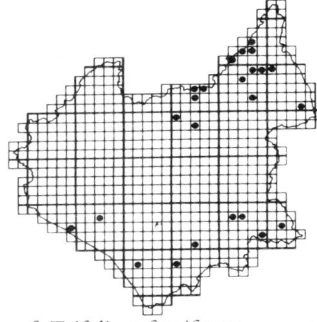

f. Trifolium fragiferum

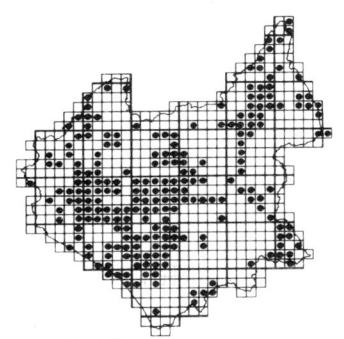

g. Trifolium campestre

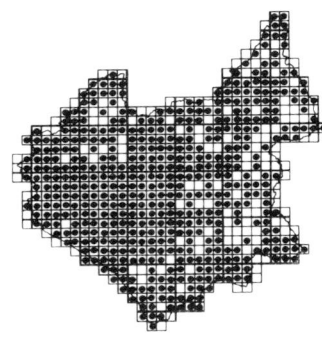

h. Trifolium dubium

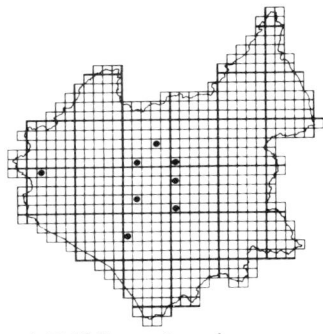

i. Trifolium micranthum

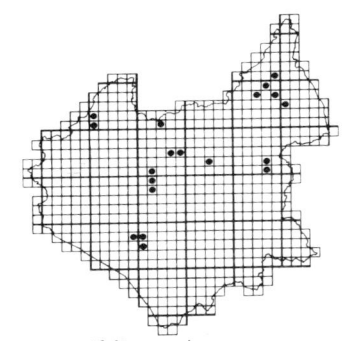

j. Trifolium striatum

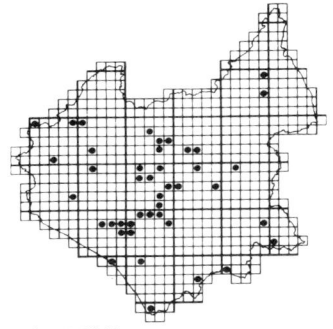

k. Trifolium arvense

l. Trifolium pratense

Leguminosae - Geraniaceae

Maps 35a-l

a. Trifolium medium

b. Trifolium subterraneum

c. Lotus corniculatus

d. Lotus uliginosus

e. Anthyllis vulneraria

f. Ornithopus perpusillus

g. Oxalis corniculata

h. Oxalis europaea

i. Oxalis acetosella

j. Geranium pratense

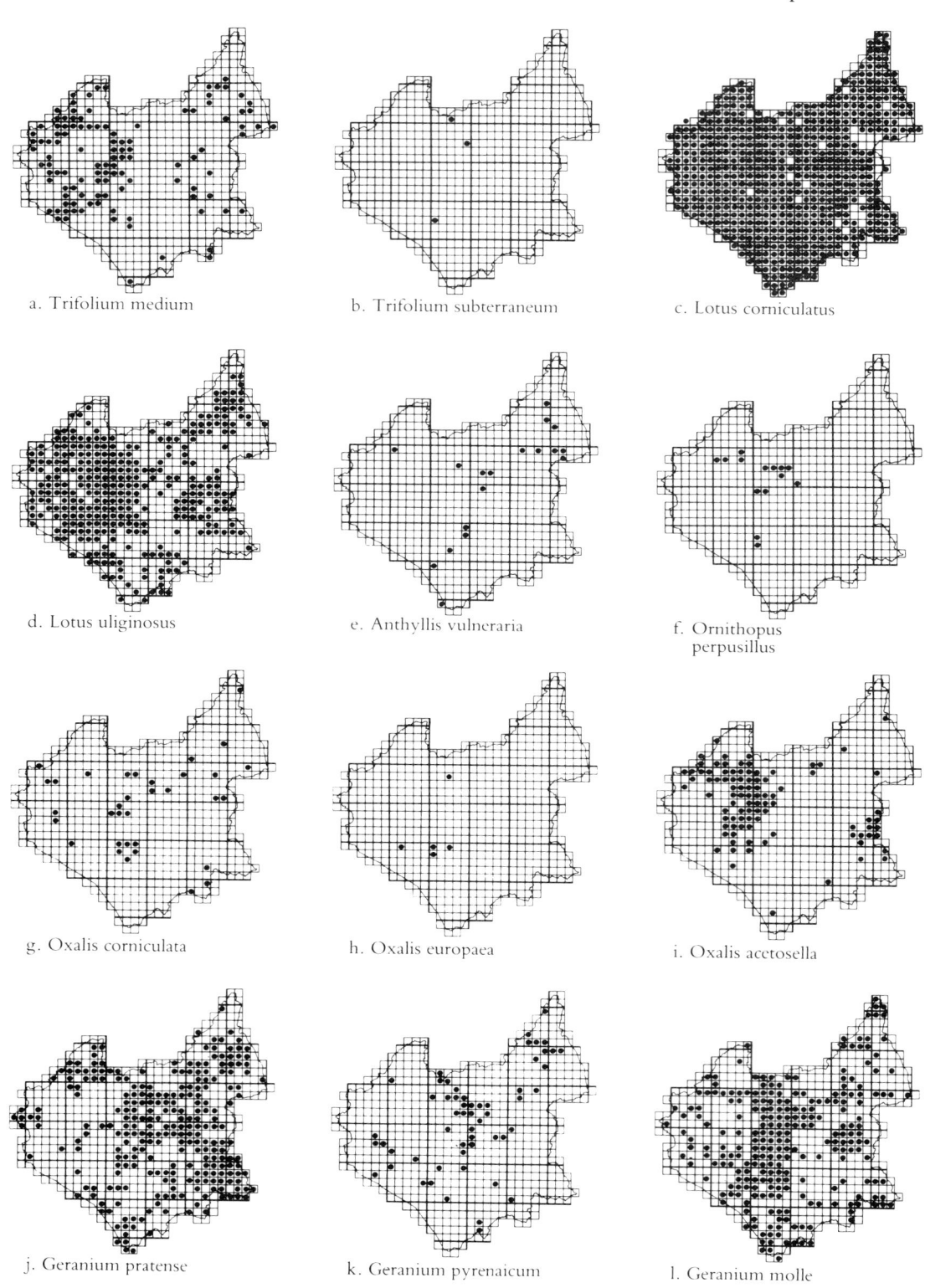

k. Geranium pyrenaicum

l. Geranium molle

Geraniaceae - Euphorbiaceae   Maps 36a-l

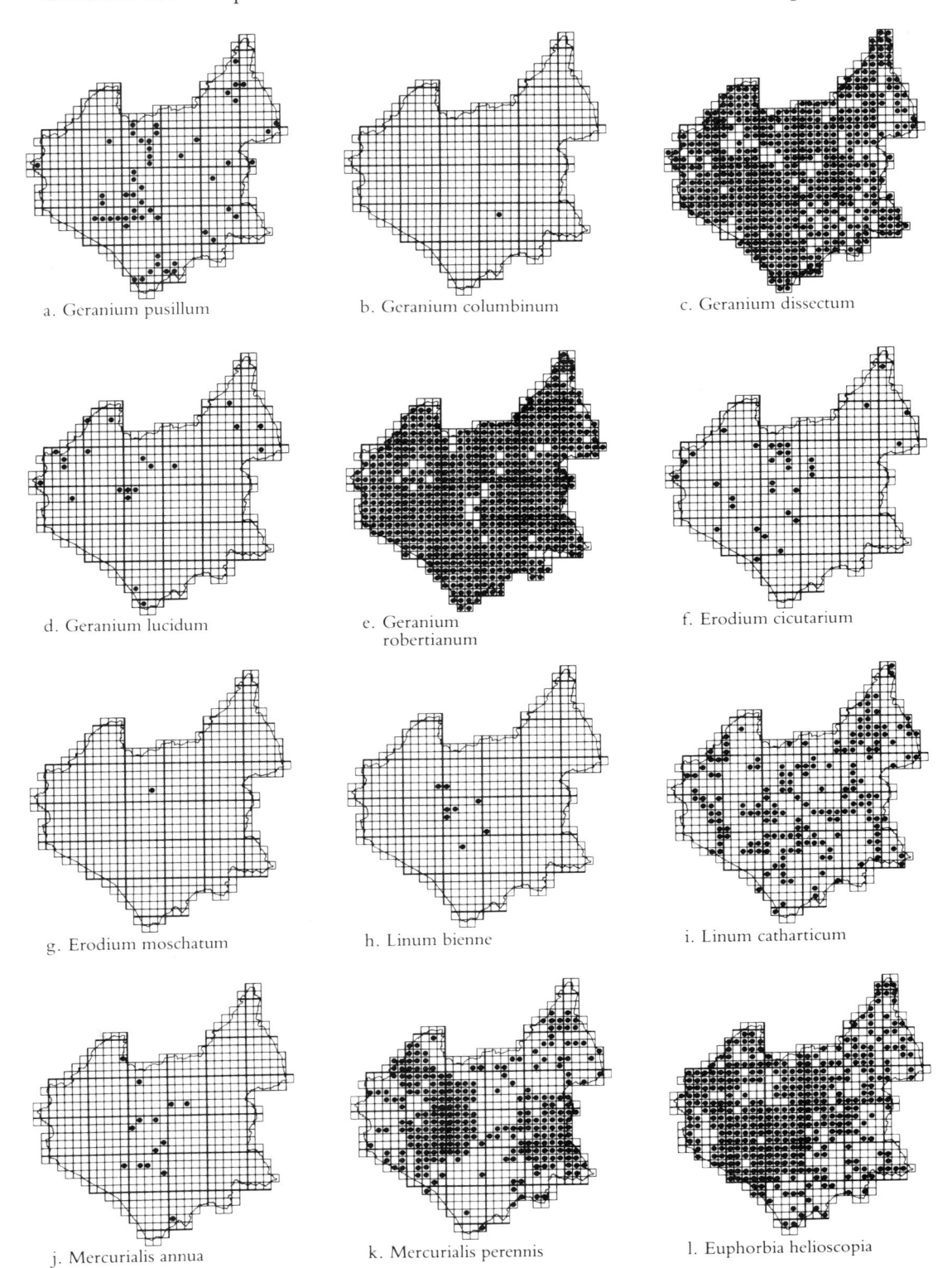

a. Geranium pusillum

b. Geranium columbinum

c. Geranium dissectum

d. Geranium lucidum

e. Geranium robertianum

f. Erodium cicutarium

g. Erodium moschatum

h. Linum bienne

i. Linum catharticum

j. Mercurialis annua

k. Mercurialis perennis

l. Euphorbia helioscopia

Euphorbiaceae - Balsaminaceae

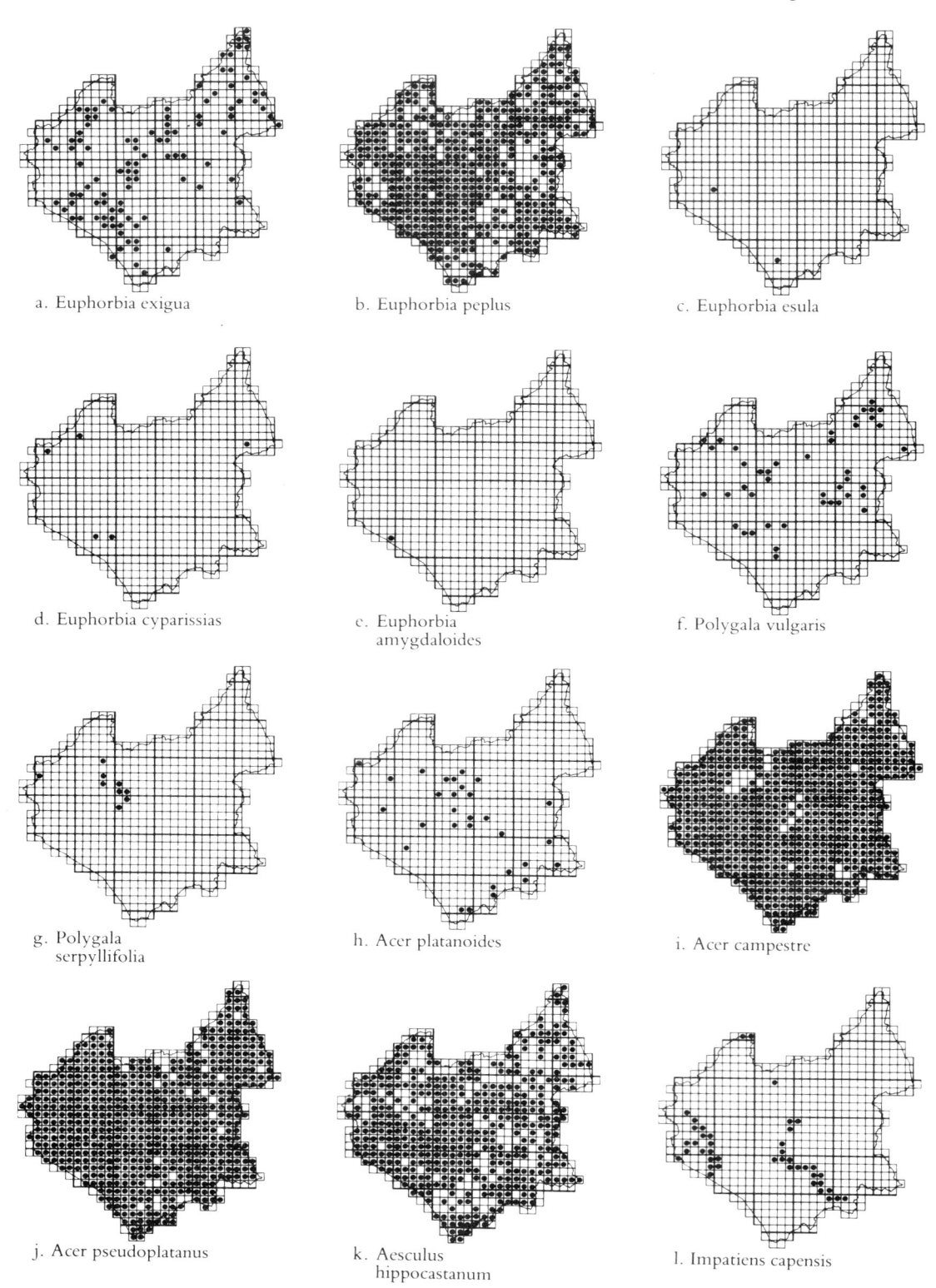

Maps 37a-l

a. Euphorbia exigua

b. Euphorbia peplus

c. Euphorbia esula

d. Euphorbia cyparissias

e. Euphorbia amygdaloides

f. Polygala vulgaris

g. Polygala serpyllifolia

h. Acer platanoides

i. Acer campestre

j. Acer pseudoplatanus

k. Aesculus hippocastanum

l. Impatiens capensis

Balsaminaceae - Malvaceae   Maps 38a-l

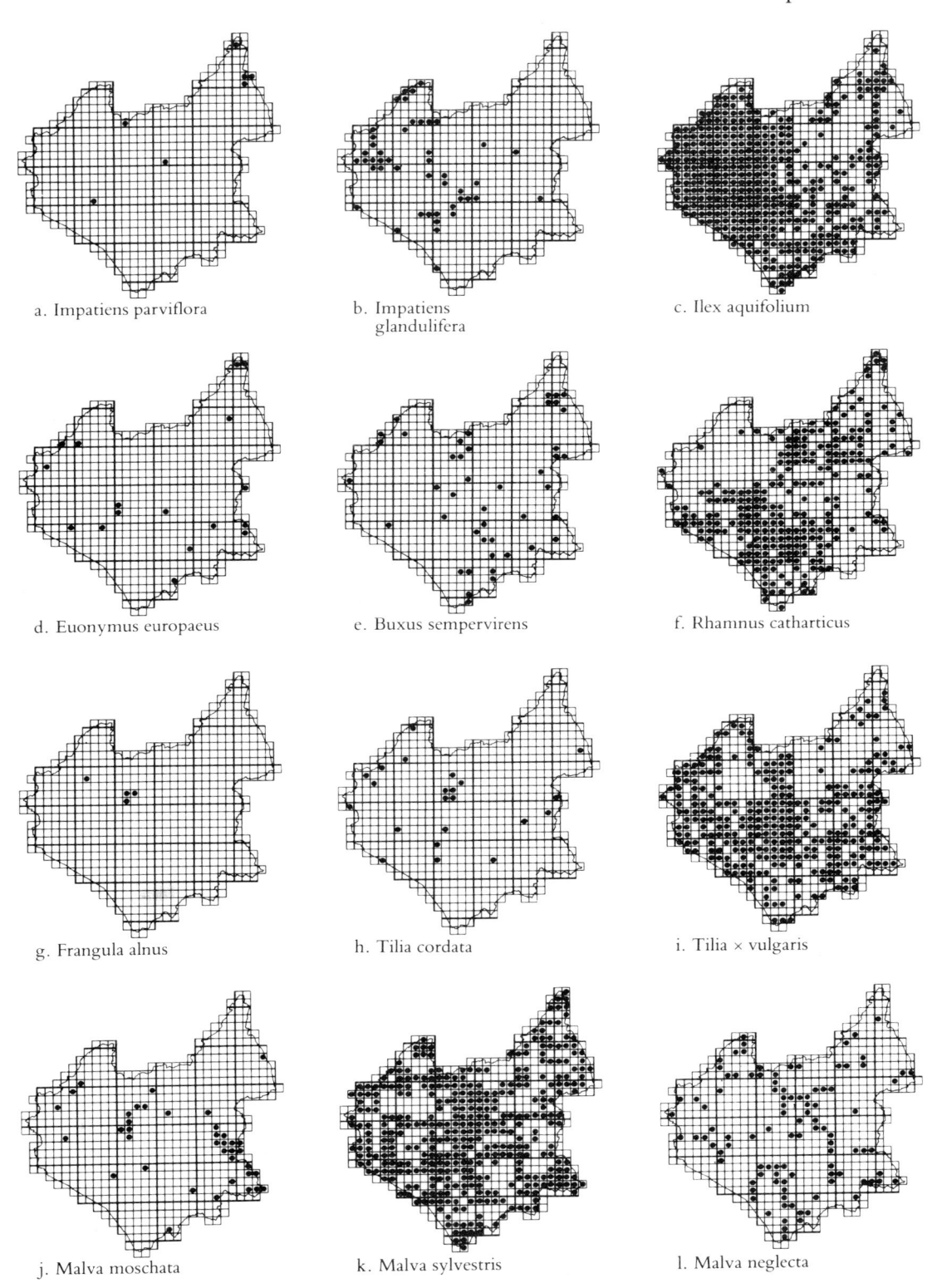

a. Impatiens parviflora

b. Impatiens glandulifera

c. Ilex aquifolium

d. Euonymus europaeus

e. Buxus sempervirens

f. Rhamnus catharticus

g. Frangula alnus

h. Tilia cordata

i. Tilia × vulgaris

j. Malva moschata

k. Malva sylvestris

l. Malva neglecta

Thymelaeaceae – Violaceae
Maps 39a-l

a. Daphne laureola

b. Hypericum hirsutum

c. Hypericum pulchrum

d. Hypericum humifusum

e. Hypericum tetrapterum

f. Hypericum maculatum

g. Hypericum perforatum

h. Viola odorata

i. Viola hirta

j. Viola hirta × odorata

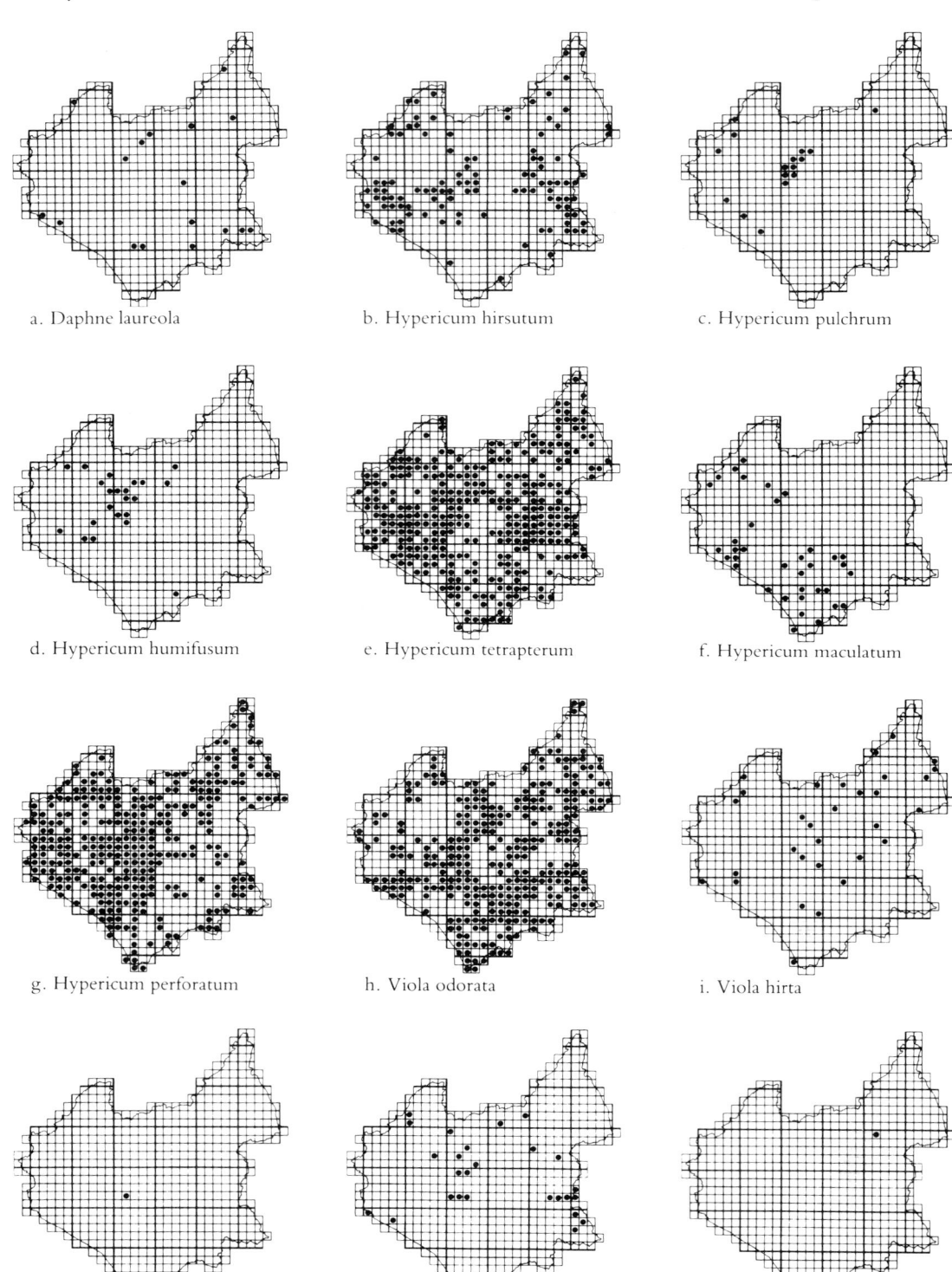

k. Viola reichenbachiana

l. Viola reichenbachiana × riviniana

Violaceae – Onagraceae

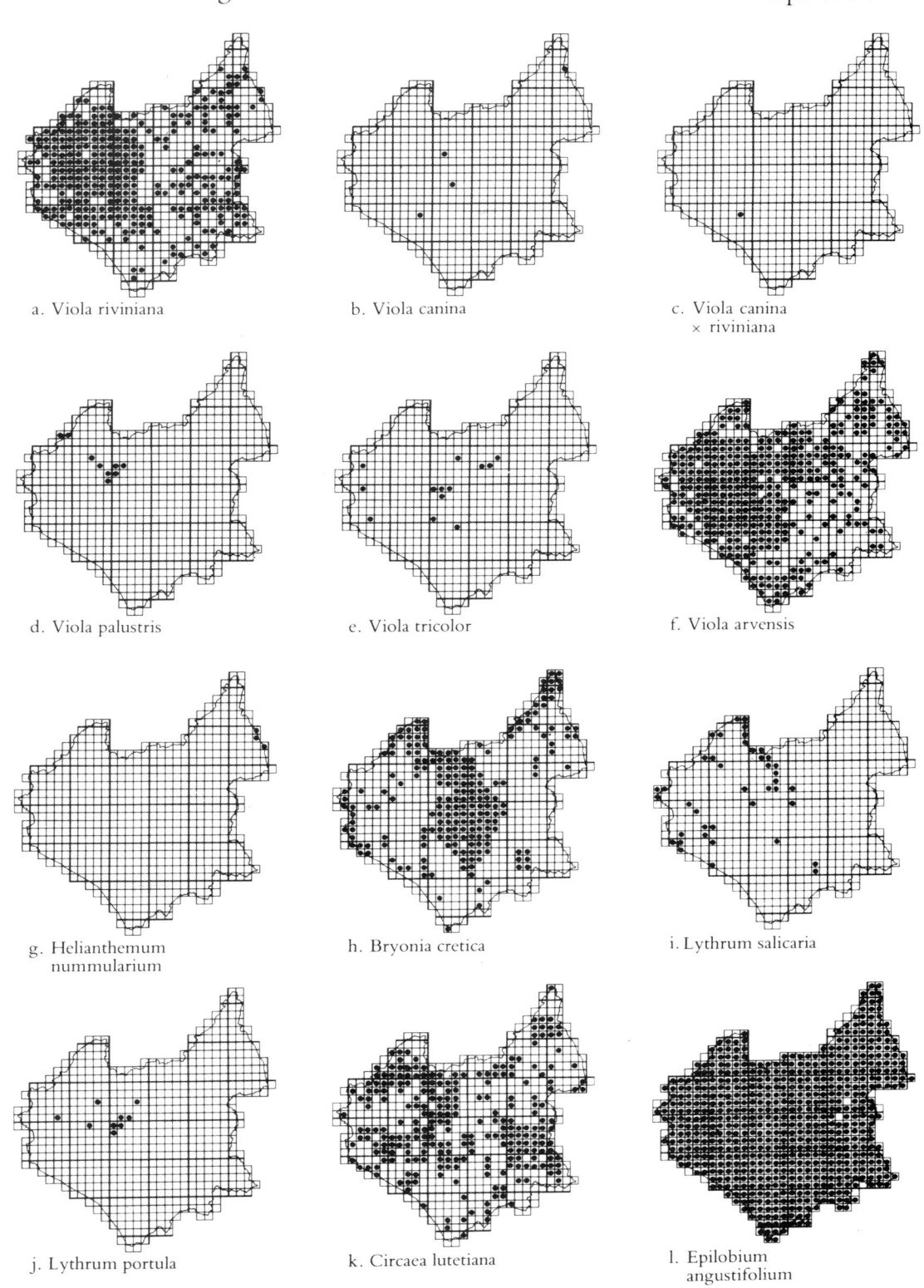

Maps 40a–l

a. Viola riviniana
b. Viola canina
c. Viola canina × riviniana
d. Viola palustris
e. Viola tricolor
f. Viola arvensis
g. Helianthemum nummularium
h. Bryonia cretica
i. Lythrum salicaria
j. Lythrum portula
k. Circaea lutetiana
l. Epilobium angustifolium

Onagraceae - Haloragaceae Maps 41a-l

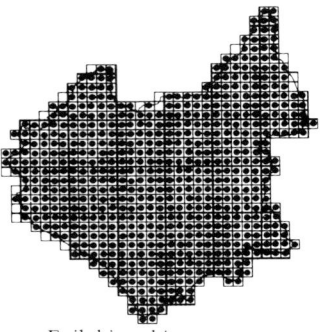

a. Epilobium hirsutum

b. Epilobium parviflorum

c. Epilobium montanum

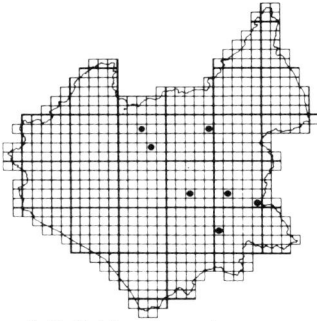

d. Epilobium lanceolatum

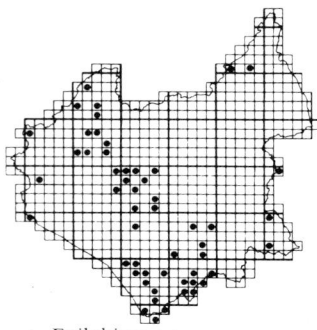

e. Epilobium tetragonum ssp. tetragonum

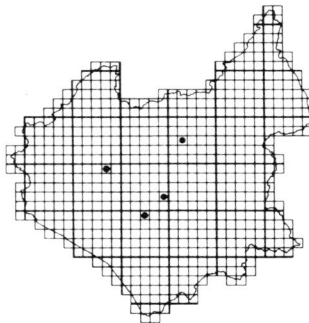

f. Epilobium tetragonum ssp. lamyi

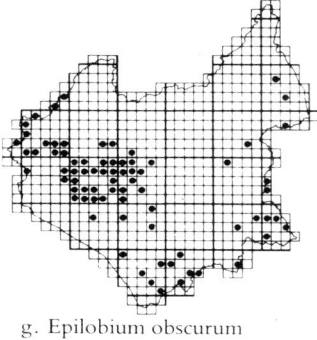

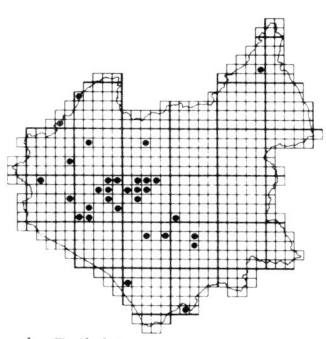

g. Epilobium obscurum

h. Epilobium roseum

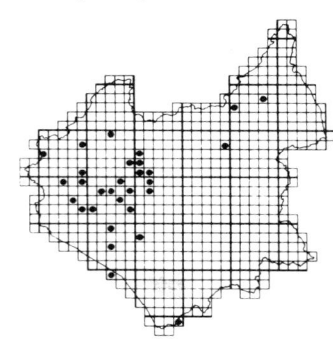

i. Epilobium palustre

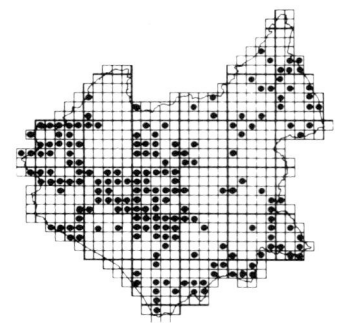

j. Epilobium adenocaulon

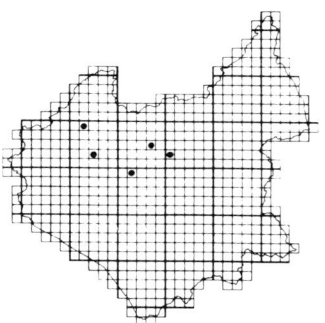

k. Epilobium nerterioides

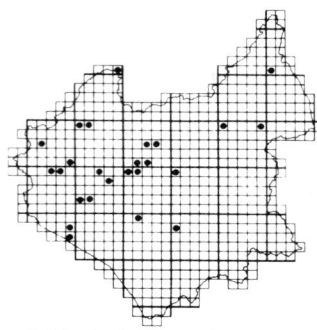

l. Myriophyllum spicatum

Hippuridaceae - Umbelliferae

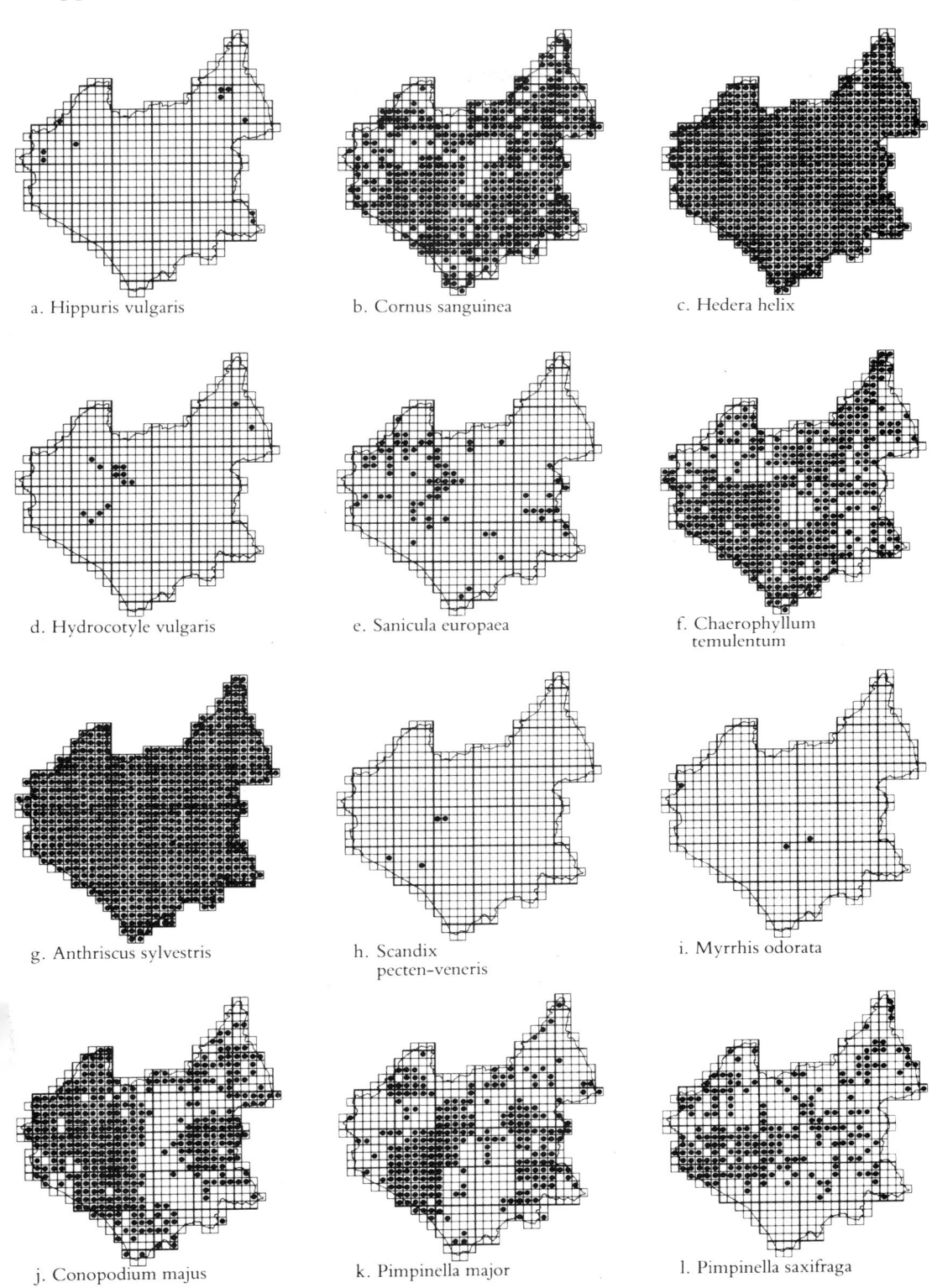

Maps 42a-l

a. Hippuris vulgaris

b. Cornus sanguinea

c. Hedera helix

d. Hydrocotyle vulgaris

e. Sanicula europaea

f. Chaerophyllum temulentum

g. Anthriscus sylvestris

h. Scandix pecten-veneris

i. Myrrhis odorata

j. Conopodium majus

k. Pimpinella major

l. Pimpinella saxifraga

Umbelliferae

Maps 43a–l

a. Aegopodium podagraria

b. Berula erecta

c. Oenanthe fistulosa

d. Oenanthe silaifolia

e. Oenanthe fluviatilis

f. Oenanthe aquatica

g. Aethusa cynapium

h. Silaum silaus

i. Conium maculatum

j. Apium nodiflorum

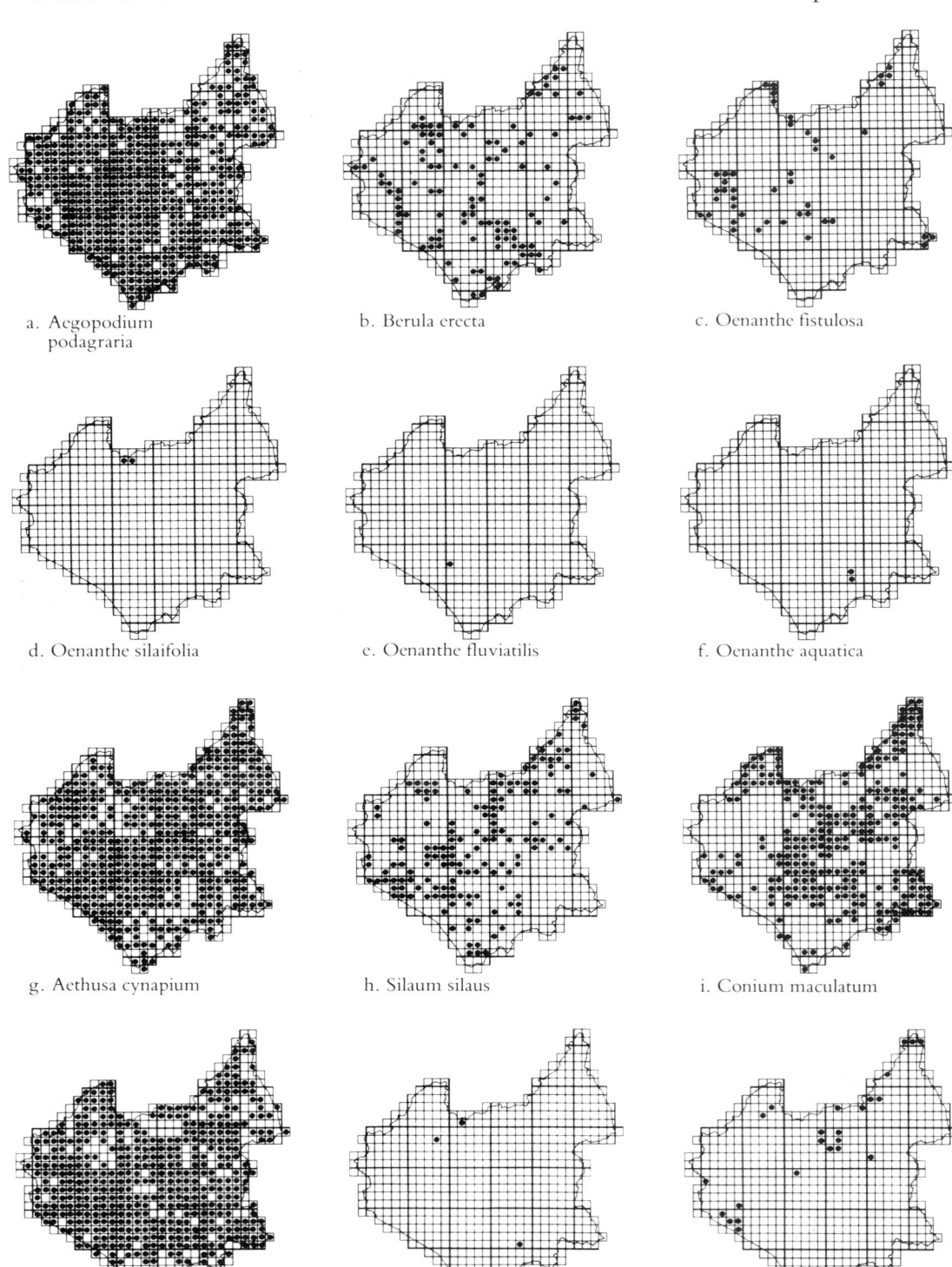

k. Apium inundatum

l. Sison amomum

Umbelliferae - Ericaceae

Maps 44a-l

a. Angelica sylvestris

b. Angelica archangelica

c. Pastinaca sativa

d. Heracleum sphondylium

e. Heracleum mantegazzianum

f. Torilis nodosa

g. Torilis japonica

h. Daucus carota

i. Erica tetralix

j. Calluna vulgaris

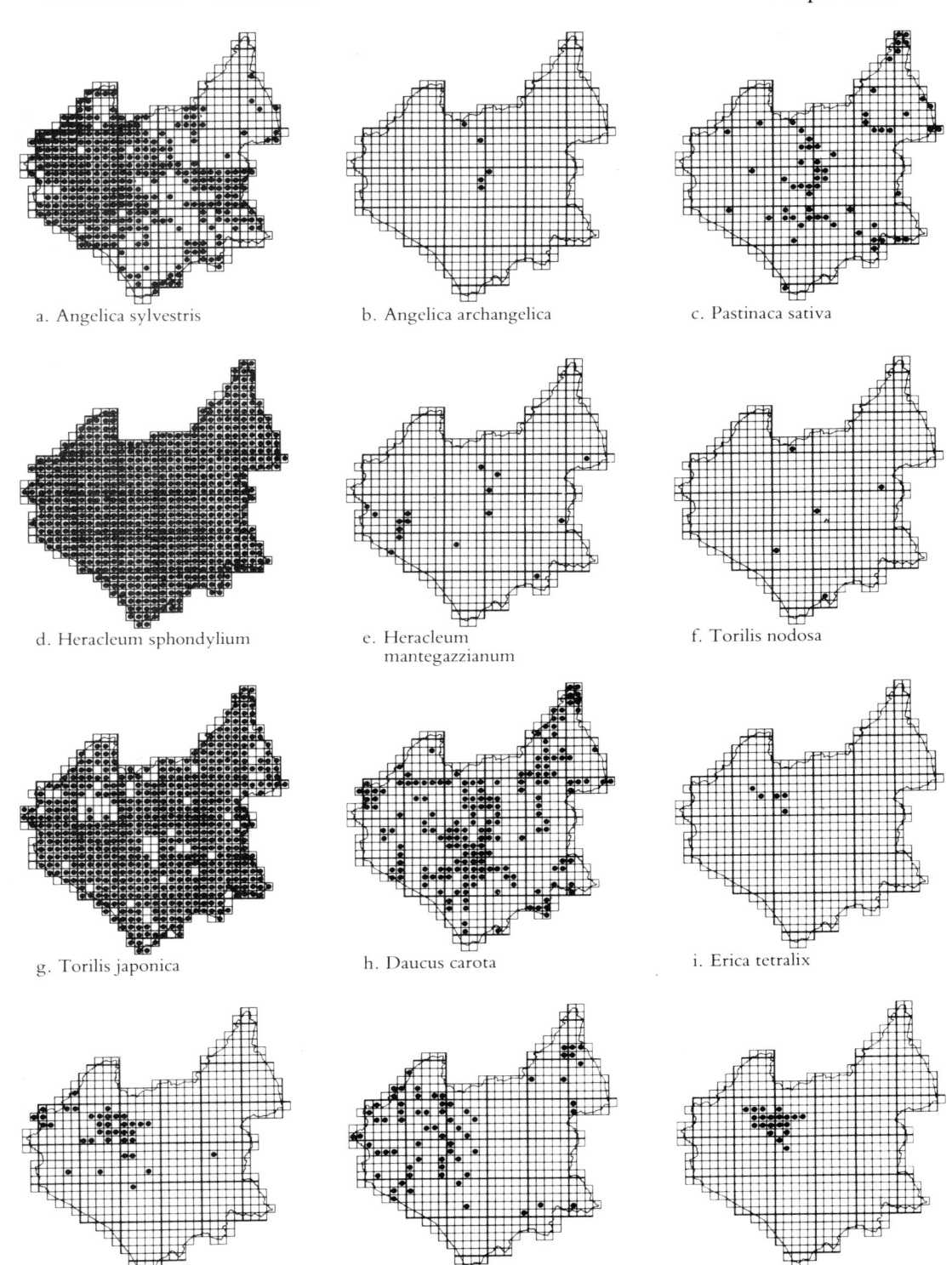

k. Rhododendron ponticum

l. Vaccinium myrtillus

Empetraceae - Primulaceae

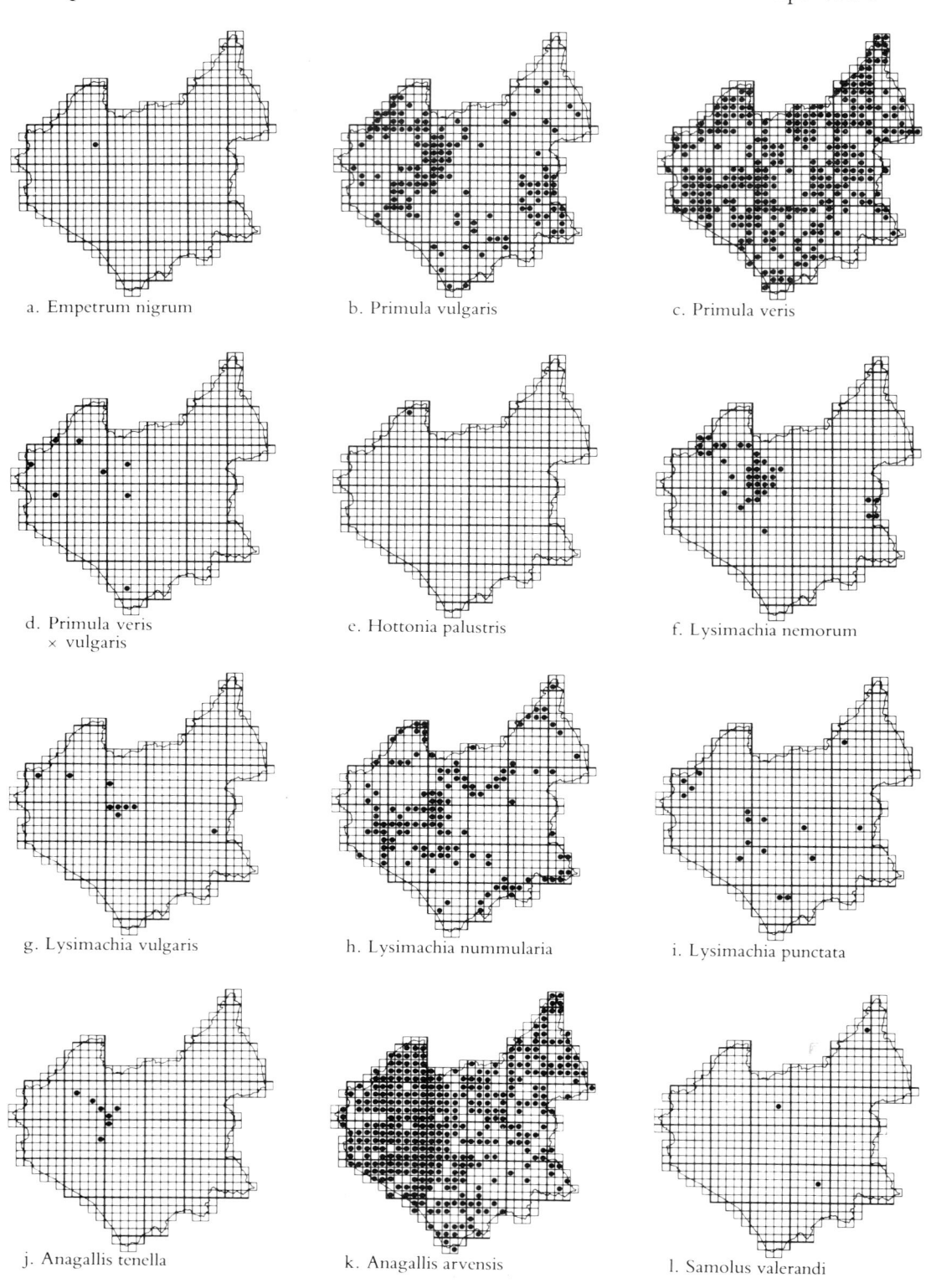

Maps 45a-l

a. Empetrum nigrum

b. Primula vulgaris

c. Primula veris

d. Primula veris × vulgaris

e. Hottonia palustris

f. Lysimachia nemorum

g. Lysimachia vulgaris

h. Lysimachia nummularia

i. Lysimachia punctata

j. Anagallis tenella

k. Anagallis arvensis

l. Samolus valerandi

Oleaceae - Rubiaceae

Maps 46a-l

a. Fraxinus excelsior

b. Syringa vulgaris

c. Ligustrum vulgare

d. Blackstonia perfoliata

e. Centaurium erythraea

f. Gentianella amarella

g. Menyanthes trifoliata

h. Nymphoides peltata

i. Vinca minor

j. Vinca major

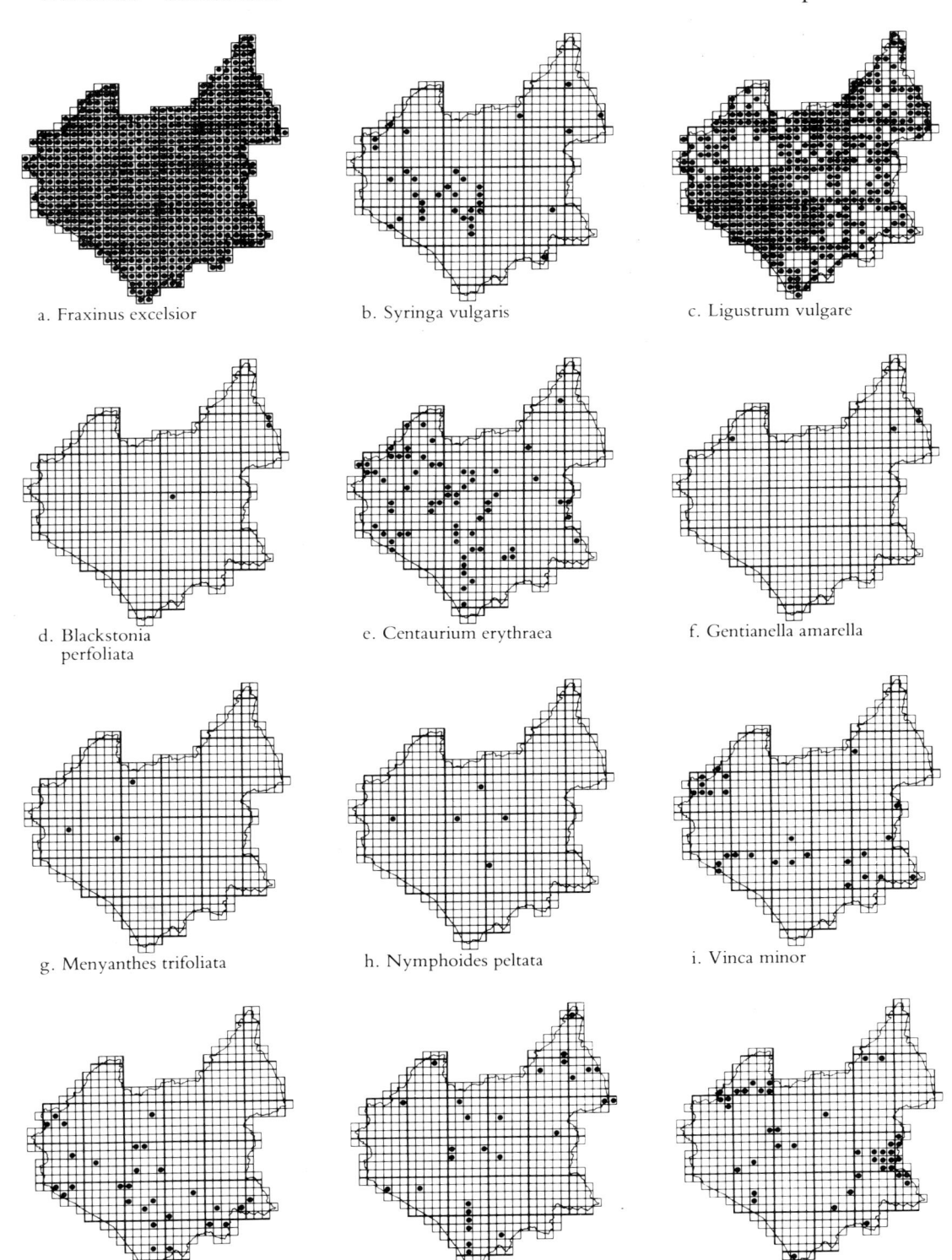

k. Sherardia arvensis

l. Galium odoratum

Rubiaceae - Convolvulaceae   Maps 47a-l

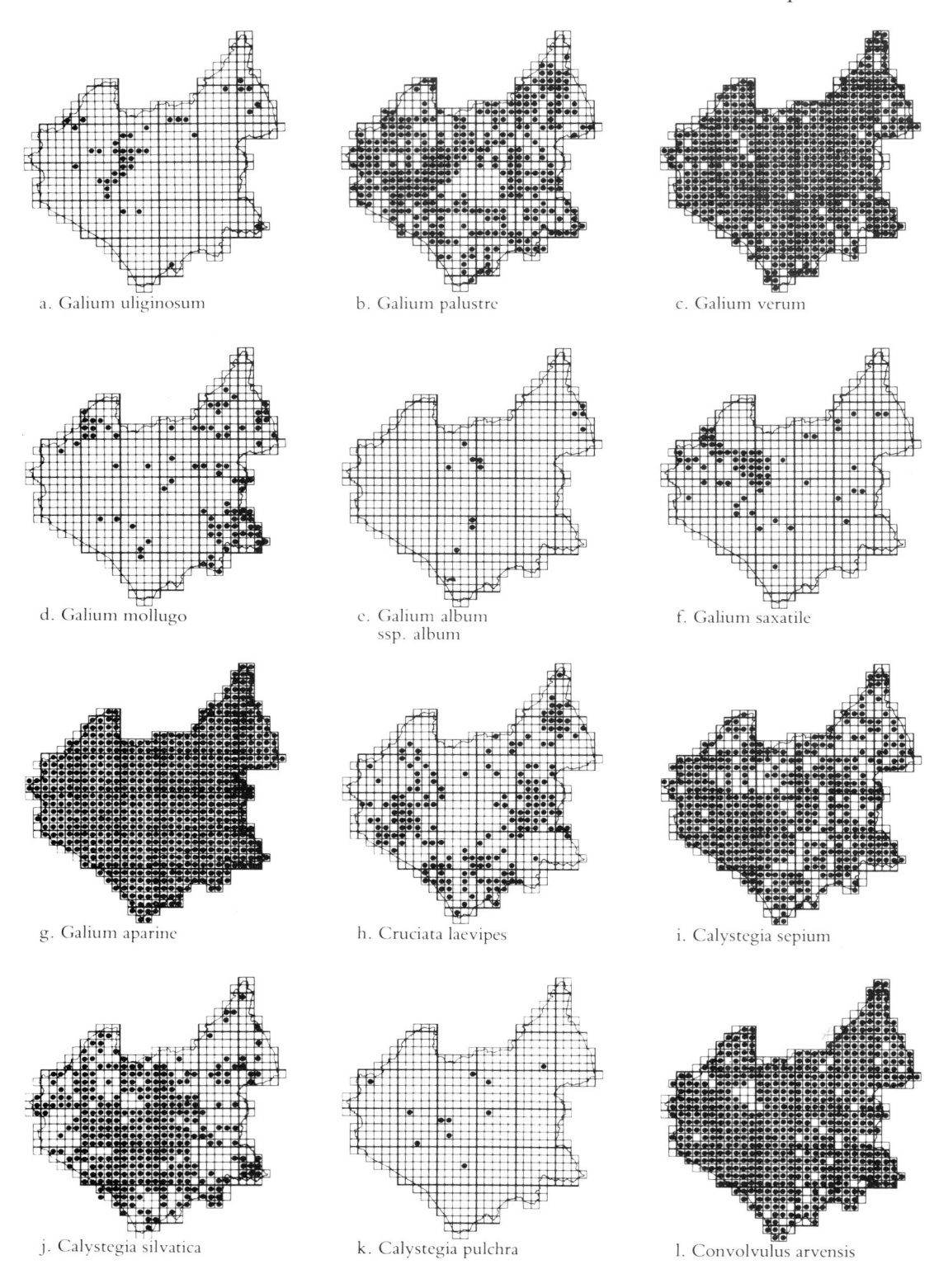

a. Galium uliginosum
b. Galium palustre
c. Galium verum
d. Galium mollugo
e. Galium album ssp. album
f. Galium saxatile
g. Galium aparine
h. Cruciata laevipes
i. Calystegia sepium
j. Calystegia silvatica
k. Calystegia pulchra
l. Convolvulus arvensis

Boraginaceae

Maps 48a-l

a. Buglossoides arvensis

b. Echium vulgare

c. Symphytum officinale

d. Symphytum × uplandicum

e. Symphytum tuberosum

f. Anchusa arvensis

g. Pentaglottis sempervirens

h. Myosotis arvensis

i. Myosotis ramosissima

j. Myosotis discolor

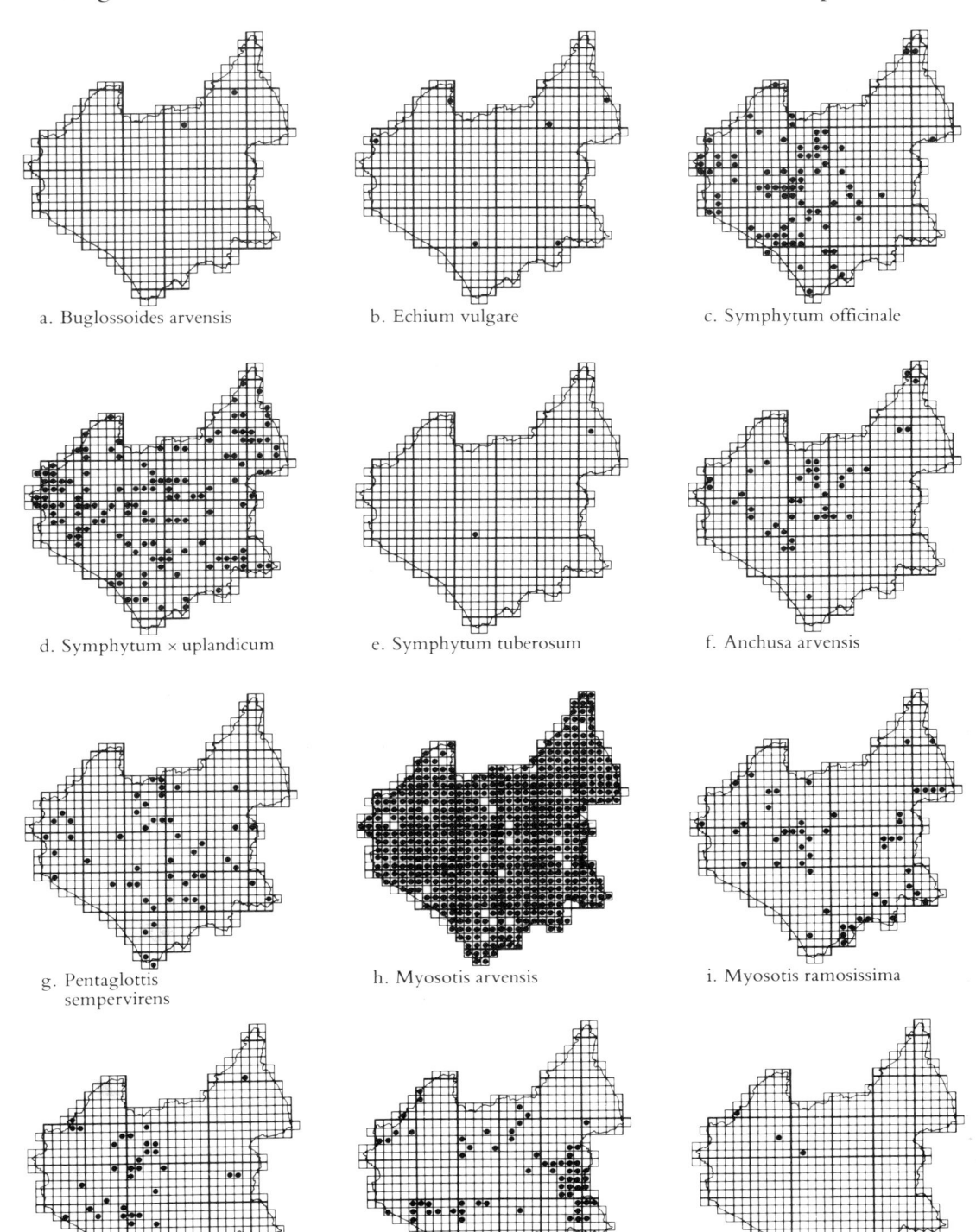

k. Myosotis sylvatica

l. Myosotis secunda

Boraginaceae - Labiatae

Maps 49a-l

a. Myosotis laxa ssp. caespitosa

b. Myosotis scorpioides

c. Cynoglossum officinale

d. Callitriche truncata

e. Callitriche stagnalis

f. Callitriche platycarpa

g. Callitriche intermedia ssp. hamulata

h. Ajuga reptans

i. Teucrium scorodonia

j. Scutellaria galericulata

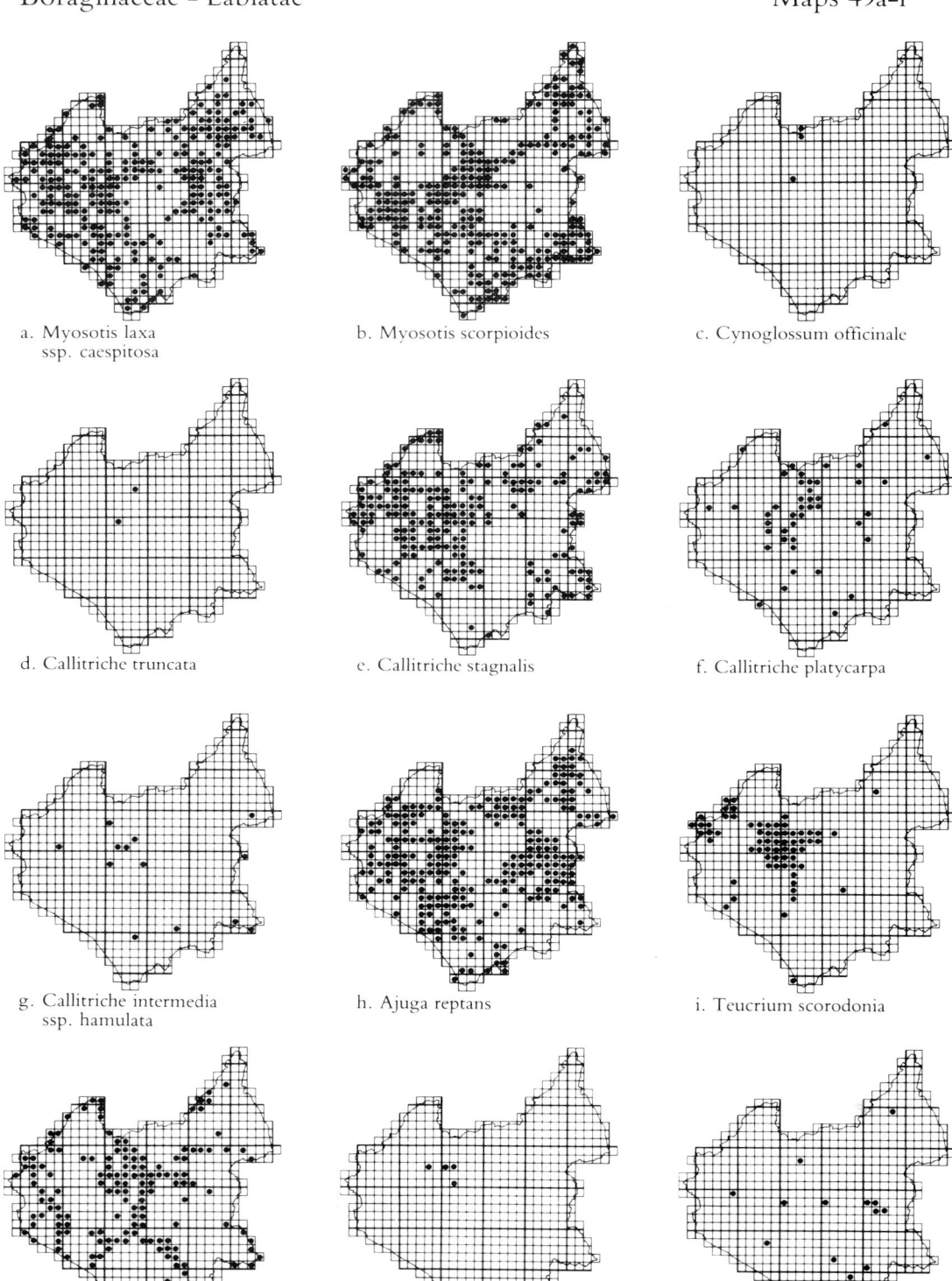

k. Scutellaria minor

l. Galeopsis speciosa

Labiatae   Maps 50a-l

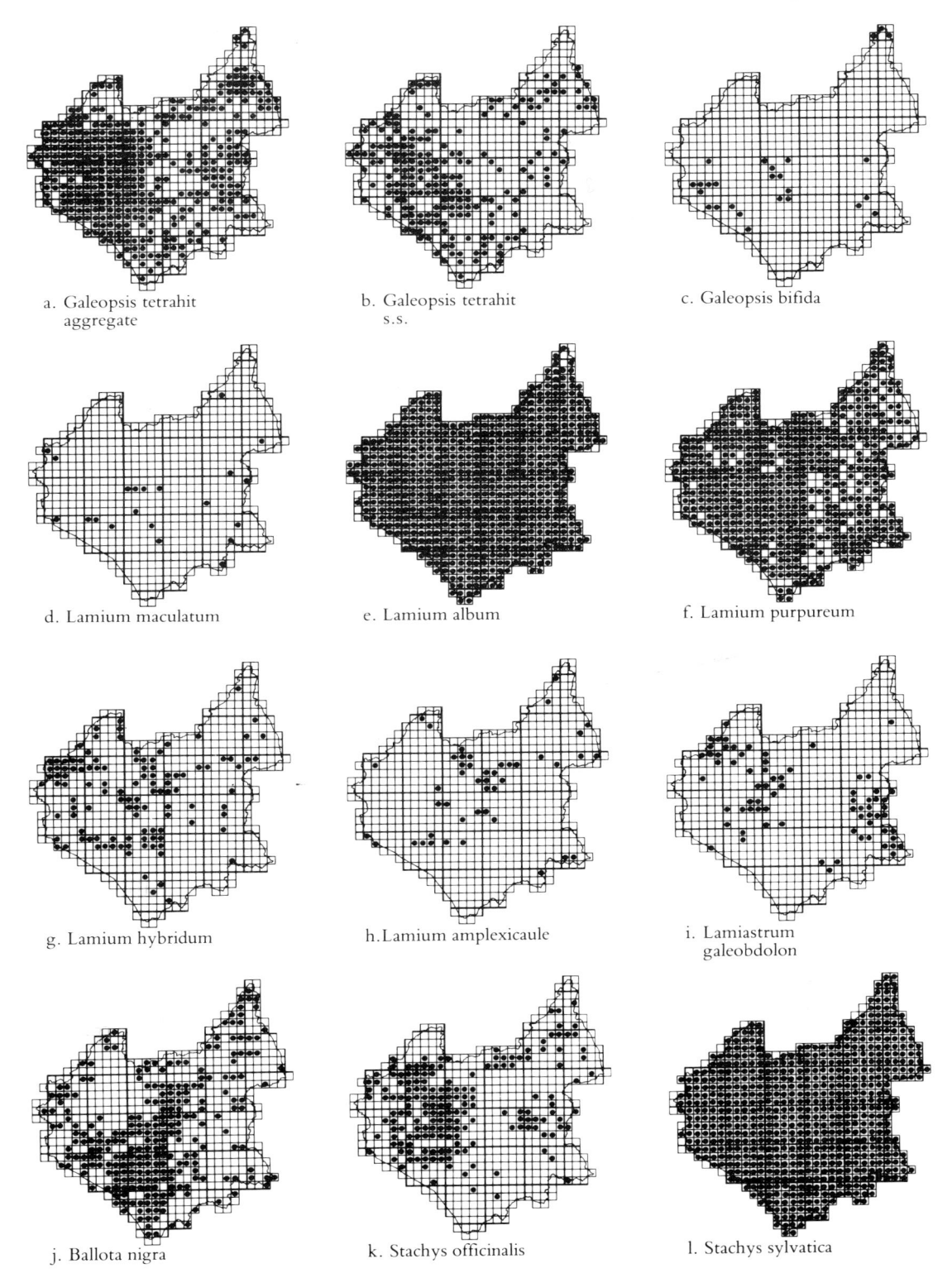

a. Galeopsis tetrahit aggregate

b. Galeopsis tetrahit s.s.

c. Galeopsis bifida

d. Lamium maculatum

e. Lamium album

f. Lamium purpureum

g. Lamium hybridum

h. Lamium amplexicaule

i. Lamiastrum galeobdolon

j. Ballota nigra

k. Stachys officinalis

l. Stachys sylvatica

Labiatae

Maps 51a-l

a. Stachys palustris

b. Stachys palustris × sylvatica

c. Stachys arvensis

d. Glechoma hederacea

e. Prunella vulgaris

f. Acinos arvensis

g. Clinopodium vulgare

h. Origanum vulgare

i. Thymus praecox

j. Thymus pulegioides

k. Lycopus europaeus

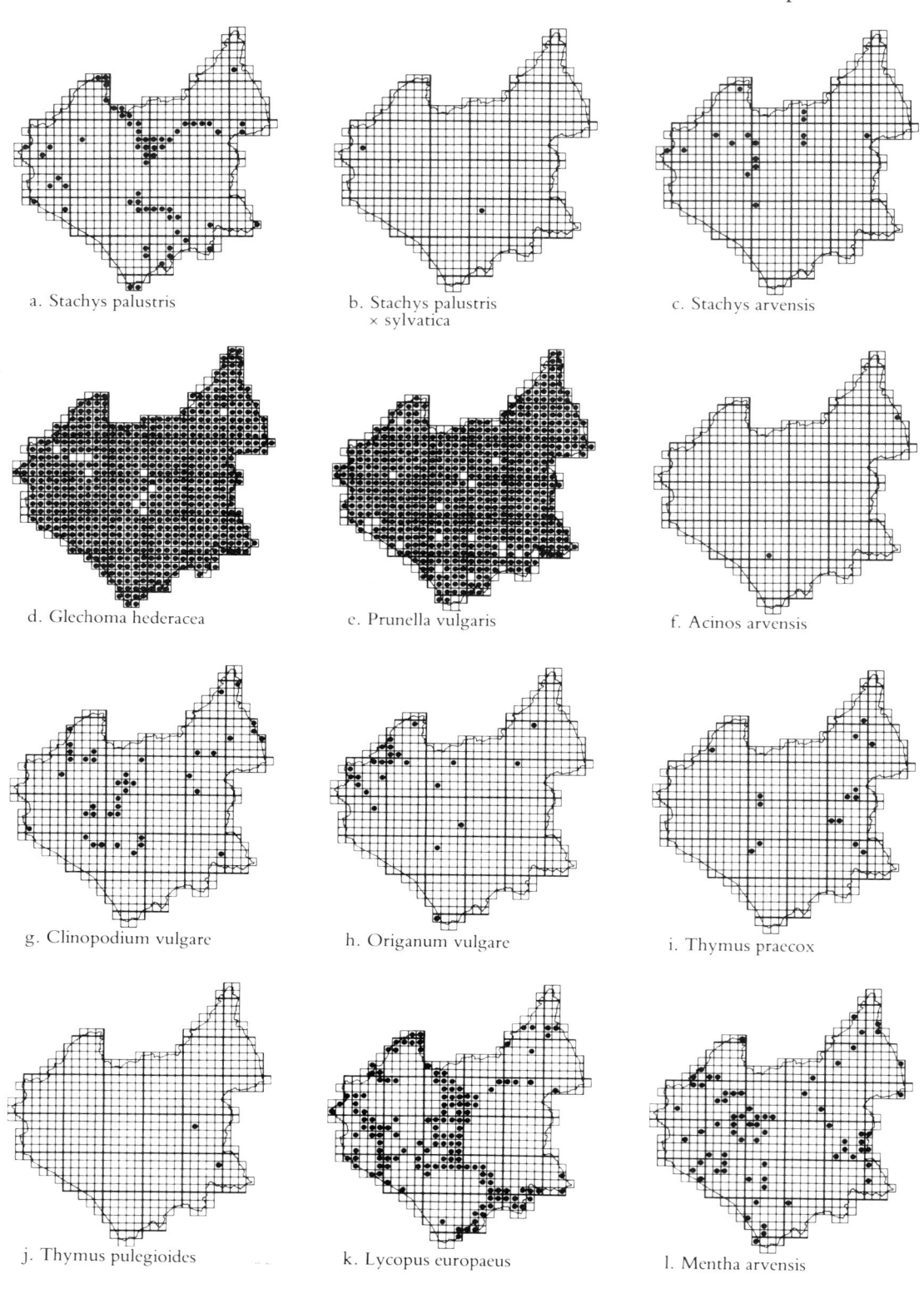

l. Mentha arvensis

Labiatae - Solanaceae

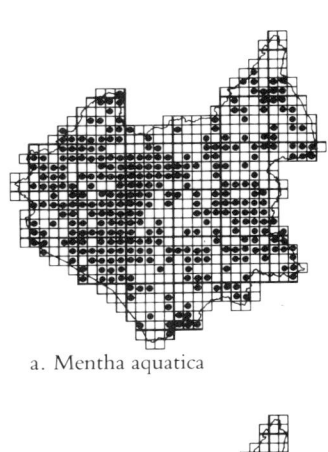

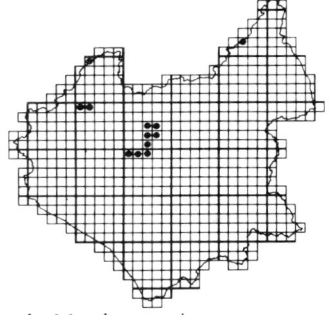

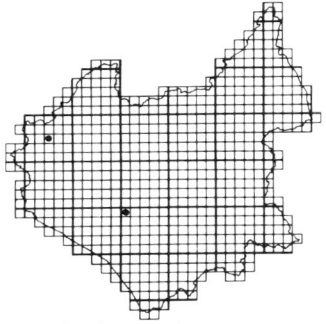

Maps 52a-l

a. Mentha aquatica

b. Mentha aquatica × arvensis

c. Mentha aquatica × arvensis × spicata

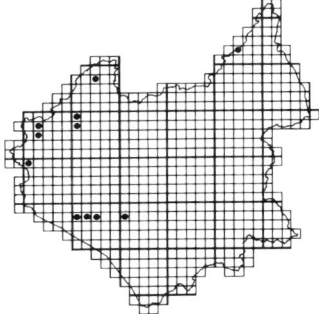

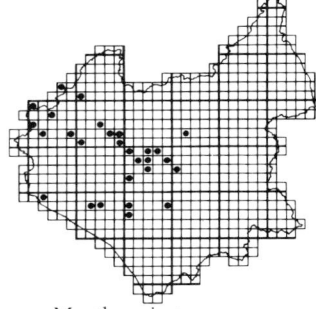

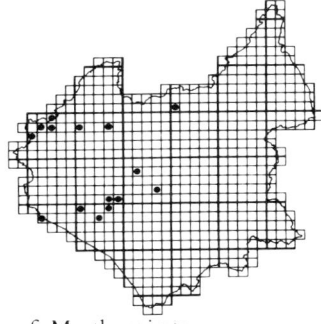

d. Mentha aquatica × spicata

e. Mentha spicata

f. Mentha spicata × suaveolens

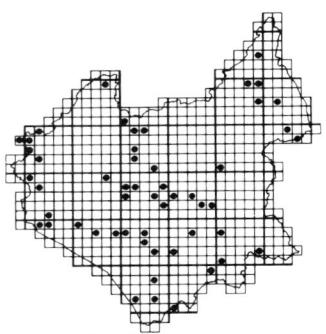

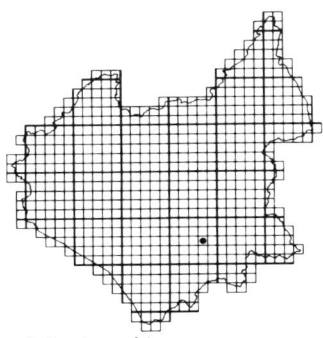

g. Salvia verbenaca

h. Lycium barbarum

i. Lycium chinense

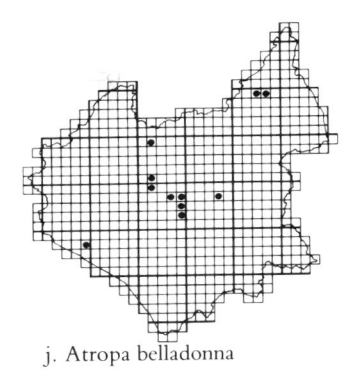

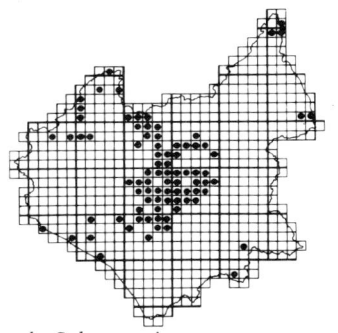

j. Atropa belladonna

k. Solanum nigrum

l. Solanum dulcamara

Scrophulariaceae

Maps 53a-l

a. Limosella aquatica

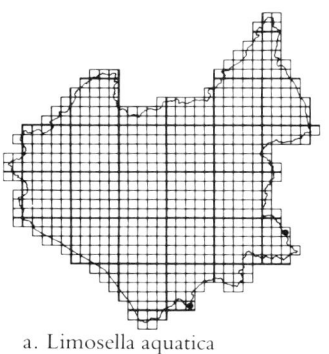

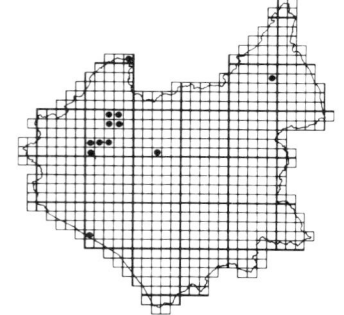

b. Mimulus guttatus

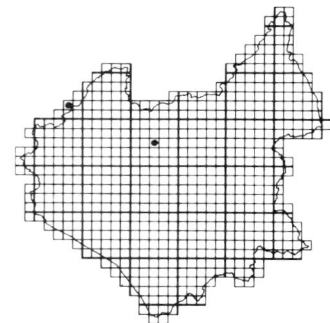

c. Mimulus moschatus

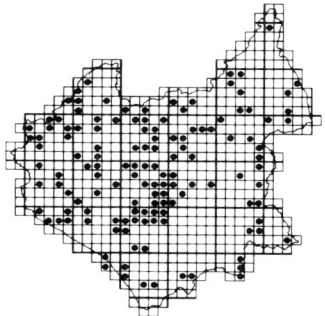

d. Verbascum thapsus

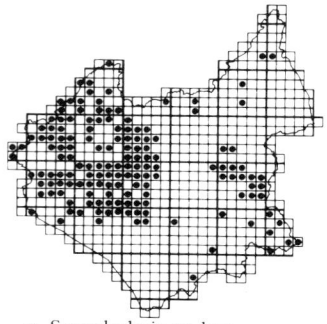

e. Scrophularia nodosa

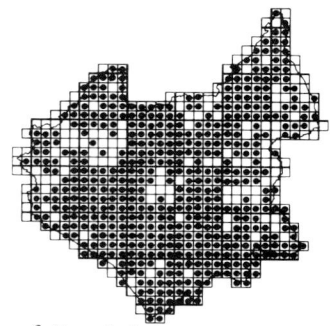

f. Scrophularia auriculata

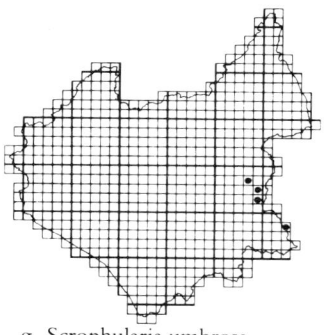

g. Scrophularia umbrosa

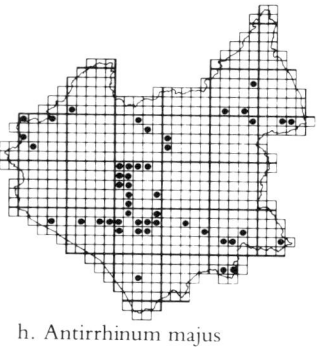

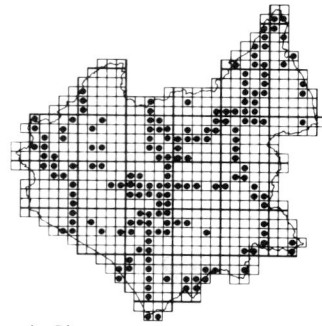

h. Antirrhinum majus

i. Chaenorhinum minus

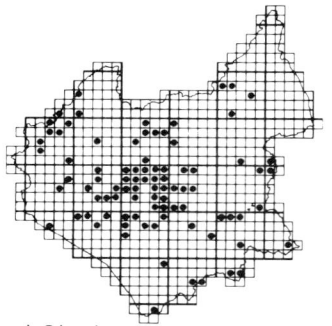

j. Linaria purpurea

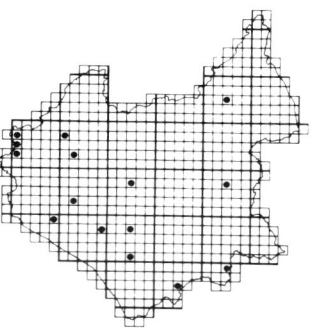

k. Linaria repens

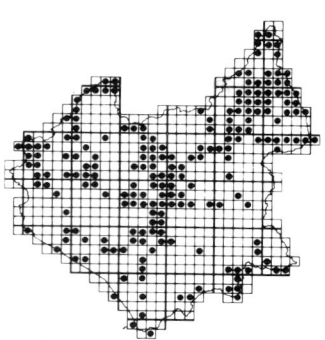

l. Linaria vulgaris

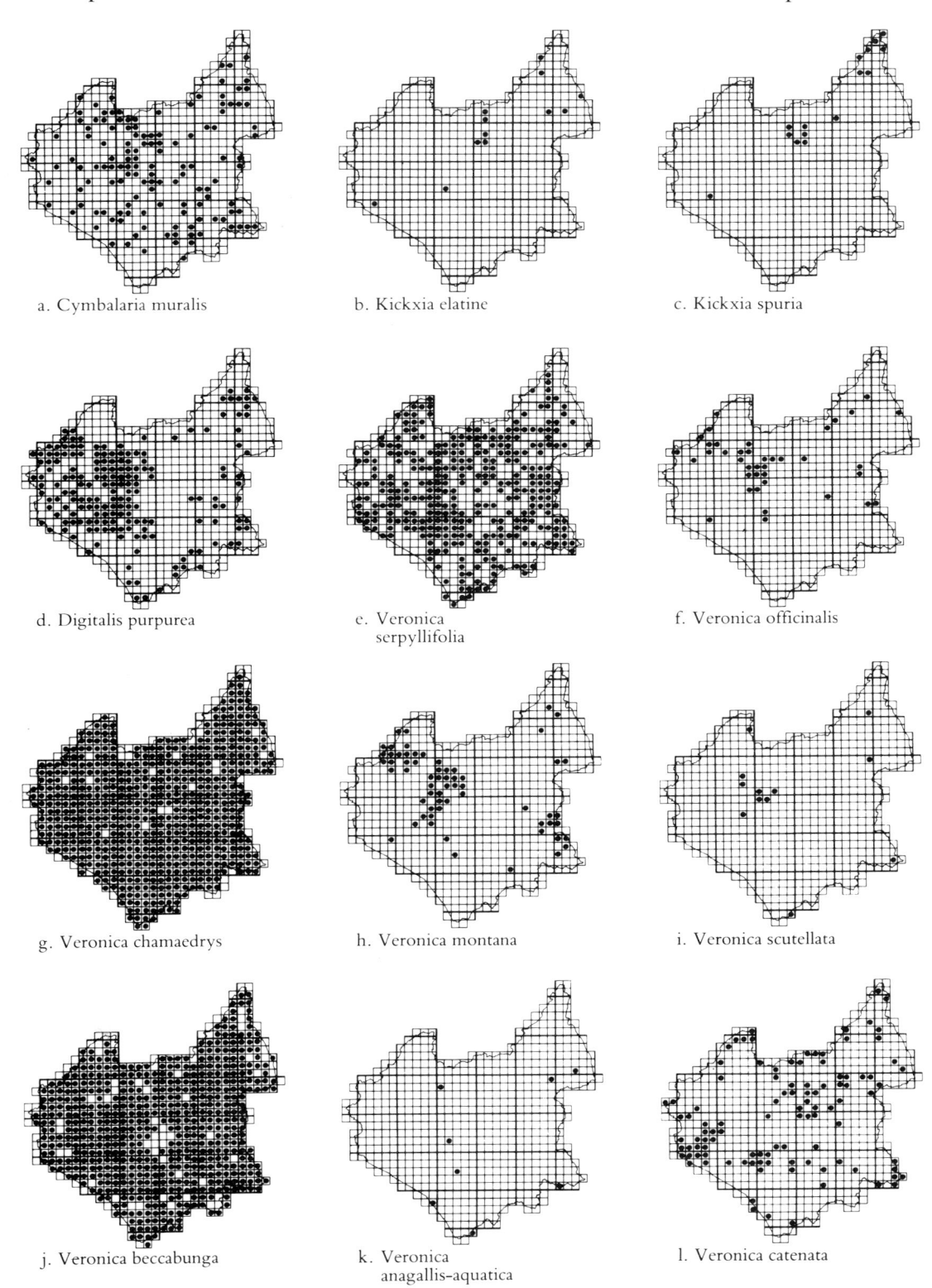

Scrophulariaceae

Maps 54a-l

a. Cymbalaria muralis

b. Kickxia elatine

c. Kickxia spuria

d. Digitalis purpurea

e. Veronica serpyllifolia

f. Veronica officinalis

g. Veronica chamaedrys

h. Veronica montana

i. Veronica scutellata

j. Veronica beccabunga

k. Veronica anagallis-aquatica

l. Veronica catenata

Scrophulariaceae

Maps 55a-l

a. Veronica arvensis

b. Veronica agrestis

c. Veronica polita

d. Veronica persica

e. Veronica filiformis

f. Veronica hederifolia

g. Melampyrum pratense

h. Euphrasia anglica

i. Euphrasia arctica ssp. borealis

j. Euphrasia nemorosa

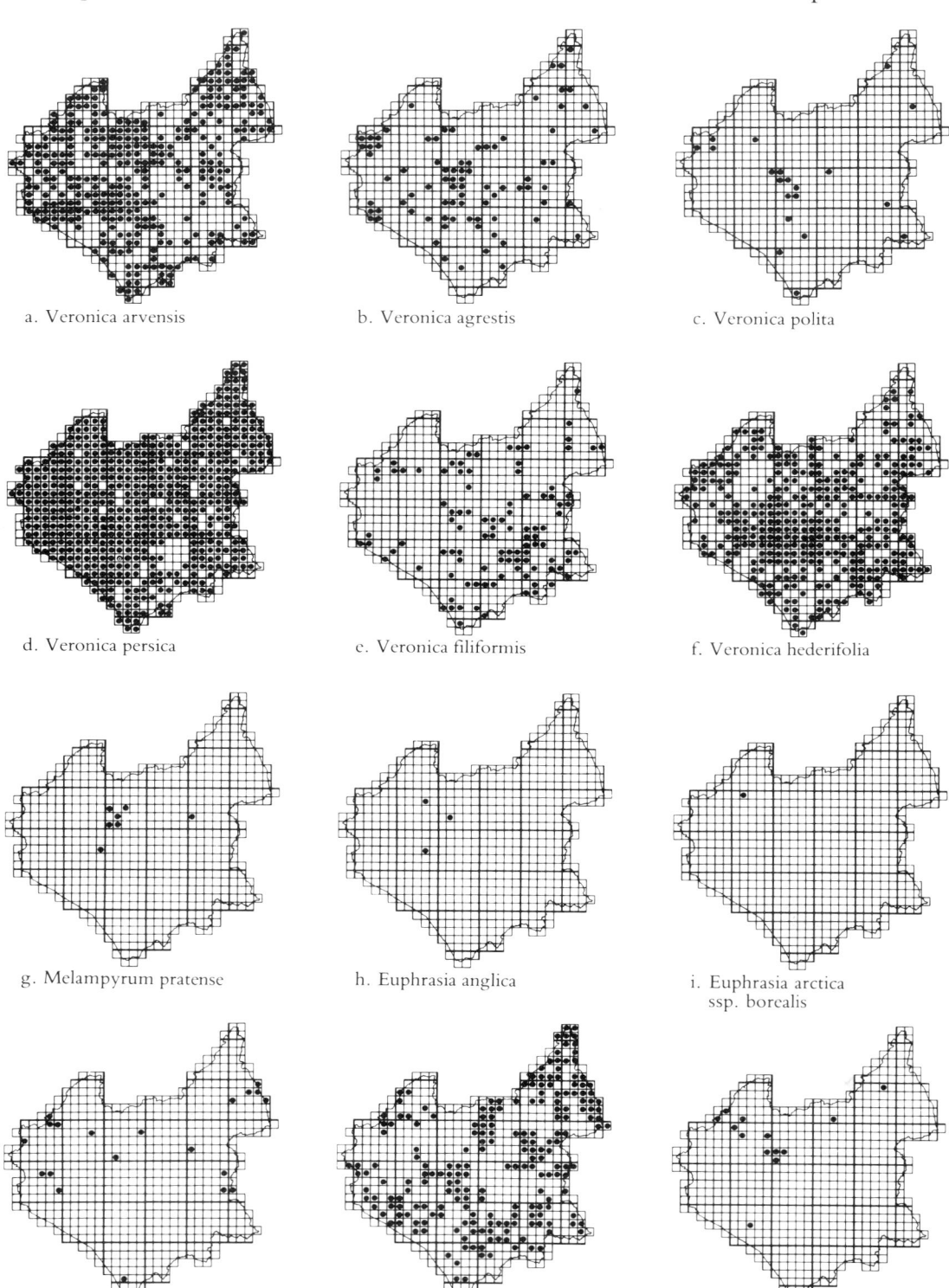

k. Odontites verna

l. Pedicularis sylvatica

Scrophulariaceae - Caprifoliaceae

Maps 56a-l

a. Rhinanthus minor

b. Lathraea squamaria

c. Orobanche minor

d. Plantago major

e. Plantago coronopus

f. Plantago media

g. Plantago lanceolata

h. Littorella uniflora

i. Sambucus ebulus

j. Sambucus nigra

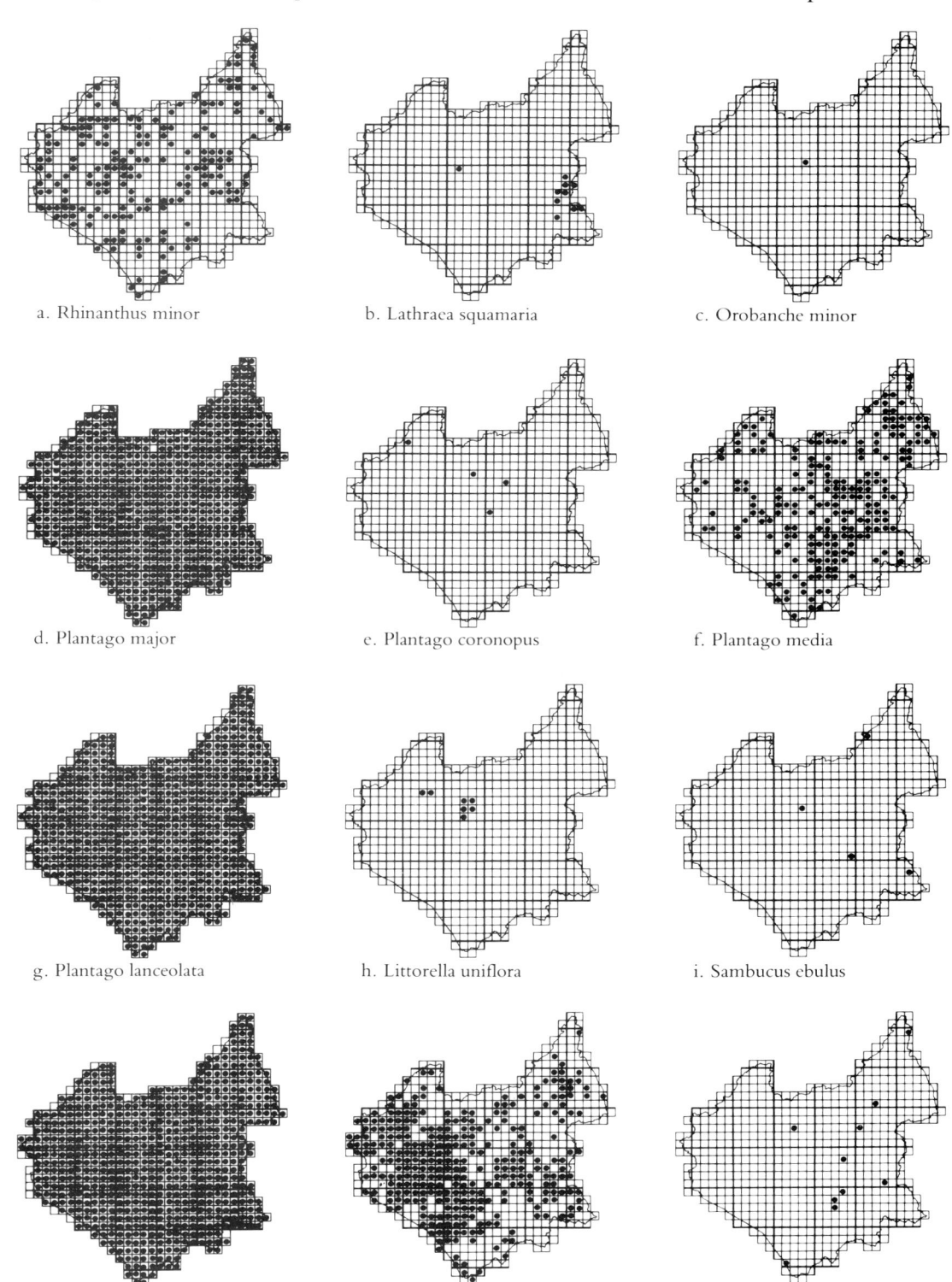

k. Viburnum opulus

l. Viburnum lantana

Caprifoliaceae - Dipsacaceae

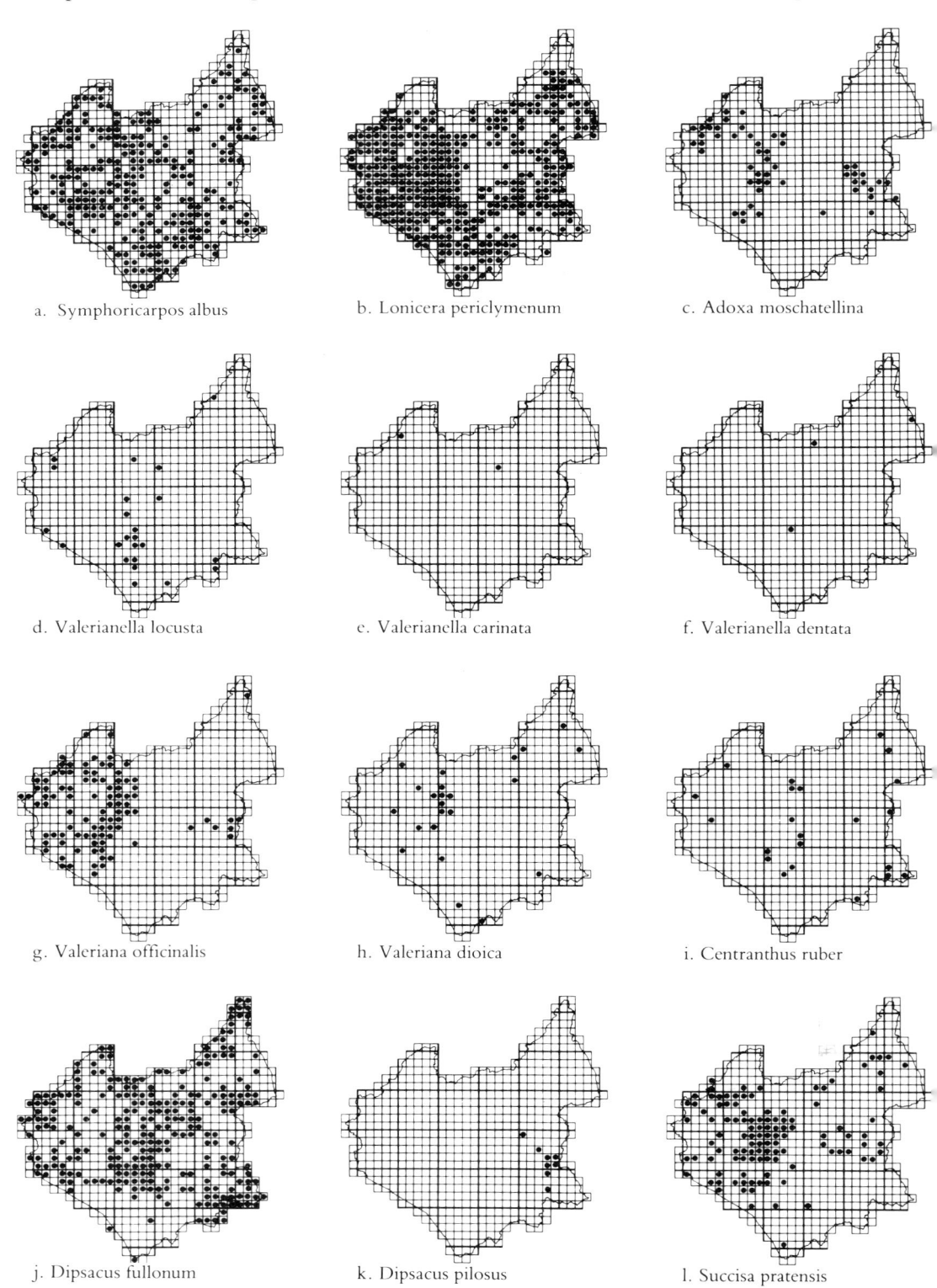

Maps 57a-l

a. Symphoricarpos albus
b. Lonicera periclymenum
c. Adoxa moschatellina
d. Valerianella locusta
e. Valerianella carinata
f. Valerianella dentata
g. Valeriana officinalis
h. Valeriana dioica
i. Centranthus ruber
j. Dipsacus fullonum
k. Dipsacus pilosus
l. Succisa pratensis

Dipsacaceae - Compositae

Maps 58a-l

a. Knautia arvensis

b. Scabiosa columbaria

c. Campanula glomerata

d. Campanula latifolia

e. Campanula trachelium

f. Campanula rapunculoides

g. Campanula rotundifolia

h. Legousia hybrida

i. Eupatorium cannabinum

j. Solidago virgaurea

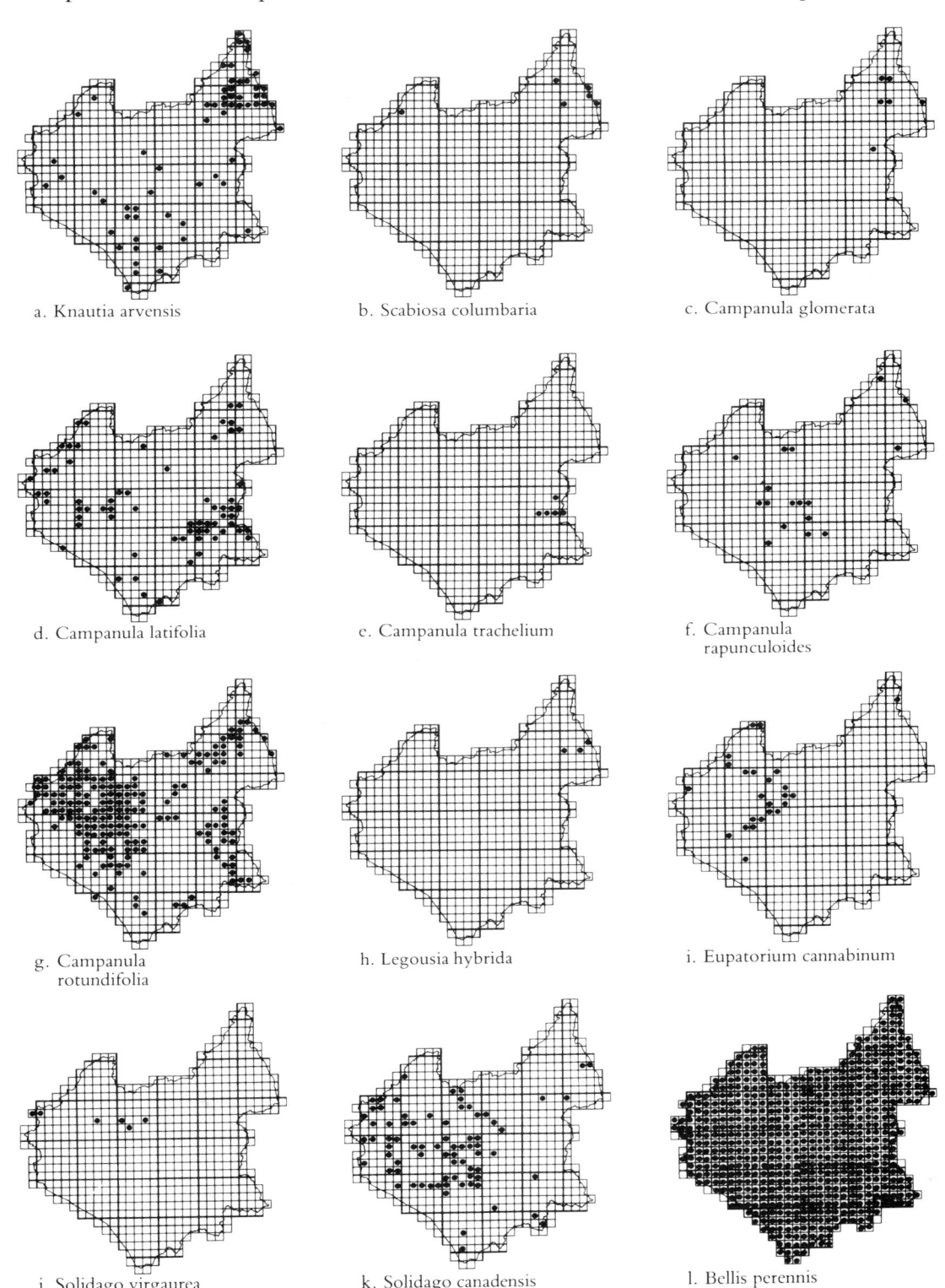

k. Solidago canadensis

l. Bellis perennis

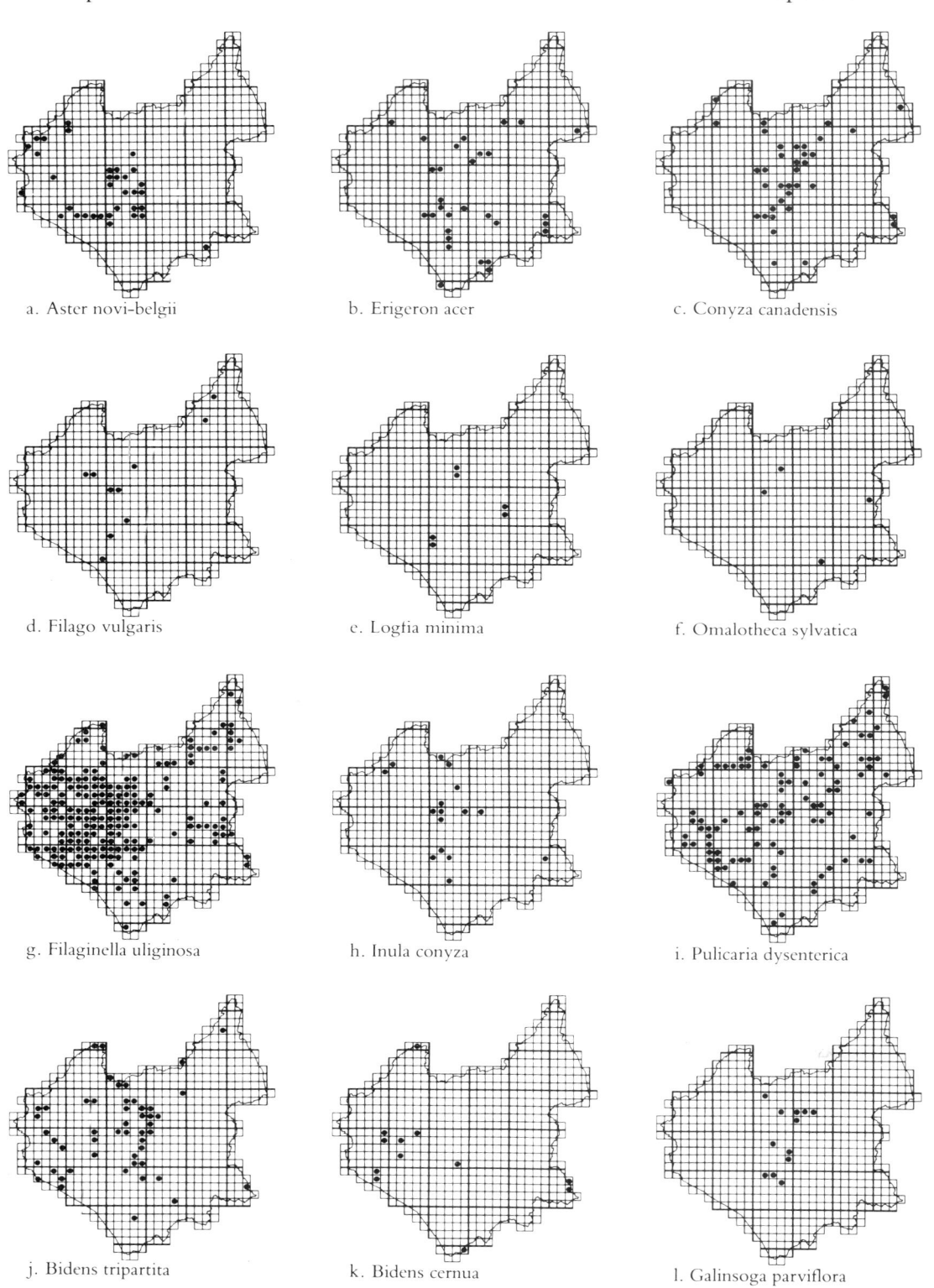

Compositae

Maps 59a-l

a. Aster novi-belgii

b. Erigeron acer

c. Conyza canadensis

d. Filago vulgaris

e. Logfia minima

f. Omalotheca sylvatica

g. Filaginella uliginosa

h. Inula conyza

i. Pulicaria dysenterica

j. Bidens tripartita

k. Bidens cernua

l. Galinsoga parviflora

Compositae

Maps 60a-l

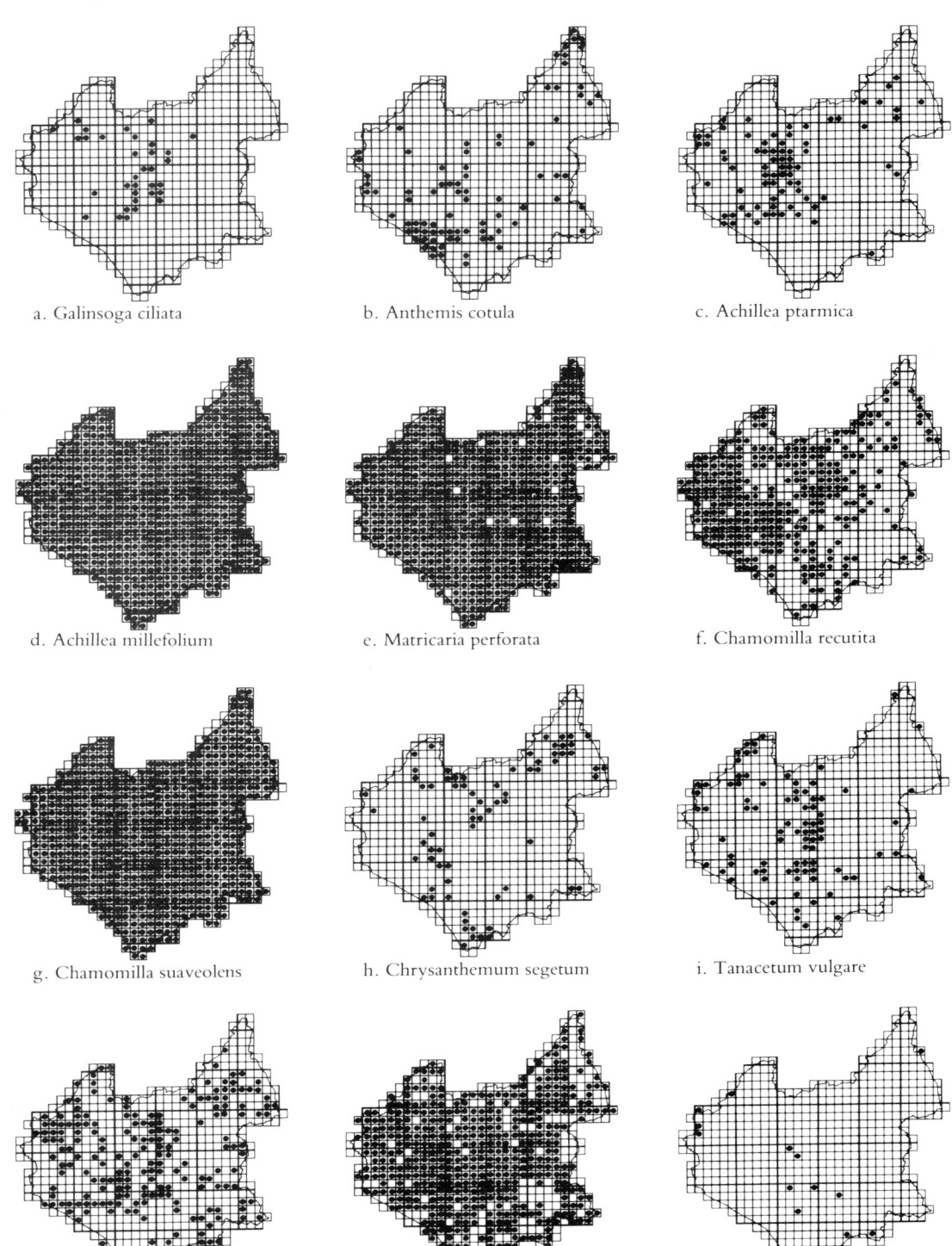

a. Galinsoga ciliata

b. Anthemis cotula

c. Achillea ptarmica

d. Achillea millefolium

e. Matricaria perforata

f. Chamomilla recutita

g. Chamomilla suaveolens

h. Chrysanthemum segetum

i. Tanacetum vulgare

j. Tanacetum parthenium

k. Leucanthemum vulgare

l. Leucanthemum maximum

Compositae   Maps 61a-l

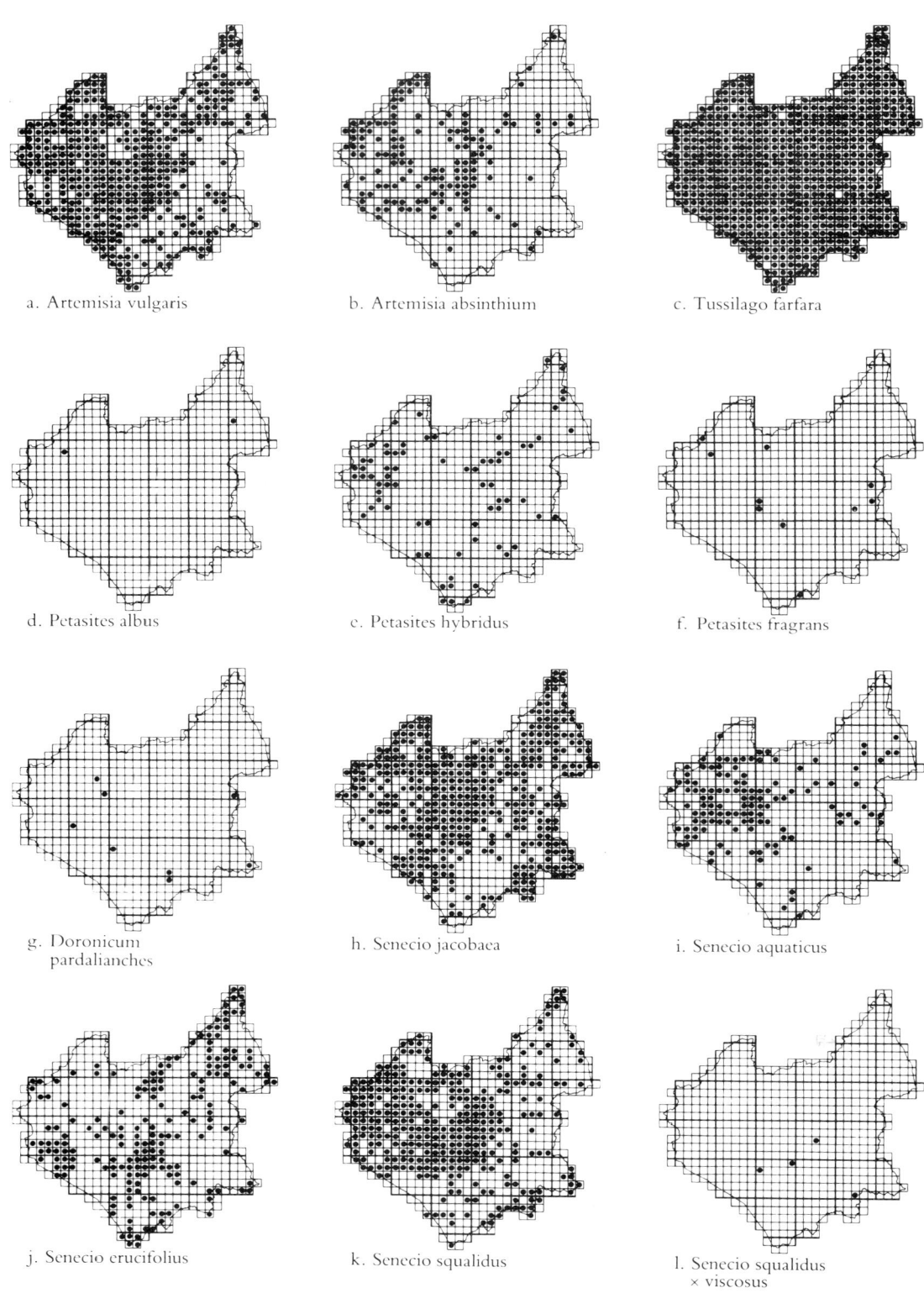

a. Artemisia vulgaris   b. Artemisia absinthium   c. Tussilago farfara

d. Petasites albus   e. Petasites hybridus   f. Petasites fragrans

g. Doronicum
   pardalianches   h. Senecio jacobaea   i. Senecio aquaticus

j. Senecio erucifolius   k. Senecio squalidus   l. Senecio squalidus
                                                                     × viscosus

Compositae

Maps 62a-l

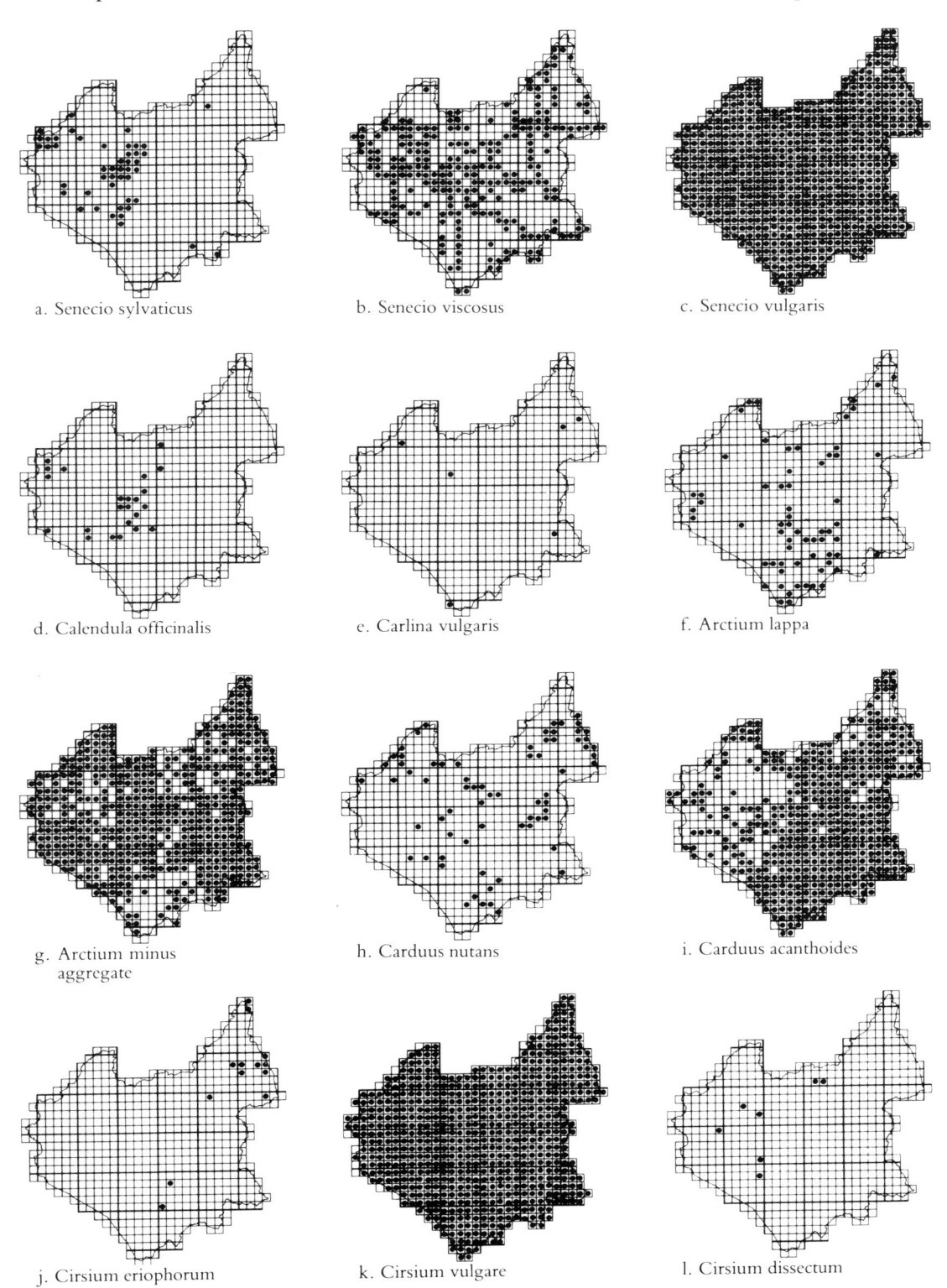

a. Senecio sylvaticus

b. Senecio viscosus

c. Senecio vulgaris

d. Calendula officinalis

e. Carlina vulgaris

f. Arctium lappa

g. Arctium minus aggregate

h. Carduus nutans

i. Carduus acanthoides

j. Cirsium eriophorum

k. Cirsium vulgare

l. Cirsium dissectum

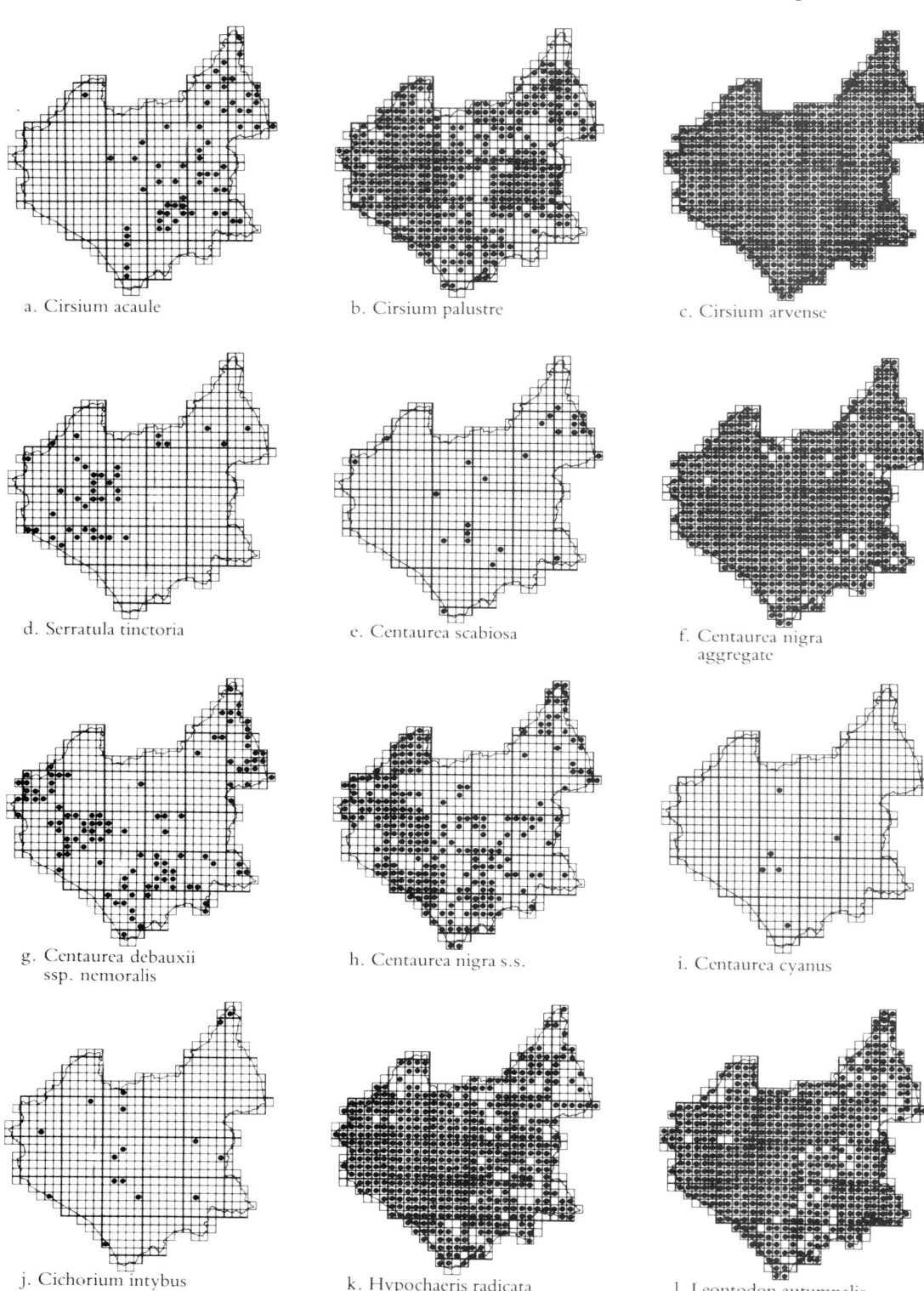

Compositae

Maps 63a-l

a. Cirsium acaule

b. Cirsium palustre

c. Cirsium arvense

d. Serratula tinctoria

e. Centaurea scabiosa

f. Centaurea nigra aggregate

g. Centaurea debauxii ssp. nemoralis

h. Centaurea nigra s.s.

i. Centaurea cyanus

j. Cichorium intybus

k. Hypochaeris radicata

l. Leontodon autumnalis

Compositae

Maps 64a-l

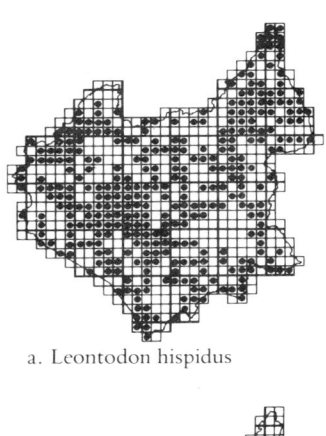

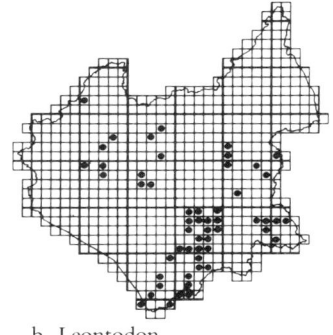

a. Leontodon hispidus

b. Leontodon taraxacoides

c. Picris echioides

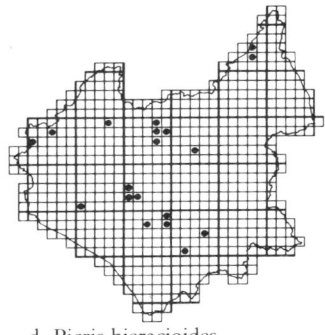

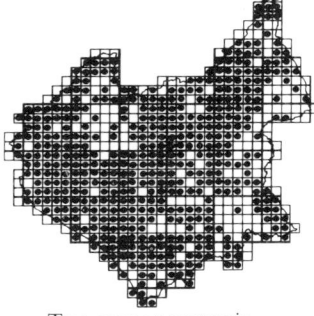

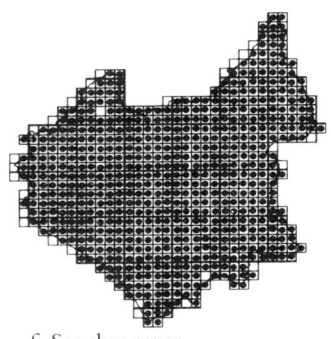

d. Picris hieracioides

e. Tragopogon pratensis

f. Sonchus asper

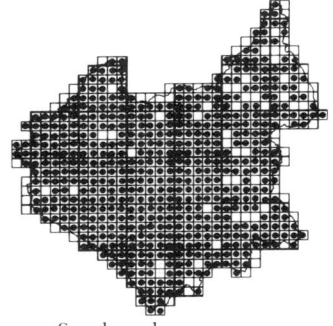

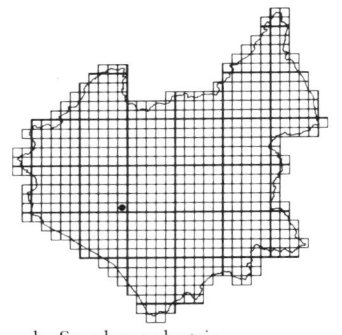

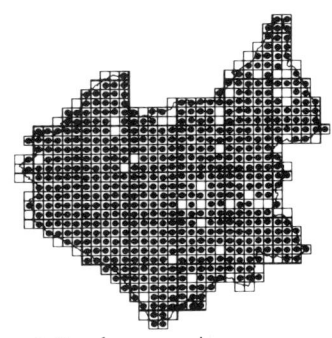

g. Sonchus oleraceus

h. Sonchus palustris

i. Sonchus arvensis

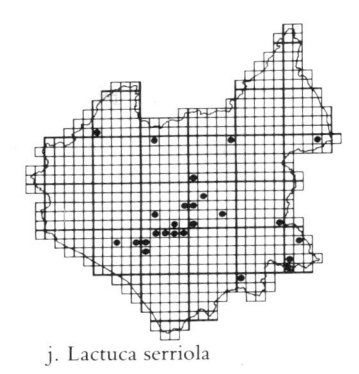

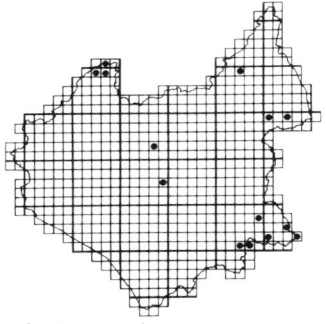

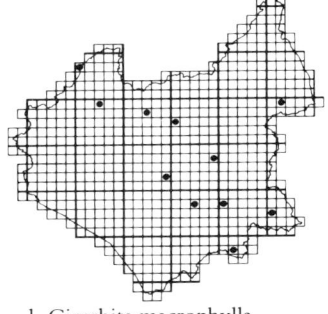

j. Lactuca serriola

k. Lactuca virosa

l. Cicerbita macrophylla

Compositae

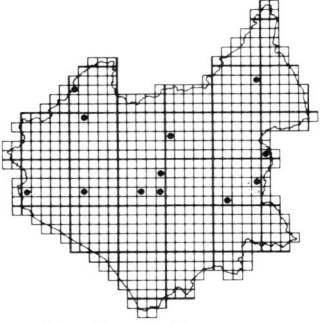

Maps 65a-l

a. Mycelis muralis

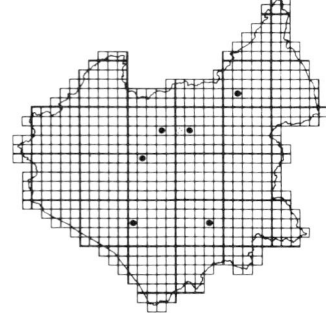

b. Taraxacum brachyglossum

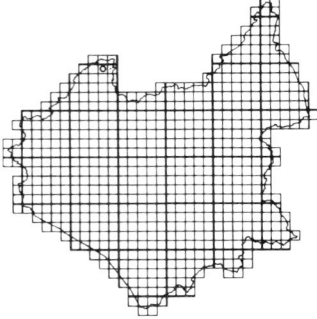

c. Taraxacum canulum

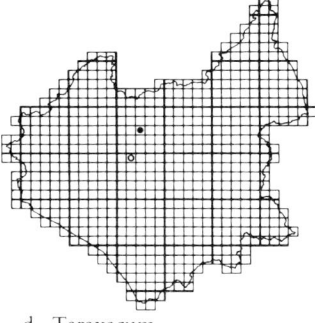

d. Taraxacum fulviforme

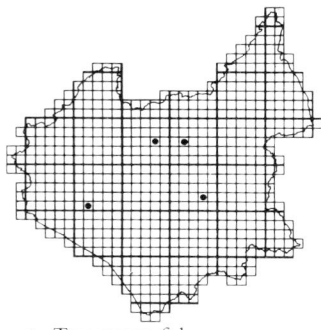

e. Taraxacum fulvum

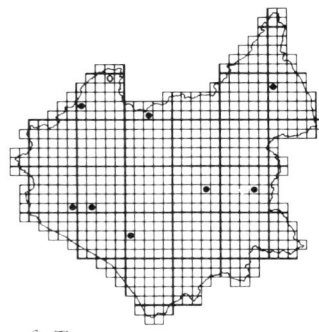

f. Taraxacum lacistophyllum

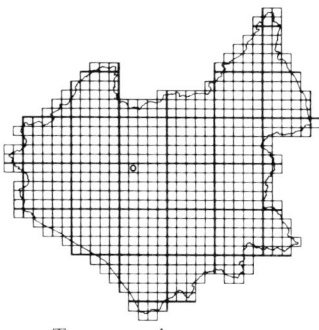

g. Taraxacum laetum

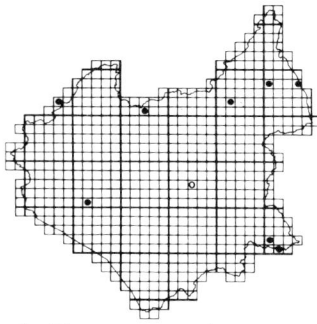

h. Taraxacum oxoniense

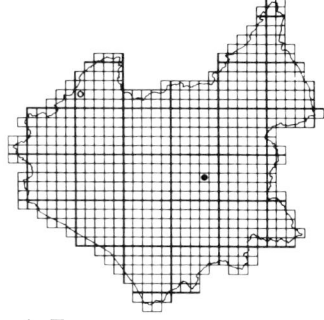

i. Taraxacum pseudolacistophyllum

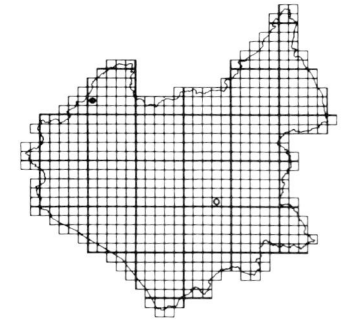

j. Taraxacum rubicundum

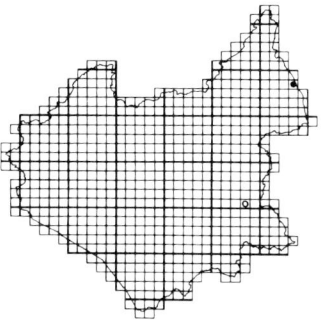

k. Taraxacum silesiacum

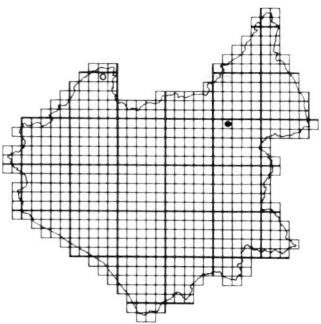

l. Taraxacum simile

Compositae

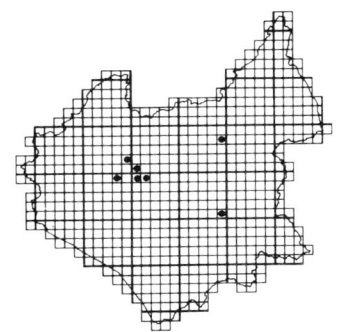

Maps 66a-1

a. Taraxacum faeroense

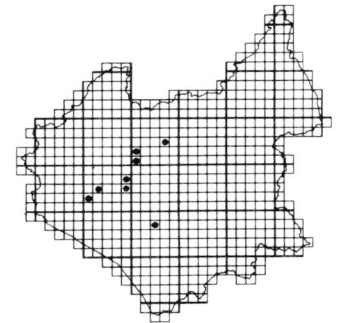

b. Taraxacum spectabile

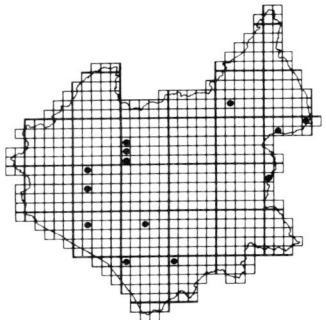

c. Taraxacum euryphyllum

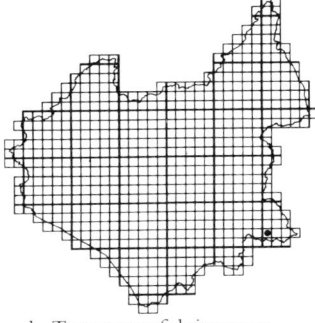

d. Taraxacum fulvicarpum

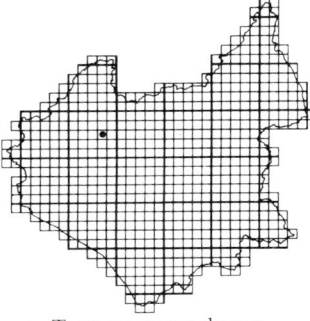

e. Taraxacum maculosum

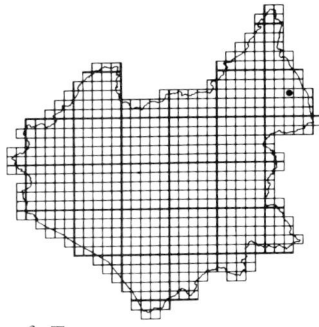

f. Taraxacum naevosiforme

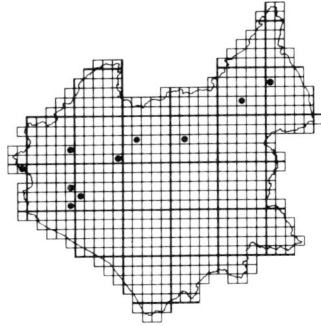

g. Taraxacum adamii

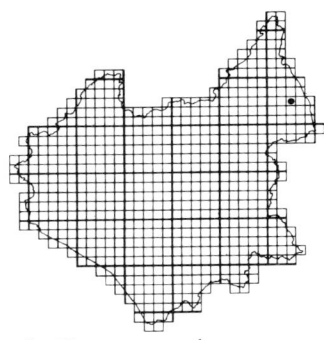

h. Taraxacum explanatum

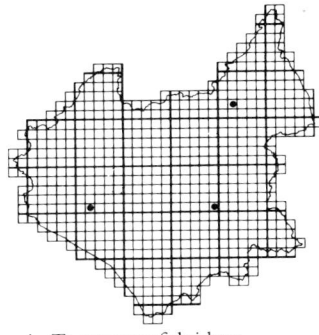

i. Taraxacum fulgidum

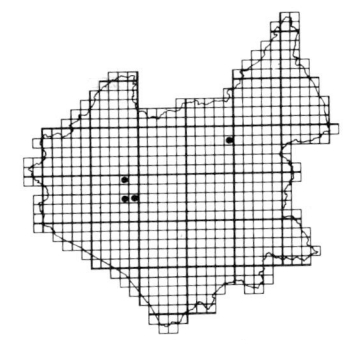

j. Taraxacum haematicum

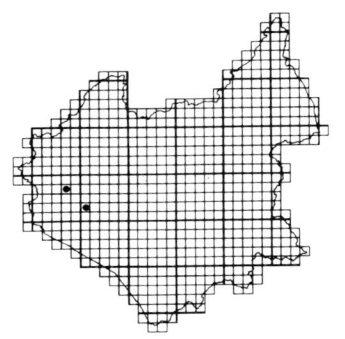

k. Taraxacum laetifrons

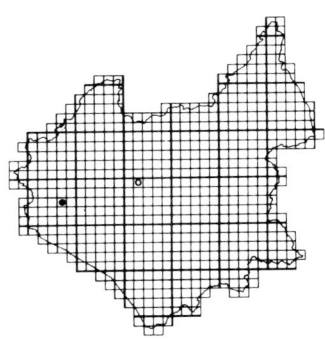

l. Taraxacum landmarkii

Compositae

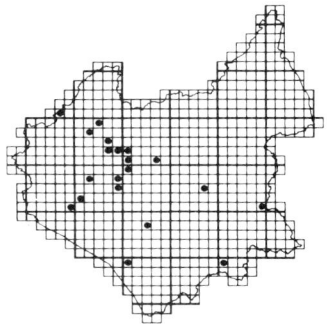

Maps 67a-l

a. Taraxacum nordstedtii

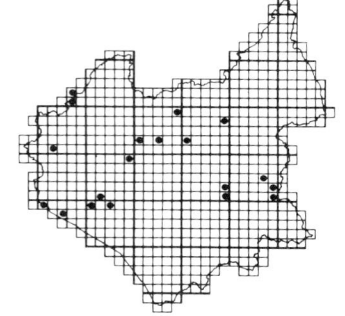

b. Taraxacum raunkiaerii

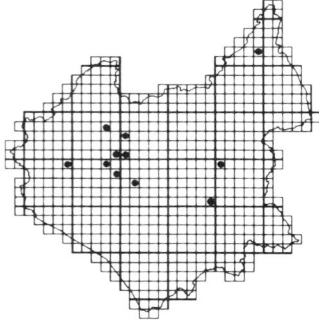

c. Taraxacum subbracteatum

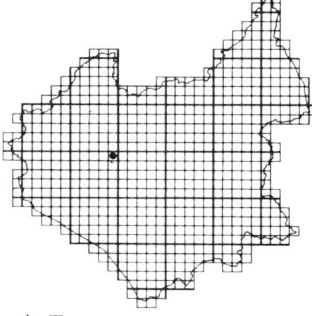

d. Taraxacum tamesense

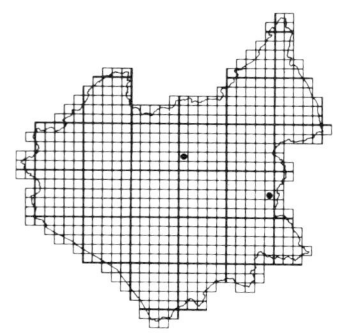

e. Taraxacum tenebricans

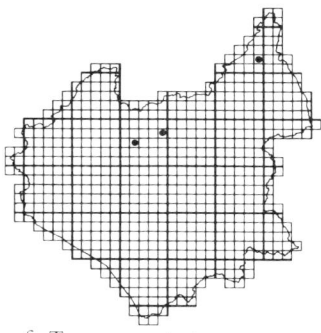

f. Taraxacum atactum

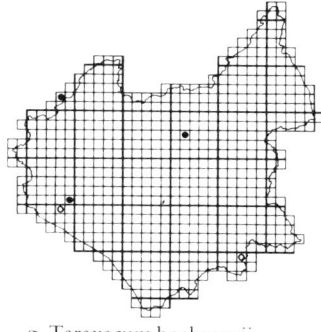

g. Taraxacum boekmanii

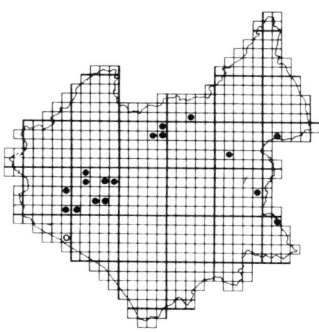

h. Taraxacum bracteatum

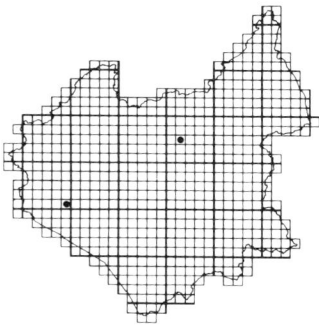

i. Taraxacum hamatiforme

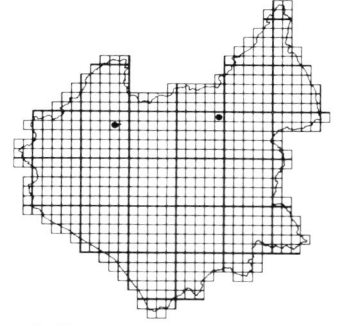

j. Taraxacum hamatulum

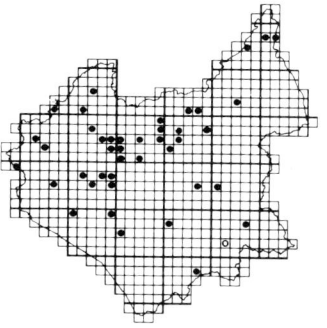

k. Taraxacum hamatum

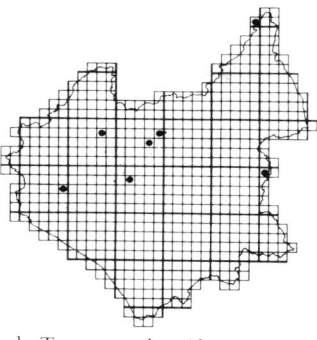

l. Taraxacum hamiferum

Compositae

Maps 68a–l

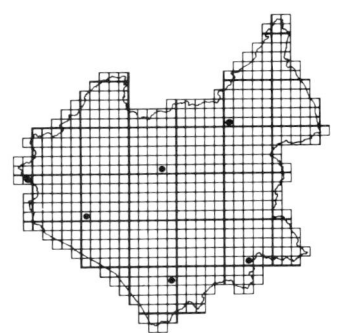

a. Taraxacum lamprophyllum

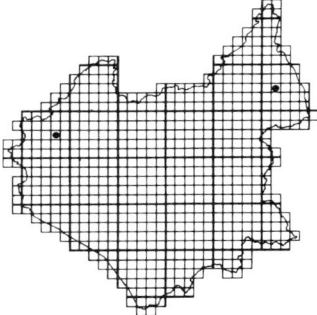

b. Taraxacum polyhamatum

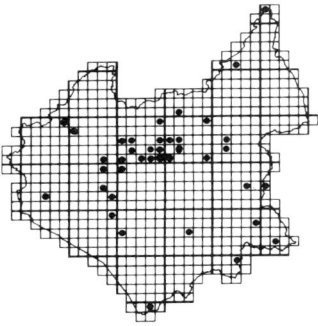

c. Taraxacum pseudohamatum

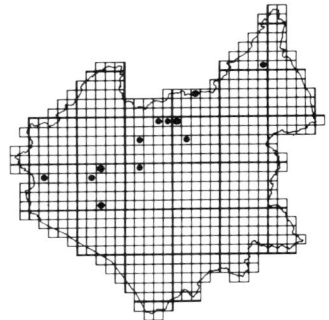

d. Taraxacum quadrans

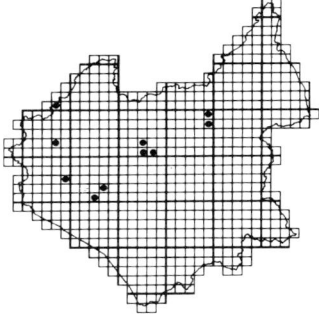

e. Taraxacum subhamatum

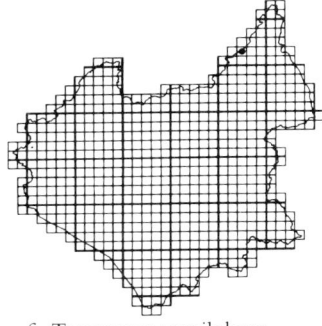

f. Taraxacum aequilobum

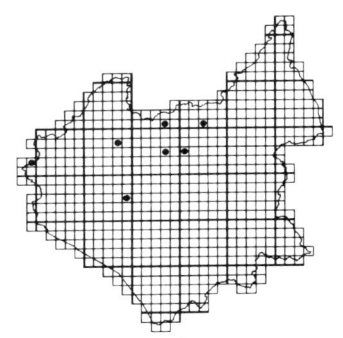

g. Taraxacum alatum

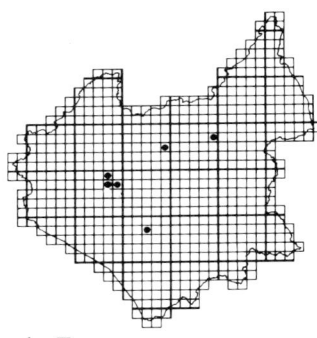

h. Taraxacum ancistrolobum

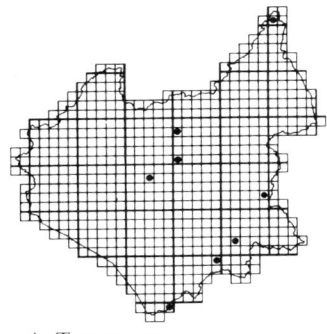

i. Taraxacum cophocentrum

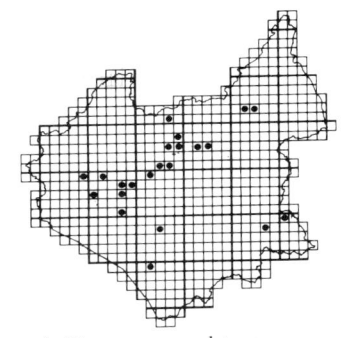

j. Taraxacum cordatum

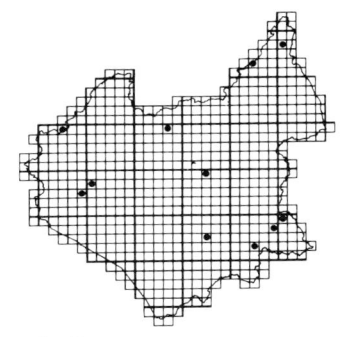

k. Taraxacum croceiflorum

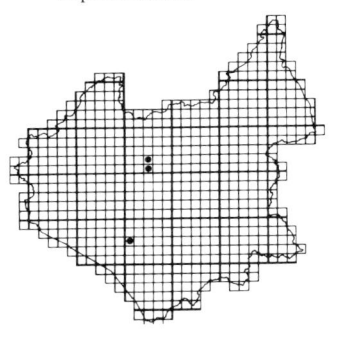

l. Taraxacum cyanolepis

Compositae

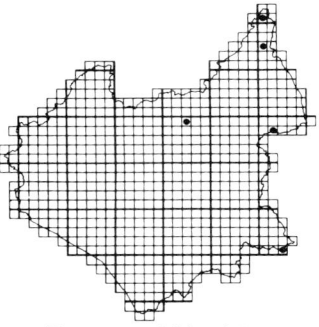

Maps 69a-l

a. Taraxacum dahlstedtii

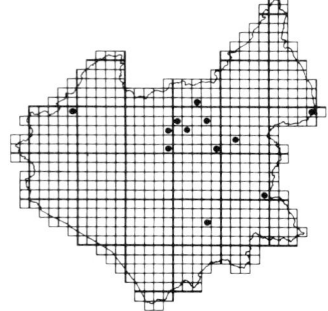

b. Taraxacum ekmanii

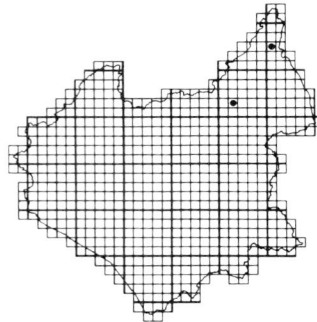

c. Taraxacum excellens

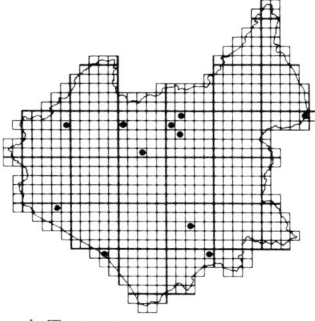

d. Taraxacum expallidiforme

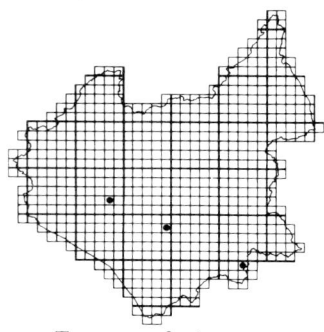

e. Taraxacum fasciatum

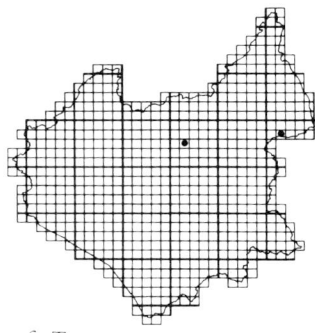

f. Taraxacum grossum

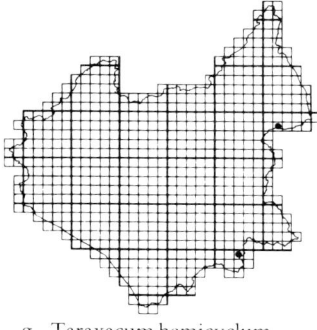

g. Taraxacum hemicyclum

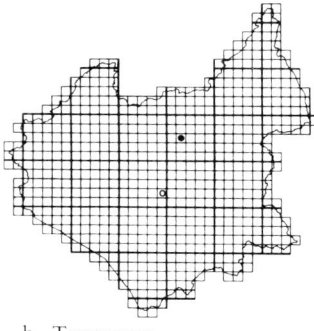

h. Taraxacum hemipolyodon

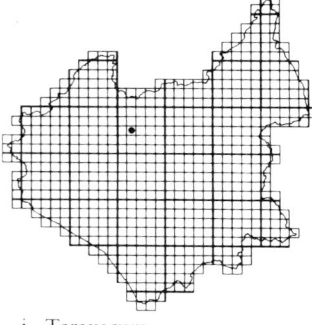

i. Taraxacum horridifrons

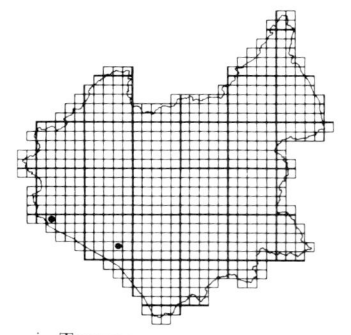

j. Taraxacum huelphersianum

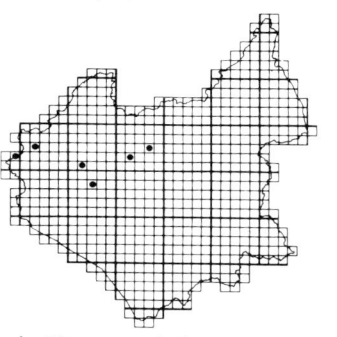

k. Taraxacum insigne

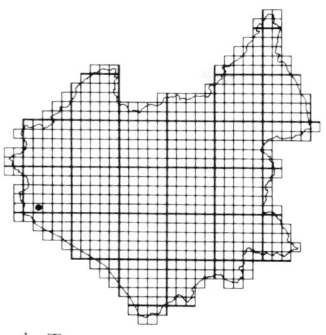

l. Taraxacum laciniosifrons

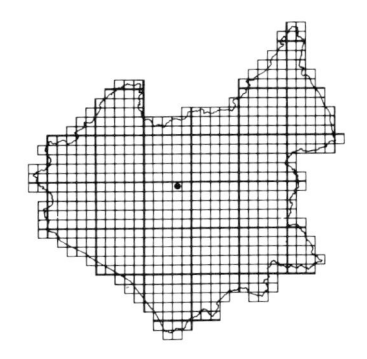

Compositae

Maps 70a-l

a. Taraxacum lacinulatum

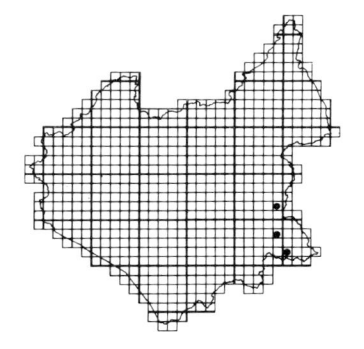

b. Taraxacum latissimum

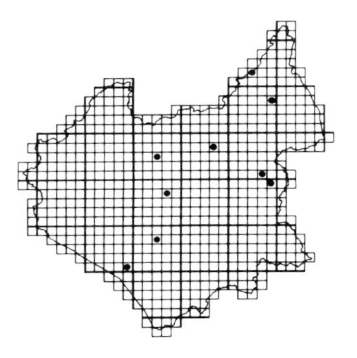

c. Taraxacum lingulatum

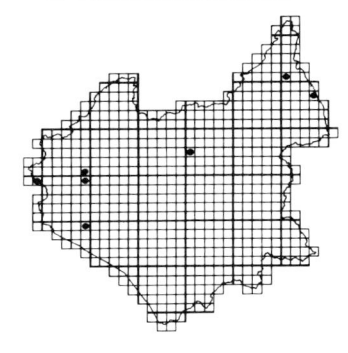

d. Taraxacum longisquameum

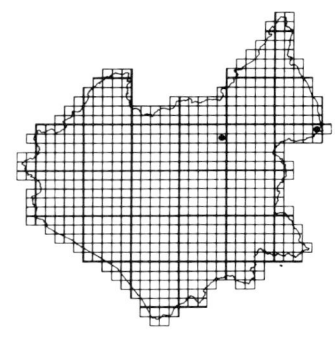

e. Taraxacum melanthoides

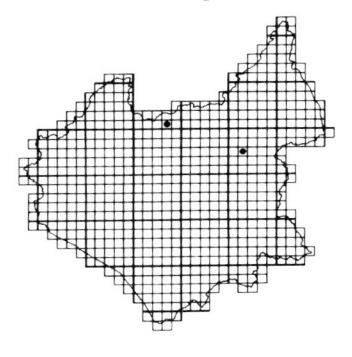

f. Taraxacum obliquilobum

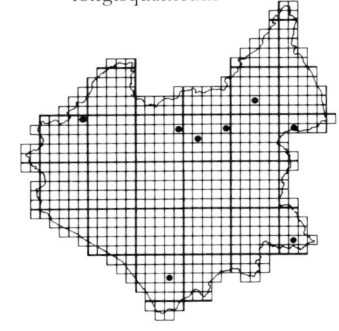

g. Taraxacum oblongatum

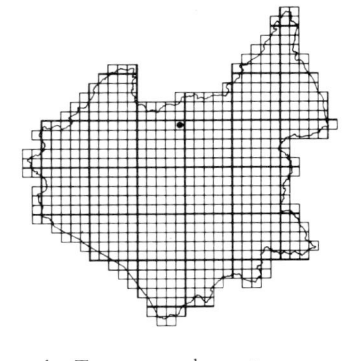

h. Taraxacum obscuratum

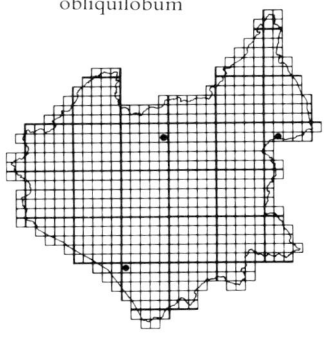

i. Taraxacum pachymerum

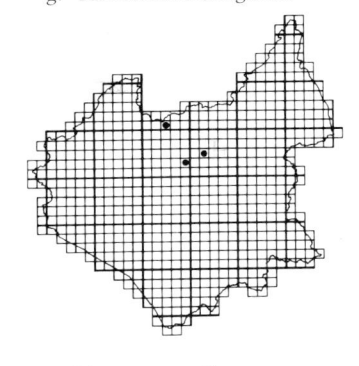

j. Taraxacum pallescens

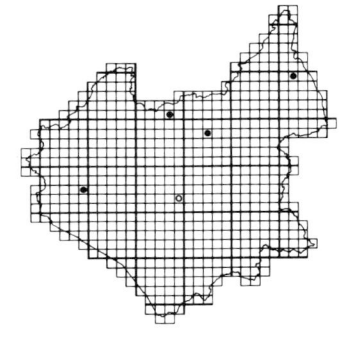

k. Taraxacum pannucium

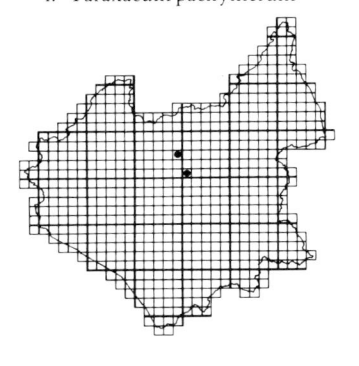

l. Taraxacum piceatum

Compositae

Maps 71a-l

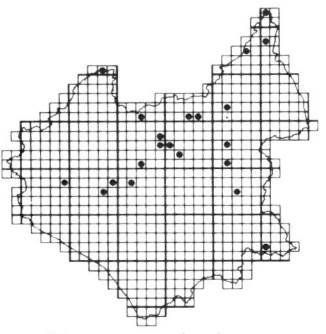

a. Taraxacum polyodon

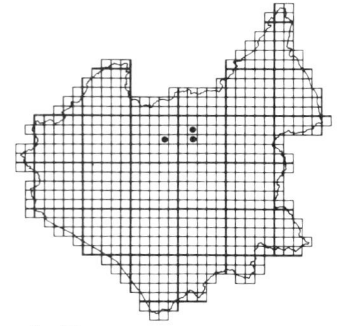

b. Taraxacum porrectidens

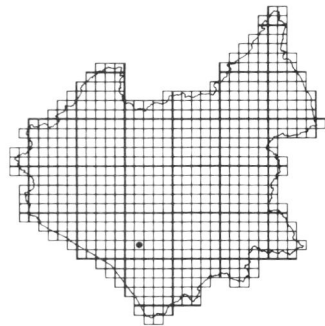

c. Taraxacum praeradians

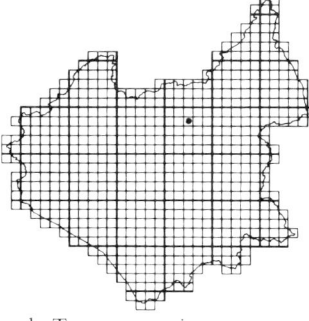

d. Taraxacum privum

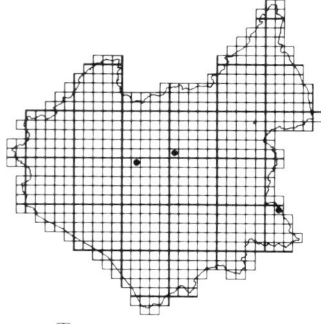

e. Taraxacum procerisquameum

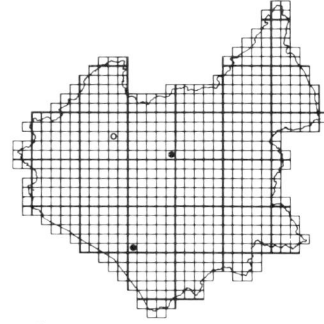

f. Taraxacum reflexilobum

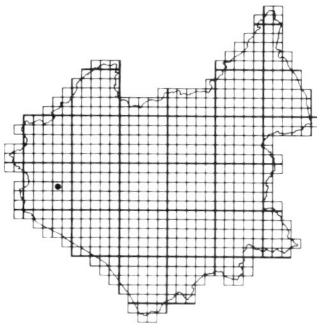

g. Taraxacum rhamphodes

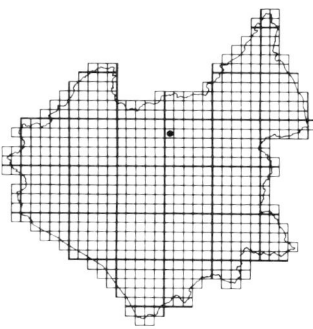

h. Taraxacum sagittipotens

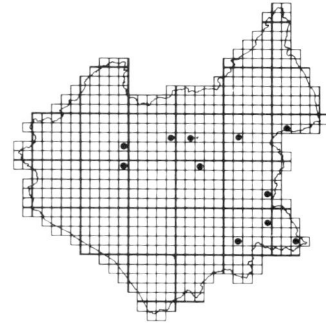

i. Taraxacum sellandii

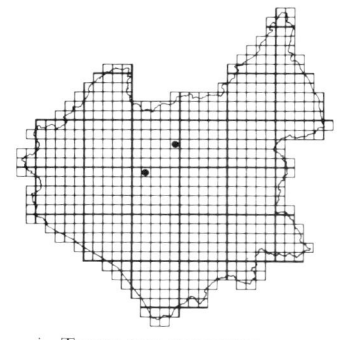

j. Taraxacum stenacrum

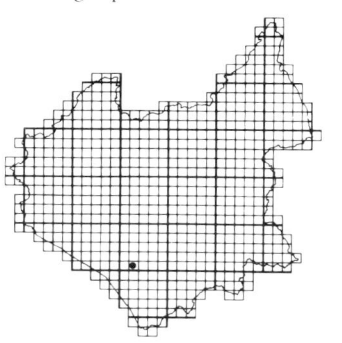

k. Taraxacum subcyanolepis

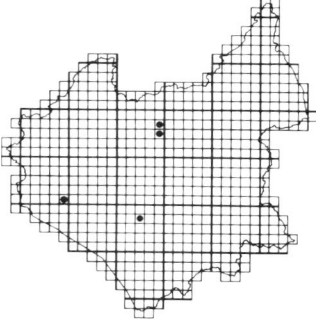

l. Taraxacum sublaeticolor

Compositae

Maps 72a-l

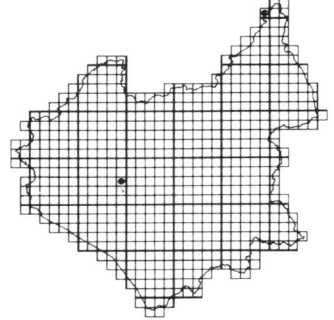

a. Taraxacum subundulatum

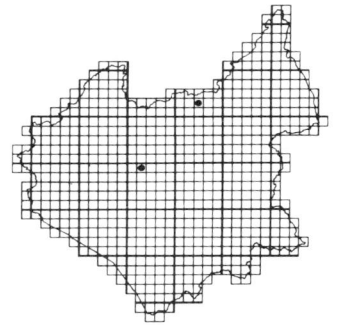

b. Taraxacum trigonum

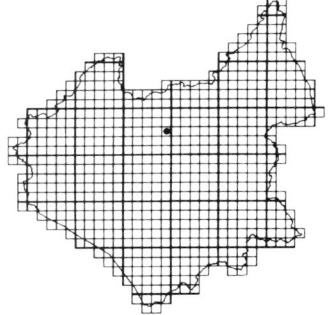

c. Taraxacum tumentilobum

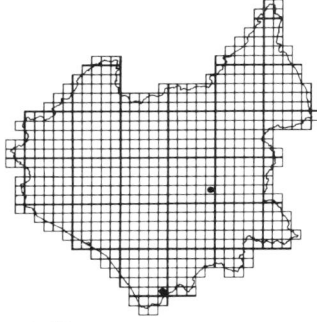

d. Taraxacum uncosum

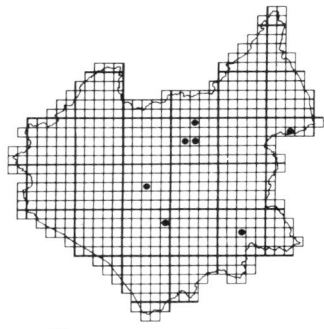

e. Taraxacum undulatiflorum

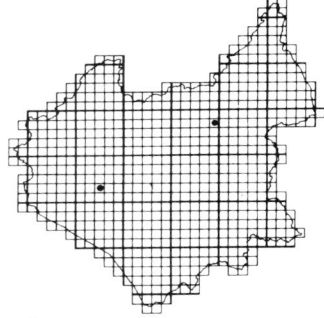

f. Taraxacum valdedentatum

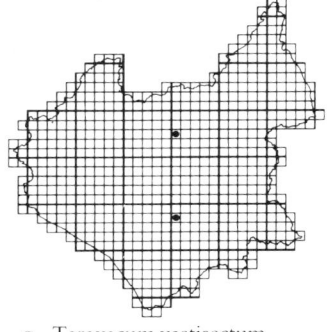

g. Taraxacum vastisectum

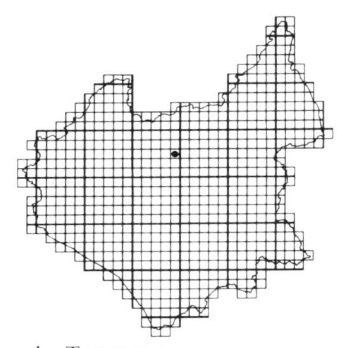

h. Taraxacum xanthostigma

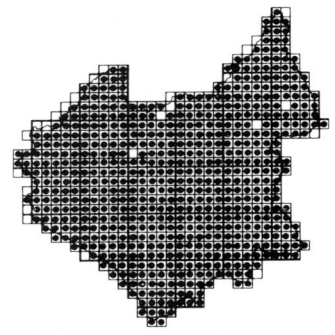

i. Lapsana communis ssp. communis

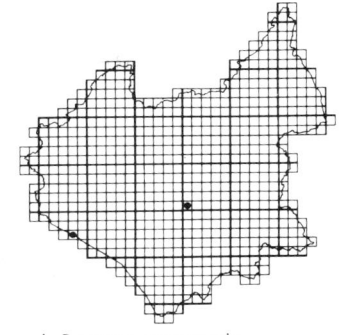

j. Lapsana communis ssp. intermedia

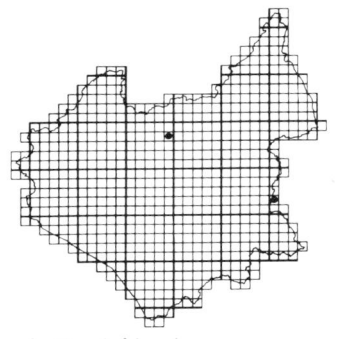

k. Crepis biennis

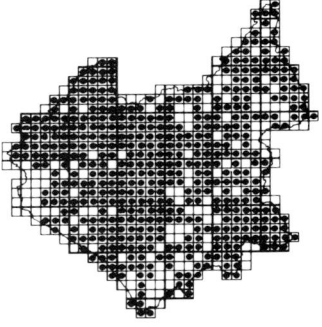

l. Crepis capillaris

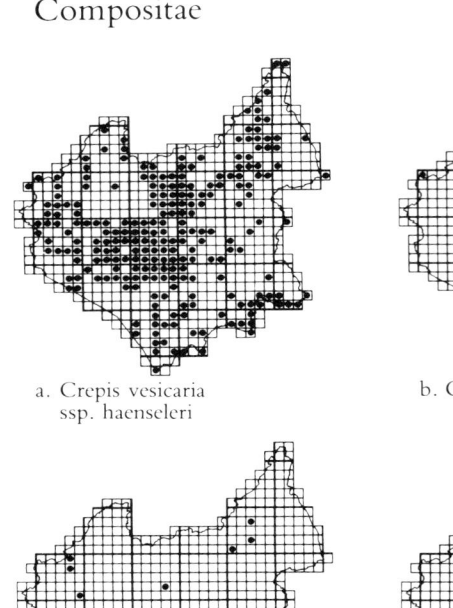

Compositae

Maps 73a-l

a. Crepis vesicaria ssp. haenseleri

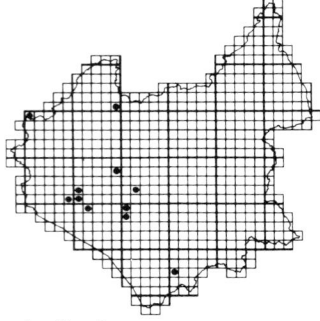

b. Crepis setosa

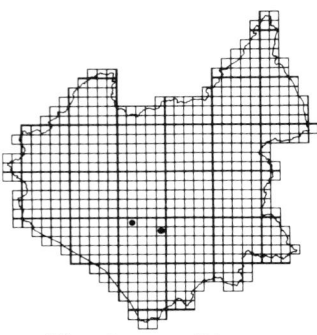

c. Hieracium grandidens

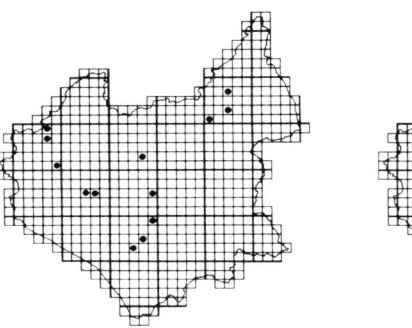

d. Hieracium vulgatum

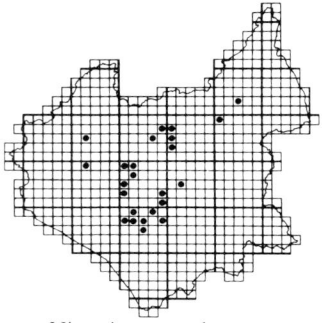

e. Hieracium maculatum

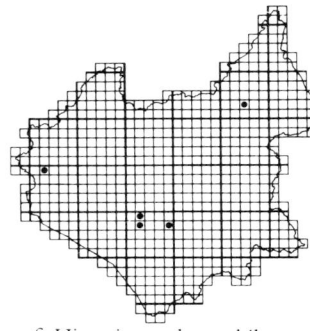

f. Hieracium submutabile

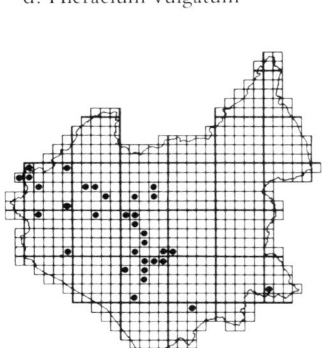

g. Hieracium diaphanum

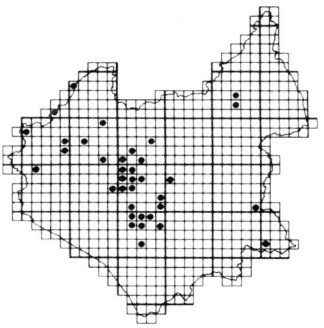

h. Hieracium strumosum

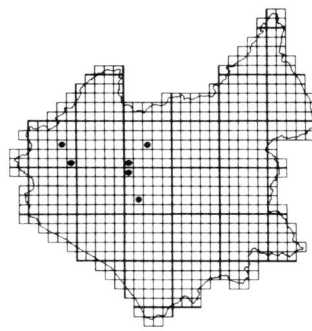

i. Hieracium eboracense

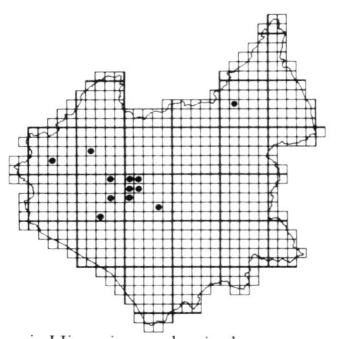

j. Hieracium calcaricola

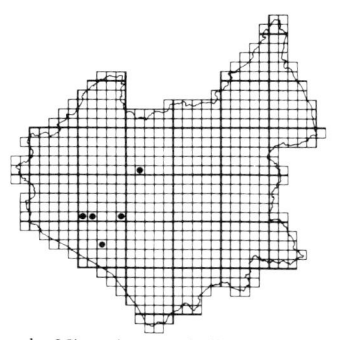

k. Hieracium umbellatum

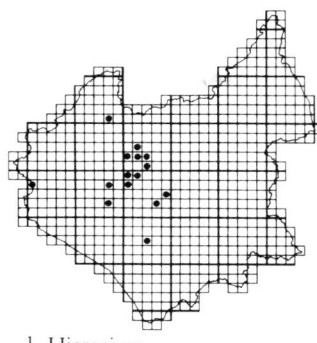

l. Hieracium perpropinquum

Compositae - Hydrocharitaceae

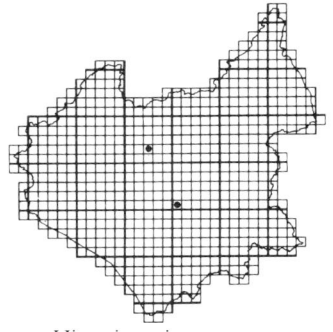

Maps 74a-l

a. Hieracium rigens

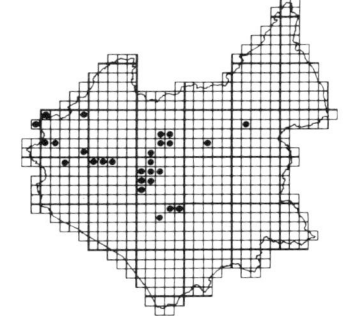

b. Hieracium salticola

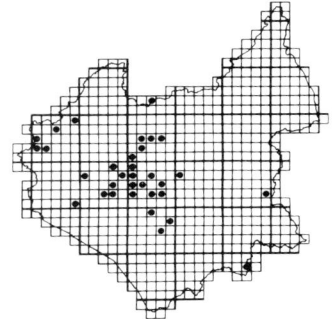

c. Hieracium vagum

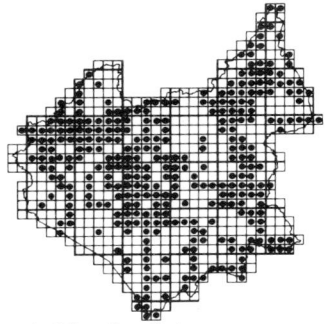

d. Pilosella officinarum

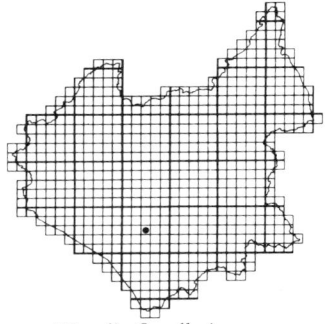

e. Pilosella flagellaris ssp. flagellaris

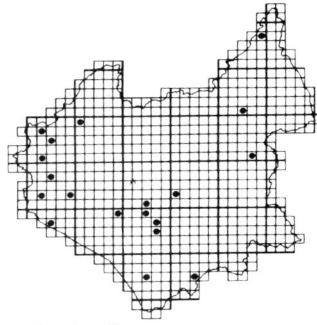

f. Pilosella aurantiaca

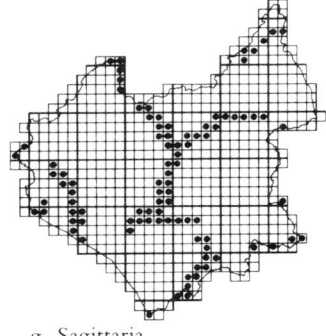

g. Sagittaria sagittifolia

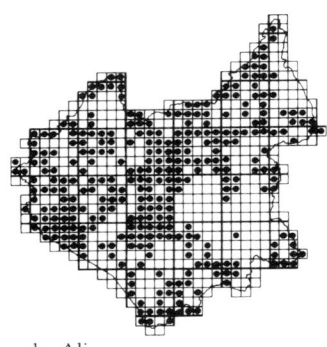

h. Alisma plantago-aquatica

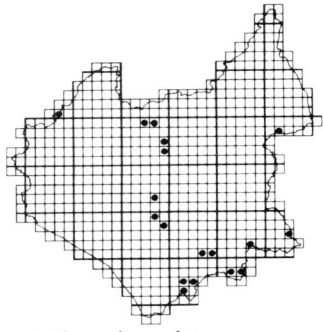

i. Alisma lanceolatum

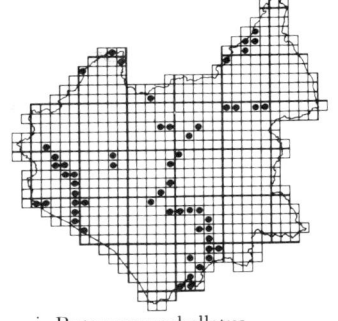

j. Butomus umbellatus

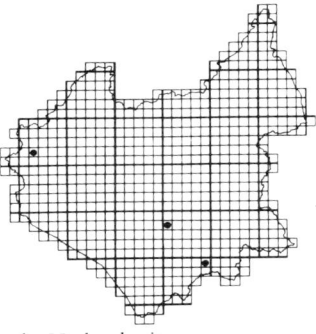

k. Hydrocharis morsus-ranae

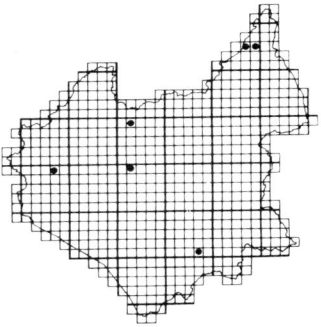

l. Stratiotes aloides

Hydrocharitaceae – Potamogetonaceae

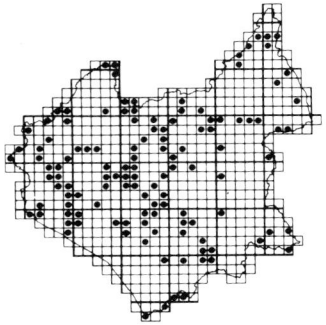

Maps 75a-l

a. Elodea canadensis

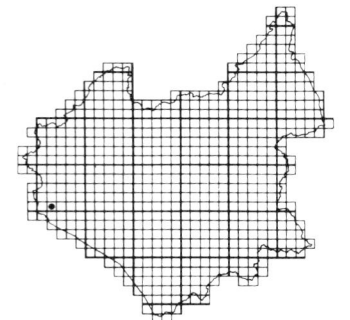

b. Elodea nuttallii

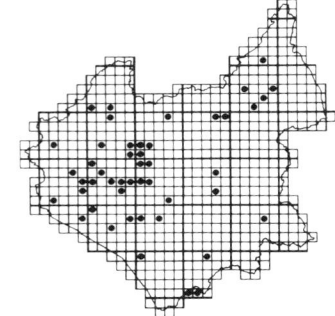

c. Triglochin palustris

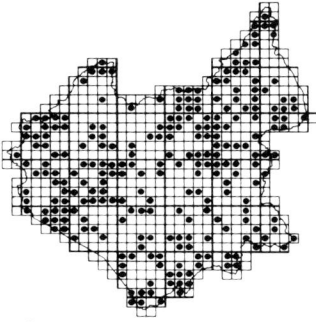

d. Potamogeton natans

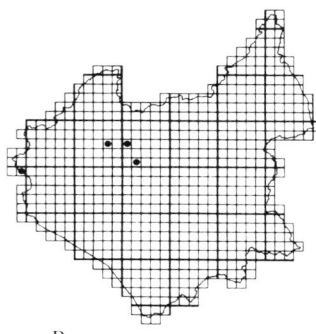

e. Potamogeton polygonifolius

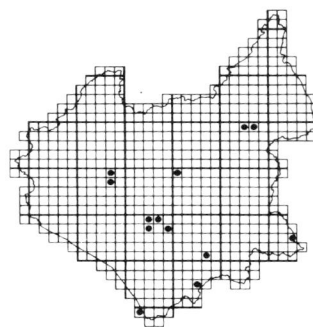

f. Potamogeton lucens

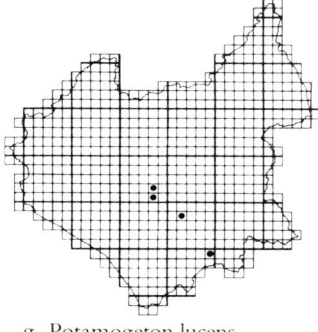

g. Potamogeton lucens × perfoliatus

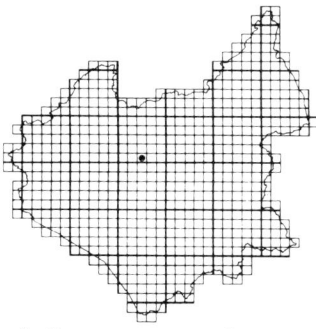

h. Potamogeton gramineus

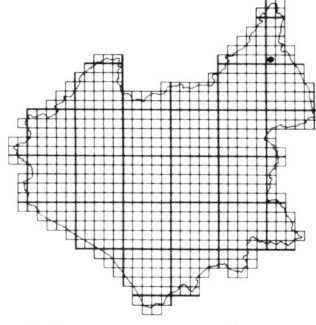

i. Potamogeton gramineus × lucens

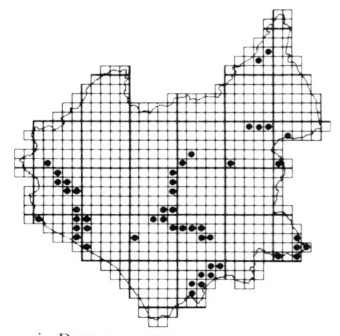

j. Potamogeton perfoliatus

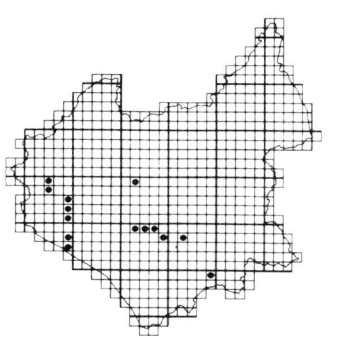

k. Potamogeton friesii

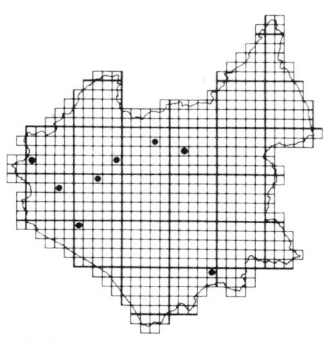

l. Potamogeton pusillus

Potamogetonaceae - Liliaceae   Maps 76a-l

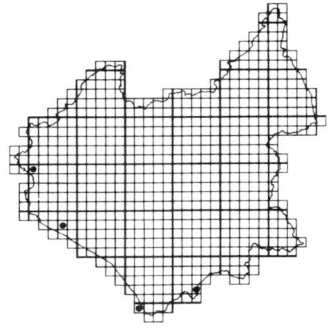

a. Potamogeton obtusifolius

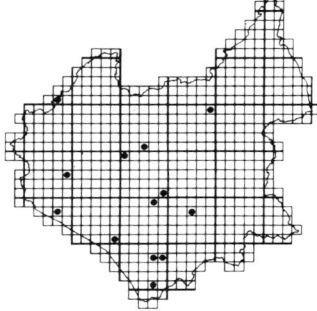

b. Potamogeton berchtoldii

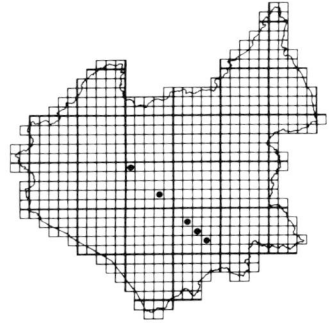

c. Potamogeton trichoides

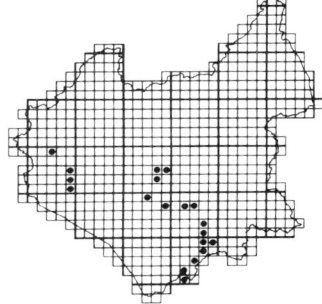

d. Potamogeton compressus

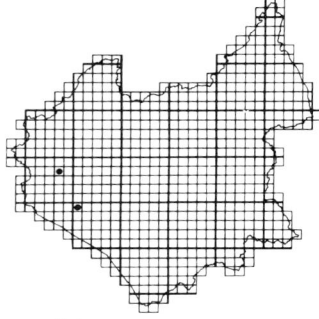

e. Potamogeton acutifolius

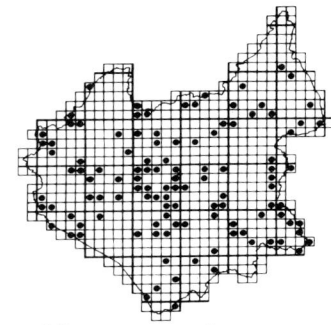

f. Potamogeton crispus

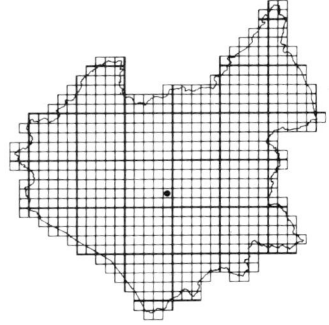

g. Potamogeton crispus × friesii

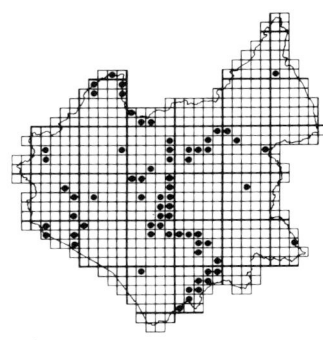

h. Potamogeton pectinatus

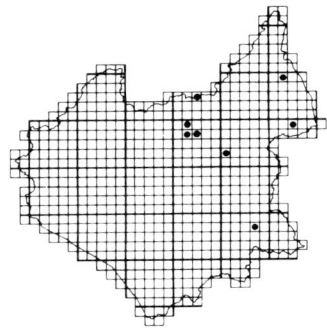

i. Groenlandia densa

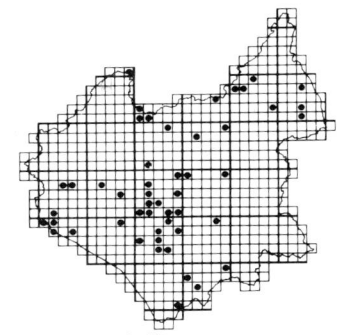

j. Zannichellia palustris

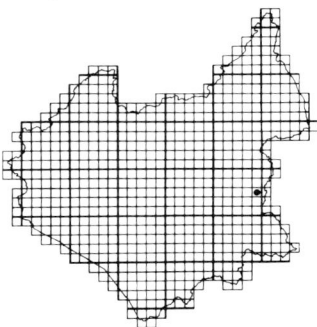

k. Gagea lutea

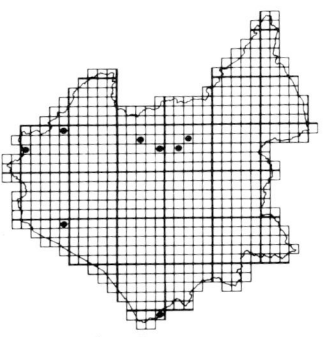

l. Ornithogalum umbellatum

Liliaceae - Iridaceae
Maps 77a-l

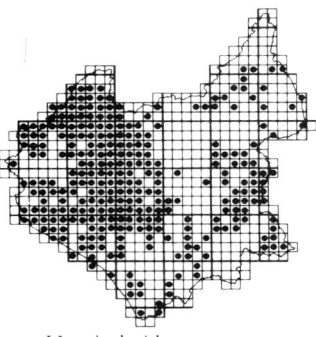

a. Hyacinthoides non-scripta

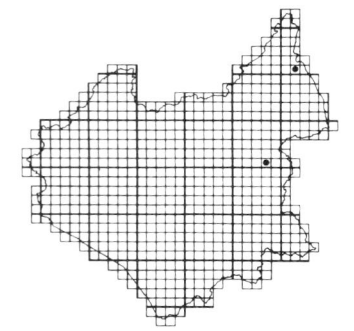

b. Allium paradoxum

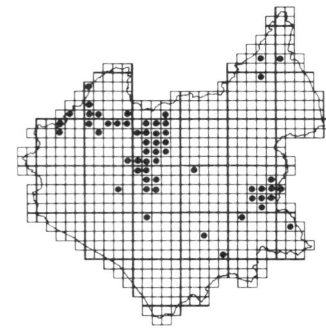

c. Allium ursinum

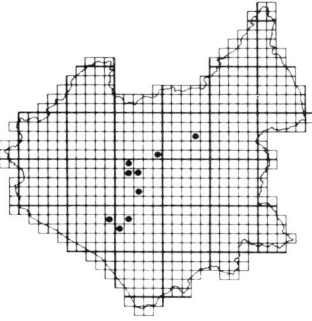

d. Allium vineale

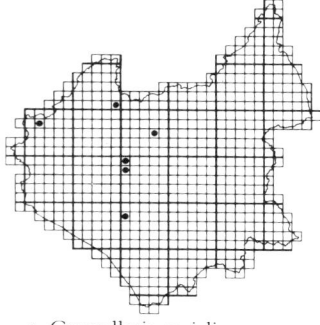

e. Convallaria majalis

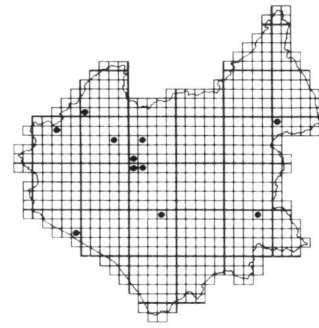

f. Polygonatum multiflorum

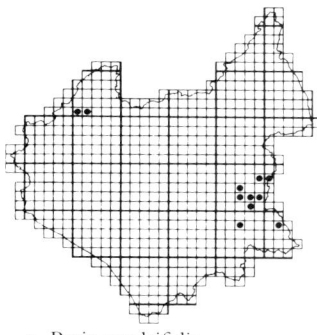

g. Paris quadrifolia

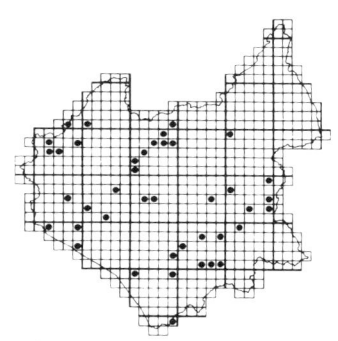

h. Galanthus nivalis

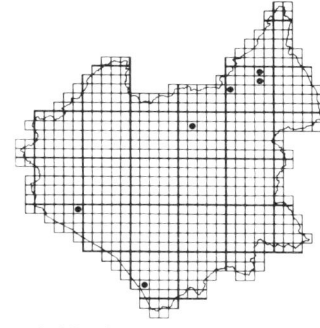

i. Narcissus pseudonarcissus

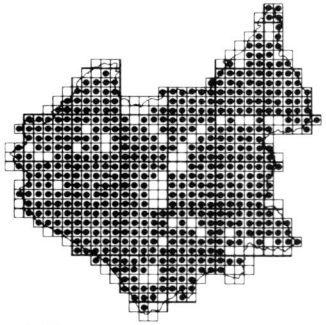

j. Tamus communis

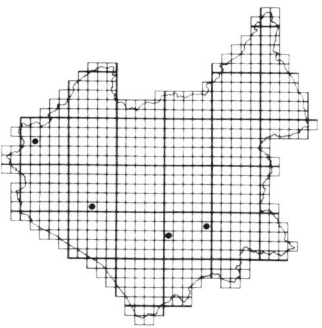

k. Iris foetidissima

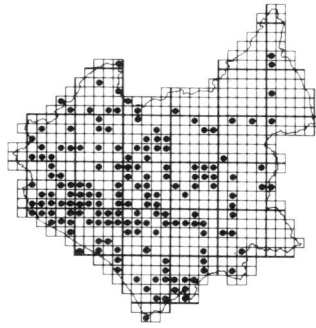

l. Iris pseudacorus

Iridaceae - Juncaceae

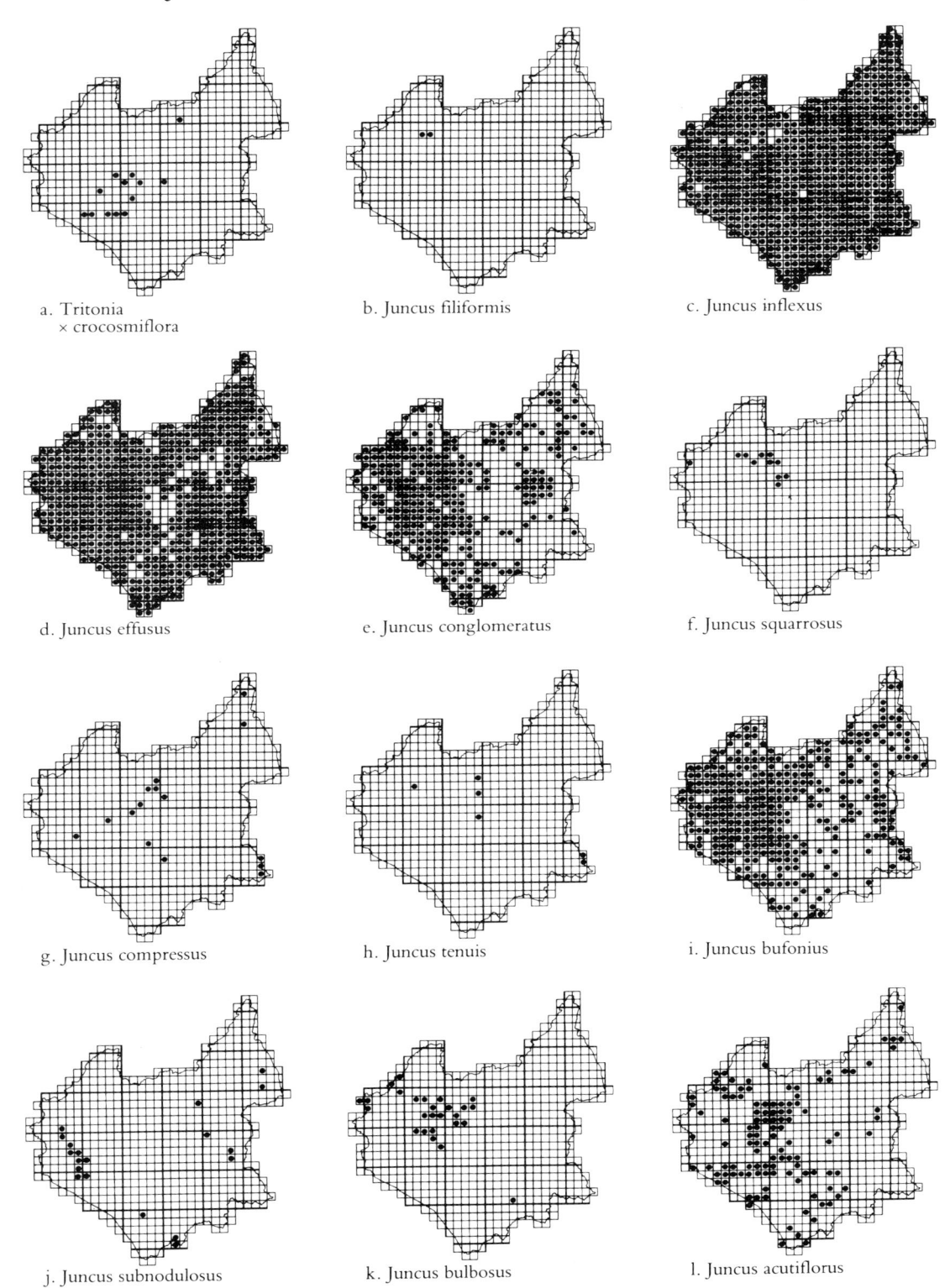

Maps 78a-l

a. Tritonia × crocosmiflora

b. Juncus filiformis

c. Juncus inflexus

d. Juncus effusus

e. Juncus conglomeratus

f. Juncus squarrosus

g. Juncus compressus

h. Juncus tenuis

i. Juncus bufonius

j. Juncus subnodulosus

k. Juncus bulbosus

l. Juncus acutiflorus

Juncaceae - Gramineae

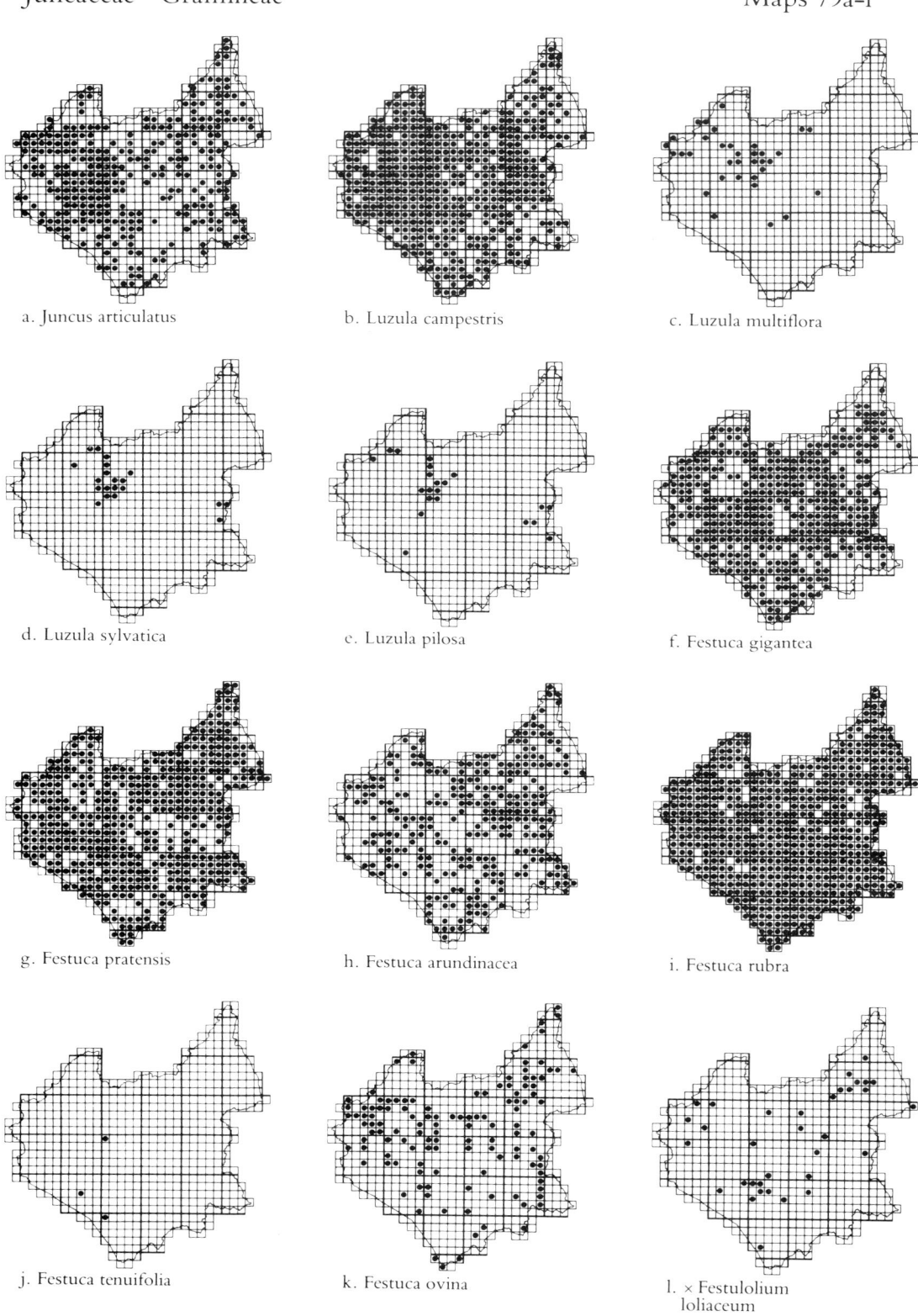

Maps 79a-l

a. Juncus articulatus

b. Luzula campestris

c. Luzula multiflora

d. Luzula sylvatica

e. Luzula pilosa

f. Festuca gigantea

g. Festuca pratensis

h. Festuca arundinacea

i. Festuca rubra

j. Festuca tenuifolia

k. Festuca ovina

l. × Festulolium loliaceum

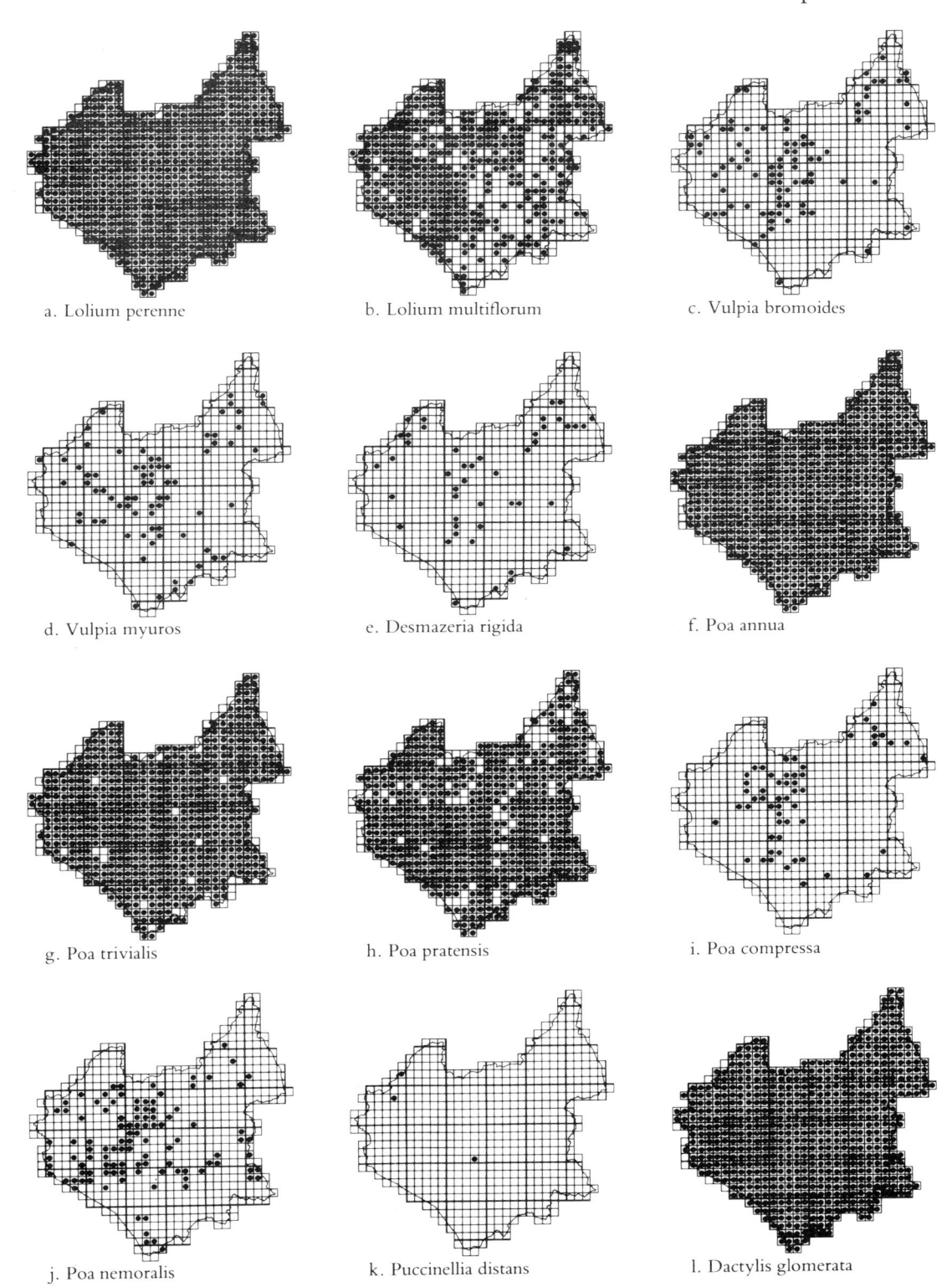

Gramineae

Maps 80a-l

a. Lolium perenne

b. Lolium multiflorum

c. Vulpia bromoides

d. Vulpia myuros

e. Desmazeria rigida

f. Poa annua

g. Poa trivialis

h. Poa pratensis

i. Poa compressa

j. Poa nemoralis

k. Puccinellia distans

l. Dactylis glomerata

Gramineae

Maps 81a-l

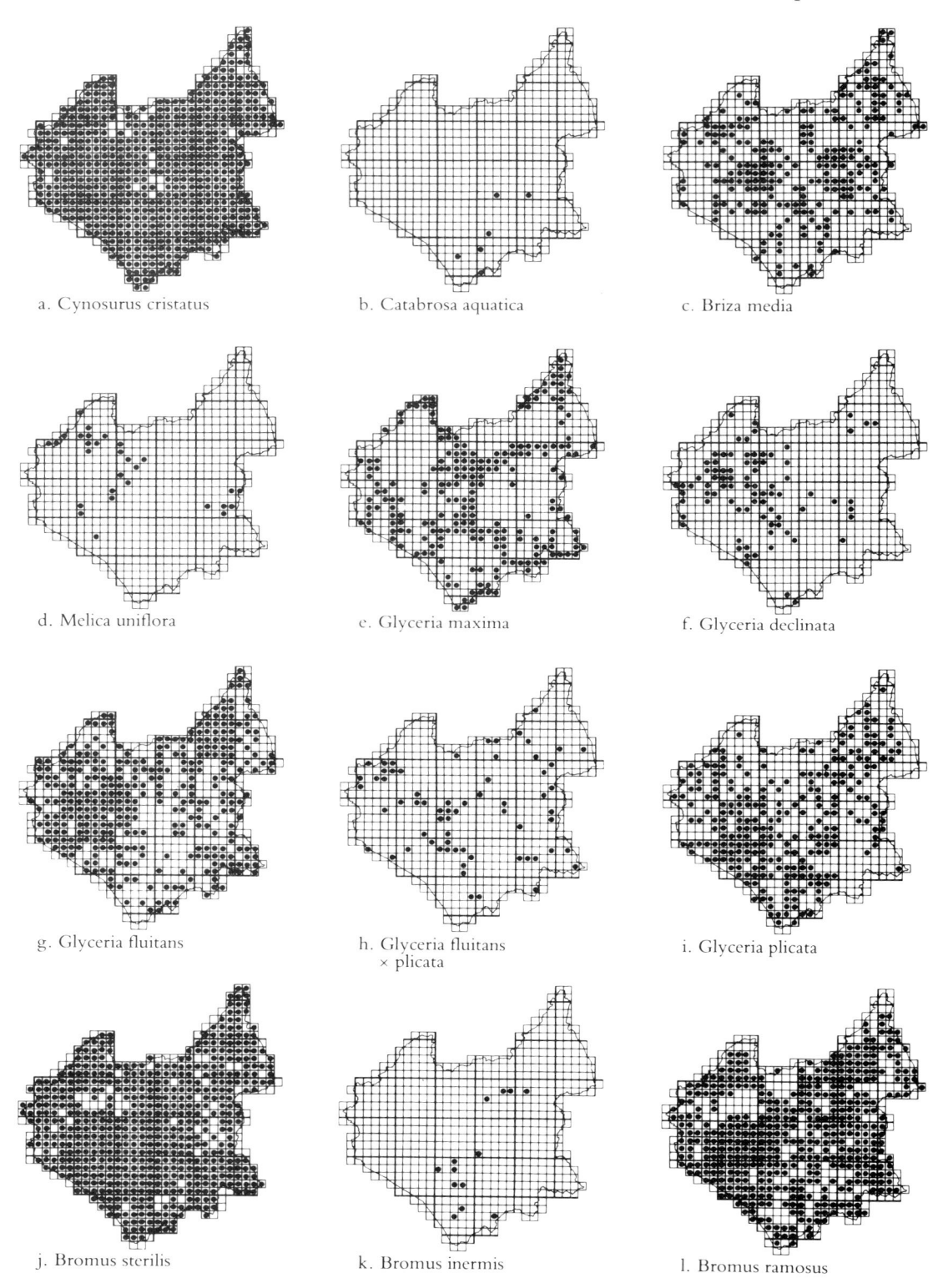

a. Cynosurus cristatus

b. Catabrosa aquatica

c. Briza media

d. Melica uniflora

e. Glyceria maxima

f. Glyceria declinata

g. Glyceria fluitans

h. Glyceria fluitans × plicata

i. Glyceria plicata

j. Bromus sterilis

k. Bromus inermis

l. Bromus ramosus

Gramineae

Maps 82a-l

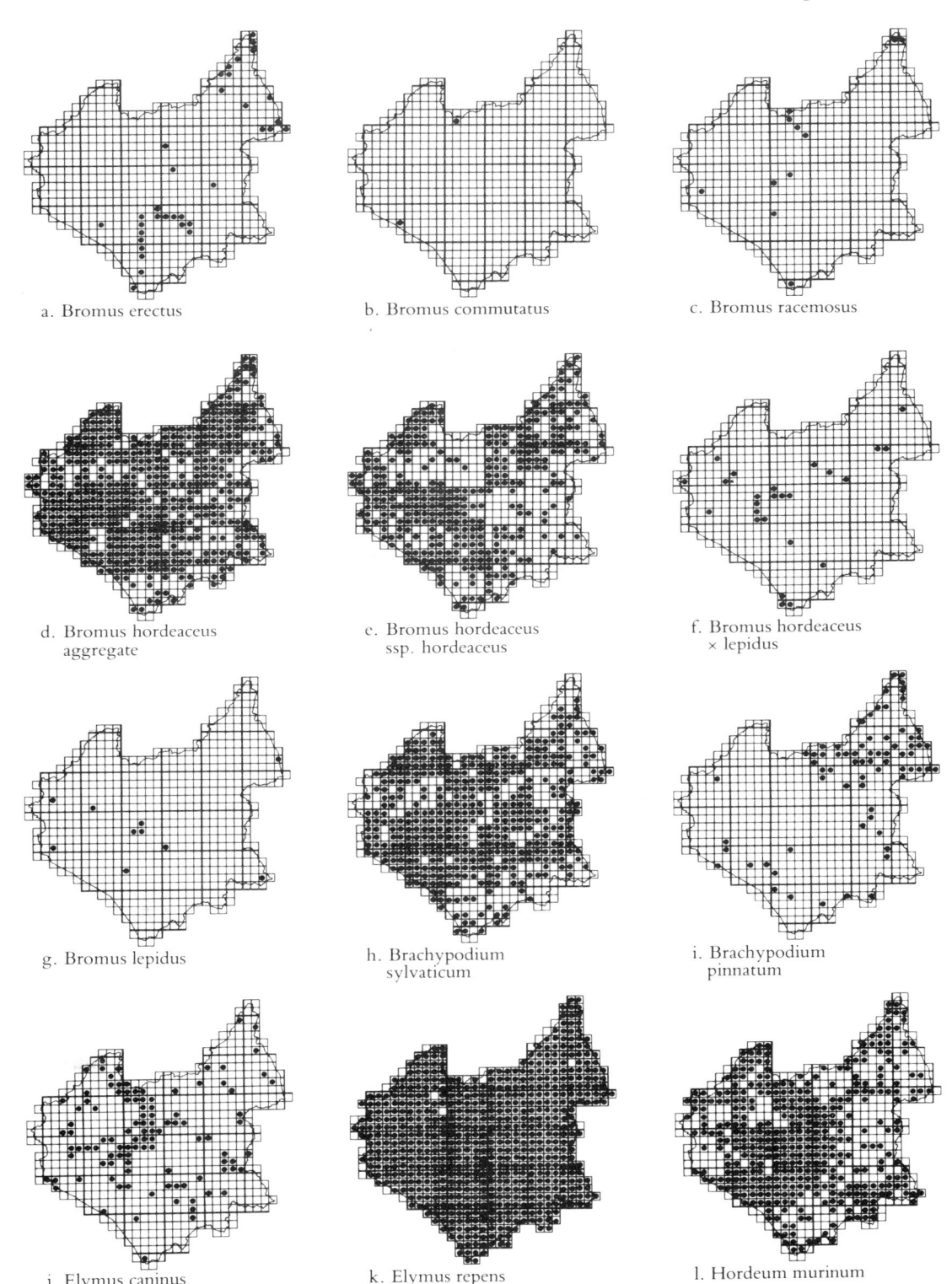

a. Bromus erectus

b. Bromus commutatus

c. Bromus racemosus

d. Bromus hordeaceus aggregate

e. Bromus hordeaceus ssp. hordeaceus

f. Bromus hordeaceus × lepidus

g. Bromus lepidus

h. Brachypodium sylvaticum

i. Brachypodium pinnatum

j. Elymus caninus

k. Elymus repens

l. Hordeum murinum

Gramineae

Maps 83a-l

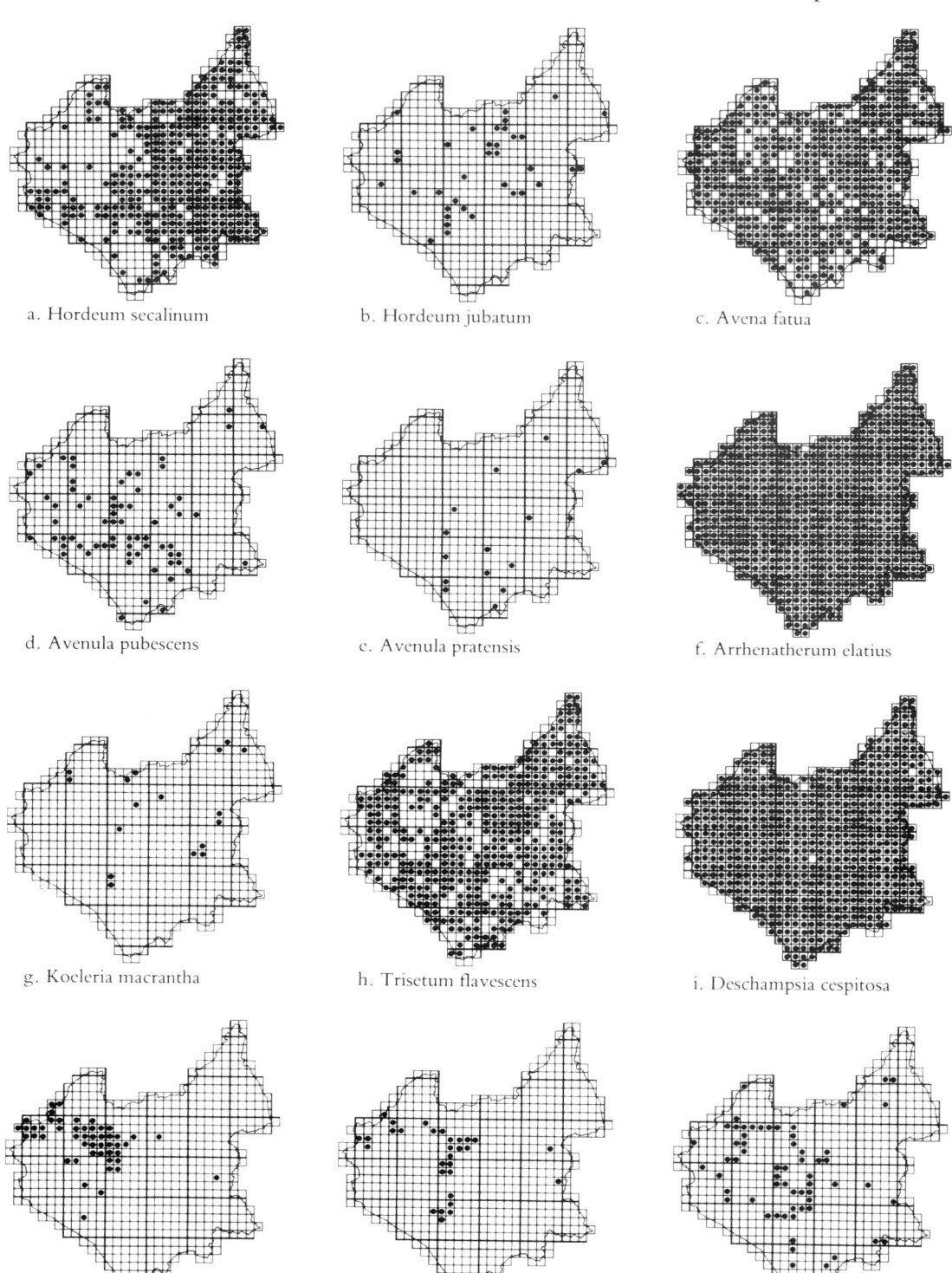

a. Hordeum secalinum

b. Hordeum jubatum

c. Avena fatua

d. Avenula pubescens

e. Avenula pratensis

f. Arrhenatherum elatius

g. Koeleria macrantha

h. Trisetum flavescens

i. Deschampsia cespitosa

j. Deschampsia flexuosa

k. Aira praecox

l. Aira caryophyllea

Gramineae

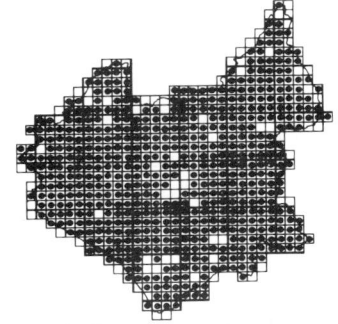

Maps 84a-l

a. Anthoxanthum odoratum

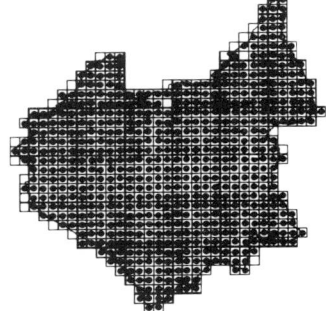

b. Holcus lanatus

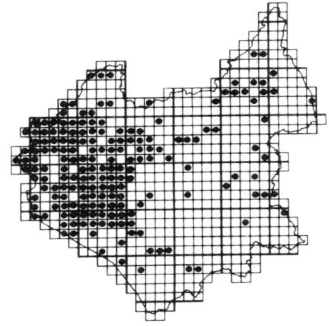

c. Holcus mollis

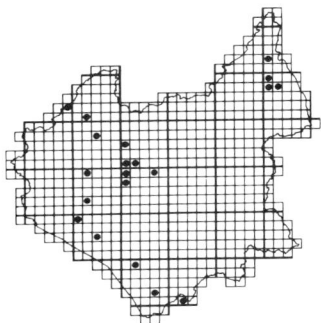

d. Agrostis canina

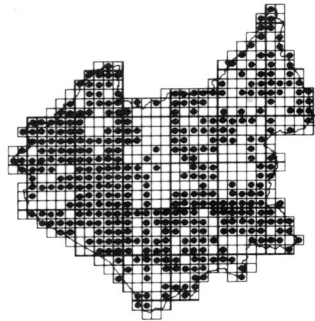

e. Agrostis capillaris

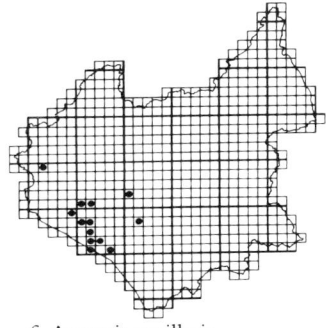

f. Agrostis capillaris × stolonifera

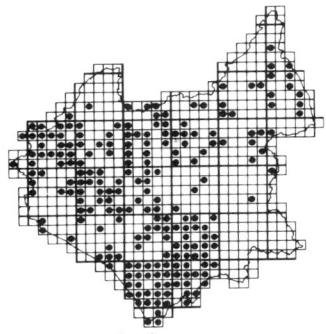

g. Agrostis gigantea

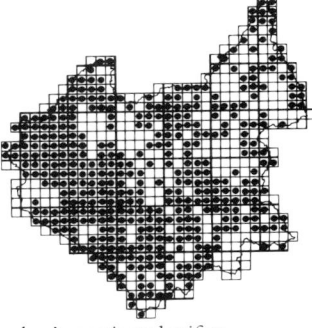

h. Agrostis stolonifera

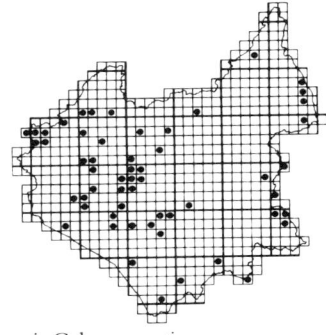

i. Calamagrostis epigejos

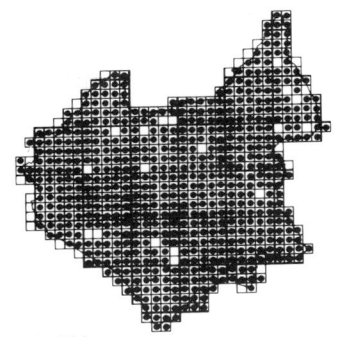

j. Phleum pratense ssp. pratense

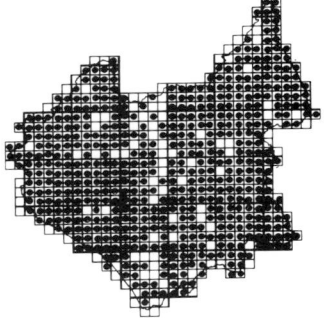

k. Phleum pratense ssp. bertolonii

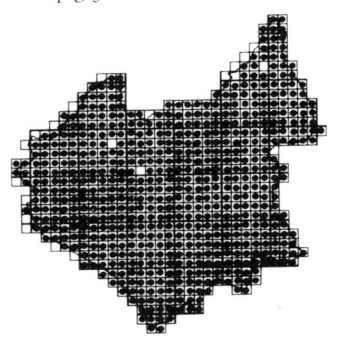

l. Alopecurus pratensis

Gramineae - Araceae

Maps 85a-l

a. Alopecurus geniculatus

b. Alopecurus aequalis

c. Alopecurus myosuroides

d. Phalaris arundinacea

e. Phalaris canariensis

f. Milium effusum

g. Phragmites australis

h. Danthonia decumbens

i. Molinia caerulea

j. Nardus stricta

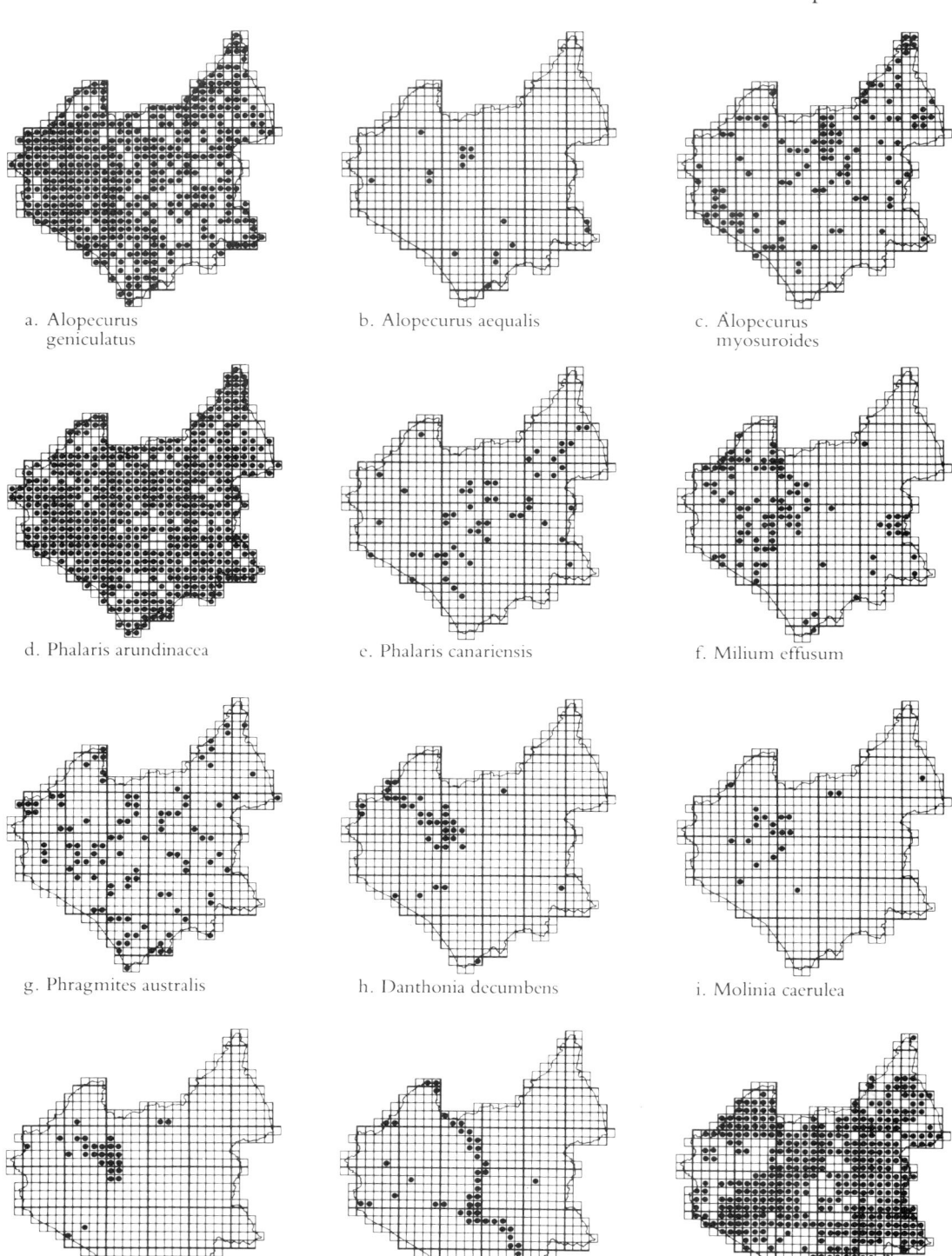

k. Acorus calamus

l. Arum maculatum

Lemnaceae - Cyperaceae

Maps 86a-l

a. Lemna trisulca

b. Lemna gibba

c. Lemna minor

d. Spirodela polyrhiza

e. Sparganium erectum

f. Sparganium emersum

g. Typha angustifolia

h. Typha latifolia

i. Scirpus sylvaticus

j. Scirpus lacustris ssp. lacustris

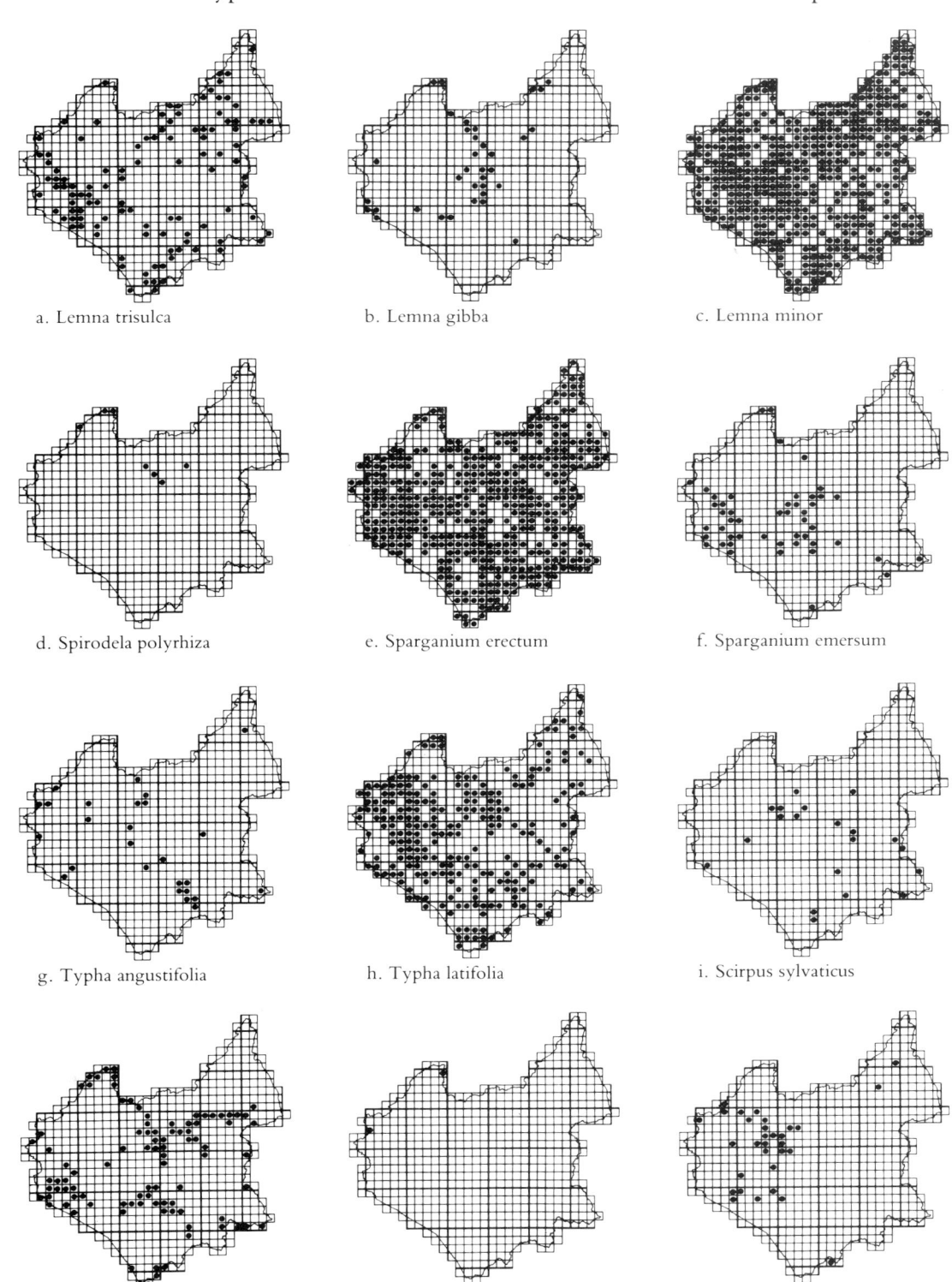

k. Scirpus lacustris ssp. tabernaemontani

l. Scirpus setaceus

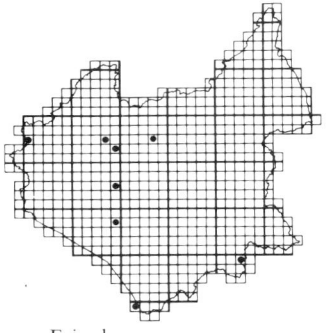

Cyperaceae   Maps 87a-l

a. Eriophorum angustifolium
b. Eleocharis acicularis 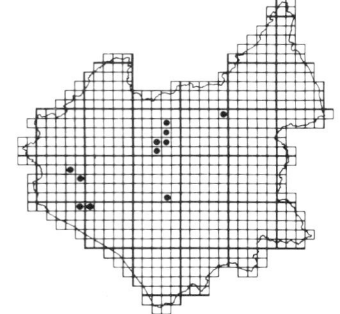
c. Eleocharis palustris 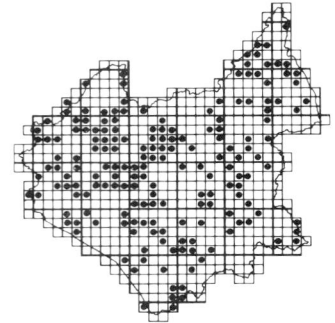
d. Schoenus nigricans 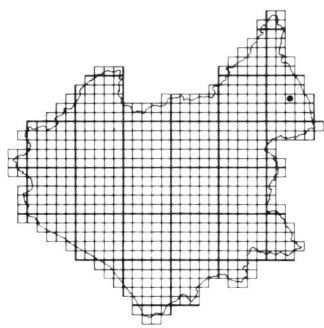
e. Carex paniculata 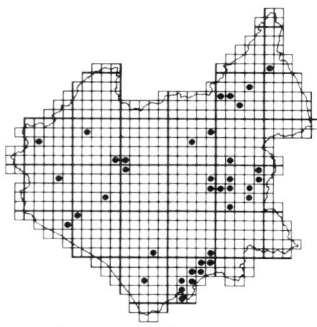
f. Carex otrubae 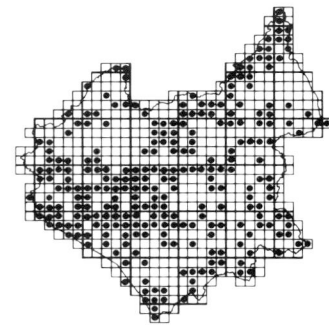
g. Carex spicata 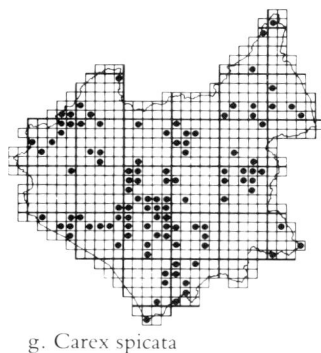
h. Carex muricata 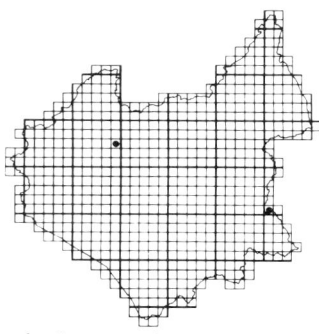
i. Carex disticha 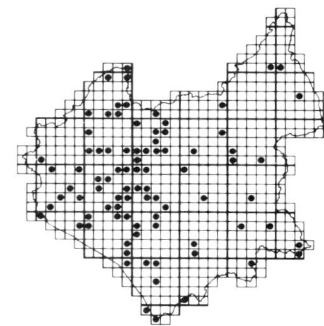
j. Carex remota 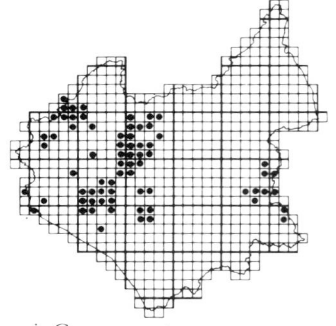
k. Carex ovalis 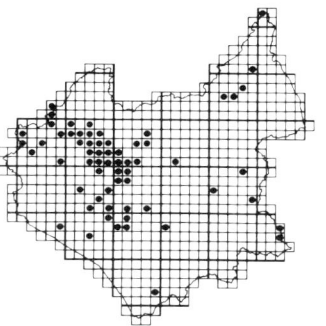
l. Carex echinata

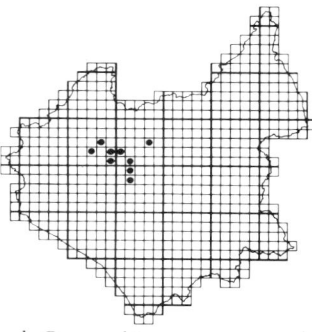

Cyperaceae

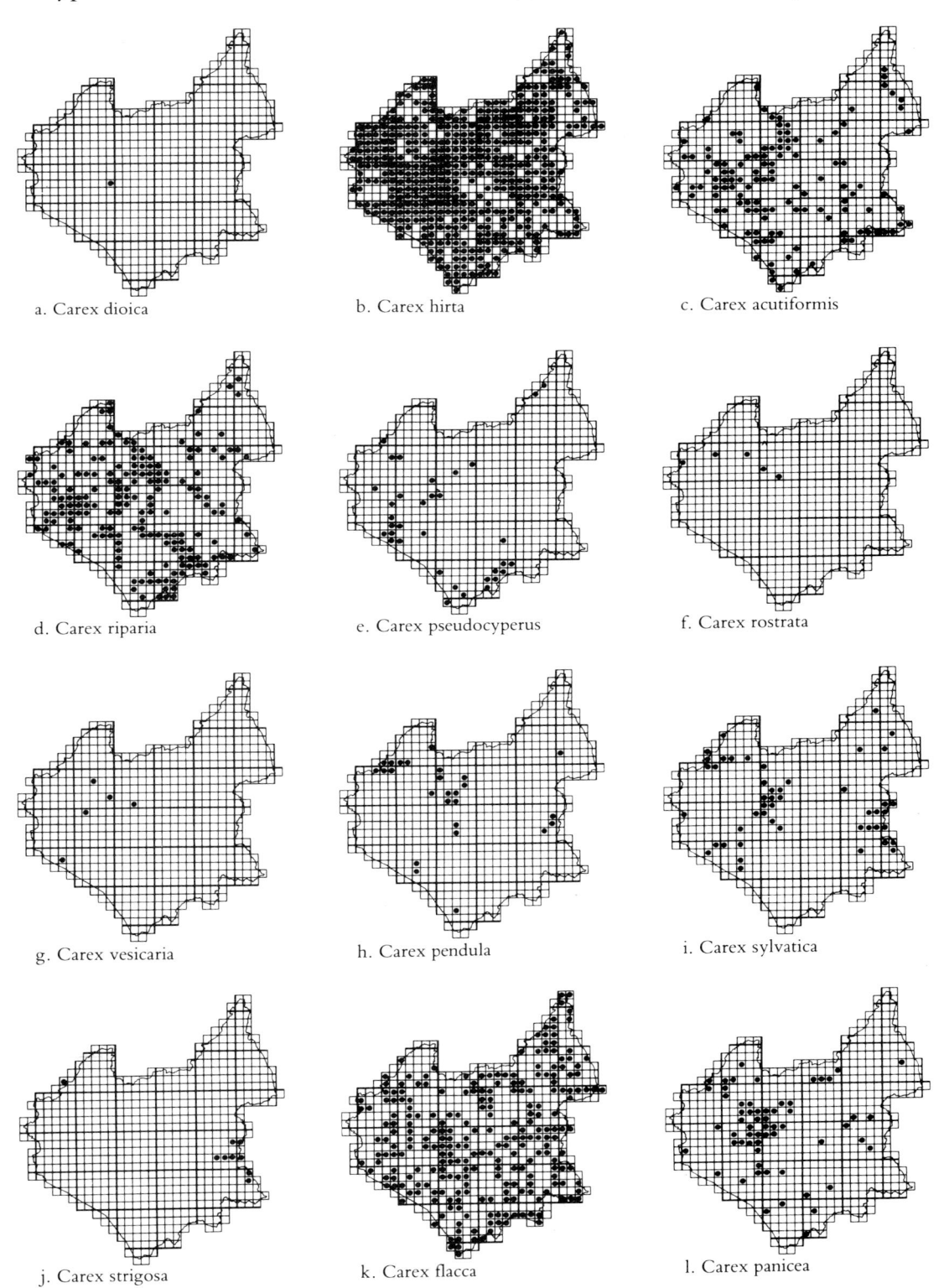

Maps 88a-l

a. Carex dioica

b. Carex hirta

c. Carex acutiformis

d. Carex riparia

e. Carex pseudocyperus

f. Carex rostrata

g. Carex vesicaria

h. Carex pendula

i. Carex sylvatica

j. Carex strigosa

k. Carex flacca

l. Carex panicea

Cyperaceae - Orchidaceae

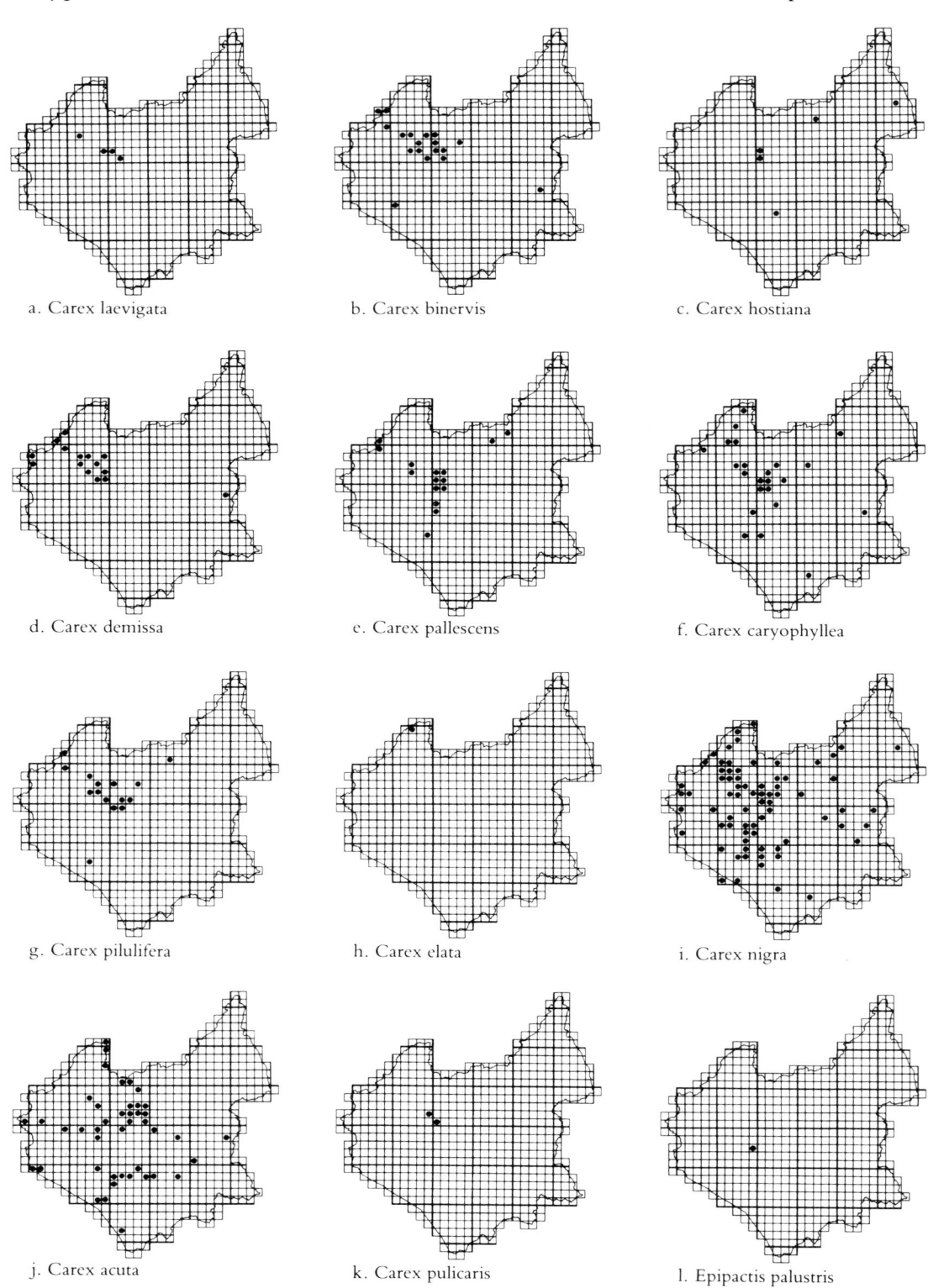

Maps 89a-l

a. Carex laevigata

b. Carex binervis

c. Carex hostiana

d. Carex demissa

e. Carex pallescens

f. Carex caryophyllea

g. Carex pilulifera

h. Carex elata

i. Carex nigra

j. Carex acuta

k. Carex pulicaris

l. Epipactis palustris

Orchidaceae

Maps 90a-l

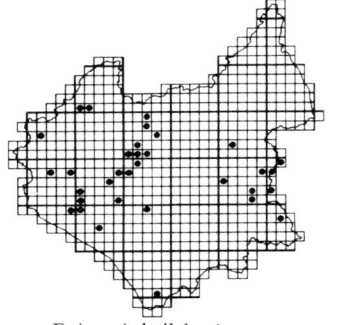

a. Epipactis helleborine

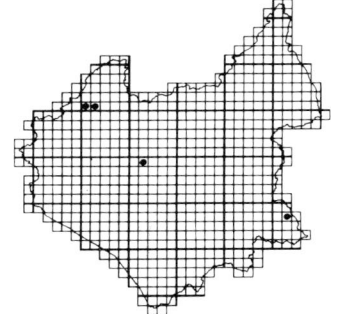

b. Epipactis purpurata

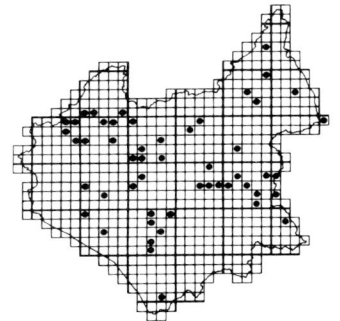

c. Listera ovata

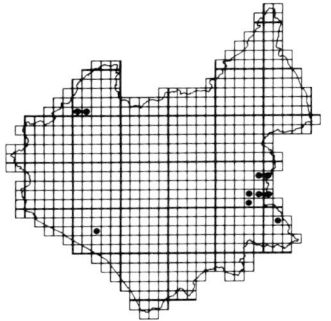

d. Platanthera chlorantha

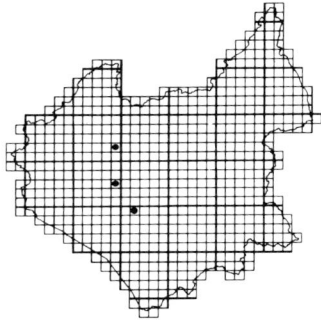

e. Gymnadenia conopsea

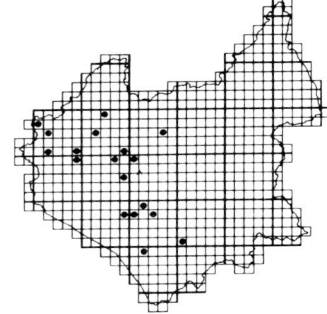

f. Dactylorhiza majalis ssp. praetermissa

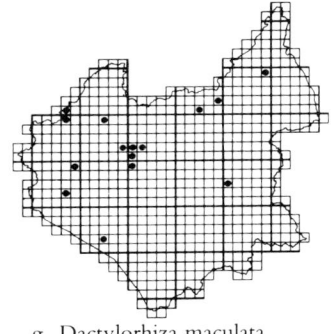

g. Dactylorhiza maculata ssp. maculata

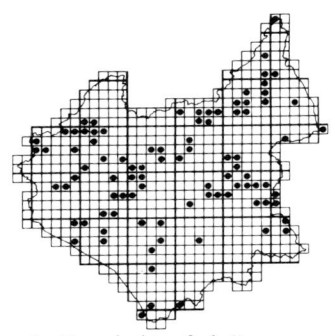

h. Dactylorhiza fuchsii

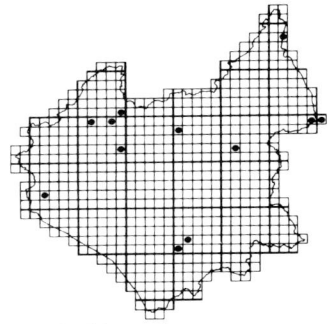

i. Orchis morio

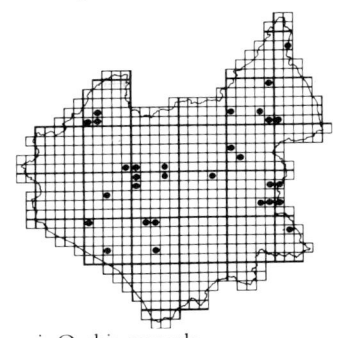

j. Orchis mascula

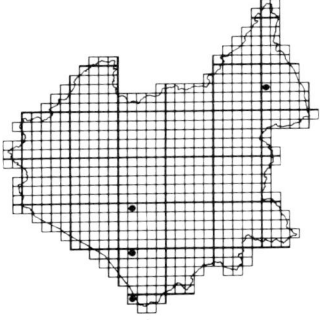

k. Anacamptis pyramidalis

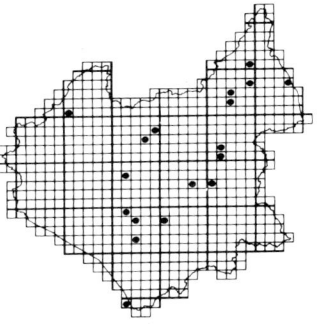

l. Ophrys apifera

GAZETTEER

Purpose and scope

This gazetteer is designed to enable readers to locate landmarks, features and sites of botanical interest, both past and present, in the area covered by the Flora. As in the rest of this work, 'Leicestershire' or 'the county' refer to the administrative county prior to local government reorganisation in 1974, therefore excluding Rutland (see figure 1). The gazetteer was compiled by K. G. Messenger, with the help of information supplied by field workers. Because of the number of people involved and the different degrees of familiarity with their areas, and because some areas were not the responsibility of any one person, there is some unevenness of treatment. It is also known that some areas that were of botanical interest at the time of compilation are no longer so. However an effort has been made to render the gazetteer as comprehensive and useful as possible from both botanical and topographical points of view.

Historical background

The history of the civil and ecclesiastical units into which Leicestershire has been divided for administrative purposes is very complicated. They include parishes, extra-parochial places, districts and boroughs. Changes in their respective boundaries have occurred on numerous occasions over the last two hundred years and are still going on. For an authoritative account of changes between 1801 and 1951 the reader is referred to the chapter on 'Population' by C. T. Smith in volume 3 of *The Victoria History of the County of Leicester*, published in 1955. Particularly useful are the footnotes to the Population Table (pp. 179-203).

For the purposes of this gazetteer the term 'parish' has been used for the present day civil parishes, as defined in 1974. The term 'former parish' refers to those civil parishes, mostly in rural areas, but some on the edge of expanding urban areas, which have either suffered major boundary changes or lost their separate identities during the last hundred years, mostly as a result of the reorganisations of the eighteen-nineties, the decade during which the county was first mapped at the scales of 25" and 6" to the mile.

The gazetteer has been set out to give equal prominence to present-day and former parishes. This is because many of the records in the earlier Floras of 1850, 1886 and 1933 relate to the former parishes, and local patriotism or conservatism has resulted in the continued use, even today, of the former parish names. The former parishes also provide a more even and detailed division of the county for such readers as are not concerned with the more precise location provided by a National Grid reference. The relationship of the boundaries of the former parishes to those of the present-day ones is shown in figures 28-34.

Layout and content of entries

There are three classes of entry:
  a. Parishes. The name of the parish is followed by a list of the tetrads in which it lies. If it was formed by the amalgamation of two or more former parishes, these are listed. Features the names of which begin with the name of the parish are listed under the heading of the parish, as also are sites of outstanding botanical interest not so named. Un-named sites of botanical interest are also included in the entry for the parish. Such features and sites are given a grid reference except where they are of sufficient importance to merit a separate entry.

b. Former parishes. The entries for these follow the same general pattern as for the parishes. Where a parish was formed by the amalgamation of two or more former parishes, features of botanical interest are listed under the appropriate former parish.
c. Named settlements, localities, landscape features, reference points and features of botanical interest which are mentioned in this Flora, or in Horwood and Gainsborough, or have been selected from the 1:25000 Ordnance Survey maps. There are also some local or popular names not found on the maps.

The grid reference given for a village or other settlement is that of the church if there is one, otherwise it is that of the approximate centre of the settlement.

Features which are known to have been of botanical interest during the period of the recent survey are in **bold** type. Features mentioned in Horwood and Gainsborough, the botanical interest of which appears to have been reduced or destroyed, are in *italic* type. Other features are in normal type.

A

Abbey Grounds, Leicester, 583058 (50X). **Ruins** and gardens.

Abbey Park, Leicester, 586056 (50X). Bounded by the Grand Union Canal and the River Soar; the latter is crossed by Abbey Corner Bridge at 586060. There is a lake in the Park.

Abbey Wood, Noseley, 729975 (79I).

Abbot Lodge, Garthorpe, 829221 (82G).

Abbot's Oak, Coalville (Whitwick), 464142 (41S). Large house and garden.

Abbot's Spinney, Barkby Thorpe, 633087 (60J). Contains pond.

Abell's Wood (or *Plantation*), Newtown Linford, 517128 (51B).

Ab Kettleby, parish, 62W, 72ABCFGHLMN.
Consists of the former parishes of Ab Kettleby, Holwell and Wartnaby.
The former parish, 72ABCFGH, contains: village at 724228, part of **Holwell Mouth** SSSI; **marlstone escarpment** coinciding with the northern parish boundary.

Abraham's Bridge, Hinckley (Barwell), 432976 (49I). Carries A447 over River Tweed; **marsh** and **meadow** nearby.

Acresford, Oakthorpe and Donisthorpe, 299131 (21W). Hamlet. **Acresford Plantation**, 301133, is on the Leicestershire side of the Hooborough Brook; **disused gravel pits**, partly restored, are at 304136.

Adam's Gorse, Twyford and Thorpe (Thorpe Satchville), 739113 (71F).

Agar Nook, Coalville (Whitwick), 455143 (41M). Now a housing estate.

Ainsloe Spinney, North Kilworth, 597835 (58W, 68B).

Albert Village, Ashby Woulds, 303183 (31E). Hamlet.

Alderman's Haw, Woodhouse, 502145 (51C). Farm.

Allexton, parish, 80AF, 89EJ.
Contains: village at 817003; Allexton Hall at 813004; lake at 813004; converted mill at 809004; **Allexton Wood** SSSI at 820994. The **Eye Brook** forms the former county boundary on the north and east sides of the parish.

Allotment Covert, Market Bosworth, 403038 (40B).

Allsopps Lane, Loughborough (Loughborough), 548202 (52K). Adjacent **wet meadows.**

Altar Stones, The, Markfield, 485108 (41V). Rock outcrop.

Alton Grange, Ravenstone with Snibston, 390147 (31X). Several small woods and plantations nearby.

Alton Wood, Ravenstone with Snibston, 382144 (31X).

Amberdale Spinney, Kilby, 632953 (69H).

Ambion Hill, 402002 (40A). Farm nearby has been developed as interpretive centre for the Battle of Bosworth (1485).

Ambion Wood, Sutton Cheney (Sutton Cheney), 403995 (39Z, 49E), formerly Sutton Ambien Wood. Part is a county trust nature reserve.

Ambro Hill, Breedon on the Hill, 416251 (42C).

Ambro Mill, Breedon on the Hill, 411247 (42C).

America Wood, Nevill Holt, 823942 (89H).

Anker, River, 29Z, 39DE, 49F. Forms part of the south-western boundary of the county on the edges of Sheepy and Witherley parishes. Rises just within Leicestershire, south of Hinckley, flows through Warwickshire to the crossing of Watling Street at the corner of Witherley parish, and thence along the county boundary, eventually diverging from it to join the River Tame at Tamworth.

Anstey, parish, 50JNPU, 51KQ.
Contains: village at 549085; **Anstey Pastures** at 553078; the ancient road known as *Anstey Gorse* at 556081; a length of the **Rothley Brook**, with Anstey Mill at 551081.

Anstey Frith, Glenfields, 551068 (50N). Former extraparochial place, now the site of County Hall.

Anstey Lane, Leicester, 561072 (50T). Retains some old grassland and rich hedges.

Appleby Magna, parish, 20Z, 30DEIJ, 31AF.
Contains: village at 315098; Appleby Parva hamlet at 308088; Appleby Park at 312086 and Appleby Hall gardens at 310088. **Hedges** along the A444 between Appleby Parva and Twycross are a classic site for *Rubus* species first recorded by Bloxam prior to 1850.

Aqueduct Spinney, Great Glen, 650959 (69MN). Beside **Grand Union Canal.**

Arnesby, parish, 69 ABCFGH.
Contains village at 617921.

Asfordby, parish, 61YZ, 71EJ, 72AFG.
Consists of the former parishes of Asfordby and Welby.
The former parish, 61YZ, 71EJ, 72A, contains: village at 708189; Asfordby Hill hamlet at 726192; extensive flooded **gravel pits** at 6918; former Melton to Nottingham railway now used as a test track, including Asfordby Tunnel West **cutting** at 724193; Asfordby Weir at 707188 on the River Wreake.

Ashby Canal (Ashby de la Zouch Canal), 39XYZ, 30NPSTVWX, 31CDFGHQ, 49BCE.
Once fed from the former Swains Park Reservoir (Moira Reservoir). Traces of the canal remain in the mining area of Ashby Woulds, but the 9.5 km. south to just north of Snarestone Tunnel at 346099 has been drained because of mining subsidence and much of it filled in with fly-ash since the nineteen-forties. The length that remains in water provides considerable botanical interest in an area of intensively cultivated countryside. Habitat study 66 was made near Carlton Bridge at 386043.

Ashby de la Zouch, town and parish, 31HIJLMNPS TUYX, 32KQ.
Consists of the former parishes of Ashby de la Zouch, Blackfordby, Willesley and part of Packington.
The former parish, 31HIJMNPSTUXY, 32KQ, contains: town at 355165; hamlet of Shellbrook; **Ashby Castle ruins** at 361166; a working railway, with the site of Ashby Station at 355162 and traces of the long dismantled Derby line; interesting woodland, including **South Wood** and **The Coppice**, and species-rich **grassland**, in the extreme north of the parish. Habitat study 21 is at 377189.

Ashby Folville, former parish, Gaddesby, 61VW, 71ABCFG.
Contains: village at 707119; Ashby Grange at 717125; Ashby Lodge at 705138; **Ashby Pastures** (woodland).

Ashby Folville Brook, 61LRW, 70J, 71ABF. Formed from tributaries rising in Lowesby, Marefield, and Owston and Newbold parishes; flows west through Twyford and Ashby Folville, between Queniborough and Rearsby and joins the Queniborough Brook at the edge of East Goscote parish; **old pasture** adjacent in 61R.

Ashby Magna, parish, 58PU, 59QRZ.
Contains: village at 563904; lengths of the M1 motorway and **dismantled railway**; a few small woods, one at 557913 with pools in it.

Ashby Parva, parish, 58DEIJNP.
Contains: village at 525886; two small woods.

Ashby Pastures, Gaddesby (Ashby Folville), 717137 (71B). Wood managed by Forestry Commission, possibly ancient.

Ashby Woulds, parish, 21Y, 31CDEHIJ.
Contains mining hamlets of Albert Village, Boothorpe, Littleworth, Moira, Norris Hill and Sweet Hill; areas of **heath, heath grassland** and other habitats which have developed on disused railway lines; numerous subsidence ponds; opencast workings for clay and coal; coal mines and derelict sites of former coal mines; derelict **Newfield Colliery site** at 320154, with an important relict heath and bog flora; three important aquatic habitats now lost, *Moira (Swains Park) Reservoir*, **Barratt Pool** and the upper three kilometres of the Ashby Canal.

Ashfield House Quarry, Dane Hills, Leicester, 50S. Disappeared in the suburban expansion of Leicester.

Ash Hill Plantation, Withcote, 798048 (70X).

Ash Spinney, Kings Norton, 682017 (60V).

Ash Spinney, Long Whatton, 476241 (42M).

Ashlands Plantation. Could refer to Long Plantation, Illston on the Hill, 717004.

Ashpole Spinney, Cosby, 535959 (59HI).

Ashpole Spinney, Somerby, 794144 (71X).

Ashpole Spinney, Witherley, 366971 (39T).

Ashpole Spinneys, Peckleton, 448014 (40K). Several small spinneys and a brook.

Asplin Wood, Breedon on the Hill, 430218 (42F). Ancient woodland, SSSI; also known as Belton Asplands or Belton Asplin Wood. Habitat study 5 is at 431218.

Aston Firs, Aston Flamville, 455940 (49LM). Ancient woodland, part of the Burbage Wood and Aston Firs SSSI. Habitat study 7 is at 457938.

Aston Flamville, parish, 49KLMQR.
Contains: village at 463927; a length of the M69 motorway; **Aston Firs.**

Atkins' Great Closes, Stoney Stanton, SP 49. Location not determined.

Atterton, former parish, Witherley, 39IJNPTU.
Contains: hamlet at 352983.

419

Austen Dyke, 61YZ. Brook; Austen Dyke Bridge is at 681175.

Avon, River, 57NTUZ, 58V, 68ABFGK.
Forms the county boundary from near Welford (Northants.), to Catthorpe; dammed in 58V to form Stanford Reservoir.

Aylestone, former parish, Leicester, 50QRVW, 59UZ.
Contains: former village at 572010; a weir on the Grand Union Canal at Aylestone Mill at 578016; **Aylestone Meadows** at 563005 which retain some of their former interest. *Aylestone Sandpit* has disappeared.

B

Baggrave, Hungarton, site of former village, 698087 (60Z). Baggrave Hall 698090, has a *lake*, former water gardens and a park with spinneys and fine trees.

Bagworth, parish, 40IJMNPSTUYZ.
Consists of the former parishes of Bagworth and Thornton.
The former parish, 40IJMNPSTU, contains: village at 449079; dismantled railway line; collieries and factories; the site of Bagworth and Ellistown Station and sidings at 442093; **Bagworth Park Moat** at 453087; a **mining subsidence pool** at 450087 which is the subject of habitat study 57. The location of *Bagworth Fields* is not known.

Bailey's Plantation, Loughborough (Thorpe Acre and Dishley), 505208 (52A).

Bardon, parish, 41KLMQRW.
Contains: **Bardon Hill** at 459132, SSSI; **Bardon Hill Quarries** at 454132; **Bardon Hill Wood** at 461132, largely quarried away; castle mound at 472131; moat at Old Hall Farm, 461122. The village and station were transferred to Coalville when that town's boundaries were enlarged. The parish boundary is coincident with that of the mediaeval deer park.

Bardon, Coalville, 449125 (41L). Village, consisting of a few houses and a church; the site of Bardon Hill Station is at 443126.

Barkby, parish, 60JPU, 61FKQ.
Contains: village at 636098; a number of small plantations; **Barkby Holt** at 672096; the Barkby Brook flowing through the parish from east to west.

Barkby Brook, 60 JPTUY, 70DEI, 61AF. Formed by the union of brooks rising near Cold Newton, Quenby and Baggrave; flows west through Beeby, Barkby and Syston to join the River Wreake at the A46 (Syston By-pass) bridge.

Barkby Spinney, Syston, 605114 (61A).

Barkby Thorpe, parish, 60IJNP.
Contains: hamlet at 636091, and the site of the deserted mediaeval village of Hamilton (Hamilton Old Town); spinneys at 647083 and 642087, which are the only woodland.

Barkestone, former parish, Redmile, 73RSTWX, 83B.
Contains: village at 777349; **Barkestone Wood** at 793324; **Church Thorns**; **grassland** at 789347, 774349 and 777353; **grassland** at 771350, since ploughed; 2km. of **Grantham Canal** with Barkestone Bridge at 772349 and Barkestone Wharf at 775352; 2 km. of **dismantled railway**; **lane** leading to Barkestone Bridge.

Barlestone, parish, 40DHIMN.
Contains: village at 428058; a small brook which crosses the parish from north-east to south-west, passing through **marshy meadows** at 428053 and 430056.

Barlow's Lodge, Clawson and Harby, 727316 (73F).

Barnes Hill Plantation, Grimston, 694229 (62W).

Barn Pool, Oadby and Wigston (Wigston Magna), 612971 (69D). Flooded limestone quarry.

Barratt Pool, Ashby Woulds, 305153 (31C). Straddles the county boundary south-west of Moira; formerly a large expanse of open water, drained and filled following subsidence damage, now vegetated right over, still of botanical importance.

Barrow upon Soar, parish, 51STUXYZ, 61D.
Contains: village at 576175; the **River Soar** and **Grand Union Canal**, which in places bound the parish and in others cross it; working railway crossing the parish, with the site of a station at 574174; **Barrow Gravel Pits** at 569167, one of oldest surviving valley pits in the county and an SSSI; derelict **osier bed** at 580158; Barrow Hill at 582195; disused limepits at 584183 and 598160. Habitat study 88 is of **hedgerow** at 583157.

Barrow Hill, Worthington, 421200 (42F). Carboniferous Limestone outcrop, formerly quarried, of some botanical interest.

Barrow Hill Quarry, Hinckley (Earl Shilton) 488971 (49Y). Variously referred to in HG as Earl Shilton or Potters Marston Quarry, formerly of considerable botanical interest, but much of this lost recently owing to tipping.

Barrowcliffe Spinney, Gaddesby, 674119 (61Q).

Barsby, former parish, Gaddesby, 61QRVW, 70V, 71A.
Contains: village at 698113; Barsby Spinney at 678118; *Barsby Windmill* at 697124; the Gaddesby Brook, which rises to the south-east of the village.

Barton in the Beans, former parish, Shackerstone, 39XY, 40CD.
Contains: village at 396063; pool and **willow holt** at 401070. In HG as Barton in Fabis but this name more correctly belongs to a village in Nottinghamshire.

Barwell, former parish, Hinckley, 49HIJMNPU.
Contains: the much enlarged village at 444968; the Tweed River forming part of the northern boundary. Habitat study 104 is of **refuse tip** at 438973.

Basin Bridge, Higham on the Hill, 393960 (39Y). Over **Ashby Canal**. Habitat study 86 is of **hedgerow** at 394958.

Bassett Farm, Thurlaston, 485996 (49Z). *Knoll and Bassett House* was formerly an extra-parochial place and civil parish in 40V and 49Z.

Bates Bridge, Shackerstone, 374057 (30S). Over **Ashby Canal**.

Bates Wharf Bridge, Shackerstone, 375060 (30T). Over **Ashby Canal**.

Bath Hotel, Shearsby, 621901 (69F). On site of former spa.

Bath Spinney, Gaulby, 694005 (60V).

Battlefield, Medbourne. Location not determined.

Battle Flat, Coalville (Hugglescote) and Markfield, 41K. An area including Battle Flat Lodge at 445108 where there was an interesting elm population.

Bawdon Castle, Ulverscroft, 497140 (41X). Rock outcrop; SSSI (geological).

Beacon Hill, Bottesford, 812396 (83E).

Beacon Hill, Woodhouse, 509148 (51C). Extensive public open space containing important outcrop of Pre-cambrian rock at over 800 ft.; SSSI; **pool** and **drain** at 520150 is site of habitat study 59.

Beanfield House, Shackerstone, 354085 (30P). Pond and coverts nearby.

Beaumanor Brook, 51HIMN. Of little interest in Beaumanor Park, but becoming more interesting downstream, before it joins the Buddon Brook in Quorn village.

Beaumanor Park, Woodhouse, 537157 (51H).

Beaumont Hall, Oadby and Wigston (Oadby), 616016 (60A). University of Leicester Botanic Garden.

Beaumont Leys, former parish, Leicester, 50IJNP, 51Q. Transferred to Leicester County Borough in 1935, together with parts of Belgrave and Leicester Abbey parishes that had been added to it in 1892. The sewage farm was of interest for a considerable time, but is now being built over.

Beeby, parish, 60NPTUZ.
Contains: village at 664083; Beeby Spring at 663083; two brooks which unite in the village to form the **Barkby Brook**.

Beech Spinney, Peckleton, 451023 (40L).

Beedle's Gravel Pit, East Goscote, 630134 (61G). Water-filled.

Beehive, The, Ashby Woulds, 327158 (31H). Site of former building.

Beeson's Barn, Rearsby, 670137 (61R); now called Topfield Farm.

Belcher's Bar, Nailstone, 408085 (40E).

Belgrave, former parish, Leicester, 50UYZ, 60CD.
Transferred to Leicester County Borough in two stages, in 1892 and 1935. Contains: parish church at 592071; some public open spaces and meadows along the River Soar with habitat study 74 at 591075; lock and weir at 590066; bridges at 590073 and 591076; botanic gardens of the Leicestershire Museums Service attached to Belgrave Hall at 593072; dismantled railway, crossing the area on a high embankment at Mowmacre Hill and forming part of the western boundary of the former parish.

Bellevue Hill, Loughborough (Dishley), 503212 (52A).

Bellows Clump, Shackerstone, 362070 (30T).

Belton, parish, 41IJNP, 42FGKLQR.
Contains: village at 447208; **Grace Dieu Wood**; Grace Dieu Manor; the **Westmeadow** and **Grace Dieu Brooks** crossing the parish from south to north. **Belton Asplin** or **Asplands Wood** is in Breedon on the Hill parish.

Belvoir, parish, 83ABCFGHKL.
Consists of the former parishes of Belvoir, Harston, and Knipton. Part of the Lincolnshire parish of Woolsthorpe was added in 1966, thus including the Lower Fish Pond in Belvoir parish.
The former parish, 83BCGH contains: Belvoir Castle at 820337; **woodland** surrounding the castle and other isolated **woods**.

Benn Hills, Sheepy (Sheepy Magna), 307019 (30A).

Benscliffe Wood, Newtown Linford, 515125 (51B). Former sessile oak-wood, now largely planted with conifers; part is an SSSI (lichenological).

Berry Covert, Freeby, 793157 (71X).

Bescaby, former parish, Sproxton, 82CDEHI.
Contains: farm, on site of former settlement at 822263, with moat of some interest; **Bescaby Oaks** at 831267, which has been partly cleared and replanted, but still has an interesting ground flora.

Betty Henser's Lane, Mountsorrel, 51ST.

Biam, River, 50Q. Formed by the union of brooks rising in the parishes of Lubbesthorpe and Enderby and flowing through Aylestone to join the River Soar.

Big Lane, Seagrave, 600190 to 614176 (61DE).

Big Ling Spinney, Prestwold, 569216 (52Q).

Billa Barra Hill, Markfield, 466114 (41Q). Also called Billa Barrow.

Billesdon, parish, 60W, 70ABCFGH.
Contains: village at 719025; **woodland** on Billesdon Coplow at 708044; marlstone **grassland** near Life Hill 717044; the Billesdon Brook, which forms the south-west boundary of the parish and unites with the Coplow Brook to form the eastern River Sence.

Billington Rough, Elmesthorpe, 461957 (49S). Former fish pond, now pasture and marsh.

Bilstone, former parish, Shackerstone, 30LMNRST.
Contains: hamlet at 363053 and a small part of Gopsall Park.

Birch Coppice, Coleorton, 393186 (31Z).

Birch Hill, Charley, 478136 (41R). Rock outcrop; SSSI (geological).

Birchnall Spinney, Markfield, 497090 (40Z).

Birchwood Plantation, Coleorton. This is perhaps the wood now known as Birch Coppice, 393186.

Bird Hill, Woodhouse, 528142 (51H).

Birstall, parish, 50UYZ, 51V, 60E, 61A.
Rather less than half the former parish was transferred to Leicester in 1935, but a small section was returned in 1966. The present parish is largely built up. Contains: village at 596088; **unimproved grassland** alongside the golf-course and around the gravel-pits in the Soar valley; the **River Soar** which forms much of the eastern boundary, with weirs at 607093 and 608099 and a **water-filled gravel pit** at 603095; a **dismantled railway** on the western boundary with **station site** at 587083. *Birstall Hill* at 588087 and *Birstall Gorse,* probably at 579098, are now in Leicester.

Bishop Meadow Bridge, Loughborough, 524217 (52F). Field bridge over **Grand Union Canal**. Bishop Meadow Lock is at 529216.

Bittesby, parish, 48XY, 58CD.
Contains: site of deserted mediaeval village at 501859; part of Bittesswell Airfield; several small brooks; a dismantled railway which crosses the parish from north to south; Watling Street which runs parallel to the south-western parish and county boundaries.

Bitteswell, parish, 58CDHIMNP.
Contains: village at 538858; most of Bitteswell airfield; Bitteswell Hall at 537872 and most of the park with a lake and spinneys.

Blaby, parish, 59STUXY.
Contains: village at 570978, with much new housing to the south; the eastern River Sence which forms the northern parish boundary; *Blaby Mill* at 580977; dismantled railway running in a **cutting** along the eastern parish boundary. *Blaby Meadows* are mentioned in HG.

Black-a-Moors Spinney, Hoton, 555229 (52L).

Blackberry Hill, Belvoir, 818328 (83BG).

Blackbird's Nest, Loughborough (Charley), 513155 (51C). House which has taken its name from an area. The latter was an important reference point for botanical records.

Black Brook, 41NPSTU, 42Q, 52AF. Formed by the union of brooks rising near Lower Bawdon, Charley Hall, Cat Hill and Charnwood Lodge. It flows past Oaks in Charnwood, where there are adjacent **marshes**, into **Blackbrook Reservoir**, passes west and north of Shepshed where there were until recently **good meadows** adjacent, between Thorpe Acre and Dishley, under the A6 at Dishley Mill, and joins the River Soar at 521219.

Blackbrook Reservoir, Loughborough (Shepshed) and Charley, 458174 (41NT). SSSI.

Black Ditches, Ashby de la Zouch, 369194 (31U). Woodland on the site of former fish-ponds.

Blackfordby, former parish, Ashby de la Zouch, 31IJ.
The former parish was divided equally between Ashby de la Zouch and Ashby Woulds when these were created Urban Districts in 1936. Contains: village at 330181; the hamlet of Boundary; a considerable amount of **mining dereliction**, much of which is of botanical interest; the Shell Brook, which rises near the village and flows south-east.

Black Friars, Leicester, 581046 (50X). Area of former All Saints parish adjacent to the River Soar.

Black Hill, Ulverscroft, 506135 (51B).

Black Holt, Belvoir, 848312 (83K). An example of how botanically barren a wood dominated by sycamore can become.

Black Pool, East Goscote, 633137 (61G). Part of Beedles gravelpit, now restored.

Black Spinney, Kings Norton, 692996 (69Z).

Blakeshay Wood, Newtown Linford, 514114 (51A).

Blaston, parish, 79WXY, 89BCDHI.
Contains: village at 802954, in which the **churchyard** is of interest; **Blaston Pastures** at 816957, with meadows and a spinney; a dismantled **railway**, which crosses the western end of the parish on an embankment.

Bleak Hills, Broughton and Old Dalby, 722245 (72H).

Bleak Moor, Rearsby, 657155 (61M). Wet woodland. The adjacent railway **cutting** is also of botanical interest.

Blobbs, The, Appleby Magna, 309081 (30E). Woodland.

Blood's Hill, Kirby Muxloe, 526047 (50H). Adjacent to Kirby Muxloe Castle.

Blowpool Spinney, Burton and Dalby, 790184 (71Z).

Blue Point, Wymondham, 894195 (81Z). Farm, formerly drovers' inn.

Boathouse Walk Plantation, Castle Donington, 415272 (41J). On a steep slope above the River Trent.

Boden Brook, 42ABG. Rises in Worthington and flows past Breedon Cloud Quarry and Tonge to Ambro Mill 411247; it enters Derbyshire as the Ramsley Brook and flows into the Trent north of Kings Newton.

Bolt Wood, Stockerston, 826969 (89I). Ancient woodland; SSSI.

Bondman Hays, Ratby, 492071 (40Y). Wood.

422

Boothorpe, Ashby Woulds, 319175 (31D). Hamlet. The area contains working, derelict and restored **claypits**. Habitat study 78 is of the **green lane** at 311166.

Booth Wood, Loughborough (Garendon), 507192 (51E).

Bosworth Field, Sutton Cheney, 406003 (30V, 40A). The battle was fought on and around Ambion Hill, where there is now an interpretive centre run by the County Council.

Bosworth Gorse, Husbands Bosworth, 642862 (68N).

Bosworth Mill, Husbands Bosworth, 629823 (68G). Between Welford arm of **Grand Union Canal** and **River Avon**.

Bosworth Park, Market Bosworth, (40AB). Public open space containing **Bow Pool**, **The Duckery** and other **pools** of interest.

Bosworth Wharf Bridge, Market Bosworth, 391032 (30W). Carries B585 over Ashby Canal.

Botany Bay, Broughton and Old Dalby, 720243 (72CH). Woodland and stream source at Bleak Hills, with interesting scrub and grassland.

Botany Bay, Houghton/Hungarton, 704046 (70C). Fox covert to west of Billesdon Coplow.

Botany Bay Spinney, Newbold Verdon, 437031 (40G).

Botany Spinney, Sheepy (Sibson), 378030 (30R).

Botcheston, Desford (Ratby), 482049 (40X). Hamlet. **Botcheston Bog** is an important wetland SSSI beside the Thornton Brook at 485047. Habitat study 50 is of the Bog and 81 of a roadside **verge** at 478051.

Bottesford, parish, 73TYZ, 74V, 83CDEHIJ, 84ABFG. Consists of the former parishes of Bottesford and Muston.
The former parish, 73TYZ, 74V, 83CDE, 84ABFG, contains: village at 806391; hamlets of Easthorpe at 810386 and Normanton at 811406; the **River Devon** flowing through Bottesford village; the **Grantham Canal** crossing the south of the parish; Bottesford Wharf at 804373; a **working railway** crossing from east to west; Bottesford Station at 810392; a **dismantled railway** running from north to south through the parish; the site of **Bottesford Junction** at 799396.

Bottle Neck, Market Bosworth, 397046 (30X). Woodland.

Bottom Plantation, Buckminster, 875240 (82RS).

Boundary, Ashby de la Zouch (Blackfordby), 337188 (31J). Hamlet on the A50, the county boundary follows the north side of the road.

Bow Bridge, Leicester (St. Mary's), 581040 (50X). Over River Soar.

Bow Pool Covert, Market Bosworth, 411027 (40B).

Bradfields Bridge, Sutton Cheney, 394001 (30V). Road bridge over Ashby Canal.

Bradgate House, Groby, 508090 (50E). Site of house in woodland with only stable block remaining; lake nearby.

Bradgate House, Newtown Linford, 534102 (51F). House in middle of Bradgate Park; in ruins apart from chapel.

Bradgate Park, Newtown Linford, 50J, 51FK. Deer park and public open space. SSSI.
Consists of heathland and grassland with plantations; contains the **ruins** of Bradgate House; **Cropston Reservoir** now forms its eastern boundary; Old John tower is at the highest point. Habitat study 45 is of **pool** and **wet heath** at 529115 and 46 is of **dry heath** at 536110.

Bradshaws, The, Peckleton (Stapleton), 429979 (49I). Site of Richard III's encampment on the eve of the Battle of Bosworth. Now known as Barn Farm.

Brand, **The**, Woodhouse, 535132 (51G). Country house and estate with **heathland**, **woodland** and **disused slate quarries**; SSSI. **Brand Hills** are at 534132.

Bran Hills, Frisby on the Wreake, 706166 (71D).

Branston, former parish, Croxton Kerrial, 73VW, 82DE, 83AB.
Contains: village at 809295 with **churchyard** and **wall flora**; part of **Knipton Reservoir**; former ironstone workings, now largely restored to agriculture.

Brascote, Newbold Verdon, 441025 (40L). Hamlet. The un-named **wet woodland** at 448026 is of botanical interest, but Brascote Covert at 447022 is not. There are currently extensive gravel workings to the west.

Braunstone, parish, 50GHKLMQR.
About half the parish was transferred to Leicester County Borough in 1935, the rest retaining its parochial status. Contains: village centre at 555028, but the parish is now largely built up. *Braunstone Park* at 559032, a public open space, with *lake* at 557029, is now in Leicester.

Braunstone Frith, former parish, transferred to Leicester in 1935, 50CH. Most of Leicester Golf Course and a small part of Western Park lie in the area.

Brazil Wood, Swithland, 558136 (51L). Former extensive woodland destroyed by the construction of Swithland Reservoir; an island carrying the Great Central Railway over the reservoir is all that remains of it.

Breach Hill, Packington, 378153 (31S).

Breach Pond, Hinckley (Earl Shilton), 470971 (49T). This and another **pond** at 470963 are both of interest.

Breakback Plantation, Woodhouse, 524148 (51H).

Breedon on the Hill, parish, 32WX, 42ABCFGH.
Contains: village below the south side of Breedon Hill; Tonge hamlet; Wilson hamlet; **Breedon Hill**, SSSI, with habitat study 22 of **limestone grassland** at 404233; church on top of Hill at 405233; Breedon Lodge with **moat** at 418222; **Breedon Brook**

423

flowing through the village and joining the Boden Brook north of Tonge; **Cloud Hill Quarry**; **Cloud Wood**; **Pasture** and **Asplin Woods**. Although much of Breedon Hill has been removed by quarrying, the church which surmounts it remains a conspicuous landmark.

Brentingby and Wyfordby, former parish, Freeby, 71UZ, 72 QVW.
Contains: villages of Brentingby at 784188 and Wyfordby at 793189; **working railway** which crosses the southern part of the parish from east to west; the **River Eye**, which forms its southern boundary; traces of the former Oakham Canal which remain, although much of its course lies under the railway; **Brentingby Wood** at 788214. The interest of grassland and marsh site known as *Brentingby Field* at 794219 has been entirely destroyed by agriculture.

Brickfield House, Somerby, 791152 (71X). **Old grassland**, and water-filled **brickpit** at 794151.

Brickfield Spinney, Husbands Bosworth, 634838 (68G).

Brick Kiln Farm, Sutton Cheney, 422014 (40F). **Grassland** in disused brickpit.

Brick Kiln Plantation, Sheepy/Twycross, 297034 (20W). **Pool** and **rough grassland** in disused claypit.

Brickyard Farm, Ashby Magna, 557917, (59K). Spinney and pools at 557914.

Brickyard Farm, Peckleton, 430009 (40F). **Grassland** on site of old brickyard at 429009.

Bridge Farm, Clawson and Harby (Long Clawson), 707267 (72D). The **River Smite** here is of some interest.

Bridge over the Brook, Desford (Newtown Unthank), 487041 (40X).

Bridget's Covert, Broughton and Old Dalby (Old Dalby), 672223 (62R).

Briery Leys Spinney, Tugby and Keythorpe, 747006 (70K).

Briery Wood, Belvoir, 822330 (83G). SSSI (heronry).

Bringhurst, parish, 89FGKL.
Contains: village at 841921 which is a hilltop landmark; **dismantled railway** line which crosses the parish from east to west; the **River Welland** which forms its southern boundary.

Broad Hill, Coalville (Whitwick), 435171 (41I). Habitat study 23 is of **rough grassland** at 425171.

Broad Hill, Kegworth, 479261 (42T).

Broad Hill, Mountsorrel, 573147 (51S).

Broadnook Spinney, Wanlip, 585114 (51V).

Brockey, The, Hinckley (Barwell), 448983 (49P). Area of land which is distinct from Stapleton Brockey at 439994. The site of *Brockey Hills* in Stoney Stanton is not known.

Brock Hill, Clawson and Harby, 745264 (72N). **Cuttings** at either end of railway tunnel under hill are important sites.

Brook Copse, Hinckley (Earl Shilton), 470989 (49U).

Brooksby, former parish, Hoby with Rotherby, 61HSTX.
Contains: church at 671160 and the Hall which is now part of the Agricultural College; working railway line with station site at 669163; **River Wreake** with a complex of weirs etc. close to the station; Brooksby Spinney at 681151; part of an ox-bow at 672164.

Brook Spinney, Stoughton, 638017 (60F).

Broombriggs Hill, Woodhouse, 515140 (51C). Part of the hill is a nature reserve.

Broome Lane Crossing, East Goscote, 639140 (61G). The **railway verge** near this level crossing has a relict sandy grassland flora.

Broughton and Old Dalby, parish, 62KLMQRSWXY, 72BCDH.
Consists of the former parishes of Nether Broughton and Old Dalby.

Broughton Astley, parish, 59ABCFGHKL.
Contains: village at 526926; hamlets of Primethorpe and Sutton in the Elms; the Fosse Way which forms much of the western boundary; a stretch of the **River Soar**; the **dismantled railway** which crosses the parish from north to south with a station site at 534919; a **quarry pool** at 535918, with interesting surrounding **grassland**.

Brown's Hill Quarry, Ab Kettleby (Holwell), 742234 (72L). County trust nature reserve and part of the Brown's Hill Quarry and Railway Cuttings SSSI.

Browns Wood, Bagworth, 484073(40Y). Clear felled.

Brown's Wood, Skeffington, 758022 (70LR). Ancient woodland, part of the Leighfield Forest SSSI.

Bruntingthorpe, parish, 58Z, 59V, 68DEIJ, 69A.
Contains: village at 600897; part of the disused Bruntingthorpe Airfield.

Buck Hill, Loughborough (Nanpantan), 508164 (51D).

Buckminster, parish, 81UZ, 82QRSVWX.
Consists of the former parishes of Buckminster and Sewstern
The former parish, 81U, 82QRSWX, contains: village at 879230; Buckminster Hall at 881233; large conifer and mixed plantations round Buckminster Park; marsh and aquatic vegetation round the **gravel pits** at 875225. Much of the parish has been opencast mined for ironstone and restored.

Buddon Brook, 51MNT. Otherwise known as the Quorn Brook; a continuation of the River Lin, it flows from Swithland Reservoir, through Quorn village to the River Soar at 565165.

Buddon Wood, Quorn, 562150 (51MS). SSSI. Former sessile oakwood, felled during the Second World

War. It was regenerating satisfactorily until reopening of the quarry in the nineteen-seventies, which has destroyed a large proportion of this important site.

Bufton, Carlton, 400057 (40C). Hamlet.

Bufton Lodge Spinney, Desford, 487045 (40X).

Bullacre Spinney, Peckleton, 458024 (40L).

Bull in the Oak, Market Bosworth, 422032 (40G). Farm at crossroads of A447 and B585.

Bulwarks, The, Breedon on the Hill, 406234 (42B). Iron Age hill fort, now largely destroyed by quarrying.

Bulwell Barn, Normanton le Heath, 385134 (31W).

Bunkers Wood, Belvoir/Croxton Kerrial, 813312 (83A). Plantation, mainly of mixed hardwoods.

Bunny's Spinney, Burton and Dalby, 770148 (71S).

Burbage, former parish, Hinckley, 48PU, 49ABFGHKLMQ.
Contains: villages of Burbage at 442927 and Sketchley, both joined to Hinckley by development; **Burbage Wood** at 450941, part of the Burbage Wood and Aston Firs SSSI; **Burbage Common** at 447951, part of which is now a golf-course, with habitat study 25 at 446950; Watling Street which runs parallel to the south-west parish and county boundary and which formerly coincided with it; the M69 which crosses it with a junction at 436911; a working **railway line** running across the northern apex of the parish; the Sketchley Brook marking the former boundary between Burbage and Hinckley; the Soar Brook crossing the southern part of the parish; large **pools** at 422928.

Burleigh Brook, 51DEJ. Formed by the union of brooks rising near Holywell Hall, Snell's Nook and Shortcliff, it forms the boundary between the former parishes of Garendon and Nanpantan, and joins the Wood Brook where it crosses the Grand Union canal at Loughborough.

Burleigh Wood, Loughborough (Nanpantan), 507176 (51D). Ancient woodland.

Burney Rough, Breedon on the Hill, 389227 (32W).

Burrough Hill, Somerby, 71KLQR. Site of Iron Age hill fort on spur of marlstone scarp, with grassland and scrub on adjacent slopes; habitat study 35 is at 761119; Burrough Hill Covert is at 760122.

Burrough on the Hill, former parish, Somerby, 70PU, 71FKLQR.
Contains: village at 757107; Burrough Court Farm at 752096; **Burrough Hill**; brooks rising on the high ground to the east and south of the parish which eventually join to form the Gaddesby Brook.

Burrow Spinney, Cotesbach, 524821 (58G).

Burrow Wood, Charley, 476145 (41S).

Burton and Dalby, parish, 71GHILMNRSTUWXYZ.
Consists of the former parishes of Burton Lazars, Great Dalby and Little Dalby.

Burton Bandalls, Burton on the Wolds, 567203. Farm.

Burton Brook, 69MSTUZ, 79E. Formed from several small brooks in the parish of Illston on the Hill. Forms the parish boundary between Burton Overy and Carlton Curlieu and Burton Overy and Kibworth Harcourt. It joins the eastern River Sence at 658950.

Burton Brook, 71RSTUY. Rises above Little Dalby Lakes and flows through Burton Lazars former parish to join the River Eye at 778181.

Burton Lazars, former parish, Burton and Dalby, 71MNSTUXYZ.
Part of the parish was incorporated into Melton Mowbray when the remainder was united with Great and Little Dalby.
Contains: village at 767169; the site of St Mary and St Lazarus Hospital at 736166; a little **old grassland** at 764169; several spinneys and ponds; a short length of the **River Eye** which forms the north-eastern parish boundary.

Burton on the Wolds, parish, 51PUZ, 52KQV, 62ABFGKL.
Contains: village at 592212; a few small woods; the **River Soar** which is its western boundary; the Fosse Way (A46) which is its eastern boundary; the Willoughby Brook which rises at Six Hills and flows north across the parish; **Twenty Acre**.

Burton Overy, parish, 69MSTUYZ, 79DE.
Contains: village at 678982; Burton Brook which is the eastern parish boundary; about 1 km each of the **Grand Union Canal** and a working railway which cross the southern tip of the parish; traces of earthworks at The Banks at 676982.

Bushby, former parish, Thurnby, 60LM.
The former parish was united with Thurnby when half of the latter was incorporated into Leicester County Borough.
Contains: village at 653039; Bushby Spinney at 660034 which is a fox covert planted in 1885.

Bushy Field Wood, Ulverscroft, 498111 (41V).

Buttermilk Hill Spinney, Burton and Dalby (Little Dalby), 774125 (71R).

C

Cadeby, parish, 40ABFG.
Contains: village at 426023; the **churchyard**; **The Gorse**; **Spring Wood**; extensive **gravel workings**, some now restored, in the east of the parish around Naneby Farm at 434025; roadside **verge** along Deeping Lane at 418035. There are several ponds and small spinneys in this largely arable parish. The former *toll gate* is at 430011.

Cademan Wood, Coalville (Whitwick), 438169 (41IN). Part of Grace Dieu and High Sharpley SSSI.

Caldicote Spinney, South Kilworth, 596834 (58W).

Callan's Lane, Staunton Harold, (32Q).

Cant's Thorns, Ab Kettleby (Wartnaby), 717218 (72A).

Cap's Spinney, Tugby and Keyworth, 774001 (70Q).

Captain's Gorse, Castle Donington, 425271 (42I).

Carington Spinney, Gaddesby, 710135 (71B).

Carland Spinney, Husbands Bosworth, 653832 (68L).

Carlisle Wood, Belvoir, 816325 (83B).

Carlton, parish, 30RSWX, 40C.
Contains: village at 396049; Carlton Bridge at 386043 carrying a minor road over the **Ashby Canal** where habitat study 66 has been made; a privately operated **railway** line, entering the parish at 382049; Friezeland.

Carlton Curlieu, parish, 69XYZ, 79CDE.
Contains: village at 693972; Carlton Clump at 696955; Carlton Curlieu Manor House at 700981 is in Illston on the Hill parish.

Carlton Hayes Hospital, Narborough, 536984 (59J).

Carr Bridge, Lowesby/Hungarton, 706085 (70E). Bridleway bridge over Queniborough Brook. Carr Bridge Spinney and **marshy meadow** nearby.

Carr Bridge, Shepshed, 474201 (42Q). Minor road bridge over Black Brook.

Carthagena Farm, Cossington, 626156 (61H). Former name of farm now called Ratcliffe Farm.

Castle Donington, parish, 42CDHIJMNP, 43K.
Contains: village at 446273; **Donington Park** with woodland to the south; an area in the extreme north of the parish with **ponds**, **oxbows** and **ditches**; part of the East Midlands Airport in the south-east corner of the parish; the power station at 433282 served by a working railway; the **River Trent** forming the north and north-west boundaries of the parish.

Castle Hill, Hallaton, 779967 (79TY).

Castle Hill, Leicester (Beaumont Leys), 565092 (50P).

Castle Hill, Mountsorrel, 582149 (51X). Habitat study 107.

Castle Rock, Coalville (Whitwick), 456148 (41M).

Cat Hill Wood, Charley, 475152 (41S).

Catsick Hill, Barrow upon Soar, 569184 (51U). *Catsick Lane* is mentioned in HG.

Catthorpe, parish, 57NPTU.
Contains: village at 552781; a short length of the M6 motorway and also of **dismantled railway**; the **River Avon** forming part of the southern boundary and entering the parish for a short distance. A length of the Watling Street formerly in the parish was transferred to Warwickshire in 1966.

Cavendish Bridge, Castle Donington, 447299 (42P). Hamlet at A6 crossing of **River Trent**; a **pond** at 445294.

Caves Inn (Warwickshire), 535792 (57J). Formerly in Shawell parish until this section of Watling Street was transferred to Warwickshire. In Leicestershire the name Caves Inn now only refers to **old gravel workings**, Cave's Inn Pits, SSSI, where habitat study 100 was made at 539795.

Cedar Hill, Croxton Kerrial, 828301 (83F).

Chadwell, Scalford (Wycomb and Chadwell), 782246 (72X). **Churchyard**, **walls** and **verges** in the village are of some botanical interest.

Chalk Pool Hill, Frisby on the Wreake, 704175 (71D).

Chamberlains Nether Close, Gaulby, 685022 (60W). Woodland.

Chaplain's Rough, Newtown Linford, 530092 (50J).

'**Charity Meadow**', Desford, 489039 (40W). Property of Desford Town Charity.

Charley, parish, 41MNRSTWXY, 51CD.
The parish was considerably enlarged in 1936 by the addition of parts of Woodhouse and Whitwick parishes. It is sparsely populated. Contains: many sites of botanical importance including those now contained within the Charnwood Lodge Nature Reserve; small settlements at Oaks in Charnwood and around Mount St. Bernard Abbey; the M1 motorway which crosses the parish from south to north; about half of **Blackbrook Reservoir**; Charley Knoll at 490157; **Charley Mill** at 475141; Charley Hall at 479147.

Charnock Hill, Isley cum Langley, 446251 (42M).

Charnwood Forest, SK40, 41, 50, 51. An area of high ground north-west of Leicester with outcrops of Precambrian rock. It still contains some **heath**, and much **woodland** and is the most intensively studied area of the county both botanically and geologically.

Charnwood Forest Canal, 41DIJPUYZ, 51D. Only short stretches remain of the former canal which is still of botanical interest at several points.

Charnwood Granite Quarries, Shepshed, 487179 (41YZ).

Charnwood Lodge Nature Reserve, Charley, 41MS. SSSI. 245 hectares of **heath**, **woodland**, **pasture** and **wetland** in which habitat study 43 was made at 466154 (**wet heath**) and 44 at 473149 (**wet heath**). Colony Reservoir is at 464153; there are several small **ponds**.

Chater, River, 70SX, 80C. Rises near Halstead Lodge at 771054 and flows east in a **deep valley** past Sauvey Castle and Launde Abbey to enter Rutland at 809044. It is of botanical interest in 70X and 80C and there are **rich meadows** and **spring-fed marshes** between the river and Launde Park Wood in 80C.

Cherry Copse, Cossington, 621153 (61H).

Cheseldyne Spinney, Knossington, 808063 (80D). Known also as Tampion's Coppice.

Chilcote, parish, 20YZ, 21QRVW.
Contains: village at 284114; **pools** at 282115; Horseley Plantation at 287108; three other woodlands of less interest; the River Mease forming the northern boundary of the parish.

Chitterman Hills (Farm), Ulverscroft, 496115 (41V).

Church Hill, Cranoe, 762956 (79S).

Church Langton, East Langton, 724934 (79G). Village.

Church Thorns, Redmile, 803330 (83B). Woodland.

Church Town, Coleorton, 396169 (31Y). One of the two principal hamlets in the parish.

City of Dan, Coalville (Whitwick), 437160 (41HI). Former hamlet, now part of Whitwick.

City of Three Waters, Coalville (Whitwick), 432166 (41I). Former hamlet, now part of Whitwick.

Clack Hill, Market Harborough (Little Bowden), 752866 (78N).

Clarendon Park, Leicester (Knighton), 596023 (50VW). Residential suburb.

Clarke's Bush, Stoughton, 652022 (60L). Woodland.

Clarke's Spinney, Cosby, 524956 (59M).

Clawson and Harby, parish, 72CDEHIJNPTU, 73AFGKL.
Consists of the former parishes of Long Clawson, Hose and Harby.

Claybrooke Magna, parish, 48UZ, 49V, 58E.
Contains: village at 490888; the Fosse Way (as a green lane) forming the north-western parish boundary, meeting Watling Street at High Cross; marsh at 498891 now ploughed.

Claybrooke Parva, parish, 48TUYZ, 58DE.
Contains: village at 496879; Watling Street running parallel to and just inside the south-west parish boundary.

Cliff by Trent, Castle Donington. Could refer to Boathouse Walk Plantation at 415271. Cliff Hill Plantation, to the west, is in Derbyshire.

Cliffe Hill Quarry, Markfield, 475106 (41Q). Older workings, formerly of botanical interest, were lost in expansion of quarry during nineteen-seventies.

Cliff Farm, Lockington and Hemington (Lockington), 489308 (43V), now ruined. **Wetland** to the south-west is an SSSI.

Cliff Spinney, Birstall, 584094 (50Z).

Cliff Spinney, Ratcliffe on the Wreake, 627138 (61G). Steep wooded Keuper Marl slope by the River Wreake.

Clifton's Bridge, Oadby and Wigston (Wigston Magna), 617967 (69D). Field bridge over the **Grand Union Canal**; there is **marshy ground** and a **pool** called Kilby Pit at 616968.

Cliftonthorpe, Ashby de la Zouch, 357183 (31P). House, pools and the **embankment** of a disused railway.

Clock Mill, Swepstone, 357116 (31K). Farm with pond.

Cloud Hill Quarry, Breedon on the Hill, 412214 (42A). Limestone quarry with botanically interesting **spoil heaps**.

Cloud Wood, Breedon on the Hill, 417214 (42A). About a third of the wood has been quarried away. The wood is an SSSI and habitat study 3 is at 415214.

Clump Hill, Broughton Astley, 533914 (59F).

Coalbourn Wood, Ulverscroft, 488121 (41VW).

Coalpit Lane Fox Covert, Wistow, 635945 (69H).

Coalville, parish, 41ABCFGHIKLMNQRS. Abolished in 1974.
The parish was created in 1894 by the amalgamation of parts of Hugglescote, Swannington, Ravenstone with Snibston and Whitwick. The changes of boundary of Coalville civil parish since then have been so extensive that to catalogue them is beyond the scope of the present work. The parish of Coalville to which former parishes such as Hugglescote are referred in this gazetteer is the unit which was defined in 1936. It persisted until 1974 when the administrative units of Coalville parish and Coalville Urban District were abolished. The area now has no parish status and is administered by North West Leicestershire District Council.
Coalville parish as delineated in 1936 consists of parts of the former parishes of Bardon, Hugglescote, Ravenstone with Snibston, Swannington, Thringstone and Whitwick. It contains: the town of Coalville, 41GH; the villages and hamlets of Bardon, Donington le Heath, Hugglescote, Snibston, Thringstone and Whitwick. **Coalville Meadows** SSSI is in Whitwick former parish.

Cocklow Wood, Quorndon, 568150 (51S). The **wood** and **quarries** were both of botanical interest prior to the re-opening of quarrying on Buddon Hill.

Cockspur Bridge, Oakthorpe and Donisthorpe, 321139 (31G). Over former **Ashby Canal**, the bed of which has **marsh** and **pools**.

Colborough Hill, Tilton, 761051 (70S).

Cold Newton, parish, 70CDEHI.
Contains: deserted mediaeval village site close to the Manor House at 717067; a ridge of high ground (over 200m.) at the southern end of the parish which has rough pasture and woods along the marlstone escarpment; interesting and diverse elm populations in various places in the parish which may now have died; a **dismantled railway** which crosses the parish from east to west.

Cold Overton, former parish, Knossington, 70Z, 71V, 80EJ, 81ABF.

Contains: village at 810101; Cold Overton Grange at 809110 on a brook which flows into an overgrown *lake* at 814115, still with surviving remnants of introduced ornamental plant species.

Cold Overton Park Wood, Knossington, 822086 (80EJ). The ground vegetation in a large part of the wood was severely damaged when it was used as a pig run in the early and late seventies.

Coleorton, parish, 31STUXYZ, 41CDE.
Small parts of Thringstone and Swannington were added to the parish when Coalville Urban District was formed.
Contains: scattered settlements, which developed in association with the coal mining industry, around the park-land belonging to Coleorton Hall at 391173, the most important of which are called Church Town, Farm Town, **Coleorton Moor**, Pegg's Green and St George's; numerous woods, including three of considerable botanical interest, **Birch Coppice**, **Rough Park** and **Spring Wood**; **The Paddock** at 399175, an area of **heath grassland** and **scrub**; a **railway line** which crosses the southern end of the parish; habitat study 55 of the **fishpond** at 399171. *Coleorton Wood* is mentioned several times in HG but it is not clear which wood is referred to. Extensive opencast operations have destroyed the interest of some land on the northern edge of the parish and threaten more.

Collier Hill, Charley, 470157 (41S). Woodland, now part of Charnwood Lodge Nature Reserve and SSSI.

Colony Reservoir, Charley, 464153 (41S). In Charnwood Lodge Nature Reserve and SSSI.

Combs Plantation, Stathern, 780305 (73QV). Habitat study 65 was made of a **pond** at 782301 (in Eaton parish), to the east of the plantation.

Conduit Spinney, Hungarton, 691054 (60X).

Conduit Spinney, Shangton, 714970 (79D).

Coneygear Wood, Croxton Kerrial, 844298 (82P).

Coney Hill Plantation, Noseley, 730984 (79J).

Congerstone, former parish, Shackerstone, 30NST.
Contains: village at 367054; sections of **dismantled railway line**; **Ashby Canal**; the **River Sence** which flows along the southern parish boundary.

Cooper's Plantation, Croxton Kerrial, 860281 (82PU). The eastern end is **scrub** developing over **limestone grassland**. Part of King Lud's Entrenchments and the Drift SSSI.

Coplow Brook, 60W, 70BG. Rises near Lodge Farm, Billesdon and flows west and south into Houghton parish, uniting with other unnamed brooks to form the eastern River Sence.

Coppice, The, Ashby de la Zouch, 368208 (32Q).

Coppice, The, Quorndon, 560156 (51S).

Coppice Plantation, Newtown Linford, 540111 (51FK).

Coppice Wood, Castle Donington, 427262 (42I).

Copt Oak, Markfield/Ulverscroft/Charley, 481129 (41W). Hamlet with church, inn and school at the junction of three parishes, with two others (Bardon and Woodhouse) within a very short distance. *Copt Oak Wood*, which is in Ulverscroft parish at 484130 has been largely cleared and replanted.

Cord Close, Stoughton, 654025 (60L). Spinney.

Cord Hill, Wymondham, 837170 (81I).

Cosby, parish, 59GHILMNR.
Contains: village at 548948; golf course 59LM; two lengths of **dismantled railway**, crossing at 555938; a few spinneys including Ash Pole Spinney; about 3km of M1 motorway.

Cossington, parish, 51W, 61BCHI.
Contains: village at 603136; a working railway crossing the parish from north to south; the **River Soar** forming part of the western boundary and the **River Wreake** the southern boundary; the **canalised River Wreake** from 609121 to its junction with the River Soar at 595127; **gravel pits** at 598129; 'Church Pond' at 603137; **Cossington Mill** at 595129. *Cossington Gorse* at 630162 has been cleared and ploughed.

Coston, former parish, Garthorpe, 81PU, 82FGKLQR.
Contains: village at 847221; the **River Eye** flowing across the parish from north to south with **grassland** adjacent to it near the village; **ponds** near Hall Farm at 851213.

Cotes, parish, 52KLQ.
Contains: **mediaeval village site** with habitat study 105 at 552209; the **River Soar** which forms the southern parish boundary from Cotes Bridge at 554205 almost to Loughborough Viaduct; habitat study 106 at 553208 of **grassland** on bank by river. **Cotes Weirs** and Cotes Lower Mill (with adjacent **ditches**) are in Loughborough parish.

Cotesbach, parish, 58FGKL.
Contains: village at 539824; about half a kilometre of the River Swift; one kilometre of **dismantled railway**; a short length of M1 motorway; a few plantations and two small brooks; extensive **gravel pits** at 525816. The south-west boundary formerly ran along Watling Street from Gibbet Hill to the River Swift but during boundary revision in 1935 was moved about 75 metres to the north east of the road.

Cotes de Val, Gilmorton, 553885 (58P). Farm with spinney and former moat, close to M1 motorway and **bridge** over dismantled railway cutting.

Coton Bridge, Market Bosworth, 388024 (30W). **Ashby Canal** and spinney beside it.

Cottage Plantation, Freeby (Stapleford), 820178 (81DEIJ).

Cotterill Spinney, Little Stretton, 673019 (60Q).

Countesthorpe, parish, 59STXY.
Contains: village at 585954, much enlarged with new

housing; a dismantled railway crossing the parish from north to south; **station site** at 577955.

Cover Cloud Wood, Newtown Linford, 498102 (41V).

Cow Pastures Spinney, Market Bosworth, 415037 (40B). Habitat study 4 is at 416037.

Cowpen Spinney, Groby, 521068 (50I).

Cow Ponds at 31EJ, formerly in Blackfordby, were transferred from Leicestershire to Derbyshire (Woodville) in 1897.

Crackbottle Spinney, East Norton, 775993 (79U).

Cradock's Ashes, Walton on the Wolds, 637205 (62F). Sometimes called Ash Plantation.

Cradock's Covert, Shackerstone (Congerstone), 376050 (30S).

Crane's Lock, Wistow, 657955 (69M).

Cranoe, parish, 79LMNST.
  Contains: village at 761953; part of the east-facing slope of Langton Caudle; a few ponds.

Craven's Rough, Newtown Linford, 508119 (51A).

Cream Gorse, Frisby-on-the-Wreake, 705145 (71C). Habitat study 14 is at 705145.

Cream Lodge Quarry, Barrow on Soar, 591186 (51Z).

Cribb's Lodge, Wymondham, 893186 (81Z). Close to a **dismantled railway** and the site of the junction of the **mineral line** leading to Market Overton quarries (Rutland).

Cribb's Meadow Nature Reserve, Wymondham, 899188 (81Z, 91E). Former glebeland comprising two sections of species-rich **pasture** separated by the embankment of the **dismantled railway**, and containing two small ponds; SSSI with habitat study 40 at 899189; headwaters of the River Witham run along the northern margin.

Croft, parish, 59CDHI.
  Contains: village at 510959; a length of the **River Soar** passing through the village at Croft Bridge 511959; **Croft Hill** at 510966 with habitat study 29; **quarry** at 512963; a length of railway with *Croft station site* at 512957; **Croft Pasture** SSSI, a county trust nature reserve, with habitat study 28 at 509959 (meadow), and habitat study 73 at 504987 (river).

Crompton's Plantation, Broughton and Old Dalby (Old Dalby), 686236 (62W).

Cropston, former parish, 50P, 51KQ.
  Contains: village at 553109, with an interesting **wall** flora; a small part of **Cropston Reservoir** at 545109 including the **dam** and outfall; water-filled claypit near the reservoir at 547116, known as 'Puddledyke' or **Cropston Brickpit** and mentioned in HG under the latter name.
  The spelling Cropstone is used on old OS maps.

Cropston Brook, 51KQR. Flows out of **Cropston Reservoir** and runs into **Swithland Reservoir**.

Crossburrow Hill, Glooston, 748951 (79M).

Cross Hands, The, Sheepy (Sheepy Magna), 330024 (30G). Cross roads on B4116. **Sheepy Fields** SSSI is immediately adjacent.

Crossing Covert, Wymondham, 828174 (81I).

Cross in Hand, Lutterworth, 508839 (58B). Road junction of the A5, A427 and B4455.

Crow Mill Bridge, Oadby and Wigston (Wigston Magna), 589977 (59Y). Road bridge over the Grand Union Canal.

Crown Hill, Burton and Dalby (Great Dalby), 748146 (71M).

Crown Hills, Leicester (Evington), 621044 (60H). Formerly farmland, now allotments, schools and houses.

Crown Inn, Snarestone, 357098 (30P). Crossroads.

Crow Spinney, Little Stretton, 665006 (60Q).

Crow Wood, Bagworth, 484064 (40Y)].

Crow Wood, Rolleston, 737002 (70F).

Crow Wood, Skeffington, 757028 (70L). Part of Leighfield Forest SSSI.

Crow Wood, Swithland, 561125 (51R).

Croxfield Spinney, Tur Langton, 727946 (79H).

Croxton Kerrial, parish, 72Z, 73VW, 82DEIJNPTU, 83ABFK.
  Consists of the former parishes of Croxton Kerrial and Branston
  The former parish, 82DEIJNPTU, 83AFK, contains: village at 835295 with limestone **wall** flora; **pond** at 833290; site of Croxton Abbey and **parkland**; several woods including Croxton Banks at 832301.

Croxton Park, Croxton Kerrial, 822276 (82I). Part is an SSSI. Contains the site of Croxton Abbey, 823276, with some **old grassland**. **Lakes** and an **osier holt** in the middle of the park are drained by a brook flowing north. *Croxton Thorns* is mentioned in HG.

Culloden Farm, Twycross, 336082 (30J).

Cumberland Lodge, Scalford, 768230 (72R).

D

Dadlington, former parish, Sutton Cheney, 39YZ, 49DE.
  Contains: village at 403980; lengths of the **Ashby Canal** and of **dismantled railway**.

Daisy Plantation, Ravenstone with Snibston, 383146 (31X). One of the woods near Alton Grange.

Dakins Bridge, Congerstone, 371052 (30S). Field bridge over **Ashby Canal**.

429

Dalby Brook, 62RS. Rises on the Dalby Wolds and flows north to the Nottinghamshire border.

Dalby Wolds, Broughton and Old Dalby, 62KL.

Dale Hill Covert, Gaddesby, 682140 (61X).

Dale Spinney, Newtown Linford, 536109 (51F). In Bradgate Park.

Dam Dyke, 72IJ. Brook rising in Long Clawson, flows north-east to join Hose Brook and form Wash Dyke which enters Nottinghamshire at 722317.

Dam's Spinney, Stoughton, 632023 (60G).

Dane Hills, Leicester (St. Mary), 568047 (50S). Now a suburb, formerly had sandstone and sand quarries.

Day's Plantation, Wymondham, 833173 (81I). Part of Wymondham Rough Nature Reserve.

Deakin's Bridge, Market Bosworth, 386023 (30W). Field bridge over **Ashby Canal**.

Debdales Farm, Bottesford, 783377 (73Y).

Debdale Spinney, Burton and Dalby (Little Dalby), 780126 (71W). On the scarp east of Burrough Hill; Debdale Lodge is at 783130, in Somerby parish.

Debdale Wharf, Gumley/Smeeton Westerby, 695915 (69V). Minor road bridge here crosses Grand Union Canal by marina.

Deer Park Spinney, Newtown Linford, 538105 (51F). In Bradgate Park.

Demoniac Plantation, Ravenstone with Snibston, 384152 (31X). Frequently mentioned in HG; one of a number of plantations and spinneys near Alton Grange.

Desford, parish, 40LMQRSTVXY, 50BC.
Consists of Desford former parish and a large area including Botcheston transferred from Ratby in boundary revisions of 1935.
The former parish, 40RSVWX, 50B, contains: village at 478034; a railway crossing the parish from east to west with Desford Station site at 487041 and Desford sidings, now dismantled; mining and industrial sites; Thornton Brook which marks the former parish boundary; interesting botanical sites at **Lindridge**.

Devon, River, 74Y, 83EFGHIJ, 84AB. Rises above **Knipton Reservoir** in Croxton Kerrial parish and flows through **Belvoir Lakes**, below which it enters Lincolnshire for two miles before re-entering Leicestershire at Muston. It passes through Bottesford village and eventually joins the River Smite in Nottinghamshire.

Diamond Spinney, Lowesby, 713076 (70D).

Dicken Bridge, Whetstone, 556971 (59N).

Dimmingsdale Spinney, Newtown Linford, 539099 (50JP). Just outside Bradgate Park.

Dimminsdale, Staunton Harold, 376217 (32Q). **Woodland, flooded limestone quarries, brook** and **siliceous grassland**; a county trust nature reserve; part of the Staunton Harold Reservoir SSSI. Straddles the Leicestershire/Derbyshire county boundary.

Diseworth, former parish, Long Whatton, 42GKLMRST.
Contains: village at 453245; two small woods of no special botanical interest; several brooks including West Meadow Brook which forms most of the boundary between Diseworth and Long Whatton; about 1 km. of the M1 motorway.

Diseworth Brook, 42HLMRS. Formed from the union of two brooks rising in Isley cum Langley, flows eastwards through Diseworth village to merge with West Meadow Brook forming Long Whatton Brook which eventually flows into the River Soar at Zouch Bridge.

Dishley, Loughborough, (Thorpe Acre and Dishley), 515199 (51E). Hamlet.
Dishley Mill formerly stood by the A6 road bridge over the **Black Brook** at 516210.

Dishley Farm, Swepstone, 365111 (31Q). Marshy ground near the road has been drained.

Dog Kennel Plantation, Noseley, 737980 (79IJ).

Dog Kennel Pool, Staunton Harold, 379204 (32Q). Now part poplar plantation and part willow holt, no open water.

Dog Kennel Spinney, Osbaston, 418045 (40C).

Doles Farm, Ashby de la Zouch, 379176 (31T).

Donington le Heath, Coalville (Hugglescote), 419125 (41BG). Hamlet, now continuous with Hugglescote.

Donington Park, Castle Donington, 42DHI. Contains Donington Hall at 420269 and a deer park south-west of the Hall; the part of the river is an SSSI. There is a motor racing circuit south of the Hall.

Donisthorpe, Oakthorpe and Donisthorpe, 315139 (31B). Village. The railway still carries mineral traffic and the **station site** is at 316138. In the early nineteenth century the boundary between Leicestershire, Staffordshire and Derbyshire was extremely ill defined in the neighbourhood of Donisthorpe, Appleby Magna, Packington and Measham. Donisthorpe itself was never a clearly defined Leicestershire parish and after various vicissitudes the whole of Oakthorpe and Donisthorpe was transferred to Leicestershire in 1897.

Dovecot Nook Hill, Freeby (Brentingby and Wyfordby), 793192 (71Z).

Drayton, parish, 89FGKL.
Contains: village at 830922 with interesting elm populations; the **River Welland** which forms part of the southern boundary; a short section of **dismantled railway** crossing the southern boundary of the parish. Drayton Wood seems to have been destroyed.

Drayton Grange Farm, Witherley (Fenny Drayton), 342971 (39N).

Drift, The, 81Z, 82PTUVWX, 83KL, 91E. Ancient pre-Roman trackway leaving Ermine Street near Stretton in Rutland and forming the county boundary between Leicestershire and Lincolnshire for about 14km. Its continuation north past Woolsthorpe is lost but it reappears as Sewstern Lane near Stenwith and rejoins the Great North Road at Long Bermington; this northern section is wholly in Lincolnshire. Botanically it is of special interest where it crosses the former Saltby Heath, and near Harston. Unfortunately minor boundary revision in 1965 moved the actual county boundary from the centre of the track, to the wall or hedge on one side or the other, and most of the best limestone vegetation along its verges is now in Lincolnshire.

Drybrook Plantation (probably the same as Drybrook Wood), Charley, 455165 (41N).

Duchess's Garden, Belvoir, 819332 (83BG). Semi-wild garden in the wooded grounds of Belvoir Castle.

Duckery, The, Market Bosworth 414014 (40A). Pool and spinney.

Duck Pond Covert, Twycross (Gopsall), 361068 (30NT).

Duck's Nest, Castle Donington, 433255 (42M).

Dumps, The, Lockington-Hemington, 464274 (42T). Woodland formerly of botanical interest.

Dumps Plantation, The, Isley cum Langley, 426242 (42H).

Dunn's Lock, Glen Parva, 572985 (59V). Grand Union Canal.

Dunton Bassett, parish, 58JP, 59FKL.
Contains: village at 547904; a section of **dismantled railway**; a section of the M1 motorway; pools and old diggings at Dunton Mill 537896; marsh at 543891, ploughed 1982; **gravel pits** at 537899.

Durrells, The, Huncote, 510977 (59D). Formerly marshland, with flooded sandpit adjacent, now largely destroyed.

E

Earl Shilton, former parish, Hinckley, 49NPTUYZ.
Contains: Earl Shilton town at 471982; Thurlaston Brook rising north of the town; an unnamed brook flowing eastwards; **Barrow Hill Quarry**, the most important botanical site in the parish. Marshy ground, known as 'Potters Marston Bog' formerly of botanical importance, lay astride the parish boundary with Potters Marston at 483965, but disappeared in the construction of the M69 motorway.

East Goscote, parish, 61GHL.
Formed in 1968 from parts of Queniborough and Rearsby. Contains: the recently constructed village of East Goscote at 639135; a working railway crossing the parish from south-west to north-east, almost entirely on level ground without embankment or cuttings; extensive **flooded gravel workings** north-west of the railway; the **River Wreake** which forms a short section of the northern boundary; **waste places** and **rough grassland** in the village.

Easthorpe, Bottesford, 810385 (83E). Hamlet on outskirts of Bottesford village; there is a **millpool** at Easthorpe Mill, 812388. The site of *Easthorpe Sands* is not known.

East Langton, parish, 79ABCFGH.
Contains: village at 727926; village of Church Langton; a section of working railway on an embankment; a section of the Langton Brook.

East Langton Station site, West Langton, 720919.

East Midlands Airport, 42MNST. Former Second World War airfield, now somewhat expanded. Occupies a considerable area in the south-east corner of Castle Donington and in the southern part of Lockington-Hemington.

East Norton, parish, 70QV, 80A, 79UZ, 89E.
Contains: village at 782004; a section of dismantled railway, including the station site at 792003 and the north tunnel cutting, both formerly of botanical interest but now entirely filled in; the **Eye Brook** which forms the northern parish boundary and the railway viaduct spanning it.

Easton Crossing, Great Easton, 856924 (89L). Level crossing of dismantled railway west of Rockingham Station formerly of some botanical interest, but gradually deteriorating.

Eastwell, former parish, Eaton, 72NTUYZ.
Contains: village at 774285; a **pond** at 777283; traces of former mineral workings and a mineral railway which have been almost completely obliterated. The north west parish boundary runs along the top of the **Harby Hills** scarp.

East Wigston, former parish, Oadby and Wigston, 69DEIJ.
Former rural parish between Newton Harcourt and Wigston Magna which existed between 1894 and 1930, containing a few farms, lengths of the **Grand Union Canal** and a working railway.

Eaton, parish, 72MNSTUXYZ, 73QV, 82DE.
Consists of the former parishes of Eastwell, Eaton and Goadby Marwood.
The former parish, 72UYZ, 73QVW, 82DE, contains: village at 798290; former ironstone quarries and associated mineral lines now largely obliterated and returned to agricultural use; habitat study 36 of **pasture** at 795309. The north-west parish boundary runs for some distance within the borders of **Stathern Wood** and **Plungar Wood**.

Edmondthorpe, former parish, Wymondham, 81INTUYZ, 91E.
Contains: village at 858176 (81NT); **Cribb's Meadow** which is an SSSI and county trust nature reserve; a section of the former **Oakham Canal** still of some botanical interest at several points; two short sections of **dismantled railway; Whissendine Station site** on a working railway; a section of one of the **brooks** forming the **River Eye; Woodwell Head**.

Egypt Plantation, Sproxton (Saltby), 869279 (82T). Part of the King Lud's Entrenchments and The Drift SSSI. Habitat study 39 at 868279 is of limestone grassland adjacent.

Eightlands Farm, Witherley (Ratcliffe Culey), 351003 (30K).

Elder Plantation, Newtown Linford, 527105 (51F). In Bradgate Park.

Eleven Acre Covert, Sutton Cheney (Shenton), 378993 (39U).

Ellaby's Spinney, Burton & Dalby (Little Dalby), 783156 (71X).

Ellistown, Ibstock, 429110 (41F). Mining settlement.

Elmesthorpe, parish, 49MNSTXY.
  Contains: village at 460964; Billington Rough at 460957; a number of irrigation ponds; several **hedgerows**; **Elmesthorpe Plantation** at 454942, part of the Burbage Wood and Aston Firs SSSI; a section of the M69 motorway; nearly 4km. of **working railway** including the station site at 470958. *Elmesthorpe Gorse* has disappeared. Elmesthorpe Meadow, a former SSSI, has been ploughed.

Elms, The, Hoby with Rotherby. 659168 (61N). Few of the elms for which the house were named were still surviving in 1977.

Elms Farm, Narborough, 523971 (59I). There is a **water-filled quarry** at 524975.

Enderby, parish, 50FKQ, 59JPU.
  Contains: village at 537994; **quarries** at 50FK and 59JP; a section of the M1 motorway running north and south near the eastern margin of the parish; a **pool** near Enderby Hall at 538996; **River Soar** with Enderby Bridge at 551985.

Ervin's Lock, Oadby and Wigston, (Wigston Magna) 594977 (59Y). On **Grand Union Canal.**

Evington, former parish, Leicester, 60ABCFGHL.
  Transferred to Leicester County Borough in two stages, in 1892 and 1935, with the exception of a small section south of the Gartree Road which became part of Oadby. Contains: village at 627027; Evington Golf Course 60BG; a section of Evington Brook 60BG; Evington Park at 624034; Crown Hills (large area of allotments) 60H; General Hospital at 622039; much new residential building around Evington village; the inner suburban areas of Spinney Hills, Horston Hall and North Evington; Thurnby Brook marking the northern boundary; *Evington Fields* and *Evington Grange* which are mentioned in HG.

Eye, **River**, 71PUZ, 81EIJ, 82FIKLMN. Rises at Bescaby, flows eastwards towards Saltby, past **Saltby Bog**, and then south and south-west through Sproxton, Coston, Garthorpe and Saxby to Stapleford Park and through **Stapleford Park lake**. Thence it flows west between Brentingby and Burton Lazars to Eye Kettleby Mill, after which it is renamed the Wreake. Habitat study 77 is at 800183. The river is an SSSI between Ham Bridge at 801186 and Swan's Nest weir on the eastern edge of Melton Mowbray at 764188.

Eye Brook, 70HLMQRY, 80AF, 89IJLMNR. Rises south-west of Tilton on the Hill and flows south-east. From 799004 onwards to its junction with the River Welland it marks the former county boundary between Leicestershire and Rutland. It is of botanical interest for most of its length. It was dammed in 1940 at 854943 to form the **Eye Brook Reservoir**.

Eye Brook Reservoir, Great Easton/Stockerston (Rutland), 89MN. Only the western part of the reservoir is in the county. Formed in 1940 to provide water for the growing steel works at Corby (Northamptonshire). Both Rutland and Leicestershire shores of the reservoir are of botanical interest.

Eye Kettleby, former parish, Melton Mowbray, 71IJNP.
  Contains: a farm on the site of the mediaeval village at 733167; the **River Eye** and the **River Wreake** marking the northern boundary (the name changes at Eye Kettleby Mill, 737181); **dismantled railway** running from north to south; the **junction** of two working railways at 742184; a small part of the former Melton Mowbray Airfield.

F

Fairfield Bridge, Shackerstone (Congerstone), 377044 (30S). Over Ashby Canal.

Fairham Brook, Broughton and Old Dalby (Old Dalby), 662251 (62S).

Far Coton, Market Bosworth, 386021 (30W). Hamlet.

Farm Town, Coleorton, 391165 (31Y). One of the two principal hamlets in the parish.

Farnham Bridge, Rothley, 590133 (51W). Carries A6 road over Rothley Brook; **marshy meadows** adjacent.

Felstead's Spinney, Burton and Dalby (Burton Lazars), 790170 (71Y).

Fenny Drayton, former parish, Witherley, 39IMNT.
  Contains: village at 350971; the Roman road from Leicester to Mancetter which runs the length of the parish. Watling Street, now transferred to Warwickshire, was formerly its south-west boundary.

Field Head, Markfield, 496100 (41V). Road junction and hamlet; Field Head Lane is still of some botanical interest.

Fieldon Bridge, Sheepy (Sheepy Magna), 307994 (39E). Carries B4116 road over River Anker at the county boundary.

Finchley Bridge, East Norton, 801003 (80A). Carries A47 road over Eye Brook at the former county boundary.

Finney Hill, Loughborough (Shepshed), 465181 (41U). Grassland with a little outcrop rock and scrub; Finney Spring at 464178 is no longer of botanical interest.

First Hill, Burton and Dalby (Great Dalby), 752143 (71M).

Fishpond Spinney, Cotes, 558214 (52K). Spinney and pond.

Fishpond Spinney, Waltham (Waltham on the Wolds), 801239 (82B).

Fishponds Spinney, Asfordby (Welby), 728211 (72F).

Fishpool Brook, 51TYZ, 61E, 62AF. Rises near Walton Thorns, flows south-west to the outskirts of Barrow upon Soar and then south to join the River Soar at 579168.

Fishpool Grange, Loughborough (Shepshed), 454194 (41P). **Marsh** immediately to the west.

Fishpool Spinney, Enderby, 544002 (50K).

Flat Hill, Charley, 465160 (41ST).

Fleckney, parish, 69GHLM.
  Contains: village at 647934; former **brick pit** at 628938.

Fleckney Tunnel, Saddington, 69LR. **Grand Union Canal**; the **cutting** at the south entrance to the tunnel at 664926 is of considerable botanical interest.

Folly Farm Rough, Peckleton, 471997 (49U).

Forest Rock Quarry, Coalville, 444159 (41MN). Disused quarry containing pool.

Fosse (Foss) Way, SP48, 49, SK50, 60, 61, 62. Crosses the county from south-west to north-east; the first two miles from High Cross survive as a green lane; it then becomes the B4114 as far as Narborough village. It is partly lost in Narborough and Enderby parishes, but the A46 joins it as it enters Braunstone and continues to mark its course almost to the crossing of the River Soar close to the site of the Roman town of Ratae (by St. Nicholas church); north-east of this, the A46 marks the line of the Roman road all the way to the county boundary near Willoughby on the Wolds (Notts.) except where bypasses have been built round Thurmaston and Syston; parish boundaries follow it both to the north and south of Leicester. Fosse Road, Leicester (St Mary's) does not follow the line of the Roman road.

Foston, former parish, Kilby, 59XY, 69BCD.
  Contains: mediaeval village site around Foston Hall Farm at 605950; the church and Foston House at 603950; the River Sence forming the northern boundary of the parish.

Four Acre Wood, Burton on the Wolds, 601205 (62A).

Fox Bridge, Market Bosworth, 384021 (30W). Field bridge over Ashby Canal; there is a spinney at 382022.

Fox Cover Farm, Higham on the Hill, 380973 (39Y). The farm has several ponds.

Fox Covert, Leicester (Thurcaston), 577104 (51Q).

Fox Covert, Ulverscroft, 487130 (41W). Part of Ulverscroft Nature Reserve.

Fox Croft Spinney, Hoton, 558217 (52K).

Fox Holes, Ab Kettleby, 724244 (72H). Woodland. Part of Holwell Mouth SSSI.

Fox Holes Spinney, Braunstone, 547028 (50L).

Fox Holes Spinney, Hungarton, 686065 (60Y).

Foxton, parish, 68Z, 69V, 78EJ, 79ABF.
  Contains: village at 699897; Langton Brook, which forms the northern parish boundary; the **Grand Union Canal** joining its Market Harborough branch at the bottom of Foxton Locks.

Foxton Locks, Foxton, 692895 (68Z). A flight of locks on the Grand Union Canal rising from 103m. to 124m. with associated **side ponds**. The site of the former barge lift a short distance to the east is now covered with scrub; the area is preserved primarily for recreational purposes. Recent development as a marina and for recreation has seriously diminished the botanical interest.

Freaks Ground, Leicester, 50MS. A small former extraparochial place built over before the end of the 19th century, and situated between New Found Pool and the eastern end of the Groby Road; the West Bridge branch of the Midland Railway ran along its northern edge.

Freeboard Spinney, Lubbesthorpe, 544005, (50K). Now destroyed by road improvements.

Freeby, parish, 71UXYZ, 72QVW, 81CDEIJ, 82ABF.
  Consists of the former parishes of Brentingby and Wyfordby, Freeby, Saxby, and Stapleford.
  The former parish, 71Z, 72AB, 81E, 82B, contains: village at 803201 where the **churchyard lawns** are of interest; **meadows** alongside the brook north-east of the village; sections of **working railway** and former **Oakham Canal** which cross the southern end of the parish; the **River Eye** which, downstream of Ham Bridge, is an SSSI with habitat study 77 at 800183; **Freeby Wood** at 802222; spinney by Freeby Lodge at 803215 with a **pasture** to the south.

Freeholt Wood, Sapcote, 460940 (49MRS).

Freeman's Meadow, Newbold Verdon, 458045 (40M).

Freemen's Common, Leicester (St. Mary's), 586025 (50W). Formerly common grazing land outside the mediaeval town walls; became allotments and market gardens in the 19th and early 20th centuries; now largely built over. Freemen's Weir at 579029 is on the River Soar.

Friezeland, Carlton, 384035, (30W). Area of fields.

Frisby, parish, 60VW, 70ABF.
  Contains three farms one of which, Frisby House, stands on the site of the deserted mediaeval village at 703016.

Frisby Gravel Pits, Asfordby, 6918 (61Z).

Frisby on the Wreake, parish, 61XYZ, 71BCDEGHIJ.
Consists of the former parishes of Frisby on the Wreake and Kirby Bellars.
The former parish, 61XYZ, 71CDE, contains: village at 695177; an important **marshland** site at The Wailes; a section of the **River Wreake**; **Kirby Bellars Gravel Pits** at 7018 and 7118; a section of **working railway** with station site at 693178.

Froanes Hill, Enderby, 535998 (59J).

Frog Hole, Groby, 518083 (50E). Once a rich marsh, now overgrown by trees.

Frog Hollow Pond, Belvoir, 818321 (83B). In Granby Wood.

Frog Island, Leicester, 580051 (50X).

Frolesworth, parish, 48Z, 49VW, 58E, 59AB.
Contains: village at 503906; Frolesworth Hill at 499898; the **Fosse Way** forming the western boundary of the parish; two small brooks which rise in the parish and flow westwards to join the River Soar.

Furze Hill, Owston and Newbold, 791078, (70Y).

G

Gaddesby, parish, 61QRSVWX, 70E, 71ABCFG.
Consists of the former parishes of Ashby Folville, Barsby and Gaddesby.
The former parish, 61RSWX, 71BC, contains: village at 689130; Gaddesby Hall and park; a section of Ashby Folville Brook with species-rich **grassland** alongside it.

Galby (see Gaulby).

Garendon, former parish, Loughborough, 41Z, 42V, 51DE, 52A.
Contains: the site of Garendon Hall now demolished, (on the site of the former Abbey), at 501198; **lake** at 500199. The Park is now largely arable but some woodland remains.

Garthorpe, parish, 81JPU, 82BFGKLQR.
Consists of the former parishes of Coston and Garthorpe.
The former parish, 81JP, 82BFGK, contains: village at 831209, with a road **verge** of interest outside the churchyard wall; Garthorpe Race Course at 840210; a section of the **River Eye** passing across the parish.

Gartree Hill Covert, Burton and Dalby (Little Dalby), 763144 (71S).

Gartree Road, 60BFGK, 69UZ, 79DEHIMRSW, 89B.
See also Via Devana; the modern name given to that part of the Roman road from Leicester to Godmanchester which lies in Leicestershire. Green lane in places, elsewhere metalled.

Gaulby, parish, 60VW, 69Z, 70A, 79E.
Contains: village at 694010; some small spinneys and brooks; a little **rough grassland** at 699006 by a brook.

Gee's Lock, Glen Parva, 557990 (59P). On **Grand Union Canal.**

Gelscoe Plantation, Isley cum Langley, 430228 (42G).

Gelsmoor, Worthington, 416183 (41E). An area of small fields of interesting **grassland**.

Gibbet Hill, formerly in Cotesbach, 528807 (58F). Junction of roads A5 and A426, now in Warwickshire.

Gilmorton, parish, 58NPSTUYZ.
Contains: village at 570878; section of **dismantled railway**; M1 motorway; part of former Bruntingthorpe airfield; **flooded gravel pit** at 582885.

Gilroes, 50MST. Former extraparochial place, transferred to Leicester County Borough in 1933. The area is now mainly occupied by a cemetery, to which it has given its name, and a hospital.

Gilwiskaw Brook, 31FKLMNPR. Rises in the Pistern Hills (Derbyshire) and flows south past Smisby, entering Leicestershire near Old Parks House; passes through Ashby de la Zouch and Packington, under the former course of the Ashby Canal near Ilott's Wharf and joins the River Mease at 336100.

Gisborne's Gorse, Charley, 470154 (41S). Nineteenth century planting forming an important part of Charnwood Lodge Nature Reserve and SSSI.

Glebe Farm, Hoby with Rotherby, 668172 (61T). **Fishpond** at 667173.

Glenfield Frith, 50ST. Formerly an extraparochial place, incorporated into Glenfields parish in 1936.

Glenfields, parish, 50HIMN.
A new parish created in 1936 from Glenfield, Glenfield Frith, Kirby Frith and parts of adjacent parishes. Contains: former village centre at 538060; some open country to the west; Rothley Brook flowing from south-west to north-east past Glenfield Mill at 535061; a section of dismantled railway, with station site at 543064.

Glen Gorse, Oadby and Wigston (Oadby)/Great Glen, 643995 (69P). Glen Gorse Golf Course is at 633986. Curiously, named Glen Course on some OS maps.

Glen Parva, parish, 59UYZ.
Urban fringe parish about one third of which was transferred to Leicester in 1966. Contains: much urban development centred round the old hamlet; the **Grand Union Canal** which runs parallel to the southern boundary; the **River Sence** which actually forms the boundary, joining the River Soar on the western boundary; **grassland** near the canal and the River Soar; a working railway crossing the south-east corner and a **dismantled railway** crossing the western end of the parish.

Glooston, parish, 79MNT.
Contains: village at 748957; **Glooston Moat** at 747959 which is a partly overgrown swamp; course of **Gartree Road** close to the village; **Glooston Wood** at 754967, partly clear felled and containing areas of scrub. Marsh referred to as 'Glooston Bog' at 751966 has now gone.

Goadby, parish, 79NPTU.
Contains: village at 750988; **Keythorpe Wood** at 760985.

Goadby Marwood, former parish, Eaton, 72MNSTXY.
Contains: village at 769263 with interesting **wall** flora; Goadby Hall at 779264 with **lakes, spinney** and **brook**; Goadby Gorse at 785257.

Gopsall, former parish, Twycross, 30IMNT.
Contains: site of Gopsall House now demolished at 346057; *Gopsall Wood* at 338064; most of **Gopsall Park** including various spinneys and coverts, a lake at 359065, duck pond at 360067 and former kennels at 343068. It was well known to Rev. A. Bloxam in the 19th century and remains of some botanical importance.

Gopsall Wharf, Snarestone/Shackerstone, 346082 (30P). On **Ashby Canal**.

Gorse, The, Cadeby, 417024 (40B).

Gorse Covert, Loughborough (Thorpe Acre and Dishley), 512207 (52A).

Gorse Farm, Hoton, 588228 (52W).

Gorse Spinney, Chilcote, 279117 (21QV).

Gorse Spinney, Wistow (Newton Harcourt), 634977 (69I).

Grace Dieu Brook, 41HIJP, 42KQ. Formed at City of Three Waters, Whitwick, from brooks rising on Bardon Hill and north of Coalville. It flows north past Thringstone, **Grace Dieu Wood** and **Grace Dieu Priory** to Belton and then turns east, joining the **Black Brook** near Shepshed.

Grace Dieu Manor, Belton (Thringstone) 437179 (41I). Now a school; some of the **grassland** is moderately rich. **Marsh** at 440181 is part of the Grace Dieu and High Sharpley SSSI.

Grace Dieu Manor Farm, Belton (Thringstone), 436181 (41J). **Pond** at 437182.

Grace Dieu Priory, Belton (Thringstone), 435183 (41I). Ruins.

Grace Dieu Wood, Belton (Thringstone), 434176 (41I). Part of the Grace Dieu and High Sharpley SSSI. Habitat study 6 is at 435175.

Granby Wood, Belvoir (Knipton), 819320 (83ABFG). Both the **wood** and the **marsh** at its northern end are of botanical importance.

Grand Union Canal, SP 68, 69, 78, 79, SK 51, 52, 53, 61. Strictly speaking the name belongs to the section of the canal system joining Foxton Locks with the Grand Junction Canal at Norton Junction (Northamptonshire), but the name is loosely used for the whole system including the Market Harborough branch, the Leicestershire and Northamptonshire Canal (Foxton to Leicester) and the canalised River Soar (Leicester to the River Trent) in addition to the section of the Grand Union Canal which enters the county north of Welford (Northamptonshire), and the short Welford Branch. Much of the system is of high botanical interest and many individual sites are mentioned in the gazetteer under the features such as bridges, locks and wharfs; between Kilby Bridge and Foxton Locks the canal is an SSSI. Habitat study 67 is in Loughborough at 529213 and 68 is near Saddington at 669922.

Grange Farm, Oadby & Wigston, 619996, (69E). **Marshy ground** nearby at 621997 is a county trust nature reserve known as Lucas's Marsh.

Grange Lane, Garthorpe (Coston), 854217 (82K). Verges of interest as far as entrance to Hall Farm.

Grange Wood, Netherseal, Derbyshire. The parish, formerly in Leicestershire, was transferred to Derbyshire in 1897.

Granitethorpe Quarry, Sapcote/Stoney Stanton, 495936 (49W). Flooded disused diorite quarry.

Grantham Canal, SK 72, 73, 83. Abandoned for navigation in 1933; the section in Leicestershire which traverses the parishes of Bottesford, Redmile, Stathern, and Clawson and Harby retains water along much of its length and is of botanical interest throughout; special points of interest are referred to in the gazetteer under the names of features such as bridges and wharfs; habitat study 69 at 756328. The section between Harby and Redmile is an SSSI.

Gravel Hole Spinney, Burton and Dalby (Burton Lazars),784182 (71Z).

Gravel Pit Spinney, Husbands Bosworth, 651840 (68LM).

Great Bowden, former parish, Market Harborough, 78IJNPU, 79AFKQ.
The former parish formed much the largest part of what is now Market Harborough parish. Contains: village at 746888, with **walls** of botanical interest; about 6km. of the Market Harborough branch of the **Grand Union Canal**; lengths of railway line both in use and dismantled; about 8km. of the River Welland; **Great Bowden Pit**; Market Harborough Sewage Works at 758888 with adjacent marshy fields.

Great Bowden Pit, Market Harborough (Great Bowden), 743898 (78P). County trust nature reserve and SSSI; excavated during construction of nearby Peterborough railway line as a borrow pit; contains some standing water throughout the year.

Great Central Railway, 51LMNP. The last main line to be constructed in Leicestershire, opened in 1899, closed in 1969. The section from Loughborough to Rothley is privately operated but the remainder of the former line, north of Loughborough to the county boundary at Stanford on Soar (Nottinghamshire) and south through Leicester and Lutterworth, is dismantled; sites of botanical interest are referred to under the names of features such as stations, tunnels, cuttings and bridges.

Great Dalby, former parish, Burton and Dalby, 71GHILMNR.
Contains: village at 742144; Great Dalby **station site** at 734143 and a length of **dismantled railway line** which is of botanical interest in places; remains of a wartime airfield north of the village.

Great Easton, parish, 89GHIKLMNR.
Contains: village at 849933; the **River Welland** forming part of its southern border; the extreme south-western corner of **Eye Brook Reservoir**; **Great Merrible Wood**. There was formerly extensive park and heathland at the northern end of the parish, of which Great Merrible Wood is the only remnant.

Great Fenny Wood, Quorndon, 562175 (51T). This small willow holt close to the River Soar is a county trust nature reserve.

Great Fox Covert, Bagworth (Thornton), 462056 (40S). The stream and wood to the south-west at 458053 are also of interest.

Great Glen, parish, 69JMNPSTU.
Contains: village at 652977, sections of the **Grand Union Canal** and a **working railway**; site of station at 650963; the River Sence crossing the parish from north to south; *Glen Gorse* partly in the parish.

Great Merrible Wood, Great Easton, 835963 (89I). SSSI. County trust nature reserve; formerly heathland adjacent to Great Easton Park, first appears on maps as enclosed woodland in 1824, although this is open to more than one interpretation.

Great Peatling Covert, Peatling Magna, 604939 (69BC).

Great Stretton, Stretton Magna, 657004 (60K). Mediaeval village site with disused church still standing and a farm near by.

Great Wood, Bagworth (Thornton), 487073 (40Y).

Green Hill, Broughton and Dalby (Old Dalby), 695235 (62W).

Green Hill, Ulverscroft, 508130 (51B). **Scrub** and **open woodland**.

Green Hill Covert, Sutton Cheney, 388992 (39Z).

Green Lane Spinney, Shawell, 534807 (58F).

Griffydam, Worthington, 412185 (41E). Hamlet.

Grimston, parish, 61UZ, 62QRVW, 72AB.
Consists of the former parishes of Grimston, Saxelby and Shoby.
The former parish, 62QRVW, 72A, contains: village at 685219; Saxelbye Park at 690209. Grimston Gorse at 689223 is in Saxelby former parish. Grimston Tunnel is in Saxelby and Old Dalby former parishes.

Groby, parish, 40YZ, 41V, 50DEIJNP.
Contains: village at 523076; Bradgate House; **Groby Pool**; several **quarries**, both working and disused; interesting **woodland** and **grassland** sites; *Groby Mill* at 522082; *Groby Lodge* at 504077; *Groby Park Farm* at 501088; habitat study 30 at 518082 (**pasture**).

Groby Old Quarry, Groby, 519076 (50DI).

Groby Pool, Groby, 521082 (50EJ). The pool and its surroundings form an important complex of habitats which include **aquatic**, **marsh**, **siliceous grassland** and **woodland**; habitat study 9 is at 520084 (wet woodland) and 60 at 521082 (aquatic); much of the area is an SSSI.

Groby Quarries, Groby/Newtown Linford, 526083 (50J). Disused quarry.

'Groby Rifle Range', Groby, 526078 (50I). Disused **quarry**.

Gumley, parish, 68UZ, 69QV.
Contains: village at 679902; Gumley Hall, now demolished, and parkland, including several woods mentioned in HG; about 1.5km. of the **Grand Union Canal**; the southern part of **Saddington Reservoir**; a small **lake** called The Mot at 679897 to the east of Gumley Covert; **fish pond** at 681901; Gumley Wood at 682904.

Gun Hill, Charley, (Whitwick), 452169 (41N). Ruined house on **rocky outcrop** south-west of Blackbrook Reservoir. Part of Grace Dieu and High Sharpley SSSI.

Gunneries, Loughborough. Mentioned in HG. Location not determined. Possibly a corruption of 'cunneries'.

Gunsels, The, Tugby and Keythorpe/East Norton, 70Q. Spinneys named Big Gunsel at 770013, Little Gunsel at 765015 and Hardy's Gunsel at 773012.

Gwash, River, Knossington, 80D. Formed from several small brooks arising around Knossington, flows eastwards into Rutland Water.

Gwen's Gorse, Ashby Magna, 569908 (59Q).

H

Hallam's Wood, Croxton Kerrial, 846303 (83K).

Hallaton, parish, 79VWXY, 89YZ.
Contains: village at 786965; nearly 5km. of **dismantled railway**, some of which has been levelled and ploughed; old grassland at **Castle Hill** at 780967; **Hallaton Wood** at 766977 now felled and grubbed. Hallaton Spinneys at 775984 are in Tugby/Keythorpe.

Hall Farm, Garthorpe, 850213 (82K). **Pond** at 852213.

Hallgates, Newtown Linford, 541114 (51F). House. *Hallgate Wood* mentioned in HG is perhaps **Hallgate Hill Spinney**, 536114.

Hall Spinney, Hungarton, 704088 (70E).

Halstead, former parish, Tilton, 70IMNST.
 Contains: hamlet at 750057; length of **dismantled railway** with Tilton Station site at 760057 and **Tilton Cutting**, county trust nature reserve and geological SSSI, at 762053.

Ham Bridge, Freeby, 801186 (81E). **River Eye** and **marshy ground** nearby.

Hamilton, Barkby Thorpe, 643073 (60N). Deserted village site of Hamilton Old Town; Hamilton Grounds at 644070.

Hammercliffe Wood, Ulverscroft, 493123 (41W). Woodland adjacent to Hammercliffe Lodge at 495122. Part of the Ulverscroft Valley SSSI and of county trust nature reserve.

Hanging Hill, Ashby Woulds, 313167 (31D). **Old clay pit** at 312165.

Hanging Rocks. Charnwood outcrop mentioned in HG, perhaps a misnomer for Hanging Stone.

Hangingstone Hills, Woodhouse, 522152 (51H). Part of the Beacon Hill and Hangingstone SSSI. Much of the ground is a **golf course** with the Hanging Stone a named feature in the centre. There are **old quarries** at 522155 and 525150.

Hangman's Hall, Sutton Cheney, 422994 (49J). There is a spinney at 419977.

Harby, former parish, Clawson and Harby, 72NPTU, 73FGKLQ.
 Contains: village at 747313; the **Grantham Canal** which is accessible at Harby Colston Bridge at 738313 and Langar Bridge; **dismantled railway line**; **Harby Hills** at 72TU which have steeply sloping **woods** of considerable botanical interest. **Harby and Stathern Station site** is in Stathern parish.

Hardwick, Shangton, 723973 (79I). Deserted village site; the Gartree Road (Roman) crosses Stonton Brook at Hardwick Bridge 728964; Hardwick Wood is at 722978.

Hare Pie Bank, Hallaton, 784959 (79X).

Harper's Hill, Higham on the Hill, 385946 (39X).

Harrington Bridge, Lockington-Hemington, 471311 (43Q). Carries B6540 over **River Trent**; nearby **lock** and **Sawley Cut** are also of botanical interest; referred to in HG as *Sawley Bridge.*

Harris Bridge, Sheepy/Twycross, 351031 (30L). Carries A444 over **River Sence**; there is marshy ground on both sides of the bridge.

Harrow Brook, Hinckley, 39W, 49B. Formed by the union of Battling Brook and another, unnamed, and flows south-westwards across the county boundary near Harrow Farm, 398931.

Harston, former parish, Belvoir, 83FGKL.
 Contains: village at 838318; interesting **roadside verge** north of the village; quarries to the south of the village now filled in and ploughed; remains of **mineral railway** to the east at 846323; **Harston Wood** at 838302; part of the Upper and Middle **Fish Ponds** in 83G.

Hathern, former parish, Loughborough, 42VW, 52AB.
 Contains: village at 502224; **grassland** off Pasture Lane in 52B; Leicestershire bank of the canalised **River Soar**.

Hawcliff Hill, Mountsorrel, 571150 (51S). **Scrub** around Hawcliff Quarry at 572151 and the **quarry** itself are of botanical interest.

Hayhill Lane, Barrow on Soar, 51Y.

Heath End, Staunton Harold, 368212 (32Q). Hamlet.

Heather, parish, 30UZ, 31QVW, 41A.
 Contains: village at 390108; a **dismantled railway** crossing the parish from north to south, with partly flooded remains of old workings alongside it; an extensive opencast coal site; **grassland** at 388096 and 391098.

Heath Wood, Higham on the Hill, 375970 (39T). Nearby **ponds** at 373970.

Hemington, former parish, Lockington/Hemington, 42MNPTU, 43KQ.
 Contains: village at 467277; the **River Trent** bounding the parish on north side; **flooded gravel pits** in 42PU; **grassland** in old quarries at 455272. *Hemington Hole* at 458287 is now filled in.

Herbert's Meadow, Ulverscroft, 492134 (41W). County trust nature reserve; part of the Ulverscroft Valley SSSI; habitat study 27 is at 494133.

Hermitage, Ashby Woulds. Location unknown.

Hermitage Brook, Loughborough, 51JP, 52K. Rises near the Grammar School and flows east and then north into the River Soar.

Herring Gorse, Sproxton (Saltby), 865271 (82T).

Heyday Hays, Newtown Linford, 502101 (51A). Mixed woodland.

Heywoods Pit, Ashby Woulds. Derelict mining area at Moira, location unknown.

Higham on the Hill, parish, 39MNSTUWXZ, 49C.
 Contains: village at 382955; disused airfield (now a motor vehicle proving ground); about 4km. of **dismantled railway** and 2.5km. of a line which was never used; about 3.8km. of the **Ashby Canal** which forms part of the eastern parish boundary; **Higham Gorse** at 388943; Higham Grange **pools** at 392945; Higham Thorns at 397945. The parish has been subjected to considerable boundary changes since the publication of HG, losing a 200 metre wide strip to Warwickshire along the A5 from the A45 junction to the A444 junction (about 5km.) and gaining nearly 2 sq. km. of Stoke Golding parish when the latter was absorbed into Hinckley.

High Bridge, Wistow, 643966 (69N). Farm bridge over **Grand Union Canal**.

High Cademan, Coalville (Whitwick), 441169 (41N). Highest point in **Cademan Wood**. Part of the Grace Dieu and High Sharpley SSSI.

High Cross (Venonae), Sharnford, 472886 (48U). Site of Roman settlement at junction of Watling Street and Fosse Way.

Highfields, Leicester (St Margarets), 50W. Long since built over.

Highfield Spinneys, Tugby and Keythorpe, 768984 and 769986 (79U).

Highfields Spinney, Burton and Dalby, (Burton Lazars), 786168 (71Y).

Highland Farm, Sharnford, 475903 (49Q). There are spinneys at 472901.

Highland Spinney, Blaston, 818963 (89D).

High Leys, Belvoir (Knipton), 813325 (83B). Former heathland, quarried and restored to arable farming.

High Sharpley, Coalville (Whitwick), 447170 (41N). **Rocky outcrop**, one of the most extensive remaining areas of **heathland** in Charnwood Forest. SSSI. Habitat study 42 is at 448171.

High Tor, Charley, 41M. Presumably in the vicinity of High Tor Farm, 459155; *High Towers* is also mentioned in HG.

Hill Hole, Markfield, 485103 (41V). Water filled quarry.

Hill's Bridge, Shackerstone, 363073 (30T). Field bridge over **Ashby Canal**.

Hill Tamborough, Gaulby/Frisby, 708004 (70A). Woodland.

Hill Top Quarry, Bardon. Presumably part of the Bardon Hill quarry complex, 41L.

Hinckley, parish, 39WXY, 49ABCDEFGHIJKLMNPT UYZ.
Consists of the former parishes of Hinckley, Burbage, Barwell, Earl Shilton and most of Stoke Golding. The former parish, 39WX, 49BCDGHIMN, contains: town at 427938; **Ashby Canal**; short length of dismantled railway; working railway with Hinckley Station at 427932.

Hoby, former parish, 61NPSTUYZ.
Contains: village at 669173; site of **old mill** at 674173; **fish pond** at 667173; c. 2 km. of the **River Wreake**; parts of **two ox-bows** at 672164 and 673166.

Hoby with Rotherby, parish, 61MNPSTUXYZ, 62KQ.
Consists of the former parishes of Hoby, Rotherby, Brooksby and Ragdale.

Hogue Hall Spinney, Hinckley (Burbage), 460910 (49KQ).

Holling Hall Wood. One of the Charnwood woods mentioned by Pulteney in the 18th century. Thought to be the wood now known as **Holywell Wood**.

Holloway Spinney, Gumley, 672902 (69Q). *Holloway's Plantation* is mentioned in HG.

Holly Hayes Wood, Coalville (Whitwick), 443154 (41M). Habitat studies 56 of nearby **pond** at 444152 and 26 of **grassland** at 447151 which is Coalville Meadows SSSI.

Holly Walk Spinney, Cotesbach, 541825 (58L).

Holt Wood, Nevill Holt, 821938 (89BH).

Holwell, former parish, Ab Kettleby, 72GHLMN.
Contains: village at 735236; **grassland** adjacent to Landyke Lane at 743251; **dismantled mineral line** crossing the parish from north to south, part of which is a county trust nature reserve; **Brown's Hill Quarry** and **Railway Cuttings** SSSI; North Quarry. Holwell Mouth is in Ab Kettleby former parish.

Holwell Mouth, Ab Kettleby (Ab Kettleby), 725243 (72H). Contains the **spring source** of the **River Smite**, **marsh** and **scrub**. It is part of the Holwell Mouth SSSI, which also contains: adjacent **grassland**; **Fox Holes**; Holwell Mouth Covert at 723247 which is in Clawson and Harby.

Holwell Works, Asfordby (Welby), 726200 (72F). **Spoil heaps** at 730207.

Holywell Wood, Loughborough (Garendon), 507182 (51E). Habitat study 8 is at 507182. *Holywell Haw,* location unknown, is mentioned in HG.

Home Barn Farm, Sapcote, 489919 (49V). There are **marshy fields** between the farm and the Fosse Way.

Home Covert, Loughborough (Shepshed), 492192 (41Z).

Honey Pot Plantation, Garthorpe, 831224 (82G). **Rides** are of some botanical interest, as are the **marshy fields** and **stream** between this and Strifts Plantation.

Hooborough Brook, 21W, 31BC. Forms part of the county boundary from Swains Park (Ashby Woulds) to the River Mease below Acresford; Netherseal parish to the west of the brook was in Leicestershire until the boundary revisions of 1892 and there are some records from that parish in HG.

Hoo Hills, Sheepy (Wellsborough), 375034 (30R).

Hookhill Wood, Loughborough (Shepshed), 455191 (41P). There is an un-named **spinney** beside **Black Brook** nearby at 459189.

Hook's Bridge, Market Bosworth, 388012 (30V). Field bridge over **Ashby Canal**.

Hoothill Slang, Skeffington/Tugby and Keythorpe, 767024 (70R). **Wooded banks** of tributary of the Eye Brook; part of the Leighfield Forest SSSI.

Hoothill Wood, Skeffington, 763024 (70R). Part of the Leighfield Forest SSSI.

Horninghold, parish, 79Y, 89DEIJ.
Contains: village at 807970; Horninghold Wood at 814973; several small woods.

Horseclose Spinney, Hallaton, 772960 (79ST).

Horse Leys Wood, Burton on the Wolds, 607210 (62A).

Hose, former parish, Clawson and Harby, 72IJNPT, 73AFK.
  Contains: village at 736292; **Grantham Canal**; **Hose Brook** rising at **Piper Hole** and flowing north west to join Dam Dyke near the canal; the **dismantled railway** from Melton to Bottesford emerging from the tunnel near the **site of Long Clawson and Hose station**, with habitat study 94 at 746266. The **marlstone escarpment** forms the south-eastern boundary of the parish with **grassland**, **woodland** and **scrub**.

Hose Gorse, Clawson and Harby (Hose), 720297 (72J).

Hose Hill, Freeby, 796160 (71Y).

Hoton, parish, 52KLQRW.
  Contains: village at 574225; Hoton Hills at 559223; old quarries at 570232; part of a disused wartime airfield; **King's Brook** which forms its northern boundary.

Houghton on the Hill, parish, 60LQRSWX, 70BC.
  Contains: village at 676032; a few small brooks and woods. In the 1960s there was large-scale hedge removal in an area north of the A47 and rich hay meadows in 6702 were destroyed in the 1970s.

Hovel Hill, Westrill and Starmore, 591806 (58V). In Stanford Park.

Hubbard's Spinney, Frisby, 716014 (70A).

Hugglescote, former parish, Coalville, 41ABFGKLQ.
  Contains: village at 427127; the hamlet of Donington le Heath and the village of Whitehill; mill pond at 425125; several kilometres of working and **dismantled railway** with station site at 423121; several working and disused collieries.

Humberstone, former parish, Leicester, 60CDHIJ.
  Contains: former village at 626059; the Humber Stone at 624070; Humberstone Hall is mentioned in HG.

Huncote, parish, 59DEIJ.
  Contains: village at 517973; Huncote Bridge over Thurlaston Brook 515972; **Huncote Quarry** 512969 with habitat study 99; part of M69 motorway; *Huncote Mill* 514973; working **sandpits** at 512981, one at 514977 now restored to agriculture; The Durrells.

Hungarton, parish, 60RSTUWXYZ, 70CDE.
  Contains: village at 690072; the deserted mediaeval villages of Ingarsby, Quenby and Baggrave; **dismantled railway** crossing the parish from east to west; Hungarton Spinneys at 693072; Queniborough Brook forming the north eastern parish boundary.

Hungerford's Clump, Twycross (Orton on the Hill). Location not determined.

Hunt's Lane, Desford, 459037 (40L). Hamlet.

Husbands Bosworth, parish, 68FGHIKLMNS.
  Contains: village at 644844; Husbands Bosworth Hall at 647843 with The Shrubbery and some small lakes; 8km. of the **Grand Union Canal**; about 3km. of **dismantled railway**; the **River Avon** forming the south western parish and county boundary for about 3.5km.; the **River Welland** forming about 2km. of the eastern boundary.

Hut Spinney, Osbaston, 421049 (40H).

Hydes Pastures (now in Warwickshire), 39W, 49B. Were included in Hinckley parish (Leics.) until 1936 when they were transferred to Warwickshire; they are included in VC55.

I

Ibstock, parish, 30Z, 31V, 40EJP, 41AFK.
  Contains: village at 404095; rather more than 1km. of dismantled railway leading towards Ibstock Brickworks at 413008, which is surrounded by spoil heaps; **roadside verge** at 406088; mine at 438103. **Ibstock sidings** at 440098 are in Bagworth parish.

Iliffe Bridge, Carlton, 381044 (30X). Field bridge over **Ashby Canal**.

Illston on the Hill, parish, 69Z, 70AF, 79DEJ.
  Contains: village at 706992; a **pond** at 704990; Illston Grange at 695987; crossroads of Three Gates at 717984.

Ilott's Wharf, Measham, 347112 (31K). Formerly loading point for Measham Colliery. Ashby Canal at this point has been filled. There is a **subsidence pool** at 348110.

Ingarsby, Hungarton, 685051 (60X). Deserted mediaeval village site. Ingarsby Old Hall is at 685054, overlooking **Ingarsby Hollow** to the south. A **dismantled railway** passes through Ingarsby Station site at 688056; **Ingarsby Cutting**, formerly a county trust nature reserve, is at 665046.

Ingle Pingle, Loughborough, 531191 (51J). Flooded claypit in grounds of Loughborough Technical College Hostel.

Inkerman Lodge, Hungarton, 706075 (70D). Until recently of special interest because of the many varieties of elms in the vicinity, though few have survived Dutch Elm Disease.

Isley cum Langley, parish, 42CGHLM.
  Consists of the former parishes of Isley Walton and Langley with almost 2sq. km. of Castle Donington parish.

Isley Walton, former parish, 42GH.
  Contains: village at 424249.

Ives Head, Loughborough (Shepshed), 477170 (41T). Conspicuous rocky outcrop with remnant **heath grassland**; SSSI (geological).

J

Jane Ball Covert, Knaptoft, 629902 (69F).

Jeremy's Ground Spinney, Cotesbach, 530820 (58G).

Jericho Lodge, Freeby, 799153 (71X). The only access to the part of tetrad 81C in Leicestershire is by the public right of way through the farm.

John Ball Covert, Knaptoft, 635901 (69F).

John O'Gaunt Fox Covert, Lowesby, 742074 (70N). John O'Gaunt Station site on the **dismantled railway** at 741096 is in Somerby parish; John O'Gaunt Farm at 738097 is in Twyford and Thorpe parish.

John's Lee Wood, Ulverscroft, 507105 (51A). **Marshland** nearby at 505107 is part of the Ulverscroft Valley SSSI.

Jubilee Plantation, Ravenstone with Snibston, 390139 (31WX). One of the plantations near Alton House.

Judge Meadow, Leicester (Evington), 60G. Now part of a golf course. A local school and Community Centre bear its name.

K

Kaye's Plantation, Quorndon, 558162 (51NT). Mixed woodland and marsh on Quorn House estate.

Keepers Lodge Plantation, Newtown Linford, 538105 (51F). In Bradgate Park. Also known as Deer Park Spinney.

Kegworth, parish, 42STUXYZ.
Contains: village at 487267. The parish and county boundary to the east follow the canalised **River Soar**. The site of *Kegworth Station* is in Nottinghamshire.

Kelham Bridge, Coalville (Hugglescote), 405120 (41B). Carries A447 over the **River Sence**, alongside which are some **meadows**.

Kennel Wood, Belvoir, 829331 (83G). Woodland on the Belvoir Castle estate; the wetter parts near the lake are of botanical interest.

Keyham, parish, 60STXY.
Contains: village at 670065; a section of dismantled railway with **cutting** at 676055; two brooks; **The Miles Piece**, a spring-fed marsh and county trust nature reserve; Keyham Bridge crossing Barkby Brook at 679068.

Keythorpe, Tugby and Keythorpe. Mediaeval village site at Old Keythorpe, now Keythorpe Hall Farm, 766993, with a reed-choked lake; Keythorpe Hall at 767002 is surrounded by planted woodland; **Keythorpe Spinney** is at 771998. Keythorpe Grange at 774006 is in East Norton parish and **Keythorpe Wood** at 760985 is in Goadby parish.

Kibworth Beauchamp, parish, 69LMRSVW, 79AB.
Contains: Kibworth Beauchamp and part of Kibworth Harcourt villages with church at 685941; two stretches of working railway line; Kibworth Station site (dismantled) at 683939; a short length of the **Grand Union Canal** with Kibworth Top Lock at 697944.

Kibworth Harcourt, parish, 69MSTWX.
Contains: most of Kibworth Harcourt village; two lengths of working railway line; 2km. of the **Grand Union Canal** with Pywell's Lock at 661949, Second Lock at 660946, Kibworth Bridge and Taylor's Turnover Lock at 662948; Kibworth Hall (now a school) at 689952. **Kibworth Rifle Range** at 695948 is partly in Carlton Curlieu parish.

Kicklewell Spinney, Laughton, 656876 (68N).

Kidger's Pond, Worthington, 410181 (41E). Subsidence pond on un-named brook.

Kilby, parish, 59XY, 69BCDGHI.
Consists of the former parishes of Kilby and Foston. The former parish, 69CDHI, contains: village at 619956. Kilby Pit is in Oadby and Wigston.

Kilby Bridge, Oadby and Wigston, 610970 (69D). Hamlet. Kilby Bridge Lock is at 604970; there is a flooded limestone quarry, known as **Kilby Pit**, on the north side of the canal at 615968 and another, Barn Pool, on the north side of the railway. Kilby Bridge Farm is at 603973 and there is a **pond** and **marsh** at 601976.

Kilby Bridge, Oadby and Wigston/Kilby, 610967 (69D). Carries A50 over River Sence.

Kilby Canal Bridge, Oadby and Wigston, 609969 (69D). Carries A50 over the **Grand Union Canal**.

Kilworth Wharf, Husbands Bosworth, 627836 (68G). On **Grand Union Canal** and A427.

Kimcote and Walton, parish, 58STXYZ, 68CDEHI.
Contains: Kimcote village at 585865; Walton hamlet; headwaters of the **River Swift**.

Kinchley Hill, Rothley, 561139 (51R). Kinchley Lane leads from Broad Hill over Kinchley Hill to the edge of Swithland Reservoir; habitat study 33 is of **siliceous grassland** at 569145.

King Dick's Hole, Sheepy (Sheepy Magna), 314991 (39E). On **River Anker** near its junction with the **River Sence**.

King Lud's Entrenchments, Croxton Kerrial, 862280 (82PTU). Earthworks on the south side of Cooper's Plantation with **limestone grassland** and **scrub**. Habitat study 98 is of an arable field nearby at 863278. County trust nature reserve of this name is in Sproxton. Part of the King Lud's Entrenchments and The Drift SSSI.

King Richard's Well, Sutton Cheney, 401999 (49E).

King's Bridge, Hoton, 574233 (52R). **Old quarries** at 570232.

King's Bridge, Market Bosworth, 390038 (30W). Field bridge over **Ashby Canal**.

King's Brook, 52KLRW. Forms county boundary along north and north-west sides of Hoton parish, joining the River Soar at 543217.

King's Lock, Leicester (Aylestone), 567007 (50Q). On Grand Union Canal.

King's Mills, Castle Donington, 417274 (42D). Hamlet by **River Trent**; habitat study 71 at 418278.

King's Norton, parish, 60QRVW, 69UZ.
Contains: village at 689005. *Norton Gorse* is at 683011.

King Street Lane, Sproxton/Garthorpe/Buckminster. Green lane with moderately rich grassland flora, running from 822245 to 859223.

King's Wood, Belvoir, 824319 (83FG). Belvoir Estate woodland.

King William's Bridge, Anstey, 556089 (50P).

Kirby Bellars, former parish, Frisby on the Wreake, 71CDEHIJ.
Contains: village with church at 717182; mediaeval village site at Kirby Park, 720175; old Priory site at 717184; large pond at Kirby Gate, 718172. **Kirby Bellars Gravel Pit** at 7018 and 7118 is in Frisby on the Wreake former parish.

Kirby Frith, Braunstone/Glenfields, 50HM. Former extraparochial place which became part of Glenfields parish in 1936. Coincides closely with the present day golf course.

Kirby Muxloe, parish, 40VW, 50ABCGH.
Contains: village at 520046; *Kirby Muxloe Castle* at 523045; Kirby Muxloe golf course at 519036; the housing estates of Leicester Forest East; a length of **working railway**, with Kirby Muxloe Station site at 521035. The parish lost a small area to Ratby and gained most of Leicester Forest East in 1935.

Kirkby Mallory, former parish, Peckleton, 49JPU, 40FGKLQR.
Contains: village at 454003; Mallory Park (now a motor racing circuit) with a lake at 450006; **Kirkby Moats** at 453018; **woods** and **pools** at 452017; **Kirkby Wood** at 446999, largely destroyed but still retaining some botanical interest.

Kite Hill, Charley, (Whitwick) 457161 (41N). The site of Mount St. Bernard Abbey; **Kite Hill Plantation** is at 458168.

Klondyke, Thurcaston (Cropston). Formerly a local name for the road between 556110 and 560105.

Knaptoft, parish, 68EIJ, 69F.
Contains: deserted village site of Knaptoft, with the remains of the church at 627895. The pond at 635881, known as **Knaptoft Pond**, is in Mowsley parish and is a county trust nature reserve.

Knighton, former parish, Leicester, 50VW, 60ABF.
Transferred to Leicester in two stages, 1892 and 1935, with the exception of a small portion of Stoneygate which became part of Oadby UD in 1935. Between 1892 and 1935 the southern half of the parish was administered by Lubbesthorpe. Contains: former village with church at 599012; **Knighton Spinney** at 605008; both the Saffron Brook and a working railway crossing the parish from north to south; **Knighton Cutting** at 589024; *Knighton Clay Pit,* now gone; *Knighton Grange Farm*. The former parish is largely urbanised with a few open spaces and recreation grounds.

Knight's Bridge, Glen Parva, 576983 (59U). Bridge over the **Grand Union Canal**.

Knight's Bridge, Oadby and Wigston, 596972 (59Y). Bridge over the **Grand Union Canal**.

Knight's End, Market Harborough (Great Bowden), 745885 (78P).

Knight Thorpe, Loughborough (Garendon). Former parish, finally incorporated into Loughborough in 1902.

Knipton, former parish, Belvoir, (83ABFG).
Contains: village at 824311; the northern end of **Knipton Reservoir**; a large part of the Belvoir Estate, including many named **woods**; part of the **Upper** and **Middle Fishpond** in 83G.

Knipton Reservoir, Belvoir/Croxton Kerrial, 817304 (83A).

Knob Hill, Horninghold, 821980 (89J).

Knoll and Bassett House, Thurlaston, 494999 (49Z, 40V). Former parish, transferred to Thurlaston in 1909. Knoll Spinney is of little botanical interest.

Knossington, parish, 70YZ, 71V, 80DEJ, 81ABF.
Consists of former parishes of Knossington and Cold Overton.
The former parish, 70YZ, 80DE, contains: village at 800087 with interesting **walls**; Knossington Grange at 799086 (now a school) with planted woodland and two fish ponds; **Lady Wood**; **Tampion's Coppice**, sometimes known as Cheseldyne Spinney.

L

Lady Hay Wood, Groby, 516085 (50E).

Lady Wood, Knossington (Knossington), 815079 (80DE). Ancient woodland.

Lake Spinney, Burton and Dalby (Little Dalby), 765141 (71S).

Landfield Spinney, Burton and Dalby (Little Dalby), 772132 (71R).

Landyke Lane, 72BGHLM. Unfenced road on line of ancient trackway from Six Hills to Belvoir.

Lane's Hill, Stoney Stanton, 493941, 49P.

Langar Bridge, Clawson and Harby (Harby), 743315 (73K). Road bridge over **Grantham Canal**.

Langham Bridge, Narborough, 531970 (59I). Bridge over River Soar.

Langham's Bridge, Oadby and Wigston (Wigston Magna), 620962 (69I). Field bridge over **Grand Union Canal**.

Langley, former parish, Isley cum Langley, 52GHLM.
Contains: Langley Priory at 433235 with **fish ponds**; several woods. *Langley Gorse* has disappeared.

441

Langton Brook, 69QRVW, 79ABFK. Formed by the union of brooks below Saddington Reservoir dam; flows east by Smeeton Westerby and the Langtons to join the River Welland near Welham sidings; it is gradually revegetating after comprehensive 'improvement' in 1960s by Anglian Water Authority; some of its **tributaries** are of botanical interest.

Langton Brook Plantation, Foxton, 710921, 79B.

Langton Caudle, Stonton Wyville/Welham, 744937 (79GHLM). Conspicuous hill feature, botanical interest diminished in recent years.

Larch Spinney, King's Norton, 677017 (60Q).

Laughton, parish, 68NPTU, 69KQ.
Contains: village at 659891; **Laughton Hills**, 68NTU (partly in Theddingworth), a partly wooded escarpment.

Launde, parish, 70STWX, 80BC.
Contains: Launde Abbey at 797043, with surrounding parkland; **Launde Big Wood** at 786037, a county trust nature reserve and SSSI, habitat study 16 at 785035; **Launde Park Wood** at 803037, recently partly felled and replanted but still of considerable botanical interest; the **River Chater** flowing through the parish, with species-rich **marsh**, **pasture**, and **scrub** alongside it, much of it forming the Chater Valley SSSI at 804044; **fish ponds** beside the river at 799047 and 800046.

Lawn Wood, Groby, 506094 (50E). Contains a pool at 505093.

Laxton's Covert, Freeby, 807165 (81D).

Lea Grange Farm, Twycross (Orton on the Hill), 322054 (30H).

Lea Wood, Ulverscroft, 502115 (41V, 51A). Habitat study 51 was made at 505114 of **marshy ground** (county trust nature reserve) near the wood; **Lea Lane** running along the northern boundary of the wood has **verges** and **hedgerows** of botanical interest.

Leesthorpe, Somerby. Hamlet. Leesthorpe Hall at 791135 has **pools** and **willow holt**. Lower Leesthorpe is farm at 795146; Leesthorpe Hill is on A606 at 787143.

Leicester, SP59, SK50, 51, 60.
Borough established by the Local Government Act of 1972 supersedes the County Borough (1888) which was enlarged in two main stages in 1935 and 1966. The modern borough consists of (a) Leicester borough as delineated in 1835 and (b) the added parishes and extra parochial places of Anstey (part), Aylestone, Belgrave, Beaumont Leys, Birstall (part), Braunstone (part), Braunstone Frith, Evington, Freaks Ground, Gilroes, Glen Parva (part), Humberstone, Knighton, Leicester Abbey, Leicester Frith, Lubbesthorpe (part), New Found Pool, New Parks, Scraptoft (part), Thurcaston (part), Thurmaston (part) and Thurnby (part). Each of these units is considered under its own name in the gazetteer.

Leicester (Borough established in 1835) 50MQRSWXY, 60BCD.
Consisted of two large parishes, St. Mary and St. Margaret and four small ones, St. Martin, St. Nicholas, All Saints and St. Leonard. The following open spaces of varying degrees of botanical interest lie within the old borough; Abbey Park; Victoria Park; allotment gardens at 563037; cemetery at 591031; Freemen's Common, formerly of interest but now almost wholly built over; the River Soar flowing across the middle of the borough from south to north, its main channel part of the Grand Union Canal system. The borough contains about 9km. of working railway with a main station and several goods yards, and traces of two dismantled railways, mainly on viaducts. Leicester Castle Mound at 582041 is no longer of botanical interest.

Leicester Abbey, former parish, Leicester, 50STUXYZ. The parish is almost completely built up. The Abbey itself at 585060 is set in a formal garden bordering the River Soar. The **walls** are of botanical interest.

Leicester Bridge, Melton Mowbray, 750191 (71P). Carries the A607 over the **River Eye**.

Leicester Forest East, Kirby Muxloe, 50BG. Former parish, transferred to Kirby Muxloe in 1935, now largely built over.

Leicester Forest West, parish, 40VW, 50A. The smallest administrative unit in Leicestershire, containing a few farms and **Old Brake**, a small woodland nature reserve.

Leicester Frith, Leicester, 50NT. Former parish, dissolved in 1936; containing Glenfrith Hospital and Leicester Frith Farm.

Leicester Hill, Newtown Linford, 511111, (51A).

Leighfield Forest, SK70, 80. Name of SSSI along the Eye Brook valley comprising **Tilton Wood**, **Skeffington Wood**, **Tugby Wood**, **Loddington Reddish** and other smaller **woods**. It should not be confused with the former Norman hunting preserve of Leighfield or Lyfield Forest which occupied most of south Rutland and only overlapped into Leicestershire by a very small amount at Withcote.

Leire, parish, 58EJ, 59AF.
Contains: village at 525900; about 2km. of **dismantled railway** including **Leire Cutting** at 519897; several unnamed brooks which rise in the parish.

Lewin Bridge, Ratcliffe on the Wreake, 622129 (61G). Carries the Fosse Way (formerly A46) over the River Wreake. The marshy fields nearby have been drained.

Life Hill, Billesdon, 722049 (70H). Spinney at 716041.

Limby Hall, Swannington, 408162 (40D).

Limekiln Bridge, Hinckley (Hinckley), 411923 (49B). Carries A5 over Ashby Canal.

Lin, **River**, 41WX, 51ABF, 50EJ. Rises in the Ulverscroft Valley and runs through Bradgate Park into Cropston Reservoir.

Lindley House, Higham on the Hill, 365960 (30ST). **Ponds** at 361969; **lakes** at 365957; several **spinneys** west of Lindley Park; **Lindley Wood** at 363969.

Lindridge, Desford, 469048 (40S). There is a **spinney** and a **brook** at 466044; **Lindridge Wood** is at 471043; there is a **moat** at 471047.

Ling Hill, Newtown Linford, 521125 (51G).

Lings Covert, Croxton Kerrial, 809274 (82D). **Old grassland** on its eastern boundary.

Little Beeby, Beeby, 666076 (60T).

Little Bowden, former parish, Market Harborough, 78HIMNPT.
Formerly a Northamptonshire parish, transferred to Leicestershire in 1892 (hence it is in VC38). Contains: village at 740869; *Little Bowden Pool* (claypit) at 743877 now gone; River Welland (south bank); nearly 4km. of working **railway line** and 1.5km. of dismantled line.

Little Dalby, former parish, Burton and Dalby, 71RSWX.
Contains: village with the parish church in the Park at 774136; **Little Dalby Lakes** at 766162 which are mostly silted up. Habitat study 89 was made of a **hedgerow** at 763126.

Little John, Markfield, 501083 (50E). Rock outcrop.

Little Markfield, Markfield, 483098 (40Z).

Little Merrible Wood, Stockerston, 831966 (89I).

Little Moor Lane Bridge, Loughborough, 545196 (51P). Over **Grand Union Canal**.

Little Orton, Twycross (Orton on the Hill), 313059 (30C). Hamlet.

Little Owston, Owston and Newbold, 70Y, 80D. Name sometimes used for the eastern end of Owston Wood; habitat study 18 is at 798066.

Little Stretton, parish, 69PU, 60Q.
Contains: village at 668002; the eastern River Sence flowing southwards through the parish.

Littlethorpe, Narborough, 541968 (59N). Hamlet, much built up in recent years.

Little Twycross, Twycross, 337053 (30H). Hamlet.

Littleworth, Ashby Woulds, 309178 (31D). Mining hamlet and farm.

Lockington-Hemington, parish, 42MNPTUZ, 43KQV.
Consists of the former parishes of Lockington and Hemington.
The former parish of Lockington, 42MNTUZ, 43QV, contains: village at 468279; **Lockington Marshes** SSSI, 42Z, 43V; the **River Soar** and the **River Trent** marking the county boundary; railway line to Castle Donington power station; M1 motorway.

Loddington, parish, 70QRW, 80AB.
Contains: village at 786020; **Loddington Reddish** at 775022, woodland, part of Leighfield Forest SSSI; the **Eye Brook** with site of *Loddington Mill* at 780015; a length of **dismantled railway line** with **bridge** at 778034; lakes and wooded parkland round Loddington Hall at 790023.

Lodge Mill Spinneys, Lutterworth, (58BG). Osier beds alongside a tributary of the River Swift from 520823 to 526833.

Lodge Plantations, Cotesbach, 542819 (58FK).

Lodge Spinney, Belvoir. Location not determined.

Long Clawson, former parish, Clawson and Harby, 72CDEHIJN.
Contains: village at 722271; a section of the **Grantham Canal** with Long Clawson Bridge at 720298; several patches of scrub woodland including Clawson Thorns at 730256.

Longcliffe Golf Course, Loughborough (Shepshed/Nanpantan), 499175 (41Y).

Longcliffe Plantation, Loughborough (Shepshed), 492170 (41Y). Site of large quarry.

Longcliff Hill, Broughton and Old Dalby, (Old Dalby), 668242 (62S).

Long Covert, Shackerstone (Congerstone), 382057 (30XY).

Long Lane, Twycross. Location not determined.

Longore Bridge, Bottesford, 826365 (83I). Over the **Grantham Canal**.

Long Plantation, Rolleston, 736000 (70F).

Long Spinney, Burton and Dalby (Burton Lazars), 787178 (71Y).

Long Spinney, Cotesbach, 543822 (58L).

Long Spinney, Scraptoft, 657057 (60M).

Long Walk, Wistow, 639954 (69HM). Wood on the edge of Wistow Park.

Long Whatton, parish, 42FGKLMQRSTVWX.
Consists of the former parishes of Long Whatton and Diseworth together with parts of Shepshed and Hathern. **Piper Wood** and **Oakley Wood**, formerly in Shepshed, are now in this parish.
The former parish, 42LRSTWX, contains: village at 482233; a number of small woods and plantations; **Long Whatton Brook** 42LRSW, 52B flowing across the parish from west to east, into Hathern parish and joining the River Soar near Zouch Bridge; **roadside verge** on B5324 south-west of Piper Wood.

Lord Aylesford's Covert, Grimston, 674213 (62Q).

Lord Morton's Covert, Cold Newton, 723053 (70H).

Lord Wilton's Gorse, Asfordby, 719211 (72A).

Loughborough, parish, 41YZ, 42VW, 51CDEHIJNPU, 52ABFGKL.
Formerly an Urban District; the present administrative unit consists of the former parishes of Loughborough, Garendon, Thorpe Acre and Dishley, Nanpantan, and Woodthorpe, together with most of Shepshed and Hathern and a small part of Woodhouse; each of these has a separate reference in the gazetteer.

Loughborough, town and former parish, 51EIJPU, 52AFGK.
Most of the modern town lies within the boundaries of the former parish but the suburbs are spreading into Garendon and Nanpantan; the main areas of botanical interest are **Loughborough Moors** and **meadows** adjacent to the **River Soar** and **Grand Union Canal**; habitat study 32 was made at 540214 on **Loughborough Big Meadow**, an exceptionally large area of species-rich grassland which is an SSSI and part of which is a county trust nature reserve; two working railways, one belonging to BR and one privately operated, pass through the town with stations at 543205 and 543193 respectively. Habitat study 92 at 540220 is on a **disused railway** north of the town. **Loughborough Viaduct** at 542217 has an interesting flora, although parts of it are in Nottinghamshire.

Lount, Staunton Harold, 386193 (31Z). Hamlet. Lount Lodge and Lount Waste (Coleorton), 383187, are referred to in HG; **Lount Wood** at 379187 has had its south-eastern half (in Coleorton) destroyed by opencast mining; **Lount Meadows** SSSI consists of three fields, at 376189, 385191 and 393191, in Ashby de la Zouch, Coleorton and Worthington respectively.

Lovett's Bridge, Sheepy (Sheepy Magna/Sheepy Parva), 335022 (30G). Carries B585 over **River Sence**.

Lower Mill, Loughborough, 553205 (52K). **River Soar** and nearby **ditches**.

Lowesby, parish, 70DEIJM.
Contains: church at 723074; deserted village site at 725079; Hall at 722075 with park containing woodland and **fish ponds** mentioned in HG; **marsh** at 707086; Queniborough Brook forming the south western parish boundary; section of **dismantled railway** with station site at 733067. Until recently the area was of interest because of the variety of different elms in the hedgerows.

Lowes Mill, Sheepy. Location not determined.

Lubbesthorpe, parish, 50ABFGKL, 59EJ.
Between 1892 and 1935 the parts of Aylestone and Knighton parishes not incorporated into Leicester were administered by Lubbesthorpe; the parish contains no population centre, but only scattered farms, spinneys and brooks; junction between M1 and M69 motorways is at 546006; *Lubbesthorpe Abbey site* at 542010.

Lubcloud, Loughborough (Shepshed), 479162 (41W). Rock outcrop and woodland adjacent to Lubcloud Farm.

Lubenham, parish, 68TUYZ, 78DEIJ.
Contains: village at 705870; about 2km. of **dismantled railway line** with Lubenham Station site at 700870; 2km. of **Grand Union Canal**; 4.5km. of **River Welland** along the southern parish and county boundary. The former Lubenham airfield contains Gartree Prison at 704892.

'**Lucas's Marsh**', Oadby and Wigston (Oadby), 621997 (69J). County trust nature reserve.

Lutterworth, parish, 58BCGHLMN.
Contains: town at 542844; Lutterworth Golf Course at 545838; about 3km. of **dismantled railway** with the station site at 548845; 3km. of M1 motorway; **marsh** at 555851; several brooks; the River Swift flowing south-westwards close to the south-eastern parish boundary. The parish gained a strip of land on the west side of the A5 between 506842 and 515827 and lost a strip from the latter to 520822 when the county boundary along the A5 was adjusted.

M

Mallory Park, Peckleton (Kirkby Mallory), 40K. Former parkland, now a motor racing circuit, **Mallory Park Lakes** at 450006.

Mantle, River, 52W. Continuation westward of Wymeswold Brook along the parish boundary between Wymeswold and Hoton, to the county boundary. Thence to the River Soar it appears to be called **King's Brook**.

Maplewell Hall, Woodhouse, 522132 (51G).

Marefield, parish, 70JNPU.
Contains: farms at 745079 marking the site of the former village; **dismantled railway** crossing the parish.

Market Bosworth, parish, 30QRVWX, 40ABC.
Contains: town at 407032; Bosworth Park occupying about one third of the parish, with **lakes** and some **woodland**; the **Ashby Canal** crossing the parish from south to north and running more or less parallel to the **dismantled railway**; the site of Market Bosworth station at 392031, now being reconditioned.

Market Harborough, parish, 78DHIJMNPTU, 79FKQ.
Consists of the former parishes of Market Harborough, Great Bowden and Little Bowden, together with parts of East Farndon (Northamptonshire) added in 1936 and 1966. The former parish, 78IJN, contains: much **derelict railway property** north of the station at 741874; the Market Harborough terminus of the **Grand Union Canal** branch at 727879.

Markfield, parish, 40PUZ, 41KQRVW, 50DE.
Consists of the former parishes of Markfield and Stanton under Bardon.
The former parish, 40UZ, 41KQRVW, 50DE, contains: village at 487100; the hamlet of Copt Oak; **Hill Hole**; **Billa Barra**; **Rise Rocks**; rock outcrops; amenity woodland and shrubbery at former Markfield Hospital at 493090; M1 motorway crossing the parish form south to north, much of it in cutting.

444

Marriott's Bridge, Clawson and Harby (Hose), 725296 (72J). Carries a farm track over the **Grantham Canal**.

Marriott's Spinney, Broughton and Old Dalby (Old Dalby), 687233 (62W).

Martinshaw Wood, Ratby/Groby, 510072 (50D). Large wood, almost certainly ancient, extensively replanted but with **rides** of considerable botanical interest; has an interesting **Rubus** flora.

Mawbrook Lodge, Scalford, 750250 (72M). **Railway cutting** from here to Brock Hill tunnel.

Meadow Lane Bridge, Syston, 609119 (61A). Crosses **Grand Union Canal**.

Mealy Copse, Stoughton. Location not determined.

Mease, River, 21QRW, 31ABF. Formed by the union of brooks rising in Twycross parish; forms the boundary between Snarestone, Measham, and Oakthorpe and Donisthorpe to the north and east and Appleby Magna, Stretton en le Field and Chilcote parishes to the south and west; it leaves the county at Chilcote and flows westwards to the Trent.

Measham, parish, 31FGKL.
The parish was transferred from Derbyshire to Leicestershire in 1897 but has always been included in VC55. Contains: village at 335122, mines and industrial workings with associated **railways** partly dismantled; a length of the **Ashby Canal**, now drained and filled; subsidence pool at 348110, now drained; several **pools and claypits** in the south of the parish; a **duck decoy** at 333108; **River Mease** forming south-western boundary and the **Gilwiskaw Brook** the south-eastern boundary of the parish.

Medbourne, parish, 79VWX, 89ABC.
Contains: village at 799930; the **River Welland** forming its southern boundary and Medbourne Brook flowing through the village to join it; two dismantled railway tracks crossing the parish, with the former station site at 800936; some **old walls** in the village. *Medbourne Bridge* and *Medbourne Lodge* are mentioned in HG.

Medbourne Brook, 79STWX, 89AB. Formed from brooks rising in Tugby and Keythorpe, Hallaton, and Horninghold parishes; flows southwards through Medbourne village to the River Welland at 804915.

Melbourne Lodge, Staunton Harold, 383212 (32V). The main entrance to Staunton Harold Park.

Melton Brook, 50Y, 60DIJNSTWX, 70C. Rises at Botany Bay Covert, passing through the parishes of Houghton on the Hill, Hungarton, Keyham, Scraptoft, Beeby and Barkby Thorpe, and running along the north side of Hamilton deserted mediaeval village. Having entered Leicester it joins the River Soar at 597078.

Melton Mowbray, parish, 71IJMNPTU, 72FGHLQR.
Consists of the former parishes of Melton Mowbray, Eye Kettleby and Sysonby, together with parts of Burton Lazars, Thorpe Arnold and Welby. The former parish, 71JMNPTU, 72KLQR, contains: town at 752190; **River Eye** with a **canal cut** at 747190; Scalford Brook forming part of the northern parish boundary and entering the parish at 760217; **marshy meadow** at 746204; working railway line crossing the parish from west to east, with station at 753187 and junction at 742185 with the former line to Nottingham (still in use as a test track); the **dismantled line** crossing the parish from south to north with an old station site at 753195.

Melton Spinney, Melton Mowbray, 766223 (72R). Part is in Scalford parish.

Mere Lane, Oadby and Wigston (Oadby), 636987 (69J). Bridleway running south-west from the A6 at 638990 along the eastern edge of Glen Gorse Golf Course. The south western half runs along the parish boundary.

Mere Hill Spinney, Prestwold, 562209 (52Q).

Mere Road, Sproxton (Saltby). Part of the trackway known as The Drift.

Merril Grange, Belton, 445217 (42K).

Merry Lees, Bagworth (Thornton), 470058 (40S). Former colliery at 468058, with nearby **brook** and some **waste ground**; in HG as *Merrilees.*

Mickle Hill, Aston Flamville, 465917 (49K). Spinney on northern slope.

Middle Plantation, Walton on the Wolds, 597202 (52V). There are several spinneys in the vicinity.

Middlesdale, Belvoir, 823332 (83G). Part of Belvoir Estate, containing mixed woodland with some exotic species.

Middlestile Bridge, Bottesford, 810367 (83D). Carries minor road over **Grantham Canal**.

Miles Piece, The, Keyham, 671071 (60T). Spring-fed marsh, county trust nature reserve.

Mill Covert, Market Bosworth, 399042 (30X, 40C). Deciduous woodland in Bosworth Park.

Miller's Bridge, Loughborough, 551187 (51P). Carries bridleway over **Grand Union Canal**.

Miller's Dale, Great Glen, 658974 (69N). Parkland with a **lake**, beside a tributary of the River Sence south of Great Glen Hall.

Millfield Clump, Noseley, 729993 (79J).

Mill Hill Spinney, Burton and Dalby (Little Dalby), 772135 (71R).

Mill House Farm, Ashby de la Zouch (Ashby de la Zouch), 358156 (31M). With footbridge over **Gilwiskaw Brook** nearby.

'**Mill House Pools**', Gilmorton, 581885 (58UZ). Small water-filled sand pit used for fishing.

Mires, The, Charley, 475154 (41S). Woodland.

445

Misterton, parish, 58KLMNQRSWX, 68C.
Contains: church at 557839 with Misterton Hall and **fishpond** in **parkland**; Walcote village; the **River Swift** crossing the parish from east to west with **marsh** and **woodland** on the north side at 563842; **Misterton Marshes** SSSI at 557852 and **marsh** at 586850; **Thornborough Spinney**; Shawell Wood; a long narrow **shelter belt** (about 2km.) running from Swinford Corner 565818 to near Strawfield House 563835; lengths of M1 motorway and dismantled railway.

Moat Farm, Ratby. Possibly a name for Old Hays.

Moat Hill Spinney, Cotes, 550216 (52K).

Moat House, Loughborough, 525170 (51I). Formerly Moat Farm. **Moat**.

Moat Wood, Peckleton (Kirkby Mallory), 455018 (40K). Now largely cleared.

Moira, Ashby Woulds, 315156 (31H). Mining village, surrounded by **derelict mining sites** and **dismantled railways**. *Moira (Swain's Park) Reservoir* at 304170, formerly of great botanical interest, was drained soon after the Second World War.

Moor Barns, Sheepy (Sheepy Magna), 303027 (30B).

Moor Hill Spinneys, Tugby and Keythorpe, 780985 (79UZ).

Moor Lane Bridge, Loughborough, 546193 (51P). Carries lane across **Grand Union Canal**.

Moorley Hill Plantation. Location not determined.

Morley Quarry, Loughborough (Shepshed), 476179 (41T).

Motorways. Motorway verges, cuttings and embankment slopes are subject to a lenient management regime and can be expected to develop semi-natural vegetation as they mature. The M1 traverses the county from south to north, passing through 10km. squares 57, 58, 59, 50, 40, 41 and 42; the M6 crosses the corner of the county in 10km. square 57; the M69 enters the county near Hinckley in 10km. square 49 and terminates in a junction with the M1 in 59; the M42 when completed will pass through 10km. squares 39, 30 and 31.

Mott, Big and Little, 68U. Coverts in parkland at Gumley; the lake at 679897 is called The Mot.

Moult Hill, Charley, 465169 (41T).

Mount, The, Wistow, 649949 (69M). Mixed wood in parkland.

Mount Saint Bernard Abbey, Charley (Whitwick), 457161 (41N).

Mountsorrel, parish, 51RSTWX.
Contains: village at 580146; **Castle Hill**; **Mountsorrel Common** at 569146; **flood plain meadows**; a considerable length of the canalised **River Soar**; Mountsorrel Lock at 581152; **Mountsorrel Quarries** at 578148; **Hawcliff Quarry**; a length of **dismantled mineral railway**.

Mowmacre Hill, Leicester (Belgrave), 585083 (50Z). Area now largely built-up.

Mowsley, parish, 68IJNP, 69FK.
Contains: village at 647890; pond at 635881, known as **Knaptoft Pond**, which is a county trust nature reserve; Mowsley Hills, 68IN forming part of the watershed between the Rivers Soar and Welland.

Muckelborough Plantation, Horninghold, 825983 (89J).

Mucklegate Lane, Seagrave, (61E). Green lane leaving Gorse Lane, Seagrave at 610187 and running south-east to the village.

Mucklin Wood, Woodhouse, 537164 (51I).

Mundy's (Mandy's) Gorse, Burton on the Wolds, 642210 (62K). Formerly an area of scrub, now ploughed, just north of the botanically important site **Twenty Acre**.

Muston (former parish), Bottesford, 83CDEHIJ, 84F.
Contains: village at 829378; a short length of the **Grantham Canal** crossed by Muston Gorse Bridge at 818359; **Muston Meadows National Nature Reserve** lying immediately north of the Grantham Canal in 82F with habitat study 37 at 824365; rich **meadows** alongside the River Devon at 829377; a length of working railway line and a **mineral branch** which forms part of the county boundary on the east side of the parish; **Muston Gorse** at 824357 with some **rough grassland**.

Mythe, The, Sheepy (Sheepy Magna), 316994 (39E). Farm.

N

Nailstone, parish, 40CDEIJ.
Contains: village at 418071; Nailstone Gorse at 407074; **Nailstone Wiggs** at 425085 much reduced by coal mining, with a **wood, pools, mine tips** and **mineral railway**. Habitat study 80 (farm track) at 423084.

Naneby Hall Farm, Cadeby, 434025 (40G). **Gravel workings** nearby.

Nanpantan, former parish, Loughborough, 41Y, 51CDEHIJ.
Contains: village at 505172; Nanpantan Hill at 501169 and Nanpantan Reservoir at 507171.

Narborough, parish, 50F, 59DIJNP.
Contains: village at 540975; **Narborough Bog**; about 5km. of **railway line** with the *station site* at 541973; Littlethorpe hamlet, now much expanded; about 3km. of the **River Soar**, with **oxbows** at 525967 and 529970; short lengths of M1 and M69 motorways, the latter on a high embankment; *Narborough Quarry* at 524975, disused and water-filled.

Narborough Bog, Narborough, 549979 (59NT). SSSI and county trust nature reserve with **reedbed**, **willow woodland** and species-rich **marshy meadow**, on both sides of railway and bordered by **River Soar**, reduced by drainage and ploughing. Habitat study 13 at 550979 (woodland).

Navvy's Pit, Oadby and Wigston (Wigston Magna), 601976 (69D). Borrow pit.

Near Coton, Market Bosworth, 393023 (30W). Hamlet.

Nether Broughton, former parish, Broughton and Old Dalby, 62WXY, 72BCDH.
Contains: village at 695262; part of Army Depot at 687245; Broughton Hill at 712240; Dalby Brook forming the north-western parish boundary; River Smite forming the north-eastern parish boundary; **marlstone escarpment** along the southern parish boundary between Marriott's Spinney and **Bleak Hills.**

Nether Hall, Scraptoft, 644058 (60M).

Nevill Holt, parish, 89BCGHI.
Contains: private school and church at 816936 with some **stone walls**; **Nevill Holt Quarry** at 813932; a few small woods.

Newarke, The, Leicester (St. Mary), 583040 (50X).

Newbold, Owston and Newbold, 765090 (70U). Farm near site of deserted mediaeval village.

Newbold, Worthington, 401190 (31Z, 41E). Hamlet; **grassland** at 397189.

Newbold Verdon, parish, 40FGHKLMNS.
Contains: village at 443038; Newbold Heath hamlet at 444051; Brascote hamlet; **Newbold Spinney** at 454035.

New Covert, Chilcote, 278101 (21V).

New Covert, Melton Mowbray, 775181 (71U).

New Covert, South Kilworth, 598809 (58V). Flooded gravel pits to the west.

New Found Pool, Leicester, 50M. Former parish, transferred to Leicester in 1892 and built up in the late 19th century, to provide housing for families displaced by the building of the Great Central Railway.

Newhall Park, Thurlaston, 507003 (50A). Moat.

New House Grange, Sheepy (Sheepy Magna), 308023 (30B). Spinneys to north and south.

New Humberstone, Leicester (Humberstone), 617055 (60C). The railway station site and most of the dismantled railway have now been built over.

New Ingarsby, Hungarton, 669044 (60S). Farm.

New Inn, Rolleston, 723999 (79U).

New Parks, Leicester, 50MNST. Former parish most of which was transferred to Leicester in 1936.

New Plantation, Brentingby, 793214 (72V).

New Plantation, Quorndon, 567152 (51S).

New Plantation, Skeffington, 759030 (70L). Part of Leighfield Forest SSSI.

New Queniborough, Queniborough, 635125 (61G). On the north-west side of the A607, south of the Queniborough Brook.

New Spinney, Scraptoft. Location not determined.

New Swannington, Coalville (Swannington), 425159 (41M). Hamlet.

Newton Burgoland, former parish, Swepstone, 30PTUZ, 31Q.
Contains: village at 370089; Newton Nethercote hamlet at 367092; Newton Barn at 378091, with **ponds** at 378089; **Newton Burgoland Marshes** SSSI at 381089, habitat study 48 at 381090.

Newton Harcourt, former parish, Wistow, 69IJMNP.
Contains: village at 640967; about 2km. of working railway line; a length of the **Grand Union Canal**; Newton Bridge at 639967; Newton Bottom Lock at 631966; Newton Middle Lock at 634966.

Newtown Linford, parish, 40Z, 50EJP, 41V, 51ABFGKL.
Contains: village at 522097; **Benscliffe Wood** SSSI; **Bradgate Park** SSSI; **Cover Cloud Wood**; **Cropston Reservoir** SSSI; part of **Groby Quarries**; **Heyday Hays**; **River Lin**; **Roecliffe Spinney**; **Sheet Hedges Wood** SSSI; **Swithland Wood** SSSI; **Tangle-Trees Wood**.

Newtown Unthank, Desford, 490043 (40X). Hamlet; *Newtown Unthank Mill* is at 487043.

New Wood, Long Whatton (Diseworth), 445230 (42Z).

New York Spinney, Nevill Holt, 828942 (89H).

No Man's Heath, 290088 (20Z). Hamlet. The county boundary formerly followed the Austrey to Clifton Campville road through the village, but was diverted during the boundary changes of 1892; most of the hamlet, though in VC55, is now in Staffordshire.

Normanton, Bottesford, 811407 (84A). Hamlet; Normanton Fox Covert at 805407; **Normanton Thorns** at 820427, now almost impenetrable.

Normanton Ferry, Loughborough, 519228 (52B). Footpaths through **meadows**, converge here on the **River Soar** bank opposite Normanton Church (Nottinghamshire).

Normanton le Heath, parish, 31QRVWX.
Contains: village at 377127 and **Normanton Wood** at 391136.

Normanton Turville, Thurlaston, 492983 (49Z). Former mansion surrounded by *Normanton Park* with *woods* and *lakes*; **Upper Pool** at 491986; **Lower Pool** at 491982 and **Lower Pool Spinney**; other **spinneys** at 487984; the brook draining the pools becomes Thurlaston Brook by the weir at 493977.

Norris Hill, Ashby Woulds, 326164 (31I). Hamlet; Norris Hill Farm at 325173 is in Ashby de la Zouch parish.

North Bridge, Leicester (All Saints), 580053 (50X). Over River Soar.

North Kilworth, parish, 58WX, 68ABCGHI.
Contains: village at 615831; North Kilworth House at 603834 with a park, lake and caravan site; North Kilworth Sticks (covert) at 605845; North Kilworth Mill on the **River Avon** at 619821, about 2.5km. of dismantled railway line; nearly 1km. of the Grand Union Canal, with the **western end** of North Kilworth Tunnel; the River Avon forming about 1km. of the south-eastern parish and county boundary; an unnamed brook forming the eastern parish boundary.

North Quarry, Ab Kettleby, 743238 (72LM). County trust nature reserve and part of Brown's Hill Quarry and Railway Cutting SSSI.

North's Spinney, South Croxton. Location not determined.

Norton Gorse, Kings Norton, 682010 (60V). **Scrub woodland** and adjacent **marsh**.

Norton juxta Twycross, former parish, Twycross, 30DHIJN.
Contains: village at 323070; Norton House at 318067 with a large pond; pond by the A444 at 313069; *Norton Coverts* at 337070. The **hedgerows** along the A444 were an important source of Rev. A. Bloxam's *Rubus* records.

Noseley, parish, 79EIJNP.
Contains: the deserted mediaeval village site at 734987; Noseley Hall and church, 738985, surrounded by **wooded parkland** with several pools; Noseley Wood at 733980; several other woods. The unnamed **brook** forming the eastern and south-eastern parish boundaries flows by Stonton Wyville to become Langton Brook.

Nowell Spring Wood, Ulverscroft, 502121 (51AB). The eastern margin is of botanical interest; **grass heath** and **marsh** formerly along its eastern border was destroyed by ploughing in 1982.

Nunckley Hill, Rothley, 569142 (51S). Calcifuge flora around **quarry margin**.

Nutt's Bridge, Hinckley, 410929 (49B). Carries minor road over Ashby Canal close to a flooded clay or gravel pit.

O

Oadby, former parish, (Oadby and Wigston), 60AFG, 69EJP.
Former parish, incorporated into the new borough in 1974. Contains: the enlarged village of Oadby, 624004; '**Lucas's Marsh**'; *Leicester Race Course* at 613005; **grassland** at 613013; **Glen Gorse Golf Course** at 637990 and part of *Glen Gorse* (spinney) at 643995. There are numerous early botanical records from parts of the parish now built over.

Oadby and Wigston (Borough), 50V, 59YZ, 60AFGK, 69DEHIJP.
Consists of the former parishes of Oadby and Wigston together with small areas of Knighton and Evington parishes.

Oakham Canal, 71UZ, 81EIJN.
Originally joined the Wreake Navigation in Melton Mowbray and passed through the former parishes of Thorpe Arnold, Wyfordby and Brentingby, Freeby, Saxby, Wymondham and Edmondthorpe before entering Rutland near Teigh. The Leicestershire section is now almost completely dry, though there are still short, marshy sections near Brentingby and east of Whissendine Station; a section lies within the Wymondham Rough Nature Reserve.

Oakley Wood, Long Whatton (Shepshed), 485215 (42VW). SSSI.

Oaks in Charnwood, Charley, 473163 (41T). Hamlet with parish church. **Marshy fields** nearby along Black Brook.

Oak Spinney, Gumley, 673894 (68U).

Oakthorpe and Donisthorpe, parish, 21V, 31ABCFGHL.
The parish has a complex history of administrative changes, the land having been shared between Leicestershire and Derbyshire in a very complicated pattern up to about 1835; in 1892 this and neighbouring parishes were transferred to Leicestershire in exchange for Netherseal and Overseal and part of Woodville transferred to Derbyshire; however all the parishes involved in this exchange were included by Watson in VC55. Contains: the village of Oakthorpe at 321129; village of Donisthorpe; the hamlet of Acresford; lengths of **dismantled railway**; about 2km. of the **Ashby Canal bed**, mainly dry; **rough grassland** and **scrub** around the site of Oakthorpe Colliery; habitat study 95 (arable field) at 334138 (Pasture Farm); **subsidence pools** in Saltersford Brook.

Occupation Lane, Wymeswold, 62LM. Green lane as far as bridge over the brook which forms the new county boundary.

Odstone, former parish, Shackerstone, 30TYZ, 40DE.
Contains: village at 393078; about 1.5km. of dismantled railway; River Sence forming the western parish boundary.

Old Brake, Leicester Forest West, 490012 (40V). Wood, relic of Leicester Forest. Local nature reserve.

Old Covert, Market Bosworth, 411018 (40A).

Old Covert, South Kilworth, 605810 (68A).

Old Dalby, former parish, Broughton and Old Dalby, 62KLMQRSWX.
Contains: village at 674235; **Old Dalby Wood** at 682228 in which extensive Forestry Commission plantings were made in 1960's; c.2km. of **railway** still used as a test track, with **cutting** at north end of Grimston tunnel at 683234; interesting **marsh** and **grassland** to the east of the tunnel; several ponds and small woods; the **marlstone escarpment** with diminishing slope running through the eastern side of the parish. Two Roman roads, the Fosse Way and Six Hills Lane, form the western and southern boundaries of the parish respectively.

Old Hays Farm, Ratby, 490064 (40Y).

Old Hills Wood, Ab Kettleby (Holwell), 741228 (72L).

Old Ingarsby, Hungarton, 686054 (60X). Ingarsby Old Hall is at 685054.

Old John Tower, Newtown Linford, 525112 (51F). In Bradgate Park.

Old Mere, Oadby and Wigston (Wigston Magna), 69J. Otherwise referred to as Mere Lane. Bridleway running south-west from the A6 at 638990 along the eastern edge of Glen Gorse Golf Course. It marks the former parish boundaries of Oadby and of Wigston and the present parish boundary of Oadby and Wigston.

Old Parks, Ashby de la Zouch, 31U, 32Q.
Vaguely defined area at the north end of the parish; Old Parks House and Farm, 359189; **dismantled railway cuttings** at 364186 and 361183. Habitat study 21 (**grassland**) is at 377189. Old Parks includes part at least of **South Wood**.

Old Park Spinney, Market Bosworth, 406040 (40BC).

Old Park Wood, Belvoir, 810329 (83B). **Rides**.

Old Pond Wood, Rolleston, 731001 (70F).

Old Rise Rocks, Bardon, 468123 (41R).

Old Wood, Groby, 513090 (50E). Several disused quarries, some water-filled.

Old Wood, Prestwold, 584214 (52V).

One Ash Spinney, Quorndon, 551172 (51N).

One Barrow, Loughborough (Shepshed), 457178 (41NP). Plantation, farm and viaduct at inflow end of Blackbrook Reservoir.

Orange Hill Plantation, Shackerstone, 380063 (30TY).

Orton on the Hill, former parish, Twycross, 20W, 30BCGH.
Contains: village at 304039; Orton Gorse at 298043; a lake at Orton Hall, 306038; *Orton Plantation* at 305043; *Orton Wood* at 328050.

Osbaston, parish, 40BCDGHIM.
Contains: hamlet at 425043; Hall at 423045 with **lakes** and **spinneys**; **Osbaston Hollow** at 416061; *Osbaston Toll Bar* (cross roads) at 417056; **marsh** at 417037; **grassland** at 417044, 410049 and 407049; **roadside verge** on Deeping Lane.

Osgathorpe, parish, 41DEIJ, 42FK.
Consists of the former parish of Osgathorpe together with about half of Thringstone.
The former parish 41EJ, 42FK contains: village at 431195; habitat study 24 of **grassland** at 427188; traces of **former Charnwood Forest Canal** still retaining botanical interest in places.

Osiers, The, Braunstone, 551009 (50K). Woodland.

Othorpe, Slawston, 772954 (79S). Deserted mediaeval village site.

Out Woods, Loughborough (Nanpantan), 514163 (51CD). Ancient woodland, part planted with conifers; SSSI; public open space. Outwoods Farm is at 517165.

Overclose Spinney, Blaston, 809963 (89D).

Overton, Ibstock, 408095 (40E). Hamlet.

Overton Lodge, Castle Donington. Location not determined.

Owston and Newbold, parish, 70JPTUXYZ, 71Q, 80D.
Contains: Owston village at 774079; the site of Newbold deserted mediaeval village at 766090; **Owston Woods** comprising **Owston Big Wood** and **Little Owston Wood**, SSSI, 70Y, 80D, one of the largest and most botanically important woods in the county, with habitat studies 17 at 794065 and 18 at 798066; remains of **mediaeval fish ponds** at 773079; short section of dismantled railway in the extreme west of the parish.

Ox Brook, Thrussington, 61MNP. Rises in Hoby with Rotherby parish and flows southwards to the River Wreake at 655157; **marshy hollow** at 644186.

Oxey Farm, Loddington, 778034 (70R). By **dismantled railway** in cutting. **Blue brick bridge** with rich fern flora nearby.

P

Packington, parish, 31KLMRSTWX.
The parish includes nearly a dozen small enclaves which were formerly parts of Derbyshire; two areas totalling over 0.5 sq.km. were lost to Ashby de la Zouch when the Urban District was enlarged. Contains: village at 358145; **old grassland** in churchyard; **Gilwiskaw Brook** crossing the parish from north to south.

Paddy's Lane, Broughton and Old Dalby, 62KLQ. Part of A6006, with **verge** at 659219.

Papillon Hall, Lubenham, 688869 (68Y).

Park, The, Burton and Dalby, 763166 (71T). Site of Leper Hospital; some **old grassland**.

Park Farm, Ashby de la Zouch, 345138 (31L). 300m. west of the crossroads, some cottages adjacent to a **pool** at 344140.

Park Farm, Stretton en le Field, 298111 (21V).

Park Hill Lane, Seagrave, 623172 (61I). **Verges**, with habitat study 83 at 623172.

Park Wood, Stockerston, 826977 (89I). SSSI. Ancient woodland.

Park Wood Farm, Launde/Loddington, 803032 (80B). At southern margin of **Launde Park Wood**. The parish boundary runs through the farm.

Park Woods, Castle Donington, 42DI. In Donington Park; the name is used in HG for all the woodland in the Park.

Pasture Lane, Loughborough (Hathern), 509227 (52B). **Lane** and nearby **meadowland** and **ditches**.

Pasture Lodge, Wymeswold, 644231 (62L). Farm in the triangular area of land transferred in 1966 from Nottinghamshire which includes Hades Lane and two other farms.

Pasture Wood, Breedon on the Hill, 424213 (42F). SSSI.

Paudy Crossroads, Seagrave, 606186, (61E).

Peake's Covert, Somerby, 762100 (70U, 71Q).

Peatling Magna, parish, 59VWX, 69ABC.
    Contains: village at 595924.

Peatling Parva, parish, 58UZ, 59V.
    Contains: village at 589896; part of former Bruntingthorpe airfield with pool at 596890; marsh at 581903; Hall at 589898 with pools nearby.

Peckleton, parish, 40FKLQRV, 49DEIJPUZ.
    Consists of the former parishes of Peckleton, Kirkby Mallory and Stapleton.
    The former parish, 40KLQRV, 49UZ contains: village at 470008; *Peckleton Common* at 480013, now partly industrialised; **marshy fields** at 460020 with habitat study 49; **pool** at 452015.

Peggs Green, Coleorton, 413177 (41DE). Hamlet.

Peldar Tor, Coalville (Whitwick), 449157 (41M). Active quarry, which has destroyed former area of botanical interest.

Pepper's Farm, Quorndon. Probably that at 561175 now known as Poole Farm.

Pickering Wood, Coalville. Location not determined.

Pickwell, former parish, Somerby, 71QRVWX, 81AB.
    Contains: village at 785113; Leesthorpe hamlet; brook with adjacent **grassland** from Pickwell Lodge to Somerby village; brook with adjacent **grassland** at 800116; **flooded brickpit** at 794151.

Pillings Lock and Weir, Quorndon, 565182 (51U). **River Soar**; **willow holt** at 565185. Habitat study 75 of River Soar at 564185 is just north of the lock.

Pingle Plantation, Ravenstone with Snibston, 390144 (31X). One of the woods near Alton Grange.

Pinwall, Sheepy (Sheepy Magna), 308002 (30A). Hamlet.

Piper Hole, Clawson and Harby, 759276 (72NT). Recess in marlstone scarp; **grassland** and **scrub**.

Pipers Hole, Croxton Kerrial. Location not determined.

Piper Wood, Long Whatton (Shepshed), 477216 (42QR).

Pistern Hill, Smisby, Derbyshire, 31P, 32K. Referred to in HG although never part of Leicestershire, just beyond county boundary.

Plungar, former parish, Redmile, 73LMRSVW.
    Contains: village at 769340; **Plungar Wood** at 785322; a length of **dismantled railway**; a length of the **Grantham Canal** with Plungar Bridge at 766342.

Pochin Bridge, Wanlip/Syston, 603114 (61A). New bridge carrying A607 over **River Soar**.

Pochin's Bridge, Wigston Magna, 596974 (59Y). Field bridge over **Grand Union Canal**.

Pocket Gate, Loughborough, 529158 (51H). House on edge of **Out Woods**. Pocket Gate Farm at 521157 is nearby.

Polly Bott's Lane, also known as Lea Lane, Ulverscroft, 41B, 51A. **Verges** adjacent to **Lea Wood**.

Pond Bay, Westrill and Starmore, 589803 (58Z). Spinney with brooks and springs in Stanford Park.

Pond Spinney, Aston Flamville, 466917 (49R).

Pontylue Farm, Syston, 612109 (61A). Destroyed by **gravel workings**.

Pool House, Gaddesby (Ashby Folville), 706124 (71B). Pool at 707124.

Pop's Spinney, Rolleston, 737003 (70F).

Port Bridge, Billesdon, 703031 (70B). Carries A47 over Coplow Brook.

Port Hill, Slawston/Medbourne, 787934 (79W).

Pot Bottom, Kegworth. Location not determined.

Potter Hill, Asfordby (Welby), 735218 (72F). Farm at 731221; adjacent dismantled mineral line.

Potters Marston, parish, 49XY, 59CD.
    Contains: site of former village at 498964; 0.5km. of motorway M69; several tracks of former mineral railways; **Potters Marston Pipeworks** at 490960; 'Potters Marston Bog', formerly of botanical interest, straddled the parish boundary with Hinckley (Earl Shilton) at 483965 and was destroyed during the construction of the M69 motorway.

Potter's Wood, Netherseal, 274149. Now in Derbyshire.

Poultney Wood, Ulverscroft, 494130 (41W). Part of Ulverscroft Nature Reserve, part in Ulverscroft Valley SSSI.

Preston Lodge, Knossington, 803066 (80D). On south-eastern edge of Little Owston Wood.

Prestop Park, Ashby de la Zouch, 345174 (31I). *Prestop Park Farm* is at 339178.

Prestwold, parish, 52KQRVW.
    Contains: village at 577214 with church in **grounds of Prestwold Hall**; **Park Plantation** at 576214; **Prestwold Park**, partly wooded and partly arable.

Priest Hill, Skeffington, 760029 (70R). Woodland, part of Leighfield Forest SSSI.

Primethorpe, Broughton Astley, 522933 (59G). Hamlet continuous with Broughton Astley.

Prince of Wales Covert, Hungarton, 692090 (60Z).

Princethorpe, Enderby. Location not determined.

Prior Park, Packington. Location not determined.

Privets, The, Loughborough (Shepshed), 498172 (41Y). Part of the Longcliffe group of woodlands.

Proctor's Pleasure Park, Barrow on Soar, 569167 (51T). Caravan park and adjacent **flooded gravel pits**. The latter, together with adjoining grassland, make up Barrow Gravel Pits SSSI.

Providence House, Ibstock, 427095 (40J). **Pond** at 423094.

Proving Ground, Higham on the Hill, 39STY (mainly in 39T). Sited on **former airfield.**

'**Puddledyke**', Thurcaston (Cropston), 547116 (51K). Local name for Cropston Brickpit.

Pywell's Lock, Kibworth Harcourt, 661949 (69S). **Grand Union Canal.**

Q

Quaker Plantation, Ravenstone with Snibston, 386143 (31X). One of the woods near Alton Grange.

Quakesick Spinney, Leicester (Humberstone), 629069 (60I).

Quarry Hill Plantation, Castle Donington, 419275 (42D).

Quenby Hall, Hungarton, 702064 (70D). The adjacent *Quenby Park* contains the site of the deserted mediaeval village of Quenby.

Queniborough, parish, 61FGKLQR.
Contains: village at 650120, much enlarged with new housing; a working railway line; parts of **Ashby Folville** and **Queniborough Brooks** with **adjacent grassland**; a short length of the **River Wreake** forming the western parish boundary.

Queniborough Brook, 60DEI, 61KLQV. Formed from tributaries rising in Cold Newton and Lowesby; follows parish boundaries between Lowesby, Hungarton and South Croxton, entering the last and passing through Queniborough to join Ashby Folville Brook and flow into the River Wreake. There is interesting **grassland** adjacent at intervals throughout its length.

Quorndon (otherwise known as Quorn), parish, 51MNPSTU.
Contains: village at 561166; **Quorn Park** and Quorn House, 561161, with a **lake** at 559161; **Buddon Wood**, SSSI, **River Soar** forming the eastern parish boundary, canalised as far as Pilling's Lock where the Grand Union Canal branches off as a separate cut; **Buddon Brook** flowing out of Swithland Reservoir with **grassland** on its east bank, passing through **Quorndon Mill** at 556159 to join the River Soar at 565165; **pool** at 556169; **marshy field** locally called 'Tom Long's Meadow' at 557165; **Great Fenny Wood**; the **former Great Central Railway** line now operated privately crossing the parish from north to south with Quorn and Woodhouse Station at 549162.

R

'Rabbits Bridge', Woodhouse, 553143 (51M). Local name for the bridge which carries lane over privately operated **railway**.

Race Course, The, Twycross/Shackerstone, 353058 (30MN). Mixed woodland in Gopsall Park.

Ragdale, former parish, Hoby with Rotherby, 61PU, 62KQ.
Contains: village at 661199; Ragdale Hall at 653199 with parkland, now largely arable; former rifle range with **pools** at 663210.

Ragdale Wood, Thrussington, 647201 (61P, 62K). Mixed woodland in a part of the county where woodland is scarce.

Ram's Head Spinney, East Norton, 777997 (79U).

Ramsley Brook, Worthington/Breedon on the Hill, 42ABC. Formed from streams rising in Worthington parish, flows north and north-west past Cloud Hill Quarries, through Tonge, past Ambro Mill to the county boundary near Wilson.

Rancliff Wood, Burton on the Wolds, 604208 (62A).

Ratby, parish, 40XYZ, 50CDHI.
Boundary changes have led to gains from Kirby Muxloe and losses to Desford parishes. Contains: village at 513059; **Martinshaw Wood**; **woodland** and **grassland** at Ratby Burroughs, 494060; other woods; **grassland** at 499065, M1 motorway crossing the parish.

Ratchet Hill, Coalville, 446163 (41N).

Ratcliffe College, Cossington, 625150 (61H). School.

Ratcliffe Culey, former parish, Witherley, 39EJP, 30FK.
Contains: village at 326994; **River Sence** forming part of the northern parish boundary; Ratcliffe Bridge, over the River Sence at 321996. Habitat study 70 is of the **River Sence** between 325998 and 321996 and 85 of **hedgerow** at 329989.

Ratcliffe Cut, Kegworth, 491290 (42Z). Ratcliffe Locks at 491293 and Ratcliffe Weir at 492297. Name derives from Ratcliffe on Soar, Nottinghamshire.

Ratcliffe on the Wreake, parish, 61BCGHIM.
Contains: village at 631145; the Fosse Way (A46) forming part of the western boundary and the **River Wreake** the south-eastern; Ratcliffe Mill at 631141; marshland, now partly drained, near the river at Lewin Bridge; **fishponds** at and below Ratcliffe Hall, 628143.

Ravenstone with Snibston, parish, 31VWX, 41ABC.
Contains: Ravenstone village at 402139; Alton Grange with **woodlands** nearby; course of railway, now conveyor, crossing the southern corner of the parish; a working line cutting through the northern parish boundary. The hamlet of Snibston is now in Coalville.

Raw Dykes, Leicester (50W). Roman aqueduct fragment remaining at 583026.

451

Rearsby, parish, 61GHLMRS.
   Contains: village at 651146; the **River Wreake** forming the north-west parish boundary; Rearsby Mill at 641149; several **ox-bow pools**; Bleak Moor; **gravel pits**, working and restored in the west of the parish; the Leicester to Melton Mowbray **railway line** (station site at 651151) running north-east to south-west through the parish.

Rearsby Brook, 61MSXY, 71CDGH. Formed from brooks rising near Ashby Pastures and Cream Gorse, crosses Rotherby and Brooksby former parishes, and flows through Rearsby to join the River Wreake at 641147.

Rectory Wood, Waltham, 82C. Probably at 802243.

Red Hill, Birstall, 589082 (50Z).

Red Hill, Swannington, 424166 (41I).

Redma Spinney, Rolleston. Location not determined.

Redmile, parish, 73LMRSTVWXY, 83BCD.
   Consists of the former parishes of Redmile, Barkestone and Plungar.
   The former parish, 73STWXY, 83BCD, contains: village at 797355; a **dismantled railway** crossing the parish with Redmile Station site at 786361; nearly 2km. of the **Grantham Canal**; **woods** on the marlstone scarp at the south-eastern end of the parish and some **old grassland** on the flat land below.

Reedpool Spinney, Foston (Kilby), 600942 (69C). Chalybeate spring.

Reedpool Spinney, Hungarton, 694058 (60X). Spring-fed marsh.

Reservoir Covert, Shackerstone, 351079 (30NP).

Reservoir Wood, Croxton Kerrial, 815306 (83A).

Reynolds' Field, Thurcaston. Location not determined but may be at 568111.

Ridgemere Lane Spinney, South Croxton, 673098 (60U). Adjacent to Barkby Holt.

Rigget's Spinney, Hoton, 558224 (52L).

Rise Hill Spinneys, Burton and Dalby (Little Dalby), 764122 and 767123 (71R).

Rise Rocks, Markfield, 469121 (41R).

Robie's Gorse, Shackerstone (Congerstone), 373040 (30RS).

Robin-a-Tiptoe, Tilton, 773043 (70S). Hill rising to 220m.

Robin Hole, Hinckley. Location not determined.

Rockingham Station, Great Easton, 866931 (89R). The **dismantled railway** is still of some interest, but the station has been turned into a coal depot.

Rocky Plantation, Ulverscroft, 492118 (41V). National Trust property, part of Ulverscroft Nature Reserve.

Roecliffe Hill, Newtown Linford, 533127 (51G). **Roecliffe Spinney** is at 532130. Habitat study 10 is at 531131.

Rolleston, parish, 70FGK, 79JP.
   Contains: Rolleston Hall (in parkland) and church at 732004; a lake at 733002; Rolleston Wood at 740996; other small woods.

Rookery, The, Westrill and Starmore, 583797. Plantation and pond in Stanford Park.

Ross Knob Plantation, Ravenstone with Snibston, 387139 (31WX). One of the woods near Alton Grange.

Rotherby, former parish, Hoby with Rotherby, 61TXY.
   Contains: village at 675165; part of **ox-bow** at 673166; **marsh** and **pasture** at 685172; a short stretch of the **River Wreake** forming the parish boundary; a short length of the Leicester to Melton Mowbray **railway**.

Rothley, parish, 51MQRSVWX.
   Contains: village at 586126; about 0.5km. of privately owned **railway** with Rothley Station at 568122; **Kinchley Hill** and Kinchley Lane; **Nunckley Hill Quarry**; a small area of **Swithland Reservoir**; *Rothley Plain,* now a residential area; about 1km. of **dismantled mineral railway**; Rothley Park Golf Course at 571122; Rothley Temple at 576123, with parkland. **Rothley Brook** crosses the parish from west to east to join the River Soar which forms the eastern parish boundary for about 3km. Rothley Sandpit at 567023 is in Thurcaston parish.

Rothley Brook, 50CHINPU, 51KQRW. Formed from the union of several brooks rising at Thornton (Reservoir), Desford and Kirby Muxloe; flows through Glenfield, Anstey, Thurcaston and Rothley to join the River Soar below Farnham Bridge, 592132.

Rough, The, Loughborough (Shepshed), 497170 (41Y). One of the Longcliffe group of woods.

Rough, The, Thurcaston, 557122 (51L). Woodland on the edge of Swithland Hall Park; there is a large pond nearby.

Rough Heath, Staunton Harold, 372207 (32Q). Woodland.

Rough Park, Coleorton, 392183 (31Y). Woodland.

Roundabout Spinney, Peckleton, 491005 (40V).

Roundhill Spinney, Loddington, 774031 (70R).

Roundhill, Thurmaston, 616102 (61A). Gravel pits, formerly of botanical interest, now largely filled in.

Round Stye Plantation, Woodhouse, 523136 (51G).

Rowden Gorse, Higham on the Hill, 365966 (39T). **Ponds** nearby.

Rowhele Wood, Quorndon, 564157 (51S). Part of Buddon Wood and Swithland Reservoir SSSI.

Rowlatts Hill, Leicester, 619045 (60C). Now built up.

Rowley Fields, Leicester (Aylestone), 572023 (50R). Allotments.

Rye Close Spinney, Misterton, 561845 (58MS).

S

Saddington, parish, 69FGKLQR.
Contains: village at 658917; half of **Saddington Reservoir** (the rest is in Gumley parish); the **Grand Union Canal**, more than half of which is in Fleckney Tunnel; a brook flowing north-east across the parish; **feeder channel** from the reservoir joining the canal in Smeeton Westerby parish. Habitat study 63 of the **reservoir** is at 664910. Habitat study 68 of the **canal** is at 666923.

Saffron Brook, Leicester, 50VW, 60A. Continuation of Wash Brook through Knighton and along northern boundary of Aylestone into the River Soar at 577025.

St. Ann's Well, East Langton, 727938 (79G). Site of spring.

St. John Aldeby, Enderby, 553990 (59P). Site of mediaeval church by River Soar.

St. Leonard's, Leicester, 585056 (50X). One of the smaller former parishes of the old borough.

St. Margaret's, former parish, Leicester, 50WXY, 60BCD.
One of the two largest parishes of the old borough. Contains: church at 584051; Abbey Park; parts of the River Soar and the Grand Union Canal; considerable railway property; St. Margaret's Pasture, 583054, now a recreation ground.

St. Martin's, Leicester, 585044 (50X). Very small former parish surrounding St. Martin's Church (now the Cathedral).

St. Mary de Castro, former parish, Leicester, 50LMRSWX, 60B. One of the two largest parishes of the old borough. Contains: Victoria Park; Freemen's Common; a length of the Grand Union Canal; much railway property and two areas of allotment gardens.

St. Mary's Bridge, Market Harborough, 743873 (78N). Carries A427 over the River Welland.

St. Mary's Mills, Leicester (St. Mary's), 576026 (50R). On River Soar, now wholly industrialised.

St. Nicholas, Leicester, 582045 (50X). Very small former parish containing remains of the Roman town (Ratae Coritanorum).

Saltbeck, Belvoir, 819347 (83CH). Covert on Belvoir estate with interesting **rides**.

Saltby, former parish, Sproxton, 83HIMNPSTX.
Contains: village at 851265; a wartime airfield, much of which has been converted to arable, but with fragments of **scrub** and **limestone grassland** remaining; **Egypt Plantation**; **Herring Gorse**; **Saltby Bog** with habitat study 54 at 841253. **The Drift**, an ancient green lane, the west side of which was formerly in Leicestershire but which is now wholly in Lincolnshire, is on the eastern side of the parish. Saltby Heath is the old, pre-enclosure name of the area on which the airfield was built; it survives as the name of an undefined area.

Saltersford, Croxton Kerrial. Location not determined.

Saltersford Bridge, Leicester (Humberstone), 621052 (60H). Carries A47 over Thurnby Brook.

Saltersford Brook, Oakthorpe and Donisthorpe. Drains **Willesley Lake** and flows through 31BGH, between Oakthorpe and Donisthorpe into the **River Mease** at 311122. Saltersford Bridge is at 313125.

Salters Hill, Somerby (Burrough on the Hill), 748120 (71KL).

Saltway, Croxton Kerrial, 82P. Traditional name for the road from Croxton Kerrial to Three Queens.

Sandham Bridge, Thurcaston, 564109 (51Q). Carries footpath over **Rothley Brook**.

Sandhills Wood, Newtown Linford, 518112 (51A). **Cowslip meadow** close to wood at 515112.

Sandhole Spinney, Peckleton, 457011 (40K).

Sanham Hall, Frisby on the Wreake, 729154 (71H).

Sapcoat's Spinney, Burton and Dalby (Burton Lazars), 787160 (71XY).

Sapcote, parish, 49MRSVWX, 59B.
Contains: village at 488932; 1km. of M69 motorway; the Fosse Way forming 2km. of the eastern parish boundary; **Sapcote Quarry**, disused flooded diorite quarry at 497934; part of **Granitethorpe quarry**.

Sauvey Castle, Withcote, 787052 (70X). Mediaeval castle site on natural marlstone peninsula; **Sauvey Plantation** is at 790052.

Sawley Cut, Lockington-Hemington, 43Q. Canalised section of River Trent. Sawley Mill and Sawley Bridge (Harrington Bridge) at 471311.

Saxby, former parish, 81EJ, 82ABF.
Contains: village at 820200; site of Saxby Station at 813192, formerly of botanical interest; short lengths of both working and dismantled railways; part of the disused **Oakham Canal**; **grassland** at 813213 and 818217.

Saxelby (Saxelbye), former parish, Grimston, 62VW, 72AB.
Contains: village at 700209; **Saxelby Wood** at 693225 with **Barnes Hill**, **Summer Leys** and **Ten Acre Plantations** continuous with it to the north; about 3km. of **partially dismantled railway**, with the station site at 697210; **tunnel cuttings** at 690222 and 701206.

Scalford, parish, 72KLMNQRSWX.
Consists of the former parishes of Scalford and Wycomb-and-Chadwell.
The former parish contains: village at 763241; numerous sites of old quarries and mineral railways

453

(though many have been restored to agricultural use); a section of **dismantled railway** crossing the parish, with a **tunnel cutting** at 746255 and Scalford Station site at 755242; Scalford Brook with associated **wet grassland** crossing the parish from north to south. **Scalford Gorse** 748224 is in Ab Kettleby parish.

Scotland, Burton Overy, 679984 (69U). Hamlet at north end of Burton Overy village.

Scotland, Staunton Harold, 388222 (32W). Hamlet.

Scotland End, Market Harborough, 740865 (78IN). South end of Little Bowden.

Scraptoft, parish, 60HIMNST.
Contains: village at 647055; **Scraptoft Gorse** at 664053; Scraptoft Golf Course at 653065; *Scraptoft Hall Wood* (now partly destroyed); Square Spinney; **grassland** at 659063.

The western fifth of the parish was transferred to Humberstone (Leicester) in 1966 during the expansion of Leicester. Contains: village at 647055; **Scraptoft Gorse** at 664053; Scraptoft Golf Course at 653065; *Scraptoft Hall Wood* (now partly destroyed); Square Spinney; **grassland** at 659063.

Seagrave, parish, 51YZ, 61DEIJP, 62FK.
Contains: village at 619175; two brooks with associated grassland, which rise in the north of the parish and converge south of the village; habitat study 34 of **calcareous grassland** at 621178; neutral grassland of **Seagrave Meadows** SSSI at 624188; **roadside verges** of Big Lane and Park Hill Lane with habitat study 83 at 623172; a **sheepwash pond** at 621173; Seagrave Wolds in the north-east of the parish; the Fosse Way forming the eastern boundary and the Fishpool Brook forming part of the northern boundary of the parish.

Seale Pastures Wood, Oakthorpe and Donisthorpe, 31B. It is not clear whether Horwood's records for this wood are from the Leicestershire or the Derbyshire side of the county boundary near Acresford. In any case, records prior to 1897 (when Nether and Over Seal was transferred to Derbyshire) would have been for Leicestershire. Horwood seems to have used the name Seal Brook for that part of Hooborough Brook between Acresford and the River Mease.

Seal Wood, Packington (possibly an error for the previous entry).

Sence, River (eastern), 59PTUY, 69DHIMNPU, 60QRW, 70B. Formed by the union of the Billesdon and Coplow Brooks at 700031. Flows south-west between Houghton, Gaulby and Frisby, through Little Stretton and Great Glen after which it flows west to join the River Soar near Enderby Bridge at 551985.

Sence, River (western), 30FGLRSTYZ, 31VW, 39EJ, 41ABGLR. Rises between Copt Oak and Bardon Hill; flows south-westwards through Hugglescote, Heather, Shackerstone, Congerstone and Sheepy, to join the River Anker at King Dick's Hole, 39E; there are associated **marshes** at Newton Burgoland and Harris Bridge and **pools** at Hugglescote and Sheepy.

Sewstern, former parish, Buckminster, 81UZ, 82VW. Contains: village at 889215. Most of the eastern parish and county boundary follows the Drift, here a metalled road.

Shackerstone, parish, 30LMNPRSTUXYZ, 40CDE. Consists of the former parishes of Barton in the Beans, Bilstone, Congerstone, Odstone and Shackerstone.
The former parish, 30NPSTUY contains: village at 374067; about 5km. of the **Ashby Canal**; 6km. of **dismantled railway**; 0.5km. of private railway still in use with Shackerstone Station at 378065; **River Mease** forming part of the western parish boundary; Shackerstone Park Coverts (partly in Twycross) at 359074; several other coverts. Shackerstone Gorse at 383065 is in Barton in the Beans former parish.

Shangton, parish, 79CDEHIJ.
Contains: village at 715960; **reed swamp** at 716962; the Care Village at 720966; the site of Hardwick village; **Shangton Holt** at 716979; other small woods.

Sharnford, parish, 48PU, 49QRVW.
Contains: village at 483919; **grassland** at 488914; about 2km. of **Soar Brook**; the Fosse Way forming its south-eastern boundary for 2.5km, meeting Watling Street at High Cross. This sector of Watling Street (about 0.4km.) is wholly in Leicestershire.

Sharp's Covert, Twycross (Twycross), 350050 (30M).

Shawell, parish, 57JP, 58FKQ.
Contains: village at 541796; about 2.8km. of **dismantled railway line**; **disused sand and gravel pits** known as Cave's Inn Pits, SSSI, at 537795; **flooded gravel pits** at 537801; formal ponds at Shawell Hall, 542796; the M1 motorway crossing the parish from north to south and a short length of the M6 motorway from east to west. Shawell Wood, 554819 is in Misterton. The south-west county and parish boundary runs parallel to Watling Street, which here is wholly in Warwickshire.

Shearsby, parish, 68EJ, 69AFG.
Contains: village at 623909; old mill mound at 623905; numerous small brooks.

Sheepthorns (Sheephorns) Spinney, Carlton Curlieu, 701950 (79C).

Sheepy, parish, 20VW, 29Z, 30ABFGQRSW, 39EJ.
Consists of the former parishes of Sheepy Magna, Sheepy Parva, Sibson and Upton.

Sheepy Magna, former parish, 20VW, 29Z, 30FGL, 39EJ.
Contains: village at 326013; **pools** at 326017; **Sheepy Fields** SSSI, with habitat study 20 at 332024; the **River Sence** forming the eastern parish boundary with **grassland** at 317993.

Sheepy Parva, former parish, 30FGKL, 39J.
Contains: village at 332013; **pool** at 327012.

Sheepy Wood, Hinckley (Burbage), 447947 (49M).

Sheepy Wood, Twycross (Orton on the Hill), 330044 (30H).

Sheet Hedges Wood, Newtown Linford, 529087 (50J). Ancient woodland reduced in area by quarrying; SSSI.

454

Shell Brook, Ashby de la Zouch, 31HIM. Rises near Blackfordby and flows south into Willesley Lake; the hamlet of Shellbrook is at 342166.

Shelthorpe, Loughborough, 51NPTU. Former hamlet, now a suburb of Loughborough; **spinney** at 549184.

Shenton, former parish, Sutton Cheney, 30QV, 39UZ.
Contains: village at 386003; about 1km. of **dismantled railway** including Shenton Station site at 396004 and **Shenton Cutting** Nature Reserve at 397000 with habitat study 90 at 397999; Shenton Gorse at 376015 and several other coverts; *Shenton Pool*, presumably in the grounds of Shenton Hall; part of the **Ashby Canal**. Shenton Aqueduct carrying the Canal over a minor road at 392007 is in Market Bosworth parish.

Shepshed, former parish, Loughborough, 41NPSTUXYZ, 42KQV, 52A.
Contains: village at 479192; **Charnwood Quarries**; **Longcliffe Quarry**; **Morley Quarry**; other **quarries**; Ives Head SSSI; **Oakley Wood** SSSI; **Piper Wood**; **White Horse Wood**; **woodland** at 459189; the northern half of **Blackbrook Reservoir**; **Black Brook** forming the south-western parish boundary as far as the reservoir and then running north through the parish; part of the **Grace Dieu Brook**; some **heath land** and **rocky outcrops** at the northern end of Charnwood Forest; traces of the **Charnwood Forest Canal**; **dismantled railway**; 4km. of M1 motorway.

Shilton Mill, Hinckley (Earl Shilton). Location not determined.

Shoby, former parish, Grimston, 61UZ, 62QRV.
Contains: ruined remains of Priory at 683202; **Shoby Scholes** at 672208 with fox covert, **pond** and **marshland**; **woodland** by crossroads at 663217.

Shortcliff Brook, Loughborough, 41Z, 51DE. Rises near Charnwood Farm, 491174 and flows north-eastwards to join Burleigh Brook near the University.

Short Wood. Location not determined. Probably near Netherseal, now in Derbyshire.

Shoulder of Mutton Hill, Leicester, 558042 (50M). Railway cutting through sandstone. Geological SSSI.

Shovel Nook, Leicester (Knighton). Location not determined.

Sibson, former parish, Sheepy, 30KLQRSW.
Contains: the village of Sibson at 354009 and the hamlet of Wellsborough; the **River Sence** forming its north-western boundary with Sibson Mill at 344025; *Sibson Field*, *Sibson Gorse* and *Sibson Wolds*, 365039 mentioned in HG.

Sileby, parish, 51WXY, 61BCDEHI.
Contains: village at 600151; 1.5km. of the **River Soar** forming the western parish boundary with **meadows** nearby; **working railway** crossing the parish from north to south; habitat study 62 of **field pond** at 592158; **grassland** at 596156; **pool** at 598160; former Sileby Brick pits at 606148 now a **refuse tip** at 609148 with ruderal flora.

Sileby Brook, 51X, 61CDIJ, 62FK. Rises near Six Hills and flows south-west past Seagrave and through Sileby to join the River Soar near Sileby Mill, 591148.

Sinope Mine, Swannington, 402152 (41C).

Sir Francis Burdett's Covert, Burton and Dalby (Little Dalby), 760133 (71LR).

Sir Johns Wood, Belvoir, 811324 (83B).

Six Hills, 643207 (62K). Road junction at site of a Roman crossroads on the Fosse Way where six parishes meet; **common land** (Twenty Acre) at 642210 is an important SSSI; Mundy's (Mandy's) Gorse immediately to the north of this is now ploughed up.

Skeffington, parish, 70FGHKLMR.
Contains: village at 741026; **Skeffington Wood** at 758035, and **Brown's Wood**, **Crow Wood**, **Hoothill Wood**, **Priest Hill**, all part of the Leighfield Forest SSSI. **Skeffington Vale** at 740012; *Skeffington Gap* at 754020; *Skeffington Bushes*, location not determined.

Skeffington and Stonton Brook, 70FK, 79GHIQRTUVW. Not named on 1:25000 OS map; rises between Skeffington and Rolleston and flows south along parish boundaries between Noseley and Goadby, Stonton Wyville and the Langtons, to join the River Welland at Welham.

Sketchley, Hinckley, 424922 (49G). Former hamlet, now suburb of Hinckley. Sketchley Brook, 49BGLM, divides Hinckley from Burbage.

Skillington Road, Sproxton (Saltby). Running north-east out of Sproxton village to the county boundary at 882257. Good **limestone verges**.

Slade Meadow, Stoney Stanton. Location not determined.

Slash Lane, Barrow on Soar, 51XY. Joins the B5328 and B674. **Wet grassland** nearby.

Slate Brook, 40Y, 50DEJ. Rises near Whittington Roughs and flows into Groby Pool.

Slawston, parish, 79RSTWX.
Contains: village at 781945; Othorpe deserted village site at 770955; **Slawston Hill**, 783942; about 2km. of **dismantled railway**; traces of the Roman road (Via Devana).

Sludge Hall, Cold Newton, 719055 (70C).

Slyborough Hill, Clawson and Harby, 714264 (72D).

Smeeton Westerby, parish, 69QRVW, 79AB.
Contains: village at 677927; about 3km. of the **Grand Union Canal** with habitat study 68 at 670922; Smeeton Hill at 672911; *Smeeton Gorse* at 673909; the Langton Brook forming the northern parish boundary; *Smeeton Westerby Gravel Pit*, location not determined.

Smite, River, 72CDH. Arises at Holwell Mouth and flows north until it meets and follows the county boundary in 72DE.

455

Smockington, Hinckley, 454897, (48P). Hamlet at junction of A5 and A46.

Smoile, The, Worthington, 390195 (31Z). Open **birchwood** crossed by **dismantled railway**, both now mostly destroyed by opencast mining; habitat study 2 is at 390195.

Smooth Coppice, Isley cum Langley, 427237 (42G).

Snarestone, parish, 30JPU, 31FKQ.
Contains: village at 341094; nearly 2km. of **dismantled railway line** (station site at 340092); about 2.5km. of the **Ashby Canal** passing under the village in a short tunnel and now terminating at 346100.

Snibston, former parish, Coalville, 41BCGH.
Contains: church at 411131; Snibston Colliery at 418144 is no longer working.

Soar, River, SP49, SK42, 43, 50, 51, 52, 60, 61. Formed by the union of **Soar Brook** with other brooks near Sharnford and flows north-east across the county through Leicester and Loughborough to join the River Trent near Sawley Bridge. It is canalised apart from some stretches which are by-passed with cuts and navigable from its junction with the Grand Union Canal at Glen Parva. It receives tributaries which drain the southern and eastern slopes of Charnwood Forest. The **River Sence** joins it at Glen Parva, the River Biam at Aylestone and the **River Wreake** at Cossington. It is of botanical interest along much of its length but more especially in SP59, SK51 and SK52. Habitat study 73 is between 504947 and 511940, 74 is at 591075 and 75 is at 564185.

Soar Brook, 49KQV. Rises in Warwickshire and becomes the River Soar after joining an un-named brook at 487919. Soar Brook Spinney is at 451908.

Soar Mills Bridge, Broughton Astley, 509938 (59B). Carries B581 over River Soar.

Somerby, parish, 70JPUZ, 71FKLQRVWX, 81AB.
Consists of the former parishes of Somerby, Burrough on the Hill, and Pickwell.
The former parish, 70UZ, 71QRV, contains: village at 779104; part of **Burrough Hill**; some old **quarries** and **gravel pits**.

Soper's Bridge, Croft/Stoney Stanton, 504947 (59C). Field bridge over the River Soar, with a **meadow** nearby.

South Croxton, parish, 60UZ, 61QV, 70E, 71A.
Contains: village at 691103; **Queniborough Brook** forming part of the southern parish boundary and thereafter flowing across the parish from south-east to north-west.

Southfields, Leicester (St Mary's), 50W. Now wholly urbanised. Contains **Welford Road Cemetery**.

South Kilworth, parish, 58VW, 68AB.
Contains: village at 604818; the Manor House site at 604814 with remains of fish ponds and old moats; disused gravel pits at 598811; the north-western half of **Stanford Reservoir**; the River Avon forming the south-eastern parish and county boundary for 0.5km.

South Wigston, Oadby and Wigston, at 586983 (59Z). Largely suburban village. There is a large area of **railway property** at Wigston Junctions. The **Grand Union Canal** passes south of the village.

South Wood, Ashby de la Zouch, 363205 (32Q). There are ponds at South Wood Farm, 368203.

Spinney Hill Park, Leicester (Evington), 605044 (60C).

Spring Burroughs, Bardon. Location not determined.

Spring Cottage, Ashby Woulds, 305164 (31D). **Pools** and **old workings**.

Spring Hill, Coalville (Whitwick), 448158 (41M). Partly quarried by Peldar Tor workings.

Spring Plantation, Charley, 472155 (41S).

Spring Wood, Cadeby, 417017 (40A).

Spring Wood, Coleorton, 384185 (31Z).

Spring Wood, Ravenstone with Snibston, 398141 (31X). One of the woods near Alton Grange.

Spring Wood, Staunton Harold, 382225 (32RW). Habitat study 1 is at 381228.

Sproxton, parish, 82BCDEGHILMNPRSTUX.
Consists of the former parishes of Bescaby, Saltby, Sproxton and Stonesby.
The former parish, 82GHLMRSX, contains: village with church at 856249; nearly 3km. of **dismantled mineral railway**; the remains of **Sproxton Heath**; **Sproxton Quarry**, former ironstone gullet and now geological SSSI at 82S; Sproxton Thorns at 842233; **Skillington Road**; the **River Eye** crossing the parish from north to south.

Square Spinney, Great Glen, 650993 (69P).

Square Spinney, Scraptoft, 666051 (60S).

Stanford Hall, Westrill and Starmore, 587793 (57Z). The Park contains two lakes and a large pond; it is divided by the county boundary which follows the River Avon; Stanford church is in Northamptonshire.

Stanford Reservoir, South Kilworth, 604808 58V, 68A. About half the reservoir is in Northamptonshire.

Stanleys Barn. Location not determined.

Stanton Plantation, Tilton, 759059 (70M).

Stanton under Bardon, former parish, Markfield, 40PU, 41KQ.
Contains: village at 466103; **Cliffe Hill Quarry**, formerly of botanical interest but recently much extended; traces of a dismantled mineral railway.

Stapleford, former parish, Freeby, 71XYZ, 81CDE.
Contains: Hall with church at 811182; **lake** at 816182 and nearby **marsh**; most of Stapleford Park; **Laxtons Covert**; Berry Covert; **marsh** at 804185; **River Eye** forming part of the northern parish boundary; former **Oakham Canal** crossing the eastern side of the parish.

Stapleton, former parish, Peckleton, 49EIJ, 40F.
Contains: village at 434984; Manor Farm Moat at 433988; Stapleton Wood at 428001.

Starkeys Hill, Castle Donington, 420259 (42C).

Stathern, parish, 72U, 73KLQRVW.
Contains: village at 772309; **grassland** at 768293; **Stathern Wood** at 782317 running down from the marlstone escarpment; Combs Plantation with **grassland** to the north; about 1.5km. of the **Grantham Canal** with Stathern Bridge at 755323; about 2.5km. of the **dismantled Melton to Bottesford railway**.

Staunton Harold, parish, 31UZ, 32QRVW.
Contains: Staunton Harold Hall at 379209 with associated **woodland**; Park with **Upper Lake** at 380208 and **Lower Lake** at 380212; a small part of **Staunton Harold Reservoir** at 377223; about half of **Dimminsdale Nature Reserve** at 376217; **Lount Wood**; Lount hamlet at 387194; **Rough Heath**; **Spring Wood**; Staunton Lodge at 394213.

Stemborough Mill, Leire, 532910 (59F).

Stewards Hay, Groby, 509090 (50E). House now demolished and stables in ruins; lake nearby is silting up.

Stockerston, parish, 89HIJMN.
Contains: village at 834975; **Bolt Wood**; **Park Wood**; the **Eye Brook** forming part of the eastern parish boundary; the western half of **Eye Brook Reservoir**.

Stocking Farm, Woodhouse, 544148 (51M). Woodhouse Brook and pool nearby.

Stocking Wood, Leicester (Leicester Abbey), 50Z. This and *Stocking Farm* on Mowmacre Hill destroyed and built over.

Stoke Golding, former parish, Hinckley, 39Y, 49CDE. Was divided between Hinckley and Higham on the Hill in 1936. Contains: village at 397972; about 1.3km. of **dismantled railway** with station site at 391972; about 1.8km. of the **Ashby Canal** on its western boundary.

Stoke Lodge Spinney, Hinckley (Stoke Golding), 411968 (49D).

Stonebow Bridge, Loughborough (Garendon), 504205 (52A). Carries lane to Bailey's Plantation over **Black Brook**.

Stonepit Spinney, Ab Kettleby (Wartnaby), 705237 (72B).

Stonesby, former parish, Sproxton, 82BCGH.
Contains: village at 823247; *Stonesby Spinney* at 820232; **Stonesby Quarry** much of which has been filled, but part is an SSSI, with habitat study 103 at 814252. Half of Stonesby Gorse at 812242 is in Waltham parish.

Stoney Bridge, Stoney Stanton, 503929 (59B). Carries Fosse Way over River Soar.

Stoneygate, Leicester (Knighton), 60B. Residential suburb.

Stoney Stanton, parish, 49SWXY, 59BC.
Contains: village at 489948; **Stoney Cove** (water-filled deep quarry) at 493941; about 0.7km. of working railway; the Fosse Way forming about 1.3km. of its south-eastern boundary and the **River Soar** about 1km. of its north-eastern boundary.

Stoneyway Plantation, Appleby Magna, 315084 (30E).

Stoneywell Wood, Ulverscroft, 499120 (41VW). SSSI.

Stonton Wyville, parish, 79GHILMN.
Contains: village at 735950; Langton Caudle; **Stonton Wood** at 741963; the Via Devana (green lane) crossing the parish from east to west; fishponds at 734949 which have been drained.

Stony Bridge, Potters Marston/Thurlaston, 502972 (59D). Carries minor road over Thurlaston Brook; habitat study 52 is of a **marsh** at 509975; **willow holt** at 502972 is in Potters Marston parish.

Stony Stumps, Charley or Coalville. Near Abbots Oak, location not determined.

Stordon Grange, Osgathorpe, 419192 (41E). **Moat**.

Stoughton, parish, 60FGKLQR.
A very small area of the parish was transferred to Oadby and Wigston in 1974, and an even smaller area was gained from Evington. Contains: village at 640021; most of Leicester East Airfield; *Stoughton Flats,* location not determined.

Strawberry Hill Plantation, Charley, 455171 (41N).

Stretton en le Field, 20Z, 21VW, 30E, 31AB.
Contains: hamlet with a farm and parish church at 304119; **River Mease** forming the northern parish boundary with Stretton Bridge at 300123; Stretton Hill at 304103.

Stretton Magna, parish 69SP, 60KQR.
Contains: the site of the mediaeval village and the parish church at 657004; Glenfrith Hospital at 653996; about 0.5km. of the River Sence; part of Leicester East Airfield.

Strifts Plantation, Garthorpe, 829226 (82G).

Stubble Hills, Sutton Cheney (Shenton), 375007 (30Q).

Stult Bridge, Whetstone, 562953 (59S). Carries A428 over Whetstone Brook.

Sturrad Spinney, Blaston, 804963 (89D).

Sulby Reservoir, Northamptonshire, 68K. The references in HG are probably to **Welford Reservoir**, a small corner of which is in Leicestershire. Sulby Reservoir is separated from it by a weir.

Summerpool Brook, 52FK. Flows on south-west side of Loughborough Meadows and joins the Wood Brook at 531217.

Sunny Leys, Houghton on the Hill, 660028 (60LR). Woodland.

Sutton Cheney, parish, 39UYZ, 49DEJ, 30QRVW, 40AF.
Consists of the former parishes of Sutton Cheney, Shenton and Dadlington.
The former parish, 39Z, 30V, 49EJ, 40AF, contains: village at 416004; about 1.5km. of the **Ashby Canal** with Sutton Wharf Bridge at 410993; a gated road to Market Bosworth with **wide verges** at 413010; **Ambion Wood**; Bosworth Field (Country Park); short stretch of **dismantled railway**.

Sutton Hill Bridge, Stoney Stanton, 511940 (59C). Carries A46 over **River Soar**.

Sutton in the Elms, Broughton Astley, 519939 (59BG). Hamlet.

Swains Park, Ashby Woulds, 304170 (31D). Hamlet and site of reservoir, otherwise known as Moira Reservoir, now destroyed by mining.

Swallow Hole, Sproxton (Saltby), 842277 (82N).

Swannington, parish, 41CDHI.
The parish has lost about 1 sq.km. to Coalville, gained parts of Thringstone and Coleorton. Contains: village at 414162; about 2km. of railway line (station site at 411155); numerous traces of dismantled mineral lines; **Swannington Common** at 415170; *Swannington Bog* and *Swannington Aqueduct,* locations not determined.

Swannymote Rock, Charley (Whitwick), 444172 (41N).

Swan's Nest, Melton Mowbray, 764188 (71U). **River Eye** of interest here.

Sweet Hill, Ashby Woulds, 322159 (31I). Hamlet and derelict mining area with a **pool**; *Sweet Hill Oak* is mentioned in HG.

Swepstone, parish, 30PTUZ, 31KLQRV.
Consists of the former parishes of Swepstone and Newton Burgoland.
The former parish 30U, 31KLQRV, contains: village at 368105.

Swift, River, 58GLMSXY. Formed by the union of several brooks near Kimcote, flows south-west through Misterton and Lutterworth, leaving Leicestershire just short of the A5 and joining the River Avon in Warwickshire.

Swinford, parish, 57PTUZ, 58KQ.
Contains: village at 569794; M1/M6 motorway junction and over 2km. of M1; the River Avon forming the parish and county boundary for about 1.7km.; *Swinford Corner* at 562816; Swinford Covert at 580787.

Swithland, parish, 51GKLMR.
Contains: village at 555128; Swithland Hall at 559126; **Crow Wood**; part of **Swithland Reservoir** including **Brazil Wood**; a very small part of **Swithland Wood**; 1.8km. of privately operated railway; **Swithland sidings**, with habitat study 91 at 564132. *Swithland Claypit* is mentioned in HG.

Swithland Quarries, Newtown Linford, 539122 (51G). Disused slate quarry in Swithland Wood with habitat study 101 at 539122.

Swithland Reservoir, Mountsorrel, Quorndon, Swithland, Woodhouse, 51LMRS. Part of the Buddon Wood and Swithland Reservoir SSSI. Habitat study 61 at 562142.

Swithland Wood, Newtown Linford/Swithland, 51FGKL. Ancient woodland; SSSI; habitat study 11 at 537129 (woodland), 12 at 538118 (woodland), 31 at 538128 (**old grassland**).

Sysonby, former parish, Melton Mowbray, 71JP, 72FGKL.
Contains: housing estates on the western edge of Melton Mowbray; the RAVC Remount Depot at 738197; about 1.3km. of railway; the **River Eye** forming its southern boundary.

Syston, parish, 51VW, 61ABFGK.
Contains: village at 626118; about 1.3km. of the **Grand Union Canal**; about 4km. of **railway** including the station site at 621110 and **Syston Junction** at 620117; **River Wreake** forming the northern parish boundary of interest at Syston Mills 614124; extensive **gravel workings** in the west and north of the parish.

Syston Brook, 60PU, 61ABFK. Rises near Beeby and flows north-west through Syston to join the River Wreake west of Syston Mills.

T

Talbot Lane, Ashby de la Zouch. Location not determined.

Tampion's Coppice, Knossington, 808063 (80D). Sometimes known as Cheseldyne Spinney.

Tatborough Spinney, Twycross (Norton juxta Twycross), 313075 (30D).

Taylor's Bridge, Oadby and Wigston (Wigston Magna), 600970 (69D). Field bridge over **Grand Union Canal**.

Temple Mill, Sheepy (Sheepy Magna), 356034 (30R). Foot bridge over River Sence.

Tent, The, Sproxton (Saltby), 867279 (82T). Antiquity; perhaps referring to barrows in and to south of Egypt Plantation.

Terrace Bridge, Shackerstone (Congerstone), 371055 (30S). Field bridge over **Ashby Canal**.

Terrace Hills, Croxton Kerrial (Branston)/Belvoir, 73VW, 83B. Part of the **marlstone escarpment**, **wooded**. Terrace Hill Farm at 796312.

Theddingworth, parish, 68MNSTUY.
Contains: village at 667857; about 2.5km. of dismantled railway (station site at 664860); 2.7km. of **Grand Union Canal**; the lower slopes of the **Laughton Hills**; about 2.8km. of the River Welland

458

on the parish and county boundary with a tributary brook joining it at 679863; Theddingworth Lodge at 651868 is close to a road bridge over the canal. Theddingworth Hollow Spinney at 662878 is in Laughton parish.

Thistle Bridge Hill, Thurcaston, 574104 (51Q).

Thornborough, Coalville (New Swannington), 424155 (41H). Hamlet.

Thornborough Spinney, Lutterworth/Misterton, 555847 (58M). **Misterton Marshes** SSSI immediately to north.

Thornton, former parish, Bagworth, 40NPSTUYZ.
Contains: village at 468076; **Thornton Reservoir** with habitat study 58 at 477077; mining sites; nearly 2km. of railway line; several small woods; a **marsh** at 478082.

Thornton Brook, 40STWX, 50C. Starts below the dam at Thornton Reservoir and flows by Merry Lees, **Botcheston Bog** and Newtown Unthank to join Rothley Brook near Ratby.

Thorntree Cottage. Location not determined, probably in Charnwood Forest.

Thorpe Acre and Dishley, former parish, Loughborough, 51E, 52ABF.
Contains: Thorpe Acre village at 515199, now entirely urbanised; Dishley hamlet with remains of church at 512211; **River Soar** forming the northern parish boundary; **Black Brook** crossing the parish from south to north and joining the River Soar; a short length of **Summerpool Brook** joining the River Soar; a short length of **Grand Union Canal**; 200 m. of working railway; Thorpe Hill at 514195.

Thorpe Arnold, former parish, Melton Mowbray/Waltham, 71U, 72QRVW.
Contains: village at 770200; River Eye running from north to south; golf course at 780213 with some remaining old grassland. Thorpe Broom Covert mentioned in HG was at 788219.

Thorpe Langton, parish, 79FGKLQR.
Contains: village at 740924; pool at 756913; dismantled railway crossing the parish from east to west; former **Welham Sidings**; **Stonton Brook** forming much of northern parish boundary, of botanical interest at **Thorpe Langton Ford**, 743929; the **River Welland** forming the southern parish and county boundary.

Thorpe Satchville, former parish, Twyford and Thorpe, 71AFGKL.
Contains: village at 732117; about 2.5km. of **dismantled railway** mostly in a **cutting** which is an SSSI; a large pond at 729125; **Thorpe Gravel Pits** at 729112; **Thorpe Trussels** (scrub) at 727130.

Three Gates, Illston on the Hill, 718983 (79E). Crossroads and farm.

Three Queens, farm, just in Lincolnshire. The point, close by, on the county boundary, where **The Drift** crosses the Croxton to Grantham road is at 859297.

Three Shire Oak, Bottesford (Normanton), 821428 (84G). At the junction of Leicestershire, Lincolnshire and Nottinghamshire.

Thringstone, former parish, Coalville, 41DEIJ.
The former parish was divided between Coalville, Coleorton, Belton, Osgathorpe, Swannington and Worthington on the formation of Coalville Urban District. Contains: village at 425173; Peggs Green hamlet; Grace Dieu Manor; **Grace Dieu Priory**; **Grace Dieu Wood**; length of **dismantled railway** with adjacent **grassland** and **marsh**; traces of **Charnwood Forest Canal**.

Thrussington, parish, 61HIJMNPST, 62K.
Contains: village at 650157; **Thrussington Wolds Gorse** at 642197; **Ragdale Wood;** the Ox Brook flowing southwards across the parish to join the **River Wreake** which forms the southern parish boundary; Thrussington Mill on the **River Wreake** at 656157. Habitat study 97 is of an arable field at 656162. The Fosse Way forms the western parish boundary.

Thurcaston, parish, 50PU, 51KLQRV.
Consists of the former parish of Cropston together with about two thirds of the former Thurcaston parish, the remainder of which was incorporated into Leicester in 1966.
The former parish, 50PUZ, 51KLQRV contains: village at 565106; about 1.5km. of **railway**, part of which is privately operated and part of which is dismantled; about 2km. of **Rothley Brook**; Thurcaston Weir at 567113 with **pools** nearby; part of **Breech Spinney**. *Thurcaston Fox Covert,* now known as **Fox Covert**, is in Leicester.

Thurlaston, parish, 49UYZ, 59DE, 40V, 50A.
Includes the former parish of Knoll and Bassett House. Contains: village at 502990; Normanton Turville estate; about 1.5km. of M69 motorway; habitat study 82 at 501977 (disturbed **roadside verge**).

Thurlaston Brook, 49MSTY, 59DI. Rises in Elmesthorpe, flows east through Potters Marston and Huncote and joins the River Soar at 522963; a **tributary** (49UYZ) rises in Peckleton, flows through the pools in Normanton Turville and joins it by The Yennards; another, rising in Earl Shilton, joins it near Potters Marston; Feeding Brook joins it at 310975.

Thurmaston, parish, 50Z, 60EJ, 61AF.
A largely built up parish of which nearly one third was transferred to Leicester in 1935. Contains: village at 610093; **River Soar** on the western parish boundary with Thurmaston Lock at 608094 and Thurmaston Weir at 608099; derelict land now mostly restored, between **old gravel pits** at 603090 and the River Soar; a working **railway** crossing the parish from north to south with **cutting** at 620095. Thurmaston sandpit at 621080 is now in Leicester.

Thurnby, parish, 60GHLMRS.
Consists of the whole of Bushby together with about half the former parish of Thurnby, the remainder having been incorporated in Leicester in 1935.
The former parish, 60GHLM, contains: village at 647039; site of *Thurnby Court* at 645037; a length of dismantled railway partly built over with Thurnby and Scraptoft Station site at 645044.

459

Thurnby Brook, 50X, 60CHMS. Rises near New Ingarsby and flows west between Scraptoft and Thurnby, and between Evington and Humberstone before disappearing into an underground culvert which discharges into the River Soar opposite Abbey Park.

Tilton, parish, 70GHILMNRST.
Consists of the former parishes of Tilton and Halstead.
The former parish, 70GHLMRS, contains: village at 743056; Robin-a-Tiptoe Hill; **Tilton Wood** at 759041, part of Leighfield Forest SSSI; about 1.5km. of **dismantled railway**. Tilton Station site and **Tilton Cutting**, are in Halstead former parish.

Timberwood Hill, Charley, 471148 (41S). In Charnwood Lodge Nature Reserve.

Timms Bridge, Shackerstone, 352075 (30N). Field bridge over **Ashby Canal**.

Tin Meadow, Charley, 455160 (41N). Old enclosure located to the south-west of Mount St. Bernard Abbey.

Tinsel Lane, Market Bosworth. Location not determined.

Tipping's Gorse, Croxton Kerrial, 858291 (82P).

Tonge, Breedon on the Hill, 418232 (42B). Hamlet with **mineral railway**; Tonge Gorse is at 436223.

Tooley Hall, Peckleton, 475994 (49U). **Spinneys** at 478993 and 476002.

Toot Hill, Groby, 511077 (50D). Northern extremity of Martinshaw Wood. The rest of the wood is in Ratby parish.

Top Bridge, Quorndon, 562186 (51U). Field bridge over **Grand Union Canal**.

Top Spinney, Sutton Cheney (Shenton), 390997 (39Z).

Top Town Plantation, Walton on the Wolds, 597198 (51Z).

Towers Hospital, Leicester (Humberstone), 617061 (60D). Formerly the Borough Lunatic Asylum from the grounds of which there are records in HG.

Town Bridge, Shackerstone, 373070 (30D). Carries Heather Road over **Ashby Canal**.

Trent, River, 42DIJP, 43KV. The county boundary follows the north bank of the river from west of King's Mill to south of Derwent Mouth (43K); it then diverges to the south of the river for a short distance, crosses it and follows an old meander west of Sawley, rejoining the main stream west of Harrington Bridge; thence to the mouth of the River Soar the boundary is in midstream; **Sawley Cut** is wholly in Leicestershire.

Trent Hollow Lane. Could refer to Trent Lane, Castle Donington, 444282 (42P).

Triangle Spinney, Sheepy (Sibson and Wellsborough), 362032 (30R).

Trilobate Plantation (Charnwood). Location not determined.

Tryon Spinney, Tilton/Skeffington/Loddington, 769031 (70R).

Tugby and Keythorpe, parish, 70KLQR, 79PTUZ.
Contains: Tugby village at 762010; Keythorpe (village site); **Tugby Wood** at 766020; **Tugby Bushes** with habitat study 15 at 770021; **Hoothill Slang, grassland** at 769023 (all parts of the Leighfield Forest SSSI); the **Eye Brook** forming the north-eastern parish boundary.

Tur Langton, parish, 69WX, 79BCH.
Contains: village at 713945; tributary of the River Welland forming the eastern boundary.

Turn Bridge, Shackerstone, 376067 (30T). Carries Station Road over **Ashby Canal**.

Turnover Bridge, Oadby and Wigston (Wigston Magna), 623961 (69I). Over **Grand Union Canal**.

Tweed River, 30FKQV, 39JZ, 49DEI. Rises in Barwell; flows west between Dadlington and Stapleton, past Shenton and Sibson, and joins the River Sence at Ratcliffe Culey at 325998.

Twelve Bridges, Leicester. Location not determined.

Twenty Acre, Burton on the Wolds, 642210 (62K). Common land and SSSI, **rough grass** and **scrub** with habitat study 47 at 641211.

Twycross, parish, 20WX, 30BCDGHIJLMNPST.
Consists of the former parishes of Twycross, Gopsall, Norton juxta Twycross and Orton on the Hill.
The former parish, 30GHLMN, contains: village at 338049; Little Twycross; **Twycross Park** at 340057; Twycross Hill at 333047; **River Sence** forming the south-eastern parish boundary. The roadside **hedges** on the A444 from Twycross towards Appleby contained a rich variety of *Rubus* species during the nineteenth century. Twycross Zoo at 318063 is in Norton juxta Twycross former parish.

Twyford and Thorpe, parish 70EJP, 71AFGKL.
Consists of the former parishes of Twyford and Thorpe Satchville.
The former parish, 70EJP, 71AFK contains: village at 729101; about 1.2km of dismantled railway; **Ashby Folville Brook** passing through the village.

Tyburn, Newtown Linford, 523105 (51F). Hill in Bradgate Park.

Tyler Bridge, Loughborough (Shepshed), 480207 (42V). Carries minor road over **Black Brook**.

Tythorn Bridge, Oadby and Wigston (Wigston Magna), 618964 (69D). Over the **Grand Union Canal**.

Tythorn Hill, Wistow, 627964 (69I).

U

Ullesthorpe, parish, 48Y, 58CDEI.
Contains: village at 505875; earthworks at 501874;

marshy meadows at 512889; about 0.2km. of A5; 2.8km. of dismantled railway; Ullesthorpe Court at 507884 with spinneys.

Ulverscroft, parish, 41VWX, 51ABC.
Has no village, though part of Copt Oak hamlet is in the parish. The larger part of the Ulverscroft Valley SSSI lies within the parish. Contains: **Ulverscroft Cottage** at 492111; **Ulverscroft Priory** at 501127 with **moat** at 501126 and recently constructed **pools** at 500127; **Ulverscroft Pond** at 498128; **Ulverscroft Wood** at 491112; **Ulverscroft Brook** 41W, 51AB, with habitat study 72 at 501124; **marshland** at 501123; **Ulverscroft Nature Reserve**, 41W, (partly National Trust property) containing **heath, marsh, grassland** and **woodland**; **Coalbourn Wood**; **Nowell Spring Wood**; **Poultney Wood**; **Stoneywell Wood**.

Ulverscroft Lane, Newtown Linford, 51AB. Habitat study 87 is of **hedgerow** at 506120.

Upper Fields Farm, Swepstone, 373111 (31Q). **Pond** at 375111; **marsh** at 378116.

Upton, former parish, Sheepy, 30KQ, 39PTUYZ.
Contains: hamlet at 363996.

V

Vale of Belvoir, SK62, 72, 73, 74, 83, 84. Tract of country divided equally between Leicestershire and Nottinghamshire, drained by a number of streams flowing north-eastwards to join the River Smite; its south-eastern boundary is clearly marked by the **marlstone escarpment** terminating at Belvoir in the north-east and the slopes of **Dalby Wolds** in the south-west. The main botanical features are the **woodlands** of the escarpment, the **Grantham Canal**, the **dismantled Melton to Bottesford railway**, and a number of **species-rich meadows** on the floor of the Vale. The subject of a controversial proposal to mine coal on a large scale.

Valley Farm, Ashby de la Zouch (Packington), 351148 (31M).

Venonae, Sharnford/Claybrooke Magna, 473887 (48U). Roman settlement site at junction (High Cross) of Watling Street and Fosse Way.

Verney Spinney, Westrill and Starmore, 582813 (58V).

Via Devana, SK50, 60, 69, 79. Name given in the mid-eighteenth century to the Roman road from Leicester through Godmanchester to Cambridge and Colchester, between Leicester and Medbourne it is also called the Gartree Road. A metalled road covers it between Leicester (south of Evington Golf Course) and Shangton; thence it is a green lane nearly as far as Glooston, and finally its course can be traced across fields most of the way to the River Welland which it crosses south-east of Medbourne; the site of the crossing was discovered in the 1960's.

Vicary Farm, Woodhouse, 541152 (51M).

Vice's Bridge, Oadby and Wigston, 584979 (59Y). Over the **Grand Union Canal**.

Victoria Park, Leicester (St. Marys), 597031 (50W).

Vinegar Hill, Worthington, 412192 (41E).

Vowle's Gorse, Hallaton, 789994 (79Z).

W

Wailes, The, Frisby on the Wreake, 686174 (61Y). SSSI and county trust nature reserve. Contains marsh with habitat study 53 at 687173 and oxbow with habitat study 64 at 686174.

Wain Bridge, Wistow, 69I. The name of two bridges, one over the **Grand Union Canal** at 633966 and the other over the **River Sence** at 635964.

Walcote, Misterton, 567837 (58R). The main village of Misterton parish.

Walker's Farm, Shackerstone (Odstone), 386082 (30Z). Pond.

Wallis's Spinney, Thurnby. Location not determined.

Waltham, parish, 71U, 72QRVWXY, 82BCDE.
Consists of the former parish of Waltham on the Wolds and the greater part of Thorpe Arnold parish. The former parish of Waltham on the Wolds, 72WXY, 82BCDE contains: village at 802250; **old fish ponds** at 801257 and 802239; **grassland** at 787241; **marshy fields** east of Waltham Lodge at 810232; Waltham Thorns at 802225; Waltham New Covert at 813228; dismantled mineral railway crossing the north-west corner of the parish; **Waltham Quarry** with habitat study 102 at 800264. The television mast at 809233 is a conspicuous landmark. Half of Stonesby Gorse is in the parish.

Walton, Kimcote and Walton, 595869 (58Y). Hamlet.

Walton on the Wolds, parish, 51UZ, 52V, 61EJ, 62AFK. Contains: village at 591196; the **River Soar** forming its western, Walton Brook its northern, and Fishpool Brook part of its southern boundary; **Walton Thorns** at 627204; **Walton Holme Wood** at 562197; Ash Plantation (Cradock's Ashes).

Wanlip, parish, 51QVW, 61A.
Contains: village at 602109; the **River Soar** forming its eastern boundary. **'Wanlip Gravel Pits'** are in Syston parish 61A and have varying botanic interest as new ones are dug out and old ones are filled in.

Warren, The, Belvoir. Location not determined.

Warren Hill, Newtown Linford, 529118 (51F).

Warren Hills, Charley (Whitwick), 459151 (41M).

Warren Pond, Staunton Harold, 393207 (32V).

Wartnaby, former parish, Ab Kettleby, 62W, 72AB. Contains: village at 712231 with an interesting **wall flora**; **fish pond** at 708221; **marlstone escarpment** forming the northern parish boundary; **Cant's Thorns**. *Wartnaby Stonepit* was presumably near **Stonepit Spinney** at 705237.

Wash Brook, Oadby and Wigston (Oadby), 60AF. Is renamed the Saffron Brook at 605004.

Wash Dike Bridge, Houghton on the Hill, 692029 (60W). Carries a minor road over the **River Sence**.

Watling Street, 39W, 48PTUXY, 49ABFK, 57JNP, 58BCFG. The verges have rarely been stable for long enough to have more than a transient casual flora; until about 1935 the county boundary shared with Warwickshire followed the road all the way from Catthorpe to Witherley, with a diversion near Hinckley to include Hydes Pastures. It was generally deemed to be in the centre of the roadway but followed the southern boundary hedge or fence for short stretches near Claybrooke and Burbage. Then mainly for the convenience of Highway and Rating authorities it was moved about 80-200 metres to one side or the other (to the north east from Catthorpe to Cotesbach, and Hinckley to Witherley, and to the south-west from Lutterworth to Hinckley). Hyde Pastures reverted to Warwickshire and other minor adjustments were made near Hinckley.

Watson's Spinney, Hungarton, 687085 (60Z).

Welby, former parish, Asfordby, 71EJ, 72AFG.
  Contains: deserted village site, with church at 725209 still in use; Holwell Works with **pools** and **spoil heaps**; **Welby Osier Beds** at 719210 now affected by the new Asfordby Mine; traces of a dismantled mineral railway.

Welford and Kilworth Station site, North Kilworth, 624835 (68G). On dismantled Rugby to Market Harborough railway.

Welford Reservoir, Husbands Bosworth, 646810 (68K). Only half the dam and a few hundred square metres of water are in Leicestershire.

Welham, parish, 79KLMQRS.
  Contains: village at 765924; the **River Welland** forming the south-eastern parish and county boundary. Both **Welham Crossing** at 754911 and **Welham Sidings** at 769917, are on a **dismantled railway** in Thorpe Langton parish.

Welland, River, SP68, 78, 79, 89. Rises near Sibbertoft (Northamptonshire) and flows thence to the county boundary south-east of Husbands Bosworth at 655833. It forms the county boundary between Leicestershire and Northamptonshire all the way to the Rutland border except for about 3.5km. through Market Harborough (where it ceased to be the county boundary when Little Bowden was added to Leicestershire) and in short stretches where it follows old oxbows rather than the modern dredged channel. It was last dredged and 'improved' in 1969-71; some of the old vegetation is now beginning to be re-established, though many riverside trees have been lost.

Wellsborough, Sheepy (Sibson and Wellsborough), 361023 (30R). Hamlet; Wellsborough Hill is at 365025.

Welsboro Bridge, Market Bosworth, 385017 (30V). Over the **Ashby Canal**; spinney at 382017.

West Bridge, Leicester (St. Mary's), 581043 (50X). Main road bridge over River Soar; replaced and rebuilt many times in its long history.

Westcotes Park, Leicester (St. Mary's), 569034 (50R). Residential suburbs; large area of allotments at 565037.

West End, Leicester (St. Mary's), 573038 (50R).

Western Park, Leicester (New Parks/St. Mary's), 555045 (50M).

West Langton, parish, 69W, 79ABCFG.
  Contains: Langton Hall at 716931; about 2.5km. of working railway with dismantled station site at 720919; Langton Brook forming the southern boundary; **pond** at 717940.

Westmeadow Brook, 41IJ, 42FKLR. Formed from brooks rising in Swannington and Belton, and flowing through Osgathorpe, Belton and Long Whatton parishes to join Diseworth Brook at 469239, then becoming the Long Whatton Brook.

Westrill and Starmore, parish, 57Z, 58QRVW.
  Contains: Stanford Hall and most of its Park. The **River Avon** forms the south-eastern parish and county boundary. Takes its name from two deserted mediaeval villages of Westrill and Starmsworth.

Wet Wang Wood, Belvoir. Probably refers to one of the woods adjoining West Wong at 811334.

Whatborough, parish, 70STXY.
  Contains: Whatborough Hill at 769060; Whatborough Farm at 774056.

Whatton House (as *Whatton Hall*), Long Whatton, 493241 (42X).

Wheat Hill Spinney, Burton and Dalby (Little Dalby), 782138 (71W).

Whetstone, parish, 59LMNPRSTUWX.
  Contains: village at 557975; **Whetstone Gorse** at 560942; Whetstone Brook, 59NRST; **dismantled railway** running through the parish from north to south with station site at 555976; another short length of **dismantled railway** crossing the southern part of the parish; 1.5km. of **working railway**; nearly 3km. of M1 motorway; the River Sence forming the northern parish boundary and the **River Soar** forming the north-western boundary.

Whissendine Station site, Wymondham, 836166 (81I). Relict garden flora still remains.

Whitehill, Coalville (Hugglescote), 41F. The former name of the northern end of Ellistown.

White Hollows. In the north-west of the county, location not determined.

White Horse Wood, Loughborough (Shepshed), 468185 (41U).

White House, Appleby Magna, 322107 (31F). Spinneys nearby.

White Lodge, Ashby de la Zouch (Willesley), 343152 (31M). **Marsh** nearby at head of **Willesley Lake**.

White Lodge, Eaton (Goadby Marwood), 783273 (72Y). Traces of mineral railway.

White Moor Covert, Sutton Cheney, 383990 (39Z). Richard III's encampment on the eve of Bosworth Field.

Whittle Hill, Charley, 495158 (41X).

Whitwells Farm, Coton. Location not determined, probably south-west of Market Bosworth.

Whitwick, former parish, Coalville/Charley, 41GHILMNST.
The greater part was incorporated in Coalville in 1936 and the smaller part was added to Charley. Contains: village at 435162; part of Coalville town; several hamlets associated with working and abandoned collieries and quarries; habitat study 26 of Coalville Meadows SSSI, **grassland** at 447152; **Cademan Wood**, **High Sharpley**, **Holly Hayes Wood**, **Peldar Tor** and **Spring Hill** (all in Coalville) and part of **Charnwood Lodge Nature Reserve**, **Kite Hill** and part of **Blackbrook Reservoir** (now in Charley). *Whitwick Rocks* mentioned in HG, location not determined.

Wicket Nook, Ashby de la Zouch, 360202 (32KQ). House on the county boundary with Derbyshire.

Wide Bridge, West Langton/East Langton, 721919 (79F). Carries B6047 over Langton Brook close to Langton Railway Station site.

Wide Lane, Wymeswold, 62BGL. Wide **verges**.

Wignell Hill, Nevill Holt, 828939 (89GH).

Wigston Magna, former parish, Oadby and Wigston, 50V, 60A, 59YZ, 69DEHIJ.
Contains: village at 604986; **Navvy's Pit**; Kilby Bridge; **Kilby Pit** at 615968; about 5km. of working and dismantled railway (see also under South Wigston); about 5km. of the **Grand Union Canal**.

Wigston Parva, parish, 48PU, 49Q.
Contains: village at 466897; about 1.5km. of Watling Street (A5) lies within the parish.

Wilderness, The, Cadeby, 425026 (40G). **Pond**.

Wilderness, The, Stretton Magna, 651996 (69P).

Willesley, former parish, Ashby de la Zouch, 31GHLM.
Incorporated into Ashby de la Zouch except for a small area at the southern edge which was added to Oakthorpe and Donisthorpe. Contains: **Willesley Park** with church at 340147 and a **golf course** interspersed with woods and spinneys; **Willesley Lake** with a dam at 337146 and **marsh** at 343152; *Willesley Wood* at 332154, destroyed by mining. Willesley Basin on the now filled Ashby Canal 328140 is in Oakthorpe and Donisthorpe. *Willesley railway cutting* at 333159 is in Ashby de la Zouch former parish.

Willoughby Brook, 52GHKL. Rises near Six Hills and flowing north-west forms the western boundary of the part of Nottinghamshire added to Wymeswold and Burton on the Wolds in 1966. Willoughby Gorse is at 625252.

Willoughby Waterleys, parish, 59QRVW.
Contains: village at 575925; a large pond at 580923; two brooks running northwards through the parish, joining at 572933 to form the Whetstone Brook.

Wilson, Breedon on the Hill, 405247 (42C). Hamlet.

Windmill Hill, Ashby de la Zouch, 361161 (31T). Now a housing estate.

Windmill Hill, Potters Marston. Location not determined.

Windmill Hill, Woodhouse, 526142 (51H).

Windmill Lodge, Knossington, 805076 (80D).

Windsor Hill, Belvoir, 815322 (83B).

Winkadale, Thurnby (Bushby), 662041 (60S). House.

Winterfield Spinney, Misterton, 564844 (58S).

Wistow, parish, 69HIJMNP.
Consists of the former parishes of Wistow and Newton Harcourt.
The former parish, 69HIMN, contains: church and deserted mediaeval village site at 643959; *Wistow Hall* at 641958 with park and lake; the River Sence forming the northern parish boundary.

Withcote, parish, 70XY, 80CD.
Contains: church at 795057; Withcote Hall at 797058; **lake** at 797056; Withcote Lodge at 807048; **Sauvey Castle**; Sauvey Plantation; the **River Chater** forming the southern parish boundary.

Witherley, parish, 39EIJMNPTU, 30FK.
Consists of the former parishes of Witherley, Atterton, Fenny Drayton and Ratcliffe Culey. The south-west boundary formerly coincided with Watling Street but is now about 70m. from it on the Leicestershire side.
The former parish, 39EIJ, contains: village at 325973; the site of the Roman town of Manduessedum (Mancetter) straddling the county boundary at 327968; the **River Anker** forming the parish and county boundary to the west.

Wood Bridge, Lutterworth, 525841 (58H). Carries A427 over a tributary of the River Swift.

Wood Brook, 51CDIJ, 52F. Rises at Buck Hill and flows north east across Nanpantan and Loughborough to join **Summerpool Brook** and thence into the **River Soar**.

Wood Close Plantation, Lubbesthorpe, 547016 (50K).

Woodcote, Ashby de la Zouch (Blackfordby), 353186 (31P). The county boundary crosses A50 here and passes through the house.

Wood Farm, Freeby (Brentingby), 789214 (72V). A small patch of **relict grassland** at 793216 appears to be all that remains of the important *Brentingby Field* frequently mentioned in HG.

Woodhouse, parish, 41X, 51BCGHILMN.
Formerly included about 4 sq.km. round Bawdon Lodge transferred to Charley parish and 1 sq.km. round Buck Hill transferred to Loughborough when the present boundaries of Coalville, Charley and Loughborough were established. Contains: the villages of Woodhouse at 538151 and Woodhouse Eaves at 531140; numerous important botanical sites including **The Brand, Broombriggs, Beacon Hill, Beaumanor Park, Hangingstone Hills**, a small part of **Swithland Reservoir, Mucklin Wood**, a short length of **railway**.

Woodside Cottage, Quorndon, 568154 (51S). **Quarry** at 566155.

Woodthorpe, former parish, Loughborough, 51INPU.
Contains: hamlet at 543173; about 200m. of **Grand Union Canal** with Woodthorpe Bridge at 558187; privately operated **railway** crossing the parish; 100m. of working railway; 100m. of the River Soar forming the eastern parish boundary.

Woodville, Derbyshire. About 1 sq.km. of Blackfordby was transferred in 1897 from Leicestershire to Derbyshire, as part of the new parish of Woodville in the latter county and is therefore in vc55. *Woodville House* is mentioned in HG.

Woodwell Head, Wymondham (Edmondthorpe), 879176 (81TYZ). Locally called 'Woodlehead'.

Woolrooms, Worthington, 406178 (41D). Hamlet.

Worthington, parish, 31YZ, 32VW, 41DEJ, 42ABF.
When Coalville's present boundaries were established, Worthington gained nearly 2 sq.km. from Osgathorpe and Thringstone and lost a very small area to Coleorton. Contains: village at 406206; Newbold and Griffydam hamlets; **The Smoile**; nearly 4km. of **dismantled railway** (station site at 408210); **Barrow Hill Quarry**; a small part of **Cloud Wood**; numerous small fields of **species-rich grassland** in the southern part of the parish; brooks flowing north through the parish which eventually merge to form Ramsley Brook. *Worthington Rough* and *Worthington Field Farm* at 407200 are mentioned in HG.

Wreake, River, 51W, 61BGMSTYZ, 71EJ.
Continuation of the **River Eye** west of Eye Kettleby. Formerly the boundary between parishes along much of its length, but with parish mergers, no longer so. Flows south through Kirby Bellars, Asfordby, Frisby on the Wreake, Hoby, Rotherby and Ratcliffe on the Wreake. The **Grand Union Canal** joins it at 609122 and the last 1.5km. before its junction with the **River Soar** at 595127 is canalised with a lock at 603124.

Wycomb and Chadwell, former parish, Scalford, 72RSWX.
Contains: Wycomb hamlet at 774248; Chadwell village; traces of about 2km. of dismantled mineral railways; Scalford Brook crossing the parish from north to south.

Wyfordby and Brentingby, former parish, Freeby, 71UZ, 72QVW.
Contains: the village of Wyfordby at 793189; Brentingby; the **River Eye** forming its southern boundary; a working **railway** crossing it from east to west; traces of the **Oakham Canal** near the railway.

Wykin, Hinckley, 407953 (49C). Hamlet. **Wykin Spinney** at 398953 is partly in Higham on the Hill.

Wymeswold, parish, 52VWX, 62ABCFGHKLM.
The parish was enlarged in 1966 when that part of Willoughby on the Wolds parish which lies between Willoughby Brook and the Fosse Way was transferred from Nottinghamshire to Leicestershire. Contains: village at 603234; **Wymeswold Brook** flowing westward across the parish with adjacent **grassland** south of Narrow Lane; county trust nature reserve (**grassland**) at 611231; **grassland** at 590238; **verges** of Wide Lane from 641228 westward to the village. Wymeswold Airfield is mainly in Hoton and Prestwold.

Wymondham, parish, 81IJNPTUYZ, 82V, 91E.
Consists of the former parishes of Wymondham and Edmondthorpe.
The former parish, 81IJNPUZ, 82V, contains: village at 851186 with **wall flora**; about 7km. of **dismantled railway** with station site at 851190; Wymondham Rough SSSI which is part of **Wymondham Rough Nature Reserve**, 832173, including **wetland, woodland, species-rich pasture** with habitat study 38 at 833175 and a short length of the **disused Oakham Canal** alongside a length of working railway.

Wymondham Brook, 81DEIPU, 82QR. Rises in Buckminster and flows south-west through **species-rich meadows** north of the railway viaduct at 858190, into Wymondham village; it passes along the south-eastern edge of Day's Plantation, and joining other brooks rising in Rutland, passes through Stapleford Park to join the River Eye near Saxby Station. It is of botanical interest at several points.

Y

Yennards, The, Hinckley (Earl Shilton), 493971 (49Y). Farm. Yennards Quarry is an alternative name for Barrow Hill Quarry.

Yenwoods, The, Lubbesthorpe, 529006 (50F). House.

Z

Zouch Bridge, Loughborough (Hathern), 502232 (52B). Carries A6006 over the original course of the **River Soar** which marks the county boundary. Zouch Mills and Zouch Cut are in Nottinghamshire.

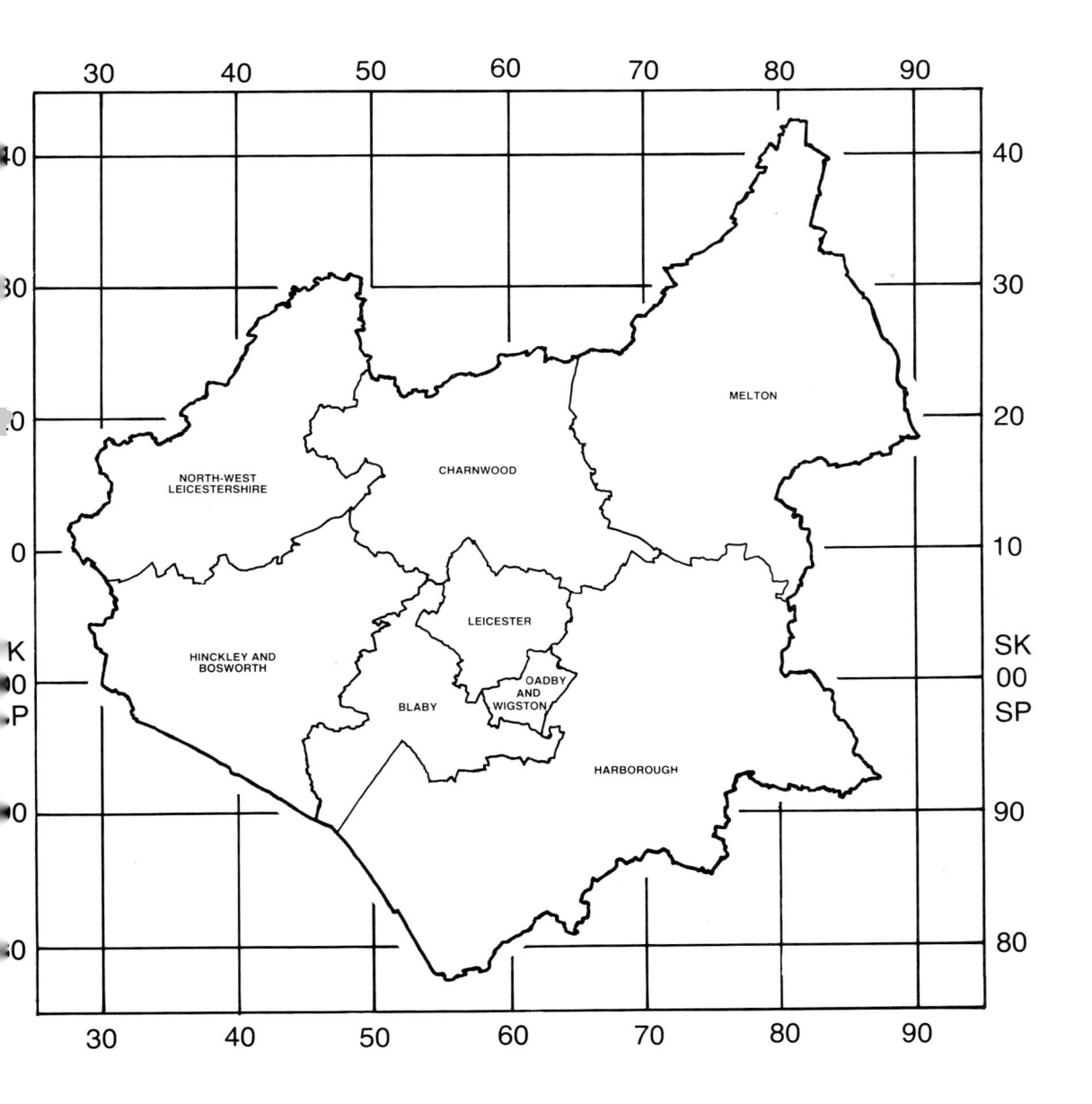

Figure 27. Relationship of the former county to the present districts.

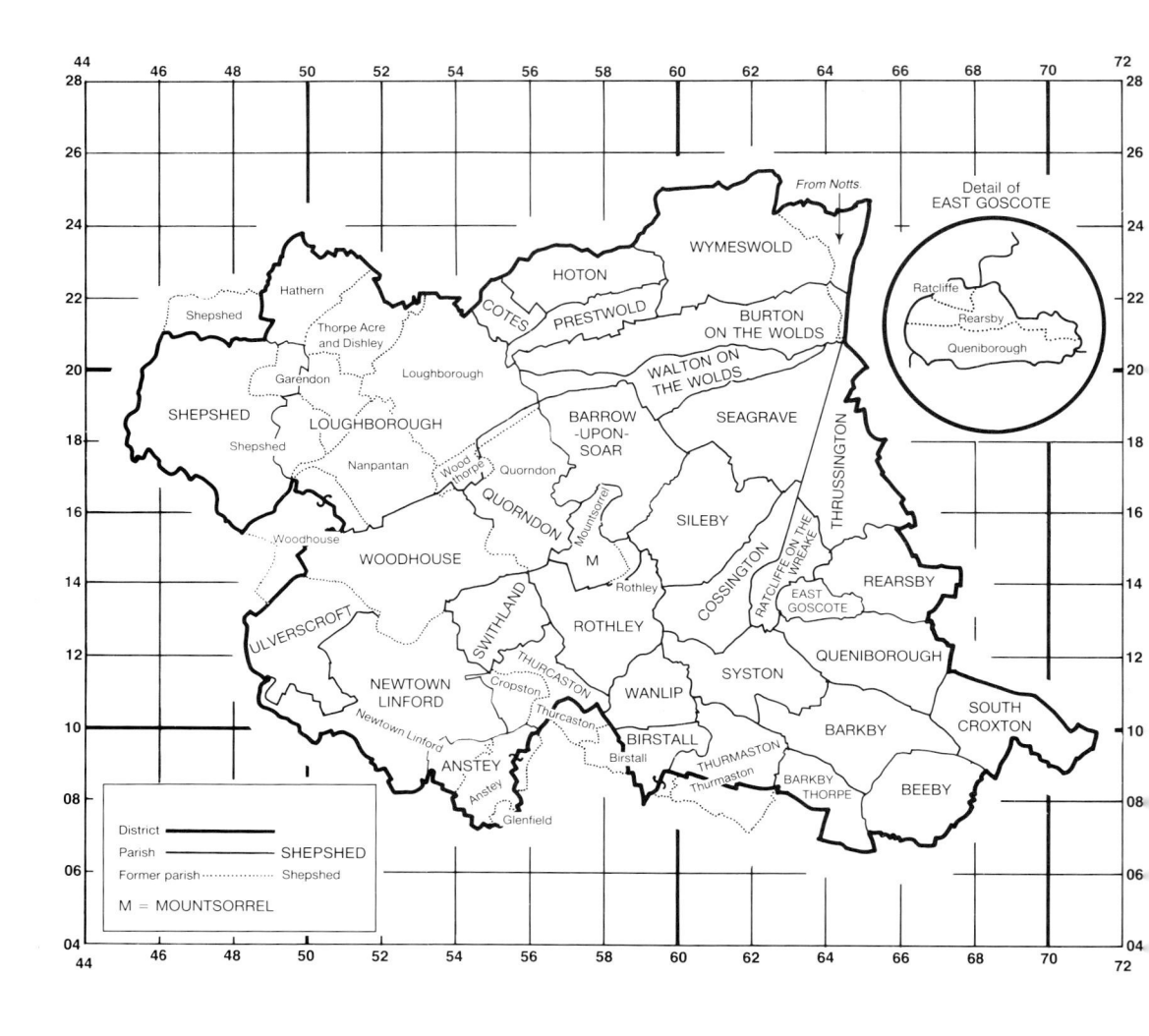

Figure 28. Parishes and former parishes in the Borough of Charnwood.

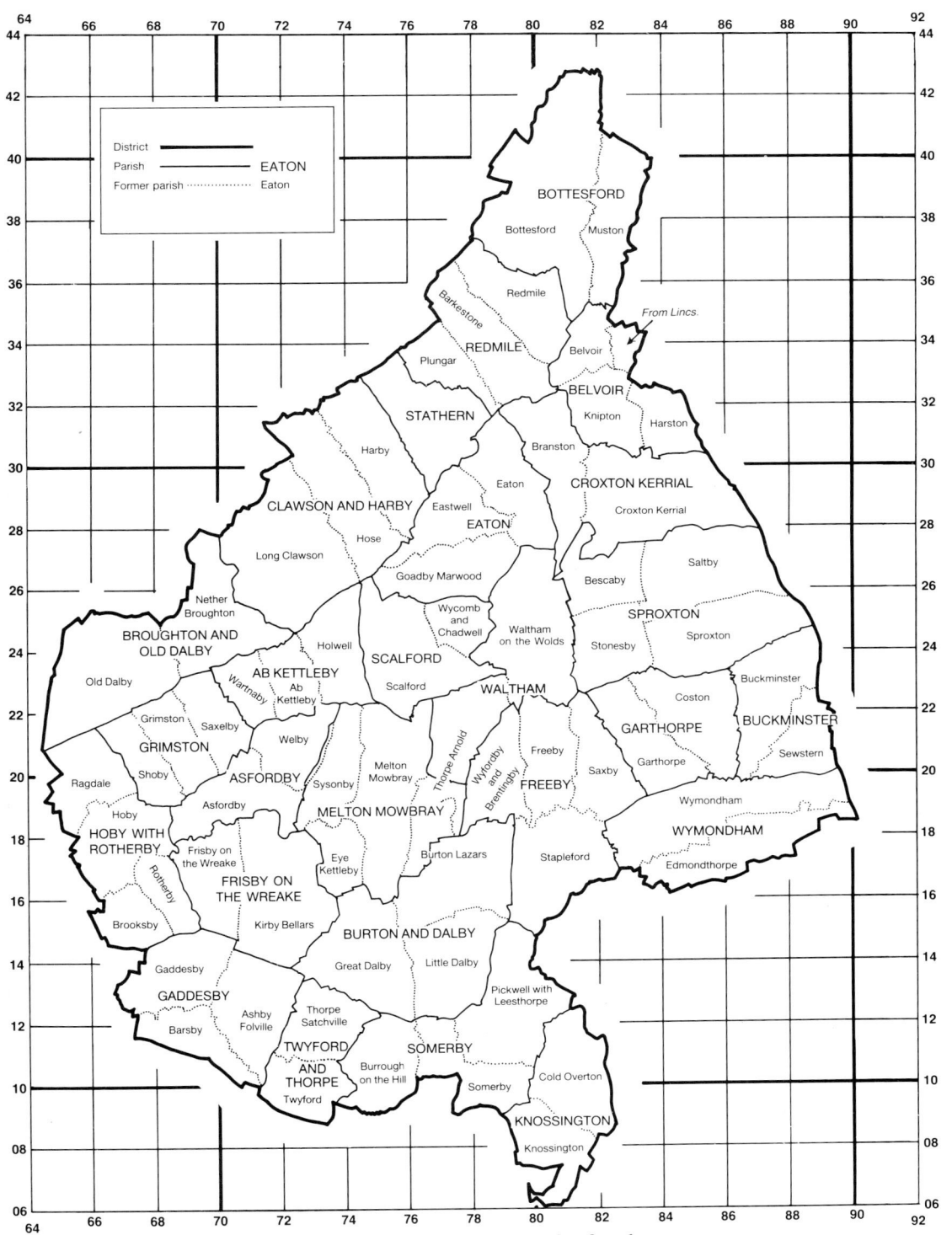

Figure 29. Parishes and former parishes in the Borough of Melton.

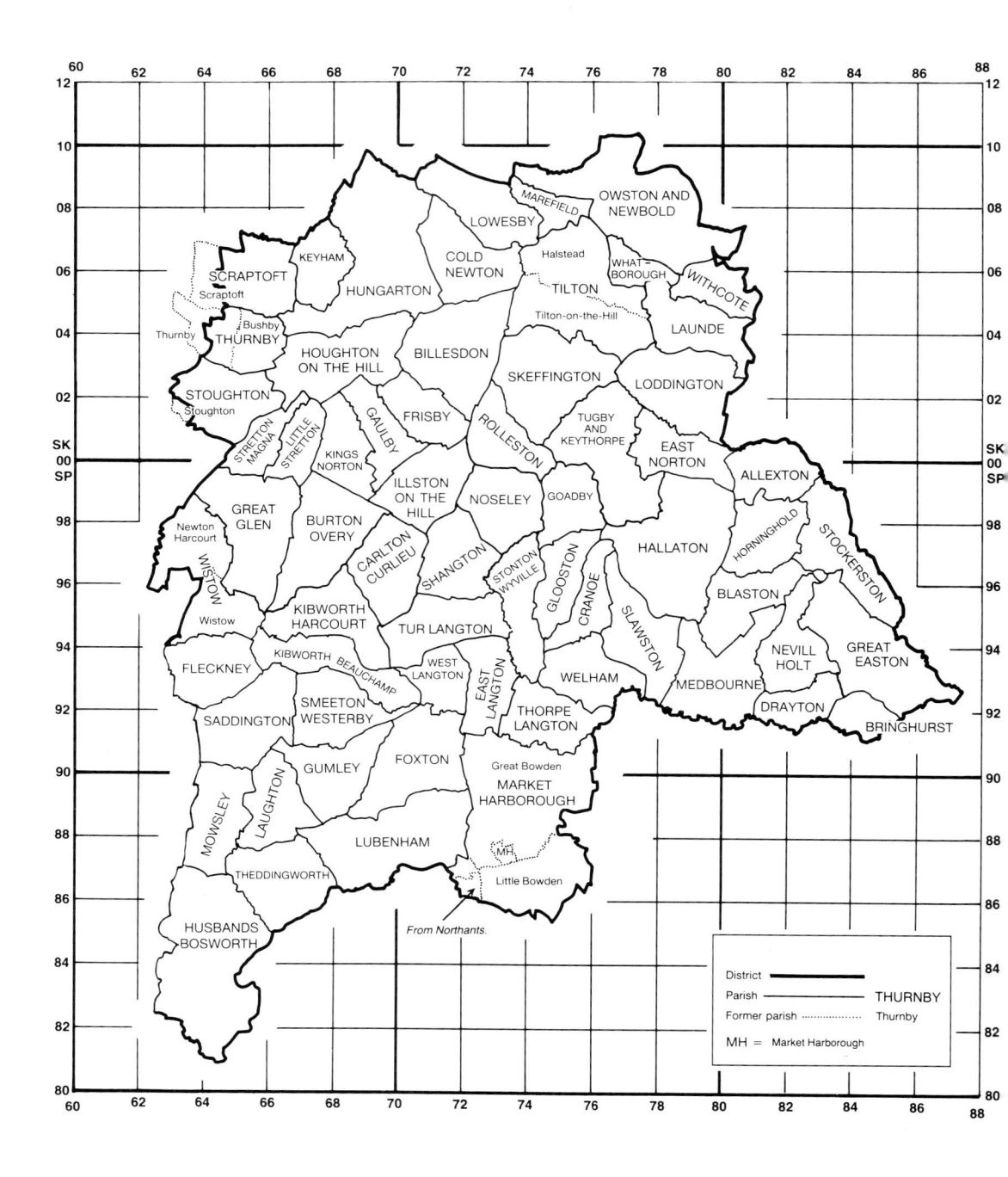

Figure 30. Parishes and former parishes in the eastern part of Harborough District.

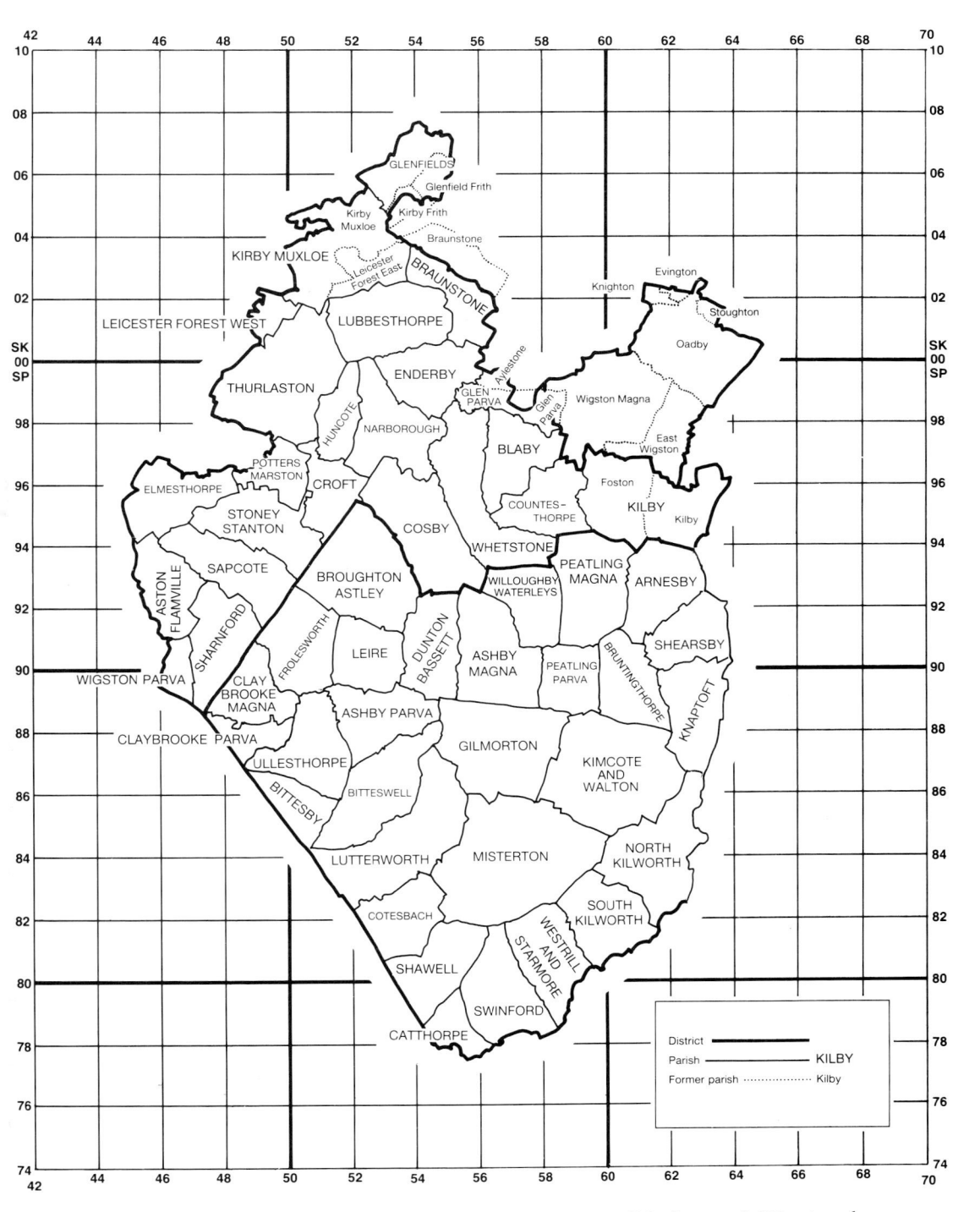

Figure 31. Parishes and former parishes in the western part of Harborough District, the Borough of Oadby and Wigston, Blaby District.

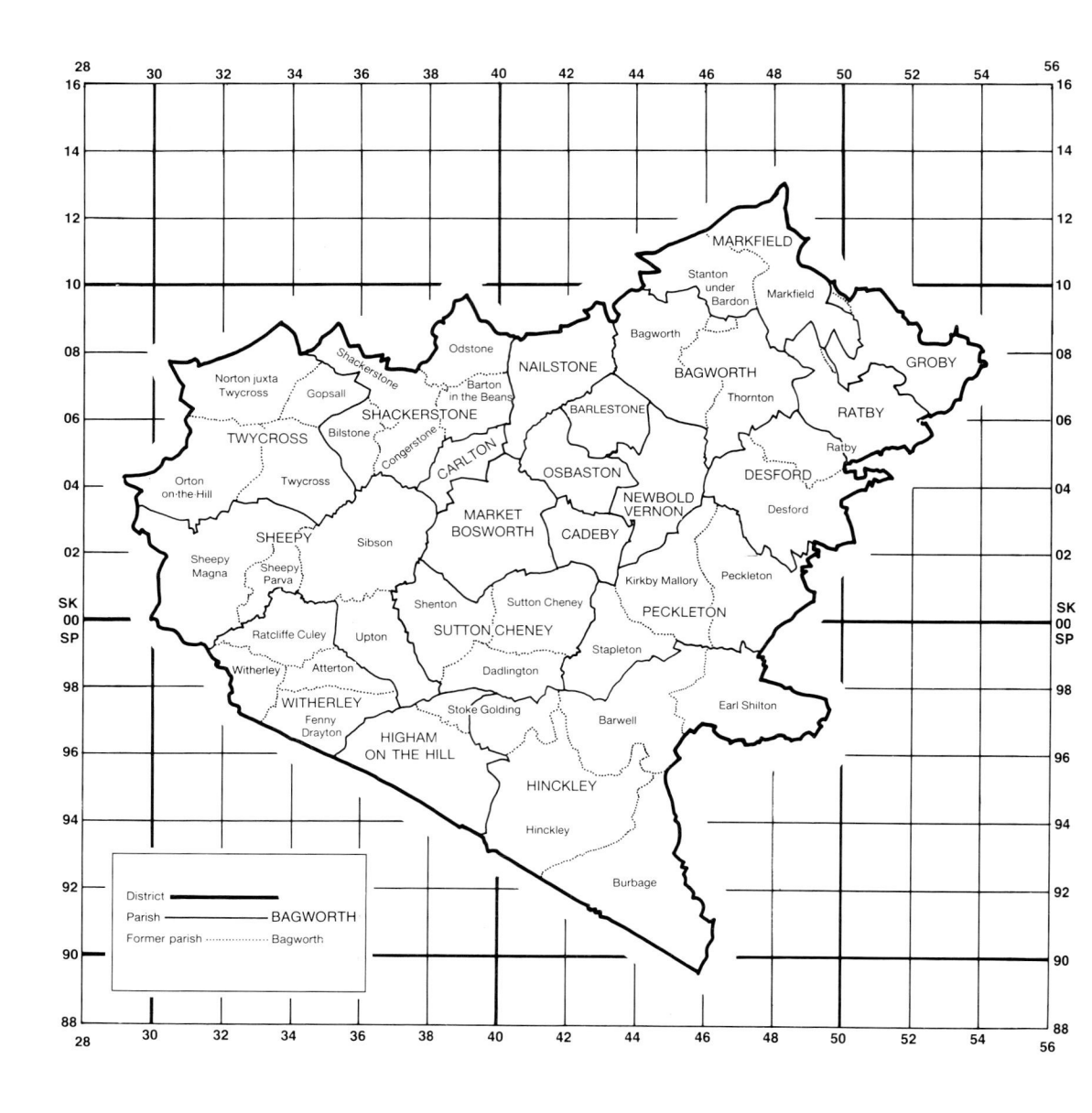

Figure 32. Parishes and former parishes in the Borough of Hinckley and Bosworth.

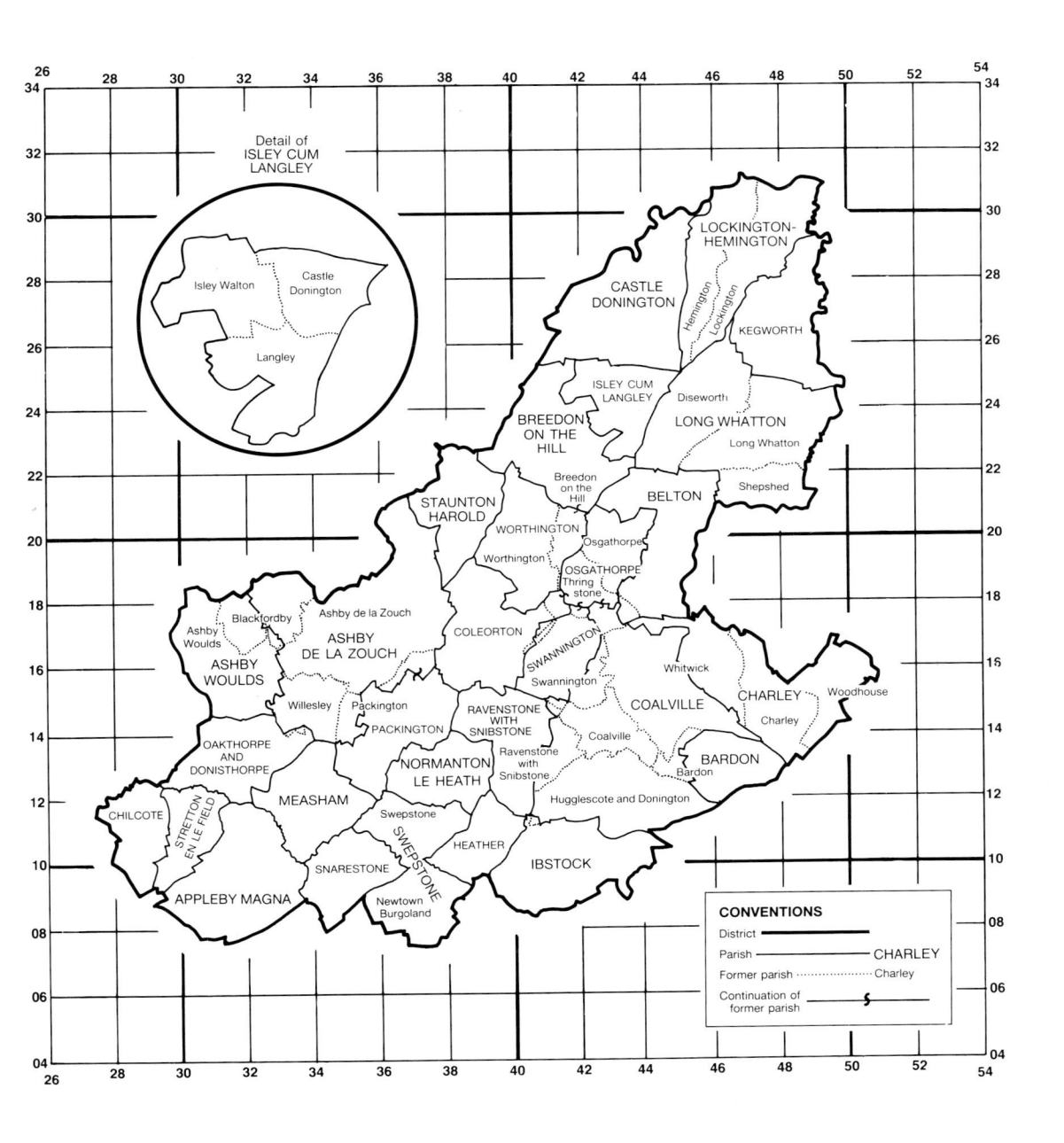

Figure 33. Parishes and former parishes in North-West Leicestershire District.

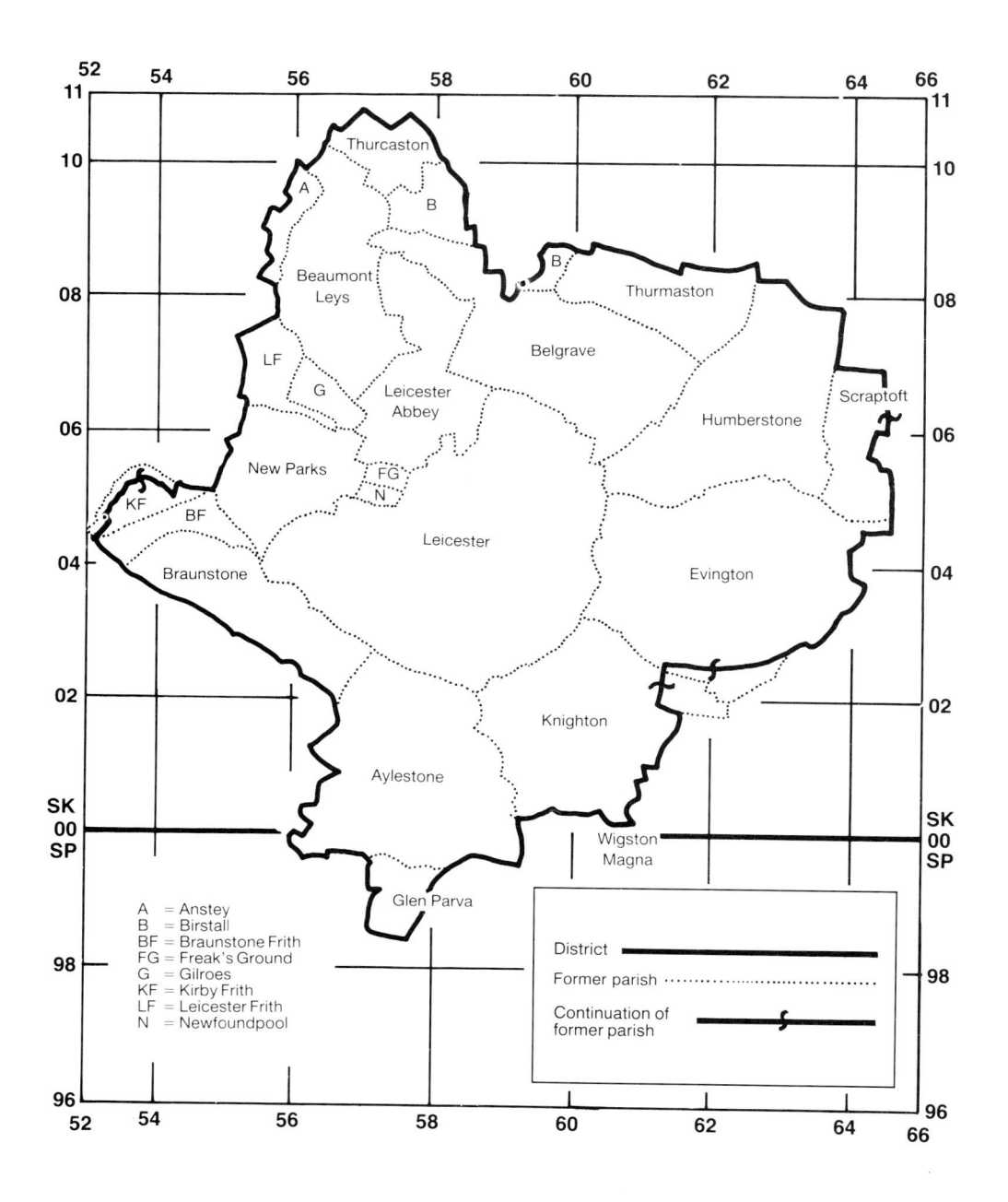

Figure 34. Former parishes in the City of Leicester.

NOTES ON CONTRIBUTORS

The biographical details included in this Flora of some of those who contributed to Horwood and Gainsborough (1933) proved difficult to ascertain after a lapse of over fifty years. For the benefit of posterity, it was therefore decided to incorporate in this Flora brief notes on contributors, compiled by themselves. Any selection of those to whom this doubtful privilege might be extended was certain to be invidious, but it was agreed that it should include members of the Flora Committee, together with other contributors to the text of the Flora. There are many others who gave generously of their time and effort, but their contribution is, we hope, adequately acknowledged elsewhere in the text.

Stephen H. Bishop
Member of the Flora Committee, with a special interest in the botany of the west and north-west of the county. Born at St Albans 27.7.1946; educated at Mill Hill Boys School and trained with the Royal Horticultural Society at Wisley. Head Gardener to the Leicestershire Museums Service 1974-80 and latterly Head Gardener at Sudeley Castle, Gloucestershire. Scientific Officer to the Leicestershire and Rutland Trust for Nature Conservation 1976-80. Currently working on a new Flora of Gloucestershire and North Avon.

Ian M. Evans, M.A., P.G.C.E., F.M.A.
Chairman of the Flora Committee. Born in north London 25.7.1936; educated at Mercers' School and Magdalene College, Cambridge. Keeper of Biology at Leicester Museum 1959-1972 and latterly Assistant Director (Natural Sciences) and County Ecologist with the Leicestershire Museums Service. Scientific Officer to the Leicestershire and Rutland Trust for Nature Conservation 1963-1976. Hon. Secretary (1960-1970) and President (1970 to present) of the Natural History Section of Leicester Literary and Philosophical Society.

Patricia A. Evans, B.A. (Mod.) (formerly Candlish, née Padmore)
Member of the Flora Committee and Joint Editor of this Flora. Born in Leicester 5.5.1929; educated at Loughborough High School and Trinity College, Dublin. Researched *Ranunculus scoticus* under Prof. T. G. Tutin at Leicester University. Author of weekly natural history column in *Leicester Mercury* since 1975. Ecological survey worker since 1978 with the Leicestershire Museums Service. Founder member, former Hon. Secretary and Editor of the Loughborough Naturalists' Club.

Peter H. Gamble
Member of the Flora Committee. Born at Quorn 20.3.1927; educated at Quorn, where he has lived all his life. Apprenticed and worked as a bricklayer for 19 years. Technician and latterly Senior Technician at Loughborough Technical College from 1961 until his retirement in 1987. Entomologist, ornithologist and natural history photographer. Adult education lecturer in natural history for over twenty years. Founder member and former Editor of the Loughborough Naturalists' Club.

F. Robert Green, M.A., Ph.D., Dip. Agric. (Cantab.)
Member of the Flora Committee. Born at Loughborough 26.6.1932; educated at Loughborough Grammar School and Jesus College, Cambridge. Lecturer in Agriculture at Nottingham University, where he took his Ph.D., 1956-1960. Farmed in Leicestershire until 1971, since then in Dorset. Former Chairman and Botanical Recorder of the Loughborough Naturalists' Club. Presently Vice-Chairman of the Dorset Farming and Wildlife Advisory Group.

Edward K. Horwood
Member of the Flora Committee; died 16.12.1977. See biographical note on page 85.

June M. Horwood, B.A., B.Sc. (Hons.) (née Hartshorn)
Member of the Flora Committee. Born in Leicester 16.6.1932; educated at Alderman Newton's Girls' School; B.Sc. in Botany (1961) as an external student of London University, B.A. in Biology (1980) with the Open University. Teacher of Biology at Wyggeston Girls' School (now a Sixth-form College, Wyggeston Collegiate) from 1964 to 1987.

John G. Martin, B.Sc. (Hons.), A.M.A.
Author of the section on geology and soils in the Flora. Born in east London 26.4.1947; educated at Haberdashers' Aske's, Hatcham, and Bristol University. Worked for Ordnance Survey and British Petroleum 1965-1970. Appointed in 1974 Assistant Keeper of Geology with the Leicestershire Museums Service and in 1979 Keeper of Earth Sciences. Interests include ornithology, Leicestershire geology, the palaeobiology of fossil marine reptiles and sauropod dinosaurs.

K. Guy Messenger, M.A., F.L.S.
Member of the Flora Committee, with special responsibility for the gazetteer and the accounts of *Hieracium*, *Rubus*, *Taraxacum* and *Ulmus*. Born at Hampstead 26.2.1920; educated at Felsted School and Emmanuel College, Cambridge. Served with Royal Corps of Signals in India and Ceylon, 1941-1945. Head of Biology at Uppingham School 1949-1968 and continued to teach there until his retirement in 1980. Cartobibliographer, with a particular interest in the 3rd series of the 1" Ordnance Survey. Author of the *Flora of Rutland*, published in 1971 and based on survey work 1958-1970. B.S.B.I. Recorder for Rutland 1958 to present.

Anthony L. Primavesi, C. Biol., M.I. Biol., F.L.S.
Hon. Secretary of the Flora Committee and Joint Editor of this Flora. Born in Northampton 18.12.1917; educated at Ratcliffe College, Leicestershire and Senior Biology Master there from 1947 until his retirement in 1982. Also taught chemistry and metalwork. Member of the Roman Catholic Order of the Institute of Charity. B.S.B.I. Recorder for Leicestershire since 1968 and Referee for *Rosa* since 1986.

Michael Walpole, F.C.A., F.L.S.
Hon. Treasurer of the Flora Committee. Born at Loughborough 13.2.1933; educated at Loughborough Grammar School. Qualified as a Chartered Accountant in 1955 and in 1959 joined Towles Ltd., a Loughborough hosiery company, of which he is now Financial Director. Founder member of the Loughborough Naturalists' Club. Hon. Secretary (1964-1973) and Chairman (1975-1986) of the Leicestershire and Rutland Trust for Nature Conservation. Bryologist and natural history bibliophile, with a special interest in British Floras. Hon. Treasurer of the B.S.B.I.

Alan W. Wildig, B.Sc. (Eng.), C. Eng., M.I. Mech. E.
Author of the section on climate in the Flora. Born in Coventry 3.5.1929; educated at King Henry VIII Grammar School, Coventry and London University. Indentured with Humber-Hillman Ltd., and presently Lecturer in Automotive Design in the Department of Transport Technology, Loughborough University. Interest in climate resulted from helping his daughter with her mathematics homework. Weather Recorder for the Loughborough Naturalists' Club since 1973.

INDEX

This index lists the scientific names of families and genera and the English names of species and groups of species. Each name is followed by the relevant page number in the systematic account and, where appropriate, by the number of the distribution map. References to taxa in other sections of the Flora are not indexed. Current scientific names are in **bold** and those synonyms used in *italic*. English names are in normal type. The nomenclature followed is defined on p. 167 in the Plan of the Systematic Account.

Abele 179, M5i
Acacia 223
Acer 231-232, M37h-j
Aceraceae 231-232
Achillea 270, M60c-d
Acinos 254, M51f
Aconite, Winter 193, M13k
Aconitum 194, M14a
Acorus 308, M85k
Adder's-tongue 174, M1h
Adonis 194
Adoxa 265, M57c
Adoxaceae 265
Aegopodium 241, M43a
Aesculus 232, M37k
Aethusa 241, M43g
Agrimonia 218, M28g-h
Agrimony 218, M28g-h
  Hemp- 267, M58i
Agropyron 303
Agrostemma 192
Agrostis 305, M84d-h
Aira 305, M83k-l
Ajuga 252, M49h
Alchemilla 220, M291-30a
Alchemilla 220
Alder 180, M6d
Alder Buckthorn 233, M38g
Alexanders 240
Alisma 291, M74h-i
Alismataceae 291
Alkanet 250, M48g
Alliaria 198, M17b
Allium 295-296, M77b-d
Allseed 230
Alnus 180, M6d
Alopecurus 306, M84l-85c
Alyssum 201
Amaranthaceae 189

Amaranthus 189
Amaryllidaceae 296
Ambrosia 269
Ammi 243
Amsinckia 250
Anacamptis 315, M90k
Anagallis 245-246, M45j-k
Anchusa 250, M48f
Anchusa 250
Anemone 194, M14b
Anemone 194
Angelica 243, M44a-b
Anisantha 302
Annual Mercury 230, M36j
Antennaria 268
Anthemis 270, M60b
Anthoxanthum 305, M84a
Anthriscus 240, M42g
Anthyllis 228, M35e
Antirrhinum 259, M53h
Antirrhinum 259
Apera 306
Aphanes 220, M30b-c
Apium 242, M43j-k
Apocynaceae 247
Apple 221, M30e
Aquifoliaceae 232
Aquilegia 196
Arabidopsis 199, M17c
Arabis 201
Araceae 308
Araliaceae 239
Archangel, Yellow 253, M50i
Archangelica 243
Arctium 273, M62f-g
Arenaria 189, M10i-j
Arenaria 189
Aristolochia 185
Aristolochiaceae 184-185

Armeria 246
Armoracia 200, M18c
Arnoseris 276
Arrhenatherum 304, M83j
Arrowgrass 292, M75c
Arrowhead 291, M74g
Artemisia 271, M61a-b
Artichoke, Jerusalem 269
Arum 308, M85l
Asarabacca 184
Asarum 184
Ash 246, M46a
Asparagus 296
Aspen 179, M5k
Asperugo 250
Asperula 247
Asphodel, Scottish 294
Aspidiaceae 175
Aspleniaceae 174-175
Asplenium 174, M11-2c
Aster 267-268, M59a
Astragalus 224, M32g-h
Astrantia 240
Athyriaceae 175
Athyrium 175, M2f
Atriplex 188, M10d-e
Atropa 257, M52j
Aubrieta 201
Avena 304, M83c
Avena 304
Avens 219, M28l-29b
Avenula 304, M83d-e
Azolla 176, M3e
Azollaceae 176

Baldellia 291
Ballota 253, M50j
Balm 254

477

Balsam   232, M37l-38b
Balsaminaceae   232
Barbarea   199-200, M17g-j
Barberry   197, M15l
Barley
   Foxtail   304, M83b
   Meadow   303, M83a
   Wall   303, M82l
Barren Strawberry   220, M29i
Bartsia, Red   262, M55k
Bartsia   262
Basil, Wild   255, M51g
Basil Thyme   254, M51f
Bastard Cabbage   205
Beak-sedge   310
Bedstraw   247-248, M47a-f
Beech   180, M6g
Bee Orchid   315, M90l
Bellflower   266, M58c-f
Bell Heather   244
Bellis   267, M58l
Bent   305, M84d-h
   Silky-   306
Berberidaceae   197
Berberis   197, M15l
Berberis   197
Berula   241, M43b
Betonica   253
Betony   253, M50k
Betula   180, M6b-c
Betulaceae   180
Bidens   269, M59j-k
Bilberry   244, M44l
Bilderdykia   186, M8i
Bindweed
   Black-   186, M8i
   Field   249, M47l
   Hairy   249, M47k
   Hedge   248, M47i
   Large   249, M47j
Birch   180, M6b-c
Bird's-foot   228, M35f
Bird's-foot-trefoil   228, M32c-d
Bird's-nest Orchid   314
Birthwort   185
Bistort   185-186, M8e-g
Bitter-cress   201, M18h,18j-k
Bittersweet   257, M52l
Bitter Vetch   225, M33d
Blackberry   208-213, M22h-26j
Black-bindweed   186, M8i
Black Bryony   297, M77j
Black-grass   306, M85c
Black Horehound   253, M50j
Black Mustard   204
Black Nightshade   257, M52k
Blackstonia   246, M46d
Blackthorn   222, M31b
Black Twitch   306, M85c
Bladder-fern, Brittle   175, M2g
Bladder-senna   223, M32f
Bladderwort   263

Blechnaceae   176
Blechnum   176, M3c
Blinks   189, M10f
Blood-drop-emlets   258
Bluebell   295, M77a
Blue Fleabane   268, M59b
Blue Pimpernel   246
Blue Sow-thistle   277, M64l
Blysmus   310
Bogbean   247, M46g
Bog Pimpernel   245, M45j
Bog-rush   310, M87d
Borage   250
Boraginaceae   249-251
Borago   250
Botrychium   174, M1i
Box   233, M38e
Brachypodium   303, M82h-i
Bracken   174, M1j
Bramble   208-213, M22h-26j
Brassica   204, M20d-e
Brassica   204
Bridewort   207, M22b
Bristle-grass   308
Brittle Bladder-fern   175, M2g
Briza   301, M81c
Broad Bean   225
Brome   302-303, M81j-82g
   False   303, M82h
Bromus   302-303, M81j-82g
Brooklime   261, M54j
Brookweed   246, M45l
Broom   223, M31k
Broom, Butcher's-   296
Broomrape   263, M56c
Bryonia   236, M40h
Bryony
   Black   297, M77j
   White   236, M40h
Buckler-fern   175, M2l-3a
Buckthorn
   Alder   233, M38g
   Purging   233, M38f
Buckwheat   186
Buddleja   258
Buddlejaceae   258
Bugle   252, M49h
Bugloss   250, M48f
   Viper's-   249, M48b
Buglossoides   249, M48a
Bullwort   243
Bulrush   309, M86g-h, 86j
Bunias   199
Bupleurum   242
Bur Chervil   240
Burdock   273, M62f-g
Bur-marigold   269, M59j-k
Burnet
   Fodder   219, M28k
   Great   218, M28i
   Salad   218, M28j
Burnet-saxifrage   241, M42k-l

Burnt orchid   315
Bur-reed   309, M86e-f
Bursa   202
Butcher's-broom   296
Butomaceae   291
Butomus   291, M74j
Butterbur   272, M61d-e
Buttercup   194-195, M14d-j
Butterfly-bush   258
Butterfly-orchid   314, M90d
Butterwort   263
Buxaceae   233
Buxus   233, M38e

Cabbage   204
   Bastard   205
   Warty   199
Calamagrostis   306, M84i
Calamint   254
Calamintha   254
Calendula   273, M62d
Californian Poppy   197
Callitrichaceae   251-252
Callitriche   251-252, M49d-g
Calluna   244, M44j
Caltha   194, M13l
Calystegia   248-249, M47i-k
Camelina   202
Cammarum   194
Campanula   266-267, M58c-g
Campanulaceae   266-267
Campion   192, M12k,13a-c
Canadian Fleabane   268, M59c
Canadian Waterweed   292, M75a
Canary-grass   307, M85d-e
Candytuft   202, M19g
Cannabaceae   184
Cannabis   184
Cape-gooseberry   257
Capnoides   198
Caprifoliaceae   264-265
Capsella   202, M19d
Caraway   243
Cardamine   201, M18g-k
Cardaria   203, M19l
Carduus   274, M62h-i
Carex   310-314, M87e-89k
Carlina   273, M62e
Carline Thistle   273, M62e
Carpinus   180, M6e
Carrot   244, M44h
Carum   243
Carum   242
Caryophyllaceae   189-193
Castalia   193
Castanea   180, M6h
Catabrosa   301, M81b
Catapodium   300
Catchfly   192,193, M12l
Cat-mint   254
Cat's-ear   276, M63k

478

Cat's-tail, Smaller   306, M84k
Caucalis   243
Celandine
  Greater   197, M16g
  Lesser   195, M14k
Celastraceae   232
Celery   242
Centaurea   275, M63e-i
Centaurium   246, M46e
Centaury   246, M46e
Centranthus   265, M57i
Cerastium   190-191, M11h-12a
Cerastium   191
Ceratochloa   303
Ceratophyllaceae   193
Ceratophyllum   193, M13g-h
Ceterach   174, M2d
Chaenorhinum   259, M53i
Chaerefolium   240
Chaerophyllum   240, M42f
Chamaemelum   270
Chamaenerion   237
Chamomile   270
  Corn   270
  Stinking   270, M60b
  Yellow   270
Chamomilla   270, M60f-g
Charlock   204, M20f
Cheiranthus   199, M17f
Chelidonium   197, M16g
Chenopodiaceae   187-189
Chenopodium   187-188, M9j-10c
Cherry   222, M31e-g
Cherry Laurel   222, M31i
Chervil
  Bur   240
  Garden   240
  Rough   240, M42f
Chestnut
  Horse-   232, M37k
  Sweet   180, M6h
Chickweed   189-190, M11a-c
  Upright   191, M12b
  Water   191, M12c
Chicory   276, M63j
China Teaplant   257, M52i
Chrysanthemum   271, M60h
Chrysanthemum   271
Chrysosplenium   206-207, M21h-i
Cicerbita   277, M64l
Cichorium   276, M63j
Cinquefoil   219, M29d; 220, M29h
Circaea   236, M40k
Cirsium   274, M62j-63c
Cistaceae   236
Clary   256, M52g
Claytonia   189
Cleavers   248, M47g
Clematis   194, M14c
Clinopodium   255, M51g
Clover   226-228, M34d-f,j-l; 35a-b
Clubmoss   173, M1a

Club-rush   309, M86i-l
Cochlearia   202, M19c
Cochlearia   200
Cock's-foot   301, M80l
Cockspur   307
Coeloglossum   314
Colchicum   294
Colt's-foot   271, M61c
Columbine   196
Colutea   223, M32f
Comfrey   249-250, M48c-e
Common Cudweed   268, M59d
Common Fleabane   269, M59i
Common Gromwell   249
Compositae   267-291
Conium   242, M43i
Conopodium   241, M42j
Conringia   203
Consolida   194
Convallaria   296, M77e
Convolvulaceae   248-249
Convolvulus   249, M47l
Conyza   268, M59c
Coralroot   201, M18g
Coriandrum   240
Cornaceae   239
Corn Chamomile   270
Corncockle   192
Cornflower   275, M63i
Corn Marigold   271, M60h
Corn Parsley   242
Cornsalad   265, M57d-f
Corn Spurrey   192, M12h
Cornus   239, M42b
Coronilla   228
Coronopus   203, M20a-b
Corydalis   197-198, M16h-i
Corylaceae   180
Corylus   180, M6f
Cottongrass   310, M87a
Cotton Thistle   274
Cotyledon   205
Couch   303, M82j-k
Cowberry   244
Cowherb   193
Cow Parsley   240, M42g
Cowslip   245, M45c
Cow-wheat   262, M55g
Crab Apple   221, M30e
Crab Grass   308
Cranesbill   229-230, M35j-36d
Crassula   205
Crassulaceae   205-206
Crataegus   221-222, M30k-l
Creeping Jenny   245, M45h
Creeping Soft-grass   305, M84c
Crepis   287-288, M72k-73b
Cress
  Bitter-   201, M18h,18j-k
  Fool's Water-   242, M43j
  Garden   203

  Hoary   203, M19l
  Penny-   202, M19f
  Rock-   201
  Shepherd's   202, M19e
  Swine-   203, M20a-b
  Thale   199, M17c
  Tower   201
  Water-   200, M18d-f
  Winter-   199-200, M17g-j
  Yellow-   200, M17k-18b
Crested Dog's-tail   301, M81a
Crested Hair-grass   304, M83g
Crocosmia   297
Crosswort   248, M47h
Crowberry   244, M45a
Crowfoot   195-196, M15b-c
  Water   196, M15d-i
Crow Garlic   296, M77d
Crown Vetch   228
Cruciata   248, M47h
Cruciferae   198-205
Cuckooflower   201, M18i
Cucurbitaceae   236
Cudweed
  Common   268, M59d
  Heath   268, M59f
  Marsh   268, M59g
  Small   268, M59e
Currant   207, M21k-l
Cuscuta   248
Cymbalaria   260, M54a
Cynoglossum   251, M49c
Cynosurus   301, M81a
Cyperaceae   309-314
Cystopteris   175, M2g
Cytisus   223, M31k

Dactylis   301, M80l
Dactylorchis   314-315
Dactylorhiza   314-315, M90f-h
Daffodil   296, M77i
Daisy   267, M58l
  Michaelmas   267, M59a
  Oxeye   271, M60k
  Shasta   271, M60l
Damasonium   291
Dame's-violet   199, M17e
Dandelion   277-287, M65b-72h
Danewort   264, M56i
Danthonia   307, M85h
Daphne   234, M39a
Darnel   300
Datura   258
Daucus   244, M44h
Day Lily   294
Deadly Nightshade   257, M52j
Dead-nettle   253, M50d-h
Deer-grass   309
Delphinium   194
Dentaria   201

Deschampsia 304, M83i-j
Descurainia 198
Desmazeria 300, M80e
Devil's-bit Scabious 266, M57l
Dewberry 208, M22f
Dianthus 193
Digitalis 260, M54d
Digitaria 308
Dioscoreaceae 297
Diplotaxis 204, M20c
Dipsacaceae 265-266
Dipsacus 265-266, M57j-k
Dittander 203, M19k
Dock 186-187, M9b-i
Dodder 248
Dog's Mercury 230, M36k
Dog's-tail
   Crested 301, M81a
   Rough 301
Dog's-tooth-violet 294
Dogwood 239, M42b
Doronicum 272, M61g
Dotted Loosestrife 245, M45i
Downy Oat-grass 304, M83d
Draba 202, M19a
Dropwort 207, M22c
   Water- 241, M43c-f
Drosera 205
Droseraceae 205
Dryopteris 175, M2j-3a
Dryopteris 174
Duckweed 308-309, M86a-d
   Greater 308, M86d
Duke of Argyll's Teaplant 257, M52h
Dutch Rush 173, M1b
Dwarf Elder 264, M56i
Dyer's Greenweed 223, M31l

Early Hair-grass 305, M83k
Early Purple Orchid 315, M90j
Echinochloa 307
Echium 249, M48b
Egeria 292
Elatinaceae 236
Elatine 236
Elder 264, M56i-j
   Ground- 241, M43a
Elecampane 268
Eleocharis 310, M87b-c
Eleogiton 309
Elisma 291
Elm 181-184, M6k-7e
Elodea 292, M75a-b
Elymus 303, M82j-k
Empetraceae 244
Empetrum 244, M45a
Enchanter's Nightshade 236, M40k
Endymion 295
Epilobium 237-239, M40l-41k
Epipactis 314, M89l-90b
Equisetaceae 173

Equisetum 173, M1b-g
Eranthis 193, M13k
Erica 244, M44i
Ericaceae 244
Erigeron 268, M59b
Erigeron 268
Eriophorum 310, M87a
Erodium 230, M36f-g
Erophila 202, M19b
Eruca 204
Erucastrum 204
Erysimum 199, M17d
Erythronium 294
Eschscholzia 197
Euonymus 232, M38d
Eupatorium 267, M58i
Euphorbia 231, M36l-37e
Euphorbiaceae 230-231
Euphrasia 262, M55h-j
Eupteris 174
Evening-primrose 237
Everlasting-pea 225, M33f
Eye-bright 262, M55h-j

Fagaceae 180-181
Fagopyrum 186
Fagus 180, M6g
False Brome 303, M82h
False Oat-grass 304, M83f
False Oxlip 245, M45d
Fat Hen 188, M10c
Fennel 242
Fern
   Brittle Bladder- 175, M2g
   Buckler- 175, M2l-3a
   Hard 176, M3c
   Lady- 175, M2f
   Lemon-scented 174, M1k
   Limestone 175, M3b
   Male- 175, M2j-k
   Royal 174
   Shield- 175, M2h-i
   Water 176, M3e
Fern-grass 300, M80e
Fescue 299, M79f-k
   Squirrel-tail 300, M80c
   Rat's-tail 300, M80d
Festuca 299, M79f-k
Festuca 300
Festulolium 299, M79l
Feverfew 271, M60j
Field Bindweed 249, M47l
Field Gromwell 249, M48a
Field Madder 247, M46k
Field Marigold 273
Field Scabious 266, M58a
Field-speedwell 261, M55b-d
Figwort 259, M53e-g
Filaginella 268, M59g
Filago 268, M59d
Filago 268

Filipendula 207, M22c-d
Fine-leaved Sandwort 189, M10l
Fireweed 237, M40l
Flag
   Sweet- 308, M85k
   Yellow 297, M77l
Flat-sedge 310
Flax 230, M36h-i
Fleabane
   Blue 268, M59b
   Canadian 268, M59c
   Common 269, M59i
   Irish 268
   Small 269
Fleawort 272
Flixweed 198
Floating Water-plantain 291
Flowering-rush 291, M74j
Fluellen 260, M54b-c
Fodder Burnet 219, M28k
Foeniculum 242
Fool's Parsley 241, M43g
Fool's Watercress 242, M43j
Forget-me-not 250-251, M48h-49b
Fox and Cubs 290, M74f
Foxglove 260, M54d
Foxtail 306, M84l-85b
Foxtail Barley 304, M83b
Fragaria 220, M29j-k
Fragrant Orchid 314, M90e
Frangula 233, M38g
Fraxinus 246, M46a
Fringed Water-lily 247, M46h
Fritillaria 295
Fritillary 295
Frog-bit 291, M74k
Frog Orchid 314
Fumaria 198, M16j
Fumitory 198, M16j
   White Climbing 197, M16h
Furze 223, M32b

Gagea 294, M76k
Galanthus 296, M77h
Galega 223, M32e
Galeobdolon 253
Galeopsis 252-253, M49l-50c
Galinsoga 269-270, M59l-60a
Galium 247-248, M46l-47g
Galium 248
Gallant Soldier 269, M59l
Garden Chervil 240
Garden Cress 203
Garlic 295-296, M77c-d
Garlic Mustard 198, M17b
Gaultheria 244
Gean 222, M31e
Genista 223, M31l-32a
Gentian 246-247, M46f
Gentiana 246-247
Gentianaceae 246-247

Gentianella 246-247, M46f
Geraniaceae 229-230
Geranium 229-230, M35j-36e
Geum 219, M28l-29b
Giant Knotweed 186, M8k
Gipsywort 255, M51k
Gladdon 297, M77k
Glechoma 254, M51d
Glyceria 301-302, M81e-i
Glyceria 301
Gnaphalium 268
Goat's-beard 276, M64e
Goat's-rue 223, M32e
Golden-rod 267, M58j-k
Golden-saxifrage 206-207, M21h-i
Goldilocks Aster 268
Goldilocks Buttercup 195, M14i
Gold-of-pleasure 202
Good-King-Henry 187, M9j
Gooseberry 207, M22a
Goosefoot 188, M9k-10b
Goosegrass 248, M47g
Gorse 223, M32b-c
Gramineae 299-308
Grape Hyacinth 295
Grass-of-Parnassus 207, M21j
Grass Poly 236
Great Burnet 218, M28i
Greater Celandine 197, M16g
Green-winged Orchid 315, M90i
Groenlandia 294, M76i
Gromwell
    Common 249
    Field 249, M48a
    Purple 249
Grossulariaceae 207
Ground-elder 241, M43a
Ground-ivy 254, M51d
Groundsel 273, M62a-c
Guelder-rose 264, M56k
Guizotia 269
Guttiferae 234-235
Gymnadenia 314, M90e
Gymnocarpium 175, M3b

Habenaria 314
Hair-grass
    Crested 304, M83g
    Early 305, M83k
    Silver 305, M83l
    Tufted 304, M83i
    Wavy 304, M83j
Hairy Bindweed 249, M47k
Haloragaceae 239
Hard Fern 176, M3c
Hard-grass 306
Harebell 267, M58g
Hare's-ear Mustard 203
Hart's-tongue 175, M2e
Hawkbit 276, M63l-64b
Hawk's-beard 287-288, M72k-73b

Hawkweed 288-290, M73c-74c
    Mouse-ear 290, M74d-e
Hawthorn 221-222, M30k-l
Hazel 180, M6f
Heath 244, M44i
Heath Cudweed 268, M59f
Heather 244, M44j
    Bell 244
Heath-grass 307, M85h
Hedera 239, M42c
Hedge Bindweed 248, M47i
Hedge Mustard 198, M17a
Hedge-parsley 243, M44f-g
Helianthemum 236, M40g
Helianthus 269
Helictotrichon 304
Heliotrope, Winter 272, M61f
Hellebore 193, M13i-j
Helleborine 314, M89l-90b
Helleborine 314
Helleborus 193, M13i-j
Helxine 184
Hemerocallis 294
Hemlock 242, M43i
Hemp-agrimony 267, M58i
Hemp-nettle 252-253, M49l-50c
Henbane 257
Heracleum 243, M44d-e
Herb-Bennet 219, M29b
Herb-Paris 296, M77g
Herb-Robert 230, M36e
Heriff 248, M47g
Herniaria 192
Hesperis 199, M17e
Hieracium 288-290, M73c-74c
Hieracium 290
Hierochloe 305
Himalayan Knotweed 186, M8h
Hippocastanaceae 232
Hippocrepis 228
Hippuridaceae 239
Hippuris 239, M42a
Hirschfeldia 204
Hoary Cress 203, M19l
Hoary Mustard 204
Hogweed 243, M44d-e
Holcus 305, M84b-c
Holly 232, M38c
Holy-grass 305
Honesty 201, M18l
Honeysuckle 264-265, M57b
Hop 184, M7f
Hop Trefoil 227, M34g
Hordeum 303-304, M82l-83b
Horehound
    Black 253, M50j
    White 252
Hornbeam 180, M6e
Horned Pondweed 294, M76j
Hornwort 193, M13g-h
Horse-chestnut 232, M37k
Horse-radish 200, M18c

Horseshoe Vetch 228
Horsetail 173, M1b-g
Hottonia 245, M45e
Hound's-tongue 251, M49c
House-leek 206
Humulus 184, M7f
Huperzia 173
Hyacinthoides 295, M77a
Hydrocharis 291, M74k
Hydrocharitaceae 291-292
Hydrocotyle 240, M42d
Hyoscyamus 257
Hypericum 234-235, M39b-g
Hypochoeris 276, M63k
Hypolepidaceae 174

Iberis 202-203, M19g
Ilex 232, M38c
Impatiens 232, M37l-38b
Inula 268-269, M59h
Iridaceae 297
Iris 297, M77k-l
Irish Fleabane 268
Isatis 199
Isolepis 309
Ivy 239, M42c
    Ground- 254, M51d
Ivy-leaved Toadflax 260, M54a

Jacob's-ladder 248
Japanese Knotweed 186, M8j
Jasione 267
Jerusalem Artichoke 269
Juglandaceae 180
Juglans 180
Juncaceae 297-299
Juncaginaceae 292
Juncoides 298-299
Juncus 297-298, M78b-79a

Keck 240, M42g
Kedlock 204, M20f
Kickxia 260, M54b-c
Kidney Vetch 228, M35e
Kingcup 194, M13l
Knapweed 275, M63e-h
Knautia 266, M58a
Knawel 191, M12g
Knotgrass 185, M7k-8a
Knotweed 186, M8h,8j-k
Koeleria 304, M83g

Labiatae 252-256
Laburnum 222, M31j
Lactuca 277, M64j-k
Lactuca 277
Lady-fern 175, M2f

Lady's-mantle 220, M29l-30a
Lady's-smock 201, M18i
Lady's-tresses 314
Lamb's Lettuce 265, M57d
Lamb's Succory 276
Lamiastrum 253, M50i
Lamium 253, M50d-h
Lamium 253
Lappula 251
Lapsana 287, M72i-j
Larch 176, M3g
Large Bindweed 249, M47j
Larix 176, M3g
Larkspur 194
Lathraea 262, M56b
Lathyrus 225, M33d-g
Laurel
 Cherry 222, M31i
 Portugal 222, M31h
 Spurge- 234, M39a
Lavatera 234
Leek, Few-flowered 295, M77b
Legousia 267, M58h
Leguminosae 222-228
Lemna 308, M86a-c
Lemna 308
Lemnaceae 308-309
Lemon-scented Fern 174, M1k
Lentibulariaceae 263
Leontodon 276, M63l-64b
Leonurus 253
Leopard's-bane 272, M61g
Lepidium 203, M19h-k
Lepidium 203
Lepidotis 173
Lesser Celandine 195, M14k
Lesser Snapdragon 259
Lesser Trefoil 227, M34h
Lesser Water-plantain 291
Lettuce 277, M64j-k
 Lamb's 265, M57d
 Wall 277, M65a
Leucanthemum 271, M60k-l
Ligustrum 246, M46c
Lilac 246, M46b
Liliaceae 294-296
Lilium 295
Lily, Martagon 295
Lily-of-the-valley 296, M77e
Lime 233, M38h-i
Limestone Fern 175, M3b
Limosella 258, M53a
Linaceae 230
Linaria 259-260, M53j-l
Linaria 259, 260
Ling 244, M44j
Linum 230, M36h-i
Liquorice, Wild 224, M32g
Listera 314, M90c
Lithospermum 249
Lithospermum 249
Littorella 264, M56h

Lobularia 201
Loganberry 208
Logfia 268, M59e
Lolium 299-300, M80a-b
London-pride 206
Lonicera 264-265, M57b
Loosestrife
 Dotted 245, M45i
 Purple- 236, M40i
 Yellow 245, M45g
Loranthaceae 184
Lords-and-ladies 308, M85l
Lotus 228, M35c-d
Lousewort 262, M55l
Lucerne 226, M34b
Lunaria 201, M18l
Lungwort 249
Lupin 223, M32d
Lupinus 223, M32d
Luronium 291
Luzula 298-299, M79b-e
Lychnis 192, M12j
Lychnis 192, 193
Lycium 257, M52h-i
Lycopersicon 258
Lycopodiaceae 173
Lycopodium 173, M1a
Lycopodium 173
Lycopus 255, M51k
Lysimachia 245, M45f-i
Lythraceae 236
Lythrum 236, M40i-j

Madder, Field 247, M46k
Madwort 250
Mahonia 197, M16a
Maiden Pink 193
Malcolmia 199
Male-fern 175, M2j-k
Mallow 233-234, M38j-l
Malus 221, M30e
Malva 233-234, M38j-l
Malvaceae 233-234
Maple 231-232, M37h-i
Mare's-tail 239, M42a
Marigold
 Bur- 269, M59j-k
 Corn 271, M60h
 Field 273
 Marsh- 194, M13l
 Pot 273, M62d
Marjoram 255, M51h
Marrubium 252
Marsh Cudweed 268, M59g
Marsh-marigold 194, M13l
Marsh Orchid 314-315, M90f
Marsh Pea 225
Marsh Pennywort 240, M42d
Marshwort 242, M43k
Marsiliaceae 176
Martagon Lily 295

Mat-grass 307, M85j
Matricaria 270, M60e
Matricaria 270
Mayweed
 Scented 270, M60f
 Scentless 270, M60e
Meadow Barley 303, M83a
Meadow-grass 300-301, M80f-j
Meadow Oat-grass 304, M83e
Meadow-rue 196, M15k
Meadow Saffron 294
Meadowsweet 207, M22d
Meconopsis 197
Medicago 226, M34a-c
Medick 226, M34a, c
Melampyrum 262, M55g
Melica 301, M81d
Melick, Wood 301, M81d
Melilot 225-226, M33j-l
Melilotus 225-226, M33j-l
Melissa 254
Mentha 255-256, M51l-52f
Menyanthaceae 247
Menyanthes 247, M46g
Mercurialis 230, M36j-k
Mercury 230, M36j-k
Michaelmas-daisy 267, M59a
Mignonette 205, M20i
Milfoil 270, M60d
 Water- 239, M41l
Milium 307, M85f
Milk Thistle 275
Milk-vetch 224, M32g
Milkwort 231, M37f-g
Millet 308
 Wood 307, M85f
Mimulus 258, M53b-c
Mind-your-own-business 184
Mint 255-256, M51l-52f
Minuartia 189, M10l
Misopates 259
Mistletoe 184, M7j
Moehringia 189, M10k
Moenchia 191, M12b
Molinia 307, M85i
Monkeyflower 258, M53b
Monk's-hood 194, M14a
Montbretia 297, M78a
Montia 189, M10f-h
Moonwort 174, M1i
Moor-grass, Purple 307, M85i
Moschatel 265, M57c
Motherwort 253
Mountain Everlasting 268
Mouse-ear 190-191, M11i-12a
Mouse-ear Hawkweed 290, M74d-e
Mousetail 196, M15j
Mudwort 258, M53a
Mugwort 271, M61a
Mullein 258-259, M53d
Muscari 295
Musk 258, M53c

Mustard
    Black 204
    Garlic 198, M17b
    Hare's-ear 203
    Hedge 198, M17a
    Hoary 204
    Treacle 199, M17d
    White 204
Mycelis 277, M65a
Myosotis 250-251, M48h-49b
Myosoton 191, M12c
Myosurus 196, M15j
Myriophyllum 239, M41l
Myrrhis 240, M42i

Narcissus 296, M77i
Nardus 307, M85j
Nasturtium 200, M18d-f
Navelwort 205, M20j
Neottia 314
Nepeta 254
Nepeta 254
Nettle 184, M7g-h
    Dead- 253, M50d-h
    Hemp- 252-253, M49l-50c
Nicandra 256
Nightshade
    Black 257, M52k
    Deadly 257, M52j
    Enchanter's 236, M40k
    Woody 257, M52l
Nipplewort 287, M72i-j
Nuphar 193, M13f
Nymphaea 193, M13e
Nymphaea 193
Nymphaeaceae 193
Nymphoides 247, M46h

Oak 180, M6i-j
Oat, Wild- 304, M83c
Oat-grass
    Downy 304, M83d
    False 304, M83f
    Meadow 304, M83e
    Yellow 304, M83h
Odontites 262, M55k
Oenanthe 241, M43c-f
Oenothera 237
Oleaceae 246
Omalotheca 268, M59f
Onagraceae 236-239
Onion, Wild 296, M77d
Onobrychis 228
Ononis 225, M33h-i
Onopordum 274-275
Ophioglossaceae 174
Ophioglossum 174, M1h
Ophrys 315, M90l
Opposite-leaved Pondweed 294, M76i
Orache 188, M10d-e

Orchid
    Bee 315, M90l
    Bird's-nest 314
    Burnt 315
    Butterfly 314, M90d
    Early-purple 315, M90j
    Fragrant 314, M90e
    Frog 314
    Green-winged 315, M90i
    Marsh 314-315, M90f
    Pyramidal 315, M90k
    Spotted 315, M90g-h
Orchidaceae 314-315
Orchis 315, M90i-j
Orchis 314-315
Oregon-grape 197, M16a
Origanum 255, M51h
Ornithogalum 295, M76l
Ornithopus 228, M35f
Orobanchaceae 263
Orobanche 263, M56c
Orpine 206
Osier 179, M5e
Osmunda 174
Osmundaceae 174
Oxalidaceae 228-229
Oxalis 228-229, M35g-i
Oxeye Daisy 271, M60k
Oxlip, False 245, M45d
Oxtongue 276, M64c-d

Panicum 307
Panicum 307, 308
Pansy 235-236, M40e-f
Papaver 197, M16b-f
Papaveraceae 197-198
Parapholis 306
Parietaria 184, M7i
Paris 296, M77g
Parnassia 207, M21j
Parnassiaceae 207
Parsley 242
    Corn 242
    Cow 240, M42g
    Fool's 241, M43g
    Hedge- 243, M44f-g
    Stone 242, M43l
Parsley-piert 220, M30b-c
Parsnip 243, M44c
    Water- 241, M43b
Pasqueflower 194
Pastinaca 243, M44c
Pea
    Everlasting- 225, M33f
    Marsh 225
Pear 221, M30d
Pearlwort 191, M12d-f
Pedicularis 262, M55l
Pellitory-of-the-wall 184, M7i
Penny-cress 202, M19f
Pennyroyal 255

Pennywort, Marsh 240, M42d
Pentaglottis 250, M48g
Peplis 236
Peppermint 256, M52d
Pepper-saxifrage 242, M43h
Pepperwort 203, M19h-j
Periwinkle 247, M46i-j
Persicaria 185, M8c-d
Petasites 272, M61d-f
Petroselinum 242
Petty Whin 223, M32a
Peucedanum 243
Phalaris 307, M85d-e
Pheasant's-eye 194
Phleum 306, M84j-k
Phragmites 307, M85g
Phyllitis 175, M2e
Physalis 257
Picea 176, M3f
Picris 276, M64c-d
Pignut 241, M42j
Pillwort 176
Pilosella 290, M74d-f
Pilularia 176
Pimpernel
    Blue 246
    Bog 245, M45j
    Scarlet 245, M45k
    Yellow 245, M45f
Pimpinella 241, M42k-l
Pinaceae 176-177
Pine 176-177, M3h-i
Pineappleweed 270, M60g
Pinguicula 263
Pink Purslane 189, M10h
Pinus 176-177, M3h-i
Plantaginaceae 263-264
Plantago 263-264, M56d-g
Plantain 263-264, M56d-g
    Floating Water- 291
    Lesser Water- 291
    Water- 291, M74h-i
Platanthera 314, M90d
Ploughman's-spikenard 269, M59h
Plum 222, M31a, 31c
Plumbaginaceae 246
Poa 300-301, M80f-j
Polemoniaceae 248
Polemonium 248
Policeman's-helmet 232, M38b
Polygala 231, M37f-g
Polygalaceae 231
Polygonaceae 185-187
Polygonatum 296, M77f
Polygonum 185-186, M7k-8h
Polygonum 186
Polypodiaceae 176
Polypodium 176, M3d
Polypody 176, M3d
Polystichum 175, M2h-i
Pondweed 292-294, M75d-76h
    Horned 294, M76j

483

Opposite-leaved 294, M76i
Poplar 179-180, M5i-6a
Poppy 197, M16b-f
   Californian 197
   Welsh 197
Populus 179-180, M5i-6a
Portugal Laurel 222, M31h
Portulaceae 189
Potamogeton 292-294, M75d-76h
Potamogeton 294
Potamogetonaceae 292-294
Potato 257
Potentilla 219-220, M29c-i
Poterium 218-219
Pot Marigold 273, M62d
Primrose 244, M45b
Primula 244-245, M45b-d
Primulaceae 244-246
Privet 246, M46c
Prunella 254, M51e
Prunus 222, M31a-i
Pseudotsuga 176
Pteridium 174, M1j
Puccinellia 301, M80k
Pulicaria 269, M59i
Pulmonaria 249
Pulsatilla 194
Purging Buckthorn 233, M38f
Purple Gromwell 249
Purple-loosestrife 236, M40i
Purple Moor-grass 307, M85i
Purslane
   Pink 189, M10h
   Water- 236, M40j
Pyramidal Orchid 315, M90k
Pyrola 244
Pyrolaceae 244
Pyrus 221, M30d
Pyrus 221

Quaking-grass 301, M81c
Quercus 181, M6i-j

Radiola 230
Radish 205, M20g
Ragged-Robin 192, M12j
Ragweed 269
Ragwort 272-273, M61h-l
Ramping-fumitory 198
Ramsons 295, M77c
Ranunculaceae 193-196
Ranunculus 194-196, M14d-15i
Rape 204, M20d
Raphanus 205, M20g
Rapistrum 205
Raspberry 208, M22e
Rat's-tail Fescue 300, M80d
Rattle, Yellow 262, M56a
Red Bartsia 262, M55k
Redshank 185, M8c

Red Valerian 265, M57i
Reed
   Bur- 309, M86e-f
   Common 307, M85g
   Small- 306, M84i
Reed Canary-grass 307, M85d
Reedmace 309, M86g-h
Reed Sweet-grass 301, M81e
Reseda 205, M20h-i
Resedaceae 205
Restharrow 225, M33h-i
Reynoutria 186, M8j-k
Rhamnaceae 233
Rhamnus 233, M38f
Rhamnus 233
Rhinanthus 262, M56a
Rhododendron 244, M44k
Rhynchospora 310
Ribes 207, M21k-22a
Robinia 223
Rock-cress 201
Rocket 198, M16k-l
   Wall- 204, M20c
Rockrose 236, M40g
Rorippa 200, M17k-18b
Rorippa 200
Rosa 213-218, M22k-28f
Rosaceae 207-222
Rose 213-218, M22k-28f
Rose-of-Sharon 234
Rough Chervil 240, M42f
Rough Dog's-tail 301
Rowan 221, M30f
Royal Fern 174
Rubiaceae 247-248
Rubus 208-213, M22e-26j
Rumex 186-187, M8l-9i
Ruscus 296
Rush 297-298, M78b-79a
   Bog- 310, M87d
   Club- 309, M86i-l
   Dutch 173, M1b
   Flowering- 291, M74j
   Spike- 310, M87b-c
   Wood- 298-299, M79b-e
Russian-vine 186
Rustyback 174, M2d
Rye-grass 299-300, M80a-b

Saffron, Meadow 294
Sage, Wood 252, M49i
Sagina 191, M12d-f
Sagittaria 291, M74g
Sainfoin 228
St. John's-wort 234-235, M39b-g
Salad Burnet 218, M28j
Salicaceae 177-180
Salix 177-179, M3k-5h
Salsify 276
Salsola 189
Saltmarsh-grass 301, M80k

Saltwort 189
Salvia 256, M52g
Sambucus 264, M56i-j
Samolus 246, M45l
Sand Spurrey 192, M12i
Sandwort
   Fine-leaved 189, M10l
   Slender 189, M10j
   Three-nerved 189, M10k
   Thyme-leaved 189, M10i
Sanguisorba 218-219, M28i-k
Sanicle 240, M42e
Sanicula 240, M42e
Saponaria 193, M13d
Saponaria 193
Sarothamnus 223
Satureja 254
Saw-wort 275, M63d
Saxifraga 206, M21e-g
Saxifragaceae 206-207
Saxifrage 206, M21e-g
   Burnet- 241, M42k-l
   Golden- 206-207, M21h-i
   Pepper- 242, M43h
Scabiosa 266, M58b
Scabiosa 266
Scabious
   Devil's-bit 266, M57l
   Field 266, M58a
   Small 266, M58b
Scandix 240, M42h
Scarlet Pimpernel 245, M45k
Scented Mayweed 270, M60f
Scentless Mayweed 270, M60e
Schoenoplectus 309
Schoenus 310, M87d
Scilla 295
Scirpus 309, M86i-l
Scirpus 310
Scleranthus 191, M12g
Scottish Asphodel 294
Scrophularia 259, M53e-g
Scrophulariaceae 258-263
Scurvygrass 202, M19c
Scutellaria 252, M49j-k
Sedge 310-314, M87e-89k
   Beak- 310
   Flat- 310
Sedum 206, M20k-21d
Selfheal 254, M51e
Sempervivum 206
Senecio 272-273, M61h-62c
Serratula 275, M63d
Service-tree 221, M30g
Setaria 308
Shaggy Soldier 270, M60a
Shallon 244
Shasta Daisy 271, M60l
Sheep's-bit 267
Sheep's-fescue 299, M79j-k
Shepherd's Cress 202, M19e
Shepherd's-needle 240, M42h

484

Shepherd's-purse   202, M19d
Sherardia   247, M46k
Shield-fern   175, M2h-i
Shoreweed   264, M56h
Sideritis   252
Sieglingia   307
Silaum   242, M43h
Silene   192-193, M12k-13c
Silky-bent   306
Silver Hair-grass   305, M83l
Silverweed   219, M29c
Silybum   275
Sinapis   204, M20f
Sison   242, M43l
Sisymbrium   198, M16k-17a
Sisymbrium   198-199
Sium   241
Sium   241
Skullcap   252, M49j-k
Slender Sandwort   189, M10j
Slender Trefoil   227, M34i
Sloe   222, M31b
Small Cudweed   268, M59e
Small Fleabane   269
Small-reed   306, M84i
Small Scabious   266, M58b
Small Toadflax   259, M53i
Smyrnium   240
Snapdragon   259, M53h
   Lesser   259
Sneezewort   270, M60c
Snowberry   264, M57a
Snowdrop   296, M77h
Snow-in-summer   190, M11h
Soapwort   193, M13d
Soft-grass, Creeping   305, M84c
Solanaceae   256-258
Solanum   257, M52k-l
Solanum   258
Soleirolia   184
Solidago   267, M58j-k
Solomon's-seal   296, M77f
Sonchus   276-277, M64f-i
Sorbus   221, M30f-j
Sorrel   186, M8l-9a
   Wood-   229, M35i
Sow-thistle   276-277, M64f-i
   Blue   277, M64l
Sparganiaceae   309
Sparganium   309, M86e-f
Spearwort   195, M14l-15a
Speedwell 260-262, M54e-i,55a,55e-f
   Field-   261, M55b-d
   Water-   261, M54k-l
Spergula   192, M12h
Spergularia   192, M12i
Spike-rush   310, M87b-c
Spindle   232, M38d
Spiraea   207, M22b
Spiraea   207
Spiranthes   314
Spirodela   308-309, M86d

Spleenwort   174, M1l-2b
Spotted Orchid   315, M90g-h
Springbeauty   189, M10g
Spruce   176, M3f
Spurge   231, M36l-37e
Spurge-laurel   234, M39a
Spurrey
   Corn   192, M12h
   Sand   192, M12i
Squinancywort   247
Squirrel-tail Fescue   300, M80c
Stachys   253-254, M50k-51c
Starfruit   291
Star-of-Bethlehem   295, M76l
   Yellow   294, M76k
Star-thistle   275
Statice   246
Stellaria   189-190, M11a-g
Stellaria   191
Stinking Chamomile   270, M60b
Stitchwort   190, M11d-g
Stonecrop   206, M20k-21d
Stone Parsley   242, M43l
Stork's-bill   230, M36f-g
Stratiotes   292, M74l
Strawberry   220, M29j-k
   Barren   220, M29i
Succisa   266, M57l
Sundew   205
Sunflower   269
Sweet Alison   201
Sweet Chestnut   180, M6h
Sweet Cicely   240, M42i
Sweet-flag   308, M85k
Sweet-grass   301-302, M81e-i
Sweet Vernal-grass   305, M84a
Swine-cress   203, M20a-b
Sycamore   232, M37j
Symphoricarpos   264, M57a
Symphytum   249-250, M48c-e
Syringa   246, M46b

Tamus   297, M77j
Tanacetum   271, M60i-j
Tansy   271, M60i
Taraxacum   277-287, M65b-72h
Tare   224, M32k-l
Taxaceae   177
Taxus   177, M3j
Teaplant   257, M52h-i
Teasel   265-266, M57j-k
Teesdalia   202, M19e
Teucrium   252, M49i
Thale Cress   199, M17c
Thalictrum   196, M15k
Thelycrania   239
Thelypteridaceae   174
Thelypteris   174, M1k
Thelypteris   175

Thistle
   Carline   273, M62e
   Cotton   274
   Creeping   274, M63c
   Dwarf   274, M63a
   Marsh   274, M63b
   Meadow   274, M62l
   Milk   275
   Musk   274, M62h
   Sow-   276, M64f-i
   Spear   274, M62k
   Star-   275
   Welted   274, M62i
   Woolly   274, M62j
Thlaspi   202, M19f
Thorn-apple   258
Thorow-wax   242
Three-nerved Sandwort   189, M10k
Thrift   246
Thuja   176
Thyme   255, M51i-j
   Basil   254, M51f
Thymelaeaceae   234
Thyme-leaved Sandwort   189, M10i
Thymus   255, M51i-j
Tilia   233, M38h-i
Tiliaceae   233
Timothy   306, M84j
Toadflax   259-260, M53j-l
   Ivy-leaved   260, M54a
   Small   259, M53i
Tofieldia   294
Tomato   258
Toothwort   262, M56b
Tor-grass   303, M82i
Torilis   243, M44f-g
Tormentil   219, M29e-g
Tower Cress   201
Tragopogon   276, M64e
Traveller's-joy   194, M14c
Treacle Mustard   199, M17d
Trefoil
   Bird's-foot-   228, M35c-d
   Hop   227, M34g
   Lesser   227, M34h
   Slender   227, M34i
Trichophorum   309
Trifolium   226-228, M34d-35b
Triglochin   292, M75c
Tripleurospermum   270
Trisetum   304, M83h
Tritonia   297, M78a
Tsuga   176
Tufted Hair-grass   304, M83i
Tulip   295
Tulipa   295
Turnip   204, M20e
Tussilago   271, M61c
Tutsan   234
Twayblade   314, M90c
Twitch, Black   306, M85c
Typha   309, M86g-h
Typhaceae   309

485

Ulex 223, M32b-c
Ulmaceae 181-184
Ulmus 181-184, M6k-7e
Umbelliferae 240-244
Umbilicus 205, M20j
Upright Chickweed 191, M12b
Urtica 184, M7g-h
Urticaceae 184
Utricularia 263

Vaccaria 193
Vaccinium 244, M44l
Valerian 265, M57g-h
  Red 265, M57i
Valeriana 265, M57g-h
Valerianaceae 265
Valerianella 265, M57d-f
Venus'-looking-glass 267, M58h
Verbascum 258-259, M53d
Verbena 251
Verbenaceae 251
Vernal-grass, Sweet 305, M84a
Veronica 260-262, M54e-55f
Vervain 251
Vetch 224, M32i-j,33a-c
  Bitter 225, M33d
  Crown 228
  Horseshoe 228
  Kidney 228, M35e
  Milk- 224, M32g
Vetchling 225, M33e,g
Viburnum 264, M56k-l
Vicia 224-225, M32i-33c
Vinca 247, M46i-j
Viola 235-236, M39h-40f
Violaceae 235-236
Violet 235, M39h-40d
  Dame's- 199, M17e
  Dog's-tooth- 294
  Water- 245, M45e
Viper's-bugloss 249, M48b
Virginia Stock 199
Viscum 184, M7j
Volvulus 249
Vulpia 300, M80c-d

Wall Barley 303, M82l
Wallflower 199, M17f
Wall lettuce 277, M65a
Wall-rocket 204, M20c
Wall-rue 174, M2c
Walnut 180
Warty Cabbage 199
Water Avens 219, M28l
Water Chickweed 191, M12c
Water-cress 200, M18d-f
  Fool's 242, M43j
Water crowfoot 196, M15d-i
Water-dropwort 241, M43c-f
Water Fern 176, M3e

Water-lily 193, M13e-f
  Fringed 247, M46h
Water-milfoil 239, M41l
Water-parsnip
  Greater 241
  Lesser 241, M43b
Water-pepper 185, M8b
Water-plantain 291, M74h-i
  Floating 291
  Lesser 291
Water-purslane 236, M40j
Water-soldier 292, M74l
Water-speedwell 261, M54k-l
Water-starwort 251-252, M49d-g
Water-violet 245, M45e
Waterweed, Canadian 292, M75a
Waterwort 236
Wavy Hair-grass 304, M83j
Wayfaring-tree 264, M56l
Weld 205, M20h
Welsh Poppy 197
Whitebeam 221, M30h-j
White Bryony 236, M40h
White Horehound 252
White Mustard 204
Whitlowgrass 202, M19b
  Wall 202, M19a
Whorl-grass 301, M81b
Wild Basil 255, M51g
Wild Garlic 295, M77c
Wild Liquorice 224, M32h
Wild-Oat 304, M83c
Willow 177-179, M3k-5h
Willowherb 237-239, M40l-41k
Winter Aconite 193, M13k
Winter-cress 199-200, M17g-j
Wintergreen 244
Winter Heliotrope 272, M61f
Woad 199
Wood Avens 219, M29b
Wood Melick 301, M81d
Wood Millet 307, M85f
Woodruff 247, M46l
Wood-rush 298-299, M79b-e
Wood Sage 252, M49i
Wood-sorrel 229, M35i
Woody Nightshade 257, M52l
Wormwood 271, M61b
Woundwort 254, M50l-51c

Yarrow 270, M60d
Yellow Archangel 253, M50i
Yellow Chamomile 270
Yellow-cress 200, M17k-18b
Yellow-flag 297, M77l
Yellow Loosestrife 245, M45g
Yellow Oat-grass 304, M83h
Yellow Pimpernel 245, M45f
Yellow Rattle 262, M56a
Yellow Star-of-Bethlehem 294, M76k
Yellow-wort 246, M46d

Yew 177, M3j
Yorkshire-fog 305, M84b

Zannichellia 294, M76j
Zannichelliaceae 294
Zerna 302